6th International Conference on Metamaterials and Nanophotonics (METANANO 2021)

Journal of Physics: Conference Series Volume 2015

Tbilisi, Georgia
13 - 17 September 2021

ISBN: 978-1-7138-4986-5
ISSN: 1742-6588

Printed from e-media with permission by:

Curran Associates, Inc.
57 Morehouse Lane
Red Hook, NY 12571

CURRAN ASSOCIATES INC. proceedings.com

Some format issues inherent in the e-media version may also appear in this print version.

Institute of Physics
Dirac House, Temple Back
Bristol BS1 6BE UK

Phone: 44 1 17 929 7481
Fax: 44 1 17 920 0979

techtracking@iop.org

Additional copies of this publication are available from:

Curran Associates, Inc.
57 Morehouse Lane
Red Hook, NY 12571 USA
Phone: 845-758-0400
Fax: 845-758-2633
Email: curran@proceedings.com
Web: www.proceedings.com

TABLE OF CONTENTS

PREFACE

This issue of IOP Conference Proceedings contains extended contributions presented at the **6th International Conference on Metamaterials and Nanophotonics METANANO 2021** held **online on 13-17 September 2021**. We have been hoping to hold a hybrid conference with the offline part in Tbilisi, Georgia, but, after a thorough consideration we have decided to move METANANO 2021 fully online. The situation in the world remains unstable and to ensure the safety of all of you we believe that it is better to postpone the real-life meeting to next year and put all our efforts into making the online conference useful and pleasant for you.

For the last several years the METANANO conference has acquired a strong reputation among the scientists and is attracting more and more participants every year from different scientific fields from fundamental research in photonics and plasmonics to business-oriented projects in RF technologies, bionanotechnologies and solar energy.METANANO is a hub for researchers working in these fields and studying nanoscience, photonics, nonlinear optics and related topics. The aim of the conference is to encourage discussion and sharing of knowledge and experience between the scientific community, engineers, industry, and academic members all around the world.

The first METANANO in Russia was held in Anapa in 2016 and accumulated nearly 100 speakers, in 2017 in Vladivostok we gathered over 230 participants from Europe, Asia, and Russia. METANANO 2018 in Sochi had already over 330 participants. In 2019 more than 400 participants gathered in St. Petersburg. With the new reality brought on us last year we have had almost 500 participants from all over the world joining us. Despite the online format the conference brought together and united speakers and we made the most out of a new format.

This year we received more than 550 applications and we would like to express enormous gratitude to everyone who supported us and stayed with us for METANANO 2021 after switching to an online format again.

We acknowledge all the speakers of the conference, members of the Technical Program Committee and Advisory Board who have conducted enormous work reviewing the conference papers and forming its program. We are also thankful to the conference sponsors and co-organizers who have made an invaluable input to the growing quality of this conference.

We hope you will enjoy reading these Proceedings and discovering new and original research results.

And we welcome you to attend the forthcoming METANANO conferences.

Prof. Pavel Belov (General Chair) and Dr. Alexander Poddubny (Technical Program Chair)

METANANO conference

ACKNOWLEDGMENTS

The organizers of the conference are thankful to all the supporters who have made valuable impact to METANANO 2021 conference.

Informational support was provided by the Metamorphose VI, The Virtual Institute for Artificial Electromagnetic Materials and Metamaterials (metamorphose-vi.org).

meta morphose VIRTUAL INSTITUTE

ITMO UNIVERSITY

ITMO University is one of the Russia's leading higher education and research institutions, specializing in Information Technology, Optical Design and Engineering. Founded in 1900, one hundred years later tireless work of several generations of professors, lecturers and staff have brought our university a well-deserved success.

In 2011 we were named St. Petersburg National Research University of Information Technologies, Mechanics and Optics and today we are home to 11 125 students blending the culture of innovation and discovery with world-class education. Our focus is on preparing elite scientists, engineers and programmers, notably in the areas of IT, Photonics, Optics, Mechatronics, Robotics, Smart Technologies, Art & Science, and Urban Studies.

ITMO University's mission is to provide opportunities for the holistic development of individuals and to inspire them to tackle global challenges, as the university's philosophy is to be "More than a University".

COMMITIES

General Chair
Pavel Belov, Russia

Technical Program Committee Chair
Alexander Poddubny, Russia

Technical Program Committee
Alexey Akimov, USA
Redha Abdeddaim, France
Georgy Astakhov, Germany
Miguel Beruete, Spain
Nicolas Bonod, France
Christopher Collins, USA
Costantino De Angelis, Italy
Andrey Evlyukhin, Germany
Pavel Ginzburg, Israel
Stanislav Glybovski, Russia
Maxim Gorlach, Russia
Ivan Iorsh, Russia
Ido Kaminer, Israel
Polina Kapitanova, Russia
Alexander Khanikaev, USA
Yuri Kivshar, Australia
Mikhail Lapine, Australia
Daniel Leykam, Korea
Sergey Makarov, Russia
Andrey Miroshnichenko, Australia
Ivan Mukhin, Russia
Alexey Nikitin, Spain
Nikita Olekhno, Russia
Alexander Petrov, Germany
Alexander Poshakinsky, Russia
Carsten Rockstuhl, Germany
Andrey Rogach, China
Charles Roques-Carmes, USA
Alexander Shalin, Russia
Alena Shchelokova, Russia
Maxim Shcherbakov, USA
Timur Shegai, Sweden
Alexandra Sheremet, Russia
Mário Silveirinha, Portugal
Andrei Slavin, Russia
Alexey Slobozhanyuk, Russia
Mingzhao Song, Russia
Isabelle Staude, Germany
Nicolas Stenger, Denmark
Anton Souslov, UK
Andreas Tittl, Germany
Sergei Tretyakov, Finland
Valentyn Volkov, Denmark

Journal of Physics: Conference Series **2015** (2021) 011001 doi:10.1088/1742-6596/2015/1/011001

Andrew Webb, the Netherlands
Anvar Zakhidov, USA
Pavel Zemanek, Czech Republic
George Zograf, Russia
Dmitry Zuev, Russia

Conference Secretary
Anastasia Kaptsova, Russia

PR Manager
Julia Kuznetsova, Russia

Peer Review Declaration Form Guidance

On the following page you will find the declaration form.

- Please answer each question.
- You should submit the form along with the rest of your submission files.
- The deadline is the submission date written in your publishing agreement.

All conference organisers/editors are required to declare details about their peer review.

We will published the information you provide as part of your proceedings.

Journal of Physics: Conference Series **2015** (2021) 011002 doi:10.1088/1742-6596/2015/1/011002

Peer review declaration

All papers published in this volume of Journal of Physics: Conference Series have been peer reviewed through processes administered by the Editors. Reviews were conducted by expert referees to the professional and scientific standards expected of a proceedings journal published by IOP Publishing.

- **Type of peer review: Single-blind / Double-blind / Triple-blind / Open / Other (please describe)**

All submitted papers will be reviewed by at least two independent reviewers, who are encouraged to give constructive comments and suggestions to the authors. The review is single-blind – the author doesn't know the names of the reviewers assigned to the review of the paper. The authors of submitted papers will receive the review comments and recommendations and will have a possibility to amend their papers based on the review comments and suggestions and upload the revised versions. If both reviewers indicate that the paper can't be published then it will not be accepted for publication, however, the talk might be delivered at the Conference if the reviewers specify that in the review.

The publication of the papers in the Conference Proceedings is possible only in case the talk is presented at the Conference.

- **Conference submission management system:**

Self-made submission management system

- **Number of submissions received:**

Number of full papers submitted for publication in IOP Conference Proceedings - 232

- **Number of submissions sent for review:**

232

- **Number of submissions accepted:**

172

- **Acceptance Rate (Number of Submissions Accepted / Number of Submissions Received X 100):**

74.13%

Journal of Physics: Conference Series **2015** (2021) 011002 doi:10.1088/1742-6596/2015/1/011002

- **Average number of reviews per paper:**

1.9

- **Total number of reviewers involved:**

48

- **Any additional info on review process:**

-

- **Contact person for queries:**
Name : Anastasia Kaptsova
Affiliation:ITMO University, Russia
Email :anastasia.kaptsova@metalab.ifmo.ru

Focusing grating couplers for radio-frequency surface ion traps

Timur Abbasov[1], Ivan Kazakov[1], Ivan Sherstov[1], Sergey Kontorov[1], Arkadi Shipulin[1], Franko Kueppers[1] and Kirill Lakhmanskiy[2]

[1] Skolkovo Institute of Science and Technology, 121205 Moscow, Russia

[2] Russian Quantum Center, 121205 Moscow, Russia

E-mail: timur.abbasov@skoltech.ru

Abstract. We present a photonic integrated circuit design with multiple focusing grating couplers that can be used in a surface ion trap. This system allows transferring laser radiation from different laser sources to the ion trapped 240 μm above the surface for further manipulations.

1. Introduction

Grating couplers are widely used in photonic integrated circuits for coupling laser light between optic fibers and waveguides [1–3]. Here we demonstrate a novel grating coupler system that allows laser light transfer from a radio-frequency (RF) ion trap surface to the stored ion.

Typically, in RF surface ion traps, electrodes are located on one surface, while stored ion is held above this surface at some altitude [4]. Such ion traps can be used for different applications: from frequency standards to quantum computers [4,5]. Laser light is used for manipulations with the ion energy levels.

2. Methods

The grating coupler system was calculated using the Wave Optics module of COMSOL Multiphysics to be used with the $^{171}Yb^{+}$ ion trap, where an ion is held 240 μm above the surface of electrodes. As can be clearly seen from figure 1, the grating couplers are located around the electrode and focus the laser light on the stored ion.

The laser light propagates through the optical waveguide to the grating couplers, where it goes out and focuses on the ion. Five lasers with different wavelengths (369, 399, 435, 745, and 935 nm) are used for manipulations with the $^{171}Yb^{+}$ ion [6]. Silicon nitride (Si_3N_4) is used as a material for integrated optical waveguides and grating couplers due to its ability to propagate light in a wide optical range and low losses.

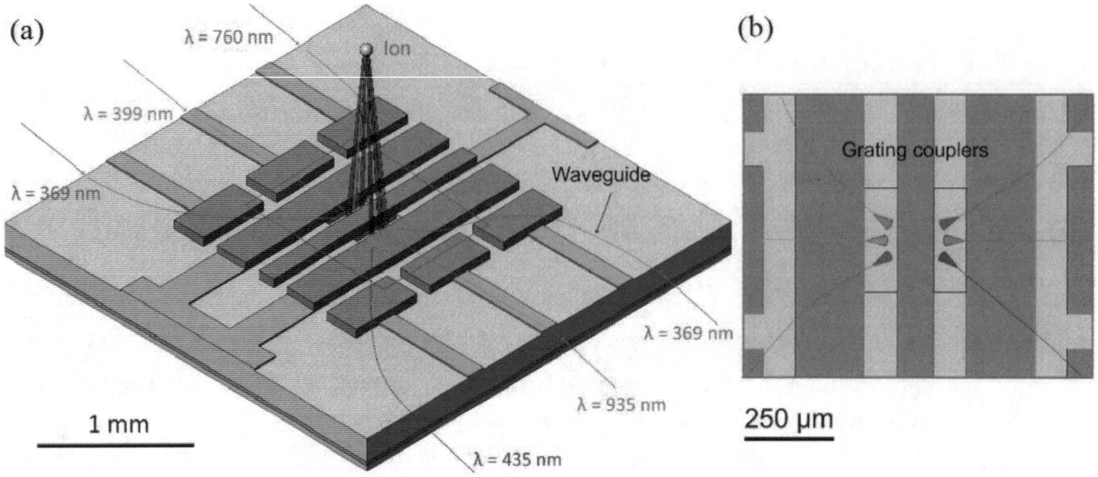

Figure 1. Schematic representation of the RF surface trap with grating couplers designed to transfer laser light from grating couplers to the ion (a), the grating region near the central electrode (b).

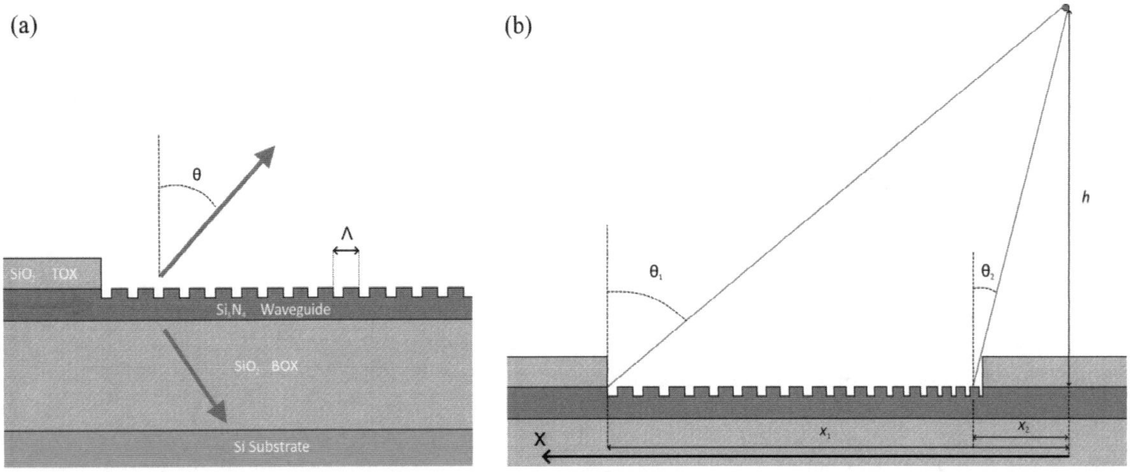

Figure 2. Schematic picture of a grating coupler (a) and its focusing configuration (b).

The main objective of calculating a focusing grating coupler for an ion trap is to create the smallest possible waists of the laser beams for all required wavelengths at an altitude of the stored ion. The focusing configuration of a grating coupler was obtained by changing an angle θ of emitted light along the grating coupler to focus it into one spot, as shown in figure 2. Bragg condition for an angle θ can be written as $n_{eff} - n_1 \cdot \sin(\theta) = {}^{\lambda}/_{\Lambda}$, where n_{eff} is an effective refractive index of a grating, n_1 is a refractive index of the medium above the grating, λ is a wavelength of the laser light and Λ is a grating period. It also can be rewritten for the coordinate x as $\Lambda(x) = \frac{\lambda}{n_{eff} - \frac{x}{\sqrt{x^2+h^2}}}$, x and h are given in figure 2. It also can be linearly approximated through Λ_{n+1} and Λ_n: $\Lambda_{n+1} = k \cdot \Lambda_n + c$, where Λ_n is the n-th period of the grating, k and c are coefficients of the linear approximation. Therefore, we need to find coefficients k and c, which give us the smallest waists for the studied wavelengths.

3. Results and Discussion

Optimized values of *k*, *c*, corresponding laser light and grating parameters are presented in table 1. We also calculated "coupling efficiency" for each grating coupler as a ratio of the power in the focusing spot to the initial power of the laser light in the waveguide before the grating coupler. Power is calculated by the standard COMSOL functions by integrating the Poynting vector over the area around the waveguide and focusing spot.

Typical wavefront and beam profile near the focus are presented in figure 3. The wavefront near the focus is flat and the beam profile is almost Gauss-shaped. It makes possible to conclude that we created a beam waist in the desired area. The waist of the beam is calculated on the power level $1/e^2$.

Table 1. Results of focusing grating couplers optimizations.

Wavelength (λ), nm	k	c, nm	First grating period (Λ_1), nm	Waist of the beam (w), μm	Coupling efficiency
369	0,99972	1,9e-4	230	5	0.25
399	0,99941	1,63e-6	260	6	0.31
435	0,99954	5,7e-5	278	8	0.28
760	0,99966	2,1e-3	515	10	0.31
935	0,99983	7e-6	614	10	0.3

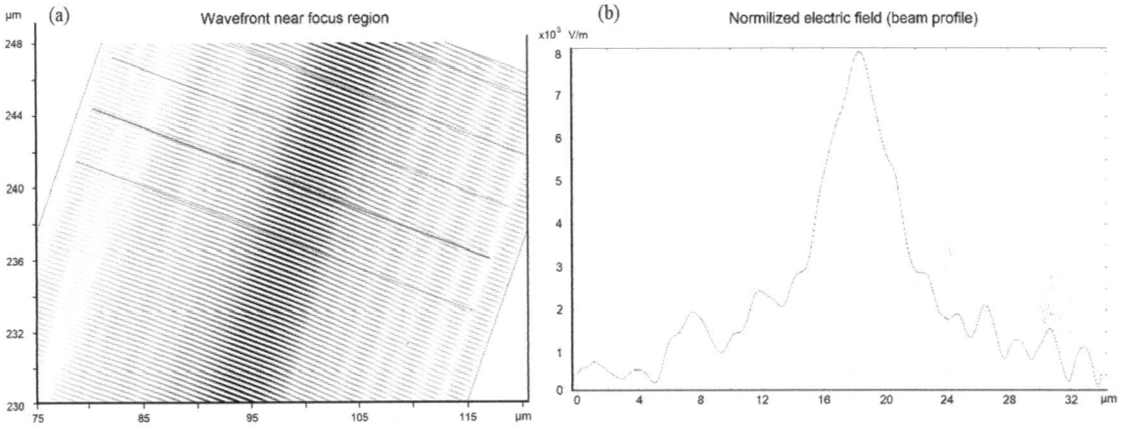

Figure 3. An example of simulated grating coupler: wavefront near focus (a) and beam profile in the focus (b). The red line represents a cut plane of the diffracted beam on the focus. Presented figures are obtained from an optimized 935 nm grating coupler.

Transverse focusing (perpendicular to the plane in figure 2) could be performed by adding curvature to the grating stripes. Such a process is described in [7]. Calculations and optimizations in the third dimension require a lot of computing power or require new approaches since the focusing region is more than 200 μm above the surface. It will be done as the next step in our research.

4. Conclusion

In conclusion, we have designed and optimized a waveguide system with the focusing grating couplers for usage in the RF surface ion trap, where the ion is held 240 μm above the surface of electrodes. This system allows delivering laser light within a range of wavelengths from 369 nm to 935 nm, with the beams waists in focus in a range from 5 to 10 μm for every wavelength. Coupling efficiency for all grating couplers is higher than 25%.

References

[1] Marchetti R, Lacava C, Carroll L, Gradkowski K and Minzioni P 2019 Coupling strategies for silicon photonics integrated chips [Invited] *Photonics Res.* **7** 201

[2] Vermeulen D and Poulton C V 2018 Optical Interfaces for Silicon Photonic Circuits *Proc. IEEE Inst. Electr. Electron. Eng.* **106** 2270–80

[3] Cheng L, Mao S, Li Z, Han Y and Fu H Y 2020 Grating couplers on silicon photonics: Design principles, emerging trends and practical issues *Micromachines (Basel)* **11** 666

[4] Niedermayr D 2015 *Cryogenic surface ion traps* (Innsbruck: University of Innsbruck)

[5] Brown K R, Kim J and Monroe C 2016 Co-designing a scalable quantum computer with trapped atomic ions *Npj Quantum Inf.* **2**

[6] Huntemann N 2014 *High-Accuracy Optical Clock Based on the Octupole transition in 171Yb+* (Hannover: University of Hannover)

[7] Mehta K K 2017 *Integrated optical quantum manipulation and measurement of trapped ions* (Cambridge: Massachusetts Institute of Technology)

Investigation of directivity of the nanoantenna by its inherent resonant states

Rajan Agrahari[1], Hadi K. Shamkhi[2]

[1] Electronics and Communication Engineering Department, National Institute of Technology, Patna 800005 India
[2] ITMO University, Kronverksky 49, 197101, St. Petersburg, Russia

rajan.ec@nitp.ac.in

Abstract. We present a quasinormal mode (QNM) approach for modeling the nanoantenna and describe the response of localized dielectric cylinder resonators. The inherent resonant states of the dielectric cylinder nanocavity are investigated for modified and reduced geometrical symmetry. We find some modes contributing mainly to the directivity and have a high-quality factor. The variation of the eigenmodes with cylinder height and substrate refractive index has also been investigated.

Nanoantennas are of great importance for the development of wireless communication systems for future photonic chips [1,2] as semiconductor chips are currently limited by the low data transfer rate due to high temperature rising [3]. As the electromagnetic fields of surface-plasmon-polariton (SPP) waves are highly localized to dielectric/metal interfaces, SPP-wave-based photonic chips can overcome the limitation [4]. However, a metal nanostructure possesses high ohmic losses which limit the propagation distance [5]. Therefore, integrating all-dielectric optical nanoantennas in a photonic chip is one of the most promising for the integrated platform.

All-dielectric nanoantennas usually consist of several dielectric nanoparticles of various shapes, e.g., nanospheres [6], nanoblocks [7], nanodisks [8], nanocubes [9], and V-shaped nanoparticles [10]. Forward and backward scattering characteristics of individual nanoparticles can be controlled by engineering the electric and magnetic resonances [11-13]. A dielectric cylinder resonator has been studied extensively both theoretically and experimentally to investigate its optical properties. Several studies examine the multipole decomposition of the dielectric cylinder [14–16] and its effect on optical emission. Cylinders can be optimized to enhance scattering by constructively overlapping several multipoles [17], to transfer power from lower-order multipoles to higher ones, and reproduce the quasi-bound state in the continuum (BICs) [18], or to suppress scattering as traditionally known as the anapole state [19]. However, few studies only analyze the inherent resonant states of a dielectric cylinder nanocavity and its influence on the directivity in order to obtain superdirective nanoantenna.

With this motivation, we investigate the inherent resonant states of a dielectric cylinder nanocavity by employing the quasinormal modes (QNMs) method and reveal how the QNMs contribute to the directivity. First, the effect of the height of the nanocylinder and refractive index of the substrate on the complex eigenfrequency will be determined together with the contributions of major modes to the directivity. Next, we determine how the directivity of a Si-cylinder nanoantenna is tailored when the nanostructure height is varying. Finally, we demonstrate the nanoantenna with substrate providing very high directivity.

1. Inherent resonant states of a dielectric nanocylinder

We consider a geometry consisting of a Si-dielectric cylinder ($n = 3.6$) with height h_{cyl} and radius r_{cyl}. For all the numerical results presented hereinafter, we choose the radius of the cylinder $r_{cyl} = 100$ nm and the cylinder is shined by x-polarized light.

We identify the inherent resonant mode of the dielectric nanocylinder by using the quasinormal mode (QNM) expansion. The natural response of resonance modes of the plasmonic and dielectric resonators is called QNMs. The QNMs are the time-harmonic solution of Maxwell's equations. It is used to describe the resonances of the non-conservative system which is not possible by normal modes. The eigenfrequency ω_m of the QNMs is complex with the quality factor $Q_m = \frac{1}{2}\ \mathrm{Re}\,(\omega_m)/\mathrm{Im}\,(\omega_m)$. The electric field E_m and eigenfrequency ω_m of the mth QNM are determined by COMSOL Multiphysics with its eigenmode solver [20,21]. Once the E_m and ω_m are known, we determine the scattered field $E_s(r,\omega)$ due to interaction of light with dielectric cylinder resonator by the QNM expansion of the scattered field, i.e.

$$E_s(r,\omega) = \sum_m \alpha_m(\omega)\ E_m(r),$$

where $\alpha_m(\omega)$ is the excitation coefficient, which depends on the incident fields. More details to determine the $\alpha_m(\omega)$ are given elsewhere [21].

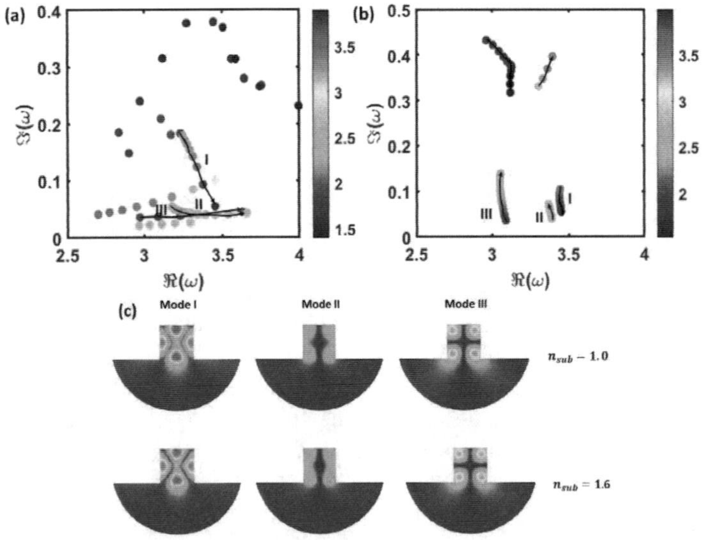

Figure 1. Variation of the quasinormal modes (QNMs) of a Si-cylinder nanoparticle with; (a) the height of the cylinder $h_{cyl} \in [100, 400]$ nm and (b) refractive index of the substrate $n_{sub} \in [1.0, 4.0]$; when the radius $r_{cyl} = 100$ nm. Arrows show how the complex eigenfrequency moves with the increase in h_{cyl} and n_{sub}. The colors of the dots show the directivity of the cylinder nanoantenna at that QNMs frequency. Three main calculated modes that contribute most in the scattering are numbered. Other modes (blue dots) are background modes that do not contribute to the scattering and directivity; (c) E-field profile inside the cylinder for three most relevant mode when $h_{cyl} = 200$ nm and $n_{sub} = \{1, 1.6\}$.

Figures 1 (a) and (b) show the variation of the QNM modes eigenfrequencies with cylinder height (h_{cyl}) and refractive index (n_{sub}) of the substrate, respectively. Arrows in the figures describe how the eigenmodes change with an increase in h_{cyl} and n_{sub}. Further, the color of the dots in the figures shows the directivity of the nanoantenna at that eigenfrequency. It can be observed from Fig. 1(a), as the height of the cylinders increases, Re (ω_m) increases, i.e. Q_m increases too, which enables the electromagnetic fields are trapped for sufficiently long times in a small space. Similarly, Q_m decreases when n_{sub} increases (see Fig. 1(b)). Furthermore, Modes I, II, and III have high directivity and contribute mostly to the directional scattering. Other modes (with blue dots) do not significantly contribute to the directivity and we call them background modes. Figure 1(b) also shows the directivity of the nanoresonator increases, when n_{sub} increases. The electric field profile inside the cylinder for Mode I, II and III are shown in Fig. 1(c) when $h_{cyl} = 200$ nm and $n_{sub} = \{1,1.5\}$.

The directivity of the Si-cylinder nanoantenna is investigated numerically based on the COMSOL by the finite-element method. Figure 2(a) shows the directivity of the nanoantenna in the visible spectrum $\lambda \in [450, 800]$ nm, when the height of the cylinder varies in the range $h_{cyl} \in [140, 400]$ nm. We observe that high directivity is obtained when the height of the cylinder is also high. Maximum directivity 6.1 is obtained at the wavelength $\lambda = 450$ nm when $h_{cyl} = 400$ nm. Further, when we introduce a substrate, indirectly we are reducing the structure overall symmetry, which results in higher directivity as can be seen by Fig. 2(b). By introducing the asymmetricity, significant higher modes are excited (as can be seen from Fig. 1(b)), which, in its turn, results in higher directivity of scattering. In Fig. 2(b) it is clearly seen that directivity increases sharply when the nanocylinder is placed on a substrate.

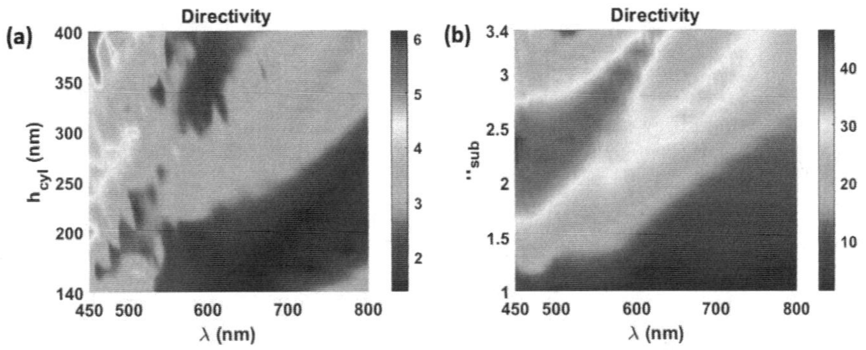

Figure 2. The dependency of the directivity on wavelength and (a) the height of the cylinder (h_{cyl}); (b) substrate refractive index (n_{sub}); h_{cyl} is fixed to be 200 nm.

2. Conclusion

In summary, we have investigated the directivity of all-dielectric nanoantenna with its inherent resonant states by using the QNMs method. We demonstrated that the directivity of a nanoantenna can be controlled by the height of the cylinder and the refractive index of the substrate. The strong enhancement of the directivity has been observed when we break the symmetry by introducing the substrate. We showed directivity and eigenmodes frequency evolution of the dielectric nanoantenna by QNMs expansion method and found a directivity with the increase of the substrate refractive index. We showed three different QNMs contributing most to the directivity and other modes called background modes do not contribute.

References

[1] Dong, P, Chen, Y K, Duan, G H and Neilson, D T 2014 *Nanophotonics* **3** 215-228.
[2] Bogaerts, W and Chrostowski, L 2018 *Laser Photonics Rev.* **12** 1700237.

[3] Ozbay, E, 2006 *Science* **311** 189-193.
[4] Agrahari, R, Lakhtakia, A and Jain, P K 2018 *J. Appl. Phys.* **124** 053104.
[5] Kostina, N, Petrov, M, Ivinskaya, A, Sukhov, S, Bogdanov, A, Toftul, I, Nieto-Vesperinas, M, Ginzburg, P and Shalin, A 2019 *Phys. Rev. B* 99 125416
[6] Fu, Y H, Kuznetsov, A I, Miroshnichenko, A E, Yu, Y F and Luk'yanchuk, B 2013 *Nature communications* **4** 1-6.
[7] Zhang, T, Li, X, Xu, J, Zhang, X, Deng, Z L and Li, X 2020 *Nanomaterials* **10** 1242.
[8] Staude, I, Miroshnichenko, A E, Decker, M, Fofang, N T, Liu, S, Gonzales, E, Dominguez, J, Luk, T S, Neshev, D N, Brener, I and Kivshar, Y 2013 *ACS nano* **7** 7824-32.
[9] Sikdar, D, Cheng, W and Premaratne, M 2015 *J. Appl. Phys.* **117** 083101.
[10] Li, J, Verellen, N, Vercruysse, D, Bearda, T, Lagae, L and Van Dorpe, P 2016 *Nano letters* **16** 4396-03.
[11] Kozlov, V, Filonov, D, Shalin, A S, Steinberg, B Z and Ginzburg, P 2016 *Appl. Phys. Lett.* **109** 203503.
[12] Terekhov, P D, Shamkhi, H K, Gurvitz, E A, Baryshnikova, K V, Evlyukhin, A B, Shalin, A S and Karabchevsky, A 2019 *Opt. express* **27** 10924-35.
[13] Terekhov, P D, Baryshnikova, K V, Greenberg, Y, Fu, Y H, Evlyukhin, A B, Shalin, A S and Karabchevsky, A 2019 *Sci. Rep.* **9** 1-9.
[14] Shamkhi, H K , Baryshnikova, K V, Sayanskiy, A, Kapitanova, P, Terekhov, P D, Belov, P, Karabchevsky, A, Evlyukhin, A B, Kivshar, Y and Shalin, A S 2019 *Phys. Rev. Lett.* **122** 193905.
[15] Shamkhi, H K, Sayanskiy, A, Valero, A C, Kupriianov, A S, Kapitanova, P, Kivshar, Y S, Shalin, A S and Tuz, V R 2019 *Phys. Rev. Mater.* **3** 085201.
[16] Gurvitz, E A, Ladutenko, K S, Dergachev, P A, Evlyukhin, A B, Miroshnichenko, A E and Shalin, A S 2019 *Laser Photonics Rev.* **13** 1800266.
[17] Shcherbinin, V I, Fesenko, V I, Tkachova, T I and Tuz, V R 2020 *Phys. Rev. Appl.* **13** 024081.
[18] Rybin, M V, Koshelev, K L, Sadrieva, Z F, Samusev, K B, Bogdanov, A A, Limonov, M F and Kivshar, Y S 2017 *Phys. Rev. Lett.* **119** 243901.
[19] Baryshnikova, K, Filonov, D, Simovski, C, Evlyukhin, A, Kadochkin, A, Nenasheva, E, Ginzburg, P and Shalin, A S 2018 *Phys. Rev. B* **98** 165419.
[20] Yan, W, Faggiani, R and Lalanne, P 2018 *Phys. Rev. B* **97** 205422.
[21] Lalanne, P, Yan, W, Vynck, K, Sauvan, C and Hugonin, J P 2018 *Laser Photonics Rev.* **12** 1700113.

Highly Directive All-Dielectric Nanoantenna

Rajan Agrahari[1], Hadi K. Shamkhi[2]

[1] Electronics and Communication Engineering Department, National Institute of Technology, Patna 800005, India
[2] ITMO University, Kronverksky 49, 197101, St. Petersburg, Russia

rajan.ec@nitp.ac.in

Abstract. A highly directive dielectric nanoantenna in an integrated chip may enable faster communication as their low losses and small size overcome the limitation of temperature enhancement and low data transfer rate. We optimize nanoantenna consist of Si-nanoblock in the near-infrared region to efficiently transfer a point dipole light to a highly directive light in the far-field region. We engineer the intrinsic electric and magnetic resonances of a Si-block nanoantenna by modifying and reducing its geometrical symmetry. We realize a pronounced enhancement of directivity by systematically inducing perturbation in the Silicon block so that both its reflection and rotational symmetries are broken. Finally, we retain the traditional method to increase resonance's coupling to outer space by introducing substrate with an increasing refractive index. We find that the directivity has boosted rapidly.

The study of the nanoantenna has been one of the hottest areas in the optical range due to their application in integrated optoelectronic semiconductor chips [1,2]. Currently, semiconductor chips are limited by high enhancement of temperature which results in a low data transfer rate [3]. The surface-plasmon-polariton wave can overcome the limitation as they are highly localized near the metal/dielectric interface [4]. However, a metal nanostructure possesses high ohmic losses which limit the propagation distance [5]. Therefore, integrating the all-dielectric optical nanoantenna in a semiconductor chip is one of the most promising for the integrated chip. It will not only increase the data transfer rate but also reduce the size of a chip.

A nanoantenna has the ability to bridge the size and impedance mismatch between nano-emitters and free-space radiation. It must have high radiation efficiency and be highly directive for wireless optical communication in photonic chips. But, existing optical nanoantennas usually do not radiate efficiently and have low directivity if they are far from their self-resonance. An all-dielectric nanoantenna with a high refractive index has the ability to excite both electric and magnetic dipole resonances, thus adding an extra degree of freedom to tailor those resonances at a certain wavelength [6]. Such all-dielectric nanoantenna could have high radiation efficiency with low dissipative losses and high magnetic response. Forward and backward scattering characteristics of individual nanoparticles could be controlled by engineering the electric and magnetic resonance [7-11].

Because of the large refractive index and negligible absorption losses in the visible spectrum region, dielectric nanoantenna (Si or Ge-nanoparticle) has been investigated for various shapes [12-15]. Over the last few decades, several all-dielectric metasurfaces have been designed to manipulate the light for possible applications in polarization, absorptance, optical nanoantenna, transmission, and many others [16-21]. In order to achieve high directive, several studies combined plasmonic and dielectric

Journal of Physics: Conference Series **2015** (2021) 012003 doi:10.1088/1742-6596/2015/1/012003

nanostructures to use their respective advantages [22,23]. However, high directive all-dielectric nanoantenna excited by point dipole in the near-infrared region was not much explored.

With this motivation, here we study the directivity of an all-dielectric nanoantenna in the near-infrared region. First, we demonstrate how the directivity of a Si-block nanoantenna is tailored when the nanostructure is excited by the point dipole. Next, we reveal that directivity of the such type of nanoantenna can be controlled by engineering symmetry breaking, as well as making a cluster of them. Finally, we demonstrate the nanoantenna with substrate providing very high directivity. Our main goal is to show how one can change the directivity of nanoantenna by engineering its structure.

1. High Directive Nanoantenna

The Directivity of an antenna is a measure of the concentration of the maximum radiated power in a particular direction (θ, ϕ), that is, $D_{max} = (4\pi U(\theta, \phi)_{max})/P_{rad}$, where θ and ϕ are the polar and azimuthal angles in a spherical coordinate system, respectively. $U(\theta, \phi)_{max}$ is the maximum radiation intensity (power per unit solid angle) in (θ, ϕ) direction and P_{rad} is the total radiated power. The directivity is investigated numerically based on the COMSOL by the finite-element method for the geometry shown in Fig. 1. We consider several dielectric nanoantennas shown schematically in Figure 1a—c, composed of Si-nanoblock ($n = 3.5$). For all numerical results presented here, we choose $l =$ 380 nm, which is much smaller than the operating wavelength.

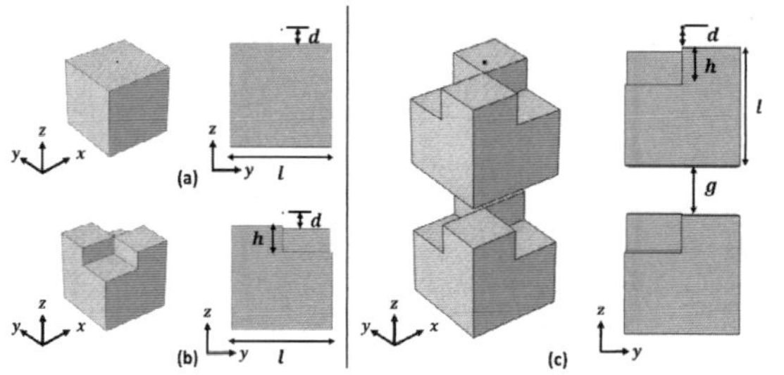

Figure 1: Perspective and side view of the nanoantenna composed of Si and excited by the point dipole which is placed at a distance d from the top surface of the nanostructure: (a) consist of a nanoblock of sides l; (b) an asymmetric block created by removing the bricks of size $l/2 \times l/2 \times h$ from the block top surface; (c) cluster of asymmetric blocks having gap g from each other.

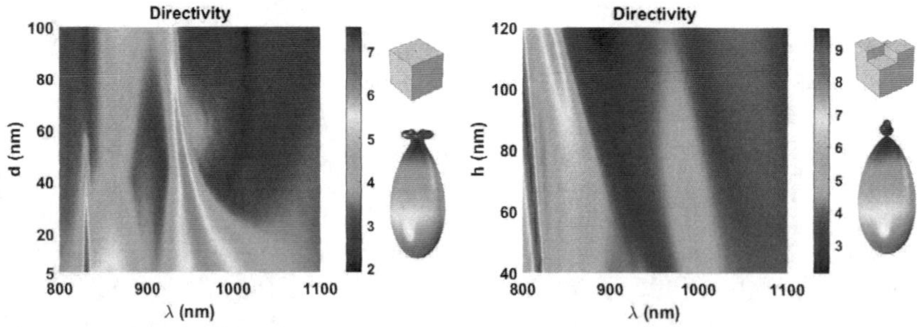

Figure 2: The dependency of the directivity on wavelength and (a) the distance between the dipole source and top surface of the block; (b) the height of the bricks (h) removed from the top surface of the block. The three-dimension radiation pattern for maximum directivity is also shown.

In Figure 2, we study the directivity of the nanoantenna for the schematic shown in Fig. 1(a) and (b). It can be seen in Fig. 2(a) that the directivity of the nanoantenna can be tailored by the position of the dipole (d). Higher directivity is obtained when the dipole is placed near the nanoantenna. Maximum directivity $D_{max} = 7.7$ is observed at a wavelength $\lambda = 830$ nm when $d = 5$ nm. Now, we break the symmetry of the block by removing the bricks of size $l/2 \times l/2 \times h$ from the top surface of the block as shown in Fig. 1(b). The dependency of the directivity on the height of the brick (h) is shown in Fig. 2(b). Maximum directivity $D_{max} = 9.6$ is achieved at a wavelength $\lambda = 805$ nm when $h = 85$ nm. The radiation pattern corresponds to $D_{max} = 7.7$ and 9.6 are also shown in Fig. 2(a) and (b), respectively.

Here, the directivity is higher than the directivity of the point dipole (1.5) because when a local source (electric or magnetic emitter) is placed close to the antenna, the near-field coupling between the emitter and the dipole resonance (ED and MD) inside the antenna offers the highly unidirectional emission. It is important to note that, a higher directivity is obtained when we introduced the asymmetricity in the block. By introducing the asymmetricity, significant higher-order multipole is excited which results in more contribution in scattering.

Next, we assume that nanoantenna showed in Fig.1 (a) and (b) are placed on a substrate (n_{sub}). Figure 3 shows the variation of directivity with the refractive index of the substrate n_{sub}. As observed from Fig. 3, directivity increases sharply when the nanoblock is placed on the substrate which is due to the coupling of the light into the substrate when the nanoblock embedded on a substrate and excited by the local source such as dipole emitters. The coupling mechanism plays an important role in achieving directional light scattering and emission using nanoantenna.

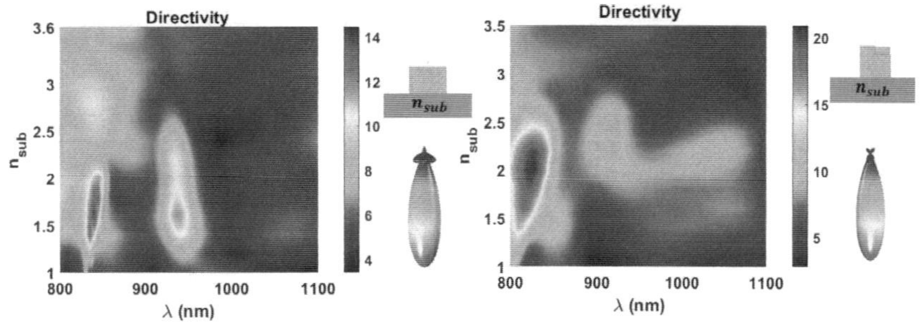

Figure 3. Directivity of the nanoantenna tailored by the refractive index of the substrate: (a) simple block on substrate leads to maximum directivity of 14.4 with $n_{sub} = 1.6$ at $\lambda = 840$ nm when $d = 5$ nm; (b) asymmetric block on substrate gives maximum directivity of 20.8 with $n_{sub} = 2.0$ at $\lambda = 820$ nm when $d = 5$ nm and $h = 85$ nm. The three-dimension radiation pattern for maximum directivity is also shown.

Finally, we demonstrate how the directivity is affected by the cluster of the nanoblock. We consider an arrangement of two asymmetric blocks having a gap g excited by the point dipole (see Fig. 1(c)). The variation of the directivity with the wavelength and the gap g is shown in Fig. 4. While comparing Fig. 2(b) and Fig. 4, we found higher directivity for the cluster of nanoblocks due to the enhance in the amplitude of higher-order multipole moments, which is not achievable with a single nanoblock. When the cluster of blocks is placed at the optimal distance, the external field is so inhomogeneous that the magnetic and electric dipole components of the polarization current in the cylinder are no longer dominant, as they would be for the case of a single block.

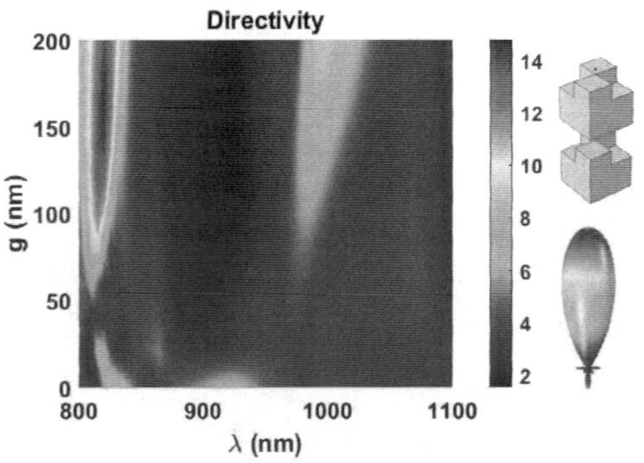

Figure 4. Directivity of the nanoantenna tailored by the gap between the two asymmetric blocks (cluster). Maximum directivity of 14.7 is observed with g=160 nm at λ=820 nm when d=5 nm, and h=120 nm. The three-dimension radiation pattern for maximum directivity is also shown.

2. Conclusions

In summary, we have studied the high directive all-dielectric nanoantenna in the near-infrared region for several nanoparticles consist of Si-nanoblock. We demonstrated that the directivity of the nanoantenna can be controlled by the position of the point dipole and it was maximum when the dipole was placed near to the nanoblock. An enhancement in the directivity has been observed when we break the symmetry by removing the brick shape from the Si-nanoblock. The directivity of the nanoantenna has been further increased due to the coupling of light into the substrate when the nanoblock was embedded on the substrate. Finally, to observed the effect of a cluster on directivity, we considered two asymmetric nanoblock. We showed that the cluster of the nanoblock enhances the directivity by increasing the amplitude of the higher-order multipole moment.

References

[1] Dong, P, Chen, Y K, Duan, G H and Neilson, D T 2014 *Nanophotonics* **3** 215-228.
[2] Bogaerts, W and Chrostowski, L 2018 *Laser Photonics Rev.* **12** 1700237.
[3] Ozbay, E, 2006 *Science* **311** 189-193.
[4] Agrahari, R, Lakhtakia, A and Jain, P K 2019 *Sci. Rep.* **9** 1-9.
[5] Kostina, N, Petrov, M, Ivinskaya, A, Sukhov, S, Bogdanov, A, Toftul, I, Nieto-Vesperinas, M, Ginzburg, P and Shalin, A 2019 *Phys. Rev. B* **99** 125416.
[6] Evlyukhin, A B, Fischer, T, Reinhardt, C and Chichkov, B N 2016 *Phys. Rev. B* **94** 205434.
[7] Baryshnikova, K, Filonov, D, Simovski, C, Evlyukhin, A, Kadochkin, A, Nenasheva, E, Ginzburg, P and Shalin, A S 2018 *Phys. Rev. B* **98** 165419.
[8] Kozlov, V, Filonov, D, Shalin, A S, Steinberg, B Z and Ginzburg, P 2016 *Appl. Phys. Lett.* **109** 203503.
[9] Canós Valero, A, Kislov, D, Gurvitz, E A, Shamkhi, H K, Pavlov, A A, Redka, D, Yankin, S, Zemánek, P and Shalin, A S 2020 *Adv. Sci.* **7** 1903049.
[10] Barhom, H, Machnev, A A, Noskov, R E, Goncharenko, A, Gurvitz, E A, Timin, A S, Shkoldin, V A, Koniakhin, S V, Koval, O Y, Zyuzin, M V and Shalin, A S 2019 *Nano Lett.* **19** 7062-71.
[11] Kozlov, V, Filonov, D, Shalin, A S, Steinberg, B Z and Ginzburg, P 2016 Appl. Phys. Lett. 109 203503.
[12] Krasnok, A E, Simovski, C R, Belov, P A and Kivshar, Y S 2014 *Nanoscale* **6** 7354-61.
[13] Ee, H S, Kang, J H, Brongersma, M L and Seo, M K 2015 *Nano Lett.* **15**, 1759-65.
[14] Moitra, P, Slovick, B A, Gang Yu, Z, Krishnamurthy, S and Valentine, J 2014 *Appl. Phys. Lett.* **104** 171102.
[15] Forouzmand, A and Mosallaei, H 2017 *Adv. Opt. Mater.* **5** 1700147.
[16] Kruk, S, Hopkins, B, Kravchenko, I I, Miroshnichenko, A, Neshev, D N and Kivshar, Y S 2016 *APL Photonics* **1** 030801.
[17] Terekhov, P D, Baryshnikova, K V, Greenberg, Y, Fu, Y H, Evlyukhin, A B, Shalin, A S and Karabchevsky, A 2019 *Sci. Rep.* **9** 1-9.
[18] Terekhov, P D, Shamkhi, H K, Gurvitz, E A, Baryshnikova, K V, Evlyukhin, A B, Shalin, A S and Karabchevsky, A 2019 *Opt. Express* **27** 10924-35.
[19] Fu, Y H, Kuznetsov, A I, Miroshnichenko, A E, Yu, Y F and Luk'yanchuk, B 2013 *Nat. Commun.* **4** 1-6.
[20] Shalin, A S 2010 *JETP Lett.* **91** 636-642.
[21] Zhao, W, Jiang, H, Liu, B, Song, J, Jiang, Y, Tang, C and Li, J 2016 *Sci. Rep.* **6** 1-7.
[22] Sugimoto, H, Hinamoto, T and Fujii, M 2019 *Adv. Opt. Mater.* **7** 1900591.
[23] Rusak, E, Staude, I, Decker, M, Sautter, J, Miroshnichenko, A E, Powell, D A, Neshev, D N and Kivshar, Y S 2014 *Appl. Phys. Lett.* **105** 221109.

Effect of the lattice mismatch on the efficiency of the GaAs nanowire/Si substrate solar cell

P A Alekseev[1,*], V A Sharov[1,2], B R Borodin[1], M S Dunaevskiy[1], R R Reznik[2], and G E Cirlin[2]

[1]Ioffe Institute, Saint-Petersburg, 194021, Russia

[2]Alferov University, Saint-Petersburg 194021, Russia

E-mail: *npoxep@gmail.com

Abstract. The effect of the crystal lattice mismatch between single p-GaAs nanowire grown on p-Si substrate on the solar cell efficiency is studied. The study is performed by measuring the I-V curves under red (wavelength=650 nm) laser illumination. The measurement of the single nanowire was done by conductive atomic force microscopy (C-AFM). The measured curve was reproduced by numerical simulations accounting piezoresistance and piezoelectric effects. The analysis demonstrated the presence of the tensile (2%) zinc blend insert at the interface between nanowire and substrate induced by crystal lattices mismatch. Strained insert at the interface changes the polarity of the photogenerated current and increases the efficiency by 2 times.

1. Introduction

GaAs nanowires (NWs) grown on low-cost lattice-mismatched substrates provide a possibility of creating solar cells with high absorption of solar light with low consumption of GaAs material[1,2]. To increase the efficiency of the cells, additional approaches are demanded. Tandem solar cell consisting of a silicon substrate and GaAs nanowires array[3,4] is among these approaches currently being investigated.

Recently it was shown that growth of the GaAs NWs on the (111) Si substrate leads to the formation of the interface energy barrier which affects the solar cell efficiency[5]. The effect of the induced barrier in the numerical model for simplicity has been accounted by adding the fixed positive electric charge to the NW/substrate interface. However, for proper modeling piezoresistance[6] and piezoelectric[7] effects should be considered. The aim of this work was to study the effect of the lattice mismatch on the efficiency of the wurtzite GaAs NW/Si substrate solar cell.

2. Samples and methods

Wurtzite GaAs nanowires were grown in (111)B direction under 20 nm Au nanoparticles on p-Si (111) substrate. The growth was done in a molecular beam epitaxy chamber with Be p-type doping (10^{17}-10^{18} cm^{-3}). The substrate also was p-type doped with an impurity level of 10^{16} cm^{-3}. The length of nanowires was 6 μm and a diameter d=100±20 nm [8]. NWs were passivated by covering a 7-nm AlGaAs layer to reduce the density of surface non-radiative recombination centers. Conventionally, a zinc blend phase forms at the first stage during the growth of the wurtzite GaAs nanowires on the Si substrate [9]. Thus, a zinc blende (ZB) insert with a thickness of ~10 nm was anticipated at the bottom of the NW. Details of the NW growth are presented elsewhere [5].

Measurement of the I-V curves on a single NW was performed by conductive atomic force microscopy (C-AFM)[10]. The experiment was done using NTEGRA AURA (NT-MDT, Russia) scanning probe microscope equipped with conductive HA_NC/W2C + (NT-MDT) probes. For the I–V curves measurement, the grounded probe was contacted to the NW top (see figure 1 (a)) and bias voltage was applied to the second contact created to the silicon substrate. Since the system detecting the probe position is equipped wth a red laser with a wavelength of ~650 nm and a power density of 10^4W/m^2, the NW also was illuminated by this laser during the measurement.

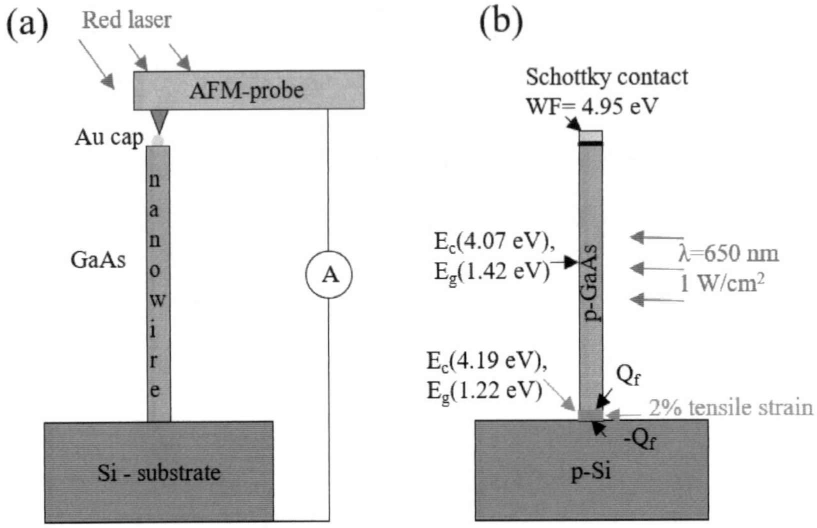

Figure 1. **(a)** Measurement of the I-V curve of single GaAs NW by C-AFM **(b)** Scheme of the positioning of the parameters implemented in the model.

Modeling of the I–V curves under red laser illumination was performed in 2D with cylindrical symmetry using the Atlas Silvaco software package[11]. The model replicated cylindrical GaAs with a diameter of 100 nm and a length of 6 μm placed on the Si substrate. Si substrate in the model had an affinity (conduction band minimum CBM, E_c) of 4.05 eV and a p-type doping level of 10^{16} cm^{-3}. The NW was also p-type doped with a level of 10^{17} cm^{-3}. The bottom contact between the Si substrate and a sample holder was set as Ohmic and the top contact with the NW was a Schottky-type with a barrier height of 0.54 eV [12]. Universal Schottky tunneling model accounting for tunneling across the Schottky barrier, recombination, and thermionic emission was used. The current was calculated accounting band-to-band and surface recombination. The mobility of holes was 0.1 cm^2/(V·s) [13].

Illumination by the red laser with a wavelength of ~650 nm and a power density of 10^4W/m^2 was set with uniform space distribution and normal incident on the side surface of the NW.

Lattice constants mismatch of GaAs and Si (~4%) leads to the compression of the GaAs lattice in the parallel to substrate direction at the NW/substrate interface. However, in perpendicular to the substrate direction, the NW is exposed to the tensile strain which can be estimated as 2%, according to the doubled ratio of the C_{13} and C_{33} GaAs elastic constants. Tensile strain exponentially decreased along the NWs at the length *z* (10-30 nm) [14]. Both piezoresistance and piezoelectric effects caused by lattice mismatch were included in the model during the calculations of I-V curves. The piezoelectric effect induces fixed charges of opposite polarities. These charges ($-Q_f$ and Q_f) were placed at the top and bottom planes of the ZB insert with a *z*-thickness at the NW/substrate interface (figure 1(b)). The polarity of the charges was set according to the tensile deformation in the z-direction of the insert and the (111)B direction of the NW growth [15]. The Q_f charges per 1% strain for the ZB

Journal of Physics: Conference Series **2015** (2021) 012004 doi:10.1088/1742-6596/2015/1/012004

GaAs was $1\cdot10^{12}$ cm^{-2} due to the piezoelectric constant (e_{14} = -0.16 C/m^2 [7]. The piezoresistance effect was included by changing the bandgap E_g=1.22 eV and the CBM E_c =4.19 eV of the insert under tensile strain into the model [6]. CBM of the unstrained ZB and WZ GaAs was 4.07 eV. During the modeling, the thickness z of the insert was varied in the range 3-30 nm to obtain the best fit with the experiment.

3. Results and discussion

I–V curve (black solid) measured on single GaAs NW by C-AFM is presented in figure 2. The bias voltage was applied to the substrate and a negative open-circuit voltage (V_{oc}) is detected. I–V curve calculated by a model which does not include an effect of the crystal lattice mismatch (without strained ZB insert at the NW/substrate interface) is represented as a red dashed line in figure 2. The sign of the open-circuit voltage (V_{oc}) for the red curve is positive. The sign of the open-circuit voltage depends on the polarity of the photocurrent, which is controlled by the balance of the Schottky-barrier and the hetero barrier (NW/substrate). For proper modeling, a strained ZB insert was added to the model. The insert with the thickness z=20 nm gives good agreement between the experimental and calculated I-V curve (black dashed curve). It is worthy to note that the strained ZB insert increases the efficiency of the solar cell by ~2 times (areas under black and red curves). This is due to the low solar efficiency of the Schottky barrier which dominates in the model represented by the red curve. Strain at the NW bottom increases the hetero barrier (Si/GaAs NW) height and changes the sign of the photocurrent and the efficiency of the whole solar cell.

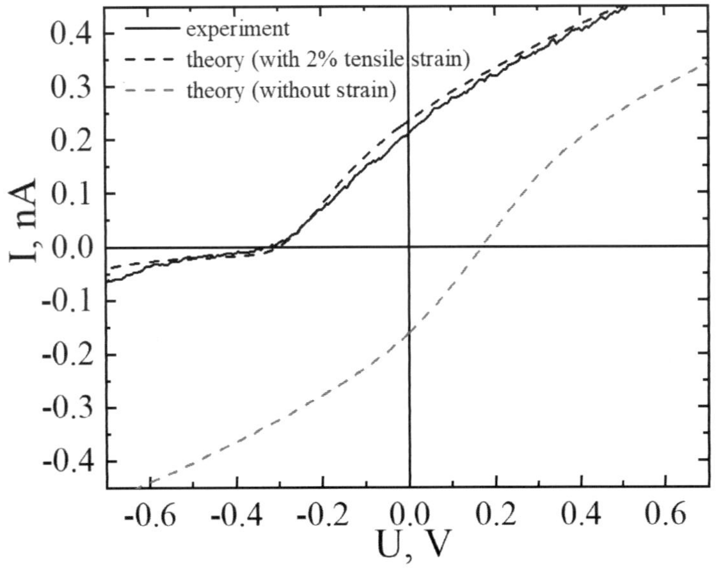

Figure 2. I-V curves of the p-GaAs NW grown on p-Si substrate. Measured curve of the NW (black solid) and calculated curves with accounted tensile strain at the NW bottom induced by lattice mismatching (black dashed) and without tensile strain (red dashed curve).

4. Conclusion

Thus, p-GaAs NW/p-Si substrate solar cell was studied by measuring the I-V curves under red laser illumination by C-AFM. Observed open-circuit voltage of the NW-substrate solar cell is governed by the balance of the Au cap/GaAs NW and GaAs NW/Si barriers. Sign of the photogenerated current indicates an impact of the GaAs NW/Si barrier increased by a strain which is induced due to GaAs/Si lattice mismatch. Numerical calculation accounting piezoresistance and piezoelectric effects shows a presence of the 2% tensile along NW in the 20 nm region at the NW/substrate interface. The calculation shows ~2 times increasing of the solar cell efficiency due to the presence of the strained GaAs NW/Si substrate interface.

References

[1] Green M A 2001 Third generation photovoltaics: Ultra-high conversion efficiency at low cost *Progress in Photovoltaics: Research and Applications* **9** 123–135
[2] Krogstrup P, Jørgensen H I, Heiss M, Demichel O, Holm J V, Aagesen M, Nygard J and i Morral A F 2013 Single-nanowire solar cells beyond the Shockley–Queisser limit *Nature Photonics* **7** 306
[3] LaPierre R 2011 Theoretical conversion efficiency of a two-junction III-V nanowire on Si solar cell *Journal of applied physics* **110** 014310
[4] LaPierre R, Chia A, Gibson S, Haapamaki C, Boulanger J, Yee R, Kuyanov P, Zhang J, Tajik N, Jewell N and others 2013 III–V nanowire photovoltaics: review of design for high efficiency *physica status solidi (RRL)–Rapid Research Letters* **7** 815–830
[5] Alekseev P A, Sharov V A, Borodin B R, Dunaevskiy M S, Reznik R R and Cirlin G E 2020 Effect of the Uniaxial Compression on the GaAs Nanowire Solar Cell *Micromachines* **11** 581
[6] Signorello G, Karg S, Björk M T, Gotsmann B and Riel H 2013 Tuning the light emission from GaAs nanowires over 290 meV with uniaxial strain *Nano letters* **13** 917–924
[7] Arlt G and Quadflieg P 1968 Piezoelectricity in iii-v compounds with a phenomenological analysis of the piezoelectric effect *physica status solidi (b)* **25** 323–330
[8] Alekseev P A, Sharov V A, Geydt P, Dunaevskiy M S, Lysak V V, Cirlin G E, Reznik R R, Khrebtov A I, Soshnikov I P and Lähderanta E 2018 Piezoelectric current generation in wurtzite GaAs nanowires *physica status solidi (RRL)–Rapid Research Letters* **12** 1700358
[9] Biermanns A, Breuer S, Trampert A, Davydok A, Geelhaar L and Pietsch U 2012 Strain accommodation in Ga-assisted GaAs nanowires grown on silicon (111) *Nanotechnology* **23** 305703
[10] Geydt P, Alekseev P, Dunaevskiy M, Haggrén T, Kakko J-P, Lähderanta E and Lipsanen H 2016 Influence of surface passivation on electric properties of individual GaAs nanowires studied by current–voltage AFM measurements *Lithuanian Journal of Physics* **56**
[11] Anon *Silvaco Atlas; https://www.silvaco.com/products/tcad/device_simulation/atlas/atlas.html*
[12] Alekseev P, Geydt P, Dunaevskiy M, Lähderanta E, Haggrén T, Kakko J-P and Lipsanen H 2017 I–V curve hysteresis induced by gate-free charging of GaAs nanowires' surface oxide *Applied Physics Letters* **111** 132104
[13] Lin A, Shapiro J N, Senanayake P N, Scofield A C, Wong P-S, Liang B and Huffaker D L 2012 Extracting transport parameters in GaAs nanopillars grown by selective-area epitaxy *Nanotechnology* **23** 105701
[14] Zhang X, Dubrovskii V G, Sibirev N V and Ren X 2011 Analytical study of elastic relaxation and plastic deformation in nanostructures on lattice mismatched substrates *Crystal growth & design* **11** 5441–5448
[15] Calahorra Y, Husmann A, Bourdelain A, Kim W, Vukajlovic-Plestina J, Boughey C, Jing Q, i Morral A F and Kar-Narayan S 2019 Highly sensitive piezotronic pressure sensors based on undoped GaAs nanowire ensembles *Journal of Physics D: Applied Physics* **52** 294002

Journal of Physics: Conference Series **2015** (2021) 012005 doi:10.1088/1742-6596/2015/1/012005

Wide-angle Reconfigurable Refraction by Silicon Fourier Metasurfaces

A A Antonov and M V Gorkunov

E-mail: antonov.wasd@yandex.ru, gorkunov@crys.ras.ru
Shubnikov Institute of Crystallography FSRC "Crystallography and Photonics", Russian Academy of Sciences, 119333, Moscow, Russia

Abstract. We design metasurfaces based on silicon films with smooth relief described by several Fourier harmonics and study their ability to redirect the refracted light over a wide angular range controlled by subtle variations of the optical setup. We use semi-analytical approach based on the Rayleigh hypothesis as well as full-scale numerical solutions to optimize the relief shape. To illustrate the reconfigurability potential, we design metasurfaces efficiently redirecting the refracted light from 83° to -73° with respect to the normal, when the angle of incidence is varied from 0° to 2°, and from 80° to -74°, when the substrate permittivity is altered from 2.3 to 2.2.

1. Introduction

The key advantage of metamaterials and metasurfaces is their ability to precisely manipulate electromagnetic waves upon careful design of structural elements and their arrangement. While typical building blocks of the optical dielectric metasurfaces are nanoridges, nanorods and nanodisks carved from thin high-index transparent layers [1], a novel type–Fourier metasurfaces based on smooth periodic reliefs–has been recently introduced. Such metasurfaces can be fabricated by the thermal scanning probe lithography [2] or by the focused ion beam milling [3]. Importantly, the periodic relief described in terms of amplitudes of several Fourier harmonics allows fast optimization. As has been exemplified for the efficient anomalous refraction in near-grazing directions [4], Fourier metasurfaces can outperform even the most efficient to date analogues which counterintuitive complex unit cell shapes are obtained by a topology optimization requiring substantial computational resources [5].

Efficient anomalous refraction is a basic metasurface functionality necessary for creating versatile meta-lenses and beam steering devices [6]. It critically relies on the formation of a single dominating diffraction order collecting the most part of the outgoing light. For the refraction in near-grazing directions, the light wavelength has to be close to the cutoff of the corresponding diffraction order. Accordingly, subtle modifications of the system shifting the cutoff wavelength can trigger a drastic redistribution of the outgoing energy from the previously prevailing diffraction order to other channels. Ensuring that this redistribution occurs to another single dominant channel, one can achieve highly efficient optical reconfigurability in a wide range of deflection angles. Here we theoretically analyze the potential of reconfigurable Fourier metasurfaces and, for definiteness, focus on those redirecting the incident green light upon small variations of the angle of incidence and of the substrate permittivity.

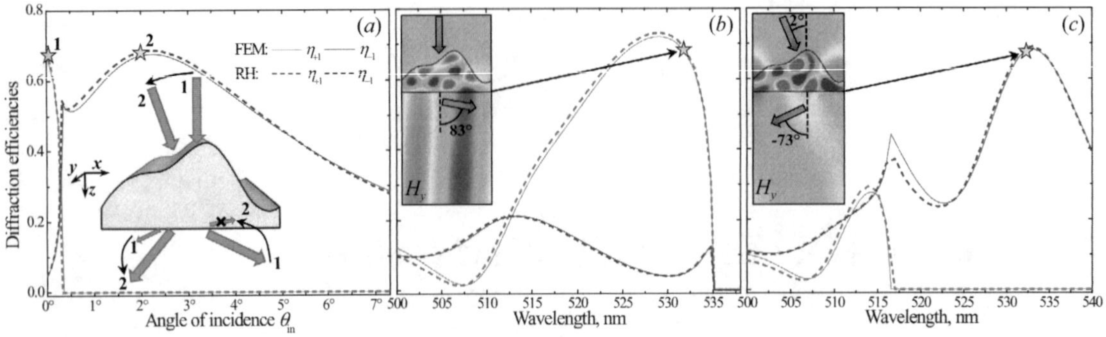

Figure 1. Silicon Fourier metasurface optimized using the RH-based (dashed lines) and FEM-based (solid lines) approaches for two dominant diffraction channels activated by a subtle variation of the angle of incidence: (a) dependencies of the diffraction efficiencies $\eta_{\pm 1}$ at the wavelength $\lambda = 532$ nm on the angle of incidence θ_{in} with the stars marking the maximized efficiency values. The inset schematically shows the reconfigurability scheme. (b) Spectra of the diffraction efficiency $\eta_{\pm 1}$ of the optimized metasurface for the normal incidence with the inset showing the distribution of the magnetic field component H_y at $\lambda = 532$ nm; (c) same as (b) for the light incident obliquely at $\theta_{in} = 2°$.

2. Results

For the light of a wavelength λ incident at an angle θ_{in} on a metasurface of a period Λ supported by a substrate with permittivity ε, the in-plane component of the wavevector of m-th transmission diffraction order reads as: $k_m = 2\pi\lambda^{-1}\sqrt{\varepsilon - (\sin\theta_{in} + m\lambda/\Lambda)^2}$. Thus by increasing the positive angle θ_{in} or by decreasing ε one can transform k_{+1} from real to imaginary while sustaining k_{-1} real. As there always exist other open outgoing channels (the direct transmission, specular and diffracted reflections) one has to precisely optimize the structure to ensure that the outgoing light is efficiently collected into a single desired direction in the both regimes.

We consider metasurfaces based on layers of monocrystalline silicon possessing the experimental permittivity values [7] patterned with 1D periodic profiles described by two Fourier harmonics and focus on controlling the propagation of TM-polarized light. The optimization goal is to maximize the transmission diffraction efficiencies η_{+1} and η_{-1} for the corresponding combinations of the angle of incidence and substrate permittivity. Technically, we employ two approaches: the semi-analytical model based on the Rayleigh hypothesis (RH) [4] and more extensive full-scale numerical solution using the finite element method (FEM) by COMSOL Multiphysics. Comparing the results also allows us to clarify the otherwise uncertain limits of applicability of the RH to dielectric Fourier metasurfaces.

2.1. *Reconfigurability by angle of incidence*

Here we maximize the diffraction efficiency η_{+1} for the light incident normally at $\theta_{in} = 0$ and η_{-1} for the slightly oblique incidence at $\theta_{in} = 2°$. We set the metasurface period to $\Lambda = 535$ nm and assume it to be suspended in the vacuum ($\varepsilon = 1$). For the green light with $\lambda = 532$ nm, the anomalous refraction into the $+1$ diffraction order occurs at an angle of 83° from the normal. Inclining the incident wave by 2° cancels the $+1$ diffraction order while the waves diffracted into the -1 order propagate at $-73°$ from the normal, see the sketches in the insets in Figs.1(a-c). Both the RH-based and FEM-based optimizations yield the same metasurface relief. The corresponding dependencies of the efficiencies $\eta_{\pm 1}$ on the angle of incidence θ_{in} in Figs. 1(a) show that the 67%-efficient refraction at an angle of 83° is reconfigured into similarly strong

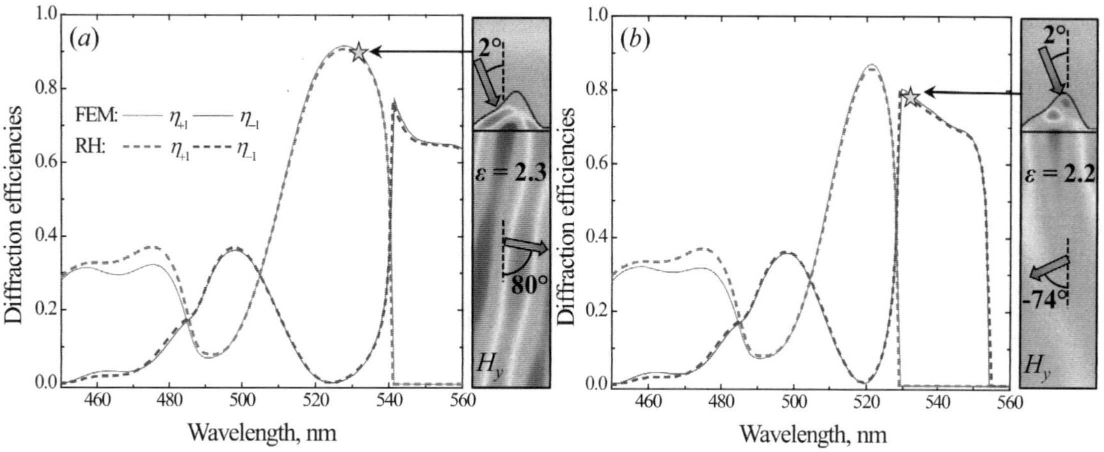

Figure 2. Silicon Fourier metasurface optimized using the RH-based (dashed lines) and FEM-based (solid lines) approaches for different dominant diffraction channels activated by a subtle variation of the substrate permittivity ε. Spectra of the diffraction efficiencies $\eta_{\pm 1}$ are shown for $\varepsilon = 2.3$ in (a) and for $\varepsilon = 2.2$ in (b) with the insets illustrating the distribution of the magnetic field component H_y at $\lambda = 532$ nm. The stars mark the maximized efficiency values.

refraction at an angle of $-73°$ by a very subtle alteration of the angle of incidence. Figs. 1(b) and (c) present the diffraction efficiencies spectra for the both cases. As seen in the insets in Figs. 1(b) and (c), the spatial distribution of the only magnetic field component H_y clearly indicates the formation of plane waves directly under the metasurface and their propagation in the corresponding directions.

2.2. Reconfigurability by substrate permittivity

Our next goal is to achieve similarly efficient wave deflection controlled by weak variations of the substrate permittivity ε in the case of slightly oblique incidence at $\theta_{in} = 2°$. For definiteness, we choose the substrate permittivity range characteristic to standard transparent optical materials (e.g. glasses, polymers or liquid crystals) and set $\varepsilon = 2.3$ for the case with maximum η_{+1}, and $\varepsilon = 2.2$ for the case when the +1 order is an evanescent wave and the efficiency η_{-1} is to be maximized. The corresponding value of the metasurface period is $\Lambda = 365$ nm, which provides the refraction angles 80° and $-74°$ from the normal for the +1 and -1 diffraction orders in the corresponding cases. Both the RH-based and FEM-based optimizations yield the profile, schematically sketched in the insets in Fig. 2. The spectra of diffraction efficiencies $\eta_{\pm 1}$ shown in Fig. 2 for $\varepsilon = 2.3$ and $\varepsilon = 2.2$ illustrate how the optimized maximum values, $\eta_{+1} \approx 88\%$ and $\eta_{-1} \approx 78\%$, are provided by the same metasurface supported by a substrate with slightly different permittivity.

3. Discussion

We have shown how dielectric Fourier metasurfaces relatively simply described by two harmonics of their periodic relief can be optimized for efficient light deflection in the directions different by more than 154° controlled by subtle variations of the optical setup. However remarkable such reconfigurability might seem per se, its true practical potential will be fully revealed in combination with tunable metasurfaces. The latter are being developed based on very different physical principles and offer the options of convenient control by voltage, external fields or laser irradiation [8]. As a trade-off, they often produce relatively weak direct effects on light. For example, liquid-crystal metasurfaces are capable of efficient low-voltage-driven millisecond-fast deflection of visible light, but their angular range is limited by several degrees [9]. Apparently, the above Fourier metasurfaces reconfigurable by the angle of incidence can be used to amplify this range by orders of magnitude.

On the other hand, the sensitivity to substrate properties can be directly employed to create tunable configurations provided that the substrate allows controlling its permittivity if even in a narrow specific range. In this regard, the required effect can be also achieved by realigning the liquid crystal adjacent to the metasurface.

Finally, it is worth noting that the material of Fourier metasurfaces supporting anomalous refraction in near grazing directions is not restricted to silicon [10]. We expect that similarly designed reconfigurable metasurfaces can be produced from other transparent high-index materials such as gallium phosphide or titanium dioxide.

Acknowledgments

The research is funded by the Russian Science Foundation project 18-12-00361, https://rscf.ru/project/18-12-00361/.

References

[1] Jahani S and Jacob Z 2016 *Nature Nanotechnology* **11** 23–36

[2] Lassaline N, Brechbühler R, Vonk S J W, Ridderbeek K, Spieser M, Bisig S, le Feber B, Rabouw F T and Norris D J 2020 *Nature* **582** 506–510

[3] Gorkunov M V, Rogov O Y, Kondratov A V, Artemov V V, Gainutdinov R V and Ezhov A A 2018 *Scientific Reports* **8**

[4] Antonov A A and Gorkunov M V 2019 *Journal of the Optical Society of America B* **36** 2118

[5] Sell D, Yang J, Doshay S, Yang R and Fan J A 2017 *Nano Letters* **17** 3752–3757

[6] Yu N, Genevet P, Kats M A, Aieta F, Tetienne J P, Capasso F and Gaburro Z 2011 *Science* **334** 333–337

[7] Vuye G, Fisson S, Van V N, Wang Y, Rivory J and Abelès F 1993 *Thin Solid Films* **233** 166–170

[8] He Q, Sun S and Zhou L 2019 *Research* **2019** 1–16

[9] Gorkunov M V, Kasyanova I V, Artemov V V, Ezhov A A, Mamonova A V, Simdyankin I V and Palto S P 2020 *ACS Photonics* **7** 3096–3105

[10] Antonov A A and Gorkunov M V 2020 Corrugated dielectric metasurfaces for anomalous refraction in near-grazing directions *AIP Conference Proceedings* vol 2300 (AIP Publishing) p 020003

Enhancement of second harmonic generation in a layered MoS_2 nanoresonator

I.M. Antropov[1], A.A. Popkova[1], G.I. Tselikov[2], V.S. Volkov[2], V.O. Bessonov[1,3], and A.A. Fedyanin[1]

[1] Faculty of Physics, Lomonosov Moscow State University, Moscow 119991, Russia
[2] Center for Photonics and 2D Materials, Moscow Institute of Physics and Technology, Dolgoprudny 141700, Russia
[3] Frumkin Institute of Physical Chemistry and Electrochemistry, Russian Academy of Sciences, Moscow, 119071 Russia

E-mail: antropov@nanolab.phys.msu.ru

Abstract. Molybdenum disulfide (MoS_2) is a layered material with a high refractive index in the visible and infrared spectral range. In this work, we theoretically and experimentally demonstrate Mie-resonant MoS_2 nanodisks. We show enhanced second harmonic generation from MoS_2 nanodisk resonators due to the overlap of Mie-type resonances at the fundamental wavelength with the C-exciton resonance at the second-harmonic wavelength.

1. Introduction

Unique optical properties of 2D TMDCs materials such as strong photoluminescence [1, 2], valley polarization [3], excitonic valley coherence [4], high refractive index [5], and giant non-linear response [6], enable these materials to be used as building blocks for integrated photonics. Molybdenum disulfide is a prominent representative of TMDCs. It is a semiconductor with indirect bandgap in bulk form, but monolayer form is a direct bandgap semiconductor [1]. Theoretical and experimental studies show that optical absorption are dominated by excitonic transitions in MoS_2, where the exciton binding energy can be close to 1 eV due to dramatically enhanced electron–hole and electron-electron interactions in monolayers [7]. Nonlinear optical (NLO) properties of MoS_2 layers are studied well [8–11]. MoS_2 has a centrosymmetric lattice structure in bulk form [12] and for an even number of layers, which leads to the elimination of SHG [11]. On the other hand, the strong SHG arises in the MoS_2 monolayer or multilayers with an odd number of layers [10,11]. Second harmonic (SH) intensity can be strongly increased by orders of magnitude when excitation photons or SH photons are resonant with the A-, B-, C-excitons in MoS_2 monolayers [12, 13]. Another way to enhance SHG is to employ Mie-type resonances for structured objects based on MoS_2. MoS_2 Layers possess high and anisotropic dielectric constants [14, 15], which opens up the possibility of utilizing them to construct nanoantennas supporting distinct Mie-type resonances [16, 17] and consider them as a novel platform for high-refractive-index all van der Waals nanophotonics [18]. Nanoresonators based on TMDCs with optical response governed by Mie-type resonances can enhance the efficiency of NLO processes including SHG [17].

In this work, we study the SHG response of an individual MoS_2 disk-like nanoresonator. We observed the strong enhancement of SHG near 900-nm pump wavelength corresponding to the C-exciton resonance at the SH wavelength. Numerical calculations show that this enhancement is associated with the localization of the pump field inside the resonator due to the excitation of the magnetic dipole resonance.

2. Results and discussion

First, we estimated the parameters of nanoresonators so that the magnetic dipole (MD) Mie-type resonance at the excitation wavelength matched with C-exciton resonance in a monolayer of MoS_2 at SH wavelength. Then we fabricated an array of disk resonators by electron beam lithography followed by reactive ion etching. The cleaved silicon wafer with a 290-nm thermally grown oxide was used as a substrate. MoS_2 flakes were deposited on substrates using mechanical exfoliation from highly oriented synthetic 2H-phase MoS_2 crystal. Fig.1(a) shows a sketch of experimental sample. The real sizes of the nanoresonators were determined by scanning electron microscopy (SEM) and atomic force microscopy (AFM). The diameter and height of resonators are 550 nm and 110 nm, correspondingly. Fig.1 (b),(c) demonstrate SEM and AFM images of the nanoresonators, respectively. Non-etched remnants of MoS_2 flake approximately 5 nm thick between the nanoresonators was used as a reference.

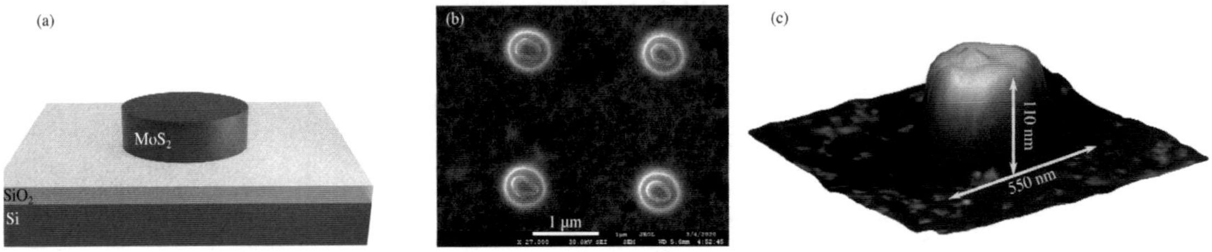

Figure 1. Images of experimental sample (a) Sketch of the sample. (b) SEM image of the sample. (c) AFM profile of a single nanodisk.

SHG was measured using home-built multiphoton microscope designed for optical harmonics investigation in reflection scheme. The 150-fs laser pulses with 80 MHz repetition rate and wavelength tuning in the 680-1080 nm range were used as a pump radiation. The laser beam, passing through the system of the power and polarization control, was focused in the back focal plane of an objective lens (NA=0.95, 50x), which corresponds to the Kohler illumination scheme with a 12 μm beam size at the sample. Reflected SHG signal from a large area formed a SH image of the nanoresonators on a CMOS camera (Photometrics Prime). Fig.2 (a) shows a SH image of the nanoresonators for Kohler illumination when the excitation wavelength is 900 nm. Each bright spot corresponds to one MoS_2 nanoresonator. The power dependences of the nonlinear response were measured to determine the nature of measured signal. For this the signal from sample area corresponded to a single disk was integrated for various pump powers. The obtained dependence is depicted in Fig. 2 (b) in double logarithmic scale and demonstrates the quadratic dependence on the pump power. The spectra of the nonlinear signal do not contain any features that confirms the absence of the contribution of luminescence and other multi-photon processes.

To determine the effect of Mie-type resonances on SHG, we measured the SH signal dependence on the pump wavelength (Fig.2 (c)). The intensity of SHG has a sharp resonance near 445-nm wavelength that corresponds to the C-exciton resonance position [13]. A similar signal behavior is also observed for the reference sample with the response value being approximately 50 times smaller. The numerical simulation of scattering cross-section shows the MD resonance in the disk plane near the resonant wavelength for the experimental sample parameters (see Fig. 3 (a)). The corresponding distributions of the components of the electric

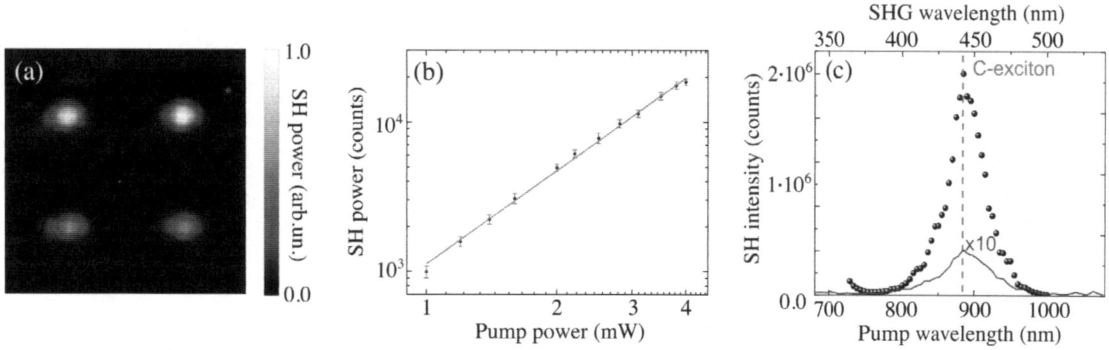

Figure 2. SH characterization (a) SHG image of the sample. (b) Power-dependent nonlinear optical response of MoS_2 disk (black dots), plotted in double logarithmic scale, and its approximation by power function $y = y_0 + Ax^p$ with $p = 2.06 \pm 0.03$. (c) SHG spectrum (black dots) of the MoS_2 nanodisk and the reference sample (black line). The red dashed line shows the position of the C-exciton peak.

and magnetic fields in the resonator are shown in Fig.3 (b). The field localization in the disk volume leads to an increase in the efficiency of nonlinear effects which in our experiment is manifested in the form of SHG enhancement.

Conclusion

We have demonstrated that the MoS_2 nanodisk resonators fabricating by electron beam lithography followed by reactive ion etching enhance the nonlinear optical response. This enhancement is associated with the excitation of the Mie-type resonance modes at the pump wavelength corresponding to the doubled C-exciton resonance wavelength. The maximum value of the SHG intensity for the MoS_2 nanoresonators is 50 times of magnitude higher than that for the unstructured MoS_2 film. This work opens a new route to control and modulate nonlinear response of Mie-resonant nanostructure based on 2D TMDCs materials by tailoring

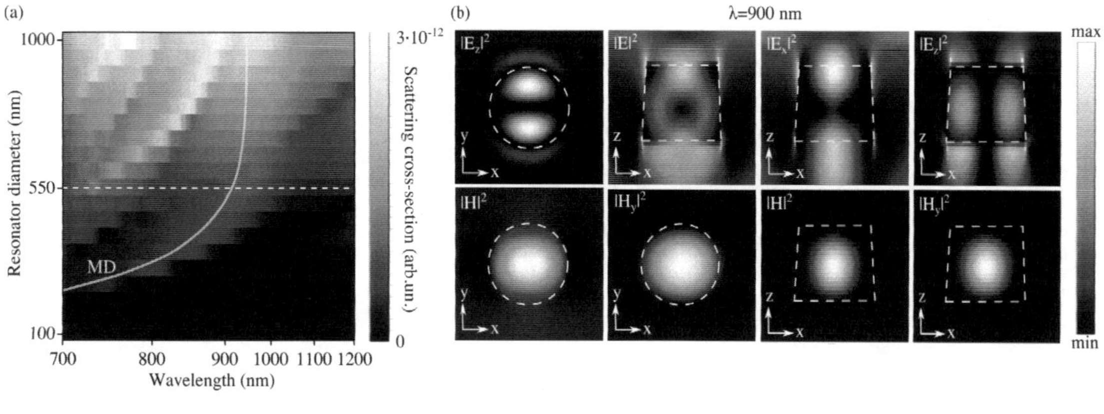

Figure 3. Numerical simulations (a) Scattering cross-section for different resonator diameters. The yellow curve shows the position of MD resonance. The dashed line indicates the size of the experimental nanodisk. (b) Electric (up row) and magnetic (down row) fields distributions in the nanoresonator for SH-resonance case (λ=900 nm, d=550 nm). Dashed curve indicates the resonator boundary.

their excitation conditions. It may be applied for nonlinear 2D photonics and holographic imaging applications.

Acknowledgements

The experimental work was supported by the Russian Science Foundation, grant No. 19-79-00350. The numerical work was performed under financial support of the Russian Science Foundation (Grant No. 20-12-00371) and according to the Development program of the Interdisciplinary Scientific and Educational School of Lomonosov Moscow State University "Photonic and Quantum technologies. Digital medicine". I.M.A. and V.O.B. thank MSU Quantum Technology Centre for the support. A.A.P. acknowledges support by Foundation for the Advancement of Theoretical Physics and Mathematics "BASIS" (Grant No. 19-2-6-28-1). V.S.V. gratefully acknowledge financial support from the Ministry of Science and Higher Education of the Russian Federation (No. 0714-2020-0002).

References

[1] Mak K F, Lee C, Hone J, Shan J and Heinz T F 2010 *Phys. Rev. Lett.* **105**
[2] Splendiani A, Sun L, Zhang Y, Li T, Kim J, Chim C Y, Galli G and Wang F 2010 *Nano Letters* **10** 1271–1275
[3] Mak K F, He K, Shan J and Heinz T F 2012 *Nature Nanotechnology* **7** 494–498
[4] Jones A M, Yu H, Ghimire N J, Wu S, Aivazian G, Ross J S, Zhao B, Yan J, Mandrus D G, Xiao D, Yao W and Xu X 2013 *Nature Nanotechnology* **8** 634–638
[5] Green T D, Baranov D G, Munkhbat B, Verre R, Shegai T and Käll M 2020 *Optica* **7** 680
[6] Wang G, Marie X, Gerber I, Amand T, Lagarde D, Bouet L, Vidal M, Balocchi A and Urbaszek B 2015 *Phys. Rev. Lett.* **114**
[7] Qiu D Y, da Jornada F H and Louie S G 2013 *Phys. Rev. Lett.* **111**
[8] Yin X, Ye Z, Chenet D A, Ye Y, O'Brien K, Hone J C and Zhang X 2014 *Science* **344** 488–490
[9] Li Y, Rao Y, Mak K F, You Y, Wang S, Dean C R and Heinz T F 2013 *Nano Lett.* **13** 3329–3333
[10] Kumar N, Najmaei S, Cui Q, Ceballos F, Ajayan P M, Lou J and Zhao H 2013 *Physical Rev. B* **87**
[11] Wang Y, Cong C, Qiu C and Yu T 2013 *Small* **9** 2857–2861
[12] Trolle M L, Tsao Y C, Pedersen K and Pedersen T G 2015 *Phys. Rev. B* **92**
[13] Malard L M, Alencar T V, Barboza A P M, Mak K F and de Paula A M 2013 *Phys. Rev. B* **87**
[14] Ermolaev G A, Stebunov Y V, Vyshnevyy A A, Tatarkin D E, Yakubovsky D I, Novikov S M, Baranov D G, Shegai T, Nikitin A Y, Arsenin A V and Volkov V S 2020 *npj 2D Materials and Applications* **4**
[15] Ermolaev G A, Grudinin D V, Stebunov Y V, Voronin K V, Kravets V G, Duan J, Mazitov A B, Tselikov G I, Bylinkin A, Yakubovsky D I, Novikov S M, Baranov D G, Nikitin A Y, Kruglov I A, Shegai T, Alonso-González P, Grigorenko A N, Arsenin A V, Novoselov K S and Volkov V S 2021 *Nature Communications* **12**
[16] Ding S J, Luo Z J, Xie Y M, Pan G M, Qiu Y H, Chen K, Zhou L, Wang J, Lin H Q and Wang Q Q 2018 *Nanoscale* **10** 124–131
[17] Busschaert S, Reimann R, Cavigelli M, Khelifa R, Jain A and Novotny L 2020 *ACS Photonics* **7** 2482–2488
[18] Ling H, Li R and Davoyan A R 2021 *ACS Photonics* **8** 721–730

Effective electrodynamical parameters and microwave heating of radially heterogeneous pellets containing EAF dust and biochar

Anton P Anzulevich[1], Leonid N Butko[1], Dmitry A Pavlov[1], Dmitry A Kalganov[1], Valentin A Tolkachev[1], Svetlana N Anzulevich[1], Vasiliy D Buchelnikov[1], Sergey G Moiseev[2], and Zhiwei Peng[3]

[1] Chelyabinsk State University, 129 Br. Kashirinykh Str., Chelyabinsk 454021, Russia
[2] Ulyanovsk State University, 42 L. Tolstoy Str., Ulyanovsk 432017, Russia
[3] Central South University, Changsha, Hunan 410083, China

E-mail: anzul@list.ru

Abstract. In this work, we investigated the effect of microwave radiation with radially heterogeneous pellets consisting of electric arc furnace (EAF) dust and biochar. We reviewed the possible content of EAF dust in terms of permittivity and permeability of its components and calculated effective permittivity and permeability of EAF dust by an effective medium approach. Using obtained values we calculated dependencies of effective permittivity and permeability of EAF dust - biochar composite mixture on the volume fraction of EAF dust and conductivity of biochar. Taking into account these dependencies we simulated electromagnetic field and temperature distribution within pellet with a radial dependency of volume fraction of EAF dust and effective permittivity correspondingly.

1. Introduction

Dust formation in electric arc furnaces can reach up to 30 kg per 1 ton of steel. The need to reduce dust formation and dust disposal technology is due to the negative impact of dust on the environment and human health [1]. Under these conditions, it is necessary to find ways to reduce dust formation and use dust as a secondary raw material in order to extract valuable components, primarily Fe, Zn, Pb.

In addition, millions tons of products of the pulp and paper industry are manufactured, waste in the form of hydrolytic lignin or biochar is at least 35% of this mass [2].

Also, in recent years, much attention has been attracted by the technology of direct reduction of iron from iron-containing compounds, the advantages of which are temperatures below the melting point, reduced slag formation, low requirements for fuel [3]. The use of microwave energy makes it possible to carry out all stages of this process, and to regulate the course of the process by changing the power of the microwave generator.

The chemical composition of the dust of electric arc furnaces varies greatly from specific manufacturer, but over 90% by weight are 4 substances: Fe_3O_4, Fe_2O_3, ZnO, $ZnFe_2O_4$ [4, 5]. The main chemical processes occurring during the direct reduction of iron with solid carbon C depend on the initial composition of the EAF dust, but generally correspond to reduction reactions.

Journal of Physics: Conference Series **2015** (2021) 012007 doi:10.1088/1742-6596/2015/1/012007

2. Calculation of effective permittivity and permeability

We used effective medium approximation to calculate effective permittivity and permeability of EAF dust itself as well as of mixture of EAF dust and biochar. For this purpose Bruggeman equation was generalized for a mixture of N types of spherical particles [6]:

$$\sum_{i=1}^{N} p_i \frac{\varepsilon_i - \varepsilon_{eff}}{\varepsilon_i + 2\varepsilon_{eff}} = 0 \tag{1}$$

Bruggeman equation was derived in quasi-static approximation. However it is valid for our case because particle size (around 3-30 μm) is much less than wavelength.

For the model volume fractions and electrodynamic parameters of components of EAF dust on table 1, and according to Bruggeman equation calculation, effective permittivity and permeability of EAF dust are $\varepsilon_{eaf} = 10.69 + 2.6022i$ and $\mu_{eaf} = 0.8723 + 0.0975i$. Taking into account these values of effective electrodynamic parameters of EAF dust, dependencies of effective permittivity and permeability of EAF dust - biochar mixture on volume fraction of EAF dust and conductivity of biochar were calculated and are shown on figure 1. In fact, these are typical curves of the percolation transition [7, 8]. In addition, these dependencies are qualitatively consistent with experimental results in [9] for particular type of biochar.

Table 1. Characteristics of components of EAF dust [10–15].

	ε'	$\sigma, \frac{Sm}{m}$	μ'	μ''	Weight fraction	Model volume fraction
$ZnFe_2O_4$	3.7	0.17	1.25	0.55	$1-20\%$	10%
ZnO	5.822	2×10^{-5}	1	1	$1-20\%$	10%
Fe_3O_4	57.355	1.2	0.2	0.2	$10-23\%$	15%
Fe_2O_3	14.922	1×10^{-5}	1	1	$10-23\%$	15%
SiO_2	6	1.125×10^{-9}	1	1	$6-34\%$	10%
$CaMgSiO_4$	8.5	7.7×10^{-11}	1	1	$22-60\%$	30%

3. Electromagnetic fields and temperature distribution

It is possible to prepare pellets with radial distribution of EAF dust volume fraction so that permittivity and permeability within pellet can depend on radia. This approach may allow to achieve better impedance matching and optimal microwave heating of pellet under investigation. Electromagnetic field and temperature distribution as well as temperature curves were simulated by finite element method and are demonstrated on figures 2 and 3.

The following model parameters have been considered: electromagnetic wave frequency - 2.45 GHz, microwave power - 450 W, heat capacity of EAF dust - 1050 J/(kg*K), thermal conductivity of EAF dust - 0.3 W/m/K*(T/300K). Maxwell equations and heat equation were simulated by finite element method for the law of conservation of energy in a differential form when the heat source depends on time and taking into account thermal conductivity and convection. Electromagnetic heat sources were taken as following:

$$Q = \frac{\omega}{8\pi}(\varepsilon''_{eff}|E|^2 + \mu''_{eff}|H|^2) \tag{2}$$

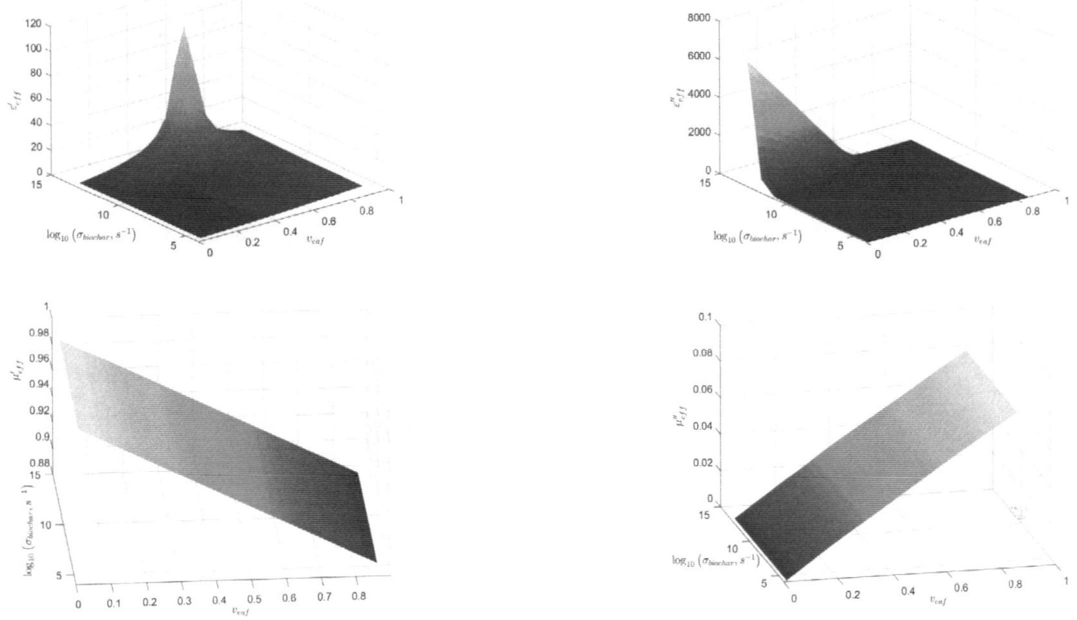

Figure 1. Real and imaginary part of permittivity and permeability of EAF dust - biochar powders mixture on volume fraction of EAF dust v_{eaf} and conductivity of biochar $\sigma_{biochar}$.

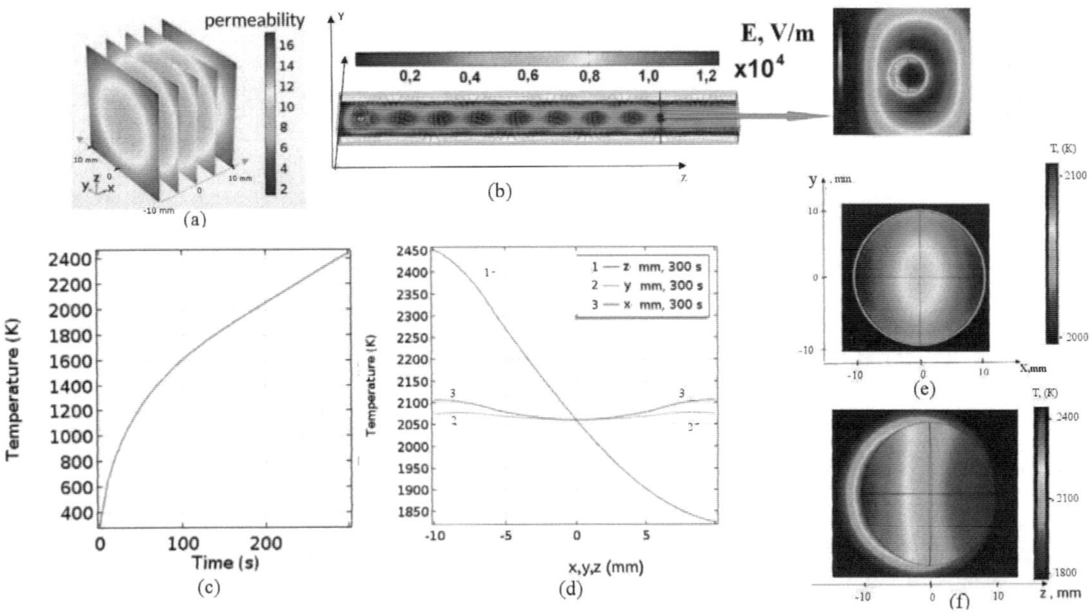

Figure 2. Permittivity and permeability linearly increases from 1 to 10 on radius. a - spacial distribution of imaginary and real parts inside pellet; b - electric field distribution in rectangular waveguide (mode $H10$); c - time dependence of maximum temperature inside pellet with radially heterogeneous distribution of EAF dust; d, e, f - spacial distribution of temperature inside pellet.

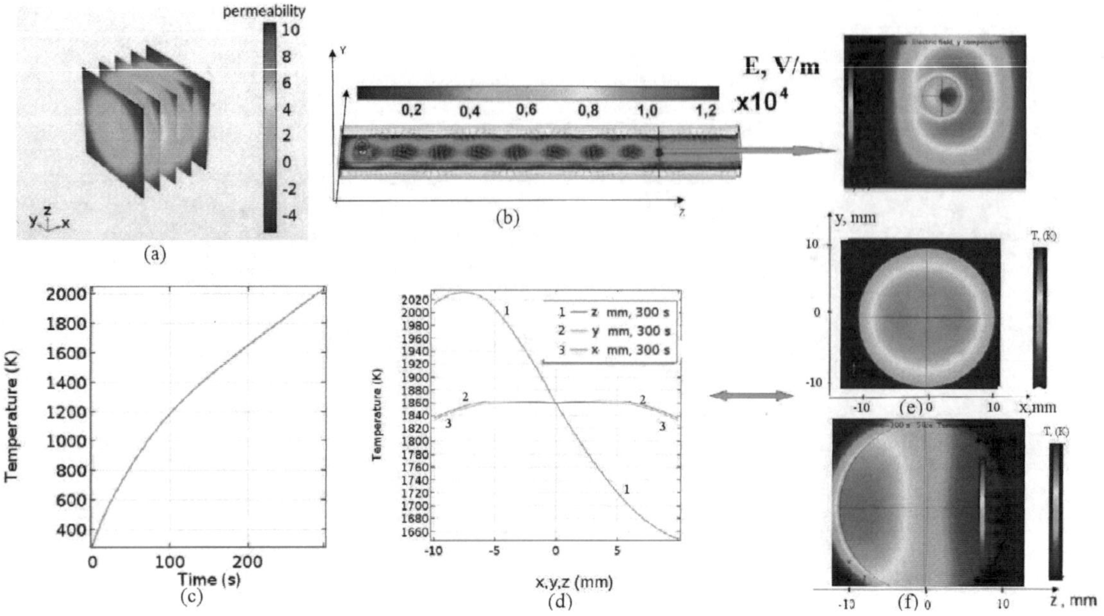

Figure 3. Permettivity and permeability linearly decreases from 10 to 1 on radius. *a*, *b*, *c*, *d*, *e*, *f* - the same as on figure 2.

4. Conclusion

On the figure 2 we can see that if EAF dust is concentrated in the core of pellet than it is possible to achieve higher temperature. However the temperature distribution is less uniform and pellet better heats in the shell layer. On the figure 3 the EAF dust is concentrated in the shell of pellet and temperature distribution more uniform. However, maximum temperature is less than on figure 2.

Thus, spacial distribution of EAF dust inside pellet drastically affects on process of microwave heating of pellet. Adjusting pelletization process so as to achieve heterogeneous distribution of EAF dust on radius of pellet decreasing from core to shell will lead to more optimal microwave heating and reduction of EAF dust.

Acknowledgments

The reported study was funded by RFBR and NSFC, project number 21-58-53044, and was supported by the National Natural Science Foundation of China under Grant 52111530046.

References

[1] Adon'ev S.M., Filip'ev O.V. Dust and gas emissions from ferrous metallurgy enterprises – M.: Metallurgy, 1973. – 200 p.

[2] Manturov D. Prospects of the forest complex in the modern economy // Pulp and Paper Industry. 2017. 2/3. pp. 16–17.

[3] Yusfin Yu.S., Gimmelfarb A.A., Plashkov N.F. Modern metal production processes // reviewers: Loginov V.I., Lazutkin S.E. - M.: Metallurgy, 1994. - 320 p.

[4] Q. Ye, G. Li, Zh. Peng, R. Augustine, M.D. Pérez, Y. Liu, M. Liu, M. Rao, Y. Zhang, T. Jiang // Powder Technology, 2020. V. 362. pp. 781–789.

[5] L.M. Simonyan, N.V. Demidova Selective extraction of carbon-free zinc and lead from EAF-dust // Izvestiya VUZov. Chernaya Metallurgiya = Izvestiya. Ferrous Metallurgy. 2020. V. 63. No. 8. pp. 631-638.

[6] A.P. Anzulevich, I.V. Bychkov, V.D. Buchelnikov, S.N. Anzulevich, D.A. Kalganov, Zh. Peng Calculation of effective permittivity and permeability for iron ore - biochar - bentonite binder powders mixture // Materials Research Proceedings, Materials Research Forum LLC, 2018. V. 9. pp. 136-139.

[7] A. Bunde, W. Dieterich Percolation in Composites // Journal of Electroceramics, 2000. 5:2. pp. 81-92.

[8] A.A. Snarskii, D. Zorinets, M. Shamonin and V.M. Kalita Theoretical method for calculation of effective properties of composite materials with reconfigurable microstructure: Electric and magnetic phenomena // Physica A: Statistical Mechanics and its Applications, 2019. 535. p. 122467.

[9] Q. Ye, G. Li, Z. Peng, R. Augustine, M.D. Perez, Y. Liu, M. Liu, M. Rao, Y. Zhang, T. Jianga Microwave-assisted self-reduction of EAF dust-biochar composite briquettes for production of direct reduced iron // Powder Technology, 2020. 362. pp. 781–789.

[10] L. Wang, Zh. Peng, L. Yang, L. Zheng, J. Wang, W. Shang, A. Anzulevich, M. Rao, G. Li and T. Jiang Self-reduction of core-shell EAF dust-biochar composite pellets under microwave irradiation // 11th International Symposium on High-Temperature Metallurgical Processing, The Minerals, Metals Materials Series, 2020. pp. 405-415.

[11] J. Huang, Y. Liu, Y. Li and X.X. Liu Microwave electromagentic and absorption properties of AFe2O4(A=Ni,Mn,Zn) ferrites // International Conference on Manufacturing Construction and Energy Engineering (MCEE), 2016. pp. 190-196.

[12] J.K. Srivastava, M. Prasad, and J.B. Wagner, Jr. // J. Electrochem. Soc., 1985. V. 132. pp. 955-963.

[13] M. Hotta, M. Hayashi, A. Nishikata and K. Nagata // ISIJ International, 2009. V. 49. No. 9. pp. 1443–1448

[14] L.P. Martin, D. Dadon, M. Rosen, D. Gershon, K.I. Rybakov, A. Birman, J.P. Calame, B. Levush, Y. Carmel, and R. Hutcheon // Journal of Applied Physics, 1998. V. 83. pp. 432-437.

[15] R.A. Lunt, A.J. Jackson, A. Walsh // Chemical Physics Letters, 2013. V. 586. pp. 67-69.

Polarization degeneracy of TE and TM eigenmodes for dielectric metasurface in the microwave

Syuzanna Asadulina, Andrey Bogdanov, Stanislav Glybovski, and Oleh Yermakov

School of Physics and Engineering, ITMO University, St. Petersburg, Russia

E-mail: o.yermakov@metalab.ifmo.ru

Abstract. We analyze the TE-TM polarization degeneracy of the guided modes of a dielectric metasurface in the microwave frequency range. We find the optimum metasurface design and investigate the dependence of the degeneracy degree on the propagation direction. Finally, we simulate the possible microwave experiment demonstrating the extraction of isofrequency contours for this metasurface. The results obtained could be useful for flat photonic devices with the function of polarization control.

1. Introduction

Metasurfaces allow to engineer efficiently the properties of electromagnetic plane waves (reflection, refraction, diffraction, polarization transformation, etc.) [1, 2]. Besides, metasurfaces pave way towards controlling the properties of strongly localized light [2, 3, 4]. Nevertheless, the polarization control of surface and guided waves still remains relevant and unresolved fundamental problem requiring the polarization degree of freedom for localized light.

The spectrum of electromagnetic plane waves in any isotropic medium is twice degenerate for all frequencies in any direction, because the dispersions of TE- and TM-polarized eigenmodes are absolutely the same. The operational principle of any bulk classical polarizer is based on the removal of the polarization degeneracy by using an anisotropic slab. However, there is no polarization degree of freedom for planar photonic devices with in-plane electromagnetic wave propagation, since the TE and TM localized modes are not degenerate. It significantly limits the functionality of flat optics and planar photonics devices. In particular, the polarizer of guided waves cannot be implemented without the broadband TE-TM polarization degeneracy.

In this work, we theoretically study the dielectric metasurface based on high-index cylinders that supports the broadband degeneracy of TE and TM guided modes in the microwave frequency range. We analyze numerically the degree of degeneracy and its dependence on the propagation direction for guided waves.

2. Design optimization

We perform a numerical optimization of a metasurface design representing a square periodic array of ceramic cylinders (Fig. 1) with a fixed diameter and permittivity equal to 5.2 mm and 40, respectively. The broadband degeneracy of the TE and TM polarized modes has been achieved with the cylinders period and height equal to 10 mm and 4.85 mm, respectively.

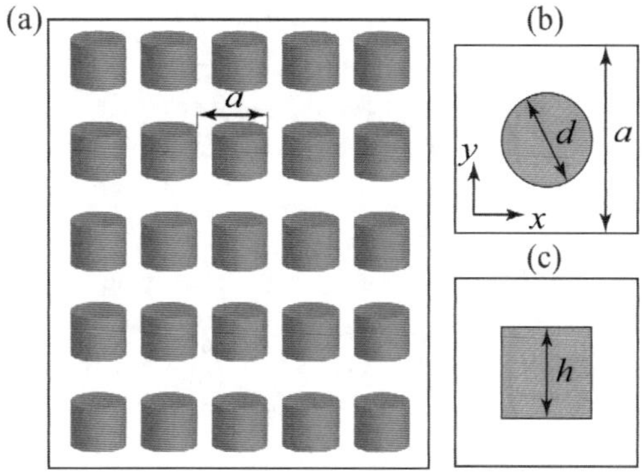

Figure 1: (a) The geometry of the metasurface with the lattice constant a surrounded by air. (b) Top and (c) side views of the unit cell of a metasurface representing a single cylinder with a diameter d and a height h.

The dispersion dependencies of guided modes of the dielectric metasurface demonstrate the TE-TM polarization degeneracy in the frequency range up to 8.2 GHz (Fig. 2), which was calculated in CST Microwave Studio using the Eigenmode Solver package. We introduce the degree of degeneracy value defined as the difference between the wave vectors of TE and TM guided modes at the same frequency (δk) or vice versa (δf). By increasing the period of the structure with fixed cylinders, the degree of degeneracy increases, but the eigenmodes become less localized, i.e. their dispersions become closer to the light line, and vice versa. For instance, the localization can be characterized by the in-plane wavevectors which are about $k_1 = 1.75k_0$, $k_2 = 1.41k_0$ and $k_3 = 1.32k_0$, where $k_0 = 2\pi f_0/c$, for the periods $a_1 = 9$ mm, $a_2 = 10$ mm

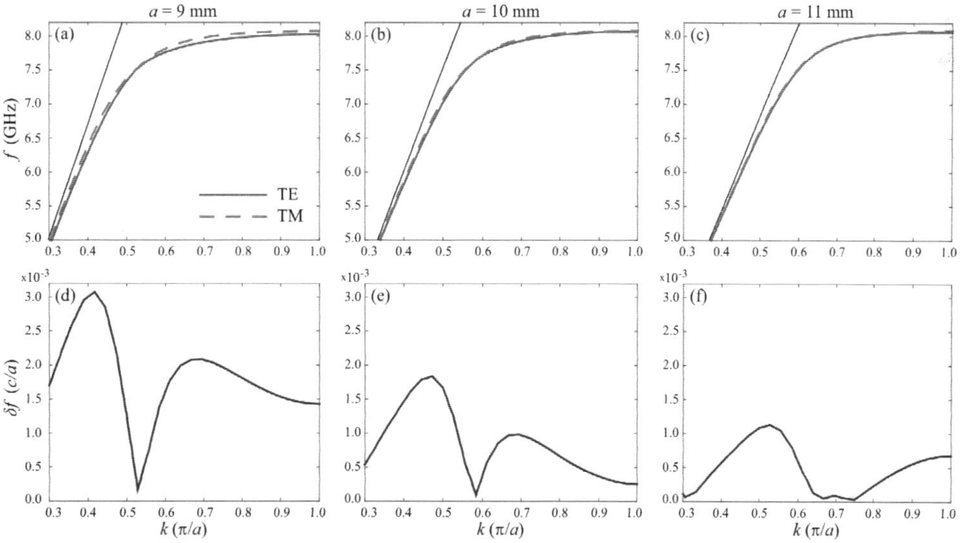

Figure 2: The dispersions of TE (blue line) and TM (red dashed line) guided modes along x-axis direction for different periods: (a,d) $a = 9$ mm, (b,e) $a = 10$ mm, (c,f) $a = 11$ mm. Black line corresponds to the light line. (d-f) Dependence of the degeneracy degree defined as the difference between the frequencies of TM and TE modes on the wavevector in the dimensionless units for corresponding lattice constants.

and $a_3 = 11$ mm, respectively, at the frequency $f_0 = 8$ GHz (Fig. 2). Hence, we have chosen a metasurface design giving the relatively high localization of guided waves ($k = 1.41k_0$) saving the degree of degeneracy not exceeding $\delta f_{max} = 0.002\, c/a$ (Fig. 2e), or in other terms $\delta k_{max} = 0.004\, \pi/a$ in the frequency range up to 8.1 GHz along ΓX direction.

3. Dependence on propagation direction

In addition, we analyze the sustainability of eigenmodes polarization degeneracy in different propagation directions. The degeneracy is removed at high frequencies in ΓM propagation direction (at angle 45° with respect to x- or y-axis), see Fig. 3. Isofrequency contours of TE and TM guided modes explicitly show that the degeneracy is achieved for all propagation directions in the frequency range up to 7.85 GHz (see the insert in Fig. 3). In the vicinity of 8 GHz one can notice the perceptible difference between TE and TM modes, especially at the propagation angle 45°. At slightly higher frequencies, only TM mode exists which completely breaks the polarization degeneracy. So, we conclude that the polarization degeneracy for the chosen metasurface design is fulfilled in the frequency range up to 7.85 GHz in all propagation directions.

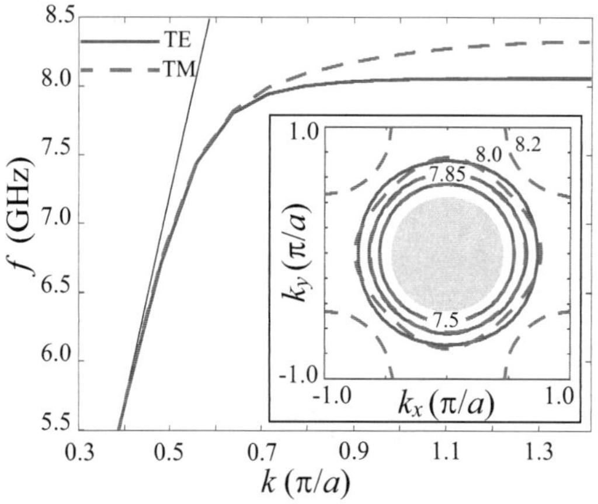

Figure 3: The dispersion of TE (blue line) and TM (red dashed line) guided modes along the propagation direction at 45° with respect to x-axis. Insert shows the isofrequency contours within the first Brillouin zone at the frequencies 7.5, 7.85, 8 and 8.2 GHz. The light cone at 7 GHz is shown by yellow circle.

We have numerically simulated the guided TE and TM modes at the dielectric metasurface under consideration in CST Microwave Studio using the Transient package. The guided waves were excited by the effective electric and magnetic point dipoles representing as coaxial probe and loop, respectively. Figure 4 shows the distribution of the normal component of electric field for guided waves excited by vertical probe-like (electric) dipole. For frequencies 7.7 GHz and 8 Ghz (Figs. 4a-4b), one can observe the conventional dipole-like field distribution corresponding to the elliptic isofrequency contours (Figs. 4d-4e). The field distribution at 8.2 GHz is shown in Fig. 4c, it corresponds to the arc-like isofrequency contour of TM mode (Fig. 3). After that we have reconstructed the isofrequency contours applying the 2D Fourier transform operation to normal components of electric and magnetic fields (see color maps in Figs. 4d-4f) following the same procedure as in Refs. [5, 6]. The extracted isofrequency contours are in good agreement with the ones calculated directly in Eigenmode solver. This retrieval technique can be used in the experimental verification of the broadband TE-TM polarization degeneracy of eigenmodes at all-dielectric metasurface.

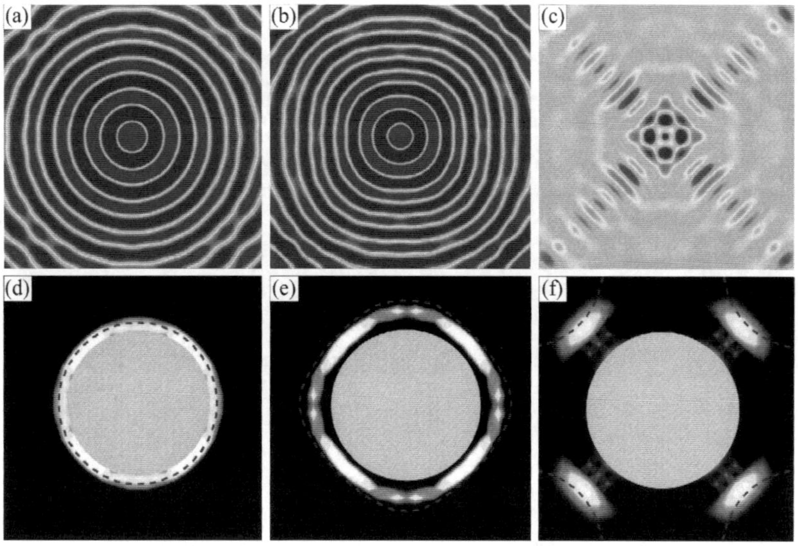

Figure 4: (a-c) The spatial distribution of the normal component of electric field for guided waves excited by a vertical point-like electric dipole (coaxial probe) located 1 mm above the surface at the frequencies: (a,d) 7.7 GHz, (b,e) 8 GHz and (c,f) 8.2 GHz. The size of the simulated region is 250×250 mm^2. (d-f) The isofrequency contours retrieved from the field distributions (color maps) and calculated numerically (dashed lines). The green circles correspond to the light cones.

4. Conclusions

We have proposed the optimum design of a metasurface consisting of dielectric high-index cylinders to achieve the polarization degeneracy of TE and TM guided modes in a wide spectral range. We have investigated the stability of degeneracy degree for this metasurface under different propagation directions by calculating the isofrequency contours. Finally, we have numerically simulated the potential microwave experiment and retrieved the isofrequency contours from the field distribution. This concept can be transferred to the optical and infrared ranges using other material platforms and opens new opportunities for the polarization control of guided waves such as metasurface-based waveguide polarizer [7].

Acknowledgements

This work was supported by the Russian Foundation for Basic Research (20-02-00636) and the Grant of the President of the Russian Federation (MK-2224.2020.2). O.Y. and A.B. acknowledge the support from the Foundation for the Advancement of Theoretical Physics and Mathematics "BASIS".

References

[1] Yu N, Capasso F 2014 *Nat. Mater.* **13(2)** 139
[2] Glybovski S B, Tretyakov S A, Belov P A, Kivshar Y S and Simovski C R 2016 *Phys. Rep.* **634** 1
[3] Kildishev A V, Boltasseva A, Shalaev V M 2013 *Science* **339** 6125
[4] Chen H T, Taylor A J, Yu N 2016 *Rep. Prog. Phys.* **79** 076401
[5] Yang Y, Jing L, Shen L, Wang Z, Zheng B, Wang H, Li E, Shen N-H, Koschny T, Soukoulis C M and Chen H 2017 *NPG Asia Mater.* **9(8)** e428
[6] Yermakov O Y, Hurshkainen A A, Dobrykh D A, Kapitanova P V, Iorsh I V, Glybovski S B and Bogdanov A A 2018 *Phys. Rev. B* **98(19)** 195404
[7] Yermakov O Y, Bogdanov A A, Lavrinenko A V 2019 *IEEE J. Sel. Top. Quant. Electron.* **25(3)** 1

Large-scale ReRAM device based on HKUST-1 thin film

Semyon V. Bachinin[1,a)], Sergey S. Rzhevskiy, Valentin A. Milichko[1,2]
1) ITMO University, St. Petersburg 197101, Russia
2) Institut Jean Lamour, Université de Lorraine, Nancy F-54011, France
a) Corresponding author: semen.bachinin@metalab.ifmo.ru

Abstract. The humanity creates more and more information every day. This aspect collides on a big problem – storage and processing. We can separate this problem on two tasks – choose some technologies which can satisfy modern requirement and searching new perspective material suitable for creating data-storage device. Herein, we demonstrate a concept of solving this problem by matching ReRAM – technology and metal-organic frameworks. We have achieved significant results and created a large-scale device with remarkable characteristics on/off ration.

1. Introduction

Metal-organic frameworks (MOFs) have been emerged as a new class of functional materials in chemistry and physics. The combination of porosity, metallic and organic nature, as well as crystallinity made possible the adaptive behavior of the frameworks in external fields [1]. This allows utilizing MOFs for controlled filtering devices, optoelectronic switches and data processing and storage. The latter, in particular, is actively developing within the framework of the ReRAM concept, where frameworks act as an active medium [2, 3, 4, 5]. Achieving low setting voltages up to 0.2 V, when using porous thin films, high signal-to-noise ratio up to 10^6, and recyclability make MOFs one of the key materials for future information recording devices. Herein, the expansion of the working MOF area from micro to a larger scales [6] and stable operation on a large number of cycles remains a challenge.
Here we demonstrate experimental realization local switching of conductivity our structure

2. Synthesis

A lower layer of gold contacts with the Thermal evaporation method was deposited on a glass treated in oxygen plasma with dimensions of 2.5x2.5 cm, after which a film of a metal-organic framework was applied to the structure with a lower contact by spin-coating at a rotation frequency of 500 rpm. The solutions were prepared in the same way as in [7], namely, 1.5 mM of $Cu_2(CH_3COO)_4 \cdot xH_2O$ in ethanol and 1,3,5-benzenetricarboxylic acid (H_3btc) ethanol. Next, an upper gold contact was applied on top of the hkust-1 film by a similar method. A photograph of the finished structure is shown in Figure 1.

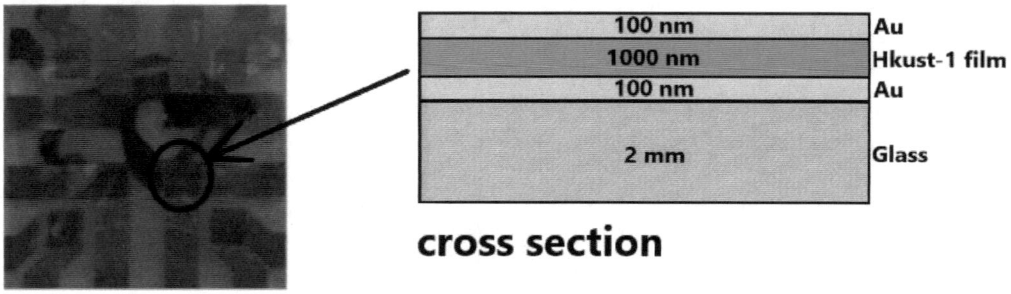

Figure 1. Our structure

3. Raman characterization

To confirm the composition of the obtained composition of the HKUST-1 film, Raman spectroscopy was carried out using a He-Ne laser with a wavelength of 633 nanometers as a source. *Objective* - 100X Mitutoyo with 0.9 numerical aperture.

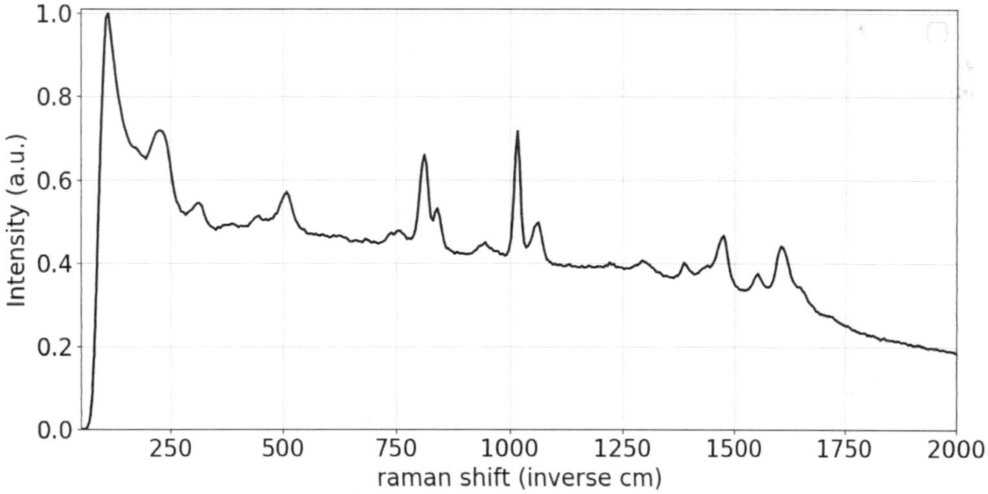

Figure 2. Raman spectra

4. Current mesearment and results

To confirm the operation of the obtained structure as a ReRAM device, current measurements of the dependence of the current strength on the applied voltage were carried out. To implement this kind of characterization, wires were soldered to the contacts using indium solder. The result of characterization can be seen in Figure 3. As a result, we can say that our structure can be used to record information by locally changing the conductivity property by applying a voltage of 5 V, while receiving a resistance jump from 2.5 MOhm to 13 Ohm. To rewrite information on our structure, you need to apply 13 volts in reverse polarity.

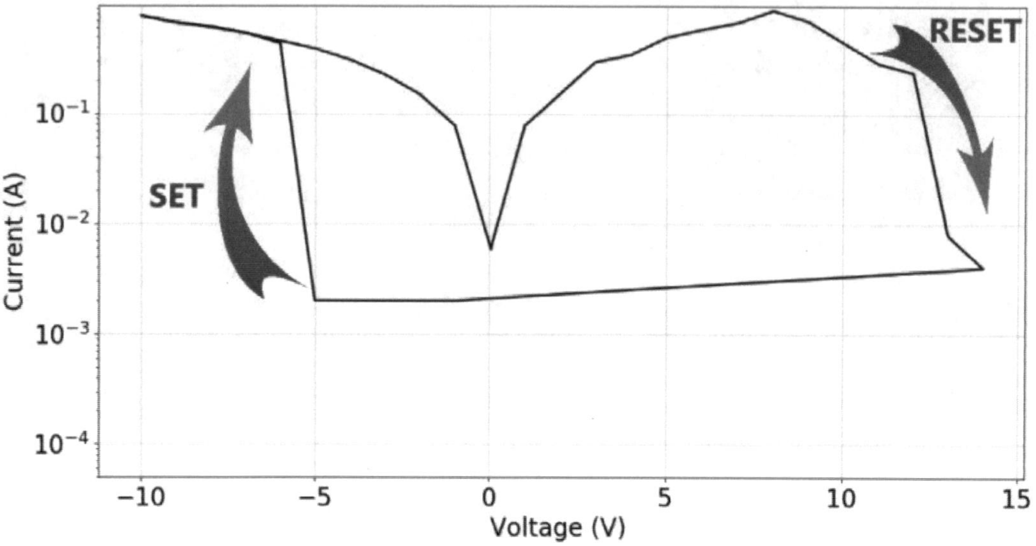

Figure 3. volt-ampere characteristics our structure

References:

[1] Y. A. Mezenov, A. A. Krasilin, V. P. Dzyuba, A. Nominé and V. A. Milichko, Advanced Science 6, 1900506 (2019).
[2] 2D Metal–Organic Framework Nanosheets with Time-Dependent and Multilevel Memristive Switching, Advanced Science 29, 1806637 (2019)
[3] The synthetic strategies of metal–organic framework membranes, films and 2D MOFs and their applications in devices, ***J. Mater. Chem. A***, 37, 21004-21035 (2019)
[4] Simultaneous implementation of resistive switching and rectifying effects in a metal-organic framework with switched hydrogen bond pathway, *Science Advances,* 5, 4515 (2019):
[5] Zeolitic-imidazole framework thin film-based flexible resistive switching memory, RSC Advances, 34, 20709 – 21359, (2017)
[6] Ambipolar Resistive Switching in an Ultrathin Surface-Supported Metal–Organic Framework Vertical Heterojunction, Nano Lett., 20, 2, 1080–1088 (2020)
[7] Advanced Fabrication Method for the Preparation of MOF Thin Films: Liquid-Phase Epitaxy Approach Meets Spin Coating Method, ACS Appl. Mater. Interfaces , 8, 31, 20459–20464 (2016)

Acknowledgments

The work was financially supported by Russian Science Foundation. Grant Number: 19-79-10241.

ITO-free Perovskite Light-Emitting Electrochemical Cell

M Baeva[1,2], D Gets[2], E Bodyago[2], A Mozharov[3], V Neplokh[4], A Nasibulin[5], I Mukhin[3] and S Makarov[2]

[1] Institute of Automation and Control Processes (IACP), Far Eastern Branch of Russian Academy of Sciences, Ulitsa Radio 5, 690041, Vladivostok, Primorsky Krai, Russia
[2] Department of Physics and Engineering, ITMO University, Lomonosova 9, 197101, St. Petersburg, Russia
[3] [1]Department of Physics, Alferov University, Khlopina 8/3, 194021, St. Petersburg, Russia
[4] Peter the Great St. Petersburg Polytechnic University, Politechnicheskaya 29, 195251, St. Petersburg, Russia
[5] Skolkovo Institute of Science and Technology, 30 Bolshoy Boulevard, bld. 1, 121205, Moscow, Russia

* Correspondence: baeva@spbau.ru

Abstract. Since Complementary metal–oxide–semiconductor technology is the conventional technology for micro- and optoelectronics, integration of emerging materials, such as halide perovskites, into the process is an important branch of perovskite technologies development. In this regard ITO free device research becomes increasingly important. The Perovskite Light-Emitting electrochemical cells are a promising alternative to conventional Perovskite Light Emitting Diodes. In this work we demonstrate green (λ_{EL} = 523 nm) $CsPbBr_3$ Perovskite Light-Emitting electrochemical cells with luminescence intensity of 50 kd/m^2 integrated with Si^{++}(111) substrate.

1. Introduction

Since first reported [1] in the early 90's PeLEDs research has been the staple research path for perovskite optoelectronic applications. Even though PeLEDs demonstrate some of the most impressive optical characteristics out of all devices based on emerging materials [2] some of the efficiency issues (low EQE) alongside with the complicated design hinder the technology introduction into industry.

Specifically, a typical PeLED device consists of photoactive perovskite layer enclosed between charge injection layers and electrodes in order to provide efficient charge injection from the low charge mobility perovskite film. Hence, while PeLED is a multilayer structure, that requires sophisticatedly balanced components in terms of their electrophysical and optical properties, PeLEC device is formed with a single multifunctional layer, that combines properties of high ion mobility and enhanced PLQY [3]. Due to the ion migration the p-i-n structure is formed, which makes for much simpler design. In addition, lowered work function requirements decrease demand for intricate electrode selection, which paves the way for novel electrode materials, such as carbon nanotubes, graphene, etc. [4] The simplicity of the design provides for scalable and cost-efficient technological solutions [5,6].

Likewise, flexibility of PeLECs processing and design opens a juncture to CMOS technology and microelectronics industry. Fabricating PeLECs directly on chip is a huge step into two technologies integration. Here, we report PeLEC manufactured directly on the Si(111) substrate with semitransparent Single Walled Carbon Nanotubes (SWCNt) top electrode [7].

Our PeLECs architecture, see Figure 1, consists of a patterned n-doped $Si^{++}(111)$ substrate ($n{\sim}10^{19}$ cm^{-3}) with thermally evaporated Al back contact (w~300 nm), single layer of photoactive perovskite material and SWCNt top contact with 80% transparency at 550 nm [7]. The $CsPbBr_3$ (E_g = 2,5 eV, λ_{EL} = 523 nm) perovskite was chosen as the photoactive material due to its enhancement ambient atmosphere and thermal stability [8] in addition to the high PLQY [2]. The SWCNt entangled network is the perfect candidate to use as a top semitransparent contact for PeLECs due to their chemical inertness, low resistivity in combination with high transparency [7].

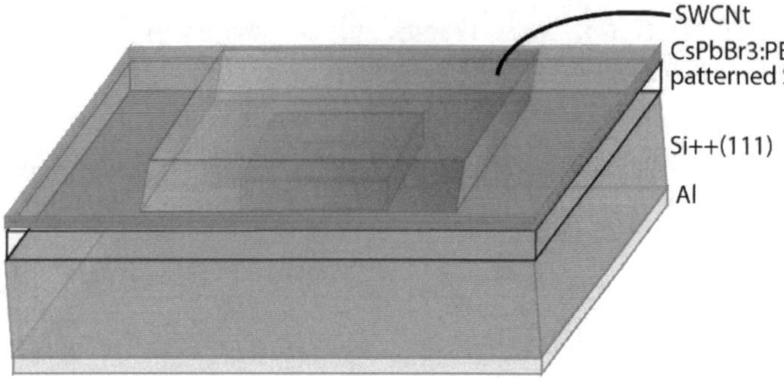

Figure 1. $CsPbBr_3$ LEC architecture scheme.

2. Experimental section

2.1 $Si^{++}(111)$ substrate patterning

The device pattern from the Figure 1 was obtain from masking $Si^{++}(111)$ surface area with dielectric epoxy-based negative photoresist. The $Si^{++}(111)$ substrates with preliminary deposited back Al contact were cleaned in a series of organic solvents (Acetone, IPA) and DE water to remove polar organic residues. Next, the samples were annealed at 200 °C for 2 h to remove water residues. After that, the SU8 epoxy-based resin was spin-coated onto the samples surface at 2000 rpm (thin film thickness calibrated to be 150 nm). In order to remove solvent from the thin film Si^{++} substrates were annealed at 95 °C on a hot plate for 2 min. The photoresist exposure was done by UV source (λ = 365 nm) through a mask with the square pattern of a size 3x3mm, see Figure 2. Exposure dose was set to 1,5 J/cm^2. The post-bake was done at 95 °C for another 2 min. The photoresist developing was done by 20 sec propylene glycol methyl ether acetate (PGMEA) treatment. In order to provide the samples chemical inertness, they were cured at 200 °C for 5 min.

Figure 2. The $Si^{++}(111)$ patterned sample.

2.2 Perovskite solution preparation

The $CsPbBr_3$ solution was prepared by adding CsBr (0,3 mmol/ml) and $PbBr_2$ (0,2 mmol/ml) in anhydrous DMSO solvent. The salts were left dissolving with moderate stirring (300 rpm) overnight without heating. The polyethylene oxide (PEO, M_w = 1000000 g/mol) DMSO solution (concentration 10 mg/ml) was stirred overnight at 60 °C, 200 rpm. Right before device fabrication the two pre-made solutions were mixed in a clean vial with the mass ratio $CsPbBr_3$:PEO = 1,25:1 and stirred for 1 hour without heating.

2.3 Device fabrication

First, the patterned Si^{++}(111) substrates surface was treated in O_2 plasma (P = 10 W) for 2min. Then, the perovskite active region (thickness ~100 nm) was fabricated through spin-coating process (10 µl, 60 sec) in dry N_2 filled glovebox. After that, a 40 nm thick layer of entangled SWCNt network was placed on top of the patterned structure and densified with dry IPA. The final device can be seen in the Figure 3,a).

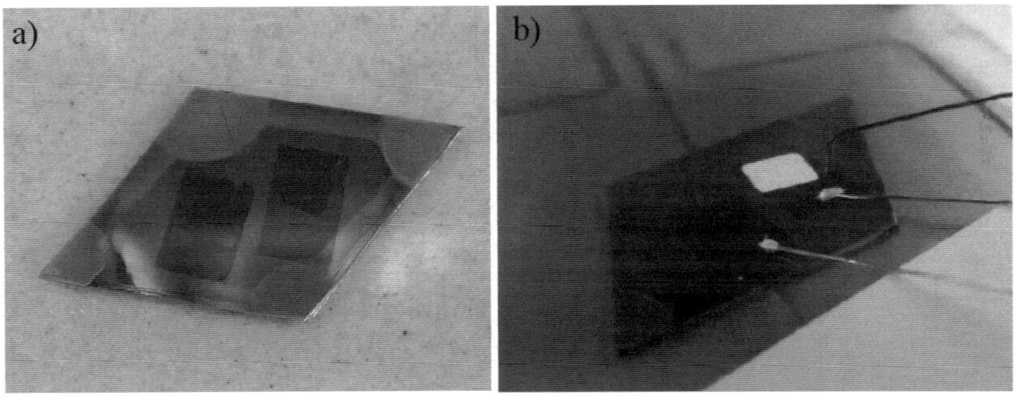

Figure 3. (a) The PeLEC on Si^{++}(111) device; (b) Our PeLEC device visible EL.

3. Results and discussion

The $CsPbBr_3$ thin film surface morphology was investigated with Atomic Force Microscopy (AFM) NT-MDT NTegra Aura set-up in tapping regime. In the Figure 4 the surface microstructure can be clearly seen: the overall homogeneous phase distribution indicates perovskite material uniformity with evident PEO inclusions only at the crystal grain boundaries. The mean perovskite composite crystal grain size is estimated to be ~230 nm.

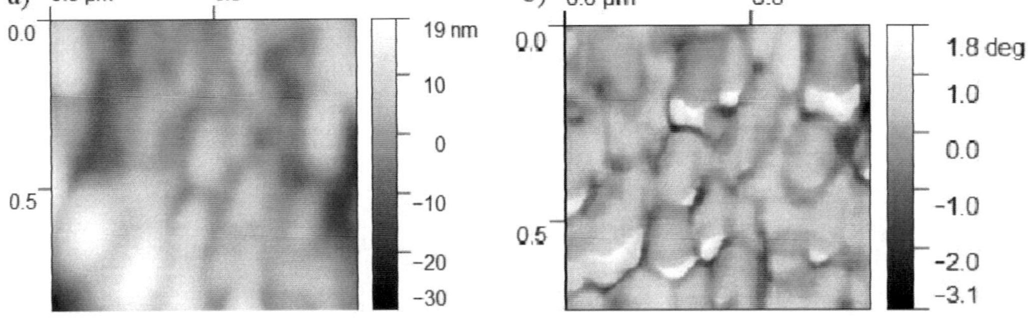

Figure 4. AFM (a) height and (b) phase scans of the $CsPbBr_3$:PEO thin film surface.

In order to measure key PeLEC characteristics, the device was connected to a Keithley 2401 source meter and a voltage 0f 0V to 3V was applied to the electrodes, see Figure 3,b). As can be seen from the

J-V curve in the Figure 5, our PeLEC opens at ~2,5 V, which corresponds well to the active material bandgap energy.

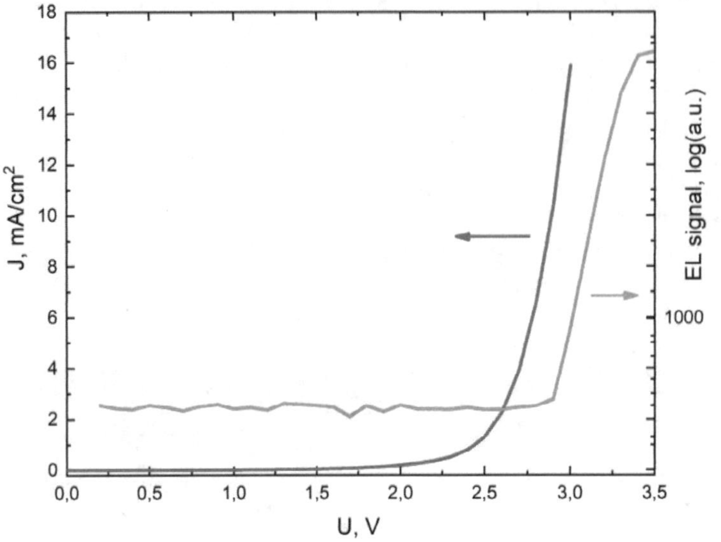

Figure 5. $CsPbBr_3$ LEC J-V characteristic (red curve) and LEC EL characteristic (green curve).

The visible electroluminescence (EL) curve of our device can be seen in the Figure 5. The EL signal emerges at voltage grater then V_{th} ~ 2,8 V. The luminescence characteristic of $CsPbBr_3$ LEC were measured using a spectroradiometer Spectrascan PR650. Our LEC demonstrates 50 kd/m^2 at 12 mA.

4. Conclusion

We have demonstrated green bright PeLEC device integrated with the key microelectronics material – silicon. The exceptional device design puts our work on the CMOS map and opens a whole route for Si and perovskites integration.

Acknowledgments

This work was supported by the Russian Science Foundation (project № 19-73-30023).

References

[1] Era, M., Morimoto, S., Tsutsui, T. & Saito, S. *Appl. Phys. Lett.* **65**, 676–678 (1994).
[2] Liu, X. K. *et al. Nat. Mater.* **20**, 10–21 (2021).
[3] Chang, S., Bai, Z. & Zhong, H. *Adv. Opt. Mater.* **6**, 1–19 (2018).
[4] Youssef, K., Li, Y., O'Keeffe, S., Li, L. & Pei, Q. *Adv. Funct. Mater.* **30**, 1–27 (2020).
[5] Xu, T. *et al. J. Mater. Chem. C* **7**, 8287–8293 (2019).
[6] Ling, Y. *et al. Adv. Mater.* **28**, 8983–8989 (2016).
[7] Tsapenko, A. P. *et al. Carbon N. Y.* **130**, 448–457 (2018).
[8] Zhang, C. *et al. APL Mater.* **7**, 071110 (2019).

Optimization of leaky-wave surface coil current using an analytical approach.

R. Balafendiev[1], V. Zalipaev[2], S. Glybovski[1], G. Solomakha[1]
[1]Department of Physics and Engineering, ITMO University, 197101 Saint Petersburg, Russia
[2]Department of Photonics, ITMO University, 197101 Saint Petersburg, Russia

Abstract. In this work, we investigate the optimal coefficients of the exponential current excited on a leaky wave surface coil. The respective functional is first derived analytically and later computed numerically using Python. The results are compared to the same problem modeled in Comsol Multiphysics.

1. Introductions

Most surface coils used in magnetic resonance imaging (MRI) nowadays work using a resonant approach. Such devices, based on resonant principles, are able to create current distributions at the surface of the phantom with constant phase. However, recently another approach has been suggested, which uses leaky-wave antennas as its underlying idea. Leaky-wave antennas work by creating an exponential current distribution (i.e. exponential amplitude and linear phase) which is depicted in Fig.1. Such leaky-wave antenna has non-resonant behavior, which offers several benefits, namely the current distribution being more configurable and the coil being significantly more broadband [1]. There are two ways to feed such a coil - unilateral feeding with the port positioned at the edge of the coil and bilateral feeding with the point in the center. In this work, we study the latter, since it realizes the possibility of focusing the field underneath the feeding point. The primary metric when determining such coil's efficiency is the ratio of the right-handed circularly polarised transverse magnetic field at the region of interest - B_1^+ to the square root of the accepted power $\sqrt{P_{acc}}$ i.e. the power dissipated in an object of investigation. Said metric has a theoretical maximum in ultimate intrinsic signal-to-noise ratio (UISNR) previously found by R. Lattanzi et. al. [2] to be realized by a specific current distribution with non-constant phase and amplitude. It is our assumption that exponential current serves as a good approximation of such UISNR distribution. For resonant coils, this ratio is determined mostly by coil length and shape. However, in leaky-wave coils realizing exponential current distribution, there are two additional degrees of freedom we can utilize: attenuation coefficient α and phase constant β. For the purpose of finding the optimal value for these two coefficients, a 2D problem has been considered. The expression for the functional giving the ratio of $B_1^+/\sqrt{P_{acc}}$ has been derived analytically and then computed numerically using Python code. The resulting values were then verified using the finite element method in the numerical simulation software Comsol Multiphysics.

2. Problem statement and Theoretical Investigation

We considered a half-space in the positive values of z filled with a dielectric representing a phantom with permittivity ϵ and conductivity σ. The exponential current flows along the boundary in both positive and negative x-directions (Fig.2). As such, the current $\mathbf{j}$ can be expressed componentwise as $j_x(x) = I_0 e^{(-\alpha+i\beta)|x|}$, $j_y(x) = j_z(x) = 0$.

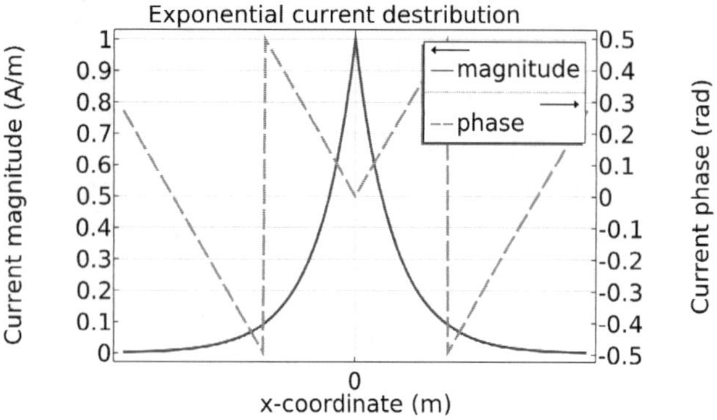

Figure 1. Plots of current magnitude and phase.

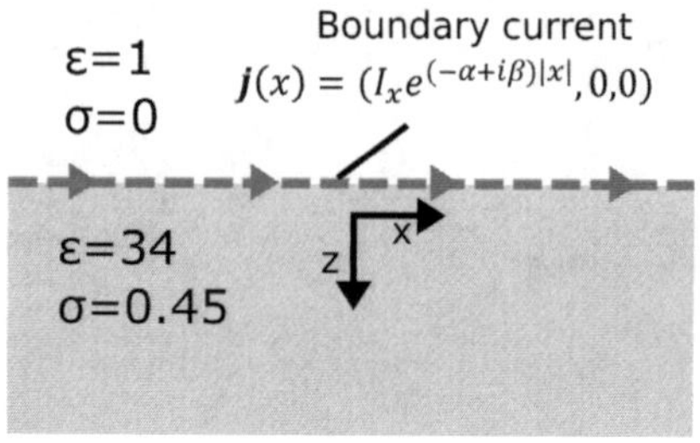

Figure 2. An exponential line current along the boundary of a semi-infinite phantom.

We start with Maxwell's equations and some other definitions [3]:

$$rot\mathbf{E} = ik_1\mathbf{H}, \qquad rot\mathbf{H} = -ik_1\mathbf{E} + \frac{4\pi}{c}\mathbf{j}, \qquad k_1 = k_0\sqrt{\epsilon}, \qquad k_0 = \frac{\omega}{c}, \qquad \epsilon = \epsilon_0 + \frac{4\pi i\sigma}{\omega}, \tag{1}$$

where k_0 is the wave number in the free space and k_1 is the wave number in the dielectric. The vector potential satisfies the inhomogeneous Helmholtz equation $(\Delta + k_1^2)\mathbf{A}(x,y,z) = -\frac{4\pi}{c}\mathbf{j}(x,y,z)$. Its solution is given via a convolution-type integral of the product of the Green function $G(R)$ and the current $\mathbf{j}$, where $R = |\mathbf{r} - \mathbf{r}'|$ and $\mathbf{r}$ and $\mathbf{r}'$ are the radius vectors to a point in the phantom and a point on the boundary respectively.
Since the current density is not dependant on the y coordinate, we can consider a 2D problem in which the only existing component of the vector potential is given by

$$A_x(x,z) = -\frac{i}{4c}I_0 \int_{-\infty}^{\infty} H_0^{(1)}(k_1\sqrt{(x-x')^2+z^2})e^{-\alpha|x'|+i\beta|x'|}dx', \tag{2}$$

with the Green function of a line current being given by a Hankel function $H_0^{(1)}$.

Using the plane wave decomposition for a Hankel function in the vector potential A_x we can express it as $A_x(x,z) = -\frac{I_0\gamma}{2\pi c}\int_{-\infty}^{\infty} e^{ik_x x + ik_z z}\frac{dk_x}{k_z(k_x^2-\gamma^2)}$, where k_x and $k_z = \sqrt{k_1^2 - k_x^2}$ are the wave numbers along x and z respectively. If we then find the field vectors using their relations to the vector potential $\mathbf{H} = rot\mathbf{A}$ and $\mathbf{E} = \frac{i}{k_1}(k_1^2 + \nabla div)\mathbf{A}$, we can finally obtain following expressions:

$$H_y(x,z) = -\frac{iI_0\gamma}{\pi c}\int_0^{\infty}\cos(k_x x)e^{ik_z z}\frac{dk_x}{k_x^2-\gamma^2}, \tag{3}$$

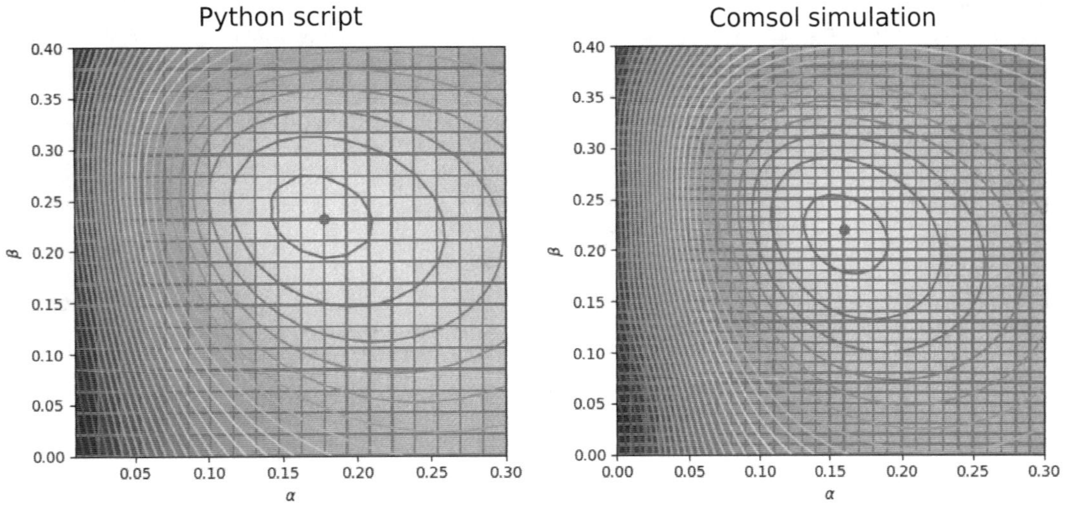

Figure 3. Comparison between the two methods used to calculate the functional $\Phi(\alpha, \beta)$. α and β are given in cm^{-1}

$$E_x(x, z) = -\frac{iI_0\gamma}{\pi c k_1} \int_0^\infty \cos(k_x x) e^{ik_z z} \frac{k_z dk_x}{k_x^2 - \gamma^2}, \tag{4}$$

$$E_z(x, z) = -\frac{I_0\gamma}{\pi c k_1} \int_0^\infty \sin(k_x x) e^{ik_z z} \frac{k_x dk_x}{k_x^2 - \gamma^2}, \tag{5}$$

where $\gamma = i\alpha - \beta$.

In our case the optimization is studied by a maximum of the deformed functional

$$\Phi(\alpha, \beta) = \frac{|H_y(0, h)|^2}{2\sigma \int_0^{x_{max}} dx \int_{z_{min}}^{z_{max}} dz |\mathbf{E}(x, z)|^2}. \tag{6}$$

The numerical computation of the integral in the denominator is complicated by the integrand diverging with z_{min} approaching 0. As such, a non-zero z_{min} had to be chosen, potentially introducing some error to the results. The functional was calculated using the `integrate` module of the Python 3.7 `scipy` package. The search of maximum was then done using the Nelder-Mead algorithm implementation from the `optimize` module yielding the optimal values of $\alpha = 17.250$ and $\beta = 23.498$ for the parameters used. Comsol Multiphysics 4.5 model was then used to verify the values found via Python with $\alpha = 15.822$ and $\beta = 21.678$ being the result (Fig.3).

3. Conclusion

In this work, we have proven the existence of an optimum of exponential current parameters by finding it numerically. The result given by numerically calculating the functional is similar to the one obtained via Comsol and is close to 8% difference for both α and β. The difference could be partially explained by the Comsol model factoring in the contribution of the layer near the boundary to P_{acc} and effectively having z_{min} be equal to zero. The fact that the functional reaches its maximum at non-zero values of α and β serves to further strengthen our assertion that such current distribution can be used to approximate the UISNR current from [2]. Knowing the optimal current distribution will be crucial in manufacturing the most effective leaky-wave coils possible.

Acknowledgments

This work was supported by the Russian Foundation for Basic Research (Grants No. 19-29-10038).

References

[1] Solomakha G, Svejda J T, van Leeuwen C, Rennings A, Raaijmakers A J, Glybovski S and Erni D 2021 *Nat. Commun.* **12** 455 ISSN 2041-1723 URL `http://www.nature.com/articles/s41467-020-20708-w`
[2] Lattanzi R, Wiggins G C, Zhang B, Duan Q, Brown R and Sodickson D K 2018 *Magn. Reson. Med.* **79** 1789–1803 ISSN 15222594
[3] Weinstein L 1988 *Electromagnetic waves [in Russian]* (Radio i svyaz', Moscow)

 2015 (2021) 012012 doi:10.1088/1742-6596/2015/1/012012

Chiral light in single-handed Fabry-Perot resonators

Alexey Taradin and Denis G. Baranov*

Center for Photonics and 2D Materials, Moscow Institute of Physics and Technology, Dolgoprudny 141700, Russia

*e-mail: denis.baranov@phystech.edu

Abstract. Chirality is a universal phenomenon that is encountered on many different length scales in nature. Interaction of chiral matter with chiral light results in the effect of circular dichroism, which underlies many techniques of discriminating molecular enantiomers. Enhancing dichroic effects is typically achieved by interfacing chiral matter with various optical resonators. In this context it is important to understand how the eigenmodes of optical cavities relate to the field states with well-defined handedness. Here, we present the model of a single-handedness chiral optical cavity supporting only an eigenmode of a given handedness without the presence of modes of other helicity. Resonant excitation of the cavity with light of appropriate handedness enables formation of a helical standing wave with a uniform chirality density, while the opposite handedness does not cause any resonant effects. Our findings expand the set of tools for investigations of chiral matter and open the door towards studies of chiral electromagnetic vacuum states.

1. Introduction

Interaction of chiral electromagnetic field with chiral matter results in the well-known effect of circular dichroism, which underlies numerous techniques of discriminating molecular enantiomers [1]. Enhancing dichroic effects is typically achieved by interfacing chiral matter with various optical resonators [2,3]. For this reason, it is important to understand how the eigenmodes of optical cavities are projected onto the field states with well-defined handedness. Considering the classical type of an optical cavity - a Fabry-Perot (FP) resonator formed by two homogeneous metallic mirrors - one could expect that excitation of such a cavity with a circularly polarized light would result in excitation of a helical FP mode. In reality, eigenmodes of a FP cavity (at normal incidence) do not have any handedness [4] - helicity flipping of the wave travelling between the two mirrors creates a standing wave with exactly zero helicity, as sketched in Fig. 1(a). In the context, the concept of duality [5] plays an important role: dual structures preserve the handedness of incident light upon scattering, thus allowing existence of eigenmodes of well-defined helicity.

This poses a quest for optical cavities supporting eigenmodes of well-defined handedness [6]. An additional requirement that one could impose is that the structure supports an eigenmode of a certain handedness only, but not that of the opposite one (at least in the given spectral range). That would result in a single-handedness chiral optical cavity that would couple efficiently to emitters of a certain handedness, and considerably weaker to the opposite enantiomer. A progress has been made in design of helicity-preserving cavities that are based on excitation of large in-plane momentum modes [7]. However, such cavities possess mirror or inversion symmetry and therefore support eigenmodes of both helicities at the same time.

In this paper, we theoretically investigate a system fulfilling this requirement - an optical resonator supporting a single eigenmode of well-defined helicity, and lacking modes of opposite helicity in the appropriate spectral range. The proposed structure is a FP cavity formed by a stack of two photonic crystal slab mirrors [8]. The handedness preserving property of the mirrors allows resonant excitation of a standing helical wave of a certain handedness inside the cavity, while at the same time completely transmitting waves of the opposite handedness. The proposed system expands the set of tools available for investigations of chiral matter and opens the door to studies of chiral electromagnetic vacuum.

2. Results

In order to implement the desired modal spectrum of the cavity, we need to have at our disposal a reflecting structure that preserves handedness of the reflected wave, but is transparent to the opposite handedness, as sketched in Fig. 1(b,c). To that end, we utilize the geometry proposed in ref. [8]. The helicity-preserving mirror is a dielectric photonic crystal slab with a square lattice and its unit cell possessing only the mirror symmetry with respect to the horizontal plane M_{xy} and two-fold rotational symmetry around the z-axis $C_{2,z}$, Fig. 1(d). M_{xy} ensures that the single mirror itself is not chiral, while $C_{2,z}$ allows conversion between clockwise and counter-clockwise polarization in reflection, necessary for handedness preservation. At the same time, $C_{2,z}$ makes the effect slightly more robust with respect to the incidence angle.

We begin by designing a single mirror made of a dielectric with refractive index $n = 4$ in air featuring the desired spectral characteristics at a fixed wavelength of 800 nm. By optimizing the structure with the use of the particle swarm algorithm (and keeping sub-diffraction lattice periodicity), we find the set of unit cell dimensions yielding near-perfect polarization conversion for normally incident light of both helicities: thickness H = 142 nm, unit cell size L = 366 nm, rc = 90 nm, rx = 89 nm, ry = 34 nm, Cx = 59 nm, Cy = 104 nm.

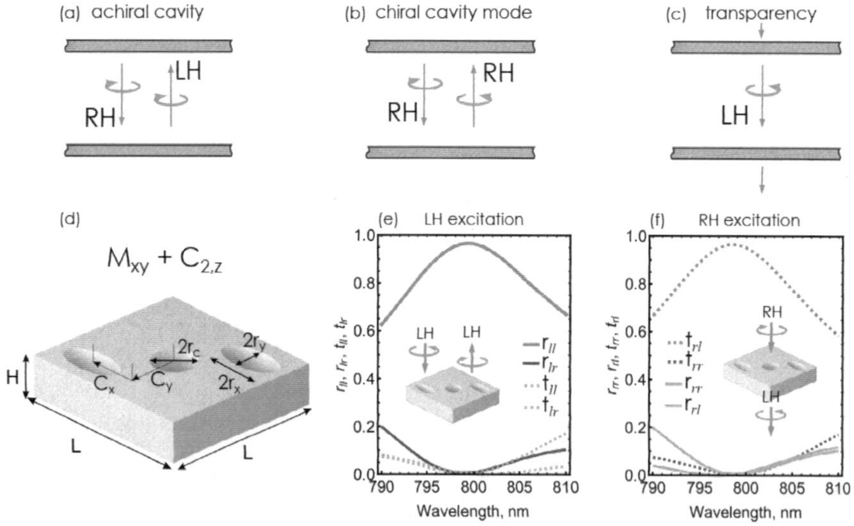

Fig. 1. (a-c). Concept of a single-handedness cavity. An ordinary achiral cavity, (a), does not support chiral eigenmodes due to handedness flipping of the standing wave inside the cavity. A single-handedness optical cavity supports a mode of a well-defined handedness, (b), and does not support a mode of the opposite handedness at the same wavelength, (c). (d) The unit cell of the photonic crystal slab mirror featuring handedness-preserving reflection at normal incidence. (e) Reflection and transmission coefficients of the optimized mirror in the circular polarization basis for the LH incidence exhibiting near-unity reflection with preserved handedness. (f) Reflection and transmission coefficients of the optimized mirror for the RH incidence exhibiting near-complete transmission.

With these dimensions, LH light normally incident from top is nearly perfectly reflected with handedness preservation, $|r_{ll}|{\sim}1$, while a RH incident plane wave is nearly perfectly transmitted into the opposite handedness, $|t_{rl}|{\sim}1$, Fig. 1(e,f).

Now we stack two such mirrors to form a cavity. In order to find the optimal cavity thickness sustaining a chiral eigenmode, we simulate volume-average electric field enhancement in the region inside the cavity and sweep the distance between the mirrors L. For thickness L=1724 nm the field enhancement spectra demonstrate handedness-selective electric field enhancement for LH excitation at the wavelength of $\lambda_{res} = 798$ nm, and much weaker enhancement for RH excitation, Fig. 2(b).

Furthermore, the electric field intensity inside the cavity plotted at λ_{res} demonstrates strikingly different spatial distributions for two incident polarizations. Field induced by the incident RH light features a typical standing wave pattern with alternating peaks and troughs caused by the non-zero reflection of the wave by each of the mirrors. The LH excitation, on the contrary, features a region of nearly uniform electric field intensity. Such a pattern is indicative of the helical standing wave, or so-called polarization standing wave formed by two counter-propagating plane waves of the same handedness.

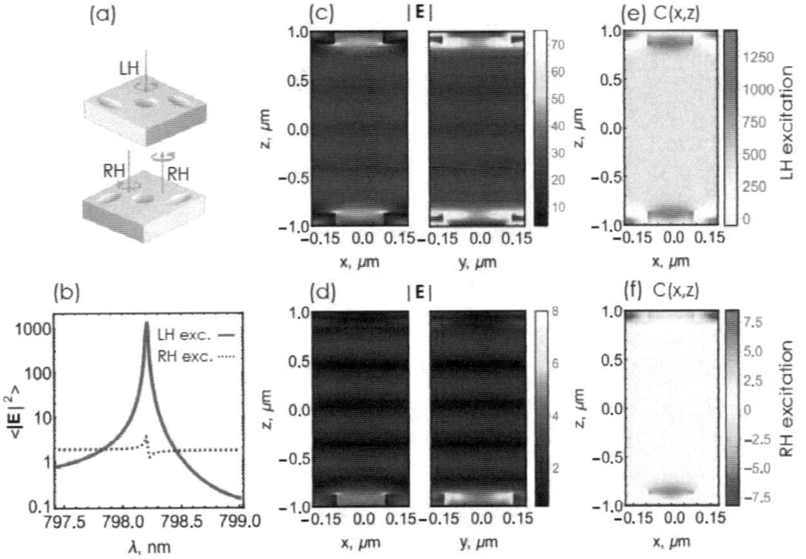

Fig. 2. (a) Illustration of the formation of a standing chiral wave in the region between two handedness-preserving mirrors. (b) Volume-average electric field enhancement in the region inside the cavity for the cavity thickness L = 1724 nm for RH and LH excitations at normal incidence. (c) Spatial distribution of electric field intensity inside the cavity induced by the LH incident wave at the resonant wavelength $\lambda_{res} = 798$ nm; the field is plotted in the two vertical middle planes of the unit cell. (d) The same as (c) for RH excitation. (e) Normalized optical chirality density C for the LH illumination at $\lambda_{res} = 798$ nm indicating excitation of a pure RH cavity mode. (f) The same as (e) for the RH excitation.

In order to confirm the excitation of a chiral cavity mode by the LH incident light, we calculate the optical chirality density C inside the cavity: $C = \frac{\omega\varepsilon_0}{2}\mathrm{Im}(\mathbf{E}\cdot\mathbf{B}^*)$. The resulting spatial distribution of C (normalized by the absolute value of the chirality density of the incident circularly polarized wave C_0) at the resonant wavelength $\lambda_{res} = 798$ nm for a LH excitation clearly demonstrates a uniform region of largely enhanced positive chirality density inside the cavity, Fig. 2(e), suggesting excitation of RH fields. RH excitation, at the same time, results in a significantly lower magnitude of chirality density, which is also positive, indicating that the RH illumination also partially couples to the RH chiral mode.

3. Conclusions

To conclude, we have introduced the concept of a single-handedness chiral optical cavity. By utilizing the design of helicity-preserving photonic crystal mirrors, we engineered a Fabry-Perot cavity that supports an optical eigenmode of a well-defined handedness, and does not support the opposite handedness in the relevant wavelength range. Upon excitation, such a cavity features a uniform region of enhanced electromagnetic field intensity and chirality density. The concept of single-handedness optical cavity adds a new tool for investigations of chiral matter and opens the door towards studies of chiral electromagnetic vacuum states.

Acknowledgments
The work was supported by the Russian Science Foundation grant 19-79-00362.

References

[1] Barron L D 2004 *Molecular Light Scattering and Optical Activity* (Cambridge University Press)
[2] Y. Tang and A. E. Cohen 2010 *Physical Review Letters* **104**, 163901
[3] A. O. Govorov, Z. Fan, P. Hernandez, J. M. Slocik, and R. R. Naik 2010 *Nano Letters* **10**, 1374
[4] I. Fernandez-Corbaton, M. Fruhnert, and C. Rockstuhl 2016 *Physical Review X* **6**, 031013
[5] I. Fernandez-Corbaton, X. Zambrana-Puyalto, N. Tischler, X. Vidal, M. L. Juan ,and G. Molina-Terriza 2013 *Physical Review Letters* **111**, 060401
[6] H. H¨ubener, U. De Giovannini, C. Schafer, J. Andberger, M. Ruggenthaler, J. Faist, and A. Rubio 2020 *Nature Materials*, https://doi.org/10.1038/s41563
[7] J. Feis, D. Beutel, J. K̈opfler, X. Garcia-Santiago, C. Rockstuhl, M. Wegener, and I. Fernandez-Corbaton 2020 *Physical Review Letters* **124**, 033201
[8] B. Semnani, J. Flannery, R. Al Maruf, and M. Bajcsy 2020 *Light: Science and Applications* **9**, 23

Fabrication of Au and Ag – Coated AFM Probes for Tip-Enhanced Raman Spectroscopy

L S Basalaeva[1], **N N Kurus**[1], **E E Rodyakina**[1,2], **K V Anikin**[1], **A G Milekhin**[1,2]
[1] Rzhanov Institute of Semiconductor Physics, Lavrentiev Ave. 13, 630090 Novosibirsk, Russia
[2] Novosibirsk State University, Pirogov 2, 630090 Novosibirsk, Russia

E–mail of the corresponding author: basalaeva@isp.nsc.ru

Abstract. In this work, approaches for fabrication metal coated probes for Tip Enhanced Raman Spectroscopy (TERS) are considered. It was proposed to use optical characterization of probes to achieve the effective TERS of semiconductor nanoobjects. The shape and size of the metal cluster at the tip apex determines the position of the localized surface plasmon resonance, the electromagnetic field enhancement and, thus, TERS performance. The possibility of optimizing the characteristics of the probes for TERS studies of nanoobjects has been investigated.

1. Introduction

Tip Enhanced Raman Spectroscopy – is a unique research technique that allows to obtain simultaneously information on the surface topology and its chemical state with a nanometer spatial resolution [1, 2]. TERS has become increasingly popular in recent years in various fields of science, including physics, chemistry of low-dimensional systems, biology, and pharmaceuticals [3, 4]. Usually TERS experimental setup combines scanning probe microscope such as scanning tunneling microscope (STM) or atomic force microscope (AFM) with a Raman spectrometer. To record the TERS signal from an object, a laser spot is focused onto the tip apex, exciting a localized surface plasmon, thus produce an “electromagnetic hotspot” at the tip apex. The TERS enhancement and spatial resolution depend strongly on the tip parameters (materials, radius, surface roughness and cone angle). At the moment, various methods for fabricating TERS tips have been proposed, including vacuum evaporation, chemical deposition of metals on silicon probes [5, 6], attachment of plasmonic nanoparticles to the tip, selective deposition [7], electrochemical etching [8, 9]. Despite remarkable progress, development effective techniques for the fabrication of stable probes for TERS measurements are still under demand. The goal of this work is to develop approaches for fabricating TERS probe that are capable of providing the high enhancement of the TERS signal and high spatial resolution for TERS mapping.

In this work, we have considered the possibility of fabricating gold and silver coatings on silicon probes for the implementation of TERS studies. Gold and silver are the two most widely used metal for TERS tips due to their high free electron density, strong LSPR effect in the visible region and relative chemical stability. Ag-coated probes allow higher field enhancement in TERS mode than Au-coated probes. However, silver sulfide or silver oxide films rapidly grow on the Ag-coated tip, thus greatly reducing the lifetimes of such a probe. The shape and size of the metal cluster at the tip apex determines the spectral position of the localized surface plasmon resonance (LSPR) and the enhancement factor of the local electromagnetic field. However, the SEM image of a probe which demonstrates the presence of a metal cluster of the required size on its tip is no guarantee of sufficient TERS enhancement. Therefore, we use optical characterization of TERS tips in combination with direct structural analysis and demonstrate TERS images of certain nanoobjects.

2. Experimental

A TERS setup consisting of the AFM AIST-NT coupled with Raman spectrometer (Xplora, Horiba) was used for the TERS experiments. Solid –state lasers with the wavelengths of 638 and 785nm were used as an excitation source. A power of the laser radiation was 1.6 and 0.16 mW for excitation of Au and Ag probe, respectively. To create TERS probes, we used standard semicontact silicon AFM probes of the Vit_P series [10] with optical access from above. The metal coating (Ti, Au, Ag) of the probe was formed by vacuum deposition process. Before metal deposition some of Si probes were covered with 5 nm thick SiO_2. The SiO_2 layer was formed by the oxidation of monosilane in a tubular reactor at a temperature of 350°C and a pressure of 0.95 Torr. To ensure better adhesion of the gold film to silicon or SiO_2, a 20 nm thick Ti layer was previously deposited on the probes. A thin layer of gold (5 nm) was deposited on an Ag coated probe to protect against a rapid degradation. SEM measurements were carried out on a Hitachi SU8220 setup.

3. Results and discussion

A silver or gold metal film with a thickness of 100-250 nm was deposited on the tip of the initial silicon probes. The morphologies of the tip of the probe were characterized by scanning electron microscopy. SEM images of the apexes of TERS probes covered with Au without intermediate layer of SiO_2 (Fig 1a), covered with Au with intermediate layer of SiO_2 (Fig 1b), and covered with Ag (Fig 1c) are shown in Figure 1. For comparison, the SEM image of the tip of a commercially available Horiba TERS probe is presented in Fig 1d. For simplicity these cantilevers are denoted as Au-Si, Au-SiO2, Ag-Si, and Au-H, correspondingly.

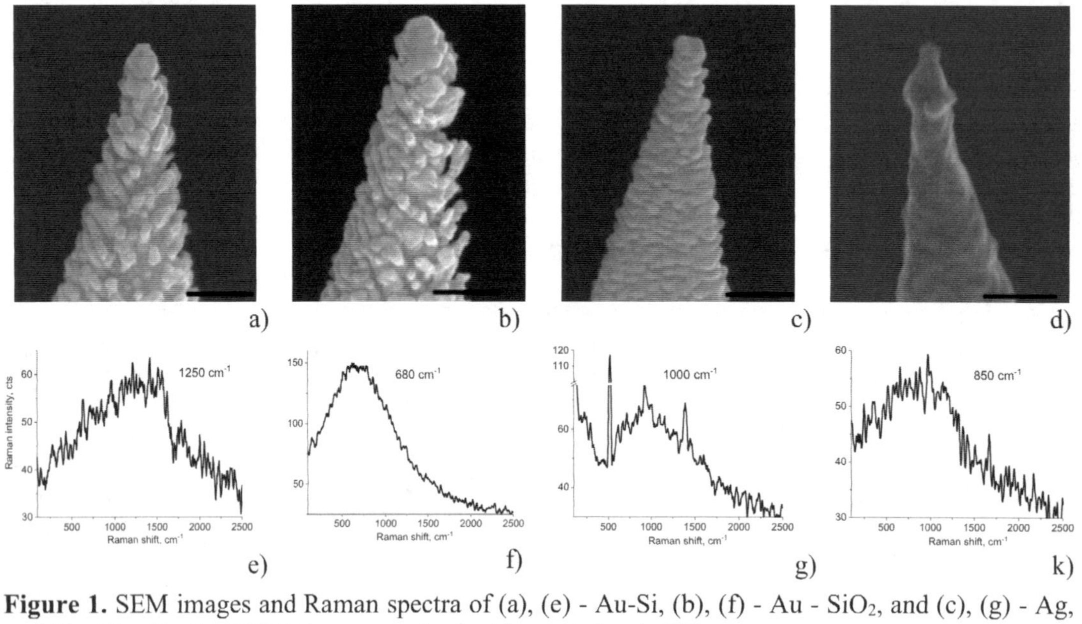

Figure 1. SEM images and Raman spectra of (a), (e) - Au-Si, (b), (f) - Au - SiO_2, and (c), (g) - Ag, and (d), (k) - Horiba TERS tips, respectively. The scale bar is 200 nm.

It can be seen from the Fig 1a-d) that a single metal nanocluster is formed at the tip apex of all probes. The size of the metal cluster is 80 nm for the Ag-Si probe, ~70 nm for the Au - Si probe, ~100 nm for the Au-SiO_2 probe, ~50 nm for the Horiba probe.

Obviously, the TERS probes with nanoclusters having different structural parameters and consisting of various metals should have different LSPR frequencies. The LSPR frequency can be shifted by changing the parameters of the metal film deposited on the silicon probe or/and the refractive index a dielectric sublayer preliminary deposited on the probe, for example, SiO_2. In this case, an increase in the magnitude of the electric field enhancement and, consequently, the TERS signal is possible. However, the SiO_2 sublayer increases the radius of the tip apex (Fig 1b), that leads to decreasing the spatial resolution.

For TERS experiments, it is necessary to excite effectively the LSPR at the tip apex by focusing the laser light. During the alignment process, the tip apex is scanned with a laser beam and the scattered light from the tip is detected. Fig 1e-k shows the Raman spectra from the tips shown in Figs 1a-d, respectively. All spectra shown in Fig 1e-k have a broad peak, which corresponds to the position of the LSPR. The LSPR energy amounts to 1.78 эВ (λ_{exc}=680nm) for the Au-Si probe, 1.85 эВ (λ_{exc}=670 нм) for the Au - SiO_2 probe, 1.815eV (λ_{exc}=683 нм) for the Ag-Si probe, 1.828 эВ (λ_{exc} =678 нм) for the Au-H probe. TERS measurements of semiconductor nanoobjects (carbon nanotubes and CdSe nanocrystals (NCs)) were obtained for all types of probes. However, the TERS intensity was different for different probes. The AFM and TERS images of a carbon nanotube (for G peak 1591 cm^{-1}) on an Au surface, measured simultaneously using Au -SiO_2 probe are presented in Fig 2 a, b) correspondingly. TERS resolution estimated from Fig.2b is about 10 nm. A typical TERS spectrum of carbon nanotube is shown in Fig.2 c.

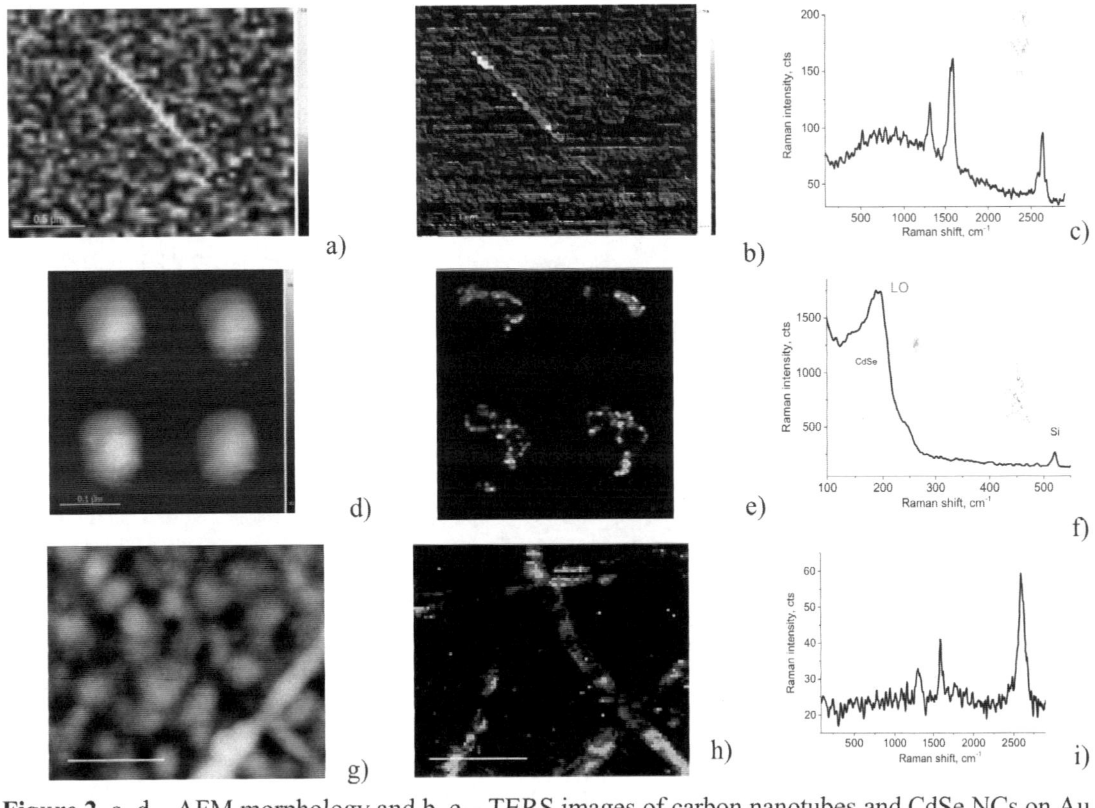

Figure 2. a, d – AFM morphology and b, e – TERS images of carbon nanotubes and CdSe NCs on Au nanodisk array, correspondingly, measured simultaneously with Au -SiO_2 tip. The TERS images of the carbon nanotubes and CdSe were obtained using the G peak (1591 cm^{-1}) and LO phonon mode of CdSe (near 200 cm^{-1}). The TERS spectra shown in (c) and (i) are taken at positions indicated by red crosses; g, h, i– AFM, TERS images, and TERS spectra of carbon nanotubes obtained by a Horiba tip are shown for comparison. TERS spectra of carbon nanotubes and CdSe NCs were excited with λ_{exc}=638 and 785 нм, respectively. The scale bar is 500 nm (a), 100 nm (d) and 200 nm (g).

Using the same probe an intense TERS signal from longitudinal optical (LO) phonons from CdSe NCs was detected at the wavelength of 785 nm (Fig 2 c-e). NCs were deposited on an array of gold nanodisks with a height of 50 nm.

For the first time, TERS images from CdSe NCs on Au nanodisk arrays were demonstrated and described in detail in [11]. As one can see from Fig.2d, TERS signal originates predominantly from CdS NCs deposited on Au nanodisks while the signal from the NCs located between Au nanodisks is significantly weaker. The intensity of the TERS map obtained by the Au-SiO_2 probe is approximately three times higher than that for the Au-Si probe (not shown in the figure).

TERS images of carbon nanotubes obtained using the Au-Horiba probe (Fig. 2f-m) have comparable intensity with those obtained by Au-SiO_2 probe (Fig. 2a-c). The Ag-Si tip also has demonstrated effective TERS enhancement from carbon nanotubes deposited on a gold substrate. However, the plasmon enhancement ability of the Ag coated TERS probe rapidly diminishes over time, despite the thin protective layer of gold. Here, the scattering signal from the tip apex can be an important factor for determining the state of the metal coating of the tip.

4. Conclusion

In this paper, Au- and Ag-coatings of AFM probes were deposited and single metal nanoclusters were formed at the tip apexes. Using these metal tips significant enhancement of Raman scattering from LO phonons in CdSe NCs and G-mode in carbon nanotubes was demonstrated and TERS images from semiconductor nanoobjects were recorded. The LSPR wavelength for tip of the probe with gold coating (formed with or without SiO_2 sublayer) and silver coating is determined from Raman spectra of the metal probes. The LSPR position of the probes is significantly changed by varying the parameters of the deposited metal film and affects the TERS performance.

Acknowledgment

The authors gratefully acknowledge financial support from RFBR (Project Nos. 19-52-12041_NNIO_a) and the Ministry of Education and Science of the Russian Federation. SEM studies were performed using the equipment of CKP "Nanostructures". The authors are thankful to Mr. A.S. Medvedev for metal deposition.

References

[1] Sartin M M, Su H-S, Wang X and Ren B 2020 *J. Chem. Phys* **153** 170901.
[2] Zhang Yao, Zhang Yang and Dong Z-C 2021 *Chinese Journal of Chemical Physics* **34** 1-14.
[3] Deckert-Gaudig T, Taguchi A, Kawata S and Deckert V 2017 *Chem. Soc. Rev* **46** 4077-4110.
[4] Lee J, Crampton K T, Tallarida N and Apkarian V A 2019 *Nature* **568** 78-82.
[5] Huang T-X, Huang S-C, Li M-H, Zeng Z-C, Wang X and Ren B 2015 *Anal Bioanal Chem* **407** 8177-8195.
[6] Capaccio A, Sasso A,Tarallo O and Rusciano G 2020 *Nanoscale* **12** 24376-24384.
[7] Yuan C C, Zhang D and Gan Y 2017 *Review of Scientific Instruments* **88** 031101.
[8] Williams C and Roy D 2008 *Journal of Vacuum Science & Technology* B **26** 1761-1764.
[9] Kim D H, Lee C, Jeong B G, Kim S H and Jeong M S 2020 *Nanophotonics* **9** 2989-2996.
[10] Tips Nano; url: https://tipsnano.com/catalog/afm-special/top-visual/vit-p/
[11] Milekhin I A et al. 2020 *Nanoscale Advances* **2** 5441-5449.

Guided exciton-polaritons in a subwavelength dielectric slab integrated with a 2D semiconductor

F. Benimetskiy[1], **A. Yulin**[1], **V. Kravtsov**[1], **A. Mikhin**[1], **I. Iorsh**[1], **A. Samusev**[1], **D. N. Krizhanovskii**[1,2]

[1]Department of Physics and Engineering, ITMO University, St. Petersburg, 197101, Russia
[2]Department of Physics and Astronomy, University of Sheffield, Sheffield S3 7RH, UK

E-mail: fedor.benimetskiy@metalab.ifmo.ru

Abstract. New-generation nonlinear planar polaritonic devices based on 2D semiconductors demonstrate great potential for a wide range of practical applications. In this work, we experimentally study strong light–matter coupling between waveguide photons and excitons in a photonic system based on dielectric slab waveguides integrated with 2D transition metal dichalcogenides.

1. Introduction

One of the challenges facing the field of nanophotonics and nanoelectronics is the development of on-chip active photonic devices, where light can be controlled by light of very weak intensity. A possible solution could be a polaritonic system, where efficient photon-photon interactions at low intensities could be achieved via strong photon-exciton coupling, resulting in the emergence of new quasiparticles, so-called exciton-polaritons. Strong light-mater coupling has been demonstrated in different designs of photonic structures based on III – V semiconductors. In particular, these types of semiconductor structures have been used to demonstrate several interesting nonlinear optical effects associated with exciton-polaritons, such as waveguide polariton lasers [1], parametric effects [2], and polariton solitons [3, 4].

Along with the study of polaritonic systems where III–V materials are used as an active medium, two-dimensional transition metal dichalcogenides (TMDs) attract the attention of the scientific community because excitons in these monolayer semiconductors have large oscillator strengths and binding energies, which makes them an excellent active medium for new polariton systems, with possibilities of room temperature operation [5].

In this work, we demonstrate strong light–matter coupling between waveguide modes and excitons in a photonic system based on slab dielectric waveguides with embedded TMD monolayers as an active medium with high optical nonlinearity.

2. Result and discussion

To realize the proposed polaritonic system, we chose an unpatterned Ta_2O_5 waveguide on a SiO_2 sublayer, where light is coupled in/out through grating couplers. After preliminary characterization of the waveguide transmission, the WSe_2 monolayer, which was fabricated by mechanical exfoliation from a commercial bulk crystal from HQ Graphene, was transferred onto the waveguide surface between the grating couplers (Fig.1).

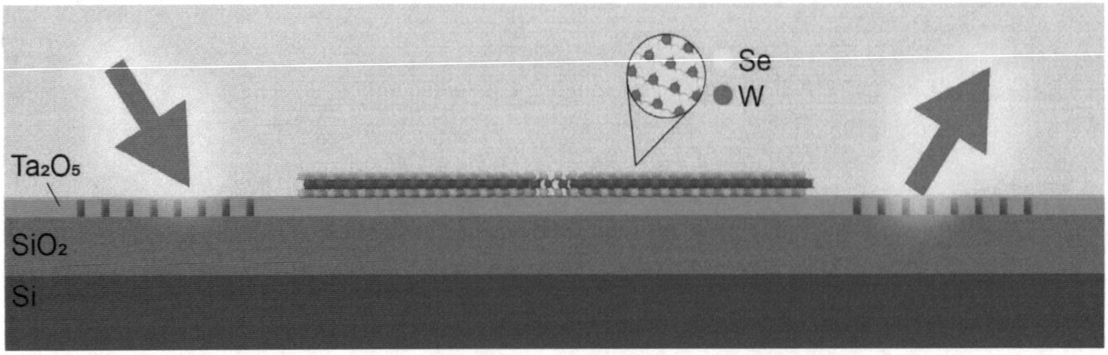

Figure 1. Schematic diagram of a waveguide integrated with a WSe_2 monolayer.

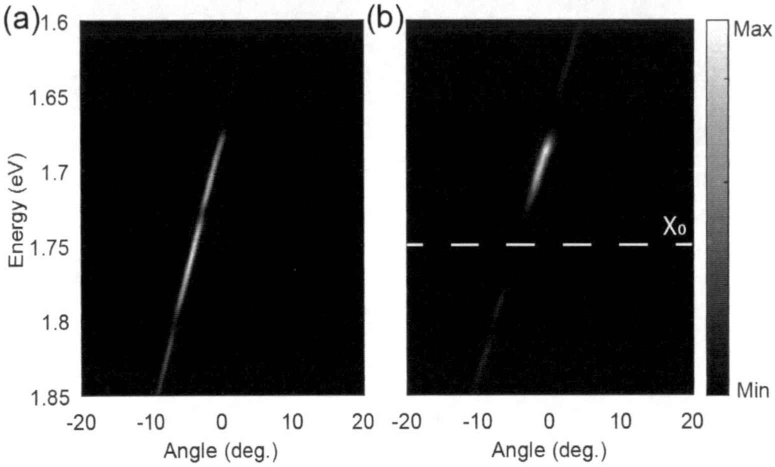

Figure 2. Angle- and photon-energy-resolved transmittance maps (a) before and (b) after transfer of a monolayer. For the transmission measurements, a halogen lamp was used as a white light source. The exciton spectral position is highlighted by a dashed white line.

In order to characterize the transmission of the fabricated hybrid structure, we used a setup for Fourier plane imaging combined with a closed-cycle helium cryostat. During the experiment, we used a halogen lamp as a white light source for the transmission measurements. The white light (polarized along the grating that corresponds to transverse-electric (TE) field mode) from a tungsten halogen lamp (Ocean Optics HL-2000) was focused onto the input grating coupler using an objective lens (Mitutoyo Objective 50x/0.65). Light coming from the sample was collected by the same microscope objective and spatially filtered. As a result, only light from the output grating was detected. The spectrally resolved signal was measured by projecting the Fourier plane of the objective onto the entrance slit of an imaging spectrometer (Princeton SP 2550) and detecting the light with a CCD detector (PyLoN 400BR eXcelon).

Figures 2a, b show the measured angle-resolved transmission maps before and after TMD monolayer transfer (the sample temperature is 7 K). We observed a dip in the transmission spectrum at the wavelength corresponding to the neutral exciton in the WSe_2 monolayer centered around 1751 meV.

Finally, we measured the evolution of ps laser pulses (OPA Light Conversion Ltd + pulse shaper) in the transmittance spectra after their propagation through the waveguide with

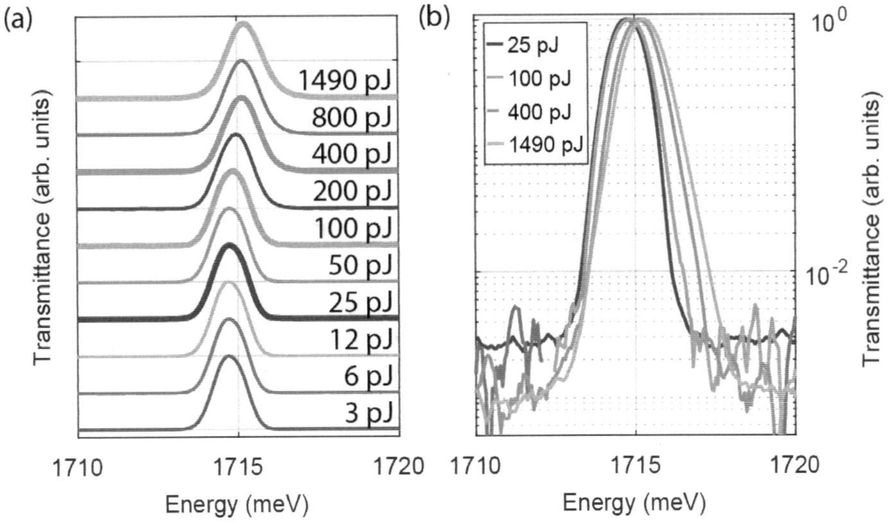

Figure 3. (a) Measured laser transmission spectra for pulse energy increasing from bottom to top. Selected spectra (bold lines) are plotted in log scale (b).

increasing pulse energy in order to verify the realization of the strong coupling of light and matter in the hybrid structure (Fig. 3 a,b). The sample is excited by pulses with a spectral full width at half maximum of 1 nm corresponding to a ~1 ps temporal duration. The pulses are coupled in through the input grating, propagate 100 um, and coupled out through the output grating. But only 50 um of the waveguide length is covered by a TMD monolayer. The detuning of the laser spectrum from the exciton resonance is approximately -36 meV, despite this, in the experiment, we observed a blue shift of the laser spectrum. Such behavior is typical for polariton systems and related to nonlinear phenomena [6].

3. Conclusion

To summarize, we have experimentally observed strong light–matter coupling between waveguide photons and excitons in a hybrid photonic system based on slab dielectric waveguides integrated with two-dimensional semiconductors. Our results suggest a new platform for future TMD-based polariton devices where various nonlinear effects can be observed.

Acknowledgments

The reported study was funded by Russian Science Foundation, grant #21-12-00218.

References

[1] Suárez-Forero D, Riminucci F, Ardizzone V, De Giorgi M, Dominici L, Todisco F, Lerario G, Pfeiffer L, Gigli G, Ballarini D *et al.* 2020 *Optica* **7** 1579–1586

[2] Suárez-Forero D G, Riminucci F, Ardizzone V, Karpowicz N, Maggiolini E, Macorini G, Lerario G, Todisco F, De Giorgi M, Dominici L *et al.* 2020 *arXiv preprint arXiv:2005.11385*

[3] Walker P M, Whittaker C E, Skryabin D V, Cancellieri E, Royall B, Sich M, Farrer I, Ritchie D A, Skolnick M S and Krizhanovskii D N 2019 *Light: Science & Applications* **8** 1–11

[4] Walker P, Tinkler L, Skryabin D, Yulin A, Royall B, Farrer I, Ritchie D, Skolnick M and Krizhanovskii D 2015 *Nature communications* **6** 1–7

[5] Schneider C, Glazov M M, Korn T, Höfling S and Urbaszek B 2018 *Nature communications* **9** 1–9

[6] Kravtsov V, Khestanova E, Benimetskiy F A, Ivanova T, Samusev A K, Sinev I S, Pidgayko D, Mozharov A M, Mukhin I S, Lozhkin M S *et al.* 2020 *Light: Science & Applications* **9** 1–8

Nonlinear imaging of whispering gallery modes in GaN microwires

Yury Berdnikov[1], Igor Shtrom[1,2], Maria Rozhavskaya[3], Wsevolod Lundin[4], Nicholas Hendricks[5], Rachel Grange[5], and Maria Timofeeva [5]

[1] St. Petersburg State University, Universitetskaya Emb. 13B, 199034, St. Petersburg, Russia
[2] Institute for Analytical Instrumentation RAS, 190103, Saint-Petersburg, Russia
[3] Soitec, Chemin des Franques, 38190, Bernin, France
[4] Ioffe Institute, Politekhnicheskaya 26, 194021, Saint-Petersburg, Russia
[5] ETH Zurich, Optical Nanomaterial Group, Institute for Quantum Electronics, Department of Physics, Auguste-Piccard Hof 1, 8093 Zurich, Switzerland

yury.berdnikov@itmo.ru

Abstract. In this work non-scanning far-field nonlinear optical microscopy is employed to study the whispering gallery modes in tapered GaN microwire resonators. We demonstrate the confinement of whispering gallery modes under near-infrared excitation with the photon energy close to half of GaN bandgap. Our results indicate the enhancement of yellow-green luminescence by whispering gallery modes in GaN microwires.

1. Introduction

Epitaxially grown III-V nano- and microwires (MW) demonstrate smooth facets and high crystalline quality, and therefore, may be employed as naturally formed optical resonators for whispering gallery modes (WGM). Generally, WGM resonators are attractive structures for photonic applications due to their ultrahigh quality factors (Q). For example, recently, lasing in high-Q WGM cavities has been successfully demonstrated in hexagonal micro- and nanorods of GaN and ZnO [1,2].

Among III-V semiconductors, gallium nitride (GaN) is promising material for applications in nonlinear photonics in view of large second and third order nonlinear coefficients, wide direct bandgap (3.4 eV) and compatibility with high operational temperatures (melting point is 2500 °C) [3,4]. Large nonlinear coefficients measured in GaN [5,6] enable various nonlinear photonic applications such as signal processing [7], imaging and others [8].

Recently, the two-photon absorption (TPA) enhancement by WGM were observed [9] in GaN structures, which attracts additional interest to GaN NWs as TPA-compatible WGM cavities. Meanwhile, several studies of epitaxial bulk and thin-film GaN reported the enhancement of TPA around one half of the bandgap which allows efficient conversion from near-infrared (NIR) to ultra-violet visible range [5,6].

Spatial information on local field enhancement and structural defects is crucial for device design and operation diagnostics the MW-based applications [10,11]. This information may be accessed by far-field nonlinear optical microscopy, which is the fast non-invasive technique for high-resolution mode mapping in nano- and microresonators [12,13]. Here we use non-scanning far-field nonlinear optical microscopy to characterize the WGMs confined in epitaxially grown GaN MW resonators under NIR excitation.

2. Methods

2.1. MW synthesis

The samples of GaN MWs were synthesized by metal-organic chemical vapor deposition (MOCVD) on sapphire substrates covered by titanium films using a system with a horizontal flow reactor [14]. Ti films with a thickness around 14 nm were thermally deposited on the substrates prior to GaN growth. The MOCVD process started with 10 min heating in a nitrogen-ammonia atmosphere. After that, the atmosphere was changed to hydrogen-ammonia and trimethylgallium was introduced into the reactor chamber. The growth temperature was fixed to 1040 °C, and the growth pressure was 200 mbar.

Morphology of the MWs was analyzed by scanning electron microscopy (SEM) in Magellan FEI 400 system. For optical characterization, the MWs were mechanically transferred on indium-tin-oxide substrates.

2.2. Optical characterization

Nonlinear optical studies of GaN MWs resonators were carried out in a home-built non-scanning transmission nonlinear optical microscope [13]. The MWs were placed in the focal plane of the incident polarized laser beam. The samples were excited with Ti:Sapphire tuneable laser source, that can operate in the region from 690 nm to 1000 nm. The laser was operated in pulsed regime with the pulse frequency of 80 MHz. The laser beam was focused with 10X objective and collected with 100X objective after passing thought the sample. Then the signal at excitation wavelength was filtered behind the sample with BG-39 filter (transmission range 300 – 700 nm). Output nonlinear signal was focused onto an CCD camera to get images. To measure the spectra from MWs excited with NIR laser source the setup was equipped with the SpectraPro imaging spectrometer. The intensity of the signal on each image was analyzed precisely in each pixel.

3. Results and discussion

In nonlinear imaging we collected the visible-range signal generated in MW resonator under NIR excitation at 740 nm. Panels (a) and (c-e) in Figure 3 show the SEM and CCD images of tapered GaN MW, respectively. Figure 1(f) characterize the MW tapering by diameter (D) dependence on the coordinate along the MW axis z.

CCD images in Figure (b-d) show the maximums of luminescence forms the three-dot patterns for specific positions along the MW axis. The intensive signal was also observed at the fractured ends of the MW. Images of the same MW in panels (b), (c) and (d) of Figure 1 correspond to three different positions of the excitation beam. By comparing CCD and SEM images we attribute the three dots of the luminescence patterns to the centers of MW facets and relate to WGMs confined at specific resonance diameters.

In tapered MW the resonant facet separations d_{WGM}^{m} for m^{th} WGM can be calculated as [15]:

$$d_{WG}^{m} = \frac{\lambda}{3n}\left(m + \frac{6}{\pi}\left(\beta\sqrt{3n^2 - 4}\right)\right), \quad (1)$$

where λ is the wavelength, n is the refractive index and the parameter $\beta = n$ for TM polarization and $\beta = 1/n$ for TE polarization.

Figure 1(e) shows the resonance diameters $D = 2d/\sqrt{3}$ for WGMs in MW resonator with hexagonal cross-section calculated using eq. (1). The calculated resonant diameters D_{WGM} match to three-dot intensity patterns in CCD images via the D(z) dependence. The D(z) was extracted from the fit of the diameters obtained from the analysis of the SEM image. We stress that matching can be obtained in the case of the second harmonic with λ = 370 nm.

Figure 1(f) shows the spectrum of the nonlinear signal collected from MW resonator. The spectrum has a broad yellow-green luminescence (YGL) band and the peak of second harmonic at 370 nm corresponding to double frequency of the excitation at 740 nm.

The YGL band in luminescence of GaN nanostructures is known to originate from the different types of the surface defect states [16,17].

Placing the 400 nm longpass FLG-400 filter between the sample and the CCD camera does change considerably the CCD images. Thus, the observed three-dot patterns can be related to YGL band. At the same time, mode matching shows that the three-dot pattern at the WGM resonance diameters. Therefore, we observe the enhancement of YGL by WGMs, as illustrated by schematic in Figure 1(g). These observations are consistent with recently reported increase of luminescence [18] and two-photon absorption [9] intensity related to confinement of WGMs.

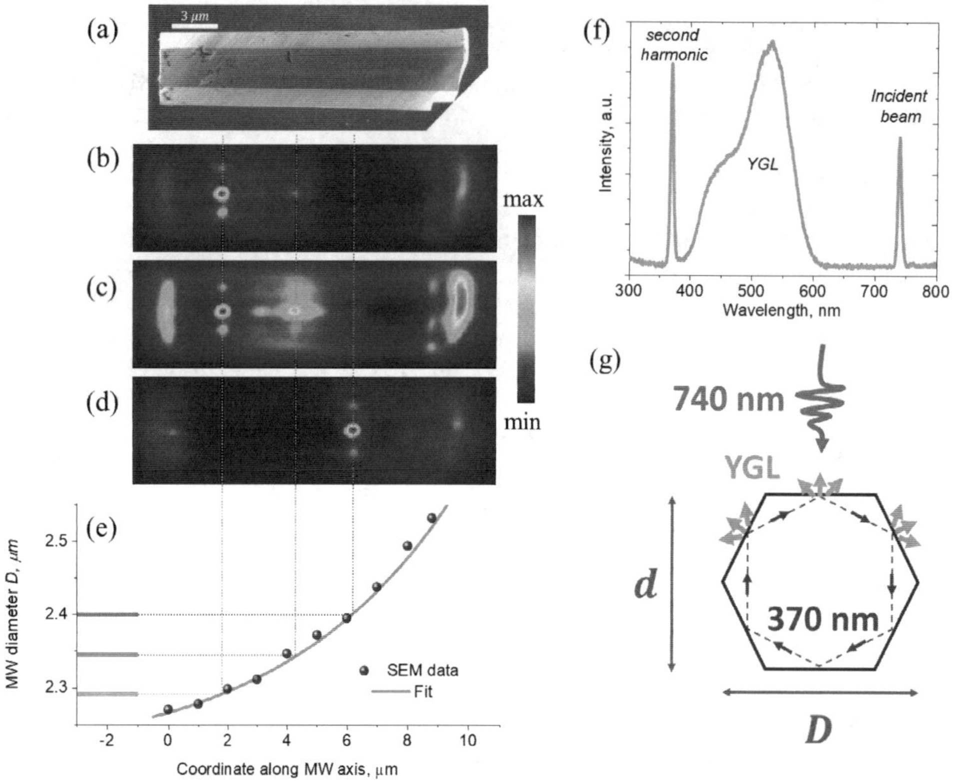

Figure 1. SEM (a) and CCD images (b-d) and spectrum (f) of the MW resonator on indium-tin oxide substrate under 740 nm excitation. The CCD images are taken at three different beam positions (using only BG-39 filter). (e) Diameter of the tapered MW as the function of coordinate z along the MW axis and matching of WGM positions to resonance diameters.

Conclusions

In conclusion, we have shown that tapered GaN MWs, grown in Ti-assisted MOCVD process can be employed as WGM resonators. In this type of nanostructures, the confinement of WGMs was demonstrated under NIR excitation with the photon energy close to half of GaN bandgap. Far-field nonlinear imaging and spectroscopy were successfully employed as the tools for mapping and characterization of the confined WGMs. The obtained luminescence patterns witness the enhancement of YGL by WGM confined at resonance diameters of the GaN MW. We believe that our method of mode mapping will be usefull for fundamental studies as well as for device diagnostics.

Aknowledgements

Y. B. and I. S. acknowledge the financial support of the Russian Science Foundation under the grant 19-72-30004. The authors thank the Scientific Centre for Optical and Electron Microscopy (ScopeM) of ETH Zurich. This work was supported by the Swiss National Science Foundation grant 150609, 163916 and Ambizione grant no 179966, by the European Union’s Horizon 2020 research and innovation program from the European Research Council under the Grant Agreement No 714837 (Chi2-nano-oxides).

References

[1] Yang Y-D, Tang M, Wang F-L, Xiao Z-X, Xiao J-L and Huang Y-Z 2019 Whispering-gallery mode hexagonal micro-/nanocavity lasers [Invited] *Photonics Res.* **7** 594

[2] Tessarek C, Heilmann M and Christiansen S 2014 Whispering gallery modes in GaN microdisks, microrods and nanorods grown by MOVPE *Phys. status solidi* **11** 794–7

[3] Stassen E, Pu M, Semenova E, Zavarin E, Lundin W and Yvind K 2019 High-confinement gallium nitride-on-sapphire waveguides for integrated nonlinear photonics *Opt. Lett.* **44** 1064

[4] Munk D, Katzman M, Westreich O, Nun M Bin, Lior Y, Sicron N, Paltiel Y and Zadok A 2018 Four-wave mixing and nonlinear parameter measurement in a gallium-nitride ridge waveguide *Opt. Mater. Express* **8** 66

[5] Krishnamurthy S, Nashold K and Sher A 2000 Two-photon absorption in GaN, GaInN, and GaAlN alloys *Appl. Phys. Lett.* **77** 355–7

[6] Sun C-K, Liang J-C, Wang J-C, Kao F-J, Keller S, Mack M P, Mishra U and DenBaars S P 2000 Two-photon absorption study of GaN *Appl. Phys. Lett.* **76** 439–41

[7] Roth J M, Murphy T E and Xu C 2002 Ultrasensitive and high-dynamic-range two-photon absorption in a GaAs photomultiplier tube *Opt. Lett.* **27** 2076

[8] Hayat A, Nevet A, Ginzburg P and Orenstein M 2011 Applications of two-photon processes in semiconductor photonic devices: invited review *Semicond. Sci. Technol.* **26** 083001

[9] Chen Z, Lou G, Zhu H, Chen A, Wu Y, Ren Y, Li J, Qiu Z, Gui X and Tang Z 2018 Enhancement of two-photon absorption photoresponse based on whispering gallery modes *Nanoscale* **10** 14047–54

[10] Usui S, Kitade S, Morichika I, Kohmura K, Kusa F and Ashihara S 2017 Near-Field Imaging of Infrared Nanoantenna Modes Under Oblique Illumination *J. Phys. Chem. C* **121** 26000–6

[11] McGehee W R, Michels T, Aksyuk V and McClelland J J 2017 Two-dimensional imaging and modification of nanophotonic resonator modes using a focused ion beam *Optica* **4** 1444

[12] Grange R, Brönstrup G, Kiometzis M, Sergeyev A, Richter J, Leiterer C, Fritzsche W, Gutsche C, Lysov A, Prost W, Tegude F, Pertsch T, Tünnermann A and Christiansen S 2012 Far-Field Imaging for Direct Visualization of Light Interferences in GaAs Nanowires *Nano Lett.* **12** 5412–7

[13] Timofeeva M, Lang L, Timpu F, Renaut C, Bouravleuv A, Shtrom I, Cirlin G and Grange R 2018 Anapoles in Free-Standing III–V Nanodisks Enhancing Second-Harmonic Generation *Nano Lett.* **18** 3695–702

[14] Rozhavskaya M M, Lundin W V., Lundina E Y, Davydov V Y, Troshkov S I, Vasilyev A A, Brunkov P N, Baklanov A V., Tsatsulnikov A F and Dubrovskii V G 2015 Gallium nitride nanowires and microwires with exceptional length grown by metal organic chemical vapor deposition via titanium film *J. Appl. Phys.* **117** 024301

[15] Kirschbrown J R, House R L, Mehl B P, Parker J K and Papanikolas J M 2013 Hybrid standing wave and whispering gallery modes in needle-shaped ZnO rods: Simulation of emission microscopy images using finite difference frequency domain methods with a focused Gaussian source *J. Phys. Chem. C* **117** 10653–60

[16] Huang P, Zong H, Shi J, Zhang M, Jiang X, Zhong H, Ding Y, He Y, Lu J and Hu X 2015 Origin of 3.45 eV Emission Line and Yellow Luminescence Band in GaN Nanowires: Surface Microwire and Defect *ACS Nano* **9** 9276–83

[17] Reshchikov M A 2014 Temperature dependence of defect-related photoluminescence in III-V and II-VI semiconductors *J. Appl. Phys.* **115** 012010

[18] Schmitt S W, Sarau G and Christiansen S 2015 Observation of strongly enhanced photoluminescence from inverted cone-shaped silicon nanostructures *Sci. Rep.* **5** 17089

High-Q two-groove resonator for all-dielectric Bloch surface wave platform

E A Bezus[1,2], E A Kadomina[1,2], D A Bykov[1,2], A A Morozov[1] and L L Doskolovich[1,2]

[1] Image Processing Systems Institute of RAS — Branch of the FSRC "Crystallography and Photonics" RAS, 151 Molodogvardeyskaya st., Samara 443001, Russia
[2] Samara National Research University, 34 Moskovskoe shosse, Samara 443086, Russia

evgeni.bezus@gmail.com

Abstract. We propose a simple integrated resonant structure for the Bloch surface wave platform, which consists of two subwavelength grooves patterned on the surface of a one-dimensional dielectric photonic crystal. We demonstrate that the investigated structure can operate in a parasitic-scattering-free regime and, in this case, provide unity transmittance and zero reflectance at resonance conditions associated with the excitation of a leaky mode of the structure localized at the central ridge formed by the grooves. The proposed structure may find application in integrated photonic devices for optical filtering and analog optical computing.

1. Introduction

In recent years, all-dielectric nanophotonics has attracted considerable research attention due to both theoretical interest and potential practical applications driven by the possibility of obtaining high-Q low-loss resonant structures [1]. One of the promising platforms of all-dielectric integrated nanophotonics is the Bloch surface wave (BSW) platform. Bloch surface waves are electromagnetic waves supported by the interfaces of photonic crystals. In the simplest case, they propagate along an interface between a one-dimensional photonic crystal and a homogeneous dielectric medium [2]. For Bloch surface waves, various integrated optical elements have been proposed, for example, refractive lenses [3] and phase-shifted Bragg gratings [4].

In the present work, we propose and numerically investigate a simple resonant structure for Bloch surface waves, which comprises two parallel grooves on the surface of a one-dimensional dielectric photonic crystal. We show that if the bandgap configuration of the photonic crystal is chosen properly, the structure can operate without any "parasitic" scattering in the case of oblique incidence of a TE-(transverse electric) polarized BSW, exhibiting high-Q resonances. The quality factor of the resonances can be tuned by changing the width of the grooves and, neglecting the absorption losses, surface imperfections and finiteness of the structure, can be made arbitrarily large. The obtained results may be useful for filtering as well as pulse- and beam-shaping applications in integrated nanophotonics.

2. Results and discussion

The geometry of the investigated resonant integrated structure is shown in Fig. 1(a). The structure consists of two grooves on the surface of a one-dimensional dielectric photonic crystal (PC). Here, we consider an example with the following parameters: free-space wavelength of the incident wave $\lambda = 630$ nm, refractive indices of the PC layers $n_1 = 2.3227$ (Nb_2O_5) and $n_2 = 1.4762$ (SiO_2), and thicknesses of the PC layers $h_1 = 63$ nm and $h_2 = 189$ nm. Thickness of the upper layer of the photonic crystal having the refractive index n_1 amounts to $h_1' = 136$ nm, and the thickness of this layer in the groove regions equals $h_1'' = 69$ nm. At the given parameters, the interface of the PC supports a TE-polarized BSW with effective refractive index (propagation constant normalized by the wavenumber) $n_{\mathrm{BSW}} = 1.95$ and a TM-polarized BSW with effective refractive index $n_{\mathrm{BSW,TM}} = 1.6$. At the same time, in the groove regions, no BSWs are supported. We will study the oblique incidence of the TE-polarized BSW on the two-groove structure (the angles of incidence θ are measured in the *yz*-plane from the negative direction of the *z* axis). The bandgap and interface of the PC are designed in such a way that at angles of incidence exceeding the "critical" angle $\theta = 66.5°$, no reflected and transmitted cross-polarized (TM-polarized) BSWs are excited, and no energy is scattered from the PC interface to the superstrate and to the "bulk" part of the PC. Thus, all the energy of the incident TE-polarized BSW is divided between the reflected and transmitted BSWs having the same polarization. In this case, the considered two-groove structure can be considered as an integrated counterpart of a resonant three-layer structure in a high-index environment, which was considered in [5] for achieving large optical field enhancement over a large volume.

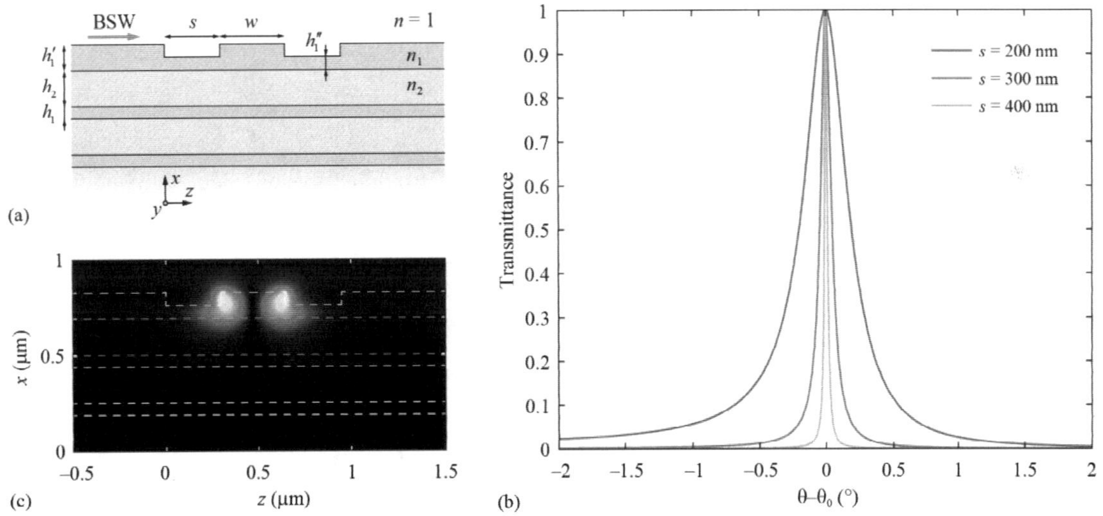

Figure 1. (a) Geometry of the two-groove resonant structure on the Bloch surface wave platform. (b) Angular transmittance spectra of three two-groove structures with groove width *s* ranging from 200 to 400 nm. (c) Distribution of the *y*-component of the electric field at resonance conditions ($s = 300$ nm, $\theta = 70.42°$).

Let us consider the optical properties of three two-groove structures with a fixed ridge width $w = 350$ nm and different groove widths *s* of 200, 300, and 400 nm. Figure 1(b) shows the dependencies of the transmittance on the angle of incidence θ for the three considered groove widths. These plots were calculated using an efficient in-house implementation of the Fourier modal method [6] adapted for the solution of integrated optics problems [7]. It is evident that the structures exhibit resonant peaks, the width of which can be changed by changing the groove width. For ease of comparison, the transmittance spectra are shown vs. the deviation from the resonance angle θ_0, which is slightly different for the three considered structures and ranges from $\theta_0 = 70.40°$ to $\theta_0 = 70.47°$. Away from the resonance, the incident wave is almost completely reflected by the structure due to an effect similar to total internal reflection. However, if the leaky mode of the central ridge is excited, which in this case acts as a ridge BSW waveguide [8], the energy is "tunnelled" through the grooves and is completely transmitted. The resonant nature of the transmittance peaks is confirmed by Fig. 1(c) showing the field distribution in the structure at resonance conditions. Note that from the theoretical point of view, the width of the transmittance peaks can be made arbitrarily small by increasing the groove width. Let us also note that similar resonant transmittance peaks are observed in the frequency spectra of the investigated structures, which are not presented here for the sake of brevity.

3. Conclusion

In this work, we presented a simple resonator for Bloch surface waves, which can be effectively used as a spatial (angular) or spectral BSW filter operating in transmission. The proposed resonator consists of two parallel grooves on the surface of a one-dimensional photonic crystal. The grooves form a central ridge, the leaky modes of which are excited in the geometry of oblique incidence of a TE-polarized BSW. The quality factor of the resulting resonance can be controlled by changing the width of the grooves. We believe that the obtained results may find application in novel integrated nanophotonic devices.

Acknowledgments

This work was funded by Russian Science Foundation (RSF) under the project 19-19-00514 (investigation of the resonant two-groove structure for BSWs) and by the Russian Federation Ministry of Science and Higher Education within the state assignment to the FSRC "Crystallography and Photonics" RAS (implementation of the numerical simulation software).

References

[1] Staude I, Pertsch T and Kivshar Y S 2019 *ACS Photon.* **6** 802–14
[2] Vinogradov A P, Dorofeenko A V, Merzlikin A M and Lisyansky A A 2010 *Phys.-Usp.* **53** 243–56
[3] Yu L, Barakat E, Sfez T, Hvozdara L, Francesco J D and Herzig H P 2014 *Light: Sci. & Appl.* **3** e124
[4] Doskolovich L L, Bezus E A and Bykov D A 2015 *Opt. Express* **23** 27034–27045
[5] Sainidou R, Renger J, Teperik T V, González M U, Quidant R and García de Abajo F J 2010 *Nano Lett.* **10** 4450–4455
[6] Moharam M G, Grann E B, Pommet D A and Gaylord T K 1995 *J. Opt. Soc. Am. A* **12** 1068–76
[7] Silberstein E, Lalanne P, Hugonin J-P and Cao Q 2001 *J. Opt. Soc. Am. A* **18** 2865–75
[8] Descrovi E, Sfez T, Quaglio M, Brunazzo D, Dominici L, Michelotti F, Herzig H P, Martin O J F and Giorgis F 2010 *Nano Lett.* **10** 2087–2091

Are there bound states in the continuum in a dielectric ring?

D V Bochek[1], N S Solodovchenko[1], K B Samusev[1,2] and M F Limonov[1,2,a)]

[1]Department of Physics and Engineering, ITMO University, Saint Petersburg 197101, Russia
[2]Ioffe Institute, St. Petersburg 194021, Russia

[a)]E-mail: M.Limonov@metalab.ifmo.ru

Abstract. Trapping and confining electromagnetic waves is important in both basic research and a variety of applications. For these purposes, various physical mechanisms are exploited including bound states in the continuum, which have been actively investigated recently. Bound states in the continuum have been observed in various objects consisting of both one and a number of dielectric structures. In particular, these photonic states were observed in high-contrast dielectric cylinders in the regime of strong eigenmode coupling, which leads to destructive interference in the far-field zone. In this article, we present the results of a study of bound states in a continuum in a dielectric ring, i.e. cylinder with coaxial air hole. The dependence of the quality factor Q on the normalized diameter of the hole is discussed.

1. Introduction

Bound states in the continuum (BIC) is a general wave phenomenon that was predicted in 1929 by von Neumann and Wigner [1] for electronic states and further considered for various types of waves such as electromagnetic, acoustic, elastic, and water waves. Photonic BIC coexist with extended electromagnetic waves and lie within the continuum, but theoretically remain completely confined without any radiation [2-4]. Theoretically, BICs are states with an infinite lifetime and can be realized if at least one dimension of the structure extends to infinity [2] or in systems of finite length, when the permittivity approaches zero [5, 6]. Unfortunately, in reality there are no such structures suitable for experiment. However, to simulate an ideal BIG one can use a resonator with a set of eigenmodes and change its geometric and dielectric parameters until the system most closely satisfies the condition required for the BIC in the corresponding infinite system. This quasi-BIC is achieved when the Q-factor no longer changes slowly as in a conventional resonator, but instead rises rapidly, following the BIC trend, before it reaches its maximum value, due to the finite size of the resonator [7].

To date, quasi-BICs have been observed in a large number of photonic objects, both single and consisting of a set of separate structures. In this case, of particular interest is the case when a quasi-BIC is observed in the simplest geometric object, for example, a single homogeneous finite-length dielectric cylinder [8, 9]. It was demonstrated that a quasi-BIC arises in a cylinder in accordance with the Friedrich-Wintgen theory [10], which implies a strong interaction of two eigenmodes of the structure. When changing the parameters of the cylinder, the damping rate of one mode increases, while the damping rate of the other mode decreases and at a minimum value a quasi-BIC is achieved. It has been demonstrated that a multitude of quasi-BIC associated with strong interactions between Mie and Fabry–Perot -like modes are observed in the cylinder at various geometric parameters [8, 9].

In this article, we continue the search for quasi-BIC in the simplest dielectric objects, which are bodies of revolution. We transform the cylinder into a ring by adding a coaxial air hole to the high-index dielectric cylinder and gradually increasing the radius of the hole. The results of numerical

Journal of Physics: Conference Series **2015** (2021) 012017 doi:10.1088/1742-6596/2015/1/012017

calculations are presented that demonstrate the regularities of the change in the quasi-BIC with the expansion of the hole in the cylinder and the change in the eigenmodes of the dielectric ring.

2. Results and discussion

We calculate the scattering cross-section spectra of the dielectric cylinder with a coaxial air hole as a function of its aspect ratio R/L for a set of hole radius r normalized to the cylinder radius r/R starting from r = 0, that is, a solid cylinder. Here we analyze the modes excited by a TE-polarized plane wave and chose the value of the dielectric permittivity $\varepsilon = 80$ following Refs. [8,9].

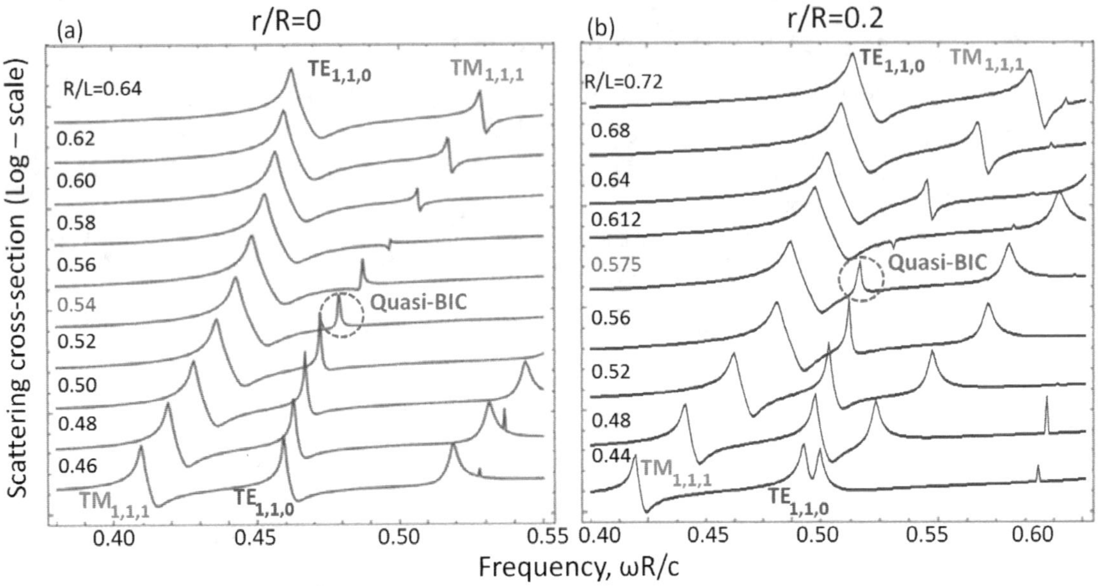

FIGURE 1. Scattering cross-section spectra of the dielectric cylinder (a) and the ring (cylinder with a coaxial air hole, r/R=0.2) (b) as a function of aspect ratio R/L in the region of the avoided resonance crossing between the modes $TE_{1,1,0}$ and $TM_{1,1,1}$. The dielectric permittivity of both structures $\varepsilon = 80$.

The modes shown in Fig. 1 have the symmetry $TE_{1,1,0}$ (slow-varying Mie-like mode) and the symmetry $TM_{1,1,1}$ (rapidly-varying Fabry-Perot-like mode) [8,9]. Here, we employ the standard mode classification [11], the first index m corresponds to the azimuthal number of the mode, whereas other two indices k and n enumerate the Mie and Fabry-Perot resonances, respectively. Due to different spectral shifts of the Mie and Fabry-Perot modes when the aspect ratio varies R/L, the modes with the same azimuthal index m could interact, and they undergo coupling each other with the avoid crossing at special values of the parameter R/L with a dramatic narrowing of the high-frequency line, which corresponds to the quasi-BIC, as shown earlier [8,9].

The calculation results show that the $TE_{1,1,0}$ and $TM_{1,1,1}$ modes do not change dramatically in the presence of a hole in a sufficiently wide range of the parameter $r/R \leq 0.2$, as can be seen from the comparison of the spectra in the panels (a) and (b) in Fig. 1, as well as from the comparison of the field distribution in the cylinder and the ring, Fig. 2. From Fig. 2, the type of the resonance mode is also well defined with a change in the parameter R/L. Below the region of the avoid crossing, the mode belongs to the Mie-type (the field changes in the xy plane), and above this region, the mode belongs to the Fabry-Perot type (the change in the field along the z axis). In the region of the quasi-BIC, the mode has a hybrid character, changing both in the xy plane and along the z axis.

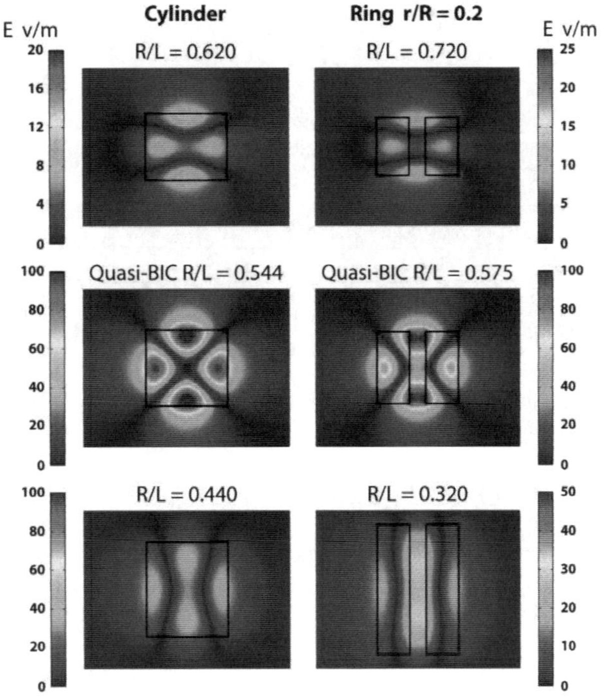

FIGURE 2. The calculated distribution of the electric field amplitude $|E|$ in the (x,z) cross sections (resonator side view) for the high-frequency branch of the interacting pair of modes $TE_{1,1,0}$ and $TM_{1,1,1}$. The left column is the field in the cylinder, the right column is the field in the ring, r/R=0.2. The parameter R/L is indicated.

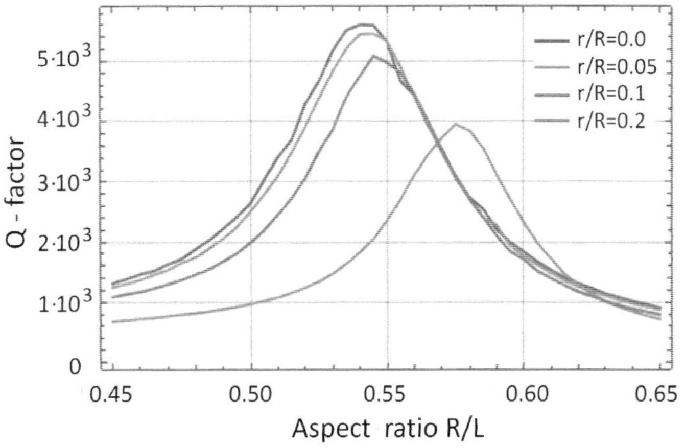

FIGURE 3. The evolution of the Q-factor for the high-frequency branch of the interacting pair of modes $TE_{1,1,0}$ and $TM_{1,1,1}$ in the region of the quasi-BIC. The dependence from the normalized hole size r/R.

Figure 3 contains the main result of this work, demonstrating the dependence of the Q-factor of the quasi-BIC on the hole size. With an increase in the parameter r/R, two effects are clearly observed: the shift in the position of the quasi-BIC and a decrease in the value of the Q-factor by only 20%. Thus, we have demonstrated a robust quasi-BIC sate in a cylinder which can endure a loss of dielectric material of 20% or more.

3. Conclusion

Our results demonstrate that we can give a positive answer to the question: are there bound states in the continuum in a dielectric ring? Based on the calculations performed, we can draw a not the most obvious conclusion. Quasi-BIC turn out to be robust when a coaxial air hole is introduced into the structure of a high-contrast cylinder in a rather wide range of values of the normalized hole radius r/R. With an increase in the parameter r/R to the value of 0.2, the Q-factor of the quasi-BIC decreases The result obtained can find application in various resonant devices, the principle of operation of which is based on the use of the quasi-BIC.

Acknowledgments

This work was supported by the Russian Science Foundation (Project 20-12-00272).

References

[1] von Neumann J and Wigner E 1929 *Phys. Z.* **30** 465
[2] Hsu C W, Zhen B, Stone A D, Joannopoulos J D and Soljačić M 2016 *Nat. Rev. Mater.* **1** 16048
[3] Monticone F, Doeleman H M, Den Hollander W, Koenderink A F and Alù A 2018 *Laser Photonics Rev.* **12** 1700220
[4] Azzam S I and Kildishev A V 2021 *Adv. Optical Materials* **9** 2001469
[5] Monticone F and Alù A 2014 *Phys. Rev. Lett.* **112** 213903
[6] Silveirinha M G 2014 *Phys. Rev. A* **89** 023813
[7] Rybin M and Kivshar Y 2017 *Nature* **541** 164
[8] Rybin M V, Koshelev K L, Sadrieva Z F, Samusev K B, Bogdanov A A, Limonov M F and Kivshar Y S 2017 *Phys. Rev. Lett.* **119** 243901
[9] Bogdanov A A, Koshelv K L, Kapitanova P V, Rybin M V, Gladyshev S A, Sadrieva Z F, Samusev K B, Kivshar Y S and Limonov M F 2019 *Adv. Photonics* **1** 016001
[10] Friedrich H and Wintgen D 1985 *Phys. Rev. A* **32** 3231
[11] Zhang K and Li D 2008 *Electromagnetic Theory for Microwaves and Optoelectronics* (Berlin: Springer)

Altering the distribution of excited-state lifetimes in aminated GFP chromophores by Ag nanohole arrays

VR Aslopovsky[1], **AM Shakhov**[2], **AA Astafiev**[2] **and VE Bochenkov**[1,*]

[1] Chemistry department of Lomonosov Moscow State University, Moscow 119991, Russia
[2] N.N. Semenov Federal Research Center for Chemical Physics of RAS, 119991 Moscow, Russia

E-mail: `boch@kinet.chem.msu.ru`

Abstract. Fluorescence of the modified GFP chromophore diethyl-ABDI-BF2 dispersed in PMMA matrix is studied on top of glass, continuous and perforated optically thin silver films. In polymer, the fluorescence decay kinetics becomes non-exponential and can be described by the distribution of rate constants. The results demonstrate shortening of the excited state lifetime in the presence of silver and broadening of the lifetime distribution caused by the nanoholes.

1. Introduction

Labelling biomolecules via chemical bonding with various fluorescent tags (usually fluorescent proteins or organic dyes) offers unique possibility to visualize molecular structures and study the numerous processes inside living cells. The performance of this technique depends on the intrinsic properties of a fluorophore, such as its fluorescence brightness and photostability. In general, two main approaches exist to enhance these characteristics. The 'chemical' one is based on altering the chemical structure of the fluorophore, whereas the 'physical' approach relies upon exploiting the amplified optical fields near plasmonic nanostructures.

One of the most widely used fluorescent tags nowadays is the Green Fluorescent Protein (GFP). At the heart of this protein lies a small chromophore molecule, which is responsible for light adsorption and emission. Without protein environment, the GFP chromophore has a very low quantum yield, which hampers its direct application as a fluorescent label [1, 2, 3]. This feature is due to existence of an efficient nonradiative de-excitation channel through the intramolecular rotation around the methine bridge coupled to double bond isomerization. To overcome this obstacle, several modifications of this chromophore have been synthesized and studied. By arresting the rotation of the arylidine fragment of the chromophore via the internal bridge, the researchers were able to drastically increase the quantum yield of the chromophore [4, 5, 6, 7].

Recently, we demonstrated that the brightness of conformationally locked aminated derivative of the GFP chromophore diethyl-ABDI-BF2 can be increased by an order of magnitude by plasmon-induced near fields of optically thin perforated Ag films [8].

Here, by using a time-correlated single-photon counting (TCSPC) technique we show how the presence of Ag nanohole arrays (NHAs) affects the excited state lifetime of the diethyl-ABDI-BF2 molecules, dispersed in thin poly(methyl methacrylate) (PMMA) films.

2. Experiment

The samples are prepared using the previously reported technique [8]. In brief, the short-range ordered NHAs with average hole diameter of 120 nm in 20 nm thick Ag films are fabricated on glass slides of 24 mm diameter using the sparse colloidal lithography approach [9]. Deposition of polystyrene nanospheres on top of triple polyelectrolyte layer is followed by 3 min UV-ozone treatment and a vacuum deposition of thermally evaporated silver. After the lift-off process, the arrays of nanoholes in Ag film are obtained.

Polymer films are produced by dissolving PMMA (MW=350000) in phenetole to a concentration of 5% wt., followed by the addition of diethyl-ABDI-BF2 up to a final concentration of 30 μM. The mixture is then deposited onto thesilv er nanohole arrays by spin-coating at 4000 rpm. The sample schematics is presented in Fig. 1.

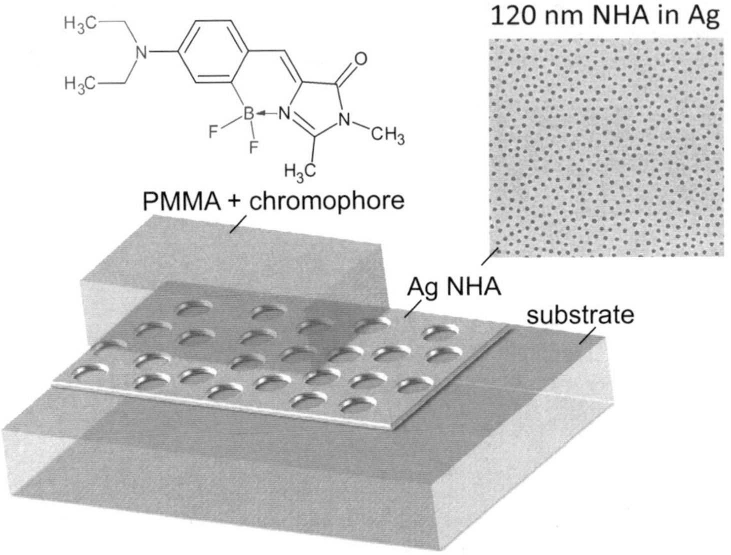

Figure 1. Schematics of the nanostructured AgNHA–PMMA sample, the structural formula of the chromophore and the SEM image of the perforated Ag flim.

The control samples are comprised of the same dye-containing PMMA films deposited on glass and on the continuous 20 nm Ag films without nanoholes.

Optical properties of the samples are studied using UV-vis-NIR spectroscopy (Jasco V-770) before and after the polymer deposition. Fluorescence spectra are registered using a custom-built optical setup. Femtosecond laser pulses with a 80 MHz repetition rate, 980 nm central wavelength, and energy up to 25 nJ are generated by a Titanium- Sapphire oscillator (Tsunami, Spectra-Physics). Frequency-doubled pulses at 490 nm are focused by a 40×0.75NA objective lens (UPlanFLN, Olympus) on the sample's top surface covered with the polymer film. The fluorescence is collected using the same objective lens and filtered using 520 nm long-pass filter and directed to SP300i monochromator connected to PI-MAX 2 CCD camera (Princeton Instruments) for recording the fluorescence emission spectra.

3. Results and discussion

In PMMA matrix, the absorption and emission spectra of diethyl-ABDI-BF2 chromophore are characterized by maxima at 497 and 522 nm, respectively. The Ag NHA samples possess a broad extinction peak between 500 and 650 nm with a maximum at 553 nm. The fluorescence decay in solution is mono-exponential, with characteristic decay time strongly depending on the dielectric constant of the solvent, changing from 0.83 ns in methanol to 3.2 ns in dioxane.

We find that polymer matrix affects the fluorescence decay kinetics of diethyl-ABDI-BF2, which cannot be adequately described by a single-exponential function. This is anticipated due to hindering of the rotation of diethylamine moiety of the chromophore in PMMA matrix leading to the distribution of the decay times of individual molecules. In such cases the fluorescence signal can be described by the following function:

$$I(t) = \int_0^\infty H(k)e^{-kt}dk, \tag{1}$$

with $I(0) = 1$.

The recovery of the shape of the distribution function $H(k)$ from the decay kinetics data is an ill-conditioned problem, which cannot be solved exactly. In the absence of a physical model, the distribution is often approximated by some continuous function, such as Gaussian of Lorentzian. This helps an analysis of the changes observed in various experimental conditions. In this work use the stretched exponential (or Kohlrausch) function:

$$I(t) = exp\left[-(t/\tau_0)^\beta)\right], \tag{2}$$

where $0 < \beta \leq 1$ and τ_0 are the parameters, describing the distribution. This simple and relatively flexible function has been successfully used for luminescence decays with underlying distributions [10].

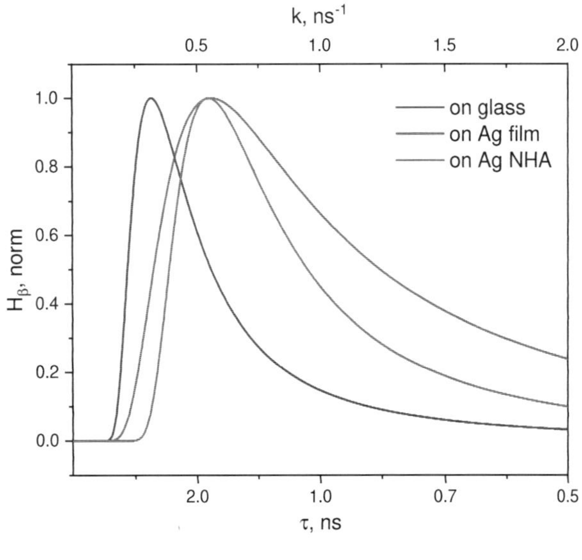

Figure 2. The normalized distribution functions of the fluorescence lifetime for diethyl-ABDI-BF2 chromophore in PMMA film on the glass (blue), on the continuous Ag film (green) and on the Ag NHAs (red).

The results of fitting the experimental decay data is presented in Fig. 2. In general, the longest excited state lifetimes are observed for the chromophore in PMMA on glass. The presence of silver results in faster decay due to the transfer of excitation to the metal. When nanoholes are present in the metal film, the fluorescence lifetimes become even shorter, and the distribution broadens, which can be explained by different positions of dye molecules in relation to nanoholes.

4. Conclusion

The non-exponential fluorescence decay kinetics observed for the diethyl-ABDI-BF2 chromophore in the PMMA film can be explained by the existence of a distribution of decay rate constants due to hindered rotation of the diethylamine moiety. The experimental data can be well fitted using the stretched exponential function. The observed plasmon-induced shortening of the excited-state lifetime can improve the photostability of the dye. In combination with increased brightness, this can be important for various applications involving the fluorescence life-time imaging.

Acknowledgments

The work is supported by the Russian Science Foundation, grant no. 17-13-01276. Fluorescence studies were performed on the equipment of Semenov FRCCP RAS CCE (no.506694).

References

[1] Svendsen A, Kiefer H V, Pedersen H B, Bochenkova A V and Andersen L H 2017 *J. Am. Chem. Soc.* **139** 8766–8771
[2] Addison K, Heisler I A, Conyard J, Dixon T, Bulman Page P C and Meech S R 2013 *Faraday Discuss.* **163**(0) 277–296
[3] Bochenkova A V and Andersen L H 2013 *Faraday Discuss.* **163**(0) 297–319
[4] Baranov M S, Lukyanov K A, Borissova A O, Shamir J, Kosenkov D, Slipchenko L V, Tolbert L M, Yampolsky I V and Solntsev K M 2012 *J. Am. Chem. Soc.* **134** 6025–6032
[5] Wu L and Burgess K 2008 *J. Am. Chem. Soc.* **130** 4089–4096
[6] Baldridge A, Solntsev K M, Song C, Tanioka T, Kowalik J, Hardcastle K and Tolbert L M 2010 *Chem. Commun.* **46**(31) 5686–5688
[7] Baleeva N S, Zaitseva S O, Gorbachev D A, Smirnov A Y, Zagudaylova M B and Baranov M S 2017 *Eur. J. Org. Chem.* **2017** 5219–5224
[8] Aslopovsky V R, Shakhov A M, Astafiev A A and Bochenkov V E 2020 *AIP Conf. Proc.* **2300** 020004
[9] Hanarp P, Sutherland D S, Gold J and Kasemo B 2003 *Colloids Surf., A* **214** 23–36 ISSN 0927-7757
[10] Berberan-Santos M, Bodunov E and Valeur B 2005 *Chemical Physics* **315** 171–182 ISSN 0301-0104

Journal of Physics: Conference Series **2015** (2021) 012019 doi:10.1088/1742-6596/2015/1/012019

Multi wall carbon nanotubes as a top electrode for perovskite light-emitting electrochemical cells

Elena Bodyago[1], Dmitry Gets[1], Maria Baeva[2,3], Ivan Mukhin[2], Sergei Makarov[1] and Anvar Zakhidov[1,4]

[1] Department of Physics and Engineering, ITMO University, Lomonosova 9, 197101, St. Petersburg, Russia
[2] Alferov University, Khlopina 8/3, 194021, St. Petersburg, Russia
[3] Institute of Automation and Control Processes (IACP), Far Eastern Branch of Russian Academy of Sciences, Ulitsa Radio 5, 690041, Vladivostok, Primorsky Krai, Russia
[4] Physics Department, Alan G. MacDiarmid NanoTech Institute, University of Texas at Dallas, Richardson, TX 75083, USA

Abstract. Carbon nanotubes are promising materials for use as flexible and transparent electrodes for a wide variety of optoelectronic devices like solar cells and light-emitting diodes. Moreover, the chemical inertness of carbon nanotubes matches with the operation of light-emitting electrochemical cells where ion migration is a main working principle especially based on halide perovskite, in which ions corrode metal electrodes. Here we demonstrate how the removal of synthesis residuals via synthesis enables the use of multiwall carbon nanotubes as a transparent electrode.

1. Introduction

Light-emitting electrochemical cells (LECs) are promising for the development of the next generation light-emitting devices owing to their numerous advantages like the use of air-stable electrodes, simple device design, scalable and cost-efficient processing, formation of p-i-n structure inside the photoactive layer [1–4]. Usually, LECs utilize simple single-layer device architecture, where one photoactive layer is sandwiched between transparent conductive oxide and metal electrodes. The photoactive layer consists of a mixture of solid polymer electrolyte (e.g., polyethylene oxide), lithium salt that is responsible for p-i-n structure formation inside the photoactive layer and luminescent material like ionic transition metal complexes, quantum dots, conjugated polymers, and halide perovskites [5–7].

Halide perovskites with the formula ABX_3, where A is usually an ion of methylammonium (MA^+), formamidinium (FA^+) or cesium (Cs^+), B is an ion of lead (Pb^{2+}), and X is a halogen ion of iodine (I^-), bromine (Br^-) or chlorine (Cl^-) are a family of promising materials for realization of next generation of different optoelectronic devices like solar cells, light-emitting diodes due to its magnificent properties like the ease of processability, a wide variety of deposition methods, narrow luminescence line, compatibility with different materials [8]. LECs based on halide perovskites have already demonstrated

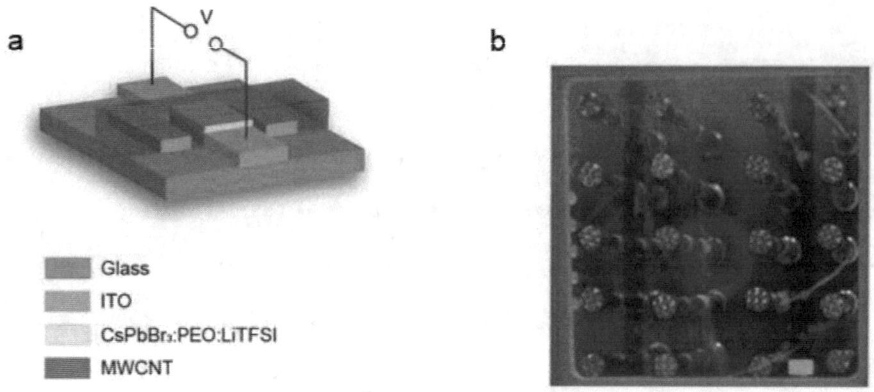

Figure 1. **a** Single-layer LEC structure **b** The image of LEC.

high luminances and quite long operating times and further improvement relates to the corrosion resistance of electrodes.

Carbon nanotubes are modern materials for the realization of chemically stable, flexible, and transparent electrodes. CNT found its application in electronics in the development of light-emitting diodes, solar cells, field-effect transistors [9,10]. The chemical inertness of CNT is the most interesting property in application in perovskite light-emitting devices. Single-wall carbon nanotubes (SWCNTs) have several advantages over multi-wall carbon nanotubes (MWCNTs) like transparency, low resistivity but the cost of SWCNTs production is much higher than MWCNTs. Therefore, finding a way of implementing MWCNTs as an electrode will allow the development of stable, cost-effective, and flexible perovskite-based LECs.

2. Experimental section

2.1. Solution Preparation.

A $CsPbBr_3$ solution was prepared by dissolving $PbBr_2$ and CsBr with a molarity of 0.2 in an anhydrous DMSO solution. A polyethylene oxide (PEO) solution was prepared by dissolving 10 mg PEO (M_w = 1 000 000 g/mol) in 1 ml DMSO, then the solution was stirred overnight at 60°C at 300 rpm. The perovskite-polymer solution $CsPbBr_3$:PEO (1:0.08 by weight) by mixing required masses of PEO and

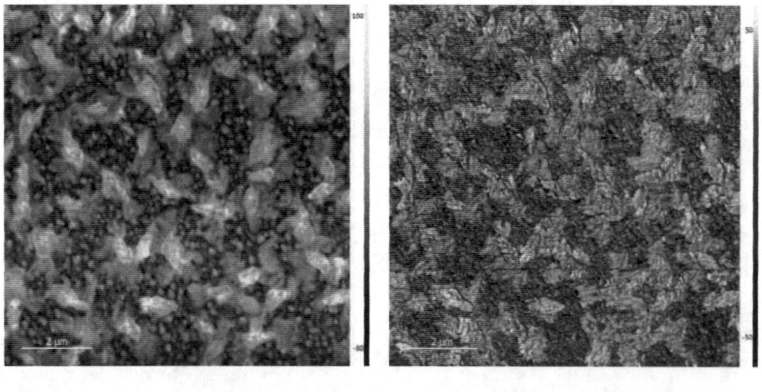

Figure 2. **a** AFM height profile **b** AFM phase profile images of perovskite-polymer composite indicating partial coverage of perovskite grains by polyethylene oxide.

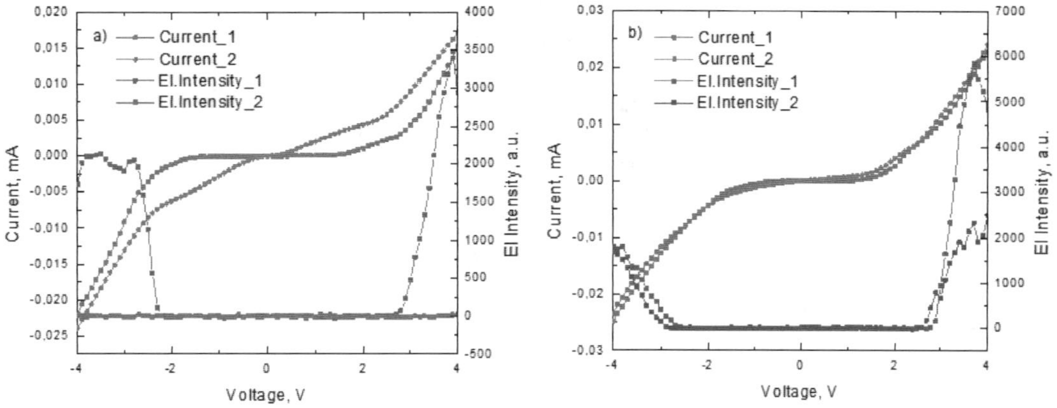

Figure 3. Current and luminance versus voltage for LEC devices with MWCNTs. **a** MWCNTs untreated acid vapors. **b** MWCNTs treated acid vapors.

$CsPbBr_3$. Next, to the perovskite-polymer solution required amount of LiTFSI solution DMSO (10mg/ml) to fulfill ratio 1:0.005 by mass. The resulting solution was stirred for 1 hour at 300 rpm.

2.2. Device fabrication.

LEC devices (Figure 1) were fabricated on a transparent, indium-tin oxide (ITO) coated glass substrate, deposition of perovskite-polymer composite was conducted in a nitrogen filled glovebox. Before deposition of the emission layer, the substrates were cleaned from various kinds of contaminants, after which they were subjected to thermal treatment and treatment in oxygen plasma for 10 minutes at 40 W. Cleansing consists of several stages. Each substrate was sequentially cleaned with a detergent solution, followed by a NaOH solution, deionized water, acetone, and isopropyl alcohol in an ultrasonication bath for 5 minutes. To obtain an emission layer with a thickness of 80 nm, the precursor solutions are centrifuged onto ITO substrates at 1000 rpm for 1 min., precursor solutions were spin-coated onto ITO substrates at 1000 rpm for 1 min. The film morphology was controlled by AFM spectroscopy (Figure 2).MWCNTs were used as a top electrode. Before deposition, MWCNTs are thermal treated at 100°C then treated with HCl vapor at 60°C, IPA was used for MWCNTs densification.

3. Results and discussion

Chemical treatment of CNT is a known method for purification and removal of different chemical impurities [11]. The removal of different residuals should affect the electronic properties of MWCNTs and the characteristics of perovskite LEC. Therefore, the study was carried out by measurement of IV, EL-V curves, and brightness for devices with MWCNT treated and non-treated with acid vapor, Figure 3 a and b shows obtained IV and EL-V curves. Application of a voltage of 0 V to 4 V corresponds to the case where ITO injects holes and MWCNTs injects electrons, applying a reverse voltage of 0V to -4 V corresponds to injection of holes from MWCNTs and electrons from ITO.

The applied bias voltage leads to an ion redistribution of perovskite and lithium salt in the perovskite film, cation ions drift to the cathode and form accumulation layer, and anions form the accumulation layer at an anode [5]. The formation of p-i-n structure and accumulation layers at the interface layers corresponds to the pinning of a Fermi level, which leads to the formation of ohmic contacts between electrodes and the perovskite-polymer layer. The formation of these ion accumulation layers leads to chemical interaction between ions and electrode material, moreover increased applied voltage can result in penetration of ions into the electrode, which is the main reason for electrode degradation [12].

The initial state of perovskite LEC shows visible electroluminescence (EL) when applied voltage higher than V_{th} = 2.4 V, with and without acid vapor treatment (Figure 3a and b). Further successive I-V sweeps of the device without HCl vapor treatment demonstrate an increase of current below V_{th} region as well as in the higher V_{th} region, which is related to higher recombination via defects and loss of diode characteristic. After several voltage sweeps, the I-V curve loses its diode characteristic and becomes more ohmic that results in the disappearance of EL (Figure 3a, gray and red curve). In contrast, the device with MWCNTs treated with acid vapors does not demonstrate a current increase in the whole voltage region and the I-V curve preserves its diode characteristic. After several repetition of voltage sweeps of both polarities, there is no degradation of the I-V characteristic and electroluminescence (Figure 3b). Figure 3b illustrated the dependences I-V and EL-V, according to which the treatment with acid vapors of MWCNTs avoids degradation of the devices and preserves electroluminescence for the subsequent application of a voltage of both polarities (Figure 3b). That allows achievement of luminance 516 cd/m^2. This difference in device operation can be related to acid cleaning of MWCNTs, which is a well-known method for MWCNTs cleaning from synthesis residuals and its functionalization.

4. Conclusion

We have shown that acid-treated MWCNTs can be used as a chemically stable top electrode of perovskite-based LECs. Initial MWCNTs contain many synthesis residuals that interact with perovskite ions and degrade device performance. Preliminary acid treatment of MWCNTs allows evading the degradation of perovskite LECs. The acid treatment removes residuals of MWCNTs production and allows preserve the diode characteristic of the device via suppression of defects development during accumulation of ions at the perovskite-MWCNTs interface. The use of acid vapor for MWCNTs treatment allows MWCNTs to be used as the top electrode for perovskite LECs.

Acknowledgments
This work was supported by the Russian Science Foundation (project № 19-73-30023).

References

1. Friend R.H. et al. Electroluminescence in conjugated polymers // Nature. 1999. Vol. 397, № 6715. P. 121–128.
2. Qibing Pei, Gang Yu, Chi Zhang, Yang Yang A.J.H. Polymer Light-Emitting // Science (80-.). 1995. Vol. 269, № 5227. P. 1086–1088.
3. Fresta E., Costa R.D. Advances and Challenges in White Light-Emitting Electrochemical Cells // Adv. Funct. Mater. 2020. Vol. 30, № 33. P. 1–25.
4. Nannen E., Frohleiks J., Gellner S. Quantum Dot Light-Emitting Diodes: Light-Emitting Electrochemical Cells Based on Color-Tunable Inorganic Colloidal Quantum Dots (Adv. Funct. Mater. 33/2020) // Adv. Funct. Mater. 2020. Vol. 30, № 33. P. 2070224.
5. Gets D. et al. Reconfigurable Perovskite LEC: Effects of Ionic Additives and Dual Function Devices // Adv. Opt. Mater. 2021. Vol. 9, № 3.
6. Youssef K. et al. Light-Emitting Conjugated Polymers: Fundamentals of Materials Selection for Light-Emitting Electrochemical Cells (Adv. Funct. Mater. 33/2020) // Adv. Funct. Mater. 2020. Vol. 30, № 33. P. 2070220.
7. Bowler M.H. et al. Circumventing Dedicated Electrolytes in Light-Emitting Electrochemical Cells // Adv. Funct. Mater. 2020. Vol. 30, № 33. P. 1–13.
8. Akin S. et al. New Strategies for Defect Passivation in High-Efficiency Perovskite Solar Cells // Adv. Energy Mater. 2020. Vol. 10, № 13. P. 1–29.
9. Habisreutinger S.N., Nicholas R.J., Snaith H.J. Carbon Nanotubes in Perovskite Solar Cells // Adv. Energy Mater. 2017. Vol. 7, № 10. P. 1–7.
10. Bansal M. et al. Carbon nanotube-based organic light emitting diodes // Nanoscale. 2009. Vol. 1, № 3. P. 317–330.
11. Balasubramanian K., Burghard M. Chemically functionalized carbon nanotubes // Small. 2005. Vol. 1, № 2. P. 180–192.
12. Ming W. et al. Formation and Diffusion of Metal Impurities in Perovskite Solar Cell Material CH3NH3PbI3: Implications on Solar Cell Degradation and Choice of Electrode // Adv. Sci. 2018. Vol. 5, № 2.

Mechanical scanning probe lithography of nanophotonic devices based on multilayer TMDCs

B R Borodin[1], F A Benimetskiy[2], V Yu Davydov[1], I A Eliseyev[1], S I Lepeshov[2], A A Bogdanov[2], and P A Alekseev[1]

[1] Ioffe Institute, Saint-Petersburg, 194021, Russia.

[2] ITMO University, Saint-Petersburg, 197101, Russia.

E-mail: brborodin@gmail.com

Abstract. In this work, we demonstrate the possibility of using mechanical Scanning probe lithography (m-SPL) for fabricating nanophotonic devices based on multilayered transition metal dichalcogenides (TMDCs). By m-SPM, we created a nanophotonic resonator from a 70-nm thick $MoSe_2$ flake transferred on Si/Au substrate. The optical properties of the created structure were investigated by measuring microphotoluminescence. The resonator exhibits four resonance PL peaks shifted in the long-wavelength area from the flake PL peak. Thus, here we demonstrate that m-SPL is a high-precision lithography method suitable for creating nanophotonic devices based on multilayered TMDCs.

1. Introduction

After graphene, Transition Metal Dichalcogenides (TMDCs) seem to be the most studied layered materials. This is due to the unique optical and electronic properties of their monolayers such as direct band gap[1], large exciton binding energy[2], valley polarization[3], etc. Moreover, such strong interest in monolayers of these materials encounters researchers to take a fresh look at the properties of bulk TMDCs. Multilayer TMDCs are attractive to nanophotonic applications owing to the high values of the refractive index and its strong anisotropy caused by the layered structure[4]. These properties allow the creation of optical waveguides and light resonators based on multilayer TMDCs. However, these materials are highly sensitive to surface conditions, and some of them can degrade even in ambient conditions[5]. Thus, the traditional methods of lithography involving resist usage and its chemical treatment are poorly suited for creating TMDC nanophotonic structures. Nevertheless, Scanning Probe Microscopy (SPM) can offer several resistless techniques of lithography suitable for creating TMDC nanophotonic devices[6]. For example, oxidative Scanning Probe Lithography (o-SPL) has established itself as an effective resistless method of lithography of layered materials with a resolution down to 10 nm. It has already been successfully used for graphene[7], $MoSe_2$[8], WSe_2[9], etc. However, we suggest that for materials that are sensitive to ambient conditions and humidity, mechanical-SPL is more suitable. Besides, applying mechanical-SPL for TMDCs is barely studied. This method consists in the mechanical pressure of a probe on a sample with force enabling partial destruction of the surface (puncture, cutting, scratching, etc.). Thus, this lithography technique does not require any resist nor chemical treatment and can be performed in vacuum conditions. Our work aimed to demonstrate the possibility of creating nanophotonic devices (resonators) based on multilayer $MoSe_2$ by mechanical-SPL.

2. Samples and methods

For the creation of nanophotonic devices, the following structure was prepared. $MoSe_2$ multilayer flakes obtained by adhesive tape exfoliation were transferred on the Si substrate covered with 50 nm of gold. The experiment was carried out on a Ntegra Aura (NT-MDT) SPM using HA_NC_DCP probes with a tip curvature radius of 100 nm, a spring constant of 35-85 N/m, and a resonant frequency of 350-650 kHz. Ntegra Aura provides a thermal drift of about 10 nm/°C and a mechanical drift of about a dozen of nm per hour, which ensures the high stability of the sample during the fabrication of a resonator. The scheme of Mechanical Scanning Probe Lithography (m-SPL) is shown in Figure 1.

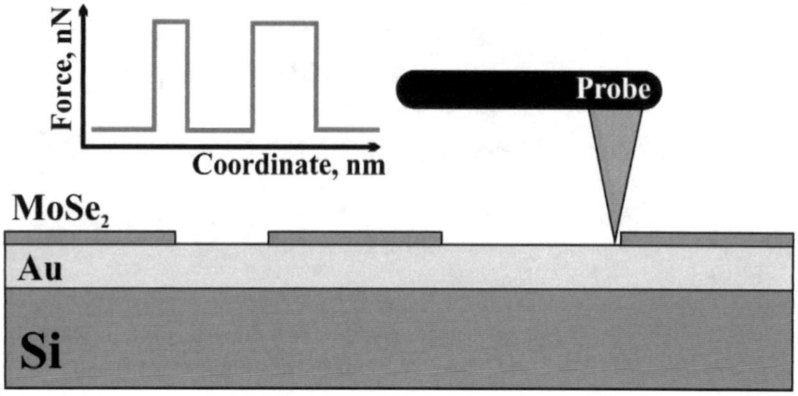

Figure 1. The scheme of Mechanical Scanning Probe Lithography.

After selecting a flake of suitable sizes (with a thickness comparable to the wavelength of visible light in $MoSe_2$), m-SPL was used to create nanophotonic structures. During m-SPL the probe passes the chosen lithographic pattern with given pressure on the surface. Mechanical interaction of the rigid diamond-coated probe with the relatively fragile surface of TMDC flake results in the erasing of surface TMDC layers. A few repetitions of such procedure can provide complete deleting of TMDC flake in chosen areas. In our experiments, the hard probes with a spring constant of 35-85 N/m were used to prevent axial twisting of the cantilever, which takes place using soft probes and leads to a decrease in lithography resolution.

Optical properties were studied by measuring microphotoluminescence (PL) spectra. The experiments were carried out at room temperature (300 K) using Horiba Jobin Yvon T64000. The measurements were performed with continuous-wave (cw) excitation using the 532 nm laser line of a Nd: YAG laser. We used a Mitutoyo 100 × NIR (NA = 0.90) long working-distance objective lens to focus the incident beam into a spot of ~ 0.7 μm diameter.

Multilayer TMDC flakes can exhibit photoluminescence at high excitation densities. An excitation power of 4 mW was used here.

3. Results and discussions

Figure 2(a-c) shows the process of investigated nanophotonic device creation. To create a resonator, a 70 nm thick flake was chosen (Figure 2a). First of all, a dozen of lithography iterations were performed to reveal a degree of interaction between the probe and the flake and to determine the suitable pressure force (Figure 2b). Then, a few hundreds of iterations of lithography provided the creation of the structure with the given size and shape. Figure 2c shows the finished resonator with a thickness of 70±2nm and a diameter of 1500±30 nm. The gap between the created structure and the flake is of 900 nm, which protects the modes in the resonator from leaking in the flake. Figure 2d demonstrates the height profile of the structure. As you can see, the level of lithography is lower than the level of the Au sublayer. This indicates that the resonator is fully separated from the flake, and there are no thin $MoSe_2$ layers on the floor of the trench. The optical properties of the created structure were investigated by Photoluminescence spectroscopy (PL).

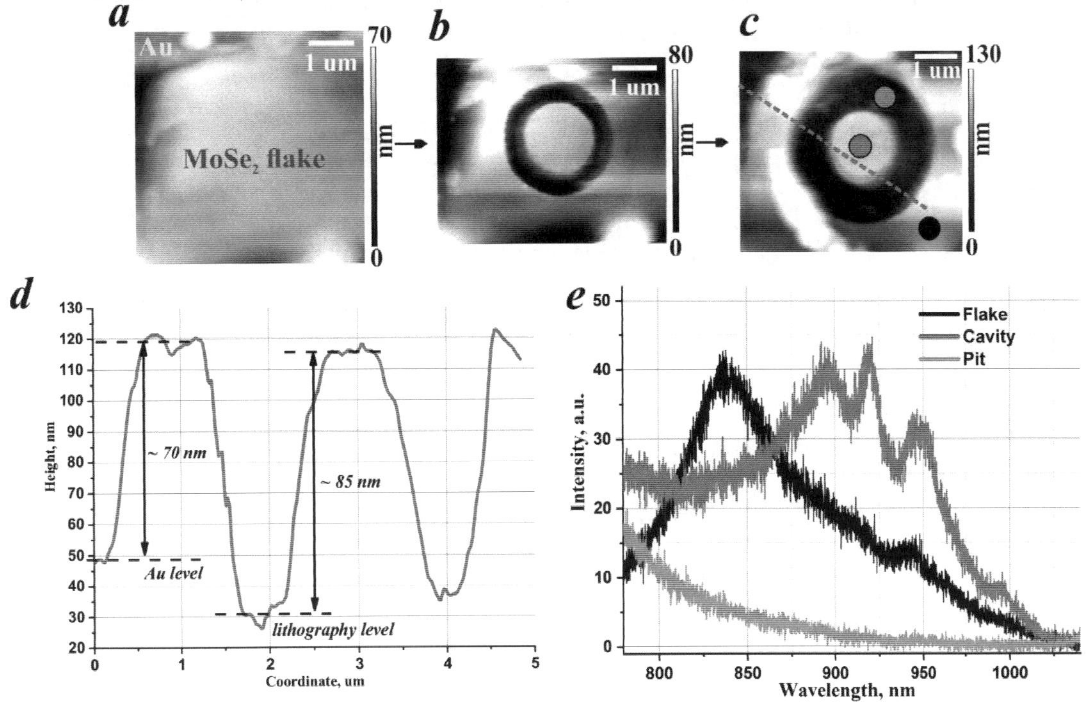

Figure 2. (a-c) Steps of the resonator creation: (a) the chosen flake; (b) determination of lithography parameters; (c) the finished structure. Colored dots show PL measurement areas. (d) The height profile of the structure; (e) Results of PL investigation of the structure.

Figure 2d shows the results of the PL investigation of the structure. The green curve (corresponding to the green dot in Figure 2c) was measured in the pit around the resonator. It can be seen that, apart from the background photoluminescence associated with the plasmon in gold[10], there are no peaks in the investigated wavelength range. The black curve (corresponding to the black dot in Figure 2c) was measured on the flake close to the resonator. This spectrum demonstrates a peak related to the photoluminescence of the multilayered $MoSe_2$ flake. The red curve (corresponding to the red dot in Figure 2c) was measured on the resonator. As can be seen, the spectrum of the resonator significantly differs from the one of the flake. Four peaks shifted in the long-wavelength area unambiguously indicate the presence of resonance in the created structure.

4. Conclusions

To conclude, in this work, the possibility of using mechanical scanning probe lithography for processing multilayered TMDCs was demonstrated. In the course of our work, we created a nanophotonic resonator with a thickness of 70 nm and a diameter of 1500 nm by m-SPL. The resonator exhibits four PL peaks shifted in the long-wavelength area from the flake PL peak that unambiguously indicates the presence of resonance in the created structure. Thus, it was demonstrated that m-SPL can be considered an effective method of creating nanophotonic devices based on multilayered TMDCs.

References

[1] Mak K F, Lee C, Hone J, Shan J and Heinz T F 2010 Atomically thin MoS 2: a new direct-gap semiconductor *Phys. Rev. Lett.* **105** 136805

[2] Wang G, Gerber I, Bouet L, Lagarde D, Balocchi A, Vidal M, Amand T, Marie X and Urbaszek B 2015 Exciton states in monolayer MoSe2: impact on interband transitions *2D Mater.* **2** 045005

[3] Gong Z, Liu G-B, Yu H, Xiao D, Cui X, Xu X and Yao W 2013 Magnetoelectric effects and valley-controlled spin quantum gates in transition metal dichalcogenide bilayers *Nat. Commun.* **4** 1–6

[4] Ermolaev G, Grudinin D, Stebunov Y, Voronin K, Kravets V, Duan J, Mazitov A, Tselikov G, Bylinkin A, Yakubovsky D, and others 2021 Giant optical anisotropy in transition metal dichalcogenides for next-generation photonics *Nat. Commun.* **12** 1–8

[5] Diaz H C, Chaghi R, Ma Y and Batzill M 2015 Molecular beam epitaxy of the van der Waals heterostructure MoTe2 on MoS2: phase, thermal, and chemical stability *2D Mater.* **2** 044010

[6] Ryu Y K and Garcia R 2017 Advanced oxidation scanning probe lithography *Nanotechnology* **28** 142003

[7] Alekseev P, Borodin B, Dunaevskii M, Smirnov A, Davydov V Y, Lebedev S and Lebedev A 2018 Local Anodic Oxidation of Graphene Layers on SiC *Tech. Phys. Lett.* **44** 381–3

[8] Borodin B R, Benimetskiy F and Alekseev P A 2021 Study of local anodic oxidation regimes in MoSe2 *Nanotechnology* **32** 155304

[9] Dago A I, Ryu Y K and Garcia R 2016 Sub-20 nm patterning of thin layer WSe2 by scanning probe lithography *Appl. Phys. Lett.* **109** 163103

[10] Sivun D, Vidal C, Munkhbat B, Arnold N, Klar T A and Hrelescu C 2016 Anticorrelation of photoluminescence from gold nanoparticle dimers with hot-spot intensity *Nano Lett.* **16** 7203–9

On the link between mean square-radii and high-order toroidal moments

Dmitrii Borovkov[1,2a)] and Adrià Canós Valero[1b)]

[1] ITMO University, Kronversky Prospect 49, 197101, St Petersburg Russia
[2] Riga Technical University, Institute of Telecommunications, Riga 1048, Latvia

[a)]Corresponding author: dmitrii.borovkov@metalab.ifmo.ru
[b)] adria.canos@optomech.ifmo.ru

Abstract. Multipole expansions of the source play an important role in a broad range of disciplines in modern physics, ranging from the description of exotic states of matter to the design of nanoantennas in photonics. Within the context of the latter, toroidal multipoles, a third group of multipoles complementing the well-known electric and magnetic ones, have been widely investigated since they lead to the formation of non-radiating sources. In the last years, however, the photonics community has brought to light the existence of a fourth type of multipoles that is commonly overlooked. Currently, different groups have provided different mathematical expressions to describe such sources, and they have been coined with different names; on the one hand mean-square radii, and on the other hand, as high order toroidal moments. Despite their clear physical similarity, a formal relation between the two has not yet been established. While explicit formulas for both types have been derived, they are not expressed in the same basis, and therefore it is not possible to draw a clear physical connection between them. In this contribution, we will bridge this gap and rigorously derive the connection between the two representations, taking as an example the cases of the nth order mean square radius of the electric dipole and the nth order electric toroidal dipole. Our results conclusively show that both types of representations are exactly equivalent up to a prefactor.

INTRODUCTION

Multipole decompositions are an essential tool in many areas of modern theoretical physics, ranging from the description of condensed matter systems [1] to the design of antennas [2]. In recent years, they have been actively exploited to understand and efficiently tailor the electromagnetic response of man-made metamaterials and metasurfaces for a broad range of applications in optics [3], and are playing an essential role in the development of the emerging field of all-dielectric nanophotonics [3,4]. Counterintuitively, high-index dielectric particles support multipolar response of both electric and magnetic character [3,5,6]. The multipolar analysis of dielectric nanoantennas sheds light on interesting optical phenomena such as Fano resonances [7], directional scattering [8,9], optical forces [10], or the transverse Kerker effect [11].

The multipole moments of a localized current distribution are commonly classified in terms of electric and magnetic multipoles, associated to oscillating charges and circulating currents, respectively. In general, they can be obtained from a small argument expansion of the vector potential, or equivalently from an expansion of the current in a series of form factors. Through this approach, a third class of multipoles having identical radiation patterns as the conventional ones can be derived, known as toroidal multipoles [12,13]. The interference of the electric dipole with the conventional electric dipole leads to the formation of nonradiating states known as 'anapoles' which have spawned intensive research efforts in the photonics community [14]. Notably, the complex current distributions arising within dielectric particles have also been shown to give rise to higher-order [15] and hybrid [16] anapole states of mixed electric and magnetic character. To understand such an exotic type of nonradiating current distributions, expressions for the so-called higher-order toroidal multipoles were developed [7]. They arise as higher order terms in the expansion of the vector potential, and therefore account for larger retardation effects arising within the source. In parallel, however, there exists an alternative conceptual background to describe these effects, based on a fourth family of multipoles known as the mean-square radii [17,18]. Interestingly, the latter have been proposed as a means to achieve new types of nonradiating current distributions without the need of the conventional electric or magnetic multipoles [18].

Despite their physical similarity, there exists no formal connection between the two electromagnetic descriptions. Mathematical expressions for the high order toroidal moments and the mean-square radii are readily available in the literature [7,17], but are expressed in different basis, and therefore cannot be straightforwardly compared. Here we perform a systematic analysis of the two representations taking as an example the toroidal dipole and demonstrate how both high-order toroidal moments and mean-square radii are in fact equivalent up to a prefactor. Consequently, they can be indifferently used to describe a localized current distribution.

THE NTH ORDER ELECTRIC TOROIDAL DIPOLE

We will derive toroidal moments starting from the expression of what is known as the exact electric dipole, hereby referred to as $\tilde{\mathbf{p}}$, which groups the contributions to radiation of the elementary electric dipole, the toroidal dipole, and higher order terms, given by [2]:

$$\tilde{\mathbf{p}} = -\frac{1}{i\omega}\int d^3\mathbf{r}\left\{\mathbf{J}(\mathbf{r},t)\,j_0(kr) + \frac{k^2}{2}\left[3(\mathbf{r}\cdot\mathbf{J}(\mathbf{r},t))\mathbf{r} - r^2\mathbf{J}(\mathbf{r},t)\right]\frac{j_2(kr)}{(kr)^2}\right\}, \tag{1}$$

where $\mathbf{J}(\mathbf{r},t)$ is the current density of the source, and $j_0(kr)$ and $j_2(kr)$ are spherical Bessel functions of the first kind. In the following we will omit the arguments of the current density and assume a time harmonic dependence of all quantities of interest. A simple way to obtain expressions for the nth order toroidal dipole is to perform a Taylor series of $j_0(kr)$ and $j_2(kr)$ in Eq.(1). In the next step we plug in the expansions for $j_0(kr)$ and $j_2(kr)$ into Eq.(1) and combine them into one sum. After some calculations one obtains an expression for the nth order **electric** toroidal dipole:

$$\mathbf{T}^{(n)} = -\frac{\alpha^{(n)}}{i\omega}\int d^3\mathbf{r}(kr)^{2n}\left[(n+1)\mathbf{J} - n(\mathbf{r}\cdot\mathbf{J})\frac{\mathbf{r}}{r^2}\right], \tag{2}$$

where

$$\alpha^{(n)} = (-1)^n\frac{6(n+1)}{(2n+3)!}. \tag{3}$$

Note that for $n = 0$ Eq.(2) yields the well-known electric dipole:

$$\mathbf{p} = \frac{i}{\omega}\int d^3\mathbf{r}\mathbf{J}. \tag{4}$$

We draw attention to the fact that with this notation all prefactors that go with electric toroidal dipole are already included in Eq.(2). A more common way to define toroidal moments is given for example in [7], where a prefactor is extracted from each term , i.e. $\mathbf{T}^{(1)} \to \frac{ik}{c}\mathbf{T}^{(1)}$, and similarly for higher order terms:

$$\tilde{\mathbf{p}} = \mathbf{p} + \frac{ik}{c}\mathbf{T}^{(1)} + \frac{ik^3}{c}\mathbf{T}^{(2)} + \ldots. \tag{5}$$

MEAN-SQUARE RADII OF THE TOROIDAL DIPOLE

The expression for the mean-square radii of an arbitrary toroidal moment in the spherical basis can be found in [17], and corresponds to the electromagnetic excitations investigated in [18]:

$$R_m^{(n)} = \frac{-1}{3c}\sqrt{\frac{4\pi}{2}}\int d^3\mathbf{r}r^{2n+2}\left[\frac{1}{2n+5}\mathbf{Y}^*_{12m}(\mathbf{n}) + \frac{\sqrt{2}}{2(n+1)}\mathbf{Y}^*_{10m}(\mathbf{n})\right]\cdot\mathbf{J}, \tag{6}$$

where $m = -1, 0, 1$, represents the components of a vector in the spherical basis. Eq. (6) is written in terms of the vector spherical harmonics $\mathbf{Y}_{ll\pm1m}(\mathbf{n})$, whose definition can also be found in [17]. They are related to the Cartesian components by

$$\begin{aligned} R_x^{(n)} &= \frac{1}{\sqrt{2}}\left(R_1^{(n)} - R_{-1}^{(n)}\right), \\ R_y^{(n)} &= \frac{i}{\sqrt{2}}\left(R_1^{(n)} + R_{-1}^{(n)}\right), \\ R_z^{(n)} &= -R_0^{(n)}, \end{aligned} \tag{7}$$

In addition, we require the expression for the Cartesian components of each vector spherical harmonic $\mathbf{Y}_{ll-1m}$:

$$\begin{aligned} \left(\mathbf{Y}_{ll-1m}\right)_x &= -\frac{b_1}{\sqrt{2}}Y_{l-1,m-1} + \frac{b_3}{\sqrt{2}}Y_{l-1,m+1}, \\ \left(\mathbf{Y}_{ll-1m}\right)_y &= -\frac{ib_1}{\sqrt{2}}Y_{l-1,m-1} - \frac{ib_3}{\sqrt{2}}Y_{l-1,m+1}, \\ \left(\mathbf{Y}_{ll-1m}\right)_z &= b_2 Y_{l-1,m}, \end{aligned} \tag{8}$$

and $\mathbf{Y}_{ll+1m}$:

$$\begin{aligned} \left(\mathbf{Y}_{ll+1m}\right)_x &= -\frac{c_1}{\sqrt{2}}Y_{l+1,m-1} + \frac{c_3}{\sqrt{2}}Y_{l+1,m+1}, \\ \left(\mathbf{Y}_{ll+1m}\right)_y &= -\frac{ic_1}{\sqrt{2}}Y_{l+1,m-1} - \frac{ic_3}{\sqrt{2}}Y_{l+1,m+1}, \\ \left(\mathbf{Y}_{ll+1m}\right)_z &= c_2 Y_{l+1,m}, \end{aligned} \tag{9}$$

where the b_i and c_i coefficients are a function of l and m [17]. Using Eqs. (8)-(9), one can find the mean-square radii of the toroidal dipole in the spherical basis from Eq. (6) and then, using Eq.(7), convert them into Cartesian. The final result is:

$$\mathbf{R}^{(n)} = \frac{1}{2c}\frac{1}{(n+1)(2n+5)}\int d^3\mathbf{r}r^{2n}\left((n+1)(\mathbf{r}\cdot\mathbf{J})\mathbf{r} - (n+2)r^2\mathbf{J}\right). \tag{10}$$

Comparing Eq.(10) with Eq.(2) we found that the nth order mean-square radius and the (n+1)th order electric toroidal dipole have an identical functional form . With Eqs.(2)-(3) and Eq.(10), the relation between them is reduced to:

$$\mathbf{T}^{(n+1)} = i\alpha^{(n)}k^{2n+1}\mathbf{R}^{(n)}. \tag{11}$$

Eq.(11) is our central result, demonstrating how the two representations are related.

CONCLUSION

Summarizing, we presented a formula for the nth order electric toroidal dipole (see Eq.(2)), and discussed the mean-square radii of the toroidal dipole. By converting the expressions of the mean-square radii into more intuitive formulas, we have explicitly revealed the connection between the two quantities; Eq.(11) clearly demonstrates that they are *equivalent up to a prefactor*, i.e. the nth order mean-square radius is nothing more than the $(n+1)^{th}$ order toroidal dipole. We emphasize that the proof presented here can be directly extended to the magnetic toroidal moments and the magnetic mean square radii.

REFERENCES

[1] Spaldin N A, Fiebig M and Mostovoy M 2008 The toroidal moment in condensed-matter physics and its relation to the magnetoelectric effect *J. Phys. Condens. Matter* **20**

[2] Alaee R, Rockstuhl C and Fernandez-Corbaton I 2018 An electromagnetic multipole expansion beyond the long-wavelength approximation *Opt. Commun.* **407** 17–21

[3] Kivshar Y and Miroshnichenko A 2017 Meta-Optics with Mie Resonances *Opt. Photonics News* **28** 24

[4] Shalin A S 2010 Broadband blooming of a medium modified by an incorporated layer of nanocavities *JETP Lett.* **91** 636–42

[5] Baryshnikova K, Filonov D, Simovski C, Evlyukhin A, Kadochkin A, Nenasheva E, Ginzburg P and Shalin A S 2018 Giant magnetoelectric field separation via anapole-type states in high-index dielectric structures *Phys. Rev. B* **98** 1–9

[6] Terekhov P D, Baryshnikova K V., Greenberg Y, Fu Y H, Evlyukhin A B, Shalin A S and Karabchevsky A 2019 Enhanced absorption in all-dielectric metasurfaces due to magnetic dipole excitation *Sci. Rep.* **9** 3438

[7] Gurvitz E A, Ladutenko K S, Dergachev P A, Evlyukhin A B, Miroshnichenko A E and Shalin A S 2019 The High-Order Toroidal Moments and Anapole States in All-Dielectric Photonics *Laser Photonics Rev.* **13** 1800266

[8] Terekhov P D, Shamkhi H K, Gurvitz E A, Baryshnikova K V., Evlyukhin A B, Shalin A S and Karabchevsky A 2019 Broadband forward scattering from dielectric cubic nanoantenna in lossless media *Opt. Express* **27** 10924

[9] Kozlov V, Filonov D, Shalin A S, Steinberg B Z and Ginzburg P 2016 Asymmetric backscattering from the hybrid magneto-electric meta particle *Appl. Phys. Lett.* **109** 203503

[10] Kostina N, Ivinskaya A, Sukhov S, Bogdanov A, Toftul I, Nieto-Vesperinas M, Ginzburg P, Petrov M and Shalin A 2017 Optical binding via surface plasmon polariton interference *Phys. Rev. B* **99** 125416

[11] Shamkhi H K, Baryshnikova K V., Sayanskiy A, Kapitanova P, Terekhov P D, Belov P, Karabchevsky A, Evlyukhin A B, Kivshar Y and Shalin A S 2019 Transverse Scattering and Generalized Kerker Effects in All-Dielectric Mie-Resonant Metaoptics *Phys. Rev. Lett.* **122** 193905

[12] Dubovik V M and Tugushev V V. 1990 Toroid moments in electrodynamics and solid-state physics *Phys. Rep.* **187** 145–202

[13] Kaelberer T, Fedotov V A, Papasimakis N, Tsai D P and Zheludev N I 2010 Toroidal Dipolar Response in a Metamaterial *Science (80-.).* **330** 1510–2

[14] Miroshnichenko A E, Evlyukhin A B, Yu Y F, Bakker R M, Chipouline A, Kuznetsov A I, Luk'yanchuk B, Chichkov B N and Kivshar Y S 2015 Nonradiating anapole modes in dielectric nanoparticles *Nat. Commun.* **6** 8069

[15] Zenin V A, Novikov S M, Yang Y, Chichkov B N, Bozhevolnyi S I, Evlyukhin A B, Malureanu R and Lavrinenko A V 2017 Direct Amplitude-Phase Near-Field Observation of Higher-Order Anapole States *Nano Lett.* **17** 7152–9

[16] Valero A C, Gurvitz E A, Benimetskiy F A, Pidgayko D A, Samusev A, Evlyukhin A B, Redka D, Tribelsky M I, Rahmani M, Kamali K Z, Pavlov A A, Miroshnichenko A E and Shalin A S 2020 Theory, observation and ultrafast response of novel hybrid anapole states *http://arxiv.org/abs/2009.00322*

[17] Radescu E E and Vaman G 2002 Exact calculation of the angular momentum loss, recoil force, and radiation intensity for an arbitrary source in terms of electric, magnetic, and toroid multipoles *Phys. Rev. E - Stat. Physics, Plasmas, Fluids, Relat. Interdiscip. Top.* **65** 47

[18] Nemkov N A, Basharin A A and Fedotov V A 2018 Electromagnetic sources beyond common multipoles *Phys. Rev. A* **98** 1–9

Directional luminescence of the diamond NV center via Bloch surface waves in one-dimensional photonic crystals

A A Bragina[1], K R Safronov[1], V O Bessonov[1,2], A A Fedyanin[1]

[1] Quantum Electronics Department, Faculty of Physics, Lomonosov Moscow State University, Moscow 119991, Russia

[2] Frumkin Institute of Physical Chemistry and Electrochemistry, Russian Academy of Sciences, Moscow 119071, Russia

E-mail: bragina@nanolab.phys.msu.ru

Abstract. In this work, we numerically study the luminescence of nanodiamonds with NV centres embedded in a polymer layer on the surface of one-dimensional photonic crystal. The interaction of NV center spontaneous emission with the Bloch surface wave (BSW) is demonstrated. The presence of a photonic crystal leads to a change in the angular distribution of the emitter radiation due to the coupling of luminescence to BSW. We show that the best coupling efficiency of 71% is observed when NV centres are located in the close proximity to the BSW field maximum.

1. Introduction

In recent years, the field of nanophotonics aimed at the control over spontaneous emission of light at the nanoscale level has been actively developed, which is widely used in optical communications, sensing and integrated optics[1]. Various systems have already been studied to control spontaneous radiation, including waveguides made of various materials and two-dimensional photonic crystals with defects[2]. One of the most promising approaches is based on surface electromagnetic waves, since they provide field localization near the surface where radiation sources are located, resulting in luminescence enhancement. The most studied class of such waves is surface plasmon-polaritons (SPP) propagating along metal-dielectric interface. SPP allows achieving significant luminescence enhancement due to the sub-diffraction localization[3] and high Purcell factor. However, they have a number of disadvantages, such as a short propagation length due to Ohmic losses in metal and fixed dispersion law determined by the material of the metal film. These shortcomings limit applications of SPP and lead to research of alternative types of surface waves. One of the most promising type is Bloch surface waves (BSW) propagating on the interface between one-dimensional photonic crystal (PC) and dielectric. Important advantages of BSW over SPP are flexibility of the dispersion law controlled by PC layers materials and thicknesses, as well as the existence of both TE- and TM-polarized waves. In addition, BSW have a large propagation length (of order of 1 mm versus several tens of micrometers for the SPP) due to the absence of Ohmic losses in the dielectric[4]. These advantages have led to the development of various applications, including enhancement of spontaneous emission near the PC surface[5, 6] and integrated photonics[7]. In this work, we study luminescent sources integrated into a slab BSW waveguide. This study reveals the BSW potential for on-chip realization of linear and quantum optical circuits.

One of the most promising candidates among sources for integrated photonics are the color centres in diamonds because they are bright stable sources even at the room temperature. Moreover, a family of diamond color centres based on group-IV elements in the periodic table, i.e. nitrogen-vacancy (NV), germanium-vacancy (GeV) and silicon-vacancy (SiV) centres, have attracted attention due to their structural symmetries, leading to high emission into zero-phonon line (ZPL)[8]. These sources have already been successfully integrated into SPP waveguides[9].

This work presents the results of 2D calculations of the emission of an NV center in diamond embedded into a polymer layer on the surface of 1D PC. Calculations were performed by the Lumerical FDTD. The NV diamond is modelled as an oscillating electric dipole. Emission wavelength is λ = 637 nm matching NV ZPL. We study dependence of coupling efficiency between source and BSW on the source position.

2. Results and discussion

2.1. Model

In 2D FDTD simulation, the 1D PC was modelled as a rectangular structure consisting of five pairs of alternating SiO_2 (n_{SiO_2}=1.46) and Ta_2O_5 ($n_{Ta_2O_5}$=2.08) layers with thicknesses of 140 nm and 98 nm, respectively, where top layer is Ta_2O_5. The PC is located on a glass substrate (n_{glass}=1.52) with a thickness of 2 μm. BSW is not sustained on a bare PC surface. In order to enable excitation of TE-polarized BSW, the PC surface is covered with a 200 nm thick polymer layer of SU-8 photoresist (n_{SU8}=1.58). We study the coupling between the NV center located inside the polymer layer and BSW. NV center was modelled as a dipole source emitting at λ = 637 nm with the dipole momentum lying parallel to the PC surface. At this λ, BSW effective refractive index, defined as the ratio of BSW propagation constant and vacuum wavenumber, is equal to 1.15. The simulation volume was enclosed by Perfectly Matched Layers (PML).

2.2. Calculations

First of all, the electric field profile of BSW was calculated (Fig. 1a). BSW is localized inside the polymer layer and attenuates inside PC with oscillations. Then coupling efficiency between the source and BSW was studied. For this purpose, we calculated the field distribution at a distance of 10 μm from the dipole source and then estimated the overlap integral of this field and the BSW field. Coupling efficiency is a fraction of source radiation transmitted into BSW mode. We calculated the coupling efficiency using Mode Expansion Monitor in Lumerical FDTD. The result showing efficiency dependence with respect to the source position in the polymer layer is shown in Fig. 1b. It demonstrates that the efficiency significantly depends on the source position. The maximum efficiency of 71% is observed when the dipole is placed at h=154 nm inside the 200 nm polymer layer. The obtained value corresponds to the position of BSW field intensity maximum. For the obtained optimal source position, the angular distribution of the dipole emission intensity in a far-field was calculated (Fig. 2a). For this purpose, the dipole radiation intensity was calculated with the use of two Frequency-domain Profile and Power Monitors parallel to the surface of the PC: one located above the PC and the other placed inside the polymer film under the dipole. The intensity maximum occurs at the angles of $\theta = -46.77°$ and $\theta = 226.77°$. These angles correspond to the effective refractive index $n_{eff} = n_{SU8} \sin\theta = 1.15$. The n_{eff} of radiation is equal to BSW n_{eff}. Thus, the radiation direction is related to the excitation of BSW. The results demonstrate that NV center coupled with BSW has directional luminescence. Fig. 2b shows field intensity distribution of NV center emission inside the PC. Almost all of the source radiation is guided inside the polymer layer, which acts as a BSW waveguide. Noteworthy, dipole with polarization perpendicular to PC surface does not couple with BSW, since only TE-polarized component of field can couple with BSW.

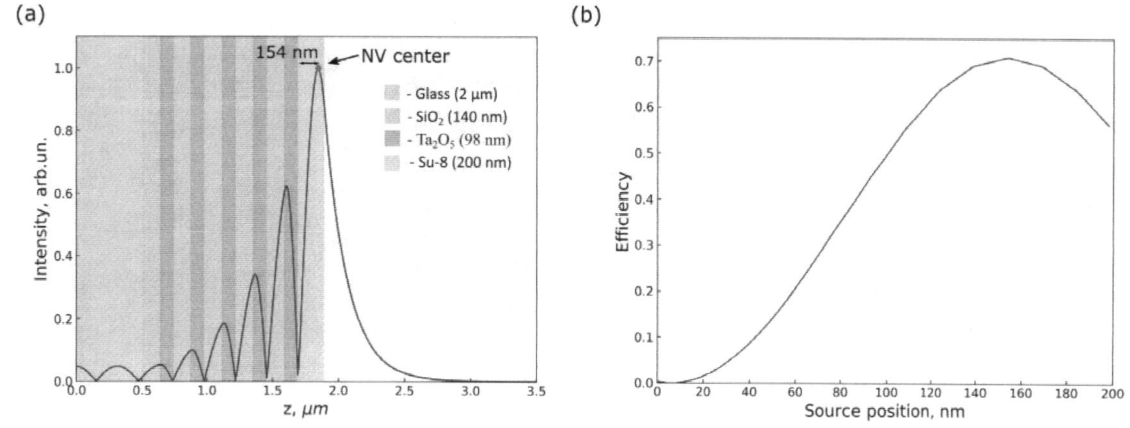

Figure 1. (a) BSW mode profile. Red dot indicates source position with maximal BSW coupling efficiency. (b) Dependence of coupling efficiency on source position in the polymer layer.

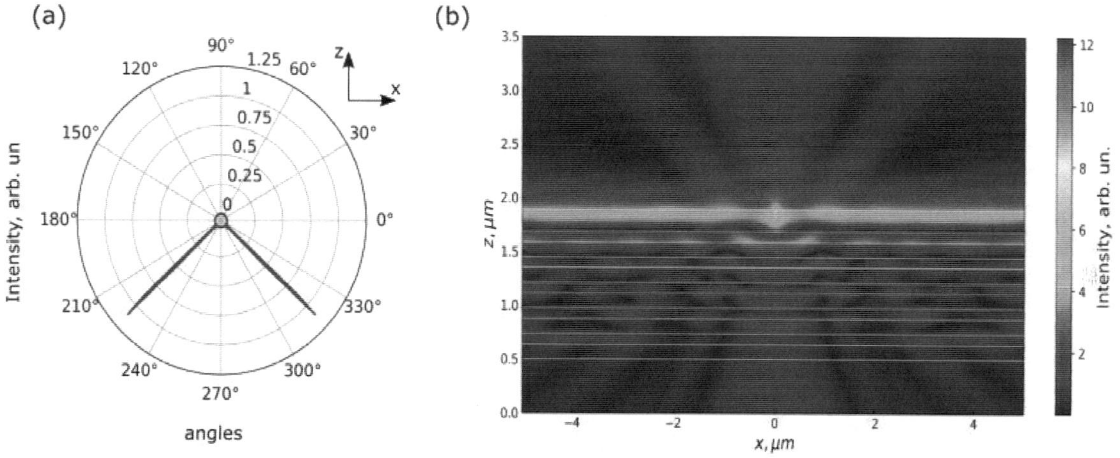

Figure 2. (a) Angular distribution of the dipole radiation intensity in the polymer layer. (b) Field intensity distribution of NV center coupled with BSW.

3. Conclusion

In this work, we performed numerical calculations of the emission from a diamond with NV center located in the polymer film on the PC surface, supporting the BSW propagation. We have shown that dipole spontaneous radiation becomes directional when coupled with BSW: the radiation intensity becomes sharply distributed at certain angles and the emission is localized in the polymer layer. In addition, the coupling efficiency between the NV center and BSW was calculated and the source position corresponding to the maximum efficiency of 71% was determined. Spontaneous emission coupled with BSW in PC is very attractive for future applications in on-chip photonic circuits.

4. Acknowledgments

This work was supported by MSU Quantum Technology Centre and the Russian Foundation for Basic Research (grant №19-32-90225). The research was performed according to the Development program of the Interdisciplinary Scientific and Educational School of Lomonosov Moscow State University "Photonic and Quantum technologies. Digital medicine".

References

[1] Dubey R, Barakat E, Häyrinen M, Roussey M, Honkanen S K, Kuittinen M and Herzig H P 2017 *Journal of the European Optical Society-Rapid Publications* **13** 1–9
[2] Aharonovich I, Englund D and Toth M 2016 *Nature Photonics* **10** 631–641
[3] Sederberg S, Firby C J, Greig S R and Elezzabi A Y 2017 *Nanophotonics* **6** 235–257
[4] Meade R D, Brommer K D, Rappe A M and Joannopoulos J 1991 *Physical Review B* **44** 10961
[5] Stella U, Boarino L, De Leo N, Munzert P and Descrovi E 2019 *ACS Photonics* **6** 2073–2082
[6] Soboleva I, Descrovi E, Summonte C, Fedyanin A and Giorgis F 2009 *Applied Physics Letters* **94** 231122
[7] Abrashitova K A, Gulkin D N, Safronov K R, Kokareva N G, Antropov I M, Bessonov V O and Fedyanin A A 2018 *Applied Science* **8** 63
[8] Siampour H, Kumar S, Davydov V A, Kulikova L F, Agafonov V N and Bozhevolnyi S I 2018 *Light: Science Applications* **7** 1–9
[9] Siampour H, Kumar S and Bozhevolnyi S I 2017 *ACS Photonics* **4** 1879–1884

Reducing "slice cross-talk" effect in metamaterial assisted fast spin-echo MRI

E Brui[1], S Rapacchi[2], D Bendahan[2], A Andreychenko[1,3]

[1]Department of Physics and Engineering, ITMO University, Saint Petersburg, 197101, Russian Federation

[2]Aix-Marseille Universite, CNRS, Centre de Résonance Magnétique Biologique et Médicale, Marseille,France

[3]Research and Practical Clinical Center for Diagnostics and Telemedicine Technologies of the Moscow Health Care Department, Moscow, 109029, Russian Federation

E-mail: e.brui@metalab.ifmo.ru

Abstract. Imaging of subtle changes in hand structures is challenged by the limited image quality, especially at 1.5 T. Reduction of specific absorption rate in metamaterial-assisted 1.5 T MRI provides an opportunity to utilize efficient pulse sequences and to improve the quality of acquired images. This work is devoted to the assessment of potential improvements of slice selectivity and reducing a "slice cross-talk" artifact, using fast spin-echo (FSE) pulse sequence together with a metamaterial-based coil. The slice selection in conventional T1-weighted FSE wrist imaging pulse sequences was modeled using a "Bloch Equations Simulator". Two types of pulses were compared: apodized SINC pulses (reference) common for clinical FSE, and optimized selective Shinnar–Le Roux (SLR) pulses constructed in the MATPULSE program. Regular and SLR-based FSE pulse sequences were tested in a phantom experiment with different gaps between slices to investigate the "slice cross-talk" artifact presence. Combining the utilization of the metamaterial-based coil with an SLR-based FSE provided 28 times lower energy deposition in a duty cycle, as compared to the regular FSE with a conventional transmit coil. When the slice gap was decreased from 100% to 0%, the "slice cross-talk" effect reduced the signal intensity by 16%-18% in the SLR-based FSE and by 23%-32% for the regular FSE. The use of SLR pulses together with the metamaterial-based coil allowed to reduce the "slice cross-talk" effect in contiguous FSE, while still being within the safe SAR limits.

1. Introduction

Diagnosis of human hand pathologies is challenged by it's anatomical complexity and by the presence of fine structures. Wrist, palm and fingers are part of the hand together with 27 bones, 27 small joints, 34 muscles, more than 100 ligaments and tendons, plenty of blood vessels and nerves. Inflammatory (eg., rheumatoid arthritis and psoriatic arthritis) and degenerative disorders (eg., osteoarthritis and post-traumatic arthritis) are associated with pathological changes in these structures. Magnetic resonance imaging (MRI) can be used to assess soft tissues and is thus a perfect tool for the detection of pathological changes occurring at extremities level. However, imaging subtle changes in hand structures, having a sub-millimeter thickness (for example, in articular cartilage, ligaments and synovial membrane), is challenged by the limited image quality, especially at 1.5 T. Taking into account the high prevalence of 1.5 T MR scanners in the world, it is important to increase their diagnostic potential, especially, for large burden associated diseases, such as osteoarthritis [1] and rheumatoid arthritis [2].

Image quality can be increased thanks to improved hardware and signal acquisition procedures of MRI procedures. Recently, a new dedicated wireless metamaterial-based radiofrequency (RF) device (wireless coil) [3] was proposed as an efficient alternative for commercial excitation-reception RF coils commonly used at 1.5T. Accordingly, wrist images recorded with this metamaterial-based device had a significantly larger signal-to noise ratio as compared to those obtained with a dedicated transmit-receive extremity coil [4]. Such a metamaterial-based coil could be used to increase resolution of the standard imaging protocols. This type of device can actually concentrate the RF magnetic field in an area of interest and shift the electric field away from the human's body. As a result, this leads to a significant reduction of the RF energy deposition in tissues when compared to conventional coils [3]. The corresponding decreased specific absorption rate (SAR) could be useful in cases where SAR-demanding pulse sequences have to be used [5], [6]. Another potential benefit of the proposed RF setup is the possibility to increase the slice selection accuracy of pulses thanks to the utilization of highly-selective RF pulses, which have a high-amplitude.

In this work, we intended to assess the potential improvement of slice selectivity and image quality using fast spin echo (FSE) pulse sequence, the most frequently used sequence in MSK MRI. We paid a particular attention to RF safety for the imaging of wrist using the metamaterial-based wireless coil previously reported [3].

2. Methods and Materials

A T1-weighted pulse sequence from the Siemens wrist imaging library for 1.5 T Magnetom Avanto scanner, was chosen. T1-weighting (T1W) refers to the use of a relatively short repetition time (TR). As compared to T2- and proton density-weighted sequences, T1W sequences are linked to a larger saturation effect and "slice cross-talk" artefact. On that basis, one can hypothesize that including highly selective pulses in T1W sequences might be of interest for image quality. Parameters of the pulse sequence were as follows: repetition time TR = 550 ms, echo time TE = 13 ms, excitation angle = 90°, refocusing angle = 180°, inter-echo time IET = 13 ms, echo train length ETL = 4.

For modeling the magnetization evolution during the pulse sequence, a "Bloch Equations Simulator" script for Matlab (The MathWorks, Natick, 2016) was used (http://mrsrl.stanford.edu/ brian /blochsim/). Spatial profiles of transverse magnetization (Mxy) along the slice selection direction were considered as slice profiles.

We initially generated selective 90° and 180° pulses, using a Shinnar–Le Roux (SLR) algorithm [7] using a Matlab-based MATPULSE program [8]. Parameters of these generated SLR pulses were optimized in order to provide a better selectivity than commonly used pulses, i.e. apodized SINC-shaped pulses, without exceeding the maximal gradient amplitude constrains. SINC-shaped pulses are commonly used in clinical FSE sequences in "Low SAR" mode. The pulse duration for both SLR and SINC pulses were similar. Then, magnetization evolution was modeled in three samples with different T1 relaxation times (329 ms, 827 ms and 992 ms), which were close to the ones reported in MSK tissues at 1.5T [9], and approximately the same T2 times (111 ms, 108 ms, 128 ms). The pulses selectivity (slice profile sharpness) was estimated for each echo using the ratio between the signal S accumulated within the desired slice thickness and the total signal S1 accumulated within a full width of the main lobe together with the first pair of side lobes of the slice profile (TH1).

The influence of pulse selectivity on "slice cross-talk" artefacts in MR images was then assessed experimentally using a metamaterial-based transmit-receive RF coil for hand and wrist MRI [3]. A corresponding pulse sequence was tested in 1/ a regular mode and 2/ with the replacement of regular pulses by the constructed SLR pulses. Tests were performed in the three

Journal of Physics: Conference Series **2015** (2021) 012023 doi:10.1088/1742-6596/2015/1/012023

Table 1. Parameters of RF pulses for simulations.

Type	FA, °	Trf, ms	BW, kHz	PBR, %	RBR, %	B1, uT	G, Gs/cm
SINC	90	2.816	1.36	-	-	7.93	0.73
SINC	180	3.840	0.68	-	-	13.68	0.53
SLR	90	2.816	1.36	1	1	8.33	0.73
SLR	180	3.840	1.057	0.5	0.5	23.99	0.82

FA - flip angle, Trf - pulse duration, BW - bandwidth, PBR - passband ripples, RBR - rejectionband ripples, B1 - pulse amplitude, G - gradient amplitude needed to excite a 3 mm slice in FSE.

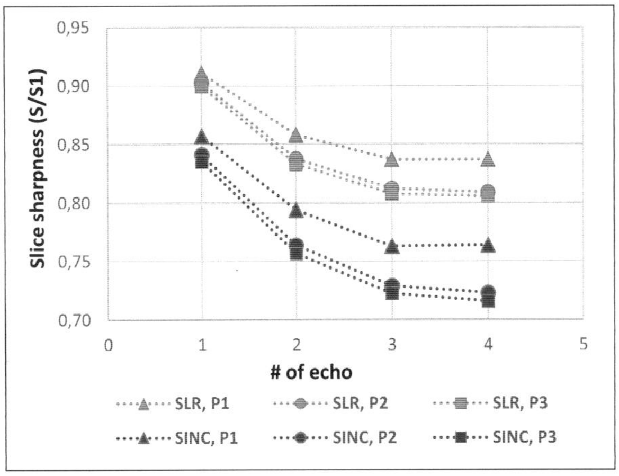

Figure 1. Modeling results. Sharpness of slice profiles (S/S1) (slice thickness - 3 mm) calculated for T1-weighted FSE in three samples with different T1 (P1 – 329 ms, P2 – 827 ms and P3 – 992 ms). The orange curves correspond to the data obtained for SLR-based sequence and the blue ones – for the regular sequence.

phantoms previously mentioned (Eurospin II, length - 11 cm, radius - 3 cm). Images were obtained in a multi-slice mode (7 slices, slice thickness - 3 mm, interleaved excitation, evenly distributed within TR) with different slice gaps of 100%, 50%, 25%, 10%, and 0%. The average signal intensity was measured in each phantom in a medial slice of each multi-slice set.

3. Results and discussion

The optimal parameters of SLR pulses are summarized in Table 1. The selected values allowed to increase the selectivity and, at the same time, not to exceed the gradient amplitude limitations (max 2.2 Gs/cm). These limitations, in fact, could allow to use an RF pulse with a wider frequency bandwidth (2.83 kHz) and, consequently, to provide higher selectivity. However, in order to leave a room for the slice thickness reduction (down to approximately 1 mm), we did not utilized this opportunity in this work. The slice profile sharpness (S/S1) of the modeled slice profiles for each of the 4 echoes was higher when the SLR pulses were used (Figure 1). The sharpness was inversely proportional to T1 values due to a larger saturation effect.

According to what has been previously reported [3], the wireless coil provided a 48-fold times SAR reduction regarding a regular wrist MRI scanning with a conventional coil. Considering the higher amplitude of the proposed SLR pulses, the corresponding SAR generated through the duty cycle of a regular FSE was 1.74 times larger. On that basis, combining the wireless coil with the SLR-based FSE pulse sequence, still provided a 28 times lower energy deposition as compared to the conventional situation. Overall, the energy deposition related to the SLR-based

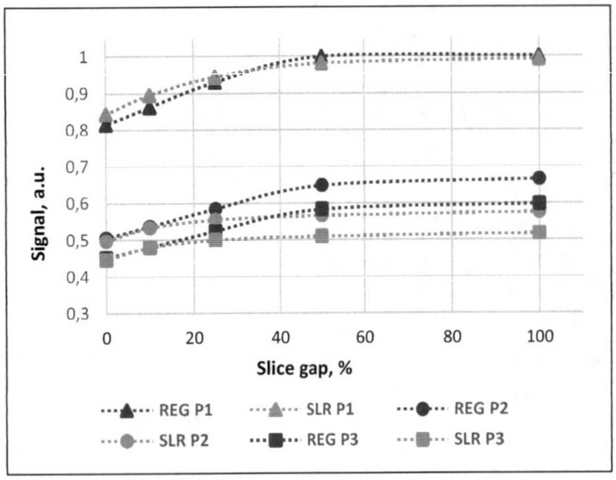

Figure 2. Experimentally measured signal intensity in MR images of three phantoms with different T1 (P1 – 329 ms, P2 – 827 ms and P3 – 992 ms) obtained in T1-weighted FSE, while reducing the gap between slices from 100% to 0% (contiguous slices). The orange curves correspond to the data obtained for SLR-based sequence and the blue ones – for the regular sequence.

FSE sequence combined with the metadevice was kept far below the initial RF safety limits. Such a reduction might be used for a further increase of the slices number which is known to be linked to a SAR raise.

Signal intensities, which were measured in central slices of experimentally obtained phantom images are summarized in a diagram in Figure 2. The “slice cross-talk” effect reduced the signal intensity by 16% -18% (depending on T1 values) in the SLR-based sequence when the slice gap was decreased from 100% to 0%. At the same time, this signal reduction was significantly higher (23% - 32%) for the regular pulse sequence and so due to a lower sharpness of the slice profiles provided by SINC pulses. The signal preservation effect was more pronounced in samples having T1 relaxation times in the range of muscle and cartilage tissues.

4. Conclusion

Our study demonstrated that SLR-based FSE sequence combined with a metamaterial-based wireless coil could reduce the "slice cross-talk" artefact in contiguous slices while remaining within the SAR limits. This combination offers an interesting alternative for MR investigations for which a reduced energy deposition is required e.g. in children, pregnant women and patients with implanted devices.

5. Acknowledgementss

This work was supported by the Russian Science Foundation (Project No. 18-79-10167).

References

[1] Fransen M, et. al., 2011 *J. Rheum. Dis.* **14(2)** 113–121
[2] Kłak A, Raciborski F, Samel-Kowalik P., 2016, *Reumatologia* **54** 73 – 78.
[3] Shchelokova A V, et. al., 2018 *Magn. Reson. Med.* **80** 1726–37
[4] Shchelokova A V, et. al., 2010 *Proc. Intl. Soc. Mag. Reson. Med.* **27** 0273
[5] Hennig J, Weigel M, Scheffler K, 2003 *Magn. Reson. Med.* **49** 527-535
[6] Borthakur A, et. al. 2003 *J Magn. Reson. Imaging* **17** 730-736.
[7] Shinnar M., Bolinger L., and Leigh J.S., 1989 *Magn. Reson. Med.* **12** 75-87
[8] Matson G.B., 1994 *Magn. Reson. Imaging* **12(8)** 1205-25
[9] Kato H, et. al., 2005 *Med. Phys.* **32(10)** 3199-3208.

Machine learning methods for Precipitable Water Vapor estimation by radiometric data in millimetre wavelength

Grigoriy Bubnov[1], Peter Zemlyanukha[1], Evgeniy Dombek[2], Vyacheslav Vdovin[1, 3]

[1]Institute of applied physics RAS, Nizhniy Novgorod, Russia

[2]University of Nizhniy Novgorod, Nizhniy Novgorod, Russia

[3]Nizhny Novgorod State Technical University, Nizhniy Novgorod, Russia

Corresponding author: bubnov@ieee.org

Abstract. This work deals with the first try to calculate the amount of Precipitable Water Vapor (PWV) in atmosphere by using machine learning and AI methods. We use the detector voltages series measured by radiometric system "MIAP-2" as the initial data for machine learning. The radiometer MIAP-2 works by "atmospheric dip method" in 2mm and 3mm atmospheric transparency windows. We also have PWV data series collected by Water Vapor Radiometer and GNSS receiver for data validation. The best convergence results were demonstrated by the independent component analysis (ICA) method with coefficient of determination R^2= 0.53 and artificial neural network method (ANN) with R^2= 0.8. These methods allow to reduce the systematic errors due to direct PWV calculation from raw radiometric data avoiding unnecessary steps opacity calculation.

1. Introduction

Studying astroclimate of a location is essential, when one chooses the most appropriate site for radio astronomical observations or space telecommunications. The suitability of a site is determined by a number of days per year, during which astronomical observations and space communication sessions are possible. In the millimeter range, astroclimatic conditions largely depend on the absorption of such waves by the atmospheric water vapor (Precipitable Water Vapor, PWV) and oxygen. This absorption depends on the elevation of the site, local and global climatic peculiarities, and weather; as the result, it may vary greatly from season to season and from day to day. The local climate is often responsible for how dry and wet air masses interchange over time, and, consequently, how the integral atmospheric absorption (or optical depth, or opacity) varies. The statistical data on the opacity are the chief astroclimatic parameter for the purposes of radio astronomy and space communications in the millimeter wavelength.

Meteorological conditions determine the astroclimate values, thus, the meteorological models and satellite data are often used for astroclimate predictions. The local climate makes the main contribution to opacity and it's the reason to measure astroclimate right on the site of interest. Finally, the satellite data and model calculations together with local measurements complement each other, and give the full information of astroclimate conditions. Machine learning methods are often used in meteorology, including PWV estimations [1][2].

Radiometric data and methods are also wide used in meteorology and climate estimation. [3] The atmospheric dip method used by us in radiometric measurements has its drawbacks. The method is based on flat-layer atmospheric model, which is not always true. Cloudiness and other inhomogeneities in the atmosphere lead to significant errors in measurements, therefore, it is necessary to improve the calculation method. The first attempts to update the methodology were associated with the use of smoothing filters and adaptive algorithms [4] and proved to be quite good [5]. In this work, we attempted to compare the classical, modified method of a flat-layered model of the atmosphere and machine learning methods.

2. Equipment and Atmospheric dip method drawback

Our task is a rough-copy 2 and 3-mm atmospheric opacity measuring. With this objective in mind, we have developed instrumentation [6] and refined our methods [4], [5] for investigation of atmospheric propagation of terahertz waves. The "tau-meter" (the above mentioned MIAP-2) which we were using, was developed by the IAP RAS as the most appropriate instrument for the astroclimate measurements in the THz band. The radiometer allows us to estimate an integral absorption by using the atmospheric dip method. The hardware includes a radiometric system comprising two self-contained radiometers operating in two different bands of 84-99 GHz ($\lambda \sim 3$ mm) and 132-148 GHz ($\lambda \sim 2$ mm), a rotary support, a control, and a firmware system.

The well-known atmospheric dip method is widely used in radiometric measurements. [7] The method is based on flat-layer assumption, allowing to approximate the dependence of the sky brightness temperature on elevation by an exponent function and opacity is the exponent index. But in real atmosphere the flat-layer assumption is not met under cloudy conditions and inhomogeneities in the atmosphere. In fact, we have the collapsed voltage shift between the records, so we are unable to differentiate it against the background of noise. Tn this case the exponent approximation appears with a huge error. Figure 1 shows an example of raw-record of the "MIAP-2" radiometer detector voltages on different elevation angles. The atmosphere is close to flat-layered model in the clear weather conditions (left) and opposite in cloudy weather (right). The detector voltage is in direct ratio to brightness temperature on the corresponding angle and has a negative sign only for technical reasons. In classical atmospheric dip method these values are approximated by exponent, which leads to catastrophic errors of approximation in cloudy conditions.

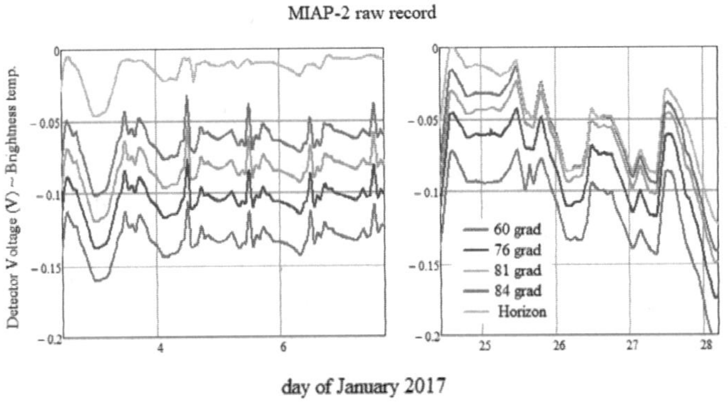

Figure 1 – The sky brightness record on different elevations was made on "Badary" observatory in January 2017. On the left is a clear sky, the atmosphere state is close to a flat-layered model. On the right - cloudy weather, unstable atmosphere.

3. A new data processing method

We are presenting a new approach for estimation of PWV by using atmospheric dip method, but without the flat-layer model. We use machine learning to produce the regression model between our data series and the PVW by Individual Component Analysis, (ICA) and k-Nearest Neighbours, (kNN) pipeline. We note that it is an only overview of an approach for the astroclimate study than the strict method. The essence of this approach is that we tried to estimate the amount of PWV without model assumption, but using independent measurements on another, more accurate device for training a neural network. The PWV data on Badary observatory is collected by two special meteorological instruments, which are well calibrated to PWV measurements and between themselves. Water Vapor Radiometer (WVR) measures the PWV at the Zenith from the sky brightness temperature at the slope of 22 GHz water line. [8] Global Navigating Satellite System receiver (GNSS) allows to calculate PWV from Zenith Path Delay by using time signals from navigation satellites. [9]. We also use the previous measurements made by atmospheric dip method in 2mm and 3mm atmospheric transparency windows on Badary observatory. Obviously, the tracks we observe carry a set of different mixed signals, wherein the contribution of water vapor absorption in the atmosphere must have nonzero covariance for the time series of detector voltages. Assuming this, we propose to use the dimension reduction procedure, which will make it possible to pass from the basis of the detector voltages to some new one, the components of which sufficiently represent the initial sample. A review of methods for choosing sufficiency is presented in the work [10]. After dimension reduction procedure, we estimate the degree of linear correlation of an individual component with a physical value – PWV by WVR and GNSS in our case.

There is a number of a dimension reduction procedure exists. To find the best one we compare three different methods: Principal Component Analysis (PCA), Individual Component Analysis (ICA) and Factor Analysis (FA) utilizing the scikit-learn (Scikit-learn: Machine Learning in Python [11]) by choosing the correlated with PVR. Since the correlation was far from linear, we use artificial neural network (ANN) and kNN as regression model and compare the results. The ANN has a hidden layer with 968 neurons with sigmoid activation function from Tensor Flow [12]. For the kNN we use 5 neighbors and m=2 metric. The detailed comparation and the analysis of the individual methods will be discussed in future papers. We use balanced ensemble divided to 2/3 training and 1/3 validation dataset.

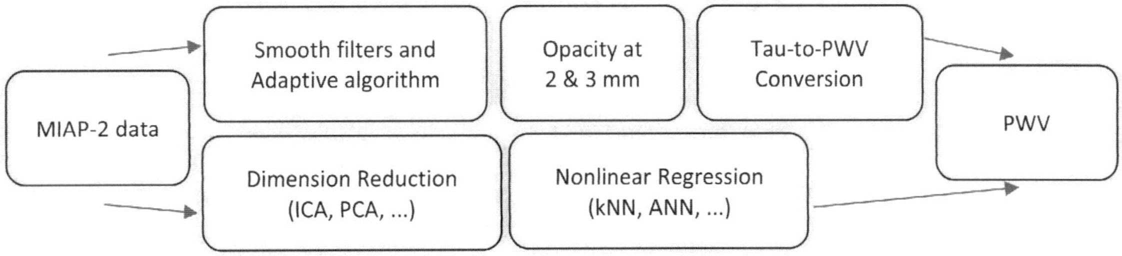

Figure 2 – PWV calculating scheme from the raw-data of MIAP-2 by the classical method (above) and by the machine learning methods (below).

4. Data sets

For the method test we took the data obtained on Badary observatory from June 2016 to June 2017. [13] Observatory is located in sharp-continental climate of Sayan mountains, which means that opacity statistics should be low. However, due to cloudiness accumulating in the Tunkinskaya valley, the measurements were not the most successful. Relatively low values of integral humidity are evidenced by the readings of the meteorological equipment of the observatory. That was the reason to research the methods to minimize the influence of cloudiness on the results of measurements by the atmospheric dip method.

We have three rows of data at our disposal:

- MIAP-2 data. It is a radiometric system operating by atmospheric dip method in atmospheric transparency windows with wavelengths of about 3mm and 2mm. In fact, the system provides a raw records of brightness temperature of the sky on 6 elevations every 10 minutes. Thus, we have 12 data series: 6 for every channel.
- Water Vapor Radiometer data (WVR). It calculates an amount of PWV in Zenith by the intrinsic radiation of the atmosphere at the slope of the water line 22 GHz. [8]
- GNSS receiver data. This is a standard tool that allows, among other things, to determine the PWV from the Zenith Path Delay by using precise time signals from navigation satellites. [9]

The time resolution of these data sets is about the same, and it provides about 50 thousand points through the year. There are clear stable atmosphere weather and cloudy weather as well. The different weather in the sample allows us to train the processing algorithm in the best way.

5. Test sample calculation results

5.1. Dimension reduction

The greatest R^2-factor (coefficient of determination of the prediction [14]) of the estimations the PWV by GNSS was obtained by ICA – 0.53 (by using the best component), FA – 0.39, PCA – 0.22. This result is a good argument that the time series we observe can be represented by independent mixed signals to a certain extent. These signals, in turn, are connected to the opacity and, probably, to clouds and other features of the real atmosphere. However, the nature of the dependence of the most correlated independent component with the PWV is nonlinear.

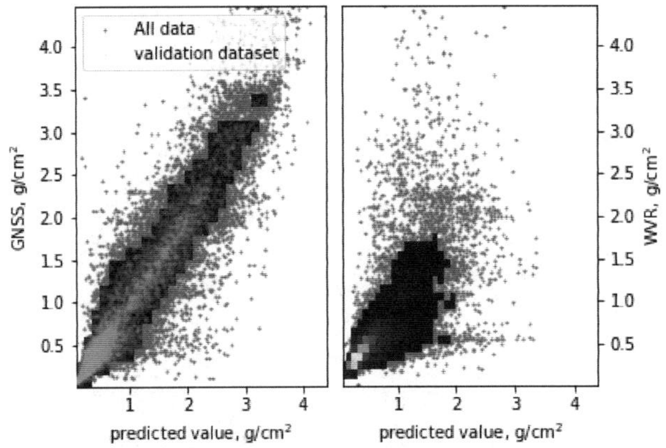

Figure 3 – The PWV regression between MIAP-2 data processed by ICA method and GNSS (left) and WVR (right). Blue dots indicate all data, and color are for predicted values.

5.2. Regression analysis

The picture below shows the data regression of PWV by ANN method with comparison to GNSS and WVR data. The WVR data was not involved in training and validation and provide a good comparation of our PWV estimation. There is no bias to the WVR data, both for the kNN and ANN regression. R^2 - factor of prediction was 0.86 for kNN and 0.8 for ANN. The kNN prediction is more linear, and the ANN are not, but we note that ANN can be better tuned. There is still the deviation from the trend line indicating presence of the stochastic noises. This deviation is much less important since we do the averaging in month bins for astroclimat analysis. But even with such data scatter, the resulting errors of PWV estimation by ANN and ICA methods are less then by old algorithm by opacity calculation. We note that the R^2 is lower if the dimension reduction procedure is excluded for kNN and ANN both.

The PWV are in a wide range of values, which suggests the possibility of using regression dependence and transformation to reduce the dimension for similar observations for other similar sessions. Finally, the methodic allows to determine the PWV bypassing the opacity calculation step and reducing systematic errors connected with uncertainty factors of specific absorption coefficients.

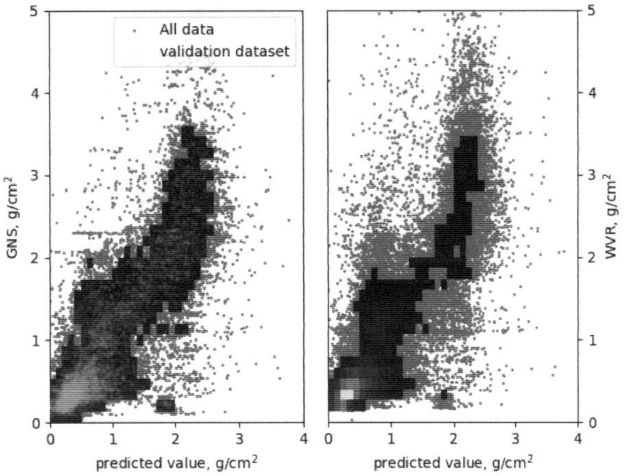

Figure 4 - The PWV regression between MIAP-2 data processed by ANN method and GNSS (left) and WVR (right). Blue dots indicate all data, and color are for predicted values.

6. Conclusion

In previous methods we used a MIAP-2 radiometric data to calculate opacity first, and then PWV values were calculated from opacity. At the first stage, errors related to the fitting of the exponent into the experimental data occurred and reached 40% in the worst conditions. [4] The calculation of PWV from opacity is connected with use of specific absorption coefficients, which can vary up to 300% depending on the site altitude, climate, calculation method and the model. [15] We offer a new algorithm which characterized by the fact that they don't need to use specific absorption coefficients and bypass opacity calculation step. This reduces the systematic errors associated with the uncertainty of the specific absorption coefficients in water vapor and oxygen.

The next step will be validation the method on other data series, obtained in different places. We have 9-year experience of astroclimate measurements in different places of Eastern Hemisphere, so the problem of PWV estimation is quite actual for our research.

7. Acknowledgments

Astroclimate measurements are supported by the Russian Science Foundation under grant №19-19-00499. Data processing was carried out by Institute of Applied Physics RAS (theme 0030-2021-0005).

References

[1] D. Diouf, A. Niang, and S. Thiria, "DEEP LEARNING BASED MULTIPLE REGRESSION TO PREDICT TOTAL COLUMN WATER VAPOR (TCWV) FROM PHYSICAL PARAMETERS IN WEST AFRICA BY USING KERAS LIBRARY," Int. J. Data Min. Knowl. Manag. Process, vol. 9, no. 6, 2019, doi: 10.5121/ijdkp.2019.9602.

[2] V. S. Kongara and D. Punyasesudu, "Forecasting the Water Vapour Distribution Over India Using Artificial Neural Networks," no. July, pp. 1–12, 2015.

[3] A. F. Almansa et al., "Column Integrated Water Vapor and Aerosol Load Characterization with the New ZEN-R52 Radiometer," Remote Sens., vol. 12, no. 9, p. 1424, Apr. 2020, doi: 10.3390/rs12091424.

[4] G. M. Bubnov, V. F. Vdovin, P. M. Zemlyanukha, and I. I. Zinchenko, "A refined method of the data processing for astroclimate measurements in mm-waves," in 2018 29th IEEE International Symposium on Space Terahertz Technology, ISSTT 2018, 2018, pp. 164–168.

[5] G. Bubnov, V. Vdovin, P. Zemlyanukha, and I. Zinchenko, "Observations of astro-climate with the broad band radiometer using the atmospheric dip method," in AIP Conference Proceedings, Dec. 2020, vol. 2300, no. 1, p. 020014, doi: 10.1063/5.0031757.

[6] V. I. Nosov et al., "A dual-wave atmosphere transparency radiometer of the millimeter wave range," Instruments Exp. Tech., vol. 59, no. 3, pp. 374–380, May 2016, doi: 10.1134/S0020441216020111.

[7] A. G. Kislyakov, "On measuring the absorption of radio waves in the atmosphere by its thermal radiation" Radiotekh Erlektron., 13, No. 7, 1161 (1968).

[8] Ilyin G.N., Troitsky A.V., "Determining the tropospheric delay of a radio signal by the radiometric method" Radiophysics and Quantum Electronics. vol. 60, no. 4, pp. 326–335, 2017, [Online]. Available: https://elibrary.ru/item.asp?id=29220875.

[9] Q. Zhao, Y. Yao, W. Yao, and S. Zhang, "GNSS-derived PWV and comparison with radiosonde and ECMWF ERA-Interim data over mainland China," J. Atmos. Solar-Terrestrial Phys., vol. 182, no. June 2018, pp. 85–92, 2019, doi: 10.1016/j.jastp.2018.11.004.

[10] R. Cangelosi and A. Goriely, "Component retention in principal component analysis with application to cDNA microarray data," Biol. Direct, vol. 2, no. 1, p. 2, Jan. 2007, doi: 10.1186/1745-6150-2-2.

[11] F. Pedregosa et al., "Scikit-learn: Machine Learning in Python," J. Mach. Learn. Res., vol. 12, pp. 2825–2830, 2011.

[12] M. Abadi et al., "TensorFlow: Large-Scale Machine Learning on Heterogeneous Distributed Systems," Mar. 2016, Accessed: Apr. 07, 2021. [Online]. Available: http://arxiv.org/abs/1603.04467.

[13] G. M. Bubnov, V. F. Vdovin, V. Y. Bukov, T. A. Makarov, G. N. Il'In, and I. I. Zinchenko, "Millimeter-wave astroclimate investigations on badary observatory near Baikal lake," 2017 32nd Gen. Assem. Sci. Symp. Int. Union Radio Sci. URSI GASS 2017, vol. 2017-Janua, no. August, pp. 1–4, 2017, doi: 10.23919/URSIGASS.2017.8105000.

[14] Altman, Naomi S. (1992). "An introduction to kernel and nearest-neighbor nonparametric regression". The American Statistician. 46 (3): 175–185. doi:10.1080/00031305.1992.10475879. hdl:1813/31637.

[15] G. Bubnov, P. Tremblin, V. Vdovin, P. Baron, and V. Khaikin, "Analysis of variations in factors of specific absorption of sub-terahertz waves in the earth ' s atmosphere," Proc. 2020 7th All-Russian Microw. Conf. (RMC), to be Publ. 2020., vol. 0, no. 1, pp. 1–4, 2020.

Journal of Physics: Conference Series **2015** (2021) 012025 doi:10.1088/1742-6596/2015/1/012025

Creation of an MRI phantom based on super-paramagnetic nanoparticles of iron oxide (SPIO) for standardization of the conversion of T2 * values in the concentration of iron in the liver

P.A. Bulanov[1,2], E.E. Nazarova[1], P.E. Menshchikov[1,3] , D.A, Kupriyanov[1,3], G.V. Tereshchenko[1] and G.A. Novichkova[1]

[1]Dmitry Rogachev National Research Center of Pediatric Hematology, Oncology and Immunology, st. Samory Machela, 1, Moscow, Russian Federation, 117997
[2]Lomonosov Moscow State University, Moscow, Russian Federation, 119991
[3]LLC Philips Healthcare, Sergei Makeev St., 13, Moscow, Russian Federation, 123022

petepeter@yandex.ru, evelina.nazarova@fccho-moscow.ru, petr.menshchikov@philips.com, dmitry.kupriyanov@philips.com, galina.tereshenko@fccho-moscow.ru, galina.novichkova@fccho-moscow.ru

Abstract. In this study the standardization method for T2* maps acquisition on various MR scanners (3T and 1.5T) is proposed. The reproducibility of the obtained T2* values is realized through the MR-compatible phantom containing paramagnetic complex iron oxide nanoparticles. The repeatability of measurements results has shown that the created phantom retains all the required characteristics (homogeneity, stability of concentrations and manifested paramagnetic properties) over a long period of time. The application of standardized T2* values allows to use previously received T2*, [ms] to iron concentrations in the dry substance of the liver (LIC), [mg/ml] conversion formulas for accurate, fast and non-invasive MRI diagnostics of liver iron overload.

1. Introduction

Iron overload is a condition in which extra iron builds up in organs and tissues causing toxic damage and consequently, organ dysfunction. Iron overload may be due to hereditary hemochromatosis or result from anemia requiring regular transfusions of donor red blood cells [1]. The classic method for determining the concentration of iron in the liver is the atomic absorption study of a biopsy sample [2]. However, biopsy, as an invasive procedure, has a number of negative factors, such as risks of infection and complications, as well as pain, which is especially relevant when working with children.

An alternative to biopsy can be the technique of noninvasive determination of iron concentration in the liver by MRI methods [3]. This method is based on the construction of T2* maps, because T2* values depend on the magnetic susceptibility of the tissue, which changes with iron overload. Then, these values are compared with biopsy data and experimental dependence between the values of iron concentrations in the dry substance of the liver (LIC), mg/ml and T2* values, ms is constructed. We obtained a similar formula earlier at the Dmitri Rogachev Pediatric Hematology Center [4]. This allowed us to introduce into clinical practice a noninvasive assessment of liver iron overload. However, the obtained formulas of T2* to LIC conversion can be reliably used only on those MRIs on which it was obtained. For the reason that T2* values are sensitive to different sequence parameters, which may not coincide in different software releases or in different MRI vendors.

For this purpose, we have created an MR-compatible phantom containing various concentrations of paramagnetic nanoparticles of complex iron oxide (SPIO).

2. Materials and methods

2.1. Phantom creation

The phantom consisted of 28 test tubes (50 ml), with various concentrations of a colloidal solution (Fig. 1) containing paramagnetic nanoparticles of complex iron oxide (Fe_3O_4). These nanoparticles were obtained as a result of the Elmore chemical reaction [5]:

$$FeSO_47H2O + 2FeCl_36H_2O + 8NH_3H_2O = Fe_3O_4 + 6NH_4Cl + (NH_4)2SO_4 + 23H_2O,$$

where 6-water iron (III) chloride and 7-water iron (II) sulfate were used as carriers of ions of two and three valence iron, ammonium hydrate was used to start the process of precipitation and formation of nanoparticles, and a surface active agent was used to control the growth and stabilization of particles. substance - citric acid. The resulting reaction solution was centrifuged at 1500 rpm for 5 minutes to precipitate large particles. The upper fraction of the solution was filtered through standard filter paper with an average pore diameter of 3-5 μm. The resulting solution was the desired colloidal suspension of iron oxide nanoparticles.

The resulting solutions with high iron concentration were diluted in order to achieve T2* values similar to those seen in a healthy liver and in the four grades of iron overload (Grade 1: >7 ms ; Grade 2: 2.5 – 7 ms; Grade 3: 1 – 2.5 ms; Grade4: 0 – 1 ms) [4].

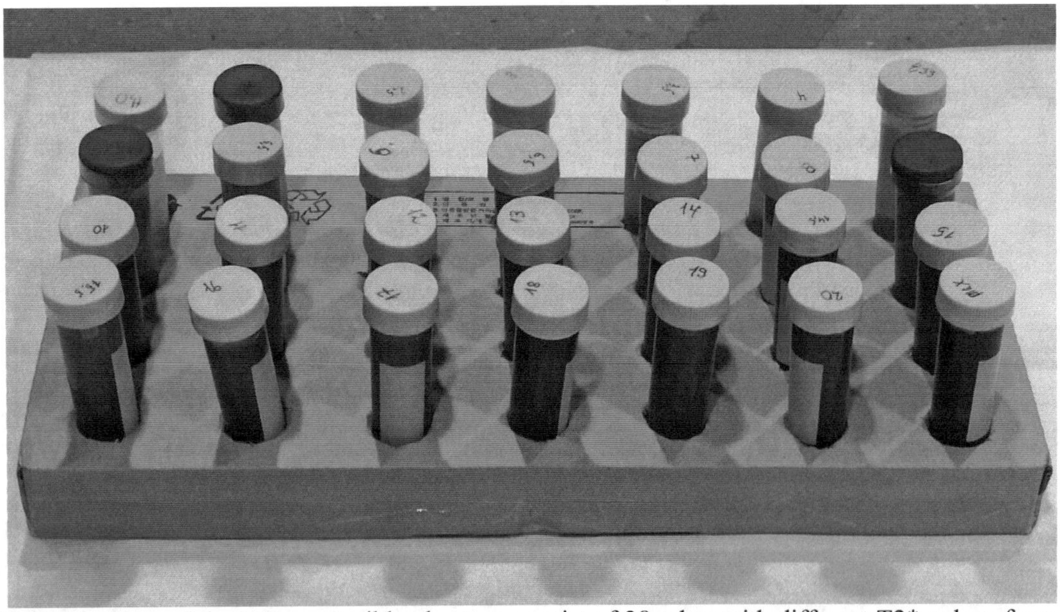

Figure 1. View of the MR-compatible phantom consist of 28 tubes with different T2* values from diapason (0.2 – 15, ms)

2.2. Atomic adsorption spectroscopy

After the creation of solutions of nanoparticles for measuring the concentration of iron in NUST MISIS, atomic adsorption spectroscopy of the samples was carried out on an Agilent 4200 apparatus (Agilent Technologies, USA). For this purpose, the method of atomic adsorption spectroscopy with inductively coupled plasma was used [6].

2.3. MRI Scanning and Post-processing

The resulting phantom was scanned (Figure 2) using 4 MR scanners from different vendors with different field strengths (Table 1). Research protocols included obtaining T2* maps using multi-phase fast gradient echo (mGRE) techniques. On all scanners, the main parameters of the sequences used corresponded to those used in in vivo studies: FA, flip angle- 45 °, TE, echo time - 1.2 ms, the number of echoes - 20 with a step $\Delta TE = 1.5$ ms, TR, repetition time - 350 ms; resolution - 1.5×1.5 mm. For MR scanners with field strengths of 1.5 and 3 T, the slice thickness was 10 and 7 mm, respectively. Relaxometric phantom maps were calculated automatically using the T2* mapping software integrated in the Philips tomograph and the commercial software package ReportCARD Functool (GE Healthcare). For each ROI, an average T2* value (ms) was automatically obtained by automatic approximation of the attenuation curve using the maximum likelihood estimation.

The name of the scanner / Magnetic field strength	Medicine center
Philips Achieva (Best, The Netherlands) / 3T	Dmitry Rogachev National Medical Research Center of Pediatric Hematology
Philips Achieva dStream(Best, The Netherlands) / 3T	Research Institute of Emergency Pediatric Surgery and Traumatology
Signa GE (Chicago, Illinois, USA) / 1.5T	Dmitry Rogachev National Medical Research Center of Pediatric Hematology
Philips Ingenia (Best, The Netherlands) / 1.5T	N.N. Priorov Scientific Research Institute of Traumatology and Orthopedics

Table 1. MRI scanners used in the research

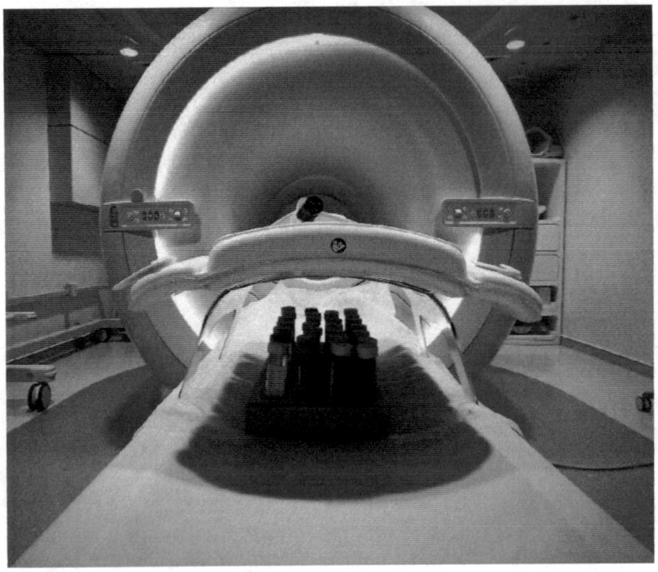

Figure 2. Example of phantom positioning in an MRI scanner

2.4. Statistical Analysis

We calculated mean T2* values and standard deviations for each tube. The statistical analysis was conducted using GraphPad Prism 8.0.1 software. The repeatability and reproducibility of T2* measurements were assessed. Repeatability was defined as the stability of data obtained during a number of scanning sessions on the same (reference) MRI scanner, over a long period of time. In order to assess repeatability, we also calculated mean coefficients of variation with the following formula: $\boldsymbol{v_x = \frac{\sigma}{\bar{x}} * 100\%}$.

Reproducibility was defined as the degree of agreement between T2* values obtained on the control MRI scanners and those obtained on the reference scanners. To test the reproducibility of the results across different MRI scanners, we estimated the correlation between measurements obtained on the reference MRI scanners and those tested for reproducibility.

Additionally, Bland-Altman plots were created to visually assess the repeatability and reproducibility of the results.

3. Results and discussion

The iron concentrations in the phantoms were measured by atomic adsorption spectroscopy. However, the structure of these nanoparticles [7] does not coincide with the structure of iron in the liver, bound to the ferritin protein and contained in the pigment hemosiderin, which have not yet been fully studied [8,9]. Therefore, the range of iron concentrations in test tubes obtained by atomic adsorption spectroscopy (0.1 - 1.5 mg / ml) did not coincide with those obtained by applying the formula for converting T2 * to LIC (1-30 mg / ml). Thus, we used the paramagnetic properties of the material to achieve the required range precisely for T2 *, ms. In addition, the study of this material can also be interesting from the point of view of contrast materials and other studies where it is used [10–12].

An example of a T2* card, as well as signal attenuation curves for two tubes with different iron concentrations are shown in the figure 3.

As a result of statistical analysis, the T2* values showed good repeatability on both our scanners, which is confirmed by the graphs of Bland Altman (Figure 4). The phantom was stable throughout, indicating that it can be used for long time.

Journal of Physics: Conference Series **2015** (2021) 012025 doi:10.1088/1742-6596/2015/1/012025

On the other scanners tested we found statistical correspondence of values, so the T2* values obtained on them can be recalculated in LIC with the help of our formulas, which is confirmed by the graphs of Bland Altman and correlations (Figure 5). In case of inconsistency of values next steps must be carried out:

1) calibration of scanning parameters and T2* mapping;
2) introducing additional calibration coefficients into the recalculation formula.

Thus, to be able to use formulas, it is enough to check the correspondence of T2* values using the phantom we created. If there are discrepancies, it is necessary to refine the scanning protocol up.

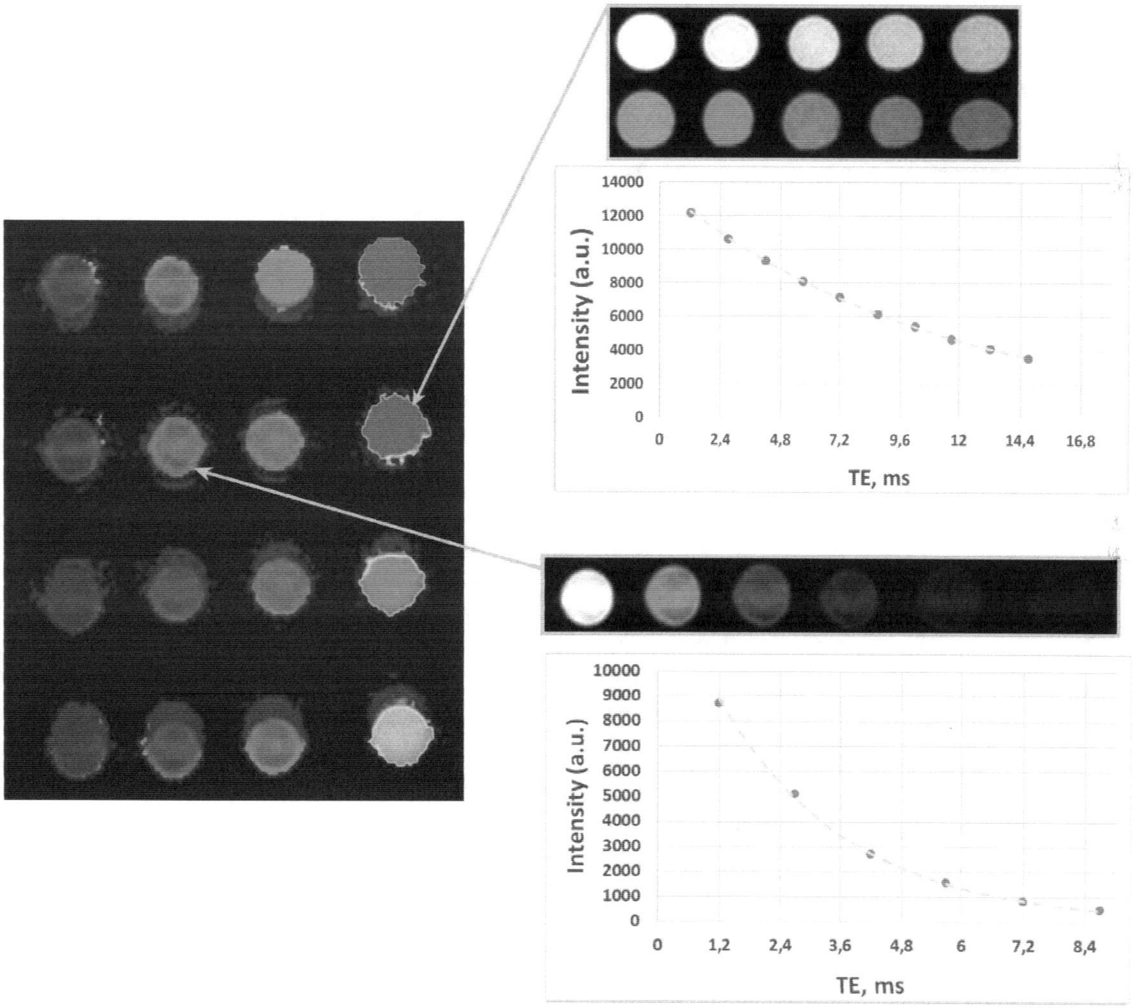

Figure 3. An example of a T2* map (on the left) obtained with commercial processing software on the Philips 3T scanner. For two tubes, the examples of T2* signal decay and the dependence of signal

intensity (relative units) on TE, ms (TE_1 = 1.2 ms, ΔTE = 1.5 ms) are given. The upper plot was constructed for the tube with the real iron concentration of 0.21 mg/mL and the T2* time = 1.2 ± 0.2 ms, the lower plot was constructed for the tube with the real iron concentration of 0.77 mg/mL and the T2* time = 9 ± 0.2 ms.

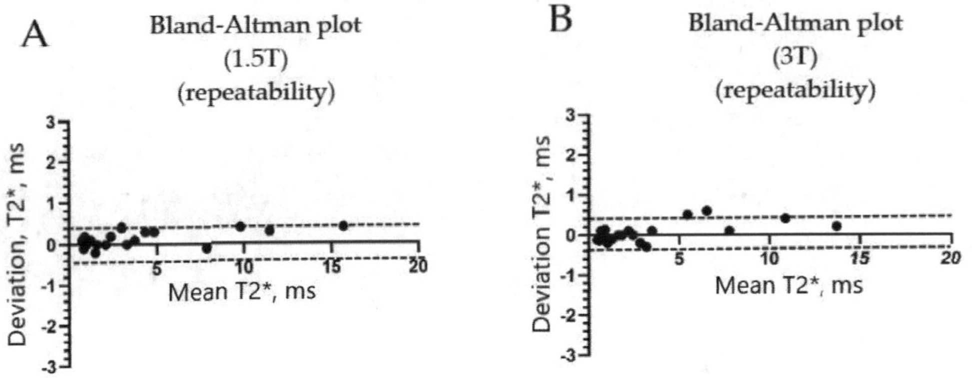

Figure 4. Testing the repeatability of the results: Bland-Altman plots for the repeatability of the results obtained on the reference scanners with field strengths of 1.5T (A) and 3T (B), showing the mean deviations of the T2* values (ms) from the average during 6 observations.

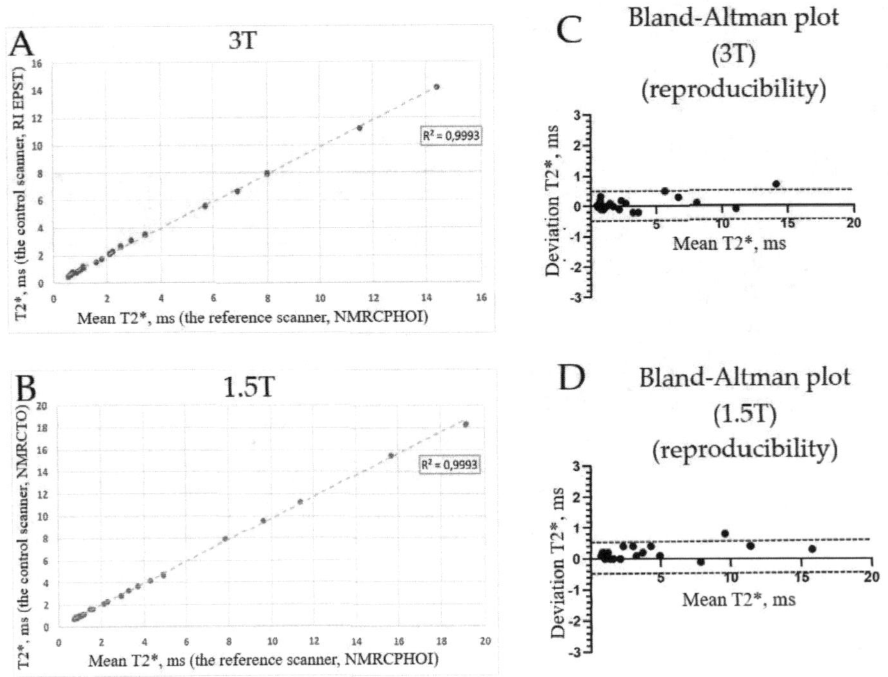

Figure 5. Testing the reproducibility of the results: A, B - correlations between the T2* values obtained on reference and other devices with field strengths of 3T and 1.5T, respectively. C, D - Bland-Altman plots for similar T2* values.

References

[1] C.W. Siah, D. Trinder, J.K. Olynyk, Iron overload, Clin. Chim. Acta. 358 (2005) 24–36. https://doi.org/10.1016/j.cccn.2005.02.022.
[2] A.C. Wortmann, P.E. Froehlich, R.B. Pinto, R.B. Magalhães, M.R. Álvares-Da-Silva, J.J. Ferreira, T.R. Silveira, Hepatic iron quantification by atomic absorption spectrophotometry: Full validation of an analytical method using a fast sample preparation, Spectroscopy. 21 (2007) 161–167. https://doi.org/10.1155/2007/761029.
[3] R. Labranche, G. Gilbert, M. Cerny, K.N. Vu, D. Soulières, D. Olivié, J.S. Billiard, T. Yokoo, A. Tang, Liver iron quantification with MR imaging: A primer for radiologists, Radiographics. 38 (2018) 392–412. https://doi.org/10.1148/rg.2018170079.
[4] E.E. Nazarova, D.A. Kupriyanov, G.A. Novichkova, G. V. Tereshchenko, Noninvasive assessment of iron overload by magnetic resonance imaging, Pediatr. Hematol. Immunopathol. 19 (2020) 158–163. https://doi.org/10.24287/1726-1708-2020-19-3-158-163.
[5] S.P. Gubin, Y.A. Koksharov, G.B. Khomutov, G.Y. Yurkov, Magnetic nanoparticles: Preparation, structure and properties, Usp. Khim. 74 (2005) 539–574. https://doi.org/10.1070/rc2005v074n06abeh000897.
[6] V. Yeung, D.D. Miller, M.A. Rutzke, Atomic Absorption Spectroscopy, Atomic Emission Spectroscopy, and Inductively Coupled Plasma-Mass Spectrometry, in: 2017: pp. 129–150. https://doi.org/10.1007/978-3-319-45776-5_9.
[7] Y. Wei, B. Han, X. Hu, Y. Lin, X. Wang, X. Deng, Synthesis of Fe3O4 nanoparticles and their magnetic properties, in: Procedia Eng., Elsevier Ltd, 2012: pp. 632–637. https://doi.org/10.1016/j.proeng.2011.12.498.
[8] H. Saito, Storage Iron Turnover from a New Perspective, Acta Haematol. (2019) 201–208. https://doi.org/10.1159/000496324.
[9] G.W. RICHTER, The nature of storage iron in idiopathic hemochromatosis and in hemosiderosis. Electron optical, chemical, and serologic studies on isolated hemosiderin granules., J. Exp. Med. 112 (1960) 551–570. https://doi.org/10.1084/jem.112.4.551.
[10] S. Gilchrist, P. Kinchesh, N. Zarghami, A.A. Khrapitchev, N.R. Sibson, V. Kersemans, S.C. Smart, Improved detection of molecularly targeted iron oxide particles in mouse brain using B0 field stabilised high resolution MRI, Magn. Reson. Imaging. 67 (2020) 101–108. https://doi.org/10.1016/j.mri.2020.01.002.
[11] W. Cheng, Y. Ping, Y. Zhang, K.H. Chuang, Y. Liu, Magnetic resonance imaging (MRI) contrast agents for tumor diagnosis, J. Healthc. Eng. 4 (2013) 23–46. https://doi.org/10.1260/2040-2295.4.1.23.
[12] C.D.G. Hines, H. Yu, A. Shimakawa, C.A. McKenzie, J.H. Brittain, S.B. Reeder, T1 independent, T2* corrected MRI with accurate spectral modeling for quantification of fat: Validation in a fat-water-SPIO phantom, J. Magn. Reson. Imaging. 30 (2009) 1215–1222. https://doi.org/10.1002/jmri.21957.

Extracting Dielectric Permittivity with a Cross-Like Stripline

Vladimir Burtsev[1, a)], Tatyana Vosheva[1], Dmitry Filonov[1]

[1]*Center for Photonics and 2D Materials, Moscow Institute of Physics and Technology, Dolgoprudny 141700, Russia*

[a)] burtsev.vd@phystech.edu

Abstract — Parametric retrieval of electromagnetic properties is important for both new materials characterization and an accurate design of devices. While quite a few techniques have been developed over the years, precise mapping of high-permittivity samples remain challenging. Here we advance a so-called micro-strip technique, where transmission coefficients of a waveguide system with an analyte on top are used to extract electromagnetic parameters of the later. Our cross-like strip line configuration has a split ring resonator on one edge and an open circuit termination on another. This design allows performing a simultaneous test of cylindrical and rectangular samples. Our new post-processing scheme was tested on a water-filled container and showed 96.3% accuracy, assessed by comparing our results with tabulated data.

1. INTRODUCTION

Material science keeps developing and provide novel composites with remarkable electromagnetic properties. Apart from the celebrated topic of metamaterials, where, in principle, electromagnetic constants can be tailored on demand [1],[2], there are other promising directions, based on conventional approaches. For example, high-quality ceramic materials, obtained by sintering of powder in a preform, become extremely important in a range of modern applications [3]. Since free space wavelength shrinks with the refractive index of an embedding medium, high index materials allow performing an efficient miniaturization of electromagnetic devices. This capability is especially important for MHz and low GHz applications, where antenna devices are relatively bulky in respect to integrated circuits, powering them. However, this promising size reduction approach comes at a price – the design become increasingly sensitive to refractive index fluctuations. In particular, a small inaccuracy in estimating the permittivity can lead to appearance of parasitic resonances, which can completely govern the performance. Consequently, a capability of an accurate parametric retrieval of high-index composites is a new challenging and highly important task.

Existing approaches to material parameters extraction include anomalous dispersion analysis method on low-GHz range[4],[5], sensor response analysis for ε-μ-extraction[6],[7], and stretchable resonators techniques [8]. In many occasions waveguide [9] and strip line [4],[5],[7] geometries are in use owing to the ease of their practical implementation.

Here we utilize the transmission line technique for special purposes of high-permittivity samples retrieval. We also develop a new algorithm for data post processing. Our approach has several key advantages, which allow an accurate assessment of both real and imaginary parts of permittivity. Typically, a real (ε') and an imaginary (ε") parts of a response function, being linked to each other by Kramers–Kronig relations, affect each other in a measurement, making their retrieval less accurate. In our method, however, ε' is predominately responsible for a frequency shift, while ε" influences the resonance quality factor. While this behaviors is quite obvious from basic physical consideration, the measurement is done by assessing both amplitude and phase information, making the data extraction significantly more accurate. The main drawbacks of our method is a demand to perform an accurate pre-calibration, though similar steps are done in waveguide-based techniques. Another issue, inherently directly linked to the high-permittivity samples retrieval is an extreme sensitivity to noise. In order to reduce the noise sensitivity we apply Savitsky-Golay signal filter[10], which increase the method robustness without a significant loss of accuracy.

2. RESULTS

The circuit, presented in the Fig. 1 (a, b), consists of a strip line, connected to feeding ports. The geometry crossed with another strip line, which has one open circuit edge, while another is functionalized with a circular split ring resonator. The geometry is placed above a ground plane and the spacer is FR-4 dielectric. The design of the scheme is an improved T-like design[4] is conditioned by the aim to gain high-quality peaks that are relatively uniformly splitted between 1 GHz and 20 GHz to achieve the permittivity on these frequencies. On the stripline there can be a sample, which can be either rectangular parallelepiped or a cylinder those are put on the simple stripline or above the SRR-unit correspondingly with their shape-factor. It is noticeable that the position of the probe brings the impact into the final result, so it is recommended to cover the SRR (or a free-end) uniformly or even leave outbounds to avoid possible discrepancies between the numerical experiment and the real one.

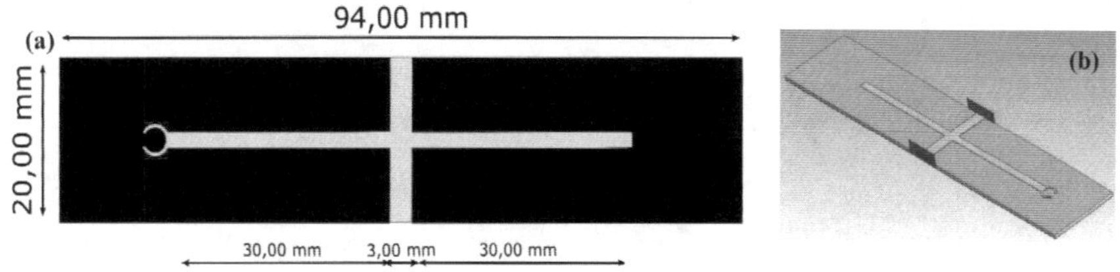

Fig.1. (a) Geometry of the circuit. (b) Picture from CST Studio Suite.

This model is put into the numerical experiment using CST Studio Suite. Its parameters are given below in the Tab. 1. below.

Tab.1. Parameters of the circuit

Parameter name	Value
PCB length	94 mm
PCB width	20 mm
Transmission line length	20 mm
Transmission line width	3 mm
Stripline length	30 mm
Stripline width	2 mm
SRR outer radius	2 mm
SRR inner radius	1.5 mm
Copper foil layer height	0.07 mm
FR-4 layer height	1 mm

Computational algorithm

The key is in sweeping two parameters inside the digital experiment: tangent delta, which is responsible for loss in the material, and real part of relative dielectric permittivity. First, we set the tangent to a relatively small value and do a sweep with ε_r'. Then we put the achieved spectra into our program that continues analyzing data. For every s_{21} spectra the algorithm converts it into module/phase view and then finds every inflection point on the phase curve and memorizes its frequency. The trend of the inflection is the same for all spectra (Fig. 2a), but their special frequencies differ. The algorithm takes first derivative of the phase curve (Fig. 2b) and finds all significant peaks in it. These peaks corresponds to inflection points in original signal. These points will be further used for finding the frequency shift. After all the inflection points for all spectra are found, one of the spectra ($\varepsilon = 2$ in the Fig.2) becomes a reference one and the program calculates for other their frequency shifts for the inflection points relatively to the reference one. Then the algorithm predicts sample ε_r', basing on the calibration values fitting curve. This means ε_r' is responsible for the relatively small (up to several hundreds of megaherz) frequency shifts of resonant peaks on the phase derivative plot. For our numerical experiment we set a task to predict water's dielectric permittivity. We made an experiment with water sample on the stripline and put its spectrum into algorithm.

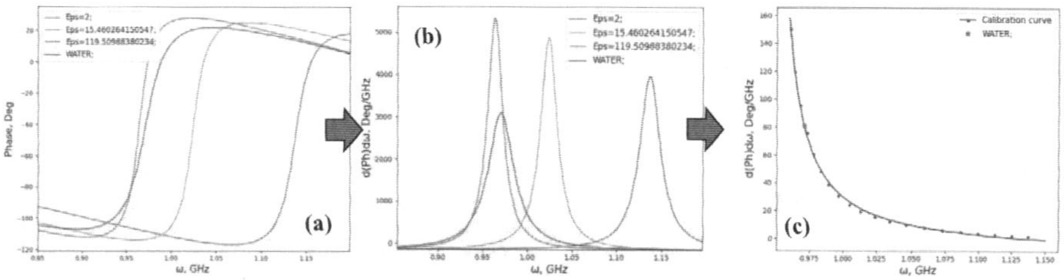

Fig.2. (a) Phases of signals and (b) Derivatives of these phases for $\varepsilon_r' = 2$, $\varepsilon_r' \approx 15.5$, $\varepsilon_r' \approx 120$ for the same tgδ and $\varepsilon_r'_{water}$ which is studied. (c) Calibration plot (yellow) and water position (blue) on the frequency axis.

As we analyzed frequency shifts for ε_r', for the imaginary part of dielectric permittivity (ε_r'') we analyze the half-width of the corresponding resonant peaks as far as with increasing losses in material the peak diffuses.

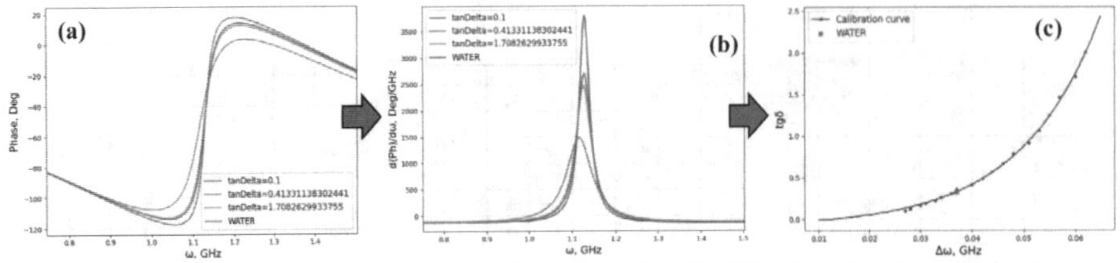

Fig.3. (a) Phases of signals and (b) Derivatives of these phases for tgδ = 0.1, tgδ ≈ 0.41, tgδ ≈ 1.70 for the same ε_r' and $tg\delta_{water}$ which is studied. (c) Calibration plot (blue) and water position (yellow) on the frequency axis.

Iterative nature of the experiment

After we gain the approximately value of $\varepsilon_r'_{sample}$ for our sample from the first attempt, there is a sense to make the second part of experiment for tangent delta using that exact value of $\varepsilon_r'_{sample}$. Thus we can get closer to the real value of $\varepsilon_{r,\ sample}$, using better calibration curve. Then we take this value of tangent delta and re-make the first part of experiment, gaining new $\varepsilon_r'_{sample} = \varepsilon_r'_{(2)}$ from the second attempt and new calibration curve for the real part of dielectric permittivity. Repeating these iterations we are able to get the precise value of $\varepsilon_{r,\ sample}$. In our experiment after the first iteration we obtained $\varepsilon_{(1)} = 80.83 + 25.87i$ on the 1.11 GHz frequency, whether the exact value for this frequency is about $78.00 + 25.59i$. That makes a total error of 3.7% after the first iteration.

Real experiment

To bring this experiment to life we built an appropriate PCB-plate with a cross-like stripline on the surface (Fig. 4a and 4b). After that we connect the circuit to a Network Analyzer, gain its phase spectrum (Fig. 4c) and implicited it into our algorithm. The in-between result is a smoothed curve for the derivative of the sample phase spectrum (Fig. 4d).

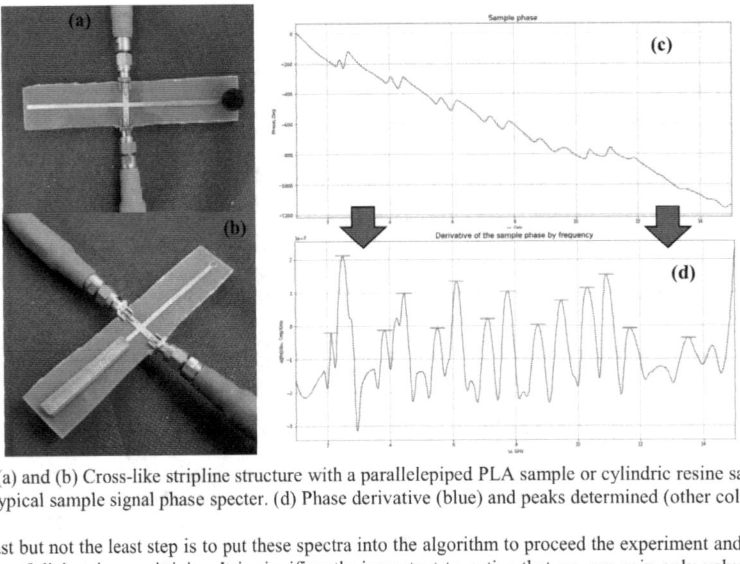

Fig.4. (a) and (b) Cross-like stripline structure with a parallelepiped PLA sample or cylindric resine sample. (c) Typical sample signal phase specter. (d) Phase derivative (blue) and peaks determined (other colors).

The last but not the least step is to put these spectra into the algorithm to proceed the experiment and gain the final value of dielectric permittivity. It is significantly important to notice that we can gain only values on the frequencies for inflection points on the phase curve. Changing the geometry of our stripline, we can measure epsilon in every particular position of inflection.

3. CONCLUSION

At the conference we are going to present the complete high-accuracy method for enumeration ε_r, using stripline circuit and our computational algorithms. There will also be discussed the impact of form-factor on the determination the exact value and the results for different humidity, densities and materials e.g. water, PLA, resine and some others in range 1 GHz to 20 GHz.

ACKNOWLEDGMENTS

The work has been supported by the Russian Science Foundation (Project 20-19-00480).

REFERENCES

[1] N. Engheta and R. W. Ziolkowski, *Electromagnetic Metamaterials: Physics and Engineering Explorations*. Wiley-IEEE Press; 1 edition, 2006.

[2] D. Filonov, A. Shmidt, A. Boag, and P. Ginzburg, "Artificial localized magnon resonances in subwavelength meta-particles," *Appl. Phys. Lett.*, vol. 113, no. 12, p. 123505, Sep. 2018, doi: 10.1063/1.5047445.

[3] D. Dobrykh *et al.*, "Long-range miniaturized ceramic RFID tags," *IEEE Trans. Antennas Propag.*, 2020, doi: 10.1109/TAP.2020.3037663.

[4] O. Siddiqui, R. Ramzan, M. Amin, and O. M. Ramahi, "A Non-Invasive Phase Sensor for Permittivity and Moisture Estimation Based on Anomalous Dispersion," *Sci. Rep.*, vol. 6, no. 1, pp. 1–9, Jun. 2016, doi: 10.1038/srep28626.

[5] R. Ramzan, O. F. Siddiqui, M. W. Arshad, and O. M. Ramahi, "A Complex Permittivity Extraction Method Based on Anomalous Dispersion," *IEEE Trans. Microw. Theory Tech.*, vol. 64, no. 11, pp. 3787–3796, Nov. 2016, doi: 10.1109/TMTT.2016.2605664.

[6] X. Lu *et al.*, "A ceramic-based microwave sensor for both permittivity and permeability characterization of materials," *J. Phys. D. Appl. Phys.*, vol. 53, no. 34, p. 345103, Aug. 2020, doi: 10.1088/1361-6463/ab8c80.

[7] Y. Abdulkarim *et al.*, "Novel Metamaterials-Based Hypersensitized Liquid Sensor Integrating Omega-Shaped Resonator with Microstrip Transmission Line," *Sensors*, vol. 20, no. 3, p. 943, Feb. 2020, doi: 10.3390/s20030943.

[8] D. S. Filonov, E. I. Kretov, S. A. Kurdjumov, V. A. Ivanov, and P. Ginzburg, "Broadband resonant calibration-free complex permittivity retrieval of liquid solutions," *J. Quant. Spectrosc. Radiat. Transf.*, vol. 235, 2019, doi: 10.1016/j.jqsrt.2019.06.020.

[9] H. B. Sequeira, "Extracting μr and ∊ r of Solids from One-Port Phasor Network Analyzer Measurements," *IEEE Trans. Instrum. Meas.*, vol. 39, no. 4, pp. 621–627, 1990, doi: 10.1109/19.57244.

[10] A. Savitzky and M. J. E. Golay, "Smoothing and Differentiation of Data by Simplified Least Squares Procedures," *Anal. Chem.*, vol. 36, no. 8, pp. 1627–1639, Jul. 1964, doi: 10.1021/ac60214a047.

Journal of Physics: Conference Series **2015** (2021) 012027 doi:10.1088/1742-6596/2015/1/012027

Conditions for a pure toroidal dipole source

Adrià Canós Valero

ITMO University, Kronversky Prospect 49, 197101, St Petersburg Russia

adria.canos@optomech.ifmo.ru

Abstract. Recently, the physical significance of dynamic toroidal multipoles in the context of electrodynamics has been put under discussion. Indeed, the latter can be shown to arise simply from a Taylor series of the exact source (Cartesian) multipole moments. The split into elementary and toroidal parts was demonstrated to lead to an unphysical result were forbidden components of the momentum transform of the current could radiate into free space. In this contribution, we elaborate the conditions that a current distribution must necessarily satisfy to be considered a 'pure' toroidal dipole source. We demonstrate for the first time that symmetry prevents such current distribution to radiate as an elementary electric dipole moment, without leading to an unphysical result. Thus, while both elementary electric dipole and toroidal dipoles are indistinguishable outside the source, they display topologically distinct characteristics within the smallest spherical surface enclosing the source itself and have different physical origin. Based on our results, a pure 'toroidal' source can be designed. We believe the outcome of our investigations will help clarify further the formal meaning of the toroidal multipoles.

INTRODUCTION

In modern physics, multipole expansions are commonly used tools to understand a broad range of phenomena in optics, nuclear physics and condensed matter [1,2]. It is nowadays well-known that the so-called 'Cartesian' or multipole expansions of the source, which arise in a Taylor series of the vector potential [3,4], require the introduction of dynamic toroidal moments [4]. The most well-known constituent of the toroidal family is the toroidal dipole, first identified [5] and observed as a static current distribution. It was later shown to naturally give rise to a different form of long-range order in ferroics violating both space and time inversion symmetries [6], and predicted to have a strong impact in the electronic properties of several molecular structures [2]. Nowadays, dynamic toroidal moments in electrodynamics are a subject of renewed scientific and technological interest in the growing field of all-dielectric nanophotonics [7–12] due to their role in the formation of the nonradiating sources known as anapoles [13].

Their physical relevance, as well as their formal meaning in electrodynamics, is still however a matter of on-going discussion [14,15]. Corbaton et.al. rigorously showed in [15] how the toroidal moments appear as high order terms of a Taylor series of the exact source multipoles. Taking as a case example the electric and electric toroidal dipole, the split into elementary and toroidal terms leads to an unphysical result if considered separately, i.e. the toroidal dipole cannot, in fact, be regarded differently from the elementary electric dipole moment.

In this work, we show that such a restriction can in fact be circumvented under a few special conditions, which effectively allow a 'pure' naturally broadband electric toroidal dipole to arise in the absence of the elementary electric dipole. With our analysis, we demonstrate that, while indistinguishable for an observer outside the source, the physical origin of the elementary electric dipole and the toroidal dipole are different and must be distinguished when investigating the effects taking place within the volume of the smallest sphere surrounding the source.

GENERAL THEORETICAL CONSIDERATIONS

We start with the dipolar vector of electric parity in the exact form [15]:

$$\mathbf{a}_1^{\omega} = -\frac{1}{i\omega}\left\{\int d^3\mathbf{r}\mathbf{J}_{\omega} j_0(kr) + \frac{k^2}{2}\int d^3\mathbf{r}\left[3(\mathbf{r}\cdot\mathbf{J}_{\omega})\mathbf{r} - r^2\mathbf{J}_{\omega}\right]\frac{j_2(kr)}{(kr)^2}\right\}. \quad (1)$$

The Bessel functions within the integral act as 'filters' for the momentum components of the current, such that only those with total momentum $|\mathbf{p}| = \omega / c$ contribute to radiation. A Taylor series of Eq.(1) leads to the well-known expressions of the elementary dipole $\mathbf{p}_1^{\omega}$ and toroidal dipole $\mathbf{t}_1^{\omega}$ moments:

$$\mathbf{a}_1^{\omega} \approx \mathbf{p}_1^{\omega} + \mathbf{t}_1^{\omega} \approx -\frac{1}{i\omega}\left\{\int d^3\mathbf{r}\mathbf{J}_{\omega} + \frac{k^2}{10}\int d^3\mathbf{r}\left[(\mathbf{r}\cdot\mathbf{J}_{\omega})\mathbf{r} - 2r^2\mathbf{J}_{\omega}\right]\right\}. \quad (2)$$

In what follows, it will be convenient to recall the expression for the momentum transform of a localized current distribution:

$$\mathbf{J}_{\omega}(\mathbf{p}) = \frac{1}{\sqrt{(2\pi)^3}}\int d^3\mathbf{r}\mathbf{J}_{\omega}\exp(-i\mathbf{p}\cdot\mathbf{r}), \quad (3)$$

comparing Eqs.(2)-(3) one immediately notices that $\mathbf{p}_1^{\omega}$ corresponds to $\mathbf{J}_{\omega}(\mathbf{p}=\mathbf{0})$, which cannot by definition radiate into free space. Therefore, the toroidal terms in the Taylor series of Eq. (1) are linked to the first order term and must necessarily contain the $\mathbf{J}_{\omega}(\mathbf{p}=\mathbf{0})$ that cancels the first one when summed together. Altogether, from the previous the authors in Ref. [15] concluded that, since each term in the expansion contains nonradiative terms, the elementary dipole and the toroidal dipole moment couple identically to electromagnetic radiation, and, as such, there is no need for separating their contributions outside the source.

Here, we focus however on the way that the elementary (electric) and toroidal terms are excited within the source itself. Within the smallest spherical volume enclosing the source, there is no restriction as to which components $\mathbf{J}_{\omega}(\mathbf{p})$ can contribute to the electromagnetic fields, and therefore the previous arguments do not apply.

RESULTS AND DISCUSSION

First, we focus on a simple way to obtain a pure 'toroidal' source, i.e. with $\mathbf{p}_1^{\omega} = \mathbf{0}$. A current distribution validating this condition would have $\mathbf{J}_{\omega}(\mathbf{p}=\mathbf{0}) = \mathbf{0}$, therefore allowing the toroidal part to radiate by itself. This idea was discussed in Ref. [15] and discarded as trivial, since in the far field pure toroidal radiation cannot be distinguished from that of the elementary dipole. However, a current distribution validating such condition has a very well-defined topology, very different from that of an elementary electric dipole. Such a current must, in fact, not only be *divergence-less* $\nabla\cdot\mathbf{J}_{\omega} = 0$ and *poloidal* ($\mathcal{L}\cdot\mathbf{J}_{\omega} = 0$, where $\mathcal{L}$ is the orbital angular momentum operator), but also it must not establish a surface charge density σ_{ω} at the boundary of the source volume. To satisfy these requirements, we propose a current distribution of the form:

$$\mathbf{J}_{\omega}(\rho,\theta) = \frac{K\mathcal{H}(\Delta\rho)}{R_0 + \rho cos\theta}\mathbf{e}_{\theta}. \quad (4)$$

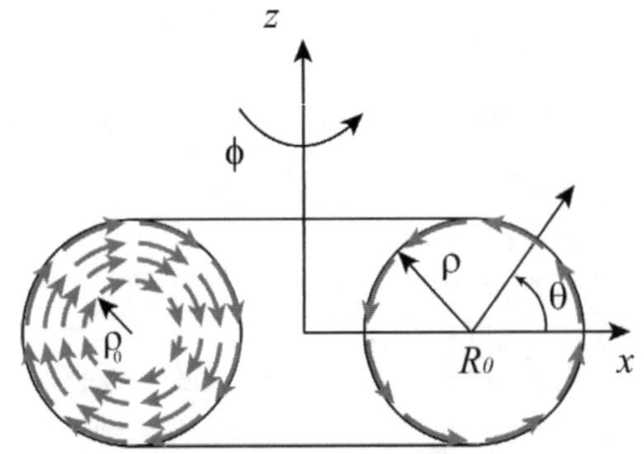

FIGURE 1. Scheme depicting a cross-section of the considered toroidal current distribution. The toroidal coordinate system utilized for the parametrization of Eq. (4) is also illustrated.

This current can be rigorously obtained from imposing the conditions above in a toroidal coordinate system as depicted in Fig.1, where K is a constant and $\mathcal{H}(\Delta\rho)$ is a step function in the coordinate ρ. Eq.(4) effectively corresponds to a current flowing along the poloidal axis $\mathbf{e}_\theta$ of a torus. We remark that this result is *independent on the extension of the torus*. Indeed, even if the dimensions of the source are larger than the wavelength, an elementary electric dipole will never occur in this system. In addition, since the current is poloidal, the system can also *never support magnetic moments*; they are proportional to $\mathcal{L}\cdot\mathbf{J}_\omega$, and therefore zero.

We now obtain analytical results for the expected behavior of the toroidal part. Indeed, integrating Eq. (2) with Eq. (4) yields $\mathbf{p}_1^\omega = 0$, while toroidal moments remain non-zero. The toroidal dipole takes the form:

$$\mathbf{t}_1^\omega = \frac{k^2\pi^2}{3} K\rho_0^3 R_0 \mathbf{e}_z \,. \tag{5}$$

CONCLUSION

We have discussed the possibility to excite the dynamic toroidal dipole in the absence of the elementary electric dipole and elaborate a sufficient set of conditions that a current distribution must satisfy in order to support it. Based on the latter, we propose a current easily imprintable on the surface of a metallic body that cannot, by definition, excite an elementary electric dipole. Outside the smallest spherical surface enclosing the source, the system cannot be distinguished from that of an elementary electric dipole. Inside, however, the topology of the fields and associated electromagnetic quantities differ significantly.

ACKNOWLEDGEMENTS

The author acknowledges the support of grant Num. 20-52-00031 of the Russian Foundation for Basic Research (RFBR). This project is inscribed within the Grant of the President of the Russian Federation.

REFERENCES

[1] Zimmermann A S, Meier D and Fiebig M 2014 Ferroic nature of magnetic toroidal order *Nat. Commun.* **5** 1–6
[2] Ceulemans A, Chibotaru L F and Fowler P W 1998 Molecular anapole moments *Phys. Rev. Lett.* **80** 1861–4
[3] Dubovik V M and Tugushev V V. 1990 Toroid moments in electrodynamics and solid-state physics *Phys. Rep.* **187** 145–202
[4] Gurvitz E A, Ladutenko K S, Dergachev P A, Evlyukhin A B, Miroshnichenko A E and Shalin A S 2019 The High-Order Toroidal Moments and Anapole States in All-Dielectric Photonics *Laser Photon. Rev.* **13**

1800266
[5] Zel'Dovich 1957 Electromagnetic interaction with parity violation *Sov. J. Exp. Theor. Phys.* **6** 1184–6
[6] Hayami S, Kusunose H and Motome Y 2014 Toroidal order in metals without local inversion symmetry *Phys. Rev. B - Condens. Matter Mater. Phys.* **90** 1–12
[7] Baryshnikova K, Filonov D, Simovski C, Evlyukhin A, Kadochkin A, Nenasheva E, Ginzburg P and Shalin A S 2018 Giant magnetoelectric field separation via anapole-type states in high-index dielectric structures *Phys. Rev. B* **98** 1–9
[8] Kostina N, Ivinskaya A, Sukhov S, Bogdanov A, Toftul I, Nieto-Vesperinas M, Ginzburg P, Petrov M and Shalin A 2017 Optical binding via surface plasmon polariton interference *Phys. Rev. B* **99** 125416
[9] Terekhov P D, Shamkhi H K, Gurvitz E A, Baryshnikova K V., Evlyukhin A B, Shalin A S and Karabchevsky A 2019 Broadband forward scattering from dielectric cubic nanoantenna in lossless media *Opt. Express* **27** 10924
[10] Kozlov V, Filonov D, Shalin A S, Steinberg B Z and Ginzburg P 2016 Asymmetric backscattering from the hybrid magneto-electric meta particle *Appl. Phys. Lett.* **109** 203503
[11] Terekhov P D, Baryshnikova K V., Greenberg Y, Fu Y H, Evlyukhin A B, Shalin A S and Karabchevsky A 2019 Enhanced absorption in all-dielectric metasurfaces due to magnetic dipole excitation *Sci. Rep.* **9** 3438
[12] Shalin A S and Moiseev S G 2009 Optical properties of nanostructured layers on the surface of an underlying medium *Opt. Spectrosc.* **106** 916–25
[13] Valero A C, Gurvitz E A, Benimetskiy F A, Pidgayko D A, Samusev A, Evlyukhin A B, Redka D, Tribelsky M I, Rahmani M, Kamali K Z, Pavlov A A, Miroshnichenko A E and Shalin A S 2020 Theory, observation and ultrafast response of novel hybrid anapole states
[14] Papasimakis N, Fedotov V A, Savinov V, Raybould T A and Zheludev N I 2016 Electromagnetic toroidal excitations in matter and free space *Nat. Mater.* **15** 263–71
[15] Fernandez-Corbaton I, Nanz S and Rockstuhl C 2017 On the dynamic toroidal multipoles from localized electric current distributions *Sci. Rep.* **7**

Exceptional points of all-dielectric nanoresonators

Adrià Canós Valero

ITMO University, Kronversky Prospect 49, 197101, St Petersburg Russia

adria.canos@optomech.ifmo.ru

Abstract. In the recent years, all-dielectric nanophotonics has been showing promising potential for biotechnology, with important progress in the development of efficient all-optical, all-dielectric nanosensing devices overcoming the ohmic losses inherently present in their plasmonic counterparts. In the quest to achieve single molecule sensitivities, a judicious design of the optical response of the nanoantennas is required. Here, we approach this problem from the perspective of non-Hermitian physics and investigate the interaction of two finite nanorods supporting Mie resonances, with the aim of maximizing the frequency detuning induced by a perturbation of the structure. We develop a simple semi-analytical technique to efficiently investigate the coupled system, and we find that Coulomb interactions, together with mutual interference induced by breaking the dimer symmetry, can effectively bring the structure towards a non-Hermitian singularity, an exceptional point, that can potentially increase the sensitivity. The results of this work are expected to lead to novel developments in all-optical single molecule detection, and merge for the first time all-dielectric nanophotonics with exceptional point physics.

INTRODUCTION

As a promising alternative to nanoplasmonics, all-dielectric nanophotonics has emerged as a means to overcome the fundamental limitation of ohmic losses at the nanoscale, while benefitting from ultrasmall light confinement [1–5]. The electric and magnetic character of dielectric nanoantennas [6–8] has made possible their implementation in Huygens metasurfaces featuring full transmission with varying phase [1], and to achieve outstanding tasks such as efficient third and second harmonic generation [9], as well as single particle lasing [10]. With the field reaching its maturity, the conceptual ground has now been set for the development of flat, all-optical, all-dielectric platforms for biosensing [11,12]. Indeed, the study of biological analytes by means of nanophotonic structures is a well-established trend of the last decade [13], due to its wide applicability in a variety of relevant biological processes such as reaction monitoring [14], drug detection [15] , and disease diagnosis [14], which are also compatible with lab-on-a-chip platforms [16,17] (LOC).

In the last few years, emerging biosensing studies exploiting some of the aforementioned effects are starting to report detection of bulk changes in refractive index and thickness variations of ultrathin biomolecular layers [18], having comparable figures of merit as that of plasmonic structures [18].

Unfortunately, the fundamental electric and magnetic dipole resonances of dielectric nanoresonators feature very low quality (Q) factors, which significantly limit the measurable frequency shift induced by variations in the refractive index of the surrounding medium. In this context, high-Q Fano resonances and quasi-bound states in the continuum (quasi-BICs) may constitute the next generation of all-dielectric biosensing devices [19]. An alternative approach that remains largely unexplored, originated, like BICs, in the field of open quantum systems, and consists in the design of non-Hermitian singularities, the so-called exceptional points [20] (EPs). In formal terms, they correspond to the collapse of two or more complex eigenfrequencies and eigenvectors of an open system that occurs when a set of degrees of freedom of the structure are carefully tuned. In photonics, they have been mostly studied in structures involving gain [20], but only a handful of investigations of EPs in passive structures have been published. Besides possessing many interesting physical properties, the eigenfrequencies near EPs display a dependence (to first order) of $\epsilon^{1/n}$ on the perturbation parameter ϵ, where n is the number of coalescing eigenvalues [20]. This peculiarity renders EPs highly attractive for biosensing schemes relying on resonance tracking [21].

In our study, we aim at designing an experimentally accessible EP with realistic all-dielectric nanoresonators. To achieve this goal, we study the interactions taking place in a dimer of dielectric nanodisks where each constituent supports a Mie resonance. Making use of a semi-analytical 2x2 Hamiltonian, we then perform careful symmetry breaking of the structure to effectively tune the Coulomb interactions between the nanorods and the radiative interference of the symmetric and antisymmetric modes. The calculations unambiguously demonstrate the presence of an EP in the investigated parameter space.

THEORY AND DESIGN

EPs in microresonators are commonly described in terms of temporal coupled mode theory (TCMT)[22]. While providing strong analytical insight on the problem, subwavelength nanoresonators typically feature large radiative losses, significantly limiting the applicability of this approach. Furthermore, fitting of a few parameters is usually required for such models. Instead, we rely on quasinormal mode perturbation theory [23], which has already proven its effectiveness for the optimization of EPs. The starting point of our analysis relies on the numerical calculation of the quasinormal modes (QNMs) of a guessed geometry with identical silicon nanorods of radius $R_0 = 140$ nm separated by a distance $g = 60$ nm, and then successively perform careful symmetry breaking of the structure by increasing the size of one of the nanorods (see inset of Fig.1a). Each silicon nanorod supports high order Mie resonances. Once the fields are known, the interaction Hamiltonian for a radial perturbation δr can be readily constructed through "exact" first order perturbation theory [23]. Considering only two QNMs, which we label as symmetric (+), and antisymmetric (-), the Hamiltonian reads:

$$\mathcal{H} = \delta r \begin{pmatrix} \langle \mathbf{E}_d^{(+)} \cdot \mathbf{E}_u^{(+)} \rangle & \langle \mathbf{E}_d^{(+)} \cdot \mathbf{E}_u^{(-)} \rangle \\ \langle \mathbf{E}_d^{(-)} \cdot \mathbf{E}_u^{(+)} \rangle & \langle \mathbf{E}_d^{(-)} \cdot \mathbf{E}_u^{(-)} \rangle \end{pmatrix}, \tag{1}$$

The Bra-Ket notation designates an integral over the perturbed resonator surface, the subscript d indicates the internal surface of the resonator, while u denotes the external one. The perturbed eigenfrequencies $\tilde{\omega}_p$ are calculated by solving the generalized eigenproblem

$$\Omega_0 |\mathbf{\Phi}\rangle = \tilde{\omega}_p [I + \mathcal{H}] |\mathbf{\Phi}\rangle, \tag{2}$$

where Ω_0 is a diagonal matrix containing the unperturbed eigenfrequencies, and $|\mathbf{\Phi}\rangle$ are the eigenvectors in the basis of the unperturbed QNMs. Making use of Eqs. (1)-(2) we are able to explore analytically the entire parameter space mapped by the radius of the nanorod (R_1), for different center-to-center separations g, reducing by half the amount of numerical computations. Essentially, to obtain an EP we need to optimize the system such that $|\mathbf{\Phi}_+\rangle \to |\mathbf{\Phi}_-\rangle$ to enforce the degeneracy of the eigenvectors.

RESULTS AND DISCUSSION

Figs.1a-b display the projection of the Riemann sheets for $g = 45$ nm in the vicinity of the EP. We clearly observe the coalescence of the real and imaginary parts of the eigenvalues of the symmetric and antisymmetric modes. Near the EP, the eigenvectors of the QNMs become degenerate, implying that the field distributions of the two modes are very similar, as can be appreciated in points A,B of Fig.1c. Slightly perturbing the system from the EP by a factor

$\epsilon \approx 0.1$ already leads to remarkably different field distributions, resonant frequencies and resonance linewidths (points C,D in Fig.1b-c). While the current results have been obtained for the dimer placed in vacuum, taking into consideration the substrate is essential for an accurate prediction of the EP. Since our formalism relies on the precomputed QNMs for one geometry, no modifications to the theory are necessary. In the future, we will make use of Eqs. (1)-(2) to carefully tune the EP of the dimer placed on a glass substrate.

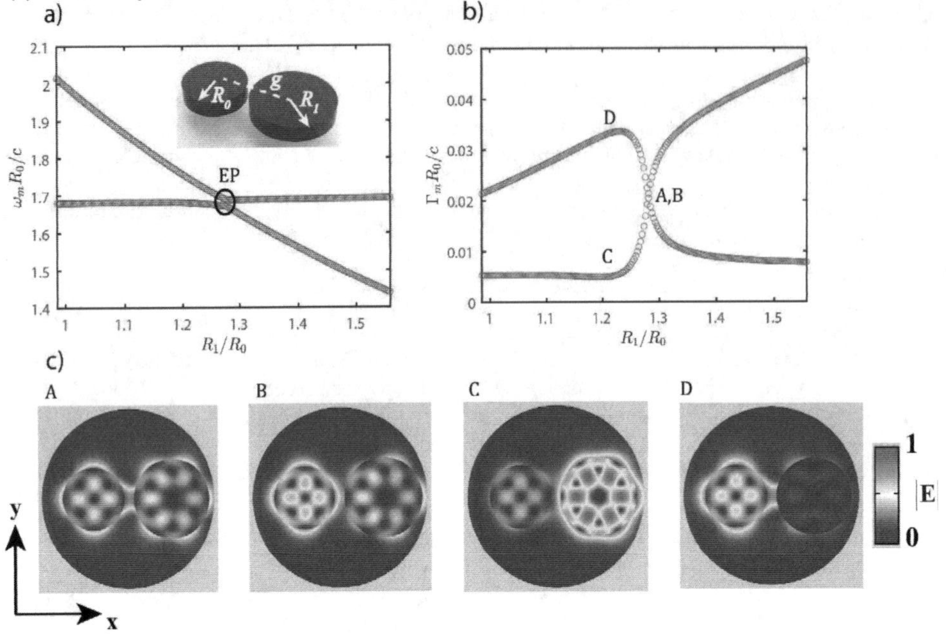

FIGURE 1. (a) Collapse of the real part of the two eigenfrequencies near the EP. Inset: investigated dimer structure consisting of two silicon nanorods with center-to-center distance g and radii R_0, R_1. The EP is achieved by simultaneously tuning two system parameters, g and the ratio R_1 / R_0. (b) Degeneracy of the radiative losses of the two modes near the EP. (c) Normalized electric field norms in the x-y plane for points A,B,C,D indicated in (b).

CONCLUSION

In summary, we have designed an exceptional point entirely based on realistic silicon resonators without the need of absorption nor gain, simply by judiciously engineering the radiative losses and the Coulomb interactions between the nanoantennas. To our knowledge, our structure constitutes the first design of an exceptional point within the framework of all-dielectric nanophotonics. We envision the possibility to realize ultrasensitive single molecule sensors without heating losses, based on the exotic properties of the investigated non-Hermitian singularity, potentially integrable in LOC platforms. In addition, our study opens an avenue towards the investigation of exceptional points at the nanoscale.

ACKNOWLEDGEMENTS

The author acknowledges the support of the Russian Science Foundation for the project 21-12-00383 and the grant 20-52-00031 of the Russian Foundation for Basic Research (RFBR).

REFERENCES

[1] Kivshar Y 2018 All-dielectric meta-optics and non-linear nanophotonics *Natl. Sci. Rev.* **5** 144–58
[2] Kruk S and Kivshar Y 2017 Functional meta-optics and nanophotonics govern by Mie resonances *ACS Photonics* **4** 2638–49

[3] Shalin A S 2010 Broadband blooming of a medium modified by an incorporated layer of nanocavities *JETP Lett.* **91** 636–42
[4] Shalin A S and Moiseev S G 2009 Optical properties of nanostructured layers on the surface of an underlying medium *Opt. Spectrosc.* **106** 916–25
[5] Baryshnikova K, Filonov D, Simovski C, Evlyukhin A, Kadochkin A, Nenasheva E, Ginzburg P and Shalin A S 2018 Giant magnetoelectric field separation via anapole-type states in high-index dielectric structures *Phys. Rev. B* **98** 1–9
[6] Kozlov V, Filonov D, Shalin A S, Steinberg B Z and Ginzburg P 2016 Asymmetric backscattering from the hybrid magneto-electric meta particle *Appl. Phys. Lett.* **109** 203503
[7] Terekhov P D, Baryshnikova K V., Greenberg Y, Fu Y H, Evlyukhin A B, Shalin A S and Karabchevsky A 2019 Enhanced absorption in all-dielectric metasurfaces due to magnetic dipole excitation *Sci. Rep.* **9** 3438
[8] Terekhov P D, Shamkhi H K, Gurvitz E A, Baryshnikova K V., Evlyukhin A B, Shalin A S and Karabchevsky A 2019 Broadband forward scattering from dielectric cubic nanoantenna in lossless media *Opt. Express* **27** 10924
[9] Koshelev K, Kruk S, Melik-Gaykazyan E, Choi J H, Bogdanov A, Park H G and Kivshar Y 2020 Subwavelength dielectric resonators for nonlinear nanophotonics *Science (80-.).* **367** 288–92
[10] Tiguntseva E, Koshelev K, Furasova A, Tonkaev P, Mikhailovskii V, Ushakova E V., Baranov D G, Shegai T, Zakhidov A A, Kivshar Y and Makarov S V. 2020 Room-Temperature Lasing from Mie-Resonant Nonplasmonic Nanoparticles *ACS Nano* **14** 8149–56
[11] Bontempi N, Chong K E, Orton H W, Staude I, Choi D-Y, Alessandri I, Kivshar Y S and Neshev D N 2017 Highly sensitive biosensors based on all-dielectric nanoresonators *Nanoscale* **9** 4972–80
[12] Spindel S and Sapsford K 2014 Evaluation of Optical Detection Platforms for Multiplexed Detection of Proteins and the Need for Point-of-Care Biosensors for Clinical Use *Sensors* **14** 22313–41
[13] Krasnok A, Caldarola M, Bonod N and Alú A 2018 Spectroscopy and Biosensing with Optically Resonant Dielectric Nanostructures *Adv. Opt. Mater.* **6** 1701094
[14] Soler M, Estevez M C, Cardenosa-Rubio M, Astua A and Lechuga L M 2020 How Nanophotonic Label-Free Biosensors Can Contribute to Rapid and Massive Diagnostics of Respiratory Virus Infections: COVID-19 Case *ACS Sensors* **5** 2663–78
[15] Frosch T, Knebl A and Frosch T 2019 Recent advances in nano-photonic techniques for pharmaceutical drug monitoring with emphasis on Raman spectroscopy *Nanophotonics* **9** 19–37
[16] Canós Valero A, Kislov D, Gurvitz E A, Shamkhi H K, Pavlov A A, Redka D, Yankin S, Zemánek P and Shalin A S 2020 Nanovortex-Driven All-Dielectric Optical Diffusion Boosting and Sorting Concept for Lab-on-a-Chip Platforms *Adv. Sci.* **7** 1903049
[17] Kostina N, Ivinskaya A, Sukhov S, Bogdanov A, Toftul I, Nieto-Vesperinas M, Ginzburg P, Petrov M and Shalin A 2017 Optical binding via surface plasmon polariton interference *Phys. Rev. B* **99** 125416
[18] Bosio N, Šípová-Jungová H, Länk N O, Antosiewicz T J, Verre R and Käll M 2019 Plasmonic versus All-Dielectric Nanoantennas for Refractometric Sensing: A Direct Comparison *ACS Photonics* **6** 1556–64
[19] Tseng M L, Jahani Y, Leitis A and Altug H 2021 Dielectric Metasurfaces Enabling Advanced Optical Biosensors *ACS Photonics* **8** 47–60
[20] Miri M-A and Alù A 2019 Exceptional points in optics and photonics *Science (80-.).* **363** eaar7709
[21] Park J-H, Ndao A, Cai W, Hsu L, Kodigala A, Lepetit T, Lo Y-H and Kanté B 2020 Symmetry-breaking-induced plasmonic exceptional points and nanoscale sensing *Nat. Phys.* **16** 462–8
[22] Lai Y-H, Lu Y-K, Suh M-G, Yuan Z and Vahala K 2019 Observation of the exceptional-point-enhanced Sagnac effect *Nature* **576** 65–9
[23] Yan W, Lalanne P and Qiu M 2020 Shape Deformation of Nanoresonator: A Quasinormal-Mode Perturbation Theory *Phys. Rev. Lett.* **125** 013901

Optical Heating Controlled With a Thermoplasmonic Metasurface

E A Chernykh[1, a)], **A N Filippov**[2], **A M Alekseev**[1,3] **M A Makhiboroda**[2] **and S S Kharintsev**[1, 3, b)]

[1] Department of Optics and Nanophotonics, Institute of Physics, Kazan Federal University, 16 Kremlevskaya str., 420008 Kazan, Russia
[2] STC NMST, Moscow Institute for Electronic Technology, Moscow, 124498, Russia
[3]Institute of Perspective Technologies, Tatarstan Academy of Science, 20 Bauman str., 420111 Kazan, Russia

[a)]Corresponding author: elenorchernykh@gmail.com
[b)]skharint@gmail.com

Abstract. We develop a photothermal technology to control optical heating of polymer and liquid crystal films through a refractory titanium nitride (TiN) metasurface. The metasurface represents an array of identical square-shaped TiN nanoantennas on a Si substrate. Upon CW laser illumination, a TiN nanoantenna experiences anomalous Joule heating at a plasmon resonance. A temperature rise provides a unique opportunity for locally probing phase transitions. In the case of heterogeneous PMMA thin films or polymeric blends, a controlled optical heating is needed to probe the glass transition temperature (T_g) of their constituents. Here, we model a controlled thermal response originating from the TiN nanoantenna under CW laser illumination by using FDTD/FEM methods.

1. Introduction

Localized heat generation using optical excited plasmonic nanoparticles is currently an active area of research. Under optical illumination, noble metal nanoparticles can play the role of an effective nanosource of heat. It has been demonstrated that the heating effect becomes especially strong when nanoparticles are illuminated under plasmon resonance conditions. This behavior became the basis for development of such areas as photothermal cancer therapy [1], drug release [2], nanosurgery [3], nanothermometers [4], thermophotovoltaics [5], nanochemistry [6]. The Covid-19 pandemic has also driven new developments in thermoplasmonics. In particular, the Qiu group proposed a version of plasmonic photothermal biosensors for accurate detection of Coronavirus [7]. Traditionally, Au and Ag have been preferred materials for demonstrating various plasmonic phenomena. However, their low melting points do not allow the implementation of practical plasmonic and photonic devices for technologically important high-temperature applications. Titanium nitride (TiN) has become a promising alternative plasmonic material due to its refractory property, chemical stability and good conductivity. In addition, TiN is biocompatible among many others plasmon materials.

The localized heat generated by nanoparticles can reach temperatures up to tens or hundreds degrees (K) while the temperature of the entire sample remains unchanged. This phenomenon can become the basis for creating plasmon sensors for nanoprobing of such processes as phase transitions in polymer and liquid crystal films. F.A.A. Nugroho et al. have demonstrated for the first time a 3D plasmonic sensor for determination glass transition temperature. The plasmon sensor represents a metasurface consisting of an array of truncated nanocones. The nanocone represents two silver disks of different diameters, separated by a SiO_2 truncated nanocone [8]. Thus, two localized surface plasmon resonances (LSPR) can be recorded. Upon heating the local refractive index decreases due to thermal expansion and thus the LSPR peak is shifted. The glass transition temperature was determined from the kink in the temperature dependence of the shift of the plasmon resonance of the disk. However, in this way, the glass transition temperature is only probed in one direction.

In connection with the aforementioned, we present a study of the controlled local heating of a poly (methyl methacrylate) (PMMA) film using a refractory TiN plasmonic metasurface which is a regular array of parallelepipeds of an identical shape on a silicon substrate using numerical simulation.

2. Simulation details

The design of our metasurface is an array of TiN parallelepipeds of size 200x200x50 nm^3 (Figure 1(a)) located on a silicon substrate. Nanoantennas are far enough that they can be considered optically independent. Therefore, further all analysis and calculations will be regarded in the context of one nanostructure. In our simulations, we used rounded edges and corners (10 nm) to avoid anomalous electric fields near them. Simulation of electric and magnetic fields around TiN nanoantennas was carried out using the Lumerical FDTD solver. The spatial cell size was set to 1 nm and perfectly matched layers (PML) were boundaries. A TiN parallelepiped was irradiated with a linearly polarized focused CW Gaussian light source (N.A. = 0.9) with a wavelength of 633 nm and an input intensity of 5 MW/cm^2, which corresponds to a power 16 mW. This size of parallelepiped is explained by the fact that the absorption and scattering cross sections at our wavelength are close to each other, which provides good field amplification (Figure 1(b)). In our simulations, the complex dielectric constant of TiN was determined by spectroscopic ellipsometry at room temperature.

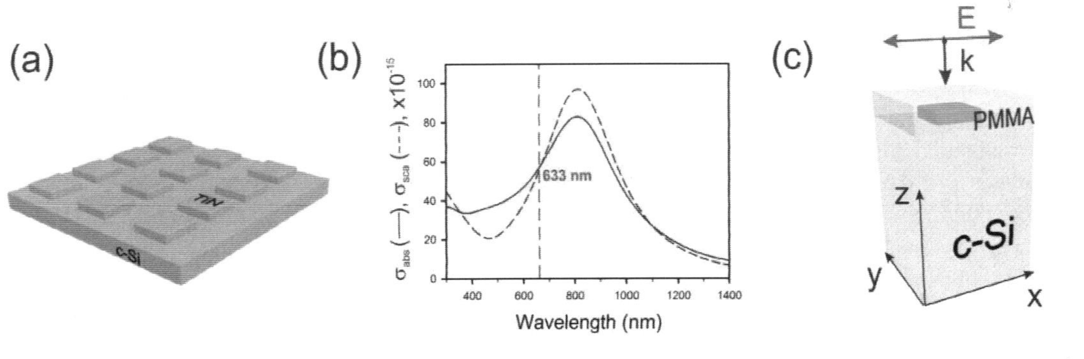

Figure 1. a) Scheme of a TiN metasurface on a Si substrate; b) FDTD calculated absorption σ_{abs} (nm^2) and scattering σ_{sca} (nm^2) cross section of a TiN antenna with a size: 200x200x50 nm; c) Schematic of the system simulated in the DEVICE Lumerical;

Simulation of steady-state heat propagation was carried out using the Lumerical FEM solver. The schematic geometry of thermal simulation is shown in Figure 1(c). A thermal boundary condition with a fixed temperature of 300 K was applied at the bottom of the simulation area. The upper part of the simulation region was covered with a PMMA film. The conditions of convective boundaries were defined at the interface with TiN and PMMA, PMMA and Si to model the heat losses of the structure through convection. We used the following thermal conductivities: k_{TiN} = 29 W/mK, k_{PMMA} = 0.19 W/mK, k_{Air} = 0.026 W/mK, k_{Si} = 148 W/mK. The optical and thermal properties of all materials were taken from the Lumerical material database. It was assumed that the thermal conductivity of all components is temperature-independent. The heat power density was calculated in the FDTD simulation.

3. Result and discussion

Heat generation in a metal nanostructure under illumination arises due to the Joule effect. The power of light absorbed by the TiN parallelepiped is a heat source inside the nanostructure and reads

$$P = \int_V p(r)dV = \frac{1}{2}\mathrm{Re}\left[\int_V J^*(r)E(r)dV\right] = \sigma_{abs} I \,, \tag{1}$$

where $p(r)$ is a power density, J is a current density, σ_{abs} is an absorption cross-section, $I = c\varepsilon_0 n_s |E_0|^2 / 2$ (ε_0 is the permittivity of vacuum, c is the speed of light, n_s is a refractive index of the environment and E_0 is intensity of internal electric field). Thus, optical absorption is proportional to the square of the internal electric field intensity.

We solve Maxwell's equations by the FDTD solver to obtain P, which we enter into the diffusive equation to determine the heating temperature of the TiN parallelepiped and PMMA film using the FEM solver. Figure 2 illustrates the profile of electric field around a TiN antenna in the case of air (a) and PMMA (b). The maximum intensity in the upper part (50 nm) of the TiN structure in the air is 11 and we conclude that the electric field, although insignificant, is amplified due to the excitation of plasmon resonance. The maximum electric field intensity falls more than twofold when TiN is coated with PMMA, while the magnetic field intensity behaves exactly the opposite (more than twofold increase) (Figure 2 (d,c)). This effect is explained by the fact that the refractive index of PMMA is higher than that of air.

In the stationary mode, the thermal diffusion equation has the form:

$$\nabla\left[k(r)\nabla T(r)\right] = -p(r) \,, \tag{2}$$

where k is the thermal conductivity of the medium and it is assumed that it does not depend on temperature. For a TiN parallelepiped dressed in PMMA we obtain the temperature increment:

$$\Delta T = \frac{q}{4\pi k_{PMMA}\beta L} = \frac{\sigma_{abs} I}{4\pi k_{PMMA}\beta L} \,, \tag{3}$$

where β is the geometric form factor and L is the lateral size of the TiN parallelepiped..

The maximum heating temperature of TiN was 630 K in air (Figure 2(e)) while it was 476 K when TiN was coated with PMMA (Figure 2(f)). Such heating is sufficient to achieve the glass transition of a polymer PMMA film (T_g=383 K). For polymers with a higher glass transition temperature, this heating will may be insufficient. In this regard, we propose a following approach to solving this problem. We offer the addition to enhance the conversion of light to heat. This can be achieved through a confined substrate effect, for example by forming c-Si pillars of some height.

Let us consider a new geometry of the experiment (inset on Figure 2(g)), where a parallelepiped is located on a silicon column with a height h. We use the same incident light conditions. Solving the thermal diffusion equation in the steady state, we get:

$$\delta T = \frac{\sigma_{abs} I}{4\pi k_{PMMA}\beta L} \cdot \frac{1}{r} \,. \tag{4}$$

Inside the TiN parallelepiped, the temperature increment corresponds to the parabolic expression:

$$\delta T = \delta T_{TiN} + \frac{\sigma_{abs} I}{6k_{TiN}}(L^2 - r^2) \,. \tag{5}$$

FEM simulation of the heating temperature is shown in Figure 2(g) for three cases of Si column heights (no column, 150 nm and 300 nm). The upper surface of the TiN parallelepiped heats up to 176 K in the absence of a column. However, as the column height increases, we observe much larger temperature increments, namely 232 K and 329 K on the top surface, respectively, for 150 nm and 300 nm in height. This means that this design and this approach are quite suitable for heating polymer films to higher temperatures. Thus, the higher the Si column, the greater the temperature rise is achieved.

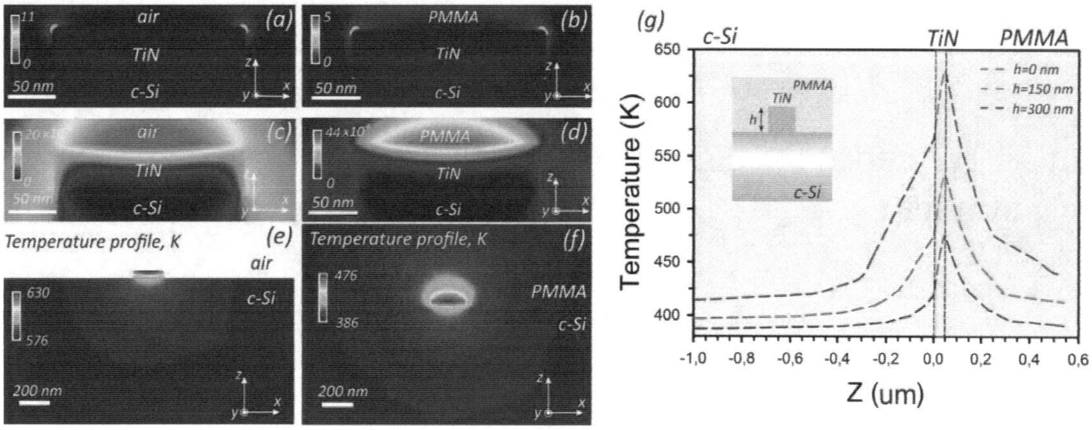

Figure 2. a), b) FDTD simulation of electric fields ($|E/E_0|^2$) around a TiN parallelepiped in air and PMMA; c), d) FDTD simulation of magnetic fields around a TiN parallelepiped in air and PMMA; e), f) FEM simulation of the temperature distribution of the TiN heating parallelepiped in air and PMMA; g) FEM calculated temperature profile of 200 nm TiN antenna for different height of Si pillars.

4. Conclusion

Summing up, in this work we have demonstrated photo-induced heating of a polymer PMMA film on a plasmonic refractory TiN metasurface under CW illumination in a steady state using FDTD/FEM simulation. We show that polymer heating can be controlled by confined Si pillar under TiN pad. We observe that the higher the Si column is, the greater the temperature rise is achieved. Thus, the idea of achieving higher temperatures using a larger column height has the prospect of creating a system of distributed heat sources with different heating temperatures under CW illumination.

5. Acknowledgments

This work was supported by grant No. 19-12-00066 of the Russian Science Foundation. E.A.C. acknowledges the help from the Russian Foundation for Basic Research (grant No. 20-32-90090) for conducting numerical FDTD simulations.

6. References

1. Gobin A M, Lee M H, Halas N J, James W D, Drezek R A and West J L 2007 *Nano Latters* **7** 1929-1934
2. Jain P K, El-Sayed I H and El-Sayed M A 2007 *Nano Today* **2** 18
3. Urban A, Pfeiffer T, Fedoruk M, Lutich A and Feldmann J 2011 *ACS Nano* **5** 3585-3590
4. Carattino A, Caldarola M and Orrit M 2018 *Nano Letters* **18** 874–880
5. Li W, Guler U, Kinsey N, Naik G V, Boltasseva A, Guan J, Shalaev V M and Kildishev A V 2014 *Advanced Materials* **26** 7959-7965
6. Jin C, Li Z, Williams R, Lee K. and Park I 2011 *Nano Letters* **11** 4818-4825
7. Qiu G, Gai Z, Tao Y, Schmitt J, Kullak-Ublick G A and Wang J 2020 *ACS Materials* **22** 4925
8. Nugroho F A A, Albinsson D, Antosiewicz T J and Langhammer C 2020 *ACS Nano* **14** 2345-2353

EBG waveguides for contactless surface impedance measurements

Sandra Rodini, Simone Genovesi, Giuliano Manara, Filippo Costa*

Dipartimento di Ingegneria dell'Informazione, 56122, Università di Pisa, Italy.

*filippo.costa@unipi.it

Abstract. A method for the estimation of sheet impedance of thin sample which does not require a direct contact with the sample under test is proposed. The surface impedance is calculated through an inversion procedure exploiting the scattering parameters obtained through a waveguide measurement setup. An inversion procedure based on the representation of the waveguide-air-waveguide section as a π junction is employed. In order to prevent the field leakage from the air gap created for hosting the thin sheet, an EBG surface is introduced on the flange of the waveguide. It is shown that the introduction of the EBG surface remarkably improves the estimation of the surface impedance of the thin sheet with respect to the case without EBG.

1. Introduction

There exist several methods to derive the surface impedance of a thin sheet. The most used method to derive the DC surface impedance of a thin sheet is the four-point method which has the disadvantage of being a contact method and can damage the material sample [1]. Microwave methods are mainly divided into resonant and non-resonant ones. Resonant methods have a good accuracy and sensitivity; however, they require an ad-hoc cavity and sample preparation can be complicated. Non-resonant methods are based on the measurement of the signal reflected and transmitted by the sample. These methods allow measurements to be made using different experimental setups and require fewer precautions [2]. In [3] a method has been proposed to derive the surface impedance of an ink deposited on a dielectric substrate using rectangular waveguide and using an inversion procedure to derive the surface impedance starting from the measured scattering parameters. A similar approach was used in [4] where the discontinuity between the two waveguides is taken into account. The aforementioned methods offer a good estimate of the surface impedance, but they require a physical contact with a waveguide. The aim of this paper is instead to develop an approach allowing the characterizations of thin sheet without a direct contact with the measuring fixture. Such a novel characterization setup would allow to estimate the surface impedance of the piezoresistive sheets under different stretching conditions and correlate this parameter with the strain in the correspondent sensor calibration curve. Piezoresistive materials are widely used in many engineering fields for fabrication of sensors, such as force sensors [5], pressure sensors [6], strain sensors [7]. Most of the works in this sector deal with high performance sensors that are wired (*i.e.,* sensor is connected to the reader via a cable). However, it could be interesting to investigate sensors able to operate *wireless* thus creating a new class of strain sensors able to provide a wireless reading of the sensed parameter could be designed by a proper use of piezoresistive materials.

Journal of Physics: Conference Series **2015** (2021) 012030 doi:10.1088/1742-6596/2015/1/012030

In this scenario it is important to consider that the presence of an air gap between the two waveguides introduces a different configuration of the field. For this reason, it is proposed to use an electromagnetic band-gap (EBG) surface to limit the harmful effects introduced by the air-gap.

2. Proposed measurement

The proposed method is based on a Transmission Line (TL) setup formed by two waveguides and a thin sheet which is placed in the middle, similarly to [5, 6]. However, unlike the former setups in which the thin sheet was in contact with the flange of the waveguide, here a contactless characterization of the sample is attempted by allowing an air gap between the two flanges. This configuration, removing the need of a contact between the waveguide and the sample, allows the wireless characterization of piezoresistive samples at radio frequencies while the sample is stretched. However, in the above-mentioned configuration there is a significant leakage of field between the flanges of the two waveguides. The field leakage is due to the presence of the air-gap and it is responsible for an inaccurate estimate of the surface impedance of the sheet under test.

The use of the EBG surface limits the leakage of electric field in the air gap and this positively impacts the estimation of the surface impedance. The proposed setup is shown in Figure 1.*a*. On one flange of the waveguide an EBG surface is applied. The top view of the EBG surface is shown in Figure 1.*b*. Figure 1.c and d show how the employment of the EBG surface modifies the electric field distribution. The field distribution is displayed at the center of the EBG stopband, that is 6.5 GHz. A similar field distribution is observed from 5.8 GHz to 7.2 GHz.

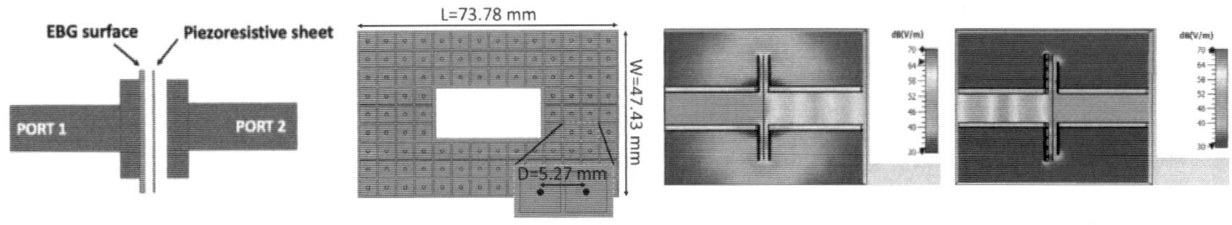

(a) *(b)* *(c)* *(d)*

Figure 1. *(a)* Modified setup with the EBG surface placed on the waveguide flange. *(b)* Top view of the EBG surface. Distribution of the electric field: (c) configuration without the EBG surface, (d) configuration with the EBG surface at 6.5 GHz.

The unit cell of the EBG surface has been suitably engineered in order to achieve bandgap properties within the operating range of the waveguide. In order to create an EBG surface that guarantees the necessary field confinement, a parametric study on the dispersion diagram has been carried out to tailor the unit cell size of the EBG. The parameters that were taken into consideration for the study of the dispersion diagram were: the size of the square patch, the radius of the vias, and the thickness of the substrate. In order to adapt the EBG surface to the waveguide flange, a periodicity D=5.27 mm was chosen. Regarding the thickness of the substrate, it is worth noting that the band-gap shifts towards lower frequencies at the increasing of the thickness. Since the range of frequencies of interest is up to 8 GHz, a substrate thickness of 1.6 mm has been chosen. A via radius equal to 0.25 mm was selected although changing the radius of the vias does not significantly affect the width of the stop band. The square patch side length has been fixed to 5 mm since larger dimensions (reduced gap size) would move the stopband to lower frequencies.

The method for determining the surface impedance, according to [6], is divided into two steps. The first step consists of an unloaded measurement aimed at calibrating the system and determining the parameters of the equivalent circuit shown in Figure 2.*a*. The two series impedances Z are assumed to be identical but the same cannot be presumed for the two parallel admittances (Y_1, Y_2), because one (Y_1) is affected by the presence of the EBG surface. In the second step of the procedure, the thin sheet is introduced between the two flanges (Figure 2.*b*). The thin sheet can be modeled as a lumped parameter Y_s that represents the inverse of the surface impedance. Once Z, Y_1, Y_2 obtained in the calibration step are known, it is possible to obtain the surface impedance of the sheet using an inversion procedure.

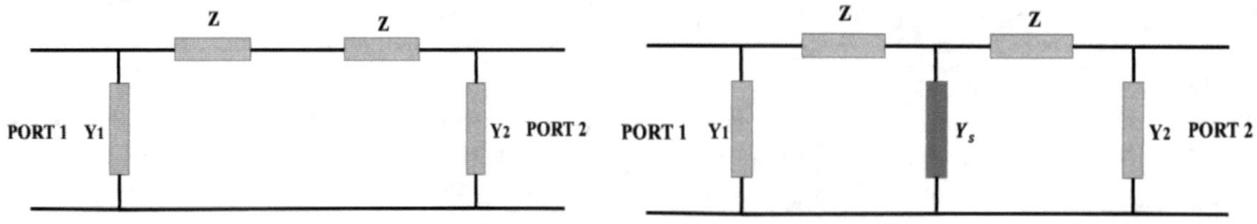

(a) *(b)*

Figure 2. Equivalent circuit models: *(a)* calibration setup and *(b)* final model with the thin sheet.

The circuit model can be described using the ABCD parameters. The ABCD matrix of the overall circuit for the first step is as follows:

$$\begin{bmatrix} A & B \\ C & D \end{bmatrix} = \begin{bmatrix} 1 & 0 \\ Y_1 & 1 \end{bmatrix} \begin{bmatrix} 1 & Z_1 \\ 0 & 1 \end{bmatrix} \begin{bmatrix} 1 & Z_2 \\ 0 & 1 \end{bmatrix} \begin{bmatrix} 1 & 0 \\ Y_2 & 1 \end{bmatrix} = \begin{bmatrix} 1 + 2ZY_2 & 2Z \\ Y_2(2ZY_1 + 1) + Y_1 & 1 + 2ZY_1 \end{bmatrix} \tag{1}$$

Starting from the matrix in (1) and exploiting the relationships between the scattering parameters and the ABCD ones, it is possible to obtain Z, Y_1 and Y_2 as follow:

$$Z = Z_0 \frac{(1 + S_{11})(1 + S_{22}) - S_{21}{}^2}{4S_{21}} \tag{2}$$

$$Y_1 = \frac{\dfrac{(1 - S_{11})(1 + S_{22}) + S_{21}{}^2}{2S_{21}} - 1}{2Z} \tag{3}$$

$$Y_2 = \frac{\dfrac{(1 + S_{11})(1 - S_{22}) + S_{21}{}^2}{2S_{21}} - 1}{2Z} \tag{4}$$

After inserting the thin sheet between the waveguides, the ABCD matrix changes as follows:

$$\begin{bmatrix} A' & B' \\ C' & D' \end{bmatrix} = \begin{bmatrix} 1 & 0 \\ Y_1 & 1 \end{bmatrix} \begin{bmatrix} 1 & Z \\ 0 & 1 \end{bmatrix} \begin{bmatrix} 1 & 0 \\ Y_s & 1 \end{bmatrix} \begin{bmatrix} 1 & Z \\ 0 & 1 \end{bmatrix} \begin{bmatrix} 1 & 0 \\ Y_2 & 1 \end{bmatrix} \tag{5}$$

Considering the expression of S'_{21} as a function of A', B', C' and D', it is possible to derive the value of Z_s as the inverse of Y_s.

3. Numerical results

The accuracy of proposed characterization methodology has been evaluated by using numerical electromagnetic simulations.

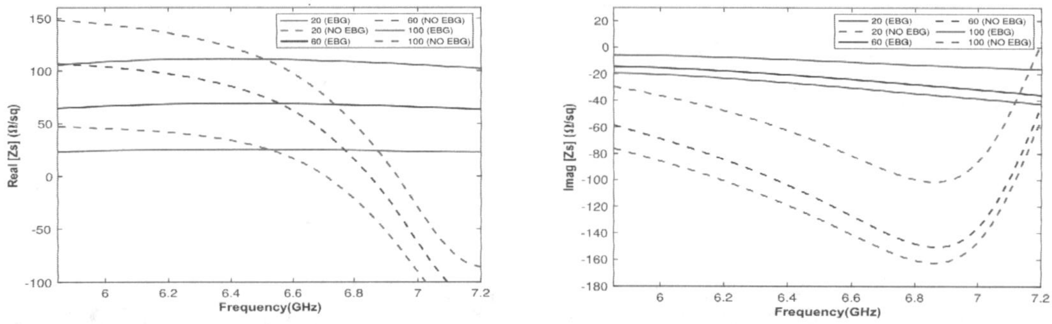

(a) *(b)*

Figure 3. *(a)* Real *(b)* imaginary part of estimated surface impedance. Comparison between configuration with and without EBG is reported.

Simulations were carried out using the CST Studio Suite. A WR137 waveguide has been used, which operates in the frequency range [5.85-8.20 GHz]. The distance between the two waveguides has been set at 5 mm to allow the insertion of the thin sheets to be characterized. The parameters Z, Y_1 and Y_2 modelling the junction have been derived according to the procedure described in section 2. Subsequently, a thin sheet of known surface impedance was inserted between the two waveguides. The surface impedance Z_s is finally estimated by using the relation obtained from the inversion procedure starting from (6). Figure 3 shows the estimates of three different real valued impedances (*i.e.* 20, 60 and 100 Ω /sq) and proves the advantage achieved through the use of the EBG surface. The imaginary part is considered to be zero as it is common in practical resistive materials such as graphene, ITO and carbon loaded materials [7]. Without the use of EBG surface the model does not offer a good estimate of the surface impedance due to leakage across the gap. The introduction of EBG leads to a significant improvement in the estimation of the surface impedance. The residual error is attributed to the assumption that Z and Y_1 and Y_2 remain constant after inserting the sheet. The inversion method allows to calculate the surface impedance starting from a correct estimation of the transmission coefficient S'_{21}.

4. Conclusion

A method for a contactless measurement of the surface impedance has been proposed for the first time. The proposed approach, due to absence of electrical contact, allows the characterization of piezoresistive materials while they are stretched. The harmful effect introduced by the air gap between the sheet and the waveguides determined by the field leakage has been prevented by using an EBG surface on the waveguide flange.

5. Acknowledgements

This work was partially supported by the Italian Ministry of Education and Research (MIUR) in the framework of the CrossLab project (Departments of Excellence) and PRIN2017 GREEN TAGS "Chipless radio frequency identification (RFID) for GREEN TAGging and Sensing" (2017NT5W7Z).

References

[1] van der PAUW, L. J. "A method of measuring specific resistivity and hall effect of discs of arbitrary shape." *Semiconductor Devices: Pioneering Papers*. 1991. 174-182.

[2] F. Costa, M. Borgese, M. Degiorgi, and A. Monorchio, "Electromagnetic Characterisation of Materials by Using Transmission/Reflection (T/R) Devices," *Electronics*, vol. 6, no. 4, p. 95, Nov. 2017, doi: 10.3390/electronics6040095.

[3] F. Costa, "Surface Impedance Measurement of Resistive Coatings at Microwave Frequencies," *IEEE Trans. Instrum. Meas.*, vol. 62, no. 2, pp. 432–437, Feb. 2013, doi: 10.1109/TIM.2012.2217661.

[4] X.-C. Wang, A. Diaz-Rubio, and S. A. Tretyakov, "An Accurate Method for Measuring the Sheet Impedance of Thin Conductive Films at Microwave and Millimeter-Wave Frequencies," *IEEE Trans. Microw. Theory Tech.*, vol. 65, no. 12, pp. 5009–5018, Dec. 2017, doi: 10.1109/TMTT.2017.2714662.

[5] X. Y. Liu, M. O'Brien, M. Mwangi, X. J. Li, and G. M. Whitesides, "Paper-based piezoresistive MEMS force sensors," in *2011 IEEE 24th International Conference on Micro Electro Mechanical Systems*, Cancun, Mexico, Jan. 2011, pp. 133–136, doi: 10.1109/MEMSYS.2011.5734379.

[6] I. Baldoli, M. Maselli, F. Cecchi, and C. Laschi, "Development and characterization of a multilayer matrix textile sensor for interface pressure measurements," *Smart Mater. Struct.*, vol. 26, no. 10, p. 104011, Oct. 2017, doi: 10.1088/1361-665X/aa644e.

[7] C. Yan *et al.*, "Highly Stretchable Piezoresistive Graphene-Nanocellulose Nanopaper for Strain Sensors," *Adv. Mater.*, vol. 26, no. 13, pp. 2022–2027, Apr. 2014, doi: 10.1002/adma.201304742.

Wave propagation control in active acoustic metamaterials

A. Bacigalupo[1], M. L. De Bellis[2], G. Gnecco[3], D. Misseroni[4]

[1]DICCA University of Genoa - Via Montallegro 1 - Genova -Italy
[2]INGEO University G. d'Annunzio Chieti-Pescara - Viale Pindaro 42 - Pescara - Italy
[3]IMT-School for Advanced Studies Lucca - Piazza S. Francesco 19 - Lucca - Italy
[4]DICAM University of Trento - via Mesiano 77 - Trento - Italy

E-mail: andrea.bacigalupo@unige.it, marialaura.debellis@unich.it, giorgio.gnecco@imtlucca.it, diego.misseroni@unitn.it

Abstract. Focus is on the design of an innovative class of tunable periodic metamaterials, conceived for the realization of high performance acoustic metafilters with settable real-time capabilities. In this framework the tunability is due to the presence of a piezoelectric phase shunted by a suitable electrical circuit with adjustable impedance/admittance. It follows that the acoustic properties of the metamaterial can be properly modified in an adaptive way, opening up new possibilities for the control of pass- and stop-bands.

1. Introduction

The study of metamaterials is increasingly emerging as an innovative interdisciplinary area, including physics, material science and engineering. Metamaterials are architectured composites, specifically conceived to exhibit outstanding constitutive properties, well above those achievable with classical materials. In this context, the present paper proposes the design of high-performance tunable acoustic filters for the passive wave propagation control. More specifically a special class of periodic metamaterials with a phase shunted by either a dissipative or non dissipative electrical circuit is proposed.

2. Active metamaterial with a piezoelectric shunting phase

Consider an infinite metamaterial with periodic microstructure characterized by three phases, as shown in the Figure 1(a). This metamaterial is made by in-plane tessellation of the centrosymmetric periodic cell $\mathscr{A}$, schematically depicted in the Figure 1(b), along two orthogonal periodicity vectors $\mathbf{v}_1 = d\mathbf{e}_1$, $\mathbf{v}_2 = d\mathbf{e}_2$, being $\mathbf{e}_1$ and $\mathbf{e}_2$ unit vectors of a given orthogonal base. The microstructure of the cell $\mathscr{A}$, with out-of-plane thickness w, is composed by a stiff and massive frame of outer edge b, embedding both an outer ring made of a soft piezoelectric material of thickness h and a inner disk of radius r, made of the same material as the frame. The geometry is completed by four flexible and light ligaments of length $l = (d - b)/2$ and width a. Each piezoelectric ring is connected in parallel to an external electrical circuit with tunable equivalent impedance/admittance, see Figure 1(c). It follows that the frequency band structure of the acoustic metamaterial can be properly controlled by modifying the equivalent impedance/admittance together with the constitutive properties of the shunted piezoelectric material. The stop- and pass-band frequency amplitudes as well as their central frequencies can be thus manipulated to exploit the performance of the periodic tunable metamaterial as acoustic filter. In the following, first the dynamic governing equations of the three phase active metamaterial, together with the constitutive relations of the shunted piezoelectric phase, are introduced. Then the capabilities of the free wave propagation control in such acoustic metamaterial are investigated and a relevant example is provided.

3. Governing equations of the three-phase active metamaterial

The in-plane governing equations of the infinite metamaterial are expressed, accordingly with [1], in the transformed Laplace space, as follows

$$\left(C^{\diamond}_{ijhl}(s)\widehat{u}_{h,l}\right)_{,j}+\widehat{b}_i=\rho s^2\widehat{u}_i, \tag{1}$$

where $C^{\diamond}_{ijhl}(s)$ are the components of the constitutive tensor, s is the complex Laplace variable, $\widehat{u}_i$ are the in-plane transformed displacement components, $\widehat{b}_i$ the transformed source term components, ρ is the mass density and the comma denotes the partial derivative with respect to x_j, i.e. the components of the in-plane position vector $\mathbf{x}=x_j\mathbf{e}_j\in\mathbb{R}^2$, with $i,j,h,l=1,2$. Due to the medium periodicity, both the constitutive tensor and the mass density are $\mathscr{A}$ periodic, so that

$$C^{\diamond}_{ijhl}\left(\mathbf{x}+n\mathbf{v}_\beta,s\right)=C^{\diamond}_{ijhl}\left(\mathbf{x},s\right), \qquad \rho\left(\mathbf{x}+n\mathbf{v}_\beta\right)=\rho\left(\mathbf{x}\right), \qquad \forall\mathbf{x}\in\mathfrak{A}, \tag{2}$$

where $n\in\mathbb{Z}$ and the index $\beta=1,2$. Note that the components $C^{\diamond}_{ijhl}(s)$ characterize the constitutive properties of each material phase pertaining to the microstructure that in Figure 1 are distinguished by color. In particular, the gray phase, referred to ligaments, is assumed as a linear elastic isotropic material. The same constitutive behavior is considered for the blue phase, pertaining both to the stiff frame and to the stiff disk. On the other hand, the red phase, i.e. the internal ring, is made by a shunted piezoelectric material. It stands to reason that the components $C^{\diamond}_{ijhl}$ of the linear elastic materials are s-independent, while those of the shunting piezoelectric phase, denoted with the apex EL, are in general s-dependent and take the following form

$$C^{EL}_{ijhl}(\lambda(s))=C_{ijhl}+\frac{e_{ij3}\widetilde{e}_{3hl}}{\beta^{EL}_{33}(\lambda(s))}-\left(C_{ij33}+\frac{e_{ij3}\widetilde{e}_{333}}{\beta^{EL}_{33}(\lambda(s))}\right)\left(\frac{C_{33hk}+\frac{e_{333}\widetilde{e}_{3hl}}{\beta^{EL}_{33}(\lambda(s))}}{C_{3333}+\frac{e_{333}\widetilde{e}_{333}}{\beta^{EL}_{33}(\lambda(s))}}\right), \tag{3}$$

where C_{ijhl} are the components of the fourth order elasticity tensor, e_{ij3} are the components of the third order stress-charge coupling tensor of the piezoelectric material and $\widetilde{e}_{pqs}=e_{qsp}$ its transpose. Moreover,

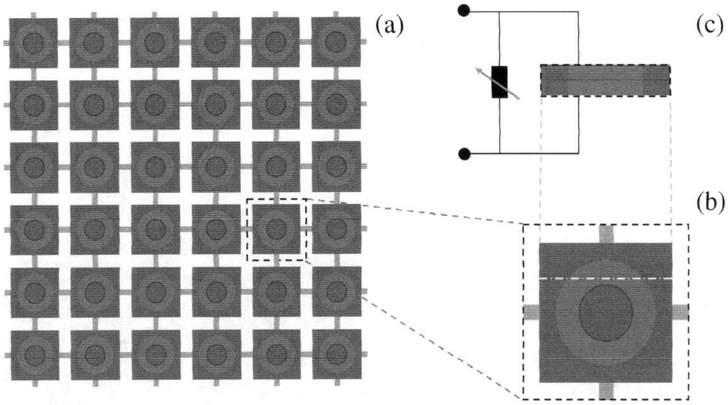

Figure 1. (a) A generic portion of the heterogeneous material; (b) scheme of the periodic cell containing the shunted piezoelectric material; (c) scheme of the tunable electric circuit.

the auxiliary s-dependent function $\beta_{33}^{EL}(s) = \beta_{33}(1+\lambda(s))$ is introduced, with β_{33} the component of the second order permittivity tensor and the so-called *tuning function* $\lambda(s) = L^{(P)}Y_{33}^{SU}(s)/(s\beta_{33}A^{(P)})$ linearly depending on the generic equivalent shunting admittance $Y_{33}^{(SU)}(s)$ that, in turn, may depend on one or more tuning parameters characterizing the specific electrical circuit. Note that $A^{(P)}$ is the area of the piezoelectric phase (red annulus in Figure 1(b)), and $L^{(P)} = w$.

4. Wave propagation modeling

With the aim of investigating the free wave propagation in the infinite periodic tunable metamaterial, accordingly with the Floquet-Bloch theory [4], the Floquet-Bloch decomposition is introduced as

$$\widehat{u}_i = \widehat{U}_i e^{Ik_jx_j} \tag{4}$$

with $\widehat{U}_i$ the $\mathfrak{A}$-periodic Bloch amplitudes of the displacement in the transformed space, k_j the wave vector components and I the imaginary unit. By plugging Eq. (4) in the field equation (1), the following field equation in terms of the Bloch amplitudes is obtained as

$$\left(C_{ijhl}^{\diamondsuit}(s)\widehat{U}_{h,l}\right)_{,j} - \left(C_{ijhl}^{\diamondsuit}(s)k_lk_j + s^2\rho\delta_{ih}\right)\widehat{U}_h + Ik_j\left[\left(C_{ijhl}^{\diamondsuit}(s)+C_{ilhj}^{\diamondsuit}(s)\right)\widehat{U}_{h,l} + C_{ilhj,l}^{\diamondsuit}(s)\widehat{U}_h\right] = 0, \tag{5}$$

where δ_{ih} is the Kronecker delta. By assuming $k_i = kn_i$ with k the wave number and $\mathbf{n} = n_ie_i \in \mathbb{R}^2$ the unit vector of propagation, the previous field equation takes the form

$$\left(C_{ijhl}^{\diamondsuit}(s)\widehat{U}_{h,l}\right)_{,j} - \left(C_{ijhl}^{\diamondsuit}(s)k^2n_ln_j + s^2\rho\delta_{ih}\right)\widehat{U}_h + Ikn_j\left[\left(C_{ijhl}^{\diamondsuit}(s)+C_{ilhj}^{\diamondsuit}(s)\right)\widehat{U}_{h,l} + C_{ilhj,l}^{\diamondsuit}(s)\widehat{U}_h\right] = 0. \tag{6}$$

By exploiting the medium periodicity, Eq. (6) can be solved by applying the well-known periodic boundary conditions on the cell boundaries. Moreover, with the aim of solving the field equation (6) through a standard Galerkin procedure, the weak formulation is obtained by integrating, over the periodic cell $\mathfrak{A}$, this equation multiplied by a properly chosen $\mathfrak{A}$-periodic test function $\widehat{\Psi}_i$, resulting in

$$\begin{aligned}\int_{\mathfrak{A}}\Big\{&\left(C_{ijhl}^{\diamondsuit}(s)\widehat{U}_{h,l}\right)_{,j} - \left(C_{ijhl}^{\diamondsuit}(s)k^2n_ln_j + s^2\rho\delta_{ih}\right)\widehat{U}_h + \\ &+Ikn_j\left[\left(C_{ijhl}^{\diamondsuit}(s)+C_{ilhj}^{\diamondsuit}(s)\right)\widehat{U}_{h,l} + C_{ilhj,l}^{\diamondsuit}(s)\widehat{U}_h\right]\Big\}\widehat{\Psi}_i d\mathbf{x} = 0.\end{aligned} \tag{7}$$

By means of the divergence theorem and observing that the boundary integrals vanish due to the $\mathfrak{A}$-periodicity of the test function, the following equation is obtained

$$\begin{aligned}&k^2\int_{\mathfrak{A}} C_{ijhl}^{\diamondsuit}(s)n_ln_j\widehat{U}_h\widehat{\Psi}_i d\mathbf{x} - Ik\int_{\mathfrak{A}}\left[\left(C_{ijhl}^{\diamondsuit}(s)+C_{ilhj}^{\diamondsuit}(s)\right)\widehat{U}_{h,l}\widehat{\Psi}_i + C_{ilhj,l}^{\diamondsuit}(s)\widehat{U}_h\widehat{\Psi}_i\right]n_j d\mathbf{x} + \\ &+\int_{\mathfrak{A}} C_{ijhl}^{\diamondsuit}(s)\widehat{U}_{h,l}\widehat{\Psi}_{i,j} d\mathbf{x} - s^2\int_{\mathfrak{A}}\rho\delta_{ih}\widehat{U}_h d\mathbf{x} = 0.\end{aligned} \tag{8}$$

In the framework of the finite elements method, the following homogeneous assembled matrix equation can be obtained by discretized the weak form (8) together with its boundary terms, in agreement with the procedure proposed by [5] and detailed in [6, 7], i.e.

$$\mathbf{H}(k,s,\mathbf{n})\widehat{\mathbf{U}} = \left(k^2\mathbf{H}^{(2,0)}(s,\mathbf{n}) + Ik\mathbf{H}^{(1,0)}(s,\mathbf{n}) + \mathbf{H}^{(0,0)}(s) + s^2\mathbf{H}^{(0,2)}(s)\right)\widehat{\mathbf{U}} = 0, \tag{9}$$

where $\widehat{\mathbf{U}}$ is the vector collecting nodal Floquet-Bloch amplitudes and $\mathbf{H}^{(2,0)}$, $\mathbf{H}^{(1,0)}$, $\mathbf{H}^{(0,0)}$ and $\mathbf{H}^{(0,2)}$ are s-dependent linear operators. More specifically, the matrix equation (9) is a generalized eigenvalue problem characterizing the free propagation of Bloch waves with angular frequency s, wave number k and polarization vector $\widehat{\mathbf{U}}$, in the infinite periodic metamaterial along the generic propagation direction $\mathbf{n}$. Aiming at investigating the temporal damping[8, 9, 7], the angular frequency s and the polarization vector $\widehat{\mathbf{U}}$ are, respectively, the eigenvalue and the eigenvector of the generalized eigenproblem (generally rational) depending on parameters $\mathbf{n}$ and k. It stands to reason that the eigenproblem (9) has non trivial solution in the case the following characteristic equation is fulfilled

$$\mathfrak{D}(k,s,\mathbf{n}) = \det[\mathbf{H}(k,s,\mathbf{n})] = 0, \tag{10}$$

determining the complex frequency band structure $s(k) \in \mathbb{C}$ in terms of the wave number $k \in \mathbb{R}$ for a fixed propagation direction $\mathbf{n}$. In the particular case of remarkable technological interest, in which the equivalent electrical circuit is characterized by a purely capacitive equivalent admittance $Y_{33}^{SU}(s) = sC$, the tuning functions turn out to be s-independent, i.e. $\lambda = CL^{(p)}/(\beta_{33}A^{(p)})$, as well as the constitutive tensor components C_{ijkl}^{EL}. In this case, as shown in [1], the complex frequency s is purely imaginary, i.e. $s = I\omega$, being $\omega \in \mathbb{R}$ the angular frequency of Bloch waves. The eigenproblem (9) takes the form

$$\mathbf{K}(k,\omega,\mathbf{n})\widehat{\mathbf{U}} = \left(k^2\mathbf{K}^{(2,0)}(\mathbf{n}) + Ik\mathbf{K}^{(1,0)}(\mathbf{n}) + \mathbf{K}^{(0,0)} - \omega^2\mathbf{K}^{(0,2)}\right)\widehat{\mathbf{U}} = 0, \tag{11}$$

where the linear operators $\mathbf{K}^{(2,0)}$, $\mathbf{K}^{(1,0)}$, $\mathbf{K}^{(0,0)}$ and $\mathbf{K}^{(0,2)}$ are s-independent so that the eigenproblem turns out to be in the polynomial form. It has non trivial solution in the case the following characteristic equation is satisfied

$$\mathfrak{D}(k,\omega,\mathbf{n}) = \det[\mathbf{K}(k,\omega,\mathbf{n})] = 0, \tag{12}$$

determining the frequency band structure $\omega(k) \in \mathbb{R}$ in terms of $k \in \mathbb{R}$ for fixed $\mathbf{n}$.

5. Conclusions

A class of tunable mechanical metamaterials characterized by periodic piezoelectric microstructure, shunted by an electrical circuit is here designed. Numerical experiments highlight that it is possible to open low-frequency band gaps and modifying on demand their amplitude and central frequencies by adjusting the tuning parameter, also starting from dense spectra. It is noted that by decreasing the tuning parameter up to about the resonant value, a stiffening of the piezoelectric material is exhibited, together with enlargments of low-frequency band gaps. Moreover, as expected the central frequencies of such band gaps tend to slightly increase.

Acknowledgements
The authors gratefully acknowledge financial support from National Group of Mathematical Physics (GNFM-INdAM), from the Compagnia San Paolo, project MINIERA no. I34I20000380007, from University of Trento, project UNMASKED 2020 and project Search for Excellence Ud'A 2019.

References

[1] A. Bacigalupo, M. L. De Bellis, and D. Misseroni. Design of tunable acoustic metamaterials with periodic piezoelectric microstructure. *Extreme Mechanics Letters*, 40:100977, 2020.
[2] L. Brillouin. *Wave Propagation and Group Velocity*. New York: Academic Press, 1960.
[3] R. S. Langley. A note on the force boundary conditions for two-dimensional periodic structures with corner freedoms. *Journal of Sound and Vibration*, 167(2):377–381, 1993.
[4] A Srikantha Phani, J Woodhouse, and NA Fleck. Wave propagation in two-dimensional periodic lattices. *The Journal of the Acoustical Society of America*, 119(4):1995–2005, 2006.
[5] F. Fantoni and A. Bacigalupo. Wave propagation modeling in periodic elasto-thermo-diffusive materials via multifield asymptotic homogenization. *International Journal of Solids and Structures*, 196:99–128, 2020.
[6] J. M. Carcione. *Wave fields in real media: Wave propagation in anisotropic, anelastic, porous and electromagnetic media*. Elsevier, 2007.
[7] A. Bacigalupo, M. L. De Bellis, and G. Gnecco. Complex frequency band structure of periodic thermo-diffusive materials by Floquet–Bloch theory. *Acta Mechanica*, 230(9):3339–3363, 2019.

Elastic Trapped Modes in Solid Acoustic Resonators of Various Shapes

Ilya Deriy, Ivan Toftul, Mihail Petrov and Andrey Bogdanov

E-mail: `ilya.deriy@metalab.ifmo.tu`
Department of Physics and Engineering, ITMO University, St. Petersburg, 197101, Russia

Abstract. Resonators are one of the main building blocks of many acoustic, photonic, and microwave devices such as metasurfaces, sensing devices, antennas, and many more. One of the main properties of any resonator, which also determines the properties of the structure, based on the resonator, is the quality (Q) factor. Q-factor of the resonator is limited due to material and radiative losses. In this paper, we propose the existence of modes of solid resonators, immersed in a nonviscious fluid, which are non-radiative, and therefore, their Q-factor is limited only by material losses.

1. Introduction

Usually, eigenmodes of the open systems such as resonators are coupled to the modes of the freespace because their energies are lying in the same region of the spectrum, which results in a finite radiative Q-factor. However, even for open systems, one can find such modes, that do not exchange energy with the radiation continuum and therefore remain purely localized inside the system. Such modes are often called in quantum mechanics and photonics as *Bound States in The Continuum* [1], [2] or, in acoustics and fluidics as *Trapped Modes* [3], [4], [5].

Trapped modes are localized solutions of the Helmholtz equation in a bounded region, whose eigenvalues are embedded into the continuous spectrum. The first analytical proof of the existence of trapped modes was made by Ursell in 1951 [6] for an oscillating fluid, bounded by fixed surfaces and by a free surface of infinite extent. Later, trapped modes were found in infinite acoustic waveguides with an obstacle [7], [3], [8]. However, as it will be shown later, such modes can be found also in open systems of a finite size e.g. resonators.

It is well-known fact, that while solid objects support both shear and pressure waves, in nonviscious fluids there are only pressure waves with longitudinal polarization. The existence of trapped modes in finite-sized objects is due to the fact, that shear waves in solids, which are purely transverse, do not couple to pressure waves in fluids when the polarization vector of shear waves is tangential to the surface of the resonator at every point of the surface. In this case, the surface of the resonator will not shrink or expand, thus creating no vibrations in the fluid [9].

2. Formulation

Consider acoustic resonator, made of isotropic solid material with density ρ, speed of pressure waves c_p and speed of shear waves c_s. Oscillations of displacement field $\mathbf{u}$ is governed by following equation [10]

$$\rho\ddot{\mathbf{u}} = \mathrm{div}\hat{\boldsymbol{\sigma}}, \tag{1}$$

where $\hat{\boldsymbol{\sigma}}$ is the Cauchy stress tensor, which is connected with displacement field via the Hooke's law

$$\sigma_{ij} = \lambda\delta_{ij}\varepsilon_{kk} + 2\mu\varepsilon_{ij}, \tag{2}$$

where $\lambda = \rho(c_p^2 - 2c_s^2)$ and $\mu = c_s^2\rho$ are the Lame parameters and $\hat{\varepsilon} = 1/2\left[\nabla\mathbf{u} + (\nabla\mathbf{u})^{\mathrm{T}}\right]$ is the infinitesimal strain tensor. It is well known fact, that Cauchy stress tensor can be decomposed into spherical and deviatoric parts [11]

$$\hat{\boldsymbol{\sigma}} = \underbrace{1/3\,\mathrm{Tr}\,\hat{\boldsymbol{\sigma}}}_{sph} + \underbrace{\hat{\boldsymbol{\sigma}}'}_{dev}, \tag{3}$$

where spherical part is connected with a change of the volume of the material. Therefore, if one can find such displacement field for which

$$\mathrm{Tr}\,\hat{\boldsymbol{\sigma}} \sim \mathrm{div}\,\mathbf{u} = 0, \quad u_n|_\Sigma = 0, \tag{4}$$

where Σ is the surface of the resonator and u_n is the component of the displacement field, normal to Σ, then one can expect, that such mode will not produce any pressure waves in the fluid outside of the resonator, and therefore will not be coupled to the radiative continuum and will have infinitely high radiative quality factor.

The so-called $\mathbf{M}$ vectorial harmonics have such properties (4) in specific geometries of the resonator, and are defined as follows [12]

$$\mathbf{M} = \mathrm{rot}\,(\mathbf{c}\psi)\,, \tag{5}$$

where ψ is the solution of the Helmholtz equation and $\mathbf{c}$ is the direction vector, which is chosen so, that $\mathbf{M}$ is a solution to the vectorial Helmholtz equation. For example, spherical vectorial harmonic

$$\mathbf{M}^{\mathrm{sph}}_{\ell m}(\mathbf{r},k) = \left[\hat{\mathbf{e}}_\theta \frac{im}{\sin\theta} j_\ell(kr) P_\ell^m(\cos\theta) - \hat{\mathbf{e}}_\varphi j_\ell(kr)\frac{\partial P_\ell^m(\cos\theta)}{\partial\theta}\right] e^{im\varphi}, \tag{6}$$

where $j_\ell(kr)$ is the spherical Bessel function of the ℓ'th order, $P_\ell^m(\cos\theta)$ is the associated Legendre polynomial, $k = \omega/c_s$ is the wavevector and ω is the frequency of oscillations, will possess properties (4) in every resonator with infinite rotational symmetry around at least one axis (D_∞ symmetry group). Indeed, one can show, by solving corresponding eigenmode problem for spherical resonator with radius R, that modes with $\mathbf{u} \sim \mathbf{M}^{\mathrm{sph}}_{\ell m}$, with eigenfrequency equation

$$(1-\ell)j_\ell(\alpha) + \alpha j_{\ell+1}(\alpha) = 0, \tag{7}$$

where $\alpha = kR$, and $j_\ell(\alpha)$ is spherical Bessel function of first kind, are purely decoupled from modes in the freespace and therefore perfectly trapped inside the resonator.

Moreover, since boundary conditions for such modes are

$$\sigma_{nn} = 0, \;\; \sigma_{n\tau} = 0, \;\; u_n = 0, \tag{8}$$

where indices n and τ represent normal and tangential components respectively, eigenmode problem can be solved analytically even in resonators with sharp corners. For example, by solving eigenvalue problem in cylinder with radius R and height h, one can obtain following equations for odd and even modes

$$\underbrace{\begin{cases} J_2(k_\rho R) = 0, \\ \cos(k_z h/2) = 0, \end{cases}}_{\text{odd mode}} \quad \underbrace{\begin{cases} J_2(k_\rho R) = 0, \\ \sin(k_z h/2) = 0, \end{cases}}_{\text{even mode}} \tag{9}$$

which leads to the eigenfrequency equation

$$\omega^2 = c_s^2 \left[\left(\frac{\alpha_n}{R} \right)^2 + \left(\frac{\pi q}{h} \right)^2 \right], \tag{10}$$

where α_n is the n'th root of the Besser function of the second order $J_2(\alpha_n) = 0$, and $q = 0, 1, 2, ...$. In this case, displacement field of trapped modes will be proportional to the cylindrical vectorial harmonics with $m = 0$

$$\mathbf{M}^{\text{cyl}\,e}_{m\ o}(\mathbf{r}, \mathbf{k}) = \left[\hat{\mathbf{e}}_\rho i \frac{m}{\rho} Z_m(k_\rho \rho) - \hat{\mathbf{e}}_\varphi \frac{\partial Z_m(k_\rho \rho)}{\partial \rho} \right] \begin{matrix} \cos(k_z z) \\ \sin(k_z z) \end{matrix} \, e^{im\varphi}, \tag{11}$$

where indices o and e represent odd and even harmonics with respect to z coordinate. Modes with odd $q = 1, 3, 5, ...$ corresponds to odd harmonics, and modes with even $q = 2, 4, 6, ...$ corresponds to even one.

3. Numerical results

To prove existence of trapped modes in such resonators, numerical simulation was used. All numerical simulations were done in COMSOL Multiphysics® [13] in 2D geometry with axial symmetry using eigenfrequency solver. Material parameters of the resonator are the following: density $\rho = 10\,\rho_0$, speed of pressure waves $c_p = 3\,c_0$ and speed of shear waves $c_s = 2\,c_0$, where $\rho_0 = 1.225$ kg/m^3 and $c_0 = 343$ m/s are the density and speed of sound of the surrounding fluid (air). Several different geometries of the resonator were investigated: sphere, hemisphere, cylinder, cylindrical shell and torus.

3.1. Sphere and hemisphere

On the figure 1 (a) one can see the Q-factor of the eigenmodes of the spherical resonator with radius $R = 5$ cm. Some of the eigenmodes have infinitely high Q-factor. One can show, that frequencies of such modes are in perfect match with ones, that can be obtained from the equation (7). Moreover, all trapped modes have $2\ell + 1$ - fold degeneracy, which can be also seen from the equation (7) because it does not depend on the azimuthal number m.

Interestingly, the fundamental mode of such a resonator is not the dipole but the quadrupole. This happens because the time-averaged angular momentum of the sphere should be equal to zero so that the resonator does not rotate as a whole. Since the displacement field of the trapped modes is proportional to $\mathbf{M}^{\text{sph}}_{\ell m}$ vectorial spherical harmonics, in the case, when azimuthal number $m = 0$, u_φ is the only nonzero component. Therefore, from the figure 1 (b), where u_φ component of the dipole, quadrupole, and octupole modes is presented it became obvious that the time-averaged angular momentum of the dipole mode is zero only in the case when the inner layer of the sphere is rotating opposite to external layer, which gives condition, that displacement field should have at least two extremes in the radial direction and therefore, $R\omega/c_s$ should be bigger than the second root of the $j_1(kR)$ spherical Bessel function. Meanwhile, for quadrupole mode and all other modes with $\ell > 1$, the condition on the time-averaged angular momentum is fulfilled automatically because, for these modes, rotation of the upper hemisphere is compensated by the lower hemisphere.

On the figure 1 (c), Q-factor of the hemispherical resonator with radius $R = 5$ cm is presented. One can see the existence of the modes with infinitely high Q-factor, which frequencies coincide with some of the frequencies of trapped modes from the figure 1 (a). The origin of this phenomenon can be understood from the figure 1 (d), where u_φ component of the field of trapped modes is presented. One can see, that the field profile is identical to one for the spherical resonator. However, for the spherical resonator, there are no modes with even ℓ. This happens because for even ℓ existence of the lower hemisphere is an important condition for

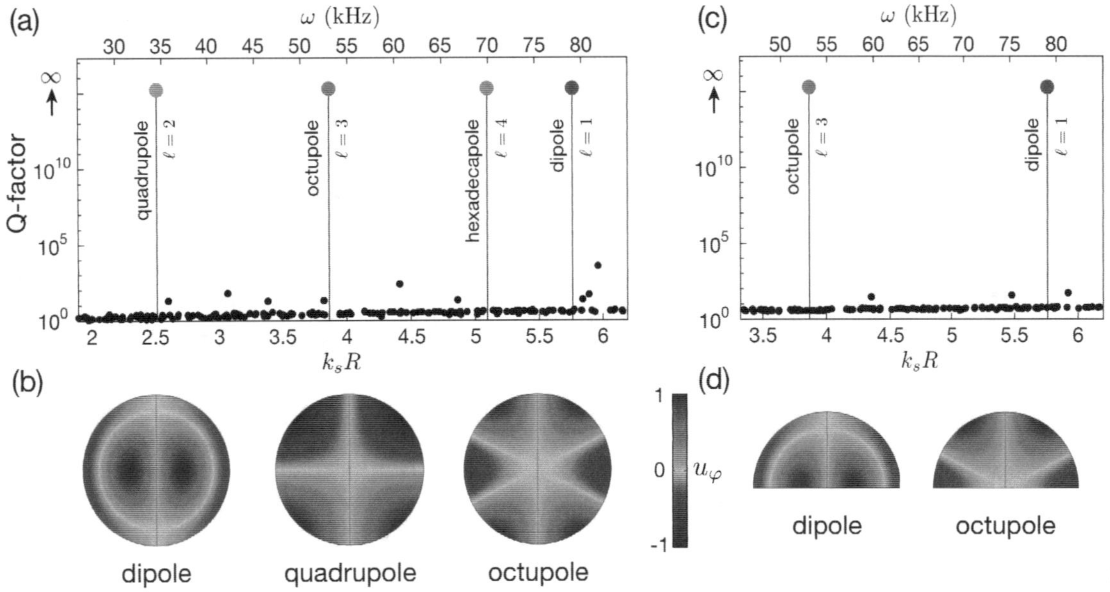

Figure 1. Spherical and hemispherical resonators with radius R. (a) and (b) - quality factor of the modes for spherical and hemispherical resonators. (b) - u_φ component of the displacement field for dipole, quadrupole and octupole modes os the spherical resonator. (c) - u_φ component of the displacement field for dipole and octupole modes of the spherical resonator. $k_s = \omega/c_s$ is the wavevector of the shear waves in the material of the resonators, $c_s = 686$ m/s is the speed of shear waves in the material of the resonator.

zeroing the time-averaged angular momentum. Another difference from the sphere is the fact that for the hemisphere trapped modes are not degenerate. Only modes with the displacement field proportional to $\mathbf{M}_{\ell 0}^{\text{sph}}$ vectorial spherical harmonics are trapped. The reason for this lies in the lower symmetry of the resonator compared to the spherical one. Another consequence of the above fact is that the displacement field has only u_φ component not equal to zero. This property holds not only for the trapped modes in the spherical and hemispherical resonator but for trapped modes in an arbitrary-shaped resonator with infinite rotational symmetry.

3.2. Cylinder

On the figure 2 (a) one can see Q-factor of the modes of the cylindrical resonator with radius $R = 2.5$ cm and height $h = 6$ cm. As in the case of the hemisphere, only modes which can be expanded in terms of $\mathbf{M}_{\ell 0}^{\text{sph}}$ vectorial spherical harmonics are trapped. On the field profiles presented on the figure 2 (b) one can see existence of the nodal lines along radial and z directions, which are due to the conditions on k_ϱ and k_z components of the wavevector (equation 9). Important to mention that there are no modes with $k_\varrho = 0$ ($\alpha_n = 0$), since in these cases $\mathbf{M}_0^{\text{cyl}} = 0$.

3.3. Cylindrical shell and torus

Like resonators of a simple form, modes can exist in more complex-shaped resonators like cylindrical shell and torus, Q-factors of whose modes are represented on the figures 2 (c) and 2 (e). The outer and inner radii of the shell and torus are equal $a = 2.5$ cm and $b = 0.1$ cm. The height of the shell is similar to the one for the cylindrical resonator. Interestingly, some modes of these resonators look similar to the ones in the cylindrical resonator. Example of such

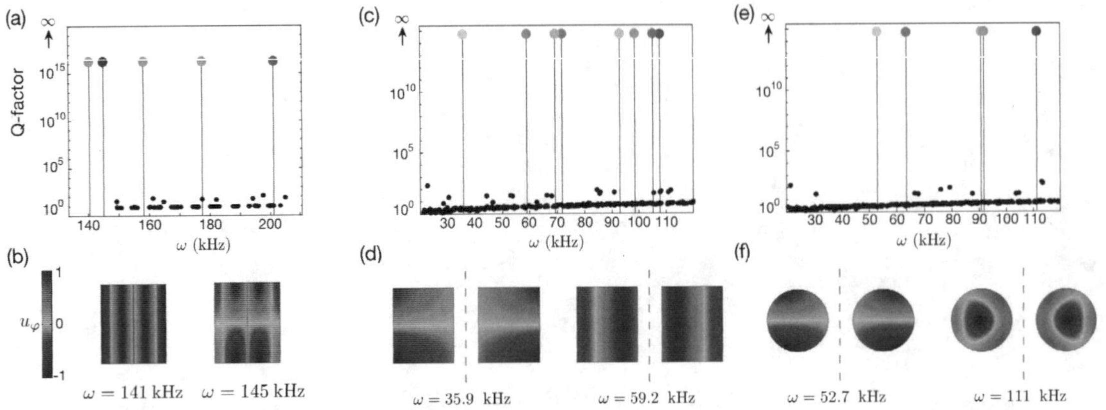

Figure 2. Cylindrical, cylindrical shell, and torus resonators. (a), (c), (e) - quality factor of the modes of the cylindrical, cylindrical shell, and torus resonators respectively. (b), (d), (f) - u_φ component of the displacement field of some of the eigenmodes of the cylindrical, cylindrical shell, and torus resonator respectively. Eigenfrequencies of the presented modes are written below the field profiles. The dotted red line represent the symmetry axis of the resonator.

modes are presented on the figures 2 (b), (d), mode with $\omega = 35.9$ kHz, and (f), mode with $\omega = 52.7$ kHz. One can see that if the width and height of the shell resonator are equal to the width and height of the cylindrical one, the eigenfrequencies of these modes are equal.

4. Conclusions

In this proceeding, we have numerically proved the existence of trapped modes in solid acoustic resonators merged in a nonviscious fluid. Such states are possible due to the polarization mismatch of the shear waves in the resonator with the pressure waves in the surrounding media. Polarization mismatch is provided by the tangential nature of the $\mathbf{M}_{\ell 0}^{\mathrm{sph}}$ vectorial spherical harmonics to the surfaces with axial symmetry, therefore, axial symmetry of the resonator is one of the main conditions of existence of such modes. We believe that our findings can be applied to many fields of acoustic engineering including high-Q acoustic metasurfaces, acoustic sensing devices, and acoustic nanoantennas.

Acknowledgments

The authors thank K. Frizyuk, K. Koshelev, and Yu. Kivshar for useful discussions and suggestions. The work is support by the Russian Science Foundation (project #20-72-10141). A. B. acknowledges a support from the "BASIS" Foundation.

References

[1] Chia Wei Hsu, Bo Zhen, A Douglas Stone, John D Joannopoulos, and Marin Soljačić. Bound states in the continuum. *Nature Reviews Materials*, 1(9):1–13, 2016.
[2] Kirill Koshelev, Andrey Bogdanov, and Yuri Kivshar. Meta-optics and bound states in the continuum. *Science Bulletin*, 64(12):836–842, 2019.
[3] D.V. Evans. Trapped acoustic modes. *IMA journal of applied mathematics*, 49(1):45–60, 1992.
[4] D.V. Evans, M. Levitin, and D. Vassiliev. Existence theorems for trapped modes. *J Fluid Mech*, 261:21–31, 1994.
[5] Lucas Chesnel and Vincent Pagneux. Simple examples of perfectly invisible and trapped modes in waveguides. *The Quarterly Journal of Mechanics and Applied Mathematics*, 71(3):297–315, 2018.
[6] F. Ursell. Trapping modes in the theory of surface waves. In *Proc. Camb. Phil. Soc*, volume 47, pages 347–358, 1951.

[7] M. Callan, C.M. Linton, and D.V. Evans. Trapped modes in two-dimensional waveguides. *Journal of Fluid Mechanics*, 229:51–64, 1991.

[8] E.B. Davies and L. Parnovski. Trapped modes in acoustic waveguides. *The Quarterly Journal of Mechanics and Applied Mathematics*, 51(3):477–492, 1998.

[9] Ilya Deriy, Ivan Toftul, Mihail Petrov, and Andrey Bogdanov. Bound states in the continuum in compact acoustic resonators. *arXiv preprint arXiv:2104.05539*, 2021.

[10] L. D. Landau and E. M. Lifshitz. *Theory of elasticity.* Pergamon, 1959.

[11] Robert Asaro and Vlado Lubarda. *Mechanics of solids and materials.* Cambridge University Press, 2006.

[12] Craig F Bohren and Donald R Huffman. *Absorption and scattering of light by small particles.* John Wiley & Sons, 2008.

[13] Comsol multiphysics® v. 5.5. www.comsol.com. comsol ab, stockholm, sweden.

Finding exceptional points in realistic systems using full-wave simulations

Alexey A Dmitriev[1] and Mikhail V Rybin[1,2]

Department of Physics and Engineering, ITMO University, St Petersburg 197101, Russia
Ioffe Institute, St Petersburg 194021, Russia

E-mail: alexey.dmitriev@metalab.ifmo.ru

Abstract. Here, we present an approach to finding exceptional points using the finite-element method. Using this method, we demonstrated exceptional points in 2D dimers of infinite cylinders and infinite parallelepipeds. The results agree well with the analytical coupled-dipole model, however a deviation due to the contribution of higher multipoles, is present. Our approach can be applied to three-dimensional particles as well.

1. Introduction

Adding gain and loss to optical systems leads to many counterintuitive effects, such as degeneration of field distributions of two or more optical modes into one at an exceptional point (EP) [1]. This leads to unique effects, such as unidirectional reflectionless propagation, asymmetric backscattering and directional emission, non-reciprocal light transmission and many others (see, e.g. [1], [2] and references therein). Moreover, the unique nature of EP degeneracy lifting found its application in sensing [3]. In optical systems, EPs have so far been intensively studied in systems larger than wavelength, such as waveguides [4], resonators supporting whispering gallery modes [5, 3] and layered structures [2]. Due to the high Q-factor of these structures, their optical modes are coupled through evanescent fields, which allows to treat such

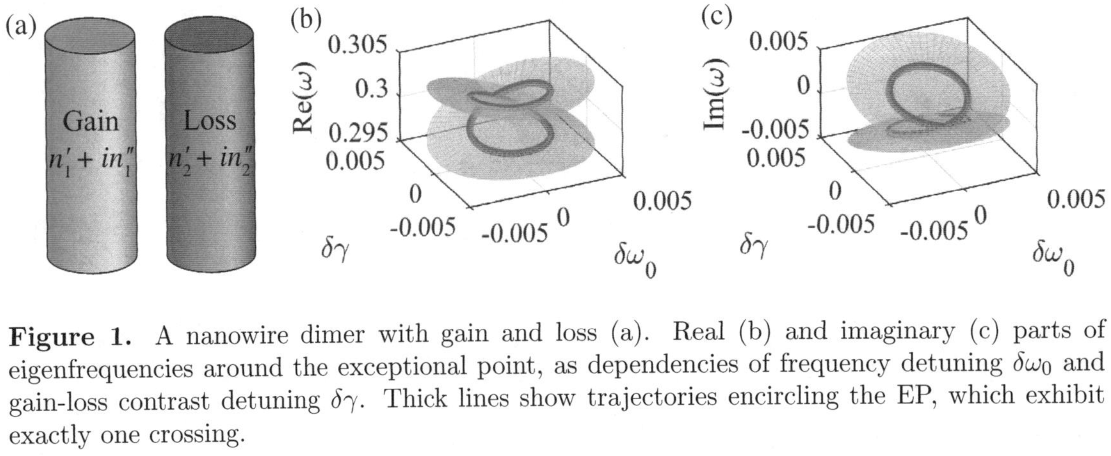

Figure 1. A nanowire dimer with gain and loss (a). Real (b) and imaginary (c) parts of eigenfrequencies around the exceptional point, as dependencies of frequency detuning $\delta\omega_0$ and gain-loss contrast detuning $\delta\gamma$. Thick lines show trajectories encircling the EP, which exhibit exactly one crossing.

systems with a simple coupled-mode theory [3]. Microparticles, in contrast, have very large radiative decay that leads to a mixed coupling of the eigenmodes consisting of an evanescent and radiative parts, which severely complicates the analysis [6, 7]. Nevertheless, realizing the exceptional point regime in a subwavelength structure or a particle is promising for creation of metasurfaces with unique properties.

Full-wave simulations are the most convenient method of finding the eigenmodes in a structure consisting of complex shapes which cannot be treated analytically, such as finite cylinders, pyramids, cubes or droplets. However, at EP the field distributions of two or more modes become degenerate, as do the corresponding frequencies. This makes a simple eigenmode analysis relying on full-wave simulations unapplicable to finding the system parameters corresponding to the EP: even if we somehow guess the right parameter values, a full-wave simulation without proper modifications would only find one mode without any proof that it is degenerated.

Here, we propose an algorithm relying on eigenmode analysis using finite-element method (FEM), that allows to find the exceptional point in a 2D space of system parameters (e.g., material gain/loss and distance between the resonators) assuming that a single second-order EP exists in the range of interest. We demonstrate our results on 2D dimers consisting of infinite cylinders and parallelepipeds, and compare the results to an analytic model.

2. Exceptional point search with full-wave simulations

To find the exceptional point, we utilize the specific topology that eigenfrequency sheets display in the parametric space around the EP, which is shown in Fig. 1. We introduce the following

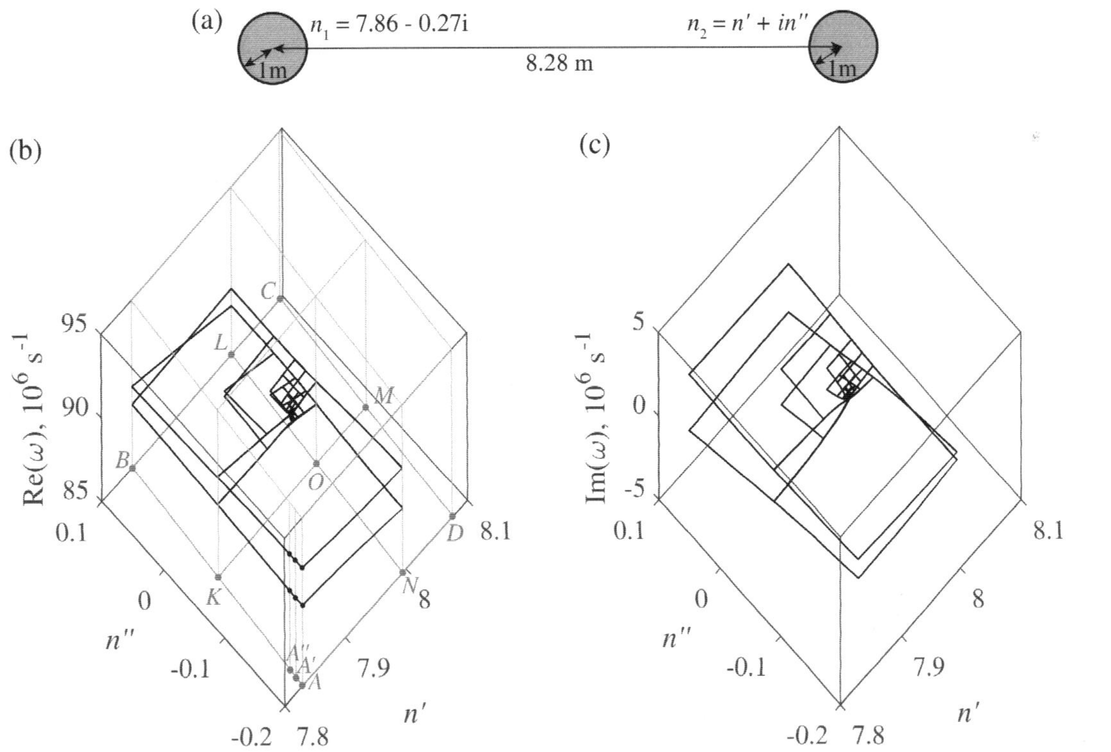

Figure 2. A dimer of infinite cylinders with gain and loss (a). Real (b) and imaginary (c) parts of eigenfrequencies around the EP, as dependencies of real and imaginary parts of the right cylinder.

crossing number criterion. Upon encircling EP twice, both the real and the imaginary part display a single crossing, as shown in Fig. 1, or possibly an odd number of crossings. If there is an even number of crossings, or no crossings at all, there is no EP inside the circle.

Using this criterion, we implement a recursive algorithm to enclose EP within the desired accuracy. First, we select a point A in the parametric space and find the two eigenmodes which are supposed to coalesce, and the corresponding eigenfrequencies. Then we choose an initial grid cell $ABCD$, which contains the exceptional point. To achieve that, we encircle the cell twice, and if the crossing number criterion indicates the abscence of EP, we extend the cell. Afterwards, the initial cell is split into four subcells $AKON$, $KBLO$, $OLCM$ and $NOMD$, as shown in Fig. 2. Then we encircle each of the subcells twice and use the crossing number criterion to check which of the subcells contains the EP. Then we take this subcell as the initial cell and continue the procedure recursively until the desired accuracy is met.

Let us now consider the EP encircling procedure in more detail. The eigenfrequencies $(\omega_{1A}, \omega_{2A})$ at the start point A in the parametric space are already known from the initial the eigenmode analysis. At a point A' which lies close to the point A, we use FEM to compute a single eigenmode using a known shift value for the eigenfrequency, searching for the mode with the eigenfrequency closest to the shift value. Using ω_{1A} as the shift value produces the new eigenfrequency $\omega_{1A'}$, and using ω_{2A} produces $\omega_{2A'}$. The set $(\omega_{1A'}, \omega_{2A'})$, which corresponds to the point A', can in turn be used to find the eigenfrequencies at A'' and so on. Provided that the points $A, A', A'', \ldots$ lie close enough to each other, the frequencies $\omega_{1A}, \omega_{1A'}, \omega_{1A''}, \ldots$ all correspond to mode 1, and the frequencies $\omega_{2A}, \omega_{2A'}, \omega_{2A''}, \ldots$ all correspond to mode 2. Our studies showed that splitting each of the segments AB, BC, CD, AD into 50 points was sufficient to prevent the modes from mixing up.

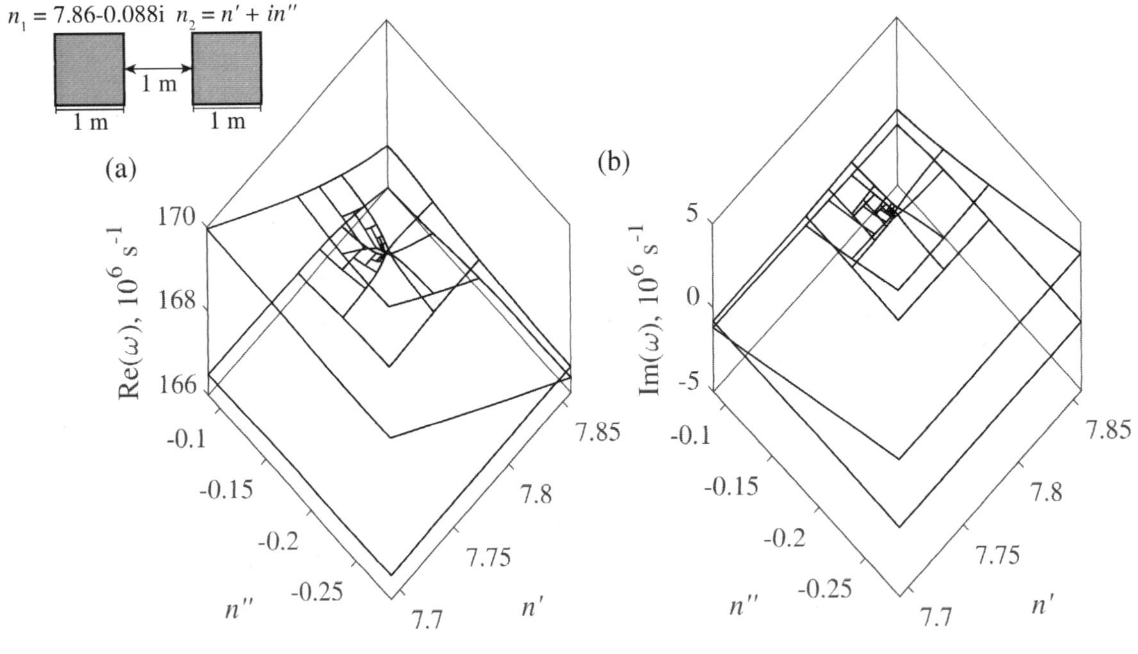

Figure 3. Real (a) and imaginary (b) parts of eigenfrequencies of a dimer of parallelepipeds (shown in the inset) around the EP, as dependencies of real and imaginary parts of the right parallelepiped.

3. Results

First, we consider a dimer of infinite cylinders shown in Fig. 2a which can be treated analytically. In our previous work [7], we have shown using the dipole approximation that EP can be achieved at $\omega_{EP} = 89.9 \cdot 10^6$ s^{-1} in a dimer made of cylinders with radius 1 m, center-to-center distance 8.28 m and refractive indices $n_1 = 7.86 - 0.27i$ and $n_2 = 7.86 - 0.088i$ in TE polarization. We set $n_2 = n' + in''$, where n' and n'' become the varied system parameters, i.e., form the 2D parameter space. Using $n_2 = 7.76 - 0.19i$ as the start point, we utilize the recursive algorithm with the exit condition $|\omega_1 - \omega_2| < 10^5$ s^{-1}. Fig 2b,c show the real and imaginary parts of eigenfrequencies ω_1 and ω_2 obtained by our algorithm. A crossing-to-anticrossing transition can be observed as we go from the upper left to the lower right quadrant through the exceptional point at $n_2 = 7.97 - 0.044i$. The obtained value for n_2 is rather close to the analytical result $n_2 = 7.86 - 0.088i$. The EP eigenfrequency $\omega_{EP} = (89.4 + 0.3) \cdot 10^6$ s^{-1} obtained with FEM has a small imaginary part, as opposed to the purely real value predicted by the coupled-dipole model. We attribute the observed deviation to the contribution of higher multipoles.

Now let us consider a dimer of infinite parallelepipeds with a 1 m square cross-section separated by an 1 m gap, as shown in the inset of Fig. 3. We choose TE polarization and set $n_1 = 7.86 - 0.088i$ and $n_2 = n' + in''$, where n' and n'' are the varied parameters. Fig 3a,b show the real and imaginary parts of eigenfrequencies obtained by our algorithm for this dimer. As in the previous case, a crossing-to-anticrossing transition can be observed as we go from the upper left to the lower right quadrant through the exceptional point at $n_2 = 7.81 - 0.14i$. The EP eigenfrequency is $\omega_{EP} = [167.0 - 0.3i] \cdot 10^6$ s^{-1}.

4. Conclusion

We have developed an approach to finding the system parameters corresponding to an exceptional point using the finite-element method. Using this method, we demonstrated exceptional points in 2D dimers of infinite cylinders and infinite parallelepipeds. The results for infinite cylinders agree well with the analytical coupled-dipole model, however a deviation due to the contribution of higher multipoles, is present. In this work we considered only two-dimensional resonators for simplicity, however our approach can be applied without any modification to three-dimensional particles as well.

Acknowledgements

We acknowledge the support from the Russian Science Foundation (grant 21-79-10190).

References

[1] Ş. K. Özdemir, S. Rotter, F. Nori, and L. Yang. Parity-time symmetry and exceptional points in photonics. *Nature Materials*, 18(8):783–798, apr 2019.

[2] Ramy El-Ganainy, Konstantinos G. Makris, Mercedeh Khajavikhan, Ziad H. Musslimani, Stefan Rotter, and Demetrios N. Christodoulides. Non-hermitian physics and PT symmetry. *Nature Physics*, 14(1):11–19, jan 2018.

[3] Hossein Hodaei, Absar U. Hassan, Steffen Wittek, Hipolito Garcia-Gracia, Ramy El-Ganainy, Demetrios N. Christodoulides, and Mercedeh Khajavikhan. Enhanced sensitivity at higher-order exceptional points. *Nature*, 548(7666):187–191, aug 2017.

[4] Mohammad-Ali Miri, Michele Cotrufo, and Andrea Alù. Anomalous optical forces in PT-symmetric waveguides. *Optics Letters*, 44(14):3558, jul 2019.

[5] Bo Peng, Şahin Kaya Özdemir, Fuchuan Lei, Faraz Monifi, Mariagiovanna Gianfreda, Gui Lu Long, Shanhui Fan, Franco Nori, Carl M. Bender, and Lan Yang. Parity-time-symmetric whispering-gallery microcavities. *Nature Physics*, 10(5):394–398, apr 2014.

[6] Alexey A Dmitriev and Mikhail V Rybin. Combining isolated scatterers into a dimer by strong optical coupling. *Physical Review A*, 99(6):063837, 2019.

[7] Alexey Dmitriev and Mikhail Rybin. Exceptional point and parity-time symmetry on dipole Mie resonances in dimer. *AIP Conference Proceedings*, 2300(1):020022, 2020.

Pump-probe spectroscopy in gold-garnet magnetoplasmonic metasurfaces

T V Dolgova[1], M A Kiryanov[1], P K Nurgalieva[1], G S Ostanin[1], A I Musorin[1], H Uchida[2], M Inoue[2], A A Fedyanin[1]

[1] Faculty of Physics, Lomonosov Moscow State University, 119991 Moscow, Russia

[2] Department of Electrical and Electronic Information Engineering, Toyohashi University of Technology, 1-1 Tempaku-cho, Toyohashi, Aichi 441-8580, Japan

dolgova@nanolab.phys.msu.ru

Abstract. Ultrafast modulation of optical response is realized in a multi-resonant magnetoplasmonic metasurface under resonant femtosecond pump. A saturation of resonant probe transient transmission is shown at dielectric quazi-waveguide mode.

1. Introduction

Active plasmonics and magnetoplasmonics rely on the resonance modes for optical response control by dielectric function tuning via laser-induced hot-carrier generation and external magnetic field [1-4]. Multi-resonance systems open a new possibility to handle the ultrafast response using resonances of different nature and their coupling. Thus, both plasmonic and high-quality-factor polaritonic magnetooptical resonances tunable by the azimuthal angle were realized in 2D hybrid metal-dielectric magnetoplasmonic metasurfaces [5]. Resonant femtosecond probe is sensitive to the ultrafast heating by non-resonant pump. The ultrafast modulation of surface-plasmon resonant transverse magneto-optical Kerr effect was shown in one-dimensional nickel magnetoplasmonic crystals [6].

In this paper a saturation of a resonant probe transient transmission is shown in a magnetoplasmonic metasurface under resonant femtosecond pump. The effect is observed within the quasi-waveguide dielectric mode.

2. Experimental

The metasurface is an array of gold nanospheres with a diameter of 100 nanometers and a period of 600 nanometers on a quartz substrate. The array is covered with a 100-nanometer layer of bismuth-substituted yttrium-iron garnet. Gold nanospheres as meta-atoms provide electromagnetic field concentration at electric dipole plasmonic resonance at 810 nm for normal incidence (pump region), and that at 735 nm with quadrupole resonance at 620 nm for the angle of incidence of 20° (probe region). The peculiarity of such structures is the possibility of excitation of the quasi-waveguide mode at 530 nm. The detailed sample description and scanning electron microscopy image can be found in [5]. The ultrafast optical responses of the magnetoplasmonic metasurface are measured using the pump-probe technique. A regeneratively amplified Ti:sapphire laser with 800 nm central wavelength, 50 fs pulse duration, and 1 kHz repetition rate is a source of laser radiation. The pulse is divided into a pump and a probe pulses with controllable delay. The 0.6 mJ/cm^2 pump pulse is directed normally to the studied sample. The probe line generates supercontinuum pulse covering the range from 450 to 750 nanometers with a total power density not more than 1 nJ/cm^2. The angle of probe pulse incidence to the sample is of 18 degrees. The spot sizes of the pump and probe pulses on the sample surface are of 800 and 400 μm, respectively.

To determine the ultrafast optical response, a differential transmittance coefficient ΔT/T was introduced, that is the normalized difference between the transmittance in the presence and in the absence of a pump pulse. Fig. 1 shows the spectrum of the differential transmittance of the supercontinuum for the 500-femtosecond delay of the probe.

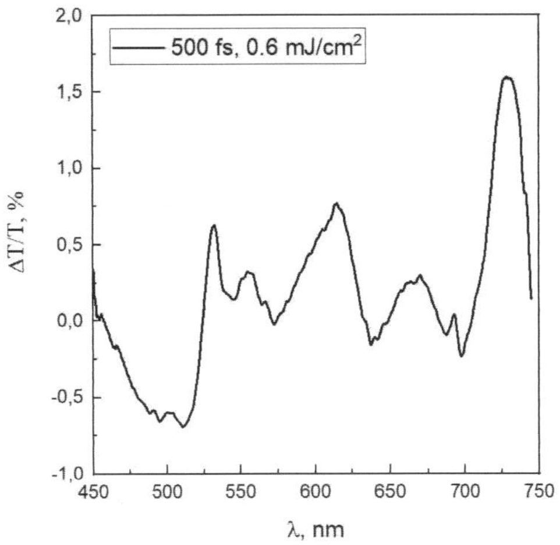

Fig 1. ΔT/T spectrum for 500 fs time delay and 0.6 mJ/cm^2 pump fluence

Maximum of differential reflectance is achieved at 730 nm associated with the excitation of surface lattice resonance. This resonance appears due to the spectral overlap of dipole surface plasmon and the Rayleigh anomaly for (-2, +1) diffraction order in the Bi:YIG layer. Maxima of differential reflectance at 615 nm and 670 nm wavelength are related to the overlap of quadrupole surface plasmon and the fading of diffraction orders in quartz (+1,0) and Bi:YIG (+2,0). The peaks at 530 nm and 555 nm wavelength are caused by waveguide propagation of (0,+-2) diffraction orders in the Bi:YIG layer. The reason of the dip at the wavelength of 500 nm is the maximum of the increase of imaginary part of the dielectric constant due to the interband transition in gold. The ultrafast modulation of the optical response of a magnetoplasmonic metasurface is associated with the interaction of the pump pulse resonant to the dipole mode with the electron and phonon systems of the sample. The presence of the resonances associated with the electronic systems of gold nanospheres and yttrium-iron garnet allows us to separate the modulation of the optical response of metal nanospheres and the dielectric layer.

The dependence of the maximum amplitude at the wavelength of 527 nanometers on the intensity of the pump pulse is not linear starting from 0.3 mJ/cm^2 (Fig.2). Saturation of the effect is observed for higher fluencies. The differential transmittance is linear with pump fluence at the other resonant features associated with the gold nanospheres. The saturation is a consequence of the nonlinear interaction of a femtosecond pulse with an yttrium-iron garnet. A saturation effect reduces during first picoseconds and disappears after electron-phonon relaxation.

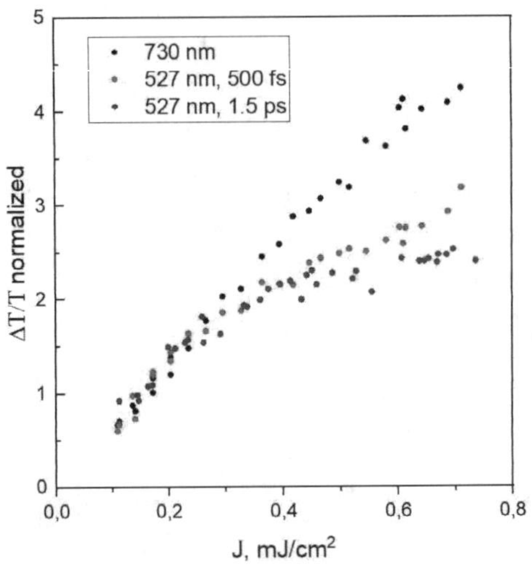

Fig 2. ΔT/T fluence dependence at 730 nm wavelength and at 527 nm wavelength for 500 fs and 1.5 ps time delay

3. Conclusions

The effect of transient transmission saturation is probed by white-light supercontinuum at the dielectric quazi-waveguide mode of an yttrium-iron garnet magnetoplasmonic metasurface. The saturation effect fades out with electron-phonon relaxation. The differential transmittance for the localized nanosphere plasmon modes shows linear behavior for the same pump fluence range.

4. Acknowledgements

T.V.D. thanks Russian Foundation for Basic Research (grant №20-02-00758, experiment). M.A.I. thanks Russian Science Foundation (grant №19-72-00168, calculations). K.M.A. thanks MSU Quantum Technology Centre for the support. This research was performed according to the Development program of the Interdisciplinary Scientific and Educational School of Lomonosov Moscow State University "Photonic and Quantum technologies. Digital medicine".

References

[1] Bossini, D. Belotelov, V. I. Zvezdin, A. K.; Kalish, A. N.; Kimel, A. V. Magnetoplasmonics and Femtosecond Optomagnetism at the Nanoscale. ACS Photonics 2016, 3, 1385.

[2] Pohl, M., Belotelov, V. I., Akimov, I. A., Kasture, S., Vengurlekar, A. S., Gopal, A. V., Zvezdin, A. K., Yakovlev, D. R., Bayer, M. Plasmonic crystals for ultrafast nanophotonics: Optical switching of surface plasmon polaritons. Phys. Rev. B: Condens. Matter Mater. Phys. 2012, 85, No. 081401.

[3] N. Rotenberg, J. Caspers, H. M. van Driel, Tunable ultrafast control of plasmonic coupling to gold films, Phys. Rev. B 80 (2009).

[4] V.V. Temnov, Ultrafast acousto-magneto-plasmonics, Nat. Photon. 6, 728 (2012).

[5] Musorin, A. I.; Chetvertukhin, A. V.; Dolgova, T. V.; Uchida, H.; Inoue, M.; Luk'yanchuk, B. S.; Fedyanin, A. A. Tunable multimodal magnetoplasmonic metasurfaces. Appl. Phys. Lett. 2019, 115, 151102.

[6] I.A. Novikov, M.A. Kiryanov, P.K. Nurgalieva, A.Yu. Frolov, V.V. Popov, T.V. Dolgova, and A.A. Fedyanin. Ultrafast Magneto-Optics in Nickel Magnetoplasmonic Crystals. Nano Lett., 2020, 20, 12, 2606.

[7] G.V. Hartland, Optical studies of dynamics in noble metal nanostructures, Chem. Rev. 111, 3858 (2011).

Spontaneous symmetry breaking in lasers with periodically modulated gain and refractive index

D Dolinina[1], K Staliunas[2,3,4] and A Yulin[1]

[1] ITMO University, Department of Physics and Engineering, Saint-Petersburg 197101, Russia
[2] Vilnius University, Faculty of Physics, Laser Research Center, Sauletekio Ave. 10, Vilnius, Lithuania
[3] ICREA, Passeig Lluís Companys 23, 08010, Barcelona, Spain
[4] UPC, Dep. de Fisica, Rambla Sant Nebridi 22, 08222, Terrassa (Barcelona) Spain

E-mail: d.dolinina@metalab.ifmo.ru

Abstract. The dynamics of light in active optical systems with periodic complex potential is considered using coupled modes approach where the field is approximated by two counter propagating waves. It is demonstrated that shifting the position of the imaginary part of the potential (effective gain) with respect to the real part of the potential (variation of the refractive index) one can control the effective gain/losses seen by the upper and the power modes. This effect can be used to control the radiation from the laser. The effect of the Kerr nonlinearity is also considered and it is shown that this can result in spontaneous symmetry breaking leading to the formation of the hybrid nonlinear states.

1. Introduction

Laser structures with periodically modulated refractive index, also referred as "distributed feedback" lasers (or "DFB" lasers), gain great attention over the past few decades [1–3]. The index grating induces rescattering of counterpropagating waves on index lattice and provides optical feedback for the laser. In such systems the working mode can be represented as two resonantly coupled counter propagating waves. At the end of the Brillouin zone the group velocity becomes equal to zero and thus this mode has the lowest radiative losses. This facilitate the selection of the modes and helps to achieve single-mode generation regime. Besides the index grating the periodical gain modulation also can couple counterpropagating waves and provide feedback [4, 5]. The combined index and gain coupling also was studied in details [6, 7] but very few works consider the effect of finite shift of the periodic modulations of the gain and the refractive index [8, 9] and, to the best of our knowledge, nonlinear effects have not been taken in account in such two-gratings laser systems. In the present paper we consider the bifurcations of the stationary states forming in the system in the absence and in the presence of the Kerr nonlinearity.

In this work we consider nonlinear active waveguide with periodically changed refractive index and periodically modulated effective gain. We would like to mention that to obtain periodically modulated gain one can use spatially uniform pump and periodic linear losses. The focus point of this paper is the symmetry breaking and the formation of the hybrid stationary states with dominating direction of the energy flow. It has been shown earlier that the similar can occur in nonlinear driven-dissipative systems where the coupling results not only in the dispersion

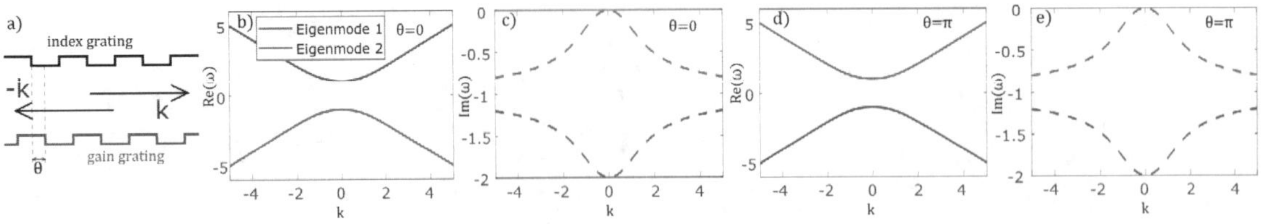

Figure 1. (a) Schematic view of a waveguide with periodical index and gain gratings. The gratings have the same lattice constant, but can have nonzero relative phase θ defined in the text. (b) Real part of dispersion characteristics with $\theta = 0$. (c) Imaginary part of dispersion characteristics with $\theta = 0$. (d) Real part of dispersion characteristics with $\theta = \pi$. (e) Imaginary part of dispersion characteristics with $\theta = \pi$.

characteristics splitting but also affects the dissipation rate of the modes so that the modes have very different Q factors [10–12]. Such states are characterized by nonzero energy flux at one of the directions of the waveguide. Here we consider the aforementioned phenomena in active optical system and consider how the relative position of the real and imaginary parts of the periodic potential affect the dynamics and the symmetry breaking bifurcation.

2. Considered system and its mathematical model

The dynamics of the system schematically shown in Fig.1(a) is described by two counter-propagating waves approach and can be expresses mathematically by following system of equations:

$$\begin{aligned}(\partial_t + \partial_x)U_+ &= (i\alpha - \gamma)(|U_+|^2 + 2|U_-|^2)U_+ + (i\sigma + \Gamma)U_- + PU_+,\\ (\partial_t - \partial_x)U_- &= (i\alpha - \gamma)(|U_-|^2 + 2|U_+|^2)U_- + (i\sigma^* + \Gamma)U_+ + PU_-,\end{aligned} \tag{1}$$

where U_+ and U_- are the slow varying complex amplitudes of the two counter-propagating waves, γ is the nonlinear losses, $\alpha = \pm 1$ is the Kerr-nonlinearity coefficient, σ is the coupling coefficient of the counter-propagating waves caused by index grating, Γ is the gain coupling caused by gain grating and P is gain in the system. It is worth noting here, that σ can be complex $\sigma = |\sigma|e^{i\theta}$, where θ is phase difference between the gratings defined as $\theta = \kappa\Delta x$, κ is the lattice constant and Δx is the shift of the real part of the periodic potential in respect to imaginary one.

The frequency of the linear guided modes in the system is given by the following dispersion characteristics:

$$\omega_{1,2} = -iP \pm \sqrt{k^2 + (\sigma - i\Gamma)(\sigma^* - i\Gamma)}. \tag{2}$$

The real part of ω is responsible for oscillatory dynamics of the modes and the imaginary part defines the rate of exponential decay or growth of the modes. From the Fig.1(b-e) it is clearly seen that one of the modes has greater gain comparing to the second one at k near zero. It is important to note here, that if gain grating has zero gain in its minima then one of the eigenmodes has zero effective gain, as it can be seen in Fig.1(c,e).

As it is clearly seen from Fig.1(b,c) the upper mode has higher gain, but situation changes if gratings are shifted on $\theta = \pi$. In this case the upper mode has lower gain comparing to the upper one, see Fig.1(d,e). A physical explanation of such difference is quite simple: if maxima of standing wave coincide with maxima of gain grating it experiences significantly higher gain comparing to the standing wave in gain grating minima. Therefore, it is possible to provide more gain for preferable mode only by changing the relative phase of the gratings.

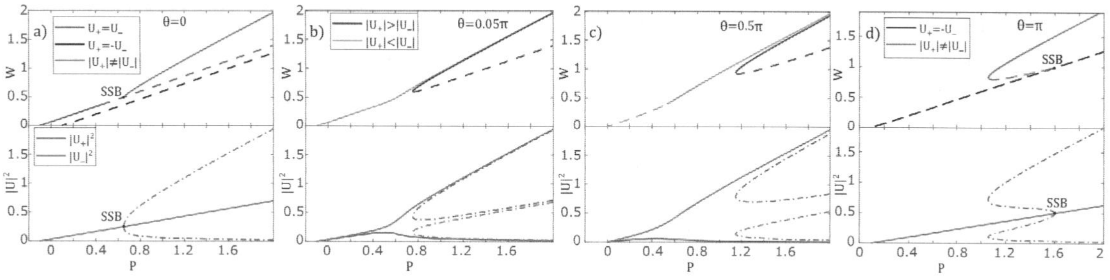

Figure 2. Bifurcation diagrams for spatially uniform states. Upper panels show dependence of the full field intensity of states on gain $W = |U_b|^2 + |U_d|^2 = |U_+|^2 + |U_-|^2$; dashed lines indicate dynamically unstable states. Lower panels demonstrate intensities of conterpropagating waves; by solid and dash-dotted different branches from corresponding upper panel are shown. (a) Upper panel shows all kinds of the uniform states with $\theta = 0$. The lower panel demonstrates field structure in terms of counterpropagating waves for symmetric and hybrid states (solid and dash-dotted lines, respectively). (b) Upper panel shows destruction of symmetry breaking bifurcation from panel (a) with $\theta = 0.05\pi$. The lower panel shows $|U_\pm|^2$ for branches from the upper panel. (c) The same as in (b), but $\theta = \pi/2$. (d) Restoring of symmetry breaking bifurcation at $\theta = \pi$. Other parameters are: $\alpha = -1$, $\gamma = 1$, $\Gamma = 0.1$, $|\sigma| = 0.3$.

We start analyzing our full system with the simplest case of real σ. When σ is real it is convenient to introduce new basis of symmetric $U_s = \frac{U_+ + U_-}{\sqrt{2}}$ and antisymmetric $U_a = \frac{U_+ - U_-}{\sqrt{2}}$ modes. In terms of variables U_s and U_a the equations have the following form:

$$(\partial_t + (\gamma - i\alpha)K - i\sigma)U_s + (\partial_x + (i\alpha - \gamma)M)U_a = (P + \Gamma)U_s \tag{3}$$

$$(\partial_t + (\gamma - i\alpha)K + i\sigma)U_a + (\partial_x + (i\alpha - \gamma)M)U_s = (P - \Gamma)U_a, \tag{4}$$

where $K = \frac{3}{2}(|U_s|^2 + |U_a|^2)$, $M = Re(U_s U_a^*)$. If the field does not depend on coordinate, the stationary solution can be found in the form $\vec{U_s} = (U_s \neq 0, U_a = 0)^T$ and $\vec{U_a} = (U_s = 0, U_a \neq 0)^T$. It can be clearly seen that the second mode has lower gain $(P - \Gamma)$ comparing to the first mode and if $P = \Gamma$ it has no gain. Such low-gain mode in linear case corresponds to the lowest dispersion branch with $\theta = 0$ and the highest one with $\theta = \pi$. In nonlinear case one more type of solutions can exist with both modes not equal to zero, such solutions will be referred as hybrid states.

3. Stationary spatially uniform states

We numerically found the full set of the stationary solutions, the bifurcation diagram showing intensity ($W = |U_b|^2 + |U_d|^2 = |U_+|^2 + |U_-|^2$) dependence on gain P for nonlinear symmetric, antisymmetric and hybrid states is shown in upper panel of Fig.2(a) by blue, black and green lines respectively. Let us note here that symmetric state exist even for negative gain because it has full gain $P+\Gamma$ while the antisymmetric state has a threshold because its effective gain is $P-\Gamma$. As it can be seen from Fig.2(a) at some point the hybrid state branches off from the symmetric state. This new mode can be interpreted as parametric excitation of the antisymmetric component. It is important to note here that because this new state appears as a result of spontaneous symmetry breaking it has non-equal amplitudes of counter-propagating waves $|U_+| \neq |U_-|$, as it can be seen in lower panel of Fig.2(a), and such states have nonzero energy flow. Since the energy flux can be directed either ot the right or to the left, the bifurcation branch of the hybrid states is double degenerated.

The hybrid states appear as a result of the pitchfork bifurcation and the points where such bifurcation takes place is easy to define. The symmetric and antisymmetric states become parametrically unstable at following values of gain:

$$P^{\pm}_{SSB} = \frac{\mp(2\Gamma^2\gamma + \alpha\Gamma\sigma + 3\gamma\sigma^2)}{\Gamma\gamma - \alpha\Gamma}, \tag{5}$$

where P^{+}_{SSB} and P^{-}_{SSB} are bifurcation points for symmetric and antisymmetric states respectively. From this expression one can verify that regardless of zero or π phase difference between the gratings and of Kerr nonlinearity sign spontaneous breaking bifurcation can take place both on symmetric or antisymmetric states.

The system dynamics becomes richer when σ is complex. In terms of the counterpropagating waves the corresponding linear eigenmodes have the form $\vec{U}_{1,2} = (\pm\sqrt{\sigma - i\Gamma}, \sqrt{\sigma^* - i\Gamma})^T$, which shows that the symmetry gets broken. Because of such symmetry breaking caused by the phase shift between the gratings the double degenerated bifurcation branches of stationary states split in two, as it can be seen in Fig.2(b). The symmetric state with nonzero phase shift θ becomes slightly asymmetric and spontaneous symmetry breaking bifurcation transforms into a fold bifurcation. This splitting becomes more significant as θ grows to $\pi/2$, see Fig.2(c). The bifurcation branches become degenerated again, as phase shift approaches $\theta = \pi$ and eigenmodes become symmetric what results in restoring of symmetry breaking bifurcation, as it can be seen in Fig2(d). It is important to note, that here we considered appearance of hybrid modes from the upper dispersion branch with higher gain at $\theta = 0$, but as $\theta \to \pi$ the upper dispersion branch becomes low-gain mode, so the threshold appears.

As it can be seen from Fig.2 both types of hybrid states with spontaneously and forced broken symmetry can be dynamically stable, what allow us to develop our research with search and studying different localized structures such as switching waves and solitons.

4. Conclusion

The rich dynamics of the electromagnetic field in the systems with complex periodic potentials and Kerr nonlinearity can be used to achieve more control over the stationary lasing states. For instance the effects discussed in the paper can potentially be used to affect the nonlinear patterns forming in the lasers, switching their radiation frequency and the direction of the emission.

5. Acknowledgments

The work was supported by Russian Fund for Basic Research (Grant “Aspiranty” No. 20-32-90227).

[1] Kogelnik H and Shank C 1972 *Journal of applied physics* **43** 2327–2335
[2] Kazarinov R and Henry C 1985 *IEEE Journal of Quantum Electronics* **21** 144–150
[3] Pourdavoud N, Mayer A, Buchmüller M, Brinkmann K, Haeger T, Hu T, Heiderhoff R, Shutsko I, Görrn P, Chen Y *et al.* 2018 *Advanced Materials Technologies* **3** 1700253
[4] Luo Y, Nakano Y, Tada K, Inoue T, Hosomatsu H and Iwaoka H 1990 *Applied physics letters* **56** 1620–1622
[5] Luo Y, Nakano Y, Tada K, Inoue T, Hosomatsu H and Iwaoka H 1991 *IEEE journal of quantum electronics* **27** 1724–1731
[6] David K, Morthier G, Vankwikelberge P and Baets R 1990 *Electronics Letters* **26** 238–239
[7] Fessant T 1998 *JOSA B* **15** 2689–2699
[8] Cardimona D, Sharma M, Kovanis V and Gavrielides A 1995 *IEEE journal of quantum electronics* **31** 60–66
[9] Kwon K Y 1996 *IEEE journal of quantum electronics* **32** 1937–1949
[10] Krasikov S, Bogdanov A and Iorsh I 2018 *Physical Review B* **97** 224309
[11] Dolinina D and Yulin A 2020 *Optics Letters* **45** 3781–3784
[12] Dolinina D and Yulin A 2021 *arXiv preprint arXiv:2102.07526*

Journal of Physics: Conference Series **2015** (2021) 012036 doi:10.1088/1742-6596/2015/1/012036

Cellular SERS structures for non-invasive study of living cells

N V Doroshina[1, a)], A A Ushkov[2], N A Brazhe[3], D A Gorin[4], M D Mokrousov[4], D I Yakubovsky[1], O O Kapitanova[1], A A Voronov[1], A V Arsenin[1], V S Volkov[1] and S M Novikov[1]

[1]Center for Photonics and 2D Materials, Moscow Institute of Physics and Technology, 141700, Dolgoprudny, Russia
[2]Univ Lyon, UJM-Saint-Etienne, CNRS, Institute of Optics Graduate School, Laboratoire Hubert Curien UMR5516, F-42023 St-Etienne, France
[3]Biophysics Department, Biological Faculty, Moscow State University, 119234, Moscow, Russian Federation
[4]Skolkovo Institute of Science and Technology, 121205, Moscow, Russian Federation
[a)]Corresponding author: doroshina.nv@phystech.edu

Abstract. Surface-enhanced Raman spectroscopy (SERS) is a potent and highly selective tool to chemically identify and determine the structure of molecules and materials. However, the large biomolecules and living cells are still difficult to study using modern SERS-substrates. Here we present a new approach to the geometry of the surface of plasmon nanostructures (cellular surfaces), which makes it possible to efficiently work with volumetric objects as cells or organelles. We use the scanning Raman microscope to map the SERS from the specially prepared microcapsules formed with a Layer-by-Layer deposition method and with built-in Raman tags (Indocyanine green). It demonstrated the possibility to detect SERS signal from Indocyanine green in microcapsules located in cavities, while the signal from the Raman dye in microcapsules located on a flat surface is not detected. The obtained results can be useful in biosensors applications.

1. Introduction

Many pathologies can be diagnosed at an early stage. However, this diagnosis requires highly sensitive analysis at the level of an individual cell/organelle or even a molecule. This requires the development of new materials and new analytical approaches. Surface-enhanced Raman scattering (SERS) is a powerful and highly selective tool to chemically identify and determine the structure of materials and molecules, based on their specific vibrational bonds [1-2]. Strong SERS effects obtained using plasmonic nanostructures/systems allow the detection of molecules at extremely low nM concentrations, thereby becoming of great interest in sensing and medical diagnostics. However, the large biomolecules and living cells are still difficult to study using modern SERS-substrates and only a few results have been reported [3-5]. The problem is that the intensity of the SERS signal decays exponentially with increasing distance between the nanostructure and the analyte. Additionally, in a flat substrate configuration, the excitation occurs in the plane of the substrate, which does not work effectively with large objects (cells/organelles). It imposes restrictions on the method since many pathologies could be diagnosed by the processes taking place in the submembrane region, and usually, the required distance is ~10-25 nm.

However, recent results have shown that a certain geometry of the surface of nanostructures and its nanostructuring (deposition of nanoparticles on its surface) can be successfully applied for SERS measurements in living organelles/cells [6]. The preliminary results indicate that the most promising types of nanostructures combine a metalized surface of a certain profile with nanoparticles of various sizes, which can provide a large contact area between organelles /cells and a substrate. In this work, we present a conceptually new approach to the geometry of the plasmonic nanostructures (cellular surfaces), which makes it possible to efficiently work with volumetric objects such as cells or organelle. The substrate was tested by special microcapsules with a layer inside of Raman dye Indocyanine green (ICG). These microcapsules were used as a model of the volumetric object. We suppose that the obtained results are very promising for further sensing applications.

2. Experimental part

Figure 1 shows a schematic diagram of the experiment. The cellular surfaces of the substrates with a period of 950 nm were patterned via the Laser Interference Lithography [7]. Photoresist S1805 was spin-coated on the cleaned surface of BK7 microscope glass slides and pre-baked at 60C for 1 min. The substrates were then homogeneously pre-exposed for 7 min with a power P=250 mW/cm^2 to switch the resist into the linear regime, and exposed twice for 2.5 min afterward with a power 2P in the interference field; finally, the photoresist was developed for 20 s in the MF-319 developer. Next, several thin layers were deposited to the cellular substrate: titanium ~5 nm (to make the surface smooth), a plasmonic layer of silver ~100 nm, a protective layer of SiO^2 ~5 nm, and a layer of gold ~7 nm to form the gold nanospheres by further annealing. The deposition was carried out using the NEE-4000 E-Beam Evaporation system under the pressure of 3×10^7 Torr in the vacuum chamber, at room temperature, with the deposition rate ≈ 2 Å/s.

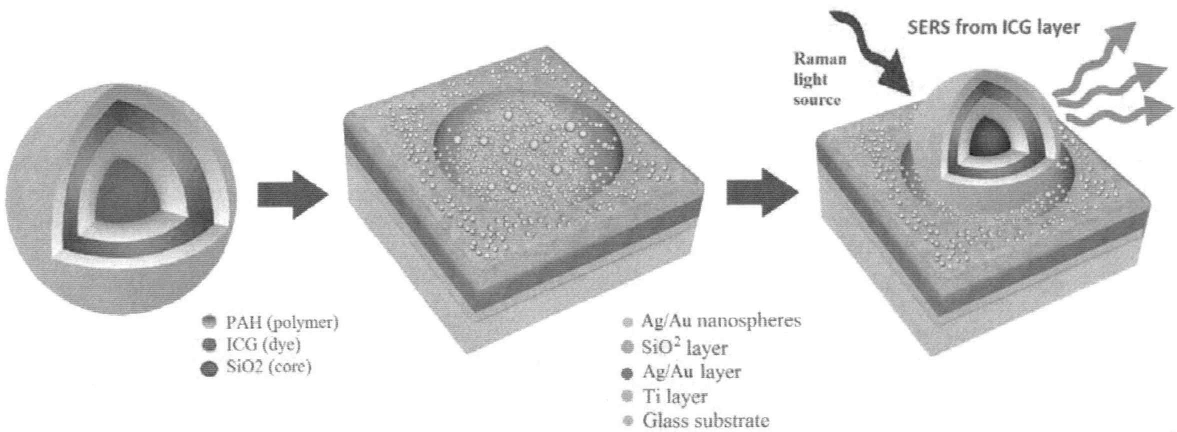

Figure 1. Schematic diagram of the experimental part with the special microcapsules.

Then, gold nanospheres with a diameter of ~50-70 nm were formed from the thin gold layer on the top of the substrates by annealing [8] (figure 2, a). At high temperatures, thin metal films begin dewetting into the metal islands, which tend to acquire a spherical shape. We optimized the annealing process parameters: the gold film thickness ~7 nm, annealing time 15 minutes, and 550° of annealing temperature. Previously, we tested the annealing method on a flat surface (figure 2, d), then on a cellular structure (figure 2, e). The specific nanostructure topography considerably enhances the near field effects due to the dipole moments, exited in the nanoparticles. A biological object, located on a flat metal surface with nanoparticles, has a limited contact area with it and the dipole moments do not penetrate deep enough into the large 3D object. In the case of our cellular structure, under normal light irradiation, an electric dipole moment directed along the electric field of the light wave resonantly excites the nanoparticles. For those particles that are located on the lateral surfaces of the cavities, these dipole moments are perpendicular to the surface, which leads to a significant increase of signal due to the reflection effect. As a result, a strong electric near field is induced at the side surface of the cellular cavity, which makes it possible to obtain a strong SERS signal from bio-objects placed in these cavities [6].

The applicability of obtained nanostructures has been investigated using a confocal scanning Raman microscope Horiba LabRAM HR Evolution (HORIBA Ltd., Kyoto, Japan). All measurements were carried out using linearly polarized excitation at a wavelength of 632.8 nm, 600 lines/mm diffraction grating, and ×100 objective (N.A. = 0.90), whereas we used unpolarized detection to have a significant signal-to-noise ratio. The spot size is ~0.43 μm. The Raman spectra were recorded with 0.8 mW incident powers and an integration time of 1 sec.

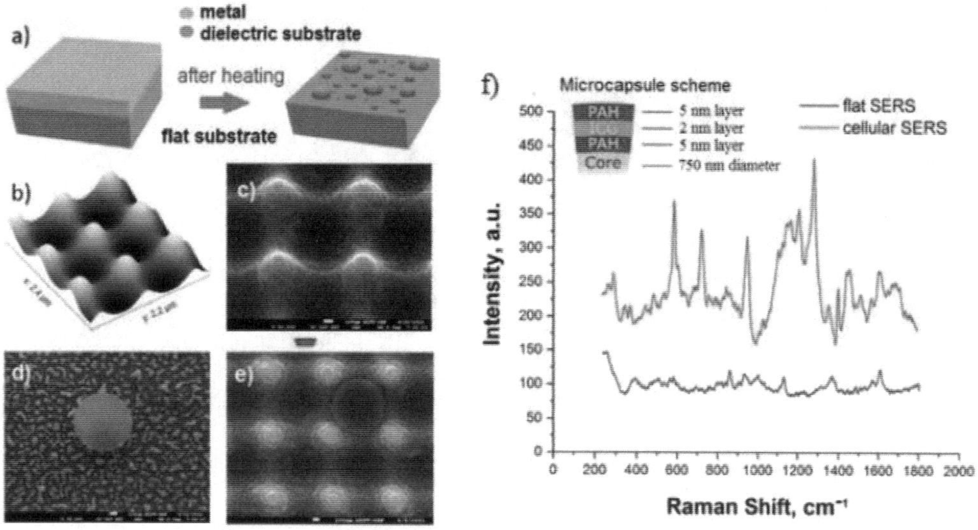

Figure 2, a) schematic annealing of the flat metal surface, b) AFM topology of the cellular substrate before nanostructuring, c) SEM image of the cellular substrate after Au nanostructuring, d) flat SERS substrate with the microcapsule, e) cellular SERS substrate with the microcapsule, f) Raman spectra from the single microcapsule located on the flat substrate (black curve) and on the cellular substrate (red curve).

As a test object, we used special microcapsules (figure 1) with a silicon dioxide core with the inflicted layer of Indocyanine green (ICG) dye on top of it and the next layers of polymer Poly (sodium 4-styrene sulfonate). The microcapsules were produced with a Layer-by-Layer deposition method [9,10]. ICG was injected as a Raman active label with concentrations that can not be detected in a flat configuration. The ICG layer was located at a depth ~10 nm in microcapsules. The ICG layer depth inside the microcapsule was varied by several polymers on top of it. These microcapsules were used as a model of the volumetric bio-object. We demonstrated the possibility to detect high SERS signal from a certain depth in a microcapsule located into a nanostructured cavity, while the analogous ICG signal from a single microcapsule on a flat nanostructured surface was detected much lower (figure 2, f). The experimental part with the single microcapsules on different substrates was provided at the similar equipment, with 632.8 nm laser excitation, 0.8 mW incident powers, and an integration time of 1 sec.

3. Conclusion

The high sensitivity of our nanostructured substrates makes it possible to use low laser powers that are safe for the bio-objects under study. Also, at the current stage, we are presented with the possibility of receiving a signal; the configuration of the substrate will be optimized in the future for the greatest reliability. In prospect, we planned to use silicon nanoparticles (they heat up slightly under laser exposure), also coating the nanostructured surface with graphene (biocompatible two-dimensional material).

In summary, we present a novel plasmonic nanostructure geometry, which allows working with volumetric objects as cells or organelle. We suppose that the obtained results pave the way toward plasmonic nanosensors for studying processes in the near-membrane regions of living organelles/cells, thus opening an opportunity to diagnose pathologies in the early stages and to analyze volumetric bio-objects by Raman.

References

[1] Langer, et al. "Sensing using plasmonic nanostructures and nanoparticles", *Nanotechnology* **26**, 322001 (2015).

[2] Langer, et al. "Present and Future of Surface-Enhanced Raman Scattering", *ACS Nano* **14**, 28-117 (2020).

[3] Shyh-Chyang, et al. "Nanofabricated SERS-Active Substrates for Single-Molecule to Virus Detection inVitro: A Review." *Biosensors and Bioelectronics*, **61**, 232–240 (2014).

[4] Ando, et al. "Dynamic SERS Imaging of Cellular Transport Pathways with Endocytosed Gold Nanoparticles." *Nano Letters*, **11**, 5344–5348 (2011).

[5] Brazhe, et al. "SERS uncovers the link between conformation of cytochrome c heme and mitochondrial membrane potential", *bioRxiv*, **14**, 322001 (2021).

[6] Brazhe, et al. "Probing cytochrome c in living mitochondria with surface-enhanced Raman spectroscopy". *Sci. Rep.* **5**, 13793 (2015).

[7] Ushkov, et al. "Systematic study of resonant transmission effects in visible band using variable depth gratings. *Sci Rep* **9**, 14890 (2019).

[8] Dong, et al. Gap-plasmon based broadband absorbers for enhanced hot-electron and photocurrent generation. *Sci Rep* **6**, 30650 (2016).

[9] Mokrousov, et al. "Amplification of photoacoustic effect in bimodal polymer particles by self-quenching of indocyanine green," *Biomed. Opt. Express* **10**, 4775-4788 (2019).

[10] Yashchenok, et al. "Multifunctional polyelectrolyte microcapsules as a contrast agent for photoacoustic imaging in blood", *J. Biophotonics* **9**, 792-799 (2016).

Acknowledgments

Authors gratefully acknowledge the financial support from the Ministry of Science and Higher Education of the Russian Federation (Agreement No. 075-15-2021-606) and the Russian Foundation for Basic Research (20-07-00475 and 20-07-00840). We thank the Shared Facilities Center of the Moscow Institute of Physics and Technology for the use of their equipment.

Journal of Physics: Conference Series **2015** (2021) 012037 doi:10.1088/1742-6596/2015/1/012037

Integrated high-contrast diffraction gratings for surface electromagnetic waves

L L Doskolovich[1,2], **E A Bezus**[1,2] **and D A Bykov**[1,2]

[1] Image Processing Systems Institute of RAS — Branch of the FSRC "Crystallography and Photonics" RAS, 151 Molodogvardeyskaya st., Samara 443001, Russia
[2] Samara National Research University, 34 Moskovskoe shosse, Samara 443086, Russia

leonid@ipsiras.ru

Abstract. We propose and theoretically and numerically investigate integrated high-contrast diffraction gratings for surface electromagnetic waves. We consider two platforms for the on-chip gratings: surface plasmon-polaritons propagating along metal-dielectric interfaces and Bloch surface waves propagating along interfaces of photonic crystals. We demonstrate that the optical properties of the studied integrated gratings are qualitatively close to the ones of the conventional high-contrast diffraction gratings. If the "parasitic" out-of-plane scattering is suppressed, the reflectance and transmittance of the on-chip gratings are not only qualitatively, but also quantitatively close to the corresponding values of the conventional "free-space" gratings. The obtained results may find application in novel integrated optical circuits.

1. Introduction

Diffraction gratings are among the most versatile and widely used photonic components due to a wide range of extraordinary optical effects exhibited under resonance conditions and a relative ease of their fabrication. Despite a long history (resonant effects in gratings were first discovered by R. W. Wood in 1902), resonant diffraction gratings remain a subject of intensive research due to both theoretical interest and their potential applications in spectrometers, chemical and biological sensors, laser resonators, and optical information processing devices, among others [1–3]. One of the promising classes of resonant diffraction gratings is constituted by the so-called high-contrast gratings (HCGs) being high-index dielectric subwavelength structures surrounded by low-index materials [3, 4]. HCGs exhibit several remarkable features, including broadband high reflectivity and high-Q resonances.

In the present work, we propose and theoretically and numerically investigate integrated high-contrast gratings for surface electromagnetic waves. We consider HCGs for two surface wave platforms: surface plasmon-polaritons (SPPs) supported by metal-dielectric interfaces and Bloch surface waves (BSWs) supported by interfaces of one-dimensional dielectric photonic crystals. The studied on-chip gratings consist of a periodic array of rectangular dielectric pillars on the propagation surface of the SPP or BSW. We show that the optical properties of the on-chip HCGs are qualitatively and, with a proper choice of the grating and incidence geometry, even quantitatively close to the ones of the conventional HCGs. The obtained results may be useful for developing new integrated optical filters and reflectors.

2. Results and discussion

The geometry of the investigated on-chip HCG is shown in Fig. 1. The integrated grating consists of periodically arranged dielectric pillars with rectangular cross-section located directly on the propagation surface of the incident SPP or BSW. The diffraction of the surface waves impinging on the gratings was simulated by a parallel in-house implementation of the aperiodic 3D Fourier modal method suitable for the solution of the integrated optics problems [5, 6].

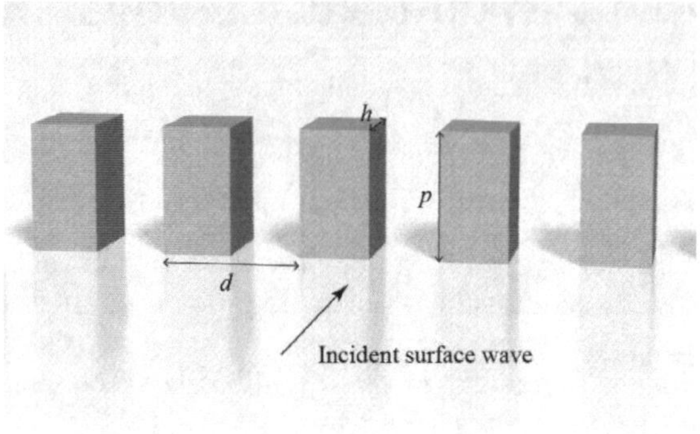

Figure 1. Geometry of the investigated integrated high-contrast gratings for surface electromagnetic waves consisting of a periodic dielectric pillar array on the propagation surface.

As a first example, let us consider a plasmonic grating with the following parameters: operating free-space wavelength range 1450–1650 nm, metal — gold (dielectric permittivity $\varepsilon_{\mathrm{Au}} = -115.11 + 11.1\mathrm{i}$ at 1550 nm), fill factor 0.5, grating pillar height $p = 4\ \mu\mathrm{m}$, dielectric permittivity of the grating pillars $\varepsilon_{\mathrm{Si}} = 12.08$ at 1550 nm. Using the approach presented in our previous work [7], a thin Si3N4 layer (dielectric permittivity $\varepsilon_{\mathrm{Si_3N_4}} = 3.99$ at 1550 nm) with the thickness $t = 11$ nm [see the inset to Fig. 2(a)] was added underneath the "main" part of the grating pillars in order to decrease the "parasitic" out-of-plane SPP scattering losses. In this case, the difference between the effective refractive indices of the incident SPP and the SPP in the pillar regions (the effective refractive index contrast) is of about 3.5, which allows us to expect that the structure will operate as a high-contrast grating.

Figure 2(a) shows the reflectance of the integrated plasmonic grating in the normal incidence case vs. the normalized parameters λ/d and h/d. It is evident from Fig. 1(a) that high- and low-reflectance regions appear in a highly regular checkerboard-like pattern in the spectrum of the studied on-chip grating, which is typical for high-contrast gratings and is caused by interference effects between the two modes supported by the grating [3, 4]. Figure 2(b) shows the reflectance, transmittance and losses spectra of three plasmonic HCG examples "extracted" from Fig. 2(a), which demonstrate that such structures can be used as efficient broadband plasmonic reflectors with a subwavelength or near-subwavelength footprint in the propagation direction and mean reflectance exceeding 85%.

As a second example, let us consider integrated gratings for transverse electric (TE-) polarized Bloch surface waves with the free-space wavelength of 630 nm propagating along the interface of a one-dimensional photonic crystal (PC) consisting of alternating gallium phosphide (refractive index $n_{\mathrm{GaP}} = 3.32$ at 630 nm) and silicon dioxide (refractive index $n_{\mathrm{SiO_2}} = 1.47$ at 630 nm) layers. Let us start with a grating designed for the case of normal BSW incidence supported by the interface of a PC with the following parameters: the PC layer thicknesses $h_{\mathrm{GaP}} = 73$ nm and $h_{\mathrm{SiO_2}} = 82$ nm, the thickness of the upper layer of the PC made of GaP $h'_{\mathrm{GaP}} = 10$ nm. At the height of the grating pillars $p = 51$ nm, which are considered to be made from the same material as the upper PC layer, the BSW effective refractive index contrast in the structure is of about 1.2, which is also relatively high.

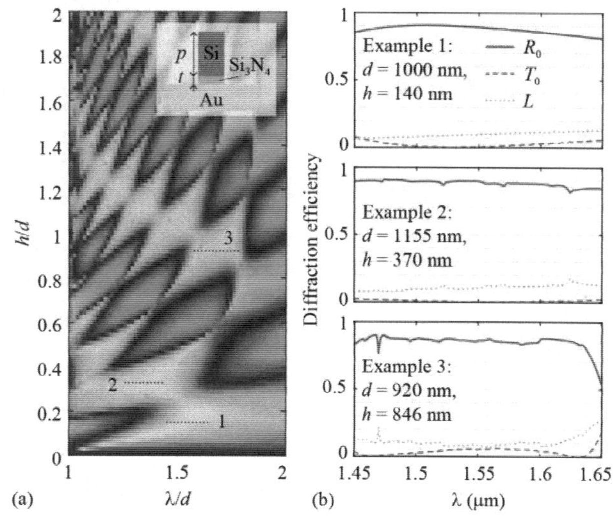

Figure 2. (a) Reflectance of a plasmonic HCG vs. the normalized parameters λ/d and h/d at normal SPP incidence. Black dotted lines correspond to the examples shown in (b). (b) Reflectance R_0, transmittance T_0 and losses L spectra of three examples of plasmonic HCGs.

Figure 3(a) shows the reflectance of the integrated HCG for BSWs in the normal incidence case vs. the normalized parameters λ/d and h/d. Although the reflectance spectrum is qualitatively close to the spectrum of a conventional HCG with the effective refractive indices equal to the effective refractive indices of the BSWs inside and outside the pillar regions (not shown for the sake of brevity), the maximum reflectance of the on-chip HCG does not exceed 0.5 due to high out-of-plane scattering losses. This hinders any potential practical applications of this particular structure.

However, with a proper choice of the photonic bandgap configuration and the incidence geometry, the "parasitic" scattering can be completely eliminated. This is illustrated by Fig. 3(b), which shows the reflectance spectrum of an integrated grating designed for the oblique BSW incidence at an angle $\theta = 55°$. In this case, the photonic crystal with the following parameters was considered: PC layer thicknesses $h_{\mathrm{GaP}} = 25$ nm and $h_{\mathrm{SiO_2}} = 164$ nm, the thickness of the upper layer of the PC (GaP) 64 nm, the height of the grating pillars $p = 31$ nm. As in the previous example, the incident BSW is TE-polarized. It is evident from Fig. 3(b) that the on-chip BSW grating exhibits a high-Q resonance with the reflectance reaching unity at the resonant peak due to the absence of the out-of-plane scattering losses. Such a structure can be used as a narrowband BSW filter.

3. Conclusion

In this work, we presented integrated high-contrast diffraction gratings for surface plasmon polaritons and Bloch surface waves, which consist of periodically arranged dielectric pillars on the propagation surface (on the metal surface in the SPP case and on the surface of a photonic crystal in the BSW case). We demonstrated that if the "parasitic" out-of-plane scattering of the incident surface wave is suppressed or completely eliminated, the studied integrated gratings can be used as efficient wideband or narrowband reflectors. We believe that the presented results may find application in novel integrated nanophotonic devices.

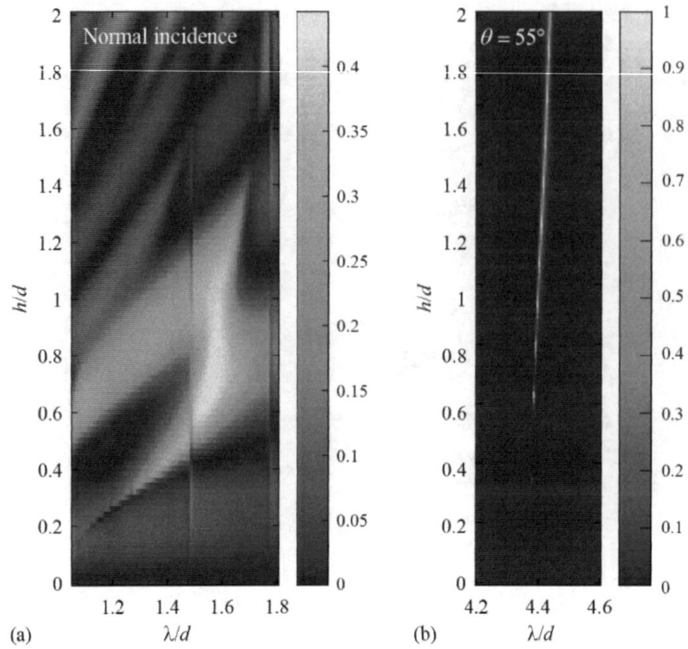

Figure 3. Reflectance of BSW HCGs designed for the cases of normal (a) and oblique (b) incidence cases vs. the normalized parameters λ/d and h/d .

Acknowledgments
This work was funded by Russian Science Foundation (RSF) under the project 19-19-00514 (investigation of the integrated high-contrast gratings for BSWs and SPPs) and by the Russian Federation Ministry of Science and Higher Education within the state assignment to the FSRC "Crystallography and Photonics" RAS (implementation of the numerical simulation software).

References

[1] Collin S 2014 *Rep. Prog. Phys.* **77** 126402
[2] Quaranta G, Basset G, Martin O J and Gallinet B 2018 *Laser Photon. Rev.* **12** 1800017
[3] Qiao P, Yang W and Chang-Hasnain C J 2018 *Adv. Opt. Photon.* **10** 180–245
[4] Chang-Hasnain C J and Yang W 2012 *Adv. Opt. Photon.* **4** 379–440
[5] Silberstein E, Lalanne P, Hugonin J-P and Cao Q 2001 *J. Opt. Soc. Am. A* **18** 2865–75
[6] Lecamp G, Hugonin J P and Lallanne P 2007 *Opt. Express* **15**, 11042–60
[7] Bezus E A, Doskolovich L L and Kazanskiy N L 2014 *Opt. Express* **22** 13547–54

Asymptotic homogenization of metamaterials elastic plates

David Faraci and Claudia Comi

Department of Civil and Environmental Engineering, Politecnico di Milano, Piazza Leonardo da Vinci 32, 20133 Milan, Italy

E-mail: david.faraci@polimi.it, claudia.comi@polimi.it

Abstract. The asymptotic homogenization technique is applied to evaluate the effective properties of thin plates with periodic heterogeneity. The effect of shear deformation in the homogenization process is evidenced and the role of cell slenderness, besides that of the plate, is clarified by several numerical analyses.

1. Introduction

The numerical simulations of thin plates with periodic heterogeneity, or metamaterials plates, have a high computational burden, especially when the size of the heterogeneity is small if compared with the global dimension of the plate. In the elastic field, asymptotic homogenization have been effectively and widely employed to obtain equivalent properties of periodic solids, both in statics and dynamics [1, 5].

For plates the procedure can be developed starting from a three-dimensional formulation [3, 4, 7, 9]. This leads to local 3D cell problems, which can be treated in a simplified manner if the unit cell can be modelled as a plate. A possible way to proceed is introducing in the 3D cell problems the kinematic constraints of plate theories [8, 9].

In this work, following a different path, we develop the homogenization procedure directly from the structural model, as done originally in [6] for the Kirchhoff-Love plate theory, and in [2] for the Mindlin-Reissner formulation and we apply it to several problems to evidence the role of shear deformation in the numerical results when the typical dimension of the heterogeneity decreases and hence the cell slenderness decreases.

2. Two-scale asymptotic homogenization

Let us consider a heterogeneous plate of mid-surface Ω characterized by an in-plane periodicity, with unit cell Y^ϵ, as depicted e.g. in Figure 1 (left), made of linear elastic isotropic materials. The typical dimension of the unit cell is denoted by $a^\epsilon = \epsilon a$, being ϵ a small parameter. Besides the macroscopic in-plane coordinate $\mathbf{x} = (x_1, x_2)$, according to the two-scale asymptotic method, we introduce the fast variable $\mathbf{y} = \mathbf{x}/\epsilon$, which lives in the in-plane re-scaled unit cell $Y = Y^\epsilon/\epsilon$, of size a, as shown e.g. in Figure 1 (middle). We assume that the thickness of the plate does not scale with ϵ, i.e. $t^\epsilon(\mathbf{x}) = t(\mathbf{x}/\epsilon)$. Similarly, we consider the transversal load $p^\epsilon(\mathbf{x}) = p(\mathbf{x}/\epsilon)$ and the periodically varying Young's modulus and Poisson's ratio, $E^\epsilon(\mathbf{x}) = E(\mathbf{x}/\epsilon)$ and $\nu^\epsilon(\mathbf{x}) = \nu(\mathbf{x}/\epsilon)$ respectively, independent from ϵ.

The homogenized (effective) properties of the plate are obtained by analyzing the asymptotic behaviour as $\epsilon \to 0$.

2.1. Homogenization of Kirchhoff-Love plates

The kinematic of a Kirchhoff-Love (K-L) plate is completely described by the out of plane displacement w^ϵ, while the equilibrium requires that:

$$\text{div div }(-\mathbb{B}^\epsilon : \nabla \odot \nabla w^\epsilon) + p^\epsilon = 0 \qquad \text{in } \Omega \qquad \text{with} \qquad \mathbb{B}^\epsilon = \frac{E^\epsilon t^{\epsilon 3}}{12(1-\nu^{\epsilon 2})}[(1-\nu^\epsilon)\mathbb{I} + \nu^\epsilon \mathbf{I} \otimes \mathbf{I}] \quad (1)$$

$\mathbb{B}^\epsilon$ is the bending stiffness tensor, div is the in-plane divergence operator and $\odot$ denotes the symmetric part of the tensorial product. The solution of the problem can be searched in the form of the asymptotic expansion:

$$w^\epsilon(\mathbf{x}) = w^0(\mathbf{x}, \mathbf{x}/\epsilon) + \epsilon w^1(\mathbf{x}, \mathbf{x}/\epsilon) + \epsilon^2 w^2(\mathbf{x}, \mathbf{x}/\epsilon) + \epsilon^3 w^3(\mathbf{x}, \mathbf{x}/\epsilon) + \dots \quad (2)$$

where the fields $w^i(\mathbf{x}, \mathbf{y})$ are defined on $\Omega \times \mathbb{R}^2$ and are $Y-$periodic with respect the fast variable. Substituting (2) into (1) one obtains a sequence of differential problems from which is possible to obtain the effective homogenized bending stiffness of K-L plate, which components can be computed as:

$$B^0_{ijhk} = \frac{1}{|Y|}\int_Y (\mathbf{e}_i \odot \mathbf{e}_j - \nabla_\mathbf{y} \odot \nabla_\mathbf{y} \alpha^{ij}) : \mathbb{B} : (\mathbf{e}_h \odot \mathbf{e}_k - \nabla_\mathbf{y} \odot \nabla_\mathbf{y} \alpha^{hk})\, d\mathbf{y} \qquad i,j,h,k \in \{1,2\} \quad (3)$$

where $\mathbf{e}_i$ is the unit vector in the $i-$th direction and $|Y|$ is the area of the re-scaled unit cell. The functions $\alpha^{ij}(\mathbf{y})$ in (3) are the $Y-$periodic solutions of the cell problems:

$$\text{div}_\mathbf{y}\text{div}_\mathbf{y}\left[\mathbb{B} : (-\nabla_\mathbf{y} \odot \nabla_\mathbf{y} \alpha^{ij} + \mathbf{e}_i \odot \mathbf{e}_j)\right] = 0 \qquad \text{in } Y \quad (4)$$

2.2. Homogenization of Mindlin-Reissner plates

In the Mindlin-Reissner (M-R) formulation, the kinematic of the plate is described by the out of plane displacement w^ϵ and the rotation φ^ϵ of the thickness segments normal to the mid-surface. The equilibrium conditions are given by:

$$\text{div}\,[\mathbf{C}^\epsilon \cdot (\nabla w^\epsilon - \varphi^\epsilon)] + p^\epsilon = 0 \qquad \text{and} \qquad \text{div}\,(-\mathbb{B}^\epsilon : \nabla \varphi^\epsilon) = \mathbf{C}^\epsilon \cdot (\nabla w^\epsilon - \varphi^\epsilon) \qquad \text{in } \Omega \quad (5)$$

where $\mathbb{B}^\epsilon$ is the plate bending stiffness defined in (1), while $\mathbf{C}^\epsilon = 5/6\, G^\epsilon \mathbf{I}$ is the plate shear stiffness tensor, G^ϵ being the shear modulus of the material. In addition to the expansion (2) adopted in the previous subsection, we assume the following asymptotic expansions of the rotation:

$$\varphi^\epsilon(\mathbf{x}) = \varphi^0(\mathbf{x}, \mathbf{x}/\epsilon) + \epsilon \varphi^1(\mathbf{x}, \mathbf{x}/\epsilon) + \epsilon^2 \varphi^2(\mathbf{x}, \mathbf{x}/\epsilon) + \dots \quad (6)$$

where the fields $\varphi^i(\mathbf{x}, \mathbf{y})$ are defined on $\Omega \times \mathbb{R}^2$ and are $Y-$periodic with respect the second variable. Substituting (2,6) in equation (5) one obtains the effective bending and shear stiffnesses of M-R plate, which components are:

$$\begin{aligned} B^0_{ijhk} &= \frac{1}{|Y|}\int_Y (\mathbf{e}_i \odot \mathbf{e}_j - \nabla_\mathbf{y} \beta^{ij}) : \mathbb{B} : (\mathbf{e}_h \odot \mathbf{e}_k - \nabla_\mathbf{y} \beta^{hk})\, d\mathbf{y} \\ C^0_{ij} &= \frac{1}{|Y|}\int_Y (\mathbf{e}_i + \nabla_\mathbf{y} \gamma^i) \cdot \mathbf{C} \cdot (\mathbf{e}_j + \nabla_\mathbf{y} \gamma^j)\, d\mathbf{y} \end{aligned} \quad (7)$$

where $\beta^{ij}(\mathbf{y})$ and $\gamma^i(\mathbf{y})$ are the $Y-$periodic solutions, respectively, of the two cell problems:

$$\text{div}_\mathbf{y}\left[\mathbb{B} : (-\nabla_\mathbf{y} \beta^{ij} + \mathbf{e}_i \odot \mathbf{e}_j)\right] = \mathbf{0} \qquad \text{and} \qquad \text{div}_\mathbf{y}\left[\mathbf{C} \cdot (\nabla_\mathbf{y} \gamma^i + \mathbf{e}_i)\right] = 0 \qquad \text{in } Y \quad (8)$$

The presented K-L and M-R homogenization are not able to take into account the real slenderness of the unit cell, since they are derived with an in-plane re-scaling. Better results are obtained by scaling also the thickness ($t^\epsilon(\mathbf{x}) = \epsilon t(\mathbf{x}/\epsilon)$) in the cell problems as in the refined M-R (R-M-R) homogenization proposed in [9].

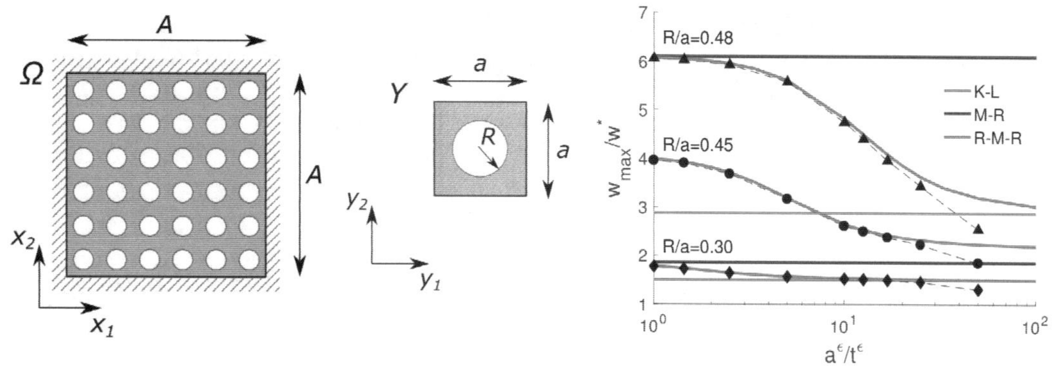

Figure 1: Periodic perforated plate with circular holes (left); re-scaled unit cell (middle); maximum deflection versus a^ϵ/t^ϵ for the holed plate (markers) and the homogenized one (continuos lines), (right).

3. Numerical analyses

To validate the above homogenized formulations and to set the level of approximation for different values of ϵ, two reference problems are considered. The results obtained of homogenized plates, with stiffness evaluated through (3) or (7), are compared with those obtained with finite element analyses on the actual heterogeneous plates.

To verify the convergence of the homogenization solution as $\epsilon \to 0$, and to investigate the effect of the unit cell slenderness, different analyses are conducted on plates with increasing numbers of cells keeping fix the global dimensions of the plate and its thickness t^ϵ.

3.1. Periodic perforated plate

As the first example, we consider a square plate, of side A and constant thickness $t = A/100$, with cylindrical holes of radius $R^\epsilon = \epsilon R$, periodically distributed in a square lattice of side a^ϵ, see Figure 1 (left). In the mid-plane of the plate, the re-scaled unit cell, shown in Figure 1 (middle), is hence a square of side a with a circular hole of radius R. The plate is subject to a uniform transversal load p and it is simply supported on its boundary.

Figure 1 (right) shows the maximum displacement w_{max} numerically computed on the holed M-R plates (markers) and the same quantity evaluated through homogenization (continuos line), for different values of R/a, as a function of the unit cell slenderness a^ϵ/t^ϵ. Green curves are obtained considering the R-M-R homogenization, while the blue and the orange ones refer to the M-R and K-L ones (for clarity these latters are only shown for $R/a = 0.3$ and 0.48). All results are normalized with the maximum displacement w^* of the uniform plate without holes ($R/a \to 0$). The results obtained with R-M-R homogenization are very close to those of the real holed plate, except when there are few cells (i.e. for high values of a^ϵ/t^ϵ). This is due to the lack of separation between macro and micro-scale, which is a fundamental hypothesis for the validity of the homogenization technique. K-L and M-R homogenization, conversely, can be used only for large and small values of the cell slenderness (respectively). In particular as the size of the holes increases (compare the curves $R/a = 0.3$ and 0.48) the shear deformation of the unit cell becomes relevant. Hence, the K-L solution significantly differ from the R-M-R one even for very slender cells.

Figure 2 shows the deformed configurations of the holed plate and of the equivalent homogeneous one for $a^\epsilon/t^\epsilon = 10$ and $R/a = 0.45$. Even though the number of cells is small (10×10), the homogenized approach is very accurate.

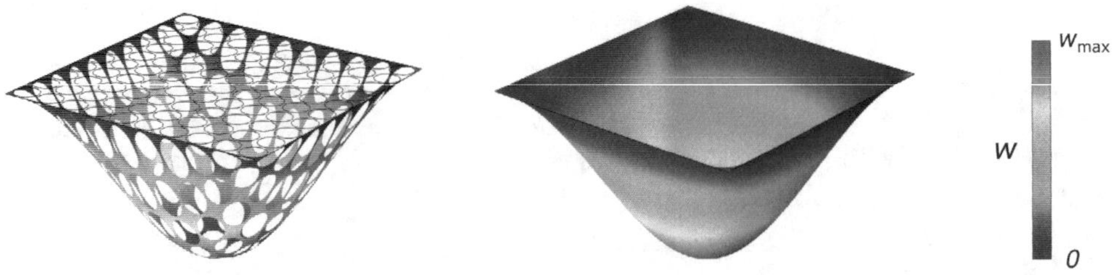

Figure 2: Contour of deflection on deformed shapes for the holed plate (left) and the R-M-R homogenized one (right), for $a^\epsilon/t^\epsilon = 10$ and $R/a = 0.45$.

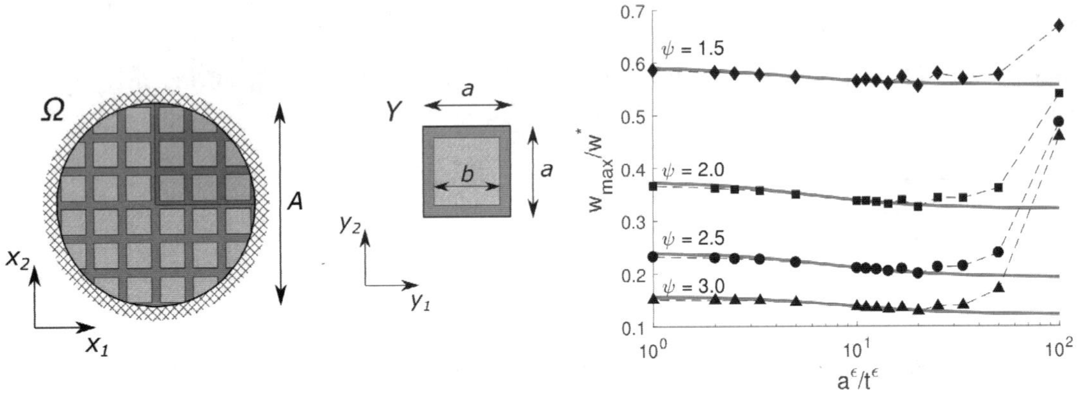

Figure 3: Two-way ribbed plate with a square inclusion (left); re-scaled unit cell (middle); maximum deflection versus a^ϵ/t^ϵ for the real problem (markers) and the R-M-R homogenized one (continuos lines), (right).

3.2. Two-ways ribbed plate

We consider a circular two-ways ribbed plate of diameter A made of a homogeneous material, as shown in Figure 3 (left). We denote by $t^\epsilon = A/100$ the thickness of the plate and, as in waffle slabs, we consider a reinforcement with a square pattern (shown in dark grey in Figure 3) of thickness ψt^ϵ, with $\psi \geq 1$, and width $a^\epsilon - b^\epsilon = 0.25a^\epsilon$. The re-scaled unit cell (Figure 3, middle) is a square of side a. The plate is clamped on its boundary and is loaded with a uniform transversal load p on the top-right quarter (the light blue region in the figure).

The maximum deflection w_{max} numerically obtained accounting for the real geometry with reinforcements and the same quantity obtained on the homogenized plate are compared in Figure 3 (right) for some values of ψ. The values are normalized with w^*, which is the maximum displacement for the case of uniform thickness plate ($\psi = 1$). As for the previous example, for few cells (i.e. for high ratios a^ϵ/t^ϵ) the R-M-R homogenized solution (continuous line) differs from the real one (markers) due to the lack of scale separation. For small values of a^ϵ/t^ϵ the shear deformability of the unit cell is no more negligible and seems rather independent on ψ.

Figure 4 shows the contour plots of the displacement w (Figure 4a and b) and of the bending moment M_{11} (Figure 4c and d) for $\psi = 2$ with 3 and 10 cells along the diameter. The first row refers to the analysis with the real geometry, the second refers to the homogenized plate. With few cells (Figure 4a), the displacement of the homogenized plate significantly differs from the real one, while with more cells (Figure 4b) the two solutions are practically equal. On the contrary, even with a lot of cells, the solution of the homogenized plate provides only average values of the bending moments, therefore it is unable

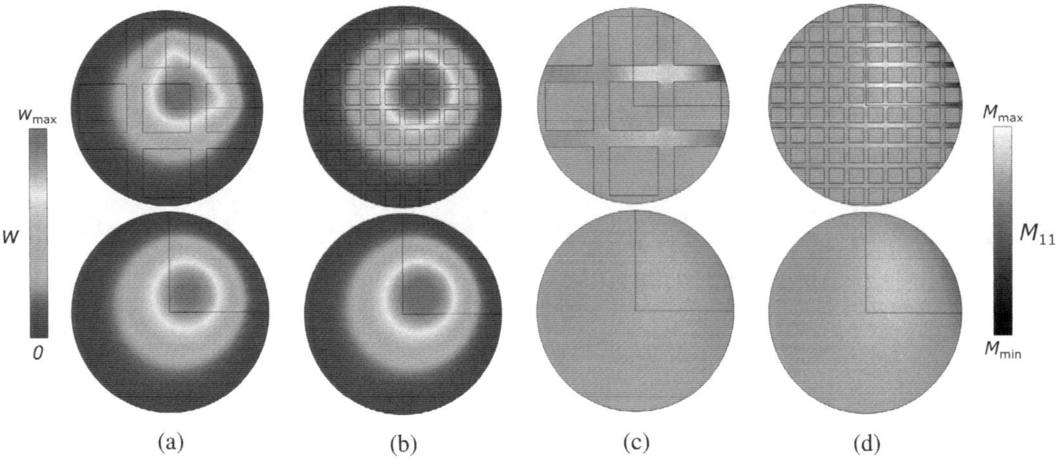

Figure 4: Top: analysis on real geometry; bottom: analysis on homogenized plate. Contours of displacement w (left) and bending moment M_{11} (right) for the cases $a^\epsilon/t^\epsilon = 33.33$ (a,c) and $a^\epsilon/t^\epsilon = 10$ (b,d) with $\psi = 2$.

to capture their maxima and minima (see Figure 4d). However, peak values can be taken into account through the so-called stress localization tensor, which is not considered in the present work.

4. Conclusions

The presented homogenization methods can be effectively adopted to simulate elastic periodic metamaterial plates when a proper scale separation (at least one order of magnitude between the lattice and the global dimensions) is guaranteed. The methods gives good results also when the load is non-uniform and the boundary of the plate does not match the periodicity of the lattice. However, one should select carefully the homogenization techinque to adopt. In particular, even for very thin plates, when the heterogeneity dimension a^ϵ is of the same order of the plate thickness t^ϵ, the shear deformation plays a significant role in the homogenization and the K-L theory cannot be used. Since, on the other hand, the separation of scales requires small values of a^ϵ, the homogenized K-L formulation has little practical use. Conversely, the homogenized R-M-R formulation, can be effectively applied in a wide range of geometric parameters, as shown by the examples here considered.

References

[1] N. Bakhvalov and G. Panasenko. *Homogenisation: Averaging Processes in Periodic Media.* Kluwer Academic Publishers, 1989.

[2] A. Bourgeat and R. Tapiéro. Homogénéisation d'une plaque mince, thermoélastique, perforée transversalement, de structure non uniformement périodique, dans le modèle de la théorie naturelle. *CR. Acad. Sci. Paris*, 297, 1983.

[3] D. Caillerie. Plaques élastiques minces à structure périodique de période et d'épaisseur comparables. *C.R. acad. Sci. Paris*, 294:159–162, 1982.

[4] D. Caillerie. Thin elastic and periodic plates. *Mathematical Methods in the Applied Sciences*, 6(1):159–191, 1984.

[5] C. Comi and J.J. Marigo. Homogenization Approach and Bloch-Floquet Theory for Band-Gap Prediction in 2D Locally Resonant Metamaterials. *Journal of Elasticity*, 139(1):61–90, 2020.

[6] G. Duvaut. Comportement macroscopique d'une plaque perforee periodiquement. In *Singular Perturbations and Boundary Layer Theory. Lecture Notes in Mathematics*, pages 131–145 (1997). Springer-Verlag, Berlin.

[7] T. Lewiński. Effective models of composite periodic plates-I. Asymptotic solution. *International Journal of Solids and Structures*, 27(9), 1991.

[8] T. Lewiński. Effective models of composite periodic plates-III. Two-dimensional approaches. *International Journal of Solids and Structures*, 27(9):1185–1203, 1991.

[9] T. Lewiński and J. J. Telega. *Plates, laminates, and shells: asymptotic analysis and homogenization.* World Scientific Publishing, 2000.

Weak plasmon modes in periodic structures for terahertz detection and amplification

D V Fateev[1, 2, *], O V Polischuk[1], M Yu Morozov[1], K V Mashinsky[1], I M Moiseenko[1] and V V Popov[1]

[1]Kotelnikov Institute of Radio Engineering and Electronics, Russian Academy of Sciences, Saratov 410019, Russia

[2]Saratov State University, Saratov 410012, Russia
[*]FateevDV@yandex.ru

Abstract. Weak plasmon modes in periodic structures with a two-dimensional electron gas without an inversion center are studied theoretically. The asymmetry of the electric field and Fourier harmonics of weak plasmon modes can lead to the excitation of a travelling plasmon by an electromagnetic wave normally incident on the structure and to the appearance of nonlinear effects leading to the rectification of the incident radiation. The low radiation damping of weak plasmon modes can be used to increase the efficiency of terahertz plasmon amplifiers.

1. Introduction

Plasma waves (plasmons) and plasmon resonances in two-dimensional electron gas (2DEG) of solid-state structures attract wide interest of researchers due to their promising properties. The strong localization of the plasmon field near the 2DEG surface and the strong deceleration of the plasmon in comparison with the electromagnetic (EM) wave make it possible to increase the interaction of the EM fields with matter. The frequencies of plasmons in modern materials fall into the terahertz range, which, together with the picosecond relaxation time of plasmons, makes it possible to observe plasmon resonances at room temperature. It is important that the classical nature of plasmons removes some of the quantum limitations characteristic of the terahertz range. For the excitation of plasmons, effective decelerating systems are needed that match the momentum and energy of EM waves and plasmons.).

The aim of the paper is to review the obtained results on using weak plasmon modes in periodical structures including new results of THz lasing with weak modes.

2. Discussions

Standard systems for exciting plasmons are short-period gratings, in which the harmonics of the scattered EM field receive momentum determined by reciprocal grating vectors. The electric field of the incident wave creates a strong dipole moment in the parts of the grating, which leads to the excitation of charge oscillations in the 2DEG. However, among all the eigenmodes of plasmons in spatially symmetric grating structures, along with modes with a large dipole moment (strong modes), there is the same number of eigenmodes with a zero dipole moment. Such weak modes are characterized by a symmetric spatial charge distribution in the plasmon resonator and an antisymmetric field distribution. The incident EM wave cannot create a charge distribution in a symmetric plasmon resonator, which is typical for weak modes. Therefore, the excitation of such "weak" plasmon modes by an incident EM wave is impossible in symmetric systems. However, weak modes can be excited in spatially asymmetric systems, in which the spatial field distribution in weak modes transforms from antisymmetric (in a symmetric system) to asymmetric (in a system without an inversion center). This report is devoted to the identification and discussion of the properties and features of the excitation of such weak plasmon modes.

Such weak modes can be excited, for example, in periodic structures in which the spatial symmetry is broken by a direct current flowing in the 2DEG [1]. Another example of a periodic structure without an inversion center, in which the excitation of weak plasmon modes is possible, is a structure with a 2DEG and a dual grating gate (figure 1). The nested gate subgratings are located in the same plane, have different gate widths, and are offset relative to each other in the grating plane to create asymmetry of the unit cell of the periodic structure. Such a double grating with an asymmetric unit cell can be located above one or several parallel sheets of 2DEG.

Excitation of weak plasmon modes due to the introducing the asymmetry into the structure leads to a doubling of the number of plasmon resonances in the spectrum [1]. Additional resonances of weak plasmon modes have a composition of Fourier harmonics different from strong modes. Due to spatial distribution of electric field in plasmon mode (figure 2) its' amplitudes of the dominant counterpropagating Fourier harmonics of weak plasmon modes are not equal to each other, while for strong plasmon modes such amplitudes are the same [2]. This feature of weak modes leads to the possibility of exciting a traveling plasma wave in 2DEG at normal incidence of an EM wave on the system [3]. Due to this property of weak modes, there is a transfer of EM energy and momentum by a traveling plasmon in 2DEG. Since, in the general case, the response of 2DEG to the acting electric field is nonlinear, then along with the excitation of a traveling plasmon, nonlinear quadratic effects will arise, such as rectification of the incident EM wave [4], for example, the effect of dragging of charge carriers in 2DEG by a plasmon. The asymmetric distribution of the plasmon field in the plasmon resonator, which is characteristic of the excitation of weak plasmon modes, leads to the appearance of the plasmon ratchet effect.

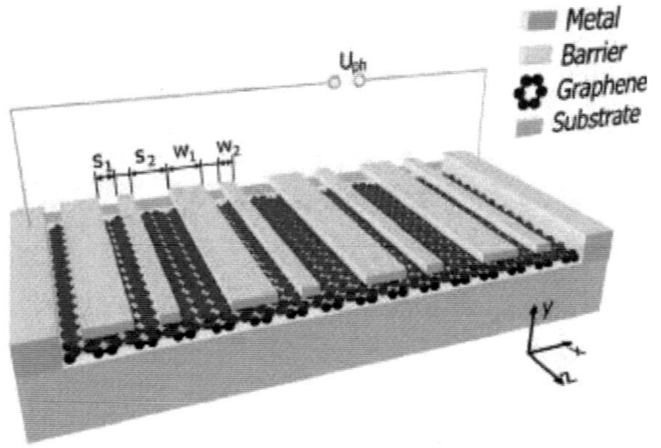

Figure 1. Schematic view of the periodic plasmon structure without center of inversion.

The indicated properties of weak plasmon modes also have disadvantages associated with weak conversion of the incident wave energy into a weak plasmon mode. Therefore, to enhance the effects on weak plasmon modes, the properties of the anti-crossing mode are used, in which strong and weak plasmon modes are excited simultaneously. In the anti-crossing mode, both modes are hybrid and have the property of good conversion of the incident wave energy into a plasmon, which is characteristic of strong modes, and the property of asymmetry of weak modes. For example, in the anti-crossing mode, it is possible to direct up to 70% of the incident wave power into a unidirectionally traveling plasma wave.

The possibility of controlling the excitation of weak modes by applying a constant voltage between the gate electrodes and 2DEG looks interesting and promising. On the one hand, using a constant voltage, it is possible to change the density of charge carriers in the 2DEG under the gate electrode, and, accordingly, the resonance frequency of plasmons, which makes it possible to switch to the anti-crossing mode of plasmon modes. On the other hand, by switching between different weak modes of different orders under different gate electrodes, one can electrically choose the direction of propagation of the traveling plasma wave [5].

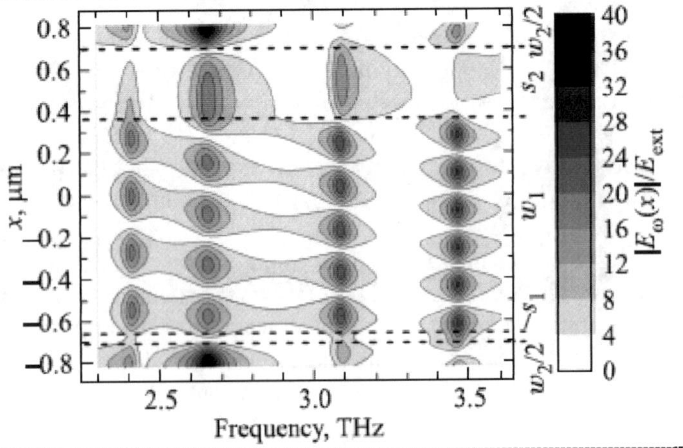

Figure 2. Spatial distribution of in-plane component of electric field in single unit cell of periodic plasmon structure with dual grating gate. Weak plasmon modes have even number of peaks at plasmon cavity, while strong modes have odd numbers of field peaks.

Another interesting property of the excitation of weak plasmon modes is their low radiative damping, and, depending on the asymmetry of the structure, the radiative damping of weak modes can be rearranged over a wide range. The most promising application of weak radiative damping is possible in structures allowing amplification of plasma oscillations [6]. The amplification mechanisms can be both drift instabilities and radiative recombination of charge carriers, for example, in graphene or cadmium-mercury-tellurium structures. Since the amplification in the active medium must compensate for the dissipative losses in the system and the radiation losses, the use of weak plasmon modes with small (sometimes almost zero) radiative damping makes it possible to significantly reduce the wave amplification thresholds [7].

3. Conclusions

This paper reviews studies of weak plasmon modes in structures without an inversion center. The use of weak modes for the plasmon detection of THz radiation, for the excitation of a unidirectionally traveling plasmon, and for lowering the plasmon threshold of THz radiation generation in structures with active graphene is considered.

Acknowledgments
This work was supported by the Russian Science Foundation, grant no. 18-79-10041.

References

[1] Aizin G R, Popov V V and Polischuk O V 2006 *Applied Physics Letters* **89** 143512
[2] Popov V V, Fateev D V, Ivchenko E L and Ganichev S D 2015 *Physical Review B* **91** 235436
[3] Fateev D V, Mashinsky K V, Polischuk O V and Popov V V 2019 *Physical Review Applied* **11** 064002
[4] Olbrich P, Kamann J, Konig M, Munzert J, Tutsch L, Eroms J, Weiss D, Ming-Hao Liu, Golub L E, Ivchenko E L, Popov V V, Fateev D V, Mashinsky K V, Fromm F, Seyller Th and Ganichev S D 2016 *Physical Review B* **93** 075422
[5] Morozov M Y, Popov V V and Fateev D V 2021 *Sci. Rep* **11**, 11431
[6] Moiseenko I M, Popov V V and Fateev D V 2020 *J. Phys. Commun.* **4** 071001
[7] Fateev D V, Polischuk O V, Mashinsky K V, Moiseenko I M, Morozov M Yu and Popov V V 2021 *Physical Review Applied* **15** 034043

High-Q states in acoustic apple-shaped resonators

Vladimir Igoshin, Anastasia Nikitina, Mariia Tsimokha, Ivan Toftul, Mihail Petrov and Kristina Frizyuk

Department of Physics and Engineering, ITMO University, St. Petersburg 197101, Russia

E-mail: k.frizyuk@metalab.ifmo.ru

Abstract. Apples play a significant role in our culture in various points of human history: starting from Adam and Eve, going on with Judgement of Paris, it also touches such great minds as Sir Isaac Newton and Alan Turing. Beyond that apples are still extremely relevant today due to Steve Jobs. In this work we study high quality (*high-Q*) resonant states of apple-shaped resonators. We have found that quasi bound states in continuum (*quasi-BICs*) are possible in the linear acoustic domain. We show that quasi-BICs are of Friedrich-Wintgen type, i.e. accompanied with avoided crossings while elongating or shrinking the apple-shaped resonator. Finally, we build a concise theory based on the group theory approach utilizing *Wigner's theorem*. We illustrate that only the resonator symmetry plays major role, but not particular resonator's shape.

1. Introduction

Studying acoustic resonators is inevitable for designing acoustic metamaterials [1, 2, 3, 4], optomechanical systems [5, 6, 7, 8], and creating the materials with specified properties [9, 10]. High-Q resonances based on observation of bound states in the continuum (BICs) are very important for engineering these properties and are extensively studied in optical [11, 12, 13, 14] and acoustical [15, 16, 17, 18, 19] systems. In our work, we focus on a Friedrich-Wintgen type BICs [20, 21] based on avoided crossings of two eigenstates decaying into the same radiation channel [22], which were recently discovered in optical resonators [23]. In solid-state physics, quantum chemistry and optics modes are classified by irreducible representations of a system's symmetry group [24, 25, 26, 27]. The classification can be applied to any structure and is defined solely by its symmetry. In this work, we obtained the high-Q resonances of Friedrich-Wintgen type in acoustic resonators of arbitrary shape, and discussed their nature in terms of the symmetry of the eigenmodes. We illustrate this with an example of an azimuthally symmetric apple-shaped resonator, to emphasize, that only the symmetry plays major role, and not particular resonator's shape.

2. Theoretical approach

The actual object of this research are eigenmodes of an acoustic resonator of *an arbitrary shape*. There is no general solution for the eigenmode problem, only specific cases can be studied fully analytically [28] such as sphere or infinite cylinder by spherical or cylindrical harmonics correspondingly. However, knowing the resonator's symmetry and applying Wigner's theorem [29, 24], one can find a set of constrains, certain rules, and properties of the modes. For the sake of con-

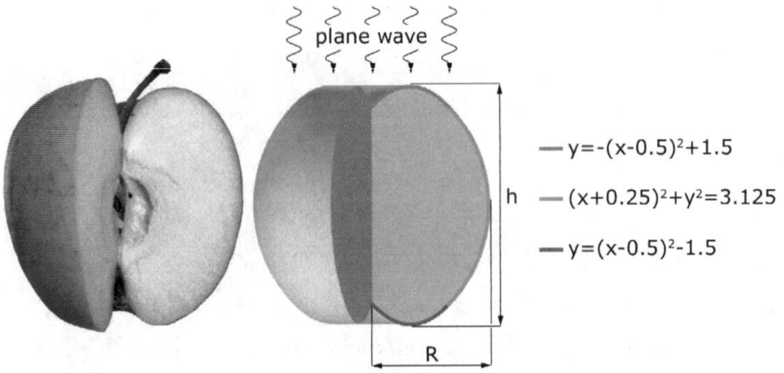

Figure 1. Geometry of the problem. Resonator boundary is represented by equations where x and y are measured in centimeters. Initial height $h_0 = 3$ cm, initial radius $R_0 = (\sqrt{3.125} - 0.25)$ cm. In this work, we scale only the height of the object by a factor h/h_0. We use density $\rho_0 = 1.225$ kg/m^3 and speed of sound $c_0 = 335$ m/s for the host and $\rho = 4.2 \cdot \rho_0$, $c = c_0/3$ for the resonator. These parameters are arbitrarily chosen while the system is scalable.

venience we explicitly state the theorem.

Wigner's theorem:

$$\mathcal{H}(\mathbf{r})\psi(\mathbf{r}) = \epsilon\psi(\mathbf{r}) \tag{1}$$

Suppose that a system (Hamiltonian) is invariant under the transformations of a symmetry group, then the eigenfunctions are transformed under irreducible representations of the group.

Namely, there are as many types of eigenmodes excited as there are irreducible representations of a symmetry group, defined by the resonator's shape. A set of modes' behavior under all symmetry transformations defines in accordance to which irreducible representation it transforms. With all of the above stated, each mode can be classified and named after its irreducible representation, which fully describes the symmetry of this mode. For the **linear acoustic domain**, Eq. (1) transforms to $-c(\mathbf{r})^2\nabla^2 p(\mathbf{r}) = \omega^2 p(\mathbf{r})$, where p is the acoustic pressure, ∇ is the nabla operator, ω is the angular frequency, and $c(\mathbf{r}) = 1/\sqrt{\beta(\mathbf{r})\rho(\mathbf{r})}$ is the speed of sound. The medium is described by the compressibility β and mass density ρ [28].

The key consequence of these considerations is that a coupling can appear only for the modes transformed in accordance with the same irreducible representation [30, 31], moreover, only modes of the same symmetry radiate in the same radiation channel. Every resonator has infinite number of modes of each type. Tuning the resonator's parameters can lead to the appearance of crossings (if there is no mode coupling or weak coupling [32]), or, on the contrary, the crossings can be avoided in case of strong coupling of the modes.

3. Results and methods

To illustrate this approach, we considered scattering by a single apple-shaped resonator with $D_{\infty h}$ symmetry [33, 34]. Incident plane wave is along the z-axis. This symmetry group contains an infinite number of elements and irreducible representations, but only two types of modes are excited by such normally incident plane wave: A_{1g} and A_{1u} modes. Both of these modes are invariant under rotations around the z-axis, as well as the incident plane wave, but they have opposite parity under reflection in the $z = 0$ plane.

We performed the numerical modelling with COMSOL Multiphysics™ simulation software. The following parameters were taken: density $\rho_0 = 1.225$ kg/m^3 and speed of sound $c_0 = 335$ m/s for the host and $\rho = 4.2 \cdot \rho_0$, $c = c_0/3$ for the resonator. We do not consider shear waves or

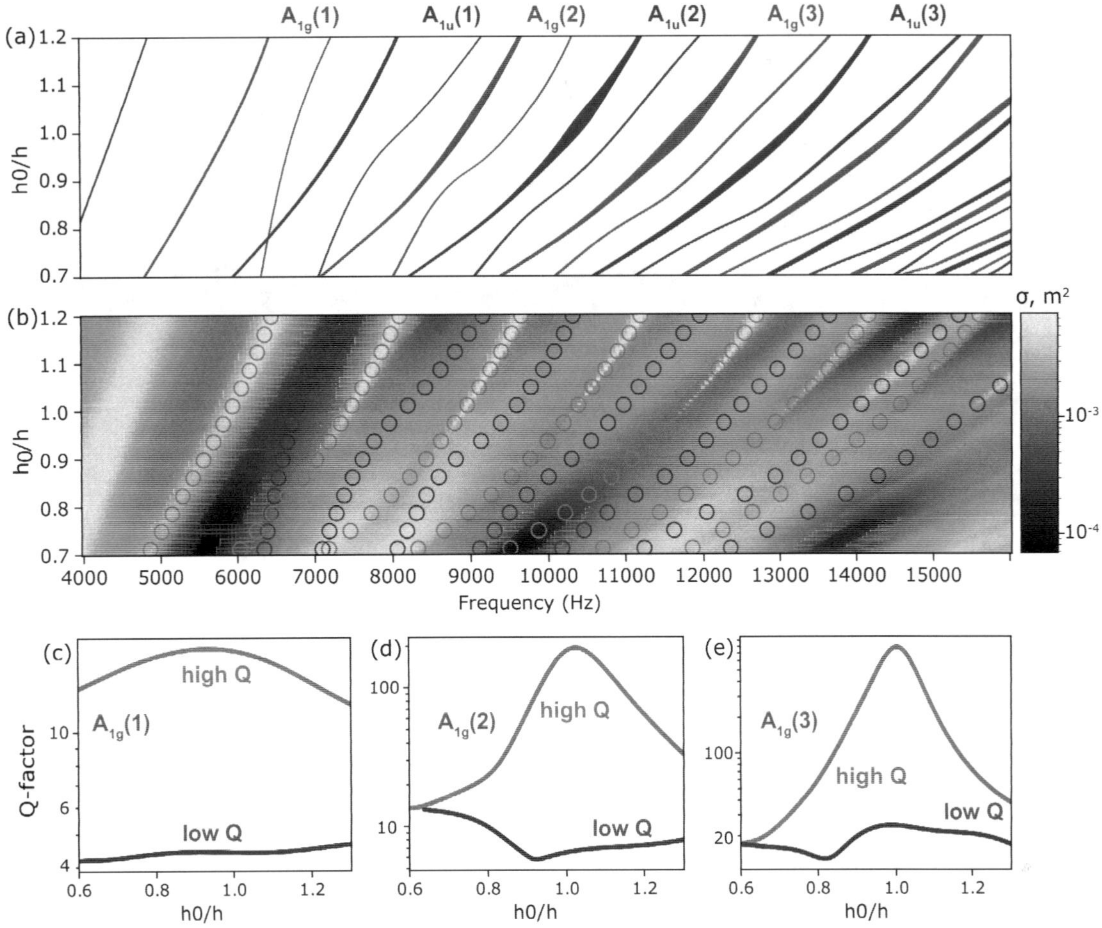

Figure 2. (a) The dependence of eigenmodes' frequency of the apple-shaped acoustic resonator on h_0/h ratio. Two types of modes shown, A_{1g} (green) and A_{1u} (purple), are named depending on their irreducible representations. Line width is proportional to the Q-factor of the mode. (b) Map of the calculated scattering cross section σ versus frequency and h_0/h ratio. High-Q (red) and low-Q (blue) modes are shown with circles. (c,d,e) Dependence of Q-factor on h_0/h ratio for three different pairs of modes is calculated.

viscosity, which lower the Q-factors, but theoretical considerations remain the same. Geometry of the problem is shown in Figure 1. We have computed scattering using the built-in 2D Pressure Acoustics physics interface as the problem possesses axial symmetry. Spherical wave radiation boundary condition was used in both models.

In Figure 2 eigenmodes and scattering maps of apple-shaped resonator are shown. It shows that between two modes transformed under the same representation an anticrossing can be observed, whereas between two modes transformed under different representations a crossing will be observed. In Figure 3 one of the eigenmodes of A_{1u} irreducible representation is shown in more detail. As one can see, at the point II (10.4 kHz frequency, $h_0/h = 1.0225$) the Q-factor increase of one of the modes takes place. Such behavior is typical for high-Q resonances of Friedrich-Wintgen type.

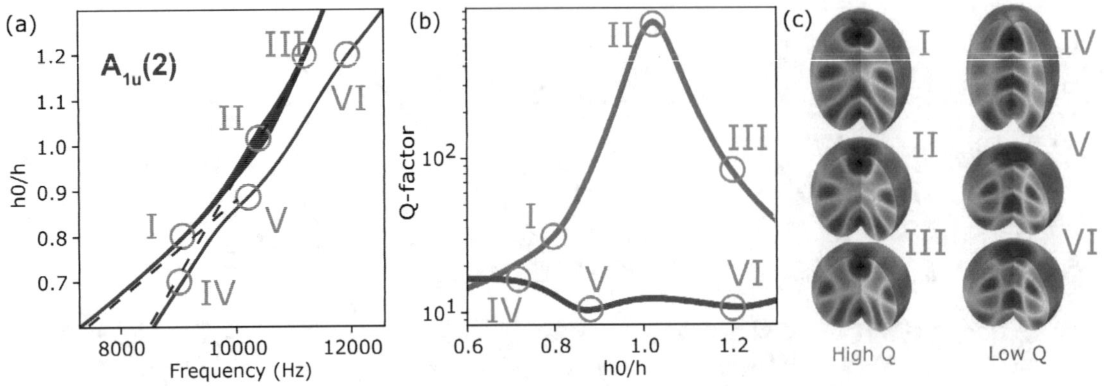

Figure 3. (a) Coupling of a pair of eigenmodes versus frequency and h_0/h ratio. Line width depends on the Q-factor. (b) Calculated dependence of Q-factor on h_0/h ratio for high Q (red) and low Q (blue) modes of the pair. (c) Field patterns in the resonator.

4. Conclusion

We have shown that scattering of an acoustic wave on an open resonator obeys the same mechanisms as light scattering on optical cavities. We have identified that for certain geometries high-Q resonant states can be observed due to the interaction of the modes with identical symmetry. The observation of these types of states is provided by the symmetry of the resonator and mode symmetry, but not the particular shape of this structure.

Acknowledgments

The work has been supported by Russian science foundation grant #20-72-10141.

References

[1] Zangeneh-Nejad F and Fleury R 2019 *Rev. Phys.* **4** 100031 ISSN 2405-4283
[2] Ma G and Sheng P 2016 *Sci. Adv.* **2** e1501595 ISSN 2375-2548
[3] Cummer S A, Christensen J and Alù A 2016 *Nat. Rev. Mater.* **1** 1–13 ISSN 2058-8437
[4] Liu J, Guo H and Wang T 2020 *Crystals* **10** 305 ISSN 2073-4352
[5] Toftul I D, Bliokh K Y, Petrov M I and Nori F 2019 *Phys. Rev. Lett.* **123** 183901 ISSN 1079-7114
[6] Carvalho N C, Bourhill J, Goryachev M, Galliou S and Tobar M E 2019 *Appl. Phys. Lett.* **115** 211102 ISSN 0003-6951
[7] MacCabe G S, Ren H, Luo J, Cohen J D, Zhou H, Sipahigil A, Mirhosseini M and Painter O 2020 *Science* **370** 840–843 ISSN 0036-8075
[8] Wang X, Qin W, Miranowicz A, Savasta S and Nori F 2019 *Phys. Rev. A* **100** 063827 ISSN 2469-9934
[9] Brunet T, Raffy S, Mascaro B, Leng J, Wunenburger R, Mondain-Monval O, Poncelet O and Aristégui C 2012 *Appl. Phys. Lett.* **101** 011913 ISSN 0003-6951
[10] Alig I and Lellinger D 1992 *J. Appl. Phys.* **72** 5565–5570 ISSN 0021-8979
[11] Hsu C W, Zhen B, Stone A D, Joannopoulos J D and Soljačić M 2016 *Nat. Rev. Mater.* **1** 1–13 ISSN 2058-8437
[12] Marinica D C, Borisov A G and Shabanov S V 2008 *Phys. Rev. Lett.* **100** 183902 ISSN 1079-7114
[13] Koshelev K, Kruk S, Melik-Gaykazyan E, Choi J H, Bogdanov A, Park H G and Kivshar Y 2020 *Science* **367** 288–292 ISSN 0036-8075
[14] Volkovskaya I, Xu L, Huang L, Smirnov A I, Miroshnichenko A E and Smirnova D 2020 *Nanophotonics* **9** 3953–3963 ISSN 2192-8614
[15] Lyapina A A, Maksimov D N, Pilipchuk A S and Sadreev A F 2015 *J. Fluid Mech.* **780** 370–387 ISSN 0022-1120
[16] Stevenson A C and Lowe C R 1998 *Appl. Phys. Lett.* **73** 447–449 ISSN 0003-6951
[17] Pillai G and Li S S 2020 *Appl. Phys. Lett.* **116** 163502 ISSN 0003-6951

[18] Sadreev A F 2020 *arXiv* (*Preprint* 2011.01221) URL https://arxiv.org/abs/2011.01221v1
[19] Huang L, Chiang Y K, Huang S, Shen C, Deng F, Cheng Y, Jia B, Li Y, Powell D and Miroshnichenko A 2021 Sound Trapping in an Open Resonator [Online; accessed 21. Mar. 2021]
[20] Friedrich H and Wintgen D 1985 *Phys. Rev. A* **32** 3231–3242 ISSN 2469-9934
[21] Lee S G, Kim S H and Kee C S 2020 *Nanophotonics* **9** 4373–4380 ISSN 2192-8614
[22] Remacle F, Munster M, Pavlov-Verevkin V B and Desouter-Lecomte M 1990 *Phys. Lett. A* **145** 265–268 ISSN 0375-9601
[23] Rybin M V, Koshelev K L, Sadrieva Z F, Samusev K B, Bogdanov A A, Limonov M F and Kivshar Y S 2017 *Phys. Rev. Lett.* **119** 243901 ISSN 1079-7114
[24] Ivchenko E L and Pikus G 1997 *Superlattices and Other Heterostructures* (Berlin, Germany: Springer-Verlag) ISBN 978-3-642-64493-1
[25] Ohtaka K and Tanabe Y 1996 *J. Phys. Soc. Jpn.* **65** 2670–2684 ISSN 0031-9015
[26] Gelessus A, Thiel W and Weber W 1995 *J. Chem. Educ.* **72** 505 ISSN 0021-9584
[27] Gladyshev S, Frizyuk K and Bogdanov A 2020 *Phys. Rev. B* **102** 075103 ISSN 2469-9969
[28] Williams G 1999 *Fourier Acoustics* (Oxford, England, UK: Elsevier Science) ISBN 978-0-08050690-6
[29] Piróth A and Sólyom J 2007 *Fundamentals of the Physics of Solids* (Berlin, Germany: Springer) ISBN 978-3-54072600-5
[30] Odit M, Koshelev K, Gladyshev S, Ladutenko K, Kivshar Y and Bogdanov A 2021 *Adv. Mater.* **33** 2003804 ISSN 0935-9648
[31] Krasikov S, Odit M, Dobrykh D, Yusupov I, Mikhailovskaya A, Shakirova D, Shcherbakov A, Slobozhanyuk A, Ginzburg P, Filonov D and Bogdanov A 2021 *Phys. Rev. Appl.* **15** 024052 ISSN 2331-7019
[32] Huang L, Xu L, Rahmani M, Neshev D N and Miroshnichenko A E 2021 *Adv. Photonics* **3** 016004 ISSN 2577-5421
[33] 2021 Character table for point group D∞h [Online; accessed 21. Mar. 2021] URL http://symmetry.jacobs-university.de/cgi-bin/group.cgi?group=1001option=4
[34] 2001 Non-Crystallographic: Dinfh [Online; accessed 21. Mar. 2021] URL http://newton.ex.ac.uk/research/qsystems/people/goss/symmetry/Dinfh.html

Journal of Physics: Conference Series **2015** (2021) 012041 doi:10.1088/1742-6596/2015/1/012041

Subwavelength probing of surface plasmons in magnetoplasmonic crystals

Aleksandr Yu Frolov[1] *, Niels Verellen[2,3], Victor V Moshchalkov[2], Andrey A Fedyanin[1]
1 Faculty of Physics, Lomonosov Moscow State University, 119991 Moscow, Russian Federation
2 Quantum Solid-State Physics, Department of Physics and Astronomy, KU Leuven, Celestijnenlaan 200D, B-3001 Leuven, Belgium.
3 IMEC, Kapeldreef 75, B-3001 Leuven, Belgium

*frolov@nanolab.phys.msu.ru

Abstract. In this work, we report on near-field studying of propagating surface plasmons (SPs) in one-dimensional magnetoplasmonic crystals (MPCs) by aperture type scanning near-field optical microscopy (SNOM). Optical near-field around the aperture probe is used to drive SPs in the MPC locally. The SNOM signal represents the scattered intensity caused by the interaction of the SNOM probe near-field with the MPC. Scanning the MPC surface with polarization resolving of the scattered radiation shows decreasing of the intensity due to the SP excitation. The observed polarization dependence of the scattered SNOM signal is associated with the selective coupling of the near-field components of the SNOM probe with SPs. Finite-difference time-domain simulations reproduce the experimental SNOM signal. It is shown the excitation of SPs with symmetric (even parity) field distribution, which is forbidden for plane wave source at normal incidence.

1. Introduction.

Surface plasmons (SP) play a crucial role in manipulating light at a subwavelength scale. It allows multifold enhancement of the electromagnetic field [1], which can boost field-dependent effects such as Raman scattering [2], nonlinear harmonic generation [3], and Purcell factor. Magnetooptical materials combined with plasmonic nanostructures open up opportunities to actively control the polarization and intensity of the light at the nanoscale by the external magnetic field [4]. Magnetoplasmonic crystals (MPCs) are one of the efficient magnetoplasmonic devices. MPCs represent either periodically corrugated metal surfaces and thin films, or the arrangement of the single nanoparticles, which can enhance the magnetooptical effects [5,6] and efficiency of sensors [7-9].

Plane wave incident from the far-field is usually employed to drive SPs in MPCs. It results in Fano-resonance in the reflectance or transmittance spectra consisting of Rayleigh and Wood anomalies. Despite significant spectral features in the far-field response, the plane wave characterization of MPCs does not provide the subwavelength visualization of SPs and excitation of SPs with even parity (symmetric SPs) forbidden for the plane wave excitation at normal incidence. In contrast, subwavelength excitation of plasmonic nanostructures with aperture or apertureless scanning near-field optical microscope (SNOM) is an excellent tool for visualization and mapping a strongly localized electromagnetic field with subwavelength resolution [10-12]. Additionally, similar to the far-field excitation, the local excitation of the single gold nanodisk with aperture SNOM probe results in the Fano-resonance [13]. However, mapping of SPs and characterization of SP-assisted subwavelength scattering in magnetooptical active nanostructure

Journal of Physics: Conference Series **2015** (2021) 012041 doi:10.1088/1742-6596/2015/1/012041

was performed only for the limited number of magnetoplasmonic devices. For instance, the nickel disk-shaped nanoantenna was probed by the apertureless type SNOM to map the dipole SP mode [14].

Here, we performed the near-field excitation of the 1D MPC made of nickel by the aperture type SNOM probe to characterize SPs with subwavelength resolution. Scanning MPC surface with polarization resolving the scattered radiation intensity allows identification of coupling with a normal near-field component. By finite-difference time-domain (FDTD) simulations of the scanning process, the simulated SNOM scattered signal reproduces the experimental one. The simulated near-field distribution shows that the SNOM probe near-field couples with symmetric (even parity) SPs at probe positions in the MPC grooves and on the MPC hills, where the scattering SNOM intensity is reduced due to the excitation SPs.

2. Results and discussions.

2.1. Experiment.

We used the commercial aperture type scanning near-field optical microscope (WiTec alpha300s) in the reflection mode to map SPs in the MPC [11,12]. The experimental scheme of the near-field scanning process and local excitation of SPs in the MPC is shown in Fig.1(a). X-polarized light was focused into the aperture SNOM probe. The aperture SNOM probe represents the hollow SiO_2 pyramid covered by an Al-layer of 100 nm thickness. The hole of 100 nm diameter was etched at the apex of the probe. The supercontinuum light (NKT Photonics, Koheras SuperK Extreme Standard) was used as a light source. The acousto-optical tunable filter (AOTF) selects simultaneously up to eight wavelength channels from the supercontinuum spectrum in the range from 550 nm to 750 nm. The spectral width of each channel is around 10−20 nm. When the probe is near the MPCs, the near-field interaction around the probe with the MPC surface leads to the light scattering to the far-field (see Fig.1(b)). Fig.1(b) demonstrates the scheme of recording the scattered radiation intensity. The scattered light is collected by the lens, which is mounted at 7 cm distance from the probe and the sample. The thin-film polarizer was placed before the lens to provide the polarization resolving of the scattered light. The collected light is registered by a spectrometer (Princeton Instruments) equipped with a Si CCD camera. SNOM cross-sections of the MPCs are obtained by scanning the surface of the MPC and registering the intensity of the scattered light at each probe position. The polarizer axis was directed along the z- or x-axis during the scan. The sample scanning was performed in atomic force microscopy (AFM) contact mode. The xyz-piezostage realized the positioning of the sample with respect to the aperture probe.

The MPC was fabricated by e-beam lithography of photopolymer and subsequent electroplating of the nickel layer. The MPCs present the 1D periodically corrugated nickel gratings with a pitch of d=600±5 nm and groove depth of h=115±5 nm (see the sketch of the MPC in Fig.1(a)).

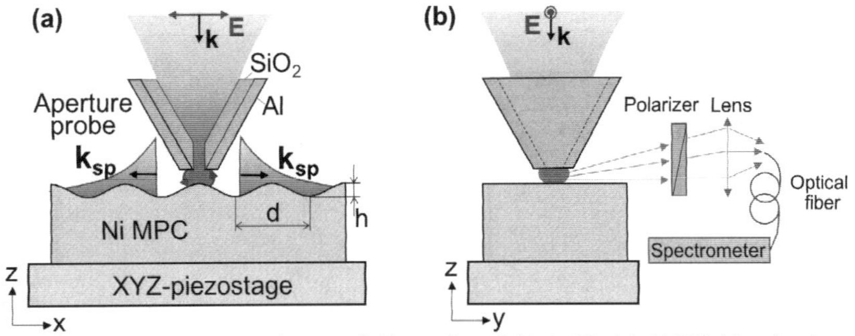

Figure 1. Schematic presentation of the scanning near-field mapping of SPs in 1D nickel MPC. (a) - view in xz- and (b) - in zy-plane.

Figure 2(a),(b) shows spatial cross-sections (red curves) of the experimental SNOM map of the MPC taken at λ=600 nm for z- and x- directions of the polarizer axis, respectively. Black curves depict AFM cross-sections of the MPC obtained with the SNOM probe. At z- and x-direction of polarizer axis, the registered signals represent the intensity of the scattered radiation with z- or x-polarization direction, respectively. SNOM spatial cross-sections consist of maximum and minimum, which correspond to enhancing or reducing the scattered radiation from the SNOM probe. The observed asymmetry of arrangement between the SNOM and AFM cross-sections is associated with unevenness of the apex of the manufactured SNOM probe. In the case of the z-axis of the polarizer, the intensity of the scattered light is reduced when the probe is located on the hills and in the grooves of the MPC. The increasing of scattered intensity is observed on the slope of the MPC hills. The change of the intensity contrast during the move of the SNOM probe along the MPC surface is associated with the excitation of SPs. Near-field light around the SNOM probe apex excites SPs, leading to decreasing the SNOM signal (in grooves and on hills). When the probe moves along the slope of the MPC profile, the coupling efficiency of the near-field around the probe with SPs is decreased, and the near-field light is efficiently scattered to the direction of the detector. It leads to the maximum of SNOM signal.

Fig.2(b) demonstrates the cross-sections of the SNOM map recorded at the x-direction of the polarizer axis, which is proportional to the $|E_x|$ component of the scattered SNOM signal. The SNOM signal possesses strong maxima at the center of hills of the MPC. These maxima correspond to minima in Fig.2(a). The maxima of the SNOM scattered intensity polarized along the x-axis is associated with non-resonant scattering of the probe near-field with x-polarization. The $|E_x|$ component of the SNOM probe at MPC hills does not couple with the SPs opposite the $|E_z|$ field component and efficiently scatters to the far-field on MPC hills. When the SNOM probe moves along slopes of the MPC and is in the grooves, the x-polarized SNOM scattered intensity is reduced due to higher distance between the SNOM probe and the MPC surface.

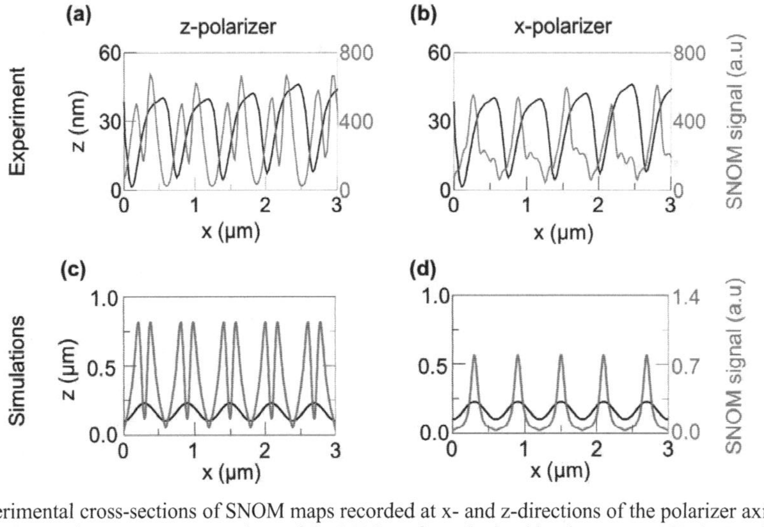

Figure 2. (a,b) Experimental cross-sections of SNOM maps recorded at x- and z-directions of the polarizer axis (red curves), respectively. Atomic force microscopy cross-sections of the MPC surface obtained by the aperture SNOM probe are shown by black curves. (c,d) Results of FDTD simulations.

2.2. Simulations.

The scanning process of the SNOM probe along the MPC surface was simulated by the FDTD technique (Lumerical solver). 3D models of the SNOM probe and MPC were used in simulations. The scattered radiation was recorded by near-to-far field projection of E_z and E_x components collected by the 2D xz-monitor placed at 350 nm distance from the SNOM probe. The E_z and E_x values were projected at the distance of 7 cm and then were integrated over the square area (area of the lens) of 2x2 cm^2 ($\int|E_z|^2dxdz$, $\int|E_x|^2dxdz$), which corresponds to experimental conditions. Simulation results of the SNOM signal are shown in Fig. 2(c) and (d), which correspond to the $\int|E_z|^2dxdz$ and $\int|E_x|^2dxdz$ values, respectively. The arrangement of minima and maxima of the SNOM signal in simulation results is in good agreement with experimental ones. It should be noted that positions of MPC hills exactly correspond to the minimum of $\int|E_z|^2dxdz$ and maximum of $\int|E_x|^2dxdz$ values in simulated spatial cross-sections of the scattered light intensity, as there is no unevenness of the apex of the SNOM probe in comparison with the manufactured one.

Figure 3(a),(b) shows FDTD simulations of field $|E_z|$ (top row) and phase $\varphi(E_z)$ (bottom row) distributions of excited SPs in the MPC taken at the hill and groove positions of the SNOM probe, respectively (λ=600 nm). $|E_z|$ and $\varphi(E_z)$ distribution was recorded in the same monitor used for the near-to-field projection (350 nm distance from the SNOM probe). It can be seen that at hill and groove positions (Fig.3(a),(b)), the SNOM probe excites symmetric (even) mode (with respect to the axis coming through the center of MPC hill, (i.e., $E_z(-x)=E_z(x)$). It has one antinode with respect to the center of MPC hills and one E_z-antinode located in grooves. Whereas, Fig.3(c) shows that the plane wave source at the normal incidence (at λ=630 nm, which corresponds to a minimum of reflectance spectrum) excites the SP with antisymmetric field distribution, $E_z(-x)=-E_z(x)$ (odd parity SP mode). It possesses two $|E_z|$ antinode symmetrically located with respect to the center of the MPC hill. The excitation of even parity SP modes is forbidden by plane wave source due to the odd symmetry of plane wave source at the normal incidence.

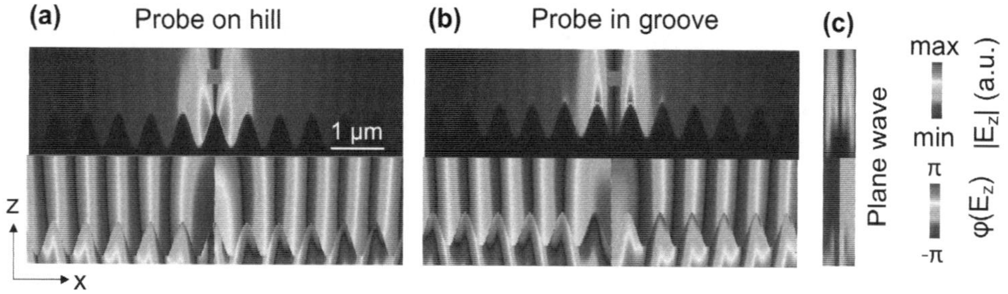

Figure 3. FDTD simulations of $|E_z|$ (top rows) and $\phi(E_z)$ (bottom rows) distribution taken at the positions of the probe on the hill (a) and in the groove (b) of the MPC surface. The purple square dot depicts positions of the SNOM probe on the MPC. (c) $|E_z|$ (top) and $\phi(E_z)$ (bottom) distribution of the SP excited by the plane wave source.

3. Conclusion

We performed the near-field studying of the surface plasmons (SPs) in opaque one-dimensional nickel magnetoplasmonic crystals (MPCs). Studying the near-field was based on the local excitation of SPs by the aperture SNOM probe and registering the scattered radiation into the far-field with polarization resolving. Analysis of the polarization of scattered light showed the presence of the electric field components normal and in-plane to the MPC. The simulations of the scanning process of the SNOM probe along the MPC surface accurately reproduced the experimental SNOM cross-sections. The near-field simulations at MPC hills and grooves positions revealed the coupling of the normal electric field component of the SNOM probe with the SP. SP excitation results in the minima of the scattered SNOM signal with normal polarization, in contrast to in-plane polarization. The excited SPs possess even parity field distribution, which is forbidden using the plane wave source at the normal incidence.

4. Acknowledgments

The authors wish to acknowledge Russian Foundation for Basic Research (grant 19-02-00876, 20-02-00758). This work is supported by MSU Quantum Technology Centre. This research was performed according to the Development program of the Interdisciplinary Scientific and Educational School of Lomonosov Moscow State University «Photonic and Quantum technologies. Digital medicine».

5. References

[1] Barnes W L, Dereux A, and Ebbesen T W, 2003, *Nature*, **424**, 824
[2] Langer J et al, 2020, *ACS Nano,* **14** 28-117
[3] Kauranen M, Zayats A V, 2012, *Nat. Photonics*, **6**, 737-48
[4] Armelles G, Cebollada A, García-Martín A. González M U 2013, *Adv. Opt. Mater.*,**1**, 10-35
[5] Frolov A Yu, Shcherbakov M R, and Fedyanin A A, 2020, *Phys. Rev. B*, **101**, 045409

[6] Musorin A I, Chetvertukhin A V, Dolgova T V, Uchida H, Inoue M, Luk'yanchuk B S, and Fedyanin A A, 2019, *Appl. Phys. Lett.*, **115** 151102
[7] Novikov I A, Kiryanov M A, Nurgalieva P K, Frolov A Yu, Popov V V, Dolgova T V, and Fedyanin A A, 2020, *Nano Lett.,* **20**, 8615-19
[8] Belyaev V K, Rodionova V V, Grunin A A, Mitsuteru I, Fedyanin A A, 2020, *Sci. Rep.*, **10**, 7133
[9] Grunin A A, Mukha I R, Chetvertukhin A V, Fedyanin A A, 2016, *J. Magn. Magn. Mater.,* **415**, 72
[10] Rotenberg N, and Kuipers L, 2014, *Nat. Photonics,* **8**, 919-26
[11] Denkova D, Verellen N, Silhanek A V, Valev V K, Van Dorpe, and Moshchalkov V V, 2013, *ACS Nano* **7**, 3168-76
[12] Frolov A Yu et al, 2017, *Nano Lett.*, **17**, 7620-37
[13] Imura K, Ueno Kosei, Misawa H, and Okamoto H, 2011, *Nano Lett.*, **11**, 960-65
[14] Bonanni V, Bonetti S, Pakizeh T, Pirzadeh Z, Chen J, Nogués J, Vavassori P, Hillenbrand R, Åkerman J, and Dmitriev A, 2011, *Nano Lett.*, **11**, 5333-38

Mesoporous perovskite solar cells with Al- and Zn-based metal-organic frameworks

A.D. Furasova[1,2], G. Hix[3], S.V. Makarov[1], A. Di Carlo[2]

[1] Department of Physics and Engineering, ITMO University, St. Petersburg, 197101, Russia

[2] C.H.O.S.E. (Centre for Hybrid and Organic Solar Energy), Department of Electronic Engineering, University of Rome-Tor Vergata, Rome, 00133, Italy

[3]Faculty of Science and Engineering, University of Wolverhampton, WV1 1LY, Wolverhampton, United Kingdom

aleksandra.furasova@metalab.ifmo.ru, furasova@inbox.ru

Abstract. The improvement of lead halide perovskites solar cells (PSC) by hydrophobic metal-organic frameworks (MOF) is one of the promising tools for modern photovoltaic technology to achieve stable and efficient thin-film devices. To show the MOF applicability for PSC, we incorporate two types of MOF: NH_2-MIL-53(Al) and basolite Z1200 in n-i-p mesoporous $MAPbI_3$ based solar cells that can add 2.2% efficiency by increasing main photovoltaic parameters. The simplicity of the proposed MOF's integration allows to use and adopt this approach to incorporate other frameworks for thin-film perovskite devices.

1. Introduction

Perovskite solar cells (PSC) present the new era for thin film photovoltaic technology due to unique properties of active material such as an easily tuneable and direct band gap, high light absorption coefficient and fabrication simplicity. At the same time, perovskite thin films prepared by chemical methods contain many charged defect centers up to $10^{15}/cm^3$ [1] due to the rapid crystal growth and a chemically active surface area.

Metal organic frameworks (MOFs) present three-dimensional crystals constructed of metal ions with a valence (II) higher and organic ligands connecting metal ions. Due to the structuring features MOFs have a high porosity and can be used as a scaffold for material crystallization. Moreover, the choice of organic chains is crucial for thin film perovskite as they can protect a contained material from chemical reactions. The perovskite film crystallization improvement by MOFs and nanoparticles were previously confirmed in a few works [2-7], whereas in our study we compare Basolite Z1200 and NH_2-MIL-53(Al) influence on the main photovoltaic properties of PSCs.

2. Discussion

Commercially available Basolite Z1200 and NH_2-MIL-53(Al) nanocrystals were carried out through a solvothermal route in a DMF–water mixed solvent system with $AlCl_3{\cdot}6H_2O$ and 2-aminoterephthalic acid as the metal source and the organic linker, respectively [8]. Then obtained crystals were dispersed in Ethanol and filtered through PTFE 220 nm pore size filter. Then, each kind of MOF nanocrystals were dropped on the planar mesoporous TiO_2 contained FTO substrates and spin coated at 1000 rpm for 10 s. After that, the $MAPbI_3$ perovskite, SPIRO-MeOTAD and gold layers were deposited by techniques properly described by Furasova [7]. The chemical formulas of NH_2-MIL-53(Al) and Basolite Z1200 compounds are presented in figure 1.

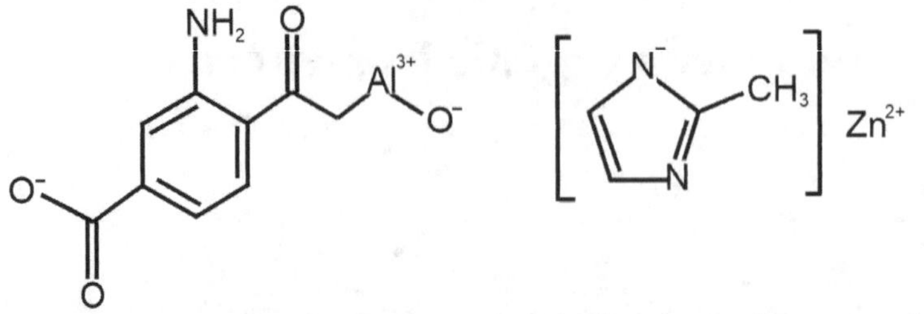

Figure 1. Chemical structure of metal organic frameworks which were used for perovskite solar cells modification. Left: NH_2-MIL-53(Al), right: basolite Z1200.

The thin film perovskite crystal growth can increase mobility for minor and major charges and their lifetime under illumination. As a result, large numbers of charges are able to get through the perovskite-transport layer interband transition, which boost almost all main PSCs parameters. Figure 2 presents J-V curves for the most efficient PSCs: pristine, containing Zn-based and Al-based MOFs. The best reference cell has 13.4% efficiency at short shunt current (J_{SC}) of 18.1 mA/cm^2 and open current voltage (V_{OC}) of 0.951 V, compared to the Basolite Z1200 with 15.4% efficiency with 20.2 mA/cm^2 J_{SC} and 0.967 V V_{OC}. The most efficient cell contains NH2-MIL-53(Al) and achieved 15.8% efficiency at 20.6 mA/cm^2 J_{SC} and 0.955 V V_{OC}.

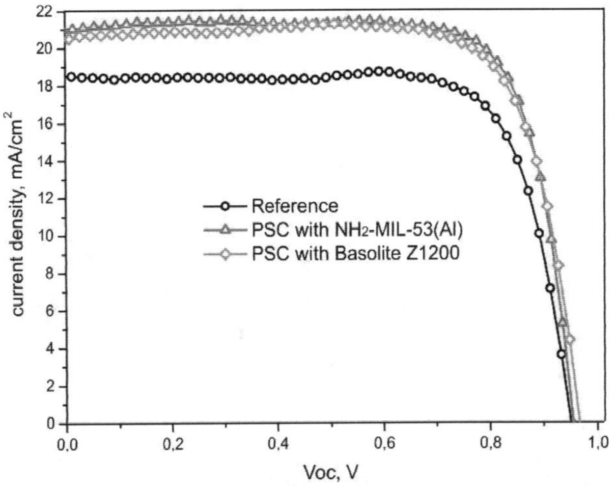

Figure 2. a) JV curves for the best reference (black circles), NH_2-MIL-53(Al) contained device (green triangles) and basolite Z1200 (orange squares) PSCs.

According to the statistical data (Figure 3), the fill factor for all kind of devices remains equal and around 77-79%, whereas the average efficiency improvement relates to the J_{SC} and V_{OC} boosting that confirms perovskite crystallisation improvement by MOF presence.

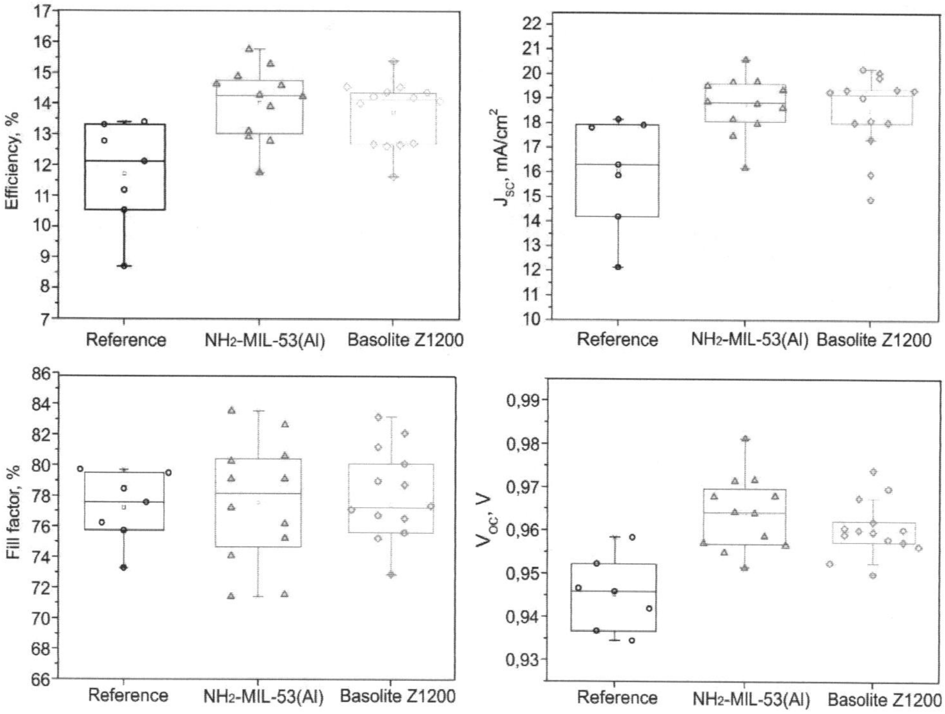

Figure 3. Statistical data for the main photovoltaic parameters (efficiency, short shunt current density - J_{sc}, fill factor, and open current voltage – V_{oc}) of different perovskite devices: black data – reference cells, green - NH_2-MIL-53(Al), orange - basolite Z1200 PSCs.

3. Conclusion

In this work, we have proved that metal-organic frameworks incorporation is a prospective approach to the crystallization improvement of perovskite layer in PSCs. We have demonstrated the high efficiency growth in 2.2% (absolute value) by using NH_2-MIL-53(Al) and in 2% (absolute value) by using basolite Z1200. The proposed PSC's improvement method can be used to prepare perovskite devices with good crystallinity and opens new opportunities for hybrid MOF-perovskite advanced materials.

Acknowledgements
The work was supported by the Russian Science Foundation (19-73-30023).

References

[1] Buin A, Pietsch P, Xu J, Voznyy O, Ip A, Comin R, Sargent E, 2014 *Nano Lett.*, **14**, 11
[2] Chang T-H, Kung C-W, Chen H-W, Huang T-Y, Kao S-Y, Lu H-C, Lee M-H, Boopathi K M, Chu C-W, Ho K-C, 2015*Adv. Mater.*, **27**, 44
[3] Lee C-C, Chen C-I, Liao Y-T, Wu K, Chueh C-C, 2019 *Adv. Sci.*, **6**
[4] Wu S, Li Z, Li M-Q, Diao Y, Lin F, Liu T, Zhang J, Tieu P, Gao W, Qi F, Pan X, Xu Z, Zhu Z, Jen A, 2020 *Nat. Nanotechnol.*, **15**
[5] Heo D Y, Do H H, Ahn S H, Young Kim S, 2020 *Polymers*, **12**, 9
[6] Shen D, Pang A, Li Y, Dou J, Wei M, 2018 *Chem. Commun.*, **54** 1253
[7] Furasova A, Voroshilov P, Lamanna E, Mozharov A, Tsypkin A, Mukhin I, Barettin D, Ladutenko K, Zakhidov A, Di Carlo A, Makarov S, 2020 *Energy Technol.*, **8**
[8] Ahnfeldt T, Gunzelmann D, Loiseau T, Hirsemann D, Senker J, Férey G, Stock N, 2009 *Inorg. Chem.*, **48**, 7

Journal of Physics: Conference Series **2015** (2021) 012043 doi:10.1088/1742-6596/2015/1/012043

Excitation of a homogeneous dielectric sphere by a point electric dipole

Roman Gaponenko,[1,*] Ilia Rasskazov,[2] Alexander Moroz,[3] Dmitry Pidgayko,[1] Konstantin Ladutenko,[1] Alexey Shcherbakov,[1] and Pavel Belov[1]

E-mail: `roman.gaponenko@metalab.ifmo.ru`
[1] School of Physics and Engineering, ITMO University, 197101, Saint-Petersburg, Russia
[2] The Institute of Optics, University of Rochester, Rochester, New York 14627, United States
[3] Wave-scattering.com

Abstract. Electrically small dielectric antennas are of great interest for modern technologies, since they can significantly reduce the physical size of electronic devices for processing and transmitting information. We investigate the influence of the resonance conditions of an electrically small dielectric spherical antenna with a high refractive index on its directivity and analyze the dependence of these resonances on the effectively excited modes of the dielectric sphere.

1. Introduction

The first antennas based on the use of a dielectric resonator appeared at the end of the last century. They have recently become in high demand during the rapid development of communication networks. This process involves a gradual increase in the quality and speed of information transfer, which inevitably forces telecom operators to move to higher frequencies, requires a reduction of the physical dimensions of devices, and also a minimization of interference effects. The development of super directional antennas can be essential for a broad variety of applications.

Dielectric resonator antenna (DRA) is currently one of the most attractive antenna types for use in the microwave and millimeter wave bands due to simplicity of design, small physical size, wide frequency range (0.7...40GHz), high radiation efficiency, varying polarizations, multiple feed systems, stable radiation patterns, ease of integration with other antennas, and high temperature tolerance. The absence of metal parts makes it possible to work more efficiently than metal antennas at high frequencies, since replacing conductive materials with dielectrics eliminates ohmic losses, as well as the lack of additional power loss into their dielectric substrates through surface waves. The absence of any metal components makes such DRA's attractive for space applications. DRA's have also its disadvantages, such as low mechanical strength (for most materials used), sensitivity to temperature changes and mechanical vibrations.

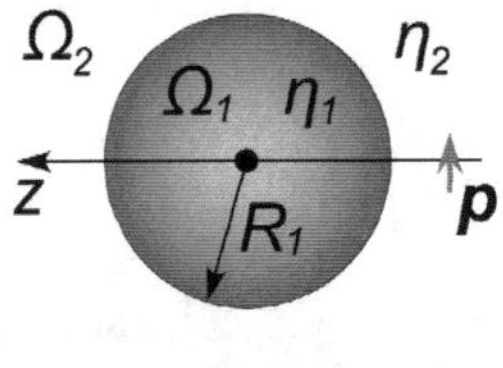

Figure 1. Sketch of the problem under study.

2. Formulation

DRA's are known to have resonant behavior [1, 2, 3, 4]. Resonances in scattering correspond to the poles of the scattering coefficients [5]. They are often called "morphology-dependent resonances", which emphasizes their dependence on the parameters of the sphere, such as the refractive index and the ratio of the radius of the sphere to the wavelength. From a physical point of view, the resonances are directly related to an efficient interaction of an incident electromagnetic wave with the sphere through the excitation the sphere's eigenmodes [6, 7]. In order to achieve a directional scattering, it is necessary that the emission frequency of an electric dipole is close to one of the resonant frequency. Usually, to obtain a directional radiation in one direction, it is considered sufficient to suppress radiation in the opposite direction using the interference of electric and magnetic modes of the scatterer [8]. While this method allows to get directional designs, it does not allow to always achieve the highest possible directivity, especially for electrically small DRA's with a high refractive index [7].

We consider an electrically small lossless antenna in the form of a homogeneous dielectric sphere located in air and excited by a *tangentially* oriented point electric dipole $\mathbf{p}$ as shown in Figure 1. The dipole can be located both inside and outside DRA. The center of coordinates is at the sphere origin. We use the following notation here: the dipole position, r_d; radius of the spherical dielectric antenna, R_1; a homogeneous isotropic concentric domain, Ω_1; refractive index of the sphere, $\eta_1 \geq 1$; refractive index of the surrounding medium, $\eta_2 = 1$ (in the domain Ω_2). We consider homogeneous, isotropic, non-magnetic materials with the permeability of a vacuum, μ_0. It is worth noting that for a resonant condition, especially for a homogeneous sphere with a high refractive index, irregularities such as roughness and bulge can be important, but in this work the case of a perfectly smooth surface is assumed.

In what follows, we make use of an exact analytical solution to the problem of scattering of an electric dipole radiation on a sphere [9]. The resonance conditions can be calculated from the poles of the scattering coefficients [9, eqs.(63),(66)], which, under some circumstances, can be similar to the conditions for electromagnetic resonances of free dielectric sphere [4]:

$$\begin{aligned}
&\text{Mode } TE_{lms}: \quad j_{l-1}(\eta_1 kR_1)/j_l(\eta_1 kR_1) = h^{(2)}_{l-1}(kR_1)/\eta_1 h^{(2)}_l(kR_1);\\
&\text{Mode } TM_{lms}: \quad \frac{l}{\eta_1 kR_1} - \frac{j_{l-1}(\eta_1 kR_1)}{j_l(\eta_1 kR_1)} = \frac{l}{kR_1} - \eta_1 \frac{h^{(2)}_{l-1}(kR_1)}{h^{(2)}_l(kR_1)}.
\end{aligned} \tag{1}$$

Here $j_l(\cdot)$ and $h^{(2)}_l(\cdot)$ are the spherical Bessel function of the first kind and the spherical Hankel function of the second kind of the order l, and k is a free space wavenumber. These equations are independent of the azimuthal mode number m because of spherical symmetry. The latter implies that all $2l+1$ modes with different m are all degenerate in frequency.

In the approximation $\eta_1 \gg 1$, the resulting zeros of the spherical Hankel function give solutions for the waves outside the sphere (kR_1-dependent "external" modes), whereas the zeros of the spherical Bessel function give solutions for the waves inside the sphere ($\eta_1 kR_1$-dependent "internal" modes). For the "external" modes the energy is concentrated on the surface or outside the sphere. The energy of "internal" modes is concentrated in the entire volume of the sphere and on its surface, but with the predominance of the stored energy inside the sphere. Since the Q factor of "external" modes is less than unity [4] and they cannot be efficiently excited in electrically small antennas, we focused only on the "internal" modes which may have significantly higher Q factor. The quality factor, Q, is a dimensionless parameter that can be roughly defined for leaky modes as the ratio of the energy stored in the resonator to the energy lost in one radian of the oscillation cycle.

Finally, the approximate resonance conditions for $TE_{(l+1)ms}$ and TM_{lms} modes in the $\eta_1 \gg 1$ approximation:

$$j_l(\eta_1 k R_1) \simeq 0 \quad (2)$$

The first several zeros ζ_{ls} of $j_l(\zeta_{ls}) = 0$ are presented in Table 1.

	$s=1$	$s=2$	$s=3$	$s=4$
$l=0$	3.14159	6.28319	9.42478	12.5664
$l=1$	4.49341	7.72525	10.9041	14.0662
$l=2$	5.76346	9.09501	12.3229	15.5146
$l=3$	6.98793	10.4171	13.698	16.9236
$l=4$	8.18256	11.7049	15.0397	18.3013
$l=5$	9.35581	12.9665	16.3547	19.6532
$l=6$	10.5128	14.2074	17.648	20.9835

Table 1. Zeros ζ_{ls} of the spherical Bessel function of the first kind $j_l(\zeta_{ls} = \eta_1 k R_1) = 0$. The subscript s denotes here the ordinal number of the zero of the l-th order spherical Bessel function.

3. Results

The directivity of an antenna is a measure of how much it concentrates power in a given direction. The maximum directivity of a homogeneous sphere in forward direction with an optimized refractive index $\eta_1 \leq \eta_{max}$ and dipole position r_d is presented in Figure 2(a). With a decrease of the R_1/λ ratio, the resonance behavior becomes stronger due to the weakening of the interference between the excited modes of the sphere. From equation (2) it is known that the approximate conditions for the $TE_{(l+1)ms}$ and TM_{lms} modes coincide. Thus the influence of these two modes affects the far-field scattering pattern at resonance condition and causes the jumps in directivity. If an electric dipole effectively excites resonances ζ_{ls} with $l \leq 3$ for a very small R/λ, then for subsequent l it becomes ineffective and a further small gradual increase in directivity occurs with an increase in the index s. Figures 2(b)-(f) show the distribution of the electric and magnetic fields for the first TE-modes of the sphere.

The impact of interference between modes on the maximum directivity in forward direction at low refractive indices can be seen in Figure 3(a), which shows the dependence of directivity on the refractive index η_1 and on the parameter $\eta_1 k R_1$. Figure 3(a) also gives a visual representation of the accuracy of the conditions (2) under the applied approximation for $\eta_1 \gg 1$. Moreover, while maintaining the value of $\eta_1 k R_1$ and decreasing R_1, the maximum directivity remains practically unchanged at all resonances. Figure 3(b) demonstrates the optimal dipole position which provides maximum directivity. If for the ζ_{01} resonance only mode TE_{101} is excited and the dipole position r_d tends to half the wavelength with increasing η_1, then for the remaining resonances the dipole position is either inside the sphere or near its surface, in order to excite two overlapping $TE_{(l+1)ms}$ and TM_{lms} modes with proper amplitudes and phases.

Figure 4 shows the dependence of the directivity on the physical size of a homogeneous sphere and on the position of the dipole for different refractive indices. With an increase in

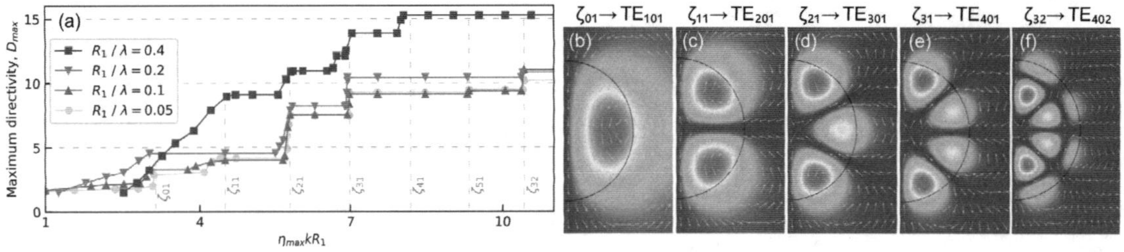

Figure 2. (a) Optimized directivity as a function of the $\eta_{max} k R_1$ value for antennas with a different R_1/λ ratio for $1 \leq \eta_1 \leq \eta_{max}$. (b)-(f) A graphic representation of the modes of the sphere corresponding to different resonance conditions. The sphere modes are mirror-symmetrical, so only half of the sphere's cross section is demonstrated. The normal component of the electric field is shown in color; magnetic field lines are shown with gray arrows.

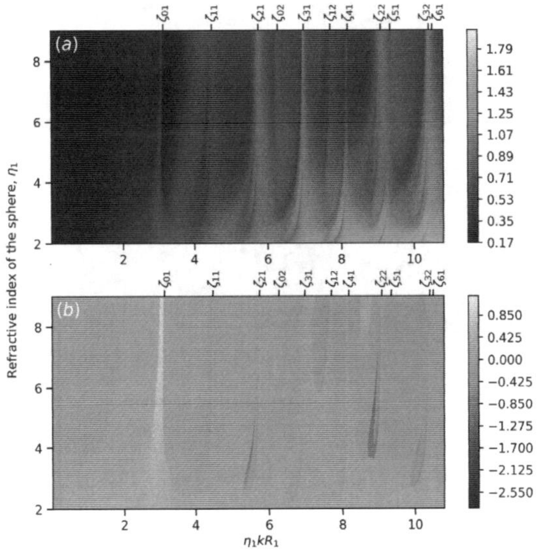

Figure 3. Optimized results for a homogeneous sphere on a logarithmic (log_{10}) scale: (a) maximum directivity in forward direction, (b) relative dipole position, r_d/R_1.

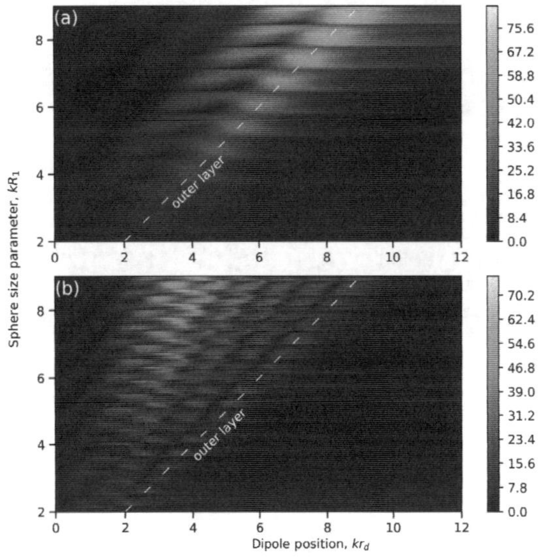

Figure 4. Dependence of the directivity of a homogeneous DRA on its geometric size and the position of the dipole: (a) $\eta_1 = \sqrt{3}$, (b) $\eta_1 = \sqrt{10}$.

the geometric size of the sphere, modes with a new index s can appear in it, and this gives characteristic stripes in the figure. It becomes possible to excite such resonances from a larger number of dipole positions located inside the sphere ($\sim s$ positions).

4. Conclusions

In this work we analyzed a directivity of an electrically small spherical DRA excited by a point electric dipole and its dependence on the effectively excited modes of the spherical DRA. We demonstrated the applicability of $j_l(\zeta_{ls} = \eta_1 k R_1) \simeq 0$ condition for $\eta_1 \gg 1$ regime to identify and characterize excited modes of a sphere. Efficient excitation of the resonances ζ_{ls} is possible by a dipole located inside the sphere or in the immediate vicinity of its surface (for $l > 0$). Maximum directivity of considered DRAs can be achieved at ζ_{3s} resonances.

Acknowledgments

This work is supported by the Russian Foundation for Basic Research (project #20-32-90243).

[1] P. Debye. Der Lichtdruck auf Kugeln von beliebigem Material. *Ann. Phys.*, 335(11):57–136, 1909.
[2] R. D. Richtmyer. Dielectric resonators. *J. Appl. Phys.*, 10(6):391–398, 1939.
[3] J. Stratton. *Electromagnetic Theory.* McGraw-Hill, 1941.
[4] M. Gastine, L. Courtois, and J. L. Dormann. Electromagnetic Resonances of Free Dielectric Spheres. *IEEE Trans. Microw. Theory Techn.*, 15(12):694–700, 1967.
[5] C. F. Bohren and D. R. Huffman. *Absorption and Scattering of Light by Small Particles.* 1998.
[6] C. Forestiere, G. Miano, G. Rubinacci, M. Pascale, A. Tamburrino, R. Tricarico, and S. Ventre. Magnetoquasistatic resonances of small dielectric objects. *Phys. Rev. Res.*, 2:013158, Feb 2020.
[7] R. Gaponenko, A. Moroz, I. Rasskazov, K. Ladutenko, A. Shcherbakov, and P. Belov. Superdirective dielectric spherical multilayer antennas. *arXiv:2104.00534*, 2021.
[8] R. Paniagua-Dominguez, B. Luk'yanchuk, and A. Kuznetsov. *Control of scattering by isolated dielectric nanoantennas.* Elsevier Ltd, 2019.
[9] A. Moroz. A recursive transfer-matrix solution for a dipole radiating inside and outside a stratified sphere. *Ann. Phys.*, 315(2):352–418, 2005.

Tip-enhanced Stokes and anti-Stokes Raman scattering in defect-enriched carbon films

Almaz R Gazizov[1,2], Myakzyum Kh Salakhov[1,2] and Sergey S Kharintsev[1,2]

[1] Institute of Physics, Kazan Federal University, Kremlevskaya St. 18, Kazan 420008, Russia
[2] Tatarstan Academy of Sciences, Baumana St. 20, Kazan 420111, Russia

E-mail: almargazizov@kpfu.ru

Abstract. Anti-Stokes Raman scattering is one of the mechanisms that lie behind an optical refrigeration due to release of photons with greater energy than of incoming photons. To achieve a cooling regime the enhancement of anti-Stokes scattering is necessary, since spontaneous Stokes scattering dominates over anti-Stokes scattering under normal conditions. Here, we investigate the opportunity of enhancement of spontaneous anti-Stokes Raman scattering in defect-enriched carbon film by means of localized plasmon resonances. In our simulations, incoherence of Raman scattering results in excess of anti-Stokes intensity over Stokes one. However, when the field is localized within the phonon coherence volume (coherent regime), the anti-Stokes intensity is lower compared to Stokes one. The provided analysis shows that plasmon-enhanced anti-Stokes Raman scattering can be achieved in highly-defective carbon films. The results are beneficial for Raman-based temperature measurements on the nanoscale.

1. Introduction

The control over the state of a quantum object, monitoring its temperature, amplification and analysis of its optical response are of important in photonics [1]. In particular, light can be used to maintain a certain local temperature using the effects of photo-induced heating and laser cooling of solids. One of the optical cooling methods is the anti-Stokes Raman scattering (ARS) [2]. When photons inelastically scatter in solid, either a creation (Stokes) of a phonon or an annihilation (anti-Stokes) of a phonon generally possible [3]. However, Stokes scattering usually dominates over anti-Stokes scattering at room temperatures due to the fact that the equilibrium population of vibrational states and phonons obey the Bose-Einstein statistics. Existing methods for inverting Stokes to anti-Stokes intensities ratio use various intrinsic and optical resonances to increase the anti-Stokes scattering cross section [2] or suppress Stokes scattering [4].

This work is devoted to a theoretical study of the possibilities of enhancing ARS in carbon films. Generally, carbon films serve as protective coatings. In addition, carbon materials are biocompatible and can be chemically functionalized by various organics. Impurities and structural defects can activate forbidden Raman lines. Moreover, moderate values of the wavenumbers of the most active lines in the Raman spectrum (> 1000 cm^{-1}) favor the use of sufficiently wide plasmon resonances in nanoparticles to enhance ARS. Graphene oxide with a thickness of 10 nm is considered as the carbon film. The optical parameters of graphene oxide fragments functionalized with hydroxyl groups were obtained from calculations based on

Journal of Physics: Conference Series **2015** (2021) 012044 doi:10.1088/1742-6596/2015/1/012044

density functional theory (DFT). Our results are beneficial in nanoscale thermometry, design of optoelectronic devices and functionalized cooling coatings.

2. The model of carbon thin film

The system under consideration consists of a homogeneous graphene oxide layer, on the surface of which functional hydroxyl groups are located, and an elongated gold nanoparticle, which plays the role of an optical nanoantenna. The nanocoating is located on the substrate; however, for the sake of computational simplicity, we take the dielectric constant of the substrate to be equal to that of the carbon coating. Due to the anisotropy of graphene oxide, the nanoantenna should be positioned obliquely to the surface in order to excite in-plane vibrations[5]. In this work, we assume the tip's tilt angle of the nanoparticle is 45 degrees.

In order to determine the scattered field $\mathbf{E}(\mathbf{r},\omega)$ at the frequency of the incident light at different points of the carbon layer, one can use the approach based on dyadic Green's function (see for details [6]). Due to the relatively large refractive index of graphene oxide [7], it is necessary to calculate the near-field interaction of the nanoparticle and the carbon layer. Elementary dipoles represent both the optical nanoantenna and the Raman-active carbon layer in the model. However, in contrast to the discrete-dipole approximation, the nanoantenna is replaced in this work with a single point dipole located at the actual tip's curvature center and that has the polarizability of a gold nanospheroid when the nanoantenna interacts with incident light. In turn, the nanoantenna follows the quasistatic theory of images for a polarizable sphere during the near-field induction from carbon film [8]. The local electric field $\mathbf{E}(\mathbf{r},\omega)$ at the point $\mathbf{r}_n$ in this model is written as:

$$\mathbf{E}(\mathbf{r}_n) = \mathbf{E}_0(\mathbf{r}_n) + \overleftrightarrow{G}(\mathbf{r}_n,\mathbf{r}_0)\hat{\alpha}_p\mathbf{E}_0(\mathbf{r}_0) + \sum_z \overleftrightarrow{G}(\mathbf{r}_n,\mathbf{r}_z)\overleftrightarrow{M}_{\mathrm{im}}(z,\mathbf{r}_0)\overleftrightarrow{R}_{\mathrm{sub}}\hat{\alpha}_p\mathbf{E}_0(\mathbf{r}_0), \tag{1}$$

where $\mathbf{E}_0$ is the incident field, $\hat{\alpha}_p$ is the polarizability of the nanoparticle at the point $\mathbf{r}_0$, $\overleftrightarrow{G}$ is the dyadic Green's function, $\overleftrightarrow{R}_{\mathrm{sub}}$ is the quasi-static reflection tensor, $\overleftrightarrow{M}_{\mathrm{im}}$ is the dressed (generalized) propagator of the problem of the image dipoles induction in the sphere-substrate system. The latter differs from bare propagator $\overleftrightarrow{M}_0$ found in [8] by the account of the self-interaction of image dipoles through the near-field reflection of dipoles [6]. One can find the dressed propagator solving the Dyson type integral equation.

One of the remarkable properties of graphene oxide is low absorption in the presence of resonant electronic transitions in the visible spectrum, which are observable in ellipsometric experiments [7]. By selecting functional groups and their surface coverage, it is possible to further enhance ARS. For resonant Raman scattering, the polarizability tensor can be estimated by the formula [9]:

$$(\alpha^R_{\rho\sigma})_{sf} = \frac{1}{\hbar}\sum_r \frac{(M_\rho)_{fr}(M_\sigma)_{rs}}{\omega_{rs} - \omega - i\Gamma_r} + \frac{(M_\sigma)_{fr}(M_\rho)_{rs}}{\omega_{rf} + \omega + i\Gamma_r} \tag{2}$$

where the subscripts "s", "r" and "f" denote the initial state, the intermediate states and the final state, respectively, M are the transition dipole moment matrix elements, Γ is the width of the excited level. Note that in this approach, we neglect the effects of electron-phonon coupling.

For the simulation, the transition energies and level widths were taken from the experimental work [7]. When the fraction of coverage of the surface of the graphene sheet with hydroxyl groups is 50-75 percent, the transition energy lies in the range 1.8-2.6 eV. Moreover, due to the anisotropy of graphene oxide, the parameters of resonance transitions do not coincide for different components of the Raman tensor. This allows us to speak about the selective electronic enhancement of Raman bands depending on the type of vibration symmetry. It should be noted

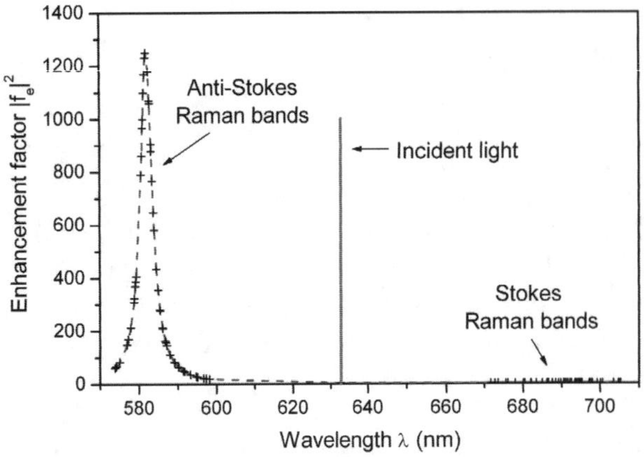

Figure 1. Dependence of the average enhancement factor of the field intensity $|f_e|^2$ on the wavelength λ. The vertical line indicates the wavelength of the incident light. The "+" signs indicate the frequencies of the scattered light.

that in equation (2) the matrix elements of the dipole operator are calculated in an unperturbed basis; therefore, they can be obtained from DFT-calculations. For the sake of simplicity we considered a graphene fragment with only hydroxyl groups and static polarizabilities were obtained at the B3LYP-D3/def2-TZVP level. The principal axes of the inertia tensor determine the orientation of the molecule. The obtained Raman tensors are then used to find the elementary dipole moments at the Stokes and anti-Stokes frequencies. Since functional groups have an arbitrary orientation, each dipole moment takes arbitrary rotation angles in space.

3. Results and discussion

Using equation (1) one can find the total gain tensor $\overleftrightarrow{F} = \mathbf{E}\mathbf{E}_0^*/|\mathbf{E}_0|^2$ at the incident frequency. Averaging the trace of its squared tensor over all elementary dipoles gives the mean enhancement factor of the field intensity $|f_e|^2 = 1/N \sum_n |\mathbf{E}(\mathbf{r}_n)|^2/|\mathbf{E}_0|^2$. In a similar way, the gain tensor $\overleftrightarrow{F}(\mathbf{r}_n, \omega_k)$ is found for each point $\mathbf{r}_n$ at which an elementary Raman-active dipole is located for each either Stokes or anti-Stokes shifted frequencies ω_k. According to the reciprocity theorem, the Hermitian-conjugate gain tensor $\overleftrightarrow{F}^\dagger(\mathbf{r}_n, \omega_k)$ will describe the field propagation back to the nanoantenna and near-field induction at the Raman-shifted frequency. Both the material dispersion parameters of the carbon layer and the nanoparticle parameters and geometry substantially affect the plasmon resonance. It is possible to tune them to achieve the plasmon resonance band covering the anti-Stokes side of the Raman spectrum. Figure (1) shows an example of such a selection. It is seen that the anti-Stokes lines experience a greater amplification averaged over all dipoles, and the enhancement factor exceeds the Stokes-anti-Stokes intensities ratio I_S/I_{aS} for the given wavenumbers at room temperature (~300 K).

The resulting Raman activity of the vibration mode taking into account the partial coherence of the dipole moment oscillations at the shifted frequency, is equal to:

$$S(\omega_k) = \iint \overleftrightarrow{F}(\mathbf{r}_n, \omega_k)\mathbf{p}(\mathbf{r}_n, \omega_k)\mathbf{p}^\dagger(\mathbf{r}_m, \omega_k)\overleftrightarrow{F}^\dagger(\mathbf{r}_m, \omega_k)g(|\mathbf{r}_n - \mathbf{r}_m|)d\mathbf{r}_n d\mathbf{r}_m, \quad (3)$$

Here $g(|\mathbf{r}_n - \mathbf{r}_m|)$ is the correlation function of phonon vibrations at the Raman-shifted frequency. Typically, spontaneous Raman scattering is considered to be an incoherent process. In various carbon materials, the phonon coherence length is tens of nanometers, which is comparable to the localization of the near field. In this regard, a coherent spontaneous Raman scattering regime becomes possible [10]. Figure (2) shows the results of the simulation for both fully coherent and incoherent enhanced Raman scattering. Applying the mean phonon number statistics gives the Raman intensities $I(\omega_k) = A(\omega_k^4/\Omega_k)S(\omega_k)/(\mathrm{e}^{-\hbar\Omega_k/k_B T} - 1)$, in which the Raman shift of k-th

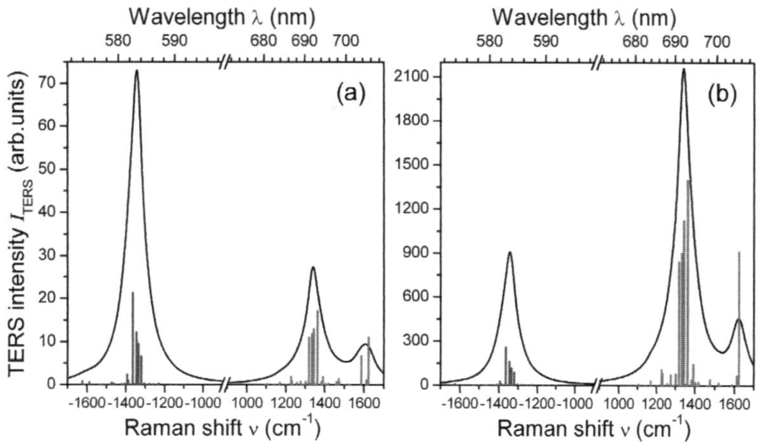

Figure 2. Intensities of Stokes and anti-Stokes Raman scattering in incoherent (a) and coherent (b) scattering regimes. The vertical strips indicate the Raman activity of the bands, the anti-Stokes Raman activity is reduced by a factor of 1000.

vibration Ω is assumed to be positive for the anti-Stokes scattering and negative for the Stokes scattering. It is worth to note that the Raman coherence gives a greater enhancement to the Stokes side of the spectrum compared to the anti-Stokes side due to the longer wavelength of light, i.e. better phase-matching of the dipoles. The absence of the G band (1600 cm^{-1}) in the anti-Stokes spectrum is associated with the rather narrow plasmon resonance band (Fig.1).

4. Conclussion

In summary, we have performed the simple analysis of tip-enhanced Raman intensities at the Stokes and the anti-Stokes frequencies. Our simulations demonstrate that in order to anti-Stokes intensity surpassed the Stokes intensity the coherence volume of a phonon should be much less than field localization. In that case, the destructive interference of secondary waves on the Raman-shifted frequency from the sources distributed over the carbon surface is suppressed. As shown, the plasmon resonances are promising to enhance anti-Stokes radiation. These findings can substantially improve measurements of both the phonon coherence volume and the local temperature in solid films using tip-enhanced Raman spectroscopy.

Acknowledgments

The work was supported by Russian Science Foundation (no. 21-72-00052).

References

[1] Aspelmeyer M, Meystre P and Schwab K 2012 *Physics Today* **65** 29–35
[2] Zhang J, Zhang Q, Wang X, Kwek L C and Xiong Q 2016 *Nat. Photonics* **10** 600–605
[3] Cardona M (ed) 1975 *Light Scattering in Solids* vol 1 (Springer-Verlag)
[4] Chen Y C and Bahl G 2015 *Optica* **2** 893–899
[5] Gazizov A R, Salakhov M K and Kharintsev S S 2020 *J. Raman Spectrosc.* **51** 442–451
[6] Kharintsev S S, Gazizov A R, Salakhov M K and Kazarian S G 2018 *Phys. Chem. Chem. Phys.* **20** 24088–24098
[7] Schöche S, Hong N, Khorasaninejad M, Ambrosio A, Orabona E, Maddalena P and Capasso F 2017 *Appl. Surf. Sci.* **421** 778–782
[8] Lindell I V, Sten J C E and Kleinman R E 1994 *J. Electromagn. Waves. Appl.* **8** 295–313
[9] Mortensen O S and Koningstein J 1968 *J. Chem. Phys.* **48** 3971–3977
[10] Cançado L G, Beams R, Jorio A and Novotny L 2014 *Physical Review X* **4** 031054

Temperature monitoring during light-induced release of cargo using polymer capsules modified with gold nanoparticles and nanodiamonds

E N Gerasimova[1,+,*], V V Yaroshenko[1,+] P M Talianov[1], O O Peltek[1], L V Mikhailova[1], M A Baranov[2], P V Kapitanova[1], D A Zuev[1], A S Timin[3,4], M V Zyuzin[1]

[1]Department of Physics and Engineering, ITMO University, Kronverksky Pr. 49, bldg. A, 197101 St. Petersburg, Russian Federation

[2]Faculty of Photonics and Optical Information, Center of Information Optical Technologies ITMO University, Kronverksky Pr. 49, bldg. A, 197101 St. Petersburg, Russian Federation

[3]Research School of Chemical and Biomedical Engineering, National Research Tomsk Polytechnic University, 634050 Tomsk, Russian Federation

[4]R.M. Gorbacheva Research Institute for Pediatric Oncology, Hematology and Transplantation, Pavlov University, 197022 St. Petersburg, Russian Federation

+equal contribution

*corresponding author: elena.gerasimova@metalab.ifmo.ru

Abstract. Application of different light-sensitive drug delivery carriers is limited due to a risk of overheating of living cells. Therefore, a real-time temperature monitoring within biological objects that controls the photothermal release of different cargos from light-sensitive carriers is highly demanded. In this work, we develop a multifunctional platform comprised of polymer microcapsules modified with nitrogen vacancies (NV) centers as nanothermometers and gold (Au) nanoparticles (NPs) as heating elements for the realization of laser-induced cargo release with a simultaneous temperature measurement inside cells. Such platform allows to prevent unwanted side effects related with the overheating of living cells and tissues.

1. Introduction

To date, micro- and nanosized drug delivery systems can be designed to realize a remote non-invasive release of bioactive compounds into cells upon different stimuli, for example, a laser-induced heating. Such photothermally responsive drug delivery platforms are able to absorb light energy and transform it into heat that further lead to the rupture of drug carriers with the consequent drug release [1]. Nonetheless, excessive heating and temperature changes strongly affect cellular functions such as cell division [2], gene expression [3], growth factor activity [4] and metabolism [5]. Thus, it is vital to provide a real-time temperature monitoring during external heating to avoid possible side effects.

The most widely used technology for temperature sensing include scanning probe microscopy [6], Raman spectroscopy [7], fluorescence-based methods using quantum dots, dyes, lanthanides or fluorescent proteins [8]. However, their applications are limited due to the low sensitivity and the influence of surrounding medium on optical properties of light-responsive carriers. However, it can be overcome by applying nanoscale thermometry based on a quantum mechanical spin associated with nitrogen vacancy (NV) centers in nanodiamonds using optically detected magnetic resonance (ODMR) [9], [10].

In this work, we develop a multifunctional system based on polymer capsules for simultaneous temperature measurements inside living cells and simultaneous drug delivery. For this, we incorporate quantum nanothermometers (NV centers) and heating agents (Au nanoparticles – Au NPs) into a single platform - polymer capsules. It is worth noticing that Au NPs were incorporated either in capsules cavity (core) or in capsules wall. Laser irradiation stimulates heating of Au NPs embedded into carriers leading to their consequent rupture, which triggers a photoinduced release of cargo. We found out a strong correlation between the localization Au NPs within a capsule (either in the capsule's cavity or attached onto the capsule's wall) and the critical laser power of capsule rupture. The developed approach allows temperature monitoring to prevent unwanted overheating of cells and tissues during photoinduced drug delivery.

2. Materials and methods

2.1. Au NPs fabrication and NV centers coating

Au NPs with 50 nm diameter were synthesized by a seed-mediated growth approach [11]. Briefly, the first step was the preparation of the initial Au seeds. The second step included Au NPs growth up to 50 nm. The final step was stabilization of Au NPs by H_2N-PEG-SH via ligand-exchange procedure [12]. Commercially available NV centers were additionally coated with a silica shell (NV@SiO_2) via a Stöber method [13].

2.2. Au NPs fabrication and NV centers coating. Synthesis of polymeric capsules with incorporated Au NPs and NV centers into capsule's core or wall, and their DAPI loading.

Micrometric capsules were synthesized as reported elsewhere with several modifications [14]. To load Au NPs and NV centers into the capsule core (CORE sample), 615 μL of calcium chloride ($CaCl_2$, 1M in water) were mixed with 2.5 mL of water, 400 μl of NV centers and 400 μl of Au NPs under magnetic stirring. Then, 615 μL of Na_2CO_3 (1 M, in water) were added to the $CaCl_2$ mixture under vigorous stirring at 1000 rpm for 30 sec at room temperature. Next, a polymer shell around CaCO3 particles was deposited using a Layer-by-Layer approach. Six monolayers of oppositely charged polyelectrolytes poly-(allylamine hydrochloride) (PAH) and poly-(sodium 4-styrenesulfonate) (PSS) were attached on the top of the $CaCO_3$ cores yielding the following architecture PAH/PSS/PAH/PSS/PAH/PSS. Finally, $CaCO_3$ cores were dissolved with a 0.2 M solution of ethylenediaminetetraacetic acid disodium salt (EDTA).

To incorporate Au NPs and NV centers into the capsule wall (WALL sample), the similar protocol described above was used. For $CaCO_3$ core-template formation, 615 μL of calcium chloride ($CaCl_2$, 1M water solution) were mixed with 2.5 mL of water under magnetic stirring. Then, 615 μL of Na_2CO_3 (1M, water solution) were added to the solution under vigorous stirring at 1000 rpm for 30 sec at room temperature. Next, after deposition of first two polyelectrolyte layers PSS/PAH, 400 μl of NV centers and 400 μl of Au NPs were added to the $CaCO_3$ particles. The obtained architecture of polymer shell was PSS/PAH/NV centers + Au NPs/PSS/PAH/PSS/PAH. Finally, $CaCO_3$ cores were dissolved by a 0.2 M EDTA.

Journal of Physics: Conference Series **2015** (2021) 012045 doi:10.1088/1742-6596/2015/1/012045

The 4',6-diamidino-2-phenylindole (DAPI) was loaded into the capsules cavity via postloading method [12]. For this, 500 μL of 5 mg/mL DAPI aqueous solution were mixed with capsules and shaken for 2 h at 65 °C. Then the loaded capsules were washed twice and redispersed in 1 mL of Milli-Q water.

2.3. Cells

Murine melanoma cell line (B16-F10 cells) was obtained from the American Type Culture Collection.

2.4. Simultaneous visualization and temperature measurements of synthesized capsules

Temperature measurements inside living cells were realized using ODMR method under continuous wave (CW) microwave excitation [15]. The experimental setup was based on a home-built confocal microscope modified with microwave (MW) and time correlated single photon counting (TCSPC) modules, developed MW resonant antenna and Mercury Arc lamp. The MW resonant antenna was specifically designed for ODMR of NV centers to provide the resonance frequency at around 2.87 GHz. It is designed as a planar ring coil with a circular hole in the middle, inside of which resonant magnetic fields are concentrated [16]. The parameters of the microwave resonant antenna were numerically optimized using CST Microwave software. During ODMR measurements, the microwave driving and laser pumping occur simultaneously. For Au NPs heating and NV centers excitation, we applied quasi-CW laser (repetition rate RR=80 MHz) at 532 nm that was selectively focused on capsules with spot size 1.2 μm.

3. Results

Two different capsule configurations (CORE and WALL) modified with Au NPs and NV centers were fabricated. Morphology of the obtained capsules was measured with Scanning Transmission Electron Microscopy (STEM) (Figure 1). As determined from STEM images, an average diameter of capsules was around 5 μm.

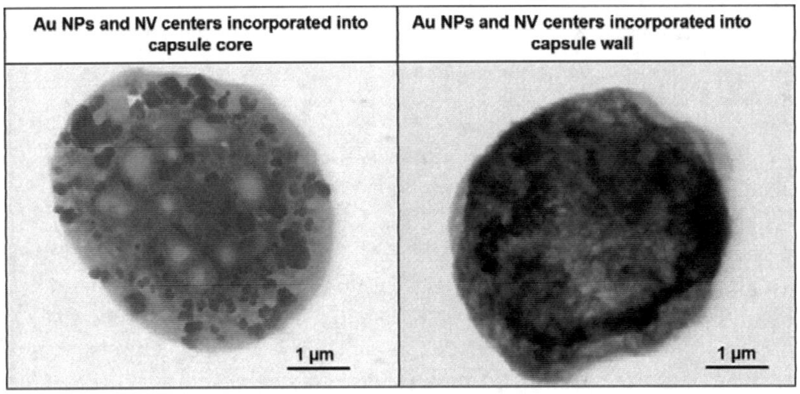

Figure 1. STEM images of capsules modified with NV-centers and Au NPs

For in vitro experiments, B16-F10 cells were seeded into confocal cell imaging dishes at amount of 5.0 x 10^4 cells per dish and left overnight. Next day, capsules loaded with DAPI were added to the cells at the ratio 1:10. Cells were washed twice with PBS to remove non-associated capsules. Afterwards, nanothermometry and photoinduced drug release was performed.

Firstly, the critical power and the temperature of capsule rupture was estimated. The temperature of capsule rupture was the same for the both sample types and it was equal to 128 ±1.12°C. Interestingly, there was a correlation between localization of Au NPs and the power, which corresponds to the capsule decomposition. For CORE sample the critical power was 1.59 mW (145 mW/cm^2). In the case of WALL sample, the critical power corresponded to 2.55 mW (230 mW/cm^2). It can be explained by the close proximity of Au NPs in the CORE sample, which caused higher heat release due to absorbance of the greater amounts of light energy. The ODMR spectra corresponding the temperature of carriers rupture and the temperature dependence on laser power are represented in Figure 2A, B. Intracellular photoinduced delivery of DAPI during temperature monitoring was observed with a fluorescence microscope. Because of the small laser spot, which was 1.2 µm, capsules could be opened selectively, without heating the whole cell. It is worth mentioning that the various applied powers of a pulse laser at 80 MHz (12.5 ns between two pulses) are low enough for the Au nanoparticles to cool down in between pulses, which allows avoiding potential side effects resulting from phototoxicity and overheating. After laser irradiation at critical power, capsule rupture led to the release of DAPI through the endo/lysosomal compartments of cells and staining of a cell nucleus (Figure 2C). It proves the ability of the developed system to deliver various cargos and perform the real-time temperature monitoring, which preserves cells from the excessive heating.

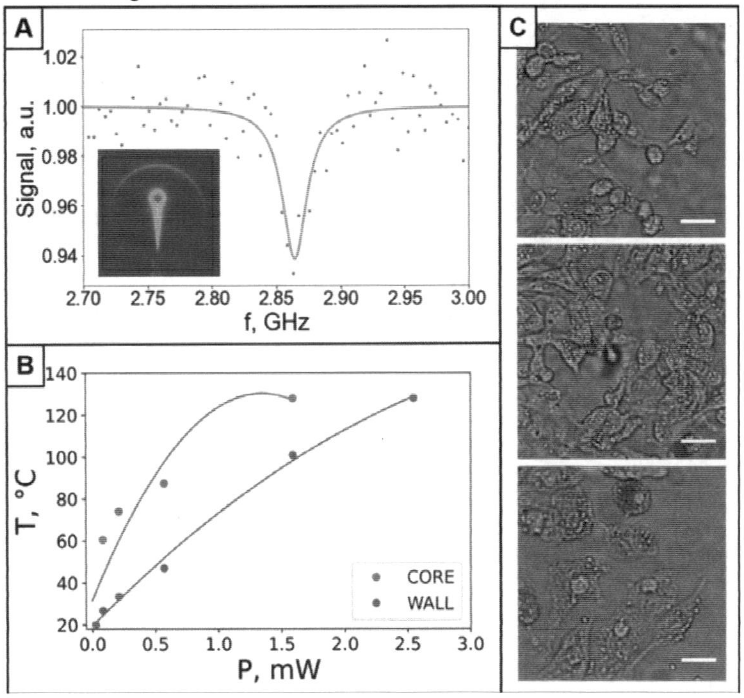

Figure 2. A – The ODMR spectra corresponding to critical temperature of capsule rupture. Points – experimental data. Line – Lorentzian fitting. Points that are greater than 1 are part of statistical error during measurements. The inset picture shows numerical simulation of the field localization in MW resonant antenna during ODMR measurements. B – Measured temperatures in capsule samples during laser treatment at various powers. C - Overlay (fluorescence + brightfield channel) images of cells with released DAPI from capsules after laser irradiation. Cells with DAPI-loaded capsules (CORE) were irradiated with laser (λ = 532 nm, 1.59 mW or 145 mW/cm^2) and DAPI staining (in blue) of cell nuclei is observed. Scale bars correspond to 5 μm.

4. Conclusion

In this work, we demonstrated the real-time temperature monitoring with an accuracy ±1.12 °C and thermally induced non-invasive cargo release from polymer capsules in living cells. Carriers were developed by loading of NV centers and Au NPs into capsules synthesized by a layer-by-layer approach. We found out the correlation between Au NPs localization and their required power for carriers rupture. Obtained results presented in the current study can be applied for the site-specific photorelease in vivo.

5. Acknowledgments
Part of this work related to the development of MW resonant antenna was supported by the Russian Science Foundation (Grant No. 19-19-00693).

References
[1] I. Koryakina, D. S. Kuznetsova, D. A. Zuev, V. A. Milichko, A. S. Timin, and M. V. Zyuzin, "Optically responsive delivery platforms: From the design considerations to biomedical applications," Nanophotonics, vol. 9, no. 1. De Gruyter, pp. 39–74, Jan. 01, 2020, doi: 10.1515/nanoph-2019-0423.
[2] J. Choi et al., "Probing and manipulating embryogenesis via nanoscale thermometry and temperature control," Proc. Natl. Acad. Sci. U. S. A., vol. 117, no. 26, pp. 14636–14641, Jun. 2020, doi: 10.1073/pnas.1922730117.
[3] Y. Kamei et al., "Infrared laser-mediated gene induction in targeted single cells in vivo," Nat. Methods, vol. 6, no. 1, pp. 79–81, Dec. 2009, doi: 10.1038/nmeth.1278.
[4] H. Yukawa et al., "A quantum thermometric sensing and analysis system using fluorescent nanodiamonds for the evaluation of living stem cell functions according to intracellular temperature," Nanoscale Adv., vol. 2, no. 5, pp. 1859–1868, May 2020, doi: 10.1039/d0na00146e.
[5] L. Shen, T. R. Xie, R. Z. Yang, Y. Chen, and J. S. Kang, "Application of a dye-based mitochondrion-thermometry to determine the receptor downstream of prostaglandin E2 involved in the regulation of hepatocyte metabolism," Sci. Rep., vol. 8, no. 1, pp. 1–12, Dec. 2018, doi: 10.1038/s41598-018-31356-y.
[6] Y. Yue and X. Wang, "Nanoscale thermal probing," Nano Rev., vol. 3, no. 1, p. 11586, Jan. 2012, doi: 10.3402/nano.v3i0.11586.
[7] G. P. Zograf et al., "All-Optical Nanoscale Heating and Thermometry with Resonant Dielectric Nanoparticles for Controllable Drug Release in Living Cells," Laser Photon. Rev., vol. 14, no. 3, p. 1900082, Mar. 2020, doi: 10.1002/lpor.201900082.
[8] T. Bai and N. Gu, "Micro/Nanoscale Thermometry for Cellular Thermal Sensing," Small, vol. 12, no. 34, pp. 4590–4610, Sep. 2016, doi: 10.1002/smll.201600665.
[9] V. M. Acosta, E. Bauch, M. P. Ledbetter, A. Waxman, L. S. Bouchard, and D. Budker, "Temperature dependence of the nitrogen-vacancy magnetic resonance in diamond," Phys. Rev. Lett., vol. 104, no. 7, p. 070801, Feb. 2010, doi: 10.1103/PhysRevLett.104.070801.
[10] G. Kucsko et al., "Nanometre-scale thermometry in a living cell," Nature, vol. 500, no. 7460, pp. 54–58, Jul. 2013, doi: 10.1038/nature12373.
[11] Y. Zheng, X. Zhong, Z. Li, and Y. Xia, "Successive, Seed-Mediated Growth for the Synthesis of Single-Crystal Gold Nanospheres with Uniform Diameters Controlled in the Range of 5-150 nm," Part. Part. Syst. Charact., vol. 31, no. 2, pp. 266–273, Feb. 2014, doi: 10.1002/ppsc.201300256.
[12] A. R. Muslimov et al., "Biomimetic drug delivery platforms based on mesenchymal stem cells impregnated with light-responsive submicron sized carriers," Biomater. Sci., vol. 8, no. 4, pp. 1137–1147, Feb. 2020, doi: 10.1039/c9bm00926d.
[13] T. Zhang et al., "Hybrid nanodiamond quantum sensors enabled by volume phase transitions of hydrogels," Nat. Commun., vol. 9, no. 1, pp. 1–8, Dec. 2018, doi: 10.1038/s41467-018-05673-9.
[14] Y. V. Tarakanchikova et al., "Layer-by-Layer-Assembled Capsule Size Affects the Efficiency of Packaging and Delivery of Different Genetic Cargo," Part. Part. Syst. Charact., vol. 38, no. 2, p. 2000228, Feb. 2021, doi: 10.1002/ppsc.202000228.
[15] G. Petrini et al., "Is a Quantum Biosensing Revolution Approaching? Perspectives in NV-Assisted Current and Thermal Biosensing in Living Cells," Adv. Quantum Technol., vol. 3, no. 12, p. 2000066, Dec. 2020, doi: 10.1002/qute.202000066.

Scattering of generalized Bessel beams simulated with the discrete dipole approximation

S A Glukhova[1, 2, 3] and M A Yurkin[1, 2]

[1] Voevodsky Institute of Chemical Kinetics and Combustion, SB RAS, Institutskaya 3, 630090 Novosibirsk, Russia
[2] Novosibirsk State University, Pirogova 2, 630090 Novosibirsk, Russia
[3] Corresponding author: stefgluhova@gmail.com

Abstract. We consider the simulation of scattering of the high-order vector Bessel beams in the discrete dipole approximation framework (DDA). For this purpose, a new general classification of all existing Bessel beam types was developed based on the superposition of transverse Hertz vector potentials. Next, we implemented these beams in ADDA code – an open-source parallel implementation of the DDA. The code enables easy and efficient simulation of Bessel beams scattering by arbitrary-shaped particles. Moreover, these results pave the way for the following research related to the Bessel beam scattering near a substrate and optical forces.

1. Introduction

Today Bessel beams are at the frontier of structured light with orbital angular momentum. One of the most prominent advantages of Bessel beams is its diffractionless, i.e., the ability to propagate while maintaining the profile near the beam axis. These beams are actively used in such areas as optical manipulation (tweezing), material proceeding, imaging, etc. In many physical problems, it is crucial to consider the scattering of Bessel beams, which is predominantly studied for spherical particles with the use of the generalized Lorenz-Mie theory (GLMT) [1] than for arbitrary particles. Moreover, various types of Bessel beams exist that only have been named [2], lacking a complete picture of how they relate to one another. In this regard, this work has two goals: the classification of various types of high-order vector Bessel beams and the development of the capability to simulate the scattering of such beams by arbitrary particle using the discrete dipole approximation.

2. Bessel beam generalization

The scalar Bessel beam was obtained by J. Durnin [3] as a solution of the scalar Helmholtz equation in a cylindrical coordinate system. Meanwhile, vector Bessel beams are vectorial solutions of the Helmholtz equation and they can be presented in several different forms or types. These types differ by their polarizations, field, and energy configurations. Among them are beams with circularly symmetric energy density (CS type), with transverse electric and magnetic fields – TE and TM types, respectively, and LE and LM – beams with linear polarizations of electric and magnetic fields, respectively.

To derive the expressions of various types, it is convenient to use Hertz vector potentials (so-called Davis approach). This approach has been applied to the Bessel beams [4], but we have extended it further, making it possible to obtain all relationships between different beam types and their polarizations. Also, we managed to relate beams of other orders using the rotation and duality operators.

2.1. *Matrix representation of Hertz vector potentials*

This section outlines the derivation of generalized Bessel beams in a homogeneous isotropic medium (with absolute permittivity ε, permeability μ, and wave impedance $\eta \overset{\text{def}}{=} \sqrt{\mu/\varepsilon}$) in SI units. The symmetry of Maxwell equations allows one to consider only the electric field, but it is more convenient to consider a pair $\{\mathbf{E}, \mathbf{H}\}$ taking into account a duality transformation $\mathbf{P}_\chi$

$$\begin{pmatrix}\mathbf{E}\\ \mathbf{H}\end{pmatrix} \to \mathbf{P}_\chi \begin{pmatrix}\mathbf{E}\\ \mathbf{H}\end{pmatrix}. \tag{1}$$

$\mathbf{P}_\chi$ in general case is a duality rotation by the angle χ, similar to the standard rotation matrix $\mathbf{R}_\chi$

$$\mathbf{P}_\chi \overset{\text{def}}{=} \begin{pmatrix}\cos\chi & -\eta\sin\chi\\ \eta^{-1}\sin\chi & \cos\chi\end{pmatrix}, \qquad \mathbf{R}_\chi \overset{\text{def}}{=} \begin{pmatrix}\cos\chi & -\sin\chi\\ \sin\chi & \cos\chi\end{pmatrix}. \tag{2}$$

Electric and magnetic fields of any beam can be obtained starting from the Hertz vector potentials of electric and magnetic types [5]. We relate the pairs $\{\mathbf{E}, \mathbf{H}\}$ and $\{\boldsymbol{\Pi}_\mathrm{e}, \boldsymbol{\Pi}_\mathrm{m}\}$ in matrix form using an operator $\mathcal{L}$

$$\begin{pmatrix}\mathbf{E}\\ \mathbf{H}\end{pmatrix} = \mathcal{L}\begin{pmatrix}\boldsymbol{\Pi}_\mathrm{e}\\ \boldsymbol{\Pi}_\mathrm{m}\end{pmatrix}, \quad \mathcal{L} \overset{\text{def}}{=} \nabla\times\begin{pmatrix}\nabla\times & ik\eta\\ -ik\eta^{-1} & \nabla\times\end{pmatrix} = \nabla\times\left(-ik\mathbf{P}_{\pi/2} + \nabla\times\right). \tag{3}$$

Note that both pairs satisfy Maxwell equations. For Bessel beams propagating along the z-axis the expressions for $\boldsymbol{\Pi}_\mathrm{e}$ and $\boldsymbol{\Pi}_\mathrm{m}$ are significantly simpler than that for the fields. The potentials have, up to the constant vector, the following concise functional form in the cylindrical coordinate system [4]

$$f_n(\mathbf{r}) = J_n(k_\mathrm{t}\rho)\mathrm{i}^n e^{in\varphi} \mathrm{e}^{\mathrm{i}k_z z}, \tag{4}$$

where J_n is the Bessel function of the first kind (n is the order of BB), $k_\mathrm{t} \overset{\text{def}}{=} k\sin\alpha_0$ and $k_z \overset{\text{def}}{=} k\cos\alpha_0$ are the transverse and longitudinal components of the wave vector k, and α_0 is the half cone angle.

The direction of Hertz vector potentials determines Bessel beam type and polarization. More specifically, a matrix $\mathbf{M}$ represents any superposition of the transverse potentials

$$\begin{pmatrix}\boldsymbol{\Pi}_\mathrm{e}\\ \boldsymbol{\Pi}_\mathrm{m}\end{pmatrix} = \begin{pmatrix}\Pi_{\mathrm{e},x} & \Pi_{\mathrm{e},y}\\ \Pi_{\mathrm{m},x} & \Pi_{\mathrm{m},y}\end{pmatrix}\begin{pmatrix}\mathbf{e}_x\\ \mathbf{e}_y\end{pmatrix} = \Pi_0\begin{pmatrix}1 & 0\\ 0 & \eta^{-1}\end{pmatrix}\mathbf{M}\begin{pmatrix}\mathbf{e}_x\\ \mathbf{e}_y\end{pmatrix} f_n. \tag{5}$$

We have shown that all known Bessel beam types: LE, LM, CS, TE, and TM are presented this way. Moreover, the corresponding matrix $\mathbf{M}$ has a simple form for most of these Bessel beams. This description clarifies relationships between various types and polarizations. It does not explicitly include longitudinal potentials, since any potential of the form $f_n\mathbf{e}_z$ is equivalent to the superposition of transverse potentials of orders $n \pm 1$.

2.2. *Relations between beam types and polarizations*

The ingenuity of the above matrix representation allows one to easily relate polarizations of one beam type and different types of Bessel beams. The expressions in Table 1 demonstrate some examples of these relationships based on polarization and duality transformations. Naturally, the polarization rotation transforms matrix $\mathbf{M}$ into the matrix of the same beam matrix but with different polarization. By contrast, duality rotation relates different beam types. For example, LE and LM Bessel beams are coupled since the electric field of one type equal to a magnetic field of the other (up to a constant)

Table 1. Relations between LE, LM, and CS beam types and their polarizations expressed in terms of matrix **M**.

Type	Field	M	Polarization rotation	Duality rotation
LE	$\mathbf{E}_{\mathrm{LE}}^{(x)}$	$\begin{pmatrix} 0 & 0 \\ 0 & 1 \end{pmatrix}$	$\mathbf{M}_{\mathrm{LE}}^{(y)}\mathbf{R}_{\pi/2} = \mathbf{M}_{\mathrm{LE}}^{(x)}$	$\mathbf{R}_{\pi/2}\mathbf{M}_{\mathrm{LM}}^{(x,y)} = \mathbf{M}_{\mathrm{LE}}^{(x,y)}$
	$\mathbf{E}_{\mathrm{LE}}^{(y)}$	$\begin{pmatrix} 0 & 0 \\ -1 & 0 \end{pmatrix}$		
LM	$\mathbf{E}_{\mathrm{LM}}^{(x)}$	$\begin{pmatrix} 0 & 1 \\ 0 & 0 \end{pmatrix}$	$\mathbf{M}_{\mathrm{LM}}^{(y)}\mathbf{R}_{\pi/2} = \mathbf{M}_{\mathrm{LM}}^{(x)}$	
	$\mathbf{E}_{\mathrm{LM}}^{(y)}$	$\begin{pmatrix} -1 & 0 \\ 0 & 0 \end{pmatrix}$		
CS	$\mathbf{E}_{\mathrm{CS}}^{(1,0)}$	$\frac{1}{2}\begin{pmatrix} 1 & 0 \\ 0 & 1 \end{pmatrix}$	$\mathbf{M}_{\mathrm{CS}}^{(0,1)}\mathrm{R}_{\pi/2} = \mathbf{M}_{\mathrm{CS}}^{(1,0)}$	$\mathbf{R}_{\pi/2}\mathbf{M}_{\mathrm{CS}}^{(0,1)} = \mathbf{M}_{\mathrm{CS}}^{(1,0)}$
	$\mathbf{E}_{\mathrm{CS}}^{(0,1)}$	$\frac{1}{2}\begin{pmatrix} 0 & 1 \\ -1 & 0 \end{pmatrix}$		

The electric field **E** of a generalized Bessel beam with matrix **M** is the following:

$$E_x = \frac{E_0}{k^2}\left\{\left(\frac{k^2+k_z^2}{2}M_{\mathrm{e},x} + kk_zM_{\mathrm{m},y}\right)f_n - \frac{k_t^2}{4}\left[\left(M_{\mathrm{e},x}+\mathrm{i}M_{\mathrm{e,y}}\right)f_{n-2} + \left(M_{\mathrm{e},x}-\mathrm{i}M_{\mathrm{e,y}}\right)f_{n+2}\right]\right\}, \quad (6)$$

$$E_y = \frac{E_0}{k^2}\left\{\left(\frac{k^2+k_z^2}{2}M_{\mathrm{e},y} - kk_zM_{\mathrm{m},x}\right)f_n - \mathrm{i}\frac{k_t^2}{4}\left[\left(M_{\mathrm{e},x}+\mathrm{i}M_{\mathrm{e,y}}\right)f_{n-2} - \left(M_{\mathrm{e},x}-\mathrm{i}M_{\mathrm{e,y}}\right)f_{n+2}\right]\right\}, \quad (7)$$

$$E_z = -\frac{E_0k_t}{2k^2}\left\{\left[k_z\left(M_{\mathrm{e},x}-\mathrm{i}M_{\mathrm{e,y}}\right) + \mathrm{i}k\left(M_{\mathrm{m},x}-\mathrm{i}M_{\mathrm{m,y}}\right)\right]f_{n+1} + \begin{bmatrix} k_z\left(M_{\mathrm{e},x}+\mathrm{i}M_{\mathrm{e,y}}\right) \\ -\mathrm{i}k\left(M_{\mathrm{m},x}+\mathrm{i}M_{\mathrm{m,y}}\right)\end{bmatrix}f_{n-1}\right\}. \quad (8)$$

Corresponding expressions for the magnetic field were obtained using duality transformation.

3. Numerical simulations

The above theoretical results were applied to generalize the Mueller calculus (involving Muller and amplitude scattering matrices) to vortex beams, i.e. beams with non-zero orbital angular momentum. The Mueller calculus is used in many implementations of the discrete dipole approximation, including a popular open-source ADDA package [6]. Besides the standard Bessel beam types (LE, LM, CS, TEC, and TMC – they are used to generate TE and TM modes [7] effectively), we have implemented a generalized Bessel beam specified by a matrix M (4 complex values) in ADDA. The latter correspond to a linear combination of 4 basis beams (x- and y-polarizations of LE and LM types). We successfully validated this implementation against the reference results of GLMT for spheres [8]. Currently, the code is available at a separate fork of ADDA: https://github.com/stefaniagl/adda .

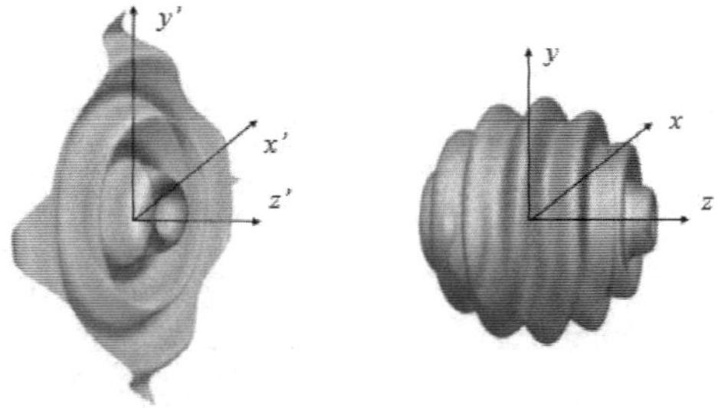

Figure 1 Geometry of scattering of high-order Bessel beam by the Chebyshev particle with $n = 8$, $\varepsilon = 0.1$.

The scattering intensity in H-plane (yz-plane) and E-plane (xz-plane) are expressed with the Mueller matrix elements as

$$I_{\perp} = S_{11} - S_{12}, \qquad I_{||} = S_{11} + S_{12}. \tag{9}$$

The results in (Figure 2) show the scattering intensities of various high-order Bessel beam types by the Chebyshev particle (expressed in spherical coordinates as $r = [1 + \varepsilon\, cos(n\theta)]$ with $\varepsilon = 0.1$ and $n = 8$). The scheme in (Figure 1) depicts particle geometry and position.

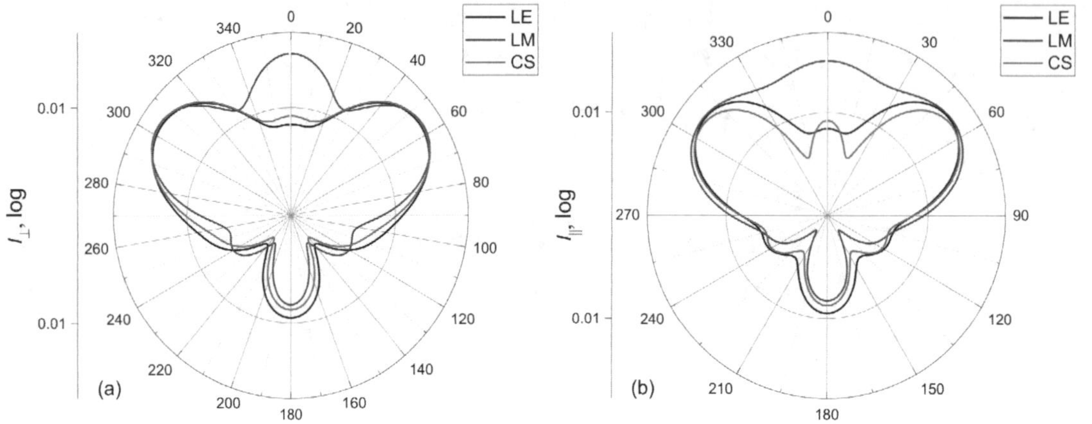

Figure 2 DDA simulations of scattering intensities (a) $I_{||}$ and (b) $I_{\perp}$ for the scattering of second-order CS Bessel beam with $\alpha_0 = 15°$ by the Chebyshev particle (refractive index $m = 1.5$). We used 15 dipoles per λ.

4. Conclusion

The theoretical results clarify the general idea and relationships between various types of Bessel beams. Similar derivations can potentially be applied to other complex light beams. The implementation of these beams in the open-source ADDA code enables one to simulate the scattering of various Bessel beams by particles with arbitrary shape and internal structure. Future research directions include scattering Bessel beams by particles near a substrate and calculating optical forces.

Acknowledgements

This work is supported by the Russian Science Foundation (Grant No. 18-12-00052).

References

[1] Lock J A and Gouesbet G 2009 Generalized Lorenz–Mie theory and applications *J. Quant. Spectrosc. Radiat. Transf.* **110** 800–7

[2] Wang J J, Wriedt T, Lock J A and Jiao Y C 2017 General description of transverse mode Bessel beams and construction of basis Bessel fields *J. Quant. Spectrosc. Radiat. Transf.* **195** 8–17

[3] Durnin J 1987 Exact solutions for nondiffracting beams. I. The scalar theory *J. Opt. Soc. Am. A* **4** 651

[4] Wang Y, Dou W and Meng H 2014 Vector analyses of linearly and circularly polarized Bessel beams using Hertz vector potentials *Opt. Express* **22** 7821

[5] Stratton J A 1941 *Electromagnetic theory* (New York: McGRAW-HILL BOOK COMPANY)

[6] Yurkin M A and Hoekstra A G 2007 The discrete dipole approximation: an overview and recent developments *J. Quant. Spectrosc. Radiat. Transf.* **106** 558–89

[7] Glukhova S A and Yurkin M A 2020 Implementation of various Bessel beams in the framework of the discrete dipole approximation *AIP Conf. Proc.* METANANO 2020 vol 2300 p 020037

[8] Chen Z, Han Y, Cui Z and Shi X 2014 Scattering of a zero-order Bessel beam by a concentric sphere *J. Opt.* **16** 055701

Principal Component Analysis Applied to Gradient Fields in Band Gap Optimization Problems for Metamaterials

Giorgio Gnecco[1], Andrea Bacigalupo[2], Francesca Fantoni[3], and Daniela Selvi[4]

[1]IMT School for Advanced Studies – AXES – Lucca, Italy
[2]University of Genoa – DICCA – Genoa, Italy
[3]University of Brescia – DICATAM – Brescia, Italy
[4]University of Florence – DIEF – Florence, Italy

[1]Corresponding author: giorgio.gnecco@imtlucca.it
[2]andrea.bacigalupo@unige.it
[3]francesca.fantoni@unibs.it
[4]daniela.selvi@unifi.it

Abstract. A promising technique for the spectral design of acoustic metamaterials is based on the formulation of suitable constrained nonlinear optimization problems. Unfortunately, the straightforward application of classical gradient-based iterative optimization algorithms to the numerical solution of such problems is typically highly demanding, due to the complexity of the underlying physical models. Nevertheless, supervised machine learning techniques can reduce such a computational effort, e.g., by replacing the original objective functions of such optimization problems with more-easily computable approximations. In this framework, the present article describes the application of a related unsupervised machine learning technique, namely, principal component analysis, to approximate the gradient of the objective function of a band gap optimization problem for an acoustic metamaterial, with the aim of making the successive application of a gradient-based iterative optimization algorithm faster. Numerical results show the effectiveness of the proposed method.

1. Introduction

The analysis and optimal design of acoustic metamaterials is of high interest to the scientific community, since such metamaterials are capable to achieve physical properties hardly found in nature, via a suitable optimization of their geometrical/mechanical model parameters. Application fields include mechanical, civil, naval, and aerospace engineering, up to biomedical and robotic engineering. Unfortunately, due to the complexity of their physical models, the analysis of acoustic metamaterials is often time-consuming, and their optimization cannot be typically obtained neither by hand, nor by applying classical optimization methods, unless in the case in which their models are relatively simple [1]. As demonstrated by recent works (see, e.g., [2, 3]), however, machine learning can make the analysis and optimization of acoustic metamaterials easier, by speeding up their technological transfer process. Moreover, due to the similarity in the respective optimization problems, machine learning techniques can be applied not only for an improved design of acoustic metamaterials, but also for the one of optical metamaterials.

In [2, 3], supervised machine learning techniques were employed to replace the objective function of a band gap optimization problem associated with an acoustic metamaterial by a suitable surrogate objective function (more specifically, by a radial basis function network [4]), which was constructed by using, respectively, a fixed/adaptively chosen interpolating training set. Then, a gradient-based iterative optimization algorithm was applied to optimize the surrogate objective function, by replacing each expensive evaluation of the original objective function by the cheaper computation of its approximation.

In this work, a related but different approach is used, based on unsupervised learning, and more specifically, on a well-known dimensionality reduction technique, Principal Component Analysis (PCA). This allows one to approximate a set of vectors belonging to a Euclidean space by projecting such vectors onto an automatically chosen lower-dimensional subspace, whose dimension is selected in such a way to achieve a desired precision in the approximation [5]. Differently from its standard application to the input vector of a machine learning problem, in the present work PCA is used to approximate the sampled gradient field of a band gap optimization problem of an acoustic metamaterial, evaluated on a subset of admissible vectors of parameters, generated by a subsequence of a quasi-Monte Carlo sequence [6]. Then, a gradient-based iterative optimization algorithm is applied to optimize the objective function, by approximating, at each iteration, the gradient of the objective function with its projection onto the subspace previously obtained by PCA. Because of that projection, the time needed for the execution of each step of the gradient-based iterative optimization algorithm is reduced, while the error in the approximation of the gradient turns out to be negligible due to the automatic choice of the dimension of the subspace. Numerical results show the effectiveness of the proposed computational method of gradient approximation.

2. Band gap optimization problem and description of the proposed method

The following band gap optimization problem for an acoustic metamaterial is considered, assuming continuous differentiability of its objective function. The problem is expressed in the constrained nonlinear programming form

$$\begin{aligned} &\underset{\mathbf{x}\in\mathbb{R}^d}{\text{maximize}} \quad && f(\mathbf{x}), \\ &\text{s. t.} && g_j(\mathbf{x}) \le 0,\ j = 1,\ldots,n_{\text{ineq}} \end{aligned} \tag{1}$$

where $\mathbf{x}$ represents a d-dimensional Euclidean vector of parameters of the acoustic metamaterial, the nonlinear objective function $f(\mathbf{x})$ models the band gap between two given dispersion surfaces of its Floquet-Bloch spectrum, and $g_j(\mathbf{x})$ are n_{ineq} (either linear or nonlinear) inequality constraints of either geometrical/mechanical nature, or expressing bounds on the range of each parameter (box constraints). It is worth noting that, in the present work, the band gap $f(\mathbf{x})$ is defined as the difference between the minimum of a dispersion surface and the maximum of the adjacent dispersion surface below. In this way, for some choices of the vector of parameters, one may obtain even a negative band gap (which corresponds effectively to the absence of a true band gap), in case the minimum of the dispersion surface above is smaller than the maximum of the dispersion surface below. Nevertheless, this definition can be preferable to setting the band gap to zero in such a situation, because it allows in principle a gradient-based iterative optimization algorithm to open a band gap, by increasing - iteration after iteration - an initially negative band gap. Otherwise, with the other definition, a zero gradient would be initially obtained, preventing any gradient-based optimization algorithm to increase the initial zero band gap.

The proposed method to solve the optimization problem (1) works as follows.

1) First, a finite number N of admissible choices $\mathbf{x}^{(n)}$ for the vector of parameters $\mathbf{x}$ is generated, by selecting the first N elements of a quasi-Monte Carlo sequence that satisfy all the constraints of the optimization problem (training set). Various choices for the sample size N are made. The final choice N° is the smallest N for which the results of the application of PCA in the successive step 3) are not significantly affected by its further increase.

2) Then, the gradient $\nabla f(\mathbf{x})$ of the objective function is computed, in correspondence of all the N° vectors $\mathbf{x}^{(n)}$ generated in step 1).

3) PCA is applied to the (sampled) gradient field $\left\{\nabla f\left(\mathbf{x}^{(n)}\right)\right\}_{n=1}^{N^{\circ}}$ of step 2), after removing from each gradient the empirical mean $\sum_{n=1}^{N^{\circ}} \nabla f\left(\mathbf{x}^{(n)}\right)/N^{\circ}$ of the gradient field on the whole training set, in such a way to obtain a centered gradient field. The number of principal components kept (or equivalently, the number of principal directions onto which the centered gradient field is projected) is chosen to be equal to the minimal number p° of projections needed to approximate the centered gradient field with a mean squared norm of the reconstruction error $MSRE\left(p^{\circ}\right)$ smaller than or equal to a desired percentage r with respect to the mean squared norm of the centered gradient field.

4) A gradient-based iterative optimization algorithm (namely, sequential linear programming with an adaptively chosen trust region [2]) is used to optimize the objective function, for a given maximum number of iterations K. At each iteration of the algorithm, the gradient of the objective function is approximated by its projection onto the

(typically $\left(p^{\circ}+1\right)$-dimensional, otherwise p°-dimensional) subspace generated by the empirical mean of the gradient field and by the p° principal directions identified in step 3).

The entire procedure is repeated N° times by choosing, as the initialization of the gradient-based iterative optimization algorithm, one of the N° admissible choices for the vector of parameters made in step 1). Details about such gradient-based iterative optimization algorithm (based on the exact gradient, not on its approximation) are in [2]. It is worth noting that a local decrease in the objective value versus the iteration number $k=1,\ldots,K$ is possible in principle, even when the gradient is not approximated. This occurs when the linear approximation of the objective function in the trust region is poor. To deal with this, the trust region size is changed adaptively, as described in [2]. The largest objective value generated during all the iterations of the various repetitions is produced as output of the proposed method. For comparison purposes, the gradient-based iterative optimization algorithm exploited in step 4) is also applied for other choices of the number $p \neq p^{\circ}$ of principal directions kept (the case $p+1=8$ refers to the exact gradient of the objective function).

By construction, at each iteration of the gradient-based optimization algorithm associated with an admissible choice for the vector of parameters that does not belong to the boundary of the admissible region of the optimization problem, the approximation of the gradient makes the algorithm update the vector of parameters according to a direction belonging to the subspace identified in step 3), at least if one ignores discretization issues associated with the choice of the trust region (which has the shape of a multi-dimensional box, likewise in [2]). Nevertheless, it is interesting to observe that, in case the current admissible choice for the vector of parameters belongs to the boundary of the admissible region, the update proceeds along a direction determined by projecting the approximated gradient onto the current linearization of that boundary. Hence, although it replaces the exact gradient with an approximation belonging to a subspace having dimension smaller than or equal to $p^{\circ}+1$, the proposed method makes the gradient-based optimization algorithm potentially able to update the vector of parameters on a subspace having even larger dimension than $p^{\circ}+1$.

3. Numerical results

The proposed method is applied to the band gap maximization of an acoustic beam-lattice metamaterial having a similar structure as the one investigated in [7]. The number of parameters is equal to $d=8$. Before applying the method, a change of variables is made (according to a standard practice in PCA), by mapping the multi-dimensional box associated with the original vector of parameters onto a d-dimensional hypercube. Then, the method is applied by considering as the vector $\mathbf{x}$ the resulting transformed vector. The initial choices $N=10,20,30,40,50,60,70,80,90,100$ are made for the number of elements in the training set. The desired upper bound on the mean squared norm of the reconstruction error, expressed as a percentage of the mean squared norm of the centered gradient field, is set to $r=5\%$.

Fig. 1 (a) shows the mean squared norm of the reconstruction error $MSRE(p)$ of the centered gradient field (expressed as a percentage of the mean squared norm of the centered gradient field) as a function of the number p of principal components kept, for different choices of the training sample size N.

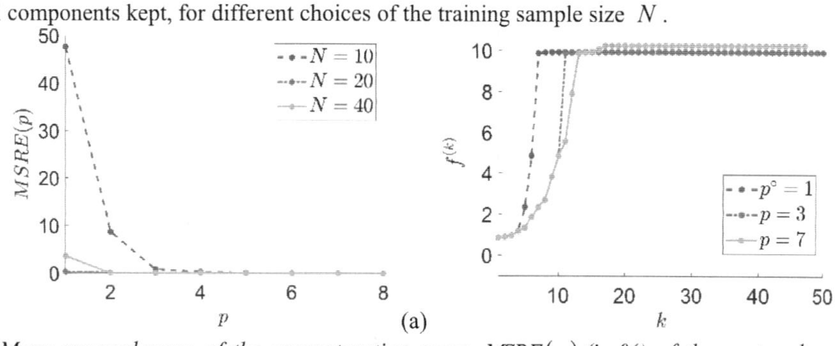

Fig. 1 *(a) Mean squared norm of the reconstruction error $MSRE(p)$ (in %) of the centered gradient field as a function of the number of principal components p kept, for different sample sizes N and (b) objective value $f^{(k)}$ versus the iteration number k for the best repetition of the proposed method with $p^{\circ}=1$, and other choices for p.*

The graph obtained for $N = 30$ overlaps with the one associated with $N = 20$. Furthermore, the graphs obtained for $N = 50, 60, 70, 80, 90, 100$ overlap with the one reported for $N = 40$. Therefore, this latter value is chosen as the training sample size N°. The gradient-based optimization algorithm is applied with a maximum number of iterations $K = 200$. The other parameters are chosen in a similar way as in [2]. Fig. 1 (b) shows the band gap $f^{(k)}$ versus the iteration number k for the best repetition of the proposed method and also, for the same repetition, analogous results obtained by selecting other choices for the number p of principal components kept . The figure shows that the proposed method can achieve a performance comparable to the one obtained when the exact gradient is used. Since $p^{\circ} + 1 = 2$, the gradient approximation takes about $1/4$ of the time required to evaluate the exact gradient.

4. Conclusions

In this work, PCA is applied to achieve dimensionality reduction in a band gap optimization problem for an acoustic metamaterial, by identifying a lower-dimensional subspace of the parameter space and projecting the gradient field of the objective function onto it. The effectiveness of the proposed method is demonstrated by the small dimension of the automatically selected subspace capable to reconstruct the centered gradient field with a given precision and, consequently, by the possibility of reducing the time required for running gradient-based optimization algorithms. This is obtained at the cost of introducing a small approximation error in the gradient, which usually does not prevent one from finding an ascent direction close to the one of the gradient. By selecting a different desired precision in the reconstruction of the centered gradient field, the proposed method could be applied to achieve a different trade-off between the precision in the gradient approximation and the time needed to find it. This is related to recent works on the trade-off between precision of supervision and number of training examples in machine learning problems [8, 9, 10]. A further research direction consists in combining the proposed method with optimization algorithms based on surrogate objective functions constructed by using adaptively chosen sets of training samples [3]. This could be achieved, e.g., by including a new sample in the current interpolation set by moving on the subspace identified by the method. Other possible extensions are the application of PCA to a subset of most recently generated samples of the gradient, and the replacement of PCA with one of its nonlinear versions (e.g., with kernel PCA [11]).

Acknowledgements

A. Bacigalupo and G. Gnecco are members of INdAM. The authors acknowledge financial support from INdAM-GNAMPA (project Trade-off between number of examples and precision in variations of the fixed-effects panel data model), from INdAM-GNFM, from the Compagnia di San Paolo (project MINIERA no. I34I20000380007), and from the University of Trento (project UNMASKED 2020).

References

[1] A. Bacigalupo, G. Gnecco, M. Lepidi, L. Gambarotta, Design of acoustic metamaterials through nonlinear programming. Lect Notes Comput Sci, 10122, 170-181, 2016.
[2] A. Bacigalupo, G. Gnecco, M. Lepidi, L. Gambarotta, Machine-learning techniques for the optimal design of acoustic metamaterials. J Optimiz Theory App, 187, 630-653, 2020.
[3] A. Bacigalupo, G. Gnecco, M. Lepidi, L. Gambarotta, Computational design of innovative mechanical metafilters via adaptive surrogate-based optimization. Comput Methods Appl Mech Eng, 375, 113623, 2021.
[4] G. E. Fasshauer, Meshfree Approximation Methods with Matlab, World Scientific, 2007.
[5] I. T. Jolliffe, Principal Component Analysis, Springer, 2002.
[6] H. Niederreiter, Random Number Generation and Quasi-Monte Carlo Methods, SIAM, 1992.
[7] F. Vadalà, A. Bacigalupo, M. Lepidi, L. Gambarotta, Free and forced wave propagation. Int J Mech Sci, 193, 106129, 2021.
[8] G. Gnecco, F. Nutarelli, On the trade-off between number of examples and precision of supervision in machine learning problems. Optim Lett, 2020, DOI: 10.1007/s11590-019-01486-x.
[9] G. Gnecco, F. Nutarelli, D. Selvi, Optimal trade-off between sample size, precision of supervision, and selection probabilities for the unbalanced fixed effects panel data model. Soft Comput, 24, 15937-15949, 2020.
[10] G. Gnecco, F. Nutarelli, D. Selvi, Optimal trade-off between sample size and precision for the fixed effects generalized least squares panel data model. Mach Learn, DOI: 10.1007/s10994-021-05976-x, 2021.
[11] G. Gnecco, M. Sanguineti. Accuracy of suboptimal solutions to kernel principal component analysis. Comput Optim Appl, 42, 265-287, 2009.

Dependence of the nonlinear optical response of CdSe nanoscrolls on coating with oleic or acetic acid

A D Golinskaya[1], A M Smirnov[1,2], V N Mantsevich[1], M V Kozlova[1], D A Kurtina[3,4], R B Vasiliev[3] and V S Dneprovskii[1]

[1]Moscow State University, Faculty of Physics, Chair of Semiconductors and Cryoelectronics and Quantum Technology Center, Moscow, Russian Federation
[2]Kotelnikov Institute of Radioengineering and Electronics of RAS, Moscow, Russian Federation
[3]Moscow State University, Faculty of Materials Science, Moscow, Russian Federation
[4]Moscow State University, Faculty of Chemistry, Moscow, Russian Federation

Abstract. The photoluminescence and nonlinear transmission of atomically thin population of colloidal CdSe rolled-up nanosheets were investigated at the one-photon stationary excitations of excitons by the third harmonic of Nd^{3+}:YAP laser pulses. It was revealed that the nonlinear optical response of the nanoscrolls is strongly influenced by multiwalled rolled-up structure. The broadening of exciton band and the red shift of the photoluminescence were observed and explained by coupling effect within the folded scroll-layers.

1. Introduction

Atomically flat nanoplatelets with precise thickness control have attracted a great deal of attention because of their remarkable optical and electronic properties, originating from strong spatial confinement of excitons in one direction and their morphological perfection [1]. Among various types of two-dimensional colloidal nanocrystals, the semiconductor CdSe nanoparticles with complicated shape, including nanoribbons, nanobelts and nanoscrolls, are promising materials for applications in polarization-enabled devices due to their strong optical activity [2].

In this work photoluminescence (PL) properties and nonlinear transmission of the 2.5-monolayer-thick CdSe nanoscrolls covered with oleic and acetic acid were investigated at the one-photon stationary excitation regime. Ligand exchange of native oleic acid molecules with shorter length acetic acid ones provides the reduced distance between layers in folded nanoscrolls.

2. Experimental details

2.1. Samples.

The synthesis of 2.5-monolayer-thick CdSe nanoscrolls followed the article [3]. Briefly, synthesis of nanoscrols was carried out by the colloidal method in the octadecene – cadmium acetate – oleic acid system. After degassing (170-190°C) a mixture of 0.13 g cadmium acetate dihydrate and 10 ml octadecene, the injection temperature (120°C) was set, 200 μl of selenium precursor (1 M solution of selenium in trioctylphosphine) was injected, after which the temperature gradually increased to the growth temperature (160°C) and maintained constant for 4 hours. At the end of the synthesis, 1 ml of oleic acid was added as a stabilizer. The resulting nanoparticles covered with oleic acid (CdSe-OA sample) were separated by centrifugation with an acetone precipitator and dissolved in hexane. To obtain a the second sample, a thermodynamically advantageous substitution of long unsaturated oleic acid for short acetic acid was performed at room temperature in hexane. In the following, the scrolls covered with oleic acid and acetic acid are denoted as CdSe-OA and CdSe-AA, respectively. The covering of nanoparticles with acids are required for passivation surface.

2.2. Experimental setup

The optical response of the samples was investigated with modification of pump and probe technique described in [4]. The difference is that the probing of transmission changing was realized by organic dye radiation. Briefly, the pump was realised by the third harmonic of Nd^{3+}:YAP laser pulses (λ=360 nm, τ=9 ns); the pump power varied from 130 kW/cm^2 to 2.1 MW/cm^2 by optical neutral filters. The radiation of BBQ organic dye ($C_{48}H_{66}O_2$, 4,4'''-Bis-(2-butyloctyloxy)-p-quaterphenyl) was used as probe signal; the wavelength of the BBQ dye PL spectrum coincides with the absorption band of heavy hole-electron (hh-e) exciton transition of the samples. All measurements were performed at room temperature.

3. Results and discussions

The PL spectra of the CdSe-OA and CdSe-AA scrolls for different pump intensities are presented in Figure 1a and Figure 1b, respectively.

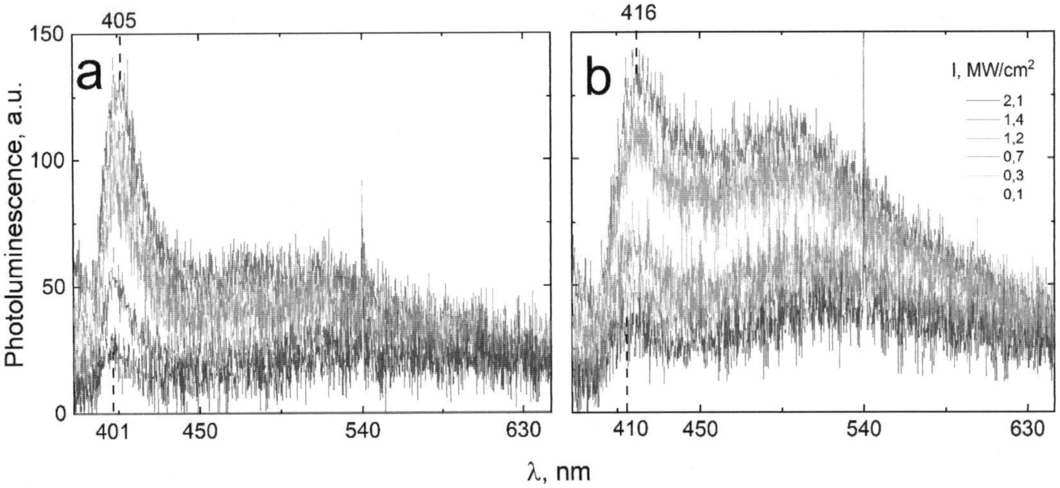

Figure 1. The PL spectra of the CdSe-OA (a) and CdSe-AA (b) scrolls for different pump intensities. Pump intensities varied from 0.1 W/cm^2 to 2.1 MW/cm^2.

For both samples PL spectra demonstrate a narrow luminescence peak (around 400-410 nm) associated with the lowest excitonic transition together with wide longwavelength luminescence band (around 520 nm) caused by the radiative carrier recombination through trap-related states. The intensities of the PL maxima of both emission bands increase with the growth of the pump intensity. Under the same excitation conditions, the excitonic PL intensities of both samples are approximately equal but trap-assisted PL intensity in CdSe-AA sample is significantly higher than trap-assisted PL intensity in CdSe-OA. Numerous induced defects in CdSe-AA sample are due to the folding effect caused a strong mechanical strain [3, 5]. For both samples the long wavelength Stokes shift of the exciton PL was found with the growth of excitation intensity. Its value increases from 8 nm (70 meV) for the lowest excitation intensity to 12 nm (100 meV) for the highest excitation intensity and from 14 nm (110 meV) to 20 nm (150 meV) for CdSe-OA and CdSe-AA, respectively. The revealed red shift of the PL spectra with increasing pump power can be the evidence of the formation of excited dimers and their radiative recombination at room temperature [6].

The measured linear and nonlinear transmission spectra of the CdSe-OA and CdSe-AA scrolls at the highest pump intensity are presented in Figure 2a and Figure 2b, respectively.

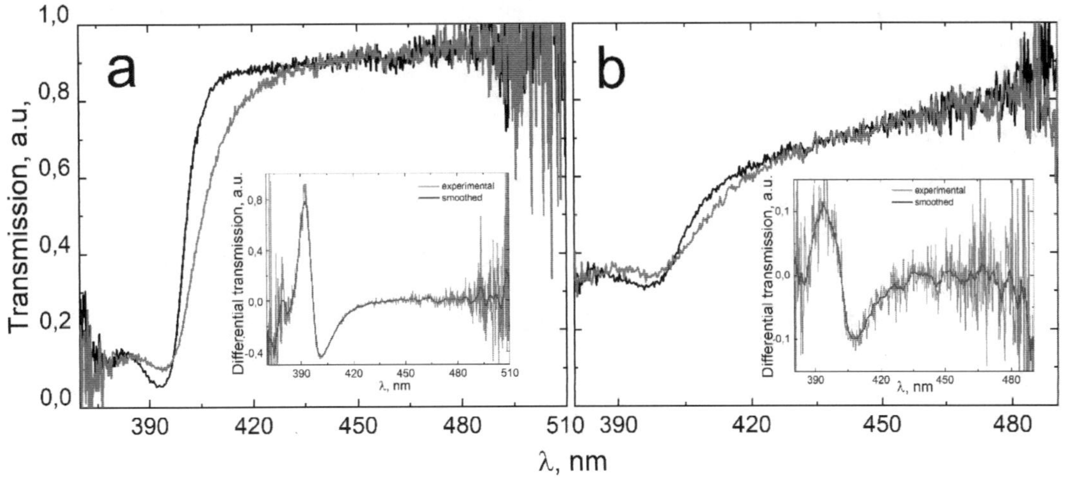

Figure 2. The linear (black line) and nonlinear (red line) transmission spectra of the CdSe-OA (a) and CdSe-AA (b) scrolls. The insets show the experimental (grey line) and smoothed (blue line) differential transmission spectra of the corresponding samples.

The broadening and slight red shift (3 nm) of exciton band was found for CdSe-AA sample, compared to the CdSe-OA one. The observed shift and broadening of CdSe-AA scrolls absorption spectrum is explained by the folding effect caused a strain appearance under acetic acid attachment [3, 5]. The modification of the exciton bands can be also assigned to the coupling effects within the rolled-up nanosheets as in quantum-dot dimer molecules due to the decreasing of interwall spacing [7]. The quantum coupling can lead to modifying of energy structure and the formation of minibands, promoting the red shift of absorption spectrum. The differential transmission (DT) spectra were measured to investigate the nonlinear properties of the scrolls (insets in Figure 2a and Figure 2b):

$$DT = (T(\lambda, I) - T_0(\lambda))/T_0(\lambda), \tag{1}$$

where $T_0(\lambda)$ and $T(\lambda, I)$ – the linear and nonlinear transmission spectrum, respectively. For both samples the significant bleaching of hh-e exciton transition was observed as a result of the phase space filling effect, energy transfer between excitons [4, 8-10] and interaction with phonons [11]. The negative DT-values were observed in in the wavelength region about 410 nm that can be caused by the broadening of absorption peaks due to exciton-exciton interaction under the action of high-intensity laser radiation [8,10].

4. Conclusion

The nonlinear optical response of oleic acid and acetic acid capped CdSe nanoscrolls were investigated by means of pump and probe technique. The role of interlayer carrier-interaction in the PL properties and transient absorption of colloidal CdSe scrolls was revealed experimentally by the observed broadening of absorption spectrum and red-shifting of PL spectra with the growth of pump intensity.

Acknowledge

This work was supported by President grant for young scientists MD-781.2021.1.2 and Russian Foundation for Basic Research (20-32-70001).

References

[1] Ithurria S, Tessier M, Mahler B, Lobo R, Dubertret B, Efros A 2011 *Nat. Mat.* **10** 936
[2] Ma W, Xu L, de Moura A, Wu X, Kuang H, Xu C, Kotov N 2017 *Chem. Rev.* **117** 8041

[3] Kurtina D, Garshev A, Vasil'eva I, Shubin V, Gaskov A, Vasiliev R 2019 *Chem. Mater.* **31** 9652.

[4] Smirnov A, Golinskaya A, Kozlova M, Zharkova E, Saidzhonov B, Vasiliev R, Dneprovskii V 2020 *Semicond.* **54** 1907

[5] Vasiliev R, Sokolikova M, Vitukhnovskii A, Ambrozevich S, Selyukov A, Lebedev V 2015 *Quantum Electron.* **45** 853

[6] Diroll B, Cho W, Coropceanu I, Harvey S, Brumberg A, Holtgrewe N., Crooker A, Wasielewski M, Prakapenka V, Talapin D, Schaller R 2018 *Nano Lett.* **18** 6948

[7] Cui J, Panfil, Koley Y, Shamalia S, Waiskopf D, Remennik S, Popov I, Oded M, Banin U 2019 *Nat. Comm.* **10** 1

[8] Smirnov A, Golinskaya A, Saidzhonov B, Vasiliev R, Mantsevich V, Dneprovskii V 2021 *J. Lum.* **229** 117682

[9] Smirnov A, Mantsevich V, Smirnov D, Golinskaya A, Kozlova M, Saidzhonov B, Dneprovskii V, Vasiliev R 2019 *Sol. St. Comm.* **299** 113651

[10] Smirnov A, Golinskaya A, Zharkova E, Kozlova M, Saidzhonov B, Vasil'ev R, Dneprovskii V 2019 *JETP Lett.* **109** 454

[11] Smirnov A, Golinskaya A, Saidzhonov B, Vasiliev R, Mantsevich V, Dneprovskii V 2019 *JETP Lett.* **109** 372

Exceptional points in composite structures consisting of two dielectric diffraction gratings with Lorentzian line shape

Nikita Golovastikov*, Dmitry Bykov and Leonid Doskolovich

Image Processing Systems Institute – Branch of the Federal Scientific Research Centre "Crystallography and Photonics" of Russian Academy of Sciences, 151 Molodogvardeyskaya st., Samara 443001, Russia
Samara National Research University, 34 Moskovskoye shosse, Samara 443086, Russia

Corresponding author: nikita.golovastikov@gmail.com

Abstract. Using scattering matrix formalism we derive analytical expressions for the eigenmodes of a composite structure consisting of two dielectric diffraction gratings with Lorentzian profile in reflection. Analyzing these expressions we prove formation of two distinct pairs of exceptional points, provide analytical approximations for their coordinates and by rigorous simulation demonstrate eigenmodes interchange as a result of encircling said exceptional points.

1. Introduction

Exceptional points (EPs) are degeneracies in non-Hermitian systems, which appear when several eigenmodes coalesce [1]. The properties of such system dramatically change in the vicinity of EPs and lead to such phenomena as enhanced optical sensing [2], loss-induced transparency [3], unidirectional transmission or reflection [4], and lasers with reversed pump dependence [5] or single-mode operation [6]. One promising feature of an EP is that adiabatically encircling an EP can result in an exchange of the eigenstate. Such behavior is expected to have applications in asymmetric mode switching [7] and on-chip non-reciprocal transmission [8] and light stopping [9].

As a rule, exceptional points are studied through analyzing the eigenvalues and eigenvectors of the proper Hamiltonian [10] or by analyzing the eigenmodes dispersion relation [11]. In this work we demonstrate the formation of EPs using ω-k_x resonant approximation of Lorentzian line shape. By obtaining analytical expressions for the eigenmodes of a composite structure consisting of two dielectric diffraction gratings (DGs) with Lorentzian line shape we show the formation of two distinct EPs which can be achieved by varying the distance between said stacked gratings l. This theoretical conclusion is supported by rigorous calculation results that show eigenmodes swapping as a result of encircling said EPs in the l-k_x parameter space.

2. ω-k_x Lorentzian line shape in composite structures

A scattering matrix **S** relates the complex amplitudes of the plane waves incident on the diffraction structure from the superstrate $\mathbf{I}_u$ and the substrate $\mathbf{I}_d$ regions with the amplitudes of the transmitted **T** and reflected **R** diffraction orders [12]. For a horizontally symmetrical subwavelength DG, which allows only the 0th reflected and transmitted diffraction orders to propagate, the **S** matrix takes the form:

$$\begin{bmatrix}\mathbf{T}\\ \mathbf{R}\end{bmatrix}=\mathbf{S}_1(k_x,\omega)\begin{bmatrix}\mathbf{I}_u\\ \mathbf{I}_d\end{bmatrix},\quad \mathbf{S}_1(k_x,\omega)=\begin{pmatrix}T_1(k_x,\omega) & R_1(k_x,\omega)\\ R_1(k_x,\omega) & T_1(k_x,\omega)\end{pmatrix}, \tag{1}$$

where $R_1(k_x,\omega)$ and $T_1(k_x,\omega)$ are the complex reflection and transmission coefficients of the DG for a unit-amplitude incident wave. It is worth noting that the scattering matrix in (1) does not describe the near-field effects associated with the evanescent diffraction orders of the DG.

In this paper we consider the elements of the scattering matrix (1) to be functions of the angular frequency ω and the in-plane wave component k_x of the incident light. Let the DG have Lorentzian line shape profile. In this case the appropriate reflection and transmission coefficients can be approximated as follows [13]:

$$R_1(k_x,\omega)=e^{i\varphi}\frac{i\,\mathrm{Im}\,\omega_{p,1}(\omega-\omega_{p,2})}{v_g^2k_x^2-(\omega-\omega_{p,1})(\omega-\omega_{p,2})},\quad T_1(k_x,\omega)=e^{i\varphi}\frac{v_g^2k_x^2-(\omega-\mathrm{Re}\,\omega_{p,1})(\omega-\omega_{p,2})}{v_g^2k_x^2-(\omega-\omega_{p,1})(\omega-\omega_{p,2})}, \tag{2}$$

As a DG under consideration we propose a dielectric structure shown in figure 1a. The agreement between its reflection spectra calculated using rigorous coupled-wave analysis and the one calculated using the approximations (2) confirms that said DG has Lorentzian reflection profile (figure 1b). See the caption of figure 1 for the parameters of the DG as well as the parameters for the approximation.

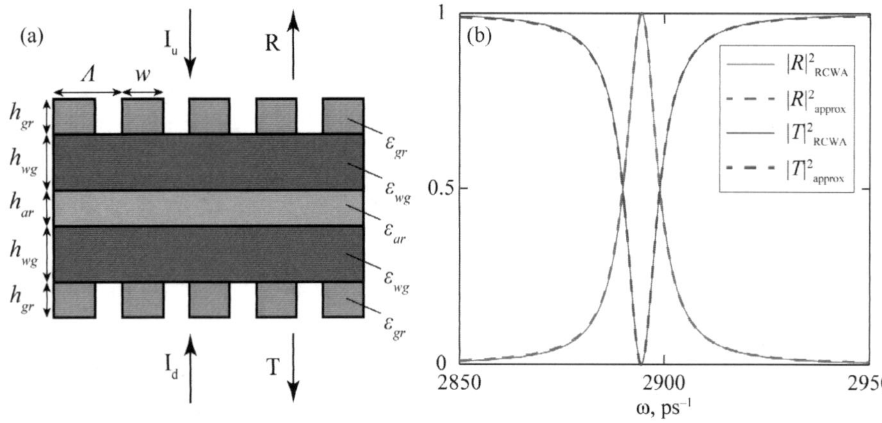

Figure 1. (a) Geometry of a DG under consideration: $\Lambda = 0.51\ \mu$m, $h_{gr} = 0.11\ \mu$m, $h_{wg} = 0.21\ \mu$m, $h_{ar} = 0.11\ \mu$m, $w = 0.27\ \mu$m, refractive indices $n_{gr} = 1.5$, $n_{wg} = 2.4$, $n_{ar} = 1.46$, $n_{env} = 1$; $\omega_{p,1} = (2894.36 - 4.57i)\cdot 10^{12}\mathrm{s}^{-1}$, $\omega_{p,2} = 2890.83\cdot 10^{12}\mathrm{s}^{-1}$, $\omega_z^T = 2894.4\cdot 10^{12}\mathrm{s}^{-1}$, $\varphi = -2.598$, $v_g = 181.64\ \mu m/ps$. (b) Reflectance $|R(\omega)|^2$ and transmittance $|T(\omega)|^2$ spectra of the DG (TE-polarization).

Let us consider a composite DG consisting of two subwavelength DGs described by scattering matrices (1) and separated by a homogeneous dielectric layer with the thickness l and the refractive index n_{env} (figure 2a). In this case, the scattering matrix of the composite DG can be expressed through the matrix $\mathbf{S}_1(\omega)$ in the form [14]

$$\mathbf{S}_2(k_x,\omega)=\mathbf{S}_1(k_x,\omega)*\mathbf{L}(l,k_x,\omega)*\mathbf{S}_1(k_x,\omega) \tag{3}$$

where the symbol $*$ denotes the Redheffer star product, $\mathbf{L}(l,k_x,\omega)=\exp(i\psi(l,k_x,\omega))\mathbf{E}$ is the scattering matrix of the intermediary layer and $\psi(l,k_x,\omega)$ is the phase shift the plane waves acquire by propagation through this layer:

$$\psi(l,k_x,\omega)=l\sqrt{(n_{env}\omega/c)^2-k_x^2(\omega)}, \tag{4}$$

where $k_x(\omega)=(\omega/c)n_{env}\sin\theta$, θ being the angle of incidence.

By substituting (2) into (3), we obtain the scattering matrix $\mathbf{S}_2$ of the composite DG with the following reflection and transmission coefficients:

$$R_2(k_x,\omega)=\frac{\mathrm{i}e^{\mathrm{i}\varphi}(\omega-\omega_{p,2})\operatorname{Im}\omega_{p,1}\left(k_x^2v_g^2-(\omega-\omega_{p,1})(\omega-\omega_{p,2})+e^{2\mathrm{i}(\varphi+\psi)}\left(k_x^2v_g^2-(\omega-\omega_{p,1}^*)(\omega-\omega_{p,2})\right)\right)}{\left(k_x^2v_g^2-(\omega-\omega_{p,2})(\omega-\omega_{\mathrm{mode},1})\right)\left(k_x^2v_g^2-(\omega-\omega_{p,2})(\omega-\omega_{\mathrm{mode},2})\right)},$$
$$T_2(k_x,\omega)=\frac{e^{\mathrm{i}(2\varphi+\psi)}\left(k_x^2v_g^2-(\omega-\operatorname{Re}\omega_{p,1})(\omega-\omega_{p,2})\right)^2}{\left(k_x^2v_g^2-(\omega-\omega_{p,2})(\omega-\omega_{\mathrm{mode},1})\right)\left(k_x^2v_g^2-(\omega-\omega_{p,2})(\omega-\omega_{\mathrm{mode},2})\right)},\qquad(5)$$

where $\omega_{\mathrm{mode},1,2}=\operatorname{Re}\omega_{p,1}+\mathrm{i}\operatorname{Im}\omega_{p,1}\left(1\pm e^{\mathrm{i}(\varphi+\psi)}\right)$.

3. Exceptional points

According to (5) the eigenmodes of the composite structure (reflection and transmission coefficients complex poles) have the following form:

$$\omega_{poles}=\left\{\begin{array}{l}\frac{1}{2}\left(\operatorname{Re}\omega_{p,1}+\omega_{p,2}+\mathrm{i}\operatorname{Im}\omega_{p,1}\left(1+e^{\mathrm{i}(\psi+\varphi)}\right)\pm\sqrt{4k_x^2v_g^2+\left[\operatorname{Re}\omega_{p,1}-\omega_{p,2}+\mathrm{i}\operatorname{Im}\omega_{p,1}\left(1+e^{\mathrm{i}(\psi+\varphi)}\right)\right]^2}\right),\\ \frac{1}{2}\left(\operatorname{Re}\omega_{p,1}+\omega_{p,2}+\mathrm{i}\operatorname{Im}\omega_{p,1}\left(1-e^{\mathrm{i}(\psi+\varphi)}\right)\pm\sqrt{4k_x^2v_g^2+\left[\operatorname{Re}\omega_{p,1}-\omega_{p,2}+\mathrm{i}\operatorname{Im}\omega_{p,1}\left(1-e^{\mathrm{i}(\psi+\varphi)}\right)\right]^2}\right)\end{array}\right\}.\qquad(6)$$

Note that expressions (6) contain square roots, which gives rise to branch points when the expressions under the radicals equal zero:

$$4k_x^2v_g^2+\left[\operatorname{Re}\omega_{p,1}-\omega_{p,2}+\mathrm{i}\operatorname{Im}\omega_{p,1}\left(1\pm e^{\mathrm{i}(\psi+\varphi)}\right)\right]^2=0.\qquad(7)$$

Let us consider the vicinity of normal incidence $\left(k_x \ll k_0 = n_{env}\operatorname{Re}\omega_p/c\right)$ so that the phase shift ψ is defined only by intermediary layer thickness l: $\psi(l,k_x,\omega)\approx\psi(l)=lk_0$. This way solving (7) with respect to ψ, we arrive at two pairs of EPs in the $\psi-k_x$ parameter space:

$$\psi=\left\{\begin{array}{l}\pm\arcsin\frac{\operatorname{Re}\omega_{p,1}-\omega_{p,2}}{\operatorname{Im}\omega_{p,1}}-\varphi+2\pi m,\quad m\in\mathbb{Z}\\ \pi\mp\arcsin\frac{\operatorname{Re}\omega_{p,1}-\omega_{p,2}}{\operatorname{Im}\omega_{p,1}}-\varphi+2\pi m,\quad m\in\mathbb{Z}\end{array}\right.\Rightarrow\quad k_x=\left\{\begin{array}{l}\pm\frac{(1+\cos\psi)\operatorname{Im}\omega_{p,1}}{2v_g}\\ \pm\frac{(1-\cos\psi)\operatorname{Im}\omega_{p,1}}{2v_g}\end{array}\right.\qquad(8)$$

The composite structure will possess exceptional points if and only if the value for ψ in (8) is real, i.e. if $\left|(\operatorname{Re}\omega_{p,1}-\omega_{p,2})/\operatorname{Im}\omega_{p,1}\right|\le 1$. This condition demands that the structure possesses low Q-factor resonance (high $\operatorname{Im}\omega_{p,1}$) as well as narrow band gap (small $\left|\operatorname{Re}\omega_{p,1}-\omega_{p,2}\right|$), and is hard to satisfy since generally speaking increasing the diffraction grating's contrast (straying away from slab waveguide) widens the band gap while simultaneously lowering the quality factor of the resonance. The necessity to meet this requirement explains the bulkiness of the structure under consideration (figure 1a).

Using the approximations (8) we derive the following estimates for the exceptional point coordinates $\psi_1=0.883$ $(l_1=0.742\mu m)$, $k_{x,1}=4.6\cdot10^{-3}\mu m^{-1}$ and $\psi_2=2.26$ $(l_2=0.885\mu m)$, $k_{x,2}=2.06\cdot10^{-2}\mu m^{-1}$. Let us choose a contour in the l-k_x parameter space centering at the analytically estimated EPs coordinates (figures 2b and 2d). After encircling the EPs counterclockwise following said contours one can notice that the two rigorously calculated using RCWA complex poles corresponding to the same square root in (6) swap places (figures 2c and 2e). This interchange of eigenmodes is an intrinsic feature of EPs and proves the existence of EPs. More accurate estimate of the EPs location can be acquired by solving (7) accounting for ψ being the function of ω and θ.

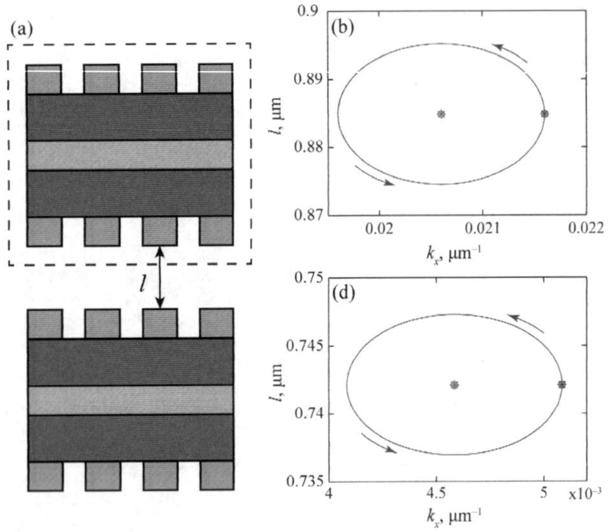

Figure 2. (a) Geometry of a composite structure containing two identical DGs in figure 1. Analytically estimated EPs positions (8) (red asterisks) and trajectories of the encircling in the l-k_x parameter space (b, d). Trajectories of the eigenmodes' complex frequencies when encircling the corresponding EPs (c, e)

4. Conclusion

By means of scattering matrix formalism we derived analytical expressions for the eigenmodes of a composite structure consisting of two dielectric diffraction gratings with Lorentzian profile in reflection. Using said approximations we formulate a criterion for the grating eigenmodes that if satisfied allows formation on two distinct pairs of exceptional points. Rigorous calculation results show eigenmodes interchange upon encircling said EPs.

5. Acknowledgments

This work was funded by the Russian Science Foundation grant no. 19-19-00514 (analytical study of the composite structure eigenmodes) and by the Russian Federation Ministry of Science and Higher Education under state contract with the "Crystallography and Photonics" Research Center of the RAS (software development and numerical simulation).

References

[1] Heiss W D 2012 *J. Phys. A: Math. Theor.* **45** 444016
[2] Chen W, Özdemir S, Zhao G, Wiersig J and Yang L 2017 *Nature (London)* **548** 192
[3] Guo A, Salamo G, Duchesne D, Morandotti R, Volatier-Ravat M, Aimez V, Siviloglou G, Christodoulides D 2009 *Phys. Rev. Lett.* **103** 093902
[4] Regensburger A, Bersch C, Miri M, Onishchukov G, Christodoulides D and Peschel U 2012 *Nature* **488** 167–171
[5] Liertzer M, Ge L, Cerjan A, Stone A, Tureci H, and Rotter S 2012 *Phys. Rev. Lett.* **108** 173901
[6] Feng L, Wong Z, Ma R, WangY and Zhang X 2014 *Science* **346** 972–975
[7] Doppler J, Mailybaev A, Böhm J, Kuhl U, Girschik A, Libisch F, Milburn T, Rabl P, Moiseyev N, and Rotter S 2016 *Nature (London)* **537**, 76–79
[8] Choi Y, Hahn C, Yoon J, Song S and Berini P 2017 *Nat. Commun.* **8** 14154
[9] Goldzak T Mailybaev A and Moiseyev N 2018 *Phys. Rev. Lett.* **120** 013901
[10] Heiss W 2012 *J. Phys. A* **45** 444016
[11] Bykov D and Doskolovich L 2018 *Phys. Rev. A* **97** 013846
[12] Gippius N, Weiss T, Tikhodeev S and Giessen H 2010 *Opt. Express* **18** 7569–7574
[13] Golovastikov N, Bykov D and Doskolovich L 2020 *AIP Conf. Proc.* **2300** 020039
[14] Li L 1996 *J.Opt. Soc. Am. A* **13** 1024–1035

Functional Photonic Elements Based on Liquid Crystal Metasurfaces

Maxim V. Gorkunov[1], **Irina V. Kasyanova**[1], **Vladimir V. Artemov**[1], **Artur R. Geivandov**[1], **Ivan V. Simdyankin**[1], **Alena V. Mamonova**[1], **Alexander A. Ezhov**[1,2] **and Serguei P. Palto**[1]

[1] Shubnikov Institute of Crystallography, FSRC "Crystallography and Photonics", Russian Academy of Sciences, 119333 Moscow, Russia
[2] Faculty of Physics, Lomonosov Moscow State University, 119991 Moscow, Russia

E-mail: gorkunov@crys.ras.ru

Abstract. We study versatile soft-matter metasurfaces based on self-assembling of nematic liquid crystal on polymer alignment layers processed with a focused ion beam. Digital control of the beam path allows imprinting patterns that induce different complex distributions of the refractive index within several micrometer thick liquid crystal layers. We optimize them to implement various optical functionalities, such as broadband anomalous refraction, wide-aperture focusing, and beam splitting in tens of channels.

1. Introduction

Alignment of liquid crystals (LCs) by surfaces belongs to the core of the design of all LC-based devices, as it is commonly employed to establish a homogeneous near-surface LC director orientation preferential for a particular functionality [1]. Recently, specially prepared surfaces supporting inhomogeneous LC alignment have attracted significant attention as the applied potential of the self-assembled modulated LC configurations has been revealed [2, 3, 4, 5]. In particular, such layers can be produced by patterning with a focused ion beam (FIB) of thin films of rubbed polyimide. While the latter are routinely employed to sustain planar orientation of the adjacent LC director, mild irradiation with the FIB locally flips the favorable director orientation to vertical. Combining micrometer-scale pristine and FIB-treated areas allows establishing precise intermediate surface LC orientations [6]. Periodic patterns of micrometer-wide stripes produce deep LC modulations sufficient for strong visible light diffraction [4]. More complex superperiodic patterns optimized for anomalous refraction establish strongly asymmetric LC configurations allowing deflecting the light in the direction of a single dominating diffraction order [5]. Importantly, all such systems as susceptible to low-voltage signals applied across the LC layer and switch between different regimes in several milliseconds [7].

Digital control of the FIB in modern scanning electron microscopy (SEM) devices allows imprinting practically arbitrary 2D patterns than can induce very different LC textures. Therefore, relying on essentially the same materials and components, it is possible to design and fabricate optical elements with very different functionalities. Here we demonstrate the broad potential of the approach and report on three optical elements optimized for specific advanced functionalities: refracting double-sided metasurfaces with the efficiency and spectral

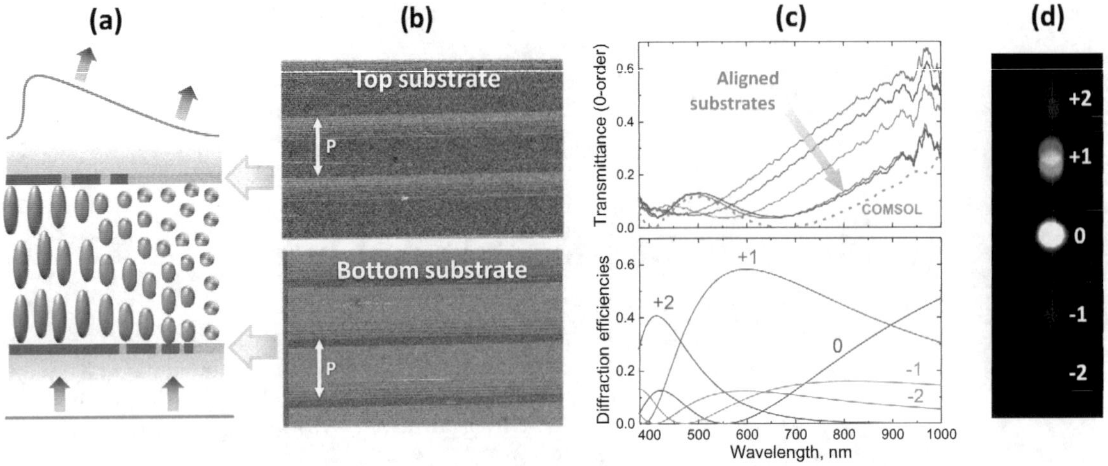

Figure 1. Double-sided superperiodic LC metasurfaces for efficient broadband anomalous refraction: (a) schematic of substrate alignment required for maximum refraction efficiency; (b) SEM-images of the fragments of exemplary substrates patterned with superperiodic stripes by the FIB; (c) transmission spectra of the 5 μm periodic metasurface measured upon different substrate alignment up to the optimum characterized by the transmission spectrum close to the modeled one (dashed line) and the diffraction spectra predicted by the optimization; (d) broadband asymmetric diffraction by a properly aligned 5 μm periodic metasurface observed by the back focal plane optical microscopy.

range enhanced due to synchronous superperiodic patterning of the both substrates (Section 2.1); arrays of cylindrical high-power short-focused LC microlenses (Section 2.2); as well as the LC Dammann-type gratings designed to evenly split the transmitted light energy between several tens of diffraction orders (Section 2.3).

2. Results

2.1. Double-Sided Metasurfaces for Anomalous Refraction

Superperiodic aligning patterns consisting of many parallel stripes of various widths have proven to induce complex LC textures giving rise to strongly asymmetric light diffraction effectively resulting in efficient anomalous refraction. Implementing such patterns on a single LC-cell substrate has allowed deflecting about 60% of the normally incident light energy into a particular oblique direction [5]. Natural factors limiting the anomalous refraction of LC layers modulated near one substrate are: the LC orientational elasticity that smooths the outgoing phase profile and prevents it from approaching the discontinuous saw-tooth shape optimum for the perfect refraction; and the moderate optical anisotropy of the LC that does not allow achieving efficient refraction at all visible light wavelengths.

Combining modulated aligning action of the both substrates allows enhancing the main functional characteristics. The main idea of the double-sided LC-metasurface design is illustrated in Fig. 1(a), where it is shown how different stripe patterns on the both substrates produce a gradual variation of the LC orientation throughout the whole layer. Note that the LC alignment is first (from left to right) varied at the top substrate and then at the bottom one. Particular optimal patterns depend on the specific LC material properties, the LC layer thickness and the overall periodicity. In order to avoid high computational costs of the design optimization based on the first-principles, we employ an approximate semi-analytical model neglecting the difference of LC elastic modules and considering the LC layer as an inhomogeneous phase grating. Such

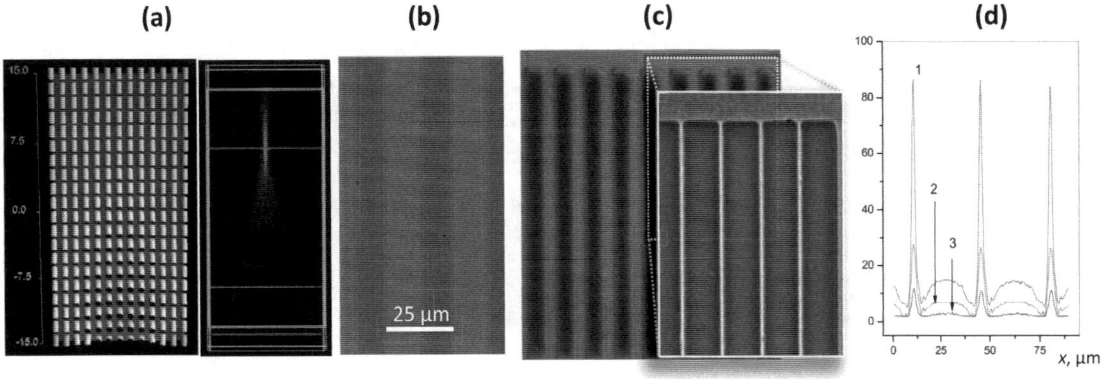

Figure 2. Cylindrical LC-lens: (a) numerical model of LC alignment and focusing of light of a 450 nm wavelength by a lens with an aperture of 15 μm; (b) SEM-image of a fragment of substrate patterned for a 25 μm periodic array of cylindrical LC-lenses; (c) optical images of a 35 μm periodic microlens array in vertically polarized light in the plane of aligning substrate and in the focal plane 35 μm above it; (d) intensity distribution (arbitrary units) in the focal plane of a 35 μm periodic microlens array, red (1), green (2) and blue (3) curves correspond to the linear RGB data of the camera sensor in red, green and blue channels, respectively.

approach has previously proven its efficiency for the single-sided design optimization [5]. Applied to the double-sided configuration, it also generates complex superperiodic stripe patterns that are to be imprinted with the FIB. Examples of the patterns optimized for a period $P = 10$ μm, for a layer of E7 nematic LC material of a thickness of $d = 4$ μm as they are imprinted in rubbed polyimide are shown in Fig. 1(b).

Precise synchronous alignment of the both patterned substrates is very important for the double-sided device functionality. As is shown in Fig. 1(c), the transmitted light spectrum substantially varies upon fine relative substrate displacement. We judge on the achievement of the optimal alignment by the transmission spectra approaching those predicted by the numerical simulations. Observing the metasurfaces using back focal plane optical microscopy (see Fig. 1(d)) confirms all theoretical predictions: the $+1$ diffraction order dominates in the whole visible range, while the -1 order remains notably suppressed. Note also the asymmetry of the ± 2 orders, as the violet light is predominantly refracted into the $+2$ order. For comparison, the optimal single-sided design of similar periodicity and thickness is only capable of a moderate asymmetry of the ± 1 diffraction orders in the blue and violet ranges [5].

2.2. Microlens Arrays

Controlling the LC by a single substrate supporting a binary (planar-vertical) director alignment profile allows creating arrays of LC microlenses. By varying the location and width of the treated and untreated micrometer- and submicrometer-wide stripes we achieve the director tilt angle distribution near the surface seen at the bottom in Fig. 2(a). As the LC elasticity transfers the alignment perturbation into the LC layer, specific spatial distribution of the LC refractive index is achieved which creates lens-like refraction of light and its focusing also shown as modeled in Fig. 2(a). The numerical simulations performed using finite-difference time-domain module of our in-house LCDTDK software predict the focal length of about 23 μm for a cylindrical lens of a 15 μm aperture, which corresponds to a remarkably low f-number of 1.5.

To implement such lenses, we fabricate them as periodic arrays by subjecting rubbed polyimide alignment layers to FIB patterning as illustrated in Fig. 2(b) and assembling 30 μm thick LC-cells filled with E7 nematic. Optical microscopy verifies that the arrays focus the light

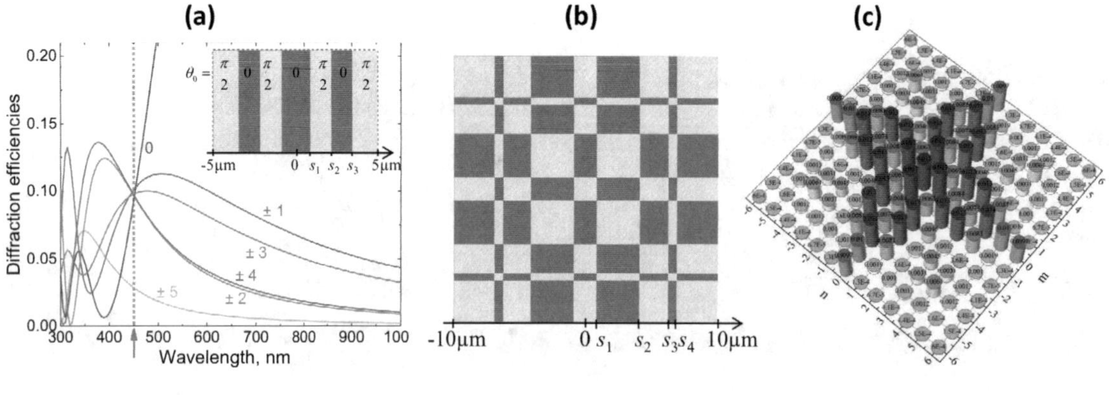

Figure 3. Optimization of LC Dammann gratings: (a) spectra of diffraction efficiencies of the first 11 orders (indicated on the lines) of a 4 μm thick LC layer assembled into a 10 μm periodic grating upon the stripe pattern (shown in the inset) optimized for even distribution of light of a 450 nm wavelength (vertical dashed line) into 9 transmission diffraction orders; (b) schematic of a 20 × 20 μm^2 unit cell of a 2D pattern optimized to induce a Damman grating in a 8 μm thick LC layer splitting the light of a 450 nm wavelength; and (c) diagram of the corresponding diffraction efficiencies.

polarized along the lens axis at about 35 μm above the aligning substrate. This determines the actual f-number equal to 2.3, i.e., higher than the predicted one, but, nevertheless, allowing classifying the microlenses as fast (wide-aperture).

By extracting the linear intensity distributions recorded by the pixels of the optical camera sensor (from a RAW file of Olympus OMD EM-5 camera) we obtain the light intensity distributions shown in Fig. 2(d) for the three color channels. In addition to a high degree of optical contrast in the focal plane, we also observe a high optical resolution just slightly below the diffraction limit. The location of the focused peaks of intensities of all colors within the same focal plane indicates a remarkably low level of potential color aberrations.

2.3. Beam-Splitting Dammann Gratings

While the efficient beam deflection described in Section 2.1 relies on the enhancement of a single outgoing diffraction channel, important optical applications spanning from parallel computing to 3D object recognition require even splitting of light beams in arrays of many bright spots. Specific diffractive elements, known also as Dammann-type gratings, are being developed for such purposes, see, e.g., the recent work [8] and references therein. Relative complexity of grating unit cells required to generate many equal output beams poses fabrication challenges. In particular, the only known so far type of electrically switchable LC-based Dammann gratings is formed upon substrates subsequently photoaligned by illumination through different masks, which allows patterning with the periodicity not smaller than 200 μm thus limiting the angle between the split channels by just 0.19° [9]. Our technique of the FIB patterning naturally avoids all such limitations and allows producing Dammann-type gratings with the diffraction angles larger by at least an order of magnitude.

To illustrate the potential possibilities, we start with optimizing 1D stripe patterns for producing a line of many equally strong diffraction spots. An example of such pattern obtained by semi-analytical optimization is shown in Fig. 3(a), along with the spectra of expected diffraction efficiencies. The symmetry of the pattern unit cell ensures the diffraction symmetry as well as simplifies the optimization by reducing the number of independent optimization variables (the coordinates s_i). One can see that the obtained pattern of a period of just 10 μm can split the light of a particular wavelength evenly in 9 directions.

Generalizing the optimization for 2D Dammann-type square patterns (see an example in Fig. 3(b)) we are able to generate structures having 20 μm periodicity and splitting the light in several tens of diffraction orders as is illustrated in Fig. 3(c). Note that restricting ourselves to the Dammann-type patterns reduces the potentially very large number of optimization variables describing a square unit cell to a small set of the coordinates s_i.

3. Conclusion

We have shown that polymer alignment layers patterned with the FIB according to various optimized designs support self-assembling of LC layers into optical elements performing qualitatively different functions: beam deflection, focusing and splitting. Naturally, the presented opportunities do not exhaust the possible LC-metasurface applications. The design flexibility, fabrication simplicity along with the options of fast electro-optical switching ensure that LC-metasurfaces are generally a very promising platform for applied flat optics.

Acknowledgments

The research is funded by the Russian Science Foundation project 18-12-00361, https://rscf.ru/project/18-12-00361/. The work is performed using the equipment of Shared Research Center of FSRC "Crystallography and Photonics" RAS supported by the Ministry of Science and Higher Education of the Russian Federation (project RFMEFI62119X0035).

References

[1] Blinov L M 2011 *Structure and properties of liquid crystals* (Springer Netherlands)
[2] Kasyanova I V, Gorkunov M V, Artemov V V, Geivandov A R, Mamonova A V and Palto S P 2018 *Optics Express* **26** 20258–20269
[3] Nys I 2020 *Liquid Crystals Today* **29** 65–83
[4] Gorkunov M V, Kasyanova I V, Artemov V V, Ezhov A A, Mamonova A V, Simdyankin I V and Palto S P 2020 *ACS Applied Materials & Interfaces* **12** 30815–30823
[5] Gorkunov M V, Kasyanova I V, Artemov V V, Ezhov A A, Mamonova A V, Simdyankin I V and Palto S P 2020 *ACS Photonics* **7** 3096–3105
[6] Gorkunov M V, Kasyanova I V, Artemov V V, Mamonova A V and Palto S P 2019 *Beilstein Journal of Nanotechnology* **10** 1691–1697
[7] Gorkunov M V, Kasyanova I V, Artemov V V, Ezhov A A, Mamonova A V, Simdyankin I V and Palto S P 2020 *AIP Conference Proceedings* vol 2300 (AIP Publishing) p 020040
[8] Wang H, Wang H, Zhang W and Yang J K W 2020 *ACS Nano* **14** 10452–10461
[9] Fan F, Yao L, Wang X, Shi L, Srivastava A, Chigrinov V, Kwok H S and Wen S 2017 *Crystals* **7** 79

Structure and properties of the low-energy deposited TiO_2 thin films: results of the molecular dynamics simulation

F.V. Grigoriev*, V.B. Sulimov and A.V. Tikhonravov

Research Computing Center, M.V. Lomonosov Moscow State University, Leninskie Gory, Moscow, 119991, Russia

**fedor.grigoriev@gmail.com*

Abstract: The classical molecular dynamics simulation of the low-energy glancing angle deposition of titanium dioxide films is performed. The deposition angle varies from 60° to 80°. It is found that the film structure consists of parallel slanted columns which lead to the anisotropy of films properties. The difference between the main components of the refractive index tensor is about 0.14, which is close to the values obtained for high-energy titanium dioxide films and larger than 0.03 obtained earlier for silicon dioxide films.

Key words: molecular dynamics, thin films, titanium dioxide, refractive index, anisotropy, glancing angle deposition.

1. Introduction

Deposition at large angles, when incoming atoms move almost parallel to the substrate, leads to the formation of glancing angle deposited (GLAD) films, consisting of separated nanostructures having different dimensions and shapes [1,2]. Due to the high porosity and anisotropy of the structure, transparent dielectric GLAD films are used in various optical devices [2,3]. The properties of GLAD films essentially depend on the deposition conditions, such as deposition energy, deposition angle, substrate temperature and others. At present for the theoretical investigation of these dependencies at the initial stage of the GLAD films formation, various methods of atomistic modelling are used [4-6].

Earlier the classical full-atomistic molecular dynamics (MD) simulation was applied to study the formation of SiO_2 [7, 8] and high-energy deposited TiO_2 [9] GLAD films. In the present work we study the formation of low-energy TiO_2 GLAD films. The differences in the properties of the high- and low-energy deposited GLAD films are discussed.

2. Simulation method

The simulation of the thin film deposition process is performed using the previously developed step-by-step procedure [10]. The energy of interatomic interactions is calculated using the pairwise force field [11, 12]:

$$U = q_i q_j / r_{ij} + A_{ij} / r_{ij}^{12} - B_{ij} / r_{ij}^{6} \quad (1),$$

where $q_{i(j)}$ is the charge of the $i(j)$-th atom, $q_O = -0.5q_{Ti} = -1.098e$, A_{ij} and B_{ij} are the parameters of Lennard-Jones potential, rewritten from the sigma-epsilon form to the *A-B* form (Table 1).

Table 1. Parameters of the Lennard-Jones potential, Eq. (1).

Interaction	A_{ij} (kJ·nm^6/mol)	B_{ij}(kJ·nm^{12}/mol)
Ti-Ti	$5.062 \cdot 10^{-4}$	$3.1195 \cdot 10^{-8}$
O-O	$3.0009 \cdot 10^{-3}$	$1.735775 \cdot 10^{-6}$
Ti-O	$1.4368 \cdot 10^{-3}$	$2.90927 \cdot 10^{-7}$

At each step of the simulation procedure, titanium and oxygen atoms with stoichiometric proportion of 1:2 are inserted randomly at the top of the simulation box. The initial value of the titanium and oxygen kinetic energies is E(Ti) = E(O) = 0.1 eV, which corresponds to a low-energy process like e-beam evaporation [13]. At each step of the simulation procedure the *NVT* (constant number of particles, volume and temperature) ensemble is used. The vertical dimension of the simulation box is increased by

0.01 nm after each injection step to compensate for the growth of film thickness. The Berendsen thermostat [14] is applied to keep the simulation cluster temperature, $T = 300$ K, constant. The duration of one injection step is 10 ps and the time step of MD modeling is 0.5 fs.

MD simulation is performed using the GROMACS package [15]. The simulation was carried out using the equipment of the shared research facilities of HPC computing resources at Lomonosov Moscow State University [16].

3. Results and discussion

Results of the simulations are presented in Figs. 1,2. Deposition at low energy of incoming atoms leads to the formation of porous structures (Fig. 1). In the case of normal deposition, $\alpha = 0$, (Fig. 1, left side), pores of different dimensions and shapes are randomly distributed over the film volume. Pore dimensions can reach several nanometers.

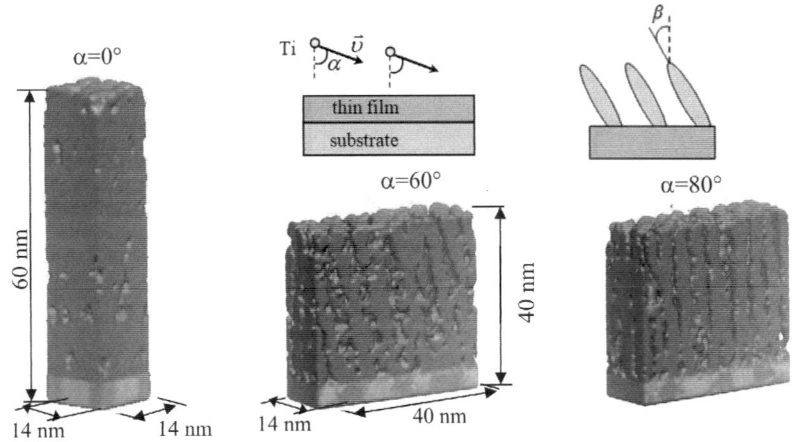

Figure 1. Structure of the deposited titanium dioxide films, α is the deposition angle, β is the tilt angle. Energy of the deposited Ti atoms is equal to 0.1 eV. Substrate is colored by yellow.

It is known that the deposition at large angles α leads to the formation of glancing angle deposited films consisting of the separate nanostructures [2]. As can be seen from the Fig. 1, in the case under consideration, these nanostructures are columns with a thickness of about several nanometers.

The deposition energy noticeably affects the structure of the GLAD TiO_2 films. According to [9], in the case of high-energy deposited films, the tilt angle β is greater than in the case of low-energy deposited films (Fig. 1) at the same values of the deposition angle α . The relationship between deposition angle α and tilt angle β, $2\tan\beta = \tan\alpha$ predicted by the ballistic model [4], performs better for high-energy deposition. This can be explained as follows. Within the framework of the ballistic model, it is assumed that the deposited atoms move along straight lines up to the moment when they reach the film [4]. This assumption performs better in the case of the high values of the kinetic energy of incoming atoms. Indeed, with the reduction of the deposition energy the influence of the interatomic interaction on the atomic trajectories grows. This leads to deviations of these trajectories from straight lines near the films.

The density profiles of the deposited films and dependence of the film porosity on the deposition angle are shown in Fig. 2. The profiles for the high-energy deposited films shown in this figure were calculated using the results from [9]. Fluctuations in the density profiles are caused by the formation of slanted columns in GLAD films. In the case of low-energy deposition, the largest fluctuations are observed in the case of $\alpha = 60°$. The thickness of the transition layer between film and vacuum is approximately 2-3 nm for all deposited films.

The film porosity $f(\alpha)$ is calculated as follows:

$$f(\alpha) = 1 - \rho(\alpha)/\rho(0) \qquad (2),$$

where $\rho(0)$ = 3.9 g/cm^3 is the density of the normally deposited film, $\rho(\alpha)$ is the density of the film deposited at an angle α. The porosity of high-energy deposited films increases by approximately threefold with an increase in the deposition angle from 60° to 80°. At the same time the porosity of low-energy deposited films varies insignificantly with increase of α.

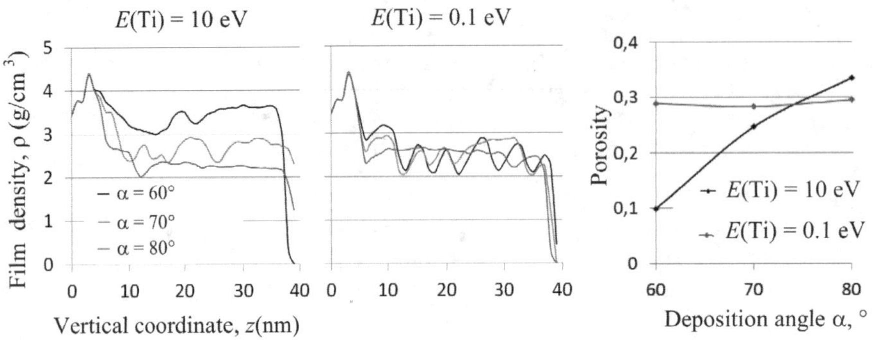

Figure 2. Density profiles and porosity of GLAD TiO_2 *films, E(Ti) – initial energy of the deposited titanium atoms.*

The structure of low-energy deposited TiO_2 film looks more ordered than the structure of low-energy SiO_2 GLAD films, as follows from a comparison with the results presented in [7, 8]. Growing SiO_2 GLAD films have a tree-like structure rather than a set of separated columns, as in the case of TiO_2.

The structural anisotropy of the GLAD TiO_2 films results in optical anisotropy. The difference Δn between the main components of the refractive index tensor is calculated in the frame of the Bruggeman effective medium approach [17] for the GLAD structures shown in Fig. 1. It is found that the maximum value of Δn reaches 0.14 and depends insignificantly on the deposition angle in the interval from 60° to 80°. This value is close to Δn = 0.13 obtained for high-energy deposited TiO_2 films at a deposition angle of 70° [9]. In contrast to low-energy deposition, the Δn value at high-energy deposition reduces significantly when the deposition angle is decreased from 70° to 60°. The Δn for the low-energy SiO_2 GLAD films simulated in [7, 8] is about 0.03 [18]. There are the following reasons for the large difference in Δn values between TiO_2 and SiO_2 low-energy GLAD films. First, the Δn value increases with an increase in the average value of the refractive index, n = 1.49 for SiO_2 [19] and n = 2.5 for TiO_2, [20]. Second, the TiO_2 films consist of the parallel separated columns (Fig. 1), which provides the anisotropy of properties, while the structure of low-energy SiO_2 GLAD films is more homogeneous [7,8].

4. Conclusion

The classical MD simulation of the low-energy GLAD TiO_2 films is performed. It is revealed that deposition at angles varying from 60° to 80° leads to the formation of anisotropic structures consisting of separated parallel columns having thickness of about 5 nm. The porosity of these films is about 0.3 and changes insignificantly when the deposition angle increases from 60° to 80°. The difference Δn of the main components of the refractive index tensor calculated using the Bruggeman effective medium approach, is about 0.14, which is close to Δn = 0.13, obtained in the case of high-energy TiO_2 films deposited at an angle of 70°. For the low-energy GLAD SiO_2 films the calculated value of Δn is about 0.03.

Acknowledgement

The work was supported by the Russian Science Foundation (grant number 19-11-00053).

References

[1].K. Robbie, M. J. Brett, A. Lakhtakia, Chiral sculptured thin films. Nature 1996, 384, 616.
[2]. M. M. Hawkeye, M. J. Brett, Glancing angle deposition: Fabrication, properties, and applications of micro- and nanostructured thin films. J. of Vac. Sci and Techn. 2007, 25, 1317-1335.
[3].A. Barranco, A. Borras, A. R. Gonzalez-Elipe, A. Palmero, Perspectives on oblique angle deposition of thin films: From fundamentals to devices. Progress in Materials Science, 2016, 76, 59-153.
[4].R. N. Tait, T. Smy and M. J. Brett. Modelling and characterization of columnar growth in evaporated films. Thin Solid Films, 226 (1993) 196-201
[5]. H. Badorreck, M. Steinecke, L. Jensen, D. Ristau, M. Jupé, J. Müller, R. Tonneau, P. Moskovkin, S. Lucas, A. Pflug, L. Grinevičiūtė, A. Selskis, T. Tolenis, T, Correlation of structural and optical properties using virtual materials analysis. Opt. Express 2019, (2019), 22209-22225, doi:10.1364/OE.27.022209.
[6].R. Tonneau, P. Moskovkin, A. Pflug, S. Lucas, TiOx deposited by magnetron sputtering: a joint modelling and experimental study. Journal of Physics D: Applied Physics, 51 (2018), 195202. https://doi.org/10.1088/1361-6463/aabb72
[7].F.V. Grigoriev, V.B. Sulimov and A.V. Tikhonravov. Structure of Highly Porous Silicon Dioxide Thin Film: Results of Atomistic Simulation. Coatings 9 (2019), 568; doi:10.3390/coatings9090568.
[8].F.V. Grigoriev, V.B. Sulimov and A.V. Tikhonravov. Atomistic simulation of the glancing angle deposition of SiO2 thin films. J. of N.-Cr. Sol. 2019, 512, 98-102, doi:10.1016/j.jnoncrysol.2019.02.016.
[9].F.V. Grigoriev, V.B. Sulimov and A.V. Tikhonravov. Application of a large-scale molecular dynamics approach to modelling the deposition of TiO2 thin films. Computational Materials Science, 2021, 188, 110202
[10]. F.V. Grigoriev, A.V. Sulimov, I.V. Kochikov, O.A. Kondakova, V.B. Sulimov, A.V. Tikhonravov, Supercomputer modeling of the ion beam sputtering process: full-atomistic level. In Proceedings of the SPIE—The international society for optical engineering, Jena, Germany, 21–24 September 2015; Volume 9627, pp. 962708-1–962708-9, doi:10.1117/12.2190938.
[11] B. Luan, T. Huynh, and R. Zhou, Simplified TiO2 force fields for studies of its interaction with biomolecules. J. Chem. Phys. 142 (2015), 234102. https://doi.org/10.1063/1.4922618
[12] M. Matsui and M. Akaogi, Molecular dynamics simulation of the structural and physical properties of the four polymorphs of TiO2, Mol. Simul. 6(4-6), 239–244 (1991). https://doi.org/10.1080/08927029108022432
[13]A.Piegari, F.Flory, Optical Thin Films and Coatings; Woodhead Publishing: Cambridge, England, 2013.
[14] H.J.C. Berendsen, J.P.M. Postma, W.F. van Gunsteren, A. DiNola, J.R. Haak, Molecular-Dynamics with Coupling to an External Bath, J. of Chem. Phys. 81 (8) (1984) 3684–3690. http://dx.doi.org/10.1063/1.448118..
[15] M. J. Abraham, T. Murtola, R. Schulz, S. Páll, J. C. Smith, B. Hess, E. Lindahl, GROMACS: High performance molecular simulations through multi-level parallelism from laptops to supercomputers, SoftwareX, 1–2 (2015) 19-25. https://doi.org/10.1016/j.softx.2015.06.001.
[16] Vl. Voevodin, A. Antonov, D. Nikitenko, P. Shvets, S. Sobolev, I. Sidorov, K. Stefanov, Vad. Voevodin, S. Zhumatiy, Supercomputer lomonosov-2: large scale, deep monitoring and fine analytics for the user community. Supercomput. Front. Innov. 6 (2019) 4–11, doi:10.14529/jsfi190201
[17] D. Schmidt, M. Schubert, Anisotropic Bruggeman effective medium approaches for slanted columnar thin films. J. Appl. Phys. 114 (2013), 083510. https://doi.org/10.1063/1.4819240
[18] F.V. Grigoriev, V.B. Sulimov and A.V. Tikhonravov. Combined Modeling of the Optical Anisotropy of Porous Thin Films. Coatings, 10 (2020), 517; doi:10.3390/coatings10060517
[19] . V. K. Leko, and O. V. Mazurin, Properties of Quartz Glass. (Nauka Publ., Leningrad, 1985). [in Russian]
[20] C. Bundesmann, T. Lautenschläge, D. Spemann, A. Finzel, M. Mensing, F. Frost, Correlation of process parameters and properties of TiO2 films grown by ion beam sputter deposition from a ceramic target. The European Physical Journal B. 90 (2017). 10.1140/epjb/e2017-80326-x

Optical properties of new hybrid nanoantenna in submicron cavity

A V Gritsienko[1,2], N S Kurochkin[1,2], P V Lega[3], A P Orlov[3], A S Ilin[3], S P Eliseev[1,2] and A G Vitukhnovsky[1,2]

[1] P. N. Lebedev Physical Institute of Russian Academy of Sciences, 53 Leninskiy Pr., 119991 Moscow, Russia
[2] Moscow Institute of Physics and Technology (State University), 9 Institutskiy Per., 141700 Dolgoprudny, Russia
[3] Kotelnikov Institute of Radioengineering and Electronics of Russian Academy of Sciences, Mokhovaya Str. 11, Build 7, 125009 Moscow, Russia

E-mail: grits_alex@rambler.ru

Abstract. An essential area of nanophotonics is the creation of efficient quantum emitters operating at high frequencies. In this regard, plasmon nanoantennas based on nanoparticles on metal (nanopatch antennas) are incredibly relevant. We have created and investigated a new hybrid nanoantenna with a cube on metal and quantum emitters. We demonstrate an increase up to 60 times for the rate of spontaneous emission and the gap-plasmon mode changing for nanopatch antenna in the metallic well. The results show the possibility of creating plasmon antennas in a controlled way by creating an array of regularly arranged nanoscale cavities-resonators.

1. Introduction

Nanoparticles on metal (nanopatch antennas, NPA) have been actively studied for more than 5 years and have shown their effectiveness for nanophotonics tasks [1]. Thus, high spontaneous emission rates (up to 10-100 ps) were obtained for various emitters, such as organic molecules [2,3], quantum dots [4–6], and nanodiamonds with NV centres [7], based on nanopatch antennas. Moreover, NPAs can provide a single-photon generation with single quantum dots and NV centres [7,8]. However, there are still several problems with this type of antenna.

The first problem is related to the uncontrolled placement of nanoparticles on the substrate since it is somewhat challenging to create nanoantennas in a given location (for colloidal nanoparticles). To solve this problem, the authors of the paper [9] propose to use the method of two-photon polymerization, fixing gold nanoparticles with emitters in a given place. The second problem is significant ohmic losses in the metal that makes up many plasmon nanoantennas [10]. Losses in metal nanoantennas can be reduced by changing the system geometry [11] or adding dielectric components [12,13]. A different approach is proposed in the research [14]. The authors show the opportunity of reducing non-radiation losses by hybridizing the nanoparticle's plasmon and the external resonator modes.

We decided to develop the idea of adding an external resonator to the nanopatch antenna. Thus, we have developed a new hybrid nanoantenna consisting of a nanocube and a round metal cavity. Previously, we showed that the nanocube absorption could decrease due to its placement

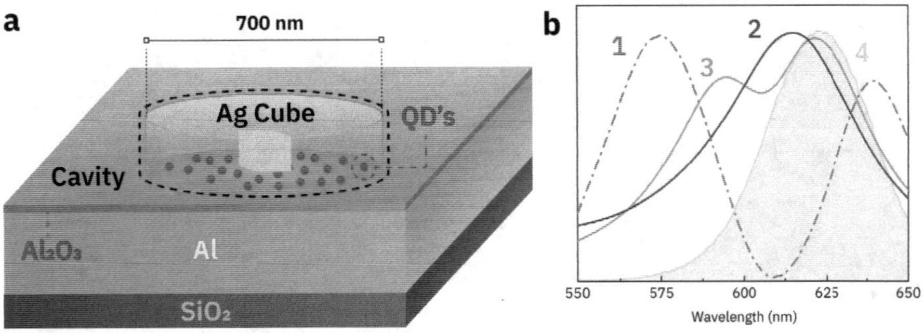

Figure 1. (a) - Diagram of a hybrid nanoantenna based on a nanocube and a circular cavity. (b) - Calculated extinction cross-section spectra of a nanocube placed in a circular cavity: the extinction cross-section spectrum of a cube located in the cavity centre (1); near the cavity wall at a distance of 55 nm (2, 3). The incident radiation is polarized in the cube-displacement direction (2) and perpendicular to the displacement relative to the cavity centre (3). Measured luminescence spectrum of CdSe/CdS quantum dots in hexane (4).

in a submicron cavity [15]. Simultaneously, the antenna's plasmon resonance could split into two resonances, independent of the exciting radiation's polarisation (Fig. 1b, curve 1) at specific parameters.

This paper experimentally shows a significant change in the QDs spontaneous emission rate and intensity placed in the hybrid nanoantenna (HNA) gap at room temperature (Fig. 1a). In this case, the studied antennas are placed in a particular structured array of a given geometry (Fig. 2).

2. Materials and Methods

Figure 1a shows a hybrid nanoantenna scheme. The nanoantenna consists of a circular aluminium cavity with a diameter of 700 nm and a depth of 300 nm. A silver cube is placed on a layer of quantum dots inside the cavity. The cube-size is about 80 nm covered by a PVP (polyvinylpyrrolidone) shell, and the QDs average diameter is about 8 nm.

Samples with an ordered array of manufactured nanoscale wells were obtained by deposition of an Al film to a Si/SiO_2 substrate by magnetron sputtering. Using the ion lithography system (FIB), the wells were etched sequentially. The Al film quality and the wells etching result were analyzed using a scanning electron microscope (SEM).

In the next step, we deposit quantum dots and silver nanocubes. CdSe/CdS quantum dots were synthesized according to the method [16]. The maximum luminescence wavelength of the synthesized quantum dots is 625 nm (Fig. 1b, curve 4).

The obtained samples were examined on a scanning fluorescent confocal microscope using an $100\times/1.40$ oil immersion objective with a numerical aperture (NA) of 1.4. The quantum dots in hybrid nanoantennas were excited by a diode laser at a wavelength of 375 nm (pulse mode). The radiation was collected on a single photon avalanche diode with pre-filtering of the reflected laser radiation.

3. Results and discussions

The study of the array of holes showed both bright and dim centres (Fig. 2d). This phenomenon must be due to the presence or absence of nanocubes in the cavities. To accurately determine cubes presence in the cavities, the sample was examined on a scanning electron microscope (Fig. 2a-c).

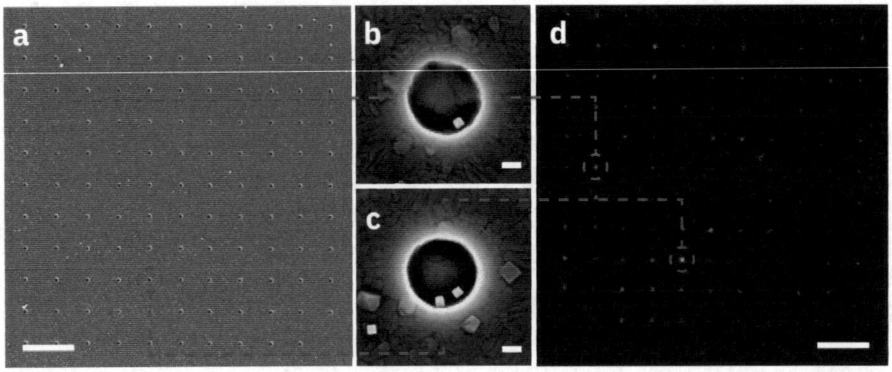

Figure 2. (a) - Scanning electron microscope image of an array of cavities and individual cavities (b, c) with applied quantum dots and cubes. The scale bars are 10 μm and 200 nm, respectively. (d) - Fluorescence-lifetime image (FLIM) of an array of cavities. The scale bar is 10 μm.

Based on the measurement of the bright points fluorescence, we obtained kinetic curves. Figure 3a shows examples of the kinetic emission spectra of quantum dots in different environments. For QDs placed on glass, the main luminescence decay time is $\tau_{glass} = 6{,}2 \pm 0{,}3$ ns. When a layer of quantum dots is placed on a metal, the radiation change properties dramatically. The spontaneous emission time decreases to $\tau_{metal} = 0{,}76 \pm 0{,}02$ ns. Simultaneously, the radiation intensity decreases that indicates the luminescence quenching effect by metal.

The emission of quantum dots in hybrid nanoantennas shows a time reduction of 1-2 orders of magnitude ($\tau_{HNA} = 0{,}13 \pm 0{,}01$ ns). In this case, the intensity in cavities with cubes is up to 5 times higher than the radiation of quantum dots in wells without nanoparticles. In Figure 3b, we show the intensity distribution of the measured cavities as a function of the luminescence decay time. The figure shows that some selected cavities luminescence time without cubes (red dots) is close to the radiation time of the layer of quantum dots on the metal.

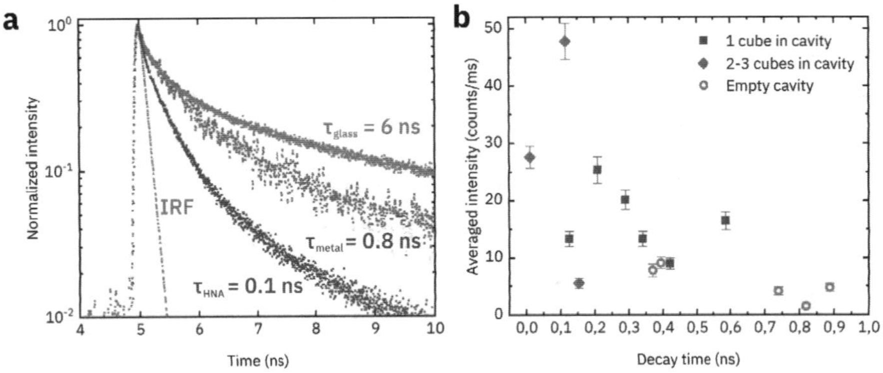

Figure 3. (a) - Comparison of the measured time-resolved fluorescence decay for quantum dots on glass (red curve), on metal (green curve), and in the HNA gap (blue curve). The grey curve is marked by the IRF detector. (b) - Comparison of the luminescence intensity of quantum dots from the characteristic time of spontaneous emission in different environments: blue squares – a cavity with one cube, green diamonds - cavities with several cubes, and hollow circles - cavities without cubes.

We also obtained brighter cavities with a short spontaneous relaxation time (about 0.4 ns) without cubes. This luminescence enhancement can be caused by the cavity optical resonance, which lies in the region of 600-610 nm. This condition can change the rate of luminescence decay, but the effect of this gain is lower compared to the gain from the plasmon resonance of the cube on the metal.

Note that the cubes were located not in the cavity centre in the resulting sample but closer to the wall. If the cube is in the centre, then two-hybrid resonances occur in this system. The resonance splitting effect weakens if the cube is located near the cavity wall (Fig. 1b, curves 2-3). We show that in this case, splitting is observed only in the case when the polarization of the incident radiation is perpendicular to the cube-displacement relative to the centre of the cavity (Fig. 1b, curves 3). Thus, the antenna resonance overlap with the QD-layer radiation in the antenna gap may change. This effect explains the difference in the intensity of individual cavities antenna radiation (Fig. 3b).

4. Conclusion

This work demonstrated a new hybrid nanoantenna with quantum emitters and showed a spontaneous emission rate increase up to 60 times. The results show the possibility of creating plasmon antennas in a controlled way by creating a cavities array. Moreover, this system can be combined with other structures, such as the bull's-eye system [17]. The proposed mechanisms open up new ways to solve the main problems of using plasmon nanostructures in photonics to create fast, ultra-bright photon sources with controlled resonance.

References

[1] Bogdanov S I, Boltasseva A and Shalaev V M 2019 *Science* **364** 532–533
[2] Akselrod G M, Argyropoulos C, Hoang T B, Ciracì C, Fang C, Huang J, Smith D R and Mikkelsen M H 2014 *Nature Photonics* **8** 835–840
[3] Gritsienko A, Kurochkin N, Vitukhnovsky A, Selyukov A, Taydakov I and Eliseev S 2019 *Journal of Physics D: Applied Physics* **52** 325107
[4] Kurochkin N, Eliseev S, Gritsienko A, Sychev V and Vutukhnovsky A 2020 *Nanotechnology* **31** 505206
[5] Hoang T B, Akselrod G M, Argyropoulos C, Huang J, Smith D R and Mikkelsen M H 2015 *Nature communications* **6** 1–7
[6] Eliseev S P, Vitukhnovsky A G, Chubich D A, Kurochkin N S, Sychev V V and Marchenko A A 2016 *JETP letters* **103** 82–86
[7] Bogdanov S I, Shalaginov M Y, Lagutchev A S, Chiang C C, Shah D, Baburin A S, Ryzhikov I A, Rodionov I A, Kildishev A V, Boltasseva A *et al.* 2018 *Nano letters* **18** 4837–4844
[8] Hoang T B, Akselrod G M and Mikkelsen M H 2016 *Nano letters* **16** 270–275
[9] Ge D, Marguet S, Issa A, Jradi S, Nguyen T H, Nahra M, Béal J, Deturche R, Chen H, Blaize S *et al.* 2020 *Nature communications* **11** 1–11
[10] Yang J, Hugonin J P and Lalanne P 2016 *ACS photonics* **3** 395–402
[11] Bogdanov S I, Makarova O A, Xu X, Martin Z O, Lagutchev A S, Olinde M, Shah D, Chowdhury S N, Gabidullin A R, Ryzhikov I A *et al.* 2020 *Optica* **7** 463–469
[12] Li W, Morales-Inostroza L, Xu W, Zhang P, Renger J, Gotzinger S and Chen X W 2020 *ACS Photonics* **7** 2474–2481
[13] Lepeshov S I, Krasnok A E, Belov P A and Miroshnichenko A E 2019 *Physics-Uspekhi* **61** 1035
[14] Gurlek B, Sandoghdar V and Martín-Cano D 2018 *ACS Photonics* **5** 456–461
[15] Gritsienko A, Eliseev S, Kurochkin N and Vitukhnovsky A 2020 Interaction effects of nano-patch antenna with external resonator *AIP Conference Proceedings* vol 2300 (AIP Publishing LLC) p 020042
[16] Vashchenko A, Lebedev V S, Vitukhnovskii A, Vasiliev R B and Samatov I 2012 *JETP letters* **96** 113–117
[17] Yang G, Shen Q, Niu Y, Wei H, Bai B, Mikkelsen M H and Sun H B 2020 *Laser & Photonics Reviews* **14** 1900213

Single photon sources based on HPHT nanodiamonds

N S Kurochkin[1,2], S P Eliseev[1,2], V V Sychev[2], A V Gritsienko[1,2], V S Gorelik[2] and A G Vitukhnovsky[1,2]

[1]Moscow Institute of Physics and Technology (State University), 9 Institutskiy Per., 141700 Dolgoprudny, Russia
[2]P.N. Lebedev Physical Institute of Russian Academy of Sciences, 53 Leninsky Pr., 119991 Moscow, Russia
E-mail: elst.conference@gmail.com

Abstract. Color centers in nanodiamonds are promising candidates for the creation of high-speed sources of single photons without blinking and degradation. Color centers in HPHT nanodiamonds have been investigated. The luminescence decay curves of color centers have been measured. Second order correlation functions were measured for nanodiamonds with sizes from 50 nm to 250 nm. Conclusions about the energy structure of color centers were made based on the correlation functions.

1. Introduction

Promising materials for creating sources of single photons are quantum dots and nanoplates [1-4], color centers in 2D materials and diamonds [5]. Ultrafast radiation sources can be obtained by coupling emitters with a nanocavity [6-8]. Among the sources of single radiation, the color centers in diamonds are the most stable. The most famous color centers in diamonds are nitrogen vacancy (NV) and silicon vacancy (SiV) luminescence centers in diamond. These centers can arise both naturally and as a result of ion implantation followed by annealing.

Most quantum-optical applications use photoluminescence of NV^- centers, which has been used to demonstrate coherent optical effects such as electromagnetically induced transparency [9] and coherent population capture [10]. A single NV center was used to implement a secure communication system at a distance of several tens of meters [11]. Also potential applications of NV centers are used in areas such as biosensing and biomarking [12], sub-diffraction imaging (for example, by STED lithography method [13, 14]) and many other scientific and technical areas (biology, neuroscience, fundamental physics [15]).

Two main methods are used to make synthetic single crystal diamonds: high pressure, high temperature (HPHT) and chemical vapor deposition (CVD). In turn, nanodiamonds can be obtained by crushing synthetic (e.g. HPHT) diamonds or by detonating carbon.

Crystal structure of HPHT nanodiamonds is preserved during their creation. This is a significant advantage of this type of nanodiamonds as sources of single photons over other types of nanodiamonds.

2. Results and discussion

In our work, we investigated the sources of single photons was carried out in HPHT nanodiamonds with characteristic sizes of 50 nm, 150 nm and 250 nm. For this, we recorded and analyzed the second-order correlation functions $g^{(2)}(t)$ according to the Hanbury Brown-Twiss scheme.

The photoluminescence was recorded using a scanning laser confocal microscope. Correlation functions of the 2nd order were measured using a CW laser source with a wavelength of 532 nm and an excitation power of about 1 mW. Photoluminescence was recorded using a broadband filter that transmits in the wavelength range of 600-800 nm. To obtain the photoluminescence kinetics, a pulsed supercontinuum laser source in the range of 500-700 nm was used.

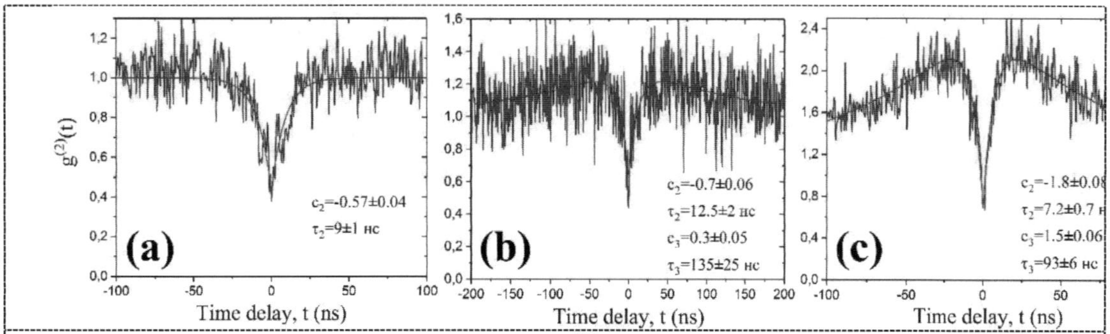

Figure 1. Second-order correlation functions $g^{(2)}$ for HPHT nanodiamonds with average sizes: 50 nm (a), 150 nm (b), 250 nm (c), located on the glass surface.

Figure 1 shows the obtained functions $g^{(2)}$ for three types of nanodiamonds with different size. The graphs show dips in the correlation function, which indicates the quantum nature of radiation with a small number (up to 5) emitters. The number of emitters n was estimated from the relation $g^{(2)}(0) = 1 - 1/n$. In this case, bunching of photons ($g^{(2)} > 1$) is observed for nanodiamonds with a size of 150 nm and 250 nm at delay times more than 10 ns. This is typical for NV centers in diamond due to the presence of a metastable level thermally coupled to the excited level. In this case, the system is not described by a simple two-level model. It is necessary to consider a three-level model of the system. The approximation of the correlation functions in figure 1 (b, c) was carried out according to the equation [16]:

$$g^{(2)}(t) = 1 + c_2 e^{-t/\tau_2} + c_3 e^{-t/\tau_3}$$

Here τ_2 and τ_3 are characteristic short and long decay times, c_2 and c_3 are exponential contribution coefficients. Coefficient $c_2 < 0$, $c_3 > 0$. The second term in the equation describes the dynamics of the $g^{(2)}(t)$ dip. The third term in the equation describes the damping of the correlation function wings. All four coefficients depend on the pump power of the NV center.

For the function $g^{(2)}(t)$ in figure 1 (a), the coefficient c_3 was set equal to 0, the system is described by a two-level model. The value $g^{(2)}(0) < 0.5$ for this nanodiamond unambiguously indicates that it is a source of single photons. For the other two nanodiamonds the number of NV centers in them can be more than one since $g^{(2)}(0) \geq 0.5$. In all three nanodiamonds from the analysis of the correlation functions the characteristic short decay time of luminescence is about 10 ns, which correlates with the luminescence time of NV^- centers in bulk diamond crystals and indicates their predominance in these HPHT diamonds. Measurement of the kinetic luminescence decay curve of a 300 μm HPHT diamond crystal (figure 2) also showes that the luminescence decay time of color centers in the bulk crystal is about 10 ns.

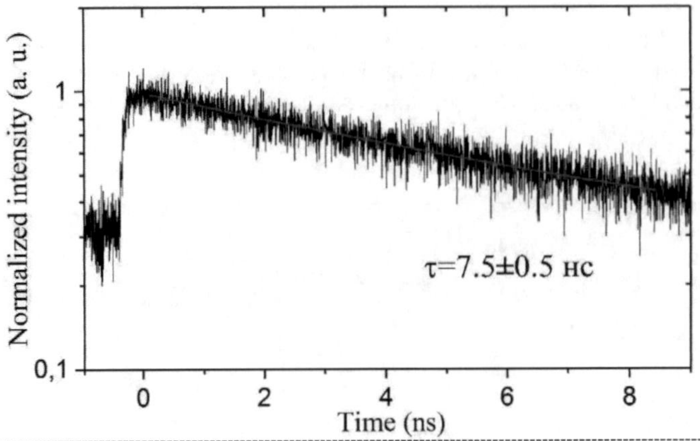

Figure 2. Normalized luminescence decay curve of HPHT nanodiamond.

3. Conclusion

HPHT nanodiamonds are promising sources of single photons due to their stable crystal structure. In the future it is possible to use them in devices for quantum-optical applications. Here we demonstrated single photon sources based on color centers in HPHT nanodiamonds with sizes from 50 nm to 250 nm.

References

[1] Eliseev S, Vitukhnovsky A, Chubich D, Kurochkin N, Sychev V and Marchenko A 2016 *JETP Lett.* **103(2)** 82
[2] Kurochkin N, Katsaba A, Ambrozevich S, Vitukhnovsky A, Vaschenko A and Tananaev P 2018 *J. Lumin.* **194** 530
[3] Kurochkin N, Eliseev S, Gritsienko A, Sychev V and Vutukhnovsky A 2020 *Nanotechnology* **31(50)** 505206
[4] Selyukov A, Danilkin M, Eliseev S, Kuznetsov A, Grafova V, Klimonsky S, Vainer Yu, Vasiliev R and Vitukhnovsky A 2020 *Quantum Electron.* **50(3)** 252
[5] Bolshedvorskii S. et al. 2017 *Opt. Mater. Express* **7(11)** 4038
[6] Eliseev S, Kurochkin N, Vergeles S, Sychev V, Chubich D, Argyrakis P, Kolymagin D and Vitukhnovsky A 2017 *JETP Lett.* **105(9)** 577
[7] Kurochkin N, Eliseev S and Vitukhnovsky A 2019 *Optik* **185** 716
[8] Gritsienko A, Kurochkin N, Vitukhnovsky A, Selyukov A, Taydakov I and Eliseev S 2019 J. *Phys. D: Appl. Phys.* **52** 325107
[9] Wilson E., Manson N and Wei C 2003 *Phys. Rev.* A **67(2)** 023812
[10] Santori C. et al. 2006 *Phys. Rev. Lett.* **97(24)** 247401
[11] Alléaume R, Treussart F, Messin G, Dumeige Y, Roch J, Beveratos A, Brouri-Tualle R, Poizat J and Grangier P 2004 *New J. of Phys.* **6(1)** 92
[12] Balasubramanian G, Lazariev A, Arumugam S and Duan D 2014 *Curr. Opin. Chem. Biol.* **20** 69
[13] Vitukhnovsky A, Chubich D, Eliseev S, Sychev V, Kolymagin D and Selyukov A 2017 *J. Russ. Laser Res.* **38(4)** 375
[14] Eliseev S, Korolkov A, Vitukhnovsky A, Chubich D and Sychev V 2016 *Nanotechnol. Russ.* **11(3)** 200
[15] Aharonovich I, Castelletto S, Simpson D, Su C, Greentree A and Prawer S 2011 *Rep. Prog. Phys.* **74(7)** 076501
[16] Kurtsiefer C, Mayer S, Zarda P and Weinfurter H 2000 *Phys. Rev. Lett.* **85(2)** 290

Journal of Physics: Conference Series **2015** (2021) 012054 doi:10.1088/1742-6596/2015/1/012054

Arrays of electrically small antennas with SINIS detectors for SubTHz astronomy and atmosphere propagation research

Aleksandra Gunbina[1, 2 a)], Mikhail Tarasov [2, b)], Sergey Lemzyakov[3, c)],

Vyacheslav Vdovin[1, d)], Grigory Yakopov[4, e)], Renat Yusupov[2, f)],

Artem Chekushkin[2, g)], Daria Nagirnaya[2, h)], Valerian Edelman[3, i)]

[1] *Institute of Applied Physics of the RAS, Nizhny Novgorod, 603155 Russia*
[2] *Kotelnikov Institute of Radio Engineering and Electronics RAS, Moscow, 125009 Russia*
[3] *Kapitza Institute for Physical Problems, Moscow 119334 Russia*
[4] *Special Astrophysical Observatory, Nizhnij Arkhyz 369167 Russia*

[a)]Corresponding author: aleksandragunbina@mail.ru
[b)] tarasov@hitech.cplire.ru
[c)] lemserj@gmail.com
[d)] vdovin@ipfran.ru
[e)] yakopov@gmail.com
[f)] yusupovrenat@hitech.cplire.ru
[g)] chekushkin@hitech.cplire.ru
[h)] darianagirnaya@mail.ru
[i)] vsedelman@yandex.ru

Abstract. An overview of different arrays of electrically small annular antennas of 350 GHz band with integrated superconductor-insulator-normal metal-insulator-superconductor (SINIS) detectors is presented in this paper. Arrays developed for both astronomical observations on the BTA 6 m telescope and measurements there of test sources for investigation of atmospheric turbulence to estimate its influence on the data rate of subTHz telecommunication channels. It is shown both numerically and experimentally that to simulate the electrodynamics of such non-phased antenna arrays it is necessary to investigate the whole array, not a single cell with Floquet ports. The first results of studying of superconducting antennas are also presented in this paper.

INTRODUCTION

The main field of application of SINIS detector arrays in the subTHz frequency range is radio astronomy [1, 2]. However, the limitations by significant atmospheric absorption of subTHz waves hinder the development of terrestrial radio astronomy in this band, which is important for astrophysics, and restrict it by space [3, 4] and high-altitude [5, 6] observatories, which is not a mass consumer of receiver developments. At the same time, the development of space and ground 7G [7] communications is also actively moving into this frequency range [8, 9]. Therefore, the task of creating highly efficient receiving systems and studying the features of the propagation in atmosphere of subTHz waves in various conditions are becoming increasingly. There are lots of works [10, 11] devoted to propagation of sub THz waves in the atmosphere or the microwave astroclimate. The Microwave

Journal of Physics: Conference Series **2015** (2021) 012054 doi:10.1088/1742-6596/2015/1/012054

astroclimate is differ from the usual astroclimate in the optical range. There are not only the effect of water [11] but other atmospheric gases, oxygen first of all.

In optics, on the other hand, an essential effect on the astroclimate is not only absorption, but also scattering by turbulences of the atmosphere has. Atmospheric turbulence can dramatically spoil the image of objects in astronomy and radically increase communication errors. Study of atmospheric turbulence in optics is carried out on the basis of an assessment of the image quality of known test astronomical sources (well-known objects on the sky like Venus, Orion, Crab nebula etc.). Taking into account the proximity of subTHz to optics, which actually represent the far-IR range, it can be assumed that turbulence in the atmosphere will also affect subTHz waves. The paper presents the results of the development of an array of antennas with SINIS detectors of subTHz waves, suitable for both astronomical observations and measurements of test sources to assess the effect of atmospheric turbulence on the data transmission rate of subTHz telecommunication channels.

MODELING

In this section we present the results of development of electrically small antenna array operating in 200-400 GHz frequency band. The first step in modeling of antennas array is choice of a unit element of the array. As it was show in our previous work [12] for arrays without polarization selectivity the best variant is to use ring antenna. The radiation pattern of such an antenna (see [12]) is pointed across the substrate and does not create substrate modes. The modeled radiation pattern of full array of 25 annular antennas on silicon lens is presented in Fig. 1. A typical metamaterial consists of a periodic array of subwavelength metallic resonators that are collectively coupled to the free-space excitation. With small antennas, the array can be designed to accommodate a wider bandwidth and occupy a much smaller physical space which allows it to be placed in the waist of a single-mode horn or in the focus of immersion lens. In the simplest case, an analytical model of lumped elements can be used to estimate parameters of such array [13-15]. Another way is to use modern software (for example, CST STUDIO SUITE or HFSS) for modeling to obtain more accurate results, an estimation of the spectral response and a radiation pattern. In our previous publications on arrays of electrically small antennas [13], [15], the calculation of a unit cell with Floquet ports that were placed at the boundaries corresponding to the minimum and the maximum of z-coordinate was used. This method allows you to quickly obtain the results, however this results are far from the real electrodynamics picture of such array (see results of modeling in ref. [13], [15]). This is due to the fact that this method can only be applied for phased antennas arrays. As an example, Fig. 2 shows the radiation patterns of non-phased array of weakly coupled to each other antennas located in different parts of the array. According this, the modeling was carried out for a full (9*9 elements) array with open boundaries and irradiation by a source from the far zone Fig. 3a, [14]. Typical dimensions of the modeled structure: diameter of antenna is 60 μm, a period is 70 μm, thickness of gold film is 200 nm, and thickness of substrate is 65 μm. Measured spectral response from [14] is presented in Fig. 3b. As a model of SINIS detector, a seriesl connection of a discrete port (50 Ω) and a lumped element (capacitance, 25 fF) was used.

SINIS detector at a glance: The sensitive element of such detector is the absorber made of normal metal. Two NIS (Normal metal – Insulator - Superconductor) junctions connected in series with an absorber serve as thermometers. Radiation absorption in the SINIS structure leads to heating of the absorber, which can be measured as the tunnel current increase in the NIS thermometers. The sensitivity of such detectors can be estimated using non-equilibrium distribution functions of electrons and phonons and the theory of quantum absorption of radiation [16]. A SINIS detector with a suspended absorber seems to be promising way for increasing of the sensitivity compared to conventional structure with absorber on the substrate [17].

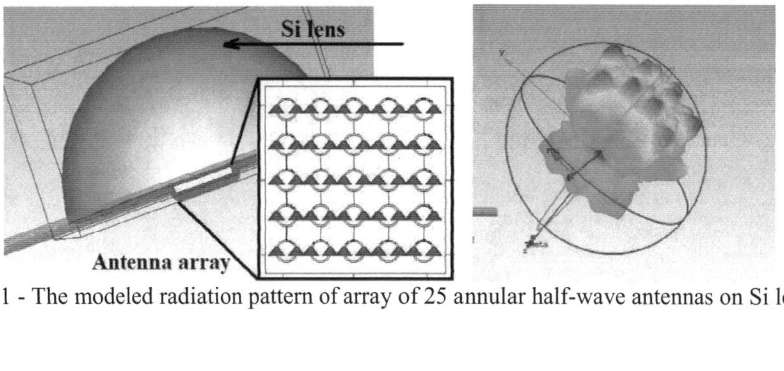

Figure 1 - The modeled radiation pattern of array of 25 annular half-wave antennas on Si lens

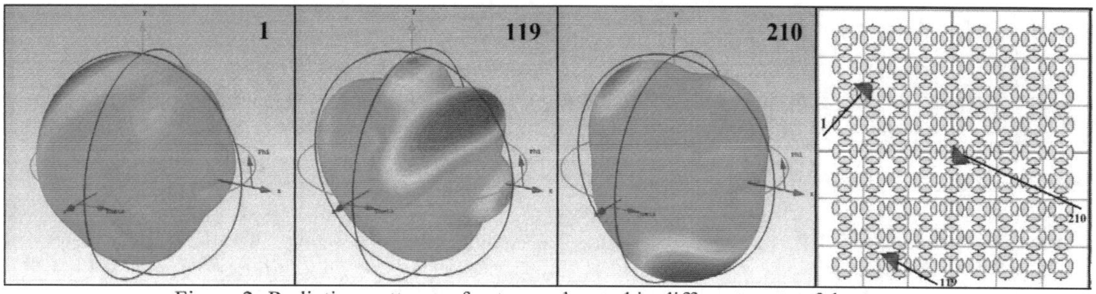

Figure 2 -Radiation patterns of antennas located in different parts of the array.

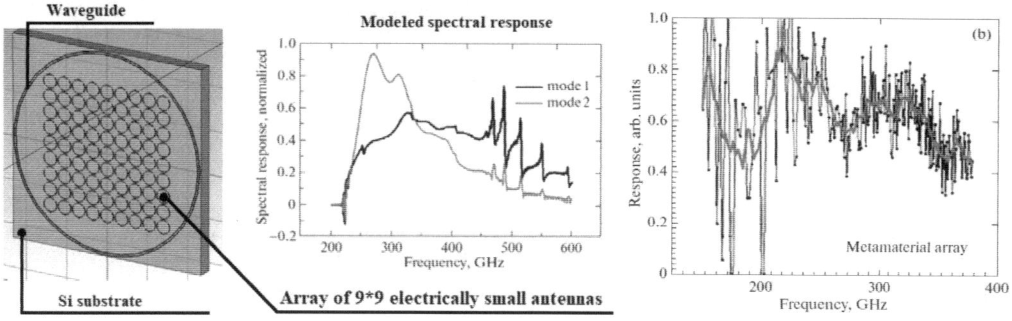

a) b)

Figure 3 – Electrically small antenna array: a) Modeling of 9*9 array of electrically small antennas: image of project (left) and modeled spectral response (right), mode 1 and 2 are two orthogonal polarizations of TE_{11} of circular waveguide, [14], b) results of measurements

SUPERCONDUCTING ELECTRICALLY SMALL ANTENNAS

Besides normal-metal antennas we also study antenna arrays made of superconducting material. The use of a superconductor antenna as prolongation of superconducting electrodes of SINIS allows both to reduce losses in the antennas and avoid overheating of the superconducting electrode of the SINIS detector by hot quasiparticles. For the first tests, we repeated the design of the array of gold electrically small antennas, replacing gold by superconducting aluminum. The advantages are presented in the measurements of the current-voltage characteristics at 300 mK. Here, the criteria are the value of the energy gap and the ratio of the resistances at zero bias and far beyond the gap. At a temperature of 300 mK and a small area of superconducting electrodes, the ratio of the resistances of the detector array did not exceed 100, whereas for a similar array with superconducting antennas, the ratio was close to 1000. And at a temperature of 100 mK, the resistance ratio exceeded of 10000 Fig. 4.

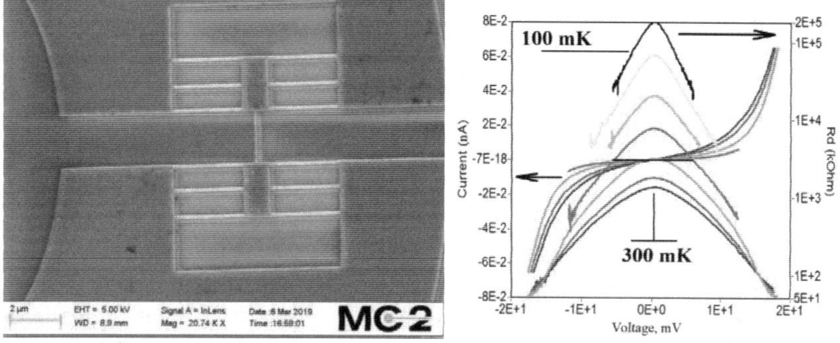

a) b)

Figure 4 – Superconducting antenna array with SINIS detector: a) SEM image of a part of single antenna with detector; b) – Measured dynamic resistance at bath temperature of 100 and 300 mK.

CONCLUSION

Different arrays (made from gold and superconducting aluminum) of electrically small annular antennas of 350 GHz band with integrated SINIS detectors are presented in this paper. It is shown that for modeling of such arrays it is necessary to simulate the whole array with open boundaries. Such arrays are matched to incoming radiation in a wide frequency range and can work in conditions of high background power (tens of pw). The spectral response was measured in frequency band from 200 to 400 GHz and we obtained relatively uniform spectral characteristic (see Fig. 3b). The first tests of superconducting aluminum electrically small antennas showed their efficiency - the measured resistance ratio exceeded 10000 at bath temperature of 100 mK. Further development of large superconducting arrays of electrically small antennas that is intended for approaching a distributed absorber model. The proposed technology of magnetron sputtering with separate lithography allows one to create tunnel structures of any shape and size and can be used in industrial production with conventional technological equipment.

ACKNOWLEDGMENTS

This work was carried out as part of the state assignment of the IRE RAS (No. 0030-2019-0003) and of the IAP RAS (No. 0030-2021-0005). The manufacture and study of samples was carried out using a unique scientific setup (USU №. 352529). Communications prospects and atmosphere problem for data rates were analyzed and development of some elements of THz detectors and creating of some devices and elements of experimental instruments for researching of developed structures were made in the framework of RSF project #19-19-00499. This work was supported at Chalmers University of Technology by the Swedish Research Council (Vetenskapsrådet, project #2016-05256), and the Knut and Alice Wallenberg.

REFERENCES

1. Sheng-Cai SHI, Terahertz Science and Technology **1**, 190-206 (2008).
2. S. Withington, Transactions of the Royal Society A: Mathematical, Physical and Engineering Sciences **362**, 395-402 (2004).
3. A. Smirnov, A. Baryshev, P de Bernardis, V. Vdovin et al., Radiophysics and quantum electronics **54**, 557-568 (2012).
4. P. A. R. Ade et al, Astron. Astrophys **536**, A1-1 – A1-16 (2011).
5. R. L. Brown, W.Wild, and C. H. Cunningham, Adv. Space Res **34**, 555-559 (2004).
6. G. Presta et al. Journal of Physics: Conference Series **1548**, 1-7 (2020).
7. http://2019.comcas.org/Portals/107/Comcas%20keynote%202019%20T%20S%20Rappaport_1.pdf
8. J.F. O'Hara, S. Ekin, W. Choi and I. Song, Technologies **7**, 1-18 (2019).
9. V. Vdovin, I. Lesnov and G. Bubnov. "Sub-terahertz Data Channels: Evaluation of Perspectives" in *2020 International Conference on Electrical, Communication, and Computer Engineering (ICECCE)*, pp. 1-3.
10. G.Bubnov, Y.Artemenko, V.Vdovin, et al, Radiophysics and quantum electronics **59**, 763-771 (2017).
11. I. E. Arsaev, V. Yu. Bykov, G. N. Il'in, E. F. Yurchuk, Technique of measurements **72**, 60-65 (2017).
12. M.A. Tarasov, A.M. Chekushkin, R.A. Yusupov, et. al., Journal of Communications Technology and Electronics **65**, 60-68 (2020).
13. A.S. Sobolev, B. Beiranvand, A.M. Chekushkin, et. al., "Wideband metamaterial-based array of SINIS bolometers", *in TERA 2018,* EPJ Web of Conferences 195, 05009-1 – 05009-2 (2018).
14. A.A. Gunbina, M.A. Tarasov, S.A. Lemzyakov, et. al., Physics of the Solid State **62**, 1604-1611 (2020).
15. M. Tarasov, A. Sobolev, A. Gunbina, et. al., Journal of Applied Physics **125**, 174501-1 - 174501-6 (2019).
16. I. A. Devyatov, P. A. Krutitskiĭ, M. Yu. Kupriyanov, Journal of Experimental and Theoretical Physics Letters **84**, 57-61 (2004).
17. M. Tarasov, V. Edelman, S. Mahashabde, et. a., "SINIS bolometer with a suspended absorber", *in 28th International Conference on Low Temperature Physics (LT28)*, Journal of Physics: Conference Series 969, pp. 1-6 (2018).

Rayleigh imaging along an optical fiber in operation

Y Haddad, J Chrétien, S Margueron, J C Beugnot and G Fanjoux

FEMTO-ST institute, University of Franche-Comté, 15B avenue des Montboucons, 25030 Besançon, France

yosri.haddad@femto-st.fr

Abstract. We present in this work a non-destructive and non-invasive imaging spectroscopic technique with a high spatial and spectral resolution to characterize the light propagation behaviour along a centimetric length and nanometric size tapered optical fiber in operation.

1. Introduction

Most of the characterization methods of an optical waveguide are limited to the output or input side of the waveguide, in order to analysis the transmitted or the backscattered field, in power, polarization, or for spectral evolution. Now, the simultaneous spatial and spectral measurements of the guided light along an optical waveguide in nominal operation represents a challenging but crucial experimental results for reverse engineering waveguide fabrication or for comparison with theoretical predictions of field propagation. A low spatial resolution in micrometric range will allow the observation of the waveguides losses, localization of hot points, scattering or defects points along the waveguide, or even spatial mode beating. Furthermore, a high spectral resolution will allow the observation of generation of spectral components along waveguides by nonlinear effects.

Several techniques are already used to longitudinally characterize an optical waveguide. Apart from the destructive cut-back technique which is used to characterize lengthwise an optical waveguide in operation, different Rayleigh or Brillouin reflectometry and derivative techniques are also generally used for long waveguides such as optical fibers. The state-of-the-art Optical Backscatter Reflectometer (OBR) allows micrometric resolution along waveguides [1]. Nevertheless, the waveguide is not in nominal operation and the system is limited to some infrared telecom wavelengths. Moreover, all these techniques cannot spectrally characterize the waveguide in nonlinear regime, except the system developed in [2] with however a worst spatial resolution.

In the present work, we demonstrate a novel experimental spectroscopic method, non-destructive and non-invasive, with a high spatial and spectral resolution, to analyze the light propagation throughout an entire optical tapered fiber with a nanometric diameter called optical nanofiber (ONF) operating in linear regime [3,4]. Tapered fiber indeed presents large interest due to its specific properties, i.e., the strong intensity of the confined field, and an evanescent field in interaction with the external medium [5]. Then, applications in both linear and nonlinear regimes go from atom trapping in the evanescent field [6] to supercontinuum generation [7], efficient evanescent Raman scattering [8], or chemical or biological sensor [9].

Our technique is based on the measurement with a confocal spectrometer of the Rayleigh scattering (RS) light radiated out of an ONF in operation. We then show that our method can spatially characterize the entire waveguide in the linear regime with a spatial resolution of 5 μm. The results are compared to the state-of-the-art OBR system (OBR 4600 ®) from Luna with a good agreement. This approach provides also direct observation of the beating of higher-order modes (HOMs), the leaking of the HOMs from the surface along the taper region and the transition from core–cladding to cladding–air guidance for the different modes (fundamental mode and HOMs) propagation along the fiber taper. Furthermore, it provides characterization of the linear loss and the localization of defects along an optical waveguide in operation. We have also analyzed the propagation of light in a single mode fiber in the visible range, which is not possible with an OBR system.

2. Experimental setup

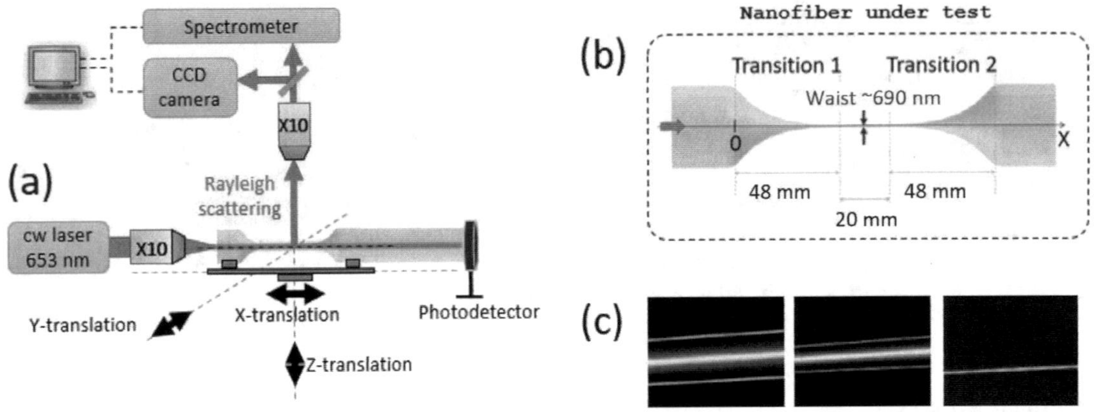

Figure 1. (a) Schematic experimental setup, (b) geometrical data of the nanofiber, (c) three pictures recorded by the CCD camera, with diameters corresponding to 51 μm, 31 μm and 2 μm, respectively.

Our measurement setup, illustrated in figure 1, consists of a confocal Raman micro-spectrometer (without Notch filter) used for Rayleigh scattering (RS) detection. A high-precision motorized stage for longitudinal X and transverse Y and Z translation (with a step size < 100 nm) of the component under study allows the generation of an accurate recordable 3D spatial trajectory point-by-point all along the waveguide. The RS of the guided light emitted to the outside of an optical waveguide is collected in the direction orthogonal to the waveguide (figure 1) with a microscope objective and recorded through a confocal spectrometer. The monochromator is equipped with three distinct diffraction gratings (300, 1500 and 2400 lines/mm), and an ultra-sensitive cooled CCD detector (−80°C) to have a low detection threshold and to increase the signal-to-noise ratio (SNR). Alternatively, a CCD image camera is also used to image the waveguide for positioning. In this work, two waveguides are studied. The first one corresponds to an ONF fabricated from a standard single mode fiber at telecom wavelength (G652D @ 1550 nm) labeled thereafter G652D-ONF. The optical losses just after manufacturing were measured at 532 nm of wavelength and equal to 0.3 dB. These losses are due to the fact that the fiber is strongly multimode at that wavelength and that the transitions are not adiabatic for all the high order modes. The second waveguide corresponds to an ONF fabricated from a standard single mode fiber in visible range (SM450) labeled SM450-ONF. The optical lost just after manufacturing were measured at 532 nm of wavelength and equal to 0.29 dB. The two ONFs are manufactured by heating and pulling the standard silica fibers (SM450 and G652D) until their diameter has reached a target value as described in [9]. After the tapering process, the two ONFs have a uniform waist of about 690 nm over around 20 mm long accurately measured by Brillouin scattering methods [2], and each transition taper is around 48 mm long (figure 1). Then, each ONF is placed inside a closed and transparent box in order to protect it from external air vibrations and to reduce any kind of contamination. For the laser source in linear regime, we use a power stabilized cw laser diode with a wavelength of 653 nm. We monitored simultaneously the transmitted power during the RS measurement and we did not observe any significant variation during the record. The duration of one measurement depends on the number of recorded points and of the exposition time for each point.

We use the CCD camera in order to make a direct link between the ONFs geometries and the light scattering process along the fiber in operation. Figure 1(c) shows three example pictures of the fiber recorded by the CCD camera for different longitudinal X positions and corresponds to externals diameters of 51 μm, 31 μm and 2 μm, from left to right respectively. These pictures allow us to measure the evolution of the fiber external diameter over the whole length of the longitudinal X scan.

3. Experimental results and discussion

Figure 2 (a-b) presents the experimental results for the G652D-ONF and figure 2(c) for the SM450-ONF. Figure 2(a) shows the backscatter trace from the OBR system at a wavelength of 1566 nm (blue curve), and the scattering trace from our experimental setup (red curve) obtained with the cw laser source at 653 nm of wavelength. We clearly observe similitudes and differences between these two curves. First, we obtain the same general scattering profiles, due to the features of the different scattering sources corresponding to the fiber core at the beginning of the transition with a large diameter, the cladding when the core no longer guides, and the fiber surface when the diameter is very small [2]. However, large differences are also visible. With our method, we clearly observe the beating of modes at the beginning of the transition (arrow 1) as well as two successive intensity dips of the scattering level due to the expansion of different optical modes (fundamental mode and high order modes (HOMs)) from the fiber core region into the cladding region. On the output side, the different optical modes couple back into the core from the cladding region at the same fiber diameters as observed at the level of the transition 1. Here, we observe again the presence of two successive decreases of the scattering level (arrow 2) situated at positions symmetrically to the positions of the first decreases (arrow 1) with respect to the microfiber.

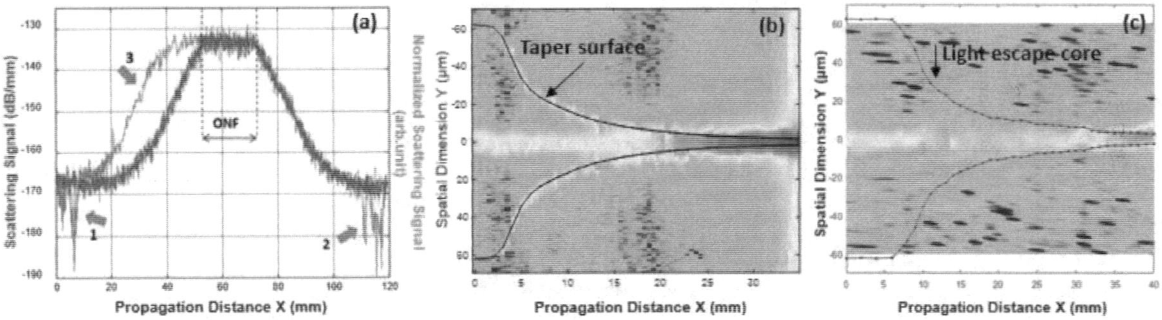

Figure 2. OBR trace (blue curve) with λ=1566 nm and scattering trace from our technique (red curve) with λ=652 nm for the G652D-ONF. (b) XY-2D map of RS intensity in the transition 1 in logarithm scale of the G652D-ONF in operation. (c) XY-2D map of RS intensity in logarithm scale of the SM450-ONF in operation. Dotted curves in Fig.1 (b-c) correspond to the ONFs surface obtained using the CCD camera and light is propagating from left to right for both 2D maps.

For lower diameter, the RS signal is strongly blurred by a large increase of the scattering level along the transition 1 (arrow 3), contrary to the RS signal in transition 2 and for the OBR measurement in transition 1. This RS feature comes from the scattering of the different HOMs which are no more guided by the cladding. To go deeper into the details, figure 2(b) represents a 2D map of the RS intensity along the waveguide, which highlights the scattering from the surface at the input transition level. It is important to note that no normalization in the intensity of these data has been realized. Therefore, the evolution of the intensity of the signal represents the evolution of the scattering process. In addition, this measurement shows that the scattered light is clearly observable all along the fiber, meaning that our

system is sensitive enough for the low-intensity scattering. We clearly observe the scattering process at the surface of the transition 1 by the radiated HOMs, which are no more guided by the clad of the fiber. Indeed, the G652D fiber before stretching is a standard single mode fiber at telecom wavelength; it is therefore strongly multimode at the wavelength of operation (653nm).
We have also analyzed the single mode SM450-ONF manufactured with adiabatic transitions. Figure 2(c) shows the corresponding XY-2D mapping of the RS intensity under the same conditions as for G652D-ONF (figure 2 (b)). The RS intensity shows the absence of the surface scattering in the taper region (transition 1). This confirms that the surface scattering obtained for the multimode G652D-ONF comes from the no longer guided HOMs, which go out of the fiber in the transition 1. Indeed, when reaching cutoff diameters, those modes cross the interface glass-air and consequently undergo large scattering by surface roughness. In addition, this measurement clearly shows that the guiding process drastically changes near X = 10 mm of propagation for the SM450-ONF (figure 2(c)), which corresponds to the transition from the core-cladding guidance to the cladding -air guidance of the fundamental mode, in perfect agreement with numerical prediction.

4. Conclusion

By correlating the spatial geometry with Rayleigh scattering intensity and the OBR measurements, preliminary results have shown that our experimental technique is a powerful tool to characterize a photonic waveguide in nominal operation. This technique provides non-destructive and non-invasive direct observation, in 1D and 2D scale, the spatial evolution of propagating fields throughout an entire optical nanofiber with a micrometric spatial resolution of 5 μm. Second, it allows us to analyze a nanofiber in operando based on a single mode fiber in the visible range that is impossible to obtain with the state-of-art system (OBR measurement). The last advantage of our technique is that provides access to distributed measurements of the nonlinear processes along the fiber due the spectral resolution of our device. We therefore can observe for example the second harmonic generation from quadratic nonlinearity along ridge waveguides, or supercontinuum generation along nonlinear waveguides such as ONF.

References

[1] Lai Y H, Yang K Y, Suh M G and Vahala K J 2017 *Opt. Express* **25** 22312-22327
[2] Hontinfinde R, Coulibaly S, Megret P, Taki M and Wuilpart M 2017 *Opt. Lett* **42** 716-1719
[3] Tong L and Sumetsky M 2010 *Subwavelength and Nanometer Diameter Optical Fibers* (China: Zhejiang University Press)
[4] Tong L, R R Gattass, J B Ashcom, S He, J Lou, M Shen and E Mazur 2003 *Nature* **426** 816–819
[5] Brambilla G 2010 *J. Opt* **12**
[6] Wuttke C, Becker M, Brückner S, Rothhardt M, and Rauschenbeutel A 2012 *Opt. Lett* **37** 1949-1951
[7] Birks T, Wadsworth W, and P S J 2000 *Opt. Lett.* **25** 1415–1417
[8] Shan L, Pauliat G, Vienne G, Tong L, and Lebrun S 2013 *Appl. Phys. Lett* **102** 201110
[9] Talataisong W, Ismaeel R, and Brambilla G *Sensors* 2018 **18** 461
[10] Godet A., Ndao A, Sylvestre T, Pecheur V, Lebrun S, Pauliat G and Huy K P 2017 *Optica* **4** 1232-1238.

Irradiation-induced Kondo resonance in two-dimensional electron systems

I. V. Iorsh[1,2], **O. V. Kibis**[2]

[1]Department of Physics and Engineering, ITMO University, Saint-Petersburg, 197101, Russia
[2]Department of Applied and Theoretical Physics, Novosibirsk State Technical University, Karl Marx Avenue 20, Novosibirsk 630073, Russia

E-mail: i.iorsh@metalab.ifmo.ru

Abstract. We demonstrated theoretically that formation of the resonant scattering states in the two-dimensional (2D) electron system irradiated by a circularly polarized electromagnetic field leads to the emergence of localized magnetic moments. As a consequence, the corresponding Kondo resonances appear. For GaAs-based quantum wells and microwave fields, we estimate the Kondo temperature around 2.5 K, which can be detected in state-of-the-art measurements.

1. Introduction

In 1964 paper [1], Jun Kondo suggested the mechanism responsible for the minimum in the temperature dependence of the resistivity of noble divalent metals, the effect which had remained a mystery for more than three decades [2]. Within the model proposed by him, the conduction electrons interact antiferromagnetically with the localized magnetic impurities leading to the $\log(T)$ corrections to the relaxation time. It has been soon demonstrated that the low energy physics of the Kondo effect shows universality and is characterized by a single energy scale T_K, Kondo temperature. Therefore, it can be effectively treated by the methods of the renormalization group theory [4]. Particularly, the Kondo physics is appropriate to describe the transformation of ground states when going from the high energy range to the low energy one. The concept of the Kondo resonance was later applied to the systems containing the dense arrays of magnetic impurities [5, 6] which triggered the research of the Kondo insulators and heavy fermion systems in condensed matter physics [7, 8].

The Kondo temperature is defined by the effective on-site Coulomb repulsion of impurity atoms, hybridization of the conduction and impurity electrons, and the relative position of their resonant energy with respect to the Fermi level. While these parameters are fixed in the bulk materials, they can be effectively tuned in nanodevices. Since the first demonstration of the Kondo effect in quantum dots [9], the tunable Kondo effect in nanostructures become an excited field of research [10, 11, 12]. While the tuning of Kondo temperature is usually achieved by the gate voltage, in the present research we show that the Kondo effect can be switched on and off by means of the off-resonant electromagnetic field. Manipulation of the properties of the condensed matter systems by an off-resonant electromagnetic field (the Floquet engineering) became an established research field with a plethora of illuminating fundamental effects [13], including numerous field-induced phenomena predicted and observed in various nanostructures [14, 15, 16, 17, 18]. Particularly, it has been recently shown that the irradiation of 2D electron systems with an off-resonant circularly polarized field leads to the formation of the quasi-stationary electron states localized at repulsive impurities [19, 20, 21]. Noteworthy, the parameters of these states can be effectively controlled by the irradiation. In the following, we explore the possibility of the formation of the Kondo effect mediated by these light-induced electron states.

2. Model

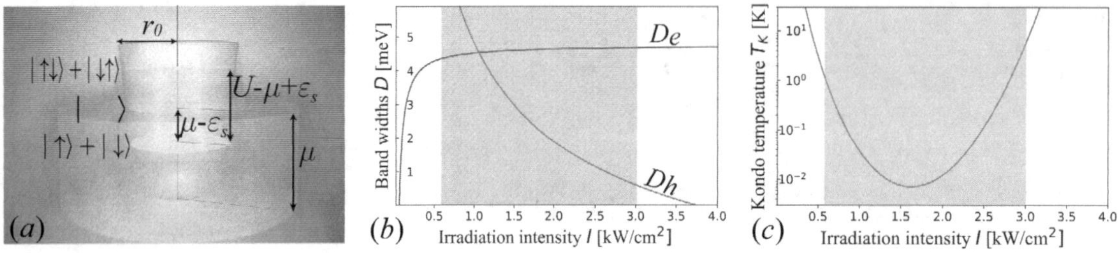

Figure 1. (a) Geometry of the problem: The irradiation leads to the transformation of the repulsive delta potential to the cylindrical potential barrier with the radius r_0. If the resonant electron energy ε_s lies below the Fermi level μ and the repulsive Coulomb energy U for electrons in the localized state is large enough, the singly occupied resonant state may become the ground state which leads to the formation of localized magnetic moment and emergence of the Kondo resonance; Effect of the circularly polarized irradiation with the frequency $\omega/2\pi = 200$ GHz and the intensity I on (b) the effective hole and electron bandwidths $D_{h,e}$ and (c) the Kondo temperature in a GaAs-based QW filled by 2D electron gas with the Fermi energy $\varepsilon_F = 5$ meV, energy broadening $\Gamma_s = 0.1\varepsilon_s$ and the electron effective mass $m_e = 0.067m_0$ (m_0 is the free electron mass) for the zero temperature. The blue shadow areas mark the validity range of the model, where the applicability conditions are satisfied for both the Kondo Hamiltonian ($\Gamma_s \ll D_e, D_h$) and the dressed potential approach ($\varepsilon_s \ll \hbar\omega$).

Assuming the parabolic dispersion of electrons with the effective mass m_e, let us consider the 2D electron system with repulsive impurities irradiated by a circularly polarized electromagnetic field (see Fig. 1a). Under action of the field, the repulsive delta potential of the impurities, $u_0\delta(\mathbf{r})$, turns into the effective cylindrical potential barrier $u_0\delta(r - r_0)/2\pi r_0$ with the radius $r_0 = eE_0/(m_e\omega^2)$, where E_0 is the field amplitude and ω is the field frequency [19, 20]. This barrier supports the quasi-stationary electron states localized inside the barrier ($r < r_0$), where the energy ε_s, the broadening Γ_s, and the wave function ψ_s of the lowest of the states read [20] $\varepsilon_s = \hbar^2\xi_0^2/2m_e r_0^2$, $\Gamma_s = 2\varepsilon_s\alpha^2/N_0^3(\xi_0)J_1(\xi_0)$ and $\psi_s(r) = \theta(r_0 - r)J_0(\xi_0 r/r_0)/\sqrt{\pi}r_0 J_1(\xi_0)$. Here $J_m(\xi)$ and $N_m(\xi)$ are the Bessel functions of the first and second kind, respectively, ξ_0 is the first zero of the Bessel function $J_0(\xi)$, and $\theta(r)$ is the Heaviside function. The dressing field approach which omits the time dependence of the Hamiltonian is valid when $\hbar\omega \gg \varepsilon_s$. Due to the semi-transparent nature of the barrier, the conduction electrons with the energy $\varepsilon_k = \hbar^2k^2/2m_e$ may tunnel in and out of the resonant state with the energy ε_s. Therefore, the electron system can be described by the tunnel Hamiltonian $H_T = \sum_{\mathbf{k},\sigma}(\varepsilon_k - \mu)\hat{c}^\dagger_{\mathbf{k},\sigma}\hat{c}_{\mathbf{k},\sigma} + \sum_\sigma(\varepsilon_s - \mu)\hat{d}^\dagger_\sigma\hat{d}_\sigma + \sum_{\mathbf{k},\sigma} T_{\mathbf{k}}[\hat{c}^\dagger_{\mathbf{k},\sigma}\hat{d}_\sigma + \text{H.c.}]$, where $\hat{c}^\dagger_{\mathbf{k},\sigma}(\hat{c}_{\mathbf{k},\sigma})$ are the production (annihilation) operators for conduction electron states, $\hat{d}^\dagger_\sigma(\hat{d}_\sigma)$ are the production (annihilation) operators for the light-induced localized electron states, $\sigma =\uparrow,\downarrow$ is the spin projection, μ is the chemical potential, $T_{\mathbf{k}}$ is the tunneling amplitude which reads $T \approx \sqrt{\Gamma_s/\pi N_\varepsilon}$ near the tunnel resonance ($\varepsilon_s = \varepsilon_k$) in the case of weak tunneling, and $N_\varepsilon = Sm_e/(\pi\hbar^2)$ is the density of conduction electron states.

To access the Kondo physics, we have to take into account the Coulomb interaction of the localized electrons by adding the Hamiltonian $H_C = U\hat{d}_\uparrow^\dagger \hat{d}_\uparrow \hat{d}_\downarrow^\dagger \hat{d}_\downarrow$ to the tunnel Hamiltonian H_T, where $U \approx \gamma e^2/(\epsilon r_0)$ is the effective Coulomb energy of the localized electrons, ϵ is the static permittivity of the material, and $\gamma \approx 0.8$. The total Hamiltonian $H = H_T + H_C$ describes the famous Anderson impurity model. Within this model, the Coulomb repulsion of impurity electrons lifts energy of the doubly occupied state and forms the spin-polarized singly occupied ground state (see Fig. 1a). Neglecting the interaction between the localized states and the conduction electron continuum, there are three eigenstates of the Hamiltonian H_T pictured schematically in Fig. 1a: (i) the empty state $|\ \rangle$ with the energy μ; (ii) the singly occupied state $|\uparrow\rangle + |\downarrow\rangle$ with the energy $\varepsilon_s - \mu$; (iii) the doubly occupied state $|\uparrow\downarrow\rangle + |\downarrow\uparrow\rangle$ with the energy $2(\varepsilon_s - \mu) + U$. The Kondo physics originates due to the emergence of the local magnetic moments at the localized electron state in the low energy range. This happens if the singly occupied resonant state ε_s has the lowest energy, i.e. for $\varepsilon_s - \mu < 0$ and $U > \mu - \varepsilon_s$. Since U and ε_s differently depend on the irradiation intensity I, there exists a range of field amplitudes for which the singly occupied state has the lowest energy. Namely, r_0 should satisfy the condition

$$\sqrt{\frac{\hbar \xi_0^2}{2m_e\mu}} < r_0 < \frac{\gamma e^2}{2\epsilon\mu r_0} + \sqrt{r_{min}^2 + \left(\frac{\gamma e^2}{2\epsilon\mu r_0}\right)^2}. \quad (1)$$

If the single occupied state has the lowest energy, we can use the Schreiffer-Wolff (SW) transformation to project the Anderson impurity model to the Kondo model, yielding [22] $H_K = \sum_{\mathbf{k},\sigma}(\varepsilon_k - \mu)\hat{c}_{\mathbf{k},\sigma}^\dagger \hat{c}_{\mathbf{k},\sigma} + J\vec{\sigma}(0)\cdot\vec{S}_s - \frac{V}{2}\hat{\psi}_\sigma^\dagger(0)\hat{\psi}_\sigma(0)$, where $\hat{\psi}_\sigma(0) = \sum_{\mathbf{k}} \hat{c}_{\mathbf{k},\sigma}$, $\vec{\sigma}(0) = \hat{\psi}^\dagger(0)\vec{\sigma}\hat{\psi}(0)$, $\vec{S}_s = \hat{d}^\dagger(\vec{\sigma}/2)\hat{d}$, and the coupling coefficients J and V read

$$J = \frac{\Gamma_s}{\pi N_\varepsilon}\left[D_e^{-1} + D_h^{-1}\right], \quad V = \frac{\Gamma_s}{2\pi N_\varepsilon}\left[D_e^{-1} - D_h^{-1}\right]. \quad (2)$$

Here $D_e = \mu - \varepsilon_s$ and $D_h = U + \varepsilon_s - \mu$ define the energies of electron-like and hole-like excitations due to the tunnelling and, therefore, should be treated as the effective bandwidth of the model. It is essential for the following that the SW transformation is applicable only if $\Gamma_s \ll D_e, D_h$. It should be noted also that the system under consideration corresponds to the Kondo model with the broken electron-hole symmetry.

3. Results and discussion

The dependence of the bandwidths $D_{e,h}$ on the irradiation intensity, I, is plotted in Fig. 1b for GaAs-based quantum well with the electron effective mass $m_e = 0.067m_0$, where m_0 is the free electron mass. The region, where the SW transformation is applicable and the Kondo effect is expected, is marked by the shadow. Since the typical Kondo temperature is essentially smaller than the conduction bandwidth $2D_0 \approx 3$eV (for GaAs), one needs to transform the initial high-energy Hamiltonian to the low-energy range in order to find the Kondo temperature. Such a transformation can be performed within the poor man's scaling renormalization approach [24, 25], which was originally proposed by Anderson. Within this approach the higher energy excitations are integrated out which changes the effective bandwidth D and the corresponding couplings $J(D), V(D)$. The renormalized coupling constant $J(D)$ diverges at some critical bandwidth $D = D_K$ associated with the Kondo temperature $T_K = D_K$.

The only difference of the considered system from the original Kondo problem [3] is the strong electron-hole asymmetry since the typical Fermi energy in GaAs-based QWs is $\varepsilon_F \ll D_0$. As a consequence, for $D > \varepsilon_F$ only the second order process involving a virtual electron should be taken into account. On the contrary, the both electron and hole processes contribute to the effective coupling rescaling for the case of $D < \varepsilon_F$. Applying the known general solution of the asymmetric electron-hole Kondo problem [26] to the considered system, we arrive at the Kondo temperature

$$T_K = \varepsilon_F \exp\left[-\frac{1}{2\pi N_\varepsilon J'(\varepsilon_F)}\right]. \tag{3}$$

where

$$J'(\varepsilon_F) = \frac{J}{\left[1 + \frac{\pi N_\varepsilon}{2}\ln\frac{D_0}{\varepsilon_F}(J+V)\right]\left[1 - \frac{\pi N_\varepsilon}{2}\ln\frac{D_0}{\varepsilon_F}(3J-V)\right]}. \tag{4}$$

In Fig. 1c we plot the dependence of the Kondo temperature on the irradiation intensity for the same GaAs-based quantum well.

4. Conclusion

It is shown theoretically that the circularly polarized irradiation of two-dimensional electron systems can induce the Kondo effect. It follows from the present analysis that the Kondo temperature reaches the value of several K in semiconductor quantum wells, which allows to observe the effect experimentally in the state-of-the-art measurements.

Acknowledgments

The reported study was funded by the Russian Science Foundation (project 20-12-00001).

References

[1] Kondo J 1964 *Progress of Theoretical Physics* **32** 37
[2] de Haas W J, de Boer J and van den Berg G J 1934 *Physica* **1** 1115
[3] Anderson P W 1961 *Physical Review* **124** 41
[4] Wilson K G 1973 *Review of Modern Physics* **47** 773
[5] Steglich F, Aarts J, Bredl C D, Lieke W, Meschede D, Franz F and Schäfer H 1979 *Physical Review Letters* **43** 1892
[6] Andres K, Graebner J E and Ott H R 1975 *Physical Review Letters* **35** 1779
[7] Tsunetsugu H, Sigrist M and Ueda K 1993 *Physical Review B* **47** 8345
[8] Stewart G R 1984 *Review of Modern Physics* **56** 755
[9] Goldhaber-Gordon D, Göres J, Kastner M A., Shtrikman H, Mahalu D and Meirav U 1998 *Physical Review Letters* **81** 5225
[10] Cronenwett S M, Oosterkamp T H and Kouwenhoven L P 1998 *Science* **281** 540
[11] Park J, Pasupathy A N, Goldsmith J I, Chang C, Yaish Y, Petta J R, Rinkoski M, Sethna J P, Abruna H D, McEuen P L and Ralph D C 2002 *Nature* **417** 722
[12] Iftikhar Z, Anthore A, Mitchell A, Parmentier F, Gennser U, Ouerghi A, Cavanna A, Mora C, Simon P and Pierre F 2018 *Science* **360** 1315
[13] Basov D N, Averitt R D and Hsieh D 2017 *Nature Materials* **16** 1077
[14] Lindner N H, Refael G and Galitski V 2011 *Nature Physics* **7** 490 (2011).
[15] Rechtsman M C, Zeuner J M, Plotnik Y, Lumer Y, Podolsky D, Dreisow F, Nolte S, Segev M and Szameit A 2013 *Nature* **496** 196
[16] Wang Y H, Steinberg H, Jarillo-Herrero P and Gedik N 2013 *Science* **342** 453
[17] Sie E J, McIver J W, Lee Y H, Fu L, Kong J and Gedik N 2015 *Nature Materials* **14** 290
[18] McIver J W, Schulte B, Stein F U, Matsuyama T, Jotzu G, Meier G and Cavalleri A 2020 *Nature Physics* **16** 38
[19] Kibis O V 2019 *Physical Review B* **99** 235416
[20] Kibis O V, Boev M V and Kovalev V M 2020 *Physical Review B* **102** 075412
[21] Kibis O V, Kolodny S A and Iorsh I V 2021 *Optics Letters* **46** 50
[22] Coleman P 2015 *Introduction to many-body physics* (Cambridge: Cambridge University Press)
[23] Riseborough P S and Lawrence J M 2016 *Reports on Progress in Physics* **79** 084501
[24] Anderson P W 1970 *Journal of Physics C: Solid State Physics* **3** 2436
[25] Anderson P W 1973 *Comments on Solid State Physics* **5** 73
[26] Zitko R and Hovrat A 2016 *Physical Review B* **94** 125138

Tunable GaAs metasurfaces for ultrafast image processing

Viacheslav Iushkov, Alexander Shorokhov, Andrey Fedyanin

Faculty of Physics, Lomonosov Moscow State University, Moscow 119991, Russia

E-mail: iushkov@physics.msu.ru

Abstract. The design and construction of optical semiconductor metasurfaces for various applications have become an important topic in the last decade. However, most metasurfaces are static; they are optimized for only one exact purpose and typically realize only one operation. In this work, we discuss the basic methods for creating dynamic metasurfaces giving special attention to ultrafast optical switching and provide numerical modeling of metasurfaces made of GaAs material realizing different amplitude-phase profiles under asymmetrical optical pumping. The metasurfaces are composed of semiconductor discs immersed in a fused silica medium. We demonstrate that based on Fourier transform and spatial filtering methods, these structures can be used for image processing and optical computing. Ultrafast switching is achieved by using an optical pump-probe scheme. The characteristic relaxation times between the pumped state and the relaxed state are on the order of several picoseconds.

1. Introduction

Vision is a sense by which we receive most of the information about the world around us. One of humanity's most desirable goals is the creation of devices that will independently watch, analyze, and draw conclusions from what they see. Thus, image processing is an essential task both for technology and science.

One of the first breakthroughs in this area was the creating of "electronic eyes" - CCD cameras around the 70s by Willard Boyle and George Smith. It opened the way for a new important field – machine vision. At the moment, with the help of computer processing, many tools have been implemented that allow to transform an image, making it more convenient for analysis: finding boundaries and edges, specific patterns, faces, etc. Currently, machine vision is a successfully developing area of IT. However, it requires significant computational power, which, besides producing drastic heat losses, also consumes an incredible amount of energy. It raises the question of searching for new devices for effective image processing.

One of the solutions can be analog methods instead of the digital-based ones [1]. Such approaches have been used earlier, but with the use of bulk optical elements. Contrary, optical metasurfaces – flat optical devices that can form an arbitrary amplitude-phase profile on demand – can provide a large number of lightweight and compact elements for different operations. These include lenses [2, 3], holograms [4, 5, 6], and phase plates [7]. In some recent studies, various mathematical operations have also been implemented on an optical signal, such as integration, differentiation, and convolution [8, 9, 10]. It has been shown that such metasurfaces are useful for applications in real-time image processing. This area includes such methods as edge detection [11] and image recognition [12, 13]. There are also the first neural networks based on a cascade of metasurfaces [14, 15] for image recognition tasks. The advantages of metasurfaces usually include CMOS compatibility and low optical losses. However, the latter advantage also sets some limitations on these dielectric and semiconductor nanostructures [16, 17, 18].

One of the most important limitations of metasurfaces is that they are generally static structures with a defined optical response. To fix this problem, one can use different mechanisms, some of which are mentioned below. Firstly, the use of silicon thermo-optical properties makes it possible to create a device with a noticeable reconfiguration of its response in a relatively small temperature range. However, it is difficult to produce it locally at the scale of individual supercells. The second method is the application of an external voltage [19, 20]. This method is faster than the first one, but it still has significant limitations on the speed of operation and difficulties in addressing individual nanoantennas. Finally, the third method is an optical injection of carriers by a femtosecond laser pulse [21], the fastest of all mentioned above.

In this work, we focus on the optical pumping mechanism under a uniform laser field irradiation to demonstrate tunable metasurface operation. Reconfiguration of the amplitude-phase profile is the result of the non-uniform absorption of the metasurface elements. Using such asymmetric optical pumping, one can switch between different amplitude-phase masks in the Fourier optical plane, providing different operations. The achieved results can be used to create reconfigurable compact imaging devices with remarkable processing time.

2. Ultrafast switching in gallium arsenide

Gallium arsenide was chosen as a material for tunable metasurfaces due to its widespread usage in semiconductor industry, well-known techniques for fabrication of micro- and nanodevices and a large number of scientific works devoted to studies of its opto-electronic properties [22]. Using the analytical description, we estimated the change in the optical response (refractive index Δn and absorption coefficient Δk), taking into account the Burstein-Moss effect, bandgap shrinkage effect, and free-carrier absorption under ultrafast optical pumping of GaAs (fig. 1).

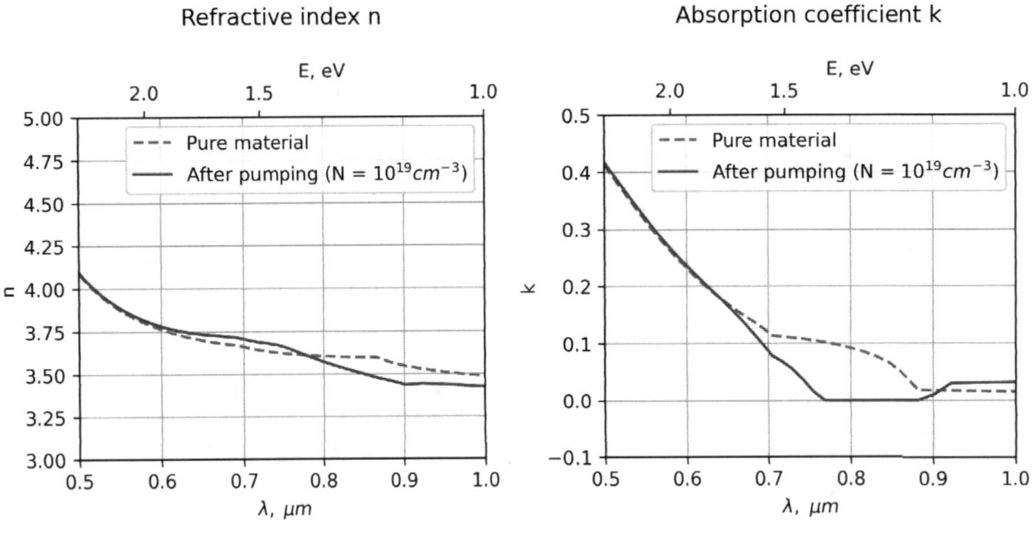

Figure 1: Change of refractive index and extinction near the band gap of GaAs ($N = 10^{19} cm^{-3}$).

The Burstein-Moss effect can be explained as following: the density of states of the conduction band is relatively small, and a small number of electrons will quickly fill it. In this case, other electrons will need to absorb more energy than the bandgap of the material. This leads to a decrease in absorption of the material. Using Kramers–Kronig relations one can obtain:

$$\Delta n_{BM}(E) = \frac{2ch}{e^2} V.p. \int_0^{+\infty} \frac{\Delta\alpha(N,P,E')}{E'^2 - E^2} dE'$$

where $\Delta\alpha(N,E,T)$ is the modulation of absorption for photon energies E lying above the band gap. To pass to the absorption coefficient, a simple relationship formula $k = \frac{\alpha\lambda}{4\pi}$ is used.

The bandgap shrinkage effect appears due to the accumulation of many electrons near the edge of the conduction band. Due to the overlapping of their wave function and, consequently, screening, their energy decreases, and the zone edge shrinks. A similar effect occurs at the edge of the valence band. As a result, the bandgap becomes narrower. Unlike bandfilling, Δn is positive for energies below the bandgap.

Free carrier absorption is the last effect included in this discussion. In contrast to those considered above, this effect is intraband: an electron can absorb a photon for transition within the boundaries of one zone. An analytical expression for the results of such an effect:

$$\Delta n_{FC} = -\frac{e^2\lambda^2}{8\pi^2 c^2 \varepsilon_0 n}\left(\frac{N}{m_e} + P\frac{m_{hh}^{0.5} + m_{lh}^{0.5}}{m_{hh}^{1.5} + m_{lh}^{1.5}}\right)$$

$$\Delta k_{FC} = \frac{1}{k^2+n^2}\frac{4\pi e^2\tau}{1+\omega^2\tau^2}\frac{N}{m}\left(\frac{n}{\omega} + k\tau\right)$$

It should be noted that the relaxation of excited carriers occurs on a very short time scale ($\tau \approx 1\ ps$), which means that devices based on the described concept can be used for ultrafast processing.

One of the essential characteristics in choosing the mechanism for tuning the optical signal was the ultra-short relaxation time. It has already been demonstrated that upon optical pumping by a picosecond laser pulse the character of relaxation of excited carriers ranges from picoseconds to several tens of picoseconds[21, 23, 24]. This phenomenon may be the basis for the creation of all-optical devices for ultrafast light control.

3. The basic concept of Fourier filtering

Image processing using Fourier filtering is a widespread practice due to the simplicity of its realisation. In general, the idea of Fourier image processing has been described previously [8, 9, 10]. The Fourier transform is implemented using a lens (fig. 2).

Assume that the image propagate along the Oz axis, the distribution of the incident electric field in the (x,y) plane is described by the function $f(x,y)$, the transmitted one is $g(x,y)$. The correspond fields in the focal plane of the lens are $F(u,v)$ and $G(u,v)$. Obviously, they are related by the dependence $f(x,y) = FT[F(u,v)]$, $G(x,y) = FT[G(u,v)]$. Placing a plate with a complex distribution of the transmission coefficient $H(u,v)$ in the focal plane, we come to the following equations:

$$F(u,v) = H(u,v) \cdot G(u,v)$$

$$f = h * g = \iint h(x-x',y-y')g(x',y')dx'dy'$$

A confocal lens is used to perform the inverse Fourier transform. Taking into account the expression $FT(FT(f(x(x,y))) \propto f(-x,-y)$, the post-processing of the final image is expressed in its reflection relative to the center of coordinates.

In this work we implement two different transformations: convolution and derivative. Convolution is required if you want to trace a given image $h(x,y)$ on the analyzed one $f(x,y)$. In this case, the role of the transmission function $H(u,v)$ is played by the amplitude-normalized Fourier transform of the function

Figure 2: Schematic representation of the proposed idea. Laser radiation propagates throw amplitude mask creating an object image, which is focused by the lens on the metasurface forming the reference amplitude and phase profile.The metasurface is optically pumped from its backside. The transmitted radiation is detected by the camera.

$h(x,y)$. If you need to find the derivative along the Ox axis, then $H(u,v) \propto iu$. Since the metasurface is a bounded structure, it is necessary to take into account its longitudinal size D. As a result, $H(u,v) = \frac{2iu}{D}$.

The main difference between the approach proposed in this work and the one, which has been used previously, is that the metasurface can be quickly reconfigurated by optical pumping, which switch it between different transmission function, for example, between a unity function and a function that implements a derivative.

4. Gradient metasurface with the tunable transmission function

We performed calculations to design metasurfaces that realize tunable analog operations. We chose a semiconductor disk immersed in a dielectric layer as a unit cell due to its polarization insensitivity and fabrication simplicity. Moreover, this geometry makes possible the excitation of electric and magnetic Mie-type resonances simultaneously for the same spectral region. It was shown that it is possible to obtain structures with reflection from 0 to 1 and a phase shift by 2π [25].

Previously, based on silicon [26], a metasurface was modeled [27] for tracing the given image on the analysed one. However, as been already mentioned, the static nature does not allow for flexibility in such devices. A similar scheme was proposed, operating for transmission and utilizing the pump-probe method. Also, the material for nanoantennas was changed to GaAs [28]. In the proposed experiment, radiation passes through a beam splitter, part of the radiation propagates as the pump, which excites the metasurface from the backside. The probe passes through the spatial light modulator, which forms

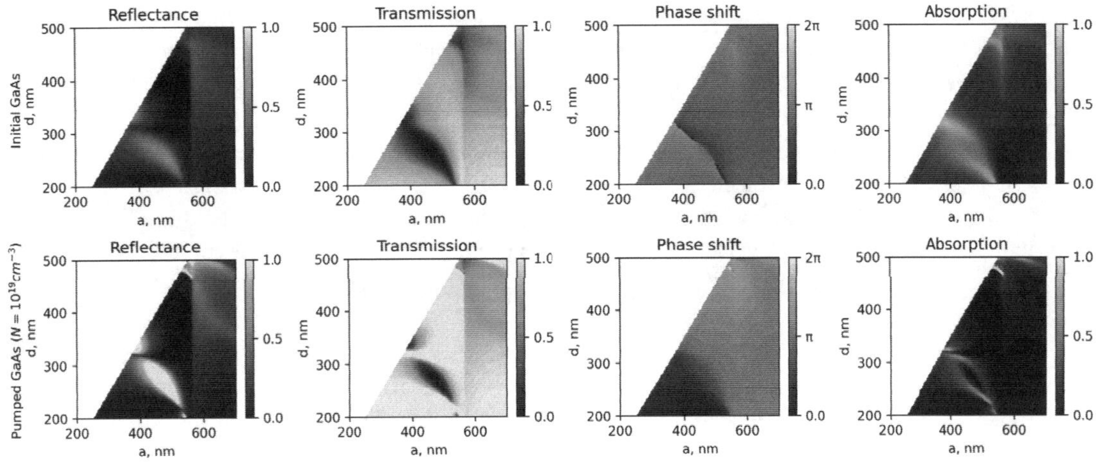

Figure 3: The dependences of reflection, transmission, phase shift, and absorption at a wavelength of 800 nm for GaAs metasurfaces without pump (top row) and under pumping (bottom row).

the analysed image, and then focuses on the metasurface. Passing through the structure is similar to multiplying the Fourier transform by the transmission function of interest. The second Fourier transform produces the output image detected by the camera.

The function of transmission without pumping equals to the function $H(u,v) = \frac{2iu}{D}$, where D is the longitudinal size of the metasurface corresponding to the derivative, which allows to find the edges on the image. The second function is $H(u,v) = 1$, corresponding to the uniform transmission. In this study the metasurface with a slowly varying period and diameter of the disks along the metasurface is proposed instead of the previously used division by pixels with defined reflection and phase.

We used Ansys Lumerical FDTD to perform numerical simulations. The dependencies of reflection, transmission, phase shift and absorption at the wavelength of 800 nm were obtained (fig. 3). Using this results, a design of the tunable metasurface was proposed (fig. 4).

Two amplitude-phase profiles were numerically demonstrated in the case of static and pumped

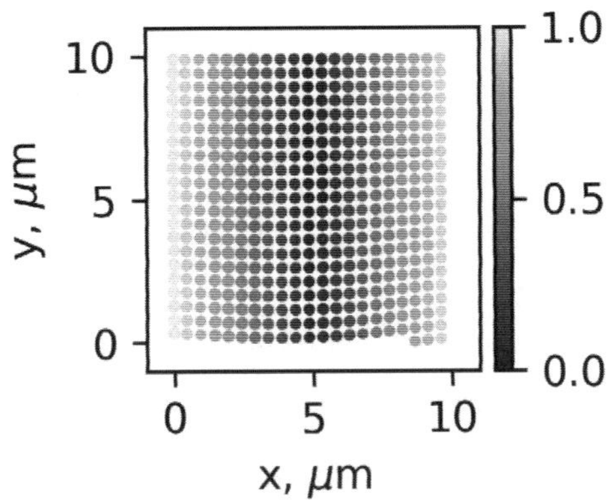

(a) Amplitude profile of the metasurface realizing $H(u,v)$ function without pumping

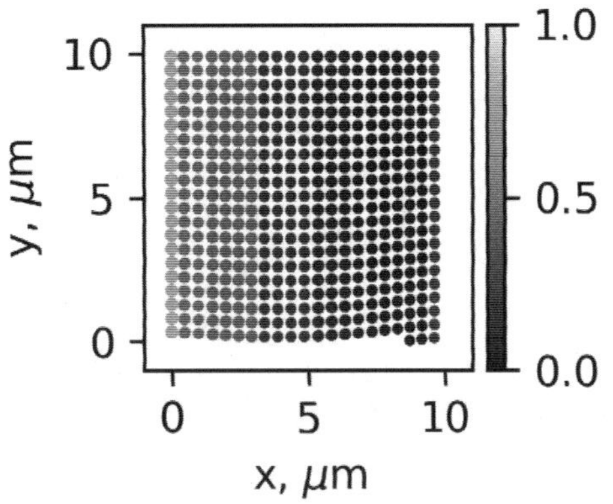

(b) Amplitude profile of the metasurface realizing $H(u,v)$ function with pumping

Figure 4: GRIN metasurface layout.

systems. They are shown in fig. 4 for initial (a) and pumped (b) cases. The spatial amplitude profile change is associated with non-uniform absorption of the metasurface elements.

5. Conclusion

A method of ultrafast spatial reconfiguration of semiconductor metasurfaces for tunable image processing is proposed. Metasurface based on GaAs material was numerically designed using the FDTD method, the approximate solution of the wave equation in the far-field, and Fourier analysis. To optimize the structure, algorithms were developed to smooth the profile of the transmission coefficient and phase shift. Tunable metasurface transmission functions are numerically demonstrated.

Acknowledgments

The authors acknowledge financial support from Russian Foundation for Basic Research (no. 20-02-00897, no. 21-52-12036). This research was performed according to the Development program of the Interdisciplinary Scientific and Educational School of Lomonosov Moscow State University "Photonic and Quantum technologies. Digital medicine". Part of the research is performed under the support of MSU Quantum Technology Center.

References

[1] Silva A, Monticone F, Castaldi G, Galdi V, Alù A and Engheta N 2014 *Science* **343** 160–163
[2] Khorasaninejad M and Capasso F 2017 *Science* **358**
[3] Lee D, Gwak J, Badloe T, Palomba S and Rho J 2020 *Nanoscale Adv.* **2** 605–625
[4] Chong K E, Wang L, Staude I, James A R, Dominguez J, Liu S, Subramania G S, Decker M, Neshev D N, Brener I and Kivshar Y S 2016 *ACS Photonics* **3** 514–519
[5] Wang B, Dong F, Li Q T, Yang D, Sun C, Chen J, Song Z, Xu L, Chu W, Xiao Y F, Gong Q and Li Y 2016 *Nano Lett.* **16** 5235–5240
[6] Genevet P, Capasso F, Aieta F, Khorasaninejad M and Devlin R 2017 *Optica* **4** 139 ISSN 2334-2536
[7] Chong K E, Staude I, James A, Dominguez J, Liu S, Campione S, Subramania G S, Luk T S, Decker M, Neshev D N, Brener I and Kivshar Y S 2015 *Nano Lett.* **15** 5369–5374
[8] Chizari A, Abdollahramezani S, Jamali M V and Salehi J A 2016 *Opt. Lett.* **41** 3451–3454
[9] Cordaro A, Kwon H, Sounas D, Koenderink A F, Alù A and Polman A 2019 *Nano Lett.* **19** 8418–8423
[10] Zhou Y, Zheng H, Kravchenko I and Valentine J 2020 *Nat. Photonics* 1–8
[11] Zhou J, Qian H, Chen C F, Zhao J, Li G, Wu Q, Luo H, Wen S and Liu Z 2019 *Proc. Natl. Acad. Sci. U.S.A.* **116** 11137–11140
[12] Manzur T, Zeller J and Serati S 2012 *Appl. Opt* **51** 4976–83
[13] Monjur M S, Tseng S, Tripathi R, Donoghue J J and Shahriar M S 2014 *Journal of the Optical Society of America A* **31** 41
[14] Shen Y, Harris N C, Skirlo S, Prabhu M, Baehr-Jones T, Hochberg M, Sun X, Zhao S, Larochelle H, Englund D and Soljacic M 2017 *Nature Photonics* **11** 441–446
[15] Feldmann J, Youngblood N, Wright C D, Bhaskaran H and Pernice W H 2019 *Nature* **569** 208–214
[16] Staude I and Schilling J 2017 *Nat. Photonics* **11** 274–284
[17] Kuznetsov A I, Miroshnichenko A E, Brongersma M L, Kivshar Y S and Lukyanchuk B 2016 *Science* **354**
[18] Yu N and Capasso F 2014 *Nat. Mater.* **13** 139–150
[19] Komar A, Fang Z, Bohn J, Sautter J, Decker M, Miroshnichenko A, Pertsch T, Brener I, Kivshar Y S, Staude I and Neshev D N 2017 *Applied Physics Letters* **110**
[20] Li S Q, Xu X, Veetil R M, Valuckas V, Paniagua-Domínguez R and Kuznetsov A I 2019 *Science* **364** 1087–1090
[21] Shcherbakov M R, Liu S, Zubyuk V V, Vaskin A, Vabishchevich P P, Keeler G, Pertsch T, Dolgova T V, Staude I, Brener I and Fedyanin A A 2017 *Nature Communications* **8** 17
[22] Bennett B R, Soref R A and Del Alamo J A 1990 *IEEE Journal of Quantum Electronics* **26** 113–122
[23] Vabishchevich P P, Vaskin A, Karl N, Reno J L, Sinclair M B, Staude I and Brener I 2021 *Applied Physics Letters* **118** 211105
[24] Della Valle G, Hopkins B, Ganzer L, Stoll T, Rahmani M, Longhi S, Kivshar Y S, De Angelis C, Neshev D N and Cerullo G 2017 *ACS Photonics* **4** 2129–2136
[25] Staude I, Miroshnichenko A E, Decker M, Fofang N T, Liu S, Gonzales E, Dominguez J, Luk T S, Neshev D N, Brener I and Kivshar Y 2013 *ACS Nano* **7** 7824–7832
[26] Palik E D 2012 *Handbook of optical constants of solids* ISBN 9780080547213

[27] Iushkov V V, Shorokhov A S and Fedyanin A A 2020 *AIP Conference Proceedings* **2300** 020048
[28] Adachi S 1989 *Journal of Applied Physics* **66** 6030–6040

Mechanical deformation of atomically thin layers during stamp transfer.

Tatiana V Ivanova, Dmitry Permyakov and Ekaterina Khestanova

Department of Physics and Engineering, ITMO University, St. Petersburg, 197101, Russia
Corresponding authors: ekaterina.khestanova@metalab.ifmo.ru

Abstract. The way transition metal dichalcogenide (TMD) strains during its transfer from one substrate to another is very interesting and holds a special place in the creation of heterostructures. In our work we observe the spectrum of photoluminescence in TMD during the transfer. For this we use a specially designed transfer system with inverted geometry. During transfer we observe a modification of exciton photoluminescence linewidth and resonance shift in atomically thin layers of TMD. We believe that our results lay grounds for the future work on the assessment of the atomically thin layer inhomogeneity introduced by the typical mechanical transfer.

1. Introduction

Nowadays, one of the most interesting directions in solid-state physics are two-dimensional materials such as graphene, transitional metal dichalcogenides (TMDs) and their heterostructures. Bulk TMD crystals are semiconductors with an indirect band gap, while monolayers are direct gap semiconductors. This is experimentally evidenced by the excitonic photoluminescence (PL) line shift and intensity increase with the decrease of sample thickness[1]. Unlike in GaAs quantum wells, excitons in TMD monolayers have large binding energies[2] and are sustained up to room temperature.

At the same time, such excitons and the corresponding photoluminescence lines are highly sensitive to external lattice perturbations via mechanical strains. The onset of stretching in a monolayer is a well-studied phenomenon. For example, monolayers were subjected to stretching in different configurations[3], strain was measured in single-, bi- and tri-layer in different TMDs[4,5,6], uniaxial strain dependence of A-exciton PL peak energy shift for monolayer TMDs was studied[7]. Other studies in chemical vapor deposition grown crystals unveiled the effect of strain on the energy and intensity of the peaks in the photoluminescence spectra.[8,9]

However, to date, TMD monolayers of the best quality are obtained by mechanical exfoliation[10]. Their combination with other two-dimensional materials by mechanical stacking with a polymer stamp makes it possible to create van der Waals (vdW) heterostructures[11] which, in turn, host a plethora of exciting phenomena such as long-lived indirect and highly localized Moiré excitons[12,13], making them promising for optoelectronics applications[14,15].

Nevertheless, there are no reported attempts to assess the strain arising in the process of stacking the layers. The aim of this work is to observe the photoluminescence of an exciton line in atomically thin layers of WS_2 during transfer onto a rigid substrate. For this purpose, we engineered an inverted transfer setup which relates to the field of precision manipulation[16] and can be used to observe the emerging strain in TMD during the transfer *in-situ*. We believe that our results lay the foundation for the next crucial technological step — the elimination of interfacial inhomogeneities caused by strains induced during typical mechanical transfer.

2. **Experiment**

Here we have studied the spatial distribution of a PL signal of WS_2 monolayers during transfer onto a glass slide from a bare polydimethylsiloxane (PDMS) stamp (Fig. 1(a)) and pick-up by a hexagonal boron nitride (hBN)/polycarbonate (PC)/PDMS stamp and subsequent transfer onto a hBN substrate (Fig. 1(b)). WS_2 monolayers were used due to their predicted high PL sensitivity to the strain.[7]

2.1 *Sample preparation and transfer setup.*

To obtain atomically thin layers of WS_2, mechanical exfoliation technology was used[17]. Blue NittoTM Scotch Tape was used to exfoliate the bulk crystal onto a commercially available (gel-film from Gelpak) viscoelastic PDMS stamp (for transfer onto glass) or a SiO_2/Si substrate (for pick-up and transfer with hBN stamp). After the thin layer was found in bright field reflection, its single layer thickness was confirmed by photoluminescence spectroscopy with the characteristic spectrum shown on Fig.3 (b). The PL characteristics can serve as both a measure of the layer thickness and a measure of its homogeneity.

Then we have used a specially designed inverted system to transfer the monolayer on the substrate. The essence of the transfer method in inverted geometry is that a transparent plate with a polymer stamp containing a layer of atomically thin material is installed in the movable stage of an inverted microscope, and the substrate on which the layer of atomically thin material is transferred is fixed on the manipulator holder. After that, the substrate is brought in contact with a certain area of the stamp.

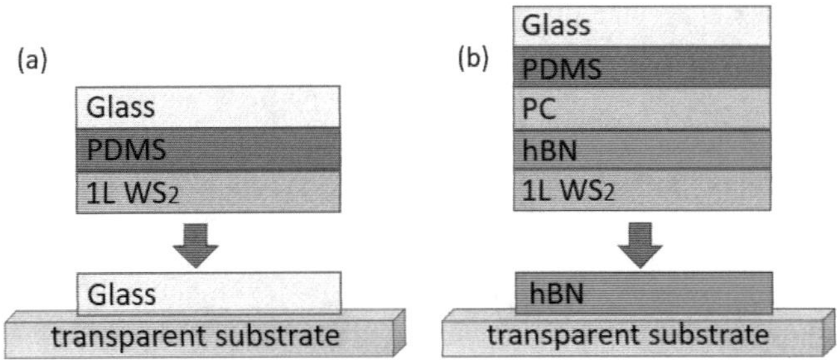

Figure 1. Schematics of transfer of monolayer WS_2 onto glass substrate using bare PDMS(a) or hBN(b) stamp. Red arrow indicates stacking direction.

2.2 *Photoluminescence measurements.*

A neodymium laser TORUS with a 532nm wavelength and a power of 15-30 μW is used to excite PL. In the experiment, the beam is focused on the sample to an approximately 1μm spot by the objective lens (water-immersion objective Olympus 60x, NA=1.2). The collected PL signal is analyzed by a slit spectrometer (Horiba LabRAM HR). The setup for PL spectroscopy mapping is shown in Fig.2.

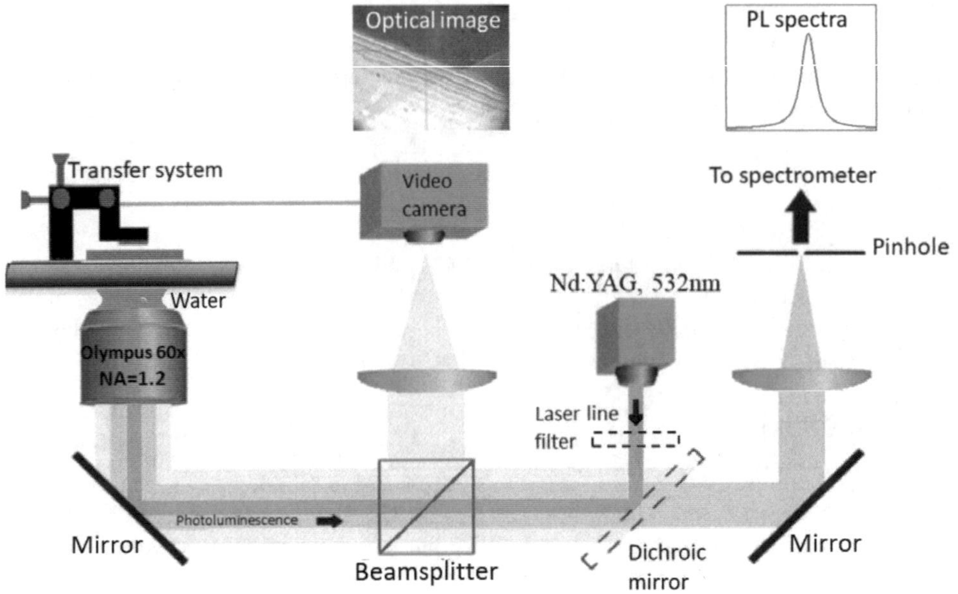

Figure 2. Setup for photoluminescence spectroscopy mapping. The optical image shows a clear transfer front of PDMS/WS_2 onto glass substrate.

As the stamp with a monolayer is slowly moved down to contact with the substrate, it develops the stamp-substrate contact front, which is clearly visible in the camera (Fig.3(a)). To measure the strain, induced during transfer, the PL signal is spatially mapped by the objective scanning with a piezo PI AIST-NT model P-517K079. The resulting PL map is presented in Fig. 3(b).
Fig. 3(b) shows the PL intensity map for a WS_2 transfer from PDMS to glass. Three regions with different photoluminescence intensities and uniformity are seen: region A, where the WS_2 monolayer is on the PDMS stamp and not yet in contact with the glass, region B, where the front of the PDMS/WS_2 contact with glass is seen and region C is the PDMS/WS_2 in full contact with glass. The corresponding spectra taken at points A, B and C are shown in Fig.3 (c). Also, Fig.3 (c) shows spectrum 0 (PDMS/WS_2/air) corresponding to the reference before transfer, featuring the main PL peak around 2.021 eV for PDMS/WS_2 reference system, in agreement with the previous findings[18]. This peak corresponds to the A-exciton, and the asymmetry of the left wing of the spectrum is associated with a closely spaced trion[19], with the intensity significantly suppressed at room temperature. To analyze the evolution of the observed peak across the transfer line, we have fitted it with a Voight function, given by the convoluted Lorenzian and Gaussian functions, which allowed us to evaluate both homogeneous (wL) and inhomogeneous (wG) broadening. The results of the fitting are presented in Table 1.

First, we note that the reference spectrum peak coincides with that from point A, indicating region A is intact despite its proximity to the transfer front. As front approaches the region B PL intensity increases by approximately 40% compared to the PDMS/WS_2/air region and then abruptly drops for the PDMS/WS_2/glass region C, also becoming spatially less homogeneous. Furthermore, the PL line spectral position redshifts across the front line: it is 2.020 eV for the A (WS_2 on PDMS) and B (front) regions and 2.015 eV in the C region (PDMS/WS_2/glass).

The article[18] reports significant changes in the band gap in TMD monolayers caused by uniaxial strain. The observed 5 meV redshift for the flake on PDMS in contact with glass can be associated with strain of 0.1-0.5%, depending on the PL shift/strain coefficient used[7]. We also note that as the monolayer is put in contact with the glass substrate, its dielectric environment changes significantly ($\varepsilon_{PDMS} = 2.5$, $\varepsilon_{glass} = 6.7$), which also can influence the PL signal. Indeed, the observed effect of a sharp change in the photoluminescence signal from area A to area C can be associated with a strong screening of the Coulomb interaction by glass as the intensity of photoluminescence I_{PL} directly depends on the dielectric constant of the environment[21]: $I_{PL} \propto \frac{|d_{cv}|^2}{\varepsilon} n_{ex}$, where $|d_{cv}|^2$ is the square of the module of the matrix element of the dipole transition for the excited electron to the ground state, n_{ex} is the concentration of excitons. Moreover, the larger dielectric constant of the glass substrate would reduce the exciton binding energy and blueshift the observed PL emission, which is the opposite of what we observe in our experiment. We, therefore, conclude, that our estimate 0.1-0.5% of transfer-induced strain is an underestimate of the real strain value.

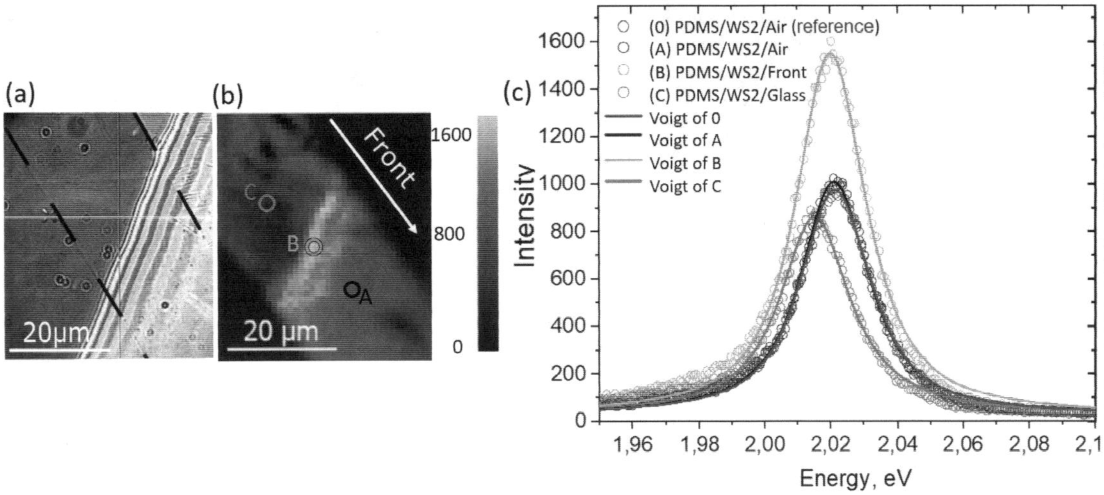

Figure 3. In-situ photoluminescence spectroscopy and spatial mapping.

(a) The optical microphotograph of the PL-mapped region of the WS_2.

(b) The spatial photoluminescence map of the WS_2 transfer from PDMS onto glass slide.

(c) Photoluminescence spectra taken at different locations, corresponding to points A ($PDMS/WS_2$/air), B ($PDMS/WS_2$/front), C ($PDMS/WS_2$/glass) of the heterostructure during transfer. Point 0 ($PDMS/WS_2$/air) corresponding to WS_2 reference before transfer was started.

Next, an hBN dielectric was used as both the stamp and the substrate layer in order to reduce the possible influence of the dielectric constant of glass (Fig.1 (b)). The hBN crystals are one of the strongest electrically insulating materials and its high interlayer integrity makes them good support as mechanical reinforcements[22] with dielectric constant value $\varepsilon = 3.5$[23].

Similar to the first experiment, the photoluminescence spectra were measured from the three distinct regions corresponding to areas A, B, C. Fig. 4 shows three regions with photoluminescence signal at 1.997 eV, in agreement with literature[24], our PL map is featuring different intensities and uniformity. First, we note that for $PDMS/PC/hBN/WS_2$ to hBN transfer the front region is much wider and does not exhibit PL intensity increase. However, similar to the first experiment, the overall PL intensity declines as $PDMS/PC/hBN/WS_2$ comes into contact with the bottom hBN, which is consistent with the increase of the effective dielectric constant as WS_2 becomes encapsulated by hBN on both sides. However, there is no spectral shift in PL within the sensitivity of our spectrometer (0.5 nm), due to the mechanical support of the WS_2 monolayer by the hBN layer and the suppression of mechanical strains.

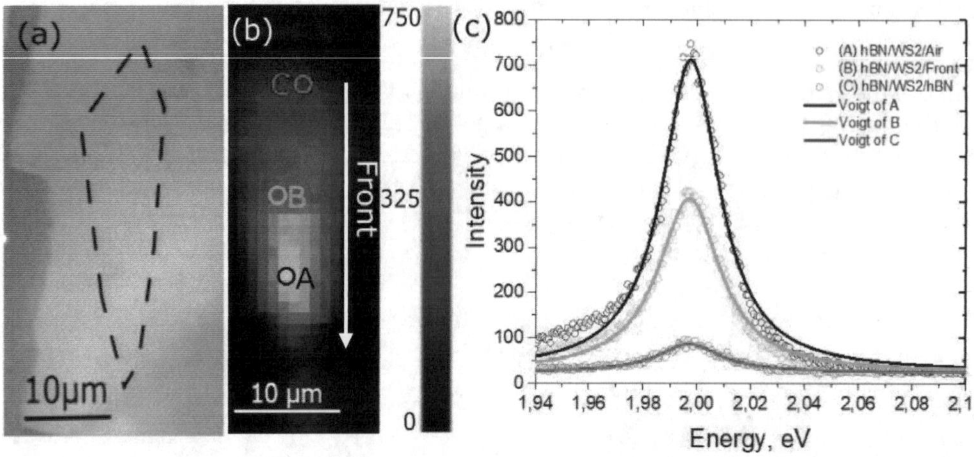

Figure 4. In-situ photoluminescence spectroscopy and spatial mapping.

(a) The optical microphotograph of the PL-mapped region of the WS_2 between two hBN.

(b) The spatial photoluminescence map of the WS_2 transfer from PDMS/PC/hBN onto hBN.

(c) Photoluminescence spectra taken at different locations, corresponding to points A (PDMS/PC/hBN/WS_2/air), B (PDMS/PC/hBN/WS_2/front), C (PDMS/PC/hBN/WS_2/hBN) of the heterostructure during transfer.

Table 1. Results of experimental data fitting

	Peak (eV)	wL (meV)	wG (meV)	Intensity (arbitrary units)
PDMS/WS_2/air(ref.)	2.021	21.3	9.8	978
PDMS/WS_2/air	2.020	21.6	9.3	1024
PDMS/WS_2/front	2.021	21.8	8.3	1550
PDMS/WS_2/glass	2.015	24.5	5.7	882
hBN/WS_2/air	1.997	25.9	0.0002	747
hBN/WS_2/front	1.997	28.0	0.0027	423
hBN/WS_2/hBN	1.997	28.4	0.0002	95

Conclusions

We have demonstrated a method to quantify transfer-induced strains and experimentally observe the photoluminescence spectral and intensity variation across the transfer front.

In the first series of experiments, PL quenching due to the increased Coulomb screening by a high-ε substrate is observed, as well as strong PL enhancement in the narrow region close to the transfer front, which is not fully understood. A redshift of 5-6 meV is observed and is consistent with the lower estimate of 0.1-0.5% strain during transfer. We emphasize that it is an underestimated value, as the increase dielectric constant of the substrate acts to blueshift PL during transfer.

In order to reduce the influence of high dielectric contrast between the substrate and the stamp in the second series of experiments, we used hBN both as a stamp and a substrate. There is no shift during encapsulation in hBN, but there is a contribution to the intensity decrease, which we relate to the increase of the effective dielectric constant during both-side hBN encapsulation.

We also strive to quantify the observed deformation and implement a method to reduce it to create homogeneous samples with a homogeneously narrow exciton line.

Acknowledgments

The study was funded by Russian Science Foundation according to project #20-72-00157.

References

1. Mak, K. F., Lee, C., Hone, J., Shan, J. & Heinz, T. F. Atomically thin MoS2: A new direct-gap semiconductor. *Phys. Rev. Lett.* **105**, 2–5 (2010).
2. Hanbicki, A. T., Currie, M., Kioseoglou, G., Friedman, A. L. & Jonker, B. T. Measurement of high exciton binding energy in the monolayer transition-metal dichalcogenides WS 2 and WSe 2. *Solid State Commun.* **203**, 16–20 (2015).
3. Benimetskiy, F. A. *et al.* Measurement of local optomechanical properties of a direct bandgap 2D semiconductor Measurement of local optomechanical properties of a direct bandgap 2D semiconductor. **101126**, (2019).
4. Carrascoso, F., Li, H., Frisenda, R. & Castellanos-Gomez, A. Strain engineering in single-, bi- and tri-layer MoS2, MoSe2, WS2 and WSe2. **12**, (2020).
5. Lee, J. H., Chung, B., Park, S. H., Moon, H. C. & Lee, D. H. Fabrication of Grid-Type Transparent Conducting Electrodes Based on Controlled Mechanical Fracture. *Macromol. Res.* **26**, 157–163 (2018).
6. Conley, H. J. *et al.* Bandgap engineering of strained monolayer and bilayer MoS2. *Nano Lett.* **13**, 3626–3630 (2013).
7. Aas, S. & Bulutay, C. Strain dependence of photoluminescence and circular dichroism in transition metal dichalcogenides: A k · p analysis. *arXiv* **26**, 28672–28681 (2018).
8. Zhang, Q. *et al.* Strain Relaxation of Monolayer WS2 on Plastic Substrate. *Adv. Funct. Mater.* **26**, 8707–8714 (2016).
9. Lloyd, D. *et al.* Band Gap Engineering with Ultralarge Biaxial Strains in Suspended Monolayer MoS2. *Nano Lett.* **16**, 5836–5841 (2016).
10. K. S. Novoselov, A. K. Geim, S. V. Morozov, D. Jiang, Y. Zhang, S. V. Dubonos, I. V. G. and A. A. F. Electric Field Effect in Atomically Thin Carbon Films. **306**, 666–669 (2016).
11. Geim, A. K. & Grigorieva, I. V. Van der Waals heterostructures. *Nature* **499**, 419–425 (2013).
12. Calman, E. V. *et al.* Indirect excitons in van der Waals heterostructures at room temperature. *Nat. Commun.* **9**, 1–5 (2018).
13. Förg, M. *et al.* Moiré excitons in MoSe2-WSe2 heterobilayers and heterotrilayers. *arXiv* 1–7 (2020) doi:10.1038/s41467-021-21822-z.
14. Novoselov, K. S. *et al.* Two-dimensional atomic crystals. *Proc. Natl. Acad. Sci. U. S. A.* **102**, 10451–10453 (2005).
15. Mueller, T. & Malic, E. Exciton physics and device application of two-dimensional transition metal dichalcogenide semiconductors. *2D Mater. Appl.* **2**, 1–12 (2018).
16. Ivanova T., Kulachenkov N., Permyakov D., Sinev I., Khestanova E., Patent of the Russian

Federation No. 2742761 dated 02/10/2021.

17. Castellanos-Gomez, A. *et al.* Deterministic transfer of two-dimensional materials by all-dry viscoelastic stamping. *2D Mater.* **1**, (2014).
18. Eizagirre Barker, S. *et al.* Preserving the Emission Lifetime and Efficiency of a Monolayer Semiconductor upon Transfer. *Adv. Opt. Mater.* **7**, (2019).
19. Wei, Ke, Yu Liu, Hang Yang, Xiangai Cheng, T. J. Large range modification of exciton species in monolayer WS2. *Appl. Opt.* **55**, (2016).
20. Wang, Y. *et al.* Strain-induced direct–indirect bandgap transition and phonon modulation in monolayer WS2. *Nano Res.* **8**, 2562–2572 (2015).
21. Chatterjee, S. *et al.* Excitonic Photoluminescence in Semiconductor Quantum Wells: Plasma versus Excitons. *Phys. Rev. Lett.* **92**, 1–4 (2004).
22. Falin, A. *et al.* Mechanical properties of atomically thin boron nitride and the role of interlayer interactions. *Nat. Commun.* **8**, 1–9 (2017).
23. Laturia, A., Van de Put, M. L. & Vandenberghe, W. G. Dielectric properties of hexagonal boron nitride and transition metal dichalcogenides: from monolayer to bulk. *2D Mater. Appl.* **2**, (2018).
24. Nagler, P. *et al.* Zeeman Splitting and Inverted Polarization of Biexciton Emission in Monolayer WS2. *Phys. Rev. Lett.* **121**, 57402 (2018).
25. Lin, Y. *et al.* Dielectric screening of excitons and trions in single-layer MoS2. *Nano Lett.* **14**, 5569–5576 (2014).

Linear optical properties of HKUST-1 metal-organic framework: Effect of the crystal size and synthetic approach

Yuliya A. Kenzhebayeva[1a)], Semyon Bachinin[1], Venera Gilemkhanova[1], Valentin A. Milichko[1,2 a)]

[1]ITMO University, St. Petersburg 197101, Russia
[2] Institut Jean Lamour, Université de Lorraine, Nancy F-54011, France

[a)]Corresponding author: y.kenzhebayeva@metalab.ifmo.ru

Abstract. We evaluated the change in the transmission and reflection spectra of single rhombic dodecahedron nanocrystals of the metal organic frameworks HKUST-1 with and without effect of near infrared laser radiation. The analysis of commercial and synthesized in the laboratory compounds was made through the optical transmittance and reflectance spectroscopy. Moreover, electronic microscopy and light scattering on the microcrystals was also utilized for revealing the size dependent colour. Such studies may find new application for MOF nanocrystals, as well as gain additional information about their usage for biological purposes or in computer sensing.

1. Introduction

In the past few years, metal organic frameworks (MOFs) have been studied in terms of their ability to change their properties under the influence of electric fields, light irradiation, or chemical transformations [1]. Therefore, recently they began to be attributed to the class of smart materials [2]. Moreover, the sizes of the MOFs can be varied in nanoscale, which is of particular interest to scientists for their further application [3]. Structural changes in nanocrystals can be associated with a change in their topology, transformation of the crystal structure, or removal of guest solvent molecules. Consequently, the transformations of such structural changes in MOFs at nanoscale open new applications for biological purposes [4], nonlinear optics [5] and computer sensing [6].

2. Experimental section

Here we present experimental realization control of the optical properties of MOFs nanocrystals with flexible structure by laser radiation. To obtain the desired results, several steps were taken, such as the synthesis of nanocrystals, optical experiments, and subsequent processing of the obtained data.

2.1. Synthesis

Single rhombic dodecahedron nanocrystals of HKUST-1 (400-700 nm) were synthesized via solvothermal method: 0,8842 g (3.66 mmol) of $[Cu(NO_3)_2*3(H_2O)]$ and 0,4203 g (2.00 mmol) of H_3BTC were mixed together in a solution of 8 ml dimethylformamide (DMF), 8 ml of isopropyl alcohol (IPA) and 8 ml of deionized water. In order to get nm-sized crystals, the synthesis was stopped after 5 minutes from the beginning. The SEM of synthesized nanocrystals of HKUST-1 is shown on the 'figure 1'.

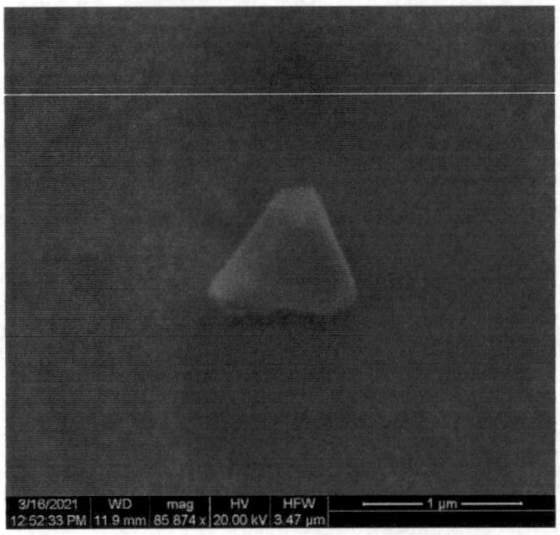

Figure 1. SEM (Scanning electron microscope) images of synthesized single rhombic dodecahedron nanocrystals of HKUST-1.

2.2. Optical experiments

Single nanocrystals of MOF were then placed on fused silica substrate on air for the optical experiments. During the experiment, the reflection and transmission spectra of commercial and synthesized HKUST-1 crystals were measured. The crystals were irradiated by a laser radiation with such technical parameters: Yb3 + laser with 1050 nm central wavelength, 150 fs pulse duration, and 80 MHz repetition rate. The radiation power was varied from 5 to 15 mW with a step of 1 mW. The spectra of irradiated and unirradiated crystals were analyzed, and in the course of the experiment, a characteristic spectral shift was observed in the switching of single MOF crystals. The transmission and reflection spectra are shown on the 'figure 2'. Turning on the laser induced the blue shift of the band gap, while turning off the laser is accompanied by relaxation of the transmission spectrum to the initial state. In addition, at certain parameters of the irradiation power, degradation of crystals was observed, which led to their destruction. Observational data and their analysis can provide additional information on the application of HKUST-1 nanocrystals in biological research, such as drug delivery, or computer sensing.

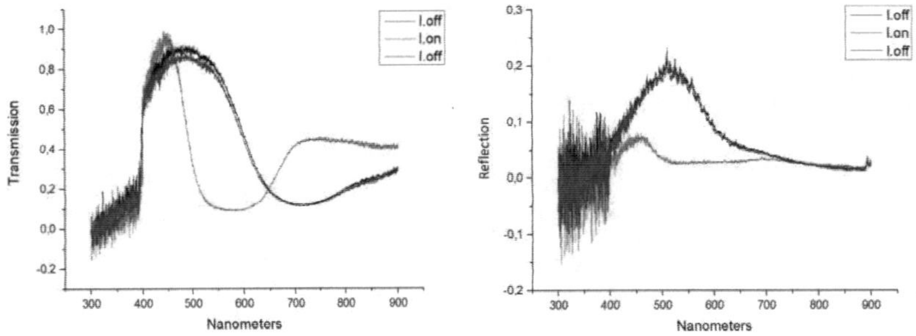

Figure 2. The transmission and reflection spectra of irradiated and unirradiated nanocrystal.

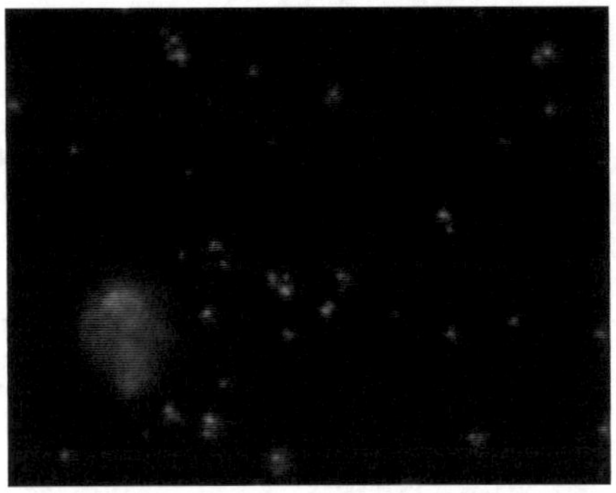

Figure 3. Image of light scattering by HKUST-1 nanocrystals depending on their size

3. Conclusion

In this work, the synthesis of HKUST-1 nanocrystals was shown as well as their optical properties, such as transmission and reflection spectra. The obtained data will make it possible to calculate the refractive index and extinction coefficient of nanocrystals depending on the effect of laser radiation. The efficiency of switching the optical properties of HKUST-1 nanocrystals makes it possible to use them in the future in medicine, microelectronics or in quantum technologies.

Acknowledgement

The work was financially supported by Russian Science Foundation. Grant Number: 19-79-10241.

References

[1] Y. A. Mezenov, A. A. Krasilin, V. P. Dzyuba, A. Nominé and V. A. Milichko, Advanced Science 6, 1900506 (2019).
[2] H. Furukawa, U. Müller and O. M. Yaghi, Angewandte Chemie Int. Ed. 54, 3417–3430 (2015).
[3] Chang B. et al. Confined Growth of ZIF-8 Nanocrystals with Tunable Structural Colors //Advanced Materials Interfaces. – 2018. – T. 5. – №. 9. – C. 1701270.
[4] Lazaro I. A., Forgan R. S. Application of zirconium MOFs in drug delivery and biomedicine //Coordination Chemistry Reviews. – 2019. – T. 380. – C. 230-259.
[5] X. Jiang, L. Zhang, S. Liu, Y. Zhang, Z. He, W. Li, F. Zhang, Y. Shi, W. Lü, Y. Li, Q. Wen, J. Li, J. Feng, S. Ruan, Y.-J. Zeng, X. Zhu, Y. Lu and H. Zhang, Advanced Optical Materials 6, 1800561 (2018).
[6] N.K. Kulachenkov, D. Sun, Y.A. Mezenov, A.N. Yankin, S. Rzhevskiy, V. Dyachuk, A. Nominé, G. Medjahdi, E.A. Pidko and V.A. Milichko, Angewandte Chemie Int. Ed. doi:10.1002/ange.202004293 (2020).

Realizing Spin-Conserved and Spin-Encrypted Hologram using Multipolar-modulated Meta-platform

Hafiz Saad Khaliq,[1] Inki Kim,[2] Kashif Riaz,[1, a)] Taimoor Naeem[1], Muhammad Zubair,[1] Junsuk Rho[2, b)] and Muhammad Qasim Mehmood[1, c)]

[1]*NanoTech Lab, Department of Electrical Engineering, Information Technology University (ITU) of the Punjab, 54600 Lahore, Pakistan.*
[2]*Department of Mechanical Engineering Pohang University of Science and Technology (POSTECH) Pohang 37673, Republic of Korea*

Corresponding author: [a)] *kashif.riaz@itu.edu.pk,* [b)] *jsrho@postech.ac.kr,* [c)] *qasim.mehmood@itu.edu.pk*

Abstract. The chiro-optical effects complemented with polarization conservation and wavefront shaping finds potential applications in advanced imaging, sorting and detection of enantiomers and quantum optics. Here, a unique design strategy proposed to manifest enormous chiro-optical effects using achiral structures (instead of conventional chiral antennas). The basic building block of the meta-platform contains a pair of achiral structures. The underlying mechanism behind the giant chiro-optical effects is explained by numerically calculating the multipolar resonances of scattering power. The designed diatomic meta-platform achieves absolute control over spin and wavefront of incident light to demonstrate the polarization-conserved and -encrypted meta-holograms.

INTRODUCTION

From stars to sub-atomic particles, the chirality phenomenon is inescapable. It is a geometrical property established in structures that do not have any rotational or mirror symmetry [1]. The chiral structures can differentiate between two spins of photons and exhibit distinct spectral responses for left-handed circularly polarized (LHCP), and right-handed circularly polarized (RHCP) light leads to chiroptical effects [2][3] such as optical activity and circular dichroism (CD) [4][5] or asymmetric transmission (AT) [6]. Recently, seen that strong chiroptical effects generated using three-dimensional (3D), multilayer, and planar nanostructures. Although these chiral structures possess outstanding abilities to yield strong optical activity and CD but due to their complex fabrication, they are far-away from state-of-the-art nano-technology, which entails miniaturized and integrable solutions. Owing to this, flat chiral meta-platforms emerges as the convenient solution to engineering the characteristics of light, leading to numerous meta-devices for applications in meta-holography [7], meta-lensing [4], and light-structuring [9-11]. Meanwhile, for chiroptical effects, several flat plasmonic meta-devices has been reported [2][8]. However, to overcome ohmic losses in plasmonic meta-devices, recently planar all-dielectric meta-devices presented [7][4]. F. Zhang et al. demonstrated all-dielectric metasurfaces for accompanying giant CD and wavefront shaping using spin-orbit interactions [6]. To the extent of our knowledge, no one has demonstrated the spin-conserved and spin-isolated giant chiroptical effect based on multipolar-modulated achiral meta-platform.

Here, we present a diatomic platform at optical frequencies based on achiral structures with extra degrees of design freedom. The proposed meta-platform yields an enormous chiroptical effect complemented with spin-conservation and spin-encrypted phase encoding. Different from the conventional design strategy of using intrinsic chiral structures, the building block of the proposed meta-platform consists of a pair of a-Si:H (hydrogenated amorphous silicon) [4] based achiral meta-atoms. The meta-atoms in the building block with distinct structural parameters and relative rotations angle of $\pi/4$ breaking rotational and mirror symmetries, thus allowing the manifestation of a strong chiroptical effect. The geometric phase manipulation is used in the meta-platform for wavefront shaping. The performance of the proposed meta-platform is much better than the already reported all-dielectric meta-devices.

Journal of Physics: Conference Series **2015** (2021) 012060 doi:10.1088/1742-6596/2015/1/012060

DESIGN METHODLOGY AND RESULTS

The schematic art of the working principle of meta-platform is depicted in Figure 1(a). The multipolar-modulated diatomic meta-platform patterned on SiO2 (silicon dioxide) substrate with a gold ground plane provides spin-conserved and –encrypted meta-hologram. The building block is carefully optimized to realize giant circular dichroism along with spin-conservation and wavefront shaping. The three-dimensional building block of the meta-platform is shown in Figure 1(b) with design parameters. Figure 1(c) demonstrates the reflectance parameters of Jones matrix for the symmetry (rotational and mirror) breaking structure of the building block for circularly polarized (CP) incidences. The reflectance parameters shows maximum reflectance for the co-polarized parameter of LHCP incident light and maximum absorption for the co-polarized parameter of RHCP illuminations at 633nm as depicted in Figure 1(c). Hence, the meta-platform yields a giant chirality in the form of circular dichroism.

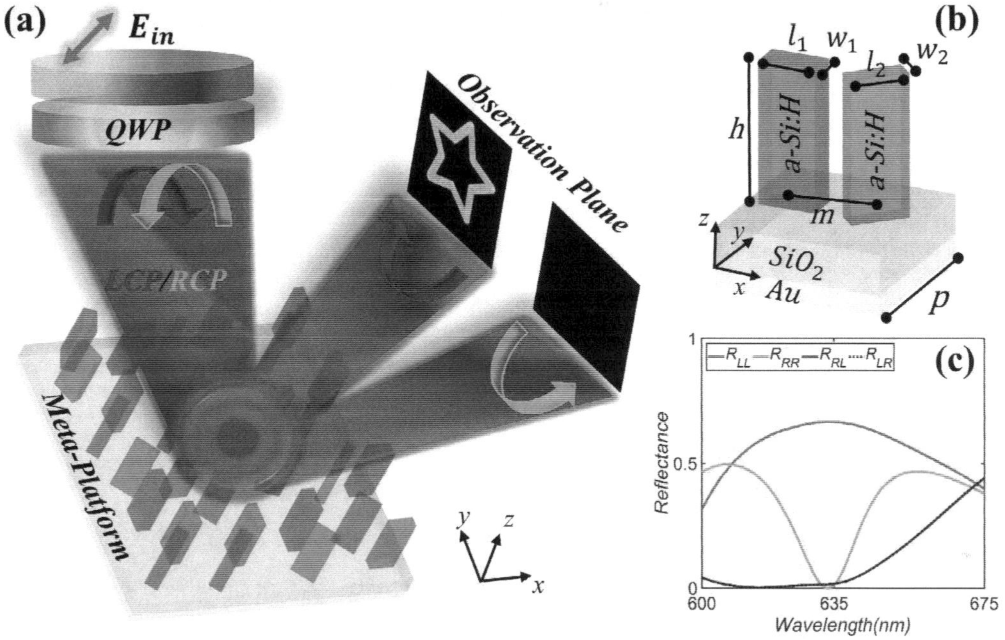

FIGURE 1. (a) Schematic art of the working principle of proposed meta-platform (b) The three-dimensional view of the building block with structural parameters ($w_1 = 80\ nm, l_1 = 145\ nm, w_2 = 60\ nm, l_2 = 150\ nm, m = 215\ nm, p = 430\ nm, h = 400\ nm$) (c) Jones matrix parameters for LHCP and RHCP illuminations.

The full geometric phase coverage of the building block is shown in Figure 2 for CP illuminations. The full (0-2π) phase modulation achieved for LHCP illumination (Figure 2(a)) but it's not the case for RHCP incident light (Figure 2(b)). To elucidate the origin of the giant chiroptical effect, we performed a multipolar decomposition of the fields and local currents excited in the center of the building block in xz-plane. Figure 3 illustrates the normalized multipolar decomposition in the form of ED (electric dipole), MD (magnetic dipole), TD (toroidal dipole), EQ (electric quadrupole) and MQ (magnetic quadrupole) modes for LHCP (Figure 3(a)) and RHCP (Figure 3(b)) illuminations. It is noticeable that TD and MD are the dominating modes for LHCP excitation, while EQ and MD show maximum contribution in the scattering for RHCP illumination at 633nm. The observed difference between the multipolar resonances for LHCP and RHCP illuminations proved the underlying mechanism of novel designed meta-platform, leading to giant circular dichroism.

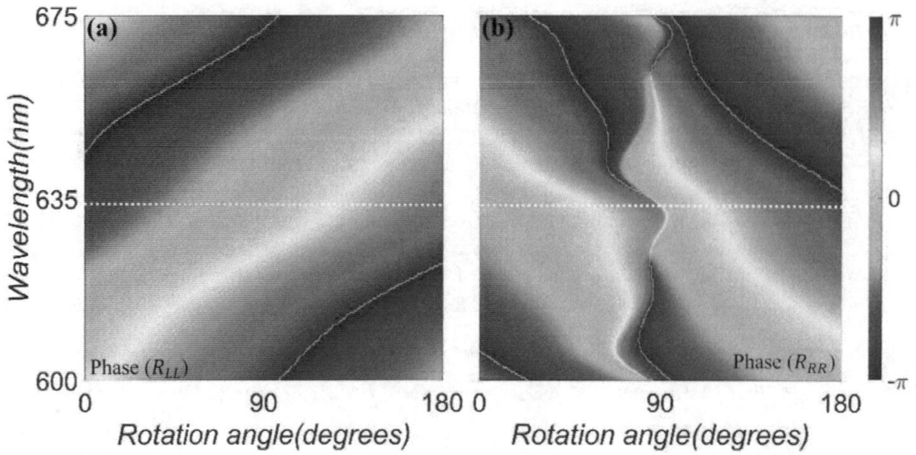

FIGURE 2. Geometric phase modulation for co-polarized reflectance parameter of (a) LHCP illumination (b) RHCP illumination. The white dotted line shows the phase coverage at working wavelength of 633 nm.

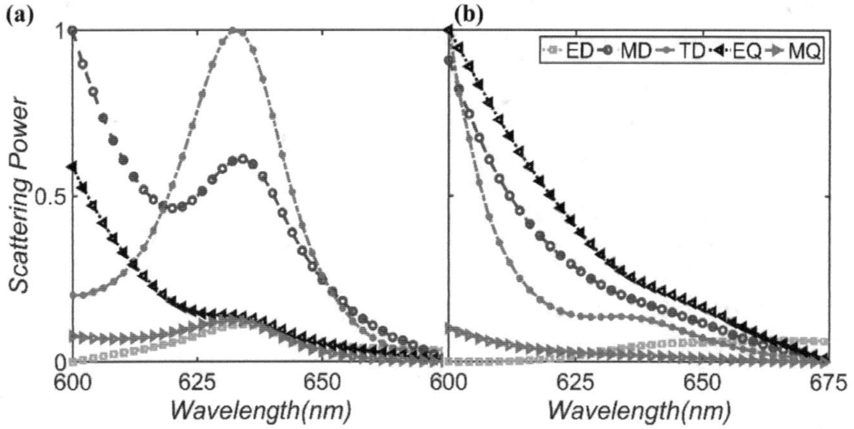

FIGURE 3. Normalized multipole decomposition of scattering power for (a) LHCP incident light (b) RHCP incident light.

To verify our aforementioned methodology, we designed meta-platform consists of a two-dimensional array of optimized building blocks patterned using the specific phase mask to reproduce a spin-conserved and -encrypted meta-hologram. The phase profile of the image ("ITU") is calculated by utilizing the Gerchberg Saxton (GS) algorithm at the focal point of 21.5 μm. The generated phase mask using the GS algorithm encoded on a meta-platform of 100 x 100 elements using FDTD solver. Figure 4(a) depicts the computer generated hologram plotted numerically in MATLAB. Figure 2(b) and 2(c) show the corresponding reproduced meta-hologram results for LHCP and RHCP illuminations, respectively. The reproduced meta-hologram for LHCP illumination shows the polarization-conservation with good image fidelity in reflection while for RHCP it shows polarization-encryption with no visible image information. The total size of the simulated meta-platform is 44 × 44 μm^2. The designed methodology can provide much better results by simulating bigger metasurface.

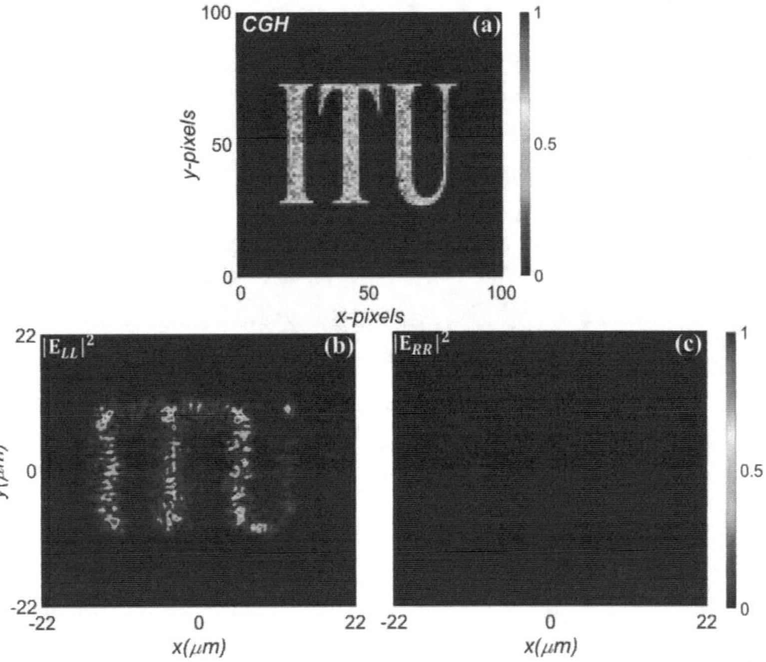

FIGURE 4. Holography as phase dictated phenomenon (a) Computer-generated hologram (CGH) includes 100 × 100 pixels. The simulated spin-conserved and –encrypted hologram for (b) LHCP illumination (c) RHCP illumination.

CONCLUSION

A single-layered all-dielectric diatomic meta-platform is proposed to generate a giant chiroptical effect verified by multipolar-decomposition. As a proof of concept, the designed meta-platform reproduces a spin-conserved and – encrypted meta-hologram. Hence, it opens a new gateway towards many applications in polarimetric imaging, CD spectroscopy, encryption, and anti-counterfeiting.

REFERENCES

1. W. Thomson and B. Kelvin, "Baltimore lectures on molecular dynamics and the wave theory of light," *C.J. Clay Sons, London, UK*, 1904.
2. V. K. Valev, J. J. Baumberg, C. Sibilia, and T. Verbiest, *Adv. Mater.*, 25, 18, pp. 2517–2534 (2013).
3. B. Frank, X. Yin, M. Schäferling, J. Zhao, S. M. Hein, P. V. Braun, and H. Giessen, *ACS Nano*, 7, 7, pp. 6321–6329 (2013).
4. H. S. Khaliq, K. Riaz, M. Zubair, M. A. Ansari, T. Naeem and M. Q. Mehmood, "All-dielectric single-layered achiral structures for simultaneous conversion circular dichroism and wavefront engineering for visible light," *Active Photonic Platforms XII*, vol. 11461, 2020.
5. H. S. Khaliq, K. Riaz, M. Zubair, M. A. Ansari, M. R. Akram, T. Naeem, J. Zhang, W. Zhu, D. Zhang, X. Wang and M. Q. Mehmood, "Highly efficient metamirror with circular dichroism and wavefront engineering," *Metamaterials XII*, vol. 11344, 2020.
6. F. Zhang, M. Pu, X. Li, P. Gao, X. Ma, J. Luo, H. Yu and X. Luo, *Adv. Funct. Mater.*, 27, 1704295 (2017).
7. M. A. Ansari, I. Kim, I. D. Rukhlenko, M. Zubair, S. Yerci, T. Tauqeer, M. Q. Mehmood and J. Rho, *Nanoscale Horizons*, 5, 1, pp. 57–64 (2020).
8. H. S. Khaliq, M. R. Akram, K. Riaz, M. A. Ansari, J. Akbar, J. Zhang, W. Zhu, D. Zhang, X. Wang, M. Zubair and M. Q. Mehmood, Optics Express, 29, 3, pp. 3230-3242 (2021).
9. M. Q. Mehmood, S. Mei, S. Hussain, K. Huang, S. Y. Siew, L. Zhang, T. Zhang, X. Ling, H. Liu, J. Teng, A. Danner, S. Zhang and C-W Qiu, *Adv. Mater.*, 28, 13, pp. 2533–2539 (2016).
10. H. S. Khaliq, I. Kim, J. Kim, D. K. Oh, M. Zubair, K. Riaz, M. Q. Mehmood and J. Rho, *Adv. Optical Mater.*, 2002002 (2021).
11. T. Naeem, A.S. Rana, M. Zubair, T. Tauqeer and M.Q. Mehmood, *Opt. Mater. Express*, *10* (12), 3342 (2020).

Broadband metasurfaces loaded with non-Foster elements

Nikita Kalmykov, Bair Buiantuev and Dmitry Kholodnyak*

Department of Microelectronics & Radio Engineering, Saint Petersburg Electrotechnical University "LETI", 5 Prof. Popov Str., Saint Petersburg, Russia

*email: Dmitry.Kholodnyak@mwlab.spb.ru

Abstract. Metasurfaces have been widely used to design low-profile antennas, thin absorbers, lenses etc. The operational frequency band of a metasurface is rather narrow due to its resonant nature. Loading metasurface unit cells with non-Foster elements allows for remarkable bandwidth extension. In this paper, design of a broadband metasurface to operate as an artificial magnetic conductor is considered. The main issues which influence the bandwidth extension such as implementation of the non-Foster load, minimization of conversion error of a negative impedance converter, and circuit stabilization are addressed.

1. Introduction

Non-Foster negative inductances and capacitances have recently attracted great attention due to the unique frequency dependence of reactance or susceptance that cannot be achieved from passive components and is therefore realized by active circuits such as negative impedance converters (NICs). A NIC converts its load impedance into a negative input impedance by flipping the sign of either load voltage or load current. As a result, the NIC input impedance mimics that of a hypothetical passive negative capacitance or inductance. Non-Foster elements (NFEs) are employed for broadband matching of electrically small antennas [1]-[6], to design broadband squint-free leaky-wave and series-fed array antennas [7]-[9], wideband amplifiers [10]-[12], broadband metamaterials [13]-[15] etc.

Metasurfaces are thin periodic two-dimensional artificial structures exhibiting extraordinary reflection or transmission properties [16]. Metasurface applications include low-profile antennas [17]-[20], lenses [21]-[23], thin absorbers [24]-[26] etc. Due to resonant behavior, metasurfaces operate within narrow frequency band. Loading metasurface unit cells with NFEs allows for remarkable bandwidth extension by compensating parasitics or reducing impedance dispersion. Bandwidth enhancement of metasurfaces loaded with NFEs was demonstrated both in simulation and experiment [14], [27]-[31]. The impact of a non-Foster load on bandwidth enhancement of the artificial magnetic conductor (AMC) metasurfaces consisting of an array of metallic patches on top of a dielectric substrate with a ground plane on the bottom was investigated in [14], [27]-[29]. It was shown that any arbitrary operational bandwidth could be achieved in theory with the aid of NFEs [28]. In practice, the bandwidth is limited due to the imperfections of realistic elements which a NIC circuit consists of.

The bandwidth of the AMC loaded with a combination of a negative capacitor and a negative inductor that were realized by Linvill's circuit NIC was shown to be 30% around the center frequency of 2 GHz, i.e. two-fold wider compared to that of the unloaded metasurface [14]. Different examples of similar non-Foster loaded metasurfaces demonstrate achievement of the operational bandwidth of 56% [29] and 80% [27] while it is limited to 15% in the unloaded case.

In [30], a mushroom-like metasurface loaded with a non-Foster circuit using negative differential resistance characteristic of a graphene-based field-effect transistor was presented. The simulation results revealed an operational bandwidth over 50% at the center frequency of 540 MHz, i.e. enhanced by a factor of 2.4 with respect to the unloaded case.

The concept of using complex non-Foster loads to enhance the bandwidth of a metasurface was also presented in [31]. The metasurface unit cell consists of two orthogonally polarized radiators connected to each other through a NIC. The approach allows for almost perfect compensation resulting in up to 200% bandwidth [31].

In order to design a broadband metasurface loaded with NFEs, implementation of a proper non-Foster load is of crucial importance for obtaining desired performance over a wide frequency band. A single non-Foster reactance or susceptance is frequently used as the load, limiting possibilities for bandwidth enhancement. Moreover, a practical NIC circuit imposes an impedance conversion error due to imperfections of realistic transistors. As a result, the desired negative capacitance or inductance value is realized with some tolerance that limits the metasurface bandwidth as well. The last but not the least, a non-Foster metasurface should be stable for most practical applications although self-oscillating non-Foster metasurfaces can offer new exciting possibilities [31].

In this paper we consider all the three issues: i) implementation of the necessary non-Foster load; ii) minimization of the NIC conversion error; iii) circuit stabilization. We focus on bandwidth extension of an AMC metasurface and demonstrate realization of a non-Foster load with necessary complex frequency dependence of reactance by using a NIC loaded with a circuit consisting of multiple passive components. We illustrate the importance of conversion error minimization. To minimize the NIC conversion error, we employ a design approach based on Linvill's circuit decomposition, a methodology we have recently introduced [32], [33] as opposed to 'blind' CAD optimization that has been commonly used for the NFE design. Besides, we present here for the first time a stabilization technique using high-pass circuits in combination with resistors connected in series.

2. Metasurface design

Fig. 1 shows the well-known mushroom-like structure of AMCs. The unit cell of such a metasurface (Fig. 1-b) consists of a square-shape metal patch and a metallized via hole connecting the center of the patch with a ground plane on the opposite side of the dielectric substrate. The substrate with $\varepsilon_r = 6$ and $h = 2.54$ mm was used. The metasurface designed for the center frequency of 2.45 GHz uses patches of the size $a = 17.6$ mm separated with the spacing $s = 1.6$ mm. The via hole diameter is $d_1 = 1$ mm.

To load the unit cell with a NFE, a lumped port was implemented by using another metallized via hole of the diameter $d_2 = 0.254$ mm going from the patch thorough the dielectric substrate. The ground plane metallization around the extra via hole was partially removed forming the circle-shape spacing of the diameter $d_3 = 2$ mm (Fig. 1-c). The via is positioned 6 mm away from the center of the patch on one of its symmetry axes. Impedance of the lumped port was chosen close to 50 Ohm.

According to results of the full-wave electromagnetic simulations, the metasurface provides full reflection with zero-degree phase at the center frequency (Fig. 1-d). The operational bandwidth measured at ±90° of the reflection coefficient phase does not exceed 7% either with or without the lumped port.

By connecting a proper non-Foster load to the lumped port, the metasurface impedance can be compensated and a broadband operation can be achieved. A necessary impedance which the unit cell should be loaded with (Fig. 2-a), can be calculated from the following equation

$$Z_L = (Z_{22}Z_{in} - \Delta Z)/(Z_{11} - Z_{in}), \quad (1)$$

where Z_{in} is the input impedance of the unit cell with regard to the waveguide port (that should obey the open circuit conditions for the AMC metasurface, i.e. $\mathrm{Re}(Z_{in}) >> 377$ Ohm and $\mathrm{Im}(Z_{in}) = 0$); $\Delta Z = Z_{11}Z_{22} - Z_{12}Z_{21}$ is the determinant of the unit cell impedance matrix; Z_{11}, Z_{12}, Z_{21}, and Z_{22} are the impedance matrix elements obtained by the electromagnetic simulation.

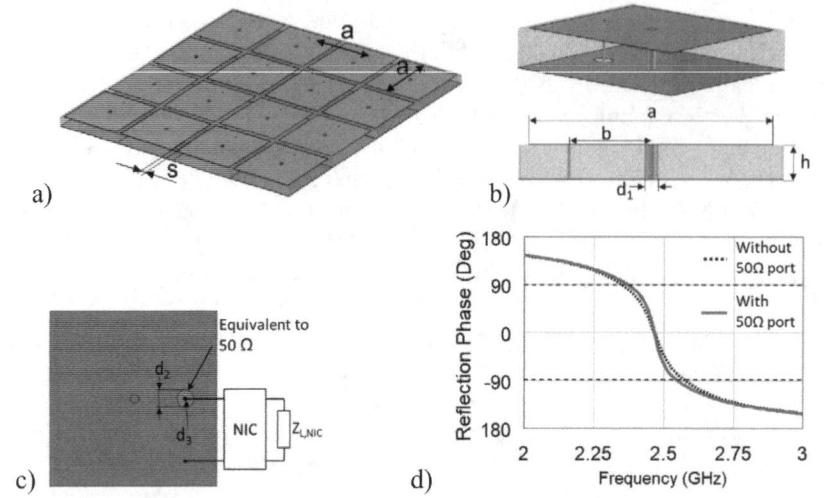

Figure 1. A mushroom-like metasurface structure: (a) bird-eye view; (b) unit cell; (c) lumped port to connect non-Foster load; (d) simulated frequency dependence of the reflection coeficient phase.

The simulated frequency dependences of the unit cell input impedance Z_{in} at the waveguide port are shown in Fig. 2-b. In turn, Fig. 2-c presents the necessary frequency-dependent load impedance Z_L at the lumped port that was calculated by (1) in case of $Z_{in} = 20$ kOhm. Obviously, the frequency dependence of Z_L corresponds to a non-Foster load to be implemented with the aid of a NIC. Hence, the unit cell load impedance is equal to the input impedance of the NIC: $Z_L = Z_{NIC}$. To realize such a non-Foster input impedance, the NIC has to be loaded with the impedance $Z_{L,NIC} = -Z_{NIC} = -Z_L$ (see Fig. 2-c). It is also clear that the target frequency characteristic of $Z_{L,NIC}$ cannot be realized with a single reactance/susceptance.

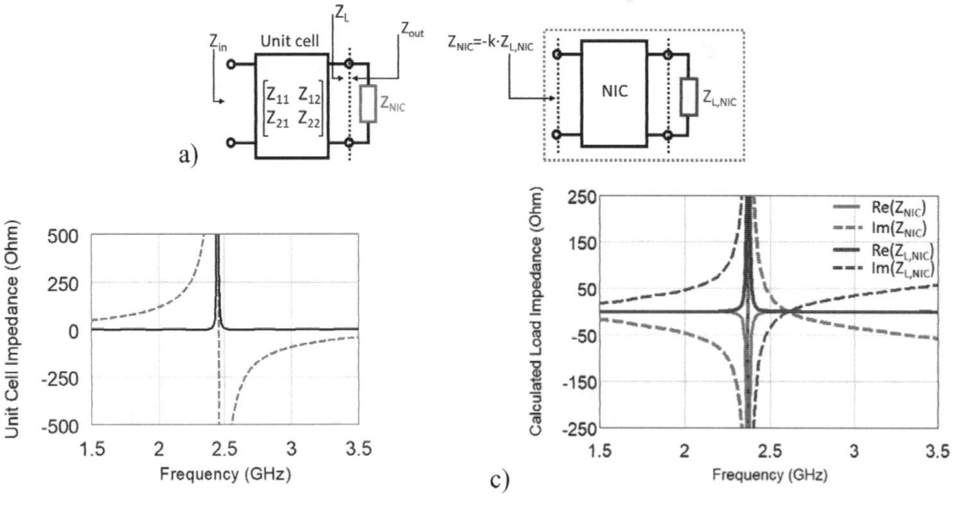

Figure 2. An equivalent diagram of a metasurface unit cell loaded with an impedance $Z_L = Z_{NIC}$ (a), frequency dependence of the real part (solid lines) and imaginary part (dashed lines) of the unit cell input impedance with regard to the waveguide port (b) and the calculated necessary load impedance at the lumped port (c).

To approximate the target frequency characteristics of the NIC load impedance, we employed the *RLC*-circuit shown in Fig. 3-a. The circuit parameters are $C_S = 0.8$ pF, $C_P = 4.2$ pF, $L_S = 7.8$ nH, $L_P = 1.1$ nH, and $R_P = 28$ kOhm. The approximation accuracy of the circuit is illustrated in Fig. 3-b.

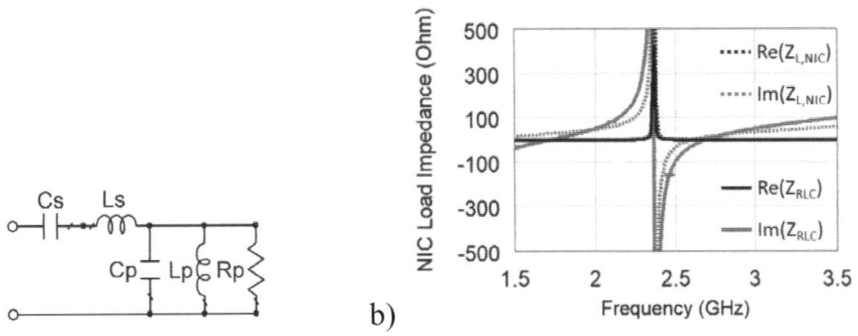

Figure 3. The RLC-circuit to serve as the NIC load (a) and its impedance Z_{RLC} compared to the calculated target impedance $Z_{L,NIC} = -Z_L$ (b).

3. Negative impedance converter design

The idealized NIC converts the load impedance into its negative counterpart with the conversion ratio $k = 1$. Due to imperfections of practical NIC circuits, a conversion error is observed in practice. The conversion error affects the tolerance which can be obtained from a NFE within a certain frequency band. Practical NIC circuits are usually optimized numerically to take the conversion error into account.

In [32], [33], we presented a simple and efficient methodology to design NFEs with predictable characteristics, based on Linvill's NIC employing bipolar junction transistors (BJTs) in the linear mode. The approach uses a decomposition of the NIC into two active and one passive two-port networks, whose frequency characteristics can be analyzed separately and compared with those of their ideal counterparts. The use of the approach simplifies NIC design and allows for conversion error minimization. The approach is much simpler than the direct analysis and optimization of a NIC behavior that is complicated due to sensitive nature of the circuit with positive feedback amplifiers. The obtained results demonstrate a possibility to design broadband NFEs with tolerable variation for GHz range without using any numerical optimization [32], [33].

The same methodology was applied to design the NIC implementing the non-Foster load for the metasurface unit cell. The Linvill's NIC circuit is presented in Fig. 4-a. A SPICE model of the BFR843EL3 SiGe:C heterojunction BJT by Infineon was employed. The DC biasing voltage was chosen as U = 12 V. The biasing resistors R_1 = 700 Ohm and DC-decoupling capacitors C = 10 nF were used.

As shown in Fig. 4-b, the NIC converts the calculated target impedance $Z_{L,NIC} = -Z_L$ with high degree of accuracy within wide frequency band. Some minor conversion error is observed in the real part of the input impedance Z_{NIC} around the resonance frequency and above 3 GHz. For comparison, Fig. 4-c illustrates the NIC input impedance when using the *RLC*-circuit (Fig. 3-a) to approximate the target impedance $Z_{L,NIC}$. The difference between the target and obtained frequency characteristics is caused by both the NIC conversion error and the approximation inaccuracy of $Z_{L,NIC}$.

The bandwidth of the metasurface unit cell loaded by the NIC with the theoretically calculated load impedance $Z_{L,NIC} = -Z_L$ is greatly improved compared to the unloaded case (Fig. 4-d). The fractional bandwidth of the reflection phase reaches $\Delta f = 53\%$ (dashed line in Fig. 4-d) demonstrating an extension by a factor of 7.6. When the real *RLC*-circuit is used as the NIC load, the bandwidth of the non-Foster loaded metasurface is only extended by a factor of 5.4 and equal to $\Delta f = 38\%$ (solid line in Fig. 4-d).

The NIC conversion error can be further reduced by using transistors with higher critical frequency or composite transistors. The ideal NIC load can be better approximated by employing more elements in the RLC-circuit. Besides, a more complex approach to identify necessary NIC load can be used which takes into account the NIC conversion error.

4. Circuit stabilization

NIC circuits are by nature amplifiers with a positive feedback loop and consequently prone to instability. Most practical applications of the non-Foster loaded metasurface require their stable operation.

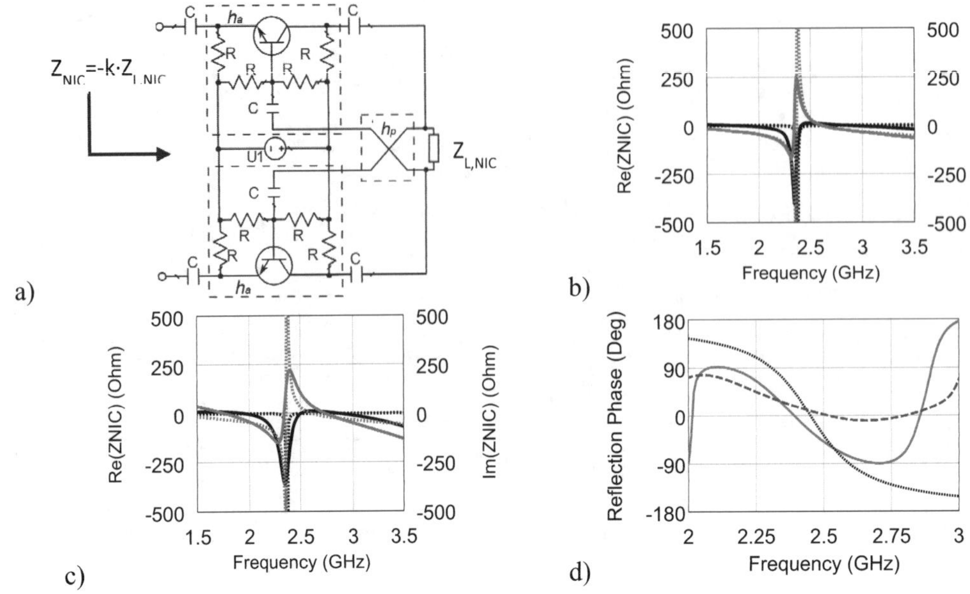

Figure 4. Linvill's NIC circuit (a), its input impedance when loaded by the calculated target impedance $Z_{L,NIC} = -Z_L$ (b) and by the real RLC-circuit (c), phase of the reflection coefficient of the unit cell loaded with the NIC (d).

In order to provide stable operation, we introduce a stabilization network in the NIC circuit as shown in Fig. 5-a. The stabilization network consists of a high-pass filter with the parameters L_{stab} = 4.3 nH and C_{stab} = 15.5 pF as well as the resistor R_{stab} = 670 Ohm. The high-pass filter is aimed at elimination of the DC-pole, which causes instability [34]. In turn, the resistor limits the current gain [33].

A combination of the high-pass filter and the current gain limiting resistor which has not been used before according to the authors' best knowledge, allows obtaining stable operation of the NIC circuit connected to the metasurface unit cell. The stability is proved by either analysis of the normalized determinant function (NDF) [35] behavior (Fig. 5-b) or the simulation of time-domain transient response of the circuit (Fig. 5-c).

Since the stabilization resistors change the NIC input impedance to some extent, the conversion error increases for a NIC with the stabilization networks. This leads to less efficient broadening of the operational bandwidth of the non-Foster loaded metasurface. Nevertheless, the bandwidth enhancement from 7% to 15%, i.e. by a factor of 2.1 was achieved by the results of simulation for the metasurface loaded by the NICs with the stabilization networks (Fig. 5-d). This is the largest increase in the bandwidth that was demonstrated for an AMC metasurface loaded with practical NICs at GHz frequencies.

The NIC performance degradation caused by the presence of stabilization networks can be mitigated by adjusting the NIC circuit parameters. On the other hand, any changes to the NIC circuit might lead to instability that will require to change parameters of the stabilization networks once again. Therefore, an iterative design approach is needed to obtain a stable NIC with suitable frequency characteristics. Furthermore, the approach should also take into account adjustment of the NIC load to compensate for residual conversion error.

5. Conclusion

Due to resonant nature, a metasurface operates within a very limited frequency band. Loading metasurface unit cells with NFEs gives possibilities for broadening the operational bandwidth. Design of a NFE for such purpose deals with implementation of the necessary frequency-dependent non-Foster load, minimization of the NIC con-version error, and circuit stabilization. All of the issues impact a metasurface bandwidth enhancement and were addressed in the paper.

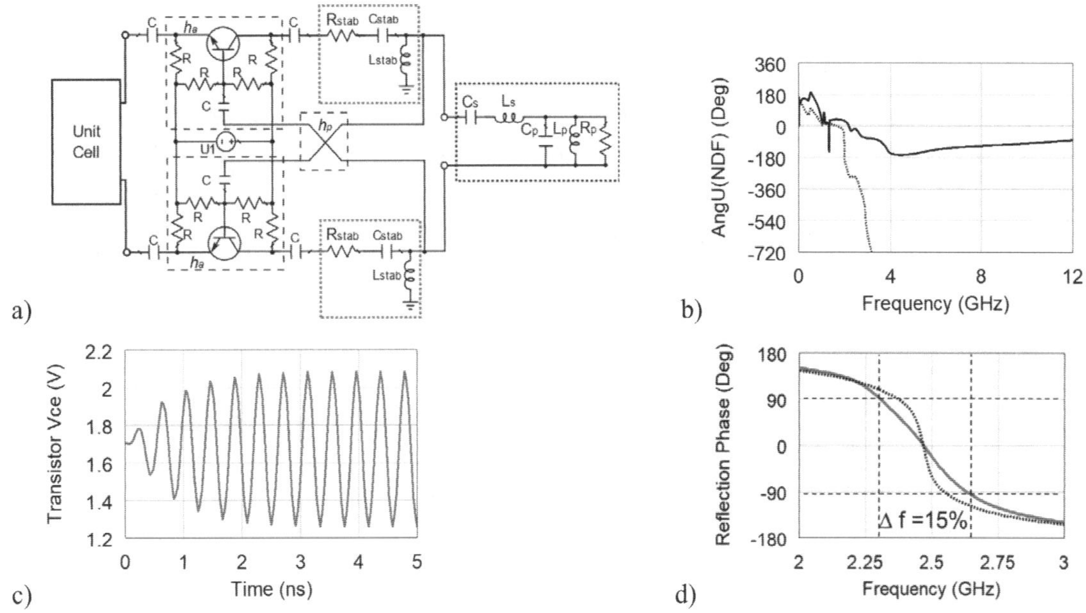

Figure 5. A metasurface unit cell loaded by the NIC with the stabilization networks: (a) circuit; (b) NDF with the stabilization network (solid line) and without the stabilization network (dotted line); (c) collector-emitter voltage of the NIC transistors indicating the entire circuit is stable; (d) reflection coefficient phase of the non-Foster loaded metasurface (solidline) and the unloaded metasurface (dotted line).

Design of a non-Foster loaded AMC metasurface with the bandwidth extended by a factor of 5.4 when using practical Linvill's BJT NIC was presented. The wide bandwidth was achieved because of low NIC conversion error and a complex NIC load used. To minimize the conversion error, the NIC was designed by using methodology based on decomposition of Linvill's circuit.

The stabilization networks consisting of a high-pass filter and a resistor limiting the current gain were presented for the first time. With the aid of such stabilization networks, a stable operation of the non-Foster loaded metasurface was achieved. Presence of the stabilization networks limits possibilities for the metasurface bandwidth enhancement. For a stable metasurface, the bandwidth extension by a factor of 2.1 compared to the unloaded case was demonstrated.

References

[1] Sussman-Fort S E and Rudish R M 2009 *IEEE Trans. Antennas Propag.* **57** 8 2230-41.
[2] Zhu N and Ziolkowski R W 2012 *IEEE Antennas and Wireless Propag. Lett.* **11** 1116-20.
[3] White C R, Colburn J S and Nagele R G 2012 *IEEE Antennas Wireless Propag. Lett.* **11** 584-7.
[4] Ivanov N, Buyantuev B, Turgaliev V and Kholodnyak D 2016 *Proc. of 2016 Loughborough Antennas and Propag. Conf. (UK)* pp 1-4.
[5] Yan Y, Liu C, Wu H and Dong Y 2016 *IEICE Electronics Express* **13** 12 1-10.
[6] Ivanov N, Turgaliev V and Kholodnyak D 2017 *Proc. of 2017 IEEE MTT-S Int. Microwave Symp. (USA)* pp 348-51.
[7] Sievenpiper D F 2011 *IEEE Antennas and Wireless Propag. Lett.* **10** 231-4.
[8] Jacob M M, Long J and Sievenpiper D F 2014 Non-Foster loaded parasitic array for broadband steerable patterns *IEEE Trans. on Antennas and Propag.* **62** 12 6081-90.
[9] Mirzaei H and Eleftheriades G V 2015 *IEEE Trans. on Antennas and Propag.* **63** 5 1997-2010.
[10] Cheng W, Annema A J, Wienk G J M and Nauta B 2012 *Proc. of 2012 IEEE Radio Frequency Integrated Circuits Symp. (Canada)* pp 13-6.
[11] Lee S, Park H, Kim J and Kwon Y 2015 *Proc. of 2015 IEEE MTT-S Int. Microwave Symp. (USA)*

pp 1-4.
[12] Akwuruoha C N and Hu Z 2018 *Proc. of 2018 Int. Conf. on Integrated Circuit Design and Tech. (Italy)* pp 149-52.
[13] Tretyakov S A and Maslovski S I 2007 *IEEE Antennas and Propag. Mag.* **49** 1 37-43.
[14] Saadat S, Adnan M, Mosallaei H and Afshari E 2013 *IEEE Trans. on Antennas and Propag.* **61** 3 1210-8.
[15] Hrabar S 2018 *EPJ Applied Metamaterials* **5** 9 1-12.
[16] Holloway C L, Kuester E F, Gordon J A, O'Hara J, Booth J and Smith D R 2012 *IEEE Antennas and Propag. Mag.* **54** 2 10-35.
[17] Feresidis A P, Goussetis G, Wang S and Vardaxoglou J C 2005 *IEEE Trans. on Antennas and Propag.* **53** 1 209-15.
[18] Foroozesh A and Shafai L 2009 *IEEE Antennas and Wireless Propag. Lett.* **8** 10-3.
[19] Cai T, Wang G, Zhang X and Shi J 2015 *IEEE Antennas and Wireless Propag. Lett.* **14** 1072-6.
[20] Zhai H, Zhang K, Yang S and Feng D 2017 *IEEE Antennas and Wireless Propag. Lett.* **16** 2692-5.
[21] Bosiljevac M, Casaletti M, Caminita F, Sipus Z and Maci S 2012 *IEEE Trans. on Antennas and Propag.* **60** 9 4065-73.
[22] Quevedo-Teruel O, Ebrahimpouri M, and Ng Mou Kehn M 2016 *IEEE Antennas and Wireless Propag. Lett.* **15** 484-7.
[23] Singh A K, Abegaonkar M P and Koul S K 2018 *Proc. of 2018 Int. Symp. on Antennas and Propag. (South Korea)* pp 1-2.
[24] Torabi E S, Fallahi A and Yahaghi A 2017 *IEEE Trans. on Antennas and Propag.* **65** 3 1464-7.
[25] Li A, Kim S, Luo Y, Li Y, Long J and Sievenpiper D F 2017 *IEEE Trans. on Microwave Theory and Tech.* **65** 8 2810-8.
[26] Zhirihin D, Simovski C, Belov P, and S. Glybovski 2017 *IEEE Antennas and Wireless Propag. Lett.* **16** 2626-9.
[27] Gregoire D J, White C R and Colburn J S 2011 *IEEE Antennas and Wireless Propag. Lett.* **10** 1586-9.
[28] Ding Y and Fusco V F 2012 *Microwave and Opt. Technol. Lett.* **54** 2111-14.
[29] Long J and Sievenpiper D F 2016 *IEEE Trans. on Antennas and Propag.* **64** 7 3003-10.
[30] Tian J, Nagarkoti D S, Rajab K Z and Hao Y 2017 *IEEE Antennas and Wireless Propag. Lett.* **16** 2655-8.
[31] Vincelj L, Krois I and Hrabar S 2020 *IEEE Trans. on Antennas and Propag.* **68** 3 1665-79.
[32] Buiantuev B, Vincelj L, Kholodnyak D and Hrabar S 2020 *Proc. of 14th Eur. Conf. on Antennas and Propag. (Denmark)* pp 1-4.
[33] Buiantuev B, Kalmykov N, Kholodnyak D, Vincelj L, Brizić A and Hrabar S 2020 *Proc. of 2020 IEEE Asia-Pacific Microwave Conf. (Hong Kong SAR PR China)* pp 272-4.
[34] Brizic A, Hrabar S, Zanic D and Krois I 2019 *Proc. of Int. Conf. on on Applied Electromagnetics and Communications (Croatia)* pp 1-3.
[35] Jimenez-Martin J L, Gonzalez-Posadas V, Parra-Cerrada A, Blanco-Campo A, Ugarte-Munoz E and Segovia-Vargas D 2012 *Proc. of 6th Eur. Conf. Antennas and Propag. (Prague, Czech Republic)* pp 135-8.

Light-induced electron pairing in two-dimensional systems

O. V. Kibis, M. V. Boev, V. M. Kovalev

Department of Applied and Theoretical Physics, Novosibirsk State Technical University, Karl Marx Avenue 20, Novosibirsk 630073, Russia

E-mail: oleg.kibis@nstu.ru

Abstract. The mechanism of electron pairing induced by a circularly polarized off-resonant electromagnetic field is suggested and examined theoretically for various two-dimensional (2D) nanostructures. Particularly, it is demonstrated that such a pairing can exist in 2D systems containing charge carriers with different effective masses. As a result of the pairing, the optically induced hybrid Bose-Fermy system appears. The elementary excitation in the system are analyzed and the possible Bose-Einstein condensation of the paired electrons and the related light-induced superconductivity are discussed.

1. Introduction

The tuning of the electronic properties of various nanostructures by an off-resonant electromagnetic field based on the Floquet theory of periodically driven quantum systems (the Floquet engineering) became the established research area of the modern condensed matter physics (see, e.g., Refs. [1, 2, 3, 4, 5, 6, 7, 8] and the references therein). Since frequency of the off-resonant field lies far from the optical absorption range, the field cannot be absorbed and only "dresses" electrons (dressing field) changing their physical properties. Among various phenomena induced by the dressing field, the dynamical localization effect [2] should be noted especially. Recently, it was demonstrated that this effect crucially modifies the Coulomb interaction in 2D systems irradiated by a circularly polarized dressing field, resulting in the attractive area in the core of the repulsive Coulomb potential [9]. As a consequence, the quasi-stationary electron states bound at repulsive scatterers appear [10, 11]. Particularly, it follows from this that the irradiation can generate the bound two-electron states (electron pairs) in 2D systems. The present study is dedicated to the theory of such a pairing and related phenomena.

2. Model

Let us consider a 2D system containing charge carriers with different effective masses, where the energy spectrum of the two ground subbands is $\varepsilon_1(\mathbf{k}) = -\Delta_0/2 + \hbar^2k^2/2m_1$ and $\varepsilon_2(\mathbf{k}) = \Delta_0/2+\hbar^2k^2/2m_2$, Δ_0 is the energy distance between the subbands, $\mathbf{k} = (k_x, k_y)$ is the momentum of charge carrier in the 2D plane, and $m_{1,2}$ are the effective masses in the subbands. Physically, such a system can be realized in a quantum well consisting of two layers (1 and 2) filled with a two-dimensional electron gas (2DEG), which are fabricated using different semiconductor materials and isolated from each other by a buffer layer with thickness d (see Fig. 1a) or in

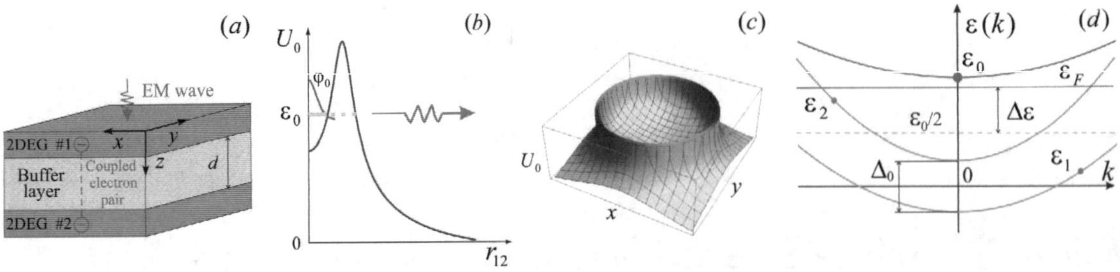

Figure 1. Sketch of the system under consideration: (a) The two-layer quantum well irradiated by a circularly polarized electromagnetic wave; (b) The dressed effective potential of electron-electron interaction, U_0, with the local minimum which confines the electron pair with the wave function φ_0 (the red curve) and the energy ε_0 (the yellow horizontal line). The paired electron state can decay by the tunneling through the potential barrier into the continuum of conduction electron states marked by the green wave arrow; (c) The dressed Coulomb repulsive potential in a hole semiconductor quantum well; (d) The energy spectrum of the subbands of conduction electrons $\varepsilon_{1,2}(k)$ (the two lowest green curves) and the coupled electrons (the upper brown curve) for the Fermi energy ε_F. The large brown circle marks the ground state of coupled electrons with the energy ε_0, whereas the two small green circles mark the states of decoupled electrons with the energies $\varepsilon_{1,2}$ satisfying the condition $\varepsilon_1 + \varepsilon_2 = \varepsilon_0$.

quantum wells based on hole semiconductor materials containing heavy and light holes. In the presence of a circularly polarized electromagnetic wave incident normally to the 2D structure, the Coulomb interaction of two electrons from the subbands $\varepsilon_1(\mathbf{k})$ and $\varepsilon_2(\mathbf{k})$ is described by the Hamiltonian $\hat{\mathcal{H}}_{12}(t) = \hat{\mathcal{H}}_1 + \hat{\mathcal{H}}_2 + U(\mathbf{r}_1 - \mathbf{r}_2)$, where $\hat{\mathcal{H}}_{1,2} = (\hat{\mathbf{p}}_{1,2} - e\mathbf{A}(t)/c)^2/2m_{1,2}$ are the Hamiltonians of free conduction electrons irradiated by the wave, $\hat{\mathbf{p}}_{1,2} = -i\hbar\partial/\partial\mathbf{r}_{1,2}$ are the plane momentum operators of the electrons, $\mathbf{r}_{1,2}$ are the plane radius vectors of the electrons, $\mathbf{A}(t) = (A_x, A_y) = [cE_0/\omega_0](\sin\omega_0 t, \cos\omega_0 t)$ is the vector potential of the wave, E_0 is the electric field amplitude of the wave, ω_0 is the wave frequency, $U(\mathbf{r}_{12})$ is the potential energy of the repulsive Coulomb interaction between the electrons, and $\mathbf{r}_{12} = \mathbf{r}_1 - \mathbf{r}_2$ is the radius vector of relative position of the electrons. If the field frequency ω_0 is high enough and lies far from characteristic resonant frequencies of the 2D structure (the off-resonant dressing field), the time-dependent Hamiltonian $\hat{\mathcal{H}}_{12}(t)$ can be reduced to the effective stationary Hamiltonian,

$$\hat{\mathcal{H}}_0 = \hat{\mathbf{p}}_1^2/2m_1 + \hat{\mathbf{p}}_2^2/2m_2 + U_0(r_{12}), \tag{1}$$

where

$$U_0(r_{12}) = \frac{1}{2\pi}\int_{-\pi}^{\pi} U_{12}(\mathbf{r}_{12} - \mathbf{r}_0(t))\, d(\omega_0 t) \tag{2}$$

is the repulsive Coulomb potential dressed by the circularly polarized field (dressed potential), $\mathbf{r}_0(t) = (-r_0\cos\omega_0 t,\, r_0\sin\omega_0 t)$ is the radius vector describing the classical circular trajectory of a free particle with the charge e and the mass $\overline{m} = m_1m_2/(m_1 - m_2)$ in the circularly polarized field, and $r_0 = |e\overline{m}|E_0/\omega_0^2$ is the radius of the trajectory [9, 10]. In the case of the layered nanostructure pictured in Fig. 1a, the dressed potential (2) can be written explicitly as

$$U_0(r_{12}) = \frac{2e^2}{\pi\sqrt{(r_{12}+r_0)^2 + d^2}}\, K\left(\sqrt{\frac{4r_0r_{12}}{(r_{12}+r_0)^2 + d^2}}\right), \tag{3}$$

whereas the dressed potential (2) for a hole semiconductor quantum well is

$$U_0(r_{12}) = \begin{cases} (2e^2/\pi r_0)K\,(r_{12}/r_0)\,, & r_{12}/r_0 \leq 1 \\ (2e^2/\pi r_{12})K\,(r_0/r_{12})\,, & r_{12}/r_0 \geq 1 \end{cases} \tag{4}$$

and $K(\xi)$ is the complete elliptical integral of the first kind. It follows from Eqs. (3) and (4) that the dressed potential $U_0(r_{12})$ turns into the usual repulsive Coulomb potential $U(r_{12})$ for $m_1 = m_2$, what leads to disappearance of the electron coupling. Thus, the different effective masses, $m_1 \neq m_2$, are crucial for the effect under consideration.

3. Results and discussion

In the following, we will assume the two subbands $\varepsilon_{1,2}(\mathbf{k})$ to be of electronic type. Since the dressed potentials (3) and (4) have the local minimum at $\mathbf{r}_{12} = 0$ (see Figs. 1b and 1c), the solution of the Schrödinger problem with the Hamiltonian (1) results in the two-electron state with the wave function of coupled electrons φ_0 and their energy ε_0 pictured schematically in Fig. 1b. This state of coupled electrons is immersed into the continuum of normal electron states of the subbands $\varepsilon_1(\mathbf{k})$ and $\varepsilon_2(\mathbf{k})$ and separated from them by the potential barrier pictured in Fig. 1b, which confines the wave function $\varphi_0(\mathbf{r}_{12})$ near the local minimum of the dressed potential $U_0(r_{12})$. Therefore, the coupled electron pair is quasi-stationary and can decay by tunneling electrons through this barrier into the normal electron states of the subbands, where the energies of the two decoupled electrons are denoted as ε_1 and ε_2 in Fig. 1c. Assuming the tunneling to be weak, the probability of this tunnel transition per unit area of the 2D system reads

$$W_{\mathbf{q}} = \frac{\hbar\Gamma_0}{m[(\varepsilon_q - \varepsilon_0)^2 + (\Gamma_0/2)^2]}, \tag{5}$$

where Γ_0 is the tunneling-induced broadening of the energy level ε_0, $\varepsilon_q = \hbar^2 q^2/2m$ is the kinetic energy of relative motion of the decoupled electrons, $\mathbf{q}$ is the momentum of the relative motion, and $m = m_1 m_2/(m_1+m_2)$ is the reduced effective mass. Taking the Pauli principle into account, the total probability of the decay of the electron pair in the presence of normal conduction electrons reads $W_0^- = \sum_{\mathbf{q}} W_{\mathbf{q}}(1-f_{\mathbf{k}_1})(1-f_{\mathbf{k}_2})$, where $f_{\mathbf{k}_{1,2}} = 1/[\exp[(\varepsilon_{1,2}(\mathbf{k}_{1,2})-\varepsilon_F)/T+1]$ are the Fermi-Dirac distribution functions for the states of normal electrons with the momenta $\mathbf{k}_1 = \mathbf{q}$ and $\mathbf{k}_2 = -\mathbf{q}$ in the subbands $\varepsilon_{1,2}(\mathbf{k})$. Correspondingly, the total probability of production of the coupled electrons from normal electrons is $W_0^+ = \sum_{\mathbf{q}} W_{\mathbf{q}} f_{\mathbf{k}_1} f_{\mathbf{k}_2}$. As a result, the Fermi gas of normal electrons is instable with respect to the pairing under the condition $W_0^+ > W_0^-$, where

$$W_0^\pm = \sum_{\mathbf{q}} \frac{W_{\mathbf{q}}}{\{\exp[\pm\varepsilon_1\{\mathbf{q}) \mp \varepsilon_F]/T+1\}\{\exp[\pm\varepsilon_2(\mathbf{q}) \mp \varepsilon_F)]/T+1\}}. \tag{6}$$

It follows from Eq. (6) that the Fermi system of normal electrons and the Bose system of paired electrons are in the equilibrium for the zero temperature, $W_0^+ = W_0^-$, if $\varepsilon_F = \varepsilon_0/2$. Assuming the Bose gas of paired electrons to be ideal, the equilibrium density of paired electrons for the temperature $T = 0$ (the density of the Bose-Einstein condensate) is $p_0 = (m_1+m_2)\Delta\varepsilon/2\pi\hbar$, where the energy difference $\Delta\varepsilon = \varepsilon_F - \varepsilon_0/2$ has the physical meaning of the energy gain of the total system per one produced pair, which corresponds to the transition of the system from the nonequilibrium state pictured in Fig. 1c to the equilibrium state with the Fermi energy $\varepsilon_F = \varepsilon_0/2$.

It follows from the aforesaid that the circularly polarized irradiation produces the hybrid Bose-Fermi system consisting of the Fermi gas of normal conduction electrons and the Bose gas of coupled two-electron molecules. Generally, the hybrid Bose-Fermi systems substantially differs from pure Bose systems since the Fermi component changes the interaction between the Bose particles. To find the dispersion of collective modes in such a system, $\Omega_{\mathbf{K}}$, let us apply the conventional condition $\zeta(\Omega, \mathbf{K}) = 0$, where $\mathbf{K}$ is the wave vector of the mode, $\zeta(\Omega, \mathbf{K}) = 1 - g_{\mathbf{K}}\Pi(\Omega, \mathbf{K})$ is the dielectric function, $\Pi(\Omega, \mathbf{K}) = \sum_{\mathbf{q}}[p_{\mathbf{q}} - p_{\mathbf{q}+\mathbf{K}}]/[\hbar\Omega + \varepsilon_0(\mathbf{q}) - \varepsilon_0(\mathbf{q}+\mathbf{K})]$ is the polarization operator, and $g_{\mathbf{K}}$ is the Fourier transform of the interaction between the Bose particles [12]. Since a small number of two-electron molecules ($p_0 r_0^2 \ll 1$) is immersed into the Fermi sea of conduction electrons, their Coulomb interaction is screened by many normal electrons and, therefore, can be described by the Fourier transform $g_{\mathbf{K}} = 8\pi e^2/(K + 2/r_s)$, where $r_s = \zeta_0 \hbar^2/(m_1 + m_2)e^2$ is the effective screening radius assumed to satisfy the condition $r_s/r_0 \gg 1$, and ζ_0 is the static dielectric constant. Substituting the single-pair energy spectrum $\varepsilon_0(\mathbf{K}) = \varepsilon_0 + \hbar^2 K^2/2(m_1 + m_2)$ and the condensate density p_0 into the polarization operator, we arrive at the energy spectrum of the collective mode, $\hbar\Omega_{\mathbf{K}} = \sqrt{2p_0 g_{\mathbf{K}} E_{\mathbf{K}} + E_{\mathbf{K}}^2}$, where $E_{\mathbf{K}} = \hbar^2 K^2/2(m_1 + m_2)$ is the kinetic energy of the two-electron molecule. This collective mode has the sound-like dispersion $\Omega_{\mathbf{K}} = v_s K$ for small wave vectors $K \ll 1/r_s$, where $v_s = \sqrt{p_0 g_0/(m_1 + m_2)}$ is the sound velocity. Therefore, the considered Bose-Einstein condensate is superfluid if the velocity of its flow, v, satisfies the Landau criterion, $v < v_s$.

For applicability of the developed theory based on the dressing field approach, the dressing field frequency, ω_0, must satisfy the two conditions [9, 10, 11]): (i) $\omega_0\tau_e \gg 1$, where τ_e is the mean free time of charge carriers; (ii) the field frequency ω_0 lies far from resonant frequencies corresponding to the optical transitions between different states of the paired charge carriers. These conditions can be met simultaneously in modern quantum wells for the frequencies ω_0 near the border of microwave range.

4. Conclusion

It is shown theoretically that a circularly polarized irradiation of two-dimensional systems containing charge carriers with different effective masses can produce electron pairs. Assuming that the reduced effective masses are $\overline{m}, m \sim 0.1m_e$ and the Fermi energy ε_F is of tens meV, we arrive at the pair radius $r_0 \sim 10$ nm, the ground energy of coupled charge carriers ε_0 around 100 meV and the energy broadening Γ_0 of meV scale for the relatively weak irradiation intensity $I \sim \mathrm{W/cm^2}$ with the frequency $\nu_0 = \omega_0/2\pi = 100\,\mathrm{GHz}$. Correspondingly, the discussed effect can be observable in modern semiconductor quantum wells.

Acknowledgments

The reported study was funded by the Russian Science Foundation (project 20-12-00001).

References

[1] Oka T and Kitamura S 2019 *Annual Review of Condensed Matter Physics* **10** 387
[2] Bukov M, D'Alessio L and Polkovnikov A 2015 *Advances in Physics* **64** 139
[3] Lindner N H, Refael G and Galitski V 2011 *Nature Physics* **7** 490
[4] Rechtsman M C, Zeuner J M, Plotnik Y, Lumer Y, Podolsky D, Dreisow F, Nolte S, Segev M and Szameit A 2013 *Nature* **496** 196
[5] Wang Y H, Steinberg H, Jarillo-Herrero P and Gedik N 2013 *Science* **342** 453
[6] Basov D N, Averitt R D and Hsieh D 2017 *Nature Materials* **16** 1077
[7] McIver J W, Schulte B, Stein F U, Matsuyama T, Jotzu G, Meier G and Cavalleri A 2020 *Nature Physics* **16** 38
[8] Kibis O V, Boev M V, Kovalev V M and Shelykh I A 2020 *Physical Review B* **102** 035301
[9] Kibis O V 2019 *Physical Review B* **99** 235416
[10] Kibis O V, Boev M V and Kovalev V M 2020 *Physical Review B* **102** 075412
[11] Kibis O V, Kolodny S A and Iorsh I V 2021 *Optics Letters* **46** 50
[12] Pitaevskii L and Stringari S 2003 *Bose-Einstein Condensation* (Oxford: Clarendon Press)

Floquet engineering of carbon nanotubes

O. V. Kibis, M. V. Boev, V. M. Kovalev

Department of Applied and Theoretical Physics, Novosibirsk State Technical University, Karl Marx Avenue 20, Novosibirsk 630073, Russia

E-mail: oleg.kibis@nstu.ru

Abstract. The Floquet theory for electrons in carbon nanotubes (CNTs) irradiated by a circularly polarized electromagnetic wave propagating along the CNT axis is developed. It is demonstrated, particularly, that the irradiation opens the gap between the conduction and valence bands of CNTs of metal type and lifts the degeneracy of electron states with mutually opposite angular momenta along the CNT axis. As a consequence, the optically-induced metal-insulator transition and the optical Zeeman effect appear in the CNTs. It follows from the theory that these light-induced phenomena can be observed in the modern experiments.

1. Introduction

The control of electronic parameters of condensed-matter structures by an electromagnetic field (the "Floquet engineering" based on the Floquet theory for periodically driven quantum systems) became the important and established research area during last decades. If the field is off-resonant, it cannot be absorbed by electrons and only "dresses" them (the dressing field), changing the band structure of irradiated solids. Particularly, many works dedicated to the Floquet engineering of electron bands were published recently for various nanostructures, including quantum rings, quantum wells, topological insulators, graphene and related 2D materials [1, 2, 3, 4, 5, 6, 7, 8, 9, 10, 11, 12]. As to such actively studied nanostructures as carbon nanotubes (CNTs) [13], the Floquet theory of them still waits for detailed analysis. The present study is aimed to fill partially this gap in modern physics at the border between the Floquet theory and the theory of CNTs.

2. Model

Let us consider the interaction between electrons in a carbon nanotube (CNT) and a circularly polarized electromagnetic wave propagating along the CNT axis with the frequency ω_0 which differs from the frequencies corresponding to optical transitions in the electron subsystem (off-resonant dressing field). Assuming that the wave length $\lambda = 2\pi c/\omega_0$ much exceeds the CNT length, the energy of electron-field interaction can be written within the dipole approximation as $-\mathbf{DE}$, where $\mathbf{D} = e\mathbf{r}$ is the electric dipole, e is the electron charge, $\mathbf{r}$ is the electron radius vector, and $\mathbf{E}$ is the electric field of the wave. Considering the problem within the conventional quantum-field approach, the classical field, $\mathbf{E}$, should be replaced with the field operator, $\hat{\mathbf{E}} = i\sqrt{2\pi\hbar\omega_0/V}\left(\mathbf{e}_+\hat{a} - \mathbf{e}_-\hat{a}^\dagger\right)$, where $\hat{a}$ and $\hat{a}^\dagger$ are the operators of photon annihilation and creation, respectively, written in the Schrödinger representation (the representation of occupation numbers), $\mathbf{e}_\pm = (\mathbf{e}_x \pm i\mathbf{e}_y)/\sqrt{2}$ are the polarization vectors, $\mathbf{e}_{x,y,z}$ are the unit vectors directed along the x, y, z-axes of the Cartesian coordinate system with the z-axis directed along the CNT axis, and V is the quantization volume. Then the total Hamiltonian of the electron-photon system reads $\hat{\mathcal{H}} = \hat{\mathcal{H}}_f + \hat{\mathcal{H}}_e + \hat{\mathcal{H}}_{\text{int}}$, where $\hat{\mathcal{H}}_f = \hbar\omega_0 \hat{a}^\dagger \hat{a}$ is the field energy operator, $\hat{\mathcal{H}}_e$ is the electron Hamiltonian of CNT, and $\hat{\mathcal{H}}_{\text{int}} = -ie\sqrt{2\pi\hbar\omega_0/V}\mathbf{r}\left(\mathbf{e}_+\hat{a} - \mathbf{e}_-\hat{a}^\dagger\right)$ is the Hamiltonian of electron-field interaction.

In the absence of the irradiation, the electron energy spectrum, $\varepsilon_\mu^{(j)}(k)$, and the electron wave function, $\psi_\mu^{(j)}(k)$, satisfy the conventional Schrödinger equation, $\hat{\mathcal{H}}_e\psi_\mu^{(j)}(k) = \varepsilon_\mu^{(j)}(k)\psi_\mu^{(j)}(k)$, where μ is the quantum number labeling different electron subbands in the CNT, $j = c(v)$ is the index marking electron states in the conduction (valence) band of the CNT, and k is the electron wave vector along the CNT axis [13]. To solve the Schrödinger problem for the irradiated CNT with the Hamiltonian $\hat{\mathcal{H}}$, we will apply the methodology [2, 3] which was developed before for various nanostructures interacting with a quantized electromagnetic field. To find the energy of the interacting electron-photon system, ε, one has to solve the secular equation,

$$\det||\mathcal{H}_{nm} - \varepsilon I|| = 0, \tag{1}$$

where $\mathcal{H}_{nm}$ is the matrix Hamiltonian of the system, and I is the unity matrix. The solution results in the energy $\varepsilon = N_0\hbar\omega_0 + \tilde{\varepsilon}_\mu^{(j)}(k)$, where N_0 is the photon occupation number of the dressing field. Correspondingly, the first term is the dressing field energy, whereas the second term, $\tilde{\varepsilon}_\mu^{(j)}(k)$, should be treated as the electron energy spectrum of CNT modified by the field. In the following, we will assume the field to be classically intensive ($N_0 \gg 1$) and, therefore, the classical field amplitude, $E_0 = \sqrt{4\pi N_0\hbar\omega_0/V}$, can be introduced. Certainly, the energy spectrum of CNT, $\tilde{\varepsilon}_\mu^{(j)}(k)$, for a classically intensive quantized field exactly coincides with the Floquet (quasi)energy which can be found as a solution of the conventional Floquet problem for the CNT driven by a classical field.

3. Results and discussion

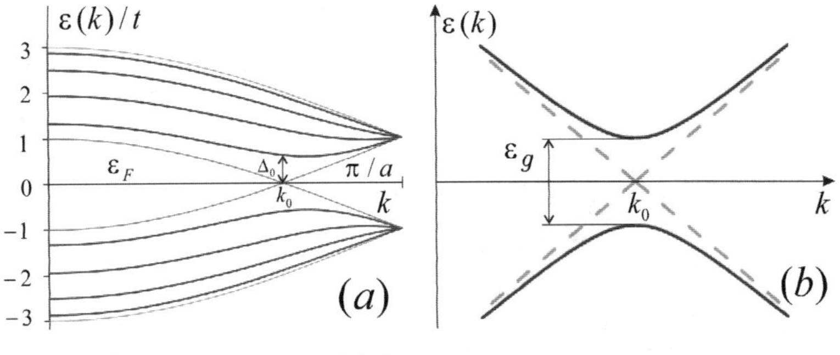

Figure 1. (a) The electron energy spectrum of CNT with the armchair crystal structure (5, 5). The Fermi energy of the electron system is $\varepsilon_F = 0$, the threshold of interband optical absorption is Δ_0 and the Dirac point wave vector is $k_0 = 2\pi/3a$; (b) The structure of electron energy spectrum of the armchair CNT near the Dirac point: in the absence of the dressing field (the dashed lines) and in the presence of the dressing field (the solid lines), where ε_g is the field-induced band gap.

For definiteness, let us apply the developed theory to CNTs with the armchair crystal structure (n, n) and the electron energy spectrum $\varepsilon_\mu^{(\pm)}(k) = \pm t[1 + 4\cos(\mu\pi/n)\cos(ka/2) + 4\cos^2(ka/2)]^{1/2}$, where $t = 3.033$ eV is the energy of interatomic electron interaction in the graphene crystal lattice, $a = 2.46$ Å is the lattice constant of graphene, $\mu = 0, 1, 2, ..., n-1$, and the signs "+" and "−" correspond to electron subbands of the conduction and valence bands, respectively [13]. This energy spectrum is plotted for the $(5, 5)$ CNT within the first Brillouin zone $(-\pi/a < k < \pi/a)$ in Fig. 1a, where the bold blue lines mark the doubly degenerate subbands corresponding to mutually opposite directions of the averaged electron momentum along the CNT axis, the thin lines (red and green) mark the nondegenerate subbands with the zero momentum, and the thin red lines correspond to the band edge subbands with $\mu = 5$. It follows from the plots that the electron energy structure of the CNT is of metal type and the Dirac (linear) electron dispersion takes place near the band edge with the Dirac point at the electron wave vectors $k_0 = \pm 2\pi/3a$. Since the selection rules forbid the optical transitions between the two band edge subbands, the threshold of optical absorption by valence electrons is $\Delta_0 = |\varepsilon^{(\pm)}_{n\pm 1}(k_0)| = 2t\sin(\pi/2n)$ (see Fig. 1a). To avoid the field absorption by the electrons, we will assume the off-resonant condition, $\hbar\omega_0 < \Delta_0$, to be satisfied. To describe the electron energy spectrum of the irradiated CNT near the Dirac point, one can apply the conventional perturbation theory to the matrix Hamiltonian $\mathcal{H}_{nm}$, considering its off-diagonal matrix elements coupling the doubly degenerate electron state at the Dirac point with other states as a perturbation. Then the electron energy spectrum near the Dirac point has the structure pictured schematically in Fig. 1b, where

$$\varepsilon_g = \left(\frac{eR_nE_0}{2}\right)^2 \left[\frac{1}{\Delta_0 - \hbar\omega_0} + \frac{1}{\Delta_0 + \hbar\omega_0}\right] \sin\left(\frac{\pi}{2n}\right) \tag{2}$$

is the field-induced gap between the conduction and valence band and $R_n = \sqrt{3}na/2\pi$ is the CNT radius. Thus, the irradiation turns the metallic CNT into insulator with the band gap (2) which depends on both the field amplitude E_0 and the field frequency ω_0. It should be noted that the field-induced band gap in graphene [2] is linear in the electric field amplitude, E_0, whereas the gap (2) in CNTs is quadratic in the amplitude. To avoid the destructive dephasing influence of scattering processes on the gap (2), the field frequency, ω_0, must meet the condition $\omega_0\tau \gg 1$, where τ is the mean free time of charge carriers in CNTs. This condition is of general character for the Floquet theory of various condensed-matter structures [2, 11] and can be satisfied for the high field frequencies starting approximately from the upper microwave range limit.

Besides the metal-insulator transition originated from the irradiation, it should be noted also the irradiation-induced splitting of the doubly generate subbands marked by the bold blue lines in Fig. 1a, which is also quadratic in the field amplitude E_0 and can be found from the numerical solving of the secular equation (1). Thus, the irradiation lifts the degeneracy of CNT electron states with respect to its angular momentum. This means, particularly, that a circularly polarized electromagnetic wave propagating along the CNT axis acts similarly to a stationary magnetic field directed along the same axis since the magnetic field also induces the gap at the Dirac point [14] and lifts the angular momentum degeneracy (the Zeeman effect). Physically, this similarity is a consequence of the broken time-reversal symmetry of the electron system, which takes place due a circularly polarized electromagnetic field as well as a stationary magnetic one. Indeed, the circularly polarized field is non-invariant with respect to the time reversal since it turns clockwise polarized photons into counterclockwise polarized ones and vice versa. Since the irradiation-induced effects discussed above follow from the broken time-reversal symmetry of the electron-photon system, they take place in CNTs with any crystal structure as well. However, CNTs with the crystal structure devoid of an inversion center (the chiral CNTs) [13] should be noted specially. It follows from the well-known Kramers theorem that the symmetric dependence of the electron energy on the electron wave vector in solids, $\varepsilon(\mathbf{k}) = \varepsilon(-\mathbf{k})$, is the direct consequence of one of two symmetries: The inversion symmetry of crystal structure and the time-reversal symmetry (see, e.g., Ref. [15]). If the both symmetries are broken, the asymmetrical energy spectrum of electrons, $\varepsilon(\mathbf{k}) \neq \varepsilon(-\mathbf{k})$, appears. Particularly, such an asymmetrical spectrum takes place in various nanostructures without an inversion center exposed to a stationary magnetic field, including asymmetric semiconductor quantum wells in the presence of an in-plane magnetic field [16], magnetic edge states in two-dimensional electron systems [17] and chiral CNTs subjected to a magnetic field directed along the CNT axis [18, 19, 20]. As a consequence of the asymmetric electron dispersion, unusual photovoltaic effect [21, 22] and the anomalous thermal effects [16, 17, 19] induced by a magnetic field appear. Since a circularly polarized electromagnetic field acts similarly to a stationary magnetic field, the corresponding optically-induced effects are expected in the irradiated chiral CNTs as well.

4. Conclusion

It is shown that an irradiation opens band gap between conduction and valence bands and lifts the degeneracy of electron states with mutually opposite angular momenta in armchair carbon nanotubes. For the field amplitude $E_0 \sim 10^6$ V/m and the resonance difference $\Delta_0 - \hbar\omega \sim$ meV, the optically induced band gap in armchair CNTs with the radius $R_n \sim 10$ nm is of meV scale and, therefore, can be detected in the modern experiments with using the angle-resolved photoemission spectroscopy (APRES).

Acknowledgments

The reported study was funded by the Russian Foundation for Basic Research (project 20-02-00084). We thank the Russian Ministry of Science and Higher Education (project FSUN-2020-0004).

References

[1] Lindner N H, Refael G and Galitski V 2011 *Nature Physics* **7**, 490
[2] Kibis O V 2010 *Physical Review B* **81** 165433
[3] Kibis OV 2011 *Physical Review Letters* **107** 106802
[4] Rechtsman M C, Zeuner J M, Plotnik Y, Lumer Y, Podolsky D, Dreisow F, Nolte S, Segev M and Szameit A 2013 *Nature* **496** 196
[5] Wang Y H, Steinberg H, Jarillo-Herrero P and Gedik N 2013 *Science* **342** 453
[6] Glazov M M and Ganichev S D 2014 *Physics Reports* **535** 101
[7] Sentef M A, Claassen M, Kemper A F, Moritz B, Oka T, Freericks J K and Devereaux T P 2015 *Nature Communications* **6** 7047
[8] Sie E J, McIver J W, Lee Y H, Fu L, Kong J and Gedik N 2015 *Nature Materials* **14** 290
[9] Basov D N, Averitt R D and Hsieh D 2017 *Nature Materials* **16** 1077
[10] Kibis O V, Dini K, Iorsh I V and Shelykh I A 2017 *Physical Review B* **95** 125401
[11] Kibis O V, Boev M V, Kovalev V M and Shelykh I A 2020 *Physical Review B* **102** 035301
[12] McIver J W, Schulte B, Stein F U, Matsuyama T, Jotzu G, Meier G and Cavalleri A 2020 *Nature Physics* **16** 38
[13] Saito R, Dresselhaus G and Dresselhaus M S 1998 *Physical Properties of Carbon Nanotubes* (London: Imperial College Press)
[14] Portnoi M E, Kibis O V and Rosenau da Costa M 2008 *Superlattices and Microstructures* **43** 399
[15] Callaway J 1964 *Energy Band Theory* (New York: Academic Press)
[16] Kibis O V 1999 *JETP* **88** 527
[17] Lawton D, Nogaret A, Makarenko M V, Kibis O V, Bending S J and Henini M 2002 *Physica E* **13** 699
[18] Romanov D A and Kibis O V 1993 *Physics Letters A* **178** 335
[19] Kibis O V 2001 *Physics of the Solid State* **43** 2336
[20] Kibis O V 2002 *Physica E* **12** 741
[21] Tarasenko S A 2008 *Physical Review B* **77** 085328
[22] Tarasenko S A 2011 *Physical Review B* **83** 035313

Electron-energy-loss spectroscopy and cathodoluminescence for particles inside substrate

Alexander A Kichigin[1,2,3] **and Maxim A Yurkin**[1,2]

[1] Voevodsky Institute of Chemical Kinetics and Combustion, SB RAS, Institutskaya 3, 630090 Novosibirsk, Russia
[2] Novosibirsk State University, Pirogova 2, 630090 Novosibirsk, Russia
[3] Corresponding author: alkichigin@gmail.com

Abstract. To simulate the interaction of a nanoparticle with an electron beam, we previously developed a theoretical description for the general case of a particle fully embedded in an infinite arbitrary host medium. The theory is based on the volume-integral variant of frequency-domain Maxwell's equations and, therefore, is naturally applicable in the discrete-dipole approximation. The fully-embedded approximation allows fast numerical simulations of the experiments for particles inside a substrate since the host medium discretization is not needed. In this work, we study how applicable the fully-embedded approach is for realistic scenarios with relatively thin substrates. In particular, we performed test simulations for a silver sphere both inside an infinite host medium and inside a finite box or sphere. For the host medium, we considered two non-absorbing cases (the denser one causes Cherenkov radiation), as well as an absorbing case. The peak positions in the obtained spectra approximately agree between substrates a few times thicker than the sphere and the infinite one. However, a much thicker substrate (of the order of μm) would be required to have a qualitative agreement for absolute peak amplitudes. The developed algorithm is implemented in the open-source code ADDA, allowing one to rigorously and efficiently simulate electron-energy-loss spectroscopy and cathodoluminescence by particles of arbitrary shape and internal structure embedded into any homogeneous host medium.

1. Introduction

Electron microscopy is an advanced experimental technique used particularly for studying the optical properties of plasmonic nanoparticles. Optical methods can only excite "bright" plasmonic modes, and the diffraction limit confines their spatial resolution. By contrast, electron microscopy excites both "bright" and "dark" plasmonic modes and has a spatial resolution less than 1 nm and energy resolution less than 0.2 eV [1]. In the experiments, relativistic electrons interacting with a particle lose energy (electron-energy-loss spectroscopy – EELS) and the particle emits light (cathodoluminescence – CL). For each position of the electron beam concerning the particle, EELS and CL spectra are obtained. To accurately interpret the data, a computer simulation of the experiment is required.

The discrete dipole approximation (DDA) is a numerically exact method for simulating the interaction of the electro-magnetic field with particles of arbitrary shape and internal structure [2]. Previous implementations of EELS and CL in the DDA [3–5] are suitable only for interaction in a vacuum when only a particle and an electron are present. However, in actual experiments, the particle is commonly placed onto (or inside) a substrate to resist gravitation and collisions with fast electrons. We previously derived a theory [6] for the general case of an arbitrary infinite host medium (refractive index m_h) even absorbing one or the one with the speed of electron exceeding that of light (Cherenkov radiation case). Moreover, this theory has been implemented in the open-source software ADDA [7].

This theory and simulation method are naturally applicable to particles inside a substrate (if the latter is large enough), but this has never been discussed in the literature except for a few cases of the most trivial (non-Cherenkov) host medium [8,9]. This work aims to study the effect of the host medium on the EELS and CL spectra. We compare the simulations for a particle fully-embedded into an infinite host medium to the same particle inside a finite box (or a sphere). The particle inside a finite medium is sometimes used as an approximation for the infinite host medium. However, it is computationally inefficient since the finite chunk of the host medium needs to be explicitly discretized. Nevertheless, in other applications, the host medium is limited (at least in one dimension). One would like to know whether the infinite-host-medium simulation would be accurate enough for this case.

2. Substrate simulation methods

In the comparisons, we use an electron with the 120 keV energy (the corresponding speed is $0.59c$, where c is the speed of light in vacuum) and a silver 15-nm-radius sphere as a test particle. We apply three approaches to account for the medium: *fully-embedded* (new theory), a sphere in a box ("*spherebox*" shape in ADDA), and a coated sphere ("*coated*" shape in ADDA) (Figure 1).

In the fully-embedded case (Figure 1, a), the sphere is discretized with the 30x30x30 dipoles grid. In both *spherebox* (Figure 1, b) and *coated* (Figure 1, c) cases, the resulting particle is twice the size of the silver sphere and is discretized with a 60x60x60 dipoles grid. The latter discretization means that the silver sphere itself is discretized with the same 30x30x30 dipoles grid as in the embedded case, while additional dipoles are used to discretize the surrounding medium. All simulations are performed using ADDA, optical properties for silver are taken from [10].

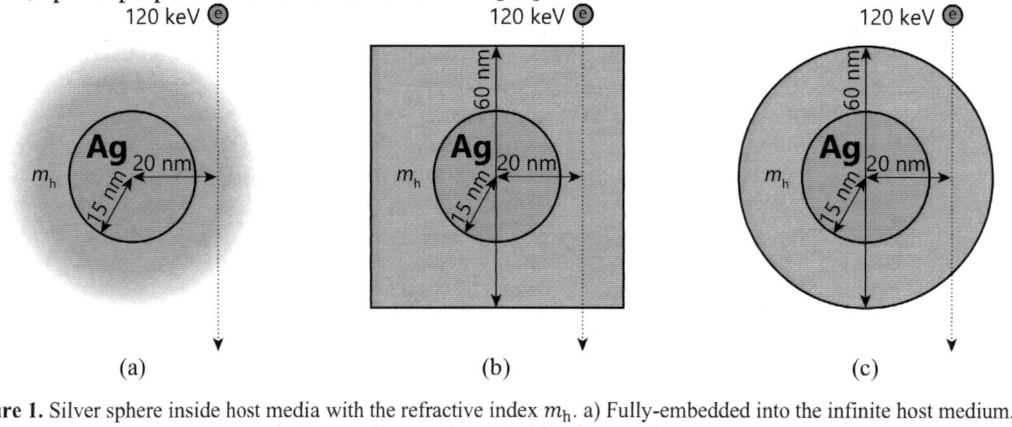

Figure 1. Silver sphere inside host media with the refractive index m_h. a) Fully-embedded into the infinite host medium. b) Inside a larger cube. c) Inside a larger sphere.

3. Results and discussion

First, we tested the accuracy of ADDA by comparing simulation results to the reference Lorenz-Mie [11] solution for the sphere in a vacuum ($m_h = 1$) (Figure 2). Simulated spectra were close to the exact solution; the error of the peak position (3.5 eV) was only 0.05 eV. Further improvement could be obtained by increasing the discretization level or using Richardson extrapolation [12].

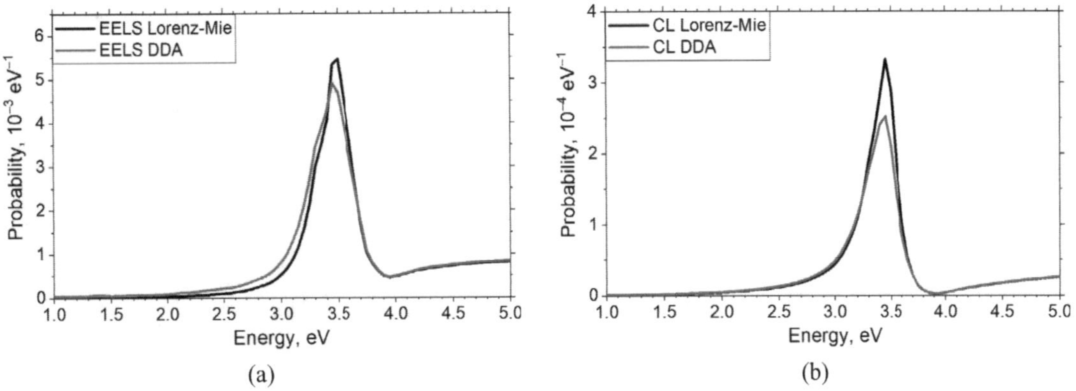

Figure 2. EELS (a) and CL (b) spectra in vacuum ($m_h = 1$) simulated with the DDA and the Lorenz-Mie theory.

EELS and CL spectra are shown for the particle in a non-absorbing host medium ($m_h = 1.5$) (Figure 3) when the electron's speed is lower than the speed of light in the medium ($0.67c$). All three peaks were shifted to the lower energy (2.9 eV). Additionally, the peak in the infinite host medium had a much smaller amplitude.

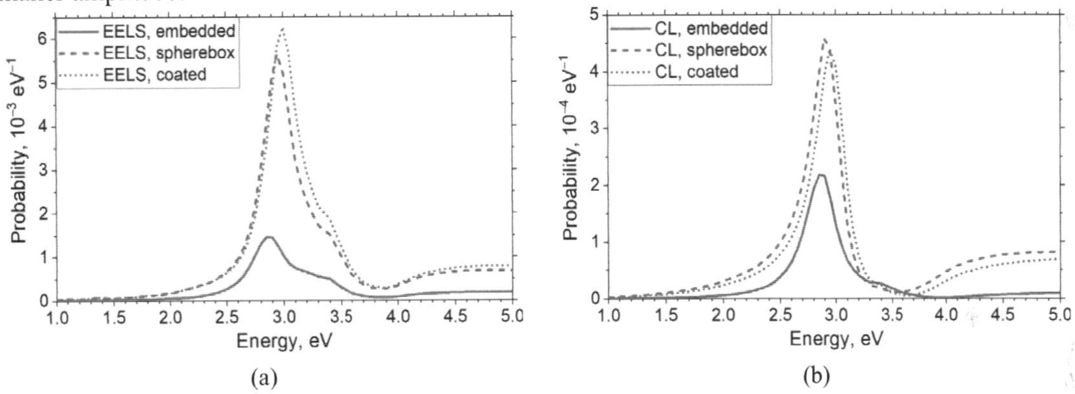

Figure 3. EELS (a) and CL (b) spectra for the case of $m_h = 1.5$, simulated with the DDA. Geometries are shown in Figure 1.

EELS and CL spectra are shown (Figure 4) for the particle in the host medium with $m_h = 2$ leading to the Cherenkov radiation since the electron's speed was greater than the speed of light in the medium ($0.5c$). Compared to the case of $m_h = 1.5$, the peaks were shifted to even lower energy (2.4 eV), but their positions remained similar for all geometries. The peak amplitude for the embedded case was further reduced, while it even increased for finite-medium. The latter can be due to the Cherenkov radiation, which is effectively accounted for in these geometries but should be added separately in the embedded case. While there is a simple Frank–Tamm formula for Cherenkov losses in a sufficiently large chunk of host medium, the studied finite systems are too small for the formula's applicability. We expect the formula to be applicable at a box size of one or several μm, depending on particular loss energy (the size should be at least several corresponding wavelengths). However, the peak position is determined by the local environment of the silver sphere and is expected to converge to the infinite-medium result for box sizes only several times larger than the sphere diameter. We will present more simulations at the conference to support this hypothesis.

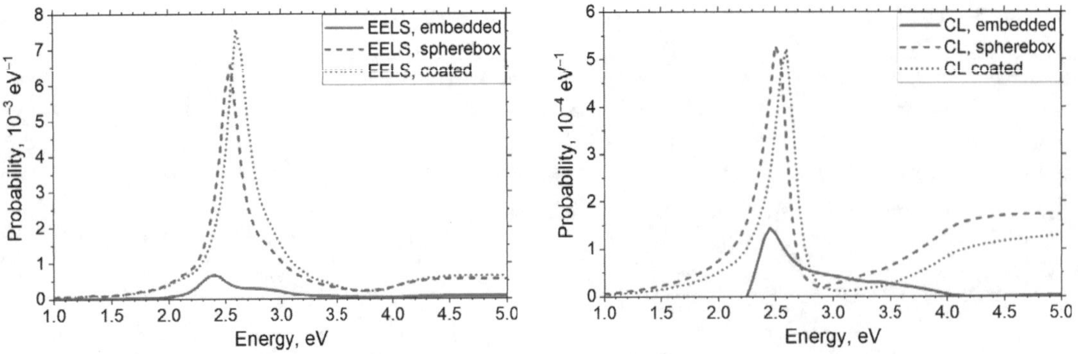

(a) (b)

Figure 4. Same as Figure 3, but for $m_h = 2$ (Cherenkov case).

For the absorbing host medium, the proper way to define CL remains an open question, so we only show the results for EELS in Figure 5. The introduced absorption did not significantly change the peak positions, determined by the real part of the refractive index (3.5 and 2.9 eV, respectively). The differences in peak amplitude result partly from the intrinsic absorption in the host medium, unaccounted for in the embedded (similar to the Cherenkov case).

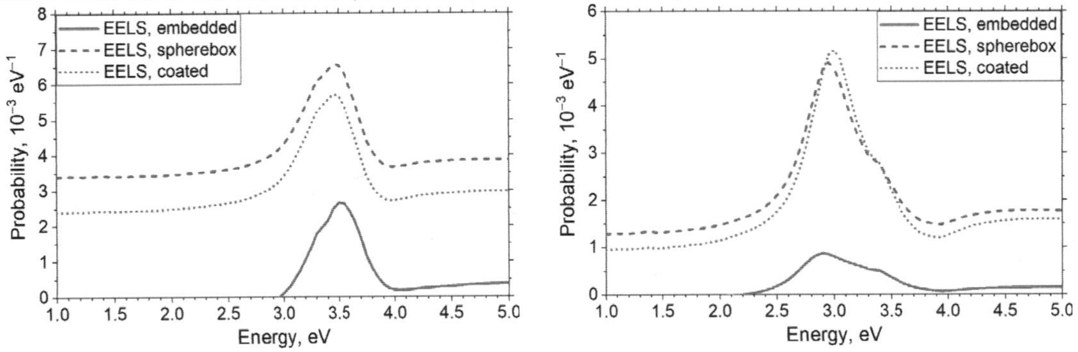

(a) (b)

Figure 5. EELS spectra in absorbing host medium, (a) $m_h = 1 + 0.1i$ and (b) $m_h = 1.5 + 0.1i$ simulated with the DDA. Geometries are shown in Figure 1.

4. Conclusion

We introduced the "embedded into the host medium" approach for simulating EELS and CL in the DDA. This approach allows more efficient calculations for EELS and CL experiments for particles inside substrate than the old-fashioned way of discretizing the surrounding medium. Test simulations for a silver sphere show that the peak positions in the obtained spectra approximately reproduce the experimental for substrates several times thicker than the sphere.

A much thicker substrate (at least 1 μm) is required to have a qualitative agreement for absolute peak amplitudes. The latter would also require manual accounting for losses in the host medium if it is absorbing or sufficiently dense to cause Cherenkov radiation. These new capabilities for simulating EELS and CL are available at https://github.com/alkichigin/adda and will be implemented in the next official ADDA release.

Acknowledgments

This work is supported by the Russian Science Foundation (Grant No. 18-12-00052).

References

[1] García de Abajo F J 2010 Optical excitations in electron microscopy *Rev. Mod. Phys.* **82** 209–75

[2] Yurkin M A and Hoekstra A G 2007 The discrete dipole approximation: an overview and recent developments *J. Quant. Spectrosc. Radiat. Transfer* **106** 558–89

[3] Geuquet N and Henrard L 2010 EELS and optical response of a noble metal nanoparticle in the frame of a discrete dipole approximation *Ultramicroscopy* **110** 1075–80

[4] Bigelow N W, Vaschillo A, Iberi V, Camden J P and Masiello D J 2012 Characterization of the electron- and photon-driven plasmonic excitations of metal nanorods *ACS Nano* **6** 7497–504

[5] Bigelow N W, Vaschillo A, Camden J P and Masiello D J 2013 Signatures of Fano interferences in the electron energy loss spectroscopy and cathodoluminescence of symmetry-broken nanorod dimers *ACS Nano* **7** 4511–9

[6] Kichigin A A and Yurkin M A 2020 Electron energy loss spectroscopy in the framework of the discrete dipole approximation *AIP Conf. Proc.* V International Conference on Metamaterials and Nanophotonics (METANANO 2020) vol 2300 (Online: ITMO University) p 020053

[7] Yurkin M A and Hoekstra A G 2011 The discrete-dipole-approximation code ADDA: capabilities and known limitations *J. Quant. Spectrosc. Radiat. Transfer* **112** 2234–47

[8] Kobylko M, Coulon P-E, Slablab A, Fafin A, Cardin J, Dufour C, Losquin A, Kociak M, Monnet I, Mailly D, Lafosse X, Ulysse C, Garcia-Caurel E and Rizza G 2018 Localized Plasmonic Resonances of Prolate Nanoparticles in a Symmetric Environment: Experimental Verification of the Accuracy of Numerical and Analytical Models *Phys. Rev. Applied* **9** 064038

[9] Sakaguchi N, Matsumoto S, Kunisada Y and Ueda M 2019 Interaction of Localized Surface Plasmons of a Silver Nanosphere Dimer Embedded in a Uniform Medium: Scanning Transmission Electron Microscopy Electron Energy-Loss Spectroscopy and Discrete Dipole Approximation Simulation *J. Phys. Chem. C* **123** 6735–44

[10] Palik E D 1985 *Handbook of Optical Constants of Solids* (Academic Press)

[11] García de Abajo F J and Howie A 1998 Relativistic Electron Energy Loss and Electron-Induced Photon Emission in Inhomogeneous Dielectrics *Phys. Rev. Lett.* **80** 5180–3

[12] Yurkin M A, Maltsev V P and Hoekstra A G 2006 Convergence of the discrete dipole approximation. II. An extrapolation technique to increase the accuracy *J. Opt. Soc. Am. A* **23** 2592–601

Nonplanar Tiled 1-bit Transmitarray with Phase Error Compensation

V V Kirillov[1] **and P A Turalchuk**[2]

Department of Microelectronics and Radio Engineering
Saint-Petersburg Electrotechnical University "LETI"
197376, St. Petersburg, Russia

[1]vvkirillov@etu.ru, [2]paturalchuk@etu.ru

Abstract. A phase compensation method for 1-bit phase quantized transmitarray is discussed. Using the tiled architecture of the transmitarray, the position of each unit cell is changed along the optical axis of the transmitarray. The spatial displacement allows changing the resulting phase distribution along the transmitarray aperture. Analytical calculations were performed to demonstrate the performance of the transmitarray radiation pattern.

1. Introduction

High-gain antennas is of practical interest for fifth-generation (5G) new radio communication systems that employ the use of both sub-6 GHz and beyond 24 GHz spectrum regions. Typically, the use of planar arrays with arbitrary control of the propagated wavefronts, and, in general, intelligent reflecting and transmitting surfaces, can be realized in conventional planar multi-layer technology using surface-mount RF components, e.g., switches and varactors allows providing an advanced functionality for communication and sensing systems. The 1-bit transmitarrays or reflectarrays are a common solution due to the simplicity of the design and fabrication. A sub-6 GHz prototype of the 1-bit tiled transmitarray were reported in [1, 2], whereby the required phase distribution across the array aperture built from standalone unit cells manufactured individually and assembled in the required pattern using a rectangular latticed plastic frame. The tiled architecture can be a viable solution for fast prototyping and teaching experiments, without significant performance deterioration, as compared with a similar single-panel transmitarray. Moreover, the possibility of replacing and adding in-dividual elements in the tiled array makes it both repairable and adjustable for a specific focal distance and feed type. Unfortunately, the effect of 1-bit phase quantization affect the antenna radiation performance. The phase quantization results in a gain reduction up to 4 dB, higher sidelobes, and a noticeable beam squint, [3]. At the same time, the effect of the phase quantization could be reduced by introducing some phase error compensation to each unit-cells. The concept proposed below is based on the theory of phase-compensated nonplanar antenna arrays, [4] and [5], as well as bears some resemblance to design of active Fresnel reflectors, [6], and adaptive millimeter-wave telescope optics, [7].

In this paper an approach based on analytical model optimization of the individual unit-cells displacement relative to the initial tiled planar array is considered. As a result of optimization of each unit-cells displacement the nonplanar tiled array provides efficient phase error compensating and more than 1 dB gain improvement.

2. Nonplanar transmitarray aproach

Essentially, phase compensation can be implemented by displacing elements of the tiled transmitarray [2] with respect to their initial positions, resulting in a nonplanar transmitarray, Fig. 1. In a basic analysis of nonplanar antenna arrays, [3], it was noticed that the element orientation, i.e., whether the element beampatterns are normal to the conformal surface of the array or all point in the boresight direction of the array, can have some effect on the sidelobes, depending on directivity of the elements. In respect to the tiled transmitarray, tile tilting seems unfeasible and it would complicate the phase-compensation model, so that in the following analysis we employ longitudinal shift of the tiles along the optical axis of the transmitarray, Fig. 1.

Considering an M x N element tiled transmitarray, Fig. 1, illuminated by a focal plane patch antenna at a distance d_f, a small longitudinal shift $d_{nm} << d_f$ along the z-axis of tile (n, m) can be represented by a corresponding phase-shift, $\delta_{mn}(\theta_0)$, in given by:

$$\delta_{mn}(\theta_0) = k\left(\cos(\theta_0) - \frac{d_f}{\sqrt{d_f^2+x_{mn}^2+y_{mn}^2}}\right)d_{mn} \quad (1)$$

The element shifts d_{nm} in (1) can be calculated for a given steering angle θ_0. An example of application of this analysis to a 10-element 1-bit linear transmitarray fed by a focal source emanating cylindrical wave is shown in Fig. 2.

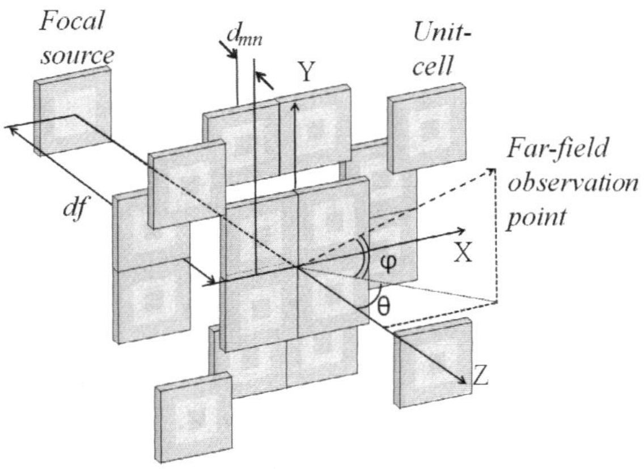

Fig. 1 Planar tiled transmitarray with shifted elements providing correction of the phase quantization error.

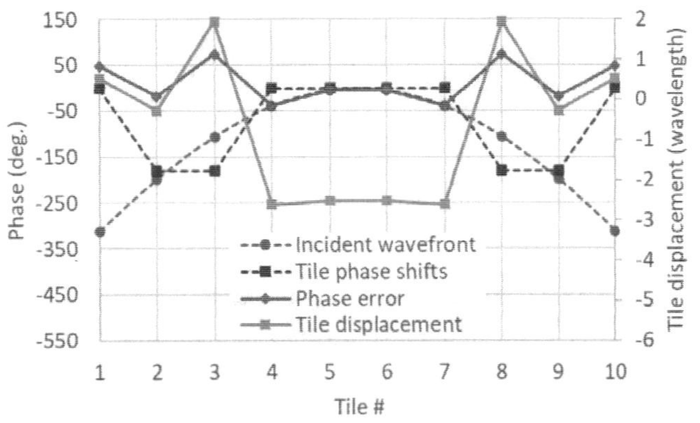

Fig. 2 Phase error and the displacement of unit cells in a linear 10-element 1-bit transmitarray emitting at the boresight and feed by an isotropic source located at 2.5λ focal distance.

As a result of calculation, in the case of unit cells close to the geometric center of the transmitarray, the shifts required for phase error compensation exceed the focal distance. Such large element shifts will not give a tangible effect, since the electromagnetic wave will flow into the space freed up between the unit cells. Also, some of the cells will shade others from the focal source irradiating the transmitarray. In order to get reasonable tile shifts, the optimization procedure with certain constraints that define maximum values of the optimized tile shifts was used.

To achieve intermediate results, an approach based on analytical model optimization was chosen instead of optimization with full-wave electromagnetic simulations due to the speed of analysis. Analytical model of the nonplanar transmitarray radiation pattern could be described as follows:

$$F(\theta,\varphi) = \sum_{m=1}^{M}\sum_{n=1}^{N} A_{mn}(\theta,\varphi)e^{j\Phi_{mn}(\theta,\varphi)}e^{ik(x_{mn}sin\theta\cos\varphi+y_{mn}sin\theta sin\varphi+d_{mn}cos\theta)} \quad (2)$$

where $A_{mn}(\theta,\varphi)$ and $\Phi_{mn}(\theta,\varphi)$ are the unit cell amplitude and phase patterns respectively, φ and θ are azimuthal and polar angles in the spherical coordinate system with the origin at the center of the transmitarray aperture and the polar direction aligned with the transmitarray optical axis.

Optimization results demonstrate more than 1 dB gain improvement compared to the flat transmitarray, Fig. 3, while different lower and upper bounds for tile shifts were used, Table 1. Gain improvement directly depends on the optimization boundary, but increasing optimization boundaries influence parasitic effects which are not described by the analytical model. More realistic results can be achieved with full-wave electromagnetic optimization. The analytical model does not take into account the interaction between the unit cells. In practice, the nonplanar transmitarrays can be easily implemented with a non-planar plastic frame, although each steering angle requires a different mounting frame. Such plastic frame was fabricated for further measurements by using a 3D printer, Fig 4.

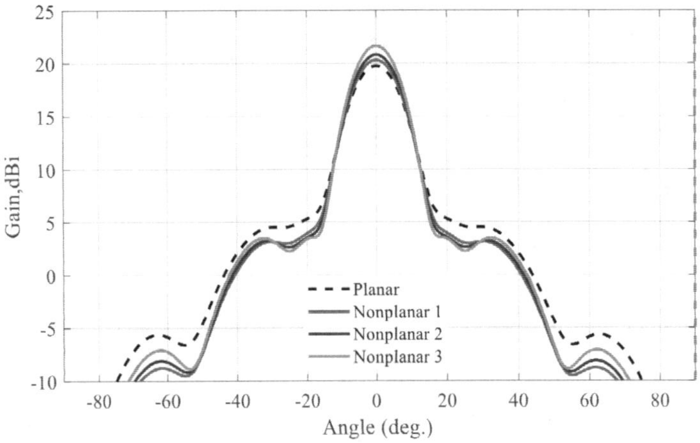

Fig. 3 H-plane beampatterns of the planar and phase-corrected nonplanar transmitarrays optimized with different bounds.

Table 1. Resulting shifted transmitarray gain for different optimization bounds.

	Upper bound, mm	Lower bound, mm	Gain, dBi
Planar	-	-	19.73
Nonplanar 1	-2.5	2.5	20.3
Nonplanar 2	-5	5	20.8
Nonplanar 3	-10	10	21.6

Fig. 4 Nonplanar tiled 1-bit transmitarray prototype.

3. Conclusion

The possibility of compensating for the phase error resulting from the 1-bit quantization of phase shifts by shifting the unit cells along the optical axis of the transmitarray was presented in this article. It was shown that satisfactory results could be achieved by using optimization procedures, and gain improvement directly depends on the optimization boundary. It is also noted that taking into account all the effects that occur due to the unit cells displaced relative to each other is necessary for optimization with the calculation of the analytical model of the transmitarray.

References

[1] I. Munina, P. Turalchuk, A. Verevkin, et al. "A study of C-band 1-bit reconfigurable dual-polarized transmitarray," in Proc. of EuCAP, 2019

[2] D. Kozlov, I. Munina, P. Turalchuk, V. Kirillov, A. Shitvov, D. Zelenchuk, "Characterization of Tiled Architecture for C-Band 1-Bit Beam-Steering Transmitarray, " Sensors, vol. 21, no. 4, pp. 1259, 2021.

[3] F. Diaby, A. Clemente, L. Di Palma, et al., "Design of a 2-bit unit-cell for electronically reconfigurable transmitarrays at Ka-band," in Proc. Europ. Radar Conf., 11–13 Oct 2017, Nuremberg, Germany, pp. 473–476, 10.23919/EURAD.2017.8249250

[4] R. L. Haupt, Antenna Arrays: A Computational Approach, Hoboken, New Jersey, USA, John Wiley & Sons, 2010, ch. 5.

[5] I. Ullah, S. Khattak, and B. D. Braaten, "Broadside Pattern Correction Techniques for Conformal Antenna Arrays," in E. Aksoy (ed.), Advances in Array Optimization, IntechOpen, Mar. 2020, 10.5772/intechopen.90957.

[6] L. Cabria, J. A. Garcia, J. Gutierrez-Rios, A Tazon, and J. Vassallo, "Active reflectors: possible solutions based on reflectarrays and Fresnel reflectors," Int. J. Anten. Propag., vol. 2009, article ID 653952, pp. 1-13, 10.1155/2009/653952.

[7] A. Greve, J.W.M. Baars, J. Peñalver, et al., "Near—focus active optics: An inexpensive method to improve millimeter—wavelength radio telescopes," Radio Science, vol. 31, no. 5, pp. 1053-1065, Se.-Oct. 1996, 10.1029/96RS01124.

Acknowledgments

The authors would like to thank Dr. Dmitry Kozlov and Dr. Alexey Shitvov for fruitful discussions of the work. This work has been supported by the Russian Science Foundation (project 17-79-20374).

Optomechanical interaction between single-walled carbon nanotubes of various structures

D Kislov[1,*], **O Kushchenko**[1] **and A S Shalin**[1,2]

[1]ITMO University, 49 Kronverksky Pr., St. Petersburg 197101, Russia
[2]Kotel'nikov Institute of Radio Engineering and Electronics of Russian Academy of Sciences (Ulyanovsk branch), Goncharova Str.48, Ulyanovsk, Russia, 432000

denis.a.kislov@gmail.com

Abstract. We consider optomechanical interaction in an asymmetric structure of a carbon nanotubes dimer of different orientations and/or different atomic structures in the field of a plane wave or a focused Gaussian beam. Here we show that optical coupling in such the system can lead to nonreciprocal interactions between the constituents. We demonstrate that a non-conservative force is applied to the center of mass of an optically coupled nanotube dimer, resulting in an unexpected lateral action. The sign and magnitude of this force depend on abrupt phase transitions in the properties of the asymmetric dimer.

1. Introduction

Carbon nanotubes are an interesting object for optics, optoelectronics, nanoelectronics, lab-on-a-chip, etc. When designing various devices, it is often necessary to manipulate individual nanotubes or several nanotubes at once.

Optical tweezers are a good tool for this. The capture of various nano- and microparticles and manipulations with them have been widely developed in various fields of science [1,2]. Recently, in optomechanics, the interaction with plasmonic particles has been studied rather well [3-6]. However, such particles have a disadvantage - significant ohmic losses. Particles made of materials with high refractive index $n > 3$ [7-12], which are a competitive alternative to resonant metal nanoparticles, are widely used, for example, in the development of metasurfaces [13] and optical or optoelectronic devices [14-17], and are free from this drawback.

Along with this, there are a few works in which the optical capture of nanotubes in optical traps and binding in optical fibers are shown experimentally [18]. There are several theoretical works in which optical binding of nanocylinders and nanowires made of glass, titanium dioxide, and other materials is considered [19,20]. However, there are still no works in which the optomechanical interaction of nanotubes, depending on their atomic structure, would be studied in detail.

2. Theoretical formalism

In this work, we consider single-wall carbon nanotubes of two types: "armchair" (x, x) and "zig-zag" (x, 0). The indices are written in brackets that specifies the coordinates of the chiral vector. In this work, x = 8, 10, 12. The imaginary part of the dielectric constant in the visible wavelength range for a given nanotubes type, simulated from first principles, was taken from the work [21]. Then, using the Kramers-Kronig relation, the real part of the dielectric constant was restored (Fig.1).

The mathematical formalism is based on the dipole approximation. To calculate the polarizability of nanotubes, we used a model for the anisotropic polarizability of a nano-needle (1)-(2) (highly elongated ellipsoid) [22]. This can be done with good accuracy since the length of the tube is much greater than its diameter.

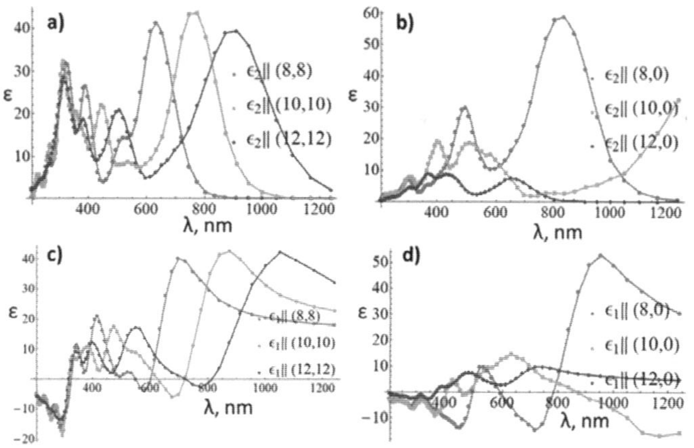

Figure 1. Real and imaginary parts of the dielectric constant of various diameters single wall carbon nanotubes: a) c) “armchair” type; b) e) “zig-zag type”.

$$\boldsymbol{\alpha}_{CNT} = \begin{pmatrix} \alpha_{\parallel} & 0 & 0 \\ 0 & \alpha_{\perp} & 0 \\ 0 & 0 & \alpha_{\perp} \end{pmatrix} \qquad \begin{cases} \alpha_{\parallel} = 4\pi\varepsilon_0 h_{CNT} r_{CNT}^2 \dfrac{\varepsilon_{CNT} - \varepsilon_m}{3\varepsilon_m + 3L_1(\varepsilon_{CNT} - \varepsilon_m)} \\ \alpha_{\perp} = 4\pi\varepsilon_0 h_{CNT} r_{CNT}^2 \dfrac{\varepsilon_{CNT} - \varepsilon_m}{3\varepsilon_m + 3L_2(\varepsilon_{CNT} - \varepsilon_m)} \end{cases} \tag{1}$$

$$L_1 = \frac{h_{CNT} r_{CNT}^2}{2} \int_0^{\infty} \frac{dq}{(h_{CNT}^2 + q)\sqrt{(h_{CNT}^2 + q)(r_{CNT}^2 + q)^2}} \qquad L_2 = \frac{h_{CNT} r_{CNT}^2}{2} \int_0^{\infty} \frac{dq}{(r_{CNT}^2 + q)\sqrt{(h_{CNT}^2 + q)(r_{CNT}^2 + q)^2}} \tag{2}$$

here h_{CNT} - is the half of the nanotube length; r_{CNT} - nanotube radius.

The force action of the nanotubes was calculated using the expression for the effective electromagnetic field based on the formalism of the Green's function for free space [23,24].

$$\begin{cases} \mathbf{E}(\mathbf{r}_1) = \left[\mathbf{I} - \omega^2\mu_0 \mathbf{G}_{21}(\mathbf{r}_1)\boldsymbol{\alpha}_{CNT\_2}\mathbf{G}_{12}(\mathbf{r}_2)\boldsymbol{\alpha}_{CNT\_1}\right]^{-1}\left[\mathbf{E}_{inc}(\mathbf{r}_1) + \omega^2\mu_0\mathbf{G}_{21}(\mathbf{r}_1)\boldsymbol{\alpha}_{CNT\_2}\mathbf{E}_{inc}(\mathbf{r}_2)\right] \\ \mathbf{E}(\mathbf{r}_2) = \left[\mathbf{I} - \omega^2\mu_0 \mathbf{G}_{12}(\mathbf{r}_2)\boldsymbol{\alpha}_{CNT\_1}\mathbf{G}_{21}(\mathbf{r}_1)\boldsymbol{\alpha}_{CNT\_2}\right]^{-1}\left[\mathbf{E}_{inc}(\mathbf{r}_2) + \omega^2\mu_0\mathbf{G}_{12}(\mathbf{r}_2)\boldsymbol{\alpha}_{CNT\_1}\mathbf{E}_{inc}(\mathbf{r}_1)\right] \end{cases} \tag{3}$$

Time averaged optical force on each nanotube in the dipolar approximation can be written as follows:

$$\mathbf{F}_{CNT\_1}(\mathbf{r}_1) = \frac{1}{2}\mathrm{Re}\left[\sum_i p_{CNT\_1}^{i*}\nabla E_i(\mathbf{r}_1)\right] \quad \mathbf{F}_{CNT\_2}(\mathbf{r}_2) = \frac{1}{2}\mathrm{Re}\left[\sum_i p_{CNT\_2}^{i*}\nabla E_i(\mathbf{r}_2)\right]; i = x, y, z \tag{4}$$

here $\mathbf{p}_{CNT\_1} = \boldsymbol{\alpha}_{CNT\_1}\mathbf{E}(\mathbf{r}_1)$, $\mathbf{p}_{CNT\_2} = \boldsymbol{\alpha}_{CNT\_2}\mathbf{E}(\mathbf{r}_2)$ - dipole moment of nanotubes.

It is known that when illuminated by an external field, identical nano- and microscopic particles can form stable structures due to their mutually acting optical forces on each other. This effect is known as optical binding (fig.2a). Potential forces act on two identical connected particles, which are equal in magnitude and oppositely directed - thereby the principle of "actio et reactio" is fulfilled. However, this is not always the case. Forces on optically interacting but dissimilar particles are not necessarily reciprocal. Dissimilar particles experience forces that do not compensate each other, which results in an overall movement of the system. In our case, a similar scenario can be obtained by considering the interaction between two tubes of the same type, but different in orientation (Fig. 2b), as well as between tubes of different types (Fig. 2c).

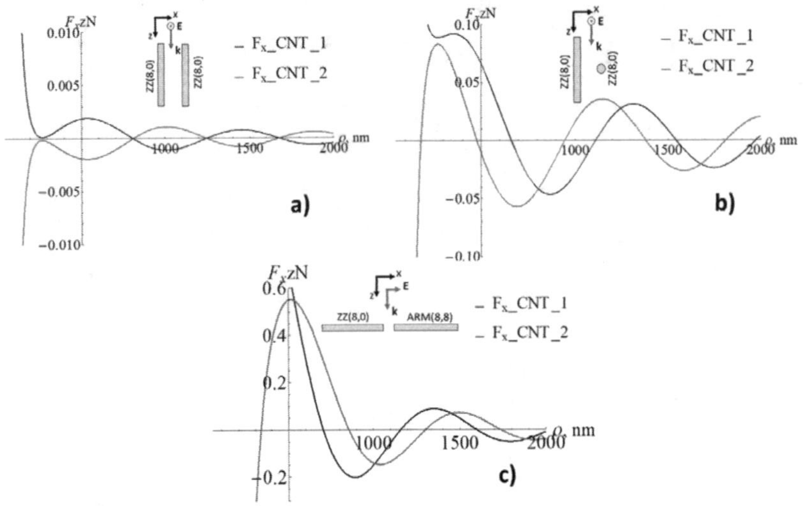

Figure 2. Transverse optical force acting on nanotubes in a plane wave field, depending on the distance between the centers of the tubes: a) tubes of the same type and the same orientation; b) tubes of the same type and different orientations; c) tubes of different types of the same orientation. The insets show the geometry of the calculation. Designations ZZ- zig-zag, ARM - armchair. Wavelength is 800 nm.

The apparent contradiction with the principle of "actio et reactio" can be resolved via taking into account that particles interacting with an external field constitute an open system, since part of the momentum is carried away by an asymmetrically scattered electromagnetic field. Optical forces between particles are mediated by their interaction with an external field and include not only conservative, but also non-conservative parts.

3. Conclusions

We have demonstrated that two dissimilar and optically interacting nanotubes experience common nonreciprocal interaction forces. This apparent violation of the principle of "action and reaction" occurs due to the presence of non-conservative optical forces that arise due to the interaction with the external field.

Acknowledgements

This work has been supported by the Russian Science Foundation (Grant No. 20-72-10141).

References

[1] Canós Valero A., Kislov D., Gurvitz E.A., Shamkhi H.K., Pavlov A.A., Redka D., Yankin S., Zemánek P. and Shalin A.S. 2020 *Adv. Sci.* **7** 1903049

[2] Favre-Bulle I.A., Stilgoe A.B., Scott E.K. and Rubinsztein-Dunlop H. 2019 *Nanophotonics* **8**, 1023

[3] Kislov D., Novitsky D., Kadochkin A., Redka D., Shalin A.S. and Ginzburg P. 2020, *Phys.Rev.B* **101** 035420

[4] Kucherenko M. G. and Kislov D. A. 2018 *J. Photochem. Photobiol. A Chem.* **354** 25

[5] Izmodenova S. V., Kislov D. A., and Kucherenko M. G. 2014 *Colloid J.* **76** 683
[6] Shalin A.S., Sukhov S.V. 2013 *Plasmonics* **8** 625
[7] Zhigunov D., Evlyukhin A.B., Shalin A.S., Zywietz U., and Chichkov B.N. 2018 *ACS Photonics* **5** 977
[8] Baryshnikova K., Filonov D., Simovski C., Evlyukhin A., Kadochkin A., Nenasheva E., Ginzburg P. and Shalin A.S. 2018 *Phys. Rev. B* **98** 165419
[9] Barhom H., Machnev A.A., Noskov R.E., Goncharenko A., Gurvitz E.A., Timin A.S., Shkoldin V. A., Koniakhin S.V., Koval O.Yu., Zyuzin M.V., Shalin A.S., Shishkin I.I. and Ginzburg P. 2019 *Nano Lett.* **19** 7062
[10] Terekhov P.D., Baryshnikova K.V., Greenberg Y. et al. 2019 *Sci Rep* **9** 3438
[11] Kozlov V., Filonov D., Shalin A.S., Steinberg B.Z. and Ginzburg P. 2016 *Applied Physics Letters* **109** 203503
[12] Terekhov P.D., Shamkhi H.K., Gurvitz E.A., Baryshnikova K. V., Evlyukhin A.B., Shalin A.S. and Karabchevsky A. 2019 *Opt. Express* **27** 8
[13] Ivinskaya A., Kostina N., Proskurin A., Petrov M.I., Bogdanov A.A., Sukhov S., Krasavin A. V., Karabchevsky A., Shalin A.S. and Ginzburg P. 2018 *ACS Photonics* **5** 4371
[14] Kostina N., Petrov M., Ivinskaya A., Sukhov S., Bogdanov A., Toftul I., Nieto-Vesperinas M., Ginzburg P. and Shalin A. 2019 *Phys. Rev. B* **99** 125416
[15] Shalin, A.S., Moiseev, S.G. 2009 *Opt. Spectrosc.* **106** 916
[16] Shalin, A.S. 2010 *Jetp Lett.* **91** 636
[17] Ivinskaya A., Petrov M.I., Bogdanov A.A., Shishkin I., Ginzburg P. and Shalin A.S. 2017 *Light Sci. Appl.* **6** e16258
[18] Li Y., Xin H., Xu X., Liu X., Li B. 2018 *Adv. Mater. Lett.* **9** 8
[19] Simpson S.H., Zemánek P., Maragò O. M., Jones P.H. and Hanna S. 2017 *Nano Lett.* **17** 6
[20] Ha S., Tang Y., van Oene M., Janissen R., Dries R. M., Solano B., Adam A. J. L. and Dekker N.H. 2019 *ACS Photonics* **6** 5
[21] Cho T. H., Su W. S., Leung T. C., Ren W. and Chan C. T. 2009 *Phys. Rev. B* **79** 235123
[22] Bohren C. F., Huffman D. R. 1998 *Absorption and Scattering of Light by Small Particles* (Wiley, New York)
[23] Kostina N.A., Kislov D.A., Ivinskaya A.N., Proskurin A., Redka D.N., Novitsky A., Ginzburg P. and Shalin A.S. 2020 *ACS Photonics* **7** 425
[24] Sukhov S., Shalin A., Haefner D. and Dogariu A. 2015 *Opt. Express* **23** 247

Second-harmonic generation enhancement in high-contrast micropillar AlGaAs resonator in bound states in the continuum regime

Stanislav Kolodny[1], **Valerii Kozin**[1,2] **and Ivan Iorsh**[1]
[1]ITMO University, Kronverkskiy prospekt 49, Saint Petersburg 197101, Russia
[2]Science Institute, University of Iceland, Dunhagi-3, IS-107 Reykjavik, Iceland

E-mail: s.kolodny@metalab.ifmo.ru

Abstract. It has been shown that the use of micropillar resonators, which comprise a cylindrical semiconductor cavity sandwiched between the Bragg mirrors can substantially increase the quality factor preserving the mode volume, and thus substantially enhance the local fields. Here, we show that these structures indeed can facilitate the significant enhancement of the SHG efficiency. We provide a specific design of the AlGaAs pillar microcavity and use the numerical modelling to directly show the resonant enhancement of the SHG efficiency. We believe that the presented results would be of high interest to the nanophotonic community, especially in nonlinear optics field.

It has been demonstrated that high-index resonant photonic nanostructures are a highly efficient tool to increase the efficiency of nonlinear optical processes [1], such as e.g. second [2], third [3] and higher harmonic generation [4]. Combining nonlinear materials and subwavelength high quality factor Q resonators, where light can be stored in a small volume V, can lead to increasing in the nonlinear conversion efficiency since it is proportional to the ratio Q/V [5], also known as Purcell factor. Photonic crystal cavities demonstrate large values of nonlinear frequency conversion efficiency [6] but creation of such structures require sophisticated fabrication techniques, since their quality factor is extremely sensitive to the geometry imperfections. On other hand, as has been recently shown, engineering of the shape of the isolated subwavelength semiconductor nanoantennae allows to create the high quality optical modes due to the destructive interference of the low order multipole modes [7, 8] and these high-Q modes are usually referred to as quasi-BIC states since the effect is similar to the bound states in the continuum (BIC) found in periodic structures [9]. It has been shown experimentally that these states in AlGaAs microresonator increase the efficiency of second harmonic generation [10].

We have recently shown, that the at certain ratios of cavity radius to cavity height, the destructive interference occurs similar to the one in the quasi-BIC state, which suppresses the side-wall leakage and resonantly increases the quality factor without strong affecting on the effective mode volume [11]. Here, we show that this quasi-BIC state occurring in pillar microcavities can be used to substantially increase the efficiency of the second harmonic generation.

We investigate the second-harmonic generation (SHG) in a pillar cavity microresonator with radius 500 nm, consisting of an AlGaAs cylinder, sandwiched between two DBR mirrors

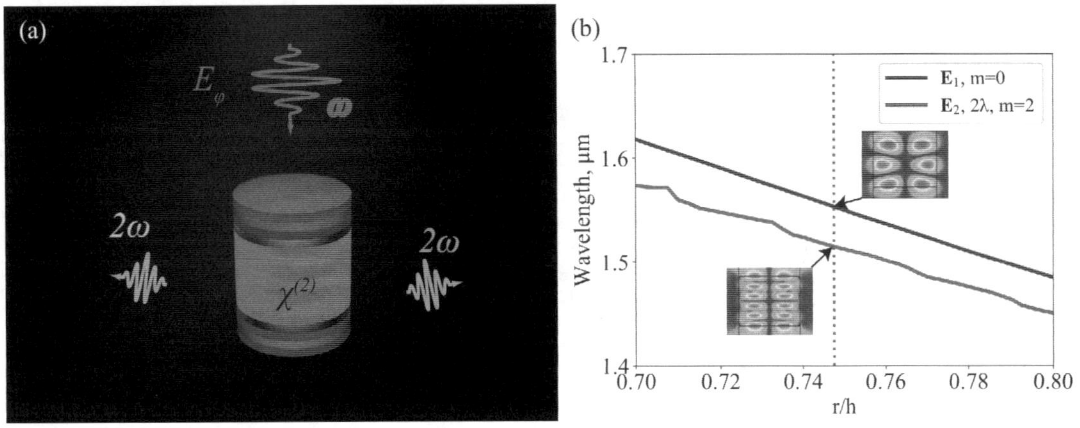

Figure 1. (a) Schematic representation of the system under consideration. Only 2 periods of Bragg reflectors on top and bottom are presented. (b) Wavelength of the modes in dependence on the r/h ratio of the cavity. The wavelength of the mode $\mathbf{E}_2$ is doubled to show that ω_2 is close to $2\omega_1$.

(AlGaAs-SiO_2, 8 layers on top and bottom), as shown in Fig. 1(a). The pillar cavity microresonator is tuned to the quasi-BIC regime [12, 7]. Refractive indices used in study are 3.48 and 1.45 for GaAs[13] and SiO_2, respectively, while their imaginary parts are neglected due to they are relative small.

The cavity is placed in the background field (pump), which is a superposition of two orthogonal linearly polarized Hermite-Gauss beams [14] and it produce an azimuthally polarized field with the azimuth number $m = 0$:

$$\mathbf{E}_{\mathrm{b}}(\omega, \mathbf{r}) = \mathrm{HG}_{01}(\omega, \mathbf{r})\mathbf{e}_x + \mathrm{HG}_{10}(\omega, \mathbf{r})\mathbf{e}_y. \tag{1}$$

The AlGaAs has a non-vanishing tensor of the second-order nonlinear susceptibility $\chi^{(2)}_{ijk}$ which contains only off-diagonal elements in the principal axis system of the zinc blende crystalline structure [15], with the components being non-zero only if $i \neq j \neq k$, $\chi^{(2)}_{xyz} \equiv \chi^{(2)} = 290\mathrm{pm/V}$.

Here we consider the two modes of the pillar cavity: $\mathbf{E}_{1,2}$ with the real and imaginary parts being equal to $\omega_{1,2}$ and $\gamma_{1,2}$, respectively. The resonator is placed in the background field with the frequency ω equals to ω_1, and the frequency ω_2 of the mode $\mathbf{E}_2$ is found to be close to 2ω as shown in Fig. 1(b). So in our simulations we consider TE_{012} mode as $\mathbf{E}_1$ since it has a confirmed quasi BIC, and the TE_{215} mode as $\mathbf{E}_2$ because of the proximity of ω_2 to 2ω (the modes field spatial distributions are shown as insets on the Fig. 1(b)). The cavity radius is fixed in our study and we vary only its height as well as the Bragg layers period to make the center of the bandgap tuned to the mode frequency.

The purpose of work is to calculate the nonlinear conversion coefficient, showing the efficiency of the second-harmonic generation, and defined as the ratio between the total SHG power and the pump power squared: $P(2\omega)/P_0(\omega)^2$. Pump peak power in our calculations was about 0.002 W. The corresponding expression for the total SHG power is given by [10]

$$P(2\omega) = \frac{8\pi}{c}\left(\frac{2\omega}{c}\right)^2 \kappa_2 Q_2 L_2(2\omega)\kappa_{12}\left[Q_1 L_1(\omega)\kappa_1(\omega)P_0(\omega)\right]^2. \tag{2}$$

Here $Q_j = \omega_j/(2\gamma_j)$ is the mode quality factor. The spectral overlap factor is defined as

$$L_j(\omega) = \frac{\gamma_j^2}{(\omega - \omega_j)^2 + \gamma_j^2}, \tag{3}$$

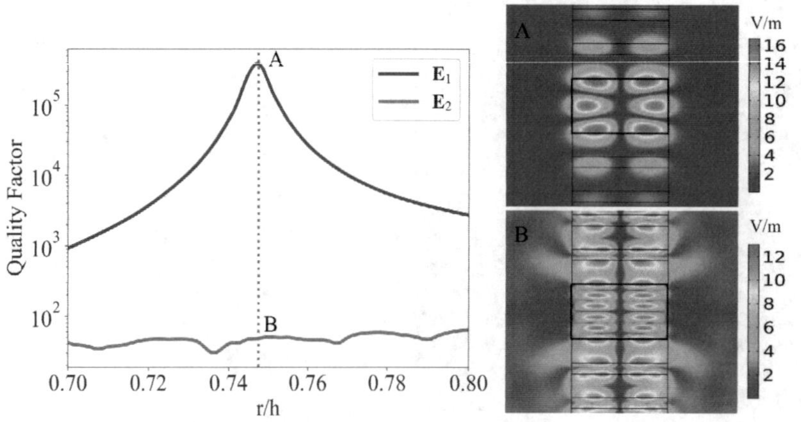

Figure 2. (a) The quality factor of the modes $\mathbf{E}_1$ and $\mathbf{E}_2$. The vertical line represents the position of the highest value of Q at r/h = 0.7475. On the right picture the modes field distributions for points A and B of plot are presented.

and the so-called coupling κ_1, cross-coupling $\kappa_{1,2}$ and decoupling κ_2 coefficients [10], where they can be defined from the spatial mode profiles $\mathbf{E}_{1,2}(\mathbf{r})$. Note that in our calculations we consider the second harmonic generation only in the AlGaAs.

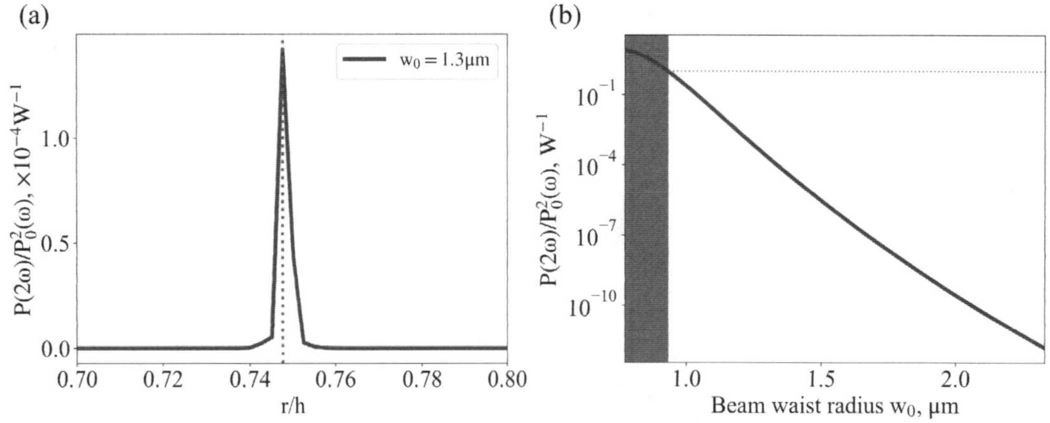

Figure 3. (a) Nonlinear conversion coefficient versus r/h of the cavity with fixed beam waist radius $w_0 = 1.3\ \mu m$. (b) Nonlinear conversion coefficient versus the beam waist radius w_0. Zone filled by red color represents the values of beam waist radius where nonlinear conversion coefficient is bigger than 1 W^{-1}(green dotted line).

The quality factor of the $\mathbf{E}_1$ mode achieves its highest value $Q \approx 4 \times 10^5$ at the aspect ratio r/h of AlGaAs cavity is approximately equal to 0.7475, see Fig. 2(a) and such large value of the quality factor allows us to neglect the material losses. Since we consider the case of high contrast Bragg reflectors the low-quality modes do not arise and the quality factor of $\mathbf{E}_1$ has a single peak [11] at the aspect ratio $r/h = 0.7475$. In the same region of the aspect ratio the mode $\mathbf{E}_2$ has a oscillating quality factor since its frequency is far away from the center of Bragg reflectors bandgap which is tuned to the $\mathbf{E}_1$ frequency. Because of this fact it can be seen from field distributions shown in Fig. 2(b) that $\mathbf{E}_2$ does not effectively decay in the reflectors. And because of this the Q factor of $\mathbf{E}_2$ remains quite low.

The dependence of second harmonic nonlinear coefficient on the aspect ratio r/h was studied with a fixed beam waist radius equal to 1.3 μm (see Fig. 3 (a)). As it can be seen from this plot, the nonlinear coefficient has a pronounced maximum at $r/h = 0.7475$, where the coefficient is at least an order of magnitude larger in comparison with the rest of the aspect ratio area, and it is about 1.4×10^{-4} $1/W$.

And as an important step we study the dependence of the nonlinear coefficient on the beam waist radius of the background field (see Fig. 3(b)). The maximal achieved value of nonlinear conversion coefficient is greater than 1 (≈ 8 $1/W$) at the value of w_0 equal to the half of wavelength of the $\mathbf{E}_1$ mode with $r/h = 0.7475$. As it can be seen, $P(2\omega)/P_0(\omega)^2$ value strongly depends on the beam waist radius of background field and simple decreasing of radius leads to significant enhancement of SHG but it requires taking into account higher order nonlinear effects such as generation of higher harmonics and difference frequency generation. It can be the next step in the study of quasi-BIC in pillar cavity microresonators since it has not been considered in the presented study.

To conclude, in this work we have studied the second harmonic generation in the pillar AlGaAs/SiO_2 cavity microresonator, and have shown its significant enhancement due to resonant increase of quality factor of first mode due to the destructive interference of two leaky modes. Results of this work may be of interest in study of nonlinear effects in micropillar resonators, especially in practical realization of sub-wavelength devices for nonlinear optics.

Acknowledgments
This work was supported by the Ministry of Science and Higher Education of Russian Federation, goszadanie no. 2019-1246. Also the authors thank Kirill Koshelev from ITMO University and Sergei Yankin from COMSOL.

References
[1] Sain B, Meier C and Zentgraf T 2019 *Advanced Photonics* **1** 024002
[2] Carletti L, Locatelli A, Neshev D and De Angelis C 2016 *ACS Photonics* **3** 1500–1507
[3] Shcherbakov M R, Neshev D N, Hopkins B, Shorokhov A S, Staude I, Melik-Gaykazyan E V, Decker M, Ezhov A A, Miroshnichenko A E, Brener I *et al.* 2014 *Nano letters* **14** 6488–6492
[4] Liu S, Vabishchevich P P, Vaskin A, Reno J L, Keeler G A, Sinclair M B, Staude I and Brener I 2018 *Nature communications* **9** 1–6
[5] Vahala K J 2003 *nature* **424** 839–846
[6] Buckley S, Radulaski M, Petykiewicz J, Lagoudakis K G, Kang J H, Brongersma M, Biermann K and Vuckovic J 2014 *ACS Photonics* **1** 516–523
[7] Rybin M V, Koshelev K L, Sadrieva Z F, Samusev K B, Bogdanov A A, Limonov M F and Kivshar Y S 2017 *Phys. Rev. Lett.* **119**(24) 243901 URL `https://link.aps.org/doi/10.1103/PhysRevLett.119.243901`
[8] Koshelev K, Favraud G, Bogdanov A, Kivshar Y and Fratalocchi A 2019 *Nanophotonics* **8** 725–745
[9] Hsu C W, Zhen B, Stone A D, Joannopoulos J D and Soljačić M 2016 *Nature Reviews Materials* **1** 1–13
[10] Koshelev K, Kruk S, Melik-Gaykazyan E, Choi J H, Bogdanov A, Park H G and Kivshar Y 2020 *Science* **367** 288–292 ISSN 0036-8075 (*Preprint* `https://science.sciencemag.org/content/367/6475/288.full.pdf`) URL `https://science.sciencemag.org/content/367/6475/288`
[11] Kolodny S and Iorsh I 2020 *Journal of Physics: Conference Series* vol 1461 (IOP Publishing) p 012067
[12] Kolodny S and Iorsh I 2020 *Opt. Lett.* **45** 181–183 URL `http://ol.osa.org/abstract.cfm?URI=ol-45-1-181`
[13] Skauli T, Kuo P, Vodopyanov K, Pinguet T, Levi O, Eyres L, Harris J, Fejer M, Gerard B, Becouarn L *et al.* 2003 *Journal of Applied Physics* **94** 6447–6455
[14] Zhan Q 2009 *Advances in Optics and Photonics* **1** 1–57
[15] Boyd R W 2003 *Nonlinear Optics* (Academic)

Silicon nanowires as multi–environment sensor elements for carbon monoxide and ammonia detection

V. M. Kondratev[1], I. A. Morozov[1, 2], E. A. Vyacheslavova[1, 2], A. S. Gudovskikh[1, 2], S. S. Nalimova[2], V. A. Moshnikov[2] and A. D. Bolshakov[1]

[1]Alferov University, [2]Saint Petersburg Electrotechnical University "LETI"

Abstract. Carbon monoxide and ammonia are inorganic agents found both in nature and in the human body, which is of great interest for modern sensing. First, in concentrations of the order of 1 ppm agents are produced by human and can be markers of changes in human health. Second, at concentrations of the order of 100 ppm, carbon monoxide and ammonia are toxic and hazardous. This work is aimed at fabrication and study of precise, technological and relatively cheap sensors compatible with a gas and liquid medium for CO and NH_3 detection, respectively.

Introduction

Low–dimensional nanostructures such as metal nanowires (NWs) [1] and metal oxide NWs [2] are the main elements for modern adsorption sensors production. High surface area and peculiar electronic properties compare to bulk material are the main reason for using of NWs in sensing. The existing works on Si NWs [3–4] consider only gaseous media and a small amounts of target adsorbates. Development of multi–environment sensors, such as for CO in gas medium and NH_3 in liquid medium can lead to qualitative breakthrough in modern adsorption sensorics, expand and combine classic applications by one multi–environment case.

1. Nanowires synthesis

Vertically oriented structures based on Si NWs with a length of about 10 μm were obtained using cryogen plasma etching of Si (001) substrate in Oxford PlasmaLab System 100 ICP380 with SF_6/O_2 flow and pressure to 5 mTorr according to the previous reports [5–6] and studied by electron microscopy. For structural characterization, Si NWs were transferred to an auxiliary substrate and studied by Raman spectroscopy. The results of the SEM and Raman studies are shown in Fig. 1 (a) and (b), respectively.

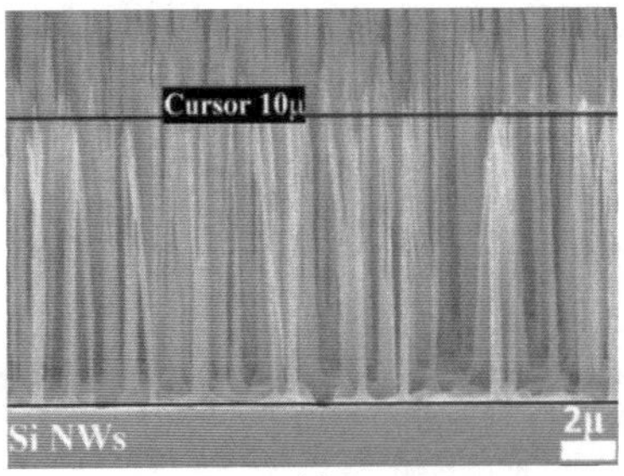

(a)

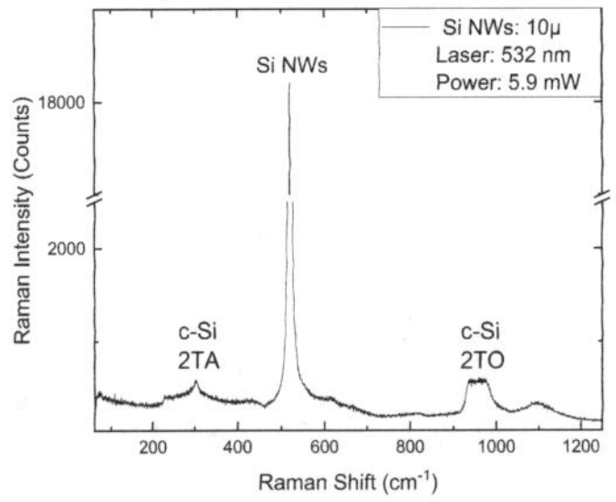

(b)

Figure 1 – SEM image of the vertically oriented Si NWs – (a) and Si NWs Raman spectrum – (b)

Fig. 1 (a) shows that the Si NWs have a length of about 10 μm and a cross section of 350–400 nm. Fig. 1 (b) demonstrates presence of transverse second order acoustic phonons 2TA (c–Si 2TA) and transverse optical phonons 2TO (c–Si 2TO) at 300 and 970 cm^{-1} [7] in the Raman spectrum, respectively, and high–intensity peak at 520 cm^{-1} (Si NWs), typical for highly crystalline Si NWs [8]. The obtained Raman spectrum also shows low–intensity peaks in the range of 400–450 cm^{-1} and 600–650 cm^{-1}, which are in a good agreement with the studies of low–dimensional Si nanostructures [9].

2. Sensor fabrication

The step–by–step protocol of the sensor development is shown in the Fig. 2.

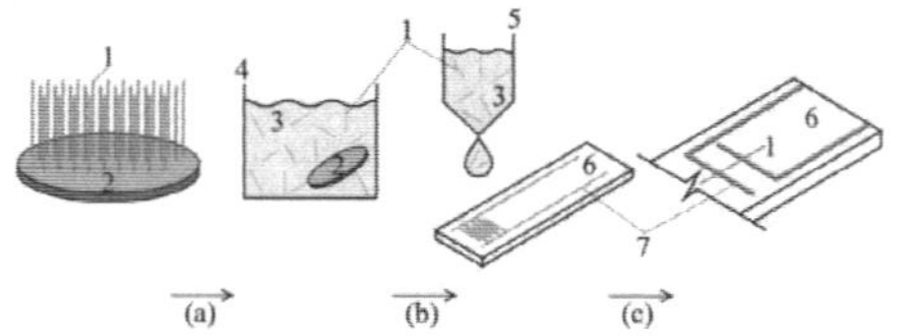

Figure 2 – The step–by–step sensor fabrication protocol: 1 –as-synthesized vertical Si NWs , 2 – Si substrate, 3 – isopropyl alcohol, 4 – ultrasonic separation tube, 5 – diffusion system, 6 – sensor platform, 7 – gold interdigital contacts; (a) – separating of NWs from the growth substrate by ultrasonic (b) – dispersion of NWs in isopropanol and transfer to the sensor platform, (c) – fabrication of sensor and formation of gold–NWs contacts by annealing

The prototype device based on Si NWs obtained using the presented protocol was used in this work as a multi–environment adsorption sensor for carbon monoxide and ammonia.

3. Sensory tests

The sensory properties of the sensor with the NWs were investigated in terms of the change in the impedance of the obtained sensors under the action of the target adsorbate in gaseous (CO) and liquid (NH_3) media and compared with the impedance in N_2 and distilled H_2O medium, respectively. The change in the impedance is represented in the form of Nyquist plot, which allows to selectively characterize the change in real and imaginary parts of the impedance of the sensor in the frequency range from 100 Hz to 500 kHz in the presence of CO and NH_3. The constant voltage signal to measure the impedance of 100mV was selected; the amplitude of the variable voltage signal is 20mV. The presented parameters for impedance measurement are optimal for working with samples with an implicit active and reactive nature of resistance change.

At the same time the concept of a multi–environment sensor is an important aspect of this work, which can be divided into 2 cases: carbon monoxide gas and liquid ammonia.

3.1. Gas case

Working with the gases allowed us to measure the sensor impedance spectrum in N_2 and CO medium not only at normal conditions, but also under heating of the sensor up to 200°C and irradiation with UV light (395nm), which excite the generation of minority charge carriers in NWs. We believe that the processes of adsorption from the gas medium occur by donor–acceptor mechanisms and minority charge carriers can make a useful contribution to the conductivity of NWs connected to the sensor configuration. At the same time, the classical idea of the NWs conductivity in this configuration is the concept of a conducting channel inside the NWs, the width of which varies depending on the adsorption processes on the surface of the low–dimensional structure, leading to the formation of depleted or enriched with the main charge carriers of the regions acting on the conduction channel. An array of conductive NWs is a sensitive part of the sensor produced according to the presented protocol. The second reason of the additional treatment of the sensor is to enhance the overall conductivity of the sensor under the influence of photoconductivity and thermal conductivity, which reduces the shunt effects. However, there no sufficient shift in the sensor impedance spectrum is observed when changing nitrogen to carbon monoxide under normal conditions and under UV irradiation. On the contrary, the corresponding spectra are very different when the sensor is heated to 200°C during measurements.

In accordance with the above, Fig. 3 shows the impedance spectra of a sensor based on Si NWs in a reference medium–nitrogen and in the medium of the target adsorbate – carbon monoxide, under conditions of heating the sensor to 200°C.

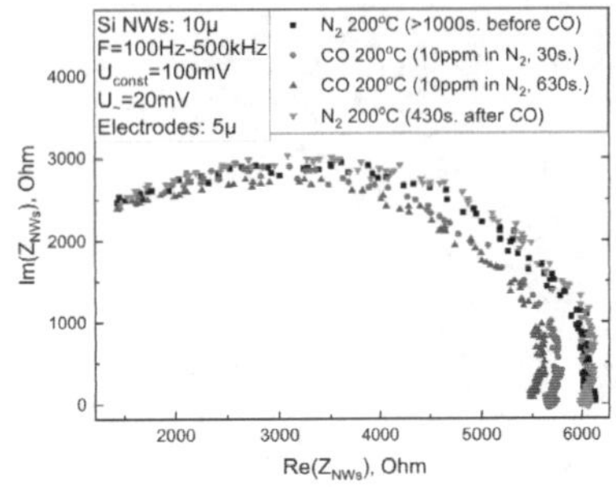

Figure 3 – The shift of the sensor impedance spectrum when changing N_2 to CO under heating of the sensor to 200°C during measurements: N_2 200°C (>1000s. before CO) – in N_2 with exposition time over 1000s.; CO 200°C (10ppm in N_2, 30s.) – in 10ppm CO with exposition time 30s.; CO 200°C (10ppm in N_2, 630s.) – in 10ppm CO with exposition time 630s.; N_2 200°C (430s. after CO) – recovery of the sensor impedance

The impedance spectra in the presence of N_2 and of CO are qualitatively different. CO presence leads to decrease of real and imaginary parts of the impedance in all of the frequency range. The latter effect may be due amorphization of the NWs surface, n – type defective conductivity and redox interaction between NWs and carbon monoxide during adsorption. The obtained data together indicate the possibility of using sensors based on Si NWs for the qualitative analysis of gaseous and vapor samples for the presence of carbon monoxide.

3.2. Liquid case

Working with the liquids allowed us to measure the sensor impedance spectrum in distilled H2O and NH_3 medium at room temperature and atmospheric pressure. We believe that the previously presented scheme of conductivity of low-dimensional nanostructures in the presence of adsorbates can be also used for liquid adsorbates, however, it is necessary to take into account the strongly polar nature of water molecules and the pronounced alkaline nature of water-ammonia solutions with reducing properties.

In analysis of the obtained results, it is necessary to take into account the contribution to the overall conductivity of the sensor from the electrochemical cell formed between the interdigital contacts when the sensor is placed in liquid medium. The impedance of the NWs and electrochemical cell in this sensor configuration are connected in parallel. Based on the latter assumption, the signal normalization algorithm, which consists in subtracting the impedance of the electrochemical cell from the total sensor impedance, is proposed further. Then the total impedance $(Z_\Sigma)_i$ of the sensor for each i-th frequency of the voltage measuring signal can be calculated using the expression 1:

$$\frac{1}{(Z_\Sigma)_i} = \frac{1}{(Z_{H_2O})_i} + \frac{1}{(Z_{NWs})_i}; \tag{1}$$

the $(Z_{H2O})_i$ is the impedance of the electrochemical cell, $(Z_{NWs})_i$ is the impedance of the NWs. To describe the sensitivity of the sensors, the real and imaginary parts of the impedance were considered:

$$Z_i = \mathrm{Re}(Z)_i + j \cdot \mathrm{Im}(Z)_i; \tag{2}$$

Z_i is the impedance, $\mathrm{Re}(Z)_i$ is the real component of the impedance, $\mathrm{Im}(Z)_i$ is the imaginary, $j = (-1)^{1/2}$.Then the expression for signal normalization, which consists in subtracting the contribution from the electrochemical cell from the sensor impedance spectrum formed between the interdigitated contacts can be represented in the form of expression 3:

$$\mathrm{Re}(Z_{NWs})_i = \frac{R}{R^2 + I^2}, \mathrm{Im}(Z_{NWs})_i = \frac{I}{R^2 + I^2}; \tag{3}$$

the ratio for the coefficients R and I is presented as an expression 4.

$$R = \left(\frac{\mathrm{Re}(Z_\Sigma)_i}{|Z_\Sigma|_i^2} - \frac{\mathrm{Re}(Z_{H_2O})_i}{|Z_{H_2O}|_i^2} \right), I = \left(\frac{\mathrm{Im}(Z_\Sigma)_i}{|Z_\Sigma|_i^2} - \frac{\mathrm{Im}(Z_{H_2O})_i}{|Z_{H_2O}|_i^2} \right); \tag{4}$$

in accordance with the above, Fig. 4 shows the impedance spectra of a sensor based on Si NWs in a comparative medium – distilled H_2O and in the medium of the target adsorbate – NH_3.

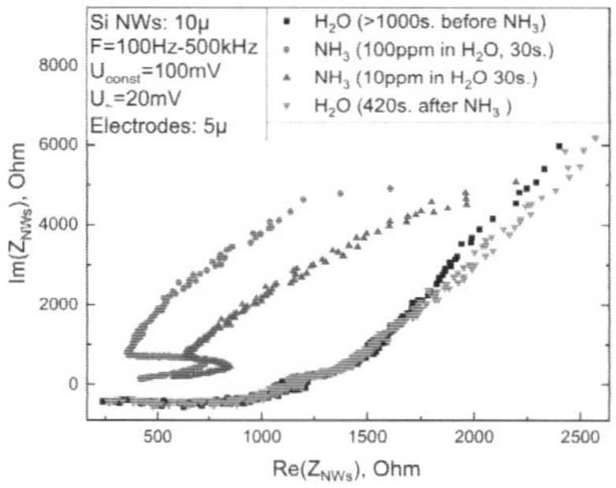

Figure 4 – The shift in the sensor impedance spectrum when changing distilled H_2O to NH_3: H_2O (>1000s. before NH_3) – in H_2O with exposition time over 1000s.; NH_3 (100ppm in H_2O, 30s.) – in 100ppm NH_3 with exposition time 30s.; NH_3 (10ppm in H_2O, 30s.) – in 10ppm NH_3 with exposition time 30s.; H_2O (430s. after NH_3) – recovery of the sensor impedance

Ammonia adsorption at NW sidewalls in liquid medium leads to decrease of imaginary part of the impedance and signal normalization allows one to evaluate the electrical properties of NWs and their correlation with adsorption properties. The impedance spectra of a sensor based on Si NWs corresponding to different concentrations of ammonia dissolved in water and distilled water are distinguishable in the entire frequency range of measurements, which indicates the possibility of using fabricated sensors according to the previously proposed protocol for the qualitative and quantitative analysis of liquid samples for the presence of ammonia.

4. Conclusion

In all the studied cases, the shift of the impedance spectrum demonstrates perspectives of Si NWs use for both quantitative and qualitative multi–environment analysis for the presence of low concentrations of carbon monoxide and ammonia.

We also demonstrate a protocol for development of multi–environment sensors based on Si NWs which can be interesting in terms of characterization of conductivity nature of low–dimensional epitaxial nanostructures and future sensory applications.

Acknowledgments

Authors thank for support the Ministry of Science and Higher Education of the Russian Federation (Grant FSRM-2020-00011).

References

[1] Nano Letters 2003, 3, 9, 1229–1233.
[2] Advanced Materials 2003, 12, 997–1000.
[3] Chemical Physics Letters 2003, 369, 220–224.
[4] Nanotechnology 2010, 21, 9, 015501.
[5] Phys. Status Solidi A 2020, 217, 1900535.
[6] Energy Procedia 2013, 38, 866–871.
[7] Physical Review B 1999, V. 59, N. 3, 1645–1648.
[8] Materials Science and Engineering C 2003, 23, 931–934.
[9] Nanotechnology 2012, 23, 275706.

Probing guided monolayer semiconductor polaritons below the light line

V. Kondratyev[1], D. Permyakov[1], V. Kravtsov[1], D. N. Krizhanovskii[1,2], A. K. Samusev[1]

[1] Department of Physics and Engineering, ITMO University, St. Petersburg, 197101, Russia
[2] Department of Physics and Astronomy, University of Sheffield, Sheffield S3 7RH, UK

E-mail: a.samusev@metalab.ifmo.ru

E-mail: valeriy.kondratiev@metalab.ifmo.ru

Abstract. In this work, we demonstrate an approach to study exciton-polaritons supported by transition metal dichalcogenide monolayers coupled to an unstructured planar waveguide below the light line. In order to excite and probe such waves propagating along the interface with the evanescent fields exponentially decaying away from the guiding layer, we employ a hemispherical ZnSe solid immersion lens (SIL) precisely positioned in the vicinity of the sample. We visualize the dispersion of guided polaritons using back focal (Fourier) plane imaging spectroscopy with the high-NA objective lens focus brought to the center of SIL. This results in the effective numerical aperture of the system exceeding an exceptional value of 2.2 in the visible range. In the experiment, we study guided polaritons supported by a WS_2 monolayer transferred on top of a Ta_2O_5 plane-parallel optical waveguide. We confirm room-temperature strong light-matter coupling regime enhanced by ultra-low intrinsic ohmic and radiative losses of the waveguide. Note that in the experiment, total radiative losses can be broadly tuned by controlling SIL-to-sample distance. This gives a valuable degree of freedom for the study of polariton properties. Our approach lays the ground for future studies of light-matter interaction employing guided modes and surface waves.

1. Introduction

Rapid development of the field of all-optical devices such as optical switches and transistors boosts the search for highly nonlinear optical systems. Those operating in the regime of strong light-matter coupling show great promise. Monolayers of transition-metal dichalcogenides (TMDs) such as WS_2, WSe_2, MoS_2 and $MoSe_2$ exhibit direct band gaps [1] and are suitable for chip integration [2]. TMD interaction with light is dominated by quasiparticles – excitons with binding energies of the order of 100 meV, large oscillator strength [3] and ability to strongly couple to optical cavities and form polaritons. TMD polaritons were studied in various optical systems, such as plasmonic cavities [4], Bragg mirrors [5] and subwavelength gratings [6, 7]. Most of the systems studied so far require relatively complicated fabrication processes, while their designs allow for radiative coupling to free-space waves with a fixed efficiency. In our work, we realize a simple hybrid planar TMD-based waveguide and demonstrate an approach for the study of polaritons intrinsically uncoupled from free-space waves propagating below the light line.

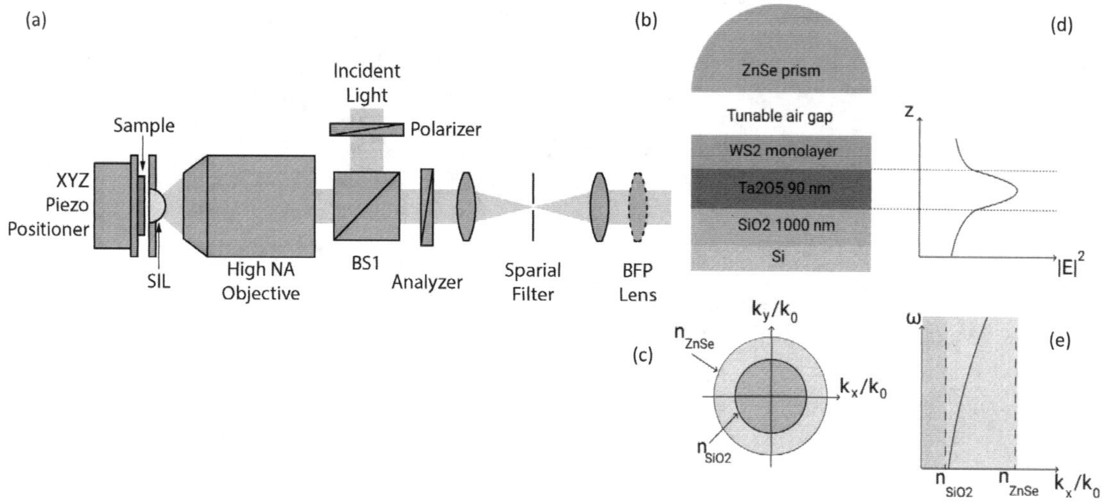

Figure 1. (a) Back focal plane spectroscopy setup with a solid immersion lens. (b) Scheme for the sample with ZnSe prism attached. The air gap can be tuned with piezo positioner. On the right side, the waveguide mode in real space and k-space. Plot under the sample represent numerical aperture in experiment. Red color corresponds to states below SiO_2 light line, blue color correspond states between SiO_2 and ZnSe light lines.

2. Results and discussion

In the experiment, we employ back focal (Fourier) plane spectroscopy setup combined with solid immersion lens (SIL) [8, 9] schematically shown in Fig. 1(a). We use a high numerical aperture objective lens (Mitutoyo, M Plan Apo HR, 100×, $\mathrm{NA}_{obj} = 0.9$) in combination with ZnSe (refractive index $n_{ZnSe} \approx 2.5$ in the visible range) prism coupled to the sample with a precisely controlled air gap in Otto geometry [10]. Such configuration enables the resulting numerical aperture of $\mathrm{NA} = \mathrm{NA}_{obj} n_{\mathrm{ZnSe}} \approx 2.25$ and allows for excitation and detection of propagating modes below the light line (see Fig. 1(e)). During the measurement, the sample was excited by a white light halogen lamp (Ocean Optics HL-2000FHSA). The angle-resolved reflectivity spectra were measured by a slit spectrometer (Princeton SP 2500, CCD camera PyLoN 400BReXcelon). The sample was attached to a piezo positioner, which allowed for precise control of the SIL-to-sample gap. The size of the gap was directly related to the radiative losses of the mode induced by SIL and thus allowed controlling mode excitation and detection efficiency.

The sample we investigated in this work consists of a WS_2 monolayer placed on a 90-nm Ta_2O_5 waveguide on top of SiO_2/Si substrate. The WS_2 monolayer was exfoliated from a bulk crystal and transferred on top of the Ta_2O_5 layer. A schematic representation of our device with a SIL (ZnSe prism) attached is shown in Fig. 1(b).

The typical field profile and dispersion of a guided mode is shown in Fig. 1(d) and (e), respectively. For a non-leaky mode, the experimentally available in-plane wavevectors $k_{||}$ satisfy the condition $n_{\mathrm{SiO_2}} < k_{||}/k_0 < n_{\mathrm{ZnSe}}$, where k_0 is the absolute value of free-space wavevector (see Fig. 1(c, e)). Since the mode field is exponentially decaying away from the waveguiding layer, it can be only coupled to free space by a high-index prism brought in a close vicinity to the sample.

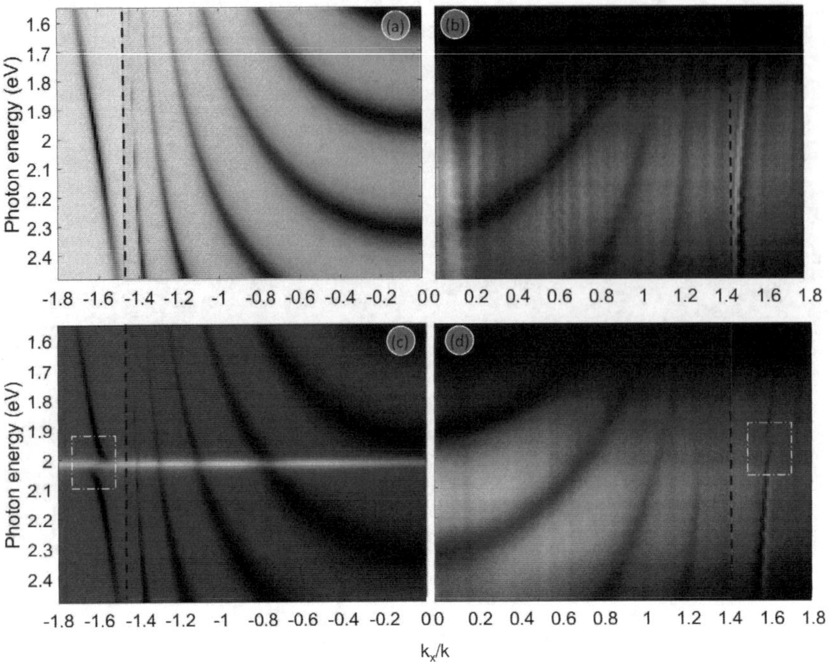

Figure 2. Simulated (left column) and measured (right column) angle-resolved TE-polarized reflectivity maps for the sample with (bottom row) and without (top row) WS_2 monolayer. Red and blue regions correspond to the states above and below the light line in SiO_2. The energy splitting between exciton resonance in TMD and waveguide mode is observed at ≈ 2.01 eV. The rectangle highlights the magnified region shown in Fig. 3.

Fig. 2 shows simulated (a, c) and measured (b, d) angle-resolved reflectivity maps for the sample without (a, b) and with (c, d) TMD monolayer obtained for a 150 nm air gap between sample and SIL. The simulations were performed using Fourier modal method [11]. The central part of the figure marked with red corresponds to small wavevectors $k_x/k_0 < n_{\mathrm{SiO}_2}$ and contains strong Fabry-Pérot interference due to reflection from Si substrate. The remaining blue regions correspond to the regime of total internal reflection, while the dips in in these regions are associated with guided TE-polarized modes.

The top row in Fig. 2 shows the dispersion of TE waveguide mode, which is efficiently coupled to free-space waves for an air gap between SIL and the bare waveguide of around 150 nm. When coupled to exciton resonance (Fig. 2, bottom row), the waveguide mode experiences Rabi splitting indicating the onset of strong light-matter coupling regime. Fig. 3 shows a magnified reflectivity map containing anticrossing of the modes. From this picture, we can estimate the Rabi splitting to be of the order of few tens of meV.

3. Conclusions

We have demonstrated a new approach for the study of guided exciton-polaritons in TMD-based planar photonic structures at room temperature. We have observed strong coupling between excitons in WS_2 monolayer and Ta_2O_5 waveguide with Rabi splitting of the order of several 10s of meV. Our results provide a basis for future investigations of radiative/non-radiative losses, lifetimes, and nonlinearities of TMD-based guided polaritons.

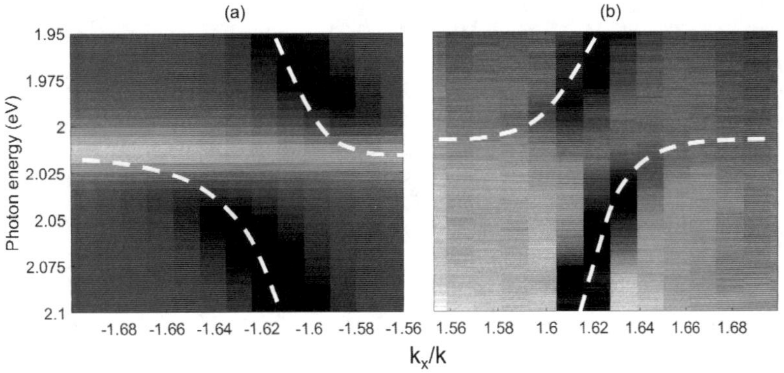

Figure 3. Magnified exciton-photon splitting region: simulation (a) and experiment (b). White dashed line is a guide for the eye. The same region is highlighted in Fig. 2.

4. Acknowledgments
The theoretical part of this study was supported by Russian Science Foundation, grant no. 21-12-00218. The experimental work was funded by RFBR according to project no. 19-52-51010.

References
[1] Mak K F and Shan J 2016 *Nature Photonics* **10** 216–226
[2] Akinwande D, Huyghebaert C, Wang C H, Serna M I, Goossens S, Li L J, Wong H S P and Koppens F H 2019 *Nature* **573** 507–518
[3] Wang G, Chernikov A, Glazov M M, Heinz T F, Marie X, Amand T and Urbaszek B 2018 *Rev. Mod. Phys.* **90**(2) 021001
[4] Munkhbat B, Baranov D G, Bisht A, Hoque M A, Karpiak B, Dash S P and Shegai T 2020 *ACS nano* **14** 1196–1206
[5] Dufferwiel S, Schwarz S, Withers F, Trichet A, Li F, Sich M, Del Pozo-Zamudio O, Clark C, Nalitov A, Solnyshkov D *et al.* 2015 *Nature communications* **6** 1–7
[6] Zhang L, Gogna R, Burg W, Tutuc E and Deng H 2018 *Nature communications* **9** 1–8
[7] Kravtsov V, Khestanova E, Benimetskiy F A, Ivanova T, Samusev A K, Sinev I S, Pidgayko D, Mozharov A M, Mukhin I S, Lozhkin M S *et al.* 2020 *Light: Science & Applications* **9** 1–8
[8] Pidgayko D, Sinev I, Permyakov D, Sychev S, Heyroth F, Rutckaia V, Schilling J, Lavrinenko A, Bogdanov A and Samusev A 2018 *ACS Photonics* **6** 510–515
[9] Permyakov D, Pidgayko D, Sinev I and Samusev A 2020 Measuring full complex dispersion of guided modes and surface waves in planar photonic structures *AIP Conference Proceedings* vol 2300 (AIP Publishing LLC) p 020096
[10] Otto A 1968 *Zeitschrift für Physik A Hadrons and nuclei* **216** 398–410
[11] Li L 1997 *JOSA A* **14** 2758–2767

Doubly excited states in a chiral waveguide-QED system: description and properties

D. Kornovan[1], **E. Vlasiuk**[1,2], **A. Poddubny**[1,3,4], **M. Petrov**[1]
[1]ITMO University, St. Petersburg 197101, Russia
[2]Moscow Institute of Physics and Technology, 9 Institutsky Lane, Dolgoprudny 141700, Russia
[3]Ioffe Institute, St. Petersburg 194021, Russia
[4]Nonlinear Physics Centre, Research School of Physics, Australian National University, Canberra ACT 2601, Australia

E-mail: d.kornovan@metalab.ifmo.ru

Abstract. In modern quantum optics chiral waveguide quantum-electrodynamical (wQED) systems are attracting a lot of attention from the perspective of fundamental science, and possible interesting applications. In our work we theoretically analyze the eigenstates in a two-excitation domain of an ensemble of two-level atoms that are periodically spaced, and asymmetrically coupled to a guided mode. We found that in a regime when all atoms emit photons in-phase, most eigenstates in such a system can be well-approximated and described through the eigenstates from a single excitation domain, while the rest present a superposition of bound states with two strongly attracting excitations, and states, for which the excitations strongly repel from each other occupying the opposite edges of the system.

1. Introduction

During last several years the topic of chiral quantum systems [1] became of high interest in a quantum optical community. The main reason for this is that emitters that are asymmetrically coupled to propagating photonic modes have already demonstrated in experiments such interesting, and unique properties as, for example, a strong coherent Bragg reflection [2], and optical isolation [3]. From a theory perspective, strongly correlated dissipation-immune photon bound states arising in chiral waveguide-QED systems are of big interest [4, 5] with a recent experimental demonstration [6].

In this work we analyze the eigenstates of a periodic chain of two-level systems that asymmetrically interact with a single guided mode (and in-phase with each other) in a two-excitation domain. We demonstrate that most states in this case can be qualitatively characterized within the rank-1 or rank-2 approximation that corresponds to the existence of two indepenent excitations in the system. We also show that there are states that can not be described within this framework, and which have a form of a superposition made of two strongly attracting excitations in a bound state, and an edge state, for which excitations prefer to occupy opposite edges of the system.

2. Theoretical framework

We consider a periodic 1D chain of identical N two-level atoms (TLAs) with interatomic spacing d that are in the vicinity of a photonic waveguide, and which are coupled asymmetrically to

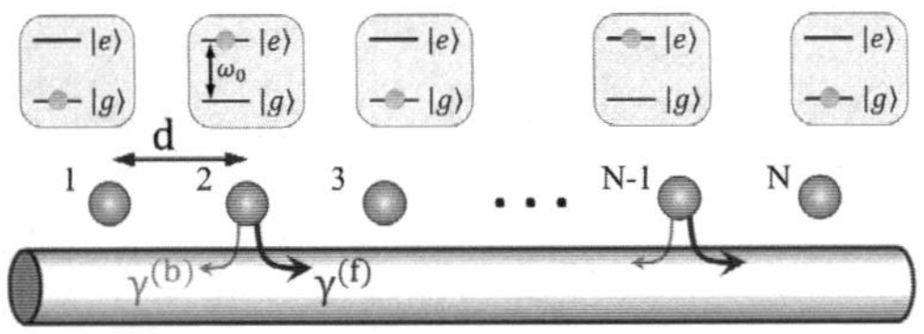

Figure 1. Basic scheme of the system. N two-level atoms are asymmetrically coupled to a single guided mode with the corresponding emission rates $\gamma^{(f)}, \gamma^{(b)}$. Out of N TLAs only 2 are excited simultaneously (atoms 2, and $N-1$ in this Figure)

forward-, and backward propagating guided modes as shown in Fig. 1. The full master equation on a reduced density matrix for such an ensemble has the following form [7]:

$$\dot{\hat{\rho}}_s(t) = i \sum_k \Delta\omega_k \left[\hat{\sigma}_k^+ \hat{\sigma}_k^-, \hat{\rho}_s(t)\right] + \hat{\mathcal{L}}\hat{\rho}_s(t),$$

$$\hat{\mathcal{L}}\hat{\rho}_s(t) = \sum_k \frac{\gamma_{kk}}{2}\left(2\hat{\sigma}_k^- \hat{\rho}_s(t)\hat{\sigma}_k^+ - \hat{\sigma}_k^+\hat{\sigma}_k^-\hat{\rho}_s(t) - \hat{\rho}_s(t)\hat{\sigma}_k^+\hat{\sigma}_k^-\right) +$$

$$\sum_{k,l}^{k\neq l} \frac{\gamma_{kl}}{2}\left(\left[\hat{\sigma}_l^- \hat{\rho}_s(t), \hat{\sigma}_k^+\right] + \left[\hat{\sigma}_k^-, \hat{\rho}_s(t)\hat{\sigma}_l^+\right]\right) + \sum_{k,l}^{k\neq l} \delta_{kl}\left(-i\left[\hat{\sigma}_l^- \hat{\rho}_s(t), \hat{\sigma}_k^+\right] + i\left[\hat{\sigma}_k^-, \hat{\rho}_s(t)\hat{\sigma}_l^+\right]\right), \quad (1)$$

where $\Delta\omega_k = -\omega_{0,k} + \delta_{kk}$, with $\omega_{0,k}$ - transition frequency for atom labeled k, $\hat{\mathcal{L}}$ is the Lindblad superoperator accounting for the description of the dissipative dynamics, $\delta_{kl} = 4\pi k_{0,l}^2 \mathrm{Re}\left[\mathbf{d}_\mathbf{k}^\dagger \mathbf{G}(\mathbf{r_k}, \mathbf{r_l}, \omega_{0,l})\mathbf{d_l}\right]/\hbar$, $\gamma_{kl} = 8\pi k_{0,l}^2 \mathrm{Im}\left[\mathbf{d}_\mathbf{k}^\dagger \mathbf{G}(\mathbf{r_k}, \mathbf{r_l}, \omega_{0,l})\mathbf{d_l}\right]/\hbar$, are responsible for the coherent, and dissipative coupling of atoms, and are described through the classical electromagnetic Green's tensor of the photonic environment $\mathbf{G}(\mathbf{r_k}, \mathbf{r_l}, \omega_{0,l})$.

We want to find the probability that a pair of atoms with numbers k, and $l, k \neq l$ that are excited at time t, given by the respective component of the density matrix $\langle e_k e_l|\hat{\rho}_s(t)|e_k e_l\rangle = \rho_{kl,kl}(t)$, where $|e_k e_l\rangle = |e_k\rangle|e_l\rangle|g\rangle^{\otimes N-2}$. Therefore, one can obtain:

$$\dot{\rho}_{kl,kl}(t) = -\left(\gamma_{k,k} + \gamma_{l,l}\right)\rho_{kl,kl}(t) + \sum_{m\neq k,l}\left(\rho_{km,kl}(t)G_{l,m} + \rho_{kl,km}(t)G_{l,m}^* + \rho_{lm,kl}(t)G_{k,m} + \rho_{kl,lm}(t)G_{k,m}^*\right), \quad (2)$$

with $G_{k,l} = i\delta_{k,l} - \frac{\gamma_{k,l}}{2}$, and by further substituting $\rho_{ij,kl}(t) = C_{i,j}(t)\cdot C_{k,l}(t)$, one arrives at:

$$\dot{C}_{k,l}(t) = -\frac{\gamma_{k,k}+\gamma_{l,l}}{2}C_{k,l}(t) + \sum_{m=k,l}\left(G_{l,m}C_{k,m}(t) + G_{k,m}C_{m,l}(t)\right), \quad (3)$$

where $C_{k,l}(t)$ is a probility amplitude that atoms k, l are excited (assumes $k \neq l$, and $k >$ (or $<$)l). As we consider a set of two-level atoms that are close to a waveguide, and positioned periodically along it, and at the same distance from it's surface, we can use the Green's tensor from [8], which gives $G_{k,l} = -\gamma^{(f)}(\gamma^{(b)})e^{ik^{(g)}|k-l|d}$, where the spontaneous emission rate into forward-propagating mode $\gamma^{(f)}$ is chosen when $k > l$, and $\gamma^{(b)}$ otherwise. Morever, from now on we take $d = 2\pi/k^{(g)}m$, $m \in \mathcal{Z}$, which corresponds to the case when all atoms emit photons in-phase as this will simplify the further analysis significantly.

3. Results

We begin with the analysis of the spectrum, which mathematically corresponds to the eigenfrequencies of normal modes of Eq. (3), and as there are 2 excited atoms out of N, then we have $N(N-1)/2$ eigenstates in general. As one can see from Fig. 2 (a), among all of the eigenstates there are the ones with a noticeable higher values of the Inverse Participation Ratio (IPR) that is given by IPR $= \sum\limits_{n>m} |C^{(j)}_{n,m}|^4 / \big(\sum\limits_{n>m} |C^{(j)}_{n,m}|^2 \big)^2$ - a common measure of a degree of localization. The probability distributions of these states can be described as a linear superposition of a bound state, when two excitations are strongly localized near each other ($|C_{n,m}| \sim e^{-|n-m|}$), and states, for which they are located at different chain edges $|C_{n,m}| \sim e^{|n-m|-N}$ (see label (1) in Fig. 2 (a), and Fig. 2 (b)). However, the states with larger frequency shifts from a bare atomic resonance $\Delta\omega_j = \omega_j - \omega_0$ have a lower IPR (more delocalized, see label (2) in Fig. 2 (a)), and for them both excitations tend to be localized on the right edge (Fig. 2 (c)).

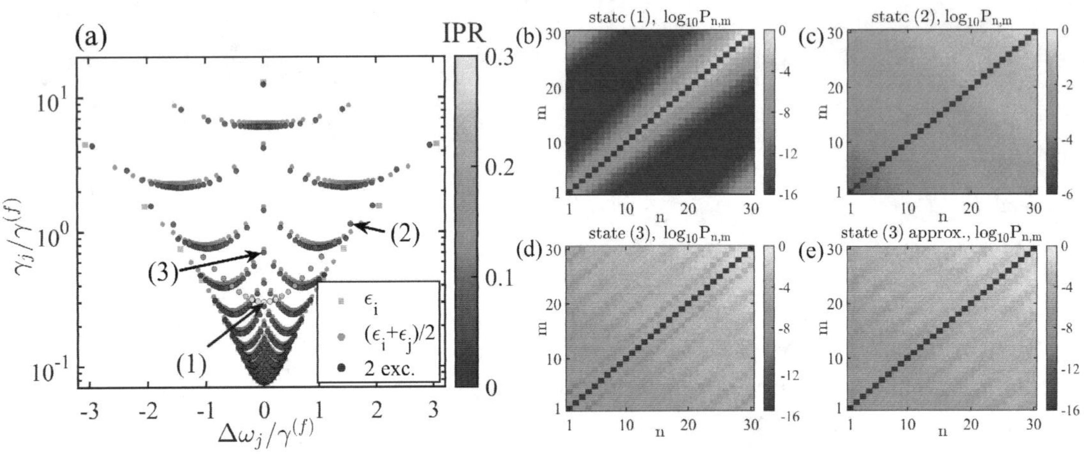

Figure 2. (a) Frequency shifts ($\Delta\omega_j/\gamma^{(f)}$), and emission rates ($\gamma_j/\gamma^{(f)}$) for 2-excitation domain (colored circles). Green squares - single particle spectrum, red circles - half-sums of all combinations of single-particle energies. The color identifies the Inverse Participation Ratio. (b)-(d) Logarithms of the probabilities for atoms n, and m to be excited for the states labeled (1)-(3) in (a). (e) - rank-2 approximation of (d) formed from the corresponding eigenstates from a single excitation domain. The parameters are $N = 30$, $\gamma^{(f)} = 1$, $\gamma^{(b)} = 10^{-2}$.

Most of other eigenenergies can be well described either by a single excitation energies ϵ_i (green squares in Fig. 2 (a)) or as an average of all possible unordered pairs of thereof $(\epsilon_i + \epsilon_j)/2$ ($i \neq j$, red circles in Fig. 2 (a)). This means that the eigenstate in this case can be approximated, roughly speaking, by the following expression $|\psi^{(k)}\rangle \sim |\psi^{(i)}\rangle\langle\psi^{(j)}| + |\psi^{(j)}\rangle\langle\psi^{(i)}|$, which is an eigenstate of the effective Hamiltonian $\hat{H}^{(1)}_{\text{eff}} \otimes \hat{I} + \hat{I} \otimes \hat{H}^{(1)}_{\text{eff}}$ with $\hat{H}^{(1)}_{\text{eff}}$ being the Hamiltonian for a single-excitation domain, which is an effective Hamiltonian describing two independent excitations in a system. However, due to the Pauli principle, one has to multiply the state $|\psi^{(k)}\rangle$ by a term $1 - \delta_{n,m}$ as two excitations can not occupy the same atom as a consequence of a photon blockade ($P_{n,n} = 0$ for any state). This corresponds to rank-2(1) approximation of a 2D representation of a given state k once $i \neq (=)j$, and as seen by comparison of (d) with (e) in Fig. 2 (state labeled (3) in Fig. 2 (a)), this approximation is qualitatevly valid.

4. Conclusions

In this work we have analyzed the doubly excited eigenstates in an ensemble of periodically spaced two-level atoms asymmetrically interacting with a single guided mode. We found that if the system period is a multiple of a wavelength of a guided mode, then most eigenstates can be qualitatively well described by making use of the eigenstates from a single excitation domain, and this can be done by constructing either rank-1 or rank-2 approximate states. We also show that some states can not be desctibed within this framework, and, depending on their frequency shift from a bare atomic resonance ω_0, they gradually morph from a superposition state of strongly attracting (bound states), and repelling (edge states) excitation pairs to states, where both excitations lie at one edge of the system due to the coupling asymmetry.

5. Acknowledgments

This work was supported by the Russian Science Foundation (Grant № 21-72-00096). D. K. also acknowledges support from the Basis Foundation.

References

[1] Lodahl, P., Mahmoodian, S., Stobbe, S., Rauschenbeutel, A., Schneeweiss, P., Volz, J., ... & Zoller, P. (2017). Chiral quantum optics. Nature, **541**(7638), 473-480.
[2] Corzo, N. V., Gouraud, B., Chandra, A., Goban, A., Sheremet, A. S., Kupriyanov, D. V., & Laurat, J. (2016). Large Bragg reflection from one-dimensional chains of trapped atoms near a nanoscale waveguide. Physical review letters, 117(13), 133603.
[3] Sayrin, C., Junge, C., Mitsch, R., Albrecht, B., O'Shea, D., Schneeweiss, P., ... & Rauschenbeutel, A. (2015). Nanophotonic optical isolator controlled by the internal state of cold atoms. Physical Review X, 5(4), 041036.
[4] Mahmoodian, S., Čepulkovskis, M., Das, S., Lodahl, P., Hammerer, K., & Sørensen, A. S. (2018). Strongly correlated photon transport in waveguide quantum electrodynamics with weakly coupled emitters. Physical review letters, 121(14), 143601.
[5] Mahmoodian, S., Calajó, G., Chang, D. E., Hammerer, K., & Sørensen, A. S. (2020). Dynamics of many-body photon bound states in chiral waveguide QED. Physical Review X, 10(3), 031011.
[6] Prasad, A. S., Hinney, J., Mahmoodian, S., Hammerer, K., Rind, S., Schneeweiss, P., ... & Rauschenbeutel, A. (2020). Correlating photons using the collective nonlinear response of atoms weakly coupled to an optical mode. Nature Photonics, 14(12), 719-722.
[7] Dung, H. T., Knöll, L., & Welsch, D. G. (2002). Resonant dipole-dipole interaction in the presence of dispersing and absorbing surroundings. Physical Review A, **66**(6), 063810.
[8] Rao, V. M., & Hughes, S. (2007). Single quantum-dot Purcell factor and β factor in a photonic crystal waveguide. Physical Review B, 75(20), 205437.

Fabrication of metal-dielectric nanoparticles from a bi-layer gold-silicon film by femtosecond laser-induced dewetting

S. V. Koromyslov, E. I. Ageev, E. Yu. Ponkratova and D. A. Zuev

Department of Physics, ITMO University, St. Petersburg 191002, Russia

E-mail: `seregj.koromyslov@metalab.ifmo.ru`

Abstract. Today the world demand for the creation of highly efficient nanoscale white light sources is growing. It happens because energy-efficient information and communication systems are being developed, in which optical signals are replacing electrical signals. For the fabrication of such devices, creating efficient nanoscale white light sources with high efficiency is very acute. Such structures obtained by current methods have a common disadvantages: a small spectral width and low efficiency. Here we demonstrate the development of metal-dielectric structures exploiting the single-step and lithography-free laser-induced dewetting of bi-layer gold silicon films and study their broadband photoluminescence.

1. Introduction

Plasmonic materials (gold and silver) are popular for the fabrication of nano-sized white light sources [1], [2], but they have significant disadvantages, for example, resistive losses. As alternative dielectric nanoparticles are proposed [3], [4] reduce resistive losses and enhance resonant response. Si can be chosen as a dielectric, due to at the wavelength of incident radiation more than 400 nm silicon has dielectric properties [5]. For metal-dielectric nanostructures, it is possible to combine the plasmonic and dielectric components in one nanostructure [6], [7], providing benefits of both components, namely a high degree of localization of electromagnetic energy (plasmonic component) and low optical losses (dielectric component), as well as strong nonlinear optical properties. Recently Makarov and al. [8] demonstrated the formation of spherical nanoparticles made from metal and dielectric with a high refractive index by laser printing. These particles are called "nanosponges," where the pores of a gold sponge are filled with silicon [9]. At the "nanosponges," gold enhances the local optical field because of maintaining localized plasmons [10], while silicon reduces optical losses. Through the nonlinear process (two-) three-photon absorption in gold, hot electrons are injected into the conductance band of silicon[8], and broadband PL is increased because of growing density of electron–hole pairs in Si. A surface patterning of a thin film by femtosecond laser pulses followed by a nanoscale hydrodynamical instability was used for functional plasmonic nanostructures [11] and crystalline silicon subwavelength nanoresonators [12] development. Such controllable femtosecond laser-induced dewetting provides lithography-free and large-scale fabrication technique. Here we apply the laser writing technique to manufacture hybrid metal-dielectric nanoparticles from a bi-layer gold-silicon film. We believe that the dewetting-based mechanism of nanoparticles formation is capable of large-scale fabrication of white light sources [13]. All this makes it possible to control light at the nanoscale, decrease the required processing time, and contribute to a significant increase in broadband photoluminescence efficiency compared with single silicon nanoparticles. Therefore, it will make it possible to use nanoscale light sources in real applications such as optical chips, nano-spectroscopy, broadband near-field sounding of nanophotonic devices or mapping their optical properties associated with local densities of states.

2. Results and discussion

In this work we use Au/Si bi-layer film supported by the glass substrate. Films was made in two stages. First, amorphous silicon layer was deposited on glass substrate by plasma-enhanced chemical vapor deposition from SiH_3 precursor gas. Next, Au layer was deposited on top of Si layer by magnetron sputtering. The thicknesses of the gold and silicon layers are 20 nm and 50 nm, respectively. At the first step, searching of the optimal mode of fabrication of the needed particles was carried out by lithography method. At this project was used Femtosecond Oscillator TiF-100F (Avesta Project). Incident fs laser pulses with wavelength of 790±5 nm, repetition rate (RR) of 80 MHz, and pulse duration of 100 fs were focused in a spot with size of 0.98 μm (this value was obtain by laser calibration). The fluence was varied from 25 μJ/cm^2 to 125 mJ/cm^2 using a combination of Glan prism and half-wave plate, the scanning speed was changed from 5 μm/s to 150 μm/s and the size of the scanned area was changed from 4 μm^2 to 25 μm^2. At the next step, we analyzed geometry of produced structures by SEM imaging and some of structures are represented on Figure 1.

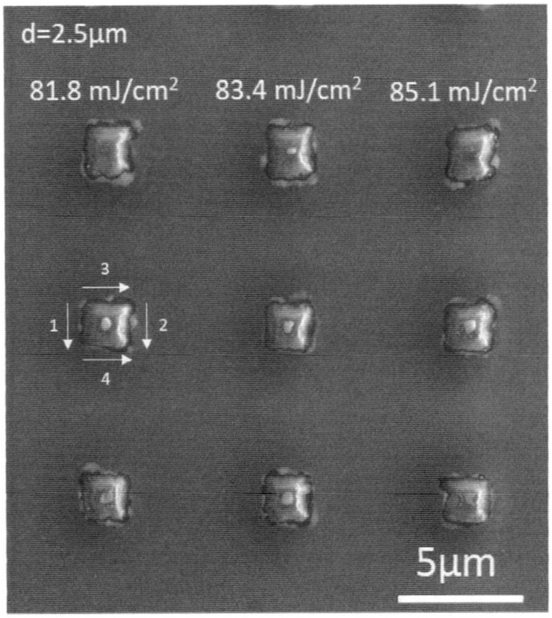

Figure 1. SEM image of obtained particles. The pulse power was changed from 81.8 mJ/cm^2 to 85.1 mJ/cm^2 horizontally, the size of the scanning area was 6.25 μm^2, and the scanning speed was varied within 7-30 μm/s vertically. White arrows and numbers show direction and priority of laser movement.

Further we investigated nonlinear optical properties of such structures. The PL signals were excited by fs laser radiation with central wavelength of 1050±5 nm, RR of 80 MHz and pulse duration of 150 fs. Laser beam was focused by Mitutoyo Plan APO NIR objective (10x, NA=0.26) in a spot with diameter of 4 μm which corresponds to laser fluence of 0.68 mJ/cm^2. Figure 2 shows a SEM image of the typical structure and its PL spectrum. For comparison, we added the spectrum of the initial film. One can see that the obtained particle has the semispherical shape, and its optical response is much higher than that of the virgin bi-layer film. Confirmation of the mechanism of photoluminescence can be seen from

Figure 3, which demonstrates the dependence of the of the intensity of photoluminescence signal from the fluence of the incident femtosecond laser pulse.

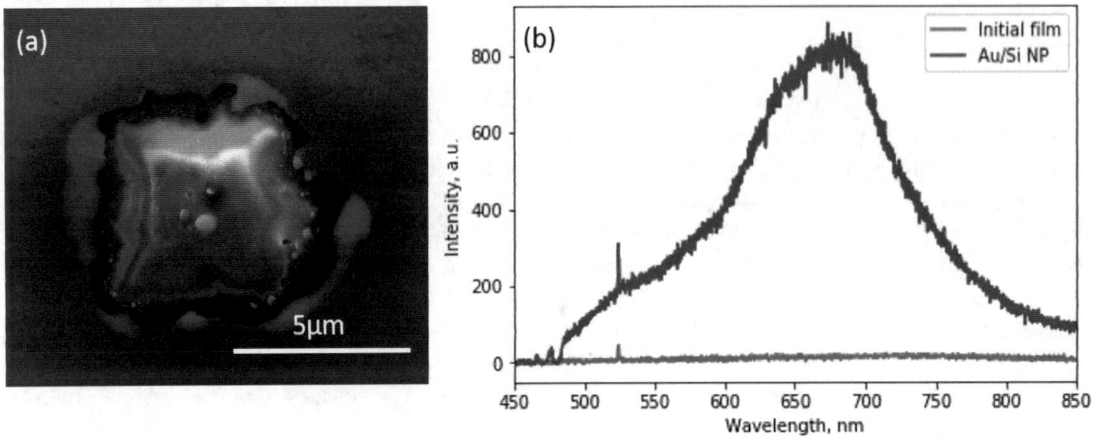

Figure 2. a) SEM image of nanoparticle, which was obtained with fluence of 300.33 mJ/cm^2, scanning speed of 30 μm/s, and a scanning period of 5 μm. b) Shows the photoluminescence of the structure (purple curve) and the intact area (green curve).

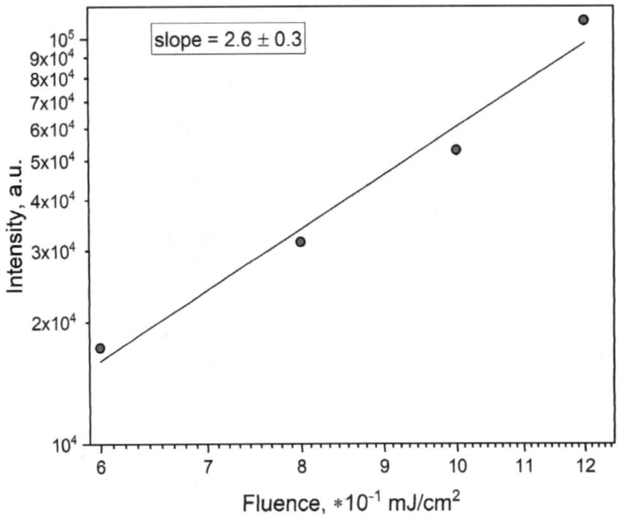

Figure 3. Dependence of the intensity of photoluminescence signal on the fluence of the incident femtosecond laser pulse in the double-log scale. Slope's value approves contribution of two- and three-photon absorption processes to the broadband PL.

3. Conclusion

Thus, it was shown that using the dewetting method, it is possible to obtain particles that are a source of broadband photoluminescence by the reason of nonlinear (two-) three-photon absorption. In the future, the optimal mode of fabrication of the desired particles will be selected, which gives the maximum response, their detailed composition will be studied and the task of manufacturing arrays of such particles will be implemented.

Acknowledgments

This work is supported by the Russian Science Foundation (Project 19-79-10259).

References

[1] Knittel V, Fischer M P, de Roo T, Mecking S, Leitenstorfer A and Brida D 2015 *ACS nano* **9** 894–900
[2] Bouhelier A, Beversluis M R and Novotny L 2003 *Applied Physics Letters* **83** 5041–5043
[3] Zhang C, Xu Y, Liu J, Li J, Xiang J, Li H, Li J, Dai Q, Lan S and Miroshnichenko A E 2018 *Nature Communications* **9** 1–7
[4] De Boer W, Timmerman D, Dohnalova K, Yassievich I, Zhang H, Buma W and Gregorkiewicz T 2010 *Nature nanotechnology* **5** 878–884
[5] Krasnok A, Makarov S, Petrov M, Savelev R, Belov P and Kivshar Y 2015 *Metamaterials X* vol 9502 (International Society for Optics and Photonics) p 950203
[6] Jiang R, Li B, Fang C and Wang J 2014 *Advanced materials* **26** 5274–5309
[7] Wang H, Liu P, Ke Y, Su Y, Zhang L, Xu N, Deng S and Chen H 2015 *ACS nano* **9** 436–448
[8] Makarov S, Sinev I, Milichko V, Komissarenko F, Zuev D, Ushakova E, Mukhin I, Yu Y, Kuznetsov A, Belov P *et al.* 2018 *Nano letters* **18** 535–539
[9] Larin A, Nominé A, Ageev E, Ghanbaja J, Kolotova L, Starikov S, Bruyère S, Belmonte T, Makarov S and Zuev D 2020 *Nanoscale* **12** 1013–1021
[10] Krasnok A E, Maksymov I S, Denisyuk A I, Belov P A, Miroshnichenko A E, Simovski C R and Kivshar Y S 2013 *Physics-Uspekhi* **56** 539
[11] Makarov S V, Milichko V A, Mukhin I S, Shishkin I I, Zuev D A, Mozharov A M, Krasnok A E and Belov P A 2016 *Laser & Photonics Reviews* **10** 91–99
[12] Dmitriev P, Makarov S, Milichko V, Mukhin I, Gudovskikh A, Sitnikova A, Samusev A, Krasnok A and Belov P 2016 *Nanoscale* **8** 5043–5048
[13] Ye J, Zuev D and Makarov S 2019 *International Materials Reviews* **64** 439–477

Journal of Physics: Conference Series **2015** (2021) 012072 doi:10.1088/1742-6596/2015/1/012072

Continuous-Flow Synthesis of Perovskite Particles for Optical Application

Irina Koryakina[1, a)], S Bikmetova[1], K Arabuli[1], A Evstrapov[2], A Pushkarev[1], S Makarov[1] and M V Zyuzin[1]

[1]School of Physics and Engineering, ITMO University, Lomonosova 9, 191002, St. Petersburg, Russia
[2]Institute for Analytical Instrumentation of the Russian Academy of Sciences, Ivana Chernykh 33A, 198095, St. Petersburg, Russia

[a)] Corresponding author: irina.koryakina@metalab.ifmo.ru

Abstract. Conventional chemical synthesis methods for nanomaterials fabrication do not always provide precise reproducible synthesis with controllable physicochemical parameters. A better control over the resultant properties of nanomaterials can be achieved by scaling down the synthesis approach to the microlevel. This can be realized by means of microfluidics. In this work, a microfluidic approach is applied for the controllable synthesis of perovskite particles with defined cubic morphology. The structural characterization of perovskite particles is performed using scanning electron microscopy and X-ray diffraction. All the synthesized particles demonstrate photoluminescence.

1. Introduction

Over the past few decades, fabrication of novel nanostructured materials with defined properties has gained a lot of attention. Today, conventional methods are challenged by novel, more convenient ones. Namely, microfluidic technology offers an alternative approach for the synthesis and characterization of nanomaterials with controllable properties.

Microfluidic synthesis of nanoparticles has a number of advantages over conventional synthesis methods, including efficient mixing, increased heat and mass transfer, precisely controlled nucleation and growth, as well as the ability to scale the synthesis outcomes. These intrinsic properties, combined with the ability to automate multiple processes on a single microchip, make microfluidics attractive as a platform for the fabrication of nanoparticles with controllable size, shape, and other properties [1].

Among known nanomaterials, perovskites are intensively investigated due to their specific properties such as superb photovoltaic [2] and optical [3] performance. The perovskites are often utilized as a component of commercially available solar cells [4]; besides, they show high performance as lasing cavities [5] and sensors [6]. Therefore, novel controllable approaches for the synthesis of perovskites with defined shapes, e.g., cubic shape [7], are highly demanded.

In this work, a microfluidic continuous-flow synthesis method is applied to demonstrate the possibility of obtaining crystalline cubic-shaped perovskite particles.

2. Materials and methods

Materials for microfluidic chip (MFC) fabrication: SU-8 photoresist, and polydimethylsiloxane (PDMS, SYLGARD™ 184 Silicone Elastomer kit, Dow Corning, USA) that consists of two parts: silicone elastomer base (PDMS base) and curing agent. Glass substrate (Menzel Microscope Coverslips).

Materials for perovskite particles synthesis: cesium bromide (CsBr, 99.999%, Sigma-Aldrich), lead (II) bromide ($PbBr_2$, 99.999%, Alfa Aesar), anhydrous dimethyl sulfoxide (DMSO, 99.8%, Alfa Aesar), isopropyl alcohol (IPA, technical grade, 95%, Vecton), and toluene (technical grade, Vecton).

Methods: to fabricate the MFC, a mold was made by photolithography method, using a silicon wafer and SU-8 photoresist. Further, uncured PDMS with a ratio 10:1 of elastomer base to curing agent was mixed and poured onto the mold, and then polymerized at 80°C for 2h. Afterwards, the inlet/outlet holes were made in the PDMS replica, and the PDMS replica was sealed to a glass substrate by O_2 plasma treatment.

The perovskite precursor was prepared by dissolving 0.3 M of CsBr and 0.3 M of $PbBr_2$ in 3 mL of anhydrous dimethyl sulfoxide (DMSO) in a N_2-filled glovebox.

The setup for synthesis consisted of a pressure controller and an optical microscope for observation (Fig. 1). The setup is described in detail in [8].

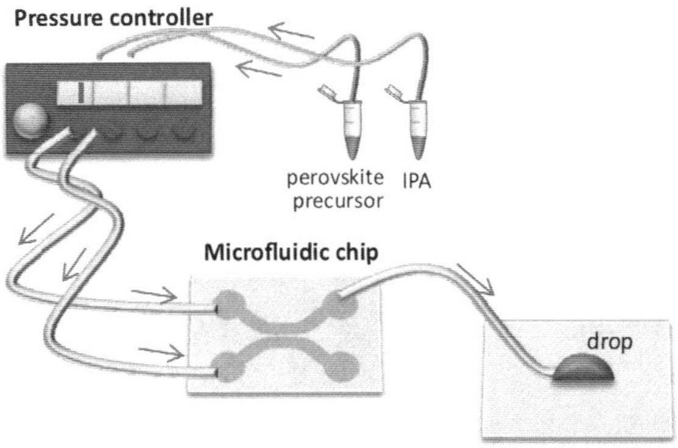

FIGURE 1. Schematic illustration of the synthesis setup [8].

To investigate the optical properties of the obtained structures, the Axio Imager A2m, Carl Zeiss microscope operating in the fluorescent regime was used.

For the structural characterization, a scanning electron microscope (SEM), Quanta 200, FEI, Netherlands, with an acceleration voltage of 10 kV, was used.

3. Results and Discussion

The perovskite structures were obtained with the microfluidic approach by changing the pressure in the MFC channels. The topology with two parallel channels described in [8] was used. In the current work, we found the following optimal settings for the synthesis: a pressure of P = 6 kPa for the IPA channel and P = 60 kPa for the perovskite precursor channel yields cubic-shaped perovskite structures.

SEM analysis was performed to examine the resulting structures' morphology (Fig. 2). The obtained structures possess a narrow variety of morphology with an average size of 4.2 ± 1 μm.

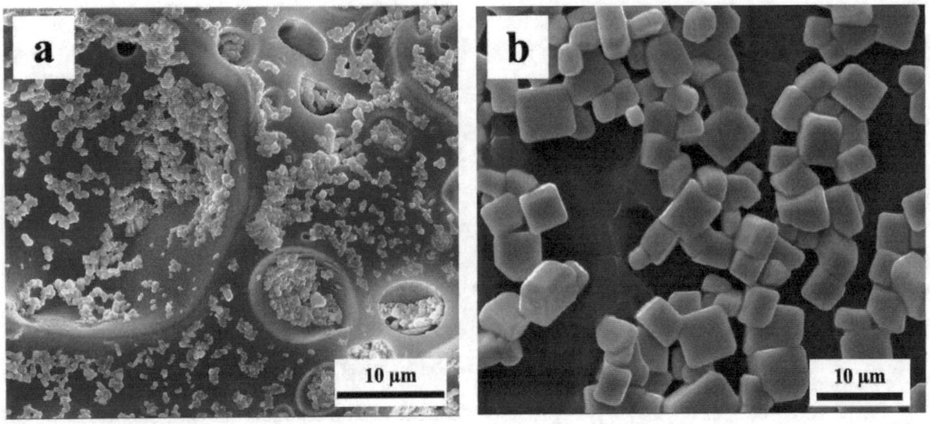

FIGURE 2. SEM image of the obtained $CsPbBr_3$ cubic-shaped perovskites. a. On a porous PDMS surface;
b. Resulting structures (P=6 kPa/60 kPa).

The powder X-ray diffraction analysis revealed an orthorhombic *Pnma* crystal structure of the obtained particles [9]. All the reflexes from the reference diffractogram (black curve in Fig.3) are present in the XRD spectrum of the obtained structures (green curve in Fig.3). The dominant peaks correspond to the (121) and (040) crystallographic planes. Intense reflexes of the lattice planes (020), (040), (101), and (202) in the diffractogram show the directional growth of the nanocrystals.

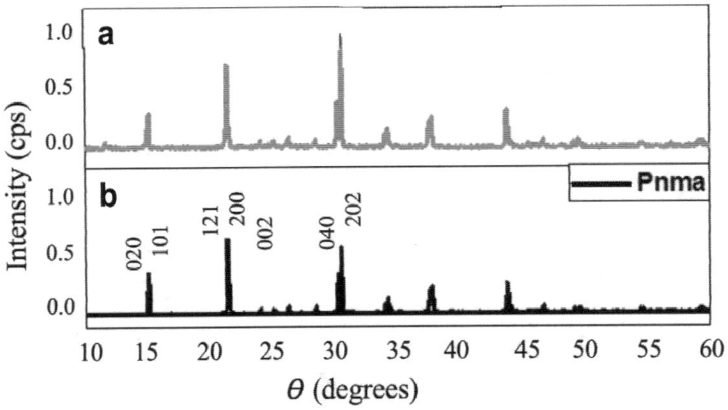

FIGURE 3. XRD of the perovskite particles.
a – experimentally obtained structures XRD, b – reference data [9].

All the obtained structures demonstrated photoluminescence when exposed to λ=351 nm (Fig.4).

The obtained results prove that microfluidic approach is suitable for the synthesis of monodisperse cubic structures that are able to demonstrate lasing properties, as shown in [7]. Besides that, microfluidics possesses a potential to realize in-line characterization of optical properties, as presented in [10].

By tuning the pressure in the MFC channels, various morphologies can be obtained. It is important to mention that the optical properties of the perovskites strongly depend on their shape and size. Thus, by varying the microfluidic synthesis conditions, such as pressure in the channels, flow rate of the reagents, and time of the reaction, it is possible to obtain structures with different optical properties.

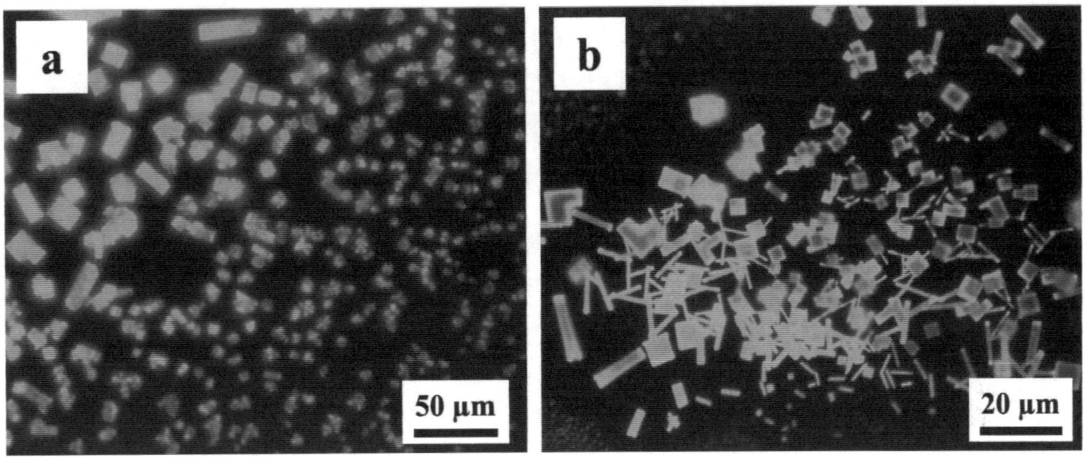

FIGURE 4. PL image of the obtained perovskite structures.

4. Conclusion

Cubic-shaped $CsPbBr_3$ particles were synthesized using microfluidics. The microfluidic synthesis approach is proven to be effective and convenient due to the low reagents' consumption, faster reaction time, and a high crystallinity of the resulting reaction product. Combined with a high rate of synthesis on the microscale, which gives a higher yield of resulting structures, the microfluidic method has a great potential for implementing the synthesis of nanoparticles with controllable properties, integrating in-line characteristics, and automating multiple processes in a single microchip.

5. References

[1] Zhao C-X & Middelberg A P J 2016 *Handbook of Nanoparticles* (Cham: Springer International Publishing) chapter 18 pp 455-473
[2] Wang Y *et al.* 2020 *Advanced Materials* **32** 2000186
[3] Li F *et al.* 2019 *Science Bulletin* **64** 698-704
[4] Park N G, Grätzel M, Miyasaka T, Zhu K, & Emery K 2016 *Nature Energy* **1** 1-8
[5] Zhou B *et al.* 2018 *Journal of Materials Chemistry C* **6** 11740-11748
[6] Kim J *et al.* 2020 *ACS Materials Letters* **2** 1368-1374
[7] Tiguntseva E *et al.* 2020 *ACS nano* **14** 8149-8156
[8] Koryakina I, Markina D, Evstrapov A, Pushkarev A, Makarov S & Zyuzin M V 2020 *AIP Conference Proceedings* **2300** 020065
[9] Stoumpos C C *et al.* 2013 *Crystal Growth & Design* **13** 2722–2727
[10] Lignos I, Stavrakis S, Kilaj A & deMello A 2015 *MRS Online Proceedings Library* **1780**1-6

Acknowledgments
This work was supported by the Russian Science Foundation (project № 21-73-20189).

Superscattering for non-spherical objects

S. D. Krasikov[1], M. A. Odit[1,2], D. A. Dobrykh[1,3], I. M. Yusupov[1], A. A. Mikhailovskaya[1,3], D. T. Shakirova[1], A. A. Shcherbakov[1], A. P. Slobozhanyuk[1], P. Ginzburg[3,4], D. S. Filonov[4], A. A. Bogdanov[1,]

[1]School of Physics and Engineering, ITMO University, St. Petersburg 197101, Russia
[2]Electrotechnical University LETI, St. Petersburg, 197376, Russia
[3]School of Electrical Engineering, Tel Aviv University, Tel Aviv 69978, Israel
[4]Center for Photonics and 2D Materials, Moscow Institute of Physics and Technology, Dolgoprudny 141700, Russia

E-mail: s.krasikov@metalab.ifmo.ru

Abstract. In this work we generalize the notion of superscattering and associate it with a symmetry group of a scattering object. Using the group theory approach we describe a way to spectrally overlap several eigenmodes of a resonator in order to achieve scattering enhancement. Importantly, this can be done by simple variation of geometric parameters of the system, implying that the symmetry is preserved. We also demonstarte that a scattering cross-section limit of a spherical object is not valid for the case of non-spherical geometries. As an example, we use finite-size ceramic cylinder and demonstrate that a dipolar scattering cross-section limit of a spherical object can be exceeded by more then 3 times. The obtained results may be promising for design of antennas and radio frequency identification systems.

1. Introduction

Electromagnetic scattering is a phenomena which plays a crucial role in a wide range of practical applications. Antenna devices, radars, and radio frequency identification (RFID) technologies are based on efficient manipulation of scattering. One of the challenging problems is the miniaturization of resonant elements, since the scattering peak bandwidth decreases with the size of the scatterer (Chu-Harrington limit [1, 2]). Possibly the only way to enhance scattering cross-section (SCS) of subwavelength particles is to spectrally overlap several resonances. The scattered field can be expanded into a series of vector spherical harmonics (multipoles), associated with the far-field of a point multipole (dipole, quadrupole, octupole, etc.) [3]. Multipoles are orthogonal and therefore represent independent scattering channels and the total SCS (C_{sct}) can be considered as the sum of partial cross-sections. For objects with spherical symmetry partial cross-sections are limited by the value [4, 5]

$$C^{\ell}_{\text{max}} = \frac{2\ell + 1}{2\pi}\lambda^2, \tag{1}$$

where ℓ is the orbital angular momentum and λ is the wavelength. The scattering cross-section reaches this limit at the Mie resonance condition. But it is possible to overcome the limit by spectral overlapping of several resonances. In this case the scattering structure is usually called a superscatterer [6, 7]. It should be highlighted that the notation of superscattering is applicable only if scattering channels can be identified. For that, there has to be a one-to-one correspondence between a resonance of a structure and a single scattering channel. This is possible only for high-symmetry objects, such as spheres and two-dimensional cylinders [6, 8, 9]. In other cases, the single-channel limit Eq. 1 defined for spherical objects is not valid, since in a non-spherical case there is a rescattering between the channels.

In this work we provide a generalization of superscattering notion for objects of non-spherical symmetry. As an illustrative example we use finite-size cylindrical resonator and demonstrate that collocation of several resonances of different symmetry can significantly enhance the SCS.

2. Generalization of superscattering for non-spherical objects

Electromagnetic scattering can be analyzed via T-matrix [10] which links incident and scattered fields via the relation

$$\mathbf{b} = \hat{T}\mathbf{a}, \tag{2}$$

where **a** and **b** are the vectors of complex amplitudes, corresponding to the incident and scattered fields, respectively. For the case of a spherical resonator $\hat{T}$ is diagonal, which means that there is no coupling between different multipoles and the field can only be scattered into specific channels. For example, a dipole harmonic of the incident field can only be scattered into the dipole channel.

The situation is different for non-spherical objects since in this case the multipoles of the scattered field are partially mixed. However, they still form independent sets linked to eigenmodes of a resonator, which can be classified with different symmetry groups. Therefore, for the case of non-sperical resonators T-matrix is block-diagonal [11]:

$$\hat{T} = \mathrm{diag}\{\hat{T}_1, \hat{T}_2, ...\}, \tag{3}$$

where the blocks $\hat{T}_s$ correspond to the modes of different symmetry. If modes are transformed differently under the operations of the resonator's symmetry group, they are affiliated with two different blocks. Similarly, transformation properties of the multipoles defines a block into which they are contributing. Therefore, mixing rules can be formulated using a group theory approach.

The multipoles contributing to the same block of $\hat{T}$ form an orthogonal basis of the *irreducible representations* (irreps) of the resonator's symmetry group [11]. The number of blocks is equal to the number of its irreducible representations [12]. Importantly, different types of modes and irreducible representations are related through a one-to-one correspondence (the Wigner theorem) [13] and therefore, the independent scattering channels are associated with irreducible representation but not with particular multipoles. This is a crucial difference between spherical and non-spherical resonators.

The total SCS can be represented as a sum of partial SCSs corresponding to the different irreducible representations:

$$C_{\mathrm{sct}} = \sum_{s=\mathrm{irreps}} \sum_{\ell,m,\sigma} C_{\ell m}^{\sigma s}. \tag{4}$$

where σ is the polarization and ℓ is the angular momentum. Therefore, scattering enhancement or superscattering for the non-sperical resonator can be achieved via spectral overlapping of several modes from different irreducible representations or, in other words, from different blocks of T-matrix.

3. Maximization of scattering cross-section for cylindrical resonator

As an illustrative example, we would like to consider a finite-size cylindrical resonator made of ceramics with permittivity $\varepsilon_{\mathrm{cyl}} = 44.8$ and and loss tangent $\tan\delta = 10^{-4}$. The cylinder is illuminated by the TE-polarized incident wave [see Fig. 1(a)]. The color map in Fig. 1(b)

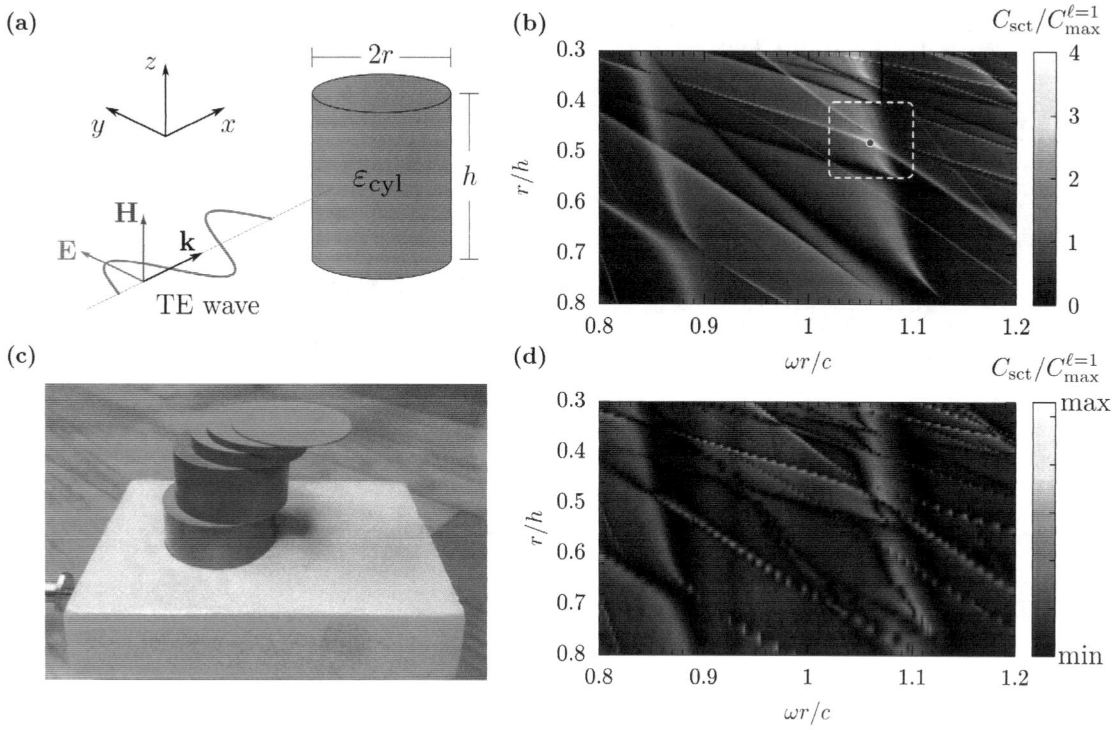

Figure 1. (a) Schematic picture of the considered system. The cylinder with height h, radius r and permittivity $\varepsilon_{\rm cyl} = 44.8$ is illuminated by a TE-polarized plane wave. Colormap (b) illustrates the numerically calculated SCS as the function of the normalized frequency $\omega r/c$ and cylinder's aspect ratio r/h. Here, values of SCS are normalized to the corresponding dipole single-channel limit [Eq. (1)]. The blue dot indicate one of the most pronounced peaks of SCS and the white dashed rectangle indicates the area around it. (c) Example of ceramic cylinders used for the experimental measurements. Each cylinder has different height but the same radius $r = 15.7$ mm. Dielectric permittivity of the sample is $\varepsilon_{\rm cyl} = 44.8$ and loss tangent $\tan\delta = 10^{-4}$. (b) The measured map of the SCS spectra.

demonstrates the evolution of modes as a function of parameters of the system, namely, the aspect ratio and the frequency. The calculated values of $C_{\rm sct}$ are normalized to the single-channel dipolar limit for spherical objects [Eq. (1), $\ell = 1$] to highlight the relative enhancement. It can be seen that SCS prevails the dipolar limit by a factor of 3-4 in the vicinity of the points where several modes of the cylinder are overlapped. Importantly, this enhancement can be achieved by tuning of only one geometric parameter of the resonator, contrary to the cases where additional coating layers or surface structuring is needed, as for example in Refs. [8, 9]. Therefore, the proposed design is very simple and can be easily implemented in practice.

The same result also was demonstrated experimentally. For that the manufactured ceramic rod with radius 16.2 mm was sliced into several sections of different heights, which allow to set the height of the resonator within the range from 0.25 mm to 15 mm with the step 0.2 mm. Figure 1(c) shows an example of several slices. The rod is manufactured by sintering ceramic powder of calcium titanate-lanthanum aluminate ($LaAlO_3$-$CaTiO_3$) into a solid. The permittivity of the rod and loss tangent are the same as those used for numerical calculations, $\varepsilon_{\rm cyl} = 44.8$ and $\tan\delta = 10^{-4}$ [14]. The rod was placed on a foam holder made of Penoplex, which is transparent in GHz region. Figure 1(d) demonstrates the measured map of the SCS.

The results are in a good agreement with the numerical calculations and the provided map also demonstrates that maximal SCS is achieved at the points where modes are overlapped.

4. Modal and multipole description of scattering enhancement

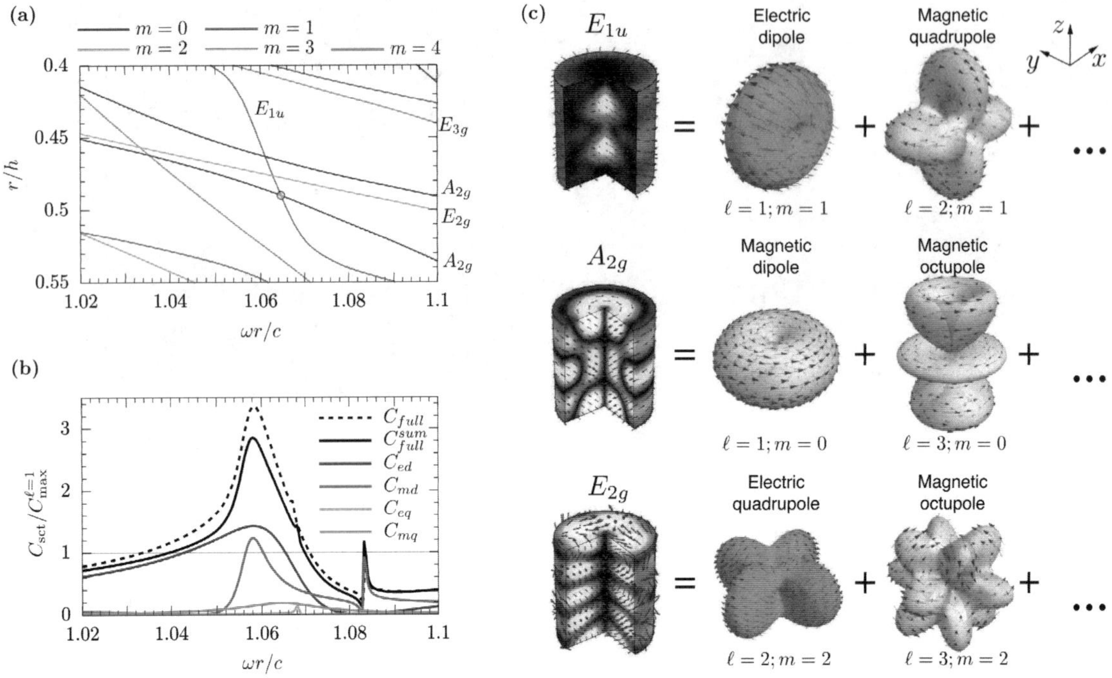

Figure 2. (a) Eigenmodes and irreducible representations of the resonator calculated in the vicinity of the SCS peak. (b) Multipole decomposition of the total SCS at $r/h = 0.48$. Solid black line corresponds to the result of summation of the partial SCSs (ED, MD, EQ, and MQ) shown by colored lines and dashed black line corresponds to full-wave numerical calculations. (c) Characteristic field distributions of eigenmodes corresponding to the irreducible representations E_{1u}, A_{2g} and E_{2g}, and their multipole far-field decompositions.

The modes of a finite cylinder can be characterized by azimuthal number $m = 0, 1, 2, ...$ and parity (odd or even) of the electric field distribution with respect to the mirror reflection $z \to -z$. However, contrary to spherical resonators, an orbital quantum number ℓ is not a good quantum number for a non-spherical resonator. To explicitly demonstrate the relation between the scattering enhancement and spectral overlapping of the modes associated to different irreducible representations, we calculate the eigenmode spectrum as a function of aspect ratio for the area marked in Fig. 1(b) by the white dashed rectangle. The spectrum obtained with COMSOL Multiphysics® software is demonstrated in Fig. 2(a). This result clearly shows that only the modes from different irreducible representation can intersect while modes from the same irreducible representations form avoid crossings. Associated examples of the electric field distributions and the multipole content for several eigenmodes from different irreducible representations are shown in Fig. 2(c).

Let us consider one of the points where SCS reaches a maximal value. In particular, when $r/h = 0.48$ [see blue dot in Fig. 1(b)] SCS prevails the dipolar single-channel limit ($\ell = 1$) for a spherical object by more than three times. This enhancement is a result of spectral overlapping of two modes from different irreducible representations, A_{2g} and E_{1u} (we use the standard notations [15]). The dominant contribution to SCSs for these modes is given by the electric and magnetic dipoles, as Fig. 2(b) demonstrates. It should be highlighted that the partial SCSs corresponding to the electric and magnetic dipole moments exceed the single-channel limit for a sphere due to rescattering between the channels corresponding to different multipoles. Therefore, the single-channel limit for a spherical resonator is not applicable for a non-spherical one. However, the question of whether the limit for low-symmetry objects exists or not, remains open and it will be the subject of further research.

5. Conclusion
In this work the concept of superscattering is generalized for the case of non-spherical resonators. The enhancement of scattering cross-section can be achieved via spectral overlapping between eigenmodes of different symmetry, which was demonstrated for cylindrical resonator with finite size. Importantly, overlapping can be achieved by simple tuning of geometrical parameters of the system, which for the case of cylindrical resonator is its aspect ratio. It is also demonstrated that the partial SCS of a non-sperical object may exceed the corresponding single-channel limit for a sphere, hence, the limit for non-spherical objects has to be introduced, which is an open question yet.

Acknowledgments
The Authors thank Kristina Frizyuk for fruitful discussions and Elizaveta Nenasheva (CEO of Ceramics Co., Ltd.) for sample fabrication. The research was supported by the Russian Science Foundation (Project 19-79-10232). A.B. acknowledges the RFBR (19-02-00419), Grant of the President of the Russian Federation (MK-2224.2020.2) and the BASIS foundation.

References
[1] Harrington R F 1960 *J. Res. Nat. Bur. Stand* **64** 1–12
[2] Chu L J 1948 *J. Appl. Phys.* **19** 1163–1175
[3] Evlyukhin A B, Fischer T, Reinhardt C and Chichkov B N 2016 *Phys. Rev. B* **94** 205434
[4] Foot C J *et al.* 2005 *Atomic Physics* vol 7 (Oxford University Press)
[5] Tribel'skii M I 1984 *JETP* **86** 915–919
[6] Ruan Z and Fan S 2010 *Phys. Rev. Lett.* **105** 013901
[7] Ruan Z and Fan S 2011 *Appl. Phys. Lett.* **98** 043101 ISSN 0003-6951
[8] Qian C, Lin X, Yang Y, Xiong X, Wang H, Li E, Kaminer I, Zhang B and Chen H 2019 *Phys. Rev. Lett.* **122** 063901
[9] Mirzaei A, Miroshnichenko A E, Shadrivov I V and Kivshar Y S 2014 *Appl. Phys. Lett.* **105** 011109
[10] Mishchenko M I, Travis L D and Mackowski D W 1996 *J. Quant. Spectrosc. Radiat. Transfer* **55** 535–575
[11] Kahnert M 2005 *JOSA A* **22** 1187–1199
[12] Tinkham M 2003 *Group Theory and Quantum Mechanics* (Mineola, N.Y: Dover Publications) ISBN 978-0-486-43247-2
[13] Wigner E P 1959 *Group Theory and Its Application to the Quantum Mechanics of Atomic Spectra* (*Pure and Applied Physics* no 5) (New York, NY: Academic Press) ISBN 978-0-323-15278-5
[14] Nenasheva Y A 1992 *MRS Online Proceedings Library Archive* **269** ISSN 0272-9172, 1946-4274
[15] Ivchenko E L and Pikus G 1995 *Superlattices and Other Heterostructures: Symmetry and Optical Phenomena* Springer Series in Solid-State Sciences (Berlin Heidelberg: Springer-Verlag) ISBN 978-3-642-97589-9

Bosonic Hofstadter butterflies in the presence of phonon modes

Yury Krivosenko, Ivan Iorsh, and Ivan Shelykh

E-mail: y.krivosenko@gmail.com
ITMO Univertisy, St. Petersburg 197101, Russia

Abstract. We have investigated the influence of lattice sites vibrations on bosonic Hofstadter's butterfly (HB) spectrum for the case of conventional square lattice. Only the pair of specific phonon modes with opposite quasimomenta has been taken into account. The study has shown that HB-type spectra can be substantially modified by the presence of phonons, depending on the ratios of such parameters of the system as site-to-site transition amplitudes, particle-phonon interaction constant, and characteristic vibration frequency.

1. Introduction

Since the invention in 1976 [1], the renowned Hofstadter's butterflies have been thoroughly studied for various lattice geometries [2–5] and in different implementations [6–11]. From this point of view, the behaviour of both fermionic and bosonic particles in two-dimensional systems subject to the magnetic or other gauge field seems to be scrupulously investigated.

On the other hand, the lattice vibrations within the second quantization formalism and tight binding approach has been recently attracted attention [12, 13]. Unfortunately, the exploration of behaviour of bosonic particles affected by both gauge field and lattice vibrations within the scope of Hofstadter butterflies can be hardly found.

In the present work, we research bosonic Hofstadter butterflies altered by the presence of lattice phonons. Tight binding approximation and the second quantization formalism are utilized throughout the paper.

2. Theory

We start with the conventional tight-binding Hofstadter Hamiltonian yielding the butterfly spectrum:

$$\widehat{H}_0 = t \sum_{n_1,n_2} \left(\hat{a}^{\dagger}_{n_1+1,n_2} \hat{a}_{n_1,n_2} + \hat{a}^{\dagger}_{n_1,n_2+1} \hat{a}_{n_1,n_2} \mathrm{e}^{-2\pi i \xi n_1} + \mathrm{h.c.} \right). \tag{1}$$

The lattice considered is square, only the nearest neighbour hoppings are taken into account. Operators $\hat{a}^{(\dagger)}_{n_1,n_2}$ are the bosonic annihilation (creation) operators at the site (n_1,n_2) (the indices 1 and 2 correspond to x and y axes, respectively). The tunneling amplitude equals t, ξ is the ratio of the magnetic field flux through the unit cell to the quantum of magnetic field flux. In the present paper, we scrutinize rational magnetic fields implying $\xi = p/q$, where p and q are coprime integers. Hereinafter, the magnetic field is treated within Landau gauge, i.e. $\vec{\mathscr{A}} = (0, Bx, 0)$.

Within the assumption of rational fields, Hamiltonian (1) can be transformed into the reciprocal space. In that case, the unit cell should be extended in the x-direction so as to possess q sites. Axis y remains, in the sense, unchanged. Redefining n_1 to enumerate the supercell, and setting s to itemize the sites within each supercell, we arrive at

$$\widehat{H}_0 = t \sum_{k_1,k_2} \left\{ \sum_{s=0}^{q-2} \hat{a}^{\dagger}_{k_1,k_2,s+1} \hat{a}_{k_1,k_2,s} + \hat{a}^{\dagger}_{k_1,k_2,1} \hat{a}_{k_1,k_2,Q-1} \mathrm{e}^{-ik_1} + \sum_{s=0}^{q-1} \hat{a}^{\dagger}_{k_1,k_2,s} \hat{a}_{k_1,k_2,s} \mathrm{e}^{-2\pi i \xi s} \mathrm{e}^{-ik_2} + \text{h.c.} \right\} \quad (2)$$

$$= t \sum_{k_1,k_2} \hat{\vec{a}}^{\dagger}_{k_1,k_2} \left\{ \widehat{I}'_q + \widehat{I}''_q \mathrm{e}^{-ik_1} + \widehat{\Xi}_q \mathrm{e}^{-ik_2} + \text{h.c.} \right\} \hat{\vec{a}}_{k_1,k_2}, \quad (3)$$

where $\hat{\vec{a}}_{k_1,k_2} = (a_{k_1,k_2,1}, a_{k_1,k_2,2}, \ldots a_{k_1,k_2,q-1})^{\mathsf{T}}$, and the rest matrices are

$$\widehat{I}'_q = \begin{pmatrix} 0 & & & \\ 1 & \ddots & & \\ & \ddots & \ddots & \\ & & 1 & 0 \end{pmatrix}_q, \quad \widehat{I}''_q = \begin{pmatrix} 0 & & 1 \\ & \ddots & \\ & & 0 \end{pmatrix}_q, \quad \widehat{\Xi}_q = \operatorname{diag}\left(\mathrm{e}^{-2\pi i \xi s}\right)\Big|_{s=0,1,\ldots(q-1)} \quad (4)$$

(All the empty positions are set to zero, the index q explicitly shows the dimension of the matrices.)

The influence of the gauge field reveals in the phase factor $\mathrm{e}^{-2\pi i \xi n_1}$ in (1) that transforms into $\mathrm{e}^{-2\pi i \xi s}$ in (3) and (4). This factor is calculated as linear integral of $\vec{\mathscr{A}}$ (with the corresponding premultipliers) along the straight path between the sites. For Landau gauge, the following expression can be derived:

$$\varphi_{1\to 2} = \frac{\pi\xi}{L^2} \left(Y_2^0 - Y_1^0 + y_2 - y_1\right) \left(X_2^0 + X_1^0 + x_1 + x_2\right) \quad (5)$$

$$\approx \underbrace{\frac{\pi\xi}{L^2} \left(Y_2^0 - Y_1^0\right) \left(X_2^0 + X_1^0\right)}_{\varphi^0_{1\to 2}} + \frac{\pi\xi}{L^2} \left[\left(Y_2^0 - Y_1^0\right)(x_2 + x_1) + \left(X_2^0 + X_1^0\right)(y_2 - y_1)\right]. \quad (6)$$

Here, the capital characters X and Y with 0 superscript stand for the equilibrium positions of the lattice sites, the lowercase characters are the corresponding displacements. The subscripts 1 and 2 respectively indicate initial and final states of tunneling. The zero-model phase factor $\varphi^0_{1\to 2}$ is that calculated for the frozen vibrations. In (6), only the terms up to the first order in x and y are kept (the vibrations are believed to be small compared to the intersite distance, L). One of the major assumptions made here is that we consider only the vibrations in y-direction. (If we assume the vibrations along one of the axes to be frozen, then the choice of the exact direction is arbitrary.) Therefore,

$$\varphi_{1\to 2} \approx \varphi^0_{1\to 2} + \frac{\pi\xi}{L^2} \left(X_2^0 + X_1^0\right)(y_2 - y_1). \quad (7)$$

Another way the vibrations can manage the tunnelings and phases is the particle-phonon interaction that can be included via the effective interaction constant, λ, and alters the tunneling amplitudes. Then, the Hamiltonian of the system can be put down as

$$\widehat{H} = \sum_{n_1,n_2} \left\{ \hat{a}^{\dagger}_{n_1+1,n_2} \hat{a}_{n_1,n_2} \mathrm{e}^{-\frac{\pi i \xi}{L}(2n_1+1)(y_{n_1+1,n_2} - y_{n_1,n_2})} \right.$$
$$\left. + \hat{a}^{\dagger}_{n_1,n_2+1} \hat{a}_{n_1,n_2} \mathrm{e}^{-2\pi i \xi n_1} \mathrm{e}^{-\frac{2\pi i \xi n_1}{L}(y_{n_1,n_2+1} - y_{n_1,n_2})} \left[1 - \lambda (y_{n_1,n_2+1} - y_{n_1,n_2})\right] + \text{h.c.} \right\} + \widehat{H}_{\text{vib}}$$
$$\approx \widehat{H}_0 + \widehat{H}_{\text{vib}}$$
$$- t \sum_{n_1,n_2} \left\{ \hat{a}^{\dagger}_{n_1,n_2+1} \hat{a}_{n_1,n_2} (y_{n_1,n_2+1} - y_{n_1,n_2}) \mathrm{e}^{-2\pi i \xi n_1} \left(\lambda + \frac{2\pi i \xi n_1}{L}\right) \right. \quad (8)$$
$$\left. + \hat{a}^{\dagger}_{n_1+1,n_2} \hat{a}_{n_1,n_2} (y_{n_1+1,n_2} - y_{n_1,n_2}) \frac{i\pi\xi(2n_1+1)}{L} + \text{h.c.} \right\} = \widehat{H}_0 + \widehat{H}_{\text{vib}} + \widehat{H}'. \quad (9)$$

The vibrations Hamiltonian $\widehat{H}_{\text{vib}}$ can be written in the form of second quantization as well ($\hbar = 1$):

$$\widehat{H}_{\text{vib}} = \sum_{\varkappa,\sigma} \omega_{\varkappa,\sigma} \hat{b}^{\dagger}_{\varkappa,\sigma} \hat{b}_{\varkappa,\sigma}. \tag{10}$$

If only the nearest neighbours interact, the spectrum of phonons can be obtained in analytic form:

$$\omega_{\varkappa,\sigma} = 4\omega_0 \sin^2\left(\frac{\varkappa}{2}\right), \tag{11}$$

where $\varkappa$ is the quasimomentum (one-dimensional), and σ denotes the phononic branch that is actually the only one, acoustic. The spectrum of phonons is q-fold degenerate, the matrix to be diagonalized is proportional to the identity matrix of shape $q \times q$. Our goal here is to consider the influence of one specific phonon mode, that is distinguished by the motion of the entire supercell as a whole. The normalized eigenvector for the mode is $(1,1,\dots 1)/\sqrt{q}$. Afterwards, the displacements can be expressed in terms of creation and annihilation operators as

$$y_{n_1,n_2} = \sum_{\varkappa} \sqrt{\frac{1}{2\omega_{\varkappa} q}} \left(\hat{b}^{\dagger}_{\varkappa} \mathrm{e}^{-i\varkappa n_2} + \text{h.c.}\right) = \sum_{\varkappa} A(\varkappa) \left(\hat{b}^{\dagger}_{\varkappa} \mathrm{e}^{-i\varkappa n_2} + \text{h.c.}\right). \tag{12}$$

This representation eliminates line (9). In line (8), we suppose that the particle-phonon interaction constant is much greater compared to the reciprocal L, $\lambda \gg 1/L$.

The next step consists in expressing operators $\hat{a}^{(\dagger)}_{k_1,k_2,s}$ in terms of operators $\hat{d}^{(\dagger)}_{k_1,k_2,s}$ diagonalizig $\widehat{H}_0$, in order to write the interaction Hamiltonian (8) and (9):

$$\hat{\vec{d}}_{k_1,k_2} = \mathscr{U}(k) \hat{\vec{a}}_{k_1,k_2}, \qquad \widehat{H}_0 = \sum_{k_1,k_2,s} \varepsilon_{k_1,k_2,s} \hat{d}^{\dagger}_{k_1,k_2,s} \hat{d}_{k_1,k_2,s}, \tag{13}$$

where $\mathscr{U}$ is q-dimensional tranisition matrix. After straightforward but lengthy algebra, we can obtain the following expression for the interaction Hamiltonian:

$$\widehat{H}' = -t \sum_{k,\varkappa} \sum_{s=0}^{q-1} \sum_{\tau,\tau'=0}^{q-1} A(\varkappa) \lambda\, \mathscr{U}^{*}_{s\tau}(k) \mathscr{U}_{s\tau'}(k+\varkappa)\, \hat{d}^{\dagger}_{k,\tau} \hat{d}_{k+\varkappa,\tau'} \left(\hat{b}_{-\varkappa} \delta_{\text{en}1} + \hat{b}^{\dagger}_{\varkappa} \delta_{\text{en}2}\right) \times$$
$$\times \underbrace{\left[\mathrm{e}^{-2\pi i \xi s - i k_2} \left(\mathrm{e}^{-i\varkappa} - 1\right) - \text{h.c.}\right]}_{\Lambda_s(\xi,k,\varkappa)}. \tag{14}$$

The symbol $\delta_{\text{en}j}$ is put here to respect the energy conservation law during the process of absorption or emission of a phonon. Finally, we restrict the analysis to the minimal set of two phonon modes withint the chosen acoustic branch, namely with quasimomenta $+\beta$ and $-\beta$, which leads to the simplification in (14) where $\varkappa$ no longer span the whole Brillouin zone but can take only these two values. Denote the occupation of these phononic states by m_+ and m_-. The eigenstate of the total Hamiltonian can be searched in the form

$$|\mu\rangle = \sum_{k,\tau,m_+,m_-} \alpha_{k,\tau,m_+,m_-} |1_{k,\tau}; m_+, m_-\rangle. \tag{15}$$

(The total Hamiltonian $\widehat{H}$ commutes with the operator of total number of bosons, $\sum_{k,s} \hat{d}^{\dagger}_{k,s} \hat{d}_{k,s}$, and we study the states with one boson and arbitrary number of phonons in the two states described above.) The equation

$$\widehat{H}|\mu\rangle = \mu\,|\mu\rangle \tag{16}$$

formulated in terms of coefficients $\alpha_{k,\sigma,m_+,m_-}$ can be treated as an eigenvalue problem, which may be more convenient for numeric calculations.

3. Numeric results

We have numerically investigated the case of $k = 0$ (Gamma-point in the first Brillouin zone). That, as can be inferred from (14), demanded diagonalization of the Hofstadter Hamiltonian in the Gamma-point and at $k_2 = \pm\beta$ (to get $\mathscr{U}(k{=}0)$ and $\mathscr{U}(k_2{=}\pm\beta)$). In the end, the eigenvalue problem for coefficients α_{k,τ,m_+,m_-} has been numerically solved. Hopping amplitude t has been taken as a unit of energy scale.

In the calculations, we studied the case when the populations of the phonon modes can be equal to either 0 or 1: $m_\pm \in \{0,1\}$.

Figure 1 demonstrates the HB spectra. Figure 1(A) shows the pristine HB without the lattice vibrations. Figure 1(B) shows the spectrum for moderate parameters ($\lambda = 1$, $\omega_0 = 1$). The spectrum may be roughly understood as the three replicas of the pristine one shifted with respect to each other, the influence of $\widehat{H}'$ in this case is present but almost negligible within the computing accuracy accepted. Figure 1(C) presents the spectrum for a huge value of vibrations frequency, $\omega_0 = 10$. No particle-phonon interaction takes place (in (14), all the $\delta_{\mathrm{en}1(2)}$ symbols equal zero). Figure 1(D) demonstrates the spectrum for small $\omega_0 = 0.01$. Such a small value of the characteristic frequency results in overall lower shifts and separation of the butterflies. On the other hand, the influence of $\widehat{H}'$ becomes more notable, a much greater amount of non-zero summands in (14) occur, which gives rise to notable alterations that smear the original spectrum.

4. Conclusion

We have investigated the behaviour of bosonic Hofstadter butterflies subject to the influence of lattice phonons. We have developed the theoretical model taking into account both phase acquisition due to the presense of magnetic (or another gauge) field and lattice vibrations. Within the second quantization formalism, the Hamiltonian of the system has been derived. Finally, we scrutinized the reduced system with only two specific phonon modes and numerically calculated the spectra.

Acknowledgements

The work was supported by Russian Science Foundation (Project No. 20-12-00224).

References

[1] Hofstadter D R 1976 *Phys. Rev. B* **14** 2239–2249.
[2] Gumbs G and Fekete P 1997 *Phys. Rev. B* **56** 3787–3791.
[3] Oh G Y 2000 *J. Korean Phys. Soc.* **37** 534–539.
[4] Oh G Y 2006 *J. Korean Phys. Soc.* **49** 672–675.
[5] Du L, Chen Q., Barr A D, Barr A R and Fiete G A 2018 *Phys. Rev. B* **98** 245145.
[6] Banerjee R, Liew T C H and Kyriienko O 2018 *Phys. Rev. B* **98** 075412.
[7] Colella E, Mivehvar E, Piazza F and Ritsch H 2019 *Phys. Rev. B* **100** 224306.
[8] Zimmerling T J and Van V 2020 *Opt. Lett.* **45** 714–717.
[9] Aidelsburger M, Atala M, Lohse M, Barreiro J T, Paredes B and Bloch I 2013 *Phys. Rev. Lett.* **111** 185301.
[10] Jaksch D and Zoller P 2003 *New Journal of Physics* **5** 56.
[11] Hafezi M, Demler E A, Lukin M D and Taylor J M 2011 *Nature Physics* **7** 907–912.
[12] Weber M, Assaad F F and Hohenadler M 2015 *Phys. Rev. B* **91** 245147.
[13] Sedov D D, Kozin V K and Iorsh I V 2020 *Phys. Rev. Lett.* **125** 263606.

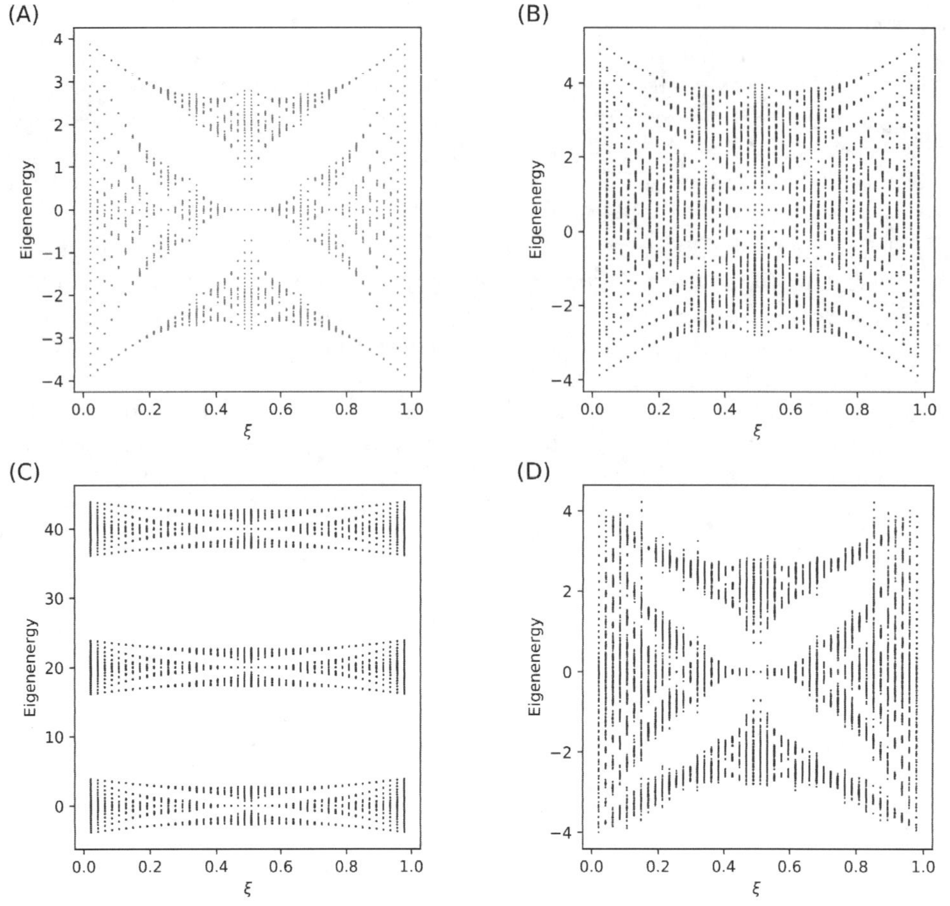

Figure 1. Hofstadter butterflies, the hopping amplitude $t = 1$ everywhere. (A) without vibrations, $\lambda = 0$; (B) $\lambda = 1$, $\omega_0 = 1$, $\beta = \pi/4$; (B) $\lambda = 0.5$, $\omega_0 = 10$, $\beta = \pi/2$; (D) $\lambda = 1$, $\omega_0 = 0.01$, $\beta = \pi/8$.

2D Perovskite Micro-optics Enabled by Direct Femtosecond-Laser Projection Lithography

A. Cherepakhin[1], A. Zhizhchenko[1,2] A. Porfirev[3], A. Pushkarev[4], S. Makarov[4] and A. Kuchmizhak[1]

[1]Institute of Automation and Control Processes, Far Eastern Branch of RAS, Vladivostok 690061, Russia
[2]Far Eastern Federal University, Vladivostok 690061, Russia
[3]Image Processing Systems Institute of the RAS-Branch of FSRC "Crystallography & Photonics" of the RAS, Samara 443001, Russia
[4]ITMO University, St. Petersburg, Russia, 197101

Abstract. Using direct femtosecond-laser projection lithography in chemically synthesized $CsPbBr_3$ perovskite microcrystals we demonstrate high-throughput fabrication of advanced binary microscale optical elements for nanofocusing as well as generation of high-order optical vortex beams. The obtained results highlight the $CsPbBr_3$ microcrystals as a promising material for realization of various complicated 2D micro-optical elements and holograms that can be directly imprinted using non-destructive and practically relevant laser technologies.

1. Introduction

Halide perovskites have recently appeared as a novel class of materials possessing intriguing combination of optical and optoelectronic properties revolutionizing such areas as solar cell technologies and light-emitting devices [1-5]. Inexpensive chemical synthesis was recently adopted to fabricate humidity-robust monocrystalline 2D microstructures (as microplates and nanowhiskers) promising for various applications ranging from nanoscale coherent light sources for next-generation optical communications to actively tunable micro-optic devices. In particular, realization of various micro-optics based on 2D perovskites hold promise considering ultra-smooth surface of chemically synthesized microcrystals as well as potential ability for in-situ variation of the material optical properties via simple vapor-phase anion exchange reaction. Convenient lithography-based nanofabrication technologies as electron- or ion-beam milling typically cause strong degradation of the perovskite photoluminescence properties being also time- and money-consuming for mass production. Recently, gentle femtosecond (fs)-laser processing appeared as a promising non-destructive approach for perovskite micro- and nano-patterning [6-9]. Additionally, fabrication of Fresnel microlenses was recently demonstrated using direct laser ablation of halide perovskite microplates by tightly focused fs-laser pulses in scanning regime. However, the quality of the obtained elements was far from being perfect, as multi-pulse scanning exposure of the perovskite surface typically leave rather deteriorated ablation layer with a surface roughness that is inconsistent with the quality required for realization of state-of-the-art micro-optical elements and devices.

Here, we reveal the superior potential of single-pulse fs-laser projection lithography for direct imprinting of advanced 2D diffraction-optical elements in chemically synthesized $CsPbBr_3$ perovskite microcrystals. In particular, we demonstrate fabrication of optical-quality micro-scale Fresnel zone plates (FZP) for generation of tightly focused laser beam, as well as binary spiral micro-axicons and binary fork-shaped gratings (FSGs) allowing generation of vortex beams in reflection mode. The observed ultra-smooth ablation of perovskite microcrystals was explained by ultrafast laser-induced thermalization rate as well as extremely low conductivity of the $CsPbBr_3$ material. The obtained results highlight the $CsPbBr_3$ microcrystals as a promising material for realization of various complicated 2D micro-optical elements and holograms that can be directly imprinted using non-destructive and practically relevant laser technologies.

2. Material and Methods

Lead (II) bromide (PbBr2, 99.999 %, Alfa Aesar), cesium bromide (CsBr, 99.999 %, Sigma-Aldrich), dimethyl sulfoxide (DMSO, anhydrous, 99.8%, Alfa Aesar), isopropyl alcohol (IPA, technical grade, 95%, Vecton) were purchased from commercial suppliers and used as received. A mixture of metal halide salts $PbBr_2$ (33 mg, 0.1 mmol) and CsBr (21 mg, 0.1 mmol) was dissolved in DMSO (1 ml) by stirring at room temperature for 30 min. The solution was filtered by using a 0.45 μm syringe filter with a PTFE membrane. The chemicals were stored and mixed inside a N_2-filled glove box with both O_2 and H_2O level not exceeding 1 ppm. The obtained perovskite solution was utilized at ambient conditions. A glass substrate was placed in the center of tilted (10 deg.) glass Petry dish (80x5 mm^2) and preheated on the hotplate up to 60 °C. A small droplet the solution was dripped onto the substrate. Immediately after that, 300 μL of $IPA{\cdot}H_2O$ azeotrope was poured in the dish and then it was sealed. The droplet was dried in the presence of the azeotropic vapor for 10 min to give isolated high-quality $CsPbBr_3$ crystals.

As-synthesized $CsPbBr_3$ microcrystals were further patterned by using second-harmonic (515 nm central wavelength) 200-fs laser pulses coming from the regeneratively amplified Yb:KGW laser system (Figure 1a). The output laser beam having Gaussian lateral intensity profile was first passed through the appropriate amplitude mask to generate the complex intensity pattern (Figure 1b). Then, the generated pattern was transferred with magnification into the focal plane of the dry microscope objective with a numerical aperture (NA) of 0.95 using 4*f*-optical system. The amplitude masks for beam shaping were fabricated using direct laser writing in a 40 nm-thick Cr films deposited onto a fused silica substrate (CLWS-200). Surface morphology of the laser-processed perovskite microcrystals was analysed using optical and scanning electron microscopy (SEM). The focusing performance of the imprinted perovskite micro-optics was studied by irradiating their surface by 515-nm CW collimated beam while the generated intensity profiles were captured in reflection at the corresponding focal plane using the calibrated CCD-camera.

3. Results and Discussions

We demonstrated the applicability of the fs-laser projection lithography by patterning the $CsPbBr_3$ microcrystal surface with several practically relevant designs of the binary optical elements: FZPs, spiral micro-axicons and FSGs (Figure 1(b)). Corresponding mm-scale amplitude masks were first produced in the glass-supported Cr films and projected into the focal plane of high-NA lens used for laser structuring. All demonstrated elements were imprinted using single-pulse irradiation of the $CsPbBr_3$ microcrystals. First, we calculated and optimized amplitude mask for imprinting of the FZP representing a well-known optical element for tight focusing of the incident plane wave into a point at the focal distance f from the element plane. It should be noted that FZP has a primary focus and an infinite number of secondary focuses. Typical FZPs are a set of alternating transparent and opaque rings (zones) having specific widths and diameters. The n-th ring has an outer radius of

$$r_n = \sqrt{n\lambda\left(f + \frac{n\lambda}{4}\right)}, \tag{1}$$

where n is an integer, λ is a wavelength of the incident laser radiation. The 10-μm diameter FZP imprinted on the surface of $CsPbBr_3$ microcrystals is shown in Figure 1(c) revealing rather smooth geometry of the produced radial grooves having the typical width between 300 and 500 nm and a depth of about 150 nm. By irradiating the fabricated optical element by CW parallel laser beam (515-nm wavelength) we found tightly focused (1/e-diameter of about 500 nm) laser spot formed few microns above the patterned surface of the $CsPbBr_3$ microcrystal that confirmed focusing performance of the imprinted FZP.

Further, using similar fabrication technique we produced two types of binary spiral axicons. Such elements allow to convert ordinary Gaussian laser beam into the quasi-non-diffracting optical vortex (OV), special type of laser beams with extended focal depth, donut-shape intensity profile and a helical wavefront [10]. The main parameter of the OV beams is their topological charge (TC) m, a number of wavefront twists within a wavelength along light propagation axis. Such structure of the OV beam leads to the presence of the orbital angular momentum in such electromagnetic fields equal to $m\hbar$ per photon ($\hbar$ is the reduced Planck's constant). A binary spiral axicon actually contains two vortex conjugated axicons ($\exp(-ik\alpha_0 r - m\varphi)$ and $\exp(ik\alpha_0 r + m\varphi)$) and has the following transmission function [11]:

$$T(r,\varphi) = \mathrm{sgn}\lfloor \cos(k\alpha_0 r + m\varphi) \rfloor, \quad (2)$$

By designing proper geometry of the binary spiral axicon, the OV with the required TC can be generated. In our work, we fabricated two spiral axicons allowing to generate the OV beam with the TC m=1 and 3. The imprinted micro-optical elements are shown in Figure 1(d,e) revealing the applicability of the fs-laser projection lithography to produce more complicated micro-optical elements. Both elements allowed to generate high-quality OV beam with m=1 and 3 at the corresponding focal plane above the $CsPbBr_3$ microcrystal surface. Noteworthy, increase of the TC of the OV beam results in corresponding increase of the donut diameter as it is illustrated by comparing focal-plane intensity patterns of the generated OVs (Figure 1d,e; bottom).

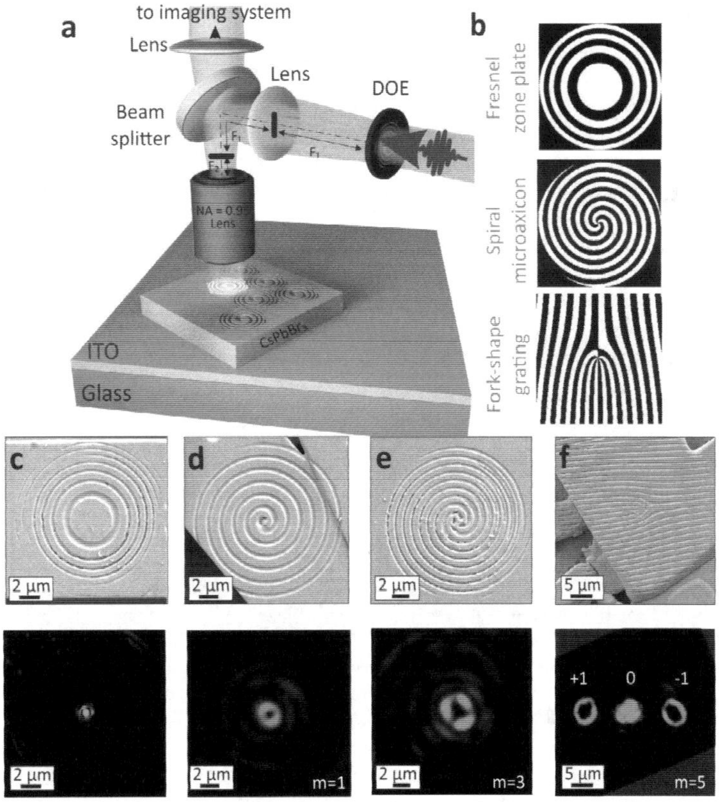

Figure 1. (a) Schematic illustration of the setup for fs-laser projection lithography. (b) Design of the amplitude masks used to generate intensity patterns for laser printing of the FZP, spiral microaxicons and fork-shaped gratings. (c) Top-view SEM images of the $CsPbBr_3$ microcrystal surface imprinted with various micro-optical elements including FZP, binary spiral axicons (b,c) and binary fork-shape grating (d) of the as well as corresponding focal-plane intensity distributions at 515 nm excitation wavelength showing focusing performance of the imprinted 2D perovskite micro-optics.

Finally, we fabricated the FSG that allows to generate OVs with the opposite TCs (±m) in opposite diffraction orders. Theoretical description of Fresnel and Fraunhofer diffraction of a Gaussian laser beam by the FSGs was presented in [12]. The general transmission function of the FSGs in the polar coordinates is defined as follows:

$$T(r,\varphi)=\sum_{p=-\infty}^{\infty} t_p \exp\left[-ip\left(\frac{2\pi}{D}r\cos\varphi-m\varphi\right)\right]=$$
$$t_0+\sum_{p=1}^{\infty} t_p \exp\left[-ip\left(\frac{2\pi}{D}r\cos\varphi-m\varphi\right)\right]+\sum_{p=1}^{\infty} t_{-p} \exp\left[ip\left(\frac{2\pi}{D}r\cos\varphi-m\varphi\right)\right], \quad (3)$$

where D is a period of its rectilinear part, and t_m are the complex (in general case) transmission coefficients. We designed and produced the corresponding amplitude mask for imprinting of the FSG that allows to generate OV beams with the m=±5. Noteworthy, the TC is defined by the number of additional grooves that start from the central bifurcation point. The imprinted perovskite fork-shaped grating is demonstrated in Figure 1f. Upon irradiation with plane wave at 515 nm produced optical element generates the two high-order (m=±5) OV beams with opposite handedness in the +1 and -1 diffraction orders as well as unconverted Gaussian beam in the main diffraction order.

Excellent performance of the imprinted micro-optics indicates the applicability of the fs-laser projection lithography for direct printing of various functional elements and devices. Here, we demonstrated few common examples of the binary diffraction optical elements that can be produced in the $CsPbBr_3$ microcrystals. However, the developed approach can be easily adopted for fabrication of more complicated elements including binary holograms. The lateral resolution of the fs-laser ablation at 515-nm wavelength of $CsPbBr_3$ can reach 250-300 nm that is generally associated with its extremely low thermal conductivity that is two orders of magnitude lower comparing to those for silicon as well as extremely fast thermalization rate of the material upon laser exposure. $CsPbBr_3$ thermalization processed at sub-picosecond timescales via Auger recombination of photo-excited carriers minimizing the heat-affected zone.

4. Conclusions

To conclude, we highlighted here remarkable performance of direct fs-laser processing for imprinting of high-quality reflection-type binary microscale optical elements – Fresnel zone plates, binary spiral micro-axicons and binary fork-shaped gratings – for laser beam focusing and generation of the vortex beams.

Acknowledgments
The work was supported by Russian Science Foundation 19-19-00177.

References

[1] Fu Y., Zhu H., Chen J., Hautzinger M. P., Zhu X. Y., and Jin S., 2019 *Nat. Rev. Mater.* **4**(3), 169–188.
[2] X. Yang, J. Wu, T. Liu, and R. Zhu, 2018 *Small Methods* **2**(10), 1800110.
[3] Berestennikov A.S., Voroshilov P. M., Makarov S.V., and Kivshar Y.S., 2019 *Appl. Phys. Rev.* **6**(3), 031307.
[4] Wang K., Xing G., Song Q., and Xiao S., 2021 *Adv. Mater.* 2000306.
[5] Wang Z., Yang T., Zhang Y., Ou Q., Lin H., Zhang Q., Chen H., Hoh H.Y., Jia B., Bao Q., 2020 *Adv. Mater.* **32**, 2001388.
[6] Zhizhchenko A, Syubaev S, Berestennikov A, Yulin A V, Porfirev A, Pushkarev A, Shishkin I, Golokhvast K, Bogdanov A A, Zakhidov A A, and Kuchmizhak A A 2019 *ACS nano* **13**(4) 4140-4147
[7] Zhizhchenko A Y, Tonkaev P, Gets D, Larin A, Zuev D, Starikov S, Pustovalov E V, Zakharenko A M, Kulinich S A, Juodkazis S, and Kuchmizhak A A 2020 *Small* **16**(19) 2000410
[8] Zhou C., Cao G., Gan Z., Ou Q., Chen W., Bao Q., Jia B. and Wen X., 2019 *ACS Appl. Mater. Interfaces* **11** 26017–26023.
[9] Zhizhchenko A.Y., Cherepakhin A.B., Masharin M., Pushkarev A., Kulinich1 S.A., Porfirev A., Kuchmizhak A.A., and Makarov S.V., 2021 *Laser Photonics Rev.* (accepted manuscript)
[10] Heckenberg N. R., McDuff R., Smith C. P., and White A. G., 1992 *Opt. Lett.* **17**, 221-223.
[11] Khonina S.N., Porfirev A.P, 2018 *Appl. Phys. B* **124**(9), 1-13.
[12] Janicijevic L. and Topuzoski S., 2008 *J. Opt. Soc. Am. A* **25**, 2659-2669.

Investigation of the Phonon Spectra in CdSe/CdS Nanoplatelets by Raman Spectroscopy

N.N. Kurus[1], A.G. Milekhin[1,2], R.B. Vasiliev[4], B.M. Saidzhonov[4], K.V. Anikin[1], R.I. Sklyar[1,2], S.A. Batsanov[1], A.K.Gutakoskii[1], A.V. Latyshev[1,2], D.R.T. Zahn[3]

[1]Rzhanov Institute of Semiconductor Physics, Lavrentiev Ave. 13, 630090 Novosibirsk, Russia

[2]Novosibirsk State University, Pirogov 2, 630090 Novosibirsk, Russia

[3]Semiconductor Physics, Chemnitz University of Technology, D-09107 Chemnitz, Germany

[4]Department of Material Science, Moscow State University, Moscow, Russia

E–mail of the corresponding author: kurus@isp.nsc.ru

Abstract. We report the phonon spectra of core/shell CdSe/CdS nanoplatelets with different shell thicknesses studied using Raman scattering. The nanoplatelets are rectangular colloidal nanocrystals, with thicknesses of core and shell layers of a few nanometers. The Raman spectra show features corresponding to the dominating longitudinal optical (LO) and surface optical (SO) phonon modes of the CdSe core in CdS shell located in the frequency regions of 200-210 and 250-290 cm^{-1}, respectively. As the shell thickness increases, the phonon modes reveal a frequency shift and a change in intensity. The frequency shift associated with a change in the stress state in the core and shell, as well as with confinement effects is discussed. The phonon mode intensities are determined by the thickness of the shell and the proximity to resonant Raman scattering conditions.

1. Introduction

Colloidal nanocrystals (NCs) of cadmium chalcogenides of various morphologies, including spherical NCs (or quantum dots (QDs), nanorods and nanoplatelets have been the object of close attention in recent decades. Interest in such nanoobjects is caused by the spatial localization of charge carriers in NCs, which makes them promising objects for various applications, such as solar cells, light-emitting diodes (LEDs), lasers and biosensors. The formation of core/shell NCs, consisting on a shell material with larger band gap than in the core, allows one to improve NC optical properties and to achieve an significant increase in the quantum yield of photoluminescence and the stability of the structure upon photooxidation [1]. Among others nanocrystals created by colloidal chemistry methods, anisotropic nanocrystals, in particular nanoplatelets, are of particular interest. Nanoplatelets are planar semiconductor structures with lateral dimensions of the order of 10 nm and a thickness of several monolayers (of the order of 1 nm). [2]. They belong to a new class of so-called 2D nanocrystals. Such geometry provides spatial confinement of charge carriers strictly in one of the directions (perpendicular to a nanoplatelet surface). The nanoplatelet thickness uniformity, provided by the specificity of the nanostructure formation method, leads to polarized ultrafast and narrow-band exciton photoluminescence with a high quantum yield and anisotropic light absorption [3]. The spectrum of lattice vibrations of a nanometer-thick structure changes significantly. It becomes quantized in one direction remaining quasi-continuous in the other two. Vibrational spectroscopy, including Raman, scattering, provides information on the phonon spectrum, electronic structure and elemental composition of a nanocrystal. It also helps to estimate structural perfection and mechanical stresses in a nanostructure, as well as to study its interaction with the environment.

Raman scattering from a bulk CdSe crystal with cubic symmetry is determined by scattering the LO phonon mode. This emerges in the spectra as a peak near 210 cm^{-1} and its overtones [4]. A decrease in the crystal thickness down to several monolayers leads to the appearance of quantum size effects. Thus, for materials with negative dispersion, this leads to a low-frequency shift of the phonon mode by several cm^{-1} [5]. Upon the formation of a heteronanostructure, *e.g.* core/shell CdSe /CdS nanoplatelets, additional features, corresponding to the SO phonon of CdSe core and CdS shell, appear in the Raman spectra between frequencies of LO and transverse optical (TO) phonons of the corresponding materials. It was previously reported that the vibrational spectra of semiconductor NCs are sensitive to the shell thickness [6]. Thus, an increase in the shell thickness leads to shift in the frequency of the LO phonon of the core due to the accumulation of mechanical stress caused by the difference in the lattice constants of the core and the shell materials. The study of effects arising at the interface between the core and the shell (stress, ordering, mixing) is an important issue from for the analysis of the optical properties of colloidal nanocrystals, including nanoplatelets.
Despite the great interest in this type of semiconductor nanocrystals, there is a limited number of works devoted to the study of their vibrational spectrum, especially nanoplatelets of the core/shell type. This work is devoted to the study of the Raman scattering by colloidal core/shell CdSe/CdS nanoplatelets with different shell thicknesses.

2. Experimental

CdSe nanoplatelets were fabricated in accordance with the method described in [7]. The core/shell CdSe/CdS heteronanostructures were prepared using the layer-by-layer colloidal shell deposition on the surface of CdSe nanoplatelets [8].
The structural parameters of CdSe-based nanoplatelets were studied with scanning electron microscopy (SEM). Typical SEM images of the CdSe/2CdS nanoplatelets (thickness of the shell is 2 monolayers (MLs)) are shown in Fig.1a. As it is seen from the images the nanoplatelets have rectangular shape with the lateral sizes of approximately 100x20 nm and the thickness depending on the shell size. The thickness of bare CdSe nanoplatelets (sample "bare CdSe") was 1.5 nm.
The samples studied were prepared by depositing the colloidal nanoplatelets on gold-coated silicon substrates using the Langmuir-Blodgett technique, described in [9].
Raman experiments were carried out using a Horiba T64000 spectrometer equipped with an optical microscope (the laser beam was focused to a spot with a size of approximately 1 μm^2) in backscattering geometry at room temperature. Ar^+ laser was used as excitation source with the wavelength of 514 nm (2.41 eV). A laser power of less than 2 mW was used in micro-Raman configuration to avoid possible effects of local heating.

3. Results and discussion

In accordance with Raman selection rules, LO phonon mode appears in the Raman spectrum of bare CdSe nanoplatelets with a cubic crystal structure reveals the LO phonon mode at 206.5 ± 0.6 cm^{-1} (Fig. 1b). Low-frequency shoulder observed at 178 cm^{-1} is probably attributed to SO vibrational mode of bare CdSe NPLs.
The LO phonon mode becomes dominant in the Raman spectra after the CdS shell formation and CdSe/1CdS nanoplatelet formation. Furthermore, the shell formation leads to the sufficient change in frequency of vibrational modes in nanoplatelets. With increasing of CdS shell thickness from 1 to 4 layers (Fig.2b) the CdS-related mode increases in frequency (from 278 to 296 cm^{-1}) and intensity.

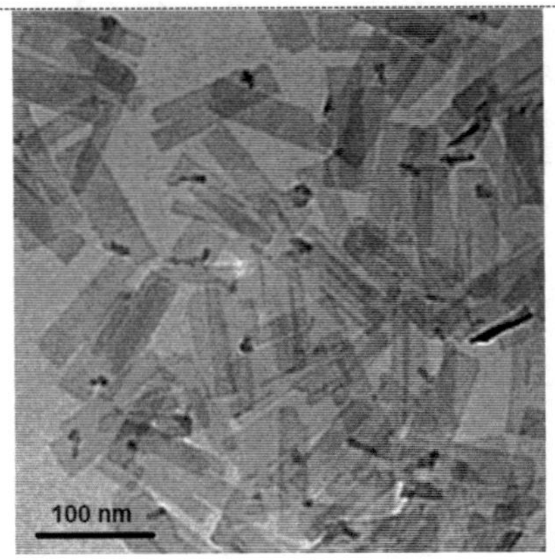

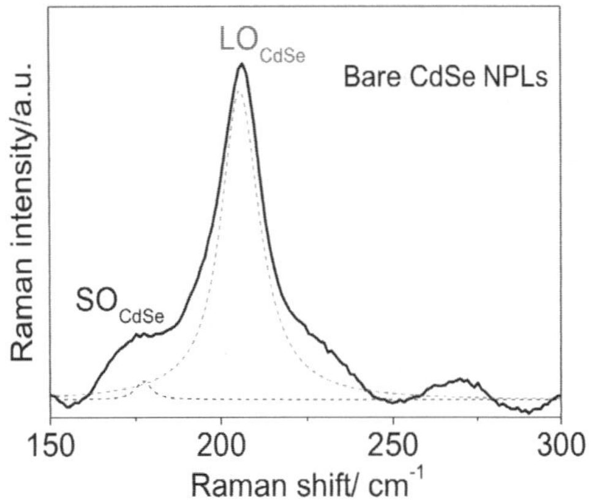

Figure 1. (a) TEM image of colloidal CdSe/2CdS nanoplatelets and (b) Raman spectra of bare CdSe nanoplatelets on a gold substrate. The dashed lines represent Lorentzian fits of the LO and SO phonon modes of CdSe nanoplatelets.

The large full-width at half maximum (FWHM) of this feature at 279 cm^{-1} (for nanoplatelets with a monolayer shell (n = 1) it amount the value of 38 cm^{-1}) indicates an SO character of the mode. The intensity increase of the mode is probably due to change in the band gap of CdS shell towards the bulk value, as a result the SO phonon mode of CdS becomes resonantly enhanced. At the same time, FWHM of the mode decreases sufficiently with increasing the shell thickness. The Raman spectrum of CdSe/4CdS nanoplatelets reveals a narrow (FWHM= 20 cm^{-1}) peak at 295 cm^{-1}, which most closely matches the LO phonon of CdS. In addition, a low-frequency shoulder (near 270 cm^{-1}) of the LO mode associated with CdS appears in the spectra of nanoplatelets with shell thicknesses n=2 and n=3 monolayers and can be assigned to the SO mode in CdS shell near CdS/CdSe interface.

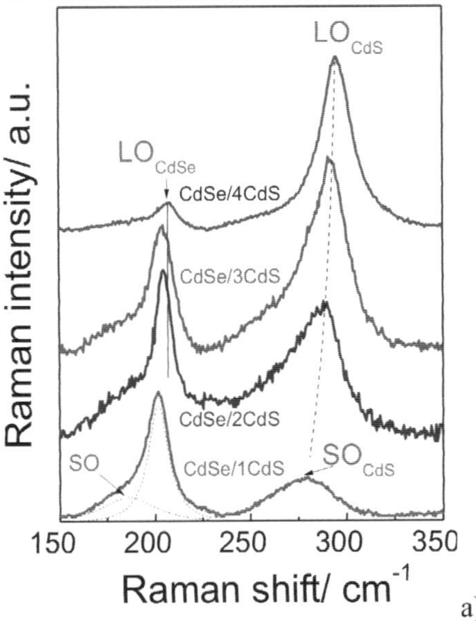

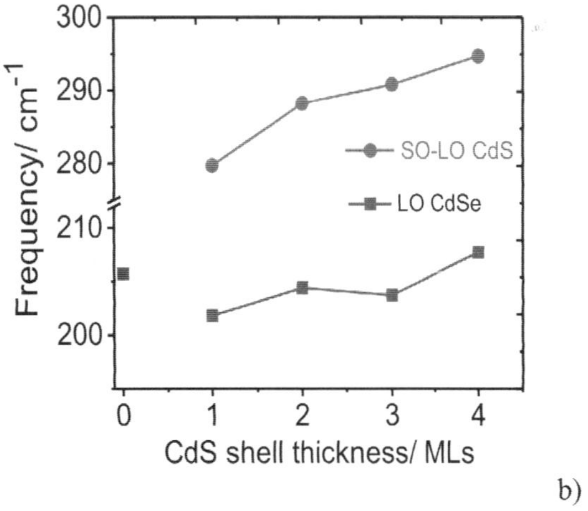

Figure 2. (a) Raman spectra of colloidal core/shell CdSe/CdS nanoplatelets and (b) the frequency dependence the LO and SO phonon modes on the CdS shell thickness.

Apparently, the features observed in this spectral region in the case of intermediate shell thicknesses have a nature of mixed SO and LO phonon modes of CdS shell. The high-frequency shift in shell-related vibrational mode with an increase of shell thickness also corresponds to the change of frequency towards the bulk LO-phonon value [10]. The high-frequency shift of LO-phonon of CdSe core, observed with increasing shell thickness is opposite to the expected red shift due to confinement effect, therefore, it is most probably caused by compressive mechanical stress. The lattice constant of CdSe ($a_{0\ CdSe}$ =6,05) sufficiently exceeds the corresponding value of CdS ($a_{0\ CdS}$ =5,81), so after the formation of the shell, the CdS layer is tensely strained, and the CdSe core is compressed. The compressive strain in the core is responsible for a blue shift in the CdSe LO phonon frequency (Fig. 2). The components of deformation tensor of core and shell materials will be quantified elsewhere. The appearance of the low-frequency shoulder of the LO phonon of CdSe near 180 cm^{-1} (Fig. 2a) can apparently be interpreted as the sum of the SO phonon of the CdSe core and the vibrational mode of the alloy at the core / shell interface of $CdSe_{alloy}$ [6].

4. Conclusion

In present paper the Raman spectra of CdSe/CdS heteronanocrystals of the core/shell type with different shell thicknesses were investigated. The formation of the CdS shell leads to an increase in the compression deformation in the CdSe core, which manifests itself in the Raman spectra in an increase in the LO-phonon frequency of the CdSe core and in the appearance of additional CdS-related vibrational modes observed in the range of 279–295 cm^{-1}. It is shown that with increasing the shell thickness from 1 to 4 monolayers, the mixed SO-LO character of the shell-related mode is changed from SO-phonon scattering to the scattering by LO-phonons localized in the CdS layer.

Acknowledgment

The authors gratefully acknowledge financial support from the Volkswagen Foundation, MERGE project (TU Chemnitz), State, DFG projects (Grants Nos.ZA 146/43-1, ZA 146/44-1), RFBR (Project No. 19-52-12041_NNIO_a), and the Ministry of Education and Science of the Russian Federation. HRTEM studies were performed using the equipment of CKP "Nanostructures." The authors are thankful to Mr. A. S. Medvedev for Au deposition.

References

[1] Dabbousi B O, Rodriguez-Viejo J, Mikulec F V, Heine, J R, Mattoussi H, Ober R and Bawendi M G 1997 The Journal of Physical Chemistry B 101, 9463
[2] Reichhelm A, Hübner R, Damm C, Nielsch K and Eychmüller A 2020 Physica status solidi (RRL)–Rapid Research Letters 14 2000282
[3] Mukhina M V, Maslov V G, Baranov A V, Fyodorov A V, Artem'ev M V and Prudnikov A V 2013 Journal of Optical Technology 80 642
[4] Hermann C and Peter Y Y 1980 Physical Review B 21 3675
[5] Klein M C, Hache F, Ricard D and Flytzanis C 1990 Physical Review B 42 11123
[6] Dzhagan V M, Valakh M Y, Milekhin A G, Yeryukov N A, Zahn D R, Cassette E and Dubertret B (2013) The Journal of Physical Chemistry C 117 18225
[7] Sokolikova M S, Vasiliev R B, and Gaskov A M 2014 Russian Journal of Inorganic Chemistry 59 413
[8] Smirnov A M, Golinskaya A D, Saidzhonov B M, Vasiliev R B, Mantsevich V N and Dneprovskii V S 2019 JETP Letters 109 372
[9] Milekhin A G, Sveshnikova L L, Duda T A, Rodyakina E E, Dzhagan V M, Sheremet E and Zahn D R T 2016 Applied Surface Science 370 410
[10] Han P and Bester G 2012 Physical Review B 85 041306

Switchable Purcell enhancement of photoluminescence by GST film

O M Kushchenko[1,a], A D Sinelnik[1], I I Shishkin[1,2], D S Gets[1], S V Makarov[1], and M V Rybin[1,3]

[1]Department of Physics and Engineering, ITMO University, St. Petersburg 197101, Russia
[2]School of electrical engineering, Tel-Aviv University, Tel-Aviv, 69978, Israel
[3]Ioffe Institute, St. Petersburg 194021, Russia

[a)] E-mail olga.kushchenko@metalab.ifmo.ru

Abstract. In the present paper perovskite radiation enhancement on crystalline GST film compared to amorphous one has been studied. The photonic local density of states has been calculated by angular spectrum representation of the dyadic Green's function. The Purcell factor has been calculated for perovskite luminescent on both amorphous and crystalline GST film. Almost 80% enhancement has been observed at wavelength 950 nm for system with perovskite thickness 25 nm, GST thickness 110 nm.

1. Introduction

Halide perovskites are widely studied as an emergent material for optoelectronic and photovoltaic applications. Comparing with analogs which are used today the perovskites offer ease of processing using solvent methods such as spin-coating or blade/slot-die coating. The energy gap of perovskites is altered by variation of halogen anion composition (Cl, Br and I), which allows tuning emission of the perovskite-based device over the whole visible spectrum [1].

Dielectric environment is known to influence the rate of emission centers. This effect was described by E.M. Purcell by considering a factor which is equal to a ratio of emission rate in environment normalized to emission rate of the same center in a homogeneous space [2]. The Purcell effect causes the luminescent enhancement in vicinity of metallic substrate because of induced surface plasmons [3]. However the interface between halide perovskite and metal leads to strong chemical degradation [4] that disables such devices in practice.

During the last decade phase change materials based on germanium-stibium-tellurium alloy (GST) attract a lot of attention in photonics due to non-volatile phase transition between crystalline (with high conductivity) and amorphous (low conductivity) phases induced by optical pulses [5, 6]. Thus, GST can be used as a substrate for halide perovskite which acts as switchable dielectric environment for the emission centers properties.

Here we show enhancement of perovskite photoluminescence by switching GST substrate from amorphous to crystalline phase. We calculate photonic local density of states (LDOS) by using angular spectrum representation of the dyadic Green's function. Further we integrate Purcell factor over distance

Journal of Physics: Conference Series **2015** (2021) 012077 doi:10.1088/1742-6596/2015/1/012077

between emitter and GST and uncover photoluminescence enhancement for perovskite on crystalline phase compared to amorphous one.

2. Results

In the present study we consider a layered structure composed of glass substrate, GST, perovskite and air. The scheme of system is represented in Fig. 1. The perovskite emits spherical waves propagating in both hemispheres. The waves reflect at boundaries perovskite-air, perovskite-GST and GST-glass. We assume the glass and air layers are semi-infinite in the normal direction. The GST thickness ranges from 50 to 200 nm and the perovskite thickness from 20 to 70 nm. The permittivity of perovskite is set to be 5.5 [7], the permittivity of GST is taken according to the literature data [6].

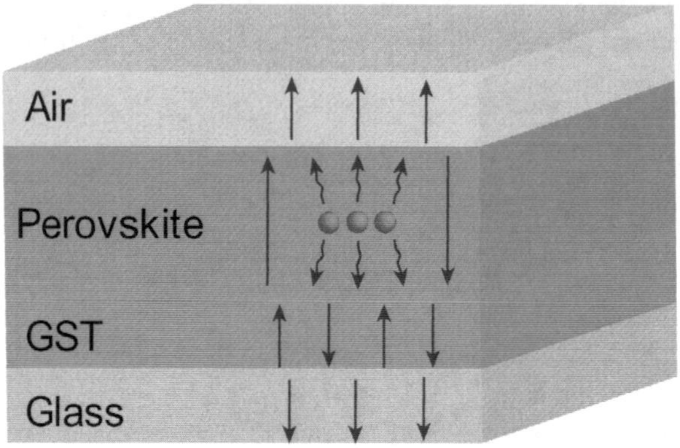

Figure 1. Schematic of the layered structure.

We calculate the local density of states as imaginary part of trace of the dyadic Green's function [3]:

$$\rho(\mathbf{r},\omega_0) = \frac{2\omega_0}{\pi c^2}\mathrm{Im}\{\mathrm{Tr}(\mathbf{G}(\mathbf{r},\mathbf{r};\omega_0))\} \quad (1)$$

where $\mathbf{G}(\mathbf{r},\mathbf{r}';\omega_0)$ is a dyadic Green function, ω_0 is a frequency, $\rho(\mathbf{r},\omega_0)$ is the photonic LDOS at position $\mathbf{r}$.

By using angular spectrum representation the problem is reduced to plane waves defined by tangential wave vector projection onto the interface plane $\mathbf{k}_\tau$. To consider all reflections at the boundaries it is convenient to solve the problem for individual plane waves by the approach based on transfer matrix [8]. After that we integrate plane waves over all $\mathbf{k}_\tau$ and obtain Green's function for layered structure. The Purcell factor is obtained by dividing the calculated photonic LSOD to one for sample with no GST layer. Figure 2 demonstrates the value of Purcell factor of perovskite on the crystalline (red curve) and amorphous (black curve) GST film versus wavelength for structure with GST layer thickness of 90 nm, perovskite layer thickness of 25 nm. The emitter is localized at a distance 45 nm off the GST layer. The Purcell factor corresponding to amorphous GST has the maximum at wavelength 525 nm while the maximum of one for the crystalline phase is shifted 15 nm to short wavelengths. Note that the specific shape of the curve depends on the position of the emitter.

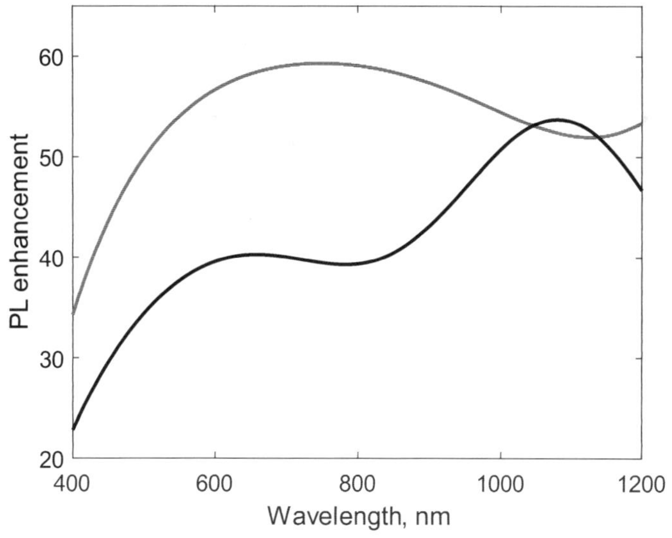

Figure 2. Purcell factor for perovskite on crystalline (red curve) and amorphous (black curve) GST layer. Perovskite layer thickness is 25 nm and GST layer thickness is 90 nm. The emitter is localized at the distance 45 nm from the interface with GST.

To evaluate the luminescence enhancement for the crystalline GST compared to amorphous one we integrate the local density of states over emitter position in the perovskite layer. Due to GST bandgap less than perovskite bandgap the near-surface charge carries leak to the GST layer and recombine in nonradiative processes. Therefore, we assume that the emission occurs due to recombination of excitons in the perovskite at a distance greater than 10 nm from the interface with GST. Figure 3 represents the results of factor Purcell calculations for perovskite thickness 25 nm and GST thickness 110 nm. For such configuration, the maximal value of Purcell factors ratio observed at the wavelength 950 nm and equal to 1.79.

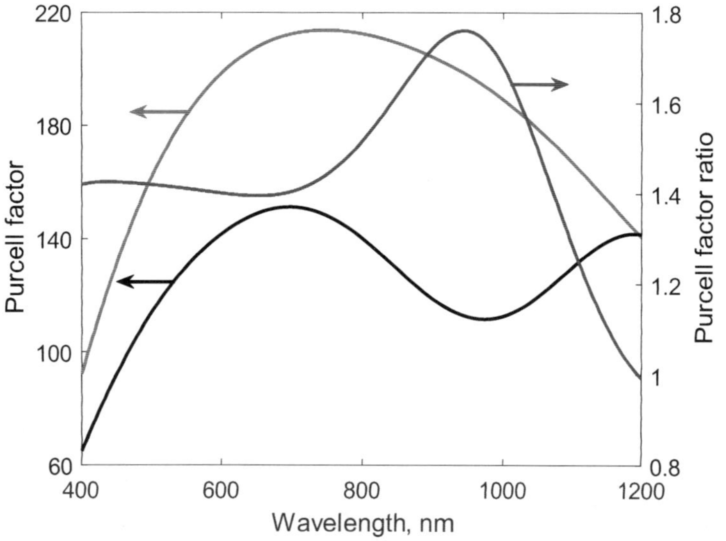

Figure 3. Purcell factor for perovskite on GST film in crystalline (red curve) and amorphous (black curve) phase and its ratio (blue curve). Perovskite thickness is 25 nm and GST thickness is 110 nm.

3. Summary

To summarize, we have revealed the switchable Purcell enhancement of perovskite photoluminescence by GST films. The Purcell factor values for perovskite radiation on amorphous and crystalline GST film have been calculated. Almost 80% value of enhancement has been obtained for the structure with GST thickness 110 nm, perovskite thickness 25 nm at the wavelength 950 nm.

Acknowledgments

This work was supported by the Russian Science Foundation (Project 20-12-00272).

References

[1] Protesescu L et al 2015 *Nano Lett.* **15** 3692
[2] Purcell E 1946 *Phys. Rev.* **69** 37
[3] Novotny L and Hecht B 2012 *Principles of nano-optics* (Cambridge: Cambridge University Press)
[4] Behrouznejad F et al. 2016 *J. Mater. Chem. A.* **4** 13488
[5] Wuttig M and Yamada N 2007 *Nat. Mater.* **6** 824
[6] Makarov S et al. 2017 *Laser Photonics Rev.* **11** 1700108
[7] Yan W et al. 2020 *Opt. Express* **28** 15706
[8] Rybin M et al. 2016 *Sci. Rep.* **6** 1

Transformation of guided modes into bound states in the continuum

Aleksandra A. Kutuzova[1, a)] and Mikhail V. Rybin[1, 2, b)]

[1)] ITMO University, St. Petersburg, 197101, Russia.
[2)] Ioffe Institute, St. Petersburg, 194021, Russia

[a)] Corresponding author: a.kutuzova@metalab.ifmo.ru
[b)] Electronic mail: m.rybin@metalab.ifmo.ru

Abstract. We study a silicon-based structure composed of parallel rods arranged in a periodic array. The structure supports guided modes which due to the deformation of structure in general become leaky modes. At the same time the modes split at the Brillouin zone surface by a pair with certain symmetry. Here we report on guided modes transformation into symmetry-protected BIC under deforming the structure, because of inherent field distribution governed by the Bragg-related mode hybridization.

INTRODUCTION

In photonics resonances are widely used for increasing light-matter interaction in vast applications and sophisticated optical devices such as biological and chemical sensors, microcavity lasers, modulators to name a few [1?]. The conventional merit of resonator efficiency is its quality (Q) factor, which determines the time of photon restraint before it leaks into free space off. For resonator modes decoupled from the radiating waves of free space Q factor tends to infinity and that resonator is perfect. Almost a century ago, in quantum mechanics a special type of electron states, bound states in the continuum (BIC), were considered [2]. These states have discrete energy embedded in the continuum of free space states, nevertheless BIC are stay localized by a certain spatial distribution of potential. Essentially the adaption of the BIC concept to optics paves the way to realization of the perfect resonators.

Unlike electrons, photonic BIC are mostly interesting because the photon energy is always positive and lies in the continuum of propagating waves. Therefore, the problem of trapping a photon near a certain dielectric structure for arbitrary time is of great practical importance. Recently, many applications of BIC in photonics were demonstrated [3, 4, 5].

We notice that perfect BIC cannot exist in finite structures, however they managed to be identified in systems that are infinitely extended in at least one direction. In extend periodic structures we distinguish resonant modes above the light line and guided mode below. Mathematically, in a homogeneous waveguide the wave number k can be combined with the frequency ω into an effective energy and then the modes above the light line $\omega/c > k$ correspond

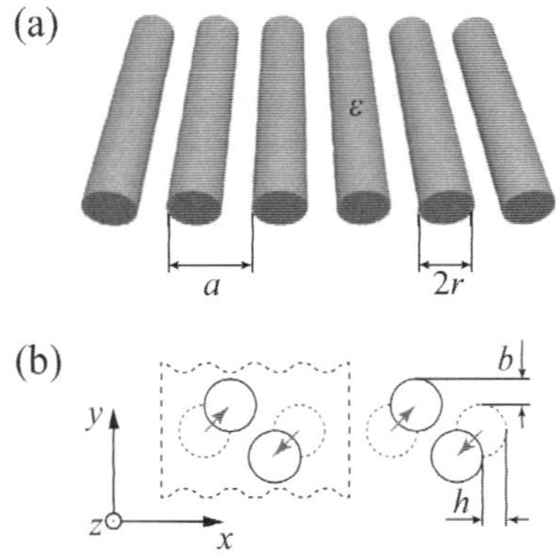

FIGURE 1. Schematic representation of the structure composed of silicon rods. (a) Periodic array of silicon rods along zx direction with radius r and dielectric index $\varepsilon = 12$. (b) Deformed structure wherein rods in each dimer are shifted at arbitrary distances in opposite directions along the x and y axes. Gray arrows specify the direction of dotted circles displacement in place of rods marked black solid circles. These shifted rods form the primitive cell depicted by black dashed lines.

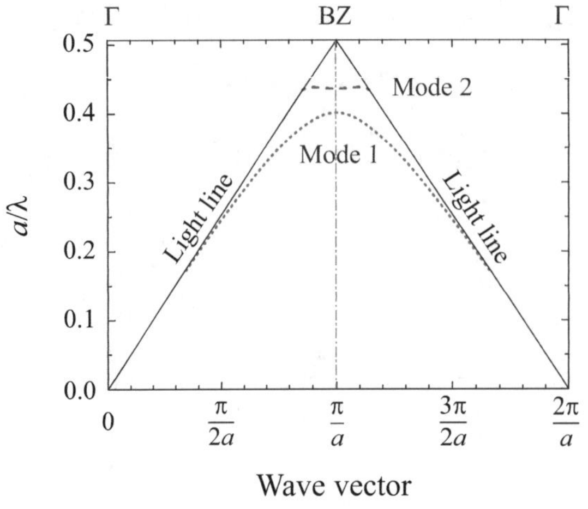

FIGURE 2. Photonic band structure of silicon structure where light cone depicted by black solid line. Guided modes below the light line are shown by pink and green dashed curves. The vertical black dashed line marks the boundary of the Brillouin zone. Wave vector $k_z = 0$.

to positive energy thus they are extend into a free space, and the modes under the light line $\omega/c < k$ correspond to negative energy. In this approach, the guided modes (below the light line) are considered as ordinary non-radiant localized states with negative effective energy. Although periodicity makes this treatment to be nonstrict it is still really illustrative. Using this approach BIC can only be modes over the light line, among other leaky modes. The periodicity leads to opening diffraction channels up, thus a portion of guided modes occur above the light line where they become leaky modes. Occasionally a structure deformation increases the period resulting to additional diffraction channels to appear in photonic band structure.

Here we show that different types of structure deformation lead to transforming guided modes into leaky ones, however Bragg interaction of two guided modes might allow one of them become BIC.

RESULTS

We consider a structure made of silicon rods with circular profile ($\varepsilon = 12$), lattice constant $a = 330\ \mu$m, radius of rods $r = 100\ \mu$m (Fig. 1). The system under study is homogeneous along the z axis and infinitely extended along the x axis. Similar finite structure is known to support BIC related high-Q supercavity modes [6].

We numerically simulated photonic band structure by using the eigenfrequency solver of COMSOL Multiphysics software. Figure 2 demonstrates modes sustained by the structure. It is important to clarify that we calculate only eigen modes of the system. Free space modes lie above the light line which is shown by black solid line. The modes of light cone occur beyond the Brillouin zone because uncertainty of the wave vector up to summation with any reciprocal lattice vector. A pair of guided modes corresponding to the opposite direction of propagation split up at the boundary of the Brillouin zone, which we refer as mode 1 (at the low frequencies) and mode 2 (at the high frequencies).

Now we deform the metasurface. So that let us consider the structure as an array of dimers consisting of two rods (Fig. 1). Parameter h is a shift of the dimerized rods in the direction to each other along the x axis; parameter b defines an offset along the normal (the y axis) such that the one rod moves upside and the other to the opposite direction. Thus one rod shifts at a vector $(h,b)^T$ and the other one has the opposite displacement $-(h,b)^T$. We consider two types of deformation and we call the horizontal deformation as the x-deformation, while vertical deformation by b parameter as the y-deformation. Both structural transformations change the primitive cell and result to open up additional light cone centered at π/a. So the both guided modes become resonant leaky modes above the light line.

Figure 3 exhibits that the modes under study are localized in different regions of structure. The magnetic field of mode 1 predominantly concentrates in site of the rod centers and localization areas protrude beyond the rod in the form of ellipses elongated in the vertical direction and the mode 1 is even relative the vertical plane passing through

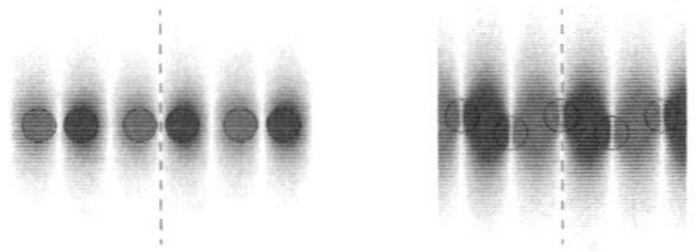

FIGURE 3. Eigen modes profiles. The intensity of blue and red colors shows, respectively, positive and negative values of the magnetic field component H_z amplitude. These modes are even or odd relative to two vertical planes: brown dashed line is symmetry element, which is located in the middle of the rod and green dashed line is symmetry element which is located between adjacent rods.

the centers of the rods and odd relative to the vertical plane passing between neighboring rods. The regions of the magnetic field localization of the mode 2, as in the case of the mode 1, have the form of ellipses elongated in the vertical direction. However, the centers of these ellipses are between the rods, so the mode 2 is opposite to the mode 1: even relative to the plane between rods and odd relative to the middle-rod plane. We notice that similar field distribution is observed for modes at edges of the fundamental gap in a one-dimension photonic crystal which are called dielectric and air bands. However the situation is opposite, namely the magnetic field is localized in the high-permittivity layers at the high-frequency band gap edge and in the low-permittivity layers at the low-frequency edge [7].

Owing to primitive cell doubling we observe appearance of additional diffraction channels whereupon modes cease to be guided. They appear above light line and propagate into free space with their Q factor reduction. We found Q-factor dependence on structure deformation thereby mode 1 stay BIC with the x-deformation while mode 2 became leaky, the opposite occurs when we deform structure with the y-deformation. Modes 1 and 2 become BIC due to even symmetry protection. We consider symmetry-protected BIC at the Γ point, although they might also exist on Brillouin zone boundary [8]. In addition we obtain that the leaky modes 1 and 2 (not protected by the symmetry) have quadratic dependence on the deformation.

CONCLUSION

We have shown that the deformation of structure leads to transformation guided modes into leaky resonances, however the field distribution at the Brilliune zone surface match the symmetry and some guided modes become symmetry-protected BIC. BIC resulting from guided modes inherit their field distribution in the material and outside the material which allows us to find different various possible applications in low-threshold lasers, high-sensitivity sensors and switchable modulators. Although we considered a structure with periodicity in one direction, similar physics is inherent in structures with more complex symmetry of systems consisting of resonant particles bounded in two directions.

ACKNOWLEDGMENTS

This work was supported by the Russian Scientific Foundation (grant 21-19-00677).

REFERENCES

1. K. J. Vahala, "Optical microcavities," nature **424**, 839–846 (2003).
2. J. Von Neumann and E. Wigner, "Uber merkwürdige diskrete eigenwerte," Phys. Z **465** (1929).
3. F. Yesilkoy, E. R. Arvelo, Y. Jahani, M. Liu, A. Tittl, V. Cevher, Y. Kivshar, and H. Altug, "Ultrasensitive hyperspectral imaging and biodetection enabled by dielectric metasurfaces," Nature Photon. **13**, 390–396 (2019).
4. S. Han, L. Cong, Y. K. Srivastava, B. Qiang, M. V. Rybin, A. Kumar, R. Jain, W. X. Lim, V. G. Achanta, S. S. Prabhu, Q. J. Wang, Y. S. Kivshar, and R. Singh, "All-dielectric active terahertz photonics driven by bound states in the continuum," Adv. Mater. **31**, 1901921 (2019).

5. A. Tittl, A. Leitis, M. Liu, F. Yesilkoy, D.-Y. Choi, D. N. Neshev, Y. S. Kivshar, and H. Altug, "Imaging-based molecular barcoding with pixelated dielectric metasurfaces," Science **360**, 1105–1109 (2018).
6. A. A. Kutuzova and M. V. Rybin, "Supercavity modes in silicon-based metasurfaces," in *AIP Conference Proceedings*, Vol. 2300 (AIP Publishing LLC, 2020) p. 020074.
7. J. D. Joannopoulos, S. G. Johnson, J. N. Winn, and R. D. Meade, *Photonic Crystals: Molding the Flow of Light*, 2nd ed. (Princeton Univ. Press, 2008) p. 304.
8. X. Gao, C. W. Hsu, B. Zhen, X. Lin, J. D. Joannopoulos, M. Soljačić, and H. Chen, "Formation mechanism of guided resonances and bound states in the continuum in photonic crystal slabs," Scientific reports **6**, 1–7 (2016).

Journal of Physics: Conference Series **2015** (2021) 012079 doi:10.1088/1742-6596/2015/1/012079

Non-Huygens transparent metasurfaces based on the novel Hybrid anapole state

A V Kuznetsov[1,2], **A Canós Valero**[1]

[1]ITMO University, Lomonosova street, 9, St. Petersburg, 191002, Russia
[2]Riga Technical University, Riga, Azenes street 12, post code 1048, Latvia

Corresponding author: alexey.kuznetsov@metalab.ifmo.ru

Abstract. In this work we studied the optical properties of the metasurfaces based on silicon nanocylinders experincing hybrid anapole. Our investigation shows the possibility of creating fully invisible metasurfaces consisting of silicon nanocylinders based on the novel state which in their properties can compete with the Huygens' metasurfaces. These results can be used in various fields of nanophotonics and may have important outcome in the development of various photonic devices.

1. Introduction

In the recent years, dielectric nanophotonics has become ubiquitous in numerous scientific and technological disciplines [1–6]. Enabling the ability to control the electric and magnetic components of light at the subwavelength scale with negligible losses, this fast-developing field holds the key to the design and fabrication of ultrasmall optical devices for flat optics. Dielectric metasurfaces have allowed the manufacturing of ultraflat lenses [7,8], hologram displays [9–12], light-managing devices [13–15], and dispersion control applications [16]. To a large extent, these developments were facilitated by Huygens metasurfaces, featuring the possibility to vary the transmitted phase with unity transmission, namely due to spectrally overlapping electric and magnetic dipole resonances of almost equal strength [17]. However, their efficiency is fundamentally limited by the near field coupling between the constituents of the metalattice [18], which has motivated the search for alternative means to achieve the desired 2π phase variation [19].

The recently observed hybrid anapole states [20] are nonradiating sources that, unlike conventional anapoles, simultaneously suppress electric and magnetic contributions to radiation (figure 1). This unique peculiarity can be exploited to suppress both far and near field coupling from the structure, unlike Huygens' sources. In this work, we explore this promising alternative to the Huygens's effect and design a fully transmissive metasurface that can control the phase, based on the hybrid anapole state of the individual metaatoms.

2. Results and discussion

Figure 1 shows the dependence of the reflection of the metasurface depending on the distance between the surfaces of the meta-atoms (a) and the distribution of the electric field norm in the particles at different distances (b). A model of a non-Huygens' metasurface consisting of silicon nanocylinders in a hybrid anapole state is schematically shown in (c).

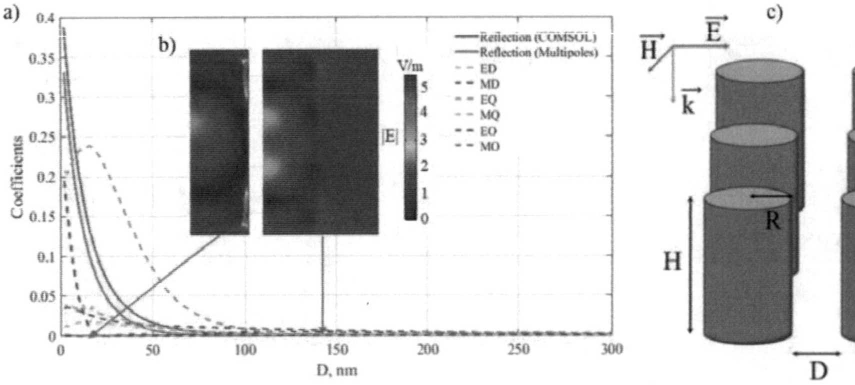

Figure 1. Dependence of the reflection of radiation with a wavelength of hybrid anapole state λ = 789 nm metasurface consisting of silicon nanocylinders with radius R = 128 nm and height H = 369 nm, on the distance between cylinders (a), fragments of the spatial electric field norm distribution in the xz-plane for distance between surfaces of metaatoms D = 20 nm, 144 nm show that near field coupling occurs only at very small periods (b) and artistic representation of the considered silicon metasurface composed of nanocylinders (c)

It is seen from the figure that even highly compact metasurfaces display minimal reflection until the near field of the metaatoms begins to influence the field inside neighboring scatterers. In addition, contrarily to other recently proposed effects inducing zero transmission, the hybrid anapole is preserved when deposited over substrates having almost any refractive index contrast. This reduces the requirements for the technical production process, allowing more design flexibility in terms of the material of the underlying substrate.

3. Conclusion

Thus, in this work, a metasurface model was developed, which, in terms of its optical properties, can become a worthy alternative to Huygens' metasurfaces. The developed non-Huygens' metsurface model based on nanocylinders in a hybrid anapole state has a number of indisputable advantages, such as full transmission of incident radiation, zero phase shift when radiation passes through it, and the possibility of deposition on various types of substrates without significant changes in the optical properties of the metasurface. All of the above properties make it possible to largely unlock the potential for creating various devices based on metasurfaces, for example, various metal lenses, sensors, holographic devices, and so on.

4. Acknowledgment

This work was supported by the Russian Scientific Foundation (grant 21-19-00677).

References

[1] Kivshar Y 2018 All-dielectric meta-optics and non-linear nanophotonics *Natl. Sci. Rev.* **5** 144–58

[2] Yan J, Liu X, Ma C, Huang Y and Yang G 2020 All-dielectric materials and related nanophotonic applications *Mater. Sci. Eng. R Reports* **141** 100563

[3] Chebykin A V., Orlov A A, Shalin A S, Poddubny A N and Belov P A 2015 Strong Purcell effect in anisotropic ε-near-zero metamaterials *Phys. Rev. B - Condens. Matter Mater. Phys.* **91**

[4] Kostina N, Petrov M, Ivinskaya A, Sukhov S, Bogdanov A, Toftul I, Nieto-Vesperinas M, Ginzburg P and Shalin A 2019 Optical binding via surface plasmon polariton interference *Phys. Rev. B* **99** 1–11

[5] Canós Valero A, Kislov D, Gurvitz E A, Shamkhi H K, Pavlov A A, Redka D, Yankin S, Zemánek P and Shalin A S 2020 Nanovortex-Driven All-Dielectric Optical Diffusion Boosting and Sorting Concept for Lab-on-a-Chip Platforms *Adv. Sci.* **7**

[6] Barhom H, Machnev A A, Noskov R E, Goncharenko A, Gurvitz E A, Timin A S, Shkoldin V A, Koniakhin S V., Koval O Y, Zyuzin M V., Shalin A S, Shishkin I I and Ginzburg P 2019 Biological Kerker Effect Boosts Light Collection Efficiency in Plants *Nano Lett.* **19** 7062–71

[7] Fathnan A A, Liu M and Powell D A 2020 Achromatic Huygens' Metalenses with Deeply Subwavelength Thickness *Adv. Opt. Mater.* **8** 1–9

[8] Chen X, Zhang X, Koten M A, Chen H, Xiao Z, Zhang L, Shield J E, Dowben P A and Hong X 2017 Interfacial Charge Engineering in Ferroelectric-Controlled Mott Transistors *Adv. Mater.* **29** 1–8

[9] Chong K E, Wang L, Staude I, James A R, Dominguez J, Liu S, Subramania G S, Decker M, Neshev D N, Brener I and Kivshar Y S 2016 Efficient Polarization-Insensitive Complex Wavefront Control Using Huygens' Metasurfaces Based on Dielectric Resonant Meta-atoms *ACS Photonics* **3** 514–9

[10] Wang L, Kruk S, Tang H, Li T, Kravchenko I, Neshev D N and Kivshar Y S 2016 Grayscale transparent metasurface holograms *Optica* **3** 1504

[11] Overvig A C, Shrestha S, Malek S C, Lu M, Stein A, Zheng C and Yu N 2019 Dielectric metasurfaces for complete and independent control of the optical amplitude and phase *Light Sci. Appl.* **8**

[12] Kozlov V, Filonov D, Shalin A S, Steinberg B Z and Ginzburg P 2016 Asymmetric backscattering from the hybrid magneto-electric meta particle *Appl. Phys. Lett.* **109**

[13] Kruk S and Kivshar Y 2017 Functional Meta-Optics and Nanophotonics Govern by Mie Resonances *ACS Photonics* **4** 2638–49

[14] Aoni R A, Rahmani M, Xu L, Zangeneh Kamali K, Komar A, Yan J, Neshev D and Miroshnichenko A E 2019 High-Efficiency Visible Light Manipulation Using Dielectric Metasurfaces *Sci. Rep.* **9** 1–9

[15] Terekhov P D, Shamkhi H K, Gurvitz E A, Baryshnikova K V., Evlyukhin A B, Shalin A S and Karabchevsky A 2019 Broadband forward scattering from dielectric cubic nanoantenna in lossless media *Opt. Express* **27** 10924

[16] Zhang X, Li Q, Liu F, Qiu M, Sun S, He Q and Zhou L 2020 Controlling angular dispersions in optical metasurfaces *Light Sci. Appl.* **9** 34–7

[17] Decker M, Staude I, Falkner M, Dominguez J, Neshev D N, Brener I, Pertsch T and Kivshar Y S 2015 High-Efficiency Dielectric Huygens' Surfaces *Adv. Opt. Mater.* **3** 813–20

[18] Olk A E and Powell D A 2019 Accurate Metasurface Synthesis Incorporating Near-Field Coupling Effects *Phys. Rev. Appl.* **11** 1

[19] Rahimzadegan A, Arslan D, Dams D, Groner A, Garcia-Santiago X, Alaee R, Fernandez-Corbaton I, Pertsch T, Staude I and Rockstuhl C 2020 Beyond dipolar Huygens' metasurfaces for full-phase coverage and unity transmittance *Nanophotonics* **9** 75–82

[20] Canos Valero A, Gurvitz E, Evlyukhin A and Redka D 2020 Theory, observation and ultrafast response of novel hybrid anapole states *arXiv*

Various multipole combinations for conical Si particles

A V Kuznetsov[1,2], A Canós Valero[1], P D Terekhov[3], H K Shamkhi[1]

[1]ITMO University, Lomonosova street, 9, St. Petersburg, 191002, Russia
[2]Riga Technical University, Riga, Azenes street 12, post code 1048, Latvia
[3]Duke University, Durham, NC, United States

Corresponding author: alexey.kuznetsov@metalab.ifmo.ru

Abstract. In this work, we investigated the possibility of creating various multipole combinations in conical silicon nanoparticles. It was found that in conical silicon particles it seems possible to create key effects for nanophotonics, such as various kinds of Kerker effects (Generalized Kerker, Transverse Kerker), Hybrid anapole state, Bound states in the continuum. This greatly simplifies the manufacturing process of photonic devices due to the easier production of nanocones in practice. Also, conical particles allow an additional degree of freedom, which opens up new horizons for obtaining previously unknown effects.

1. Introduction

In recent years, an increasing number of scientific studies have been devoted to areas in which subwavelength dielectric nanoparticles are used. The fact is that such particles can be used to create various kinds of devices, the characteristics of which either significantly exceed the characteristics of existing ones, or even open up the possibility of creating a new class of devices. Their undeniable advantage lies in their high efficiency and relatively easy production. There are a lot of examples of the use of devices based on dielectric nanoparticles: nanolasers [1], nanoantennas [2], metamaterials [3] and metasurfaces [4], ultrathin lenses [5], detectors [6] and other equally interesting applications [7,8].

A huge impetus to the development of high-index dielectric nanophotonics was made by opening the possibility of separate control of multipole excitations in scatterers and, as a consequence, combining multipole moments to obtain various effects [9–12].

Thanks to this possibility, it became possible to obtain the Kerker effect, which was originally introduced for a hypothetical magnetic sphere and did not attract the attention of researchers [13]. Now, the Kerker effect has been significantly expanded thanks to the capabilities of dielectric nanophotonics, where it became possible to obtain a magnetic moment in small particles. An excellent example for this is the generalized Kerker effect, which allows one to obtain resonant excitation and interference of various electromagnetic multipoles [14–18] or transverse Kerker, characterized by transverse isotropic scattering by subwavelength nanoparticles with simultaneous suppression of both forward and backward scattered fields [19].

Another important discovery related to multipole combinations is the possibility of obtaining a hybrid anapole state (HAS) of nanoobjects. It is possible to create a metasurface from such nano-objects, which will have many useful properties, for example, a zero phase shift of a light wave when passing through a metasurface, almost complete light transmittance, the ability to apply such a metasurface on various types of substrates [20]. This opens up incredible opportunities for creating various kinds of devices for flat optics.

Devices using bound states in the continuum (BIC) also play an important role in the development of dielectric nanophotonics. The flexibility that quasi-BICs has brought to the design and implementation of high Q resonances in nanophotonic systems has made them very useful to the nonlinear photonics community. BICs have so far been used to amplify a variety of nonlinear effects, including optical Kerr effect, laser action, second and third harmonic generation, and four-wave mixing. Thanks to BIC, lasers, harmonic generating devices, optical fibers, beamforming devices, etc. can be created [21].

All of the above cases were presented for particles with elementary geometry, for example, on spheres or cylinders, but, for example, frustoconical scatterers are currently poorly studied, although this geometry has many advantages.

Due to the additional degree of freedom in the form of the upper radius, unpredictable new effects may appear that have not been investigated before.

Another important advantage of the geometry of the cones is that the practical production of elements for photonic devices has not yet been perfected. For example, in the process of fabricating such scatterers, it is practically impossible to obtain a perfectly flat nanocylinder with the given parameters. It is much easier in practice to create a truncated cone.

Thus, this work is devoted to the study of the optical properties of the scatterer in the form of a truncated cone, the search for cases that are inaccessible to standard geometries, and testing the obtained results.

2. Results and discussion

Figure 1 shows a schematic drawing of the model for which mathematical modeling was carried out. As a scatterer, a silicon nanocylinder was taken from which, by changing the upper and lower radii, height and wavelength, various combinations of multipoles were obtained (a). Distributions for the far-field were plotted (c,e) for conical silicon particles, which were confirmed by theoretical combinations of multipoles for a point (b,d).

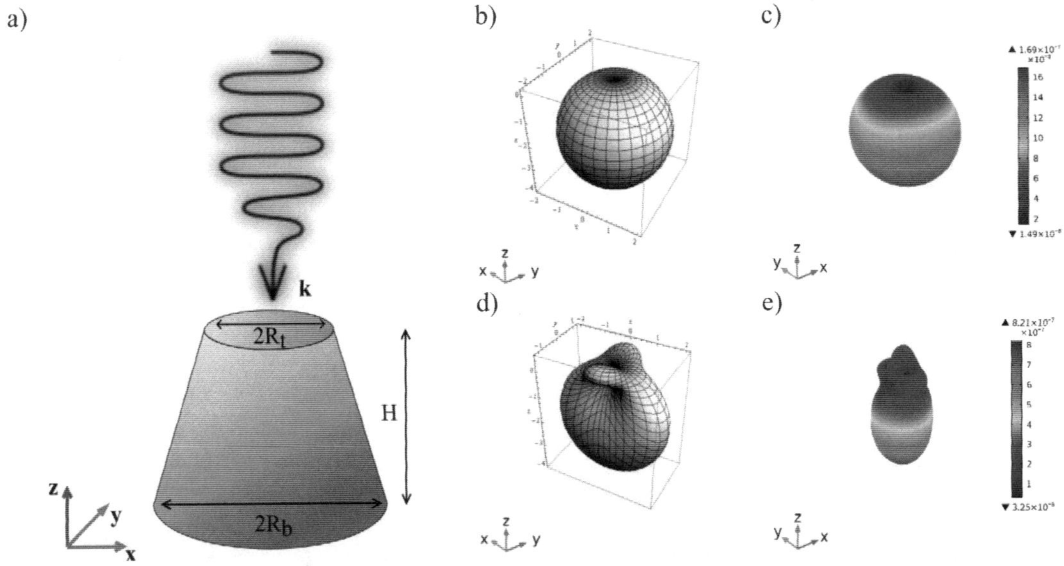

Figure 1. Artistic representation of the considered silicon particle composed of nanocylinders (infinite nanostructure illuminated with the linearly polarized plane wave) (a), theoretically far-field distributions of different combinations of multipoles for point (b,d) and simulated far-field distributions of different combinations of multipoles for conical silicon particles

It can be seen that by varying the parameters of conical silicon particles, we managed to obtain various combinations of multipoles, for example, the usual Kerker (c) and generalized Kerker effect (e). In the report we will show other possible combinations and unusual effects enabled by the truncated cone shape via separate tailoring of different multipoles.

3. Conclusion

Thus, this work is devoted to the extensive tutorial study of the optical properties of truncated conical nanoscatterers. In this work, various optical effects are obtained by tuning the parameters of the fabrication-friendly resonators - conical silicon scatterers. This opens up new possibilities for the production of photonic devices based on such shape of nanoscatterers. Also, conical particles allow an additional degree of freedom, which opens up new horizons for obtaining previously unknown effects.

4. Acknowledgement

This work was supported by the Russian Scientific Foundation (21-79-10190).

References

[1] Xu K, Fang M and Huang Z 2020 Compact unidirectional laser based on all-dielectric metasurface with high quality factor *IEEE Photonics J.* **13**

[2] Sain B, Meier C and Zentgraf T 2020 Nonlinear optics at all-dielectric nanoantennas and metasurfaces *arXiv* **1** 1–14

[3] Kozlov V, Filonov D, Shalin A S, Steinberg B Z and Ginzburg P 2016 Asymmetric backscattering from the hybrid magneto-electric meta particle *Appl. Phys. Lett.* **109**

[4] Overvig A C, Shrestha S, Malek S C, Lu M, Stein A, Zheng C and Yu N 2019 Dielectric metasurfaces for complete and independent control of the optical amplitude and phase *Light Sci. Appl.* **8**

[5] Tian Y, Li Z, Xu Z, Wei Y and Wu F 2020 High transmission focusing lenses based on ultrathin all-dielectric Huygens' metasurfaces *Opt. Mater. (Amst).* **109** 110358

[6] Mitrofanov O, Siday T, Vabishchevich P, Hale L, Harris T, Luk T S, Reno J L and Brener I 2019 Terahertz detectors based on all-dielectric photoconductive metasurfaces 36

[7] Yan J, Liu X, Ma C, Huang Y and Yang G 2020 All-dielectric materials and related nanophotonic applications *Mater. Sci. Eng. R Reports* **141** 100563

[8] Chebykin A V., Orlov A A, Shalin A S, Poddubny A N and Belov P A 2015 Strong Purcell effect in anisotropic ε-near-zero metamaterials *Phys. Rev. B - Condens. Matter Mater. Phys.* **91**

[9] Evlyukhin A B and Chichkov B N 2019 Multipole decompositions for directional light scattering *Phys. Rev. B* **100** 125415

[10] Alaee R, Rockstuhl C and Fernandez-Corbaton I 2018 An electromagnetic multipole expansion beyond the long-wavelength approximation *Opt. Commun.* **407** 17–21

[11] Kostina N, Petrov M, Ivinskaya A, Sukhov S, Bogdanov A, Toftul I, Nieto-Vesperinas M, Ginzburg P and Shalin A 2019 Optical binding via surface plasmon polariton interference *Phys. Rev. B* **99** 1–11

[12] Canós Valero A, Kislov D, Gurvitz E A, Shamkhi H K, Pavlov A A, Redka D, Yankin S, Zemánek P and Shalin A S 2020 Nanovortex-Driven All-Dielectric Optical Diffusion Boosting and Sorting Concept for Lab-on-a-Chip Platforms *Adv. Sci.* **7**

[13] Kerker M, Wang D S and Giles C L 1983 Electromagnetic Scattering By Magnetic Spheres. *J. Opt. Soc. Am.* **73** 765–7

[14] Alaee R, Filter R, Lehr D, Lederer F and Rockstuhl C 2015 A generalized Kerker condition for highly directive nanoantennas *Opt. Lett.* **40** 2645

[15] Liu W and Kivshar Y S 2017 Generalized Kerker effects in nanophotonics and meta-optics *arXiv* **26** 274–84

[16] Hesari-Shermeh M, Abbasi-Arand B and Yazdi M 2021 Generalized Kerker's conditions under normal and oblique incidence using the polarizability tensors of nanoparticles *Opt.*

Express **29** 647

[17] Terekhov P D, Shamkhi H K, Gurvitz E A, Baryshnikova K V., Evlyukhin A B, Shalin A S and Karabchevsky A 2019 Broadband forward scattering from dielectric cubic nanoantenna in lossless media *Opt. Express* **27** 10924

[18] Barhom H, Machnev A A, Noskov R E, Goncharenko A, Gurvitz E A, Timin A S, Shkoldin V A, Koniakhin S V., Koval O Y, Zyuzin M V., Shalin A S, Shishkin I I and Ginzburg P 2019 Biological Kerker Effect Boosts Light Collection Efficiency in Plants *Nano Lett.* **19** 7062–71

[19] Shamkhi H K, Baryshnikova K V., Sayanskiy A, Kapitanova P, Terekhov P D, Belov P, Karabchevsky A, Evlyukhin A B, Kivshar Y and Shalin A S 2019 Transverse scattering and generalized kerker effects in all-dielectric mie-resonant metaoptics *Phys. Rev. Lett.* **122** 193905

[20] Canos Valero A, Gurvitz E, Evlyukhin A and Redka D 2020 Theory, observation and ultrafast response of novel hybrid anapole states *arXiv*

[21] Azzam S I and Kildishev A V. 2021 Photonic Bound States in the Continuum: From Basics to Applications *Adv. Opt. Mater.* **9** 16–24

Planar Modulated Reactance Surfaces for Endfire Antenna Applications

Do-Hoon Kwon

Department of Electrical and Computer Engineering, University of Massachusetts Amherst, Amherst, Massachusetts 01003, USA

E-mail: dhkwon@umass.edu

Abstract. Modulated reactance synthesis for planar impenetrable electromagnetic surfaces for endfire radiation characteristics is presented. It is based on aperture field synthesis over the entire spectral range, where the desired radiation pattern is prescribed by a traveling-wave current over the aperture. Carefully constructed auxiliary waves in the evanescent spectrum are added and the total aperture fields are optimized such that the entire aperture surface becomes pointwise reactive. The optimized aperture fields define the modulated surface reactance that realizes the desired pattern when excited by a feed surface wave. The reactance distribution may be realized as modulated metasurfaces.

1. Introduction

A modulated surface reactance on an electromagnetically impenetrable surface can couple a bound wave and a space wave. The classical work by Oliner and Hessel established a relation between the complex propagation constant of a leaky-wave mode and the parameters of a periodic sinusoidal reactance modulation [1]. Recent active development of modulated metasurface antennas has largely built on locally periodic sinusoidal modulation, including metasurface antennas capable of amplitude, phase, and polarization control [2]. Overcoming the issues associated with locally periodic modulations, a modulated reactance design method based on complete aperture field synthesis has been reported [3] for leaky-wave antenna (LWA) applications.

In this paper, the aperture field synthesis approach in [3] is extended to treat endfire radiation. Over the aperture, a desired endfire pattern is characterized by the propagating wave component with a phase velocity matching the speed of light in free space. Evanescent spectral components are introduced and optimized such that the entire aperture surface becomes locally and globally passive and lossless. The field synthesis technique enables pattern control over the entire 180° angular range above the planar aperture, greatly expanding the limited capability in endfire pattern synthesis traditionally provided by surface wave (SW) antennas [4].

2. Complete endfire aperture field synthesis

Consider 2-D endfire radiation using a modulated reactance surface in TM polarization, as illustrated in Fig. 1(a). On an infinite perfect electric conductor (PEC) ground plane, an aperture of length $\ell = x_1 - x_0$ is characterized by an inhomogeneous surface reactance $X_s(x)$. It is desired that the modulated reactance surface realize a given endfire pattern when it is excited

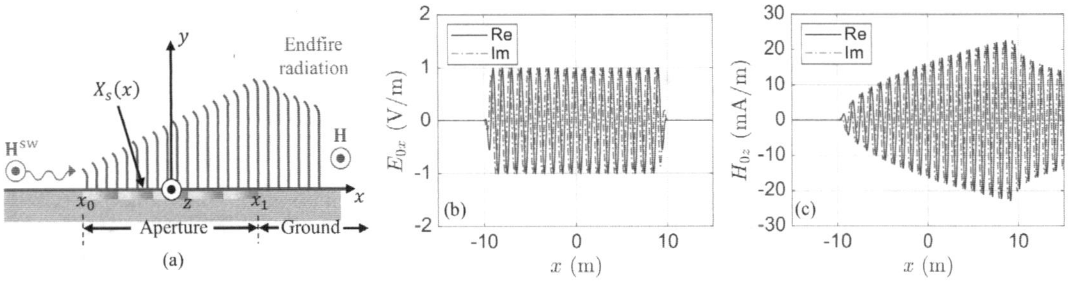

Figure 1. 2-D endfire radiation from a planar modulated reactance surface in TM polarization. (a) A planar modulated reactance surface mounted on a PEC ground excited by a feed SW. (b) The tangential E-field of a 20λ-long uniform aperture at $f = 300$ MHz. (c) The associated tangential H-field.

by a TM-polarized feed SW. Instead of $X_s(x)$, we choose the tangential E-field component over the aperture, $E_{tx}(x) = E_x(x, y = 0)$, as the quantity for synthesis. As in the LWA synthesis [3], we construct $E_{tx}(x)$ as

$$E_{tx}(x) = E_{0x}(x) + E_1(x)e^{-jk_c x} + \sum_{n=2}^{N} E_n(x)e^{j\psi_n(x)}, \tag{1}$$

where k_c is the phase constant of the feed SW. The slowly-varying functions $E_n(x)$ $(n = 1, \ldots, N)$ define the envelopes of the diminishing feed SW $(n = 1)$ and fast spatially-varying [defined via $\psi_n(x)$] higher-order components $(n = 2, \ldots, N)$. The associated tangential aperture H-field, $H_{tz}(x) = H_z(x, y = 0)$, is completely specified by $E_{tx}(x)$ via Maxwell's equations. The desired far-zone radiation pattern is specified by the field pair (E_{0x}, H_{0z}) over the aperture. For a 20λ-long aperture at $f = 300$ MHz, a uniform aperture with $E_{0x}(x) = e^{-jkx}$ V/m $(k = 2\pi/\lambda =$ free-space wavenumber) is shown Fig. 1(b), which includes a λ-wide transition at each end. The associated H-field component, $H_{0z}(x)$, is plotted in Fig. 1(c). Unlike E_{0x}, a significant tangential H-field exists beyond the aperture on the PEC ground, associated with endfire radiation.

A local condition for no loss and no gain at an impenetrable surface point is [5]

$$S_y(x) = -\frac{1}{2}\,\mathrm{Re}\{E_{tx}(x)H_{tz}^*(x)\} = 0, \tag{2}$$

i.e., the normal component for the time-average Poynting vector vanishes. The envelopes $E_n(x)$ $(n = 1, \ldots, N)$ can be numerically optimized to enforce (2) at all $x \in [x_0, x_1]$ in an average sense. To the optimization process described in [3], two pieces of refinement are added: $E_1(x)$ is allowed to be complex-valued and the feed SW strength is updated during optimization. Once the optimized tangential fields are found, the aperture surface is characterized by the surface impedance $Z_s(x) = R_s(x) + jX_s(x) = E_{tx}(x)/H_{tz}(x)$. As satisfaction of (2) implies $R_s(x) = 0$, the surface reactance $X_s(x)$ defines the aperture as a modulated reactance surface.

3. Numerical example

Consider a target endfire radiation with a main beam in the $+x$-axis direction associated with (E_{0x}, H_{0z}) in Figs. 1(b)–(c). Using $k_c = 2k$ and $N = 14$, the optimized $E_{tx}(x)$ produces the power density profile $S_y(x)$ in Fig. 2(a). Also shown is the power profile associated with the $n = 0$ term only, denoted by $S_0(x)$. It shows that an increasing amount of power leaves the surface as a

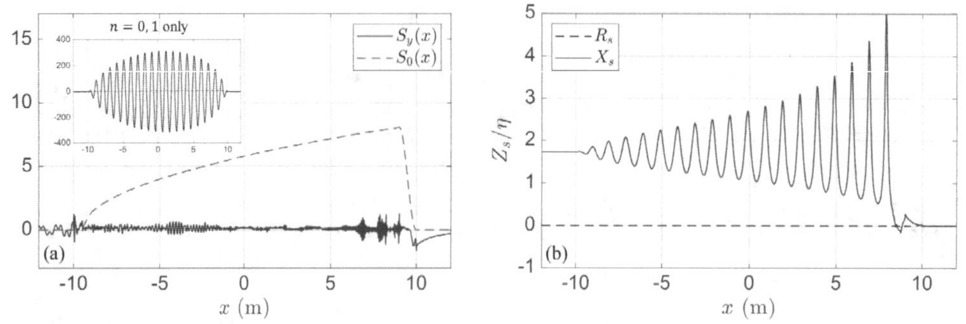

Figure 2. Field and reactance synthesis results for the 20λ-long uniform aperture. (a) The normal component of the Poynting vector in mW/m^2. (b) The retrieved surface impedance Z_s normalized by the free-space intrinsic impedance $\eta \approx 377\ \Omega$.

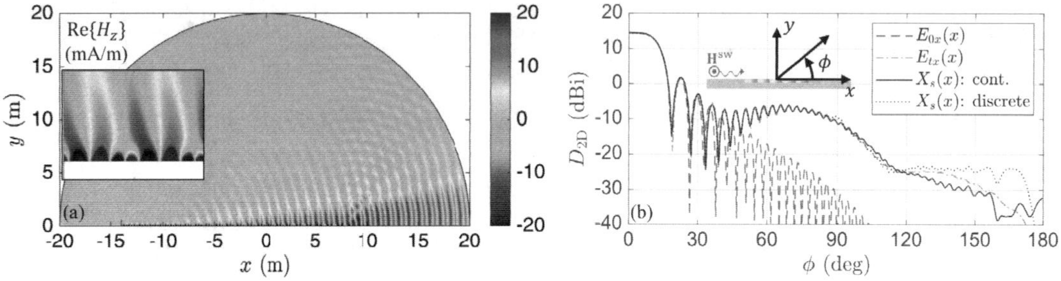

Figure 3. Characteristics of the reactance surface for a uniform-aperture endfire pattern. (a) Snapshot of $H_z(x, y)$ when the reactance surface is excited by a feed SW. The inset shows a magnified view near $x = 0$. (b) Simulated 2-D directivity patterns for the specification $[(E_{0x}, H_{0z})]$, the optimized fields $[(E_{tx}, H_{tz})]$, and the continuous as well as discretized reactance-specified aperture surfaces $[X_s(x)]$.

space wave (i.e., radiation) with increasing x. For comparison, the power density associated with the first two terms in (1) is plotted in the inset. The pattern-defining component ($n = 0$) and the diminishing SW component ($n = 1$) interfere to create a strong spatially-oscillatory interference pattern. The higher-order terms ($n = 2, \ldots, 14$) play a role of neutralizing this power interference to bring the overall profile $S_y(x)$ for the total fields, (E_{tx}, H_{tz}), close to zero everywhere. From the optimized tangential aperture fields, the surface impedance $Z_s(x)$ is retrieved, as shown Fig. 2(b). The surface resistance is essentially zero all over the 20λ-long aperture, as desired. The surface reactance has an oscillatory behavior with an increasing amplitude, starting from an unmodulated value of $X_{s0} = \eta\sqrt{(k_c/k)^2 - 1} = \sqrt{3}\eta$. The modulation profile is clearly distinct from locally sinusoidal variations.

Performance of the synthesized reactance surface in creating the desired endfire pattern is assessed using full-wave simulation. Using COMSOL MULTIPHYSICS, $X_s(x)$ in Fig. 2(b) is mathematically enforced in $x \in [-10, 10]$ m and a feed SW carrying a power of 106.5 mW/m that matches the radiated power is launched onto the aperture surface. Figure 3(a) shows a snapshot of the resulting H-field distribution. Strong endfire radiation in the $+x$-axis direction is observed. A magnified view in the inset shows a mix of an SW bound to the surface and a space wave leaving the surface. In Fig. 3(b) are plotted 2-D directivity patterns associated with the specification, the synthesized fields, and the continuous as well as discretized (20 samples per wavelength) reactance surfaces. The pattern for the synthesized fields has a broad shoulder near $\phi = 70^{\circ}$. It is associated with an abrupt change in the feed SW envelope $E_1(x)$ over a short distance that resulted from the field optimization. The patterns for both reactance surfaces have a good agreement with that associated with $E_{tx}(x)$ and they accurately reproduce the main beam and a few minor lobes of the desired pattern.

4. Conclusion

Using complete aperture field synthesis over both propagating and evanescent spectral ranges, design of modulated reactance surfaces for endfire characteristics has been presented. To the propagating wave fields that define the desired endfire pattern, auxiliary evanescent waves are added and optimized to render the entire radiating aperture passive and lossless. Starting from an appropriately defined propagating-spectrum component of the aperture fields, modulated reactance profiles for a variety of different endfire patterns can be designed. The synthesized reactance profiles may be realized using modulated metasurfaces.

Acknowledgments

This work was supported by the U.S. Army Research Office grant W911NF-19-2-0244.

References

[1] Oliner A A and Hessel A 1959 *IRE Trans. Antennas Propag.* **7** 201–208
[2] Minatti G, Caminita F, Martini E, Sabbadini M and Maci S 2016 *IEEE Trans. Antennas Propag.* **64** 3907–3919
[3] Kwon D H 2020 *IEEE Trans. Antennas Propag.* **68** 5463–5477
[4] Zucker F J 2007 *Antenna Engineering Handbook* ed Volakis J L (New York: McGraw Hill) chap 10
[5] Kwon D H 2018 *Phys. Rev. B* **98** 125137

Up-conversion photoluminescence specificity of a hybrid sponge nanostructures

Artem Larin[1], Egor Kurganov[1], Stephanie Bruyère[2], Alexandre Nominé[1,2], Eduard Ageev[1], Anton Samusev[1], Dmitry Zuev[1]

[1] Department of Physics and Engineering, ITMO University, 197101, St. Petersburg, Russia
[2] Université de Lorraine, CNRS, IJL, F-54000 Nancy, France

E-mail: `artem.larin@metalab.ifmo.ru`

Abstract.
Over the past decades, silicon is proved to be as a promising material for the development of devices in the fields of nanophotonics and optoelectronics. However, the material so popular at the current time did not find its application in nanoscale radiation sources due to the indirect bandgap of the semiconductor, which leads to low quantum efficiency. This work represents experimental results on the features of the silicon up-conversion photoluminescence enhanced by the optical resonances of the plasmonic nanosponge. The internal configuration of the nanostructure was confirmed by scanning transmission electron microscopy. The optical characterisation was provided by the dark-field spectroscopy, up-conversion photoluminescence generation and life-time measurements. The such new nanostructure type is promising for the development of nanoscale sources of broadband radiation and other applications of silicon photonics.

1. Introduction

The need for nanoscale broadband light emitters is essential for such areas as optical chip design, bioimaging, and sensing. Unfortunately, silicon has not become particularly popular in this field due to the indirect bandgap, since it gives a low quantum efficiency of radiative recombinations. In early works [1, 2, 3], which were devoted to the study of the silicon up-conversion photoluminescence (PL), the key role of resonances for a PL generation was shown. The simplest geometry for modeling properties and interpreting processes of interaction with electromagnetic radiation and for the fabrication is the nanosphere (NSph). The intrinsic resonances of the silicon NSph make it possible to enhance the field of the exciting radiation and also to increase the PL output signal. Thus, the shape of the PL spectrum curve directly depends on the shape of the enhancement factor one. Also, the intrinsic resonances of silicon are spectrally narrow, that is why silicon is not promising for the design of a nanoscale source of broadband radiation. This problem can be solved by using plasmonic nanosponges in combination with silicon. There are two key features of plasmonic nanosponges rising a special interests of special interest to such kinds of structures [4]. The first one is the high value of the surface-to-volume ratio that contributes to high electromagnetic field enhancement. The second one is the multiple localized plasmon resonance that one structure can maintain resonances at a wide spectral range. By filling plasmonic nanosponges with high refractive index semiconductor, a significant enhancement of multi-photon processes can be obtained [2, 3].

In this paper, we experimentally demonstrated the effect of the plasmonic nanosponge resonance on the output up-conversion PL of the silicon through life-time measurements. Such system representing a plasmonic nanosponge, the pores of which are filled with crystalline silicon grains, will be referred to as a hybrid sponge nanostructure (HSN).

2. Experimental setup

2.1. Fabrication

The Au-Si nanostructures were fabricated using the method of femtosecond laser ablation of a bi-layer film at the room temperature. The process of ablation was carried out with the TEMA-150 system as the source of femtosecond laser pulses with the wavelength of 1047 nm, spectral width of 7 nm, the pulse duration of 150 fs at the rate of 1 kHz and pulse energy of 50 nJ. Laser beam was focused using the Mitutoyo Plan Apo NIR Infinity Corrected Objective (10×, NA = 0.26). The Au-Si bi-layer film (15 nm and 60 nm, respectively) was used as the target (donor substrate) for laser pulses. To study the internal configuration of HSN by the scanning transmission electron microscopy (STEM) method, special metal grids were used as an acceptor substrate. Its made of copper with a thin carbon layer. Substrate of transparent silica glass was used as an acceptor substrate in the case of optical measurements.

2.2. Internal configuration characterisation

The HSN inner configuration was studied by the STEM methods: high-angle annular dark-field (STEM-HAADF) and energy-dispersive spectroscopy (STEM-EDS). A JEOL – ARM 200F Cold FEG TEM/STEM setup (200 kV operational voltage) was used with a spherical aberration probe corrector (0.078 nm resolution).

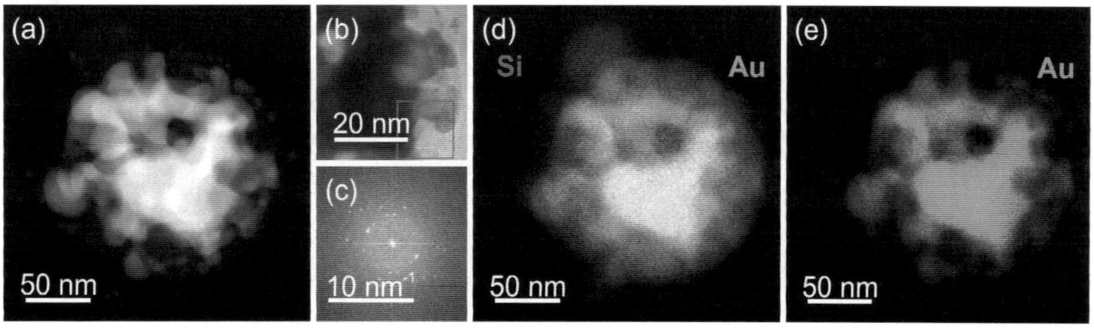

Figure 1. (a) Typical STEM-HAADF image of the HSN. (b) HRTEM image of the HSN from (a). (b) FFT of the considered area on the HSN (red square, see (b)) under electron beam incidence. (d-e) STEM-EDS map images, describing the spatial distribution of Si and Au atoms in the volume of the HSN. The presented nanostructure shows the following characteristics: particle diameter of ~186 nm, average pore size ~30 nm, 61at.%Si, and 39at.%Au.

2.3. Optical measurements

The PL measurement was performed at the room temperature on a confocal microscope system, consisting of two coaxial optical channels. The Mitutoyo Plan Apo Infinity Corrected Objectives were used to localize pump (NIR, M = ×50, NA = 0.42) and collect radiation from the sample (VIS, M = ×100, NA = 0.7). TEMA-150 system was used as the pump source at the repetition frequency of 200 kHz. The signal from the sample was spectrally decomposed with a LabRAM monochromator and collected with a CCD camera. To investigate the scattering spectra during

PL measurements, the dark-field microscopy was performed. Dark field optical channel was set at 68 ° relative to the pump optical axes. The Mitutoyo objectives were used for excitation (M = x10, NA = 0.28) and detection (M = x100 and NA = 0.70) of scattered signal. The sample was exposed with a halogen lamp and the scattered light was detected and spectrally decomposed by LabRAM monochromator.

Life-time measurements were carried out at the room temperature with time-correlated single photon counting (TCSPC) method using single-photon avalanche diode (50 ps timing resolution) to detect signal decay over time in the spectral range of 400 - 900 nm. The excitation was realised with a femtosecond laser with repetition frequency of 200 kHz. Picoharp 300 TSCPC module was used to detect the luminescence, and make synchronization with laser trigger.

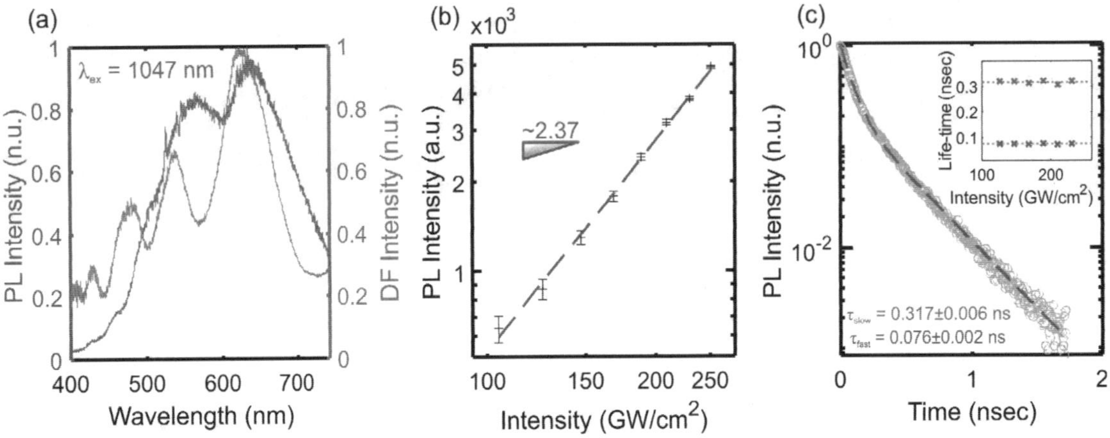

Figure 2. Optical properties of the HSN. (a) PL and DF spectra of the HSN. (b) Dependence of the PL intensity on the excitation intensity in the log-log scale. Numbers in the legend frame represent the order of the approximation polynomial (blue dashed line). (c) PL decay for the HSN. Inset: dependence of the fast and slow components of the PL decay of the HSN as a function of the pump intensity.

3. Result and discussion

The HSN were fabricated as the result of femtosecond laser ablation (Fig. 1a). High energy laser pulse induced melting of a bi-layer Si/Au film on an acceptor substrate resulting in sphere-like nanostructures with diameters varying from tens to hundreds nm in the diameter [3]. The typical content of silicon atoms, which was determined through the EDS method (Fig. 1d-e), exceeds the concentration at the point of the eutectic (~18.6% [5]). For this reason, the formed liquid droplets of the solution of Au-Si atoms starts with the process of silicon nucleation and the growth of crystalline silicon grains. These grains are uniformly and chaotically distributed in the volume of the nanostructure. The formation of the structure is completed by the Au solidification. In this case, the final structure is a plasmonic sponge, the pores of which are formed by crystalline silicon grains. The chaotic arrangement of the last leads to the formation of an Fast Fourier Transform (FFT) of the TEM image close to the structure of a polycrystalline material (Fig. 1c).

As it was shown in early work [4] plasmonic sponges having the same characteristic parameters (such as average pore size, porosity, outer diameter) possess unique linear scattering spectra. The fact that multiple localized plasmon resonance covers a wide range of wavelengths remains unchanged. For this reason, optical measurements of HSN linear scattering and their PL spectrum are demonstrated strictly for one particle.

The HSN with nanoscale size has resonances that are located in the visible spectral range (Fig. 2a). And it is possible to obtain an up-conversion PL response from HSN when it is irradiated with femtosecond laser radiation at the wavelength of 1047 nm. Obtained PL spectrum cover entire visible range and partly near-IR. Comparison of the scattering spectrum with the PL spectrum demonstrates a correlation between the shape of the PL spectrum and intrinsic resonances of the nanostructure (the coincidence of local maxima in the scattering and PL spectra). This fact is consistent with the results from [1] where the correlation of the output PL spectrum of the silicon sphere with the structure gain spectra was shown.

The dominant contribution to a broadband radiation of a HSN is a multi-photon absorption. It is lead to optically-induced electron-hole plasma caused by the plasmon-assisted transitions under high-intensity excitation [2]. The studies of the PL intensity dependence on the excitation wavelength in the log-log scale (Fig. 2b) demonstrate that the dominant processes to generate broadband PL are two- and three-photon ones. This was determined taking into account the slope of a straight line in the log-log scale ($\sim$2.37), which in turn is the degree of polynomial approximation.

The hot carriers are short-living and plasmon-assisted transitions are contributing into a sub-nanosecond ($\tau_{fast} = 0.076$ ns) exponential decay. In turn, the long-living PL ($\tau_{slow} = 0.317$ ns) we treat as the defect-supported transitions. The time scales are smaller in comparison with the work [1] ($\tau_{slow} = 1.740$ ns, $\tau_{fast} = 0.052$ ns). It can be explained by the influence of the plasmonic nanosponge, which decreases the output PL lifetime due to the Purcell effect. The considered feature is typical for the resonant plasmonic nanostructures [6]. The absence of a dependence of the PL decay on the pump intensity can be explained by relaxation through defects (see inset to Fig. 2c).

4. Conclusion

We have experimentally demonstrated the features of up-conversion PL of silicon in the HSN. It is based on two- and three-photon absorption of femtosecond laser excitation at the wavelength of 1047 nm. The geometry of the HSN was determined by STEM methods. The resonance properties of a plasmonic nano-sponge have a key influence on the PL output spectrum, which correlates with the linear scattering spectrum curve in the dark field geometry. Although, plasmonic nanosponge resonances affect the PL decay dynamics and lead to the decreasing of it in comparison with an intrinsic resonance of silicon NSph. Since the PL is based on a defect-support transitions, life-time changes are negligible and power independent. The study of HSN can open a new development designs of broadband radiation sources in nanoscale, expand the element base of structures for sensorics and catalysis.

5. Acknowledgments

This work is supported by the Russian Foundation for Basic Research (21-52-15029). A.N. is grateful to ITMO Fellowship and Professorship program. A.N. and S.B. have been supported by CNRS through International Emerging Action program.

6. References

[1] C. Zhang, Y. Xu, J. Liu, J. Li, J. Xiang, H. Li, J. Li, Q. Dai, S. Lan, and A.E. Miroshnichenko. Lighting up silicon nanoparticles with mie resonances. *Nature Communication*, 9:2964, 2018.

[2] S. V. Makarov, I. S. Sinev, V. A. Milichko, F. E. Komissarenko, D. A. Zuev, E. V. Ushakova, I. S. Mukhin, Y. F. Yu, A. I. Kuznetsov, P. A. Belov, I. V. Iorsh, A. N. Poddubny, A. K. Samusev, and Yu. S. Kivshar. *Nano Letters*, 18(1):535–539, 2018.

[3] A. Larin, A. Nomine, E. Ageev, J. Ghanbaja, L. Kolotova, S. Starikov, S. Bruyère, T. Belmonte, S. Makarov, and D. Zuev. Plasmonic nanosponges filled by silicon for enhanced white light emission. *Nanoscale*, 12:1013, 2020.

[4] D. Wang and P. Schaaf. *Advances in Physics X*, 3(1):1456361, 2018.

[5] Jin-Shyong Lina, Chien-Chon Chenb, Eric Wei-Guang Diaub, and Tzeng-Feng Liua. Fabrication and characterization of eutectic gold–silicon (au–si) nanowires. *Journal of Materials Processing Technology*, 206:425—430, 2008.

[6] Alexander E. Krasnok, Alexey P. Slobozhanyuk, Constantin R. Simovski, Sergei A. Tretyakov, Alexander N. Poddubny, Andrey E. Miroshnichenko, Yuri S. Kivshar, and Pavel A. Belov. An antenna model for the purcell effect. *Scientific Reports*, 5:12956 EP –, Aug 2015. Article.

Bound states in the continuum-induced enhancement of evanescent field confinement

S I Lepeshov[1*] **and A A Bogdanov**[1]
[1]Department of Physics and Engineering, ITMO University, Saint Petersburg, Russia

[*]s.lepeshov@metalab.ifmo.ru

Abstract. Here, the enhancement of electromagnetic field confinement in an all-dielectric metasurface is demonstrated. The enhanced confinement is achieved when the polarization singularity, corresponding to accidental bound states in the continuum, moves to the domain of evanescent fields (under the light line). Such a hybridization of the bound states and evanescent waves results in the 70-fold increase of the electric field enhancement on the top of the metasurface and boosting of the electric field localization.

1. Introduction

Metasurfaces and meta-gratings are one of the cornerstones of the modern photonics due to the high versatility of their physical and, specifically, optical properties. Since the very famous work by John Pendry in 2000 [1], the photonics of metasurfaces has got boost, and currently metasurfaces are utilized for signal modulation [2], information processing [3], color printing [4], surface-enhanced Raman scattering [5] and sensing [6]. Sensing by metasurfaces is of particular interest because of the capability of these nanostructures to confine and enhance optical fields as well as their high figure of merit [7]. Biosensing and single-molecule sensors based on metallic (plasmonic) metasurfaces have been demonstrated in several works [8], and the ability of surface plasmons to confine light makes them perspective for the practical applications. However, due to the high dissipative losses in metals in the optical range, this technology has not yet reached the industry [9].

Recently, new approaches which utilize high-Q modes in all-dielectric metasurfaces have been established [10]. The origin of such the high-Q modes has been discovered as bound states in the continuum (BIC) [11]. In theory, the Q-factors of BICs can be infinitely large, and the electric field enhancement can be respectively high [11]. Such a unique property of BIC originates from the physical symmetries of metasurfaces (parity and time) and results in cancelation of radiation in the far-field at the certain point of band diagram of the metasurface. At the same time, macroscopic dielectric structures in the waveguide regime, e.g. planar waveguides and optical fibers, are widely used in optical sensors.

Our work aims to combine unique properties of metasurface BICs and waveguides giving rise to the electromagnetic field confinement in the optical mode of a metasurface. By varying the geometrical parameters of the metasurface we achieve BIC transition under the light cone and boosting the field confinement of the waveguide mode. This approach is based on cancelation of certain Fourier term in electromagnetic field expansion of the metasurface mode.

2. Results and discussion

First, we consider a GaP meta-grating based on rectangular meta-atoms (see Fig. 1(a)) located at a quartz substrate. Notefully, such the mirror-symmetry shape of meta-atoms presumes both Γ- and off-Γ BICs.

Journal of Physics: Conference Series **2015** (2021) 012083 doi:10.1088/1742-6596/2015/1/012083

We adjust the metasurface geometrical parameters to operate in near-IR spectral range. The refractive index of the metasurface is 3.143, the refractive index of the substrate is 1.452. The height (H) of the metasurface is set to 370 nm at the first calculation step. The metasurface is surrounded by air.

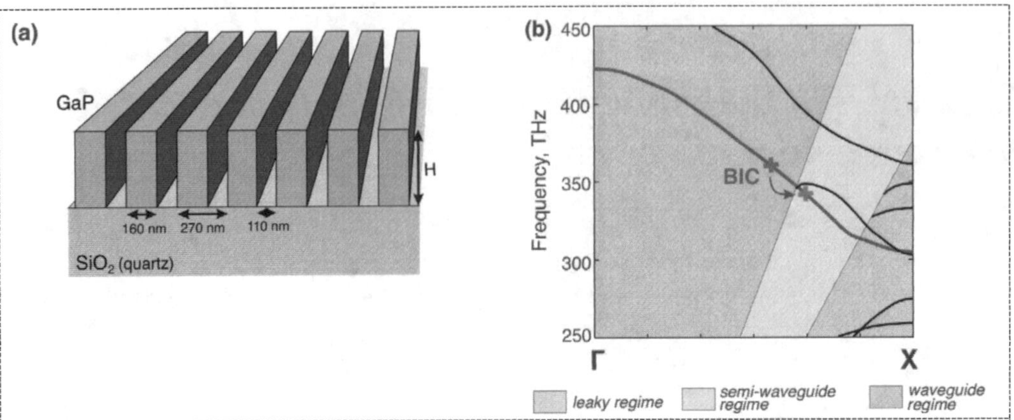

Figure 1. (a) A sketch of the metasurface operating in hybrid BIC-waveguide regime. (b) Band structure of the metasurface. H-parameter is tuned to tune the BIC under the light line of the air (yellow area).

Next, we calculate band structure of the metasurface using COMSOL Multiphysics (Fig. 1(b)). The green area corresponds to the radiation continuum above the light line of air. Yellow area is below the light line of air and above the quartz light line. Blue area designates the zone below the light line of quartz. The optical mode under consideration is distinguished by red. Initially, we observe the BIC in the green area simultaneously tracking the electric field distribution of the mode in the yellow area (Fig. 2(a)). It is clearly seen that, while the BIC is located in the radiation continuum of both air and quartz, the mode in the semi-waveguide regime gives 5-fold enhancement at the top of the metasurface and possesses long tails outside the rectangles.

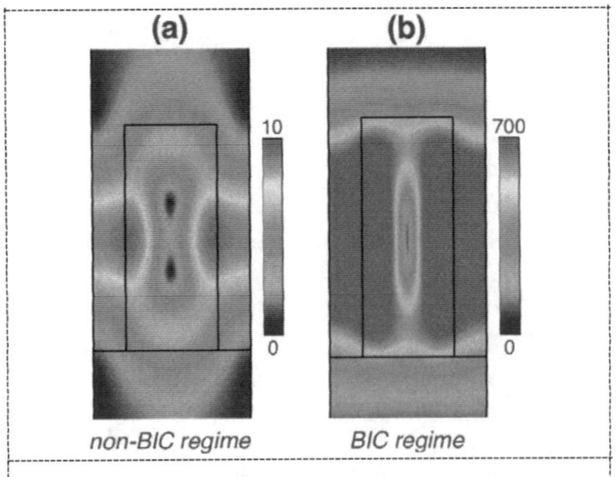

Figure 2. Electric field distributions of the optical mode of the metasurface in semi-waveguide regime at (a) H = 370 nm and (b) H = 400 nm.

Finally, we vary the height of the metasurface (H) in order to tune the BIC location in the band diagram. After several optimization cycles, we have found that the increase of H from 370 nm to 400 nm moves BIC under the light line of the air (see Fig. 1(b), blue arrow). Such a band diagram transition immediately reflects in the electric field distribution of the mode in semi-waveguide regime (Fig. 2(b)). It can be observed that the electric field enhancement in the hybrid BIC-waveguide regime is increased by 70 times, and the mode tails in the free space are shortened comparing with the Fig. 2(a). Other words, the electric field confinement of the waveguide mode gets boost after transition of BIC below the light line in the waveguiding area of the band diagram.

The presented electromagnetic confinement enhancement owes to the magnification and suppression of certain Fourier terms in the Fourier expansion of the electric field of the mode. The radiation cancelation in BIC is due to the suppression of zero Fourier term while the same term in the waveguide or semi-waveguide regime is partially responsible for the field confinement. Thus, after BIC transition below the light line, the waveguide mode obtains this expansion feature of BIC, and the zero terms gets canceled. In this case, the mode confinement is mostly determined by the first Fourier term which is characterized by the fast exponential decay of the field outside the metasurface.

3. Conclusion

To conclude, we have demonstrated the enhancement of electromagnetic field confinement in the all-dielectric metasurface. The confinement growth owes to the transition of BIC below the light line in the waveguiding area. We have shown that this transition results in 70-fold increase of the electric field and shortening of the mode tails.

References

We acknowledge support from the Russian Science Foundation (Project No. 21-72-30018) and Foundation for the Advancement of Theoretical Physics and Mathematics.

References

[1] Pendry J 2000 *Physical review letters* vol 85 p 3966
[2] Chu C H, Tseng M L, Pin J C, Wu C, Chen Y-H, Wang H-C, Chen T-Y, Hsieh W T, Wu H J, Sun G and Tsai D P 2016 *Laser & Photonics Reviews* vol 10 p 986-994
[3] Kwon H, Sounas D, Cordaro A, Polman A and Alù A 2018 *Physical review letters* vol 121 p 173004
[4] Cheng F, Gao J, Luk T and Yang X 2015 *Scientific reports* vol 5 p 1-10
[5] Narasimhan V, Siddique R and Park H 2020 *ACS omega* vol 5 p 12915-12922
[6] Lee Y, Kim S, Park H and Lee B 2017 *Sensors* vol 17 p 1726
[7] Alipour A, Farmani A and Mir A 2018 *IEEE Sensors Journal* vol 18 p 7047-7054
[8] Ding F, Yang Y, Deshpande R and Bozhevolnyi S 2018 *Nanophotonics* vol 7 p 1129-1156
[9] Boriskina S V, Cooper T A, Zeng L, Ni G, Tong J K, Tsurimaki Y, Huang Y, Meroueh L, Mahan G and Chen G 2017 *Advances in Optics and Photonics* vol 9 p 775-827
[10] Su W, Ding Y, Luo Y and Liu Y 2020 *Results in Physics* vol 16 p 102833
[11] Marinica D, Borisov A and Shabanov S 2008 *Physical review letters* vol 100 p 183902

Atmospheric Propagation Of SubTHz Waves And Space, 6G & 7G Communications

Lesnov Ilya[1,2 a)] and Vdovin Vyacheslav[1,2 b)]

[1] *Institute of Applied Physics Russian Academy of Sciences*
[2] *Nizhny Novgorod State Technical University n.a. R.E. Alekseev*

[a)]Corresponding author: ss18f@mail.ru
[b)]vdovin_iap@mail.ru

Abstract. The work is devoted to the actual problem of data rate of wireless telecommunication channels. Presented analysis of the telecommunication channel subterahertz (subTHz) frequency range - as the most promising band for the implementation of wireless telecommunications for space links and terrestrial cellular communications of high capacity. A channel considered as a combination of high effective transponder / transmitter duplex together with an open high dissipative atmospheric line. The means to achieve a high signal / noise ratio is usage of low-noise cryogenic receivers. The theoretical analysis of data rates for various atmospheric conditions and technical implementations of communication channels demonstrated reasonable limits of cooling of receivers, providing a weighty increase channel capacity, while deeper cooling is impractical due to weather restrictions in certain ranges and conditions of signal propagation, including altitude and seasonal features.

INTRODUCTION

Multimedia and data channels are like yin and yang. Multimedia pushes to increase the capacity of the data channel, and new high-speed data channels encourage the media industry to create new products that use new capacities. 5G network is already in the process of launching. The 6G standard is being formed and it considers the frequencies of 230 GHz, 340 GHz [1]. The use of a frequency of 700 GHz is considered in [2] for 7G, however, in this range atmospheric attenuation is too high and possibilities of use will be greatly limited. ~ 700 GHz frequencies definitely can be used as an alternative to Wi-Fi, operating at 2.4 GHz and 5 GHz. Whereas 230 GHz, 340 GHz can be used for outdoor communications, including in crowded places, and access points with frequencies of 230, 340 need to be placement not so close as for 700 GHz due differ atmospheric attenuation.

When the use of these frequencies in mobile phones is discussed, should not be forgotten not only the atmosphere and limitations connected with medical standards too.

An additional way to increase the capacity of communication channels is to reduce the noise of the receiver, and here we should discuss a problem of adequate cryogenic cooling systems. At the moment, there are no miniature cooling systems for use in mobile devices, and those that are relatively small (Peltier cooler) require essential power sources. However, it is not necessary to use cooling on mobile devices itself. Signal transmission from the base station to the mobile device is achieved by increasing the source power at the base station. Using a high power signal source in a mobile device is also not necessary - it is possible to install a cooled receiver at the base station to increase sensitivity [3]. But in connection with the described features, the use of the sub-THz frequency range for organizing Earth-Space communications is preferably.

CAPACITY ESTIMATION OF THZ DATA CHANNEL

The Earth-Earth ground channel, which is of particular interest [4] for mass cellular telecommunications, can be estimated using the calculation method [5] for a fixed value of atmospheric absorption, which depends on the elevation and ambient temperature. For a subTHz data channel, a heterodyne [6] circuit should be used. The altitude above the sea level of the place is chosen small with a significant absorption of THz waves, but this is typical for places of mass residence of people and the need for high-performance communication. In the case of the Earth-Space channel, the atmospheric absorption value changes with increasing altitude, which required changes in the calculation method, namely, to calculate the integral atmospheric absorption value from the altitude of the ground part of the channel to the altitude at which the space part of the channel is located. And here the task has a noticeable practical sense, since space communications are carried out both from instruments located relatively low above sea level (long-distance space communication systems in the cities of Yevpatoriya, Kalyazin, the settlement of Medvezhye Lakes), and in the mountains. Obviously, the latter for the implementation of communications in the subTHz and especially in the THz bands are much more promising.

An important parameter in the "Earth - the Earth's orbit" channel is the angle of the satellite's position above the horizon since the satellite moves through the sky (research satellites) and tracking is necessary, or they are located in a geostationary orbit (communication satellites).

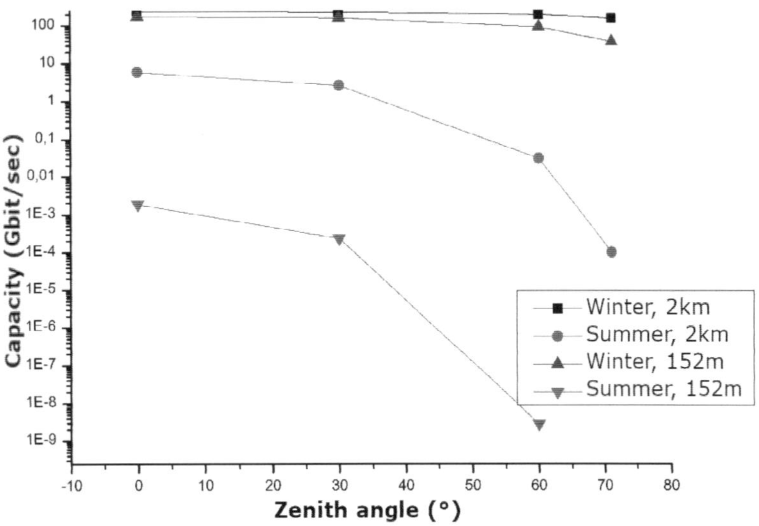

FIGURE 1. Capacity when changing the zenith angle for two seasons, 338 GHz. (0°-upright, 90°-horizontal)

The methodology is based on the MPM Liebe [7] atmospheric radio transparency calculation program, which allows one to calculate the absorption spectrum in the atmosphere under given stationary conditions - pressure, temperature, and humidity. Integral attenuation in oxygen and in water vapor at the zenith are calculated by (1).

$$\tau_{oxygen}(f) = \int_{h\_min}^{h\_max} A_{oxygen}(f, P(h), W(h), T(h))dh$$

$$\tau_{water}(f) = \int_{h\_min}^{h\_max} A_{water}(f, P(h), W(h), T(h))dh \tag{1}$$

Integral attenuation on the track with a given zenith angle are obtained from (2). For communication channels with objects located in the Earth's orbit, their support in the sky already has a significant effect, and depending on their own rotation, sometimes very fast. Even at an angle of ~ 70 °, the optical thickness will be two atmospheres - which means a significant increase in the integral attenuation for the earth-space channel. Therefore, the influence of the slope of the transmission line relative to the vertical on the transparency window capacity is analyzed below. In winter, a small effect is expected, in the summer at an altitude of 2 km at an angle of ~ 70°, the window capacity drops to ~ 0.09 Mb / s, while in winter under the same conditions the capacity increases by 3 orders of magnitude (**FIGURE 1**).

$$\tau_\theta(f) = \tau(f) \cdot \sec(\theta) \tag{2}$$

The estimates were made for a low-noise cryogenic SIS receiver with T = 4K and with a noise temperature of ~50K, and as an uncooled alternative, a Schottky Barrier Semiconductor Diode receiver at room temperature with a noise temperature of 2700K. The calculation was carried out for an altitude of 4 km, the receiver is directed vertically upward (angle 0 °).

CONCLUSION

The use of earth-space channels for data transmission with satellite constellation in orbit of the earth or long-range space missions is promising, it is also possible to use the THz earth-space and space-space channels in the three-link model “ground station” - “orbital station” - “interplanetary missions ". Also using receiver cooling can increase data link capacity. Extremely high sensitivity (low noises) of receivers are related to using superconductivity, which uses at this moment cryogenic cooling. Projects Starlink or OneWeb (so-called “global Wi-Fi”) are developing in such approach. Its propose to use base station to provide regular Wi-Fi or 4G data channel. On such base station receiver cooling (cryo cooling) can be use. At this moment we are not talking about of using cooling systems in mobile terminals because of its "not mobility". But in the future some systems can be developed, or due to advances in electronics noises of subTHz receivers will be reduced The study of the atmospheric absorption in the SubTHz range in the sense of the inhomogeneity of its transmission in the frequency and location is the separate and very popular task. It includes forming an agenda for developers of THz wave equipment. The tasks of creating highly efficient wireless communications closely related to the problems of implementing long-range space missions.

ACKNOWLEDGMENTS

This work was supported by grant RSF-19-19-00499.

References

1) M. Koziol, "It’s Never Too Early to Think About 6G," IEEE Spectrum , 2018. [Online]. Available: https://spectrum.ieee.org/tech-talk/telecom/wireless/its-never-too-early-to-think-about-6g.

2) T. Schneider, A. Wiatrek, S. Preussler, M. Grigat and R.-P. Braun, "Link Budget Analysis for Terahertz Fixed Wireless Links," *IEEE Transactions on Terahertz Science and Technology,* vol. 2, p. 250–256, 3 2012.

3) V. F. Vdovin, V. G. Grachev, S. Y. Dryagin, A. I. Eliseev, R. K. Kamaletdinov, D. V. Korotaev, I. V. Lesnov, M. A. Mansfeld, E. L. Pevzner, V. G. Perminov, A. M. Pilipenko, B. D. Sapozhnikov and V. P. Saurin, "Cryogenically cooled low-noise amplifier for radio-astronomical observations and centimeter-wave deep-space communications systems," *Astrophysical Bulletin,* vol. 71, p. 125–128, 1 2016.

4) T. S. Rappaport, Y. Xing, O. Kanhere, S. Ju, A. Madanayake, S. Mandal, A. Alkhateeb and G. C. Trichopoulos, "Wireless Communications and Applications Above 100 GHz: Opportunities and Challenges for 6G and Beyond," *IEEE Access,* vol. 7, p. 78729–78757, 2019.

5) G. M. Bubnov, I. V. Lesnov and V. F. Vdovin, "Data rates of SubTHz wireless telecommunication channels," *EPJ Web of Conferences,* vol. 149, p. 02012, 2017.

6) I. Dan, G. Ducournau, S. Hisatake, P. Szriftgiser, R.-P. Braun and I. Kallfass, "A Terahertz Wireless Communication Link Using a Superheterodyne Approach," *IEEE Transactions on Terahertz Science and Technology,* vol. 10, p. 32–43, 1 2020.

7) H. J. Liebe, G. A. Hufford and T. Manabe, "A model for the complex permittivity of water at frequencies below 1 THz," *International Journal of Infrared and Millimeter Waves,* vol. 12, p. 659–675, 7 1991.

Anion Exchange Reaction in Halide Perovskite Single Crystals Structured by Laser Pulses

A Yu Zhizhchenko[1,2], M Masharin[3], A Pushkarev[3], A A Kuchmizhak[1,2], and S V Makarov[3] and S A Kulinich[1,4]

[1]Far Eastern Federal University, Vladivostok 690061, Russia
[2]Institute of Automation and Control Processes, Far Eastern Branch, Russian
[3]ITMO University, St. Petersburg, Russia, 197101
[4]Research Institute of Science and Technology, Tokai University, Hiratsuka, Kanagawa 259-1292, Japan

Abstract. In this work, we carry out chemical processing of halide perovskite (CsPbBr3) microplates, where microlenses can be formed on surface by laser ablation by femtosecond laser ablation. Namely, we perform anion exchange reaction and observe spectral tuning of the microplates optical properties. We believe this effect will open new opportunities for tunable and reconfigurable micro-optical devices.

1. Introduction

All-inorganic lead-bromide perovskites ($CsPbBr_3$) is a promising class of materials for lasing and micro-optical applications [1-3] The refractive index of the lead-bromide perovskite is in range n = 2-2.5, which is large enough to support the various optical resonances in a single nano- or microstructure [4-5], which can enhance local electromagnetic field and provide various spectral changes in scattering and absorption of incident light [6] At the same time, large-scale and low-cost laser ablation method for fabrication of nano- and microstructures from halide perovskite crystals would enable advanced photonic designs for various applications. Moreover, combining this method with chemical post-processing of the fabricated perovskite microstructure is highly desirable for creation of tunable micro-optical elements and light-sources.

In this work, we demonstrate precise laser ablation and chemical tunability of the $CsPbBr_3$ microcrystals. The tunability is achieve via the anion exchange reaction, when the perovskite is placed in vapor of HI. We observe strong spectral tuning of photoluminescence (PL) from the perovskite microplates. Since the microplates support high-NA focusing due to microlenses on their surface, the demonstrated ability to tailor morphological, optical, and PL properties of patterned perovskites opens up a novel direction for various applications, including spectrally tunable microlasers and elements of micro-optics (lenses, phase-masks, beam-shape modulators, etc.).

2. Experimental results and Discussions

Synthesis of relatively large (10-microns scale) perovskite microplates require a specially adopted protocol. We used the following materials. Lead(II) bromide (PbBr2, 99.999%), cesium bromide (CsBr, 99.999%), dimethyl sulfoxide (DMSO, anhydrous, 99.8%), isopropyl alcohol (IPA, technical grade, 95%) were purchased from commercial suppliers and used as received. For the preparation of perovskite ink solution we used the following procedure. A mixture of metal halide salts $PbBr_2$ and CsBr was dissolved in DMSO by stirring at room temperature for 30 min to create a clear solution. Then, the solution was filtered by using a syringe filter with a PTFE membrane. The chemicals were stored and mixed inside a nitrogen-filled glove box with reduced concentration of both O_2 and H_2O level down to 1 ppm. $CsPbBr_3$ monocrystals deposition. The obtained perovskite solution was used for further deposition at ambient conditions. A glass substrate was placed in the center of a glass Petry dish and heated on the hotplate up to 60 C. A small droplet of the solution was dripped onto the substrate. Immediately after that, IPA-H_2O azeotrope was poured in the dish and, then, it was sealed. The droplet was dried in the presence of the azeotropic vapor for 10 min to give isolated microns-scale $CsPbBr_3$ single crystals.

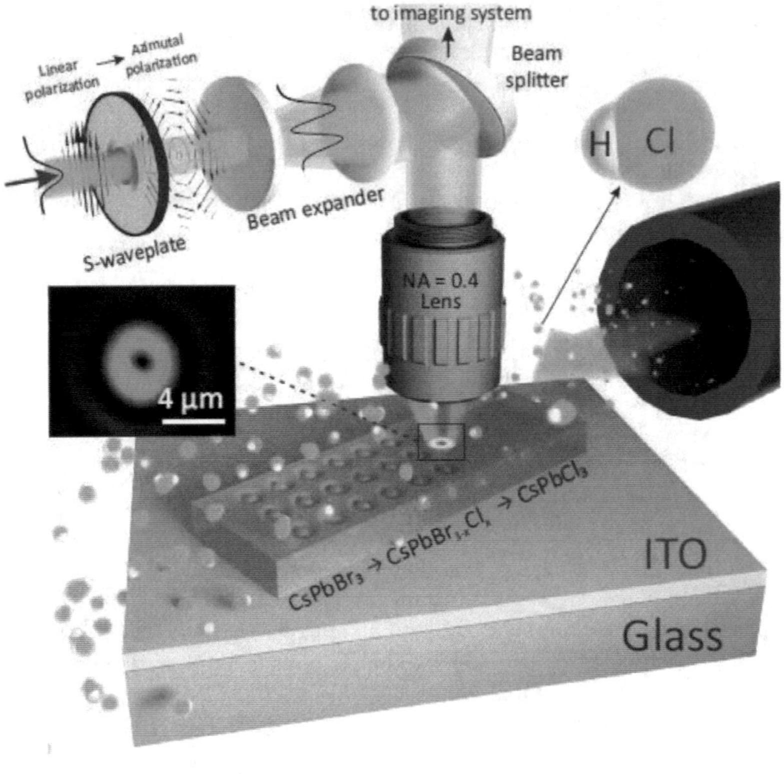

Figure 1. (a) Scheme of experimental setup used for fs-laser lithography. Inset: focal-plane intensity profile of various flat-top laser beam used for direct patterning of perovskite microplate.

To fabricate microlens designs, we use direct femtosecond laser multipulse imprinting on a perovskite microcrystal placed on a silica glass substrate with a specially designed laser beam having a donut-shaped lateral intensity profile, optical vortex (OV) beam. The proposed technology is convenient for rapid and high-throughput fabrication of microlaser arrays as well as more complicated designs [7-8]. The microplates were patterned with 180-fs second-harmonic (515 nm) laser pulses generated by a regeneratively amplified Yb:KGW laser system (Pharos, Light Conversion). The laser beam with Gaussian shape lateral intensity profile was focused onto the sample surface with a dry microscope objective (5×, NA = 0.15). We used a single-pulse regime. Incident pulse energy was measured in the sample plane by a calibrated power meter (Nova, Ophir). Samples were arranged onto the PC-driven nanopositioning system (Newport XM series) to provide their precise translation at a movement accuracy better than 100 nm along all three axes. Some additional details of the experimental setup used for laser lithography are schematically presented in Figure 1.

Figure 2a shows scanning electron microscopy (SEM) image of the fabricated array of microlenses formed on $CsPbBr_3$ microplate surface. It reveals good reproducibility of the fabrication process and absence of ablation debris on the surface. Indeed, halide perovskites have very low thermal conductivity (around 2 times lower than that of glass), which results in reduction of heat-affected zone and, thus, clean material removal. Atomic force microscopy (AFM) image shown in Fig.2b confirms high quality of the formed microaxicones (i.e. microlenses) and reveals their height around 200 nm.

The $CsPbBr_3$ material is completely transparent at the wavelengths larger than 550 nm, and have slight absorption in the range 530-550 nm allowing for light propagation through them [9]. Therefore, we irradiate the $CsPbBr_3$ microcrystal by 532-nm continuous wave laser to test focusing properties of the microlens array. Figure 2c shows far-field distribution of the transmitted light when the microcrystal surface is in imaging plane of the optical microscope, which indicates light defocusing. However, when the imaging plane of the microscope is placed 2 μm above the surface with miclenses (i.e. in their focal plane), we observe appearance of bright spots indicating tight focusing of the light passed through the microcrystal from the bottom.

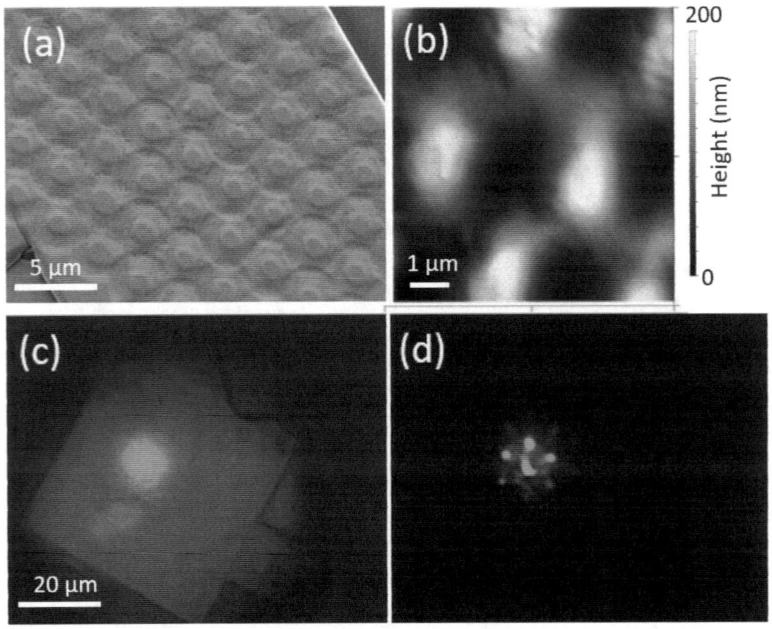

Figure 2. SEM (a) and AFM (b) images of micropatterned $CsPbBr_3$ microplate. (c) Optical image of the micropatterned microplate captured in plane of objective image and (d) in plane of the perovskite microlens focuses.

The characterized $CsPbBr_3$ microplates were subjected to the chemical vapor anion exchange reaction [9] in hydrogen iodine vapor that can be described by the following chemical reaction:

$$y\ CsPbBr_3\ (s) + z\ HCl\ (g) \leftrightarrows y\ CsPbBr_{3-x}Cl_x\ (s) + (z-xy)\ HCl\ (g) + xy\ HBr\ (g)$$

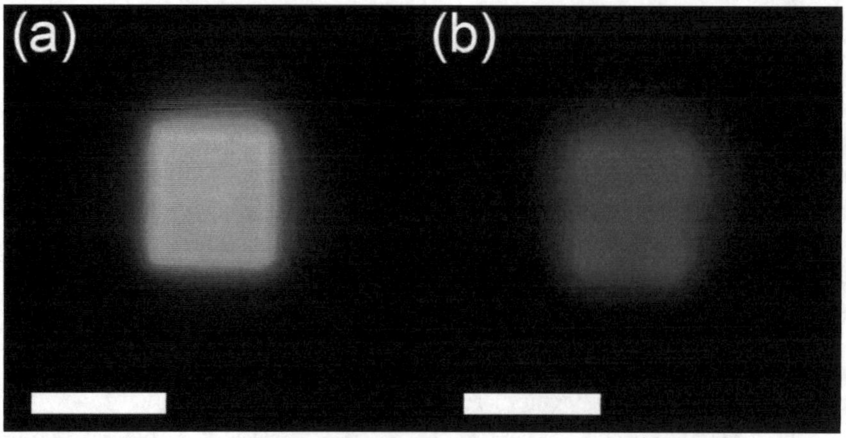

Figure 3. Photoluminescence images of perovskite microplates before and after anion exchange reaction. Scale bars are 20 μm.

Where HCl is taken in a significant excess as compared to generated HBr. For this reason, in the reaction, despite its reversibility, $CsPbBr_{3-x}I_x$ perovskite is formed. According to the position of the photoluminescence peak, the samples have a composition close to $CsPbCl_3$ [10-11]. The doping was carried out during 8 minutes in HCl acid vapor with concentration around 0.1 mg/cm^3. Our spectrometric measurements reveal spectral shift of the microplate PL from 525 nm to 421 nm. In Figure 3, one can see color change of the doped microplate.

3. Conclusion

To summarize, we have demonstrated that halide perovskite microcrystals can be efficiently structured by femtosecond laser pulses to create microlens arrays. This method is clean and does not require any additional post-processing for recovering the perovskite optical properties like in the case of focusing ion beam lithography [12]. We also have checked the ability to chemically tune the optical properties of a perovskite $CsPbBr_3$ microcrystal via anion exchange reaction in HCl vapor. As a next step, we envision further development in the direction of chemical tuning of the microlens focusing properties for transmitted light or for the light emitted from the perovskite microcystal upon external excitation to make real tunable micro-optical devices.

4. References

[1] Yang X, Wu J, Liu T, Zhu R 2018 *Small Methods* **2** 1800110

[2] Makarov S, Furasova A, Tiguntseva E, Hemmetter A, Berestennikov A, Pushkarev A, Zakhidov A, Kivshar Y 2019 *Adv. Opt. Mater.* **7** 1800784

[3] Berestennikov A S, Voroshilov P M, Makarov S V, Kivshar Y S 2019 *Appl. Phys. Rev.* **6** 031307

[4] Tiguntseva E, Zograf G P, Komissarenko F E, Zuev D A, Zakhidov A A, Makarov S V, Kivshar Y S 2018 *Nano Lett.* **18** 1185

[5] Polushkin A S, Tiguntseva E Y, Pushkarev A P, and Makarov S V 2020 *Nanophotonics* **9**(3) 599-610

[6] Berestennikov A, Li Y, Iorsh I, Zakhidov A, Rogach A, Makarov S 2019 *Nanoscale* **11** 6747

[7] Zhizhchenko A, Syubaev S, Berestennikov A, Yulin A V, Porfirev A, Pushkarev A, Shishkin I, Golokhvast K, Bogdanov A A, Zakhidov A A, and Kuchmizhak A A 2019 *ACS nano* **13**(4) 4140-4147

[8] Zhizhchenko A Y, Tonkaev P, Gets D, Larin A, Zuev D, Starikov S, Pustovalov E V, Zakharenko A M, Kulinich S A, Juodkazis S, and Kuchmizhak A A 2020 *Small* **16**(19) 2000410

[9] Tiguntseva E Y, Baranov D G, Pushkarev A P, Munkhbat B, Komissarenko F, Franckevicius M, Zakhidov A A, Shegai T, Kivshar Y S, and Makarov S V 2018 *Nano Lett.* **18**(9) 5522-5529

[10] Beal R E, Slotcavage D J, Leijtens T, Bowring A R, Belisle R A, Nguyen W H, Burkhard G F, Hoke E T, McGehee M D 2016 *J. Phys. Chem. Lett.* **7** 746

[11] Liashenko T G, Cherotchenko E D, Pushkarev A P, Pakštas V, Naujokaitis A, Khubezhov S A, Polozkov R G, Agapev K B, Zakhidov A A, Shelykh I A, and Makarov S V 2019 *Physical Chemistry Chemical Physics* **21**(35) 18930-18938
[12] Baryshnikova K., Gets D., Liashenko T., Pushkarev A., Mukhin I., Kivshar Y. and Makarov S. 2020 *Laser & Photonics Reviews* **14**(12) 2000338

Acknowledgments
The work was supported by Russian Science Foundation 19-19-00177.

Experimental study of nonlinear absorption in hyperbolic metamaterials based on ordered arrays of nanorods

I. V. Malysheva[1,2], I.A.Kolmychek[1], A. P. Leontiev[2,3], K. S. Napolskii[2,3] and T. V. Murzina[1]

[1] Department of Physics, Moscow State University, Moscow, 119991 Russia
[2] Department of Materials Science, Moscow State University, Moscow, 119991 Russia
[3] Department of Chemistry, Moscow State University, Moscow, 119991 Russia

E-mail: malysheva.iv@shg.ru

Abstract. Nanophotonics is a rapidly developing branch of physics that studies light interaction with nanoscaled objects such as metamaterials. Hyperbolic metamaterials (HMMs) based on ordered arrays of metal nanorods embedded in a dielectric matrix are of great interest due to their nontrivial optical properties and abilities to control over the parameters of light. In this article, we present the results of nonlinear absorbtion measurements in HMMs based on ordered arrays of silver nanorods. The main finding consists in the spectral vicinity of Epsilon-Near-Zero and Epsilon-Near-Pole features.

1. Introduction

Hyperbolic metamaterials (HMMs) are highly anisotropic artificial media, where the main components of the permittivity tensor are of the opposite signs. HMMs attracted considerable attention recently because of their appealing optical and nonlinear-optical properties caused by unique dispersion law unattainable in natural media [1, 2]. Special dispersion of HMMs possesses two spectral features of the effective permittivity: zero-crossing point (Epsilon-Near-Zero, ENZ) $\varepsilon_{\parallel} = 0$ and a pole (Epsilon-Near-Pole, ENP) $\varepsilon_{\perp} \to \infty$, where $\varepsilon_{\perp,\parallel}$ are the components of effective permittivity tensor perpendicular and parallel to the long axes of the nanorods. The regions where $\varepsilon_{\parallel}\varepsilon_{\perp} < 0$ are characterized by hyperbolic dispersion giving a name to the material class [3]. One of HMMs' common designs represents arrays of metal nanorods embedded in a dielectric template. In the case of nonmagnetic material $\mu = 1$, isofrequency surface for the TM mode of light is described as:

$$\frac{k_x^2 + k_y^2}{\varepsilon_{\parallel}} + \frac{k_z^2}{\varepsilon_{\perp}} = \frac{\omega^2}{c^2} \quad (1)$$

where $k_{x,y,z}$ are the components of the wave vector, ω is the frequency of the electromagnetic wave and c is the speed of light, Oz is the optical axis [1].

Light interaction with HMMs composed of metal nanorods in a dielectric mtrix leads to the excitation of transversal and longitudinal surface plasmon resonances appearing as two minima in the transmission spectra. The longitudinal one can be excited only in the p-polarized incident beam at oblique incidence. Spectral positions of resonances correspond to ENP and ENZ, respectively [4]. It is known that zero-crossing point of the $Re\varepsilon_{\parallel}$ drastically changes the light-matter interaction through transition from elliptical to hyperbolic dispersion, whereas the pole brings about the localization of the electromagnetic field. Various linear and nonlinear-optical effects have been observed in the spectral vicinities of ENZ and ENP features, such as the negative refraction of light [5], transformation of the evanescent field of an object into propagating mode [6], giant optical birefringence [1], and enhancement of such nonlinear effects as second [7] and third harmonic generation [3].

The change of the complex nonlinear refractive index can be produced by the powerful incident beam. It leads to modification of light propagation and appearance of third-order nonlinear effects such as self-focusing and two-photon absorption. An increase in the nonlinear absorption β has been studied in isotropic materials at the zero-permittivity wavelength [8] and in homogeneous Drude materials in the ENZ region [9]. The enhancement of β has been associated with the continuity of the normal component of the electric displacement field in the vicinity of $Re(\varepsilon) \rightarrow 0$ and consequent with the increase in the electric field within the structure. The enhancement of the nonlinear-optical response occurs as well in metamaterials at the zero-crossing point of the effective permittivity [10]. The nonlinear refractive index and nonlinear absorption have been meticulously studied theoretically in the hyperbolic dispersion regime, providing positive nonlinearity at elliptic regime and negative nonlinearity at hyperbolic regime [11]. At the same time, there have been no thorough experimental studies of the effect in a wide spectral range.

In this paper, experimental studies of the nonlinear absorption by means of z-scan technique in silver nanorods-based HMM are presented. Spectroscopy of the nonlinear absorption reveales the features, associated with the ENZ and ENP points.

2. Experiment

Arrays of silver nanorods were prepared by templated electrodeposition of Ag inside the pores of anodic aluminum oxide (AAO). The AAO templates were obtained by anodization of high purity aluminum in 0.3 M selenic acid electrolyte at 48 V. According to scanning electron microscopy (SEM) images, the diameter of nanorods and the distance between their centers were $D_p = (41 \pm 9)$ nm and $D_{int} = (105 \pm 14)$ nm, respectively. Thus the volume fraction of silver in the filled part of the template is about 14%. The average length of the nanorods was found to be 700 nm.

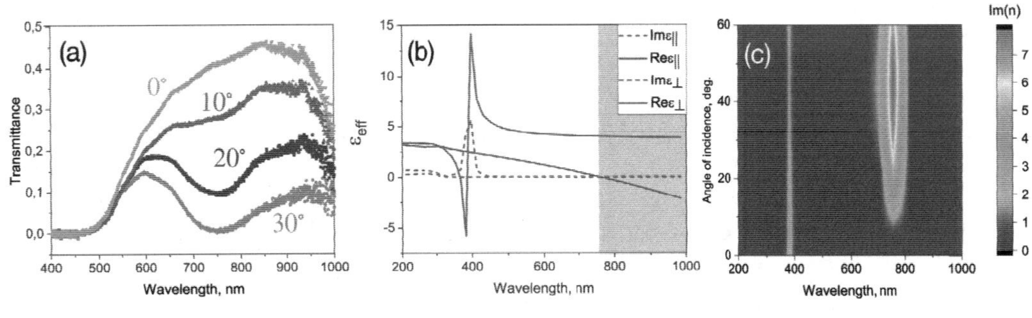

Figure 1. (a) HMM transmission spectra recorded at the angles of incidence 0°, 10°, 20°, and 30°; (b) calculated spectra of real and imaginary parts of the permittivity components with $\lambda_{ENP} = 400$ nm and $\lambda_{ENZ} = 750$ nm, hyperbolic dispersion regime is realized at $\lambda > 750$ nm (colored blue); (c) calculated wavelength-angular spectrum of p-polarized beam refractive index.

Transmission spectra of the arrays of Ag nanorods in AAO matrix measured for the p-polarized incident beam demonstrate two minima centered at $\lambda = 400$ nm and $\lambda = 750$ nm corresponding to the transversal and longitudinal resonances, respectively. It can be seen that the most efficient excitation of longitudinal resonance occurs for $\theta > 20°$ (Fig.1 (a)).

Permittivity components were calculated within the effective medium model as:

$$\varepsilon_{\perp} = \frac{(1+\rho)\varepsilon_m\varepsilon_d + (1-\rho)\varepsilon_d^2}{(1+\rho)\varepsilon_d + (1-\rho)\varepsilon_m}, \varepsilon_{\parallel} = \rho\varepsilon_m + (1-\rho)\varepsilon_d \quad (2)$$

where ρ is volume fraction of silver, ε_m, ε_d are dielectric functions of Ag and AAO [1]. Simulations, performed using $D_p = 35$ nm and $D_{int} = 105$ nm, demonstrate the spectral positions of the ENZ at $\lambda = 750$ nm and of the ENP at $\lambda = 400$ nm (Fig.1 (b)), so the hyperbolic dispersion is achieved at $\lambda > 750$ nm.

Wavelength-angular spectrum of the imaginary part of the refractive index was calculated for the p-polarized beam as $n = \sqrt{\varepsilon_{\perp} + \sin^2\theta(1 - \varepsilon_{\perp}/\varepsilon_{\parallel})}$, where θ is an incident angle [1]. In (Fig.1 (c)) one can observe a narrow increase in $Im(n)$ near ENZ and a smaller enhancement near ENP. These regions of high absorption are associated with plasmon resonances excitation. The difference in widths of the minima in the experimental transmission spectrum and calculations is associated with slightly different length of nanorods.

Z-scan technique was used for the studies of the nonlinear absorption of the HMMs near the transition from elliptical to hyperbolic dispersion law [12]. As the fundamental radiation, the p-polarized output of a Ti:Sa laser with tunable wavelength 730 $\div$ 900 nm was used, with the FWHM of the spectral line of about 10 nm, the pulse duration of 50 fs, the repetition rate of 80 MHz, and the average power of 80 mW. The incident beam with the diameter of $D = 3$ mm was focused by a simple positive-meniscus lense with the focal length of $F = 5$ cm. The sample was moved in the vicinity of the lens focal plane, thus the fundamental beam intensity irradiating the HMM was varied. The radiation at the pump wavelength was spectrally selected by color filters and detected by a photodiode (Fig. 2 (a)), so the influence of luminescence on the data is excluded. The dependence of transmission on the z-coordinate was measured, $T(z)$, where $z = 0$ corresponds to the lens focal plane. The Rayleigh length is determined as $z_0 = n_0\pi w_0^2/\lambda$, where n_0 is the air refractive index, $w_0 = 2\lambda F/D$ is the radius of the Gaussian beam in the lens focal plane. In our experiments, for example, for $\lambda = 790$ nm $z_0 \approx 3$ mm and $w_0 \approx 30$ μm.

3. Results and discussion

Typical z-scan patterns are presented in Fig. 2(b). In the spectral region with elliptic dispersion ($\lambda = 730$ nm, blue points) a gap of normalized transmission near the focus is observed, whereas in the hyperbolic dispersion range ($\lambda = 810$ nm, red points) there is a maximum of the transmitted intensity near the point z=0. The analysis $T(z)$ data obtained for various wavelengths indicated that two photon absorption occured at $\lambda < 800$ nm and saturation of absorption at $\lambda > 800$ nm.

As the obtained data $T(z)$ for all the wavelengths were well approximated by the Lorentz function with FWHM equal to $2z_0$, so the influence of higher-order nonlinear effects was excluded. The amplitude of the peaks or gaps A was obtained after the approximation. The nonlinear absorption is determined as $\beta = 2\sqrt{2}A/(w_0 L_{eff} I_0)$, where $L_{eff} = 4\pi\lambda(1 - exp^{-4\pi Im(n)L/\lambda})/Im(n)$ is the effective length, I_0 is the laser intensity, L is the thickness of the sample, i.e. the length of nanorods.

The spectrum of β for p-polarized light at $\theta = 30^\circ$ calculated from these approximation data is shown in Fig. 2 (c). One can see the change of the sign of β near $\lambda = 800$ nm. This modulation is associated solely with HMM, as the empty AAO template possesses no specific features in the spectral range under study. Also no effect was observed for the s-polarized incident beam.

The HMMs nonlinear response for the fixed incident angle is determined by:

$$\beta \propto Im\left(\frac{\chi^{(3)}_{xxzz}}{\varepsilon_{\perp}} + \frac{\chi^{(3)}_{zzzz}}{\varepsilon_{\parallel}}\right) \quad (3)$$

where $\chi^{(3)}_{xxzz,zzzz}$ are the third-order susceptibility tensor components, xOz is the incident plane [11]. As seen from equation (3), the linear anisotropy of the metamaterial can strongly enhance the nonlinear

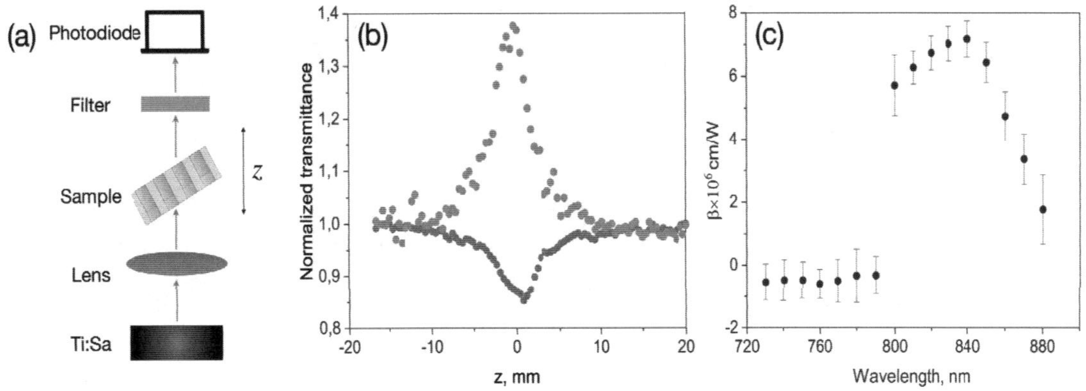

Figure 2. (a) Scheme of the z-scan setup; (b) T(z) obtained for $\theta = 30°$: $\lambda = 810$ nm (red), $\lambda = 730$ nm (blue); (c) spectrum of coefficient β.

response, particularly when the linear permittivity tensor components become small, as in the case of the ENZ regime. Near the ENZ the second term of Eq. 3 has a singularity, while $Re(\varepsilon_{\|})$ tends to zero (Fig. 1 (b)). As the contribution of first term is determined by $\varepsilon_{\perp}$ that is nearly constant in this spectral region, the $\varepsilon_{\|}$ sign-change produces the sign-change of β near the ENZ point. The sharp increase in β observed in the spectral vicinity of $\lambda = 2\lambda_{ENP} = 800$ nm can be also associated with the two-photon transversal plasmon excitation in the HMM under study.

4. Conclusion

Summing up, nonlinear absorption in HMMs based on ordered arrays of Ag nanorods in AAO template was studied experimentally in the spectral vicinity of the ENZ and ENP spectral points. A significant enhancement of β along with its sign reversal in the hyperbolic dispersion regime was observed. This effect is associated with the transition between elliptic and hyperbolic dispersion and two-photon excitation of transversal plasmon resonanse.

5. Acknowledgments

Samples preparation and characterization of their morphological properties were supported by Russian Science Foundation (grant number 18-73-10151). I.V.M. acknowledges the financial support of the Foundation for the advancement of theoretical physics and mathematics BASIS.

References

[1] Poddubny A., Iorsh I., Belov P. and Kivshar Y., 2013 *Nat. Phot.* **7** 948
[2] Litchinitser N. M. and Shalaev V. M., 2008 *Laser Phys. Lett.* **5** 466
[3] Shekhar P., Atkinson J. and Jacob Z., 2014 *Nanoconvergence* **1** 14
[4] Kolmychek I. A., Pomozov A. R., Leontiev A. P., Napolskii K. S. and Murzina T. V., 2018 *Opt. Lett.* **43** 3917
[5] Bang S., So S. and Rho J., 2019 *Scien. Rep.* **9** 14019
[6] Ferrari L., Wu C., Lepage D., Zhang X. and Liu Z., 2015 *Prog. Quantum Electron.* **40** 1
[7] Kolmychek I. A., Novikov V. B., Malysheva I. V., Leontiev A. P., Napolskii K. S., Murzina T. V., 2018 *Opt. Lett.* **45** 1866
[8] Alam, M. Z., De Leon, I. and Boyd, R. W, 2016 *Science* **352** 795
[9] O. Reshef, De Leon I., Alam M. Z. and R. W. Boyd, 2019 *Nat. Rev. Mat.* **4** 535
[10] Rashed A. R., Yildiz B. C., Ayyagari S. R. and Caglayan, H., 2020 *Phys. Rev. B* **101** 165301
[11] Neira A.D., Olivier N., Nasir M.E., Dickson W., Wurtz G.A. and Zayats A.V., 2015 *Nat. Com.* **6** 7757
[12] Sheik-Bahae M., Said A.A., Wei T.-H., Hagan D.J. and Van Stryland E.W., 1990 *IEEE Jour. of Quan. Elec.* **26** 4

Theoretical study of perovskite nanowires optical response to hydrogen halides vapor exposure

D.I. Markina[1*], A.P. Pushkarev[1], S.V. Makarov[1]
[1]Department of Physics and Engineering, ITMO University, 197101 St. Petersburg, Russia
E-mail: *daria.markina@metalab.ifmo.ru

Abstract. Highly sensitive detection of harmful to the human health and environment hydrogen halide vapors is one of the key problems for the chemical industry. The available electrochemical and optical sensors most often show no selectivity to different hydrogen halides and can be produced via costly high-tech fabrication. In contrast to them, $CsPbBr_3$ perovskite nanowires (NWs) exhibiting laser generation are capable of selective and precise detecting for HCl and HI. Exposure of a single NW to these analytes results in an anion exchange that modifies the chemical composition of the NW and therefore invokes a small spectral shift of the laser peak. Herein we propose a theoretical model describing such an optical response. Taking into account that the anion exchange occurs at the surface of the NW and initiates the formation of a core-shell structure, we perform numerical estimation of the eigenmode spectral position for different thicknesses of the chlorine- and iodine-rich shell. Calculations reveal that even a 10 nm shell causes a noticeable spectral shift of 0.81 and 0.63 nm for eigenmode in $CsPbBr_3$-$CsPbCl_3$ and $CsPbBr_3$-$CsPbI_3$ core-shell NWs, respectively.

1. Introduction

In recent decades, the chemical industry has been one of the fastest-growing productions. Currently, many branches of the chemical industry utilize hydrogen halides (HX; X = Cl, Br, I) and their by-products as most common chemicals. For instance, HCl is widely used for the production of vinyl chloride, fertilizers, rayon, and various dyes. It is also employed in electroplating, leather tanning, production of textiles and rubber. Furthermore, HCl is a major combustion product of many chlorine-containing materials. As for hydrogen iodide, it is exploited in organic chemistry for converting primary alcohols to alkyl halides. For such reactions, HI has an advantage over other hydrogen halides due to the high nucleophilicity of I^- ion. It is also utilized for the decomposition of ethers to produce alcohols and alkyl iodides. Thus, unreactive ethers can be converted into chemically active species required for the synthesis of aliphatic and alkyl aryl ethers. Nevertheless, volatile hydrogen halides are very hazardous substances affecting human health and the environment even in their low concentrations.

Common electrochemical sensors capable of HX vapor detecting possess a number of disadvantages such as a high instrumental error, narrow detection range, and no selectivity to different HX analytes. These disadvantages could be eliminated to some extent for the optical sensors operating by the following principles: i) HX molecule reacts with a detecting molecule and changes its optical properties (e.g. photoluminescence or absorption spectrum) [1, 2, 3, 4, 5, 6, 7]; ii) HX vapor changes the refractive index of the surrounding medium (n_{SM}) that interacts with an optical mode propagating in a 2D photonic crystal [8, 9] or circular microcavity resonator [10, 11, 12, 13, 14] (variation of n_{SM} changes the spectral position of the optical mode). However, the technology for producing such optical sensors remains costly. Therefore the development of highly-sensitive and inexpensive photonic designs for selective detection of different hydrogen halides in the wide concentration range is a state-of-art scientific problem since meeting all these requirements is challenging.

In the meanwhile, a new platform for optical sensing of HX vapor could be established with cesium lead halide perovskites ($CsPbX_3$; X = Cl, Br, I). They exhibit efficient and narrow-band photoluminescence whose spectrum can be tuned across the entire visible range from 420-710 nm by varying the anionic content at the X site and thus obtaining $CsPb(Cl,Br)_3$ and $CsPb(Br,I)_3$ mixed-halide structures. Recently, the good opportunity of reaching the latter has been realized via anion exchange between homo-halide perovskite crystals and HX vapor [15, 16]. Furthermore, under certain conditions, all-inorganic lead halide perovskites tend to crystalize in the form of nanowires [17] capable of generating Fabry−Pérot (F-P) laser modes due to reflective end facets, very high optical gain ($\sim 10^4$ cm^{-1} [18]), and high refractive index of the medium (n = 2-2.5). Summing up all the above-mentioned properties of perovskites, the idea for the sensing of small concentrations of HX vapor by the small spectral shift of the laser peak is straightforward and attractive. The only roadblock to measuring ultralow analyte concentrations could be an insufficient quality factor of the F-P laser mode (Q_{las}). However, recently, it has been demonstrated that integrating perovskite NWs with low-refractive-index substrates [19, 20] results in a substantial increase in Q_{las} (up to 7860) owing to the excellent optical contrast between the NW and substrate that provides the laser cavity with improved optical mode confinement.

In this work, we propose a model for optical response of a $CsPbBr_3$ nanowire laser to HCl and HI vapor exposure. According to the model, anion exchange between the NW and acid vapor produces a core-shell structure with a tribromide core and trichloride or triiodide shell. Since the refractive indices of the core and shell are slightly different the laser mode undergoes a small spectral shift. Numerical simulation shows that the dependence of the TE_{11} eigenmode spectral position on the thickness of the shell has a nonlinear behavior.

2. Results and discussion

To investigate the spectral response of $CsPbBr_3$ nanolaser on HI and HCl vapors presence in the atmosphere, we performed a computer modeling of a core-shell type structure on the low-refractive-index substrate ($n_{substrate}$ = 1.15). The choice of the substrate refractive index is based on the requirement to obtain the highest mode Q-factor, and it corresponds to the effective refractive index of the nanostructured indium tin oxide layer [21]. The core-shell structure of the NW was employed to describe the anion exchange process occurring at the surface of a perovskite crystal. From this point of view, the thickness of the shell is supposed to be proportional to the concentration of acid vapor in the atmosphere to which NW is exposed for a certain period of time. The precise relation between the concentration of the analytes and the thickness of the formed shell can be established experimentally by means of X-ray photoelectron spectroscopy depth profiling and will be reported elsewhere. The cavity volume consisting of the core and shell layer was considered to be $3 \times 0.4 \times 0.4$ μm as the optimal for stimulated emission performance from our previous experiments [20, 17] and not spending significant time cost and computing resources. The shell thickness varied in the 10-50 nm range. To analyze the mode contribution in the lasing regime, material dispersion for different perovskite compositions was acquired from ellipsometry data [15, 22] describing only the real part of a complex refractive index because extinction is supposed to be vanishing in the gain region. The eigenmode spectral position and perovskite refractive indices were estimated near the typical experimental lasing mode wavelength (530 ± 5 nm) of $CsPbBr_3$ NWs. Numerical analysis was performed by full-wave numerical simulations using a finite-element eigenmode-solver in COMSOL software.

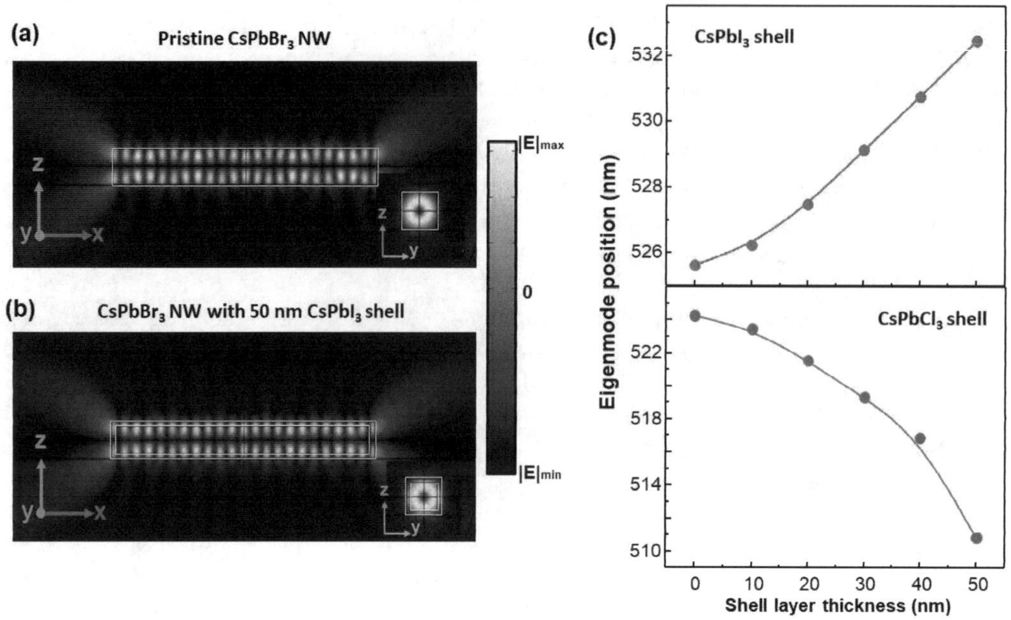

Figure 1. (a,b) Normalized electric field spatial distribution of TE_{11} eigenmode in 3×0.4×0.4 μm $CsPbBr_3$ cavity (a) and $CsPbBr_3$-$CsPbI_3$ core-shell cavity of the same size having a 50 nm shell (b). (c) Dependence of the cavity TE_{11} eigenmode spectral position on the thickness of the $CsPbI_3$ shell (top plot) and $CsPbCl_3$ shell (bottom plot).

Figure 1a,b shows TE_{11} mode distribution in the Fabry–Pérot cavity supporting a standing wave pattern by the multiple light reflections from the end facets. We provided the modeling of the core-shell structure and bare $CsPbBr_3$ NW corresponding to zero-thickness of the shell layer. The character of the field distribution in all cases remains very similar.

As it is mentioned above, the eigenmode spectral position depends on the shell thickness and its chemical composition determining the refractive index. Therefore, the eigenmode selected for the numerical simulation undergoes a red shift (Fig. 1c, top plot) in $CsPbBr_3$-$CsPbI_3$ core-shell NW having the refractive index of the shell ($n_I = 2.39$) a bit higher than that of the core ($n_{Br} = 2.3$). By contrast, for $CsPbBr_3$-$CsPbCl_3$ NW, the refractive index of the shell ($n_{Cl} = 2.14$) is a bit lower than that of the core resulting in blue shift of the eigenmode (Fig. 1c, bottom plot). Remarkably, the difference in refractive indices for $CsPbBr_3$ and $CsPbCl_3$ perovskites ($n_{Br-Cl} = 0.16$) is almost two times larger as compared to one for $CsPbBr_3$ and $CsPbI_3$ ($n_{Br-I} = 0.09$).For this reason, $CsPbBr_3$-$CsPbCl_3$ and $CsPbBr_3$-$CsPbI_3$ NWs demonstrate different rates and magnitudes of the spectral shift with increase in the shell thickness (Fig. 1c).

The minimum spectral shifts corresponding to the 10 nm shell were calculated to be 0.81 and 0.63 nm, whereas, the maximum values were equal to 13.41 and 6.84 nm for the 50 nm shell of $CsPbCl_3$ and $CsPbI_3$, respectively. Such spectral changes can be easily detected experimentally by using a spectrometer equipped with a 40 pm spatial resolution diffraction grating.

3. Conclusion

In summary, we proposed a new lead halide perovskite-based platform for the sensing of hazardous hydrogen halides vapor. To describe a working principle of the detection of the analytes by $CsPbBr_3$ perovskite nanowire laser, we introduced a core-shell structure with a tribromide core and trihalide shell enriched with halogen ions coming from the analyte molecules during the anion exchange between them and perovskite NW. Since the thickness of the shell is supposed to be correlated with the concentration of HX molecules in the atmosphere to which the perovskite NW is exposed, the numerical simulation was conducted for the different thicknesses of the shell (10-50 nm) to address the response of the NW to various concentrations of the analytes. The simulation was aimed to estimate the influence of the shell thickness on the spectral position of the TE_{11} eigenmode in the core-shell Fabry–Pérot NW. It was established that $CsPbBr_3$ NWs exhibit a more pronounced spectral shift in response to exposure to HCl vapor as compared to HI one due to the difference between refractive indices of tribromide and trichloride perovskites is a bit larger than that of between triiodide and tribromide perovskites. Furthermore, even a thin shell of 10 nm thickness has been shown to invoke a noticeable spectral shift, thus making the proposed model applicable for the detection of hydrogen halides vapor at low concentrations.

3.1. Acknowledgments
The work was financially supported by Russian Science Foundation (project no. 20-73-10183).

References
[1] Hu M, Kang W, Cheng B, Li Z, Zhao Y and Li L *Microchim. ACTA* **183** 1713
[2] Cano M, Castillero P, Roales J, Pedrosa J M, Brittle S, Richardson T, González-Elipe A R and Barranco A 2010 *Sens. Act. B: Chem.* **150** 764–769
[3] Lin F W, Xu X L, Wan L S, Wu J and Xu Z K 2015 *RSC Adv.* **5** 30472–30477
[4] Lv Y Y, Wu J and Xu Z K 2010 *Sens. Act. B: Chem.* **148** 233–239
[5] Itagaki Y, Deki K, Nakashima S I and Sadaoka Y 2006 *Sens. Act. B: Chem.* **117** 302–307
[6] Kalimuthu P and John S A 2008 *Analyt. Chim. Acta.* **627** 247–253
[7] Xu N, Wang R L, Li D P, Zhou Z Y, Zhang T, Xie Y Z and Su Z M 2018 *New J. Chem.* **42** 13367–13374
[8] Skivesen N, Têtu A, Kristensen M, Kjems J, Frandsen L H and Borel P I 2007 *Opt. Express* **15** 3169–3176
[9] Hachuda S, Watanabe T, Takahashi D and Baba T 2015 *2015 Conference on Lasers and Electro-Optics (CLEO)* (IEEE) pp 1–2
[10] Armani A M, Kulkarni R P, Fraser S E, Flagan R C and Vahala K J 2007 *Science* **317** 783–787
[11] De Vos K, Girones J, Claes T, De Koninck Y, Popelka S, Schacht E, Baets R and Bienstman P 2009 *IEEE Photon. J.* **1** 225–235
[12] Zlatanovic S, Mirkarimi L W, Sigalas M M, Bynum M A, Chow E, Robotti K M, Burr G W, Esener S and Grot A 2009 *Sens. Act. B: Chem.* **141** 13–19
[13] Chakravarty S, Hosseini A, Xu X, Zhu L, Zou Y and Chen R T 2014 *Appl. Phys. Lett.* **104** 191109
[14] Yang D, Kita S, Liang F, Wang C, Tian H, Ji Y, Lončar M and Quan Q 2014 *Appl. Phys. Lett.* **105** 063118
[15] Tiguntseva E Y, Baranov D G, Pushkarev A P, Munkhbat B, Komissarenko F, Franckevicius M, Zakhidov A A, Shegai T, Kivshar Y S and Makarov S V 2018 *Nano Lett.* **18** 5522–5529
[16] Trofimov P, Pushkarev A P, Sinev I S, Fedorov V V, Bruyere S, Bolshakov A, Mukhin I S and Makarov S V 2020 *ACS Nano* **14** 8126–8134
[17] Pushkarev A P, Korolev V I, Markina D I, Komissarenko F E, Naujokaitis A, Drabavicius A, Pakstas V, Franckevicius M, Khubezhov S A, Sannikov D A *et al.* 2018 *ACS Appl. Mater. Interfaces* **11** 1040–1048
[18] Tiguntseva E, Koshelev K, Furasova A, Tonkaev P, Mikhailovskii V, Ushakova E V, Baranov D G, Shegai T, Zakhidov A A, Kivshar Y *et al.* 2020 *ACS Nano* **14** 8149–8156
[19] Wang K, Gu Z, Liu S, Sun W, Zhang N, Xiao S and Song Q 2016 *J. Phys. Chem. Lett.* **7** 2549–2555
[20] Markina D I, Pushkarev A P, Shishkin I I, Komissarenko F E, Berestennikov A S, Pavluchenko A S, Smirnova I P, Markov L K, Vengris M, Zakhidov A A *et al.* 2020 *Nanophotonics* **9** 3977–3984
[21] Markov L, Pavluchenko A, Smirnova I and Pavlov S 2018 *Semiconductors* **52** 1349–1356
[22] Yan W, Guo Y, Beri D, Dottermusch S, Chen H and Richards B S 2020 *Phys. Status Solidi RRL* **14** 2000070

Flat-band dark polaritons in two-dimensional chiral-waveguide quantum electrodynamics

Y. Marques[1], **I. A. Shelykh**[1,2], **I. V. Iorsh**[1]
[1]ITMO University, St. Petersburg 197101, Russia
[2]Science Institute, University of Iceland, Dunhagi-3, IS-107, Reykjavik, Iceland
E-mail: yurimarques@itmo.ru

Abstract. We consider a two-dimensional extension of the one-dimensional waveguide quantum electrodynamics and investigate the nature of linear excitations in two-dimensional arrays of qubits (particularly, semiconductor quantum dots) coupled to networks of chiral waveguides. We show that the combined effects of chirality and long-range photon mediated qubit-qubit interactions lead to the emergence of the two-dimensional flat bands in the polaritonic spectrum, corresponding to slow strongly correlated light.

1. Introduction

Arrays of quantum emitters, placed in the vicinity of photonic nanostructures are now the focus of intensive research and constitute the field of waveguide quantum electrodynamics (WQED) [1, 2]. While specific material platforms of WQED span from artificial arrays of cold atoms [3] to radio-frequency circuits with superconducting qubits [4, 5], all of them possess a common feature, namely the long range inter-qubit interaction, mediated by the exchange of the propagating photons. The onset of long-range correlations enables to access the plethora of intriguing physical phenomena such as emergence of unconventional topological phases [6, 7] and collective super-radiance and sub-radiance [8, 9, 10], and it paves the way towards the design and engineering of prospective quantum networks [11].

The study of the dispersion properties of the polaritons in WQED is thus an crucial task for understanding the nature of unconventional multiphotonic states in the considered systems. In this sense, the present work address the nature of linear excitations systems consisting of one- or two-dimensional arrays of atoms or artificial atoms (semiconductor quantum dots) coupled to a single chiral waveguide or a network of chiral waveguides.

2. One-dimensional array

We start from the consideration of the 1D case. The corresponding geometry is schematically illustrated in Fig. 1. The effective real-space Hamiltonian of the system, obtained via the Schrieffer-Wolff transformation [12, 13], reads

$$\mathcal{H}_{\text{eff}} = \sum_{n=1}^{N} \delta(b^\dagger_{n,R} b_{n,L} + \text{h.c.}) - i\frac{\Gamma_0}{2} \sum_{m,n=1}^{N} \Theta(m-n) e^{iqd|m-n|} (b^\dagger_{m,R} b_{n,R} + b^\dagger_{n,L} b_{m,L}), \quad (1)$$

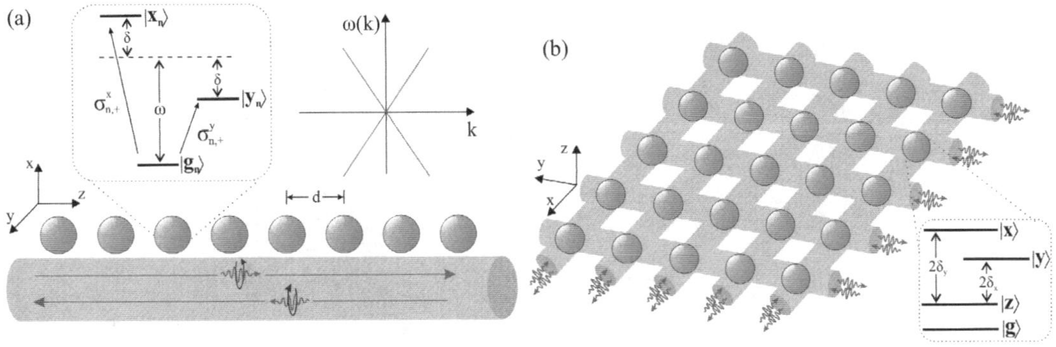

Figure 1. (a) Sketch of the considered setup consisting of an N-atomic chain placed over a waveguide supporting counter propagating modes with opposite circular polarizations. Three-level atoms in V-type configuration with dipole excited states $|x\rangle$ and $|y\rangle$, detuned by 2δ from each other are chirally coupled to right- and left-propagating modes $|R\rangle$ and $|L\rangle$ with linear dispersion relation $\omega(k)$. (b) Two-dimensional square lattice of atoms placed over a waveguide lattice hosting a pair of circularly polarized modes propagating along x and another pair propagating along y.

where the individual decay rate of a single atom $\Gamma_0 = g^2/v$ defines the characteristic energy scale of the system, $q = \omega/v$ is the wave vector of the photon, mediating the interaction between qubits equidistantly spaced by d, and Θ is the Heaviside step function defined within the half-maximum convention. Consequently, the corresponding k-space Hamiltonian for the infinite lattice is given by

$$\mathcal{H}_{\text{eff}} = \frac{\Gamma_0}{4}\sum_k \left[\cot\left(\frac{qd-kd}{2}\right) b^\dagger_{k,R} b_{k,R} + \cot\left(\frac{qd+kd}{2}\right) b^\dagger_{k,L} b_{k,L}\right] + \sum_k \delta(b^\dagger_{k,R} b_{k,L} + \text{h.c.}). \quad (2)$$

The shape of the dispersion curves ε_k, shown in Fig. 2, is determined by the phase qd and detuning δ. For the case of $\delta = 0$ illustrated by the upper panels, R and L modes are completely decoupled. The anti-Bragg structure dispersion with $qd = \pi/2$ for finite δ presents a middle slow polariton branch gapped from the upper and lower fast polariton bands [see Fig. 2 (d)]. For Bragg structures, $qd = \pi$ and $qd = 2\pi$, δ is responsible for openning a gap around $\varepsilon_k = 0$ as can be seen in Figs. 2 (e,f). The eigenmodes dispersion equation for the considered setup $\cos(kd) = \frac{1}{2}\text{Tr}(\mathcal{T})$ obtained in terms of transfer matrix ($\mathcal{T}$) over the single period of the structure composed by a layer of thickness d intersected by a scatterer is given by

$$\cos(kd) = \cos(qd)\left[\frac{(\omega-\varepsilon_k)^2 - \delta^2 - \Gamma_0^2/4}{(\omega-\varepsilon_k)^2 - \delta^2 + \Gamma_0^2/4}\right] - \sin(qd)\left[\frac{(\omega-\varepsilon_k)\Gamma_0}{(\omega-\varepsilon_k)^2 - \delta^2 + \Gamma_0^2/4}\right] \quad (3)$$

with its eigenvalues shown in Fig. 2 for different phases qd. One can see that the exact solution can diverge significantly from those obtained within the Markov approximation. In particular, it does not contain nonphysical regions with infinite group velocity around $k = 0$ for $qd = 2\pi$, where an additional gap is opened instead and substantially modifies the band structure in the vicinity of the band edges $k = \pm\pi/d$ for the case $qd = \pi$.

For finite structures, the effective Hamiltonian (1) is characterized by complex eigenenergies ε, with the real part corresponding to a frequency shift relative to ω, and the imaginary part corresponding to the radiative decay rate, $\Gamma_{1D} = -\text{Im}(\varepsilon)$ [14]. It is noteworthy that the dependence of the darkest state decay rate on the number of atoms for the considered setup, presented in Fig. 2 (g), reveals a scaling of $\Gamma_{1D} \sim N^{-1}$ in contrast to the $\Gamma_{1D} \sim N^{-3}$ scaling, characteristic for 1D waveguides hosting linear polarized modes.

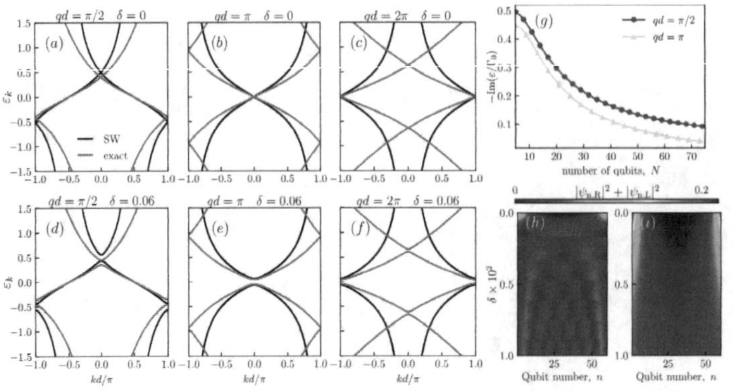

Figure 2. (a-f) Polariton dispersion for the one-dimensional chain obtained by using the Schrieffer-Wolff transformation (black) in comparison with the exact solution (red). The panels (a-c) correspond to the case of zero detuning with dispersion branches corresponding to the right and left propagating modes. The panels (d-f) illustrate the case of a finite detuning δ, when R and L modes become mixed. (g) Radiative decay rate profile scaled to Γ_0, $-\mathrm{Im}(\varepsilon)$, as a function of the array size N with $qd = \pi/2$ and $qd = \pi$. Panels (h) and (i) illustrate the effect of detuning δ on photonic distribution for $qd = \pi/2$ and $qd = \pi$, respectively.

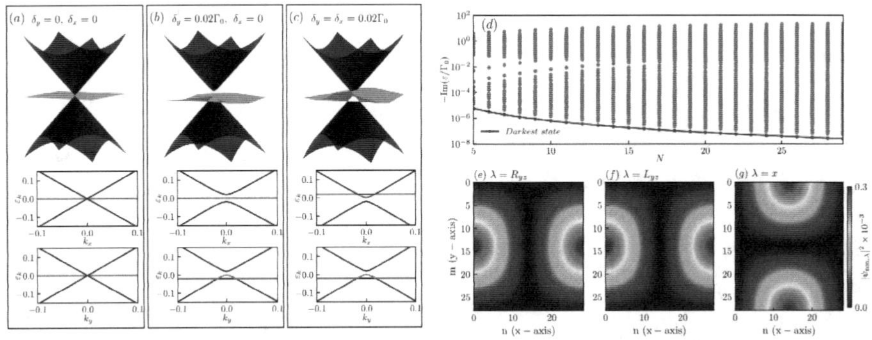

Figure 3. (a-c) Dispersions in the 2D system corresponding to low-energy effective Hamiltonian (4) and their projections corresponding to $k_y = 0$ (middle panels) and $k_x = 0$ (lower panels). (d) Radiative decay rate of the complete set of polariton modes of $N \times N$ squared lattices for $qd = \pi$ with the darkest state highlighted by the red curve. Photonic distribution in the darkest state for $qd = \pi$ with (e) right polarization ($|\psi_{nm,R_{yz}}|^2$), (f) left polarization ($|\psi_{nm,L_{yz}}|^2$), and (g) polarization along x ($|\psi_{nm,x}|^2$).

3. Two-dimensional lattice model

In the two-dimensional (2D) model, we consider an $N \times N$ square lattice of atoms at the nodes of a quadratic waveguide composed of a set of horizontal and vertical chiral waveguides in the xy plane as depicted in Fig. 1 (b). The effective 2D k-space Hamiltonian is given by $\mathcal{H}^{2D}_{\text{eff}} = \sum_{\mathbf{k}} \Psi^\dagger_{\mathbf{k}} (\mathcal{H}^x_{\text{eff}} + \mathcal{H}^y_{\text{eff}}) \Psi_{\mathbf{k}}$, with $\Psi_{\mathbf{k}} = (b_{\mathbf{k},R_{yz}} b_{\mathbf{k},L_{yz}} b_{\mathbf{k},x})^T$. Its expression in the vicinity of $k_x, k_y = 0$ for $qd = \pi$ reads

$$\mathcal{H}^{2D}_{\text{eff}} = \frac{\Gamma_0}{8} \sum_{\mathbf{k}} \Psi^\dagger_{\mathbf{k}} \begin{pmatrix} k_x d + \delta_y & 2\delta_x - \delta_y & -\frac{1}{\sqrt{2}} k_y d \\ 2\delta_x - \delta_y & -k_x d + \delta_y & \frac{1}{\sqrt{2}} k_y d \\ -\frac{1}{\sqrt{2}} k_y d & \frac{1}{\sqrt{2}} k_y d & 2\delta_y \end{pmatrix} \Psi_{\mathbf{k}} \tag{4}$$

with corresponding dispersion shown in Fig. 3 (a-c). For the case of degenerate frequencies, $\delta_x, \delta_y = 0$, the low-energy spectrum reveals a profile identical to spectra characteristic of Lieb lattices, with a pair of Dirac cones intersected by a flat band (Fig. 3 (a)). With a finite detuning in the waveguides parallel to the y axis, $\delta_y = 0.02\Gamma_0$, the spectrum shows anisotropy along k_y as well as the emergence of a gap, as it can be seen in Fig. 3 (b). Naturally, if a detuning is introduced in the waveguides parallel to the x axis the anisotropy will appear along k_x instead of k_y. In Fig. 3 (c) both detunings are finite, so that the anisotropy is present along both the x and y directions, respectively.

The diagonalization of the effective real-space Hamiltonian for $qd = \pi$ within the single occupation limit leads to the radiative decay rate and the photonic distribution shown in Fig. 3(d-g). The imaginary part of the darkest state displayed in Fig. 3 (d) reveals the collective radiative decay rate scaling $\Gamma_{2D} \sim N^{-3}$. Although the 1D chain presents a purely chiral propagation for $\delta = 0$, the 2D lattice within the condition $\delta_x, \delta_y = 0$ does not reveal purely chiral propagation due to the admixture of excitations stemming from perpendicular waveguides. This admixture can be perceived by the off-diagonal terms proportional to k_y, such as $b^{\dagger}_{\mathbf{k},R_{yz}} b_{\mathbf{k},x}$, in Hamiltonian (4). As a consequence, the radiative decay rate scaling $\Gamma_{2D} \sim N^{-3}$ is observed instead of a scaling $\sim N^{-1}$ exclusively observed for purely chiral propagation.

4. Conclusions

We analyzed dispersions and decay rates of the polariton modes emerging in 1D and 2D arrays of multilevel atoms coupled to chiral waveguides. In particular, it was demonstrated that in the 2D case the low-energy effective Hamiltonian is similar to Hamiltonians characteristic of the Lieb lattice model and thus generates polariton flat bands, which due to the large density of states may host a rich family of strongly interacting multiphotonic states.

Acknowledgments

The reported study was funded by the Russian Foundation for Basic Research (project 20-02-00084), the ITMO 5-100 Program, and supported by the Ministry of Science and Higher Education of Russian Federation, goszadanie no. 2019-1246. IAS acknowledges support from the Icelandic Research Fund (Project "Hybrid Polaritonics").

References

[1] Roy D, Wilson C M and Firstenberg O 2017 *Rev. Mod. Phys.* **89(2)** 021001
[2] Chang D E, Douglas J S, González-Tudela A, Hung C L and Kimble H J 2018 *Rev. Mod. Phys.* **90(3)** 031002
[3] Corzo N V, Raskop J, Chandra A, Sheremet A S, Gouraud B and Laurat J 2019 *Nature* **566** 359–362
[4] van Loo A F, Fedorov A, Lalumiere K, Sanders B C, Blais A and Wallraff A 2013 *Science* **342** 1494–1496
[5] Mirhosseini M, Kim E, Zhang X, Sipahigil A, Dieterle P B, Keller A J, Asenjo-Garcia A, Chang D E and Painter O 2019 *Nature* **569** 692–697
[6] Kim E, Zhang X, Ferreira V S, Banker J, Iverson J K, Sipahigil A, Bello M, González-Tudela A, MirhosseiniM and Painter O 2021 *Phys. Rev. X* **11(1)** 011015
[7] Perczel J, Borregaard J, Chang D E, Yelin S F and Lukin M D 2020 *Phys. Rev. Lett.* **124(8)** 083603
[8] Ke Y, Poshakinskiy A V, Lee C, Kivshar Y S and Poddubny A N 2019 *Phys. Rev. Lett.* **123(25)** 253601
[9] Zhang Y X, Yu C and Mølmer K 2020 *Physical Review Research* **2** 013173
[10] Albrecht A, Henriet L, Asenjo-Garcia A, Dieterle P B, Painter O and Chang D E 2019 *New Journal of Physics* **21** 025003
[11] Kimble H J 2008 *Nature* **453** 1023
[12] Schrieffer J R and Wolff P A 1966 *Phys. Rev.* **149(2)** 491–492
[13] Bravyi S, DiVincenzo D P and Loss D 2011 *Annals of Physics* **326** 2793 – 2826 ISSN 0003-4916
[14] Ke Y, Poshakinskiy A V, Lee C, Kivshar Y S and Poddubny A N 2019 *Phys. Rev. Lett.* **123(25)** 253601

Upconversion photoluminescence of perovskite nanoparticles encapsulated in porous sub-micron spheres supporting Mie resonances.

Mikhail Masharin,[1, a)] Oleksii Peltek, [1], Pavel Talianov [1], Lev Zelenkov [1], Mikhail Zuyzin [1], Sergey Makarov,[1]

[1)]Department of Physics, ITMO University, St. Petersburg 197101, Russian Federation

[a)]Corresponding author: mikhail.masharin@metalab.ifmo.ru

Abstract. Currently, halide perovskites are very perspective materials not only for photovoltaics but also for nanophotonic and especially nonlinear optics. These materials have already demonstrated high two-, three- and many- photon absorption coefficients, strong Kerr-nonlinearity, and high-efficient second harmonic generation. Easy and cheap fabrication gives halide perovskites a wide area for scientific research and engineering applications. However, to achieve the stability of perovskites is still a challenging task, which scientific community is working on. In this work, we study a new form of encapsulation of perovskite nanoparticles in sub-micron porous dielectric nanospheres. Due to small pores in such spheres, perovskites are not only protected from external factors, but also are confined in size, which brings several features in the photoluminescence emission. We also show resonant properties of spherical sub-micron particles, which can be used for enhancing upconversion photoluminescence intensity.

1. Introduction
At present, nanomaterials based on halide perovskite draw much attention of the scientific community due to their unique physical properties such as tunable spectra of photoluminescence emission, high photoluminescence quantum yield (PLQY), and defect tolerance [1] [2]. Moreover, halide perovskites are low-cost in production and the fact, that it can be synthesized by different methods, opens a lot of ways for their applications. Recently, perovskite photovoltaic devices have reached high efficiencies, comparable with their existing analogs, for example solar cells, light-emitting diodes, nanolasers, and scintillators. [3] [4] [5] [6] Also, halide perovskites demonstrate strong optically nonlinear properties such as Kerr nonlinearity, second harmonic generation, and strong two-photon absorption with upconversion photoluminescence emission. [7] [8] However, nonlinear characteristics can be not so strong for particular applications, for instance NIR laser visualizer or bioimaging, and they can be enhanced by resonance nanostructures, which can localize or direct light and enlarge nonlinear response [9].

In this work, we investigate $CsPbBr_3$ perovskite as it shows one of the highest two photon absorption coefficients and high PLQY [10] [11].We use a template-assisted method for synthesis $CsPbBr_3$ perovskite nanoparticles in $CaCO_3$ spheres with submicron sizes, which include Mie-resonances in the visible and near-infrared spectral regions [12]. Perovskite nanoparticles inserted in porous of $CaCO_3$ spheres can be encapsulated from the oxygen and humidity, which cause degradation of the material. In addition, pores in the spheres determine perovskite nanoparticles size, therefore it can be controlled by the synthesis process [13]. In our case, the pores size is around 20-30 nm, which has led to the quantum confinement in the formed perovskite nanoparticles and their pronounced excitonic optical response. Despite small sizes of perovskite nanoparticles, the PL intensity from $CaCO_3$ sphere can be high due to a large number of separated perovskite nanoparticles in one sphere. Inserted perovskite nanoparticles can be affected by Mie-resonances of $CaCO_3$ sphere, which can lead to resonant enhancement of two-photon absorption and therefore upconversion photoluminescence intensity.

2. Results and discussion

2.1. Synthesis and characterization of submicron porous spheres

Studying $CsPbBr_3$ nanoparticles embedded in submicron $CaCO_3$ spheres ($CsPbBr_3$@$CaCO_3$) were synthesized by the template-assisted method. First, 615 μL of 1 M water solution of calcium chloride ($CaCl_2$) was mixed into 2.5 mL of water. Then under vigorous stirring 615 μL of 1 M water solution of sodium carbonate (Na_2CO_3) was spilled into the mixture. The mixture was stirred for 30 seconds at 1000 rpm at room temperature. Synthesized $CaCO_3$ nanospheres were washed up via centrifugation twice and then once washed with dimethyl sulfoxide (DMSO). After this process, $CaCO_3$ nanospheres were resuspended in 2 ml of DMSO. To obtain $CsPbBr_3$ nanoparticles on $CaCO_3$ spheres in the 50 μL of $CaCO_3$ solution 140 μl of DMSO, 30 μl of 0.1 M lead bromide ($PbBr_2$), and 30 μl of 0.1M cesium bromide (CsBr) solutions in DMSO were added. The obtained solution was vigorously mixed (1400 rpm at 1 minute). After this supernatant was carefully discarded, the remaining sediment was transferred on a substrate and dried at 70 C.

Synthesized $CsPbBr_3$@$CaCO_3$ nanoparticles were transferred on ITO substrate and under a mercury lamp they showed high photoluminescence intensity. The image of photoluminescent particles is shown in **Fig 1a**. Different colors of the particles in the figure are probably associated with the quantum confinement of perovskite nanoparticles and color rendering features of the CCD camera. A scanning electron microscope image of the same particles is shown in **Fig 1b**. After the reaction we could get only $CsPbBr_3$ perovskite nanoparticles, thus, the photoluminescent wavelength can be attributed only with nanoparticles sizes and quantum confinement. It should be noted, that sizes of $CaCO_3$ nanospheres could not dictate the sizes of nanoparticles, as it is shown in **Fig 1a, Fig 1b**. In **Fig 1c** it is shown a single $CaCO_3$ nanosphere, which has a porous surface. As we could not see any perovskite nanoparticles on the surface, we suggested that $CsPbBr_3$ nanoparticles are located between $CaCO_3$ pores, which dictate their sizes. [13]

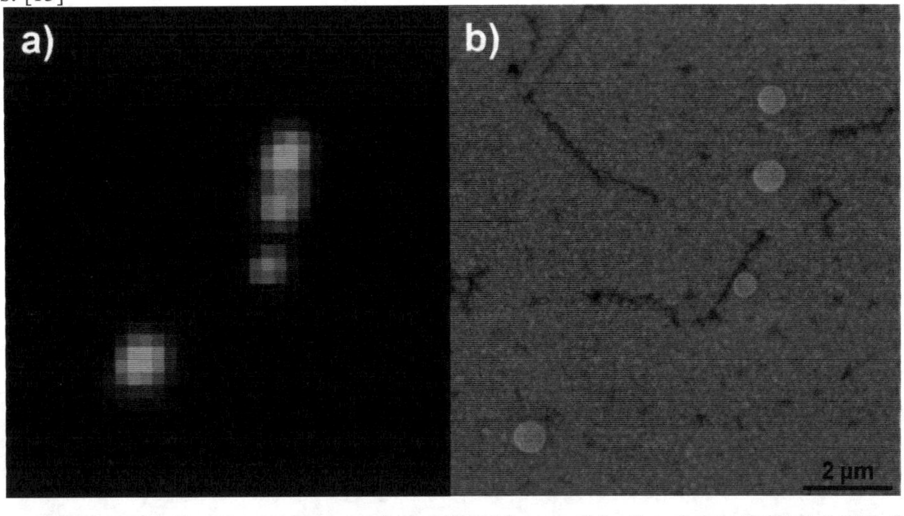

Figure 1. Photoluminescence image (a) and SEM image (b) of studying CsPbBr3@CaCO3 nanoparticles.

2.2. Scattering cross-section spectra and Mie-resonances in sub-micron spheres

SEM images showed high quality spherical particles with submicron sizes, which can have Mie resonances in scattering cross-section spectra and also can be resonators with whispering gallery modes (WGM), which can cause lasing emission from CsPbBr3 nanoparticles. Scattering cross-section spectra were experimentally measured by the dark-field (DF) spectroscopy, when white light is illuminated on the substrate with nanoparticles under an oblique incidence (67 degrees to the substrate normal) through the infinity-corrected objective (NA = 0.26), and the scattered signal is collected by the second objective from the top of the sample and sent to a spectrometer. To control nanoparticle positioning, CCD camera was used. To measure photoluminescence (PL) spectra of particular nanospheres and check possible lasing emission from the samples at room temperature, femtosecond emission from an optical parameter amplifier was used as an optical pump. Laser emission illumination from the bottom excited photoluminescence in perovskite nanoparticles, which was collected by objective from the top. Laser emission was filtered by long-pass filters, and a pure photoluminescence signal was sent to the CCD and spectrometer. Our setup allowed to conduct both experiments; the scheme of the setup is described in Fig 2 a.

Experimentally measured dark field spectra for different polarizations of incident light (Fig 2 b) showed several peaks at 556 nm and 660 nm on TE polarization and at 705 nm on TM polarization. Taking into account the sizes of the $CaCO_3$ spheres and previous works, these peaks are presumably associated with Mie resonances. As these resonances appeared lower than PL perovskite emission, such resonances can be used, for example for localization light in the red region, to enhance the two-photon absorption process and, thus, upconversion photoluminescence.

2.3. Photoluminescence of perovskite nanoparticles inserted in sub-micron spheres

Photoluminescence spectra of CsPbBr3 nanocrystals (NC) in $CaCO_3$ and bulk $CsPbBr_3$ single crystal, which was synthesized by the method mentioned above, was obtained under the femtosecond laser radiation at 405 nm and 100 kHz repetition rate. Incident laser fluence was 0.2 μJ/cm2, which caused no photodegradation of the material. PL spectrum from $CsPbBr_3$@$CaCO_3$ spheres showed a peak at 514.4 nm with a full width at half maximum (FWHM) 24 nm. From bulk $CsPbBr_3$ the peak was at 529 nm with FWHM 24 nm (Fig 3 c). The blueshift of $CsPbBr_3$ NC in comparison with the bulk is around 67 meV, it can be associated with the quantum confinement of perovskite NCs. Unfortunately, with increasing pumping power for different $CsPbBr_3$@$CaCO_3$ spheres, we did not obtain lasing emission, which could be connected with WGM. This fact can be connected with a low-quality factor for the WGM resonator of porous $CaCO_3$ spheres because the light in it can be strongly scattered on the pores and quits the sphere without resonating.

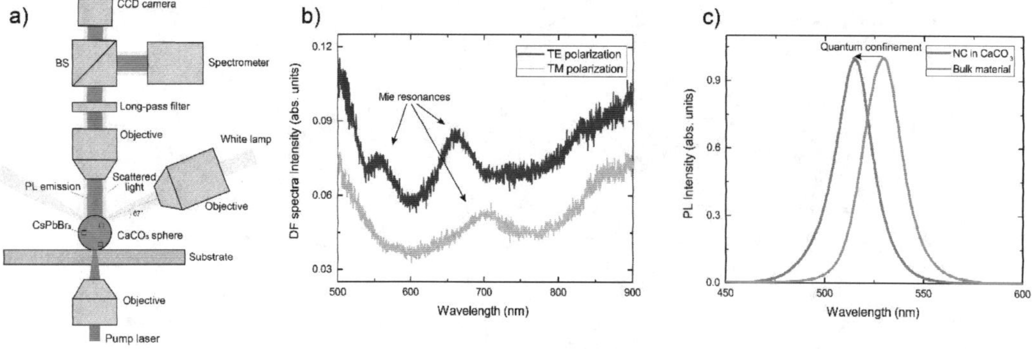

Figure 2. (a) Scheme of experimental setup for dark field and photoluminescence spectroscopy (b) measured dark field spectra with different light polarizations (c) photoluminescence spectra of perovskite nanocrystals (NC) in $CaCO_3$ spheres and bulk $CsPbBr_3$.

3. Conclusion

In summary, in this work, we have synthesized CaCO3 spheres with submicron sizes with perovskite nanoparticles inside by the template-assisted method to protect perovskite from oxygen and humidity, which cause the degradation. Moreover, our method allows to separately accumulate perovskite nanocrystals in one CaCO3 sphere, which gives a strong PL signal with an excitonic response and blueshifted photoluminescence due to quantum confinement effect. Synthesized CsPbBr3@CaCO3 submicron spheres can have Mie resonances in the red and infrared region, which localize the field inside the particle and cause enhancement of multi-photon absorption and, hence, upconversion photoluminescence. Efficient upconversion photoluminescence in such a system can be further used in infrared visualizatiors, bioimaging, and others.

4. Acknowledgements

This work was supported by The Ministry of Science and Higher Education of the Russian Federation, grant 13.2251.21.0060.

References

[1] Zhang F, Zhong H, Chen C, Wu X G, Hu X, Huang H, Han J, Zou B and Dong Y 2015 Brightly luminescent and color-tunable colloidal CH3NH3PbX3 (X = Br, I, Cl) quantum dots: Potential alternatives for display technology *ACS Nano* **9** 4533–42

[2] Huang H, Bodnarchuk M I, Kershaw S V., Kovalenko M V. and Rogach A L 2017 Lead Halide Perovskite Nanocrystals in the Research Spotlight: Stability and Defect Tolerance *ACS Energy Lett.* **2** 2071–83

[3] Tiguntseva E, Koshelev K, Furasova A, Tonkaev P, Mikhailovskii V, Ushakova E V., Baranov D G, Shegai T, Zakhidov A A, Kivshar Y and Makarov S V. 2020 Room-Temperature Lasing from Mie-Resonant Nonplasmonic Nanoparticles *ACS Nano* **14** 8149–56

[4] Polushkin A S, Tiguntseva E Y, Pushkarev A P and Makarov S V. 2020 Single-particle perovskite lasers: From material properties to cavity design *Nanophotonics* 1–12

[5] Pushkarev A P, Korolev V I, Markina D I, Komissarenko F E, Naujokaitis A, Drabavičius A, Pakštas V, Franckevičius M, Khubezhov S A, Sannikov D A, Zasedatelev A V., Lagoudakis P G, Zakhidov A A and Makarov S V. 2019 A Few-Minute Synthesis of CsPbBr 3 Nanolasers with a High Quality Factor by Spraying at Ambient Conditions *ACS Appl. Mater. Interfaces* **11** 1040–8

[6] Chen Q, Wu J, Ou X, Huang B, Almutlaq J, Zhumekenov A A, Guan X, Han S, Liang L, Yi Z, Li J, Xie X, Wang Y, Li Y, Fan D, Teh D B L, All A H, Mohammed O F, Bakr O M, Wu T, Bettinelli M, Yang H, Huang W and Liu X 2018 All-inorganic perovskite nanocrystal scintillators *Nature* **561** 88–93

[7] Zhang R, Fan J, Zhang X, Yu H, Zhang H, Mai Y, Xu T, Wang J and Snaith H J 2016 Nonlinear Optical Response of Organic-Inorganic Halide Perovskites *ACS Photonics* **3** 371–7

[8] Un J Y I, Iao L I L I M, Ie J L I, Ei W H U, Hao C H Z and En S H W 2017 Third-order nonlinear optical response of CH 3 NH 3 PbI 3 perovskite in the mid-infrared regime **7** 3894–901

[9] Makarov S, Furasova A, Tiguntseva E, Hemmetter A, Berestennikov A, Pushkarev A, Zakhidov A and Kivshar Y 2019 Halide-Perovskite Resonant Nanophotonics *Adv. Opt. Mater.* **7** 1–19

[10] Chen J, Žídek K, Chábera P, Liu D, Cheng P, Nuuttila L, Al-Marri M, Lehtivuori H, Messing M, Han K, Zheng K and Pullerits T 2017 Size- and Wavelength-Dependent Two-Photon Absorption Cross-Section of CsPbBr 3 Perovskite Quantum Dots *J. Phys. Chem. Lett.* **8** 2316–21

[11] Di Stasio F, Christodoulou S, Huo N and Konstantatos G 2017 Near-Unity Photoluminescence Quantum Yield in CsPbBr3 Nanocrystal Solid-State Films via Postsynthesis Treatment with

Lead Bromide *Chem. Mater.* **29** 7663–7

[12] Bahrom H, Goncharenko A A, Fatkhutdinova L I, Peltek O O, Muslimov A R, Koval O Y, Eliseev I E, Manchev A, Gorin D, Shishkin I I, Noskov R E, Timin A S, Ginzburg P and Zyuzin M V. 2019 Controllable Synthesis of Calcium Carbonate with Different Geometry: Comprehensive Analysis of Particle Formation, Cellular Uptake, and Biocompatibility *ACS Sustain. Chem. Eng.* **7** 19142–56

[13] Feoktistova N, Rose J, Prokopović V Z, Vikulina A S, Skirtach A and Volodkin D 2016 Controlling the Vaterite CaCO3 Crystal Pores. Design of Tailor-Made Polymer Based Microcapsules by Hard Templating *Langmuir* **32** 4229–38

Journal of Physics: Conference Series **2015** (2021) 012090 doi:10.1088/1742-6596/2015/1/012090

Stability of bound states in the continuum in low-contrast photonic structures

E. E. Maslova[1], **A. A. Bogdanov**[1], **M. V. Rybin**[1,2], **Z. F. Sadrieva**[1]

[1] Department of Physics and Engineering, ITMO University, St Petersburg 197101, Russia
[2] Ioffe Institute, St. Petersburg 194021, Russia

E-mail: ekaterina.maslova@metalab.ifmo.ru

Abstract. We consider quality (Q) factor of accidental bound states in the continuum in bilayer resonators consisting of low-index dielectric rods. The dependence of Q factor on the number of periods shows that Q factors increase with the increasing number of rods. We calculated the dependence of the resonator Q factor on the disorder by two methods: analytical (multiple scattering theory) and numerical (finite difference method) and showed that the results are in good agreement. Also, we investigated the dependence of the resonance frequency on disorder.

1. Introduction

Bound states in the continuum (BIC) are resonances with infinite radiative quality (Q) factor [1]. Although infinite Q factor is a mathematical abstraction, high-Q supercavity modes whose nature corresponds to genuine BIC can be excited in real samples [2]. In our work, we consider two types BIC: accidental and symmetry protected [3, 4]. We studied electromagnetic waves in a confined array of rods, which were calculated using the analytical multiple scattering theory (MST) [5, 6]. Also we compared results by MST with numerical simulation by using a finite difference method.

We considered a bilayer structure consisting of a lossless dielectric rods (Fig. 1a) with dielectric permittivity $\varepsilon = 2.1$. The distance between the layers along y-axis was $d = 4.7$ cm; the period along x-axis was $a = 3$ cm. A point dipole was chosen as the source, which was located at a distance of $3a$ from the left boundary of the structure. The disorder was obtained by randomly changing the coordinates of the rods along x-axis. The degree of disorder $\sigma = [0; 0.1]$. For each value of σ an ensemble of 100 structures was calculated.

Let us describe the multiple scattering theory by considering two problems in a row. As a first step, we solve the Mie problem for a single particle. Relative to the center of the particle the field can be decomposed into multipoles, which are the product of the exponential function of the angular variable and a Bessel function. Since the field has no singularities at the origin, it can be decomposed into multipoles with the Bessel function of the first kind (since the Neumann functions, i.e. Bessel function of the second kind, and the Hankel functions, i.e. ones of the third kind, have a singularity at zero). The Hankel functions correspond to outgoing cylindrical waves, therefore, the scattered field outside the scatterer region is described by multipoles with the Hankel functions $H^{(1)}(kr)$. The field also has no singularities inside the particle; therefore, it is described by the Bessel functions of the first kind $J(kr\sqrt{\varepsilon})$. The continuity of the fields

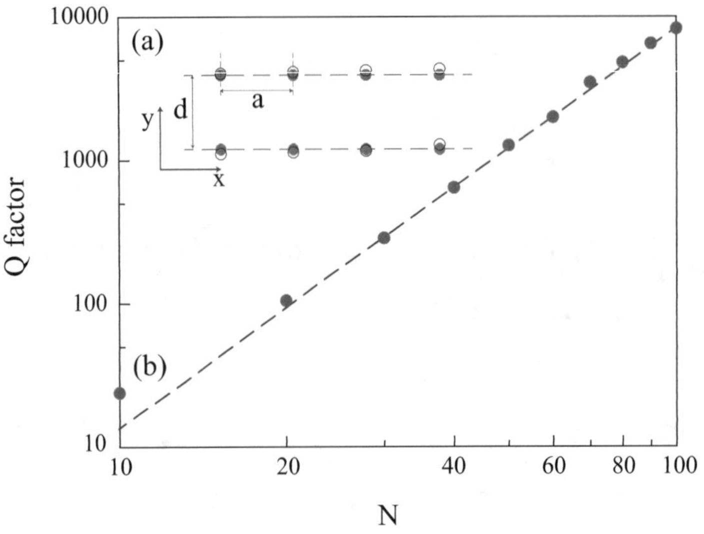

Figure 1. (a) Schematic of bilayer resonator. Red points show periodic structure, black circles show structure with a disorder along the y-axis. (b) The dependence of the Q factor on the rod-pair number of the periodic structure without material losses (solid circles). The dashed line shows the approximation.

at the particle boundary (Maxwell boundary conditions) makes it possible to write a pair of equation and, using the known multipole amplitude corresponding to the incident field, to find the unknown amplitudes of the scattered field and the field amplitude inside the particle. The amplitudes of multipoles are conveniently denoted by a column vector: $\mathbf{j}$ for the incident field; $\mathbf{h}$ for the scattered field. The correspondence between the amplitudes can be written through a T-matrix:

$$\mathbf{h} = \widehat{T}\mathbf{j}. \tag{1}$$

Moreover, the larger the particle, the greater the number of multipoles have to be taken into account for sufficient accuracy of the solution of the scattering problem. So, to find scattering on a small particle, it is sufficient to use the Rayleigh approximation, which takes into account only the dipole component of the multipole expansion. For particles comparable to the wavelength, it is often sufficient to consider only 3-5 multipoles.

Next, we move on to the problem with several particles. Let us consider 2 particles for simplicity. The incident field for the first particle consists of the sum of the external incident wave $\mathbf{j}_1^0$ and the scattered field on the second particle $\mathbf{h}_2$. The difference from the Mie problem considered above is that the scattered field from the second particle is expanded in multipoles centered on the first particle. The corresponding procedure can be written in matrix form:

$$\mathbf{j}_1 = \mathbf{j}_1^0 + \widehat{L}_{21}\mathbf{h}_2. \tag{2}$$

As a result, for each particle there is an equality connecting the amplitude of the incident field and the scattered field. There are two equations with the matrix $\widehat{L}$ and the matrix $\widehat{T}$, i.e. the number of equations exactly corresponds to the number of unknowns, and the problem can be solved by general linear algebra methods. Note that these self-consistent equations already take into account all the scattering and re-scattering events on all particles, which defines the name of the strict theory of multiple waves scattering of light. Further, the coupled multipole method is easily generalized to an arbitrary number of particles.

2. Results

We calculated the transmission spectra by MST for the vicinity of accidental bound state in the continuum. The obtained spectra were processed and the Q factor was extracted by using the Fano function fitting. We plotted the dependence of the Q factor on the number of periods N (Fig. 1b). In the absence of material losses, the Q factor of the BIC increases according to the cubic law of the number of periods N.

We plotted the dependence of the Q factor on the disorder degree (Fig. 2). The dependence decreases according to the quadratic law. Multipole scattering theory has some limits with increasing disorder parameter σ, thus, we compared MST with the finite difference method that implemented in COMSOL Multiphysics software. The results obtained by MST belong the confidence interval of the calculations obtained using COMSOL.

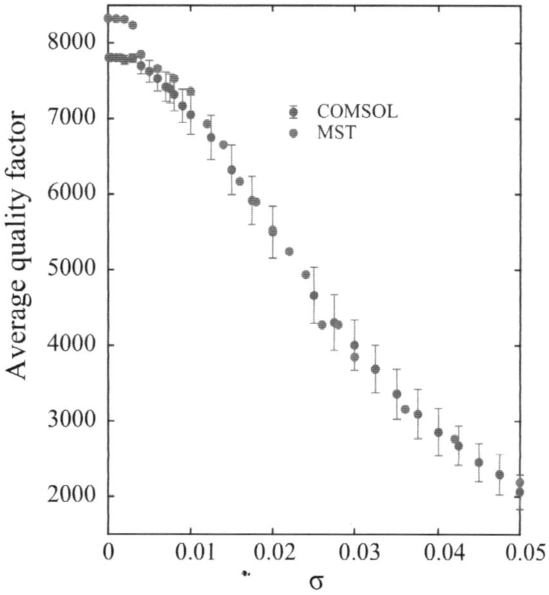

Figure 2. Dependence of Q factor on disorder degree σ. Red circles are simulation by MST and blue circles with confidence intervals by COMSOL.

In additional to our previous work, we calculated the resonance frequency for each disorder parameter σ and conducted a statistical analysis. The resonant frequency was calculated for three types of disorder: along the x-, y- and both axes (Fig. 3). Disorder for the other types was determined similarly to the disorder along the x-axis. In this part, we consider accidental and symmetry protected BIC.

For both accidental and symmetry protected BIC the average value of the resonance frequency weakly changes with increasing the disorder degree. Interestingly, that structure with disorder along the y-axis demonstrates almost a flat frequency dependence. This is associated with the fact that the symmetry protected BIC emerges at any distances between the layers d and accidental BIC depends on the rod spacing a only.

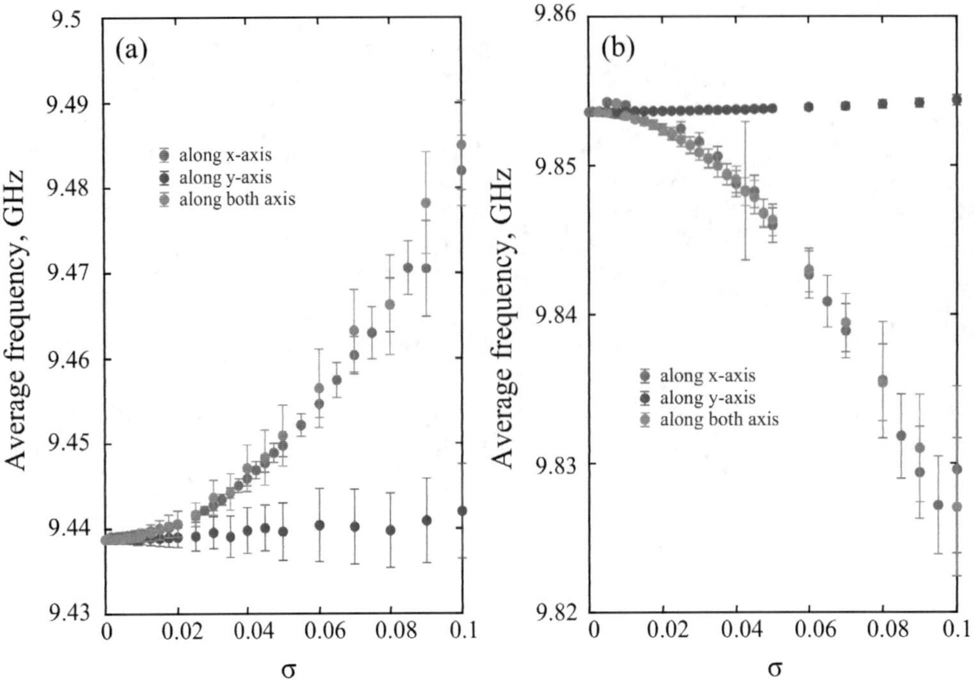

Figure 3. Dependence of frequency on the disorder degree σ for three types of disorder: along x-axis, along y-axis, along both axes. (a) accidental BIC, (b) symmetry-protected BIC.

3. Conclusion

Thus, we have shown that the multiple scattering theory and the finite difference method provide equivalent results. The resonant frequencies of both accidental and symmetry protected quasi-BIC in the disordered structure stay almost constant. The weak deviation of the resonant frequency increases with σ and that can be explained by decreasing of effective length of the structure in comparison with the ordered one.

Acknowledgments

The numerical simulations via COMSOL Multiphysics software were supported by the Russian Science Foundation 21-72-30018. The calculations via multiple scattering theory (MST) were supported by the the Russian Science Foundation 21-19-00677.

References

[1] Hsu C W et al. 2016 *Nat. Rev. Mater.* **1** 16048.
[2] Rybin M V and Kivshar Y S 2017 *Nature* **541** 164-165.
[3] Bulgakov E N and Sadreev A F 2014 *Phys. Rev. A* **90** 053801.
[4] Gao X et al. 2016 *Sci. Rep.* **6** 31908.
[5] Leung K M and Qiu Y 1993 *Phys. Rev. B* **48** 7767.
[6] Dmitriev A A and Rybin M V 2019 *Phys. Rev. A* **99** 063837.

Creation of complex composites from Metal-Organic Frameworks with pure metal nanoparticles under electronic irradiation

Yuri A. Mezenov [1*], **Stéphanie Bruyere**[2], **Valentin A. Milichko**[1,2]

[1] ITMO University, St. Petersburg 197101, Russia
[2] Institut Jean Lamour, Université de Lorraine, Nancy F-54011, France

*E-mail: iurii.mezenov@metalab.ifmo.ru

Abstract. Metal-organic frameworks (MOFs) are unique materials with high porosity and flexibility utilized widely in chemistry and physics. However, they could be used as initial materials for creation new types of composites with nanoparticles. The creation of NPs inside MOFs crystals is related with different types of outer stimuli (temperature, light, and electron irradiation). Here we report about a new approach of the creation complex composites from MOFs' crystals using the electron irradiation of the transmission electron microscope (TEM) as a highly precise method for the growing of different types of Ni and Cu nanoparticles and discuss about the dynamical process of NPs growth using the classical kinetic theory.

1. Introduction

Metal-organic frameworks are the astonishing compounds in materials science due to their high crystallinity, hierarchy, and porosity. In the last few decades these compounds became well-known in such areas as catalysis, absorption, storage, etc. Moreover, such compounds could exhibit different types of structural changes under outer stimuli as pressure, temperature, light, and magnetic or electric field. This is achieved by various types of interactions between building blocks of the compound. It plays a crucial role for observation dynamical response of MOFs' structure under different outer stimuli (temperature, pressure, light, magnetic or electric field, etc.) such as reversible/irreversible structural changes, phase transition and bistability effects. These behaviors so important for new MOFs' areas like for sensing or electronic and communication devices [1].

Moreover, the outer stimuli as temperature or light can be utilized for creation new types of complex materials from MOFs [2-4]. In this way, MOFs could be used as initial precursors. Such nanocomposites demonstrate better catalytical, optical and sorption properties. The amorphous carbon particles, pure metal or metal oxide nanoparticles, chalcogenides, phosphides, and carbides have become obtained employing these stimuli [4-6]. However, low precision of modification and thermal degradation are huge disadvantages of all these stimuli.

Here we report about the new advanced and precisely method for creation complex nanocomposites from MOFs as unitial precursors under the electronic irradiation and discuss about the dynamics of the growth process of nanoparticles. Such approach helps to modify MOFs' crystals *in situ* with high resolution and precision and with absence of thermal decomposition [7, 8].

2. Results and Discussion

In this work we utilized two compounds: HKUST-1 and Ni-BTC (BTC, benzene-1,3,5-tricarboxylate). HKUST-1 has been purchased from Sigma-Aldrich; Ni-BTC has been synthesized by solvothermal synthesis under 110°C in the oven for 24 hours and further cooling at room temperature [9]. These two compounds are composed of Cu(II) and Ni(II) ions, respectively, which are coordinated with benzene-1,3,5-tricarboxylate ligand. Recently it was found that HKUST-1 is unstable under electronic irradiation and this behavior ultimately leads to the creation of pure copper NPs within initial crystals [7, 8]. For Ni-BTC such behavior has been shown for the first time.

All major experiments have been carried out using a transmission electron microscope MET-ACCEL ARM 200F at 200 kV with variation of current density of the electron beam and magnification. Frequently, due to unstable MOFs in TEM [10], we observed absolutely amorphous crystals of HKUST-1 and Ni-BTC. However, EDX mapping revealed uniform distribution of elements in the volume of MOFs' crystals.

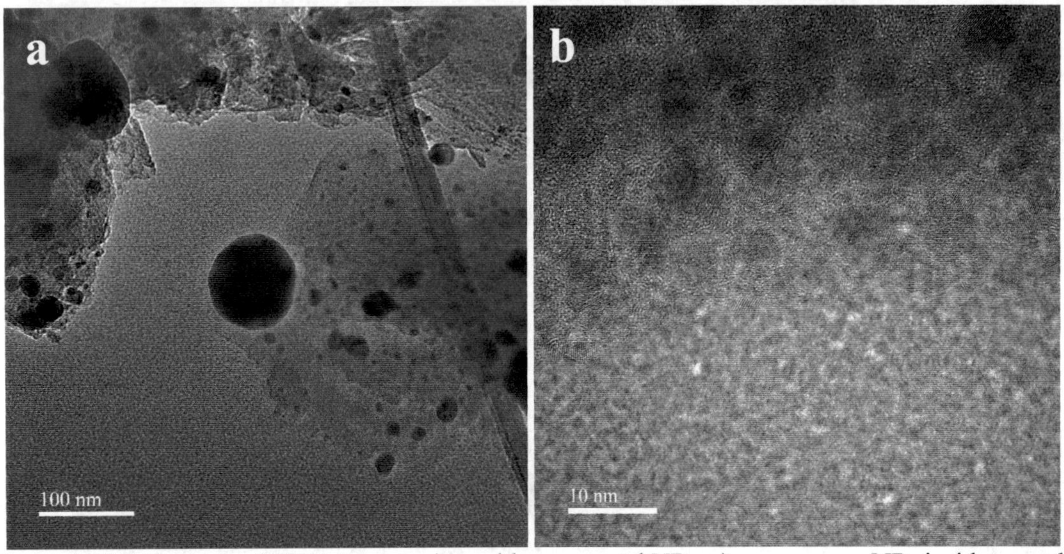

Figure 1. TEM images of MOFs' composites with pure metal NPs: a) pure copper NPs inside crystals of HKUST-1; b) pure nickel NPs inside a crystal of Ni-BTC.

Under electronic irradiation HKUST-1 and Ni-BTC exhibit the fast growing process of pure metal nanoparticles with high crystallinity (Figure 1). EDX mapping confirmed the creation of pure metal NPs inside initial crystals. However, the investigation of the growth process of NPs inside MOFs' crystals revealed several interesting dependances. Firstly, the NPs growth process for HKUST-1 and Ni-BTC has a strictly dependance with parameters of the electronic beam: pure copper NPs with the average size of 9 nm inside HKUST-1 crystals could be created in 40 s and 5.5 s under magnification of 60k and 80k, respectively, and permanent current density; pure nickel NPs with the average size of 4.5 nm could be created in 15 s and 40 s under 250k and 800k, and current density of 120 and 65 pA/cm^2, respectively. Secondly, the experiment with HKUST-1 revealed that under permanent current density, the magnification plays a crucial role for the final size of obtained pure copper NPs. It means the bigger magnification increase the electron density on a crystal, and the biggest NPs can be observed. For Ni-BTC we observed the same result but in a different way: increasing electronic density on a crystal by current density leads to have the bigger nickel NPs inside. And thirdly, we analyzed the dynamics of growth process of pure copper NPs inside HKUST-1's crystals. The experimental results of the nucleation and growth of pure copper NPs shown a high consistency with a classical model of Johnson-Mehl-Avrami-Kolmogorov (JMAK). It means that such complex process of the growth of pure copper NPs inside MOFs' crystals could be explained well by the growth theory [8].

3. Conclusions

We found a new approach for *in situ* creation of complex composites with pure metal nanoparticles inside the organic matrix from metal-organic frameworks. We selected the best way for creation such complex composites under the electron irradiation, operating with such parameters as magnification and current density. It has been revealed for the first time that Ni-BTC exhibit the same behaviour under the electron irradiation as HKUST-1. Dynamical processes of NP's growth in time revealed a good correlation with a classical model of Johnson-Mehl-Avrami-Kolmogorov (JMAK).

Acknowledgement

The work was financially supported by Russian Foundation for Basic Research (Grant Number: 20-33-90318) and the Scholarship of the President of the Russian Federation for students studying abroad.

References

[1] Mezenov Y A 2019 Metal-Organic Frameworks in Modern Physics: Highlights and Perspectives *Adv. Sci.* **6** 1900506

[2] Jagadeesh R V 2017 MOF-derived cobalt nanoparticles catalyze a general synthesis of amine *Science* **358** 326

[3] Jiang H 2019 Nanoscale Laser Metallurgy and Patterning in Air Using MOFs *J. Am. Chem. Soc.* **141** 5481

[4] Indra A 2018 Metal-Organic Framework Derived Materials: Progress and Prospects for the Energy Conversion and Storage *Adv. Mater.* **30** 1705146

[5] Kulachenkov N K 2020 Ultrafast Melting of Metal–Organic Frameworks for Advanced Nanophotonics *Adv. Funct. Mater.* **30** 1908292

[6] Mingabudinova L R 2019 Laser printing of optically resonant hollow crystalline carbon nanostructures from 1D and 2D metal-organic frameworks *Nanoscale* **11** 10155-10159

[7] Jacobs B W 2011 Electron beam synthesis of metal and semiconductor nanoparticles using metal–organic frameworks as ordered precursors *Nanotechnology* **22** 37

[8] Mezenov Y A 2020 Probing the dynamics of Cu nanoparticle growth inside metal-organic frameworks upon electron beam irradiation *Photonics Nanostructures: Fundam. Appl.* **41** 100832

[9] Fu Y 2010 Syntheses, structures and magnetic properties of Mn(II), Co(II) and Ni(II) metal–organic frameworks constructed from 1,3,5-benzenetricarboxylate and formate ligands *Inorg. Chim. Acta* **363** 645

[10] Liu L 2020 Bulk and local structures of metal–organic frameworks unravelled by high-resolution electron microscopy *Communications Chemistry* **3** 99

Miniaturized all-angle accessible RIFD tag

Anna Mikhailovskaya[1], Ildar Yusupov[2], Dmitry Dobrykh[1], Sergey Krasikov[2], Diana Shakirova[2], Andrey Bogdanov[2] and Dmitry Filonov[3] and Pavel Ginzburg[1]

[1]*School of Electrical Engineering, Tel Aviv University, Tel Aviv 69978, Israel*
[2]*Department of Physics and Engineering, ITMO University, St. Petersburg 197101, Russia*
[3]*Center for Photonics and 2D Materials, Moscow Institute of Physics and Technology, Dolgoprudny 141700, Russia*
E-mail: anna2@mail.tau.ac.il

Abstract. Radio frequency identification (RFID) is one of the commonly used approaches to a short-range wireless data exchange. Numerous passive RFID tags are available on the market, and in a vast majority of cases, their designs are based on flat meandered dipole architectures. However, besides technological advantages, those realizations suffer from polarization mismatch issues and limited spatial sectors, from which flat tags can be interrogated. Here, we demonstrate and analyze a miniature omnidirectional tag accessible from all 4π stereo angles with a commercial RFID reader.

1. Introduction

Radio frequency identification (RFID) is one of the commonly used approaches to a short-range wireless data exchange. This topic keeps advancing owing to new emerging applications, including Internet of Things (IoT) and many others. Passive RFID tags can power this technology, but need to comply with several specifications, including long reading distance, omnidirectional response, low cost realization, and small footprint. However, those requirements are challenging to accommodate on the same platform. The main electromagnetic challenge is reducing the physical sizes of the tag's antenna implies a decrease of efficiency and bandwidth. There are many approaches to reducing the antennas area, suach dipole meandering, commercial sub-centimeter designs, as a rule, have a short range and cannot be investigated from a distance of more than a meter [1], [2], [3].

Obtaining an ideal omnidirectional linearly polarized radiation pattern is fundamentally impossible [4],[5]. However, various designs have been developed to achieve ideal performance [6],[7]. Typically, typical dipole RFID tags operate with a single dipole resonance and have a donate radiation pattern. To obtain a radiation pattern close to uniform, several multipoles must constructively interfere, creating a spherical wavefront in the far zone. A possible solution in this area is to use a combination of resonances and materials with a high refractive index. These concepts have recently been explored in the field of RFID [8],[9].

Here we demonstrate a omni-RFID tag design, which complies with all of the beforehand mentioned requirements. Our experimental realization shows the capability to perform all-angle readout from a several meters distance while the overall size of the tag is 3cm.

2. Electromagnetic simulations

2.1 Design

Schematics view of the omni-RFID tag is shown on Fig.1. Design consists of several main components: folded metal strip shaped like an omega particle [10] with folded arms, placed near the split ring's gap, which control the antenna's electrical length and the T-matched configuration is optimized to impedance match the chip (integrated circuit (IC) (Impinj Monza 4 chip)) to the antenna. Overall, the geometry is similar to a recent report [6]. To reduce the area occupied by the volumetric tag, we immersed the metal strip in a ceramic sphere with a high refractive index. Omni-RFID is optimized to operate in 860-930 MHz, which is in accordance with the EPCGEN2 RFID standard, and the IC's impedance values are taken from the vendor's datesheet.

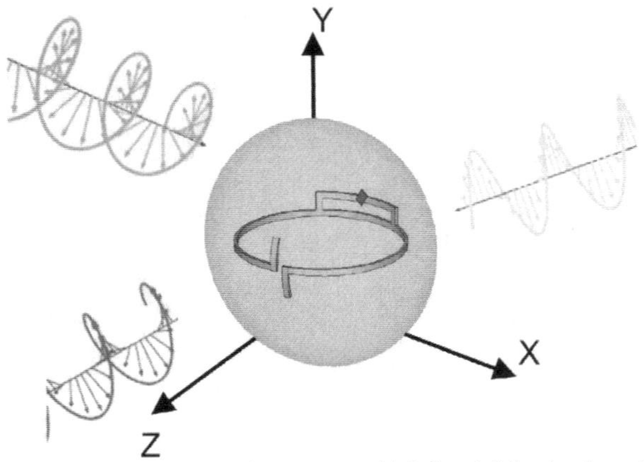

Figure. 1. Schematics of a omni-RFID tag. An antenna with left and right circular polarization is used as excitation, located at a distance of several meters from the structure.

2.2 Numerical analysis

RFID communication involves downlink (requesting a tag by the reader) and uplink (transmitting a time-modulated signal back to the reader). The structure was developed by full-wavelength numerical simulations implemented with a CST Microwave Studio frequency domain solver (FEM). To probe the uplink, the tag is considered an active antenna. To do this, instead of a microcircuit, a discrete 50 Ohm port was inserted into the gap and the far fields of such an antenna were taken into account. The geometry of the entire structure has been optimized to minimize the size of the mark and to achieve an isotropic radiation pattern. The deterioration in the uniformity of the radiation pattern was compensated for by introducing the ellipticity of the initially spherical mark. The final set of parameters is shown in Table 1.

Parameter	Axis	Value
Omni-RFID tag	x	25 mm
Omni-RFID tag	y	25 mm
Omni-RFID tag	z	18 mm
Diameter split ring		17.7 mm
Folded wires		3.4 mm
T-matched		9.5mm/1.8mm
Permittivity of ceramics		ε=9, tg=0.001

Table 1. Parameters of the omni-RFID tag.

To receive a response from any interrogation angle of omni-RFID tag, he antenna of the reader will be triggered between two states, emitting right and left circular polarization. The triggering rate is 5ms, which complies with the link standards (it takes 1-2ms slot to scan a tag).

To verify the downlink efficiency, the reciprocity principle can be applied [11]. The main optimization criterion in our case was the value of the current flowing through the microcircuit. The Impinj Monza 4 IC requires 3.7 mA RF current to activate (experimentally determined).

To make a numerical comparison our omni-RFID tag and a commercial tag were placed at 1.5 meters from the Helix antenna. Antenna radiated 0.51 W total power. Both RFID tags use the chip Impinj Monza 4 IC. Impedance of 11-143j (at 915 MHz) in one state, whereas the second is a short circuit. In numerical simulation, the tags were rotated in all planes (XY, XZ, and ZY) and were irradiated sequentially by right and left circularly polarized waves generated by a triggered Helix antenna. The RF current was calculated for all polarizations for both RFID tags (Fig.2b,c). Figure 2 (a) shows the maximal current in the logarithmic scale through the IC as a function of the tag's rotation angle. Gray area for modes below the activation threshold. This result shows close to the omnidirectional response of our design, which can collect the field from any direction of incidence and raise the IC above the activation threshold. On the other hand, the commercial dipole tag is not readable from multiple sectors.

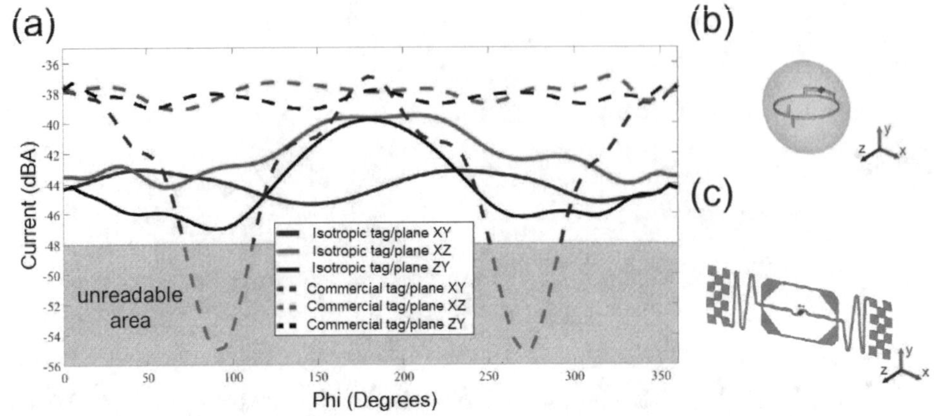

Figure 2. Numerical results of RF current flowing through the tag chip versus rotation angle in three different planes (XY, XZ and ZY). Schematics view (b) new omni-RFID tag and (c) commercial RFID. Gray shaded area on the plot demonstrates the regime, where commercial tags can't reach the activation threshold. RFID IC - Impinj Monza 4.

3. Results

An omni-RFID tag was made as follows: a metal strip was cut from a copper sheet (0.05 mm thick) and an RFID chip was soldered into the gap, detached from the commercial tag. I used an Impinj r2000 reader operating in the frequency range 865-868 MHz (European standard). The housing was 3D printed and filled with a ceramic powder with a dielectric permittivity of 9. The metal strip was embedded in the ceramic powder.

All measurements were performed in an anechoic chamber. The Impinj r2000 reader was connected to a laptop via a USB port, and the amplitudes of the received signals were monitored via the vendor's software. As an antenna of reader used a custom-made Helix antenna with right and left windings, capable of radiating left and right circularly polarized waves, operating in an 850-930 MHz frequency range. We connect two ports of our antenna, applying 0.2 kHz switching with a 50:50 duty cycle.

The tags were placed on a rotating table at 1.5 m from the reader's antenna (Fig. 3a). The tags were rotated alternately in three planes with a step of 10°, which was enough to demonstrate the omnidirectional properties. The received signal intensity was monitored for each physical position of the tags.

Figures 3 (b) and (c) show the dependence of the received signal strength on the rotation angle in three different planes (XY, XZ and ZY) for our isotropic and commercial RFID tags, respectively. You can see that our design significantly outperforms its commercial counterpart in terms of signal-to-noise ratio and angular coverage. The commercial tag is not available from the large angular sector in the case of XY plane rotation (the back reflected modulated signal is below the sensitivity of the reader (-90 dBm)), since the tag operates in dipolar resonance. Our isotropic tag, powered by a combination of an electric dipole and a magnetic quadrupole, gives us all-angle coverage and omnidirectional operation, which was the goal.

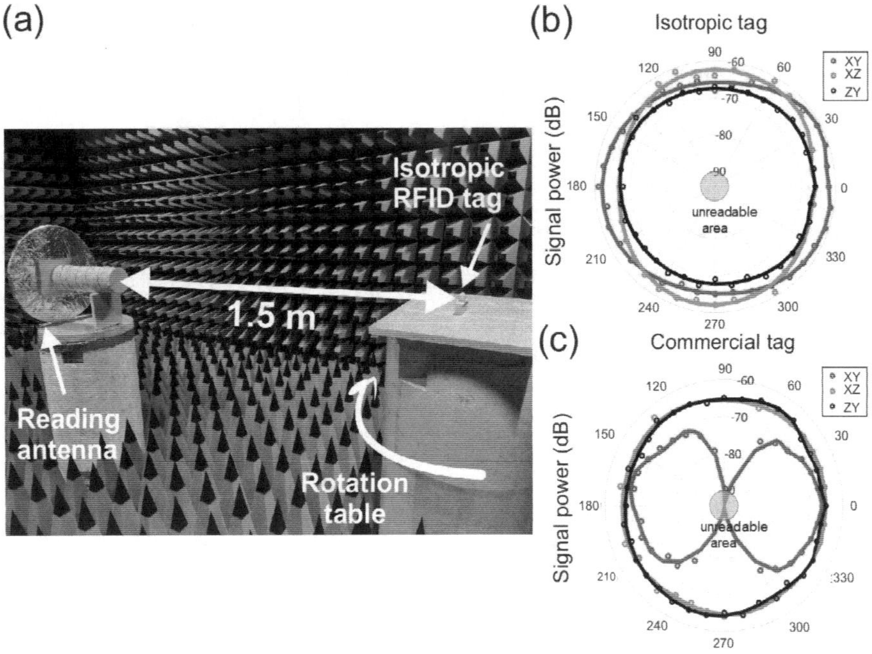

Figure 3. (a) Photo of experimental setup: RFID reader with a custom-made Helix antenna with two spiral, rotation table and the omni-tags. The power of the backscattered signal (in dB) at the reader depending on the angular orientation of the tag in three different planes (XY, XZ and ZY) (b) our omni-RFID tag and (c) a commercial RFID tag.

4 Conclusions

Demonstrated numerical simulation and experimentally setups new design of small passive RFID tag with omnidirectional response. Omni-RFID tag has a size less than 3 cm due to use high-index ceramic and is read from any angle by an antenna having left and right circular polarization. Reading from any angle is achieved by overlapping several multipole resonances. Such omni-RFID tags can be applied in many applications especially logistic, IoT, vehicle.

Acknowledgments

The work was supported by the Russian Science Foundation (Project 19-79-10232). P.G. acknowledges partial support of the ERC StG "In Motion" (802279), PAZY Foundation, and Israeli Ministry of Science and Technology (Project "Integrated 2D & 3D Functional Printing of Batteries with Metamaterials and Antennas."

References

[1] J. Rashed and C.-T. Tai, "A new class of resonant antennas," *IEEE Trans. Antennas Propag.*, vol. 39, no. 9, pp. 1428–1430, 1991, doi: 10.1109/8.99054.

[2] P. V. Nikitin and K. V. S. Rao, "Theory and measurement of backscattering from RFID tags," *IEEE Antennas Propag. Mag.*, vol. 48, no. 6, pp. 212–218, Dec. 2006, doi: 10.1109/MAP.2006.323323.

[3] K. V. S. Rao, P. V. Nikitin, and S. F. Lam, "Antenna design for UHF RFID tags: a review and a practical application," *IEEE Trans. Antennas Propag.*, vol. 53, no. 12, pp. 3870–3876, Dec. 2005, doi: 10.1109/TAP.2005.859919.

[4] L.E.J. Brouwer, "On continuous vector distributions on surfaces," *KNAW Proc.*, vol. 11, pp. 850–858, 1909.

[5] H. Matzner and K. T. McDonald, "Isotropic radiators," *AntenneX*, vol. 74, pp. 1–9, 2003.

[6] A. Bou-El-Harmel, A. Benbassou, and J. Belkadid, "Design and Development of a New Electrically Small 3D UHF Spherical Antenna with 360° of Opening Angle in the Whole Space for RFID, WSN, and RSN Applications," *Int. J. Antennas Propag.*, vol. 2016, pp. 1–18, 2016, doi: 10.1155/2016/2906149.

[7] H.-K. Ryu, G. Jung, D.-K. Ju, S. Lim, and J.-M. Woo, "An Electrically Small Spherical UHF RFID Tag Antenna With Quasi-Isotropic Patterns for Wireless Sensor Networks," *IEEE Antennas Wirel. Propag. Lett.*, vol. 9, pp. 60–62, 2010, doi: 10.1109/LAWP.2010.2043046.

[8] D. Dobrykh *et al.*, "Long-range miniaturized ceramic RFID tags," *IEEE Trans. Antennas Propag.*, 2020, doi: 10.1109/TAP.2020.3037663.

[9] D. Dobrykh *et al.*, "Multipole engineering for enhanced backscattering modulation," *Phys. Rev. B*, vol. 102, no. 19, p. 195129, Nov. 2020, doi: 10.1103/PhysRevB.102.195129.

[10] M. S. Mirmoosa, Y. Ra'di, V. S. Asadchy, C. R. Simovski, and S. A. Tretyakov, "Polarizabilities of Nonreciprocal Bianisotropic Particles," *Phys. Rev. Appl.*, vol. 1, no. 3, p. 034005, Apr. 2014, doi: 10.1103/PhysRevApplied.1.034005.

[11] C. E. Balanis, *Antenna Theory: Analysis and Design, 3rd Edition - Constantine A. Balanis*. A JOHN WILEY & SONS, INC, Pages: 1136, 2005.

Flexible Electrode Formed by Patterned Layers of Single-Walled Carbon Nanotubes for Optoelectronic Applications

D M Mitin[1,2], A A Vorobyev[1], Y S Berdnikov[3], S A Raudik[1,4], A M Mozharov[1], A G Nasibulin[4] and I S Mukhin[1,5]

[1] Saint Petersburg Academic University, 8/3 Khlopina, St. Petersburg 194021, Russia
[2] Peter the Great St. Petersburg Polytechnic University, 29 Politekhnicheskaya, St. Petersburg 195251, Russia
[3] Saint Petersburg State University, 13B Universitetskaya Emb., St. Petersburg 199034, Russia
[4] Skolkovo Institute of Science and Technology, 30 Bolshoy Boulevard, bld. 1, Moscow 121205, Russia
[5] ITMO University, 49 Kronverksky pr., St. Petersburg 197101, Russia

E-mail: mitindm@mail.ru

Abstract. The technology of creating patterned flexible electrodes based on single-walled carbon nanotubes and polydimethylsiloxane was demonstrated in this paper. A series of experiments were carried out to study whether the percolation affects the conductivity of patterned SWCNT layers. It was found that in patterns with the linewidth above 1 μm and the cell size above 50 μm the random character of SWCNT networks may be neglected. The impact of bending on the grid conductivity was studied. We observed a very moderate increase of resistance below 5% under the strain up to 4 %, which is comparable with the previous results for continuous SWCNT layers and shows the improvement in comparison with the previous reports on patterned SWCNT layers.

1. Introduction

Flexible optoelectronic device technologies have been rapidly developed in recent years. The improvement of mobile, wearable, medical, and other electronic devices imposes new requirements on their component base in terms of flexibility, stretchability, chemical stability, and service life. The flexible device technologies used today are based on organic compounds, which have fundamental limitations and some significant disadvantages: low luminosity of light-emitting diodes (LEDs), thermal and chemical instability, and relatively short service life. In turn, inorganic semiconductor compounds demonstrate high values of these parameters and are promising candidates for creating flexible and stretchable optoelectronic devices. The implementation of a flexible optoelectronic device is possible based on an array of nanowires (NWs), which are characterized by improved mechanical characteristics in comparison with bulk materials.

A typical process for flexible LEDs creating is as follows: the NWs array is filled with a layer of polymer to form a flexible matrix After that, plasma etching is used to thin the matrix and clean the upper part of the NWs. The entire layer is then separated from the growth substrate to form a composite matrix with the NWs retaining their orientation and position. Semi-transparent and fully transparent LEDs can be manufactured using this protocol. In the case of a semi-transparent device, the metallization of the lower part of the NW is used as the back contact. After that, the flexible substrate (for example, polyethylene terephthalate) is connected to the back metal contact with a silver epoxy resin. To create the upper electrode, an optically transparent and conductive material is used. For fully transparent LEDs, transparent conductive electrodes (TCEs) are used to create both the back and top contacts. The TCE is one of the components that significantly limits the flexibility and stretchability of LEDs.

The literature reports the use of conductive composites (silver nanowires/nanoparticles [1] or reduced graphene oxide penetrating an elastic polymer [2]), woven conductive fibers/threads [3], ionic hydrogels [4], and liquid metal [5] as flexible and stretchable electrodes. However, all of these solutions have their drawbacks. The resistance of silver nanowires increases dramatically when stretched due to the high Young's modulus. The reduced graphene oxide also loses its conductivity as it stretches. Some ionic hydrogels have good stretchability, but insufficient electrical conductivity can lead to low output characteristics of the device. Gallium-based liquid metal exhibits excellent stretchability and recyclability, but it is extremely susceptible to oxidation and has a high surface tension, which affects its fluidity and makes it difficult to use in systems with nanoscale objects.

Carbon nanotubes (CNTs) and especially single-walled carbon nanotubes (SWCNT) combining the advantages of graphene and silver nanowires, providing low cost, scalable production, simple application, and improved current collection [6]. These properties, together with anti-reflection properties and high stability, make SWCNTs one of the most promising transparent flexible, and stretchable contact materials.

Although the literature presents data on preliminary studies of SWCNT layers in optoelectronics, their potential has not yet been fully disclosed. This article is devoted to experimental work on the creation of a flexible SWCNT-based electrode for optoelectronic applications.

2. Experimental

Films of randomly oriented SWCNTs synthesized by the aerosol chemical vapor deposition (CVD) method were deposited on a nitrocellulose filter at the outlet of the reactor. We used the patterning procedure described earlier to overcome the tradeoff between conductivity and transparency [7]. Prior to patterning, the films of SWCNTs were transferred either onto a quartz substrate with pre-deposited metallic contacts or on quartz substrate with sacrificed layer for next transfer to flexible substrate.

Cr/Au (5/70 nm) contacts were deposited on quartz substrates before the SWCNT transfer by electron-beam physical vapor deposition (for Cr) and thermal evaporation (for Au) using Boc Edwards (UK) Auto 500 setup operating at 5×10^{-6} mbar.

After the deposition of Cr/Au contacts, the SWCNT films were transferred onto the quartz substrates, processed with isopropyl alcohol for densification and then doped in tetrachloroauric (III) acid trihydrate ($HAuCl_4{\cdot}3H_2O$ ACROS Organics) dissolved in ethanol (EtOH; 99.5%, ETAX).

The patterns of SWCNTs were produced using a combination of optical laser lithography and dry etching in an oxygen plasma. We have used the laser lithography system Heidelberg Instruments Mikrotechnik DWL 66 FS setup (Germany) with AZ MIR 701 photoresist (MicroChemicals GmbH) and AZ 726 MIF developer (MicroChemicals GmbH). The following dry etching was performed in Plasma System V 15-G (Germany) at MW-power of 400 W (O_2 flux 60 ml/min, 3 Pa) during 360 sec for 90 nm SWCNT thickness. The grids of SWCNT networks were masked by photoresist during the etching.

Unlike traditional thin films, the resistance of SWCNT stripes does not always scale linearly with the size. Thin films or narrow stripes of SWCNT networks are known to suffer from an increase of resistance due to percolation effects. Therefore, a series of experiments were carried out to study whether the percolation affects the conductivity of patterned SWCNT layers. A set of SWCNT stripes with the widths of 1-100 μm and the lengths of 10-1000 μm was prepared following the same lithography and etching procedure as described above. Figure 1 (a) shows the optical microscopy image of the SWCNT stripes with Cr/Au contacts. The resistances of SWCNT stripes were extracted from the I-V characteristics. All the measured I-V curves have shown linear or almost linear dependence revealing close to ohmic behavior of SWCNT-Au contact

Figure 1 (b) and (c) shows the normalized stripe resistances R as a function of the stripe length and width correspondingly. The $R \times w$ dependences scale almost linearly (exponent about 0.95) at the lengths above 50 μm. However, the points corresponding to the shortest measured stripes with the length of 10 μm do not follow the linear scaling law. Previously, similar observations were attributed to the transition from ballistic to diffusive transport in randomly oriented networks. The measured resistances scales as w^{-1} which is typical for SWCNT networks far above the percolation threshold. Therefore, in patterns with the linewidth above 1 μm and the cell size above 50 μm the random character of SWCNT networks may be neglected.

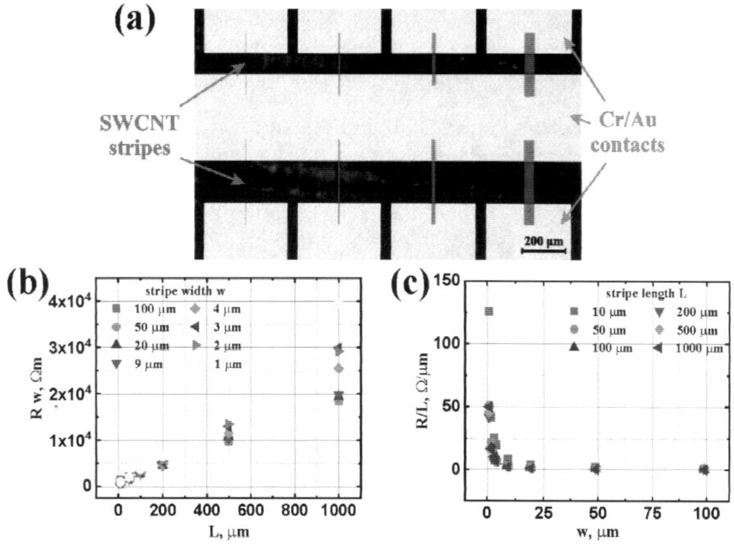

Figure 1. (a) Optical microscopy image of the set of SWCNT stripes with Cr/Au contacts (fragment); (b) the stripe resistances multiplied by its width as the functions of the stripe length; (c) the stripe resistances divided by its length as the functions of the stripe width.

To make a flexible patterned SWCNT electrode and to separate it from the quartz substrate, we used a sacrificial layer of a PMGI resist (MicroChemicals GmbH) that had been hardbaked at 200°C for 10 minutes. Hardbake was necessary to ensure PMGI persistence during further processing. The continuous SWCNT layer was transferred to the hardbaked PMGI surface. Further details of the processing, including AZ MIR photoresist spin coating, photolithography, developing, etching, and removing the photoresist are similar to those described above. Finally, we've processed patterned SWCNT electrode in geometry of square grid with cell period of 210 um and linewidth of 20 um. Additional details of fabrication and characterization were discussed in [7]. After the patterned SWCNT film was formed, the sample was poured with liquid polydimethylsiloxane (PDMS, Dow Corning Sylgard 184) and remained in the chamber drier at 60°C for 4 hours. After the PDMS dried, the substrate was placed in a dimethyl sulfoxide (DMSO) solution to dissolve the sacrificial PMGI layer. Electrical contacts to measure bending resistance were applied with silver lacquer. Figure 2 (a) shows the view of patterned electrode on flexible PDMS substrate. The final transparency of the electrode was 90% with resistance 200 Ohm/□.

Since flexibility is one of the most important advantages of SWCNT films, the impact of bending on the grid conductivity was studied. During the flexibility testing the samples were bent around quartz tubes with the diameters from 67.5 down to 22 mm. Figures 2 (b) and (c) show the sample view sketch and the resistance variation ($\Delta R/R$) in the bend-release tests correspondingly. The strain in figure 2 (c) was evaluated as film thickness divided by the quartz tube diameter. We observed a very moderate increase of resistance below 5% under the strain up to 4 %, which is comparable with the previous results for continuous SWCNT layers and shows the improvement in comparison with the previous reports on patterned SWCNT layers.

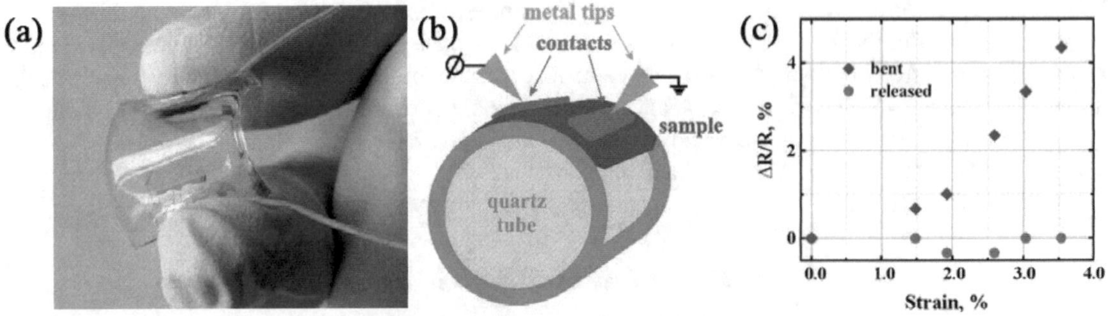

Figure 2. (a) Patterned SWCNT electrode on flexible PDMS substrate. (c) Sample view sketch and (c) variation of resistance in bend-release tests.

3. Summary

The demonstrated technology for the fabrication of flexible patterned SWCNT electrodes can be applied to a wide range of optoelectronic devices, including flexible LEDs, thin-film solar cells or displays. A series of experiments were carried out to study whether the percolation affects the conductivity of patterned SWCNT layers. It was found that in patterns with the linewidth above 1 μm and the cell size above 50 μm the random character of SWCNT networks may be neglected. The impact of bending on the grid conductivity was studied. We observed a very moderate increase of resistance below 5% under the strain up to 4 %, which is comparable with the previous results for continuous SWCNT layers and shows the improvement in comparison with the previous reports on patterned SWCNT layers.

Acknowledgments

The reported study was funded by RFBR, project number 19-38-60008, the Ministry of Science and Higher Education of the Russian Federation (grant 0791-2020-0005).

References

[1] Kang H, Zhao C, Huang J, Ho D H, Megra Y T, Suk J W ... and Cho J H 2019 *Adv. Funct. Mater.* **29** 43
[2] Kim S, Gupta M K, Lee K Y, Sohn A, Kim T Y, Shin K S ... and Kim S W 2014 *Adv. Mater.* **26** 23
[3] Kwak S S, Kim H, Seung W, Kim J, Hinchet R and Kim S W 2017 *ACS nano* **11** 11
[4] Liu T, Liu M, Dou S, Sun J, Cong Z, Jiang C ... and Wang Z L 2018 *ACS nano* **12** 3
[5] Yang Y, Han J, Huang J, Sun J, Wang Z L, Seo S and Sun Q 2020 *Adv.Funct. Mater.* **30** 29
[6] Gilshteyn E P, Lin S, Kondrashov V A, Kopylova D S, Tsapenko A P, Anisimov A S ... and Nasibulin A G 2018 *ACS Appl. Mater. Interfaces* **10** 33
[7] Mitin D, Berdnikov Y, Vorobyev A, Mozharov A, Raudik S, Koval O ... and Mukhin I 2020 *ACS Appl. Mater. Interfaces* **12** 49

Charge carriers drift induced THz amplification in dual-layer graphene structure

I.M. Moiseenko[1, a)], V.V. Popov [1, b)] and D.V. Fateev[1, 2, c)]

[1]*Kotelnikov Institute of Radio Engineering and Electronics (Saratov Branch), Russian Academy of Sciences, Saratov 410019, Russia*
[2]*Saratov State University, Saratov410012, Russia*

[a)]*Corresponding author: MoiseenkoIM@yandex.ru*
[b)]*Popov_slava@yahoo.co.uk*
[c)]*FateevDV@yandex.ru*

Abstract. The terahertz plasmon amplification in structure based on graphene with spatial dispersion of its hydrodynamic conductivity is investigated theoretically. The spatial dispersion of graphene conductivity is related to accounting of charge carriers pressure forces and direct current in graphene. It was shown that the real part of graphene conductivity becomes negative at THz frequency range due to direct electric current in graphene.

INTRODUCTION

Graphene plasmonic is an actual direction in the development of compact terahertz (THz) devices [1-3]. The possibility of creating an inverse population in the graphene does not make it a promising material for amplifying THz radiation. Various pumping methods have been proposed to create an inversion of charge carriers in graphene [4-6]. The possibility of excitation and amplification of THz plasma waves (plasmons) in a transistor structure with graphene at room temperature was shown in Ref. [7]. The possibility of plasmon instabilities and THz amplification in hydrodynamic (HD) graphene with a direct electric current (DC-current) due to the Vavilov-Cherenkov effect is shown in Ref. [8].

THEORETICAL MODEL

The amplification of THz radiation in single-layer graphene with DC-current was shown in [9]. In this work the excitation and amplification of THz plasmons in a double-layer graphene structure at room temperature was investigated theoretically. Using THz plasmons in graphene leads to strong localization of THz field near graphene and more effective THz amplification. The studied structure consists of two hydrodynamic graphene sheets separated by dielectric slab of thickness l=100 nm (Fig.1(a)). The plasmons are excited by the method of attenuated total reflection of a THz wave incident on the structure at a random angle form prism with dielectric constant ε_1, then the

x-component of wave vector of incident THz wave is $k_x = \omega\sqrt{\varepsilon_1}\sin\alpha / c$, where ω and α are the THz wave angular frequency and angle of incidence, respectively. The prism located over graphene structure at a distance d=50 μm.

The graphene layers described by hydrodynamic conductivity in the following form:

$$\sigma(k_x,\omega) = -\frac{i\,\varepsilon_F e^2 \omega\left(2(1-\beta)(\omega-k_x u)-3i\,\gamma\beta\right)}{\pi\hbar^2\sqrt{1-\beta}\left(k_x^2 V_F^2 (k_x u\chi_1+\chi_3)+\omega^2(\gamma/\omega-i)\chi_2+k_x u\chi_4\right)}, \quad (1)$$

where $\chi_1 = (2\beta-1)$, $\chi_2 = 3\gamma\beta + i\,\omega(\beta-2)$, $\chi_3 = \omega-\beta(3i\,\gamma+4\,\omega)$, $\chi_4 = \omega\left(\beta(2i\,\gamma-\omega)+(5i\,\gamma+4\,\omega)\right)$, e_0 is the electron charge, $\beta = u^2/V_F^2$, ε_F and V_F is the Fermi energy and Fermi velocity in graphene, respectively, u is the drift velocity of charge carriers in graphene, $\gamma = 1/\tau$, τ is the momentum relaxation time in graphene. The conductivity of graphene in the HD regime is obtained by solving the equations of momentum balance and energy balance of charge carriers, as well as the continuity equation [8]. The electron drift in graphene, as well as the pressure in charged liquid in graphene, leads to the appearance of a spatial dispersion of the graphene conductivity.

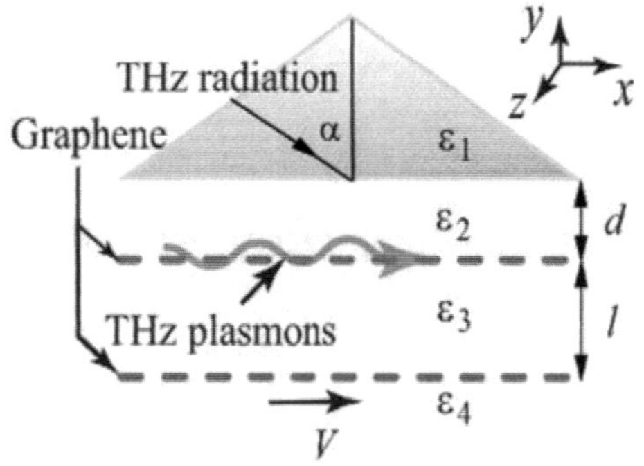

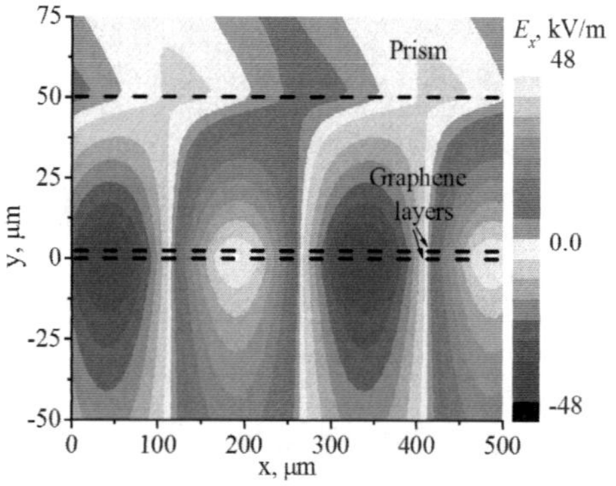

FIGURE 1. (a) Incidence of the THz wave on a graphene-based structure and (b) the electric field distribution in dual layer hydrodynamic graphene structure for angle of THz wave incidence about 38 degrees and frequency $\omega/2\pi = 0.45$ THz.

RESULTS AND DISCUSSION

The imaginary part of the conductivity of graphene with charge carrier drift (1) is negative in the case of deceleration of the incident THz wave using a prism, which does not allow for exciting plasmons in a monolayer of graphene with charge carrier drift. Therefore, the excitation of plasmons is assumed in the graphene layer without the drift of charge carriers located nearby. An electromagnetic wave at a frequency of 0.45 THz incident at an angle of 38 degrees from a prism with a dielectric constant of 12.9 excites plasmons in the structure under consideration. The characteristic distribution of the graphene-in-plane component of the electric field for plasmons is shown in Fig. 1(b). At a current density $J = -une$=68 A/cm (corresponding to charge carriers drift velocity in graphene u=0.75V_F and $n = 5.6\cdot10^{16}$ m^{-2}), the real part of the graphene conductivity is negative, which leads to amplification of plasmons, and the reflection coefficient R reached the maximal value (Fig. 2). An increase of charge carriers density (corresponds to an increase in the Fermi energy) in graphene with a drift of charge carriers leads to an increase of the screening efficiency of plasmons by graphene with a DC-current and a decrease of the plasmon frequency.

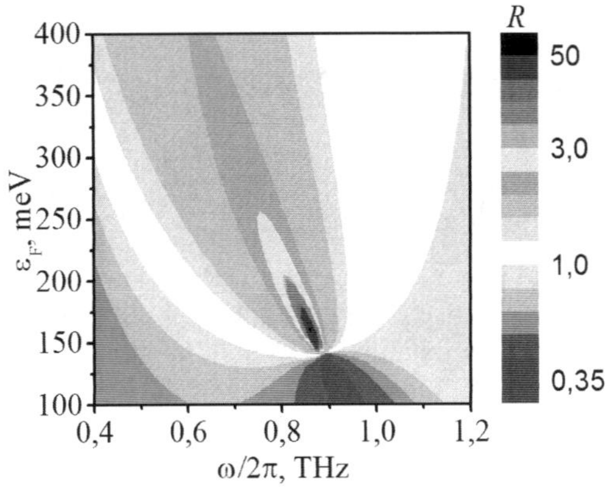

FIGURE 2. The reflection coefficient in dependence of frequency and Fermi energy in graphene with DC-current for relaxation momentum time 0.1 ps and 1ps in graphene with DC-current and without it, respectively.

CONCLUSIONS

It is shown that a dual-layer graphene structure consisting of graphene layer with DC-current and graphene-layer without it support resonant excitation and amplification of THz plasmons. The amplification occurs due to real part of hydrodynamic conductivity of graphene with DC-current becomes negative. Tuning of the resonance frequency of plasmons should be taken into account when enhancing the screening effect due to an increase in the density of charges in graphene with a constant drift.

ACKNOWLEDGMENTS

This work was supported by Russian Science Foundation through Project No. 20-79-00224.

REFERENCES

1. F. H. L. Koppens, T. Mueller, P. Avouris, A. C. Ferrari, M. S. Vitiello and M. Polini, Nature Nanotech **9**, 780 (2014).
2. X. Zhang, Quan Xu et al. Advanced Photonics **2(1)**, 014001 (2020).
3. T. Watanabe, T. Fukushima, Y. Yabe, S. A. Boubanga Tombet, A. Satou, A. A. Dubinov, V. Ya. Aleshkin, V. Mitin, V. Ryzhii, and T. Otsuji New J. Phys. **15**, 075003 (2013).
4. M. Yu. Morozov, A. R. Davoyan, I. M. Moiseenko *et al.* Appl. Phys. Lett., **106**, 061105 (2015).
5. I. V. Oladyshkin, S. B. Bodrov, Yu. A. Sergeev, A. I. Korytin, M. D. Tokman, and A. N. Stepanov Phys. Rev. B **96**, 155401 (2017).
6. V. Ryzhii, M. Ryzhii, V. Mitin, and T. Otsuji // J. of Appl. Phys., **110**, 094503 (2011).
7. S. Boubanga-Tombet et al. // Phys. Rev. X **10**, 031004 (2020).
8. D. Svintsov Phys. Rev. B. **100**, 195428 (2019).
9. I. M. Moiseenko, V. V. Popov, and D. V. Fateev AIP Conference Proceedings **2300**, 020089 (2020).

3D printed Ku band cylindrical Luneburg lens

I Grigoriev[1], I Munina[1*] and D Zelenchuk[2]

[1]Microelectronics and Radio Engineering Department, St. Petersburg Electrotechnical University "LETI", Saint Petersburg, Russia
[2]The ECIT Institute, Queen's University Belfast, United Kingdom
*ivmunina@etu.ru

Abstract. The design of the cylindrical Luneburg lens manufactured by using 3D printing facilities was presented, fabricated and experimentally verified. The two different approaches to lens parameter discretization were applied to lens design and compared. The effective dielectric permittivity of Polylactic Acid (PLA) plastic samples with different infilling percentage were estimated by a resonant method.

1. Introduction

Luneburg lens is a well-known beamformer. It consists of a sphere with a gradient refractive index and a feed placed on its surface. The main advantages of the Luneburg lens in comparison with traditional phased array antennas are high gain, wide-angle scanning without a gain drop and low side lobe level. Such systems can be used for radar and telecommunication applications. The refractive index of the lens from center to outer radius changes from 2 to 1. The classical Luneburg lens ensures 2D beam steering. 1D beam steering can be provided by the cylindrical cut of a sphere.

For the practical implementation of the lens, the dependence of the refractive index has to be discretized. There are two different approaches to discretization: using the layers with the same thickness or using the layers with the same step of refractive index or dielectric permittivity. For the fabrication of such a layered structure, additive manufacturing technologies can be used. 3D printing methods allow reducing the cost of the production process. Gradient index (GRIN) lenses can be implemented by changing the volume percentages of plastic infill. Based on this approach, it is easy to change the dielectric constant along the lens. This method was proofed by design of a flat lens [1]. The dielectric permittivity of the layer can be adjusted by using a sub-wavelength unit cell instead of percentage of filling of the layer [2]. Additive manufacturing opens new possibilities for the manufacturing of GRIN lenses.

In this paper, a 5-layer cylindrical lens is considered. Two approaches for discretising the permittivity along the lens were used: with a uniform step of the thickness of the layer and a uniform step of the permittivity of the layer. The implementation of the lenses was carried out layer-by-layer changing the percentage of filling of the PLA plastic (Tiger3D PLA). For the lens design, the permittivity for PLA samples with a different volume percentage of infill was characterized by a resonant method. The samples of the cylindrical lens were fabricated and measured using a Ku band patch antenna as a feed. The measured results are in good agreement with simulated ones.

2. The cylindrical Luneburg lens: simulated and measured results

The Luneburg lens is a graded-index (GRIN) structure with spatially varying refractive index:

$$n(r)=\sqrt{\varepsilon(r)}=\sqrt{2-(r/R)^2}\,, \qquad (1)$$

where r is the distance from the centre of the lens and R is outer radius.

It is difficult technologically to implement a continuous changing of the refractive index, and therefore the dielectric constant of the material over the entire lens profile. It is possible to resort to the use of dissimilar materials and make a single lens out of them, but it is more attractive to use the same material. This approach can be implemented using 3D printing, which ensures the relative cheapness of production and the variability of designs due to the sequential deposition of the material.

The lens can be described as concentrically arranged cylinders with a certain set of permittivity values (figure 1). The lens stratification can be performed with a uniform step along the radial coordinate or with a uniform step of the dielectric constant. Both methods are characterized by a layered structure, with each layer corresponding to its value of the dielectric constant. It was decided to use a 5-layer structure because a further increase of the number of layers does not lead to an improvement of the lens characteristics [3].

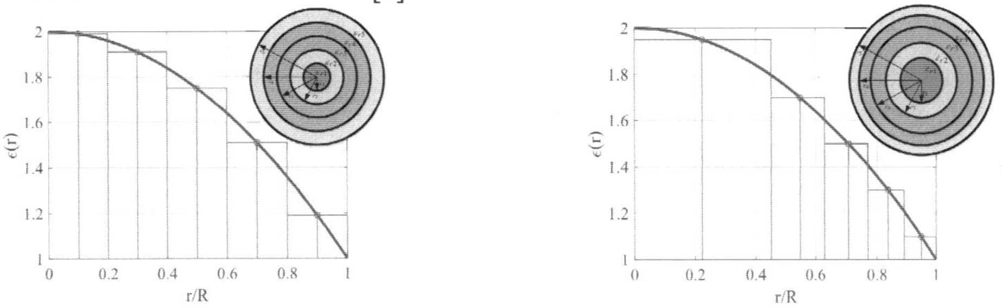

Figure 1. The permittivity discretization by a uniform layer thickness (a) and a uniform permittivity step (b).

The distance from the center of the lens r was normalized to the outer radius defined as R (figure 1). The obtained values of the permittivity of the layers were used for full-wave EM simulation using CST Studio Suite. Two models with a radius of 5 cm (3λ at 18 GHz) and a height of 18 mm were designed. The patch antenna operating at 18 GHz was chosen as a feed. For the patch antenna was used a dielectric substrate (Taconic RF35) with a permittivity of 3.5, a height of 0.762 mm, and a dielectric loss tangent of 0.0018. The operational band of the antenna according to the level of -10 dB of the reflection coefficient is 4%. It is worth noting that the Luneburg lens is frequency independent and the operational band of it depends only on the operational band of the feed.

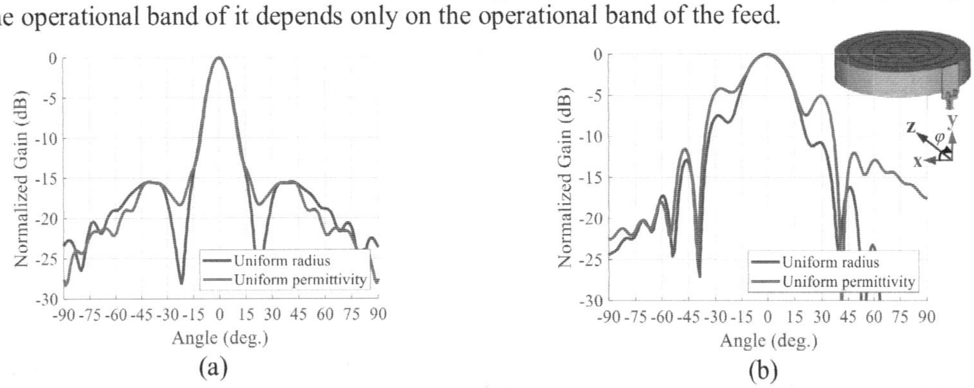

Figure 2. Normalized gain for H-(a) and E-plane (b) for two types of Luneburg lenses at 18 GHz.

The results of full wave simulation are presented in Figure 2. The simulation results show that the realized gain of the lens with a uniform thickness is 17.6 dBi at 18 GHz. The HPBW in H-plane is about 11.9° with the side lobe level about 15.5 dB below the main peak. For a lens with a uniform permittivity, the realized gain is 17.7 dBi and the HPBW is about 11.5° with the side lobe level better than 15.4 dB.

To implement a lens, it is necessary to define the effective permittivity of each layer as a percentage of the material filling of the layer. The dielectric permittivity can be recalculated to the percentage of filling of the PLA volume [1] by using (2):

$$v = \frac{\varepsilon_{\text{reff}} - 1}{\varepsilon_{r0} - 1}, \tag{2}$$

where ε_{r0} is the relative permittivity of the filament for 3D printing, $\varepsilon_{\text{reff}}$ is the effective dielectric permittivity of the layer of the lens.

For calculations of v it is important to know the permittivity of the PLA material with 100% infill. The resonant method of dielectric permittivity characterization was used. Five samples with a thickness of 3 mm were fabricated with different percentages of filling from 10% to 100%. The coplanar resonator was covered by the samples and the shift of the resonant frequency was used to determine the correspondence of the permittivity to the percentage of filling of the sample. The comparison between results obtained by (2) and the measured data is shown in figure 3. The permittivity of the 100% filled PLA sample was defined as 2.3.

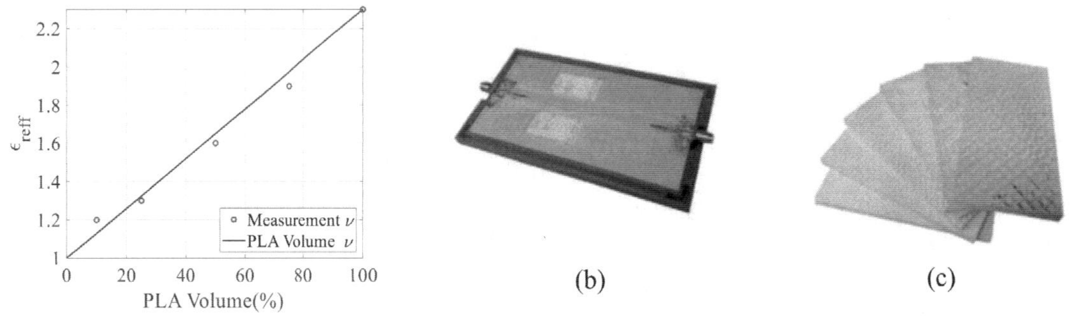

(a) (b) (c)

Figure 3. Dependence of the dielectric constant on the volume of the PLA (a) and resonator, which was used for the characterization of material (b) and the samples made of PLA plastic with different filling percentage (c).

To describe the effective volume of plastic in the total volume taking into account the outer walls (usually a minimum wall thickness is equal to the diameter of the 3D printer nozzle) of each cylindrical layer, the following expression was used [1]:

$$d = \frac{v\pi(R^2 - r^2) - 2\pi t(R + r)}{\pi(R^2 - r^2) - 2\pi t(R + r)}, \tag{3}$$

where t is the thickness of the outer wall determined by the diameter of the extruder nozzle, R and r are the outer and inner radii of the cylindrical ring respectively. The data for lens fabrication are presented in table 1.

Table 1. Lens parameters for the fabrication

Uniform thickness	$\varepsilon_{\text{reff}}$	1.99	1.91	1.75	1.51	1.19
	r/R	0.20	0.40	0.60	0.80	1.00
	d, %	74	67	54	34	7
Uniform step of dielectric const	$\varepsilon_{\text{reff}}$	1.90	1.70	1.50	1.30	1.10
	r/R	0.45	0.63	0.77	0.89	1.00
	d, %	68	50	31	11	2

For the lens excitation, 4 samples of the patch antenna (16x22 mm) were fabricated and measured. The patch antenna was chosen due to the fact that it has a symmetrical radiation pattern in the E- and H-planes. The comparison between experimental results and measured ones is shown in Figure 4. According to the required percentage of PLA infill and radii of the layers, the two experimental samples of the lens were produced (figure 5). The radiation pattern of the lenses excited by patch antennas was measured in far field zone in the H plane. The measured and simulated results are in good agreement (figure 5).

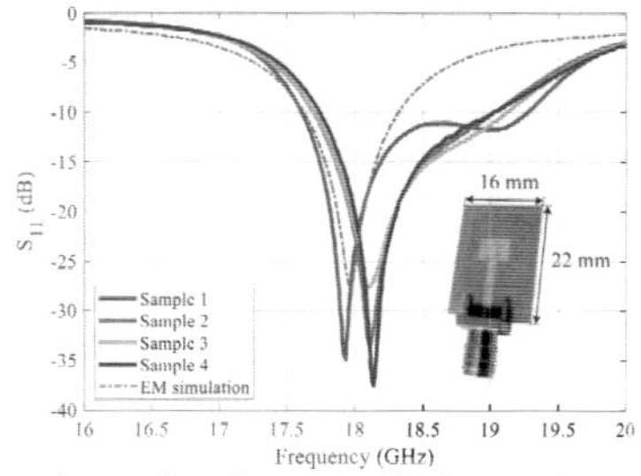

Figure 4. The reflection coefficient of patch antenna.

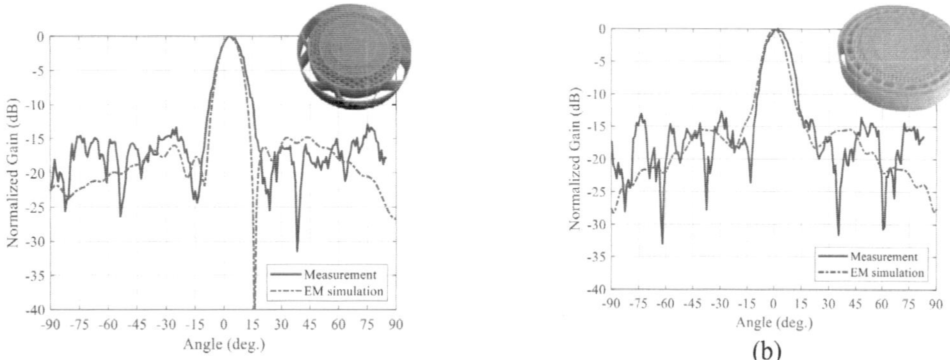

(a) (b)

Figure 5. Normalized radiation pattern of the Luneburg lenses: uniform layer thickness discretization and uniform permittivity discretization at 18 GHz.

3. Conclusion

In this paper, the 3D printed cylindrical Luneburg lens for Ku application was designed. For lens manufacturing, the samples with different infill percentage of PLA filament were fabricated and characterized by using a resonant method. Two lenses with different approaches to parameter discretization were simulated, manufactured and measured. The Ku band patch antenna was used as a feed source. The measured results correspond to the simulated ones.

Acknowledgment

The work was supported by Russian Science Foundation (project 17-79-20374).

References

[1] Zhang, Shiyu, et al. "3D-printed planar graded index lenses." *IET Microwaves, Antennas & Propagation* 10.13 (2016): 1411-1419.

[2] Norooziarab, M., et al. "Millimeter-wave 3D Printed Luneburg Lens Antenna." *2019 IEEE Radio and Antenna Days of the Indian Ocean (RADIO)*. IEEE, 2019.

[3] Korotkov, A. N., S. N. Shabunin, and V. A. Chechetkin. "The cylindrical Luneburg lens discretization influence on its radiation parameters." 2017 International Multi-Conference on Engineering, Computer and Information Sciences (SIBIRCON). IEEE, 2017.

Silicon nanowire/polymer membrane for infrared visualization via third-harmonic generation

Anna Nikolaeva,[1, a)] Vladimir V. Fedorov,[2] Alexander S. Gudovskikh,[2] Ivan A. Morozov,[2] Viktoria Mastalieva,[2] Vladimir Neplokh,[2] Ivan S. Mukhin,[2] and Sergey V. Makarov[1]

[1)] *Department of Physics ITMO University St. Petersburg 197101 Russian Federation.*
[2)] *Alferov University St. Petersburg 194021 Russian Federation.*

[a)] *Corresponding author: anna.nikolaeva@metalab.ifmo.ru*

Abstract. Nonlinear harmonic generation in nanostructures is one of the key topics in nanophotonics, as it allows infrared-to-visible light conversion at the nanoscale. This work reports on the efficient third-harmonic generation in a free-standing Si nanowire array encapsulated into a polymer membrane. High nonlinearity of Si material $\chi^{(3)}_{1111} \approx 2.45 \cdot 10^{-19} m^2/V^2$ and light coupling with optical resonances in the nanowires yield a strong third-harmonic signal and efficient infrared $(1200 - 2000\text{nm})$ to visible $(400 - 666\text{nm})$ upconversion. The fabricated membranes demonstrate high flexibility and semitransparency, which makes them convenient to use as visualizers.

INTRODUCTION

Observation and use of nonlinear optical effects became possible after the advent of powerful laser sources, but the effects themselves are rather weak, which requires new approaches for the enhancement of their generation. Recently, dielectric nanostructures with intrinsic nonlinearity appeared to be prospective objects for signal enhancement [1], where third-harmonic generation (THG) efficiency can reach values of $10^{-7} - 10^{-5}$ [2, 3] in a silicon-based structure. Third-harmonic generation from resonant nanostructures has become a powerful tool for nanoscale optical characterization [4], shaping of the directionality of nonlinear radiation [5], non-linear holography [6] as well as for the creation of nanoscale color-tunable light sources via infrared (IR)-to-visible light conversion [7].

In this work, we took advantage of the work [8], namely, that the increase of the THG conversion efficiency can be achieved with elongation of the nanoparticles in one direction, forming nanowires [9]. In such designs, a combination of Mie resonances and Fabry-Perot modes can lead to strong incident field localization and outstanding optical properties. But we use silicon material and the THG process to visualize longer wavelengths.

EXPERIMENTAL DETAILS

Sample fabrication. The optical properties have been studied in arrays of vertically oriented SiNWs encapsulated into a silicone membrane and peeled from the Si substrate. This technology enables the reuse of growth substrates and a combination of various material systems to create ultra-thin (up to several microns thickness) flexible transparent structures. "Black silicon" NWs were produced by the cryogenic process. Plasma etching of Si (100) substrates under $-140°$C temperature is used and the NW side surface is passivated due to the formation of a non-volatile SiO_xF_y compound at cryogenic temperature. It blocks the etching mechanism and the cryogenic process provides deep and anisotropic etching of silicon [10].The height of obtained NWs for X_1 sample was about $h = 9\mu$m, while diameter was in the range $d = 300 - 1350$nm and $h = 5\mu$m and $d = 200 - 300$nm for X_2 sample, what is shown in the scanning electron microscopy (SEM) images in Figure 1(a,b). Further, the NW arrays were encapsulated into a Polydimethylsiloxane (PDMS) matrix and separated from the growth substrate.

Nonlinear Optical Characterization. We perform nonlinear optical characterization of the fabricated membranes in transmission geometry with a custom nonlinear optical setup. A detailed schematic of the experiment is presented in Figure 1(c). Tunable sample excitation in the range of 1200-2000nm was performed by a femtosecond Pharos laser and Optical Parametric Amplifier (OPA) ORPHEUS-HP. In our experiments, the membranes were placed on glass coverslips and aligned perpendicular to the laser propagation axis. The laser light was focused with a 50x objective (NA = 0.42) on the sample. The signal was then collected with a 50x objective (NA = 0.55) and focused onto a scientific CMOS camera. The power was controlled via a half-wave plate placed on the excitation beam path before

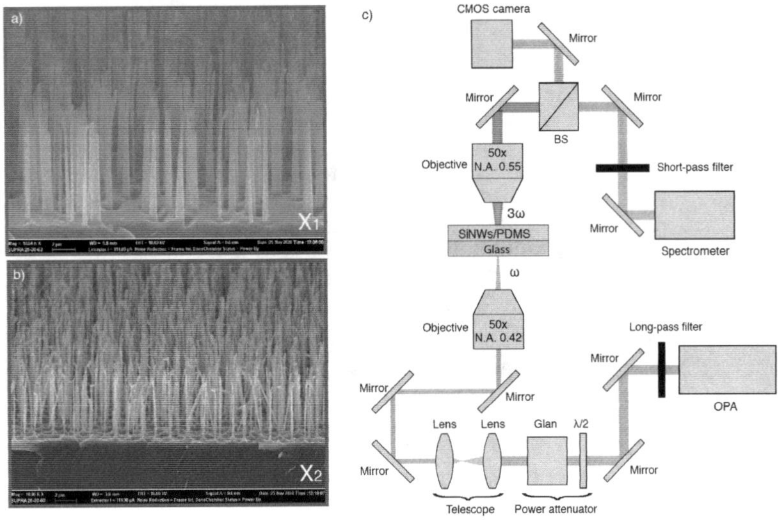

FIGURE 1. (a,b) SEM images of studied Si NW:sample X_1 highlighted in blue frame (a) and sample X_2 highlighted in red frame(b).c) Schematic of experimental setup for third-harmonic generation spectroscopy.

the Glan prism as shown in Figure 1(c). For the nonlinear optical characterization, we performed the measurements of THG intensity for different excitation wavelengths at same average power $P_{av} = 3$mW (Figure 2(b,e)) and for different average pump power at the same excitation wavelength $\lambda = 1400$nm (Figure 2(a,d)).

RESULTS AND DISCUSSION

To demonstrate efficient infrared-to-visible light visualizers based on SiNW/polymer membranes having THG signal, we studied THG from the SiNWs. First, we measured the intensity of the signal from the fluence, which we showed in Figure 2(a,d) on a double logarithmic scale. After the fitting, we got a slope of approximately 3, which ensures that we have measured the third-harmonic generation. The bend in Figure 2(d) is probably because the PDMS was warming up and the structure went out of focus, which was visible on the CMOS camera. Our structure works as an efficient visualizer over a wide range of pump wavelengths $1200 - 2000$nm, it is shown in Figure 2(b,e), which shows the dependence of the THG intensity on the pump wavelength, and in Figure2(c), showing the CMOS camera images of THG signal for different pump wavelengths.

Remarkably, optical resonances (Fabry-Perot and Mie-type) in SiNWs support strong incident light localization at certain wavelengths, resulting in the appearance of pronounced maxima in Figures 2(b,e). This effect helps to enhance THG and it is generally an advantage of resonant nanostructures, allowing for the creation of efficient and thin optical devices. Moreover, our SiNWs support the resonances in a broad spectral range due to the dispersion of the diameters and heights of nanowires, also, nanowires with varying morphologies interact differently with pump light and thus exhibit a broad spectrum of third-harmonic generation. In consequence, the conversion efficiency is homogeneously distributed over near-IR.

CONCLUSION

In this work, we have demonstrated efficient infrared-to-visible light visualizers based on SiNW/polymer membranes having THG signal. For this purpose, we have employed the strong THG process in the resonant SiNWs in a broad spectral range. Such SiNW structures enabled the development of thin 10–20μm membranes with efficiently converting IR radiation into visible light. The proposed THG-based principle of IR visualization differs in both functional and operational characteristics from most commercial analogs made of a special ceramic with anti-Stokes luminophores.

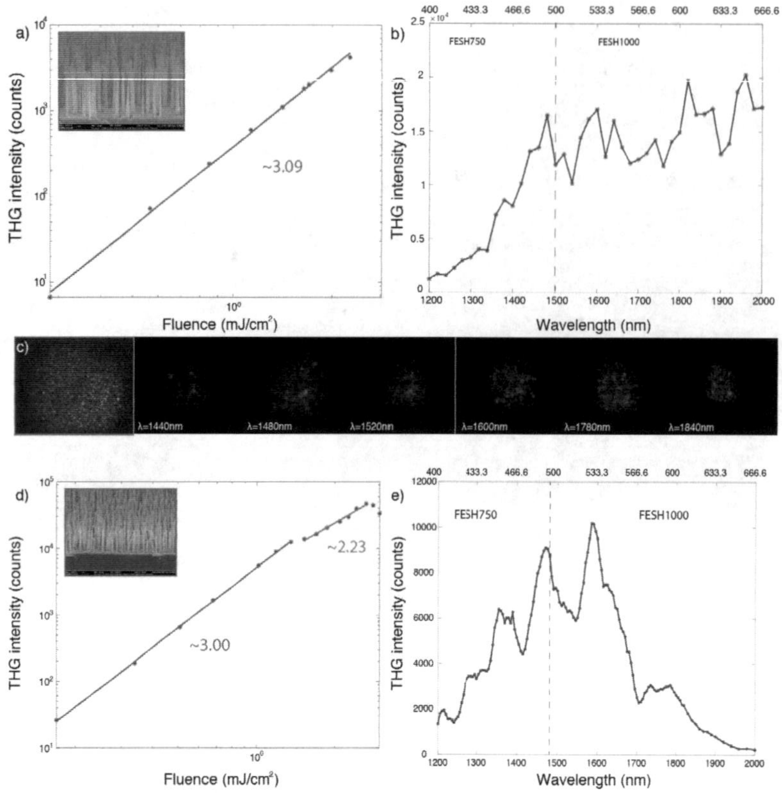

FIGURE 2. Experimental characterization of THG from SiNW/PDMS, (a,b,c) corresponds to X_1 sample and (d,e) to X_2. (a,d) Experimental dependence of the THG intensity on fluence at a wavelength $\lambda = 1400$nm and repetition rate of $RR = 1$MHz (stars); blue and orange line is the result of the fitting procedure. (b,e) Experimental dependence of the THG intensity on the pump wavelength. Up to 1500nm, we used a FESH750 short-pass filter and then we changed it to FESH1000. c) X_1 sample image captured with the CMOS camera in the light of the white lamp, as well as images of third-harmonic generation signals for different pump wavelengths $\lambda = 1440, 1480, 1520, 1600, 1780, 1840$nm.

In contrast to the anti-Stokes luminescence frequency conversion, the THG process is less temperature-dependent and makes it possible to distinguish the IR pump wavelength by the color of its second harmonic. Also, as compared to the ceramic-based IR viewers suffering from hygroscopicity problems, the SiNW array can be easily fully encapsulated in silicone elastomer, protecting them from the environment. Weak absorption of the pump radiation together with high thermal conductivity of the Si material makes the fabricated membranes suitable for high power density applications.

ACKNOWLEDGMENTS

We wish to acknowledge the support of the Presidential Grant MD-5954.2021.1.3.

REFERENCES

1. S. V. Makarov, M. I. Petrov, U. Zywietz, V. Milichko, D. Zuev, N. Lopanitsyna, A. Kuksin, I. Mukhin, G. Zograf, E. Ubyivovk, *et al.*, "Efficient second-harmonic generation in nanocrystalline silicon nanoparticles," Nano letters **17**, 3047–3053 (2017).
2. S. Sederberg, C. J. Firby, and A. Y. Elezzabi, "Efficient, broadband third-harmonic generation in silicon nanophotonic waveguides spectrally shaped by nonlinear propagation," Opt. Express **27**, 4990–5004 (2019).

3. M. R. Shcherbakov, D. N. Neshev, B. Hopkins, A. S. Shorokhov, I. Staude, E. V. Melik-Gaykazyan, M. Decker, A. A. Ezhov, A. E. Miroshnichenko, I. Brener, A. A. Fedyanin, and Y. S. Kivshar, "Enhanced third-harmonic generation in silicon nanoparticles driven by magnetic response," Nano Letters **14**, 6488–6492 (2014).
4. C. B. de Araújo, A. S. L. Gomes, and G. Boudebs, "Techniques for nonlinear optical characterization of materials: a review," Reports on Progress in Physics **79**, 036401 (2016).
5. L. Wang, S. Kruk, L. Xu, M. Rahmani, D. Smirnova, A. Solntsev, I. Kravchenko, D. Neshev, and Y. Kivshar, "Shaping the third-harmonic radiation from silicon nanodimers," Nanoscale **9**, 2201–2206 (2017).
6. E. Almeida, O. Bitton, and Y. Prior, "Nonlinear metamaterials for holography," Nature Communications **7**, 12533 (2016).
7. H. Aouani, M. Rahmani, M. Navarro-Cía, and S. A. Maier, "Third-harmonic-upconversion enhancement from a single semiconductor nanoparticle coupled to a plasmonic antenna," Nature Nanotechnology **9**, 290–294 (2014).
8. V. V. Fedorov, A. Bolshakov, O. Sergaeva, V. Neplokh, D. Markina, S. Bruyere, G. Saerens, M. I. Petrov, R. Grange, M. Timofeeva, S. V. Makarov, and I. S. Mukhin, "Gallium phosphide nanowires in a free-standing, flexible, and semitransparent membrane for large-scale infrared-to-visible light conversion," ACS Nano **14**, 10624–10632 (2020).
9. P. R. Wiecha, A. Arbouet, H. Kallel, P. Periwal, T. Baron, and V. Paillard, "Enhanced nonlinear optical response from individual silicon nanowires," Phys. Rev. B **91**, 121416 (2015).
10. P. Repo, J. Benick, V. Vähänissi, J. Schön, G. von Gastrow, B. Steinhauser, M. C. Schubert, M. Hermle, and H. Savin, "N-type black silicon solar cells," Energy Procedia **38**, 866–871 (2013).

Design and fabrication of photoactive ZnO-MgO-Ag nanocomposites for medical and environmental applications

N V Nikonorov[1,6], S K Evstropiev[1,2,3], A V Shelemanov[1], J F Podruhin[4], V M Kiselev[2] and A V Karavaeva[5]

[1] ITMO University, Saint-Petersburg, Russia
[2] JVC RPA "Vavilov State Institute", Saint-Petersburg, Russia
[3] Saint-Petersburg State Technological Institute (Technical University), Saint-Petersburg, Russia
[4] Saint-Petersburg State Polytechnical University, Saint-Petersburg, Russia
[5] Saint-Petersburg State Chemical-Pharmaceutical University, Saint-Petersburg, Russia

[6]Corresponding author e-mail address: nikonorov@oi.ifmo.ru

Abstract. Photoactive and bactericidal ZnO-MgO-Ag nanocomposites have been synthesized by liquid polymer-salt method. Polyvinylpyrrolidone plays a double role during nanocomposites synthesis: as the stabilizing component at initial solutions stages and as a fuel during materials calcination. Prepared materials demonstrate the ability to generate chemically active singlet oxygen under UV irradiation. Developed nanocomposites are promising for different medical and environmental application.

1. Introduction

Reactive oxygen species (ROS) play a key role in the photocatalytic oxidation of organic contaminants and in the environmental cleaning and disinfection. The photogeneration effectiveness depends on as the characteristics of exciting light so as on the electronic structure and the photocatalyst morphology. The main problem is that the intensity of this photogeneration often is too low for practical application at the use of the majority of common macroscopic materials.

The photogeneration process includes the generation of different ROS (singlet oxygen, superoxide radicals, hydroxyl radical formation.

ROS formation occurs on the surface and the nanomaterials, having high specific surface area, demonstrate significantly higher photocatalytic and bactericidal properties compared with macroscopic materials. Thus, the materials morphology optimization for the intensification of ROS photogeneration includes the increase of their specific surface area. This can be achieved by the decrease of the semiconductor particles size and/or by the formation of the particles having special morphology.

Solid solutions of ZnMgO, being wide band gap semiconductors (Eg ~ 3÷5 эВ) [1], are transparent in near-UV spectral range and have high thermal stability and mechanical strength. Atomic radii of Mg^{2+} and Zn^{2+} are close to each other [2] that provide the crystal structure of these solutions with small amounts of the distortions and local defects. Mixed ZnO-MgO nanostructures prepared by polymer-salt method have homogeneous structure and high specific surface area due to small crystals size [1]. This determines their effective contact with the surrounding media and high photoactive and bactericidal properties.

It was found that Ag-doped ZnO-based nanocomposites demonstrate the high ability to chemically active oxygen photogeneration and bactericidal properties [3,4]. Thus, based on the previous results [1,3,4] it is possible to conclude that ZnO-MgO-Ag nanocomposites are promising materials for photocatalytic and environmental practical application. In our work we designed and fabricated photoactive and bactericidal ZnO-MgO-Ag nanocomposites by liquid polymer-salt method with the use of polyvinylpyrrolidone.

2. Materials and Methods

Aqueous solutions of $Zn(NO_3)_2$, $SnCl_2$, $AgNO_3$ and high-molecular polyvinylpyrrolidone (PVP) (Sigma Aldrich, M_w=1300000) were used for powders synthesis.

The drying of the resultant homogeneous liquid mixtures was performed at 20°C. Thermal treatment of dried samples was carried out at atmospheric pressure at 550°C for 2 hours. Chemical compositions of prepared solutions and powders are given in Table 1.

Table 1. Chemical composition of prepared materials.

<table>
<tr><th rowspan="2">Sample</th><th colspan="6">Chemical compositions of initial solutions, wt.%</th><th colspan="3">Chemical composition of powders, wt.%</th></tr>
<tr><th>H_2O</th><th>PVP</th><th>Propanol-2</th><th>$Zn(NO_3)_2 \cdot 6H_20$</th><th>$Mg(NO_3)_2 \cdot 6H_20$</th><th>$AgNO_3$</th><th>ZnO</th><th>SnO_2</th><th>Ag[a]</th></tr>
<tr><td>*1*</td><td>53.27</td><td>0</td><td>42.04</td><td>4.18</td><td>0.46</td><td>0.05</td><td rowspan="4">90.02</td><td rowspan="4">9.00</td><td rowspan="4">0.98</td></tr>
<tr><td>*2*</td><td>52.92</td><td>0.66</td><td>41.75</td><td>4.16</td><td>0.46</td><td>0.05</td></tr>
<tr><td>*3*</td><td>52.57</td><td>1.31</td><td>41.49</td><td>4.13</td><td>0.45</td><td>0.05</td></tr>
<tr><td>*4*</td><td>51.89</td><td>2.60</td><td>40.93</td><td>4.08</td><td>0.45</td><td>0.05</td></tr>
</table>

[a]All Ag structural forms are taken into account as metallic Ag.

The diffractometer Rigaku Ultima IV was used for XRD analysis of prepared materials.
Special experimental luminescence setup was used for the study of singlet oxygen photo-generation in prepared materials. Two different LEDs (HPR40E set) (λ_{max}=370 nm; power density 0.35 W/cm^2) and (λ_{max}=405 nm; power density 0.90 W/cm^2) were used for luminescence excitation.

3. Experimental Results and Discussions

Thermal treatment of PVP-based composites at 550°C provided the full decomposition of metals nitrates and PVP and the formation of ZnO-MgO-Ag ceramic composite powders. Figure 1 demonstrates the results of XRD analysis of these powders. Numerous peaks of attributed to hexagonal ZnO crystals (JCPDS card No. 36–1451) and cubic MgO crystals (periclase) (JCPDS card No. 75–1525) are observed in these patterns. The peaks of ZnO crystals are the most intense and the ratio between the intensities of these peaks is close to the standard data (JCPDS card No. 36–1451). The absence of Ag peaks can be related to a small Ag content in prepared materials.

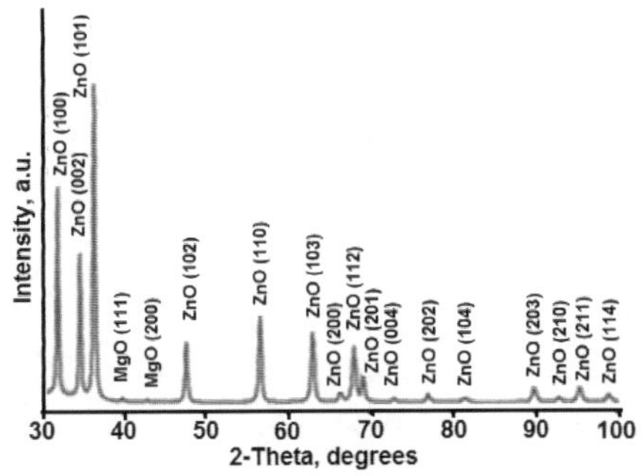

Figure 1. XRD patters of ZnO-MgO-Ag ceramic composite powder ***4***.

Average sizes d_{ZnO} and crystals cells parameters of ZnO crystals in ZnO-MgO-Ag ceramic composites were calculated using XRD data (Figure 2; Table 2). Figure 3 shows that PVP additions in the initial mixtures affect the average size of ZnO crystals in ZnO-MgO-Ag ceramic composites. The addition of even small amounts of PVP into initial metals nitrates solutions leads to the observed strong d_{ZnO} decrease.

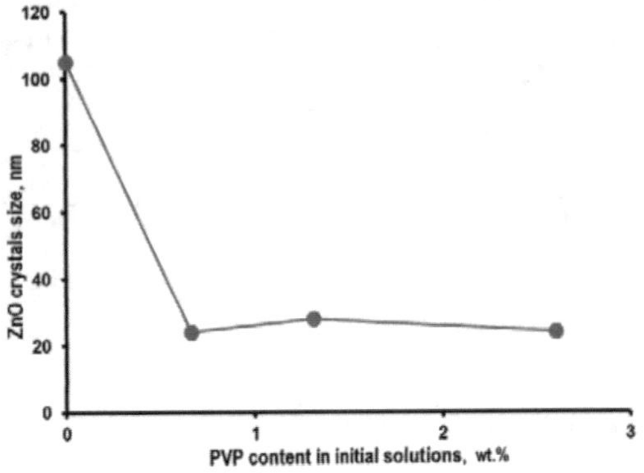

Figure 2. Dependence of the average size of ZnO crystals from the amounts of PVP in initial solutions.

It is worth to noticing that the morphology and average particle size of oxides obtained by a combustion process depends on the volume of escaping gases forming during heat treatment. Thus, the significant decrease d_{ZnO} at PVP additions into initial mixtures can be attributed to the formation of big volumes of gas products during the oxidation and decomposition of this polymer.

Table 2. ZnO crystal cells parameters in ZnO-MgO-Ag nanocomposites.

Sample	ZnO crystal cells parameters		
	a,b, Å	***c***, Å	V_{ZnO}, Å^3
1	3.2392	5.1898	47.159
2	3.2520	5.1941	47.572
3	3.2419	5.1698	47.056
4	3.2465	5.1879	47.355
JCPDS 01-070-8072	3.2465	5.2030	47.491

Figure 3 demonstrates SEM images of the morphology of ZnO-MgO-Ag powders. The powder ***1*** obtained from the solution without PVP consists of big porous particles having size ~ 10÷20 μm (Figure 3a). Some big hexagonal ZnO crystals are seen at high magnification (Figure 4b) in the powder structure.

a) b) c)

Figure 3. SEM images of ZnO-MgO-Ag nanocomposites ***1*** (a) and ***4*** (b,c).

In opposite, the structure of the powder ***4*** prepared from the solution containing significant amount of PVP (Table 1) is formed from a numerous uniform small (size about 20÷30 nm) nanoparticles

(Figure 3b,c). The size of these nanoparticles is closed to the average crystals size obtained from XRD analysis. This difference is explained by the characters of the thermal evolution of raw mixtures during powders synthesis.

The luminescent band with $\lambda_{max.}$= 1270 nm which is characteristic for singlet oxygen was observed in photoluminescence spectra of ZnO-MgO-Ag powders (Figure 4).

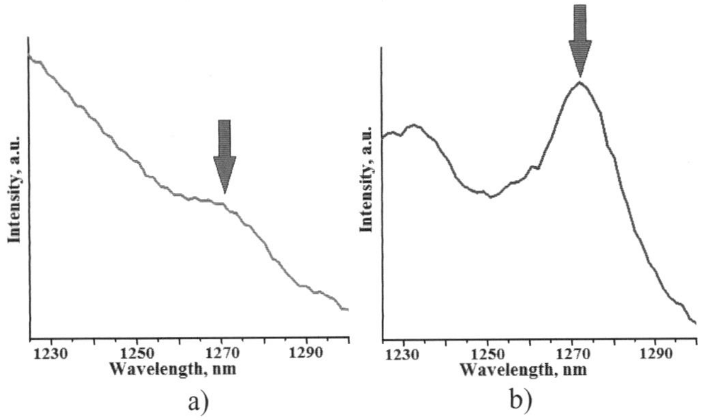

Figure 4. Photoluminescence spectra ($\lambda_{ex.}$ = 385) nm) of ZnO-SnO_2-Ag nanocomposites *1* (a) and *4* (b).

The singlet oxygen photogeneration proceeds on the surface of photoactive materials. Thus, the significant decrease of ZnO crystals size (and the corresponding growth of their specific area) at the addition of PVP into initial solutions (Figure 4) can be responsible for the observed strong increase of the intensity of singlet oxygen photogeneration. The difference in the powders morphologies is another important factor that significantly affects the intensity of singlet oxygen photogeneration.

4. Conclusion

ZnO-SnO2-Ag nanocomposites able to generate chemically active oxygen species under external lighting have been prepared by polymer-salt method. The powders morphology and crystal structure strongly affect the photoactive properties of prepared materials. The addition of polyvinylpyrrolidone into the initial mixtures decreases the average crystal sizes in prepared powders, changes the morphology of the materials, and enhance significantly their ability to generate chemically active oxygen species. The obtained results are promising for photocatalytic and environmental applications.

Acknowledgment

The work was funded by Russian Science Foundation (research project No. 20-19-00559).

References

[1] Evstropiev S K, Soshnikov I P, Kolobkova E V, Evstropyev K S, Nikonorov N V, Khrebtov A I, Dukelskii K V, Kotlyar K P, Oreshkina K V and Nashekin A V 2018 Polymer-salt synthesis and characterization of MgO-ZnO ceramic coatings with the high transparency in UV spectral range *Opt. Mater.* **82** 81-87

[2] Huheey J E, Keiter E A and Keiter R L 1993 *Inorganic Chemistry: Principles of Structure and Reactivity* (New York: HarperCollins)

[3] Evstropiev S K, Nikonorov N V, Kiselev V M, Saratovskii A S and Kolobkova E V 2019 Transparent photoactive ZnO-MgO-Ag2O films on glasses *Opt. Spectrosc.* **127** 314-21

[4] Evstropiev S K, Nikonorov N V, Saratovskii A S, Dukelskii K V, Vasiliev V N, Karavaeva A V and Soshnikov I P 2020 Photo-stimulated evolution of different structural forms of silver in solutions, composite and oxide coatings *J. Photochem. Photobiol.* A **403** 112858

[5] Bagrov I V, Kiselev V M, Evstropiev S K, Saratovskii A S, Demidov V V and Matrosova A S. 2020 Singlet oxygen generation in microcapillary optical elements with photoactive coatings *Opt. Spectrosc.* **128** 214-19

Effect of longitudinal epsilon-near-zero regime in dynamics of ultrashort laser pulses in hyperbolic metamaterials

Vladimir Novikov

Department of Physics, M. V. Lomonosov Moscow State University, Moscow, 119991, Russia

E-mail: vb.novikov@physics.msu.ru

Abstract. Optical properties of hyperbolic metamaterials (HMMs) are in stark contrast to properties of ordinary media that fuels interest to various applications of HMMs in photonics. Special attention is attributed to the epsilon-near zero regime (ENZ) of HMMs that is the spectral point in which real part of the permittivity of the HMM becomes zero. This is accompanied by the effects of field enhancement having far-reaching applications. Here we focus on the experimental and theoretical investigation of the propagation of an ultrashort laser pulse through the silver nanorod-based HMM slab in the spectral range over the ENZ. We revealed pronounced resonant change of the pulse delay in HMMs and the transition between the superluminal and slow pulse propagation at the ENZ spectral point. Observed dynamical phenomena are confirmed theoretically and attributed to unusual case when the spectral half of an ultrashort pulse has elliptical dispersion and another has the hyperbolic one. Special attention is payed to the propagation of chirped laser pulses in the HMMs.

1. Introduction

Taming of light dispersion by metamaterials benefits in a wide range of applications in photonics. Recently hyperbolic metamaterials (HMMs), which are uniaxial media with opposite signs of principle components of the dielectric tensor, have attracted much attention. Hyperbolic isofrequency surface of light in HMMs is topologically different from the conventional elliptic dispersion. It is infinite in k-space and supports high-k optical states making HMMs useful for enhancement of light emission [1] and second harmonic generation [2]. Hyperbolic dispersion is intrinsic to a number of natural materials [3] and artificial structures, for example, an array of metal nanorods in a dielectric matrix and metal/dielectric multilayers [4]. "Hyperbolic response" of artificial stuctures can be obtained in visible and near-infrared by tuning the volume fraction of metal in the composite [4]. Special interest is aimed to epsilon-near-zero (ENZ) regime of the HMMs, when real part of the principle component of the dielectric tensor of the metamaterial is zero. In the case of nanorod-based HMM, the extraordinary permittivity (ε_e) tends to zero, while ordinary one (ε_o) remains positive. In the ENZ spectral point the transition from the elliptic to the hyperbolic light dispersion takes place with the variation of the light wavelength.

Recently unusual dispersion properties of the HMMs sparked theoretical research efforts in the investigation of ultrashort laser pulse propagation in the HMMs [5]. The ENZ regime of multilayered HMMs was theoretically considered to realize soliton-like propagation of the femtosecond laser pulses in thin HMM [6]. Modulation of near-zero HMMs permittivity by the optical pump gives superior possibility for an all-optical modulation of the laser pulses that was predicted in multilayer HMMs [7] and demonstrated in nanorod-based ones [8]. In confined geometry of the HMM waveguide the coupling between forward and backward eigenmodes gives the superluminal and slow pulse propagation that was recently predicted [9]. Experimentally the slow-light effect was revealed in hexagonal boron nitride, which exhibits hyperbolic phonon-polaritons [10]. To the best of our knowledge the effect of ENZ resonance in the pulse dynamics in nanorod-based HMMs has not been considered till now, while impressive enhancement of nonlinear optical effect takes place in this regime under action of femtosecond laser radiation [11]. It makes the investigation of interaction of ultrashort pulses with the HMMs primarily important.

In this work we experimentally and numerically investigate the transmission of femtosecond laser pulses through the silver nanorod-based HMMs slab and demonstrate the pronounced resonant change of the pulse retardation in the metamaterial in the spectral vicinity of the ENZ point. The revealed phenomena are accompanied by superluminal and slow-light effects as well as the transition between these regimes. The theoretical description of the observed effects is proposed, together with the consideration of the propagation of chirped pulses in the HMMs.

2. Experiment

Experimental studies of the pulse dynamics in the HMM were performed for the structure consisted of ordered array of silver nanorods with the diameter of about $d = 34$ nm grown in anodic aluminum oxide template (AAO). The HMM thickness was about $L = 570$ nm while the thickness of AAO was 50 μm. This composite demonstrates resonant losses for p-polarized light near the wavelength of $\lambda_{ENZ} = 760$ nm, which is attributed to the ENZ regime of the considered HMM. It is confirmed by the calculation of the spectra of the principle components of the effective dielectric permittivity tensor of the HMM carried out within the effective medium approach [12]. In the ENZ regime the HMM demonstrates zero value of $Re(\varepsilon_e)$, corresponding to the longitudinal nanorod direction. Due to the sign change of $Re(\varepsilon_e)$ at λ_{ENZ}, the light with shorter wavelengths experiences the elliptic dispersion and the hyperbolic one for larger wavelengths.

The transmission of ultrashort optical pulses through the HMM was studied when using the femtosecond titanium: sapphire laser generating 80 fs pulses and tunable in the spectral range of $\lambda = 720 - 900$ nm. The retardation of the laser pulses in the HMM slab was measured using the cross-correlation method. Briefly, the laser beam was divided by a beam-splitter into two arms, the first beam passed through the HMM without focusing, the second one was retarded by a motorized delay-line. Both beams were focused by a spherical mirror to the same area of a thin nonlinear BBO crystal. The dependence of the light power of the second-harmonic beam generated under the cross-interaction of the two pump beams on the temporal delay gave the intensity cross-correlation function $C(\tau) = \int I_{HMM}(t) I_{ref}(t - \tau) dt$.

Lets consider the experimental angular dependence of the retardation of the pulse maximum in the HMM for the case of $\lambda = \lambda_{ENZ}$ [Fig. 1 (a)]. In this dependence the effect of the pulse delay in thick AAO substrate is subtracted from the total delay in the HMM+AAO structure. For this preliminary measurements of the pulse retardation in the AAO without silver nanorods were carried out. We observed a pronounced variation of the pulse delay in narrow angular ranges around 11° and 22°, where the negative retardation turns into the positive one [Fig. 1(a)]. The negative delay achieves −29 fs and indicates the superluminal pulse propagation. Positive delay of up to 42 fs corresponds to the pulse retardation. These values of the temporal effects correspond to the effective group index of about $n_g = -14$ and 24, respectively. The measured $C(\tau)$ functions for various angles of incidence (θ) are shown in Fig. 1(b) exhibiting superluminal effect, which is enhanced with the increase of θ for small angles of incidence. Then the transition to the pronounced pulse retardation takes place near $\theta = 11°$ and $\theta = 22°$. We demonstrate that the revealed effects are intrinsic to the ENZ regime of the HMM. This is illustrated in Fig. 1(c)

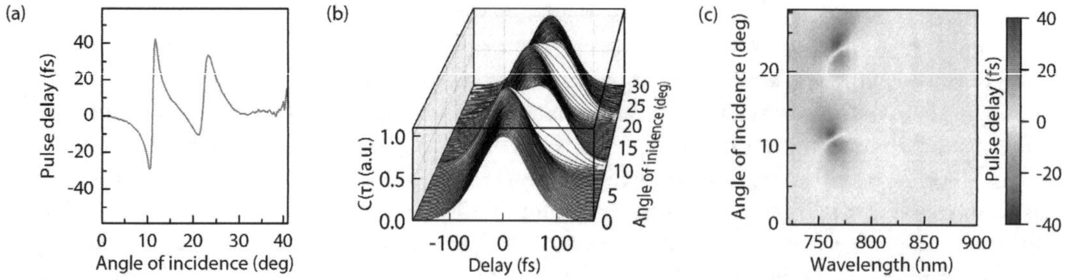

Figure 1. (a) Measured angular dependence of the pulse delay in the HMM for $\lambda = 760$ nm. (b) Measured cross-correlation functions for $\lambda = 760$ nm and various angles of incidence. (c) Experimental spectral-angular dependence of the pulse delay in the HMM.

where the measured pulse delay in the HMM with respect to the light wavelength and the angle of incidence is shown. As seen, the temporal effect is the most significant at $\lambda = \lambda_{ENZ}$.

3. Theory and discussion

To confirm the experimental findings described above, we carried out simulations of the pulse propagation though the HMM slab surrounded by air. Calculations were performed using the binding of fast Fourier transform and the transfer-matrix method assuming that the incident pulse had duration of 80 fs. We consider the chirped incident pulse $E(t) = (1/\sqrt{V})exp\left(-[1/(2V^2)]\,(t/\tau_0)^2 + i\varphi\right) exp\,(-i\omega t)$, where $V = \sqrt{1+\beta^2}$, modulation phase $\varphi = [\beta/(2V^2)]\,(t/\tau_0)^2 - (1/2)\arctan(\beta)$ and β is the chirp parameter.

The results of the calculation of the intensity profile of the transmitted p-polarized pulse with the central wavelength of $\lambda = \lambda_{ENZ}$ are shown in Fig. 2 for various β. Numerical simulation for

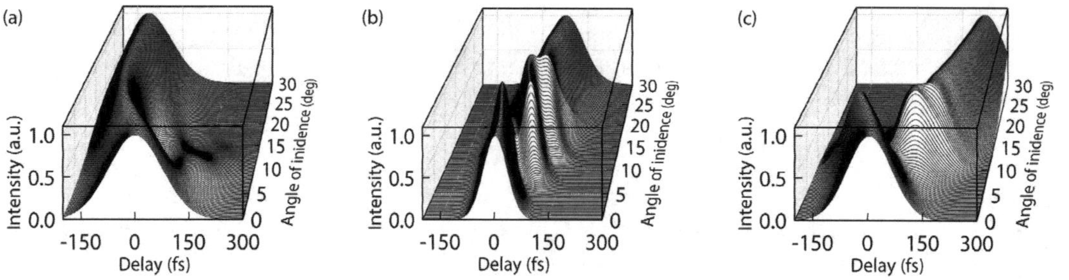

Figure 2. Calculated temporal dependencies of the normalized pulse intensity for various angles of incidence and chirp parameters: (a) $\beta = -2$, (b) $\beta = 0$, (c) $\beta = 1.5$.

the unchirped pulse ($\beta = 0$) exhibits the temporal features of the pulse in the HMMs similar to the experimental ones [Fig. 2(b)]. It demonstrates the negative pulse delay at the increase of θ up to 12° manifesting the superluminal-like effect with the pulse advance up to 40 fs. At this critical angle the pulse profile has two maxima and the pulse advance transforms to the pulse retardation. The similar feature takes place at the second angle of $\theta = 19°$ as in the experiment.

We assert that chirping of the ultrashort pulse has a significant effect in the pulse advance and its retardation in the HMM at the ENZ spectral point. For the negative $\beta = -2$ that corresponds to higher frequency at the trailing edge of the pulse we obtain a strong pulse advance in a wide range of the angles of incidence [Fig. 2(a)]. In turn, when chirp is positive calculations predict an enhancement of the pulse retardation [Fig. 2(c)]; it is accompanied by a "jump" from the pulse advance to its retardation at $\theta = 10°$ that does not appear for the negative chirp parameter.

Carried out simulations confirm the observation of the enhancement of superluminal and slow light propagation of ultrashort pulses in the spectral vicinity of the ENZ. The spectral dependence of the pulse advance/delay at inclined light incidence has a resonant form with the spectral width of about 11 nm that is in agreement with the experimental results [Fig. 1(c)].

We assert that disclosed dynamical effects are the result of (i) the beating of the pulse field inside the HMM and (ii) its negative phase velocity. Intrapulse beating originates from (i) two-band spectrum of outgoing pulse caused by enhanced optical losses near λ_{ENZ} and (ii) simultaneous elliptic and hyperbolic dispersion for various spectral components of the femtosecond laser pulse when it is tuned to the ENZ. The latter leads to significant phase difference between short- and long-wavelength spectral components of the transmitted pulse that gives superluminal fringe move inducing the effective laser pulse advance. We note that Fabry-Perot interference taken into account in simulations does not affect the revealed phenomena due to large optical losses near the ENZ. For chirped pulses the dominant mechanism of the temporal effects is the pulse reshaping caused by the absorption of leading or trailing edges of the pulse resulted from spectral asymmetry of light attenuation in the HMM near the ENZ.

4. Conclusion

Summing up, we observed superluminal and slow propagation of femtosecond laser pulses in metal nanorod-based HMM and the transition between these effects at the variation of the angle of incidence in the spectral vicinity of the ENZ spectral point of the metamaterial. We assert that these phenomena are inherent to the ENZ and disappear away from the ENZ wavelength. The proposed theoretical description of these effects is based on the intrapulse fringe move caused by the resonant absorption at the ENZ and neighborhood of the elliptic and the hyperbolic dispersion of light near the ENZ. We believe that our findings can be used in applications of the HMMs as subwavelength-thick tunable pulse delay lines. Moreover, the observed dynamical phenomena can have a significant effect in nonlinear optics of HMMs.

Acknowledgments

This work is supported by the Russian Science Foundation under grant No. 19-72-00118. V.B.N. gratefully acknowledges A.P. Leontiev and K.S. Napolskii for the fabrication of the experimental sample and thankful to Prof. T.V. Murzina for cooperation and encouraging discussions.

References

[1] Guo Y, Cortes C L, Molesky S and Jacob Z 2012 *Applied Physics Letters* **101** 131106
[2] Kolmychek I A, Malysheva I V, Novikov V B, Leontiev A P, Napolskii K S and Murzina T V 2020 *Phys. Rev. B* **102**(24) 241405
[3] Narimanov E E and Kildishev A V 2015 *Nature Photonics* **9** 214–216
[4] Poddubny A, Iorsh I, Belov P and Kivshar Y 2013 *Nature photonics* **7** 948–957
[5] Luan P G and Wu J L 2018 *AIP Advances* **8** 015106
[6] Kelly P and Kuznetsova L 2018 *Applied Physics B* **124** 1–8
[7] Kelly P and Kuznetsova L 2020 *OSA Continuum* **3** 3225–3236
[8] Nicholls L H, Rodríguez-Fortuño F J, Nasir M E, Córdova-Castro R M, Olivier N, Wurtz G A and Zayats A V 2017 *Nature Photonics* **11** 628–633
[9] Neira A D, Wurtz G A and Zayats A V 2015 *Scientific reports* **5** 1–7
[10] Yoxall E, Schnell M, Nikitin A Y, Txoperena O, Woessner A, Lundeberg M B, Casanova F, Hueso L E, Koppens F H and Hillenbrand R 2015 *Nature Photonics* **9** 674–678
[11] Kolmychek I A, Novikov V B, Malysheva I V, Leontiev A P, Napolskii K S and Murzina T V 2020 *Opt. Lett.* **45** 1866–1869
[12] Atkinson R, Hendren W R, Wurtz G A, Dickson W, Zayats A V, Evans P and Pollard R J 2006 *Phys. Rev. B* **73**(23) 235402

Long-Term Stable Structures Formed by Ion-Beam Modification of Silver Film for SERS Applications

Novikov S. M.[1, a)], Streletskiy O. A.[2], Doroshina N.V.[1], Yakubovsky D. I.[1], Mironov M. S.[1], Sychev V. V.[3], Voronov A. A.[1], Arsenin A. V.[1] and Volkov V. S.[1]

[1]Center for Photonics and 2D Materials, Moscow Institute of Physics and Technology, 141700 Dolgoprudny, Russia
[2]M.V. Lomonosov Moscow State University, Department of Physics, 119991 Moscow, Russia
[3]Lebedev Physical Institute of the Russian Academy of Sciences, 119991, Moscow, Russia

[a)]Corresponding author: novikov.s@mipt.ru

Abstract. Ensembles of silver nanoparticles (NPs) with size ~45 nm formed from the silver film using an ion beam modification are studied. The optical spectroscopy demonstrated that the fabricated ensembles of silver NPs keep stable their plasmonic properties in an ambient atmosphere for at least 39 days due to their monocrystalline nature. We use the scanning Raman microscope to map the SERS from Crystal Violet homogeneously adsorbed on these ensembles. It was found that the manufactured ensembles have a strong amplification factor, and this factor is preserved for these ensembles even after more than one month of storage in the surrounding atmosphere. Hereby, by ion beam modification of silver film, it is possible to fabricate the NPs with stable plasmonic properties and form nanostructured surfaces to be applied in sensor technologies and SERS.

1. Introduction

Controlled and reliable field enhancement (FE) effects associated with the excitation of plasmons in resonant metal nanostructures [1,2] is an essential aspect for the development of new sensors, especially for applications in surface-enhanced Raman spectroscopy (SERS) [3-5]. Noble metals nanoparticles (NPs) providing localized surface plasmon resonance (LSPR) are of increasing interest for application in optics, catalysis, electronics, sensing, for the development of ultra-bright and stable single-photon sources. The practically important issue is the stability of plasmonic properties, which often limits the use of some metals due to their chemical reactivity.

Silver is one of the metals exhibiting excellent plasmonic properties [6,7] that can be used in the above mentioned applications. There are many strategies for the fabrication of plasmonic nanostructures using silver. Different approaches such as particle formation by evaporation or sputtering, high-fluency ion implantation, chemical and photoreduction, production from colloids, etc. were suggested. Many of them allow the formation of nanomaterials with attractive efficiency. However, rapid oxidation or sulfidization of silver in the ambient atmosphere dramatically decreases all bonuses of this metal and cause complications in terms of practical applications [8, 9]. Recently, it has been shown that silver NPs formed by a cluster beam depositional technic exhibiting great stability of plasmonic properties and strong resistance to oxidization due to the monocrystalline nature [10-12]. In this work, we have demonstrated the possibility to fabricate monocrystalline silver NPs with stable plasmonic properties using a combination of dual magnetron sputtering system and ion beam modification technique [13-15]. This method allows us to control the size and surface coverage of NPs, and we anticipate that the obtained results are very promising for further sensing applications

2. Samples Fabrication

The coating process is carried out in a high-vacuum chamber with preliminary pumping up to 10^{-5} Torr, followed by the synthesis in an argon atmosphere at a residual pressure of about 10^{-4} Torr. The cover glass and Si are used as substrates. Before applying the coating, the surface of the substrates is preliminarily etched by the ion beam 1 keV energy for 5 min. A thin coating is deposited by RF magnetron with a silver target with a purity of 99.99% are used. The discharge power is approximately 30 W. The obtained coating consists of the film with a thickness of ~20 nm. The quoted "thickness" of the silver films is the average nominal coverage values measure by the quartz oscillator. After the deposition, the argon-ion irradiation of 150 eV energy and 20 mA current is applied to the deposited films to form the NPs. The ion beam is produced by Hall-effect ion source with cold hollow cathode Klan 53-M (Platar Corp.). Figure 1(a) shows schematically the sample fabrication process. Here we use electron-beam lithography (EBM) to make silver discs of about 500 nm diameter. The introduction of ion irradiation leads to the formation of discs consisting of nanoparticles, in which an average size is near 45 nm (Figure 1(b)). Thus, we show that the method allows not only to create an extended surface covered with nanoparticles, but also the ability to nanostructured the surface.

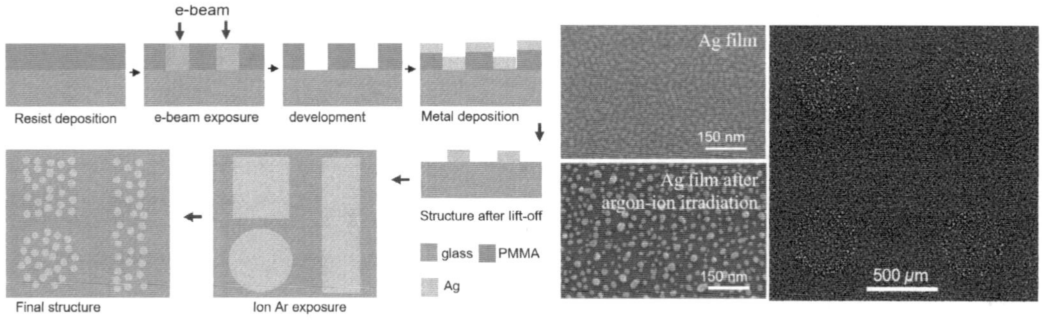

(a) (b)

FIGURE 1. (a) The schematic illustration of fabrication of sample with Ag NPs, (b) SEM images of Ag film with a thickness of ~20 nm deposited by RF magnetron on Si substrate (left-up). Formed Ag NPs after the argon-ion irradiation (left-bottom) and final structures with Ag NPs.

3. Samples Characterization

After the formation of NPs, the samples are characterized by UV-Vis-NIR spectrophotometer Agilent Technologies Cary 5000 (175-3300 nm). Optical spectra of the samples with ensembles of silver NPs are presented in Figure 2(a) and show the LSPR band with a maximum at near 418 nm. The intensity and spectral position of the plasmon band was monitored for 39 days. After 7 days, the resonance band becomes slightly red-shifted and its intensity was insignificantly decreased. In the period between 7 and 23 days, the intensity was slowly decreasing and the band became even more redshifted. After 39 days, the band intensity decreased to less than 10% from the origin intensity, thus demonstrating the excellent stability of the optical properties. The possible reason for the observed insignificant redshift and decrease in intensity can be the beginning of the formation of an oxide shell due to interaction with an ambient atmosphere. Additionally, we made the X-ray photoelectron spectroscopy (XPS) analysis. The spectra taken 0 and 14 days after NPs formations do not show any measurable difference in the peak positions or intensities (Figure 2a insert).

The Raman measurements were performed in order to estimate the applicability of the fabricated samples for SERS. We used a confocal Raman microscope Horiba LabRAM HR Evolution. The measurements were conducted with a laser wavelength of 633 nm, 600 lines/mm diffraction grating, integration time 0.5 sec., 100× objective (N.A.=0.9) and with the incident power ~0.3 mW. All samples were covered by 10^{-6} M water solution of Crystal Violet (CV) and dried under ambient conditions. The measurements were carried out on the sample kept in an ambient atmosphere for 39 days after the fabrication. The typical SERS spectra obtained from the sample with ensembles of Ag NPs (0, 14 and 39 days after the fabrication) are shown in Figure 2b. Spectra demonstrate relatively similar intensities for the samples with are kept in an ambient atmosphere for 0 and 14 days. The SERS intensity recorded for the sample after 39 days in the ambient atmosphere is a little bit less ~ 10-15% (Figure 2(b)), demonstrating high stability silver NPs even after prolonged exposure to the laboratory environment.

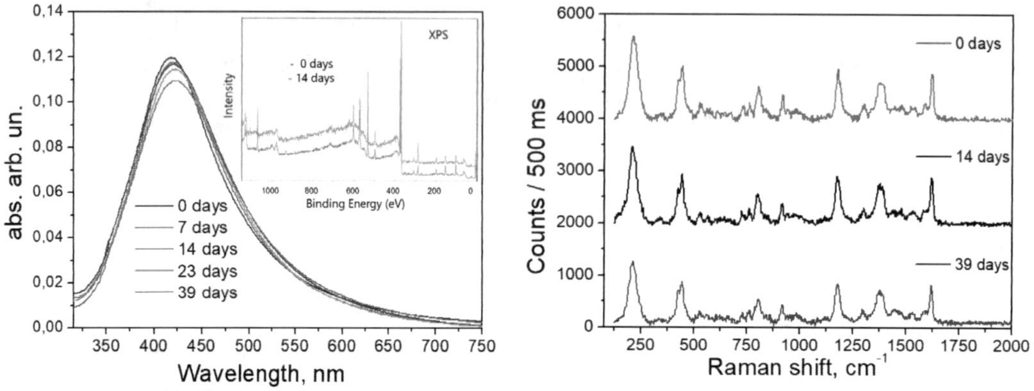

(a) (b)

FIGURE 2. (a) Evolution of the relative absorption of a sample with Ag NPs as a function of time. Insert XPS spectra obtained from the sample at the day of formation NPs and after 14 days. (b) SERS spectra of CV with concentration 10^{-6} M adsorbed on ensembles of silver NPs and obtained after 0, 14 and 39 days after the formation NPs on the glass substrate.

Finally, the analytical enhancement factor (EF) was used to compare a signal increase in SERS, with that in ordinary Raman under the same experimental parameters [16]. The average EF is estimated to be $\sim 0.82\times10^5$ for the ensembles of NPs after 0 days and $\sim 0.67\times10^5$, after 39 days, respectively. Thus, it is possible to create samples using all the advantages of silver and with the ability to use a long time.

4. Conclusion

We demonstrate the possibility of fabrication of the silver NPs with the great stability of plasmonic properties by the ion beam modification technique. The SERS characterization demonstrated that the manufactured ensembles have a strong amplification factor. The average EF is estimated to be $\sim 0.82\times10^5$ for the ensembles of NPs after 0 days and $\sim 0.67\times10^5$, after 39 days, respectively. Thus, it is possible to create samples using all the advantages of silver and the ability to use them for a long time. Considering the high stability of the plasmonic properties of our silver NPs, the presented technic can be interesting for the sensor technologies and SERS applications

Acknowledgments

We gratefully acknowledges the financial support from the Ministry of Science and Higher Education of the Russian Federation (Agreement No. 075-15-2021-606) and the Russian Foundation for Basic Research (20-07-00475 and 20-07-00840). We thank the Shared Facilities Center of the Moscow Institute of Physics and Technology for the use of their equipment.

References

[1] D. K. Gramotnev and S.I. Bozhevolnyi, Nature Photonics **4,** 83-91 (2010).
[2] S. Maier Plasmonics: Fundamentals and Applications. Springer (New York, 2007).
[3] J. Beermann, S. M. Novikov, O. Albrektsen, M. G. Nielsen, S. I. Bozhevolnyi, J. Opt. Soc. Am. B. **26,** 2370-2376 (2009).
[4] D. E. Tatarkin, D. I. Yakubovsky, G. A. Ermolaev, Y. V. Stebunov, A. A.Voronov, A. V. Arsenin, V. S. Volkov, and S. M. Novikov, Nanomaterials **10,** 164 (2020).
[5] J. Langer, D. Jimenez de Aberasturi, J. Aizpurua, R. A. AlvarezPuebla, B. Auguie, J. J. Baumberg, G. C .Bazan, S. E. J .Bell, A. Boisen, A. G. Brolo, J. Choo, D. Cialla-May, V. Deckert, L. Fabris, K. Faulds, F. J. García de Abajo, R .Goodacre, D. Graham, A. J. Haes, C. L. Haynes, et al. ACS Nano, **14**, 28-117 (2020).
[6] S. Vishnupriya, K. Chaudhari, R. Jagannathan, R. Pradeep, Part. Part. Syst. Char. **30,** 1056 (2013).
[7] N. A. Brazhe, A. B. Evlyukhin, E. A. Goodilin, A. A. Semenova, S. M. Novikov, S. I. Bozhevolnyi, B. N. Chichkov, A. S. Sarycheva, A. A. Baizhumanov, E. I. Nikelshparg, et al. Sci. Rep. **5,** 13793 (2015).
[8] W. Cao, H. E. Elsayed-Ali, Mater. Lett. **63,** 2263 (2009).
[9] M. Scuderi, M. Esposito, F. Todisco et. al. J. Phys. Chem. C **120**, 24314 (2016).
[10] S. M. Novikov, V. N. Popok, A. B. Evlyukhin, M. Hanif, P. Morgen, J. Fiutowski, J. Beermann, H.-G. Rubahn, S. I. Bozhevolnyi, J. Phys. Conf. Ser. **1092**, 012098 (2018).
[11] H. Hartmann, V. N. Popok, I. Barke I, V. von Oeynhausen and K-H. Meiwes-Broer, Rev. Sci. Instrum. **83,** 073304 (2012).
[12] S. M. Novikov, V. N. Popok, A. B. Evlyukhin, M. Hanif, P. Morgen, J. Fiutowski, J. Beermann, H.-G. Rubahn, S. I. Bozhevolnyi, Langmuir **33** 6062-6070 (2017).
[13] S. Mukherjee, CSIT **7**, 99-104 (2019).
[14] D. Dobrev, Thin Solid Films, **92**, 41-53 (1982).
[15] L. Ziqi L and C. Feng Appl. Phys. Rev**.** **4**, 011103 (2017).
[16] E. C. Le Ru, E. Blackie, M. Meyer, P. G. Etchegoin, J. Phys. Chem. C **111,** 13794-13803 (2007).

Surface ferromagnetism as a basis for the creation of planar metamaterials.

A.K. Nukhov[1,2]

[1]Moscow Institute of Physics and Technology, Institutskyi 9, 141700 Dolgoprudnyi, Russia
[2]Amirkhanov Institute of Physics of Dagestan Federal Research Centre RAS, 367015, Makhachkala, Russia

nukhov1984@mail.ru

Abstract. In this work, we study the influence of the surface on the energy of spin excitations of a ferromagnetic crystal in the Heisenberg model, with a particular focus on the possibility of using this effect to create planar metamaterials.

1. Introduction

Recently, a lot of attention has been focused on the creation of planar metamaterials and metasurfaces. With one of the successful applications being solar panels, the study of surface effects has gained particular relevance. One of the interesting research direction of creating metamaterials is to study the magnetic properties of nanostructures, superlattices and interfaces, taking into account the effect of surface energy. In metamaterials, nonopatterned and relatively thin layers of materials are used, wich makes their magnetic properties significantly different from those of a bulk material due to drastic increase of surface to volume ratio. As a result, a rigorous quantum theory for ferromagnetism which explicitly accounts for surface effects should be developed.

In [1], the effect of surface energy in the classical theory of spin waves was considered using a ferromagnetic crystal as an example, and it was shown that taking into account the local geometry of the surface leads to damping of spin waves. In [2-3], various aspects of the surface effect on the formation of magnetic and electronic properties of low-dimensional systems are considered.

In this work, using the Bogolyubov quantum mechanical hamiltonian [8] for localized electronic excitations of the crystal system, we obtained the hamiltonian of spin excitations of the Heisenberg model taking into account the surface energy.

2. Methods and results

Let's consider the derivation of the microscopic Heisenberg hamiltonian with allowance for the surface energy for a crystal, in the sites f and f' of which there are electron shells with z electrons in each with an open and local layer.

The spin Hamiltonian of such a system $\hat{H}$ can be written [6] in the form:

$$\hat{H} = \sum_{\begin{pmatrix} f,\lambda \\ f',\lambda' \end{pmatrix}, \sigma} L(f,\lambda;f',\lambda')\hat{a}^{+}_{f\lambda\sigma}\hat{a}_{f'\lambda'\sigma} + \frac{1}{2}\sum_{\begin{pmatrix} f_1\lambda_1;f_2\lambda_2 \\ f_1'\lambda_1';f_2'\lambda_2' \end{pmatrix}, \sigma_1,\sigma_2} F(f_1,\lambda_1;f_2,\lambda_2;f_1',\lambda_1';f_2',\lambda_2')\hat{a}^{+}_{f_1\lambda_1\sigma_1}\hat{a}^{+}_{f_2\lambda_2\sigma_2}\hat{a}_{f_2'\lambda_2'\sigma_2'}\hat{a}_{f_1'\lambda_1'\sigma_1'} \quad (1)$$

where λ , λ' -orbital states of electrons, σ - spin quantum number ($\sigma = 1/2$)
An additional energy operator $\hat{H}_S$ is introduced in (1) taking into account the finite dimensions of the system (crystal surface)

$$\hat{H} = \hat{H}_V + \hat{H}_S \quad (2)$$

$\hat{H}_S$ can be written:

$$\hat{H}_S = \varepsilon_{0S} - \frac{1}{2} \sum_{(\varphi_1 l_1 \neq \varphi_2 l_2)}' A(\varphi_1, l_1; \varphi_2, l_2) \hat{b}^+_{\varphi_1 l_1 \sigma_1} \hat{b}_{\varphi_1 l_1 \sigma_2} \hat{b}^+_{\varphi_2 l_2 \sigma_2} \hat{b}_{\varphi_2 l_2 \sigma_1} \quad (3)$$

Hamiltonian (3) takes into account the ground state and the interaction of only surface ferromagnons (S-magnons, SS - interaction). In addition, there is SV - interaction, then:

$$\hat{H} = \hat{H}_V + \hat{H}_S + \hat{H}_{SV} + \hat{H}_{VS} + \widetilde{H} \quad (4)$$

Where $\hat{H}_{SV}$ and $\hat{H}_{VS}$ operators SV and VS interactions, which can be written as:

$$\hat{H}^2_{VS} = \frac{1}{2} \sum_{\begin{pmatrix} f_1, \lambda_1; f_2, \lambda_2 \\ \varphi_1, l_1; \varphi_2, l_2 \end{pmatrix}, \sigma_1, \sigma_2} F_{VS}(f_1, \lambda_1; f_2, \lambda_2; \varphi_1, l_1; \varphi_2, l_2) \hat{a}^+_{f_1 \lambda_1 \sigma_1} \hat{a}^+_{f_2 \lambda_2 \sigma_2} \hat{b}_{\varphi_2 l_2 \sigma_2} \hat{b}_{\varphi_1 l_1 \sigma_1} \quad (5a)$$

$$\hat{H}^2_{SV} = \frac{1}{2} \sum_{\begin{pmatrix} \varphi_1 l_1; \varphi_2 l_2 \\ f_1, \lambda_1; f_2 \lambda_2 \end{pmatrix}, \sigma_1, \sigma_2} F_{SV}(\varphi_1, l_1; \varphi_2, l_2; f_1, \lambda_1; f_2, \lambda_2) \hat{b}^+_{\varphi_1 l_1 \sigma_1} \hat{b}^+_{\varphi_2 l_2 \sigma_2} \hat{a}_{f_2 \lambda_2 \sigma_2} \hat{a}_{f_1 \lambda_1 \sigma_1} \quad (5b)$$

$$\hat{H}^1_{VS} = \frac{1}{2} \sum_{\begin{pmatrix} f, \lambda \\ \varphi, l \end{pmatrix}, \sigma} L_{VS}(f, \lambda; \varphi, l) \hat{a}^+_{f \lambda \sigma} \hat{b}_{\varphi l \sigma} \quad (6a)$$

$$\hat{H}^1_{SV} = \frac{1}{2} \sum_{\begin{pmatrix} \varphi, l \\ f, \lambda \end{pmatrix}, \sigma} L_{SV}(\varphi, l; f, \lambda) \hat{b}^+_{\varphi l \sigma} \hat{a}_{f \lambda \sigma} \quad (6b)$$

$$\hat{H}_{VS} = \hat{H}^1_{VS} + \hat{H}^2_{VS} \quad , \quad \hat{H}_{SV} = \hat{H}^1_{SV} + \hat{H}^2_{SV} \quad (7)$$

($f\lambda$) chosen designations for V, (φl) chosen designations for S.
Condition for surface homeopolarity $\sum_{\sigma} \hat{b}^+_{\varphi l \sigma} \hat{b}_{\varphi l \sigma} = 1 \quad , \quad (l = l_1, l_2, ..l_z)$ (8)

Here it is assumed that ordinary Bose field operators commute with each other, since the surface and bulk states (wave functions) are different and the corresponding operators acting on these states must commute. We make the Bogolyubov transformation and consider separately the nodes of the surface and volume. For $\hat{H}_V$ and $\hat{H}_S$ we get the following expressions:

$$\hat{H}_V = U_{0V} - \sum_{(f_1, \lambda_1 \neq f_2, \lambda_2)} A_V(f_1 \lambda_1; f_2, \lambda_2) \, {}^V\hat{\vec{S}}_{f_1 \lambda_1} \, {}^V\hat{\vec{S}}_{f_2 \lambda_2} \quad (9)$$

$$\hat{H}_S = U_{0S} - \sum_{(\varphi_1, l_1 \neq \varphi_2, l_2)} A_S(\varphi_1, l_1; \varphi_2, l_2) \, {}^S\hat{\vec{S}}_{\varphi_1 l_1} \, {}^S\hat{\vec{S}}_{\varphi_2 l_2} \quad (10)$$

Further calculations are related to the use of similar expressions for the *SV* cross interaction terms. The members $\hat{H}^1_{VS}$ and $\hat{H}^1_{SV}$ form in *SV* interaction a kind of "basic" force and are associated with the existence of two-particle interaction, when there are interconversions between the surface and the volume. However, such terms, even if they are not equal to zero, lead only to some displacement of the constant U_0 and are not essential.

Thus, the transformed hamiltonian has the form:

$$
\begin{aligned}
\hat{H} = U_0 &- \sum_{1,2}{}' A_V(f_1\lambda_1; f_2\lambda_2)\, {}^V\hat{\vec{S}}_{f_1\lambda_1}\, {}^V\hat{\vec{S}}_{f_2\lambda_2} - \sum_{1,2}{}' A_S(\varphi_1 l_1; \varphi_2 l_2)\, {}^S\hat{\vec{S}}_{\varphi_1 l_1}\, {}^S\hat{\vec{S}}_{\varphi_2 l_2} \\
&+ \sum_{1,2}{}'' F_{VS}(f_1\lambda_1; f_2\lambda_2; \varphi_1 l_1; \varphi_2 l_2)({}_1^i\hat{\vec{S}}_{f_1\varphi_1\lambda_1 l_1})({}_1^i\hat{\vec{S}}_{f_2\varphi_2\lambda_2 l_2}) \\
&+ \sum_{1,2}{}'' F_{SV}(\varphi_1 l_1; \varphi_2 l_2; f_1\lambda_1; f_2\lambda_2)({}_1^i\hat{\vec{S}}_{f_1\varphi_1\lambda_1 l_1})^+({}_1^i\hat{\vec{S}}_{f_2\varphi_2\lambda_2 l_2})^+ \\
&+ \sum_{1,2}{}'' \widetilde{F}_1(f_1\lambda_1; \varphi_2 l_2; f_2\lambda_2; \varphi_1 l_1)({}_0^i\hat{\vec{S}}_{f_1\varphi_1\lambda_1 l_1})({}_0^i\hat{\vec{S}}_{f_2\varphi_2\lambda_2 l_2}) \\
&+ \sum_{1,2}{}'' \widetilde{F}_2(\varphi_1 l_1; f_2\lambda_2; \varphi_2 l_2; f_1\lambda_1)({}_0^i\hat{\vec{S}}_{f_1\varphi_1\lambda_1 l_1})^+({}_0^i\hat{\vec{S}}_{f_2\varphi_2\lambda_2 l_2})
\end{aligned} \tag{11}
$$

Here: $\sum_{1,2}{}' \equiv \sum_{\substack{(f_1\lambda_1 \neq f_2\lambda_2) \\ or \\ (\varphi_1 l_1 \neq \varphi_2 l_2)}} ; \sum_{1,2}{}'' \equiv \sum_{\substack{f_1\lambda_1 \neq f_2\lambda_2 \\ \varphi_1 l_1 \neq \varphi_2 l_2}}$ (11a)

Using the Holstein - Primakoff transformations, we will make the transition from spin operators to the operators of creation and annihilation of spin excitations [9], and, using some approximate expressions, we transform Hamiltonian (11) to the form:

$$\hat{H} = \hat{H}_0 + \hat{H}_{\text{int}} \tag{12}$$

Where $\hat{H}_0$ and $\hat{H}_{\text{int}}$ are respectively equal:

$$
\begin{aligned}
\hat{H}_0 = E_0 &+ \sum_j \left(\mu_0 B + L_j(0)\right) . \sum_{\xi_j} \mu^+_{\xi_j} \mu^-_{\xi_j} - \frac{1}{4} \sum_p \sum_{j,j'} \tilde{L}^p(0) \sum_{\xi_j \xi'_j}{}' \mu^+_{\xi_j \xi'_j}(p) \mu^-_{\xi_j \xi'_j}(p) - \\
&- \sum_j S_j \sum_{1,2}{}' (\xi_j) A_j\left(\vec{\xi}_{j_1} - \vec{\xi}_{j_2}\right) \mu^+_{\xi_{j1}} \mu^-_{\xi_{j2}} + \frac{1}{4} \sum_p S_p \sum_{j,j'} \sum_{1,2}{}'' (\xi_j, \xi'_j) \times \\
&\times F_p\left(\vec{\xi}_{j_1} - \vec{\xi}_{j_2}; \vec{\xi}'_{j_1} - \vec{\xi}'_{j_2}\right)\left(1 - \delta_{jj'}\right) \mu^+_{\xi_{j1}\xi'_{j1}}(p) \mu^-_{\xi_{j2}\xi'_{j2}}(p)
\end{aligned} \tag{13}
$$

$$
\begin{aligned}
\hat{H}_{int} = &- \frac{1}{2} \sum_j \sum_{1,2}{}' (\xi_j) A_j\left(\vec{\xi}_{j_1} - \vec{\xi}_{j_2}\right) \mu^+_{\xi_{j1}} \mu^-_{\xi_{j1}} \mu^+_{\xi_{j2}} \mu^-_{\xi_{j2}} + \\
&+ \frac{1}{8} \sum_p \sum_{j,j'} \sum_{1,2}{}'' (\xi_j, \xi'_j) \cdot F_p\left(\vec{\xi}_{j_1} - \vec{\xi}_{j_2}; \vec{\xi}'_{j_1} - \vec{\xi}'_{j_2}\right)\left(1 - \delta_{jj'}\right) \times \\
&\times \mu^+_{\xi_{j1}\xi'_{j1}}(p) . \mu^-_{\xi_{j1}\xi'_{j1}}(p) \mu^+_{\xi_{j2}\xi'_{j2}}(p) \mu^-_{\xi_{j2}\xi'_{j2}}(p)
\end{aligned} \tag{14}
$$

3. Discussion and conclusions

Let's consider the case of small excitations and consider the energy spectrum of an isotropic ferromagnet with a surface. In this case, $\hat{H}_{\text{int}}$ it can be represented as an excitation operator. Then, in the zeroth approximation, the energy spectrum of spin excitations can be determined by the diagonalized operator:

$$\Delta\hat{H} = \sum_j \sum_{k_j} \varepsilon_j(k) . \mu_k^{j^+} \mu_k^{j^-} + \sum_p \sum_{j,j'} \sum_{k_j} \sum_{k',j'} \widetilde{\varepsilon}^p_{jj'}(k,k') \mu_{kk'}^{jj'^-}(p) \mu_{kk'}^{jj'^+}(p) , \tag{15}$$

where $\varepsilon_j(k) = \mu_0 B + L_j(0) - L_j(k)$, $\widetilde{\varepsilon}^p_{jj'}(k,k') = -\frac{1}{4}\left[\widetilde{L}^p_{jj'}(0) - \widetilde{L}^p_{jj'}(k,k')\right]$

$$L_j(k) \equiv S_j \sum_{\xi_i \neq 0} A_j(\vec{\xi}_j) \exp(i\vec{k}_j \vec{\xi}_j) \quad , \quad L^p_{jj'}(k,k') \equiv S_p \sum_{\xi_i \neq 0} \sum_{\xi'_j \neq 0} F^p(\vec{\xi}_i, \xi'_j) \exp(i\vec{k}_j \vec{\xi}_j + i\vec{k}'_j \vec{\xi}'_j) ,$$

$$\widetilde{L}^p_{jj'}(k,k') \equiv L^p_{jj'}(k,k')(1 - \delta_{jj'}) \quad , \quad \widetilde{L}^p_{jj'}(0) \equiv \widetilde{L}^p_{jj'}(00)$$

Note that, in the zeroth approximation, the energy spectrum of spin excitations is determined here under the condition of rigid pinning of the ferromagnet ions at the lattice sites. If we take into account the possibility of their displacement from equilibrium positions, it is necessary to consider the dependence of the exchange integrals on these displacements. At small displacements of atoms, the exchange integrals are replaced by series and, after performing elementary transformations, one can obtain the following expressions for the operators of spin (pseudospin) -lattice interaction:

$$\hat{H}_{s.vib} = -\sum_{j}{}' \sum_{k,q,g}{}' (j)\left[D_j\left(\vec{k}-\vec{q}\right)-D_j(k)\right]\left(b_q^j + b_{-q}^{j+}\right)\mu_k^{j+}\mu_{k-q+g}^{j'-}$$

$$\begin{aligned}\hat{H}_{ps.vib} = &-\sum_{p}\sum_{jj'}\sum_{k,q,g}{}'(j)\sum_{k',q',g'}{}'(j).\left(1-\delta_{jj'}\right)\left[\widetilde{D}_j^p\left(\vec{k}-\vec{q}\right)-\widetilde{D}_j^p(\vec{k})\right]\\ &\times\left[\widetilde{D}_{j'}^p\left(\vec{k}'-\vec{q}'\right)-\widetilde{D}_{j'}^p(\vec{k}')\right]\left(b_q^j + b_{-q}^{j+}\right)\left(b_{q'}^{j'} + b_{-q'}^{j'+}\right)\mu_{kk'}^{jj'+}(p)\mu_{k-q+g,k'-q'+g'}^{jj'-}(p)\end{aligned} \quad (16)$$

The operator $H_{s.vib}$ in (16) is the operator of the magnon-phonon interaction, and the operator $H_{ps.vib}$ is the operator of the pseudospin-phonon interaction with the participation of two pseudomagnons and two phonons.

As we can see, taking into account the surface energy leads to the appearance of additional terms in the original Hamiltonian and has a significant effect on the energy state of the entire system.

To create metamaterials, nanostructures, including magnetic ones, are used. These are magnetic tapes, strips, filaments, nanoparticles and other low-dimensional systems.
For such systems, the effect of surface energy is of great importance. In this regard, the given theory is of great practical importance for the development of technologies for producing metamaterials

Acknowledgments
We gratefully acknowledge financial support from the Ministry of Science and Higher Education of the Russian Federation (No. 0714-2020-0002)

References

[1] A.K. Nukhov, et. al. 2011. Moscow Univ. Phys. 66, 416–421.
[2] A.K. Nukhov 2018 J. Phys.: Conf. Ser. 1092 012101.
[3] A K Nukhov et. al. 2020 J. Phys.: Conf. Ser. 1461 012122.
[4] K. Bindes, P. C.Hohenberg, 1972. Pys. Rev., B. N.9, P. 3461-3487.
[6] T. Kaneyoshi, 1996. Pys. Rev., B. V. 52, N.9, P. 7304-7310.
[7] S. Kittel, 1978. M. Mir - p. 792.
[8] N.N. Bogolyubov. 2005–2008. Collection of scientific works in 12 volumes, Nauka, Moscow.
[9] Holstein T., Primakoff H., 1940. Phys.Rev. 58, 1098.

Journal of Physics: Conference Series **2015** (2021) 012101 doi:10.1088/1742-6596/2015/1/012101

Purcell enhancement of fluorescence from silicon-vacancy color centers in Mie-resonant luminescent diamond particles

Dmitry V Obydennov[1], Ekaterina I Elyas[1], Daniil A Shilkin[1], Vitaly V Yaroshenko[2], Dmitriy A Zuev[2], Evgeny V Lyubin[1], Evgeny A Ekimov[3], Oleg S Kudryavtsev[4], Igor I Vlasov[4,5], Andrey A Fedyanin[1]

[1] Quantum Electronics Department and Quantum Technology Center, Faculty of Physics, Lomonosov Moscow State University, Moscow 119991, Russia
[2] Faculty of Physics and Engineering, ITMO University, St. Petersburg 191002, Russia
[3] Institute for High Pressure Physics, Russian Academy of Sciences, Troitsk 142190, Russia
[4] Prokhorov General Physics Institute, Russian Academy of Sciences, Moscow 119991, Russia
[5] National Research Nuclear University MEPhI, Moscow 115409, Russia

E-mail: obydennov@nanolab.phys.msu.ru

Abstract. Over the past two decades, nanosized diamond particles with various luminescent defects have found numerous applications in many areas from quantum technologies to medical science. The size and shape of diamond particles can affect drastically the luminescence of embedded color centers. Here we study diamond particles of 250–450 nm in size containing silicon-vacancy (SiV) centers. Using dark-field scattering spectroscopy, we found that fundamental Mie resonances are excited in the spectral range of interest. We then measured the fluorescence saturation curves under continuous excitation to estimate the effects of the excitation and Purcell factor enhancement on the luminescent properties of the studied particles. The results show that the saturation excitation intensity differs by several times for particles of different sizes which is well explained by the numerical model that takes into account both the Parcell factor enhancement and resonant excitation.

1. Introduction

The ability of the environment to influence the spontaneous emission rate has been well known since the middle of the last century and is the essence of the Purcell effect. Recent interest in this field was triggered by the active development of quantum optics and the growing demand for efficient single-photon sources with controllable directionality [1, 2]. Among the most promising candidates for this role are color centers in diamond providing highly stable non-bleaching luminescence at room temperature [3].

A quantitative measure of the Purcell effect, known as the Purcell factor, is defined as the ratio of the spontaneous emission rate of an emitter placed in the environment and the one of the same emitter in an infinite homogeneous medium. For subwavelength particles in the absence of an external resonator, it is usually much less than unity. A number of papers, however, predicted an increase in the Purcell factor at the excitation of fundamental morphological resonances in the hosting particles [4, 5, 6]. The experimentally observed dispersion in the emission rate of chromium [7] and nitrogen-vacancy [8] color centers in subwavelength diamond crystals was considered to be a manifestation of these resonances.

In this contribution, we report on an experimental study of the Purcell enhancement of fluorescence from silicon-vacancy color centers in Mie-resonant luminescent diamond particles. We characterize our samples using dark-field scattering spectroscopy and measure the saturation power as a function of the particle size. The saturation power, which is determined both by the resonant field enhancement at the pump frequency and the transition rate at the luminescence frequency, is additionally calculated numerically. Finally, we compare the obtained results and discuss their significance for future research.

2. Results and discussion

The studied sample of diamond particles was synthesized from a mixture of adamantane ($C_{10}H_{16}$) and tetraphenylsilane ($C_{24}H_{20}Si$) at a mass ratio of 80:1 at a temperature of 1900 K and a pressure of 8–9 GPa. After the purification process, diamonds were dispersed on a glass substrate from a water suspension.

The schematic representation of the luminescence experiments is presented in Figure 1a. For the excitation, continuous-wave laser radiation at a wavelength of 532 nm was used. The excitation beam was directed to the sample from the side of the substrate using an objective lens with a numerical aperture (NA) of 1.2 at an aperture filling factor corresponding to the resulting beam NA of 0.9. The fluorescence was collected in reflection geometry by the same objective and directed into an Ocean Optics QEPro spectrometer. A typical luminescence spectrum is shown in Figure 1b. By recording the luminescence signal at different excitation power values, we obtained the saturation curves for each particle (see Figure 1c). The saturation curves were fitted by the law $I = I_\infty P/(P + P_{\text{sat}}) + kP$, where I is the luminescence intensity, I_∞ is the saturation luminescence intensity, P is the excitation power, P_{sat} is the saturation excitation power, and kP represents the background signal. In this way, the saturation power P_{sat} was measured for 17 particles.

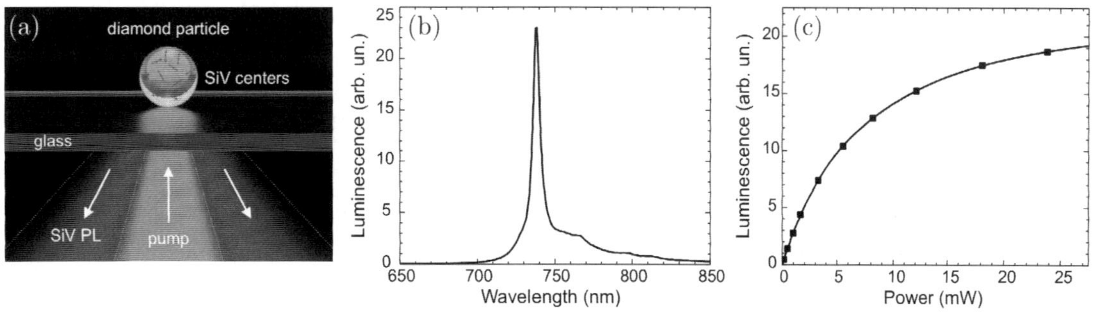

Figure 1. (a) Schematic of the experiment; SiV, silicon-vacancy; PL, photoluminescence. (b) Typical luminescence spectrum of an individual diamond particle. (c) Saturation of SiV centers luminescence. Squares — measured luminescence signal, solid line — approximation.

To reveal the fundamental modes of the studied particles, dark-field scattering spectroscopy was performed. The experimental setup is described in [9], Supporting information. After the measurements were completed, we compared the measured spectra with numerically calculated ones (see Figure 2). Based on this comparison, we have assigned an effective size for each particle (see [10] for details).

To estimate the dependence of the saturation power on particle size, we performed numerical calculations for spherical diamond particles on a glass surface. For each emitter, the saturation

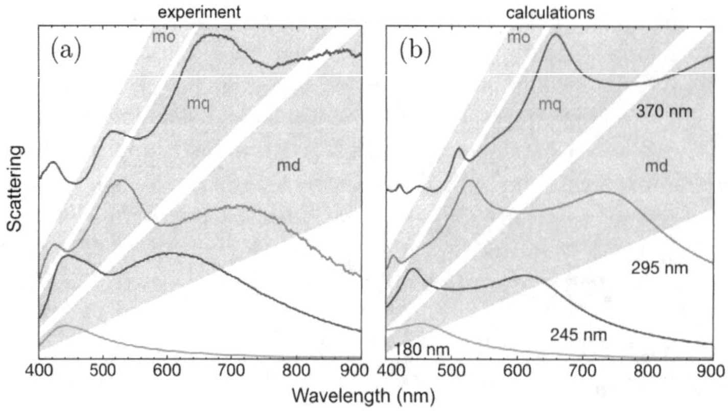

Figure 2. Scattering spectra of individual particles. Spectra are vertically shifted for better visualization. (a) Measured scattering spectra of individual particles; (b) calculated scattering spectra for spherical diamond spheres on a glass substrate. The sizes indicated on the graph correspond to effective particle diameters in the experiments. Notations 'md', 'mq', and 'mo' indicate magnetic dipole, magnetic quadrupole, and magnetic octupole resonances.

power is proportional to the spontaneous transition rate and inversely proportional to the excitation enhancement. Figure 3 shows the calculated electric field under Gaussian-beam excitation for diamond spheres of three resonant diameters.

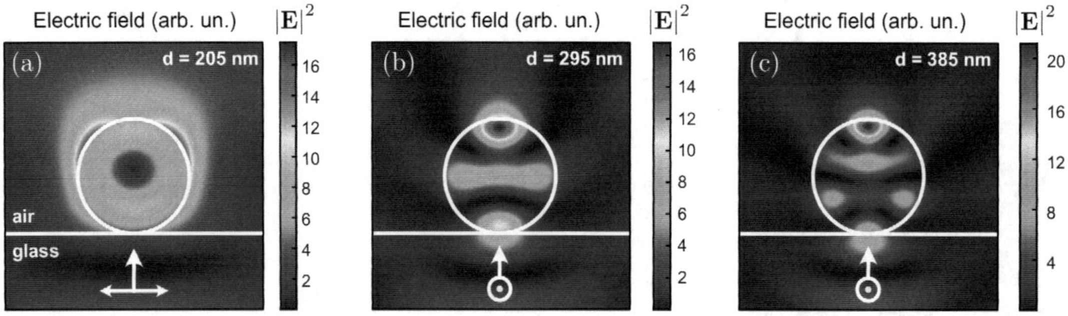

Figure 3. Resonant enhancement of the excitation field at 532 nm inside diamond spheres. (a), (b) and (c) — Electric field distribution in spherical diamond particles of diameter 205, 295, and 385 nm correspondingly. The direction and polarization of the excitation beam are shown at the bottom. The numerical aperture of the excitation beam is 0.42. The beam is focused at the center of the spheres.

The spontaneous transition rate can change due to the Purcell effect. For each sphere diameter, we have simulated an electric dipole oscillating at the frequency of SiV transition. The Purcell factor was calculated for different positions inside the sphere and three orthogonal polarization of the emitter. We have additionally calculated the integral far-field intensity directed into the experimental collection angles. The results of the simulations showing the excitation of the magnetic dipole and magnetic quadrupole resonances by an embedded dipole source are presented in Figure 4.

According to our direct measurements of the excited state lifetime in the studied particles, the spontaneous emission rate varies only for 5% that can be explained by the low quantum yield

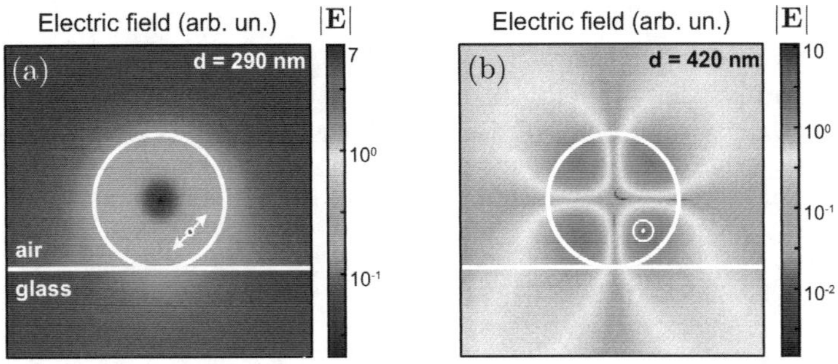

Figure 4. Dipole radiation in spherical diamond particles. (a), (b) — Electric field distribution in spherical diamond particles of diameter 290 and 420 nm. The positions and polarization of the dipoles are marked. The part of non-resonant fields is removed by apodization in the time domain.

of SiV centers. Taking it into account, we can omit the emission rate variations and estimate the saturation power of an individual emitter as $P_{\text{sat}} \propto \text{const}/\left|E\left(\lambda_{\text{exc}}\right)\right|^2$, where $\left|E\left(\lambda_{\text{exc}}\right)\right|^2$ is the excitation electric field inside the particle at $\lambda_{\text{exc}} = 532$ nm. To estimate the total saturation power of an ensemble of SiV centers, we calculated the averaged saturation power as follows:

$$\langle P_{\text{sat}} \rangle \propto \frac{\sum_{\alpha} \int_V I_{\alpha}\left(\lambda_{\text{em}}\right) / \left|E\left(\lambda_{\text{exc}}\right)\right|^2 dV}{\sum_{\alpha} \int_V I_{\alpha}\left(\lambda_{\text{em}}\right) dV},$$

where $I_{\alpha}\left(\lambda_{\text{em}}\right)$ is the far-field intensity of a dipole oscillating at $\lambda_{\text{em}} = 738$ nm and $\alpha = \{1, 2, 3\}$ represents the dipole polarization. The term $I_{\alpha}\left(\lambda_{\text{em}}\right)$ was introduced as a weighting factor.

Figure 5 shows the resulted dependence in comparison with the experimental results. Although experimental data are not consistent with the predicted dependence, the observed dispersion of the measured values is in agreement with the model. We conclude that these measurements reveal the effect of Mie resonances, which is much less apparent in life-time measurements due to the low quantum efficiency of the studied centers. This effect, however, should not be omitted when single-photon emission from isolated particles is studied and the lifetime dispersion is interpreted.

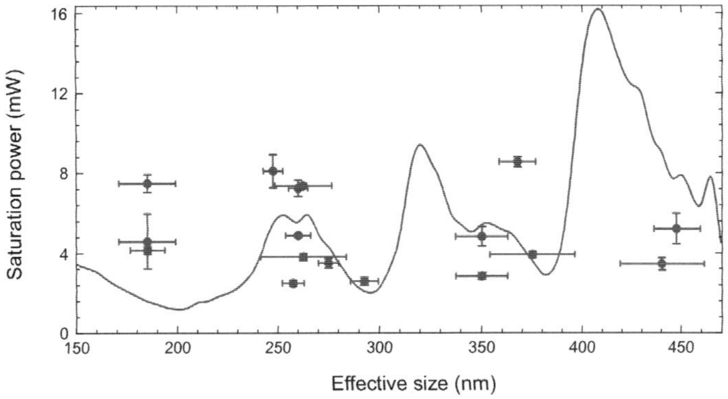

Figure 5. Dependence of the saturation power on the diamond particle size. Circles — experiment, solid line — calculations.

Conclusions

In conclusion, we have experimentally demonstrated the effect of Mie resonances of subwavelength diamond particles on the saturation of fluorescence from their SiV centres. The results demonstrate that the efficiency of diamond-based devices can be improved by choosing particles of a proper size.

Acknowledgments

This work was partially supported by the Russian Foundation for Basic Research (# 20-02-00715, calculations and spectroscopy experiments), the Russian Science Foundation (# 19-12-00407, synthesis, preparation and characterization of diamond sample), and the Development program of the Interdisciplinary Scientific and Educational School of Lomonosov Moscow State University "Photonic and Quantum technologies. Digital medicine". D.V.O., D.A.S., and E.V.L. acknowledge support from the MSU Quantum Technology Center.

References

[1] Koenderink A F 2017 *ACS Photonics* **4** 710–722
[2] Rutckaia V, Heyroth F, Novikov A, Shaleev M, Petrov M and Schilling J 2017 *Nano Lett.* **17** 6886–6892
[3] Bradac C, Gao W, Forneris J, Trusheim M E and Aharonovich I 2019 *Nat. Commun.* **10** 1–13
[4] Greffet J J, Hugonin J P, Besbes M, Lai N, Treussart F and Roch J F 2011 *arXiv preprint arXiv:1107.0502*
[5] Zambrana-Puyalto X and Bonod N 2015 *Phys. Rev. B* **91** 195422
[6] Shilkin D A, Shcherbakov M R, Lyubin E V, Katamadze K G, Kudryavtsev O S, Sedov V S, Vlasov I I and Fedyanin A A 2017 *ACS Photonics* **4** 1153–1158
[7] Castelletto S and Boretti A 2011 *Opt. Lett.* **36** 4224–4226
[8] Zalogina A, Savelev R, Ushakova E, Zograf G, Komissarenko F, Milichko V, Makarov S, Zuev D and Shadrivov I 2018 *Nanoscale* **10** 8721–8727
[9] Kroychuk M K, Shorokhov A S, Yagudin D F, Shilkin D A, Smirnova D A, Volkovskaya I, Shcherbakov M R, Shvets G and Fedyanin A A 2020 *Nano Lett.* **20** 3471–3477
[10] Obydennov D V, Shilkin D A, Lyubin E V, Shcherbakov M R, Ekimov E F, Kudryavtsev O S, Vlasov I I and Fedyanin A A 2018 *J. Phys. Conf. Series* **1092** 012102

Aldehyde Gas Detection using Nanostructured Zno-based Gas Sensor fabricated via Horizontal Vapor Phase Growth Technique

R S Olarve[1], H M Dela Torre[1], J R Foronda[1], M G Santos[1], N J Sajor[1,2], T B Lopez[1,3], K J Haygood[1], G N Santos[1]

1. Physics Department, De La Salle University, 2401 Taft Avenue, Malate, Manila, Philippines;
2. Physics Department, Central Mindanao University, Maramag Bukidnon 8710, Philippines;
3. Physics Department, Technological University of the Philippines, Taft Avenue, Ermita, Manila, Philippines;

Abstract. Detection of aldehydes such as pentanal, hexanal, octanal, and nonanal are studied with the use of nanostructured zinc oxide (ZnO) as sensing element. ZnO nanowires synthesized at optimized growth parameters using horizontal vapor phase growth (HVPG) technique was used due to its unique properties in gas sensing applications. Scanning Electron Microscope (SEM) and Energy Dispersive X-Ray (EDX) were used to verify the growth of ZnO nanowire structures. Further characterization using Source Meter was used to measure its resistance and resistivity based on the I-V graph. The sensor substrate wire set-up is connected to the Source Meter for resistance measurements as exposed to the different gas concentration of aldehydes. Gas sensing measurements were done at the static headspace gas concentration of the identified aldehydes. The sensor response of nanostructured ZnO-based gas sensor towards different gas concentrations ranges from 5.84% to 38.08%. Response time varies but it was observed that octanal gas has the longest response while pentanal has the fastest response.

1. Introduction

Aldehydes are compound commonly found in the atmosphere as air pollutants emitted by factories and industrial facilities. Aldehydes undergo a variety of reactions that lead to many different products such as organic aldehydes that are readily oxidized to give carboxylic acids. When exposed to air, aldehydes can form peroxo acids and finally carboxylic acids [1]. Aside from its presence in the atmosphere, aldehydes are also found in human body and exhaled in breath as indicators of underlying diseases [2]. Exhaled breath is largely composed of oxygen, nitrogen and carbon dioxide and contains very low concentration of various volatile organic compounds (VOC). Breath contains a few of exogenous VOC which is a result of external contamination from cigarette smoke, pollution and through ingested food or drinks. However, many of the VOC detected in exhaled breath are endogenous substances that originate from the physiological processes that occur in the body such as pentanal, hexanal, octanal, and nonanal [3,4,5].

Equipment and tools most commonly used in measuring the component of exhaled breath are gas chromatography (GC), laser-absorption spectroscopy and chemical gas sensors. GC coupled with various detection methods such as mass spectrometry (MS), flame ionization detector (FID) and ion mobility spectrometry (IMS) were used in various researches in the identification and quantification of breath components [6,7,8]. In comparison, other devices such as chemical gas sensors offer several advantages such as portability, low power consumption and low-cost production. The most common types of chemical gas sensor that are used in gas sensing are polymers, quartz microbalance sensor, acoustic wave device and metal oxide semiconductors. Most of these relies on the adsorption of gases into the sensing device which causes change in mass, vibration, conductivity or color thus a change in its output [9,10,11]. Moreover, chemical gas sensor which employ metal oxide semiconductors has had considerable impact to researchers for its wide range of sensing properties which can be synthesize using variety of techniques [12]. Among the various nanostructured metal oxide that are used in gas sensing, ZnO is widely used in gas sensing application for its excellent sensing properties [13].

Using horizontal vapor phase growth technique (HVPG), ZnO nanowires were synthesized and fabricated as sensing element to detect the identified aldehydes: pentanal, hexanal, octanal, and nonanal. Characterization method was used to identify the surface morphology of the grown nanostructured ZnO using Scanning Electron Microscopy (SEM) and to verify its elemental composition using Energy Dispersive X-Ray (EDX). Furthermore, electrical characterization using Keithley Source Meter was used to investigate the sensor's properties such as response, sensitivity, response time and recovery time through IV-curve analysis and resistance change as gaseous chemicals from the prepared VOC interact with the sensing element.

2. Methodology

ZnO nanomaterials was synthesized using HVPG technique at optimized growth parameters based on previous studies yielding dense nanowire formations at a particular zone of a quartz tube [14, 15, 16]. After the deposition process, the quartz tube is allowed to cool down at room temperature followed by mechanically cracking the walls of the tube to retrieve sections of deposited ZnO nanomaterial. The surface morphology of ZnO samples is viewed using SEM. With EDX, the captured images of ZnO nanomaterial are analyzed to determine its elemental composition. Resistance and resistivity of the substrate is measured using Keithley 2450 Source Meter as reference for gas sensing application. Resistance can be computed following Ohm's Law R=V/I considering that the I-V graph of ZnO nanomaterial shows an ohmic resistor. Resistivity of the material is computed following the Van der Pauw technique.

Pentanal (97%), hexanal (98%), octanal (99%), nonanal (98%) were purchased from Sigma-Aldrich. Amber tinted bottles of different volume capacities were used as container of the aldehyde solution to attain the desired headspace gas concentration and protect the solution from visible light radiation to prevent some form of chemical reaction due to the high volatility nature of the aldehyde. Different concentration levels of aldehyde headspace gas are attained using phase ratio of the volume of the bottle vs. volume of the aldehyde solution. For very low concentration, larger vials were used as container for small volume of aldehyde solution such that a 200ml bottle with 0.01ml of aldehyde solution produced 0.06ppm or 60ppb of headspace gas concentration. This is to reach the cut-off value for cancer recognition of 0.02ppm-1.2ppm of pentanal, hexanal, octanal, and nonanal as reported by Fuchs et al. (2010). The prepared aldehyde solution is stored for at least 1 day to allow headspace gas to saturate.

The sensor substrate with gold sputtered electrodes is connected to copper wires using silver conducting paste as contacts. Using bottle caps as the holder for the sensor substrate ensures that headspace gas does not alter drastically during the gas sensing thus retaining its concentration. The assembled gas sensor with wire setup is connected to a Keithley 2450 Source Meter for measurement of resistance during the exposure of the sensor to target gas analytes.

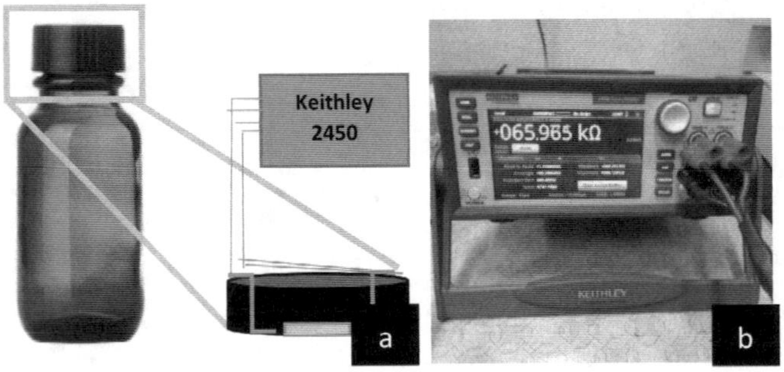

Figure 1. (a) Illustration of the ZnO sensor substrate attached to the inner wall of the cap of amber bottle and connected to the Source Meter (b) Keithley 2450 Source Meter

Gas sensing application begin by exposing the sensor to air to measure its resistance which acts as the baseline resistance of the sensor. Following the exposure of the sensor to the target gas to measure its resistance as exposed to the gas analytes. After exposure to the target gas, a purging by means of manual heater was used to remove adsorbed gas species and obtain the initial resistance reading. Assisted heating of the sensor after exposure to the target gas allows the gas species that adsorbed on the surface of the sensor to detach at a faster rate and for the sensor to return to its baseline resistance. The gas sensing for each gas is repeated for three trials. The process is also repeated as exposed to each aldehyde solution with varied concentrations.

3. Results and Discussion

Nanowire formations were found in Zone 2 section of the quartz tube measuring an average diameter of 81.12nm scattered throughout the substrate. Nanowires overlapping each other appears to be thick white areas. The atomic percentage as measured from the weight percentage is 59.27% oxygen, 37.71% zinc and 3.02% gold as shown in Fig. 2. Approximately, the ratio of zinc to oxygen is 1:2. The nanowire structures and cloud-like particles were analyzed to be zinc oxide.

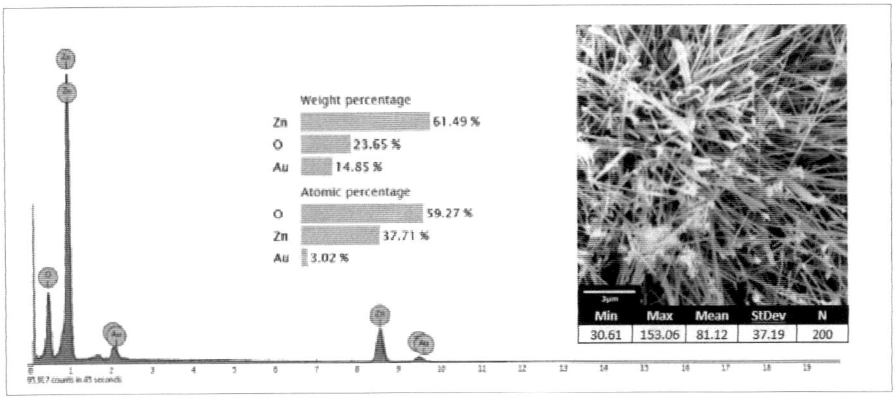

Figure 2. SEM-EDX of ZnO substrate located in Zone 2

Journal of Physics: Conference Series **2015** (2021) 012102 doi:10.1088/1742-6596/2015/1/012102

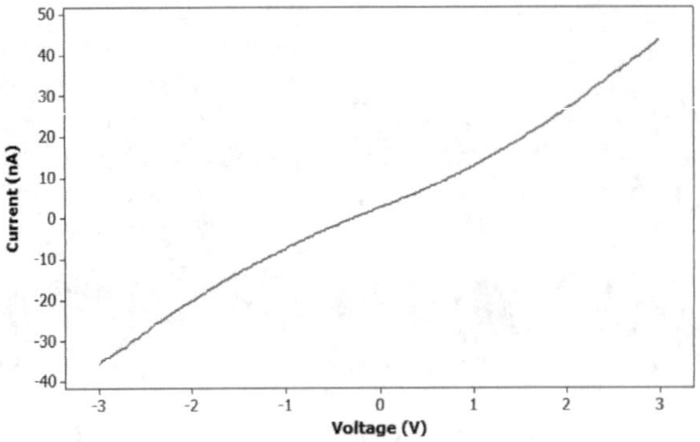

Figure 3. I-V graph of ZnO sensor substrate across a source voltage of -3V to 3V.

In Fig.3, based on the current-voltage graph, ZnO substrate has a computed resistance ($R_{AB,CD}$) of 7.0 x$10^4\Omega$ and resistivity measurement of 5.1 x 10^{-3} Ωm. The setup was done using linear sweep across a voltage from –3V to 3V resulting to current measurement of -36.0105 nA to 43.7475 nA. It can be observed that the slope follows a straight line having a linear relationship between the measured current and the applied voltage. This indicates a connection of ohmic resistor.

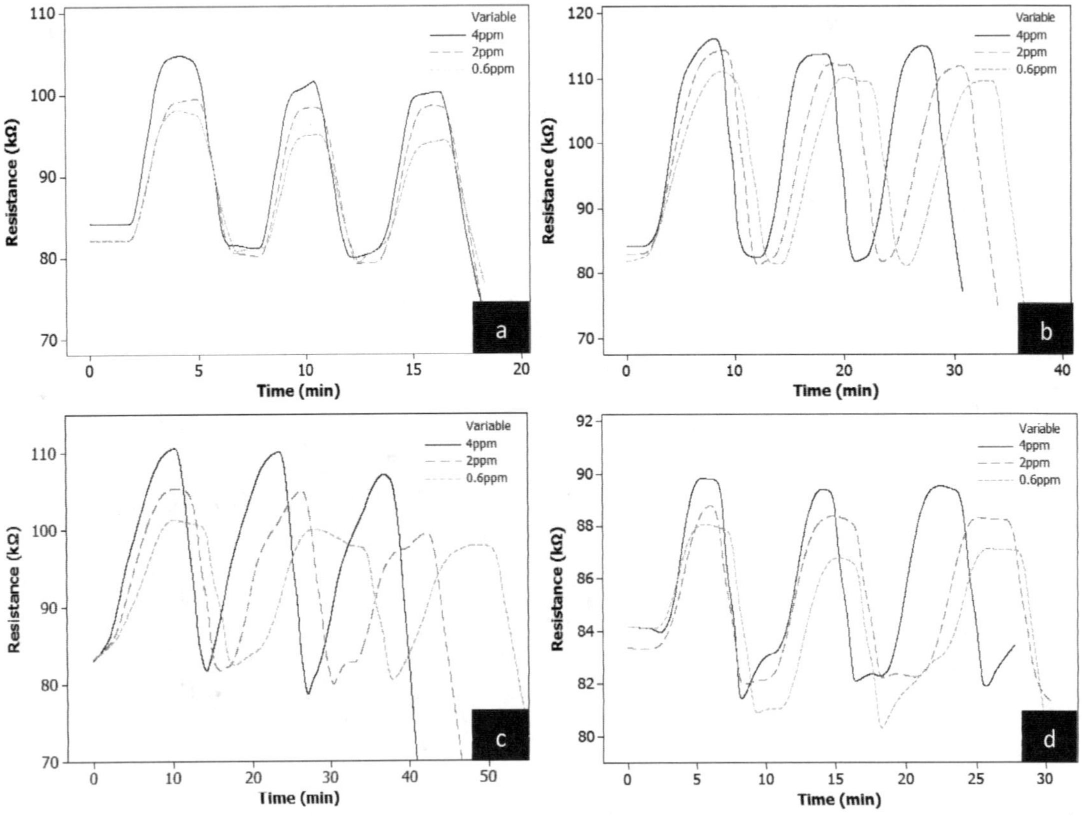

Figure 4. Dynamic sensor response (resistance vs. time plot) of ZnO towards different gas concentration of a)pentanal b)hexanal c)octanal d)nonanal at room temperature.

Table 1. Summary of sensor response, response and recovery time

		R_a (kΩ)	R_g(kΩ)	R%	Res_t (sec)	Rec_t(sec)
Pentanal	4ppm	81.91	101.86	24.35	23.07	14
	2ppm	80.58	98.54	22.3	40.27	17.33
	0.6ppm	81.33	95.85	17.86	31.87	21.33
Hexanal	4ppm	83.03	114.65	38.08	148.67	25.6
	2ppm	82.34	112.46	36.59	186.27	31.2
	0.6ppm	81.48	109.78	34.74	191.2	29.87
Octanal	4ppm	82.92	108.99	31.44	301.2	107.33
	2ppm	83.05	102.98	24	303.73	142.4
	0.6ppm	84.15	99.26	17.95	248.8	164.27
Nonanal	4ppm	83.19	89.57	7.66	94.53	48.27
	2ppm	82.61	88.45	7.07	118.8	38.8
	0.6ppm	82.48	87.3	5.84	105.87	48.93

Generally, it can be observed that the sensor response of the ZnO for each of the target aldehyde gases decreases as the concentration also decreases. It can be observed that the resistance in air averages at almost 82kΩ with uniform distribution. The resistance in gas also diminishes as the concentration is decreased from 101.86kΩ to 95.85kΩ for pentanal, 114.65 kΩ to 109.78 kΩ for hexanal, 108.99 kΩ to 99.26 kΩ for octanal and 89.57 kΩ to 87.3 kΩ for nonanal. The response and recovery time seems to be longest in octanal gas concentrations as compared to the other aldehyde gas while pentanal gas concentrations has the fastest response and recovery time.

4. Conclusion

ZnO nanowires have been successfully synthesized using HVPG technique and used as sensing substrate to detect specific aldehyde gases. The sensing parameters such as sensor response, response time and recovery time have been characterized. ZnO nanowire-based gas sensor showed high response towards hexanal and octanal for as low as 0.6ppm headspace gas concentration. Moreover, the response time is fastest with regards to pentanal gas having a response time of 23.07 seconds while the slowest response time recorded is 303.73 seconds as a response to octanal gas. This may due to the strong smell corresponding to high volatility of pentanal gas as compare to other gases. Furthermore, the recovery time of the sensor is 14.00 seconds as the fastest with regards to pentanal gas and 164.27 seconds with regards to octanal gas. The long recovery time of the sensor with regards to octanal gas may be due to the higher resistance value of the sensor as exposed to octanal gas.

Further studies on the sensor response pattern of each aldehyde gas to fabricate sensor device which can recognize the response curve of each type of gas that is exposed to the sensor and assessment of the limit of detection of ZnO nanowire sensor towards aldehyde gas are excellent progress to enhance its sensing capabilities.

Journal of Physics: Conference Series **2015** (2021) 012102 doi:10.1088/1742-6596/2015/1/012102

5. References

[1] Vanoye, L., Favre-Réguillon, A., Aloui, A., Philippe, R., & de Bellefon, C. (2013). Insights in the aerobic oxidation of aldehydes. RSC Advances, 3(41), 18931.

[2] Fuchs, P., Loeseken, C., Schubert, J.K., and Miekisch, W. (2010). Breath gas aldehydes as biomarkers of lung cancer. Int J Cancer, 126, 2663–2670.

[3] Poli, D., Goldoni, M., Corradi, M., Acampa, O., Carbognani, P., Internullo, E., Casalini, A., and Mutti, A. (2010). Determination of aldehydes in exhaled breath of patients with lung cancer by means of on-fiber-derivatisation SPME-GC/MS. J Chromatogr B Analyt Technol Biomed Life Sci, 878, 2643–2651.

[4] Horváth, I., Hunt, J., & Barnes, P. J. (2005). Exhaled breath condensate: methodological recommendations and unresolved questions. European Respiratory Journal, 26(3), 523–548.

[5] Schallschmidt, K., Becker, R., Jung, C., Bremser, W., Walles, T., Neudecker, J., Leschber, G., Frese, S., & Nehls, I. (2016). Comparison of volatile organic compounds from lung cancer patients and healthy controls—challenges and limitations of an observational study. Journal of Breath Research, 10(4), 046007.

[6] Phillips, M., Beatty, J. D., Cataneo, R. N., Huston, J., Kaplan, P. D., Lalisang, R. I., Lambin, P., Lobbes, M. B. I., Mundada, M., Pappas, N., & Patel, U. (2014). Rapid Point-Of-Care Breath Test for Biomarkers of Breast Cancer and Abnormal Mammograms. PLoS ONE, 9(3), e90226.

[7] Szulejko, J. E., McCulloch, M., Jackson, J., McKee, D. L., Walker, J. C., & Solouki, T. (2010). Evidence for Cancer Biomarkers in Exhaled Breath. IEEE Sensors Journal, 10(1), 185–210.

[8] Bannier, M., Van De Kant, K., Baumbach, J., Jobsis, Q., & Dompeling, E. (2018). Exhaled breath analysis by ion mobility spectrometry in children with asthma and cystic fibrosis. Cystic Fibrosis, 52–62.

[9] Lv, A., Pan, Y., & Chi, L. (2017). Gas Sensors Based on Polymer Field-Effect Transistors. Sensors, 17(12), 213.

[10] Kulha, P., Kroutil, J. ř., Laposa, A., Procházka, V., & Husák, M. (2016). Quartz Crystal Micro–Balance Gas Sensor with Ink–Jet Printed Nano–Diamond Sensitive Layer. Journal of Electrical Engineering, 67(1), 61–64.

[11] Länge, K. (2019). Bulk and Surface Acoustic Wave Sensor Arrays for Multi-Analyte Detection: A Review. Sensors, 19(24), 5382.

[12] Wang, C., Yin, L., Zhang, L., Xiang, D., & Gao, R. (2010). Metal Oxide Gas Sensors: Sensitivity and Influencing Factors. Sensors, 10(3), 2088–2106.

[13] Sarkar, A., Maity, S., Chakraborty, P., & Chakraborty, S. K. (2016). Synthesize of ZnO Nano Structure for Toxic Gas Sensing Application. Procedia Computer Science, 92, 199–206.

[14] Sajor, N. J., Foronda, J. R., Olarve, R. S., Dela Torre, H. M., Santos, M. G., Lopez, T. B., . . . Gratowski, S. V. (2020). Synthesis of metal oxide nanomaterials for early lung disease detection. Journal of Physics: Conference Series, 1461, 012149.

[15] Briones, J. C., Castillon, G., Delmo, M. P., & Santos, G. N. (2017). Magnetic-Field-Enhanced morphology of tin Oxide NANOMATERIALS for Gas sensing applications. Journal of Nanomaterials, 2017, 1-11.

[16] Muflikhun, M. A., & Santos, G. N. (2019). A standard method to Synthesize Ag, Ag/Ge, ag/tio2, SnO2, And Ag/SnO2 nanomaterials using the Hvpg technique. MethodsX, 6, 2861-2872.

Trapping-influenced photoluminescence intensity decay in semiconductor nanoplatelets

A A Kurilovich[1], V N Mantsevich[2], K J Stevenson[1], A V Chechkin[3,4] and V V Palyulin[5]

[1] Center for Energy Science and Technology, Skolkovo Institute of Science and Technology, 121205, Moscow, Russia
[2] Chair of Semiconductors and Cryoelectronics & Quantum Technology Center, Physics department, Lomonosov Moscow State University, 119991 Moscow, Russia
[3]Institute for Physics & Astronomy, University of Potsdam, D-14476 Potsdam-Golm, Germany
[4]Akhiezer Institute for Theoretical Physics National Science Center "Kharkov Institute of Physics and Technology", 61108, Kharkov, Ukraine
[5]Center for Computational and Data-intensive Science and Engineering, Skolkovo Institute of Science and Technology, 121205, Moscow, Russia

E-mail: V.Palyulin@skoltech.ru

Abstract. We present a diffusion-based simulation model for explanation of long time power-law decay of photoluminescence (PL) emission intensity in semiconductor nanoplatelets. In our model the shape of emission curves is an outcome of interplay of recombination, diffusion and trapping of excitons. At short times the excitons diffuse freely following the normal diffusion behaviour. The emission decay is purely exponential and is defined by recombination. At long times the transition into the subdiffusive motion happens and the emission occurs due to the release of excitons from surface traps. A power-law tail for intensity is a consequence of the release. The crossover from onelimit to another is controlled by diffusion properties. The approach reproduces the properties of experimental curves measured for different nanoplatelet systems.

1. Introduction

Among novel semiconductor materials low-dimensional structures attract significant attention due to their peculiar optical properties [1, 2, 3]. The low-dimensional semiconductor nanocrystals such as quantum dots (QDs) and nanoplatelets (NPLs) receive considerable interest due to their high tunability. The outstanding features make NPLs promising candidates for making new generation of optical devices including flexible light emitters [4] and colloidal lasers [5]. Modern chemistry is capable of precise synthesis of a wide range of QDs and NPLs of various shapes and compositions [3, 6, 7]. The emission features of QDs and NPLs are mostly defined by the behaviour of excitons. In particular, the exciton diffusion and trapping can explain the PL power-law decay dynamics of CdSe and CdS NPLs [8, 9].

The high fraction of the surface in low-dimentional materials results in its crucial importance for response of the system to an external excitation. The PL spectrum and decay statistics are among examples of such response [10]. The PL signal is caused by the electron-hole recombination. Thus, reversible or irreversible trapping of excitons by surface trap states is one of the main control factors of the decay statistics. The trapping can occur at dangling bonds which rearrange on the surface due to surface reconstruction and ligand adsorption [11].

NPL is a colloidal semiconductor nanocrystal which consists of a stuck of few atomic layers with areas of hundreds of nm^2 or more. With large surface to volume ratio these quasi-2D systems allow diffusion and trapping of excitons. For CdSe and CdS NPLs in Ref. [9] the power-law decay of PL intensity at long times was revealed in accordance with previous measurements of NPLs [12]. This decay was related with reversible trapping without radiation losses.

Most of the theoretical descriptions of PL intensity are built upon kinetic equations approach [9, 13]. The problem with purely kinetic techniques lies in the difficulty of exact interpretation of rates. This hinders a possible comparison with experiments. The exciton diffusion was also studied in free-standing and SiO_2 supported WS_2 monolayer semiconductors [14] and in 2D metal-halide perovskites [15]. In the latter case it was found to exhibit the transition from diffusive to a subdiffusive regime. This clearly points out to the existence of traps with a power-law distribution of trapping times [16]. Moreover, the subdiffusion due to trapping occurs for diffusion of copper adatoms on the graphene oxide support [17].

In our previous work [8] we have found that the power-law trapping of excitons indeed leads to power-law decay of PL instensity at long times. We supported the statement with theoretical arguments based on kinetic equations with a memory term as well as extensive simulations. Here we address the question of dependence of mean-squared displacement (MSD) of excitons on the density and trapping effects which occur at large densities of traps.

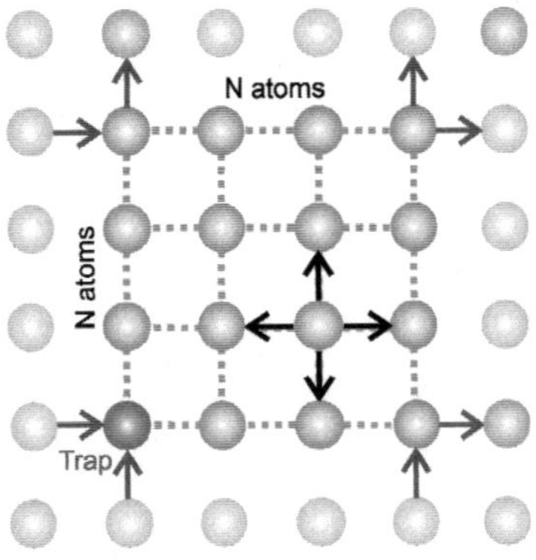

Figure 1: 2D random walk model for exciton diffusion (see main text for description).

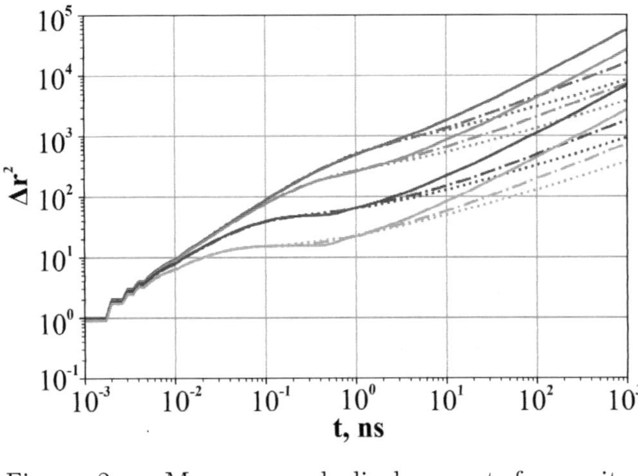

Figure 2: Mean-squared displacement for exciton diffusion as a function of time on a double logarithmic scale in the absence of recombination.

2. Simulations. Mean-squared displacement of exciton diffusion. PL intensity.

In our simulations we study the motion of excitons with and without recombination. The former model (with the recombination) allows us to study the PL decay and compare the results with the experiment from Ref. [9]. We assume that the laser pulse produces excitons which then diffuse on the surface of a nanoplatelet. They can either be trapped or recombine with an emission of a photon. We assume that in the trapped state the recombination is not possible. We also neglect nonradiative recombination or exciton dissociation assuming high quantum yield. In the previous work [8] we have shown that 2D simulations produce qualitatively similar results to the 3D models with 3 to 5 atomic layers (which is a typical thickness in experiments [9]).

The diffusion is simulated as a discrete random walk on the 2D square lattice. We apply the periodic boundary conditions for the lattice $N \times N$. At every step, the simulated particle moves to one of the nearest neighbours (see Fig. 1) with equal probability 1/4. The step duration is 1ps for all cases. The surface defects (traps) are positioned either regularly (periodically) or distributed randomly while the initial exciton position was sampled from a discrete uniform random distribution. The step duration was taken to be much smaller than the time scale of photoluminescence and trapping processes. The trapping occurs with certainty once the exciton hits the trap position. Importantly, in our model, the probability distribution of trapping times is non-exponential, having a power-law tail such that the average trapping time diverges. Such time distributions appear in the theory of transport in disordered media in the form of continuous time random walk model [16, 18]. This distribution of trapping times occurs, for example, in systems with an exponential distribution of depths of potential wells [19]. We sample the trapping times from the one-sided α-stable distribution with the Lévy index $\mu, 0 < \mu < 1$. Such distributions have long time asymptotics $\gamma(t) \sim \frac{\tau_*^\mu}{t^{1+\mu}}$. The particle could be trapped again after it escaped a trap. Thus, a recurrent trapping is possible in our approach.

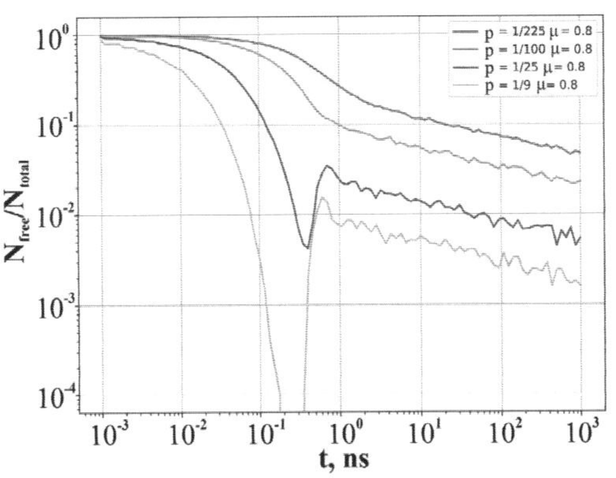

Figure 3: The fraction of free excitons for different trap densities for the case without exciton recombination (see the text for details).

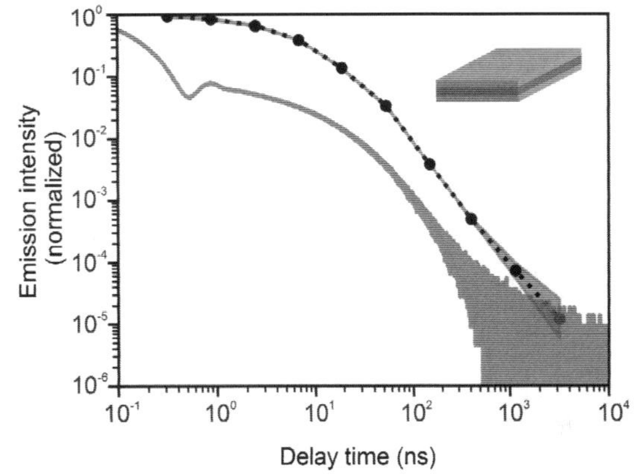

Figure 4: Comparison of photoluminescence intensities between simulations (blue data) and experiments (black dots) as reported in Ref. [9].

First we perform the simulations for the case without recombination, i.e. the number of excitons stays constant. This allows us to plot meaningful MSD in Fig. 2. The different curves correspond to different trap densities. The lattice is 4096×4096 with randomly distributed traps. The red curves are for 1 defect per 225 sites, the green are for 1/100 density of defects, the purple are for 1/25 and the cyan curves are for 1 defect per 9 cites. The continuous curves stand for the exponent of trapping times $\mu = 0.8$, the dash-doted are for $\mu = 0.6$ and $\mu = 0.5$ is pictured by dotted lines. The discreteness of the curves at short times is a lattice discretisation effect. The curves were obtained by ensemble averaging over 10000 trajectories for each curve. One can see that at short times the diffusion is Brownian, $\langle r^2(t)\rangle \sim t$ while at longer times it becomes subdiffusive $\langle r^2(t)\rangle \sim t^\mu$. For high densities of targets (1/9 and 1/25) in the transition region the MSD has a plateau at times between $0.1 - 1$ ns. This clearly points out to a caging effect. In order to see this in more detail we plot the fraction of unbound excitons in Fig. 3. In this plot in all cases $\mu = 0.8$. The curves from the upper one to the bottom one correspond to the following densities of traps $1/225, 1/100, 1/25, 1/9$. For high densities of traps corresponding to plateaus in Fig. 2 there is a dip in the number of free excitons. This fall and the following recovery of the free exciton concentration can be explained by the shorter characteristic timescale of trapping in comparison of that of the escape. Once the trapping becomes slower as for the upper two curves in Fig. 3, the dip disappears.

For the study of PL intensity decay (see Fig. 4) we assume that an exciton in a free state can recombine with a probability $p = 10^{-3}$ at each step neglecting any interactions between the excitons. As in Ref. [9] we collect the recombination data in time bins and normalise the curve by the value of the first bin. This normalisation in practice could be different in experiments, hence one could compare the curves qualitatively rather than quantitatively. In Fig. 4 the shape is the same for our simulation model and the digitised experimental data from Ref. [9] for coreshell CdSe-CdS NPLs with CdS forming the outer layers. The simulations were performed on a 10×10 periodic lattice with a single defect per lattice (i.e. 1 defect per 100 sites positioned periodically), the exponent $\mu = 0.8$, number of simulated particles (runs) $N_{sim} = 10^7$ and the binning $t_{\text{bin}} = 0.025ns$. At long times the same power law $\sim t^{-1-\mu}$ is present. At shorter times possibly due to a different binning no caging effect was found in experiments while it was produced in the simulations. The caging explains a recovery in the PL intensity in Fig. 4 at the timescale of around 1 ns. The comparison with [8] also shows that the simulations match the theory based on the non-Markovian kinetic equations with a memory term. The simulations also work for pure core NPLs as well [8].

3. Conclusions

We have shown that the simulation approach based on 2D random walk reproduces the main power-law decay features of photoluminesence in semiconductor nanoplatelets. MSD plots demonstrate a clear transition from Brownian diffusion at short times to subdiffusion at long times. At high density of traps a plateau in MSD can be observed at intermediate time scales. This happens due to caging of excitons by the traps. The subsequent release leads to a temporary increase of number of free excitons and to some recovery in PL intensity. Our simulation approach can be applied for the further analysis of exciton kinetics in semiconductor nanoplatelets to include more subtle effects such as exciton-exciton interactions.

4. Acknowledgements

V.N.M. acknowledges support from the Interdisciplinary Scientific and Educational School of Moscow University "Photonic and Quantum technologies. Digital medicine". The authors would like to acknowledge the support of the Russian Science Foundation project no. 18-72-10002.

References

[1] Ivchenko E L, *Optical spectroscopy of semiconductor nanostructures* (Alpha Science Int., Harrow, UK, 2005.).
[2] Xu M, Liang T, Shi M and Chen H 2013 *Chem. Rev.* **113** 3766–3798
[3] Ithurria S and Dubertret B 2008 *J. Am. Chem. Soc.* **130** 16504–16505
[4] She C et al. 2014 *Nano Lett.* **14** 2772 – 2777
[5] Guzelturk B, Kelestemur Y, Olutas M, Delikanli S and Demir H V 2014 *ACS Nano* **8** 6599–6605
[6] Smirnov A M et al. 2019 *J. Lumin.* **213** 29–35
[7] Smirnov A M et al. 2019 *J. Phys. Chem. C* **123** 27986–27992
[8] Kurilovich A A et al. 2020 *Phys. Chem. Chem. Phys.* **22** 24686–24696
[9] Rabouw F T et al. 2016 *Nano Lett.* **16** 2047–2053
[10] Nasilowski M, Mahler B, Lhuillier E, Ithurria S, Dubertret B 2016 *Chem. Rev.* **116**, 10934–10982
[11] Owen J 2015 *Science* **347** 615–616
[12] Olutas M, Guzelturk B, Kelestemur Y, Yeltik A, Delikanli S and Demir H V 2015 *ACS Nano* **9** 5041–5050
[13] Del Pozo-Zamudio O et al. 2015 *2D Mater.* **2** 035010
[14] Kulig M et al. 2018 *Phys. Rev. Lett.* **120** 207401
[15] Seitz M et al. 2020 *Nat. Commun.* **11** 2035
[16] Scher H and Montroll E W 1975 *Phys. Rev. B* **12** 2455–2477
[17] Furnival T et al. 2017 *Chemical Physics Letters* **683** 370–374.
[18] Metzler R and Klafter J 2000 *Phys. Rep.*, **339** 1–77
[19] Tiedje T and Rose A 1981 *Solid State Commun.* **37** 49 – 52

Ultrafast laser heating of non-plasmonic nanocylinders

Olesia Pashina, Daniil Ryabov, George Zograf, Sergey Makarov and Mihail Petrov

The Department of Physics and Engineering ITMO University, Saint-Petersburg, Russia.

E-mail: olesiya.pashina@metalab.ifmo.ru

Abstract. We develop a model describing non-equilibrium processes under the excitation of resonant semiconductor nanostructures with ultrashort laser pulses with a duration of about 100 fs. We focus on the heating effects related to pulsed excitation with account on free carriers generation, thermalization, and relaxation. The heat exchange between the electron and phonon system is treated within the two-temperature model. We applied the developed model to describing pulsed heating of silicon nanocylinder on top of a dielectric substrate. We come up with estimations of the thermal damage threshold of the considered structures which provides the limits for the experimental conditions and ensures thermal stability of the samples.

1. Introduction

The recent years have been highlighted with the rapid and extensive development of nonlinear all-dielectric photonics [1]. Compared to plasmonics, all-dielectric nanostructures posses high mode volume and large nonlinear coefficients resulting in extremely high efficiency of second [2, 3] and third [4] order nonlinear optical effects as well as time varying optical response [5, 6]. At the same time, the residual free-carrier absorption along with single and multiphoton interband absorption may result in heating of dielectric and semiconductor nanostructures which form the main platform of all-dielectric photonics [7]. The efficient optical heating has been recently observed in continuous wave regime [8] opening a way for strong thermooptical nonlinearity [9, 10, 11].

The rigorous approach of pulsed excitation heating modelling has been developed in plasmonics [12]. It allows to consider the resonant properties of metal nanoparticles for exact temperature rise estimation. However, this model is inapplicable to the semiconductor structures because of the free carriers generation during photoexcitation which changes significantly the optical properties of the investigated material [13]. The numerical simulation of ultrafast optical heating of semiconductor microscale waveguide structures has been performed with analyzing exact material properties [14]. At the same time, the thermooptical effects in the regime of pulsed excitation of resonant subwavelength structures have not been studied so far, despite of extensive studies of nonlinear optical effects under femtosecond pulse illumination. In this work we develop a semi-analytical model describing the ultrafast heating of semiconductor nanoparticles under femtosecond laser pulse excitation. We combine the exact numerical simulation with extracted key model parameters such as the rate of heat flow into the substrate and absorption cross section. Basing on our model, one can estimate the critical repetition rate of the laser pulses for different intensities which provides the limits of thermooptical damage threshold. The obtained results are highly important for describing the optical response of dielectric and semiconductor nanostructures under high-intensity laser illumination in nonlinear optical experimental setups.

2. The model

We consider cylindrical silicon nanoparticles placed on top of a glass substrate (see the inset Fig. 1(a)) and subjected to linearly polarized femto-second laser illumination. We utilize two-temperature electron-phonon model taking into account three correlated processes: i) absorption of incident femtosecond laser radiation accompanied by electron-hole plasma generation, ii) thermalization process between the electron and lattice subsystems associated with electron-phonon interaction, and iii) heat leakage into the substrate.

We simulate the process of nanocylinder heating with the following system of differential self-consistent time-dependent equations for the free carrier concentration n which corresponds to both electron and hole concentration since we have undoped semiconductor; the electron and phonon (lattice) temperatures T_e and T_{ph}:

$$\begin{cases} \dfrac{\partial n}{\partial t} = \dfrac{\alpha I(t)}{\hbar\omega} + \dfrac{\beta I(t)^2}{2\hbar\omega} - \gamma n^3, \\ C_{\mathrm{e-h}}(n)\dfrac{\partial T_{\mathrm{e}}}{\partial t} = g(T_{\mathrm{ph}} - T_{\mathrm{e}}) + Q(t), \\ C_{\mathrm{ph}}\dfrac{\partial T_{\mathrm{ph}}}{\partial t} = g(T_{\mathrm{e}} - T_{\mathrm{ph}}) - \kappa(T_{\mathrm{ph}} - T_0). \end{cases} \tag{1}$$

Here I is the incident field intensity; $\alpha = 1.0086 \cdot 10^4$ cm^{-1} and $\beta = 2$ cm/GW are the coefficients of one- and two-photon absorption respectively; $\gamma = 3.8 \cdot 10^{-31}$ cm/s is the Auger recombination coefficient [14]; $\hbar\omega$ is the energy of incident photons; $C_{\mathrm{e-h}}(n)$ and C_{ph} are the electron-hole plasma and phonon specific heat capacities; $Q(t)$ is the volume density of the absorbed electromagnetic power; g and κ are the electron-phonon coupling and heat outflow coefficients respectively.

The first equation of the system corresponds to the free carriers concentration change in the system. It increases due to the processes of one- or two-photon absorption proportional to the first and second power of the incident field intensity I respectively. The free carrier concentration decreases because of Auger recombination which is the dominant mechanism of electron relaxation for the high density plasma.

The last two equations of system (1) describe the evolution of electron-hole plasma and phonon systems temperature. The femtosecond laser pulse excitation with the volume power density $Q(t)$ generates hot electron gas which subsequently interacts with lattice phonons. Electron-phonon coupling underlies the transfer of energy from hot electrons to lattice vibrations defined. The coupling constant between the subsystems $g = C_{\mathrm{e-h}}/\tau_{\mathrm{e-ph}}$ is governed by electron-phonon relaxation time $\tau_{\mathrm{e-ph}} = 240$ fs [14].

The nanocylinder phonons are coupled to the substrate thermal reservoir, which we describe with the constant κ which depends on the material properties and on the particular configuration of the thermal contact between the nanoresonator and the substrate. We extracted the κ coefficient from exact numerical modelling of the the cylinder on the glass substrate with help of commercially available COMSOL Multiphysics software. We modeled transient cooling of nanocylinder initially heated to homogeneous temperature T_0 in the absence of optical excitation and nonequilibrium carriers. Assuming exponential dependence of phonon temperature $T_{ph} \sim \exp(-\kappa t/C_{ph})$, we obtained the relaxation constant κ from the temporal dependence of the cylinder temperature.

The absorbed density of electromagnetic power could be estimated from formula [15]:

$$Q_{\mathrm{abs}} = \frac{\omega}{2}\mathrm{Im}\varepsilon \int |\mathbf{E}|^2 dV \equiv \sigma_{\mathrm{abs}} \cdot I, \tag{2}$$

where ε is the dielectric permittivity of the nanoparticle, ω is the incident light frequency, $|\mathbf{E}|$ is the electric field amplitude inside the particle and the integral is taken over the nanocylinder volume, $\sigma_{\mathrm{abs}} = Q_{\mathrm{abs}}/I$ is the absorption cross section.

The energy absorbed during a single optical pulse can not be considered as constant since the free carriers concentration significantly changes during the excitation, which modulates the imaginary part of the dielectric constant according to the Drude model:

$$\Delta\mathrm{Im}\varepsilon(n) = \frac{\omega_p{}^2(n)\tau_{\mathrm{e-e}}}{\omega(1 + \omega^2\tau_{\mathrm{e-e}}^2)}. \tag{3}$$

Here τ_{e-e} is the characteristic electron-electron scattering time which can be as small as $\tau_{e-e} = 1$ fs for high density electron-hole plasma [16] and $\omega_p(n)$ is the plasma frequency defined as the function of generated electron-hole plasma concentration

$$\omega_p(n) = \sqrt{\frac{ne^2}{\varepsilon_0 m_{\mathrm{eff}}}}, \tag{4}$$

Journal of Physics: Conference Series **2015** (2021) 012104 doi:10.1088/1742-6596/2015/1/012104

where $m_{\text{eff}} = 0.18m_{\text{e}}$ is the effective mass of holes and electrons in crystalline silicon (c-Si), ε_0 is the permittivity of vacuum, and e is the elementary charge.

Here, for the simplicity of the model, we suppose that the real part of the refractive index and consequently the spectral position of the resonances does not change significantly during the excitation, which is a valid assumption for not very high quality resonances. Then the absorbed power as the function of the free carrier concentration can be estimated as

$$Q_{\text{abs}}(n) = \sigma_{abs}(n_0)\left(1 + \frac{\Delta\text{Im}\,(\varepsilon(n))}{\text{Im}(\varepsilon(n_0))}\right)I. \tag{5}$$

The absorption cross-section $\sigma_{abs}(n_0)$ was modelled using the COMSOL Multiphysics package. We reduced fully three-dimensional electromagnetic problem to the two-dimensional one using the axial symmetry of the structure under consideration as it was done in our previous works, for example, in [17].

3. Results

Taking into account all the equations presented above, one can solve the system of differential equations (1) with the help of the key parameters $\sigma_{abs}(n)$ and κ obtained from full wave numerical simulation. For a quantitative consideration of the model, we simulate numerically the heating of a nanocylinder with radius $r = 200$ nm and height $h = 500$ nm under normal plane wave incidence as shown in Fig. 1(a) inset. System (1) is solved numerically using MATLAB software for the experimentally measured physical constants for silicon [14].

The results of the modelling are shown in Fig. 1(a). At the initial stage of heating, we observe the sharp rise in the electron temperature under laser pulse excitation with duration of $\tau_p = 100$ fs. After the pulse is over, the electron temperature goes down and there appears slow and gradual increase in the phonon temperature. Since the characteristic exchange time between electron and phonon systems is $\tau_{\text{e-ph}} = 240$ fs, the thermal equilibrium between them is reached at the scale of ~ 2 ps.

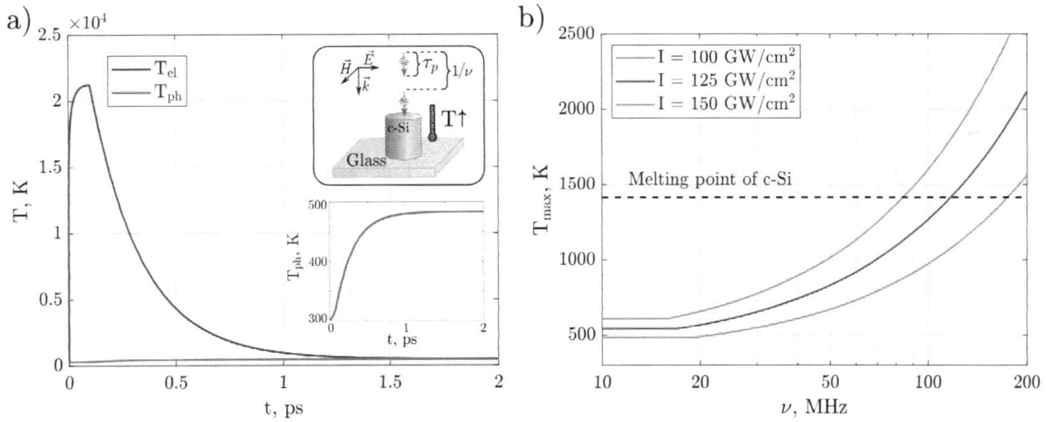

Figure 1. a) Time dependence of phonon T_{ph} and electron temperature T_{el} for a single laser pulse excitation with an intensity $I = 100$ GW/cm^2 and pulse duration $\tau_p = 100$ fs. The inset shows a schematic view of nanocylinder excitation under a series of pulses with repetition rate ν. b) Dependence of the maximal temperature on the pulse repetition rate ν for various incident intensities. The melting point of c-Si is marked with a horizontal dashed line to estimate the critical repetition rate leading to the structural damage for a given intensity.

After modelling one pulse, we proceed to study the behavior of the phonon temperature during several consequent laser pulses (Fig. 2). Simulation of a finite number of pulses was conducted by applying the results of computation for a single pulse excitation as the initial conditions for the next pulse.

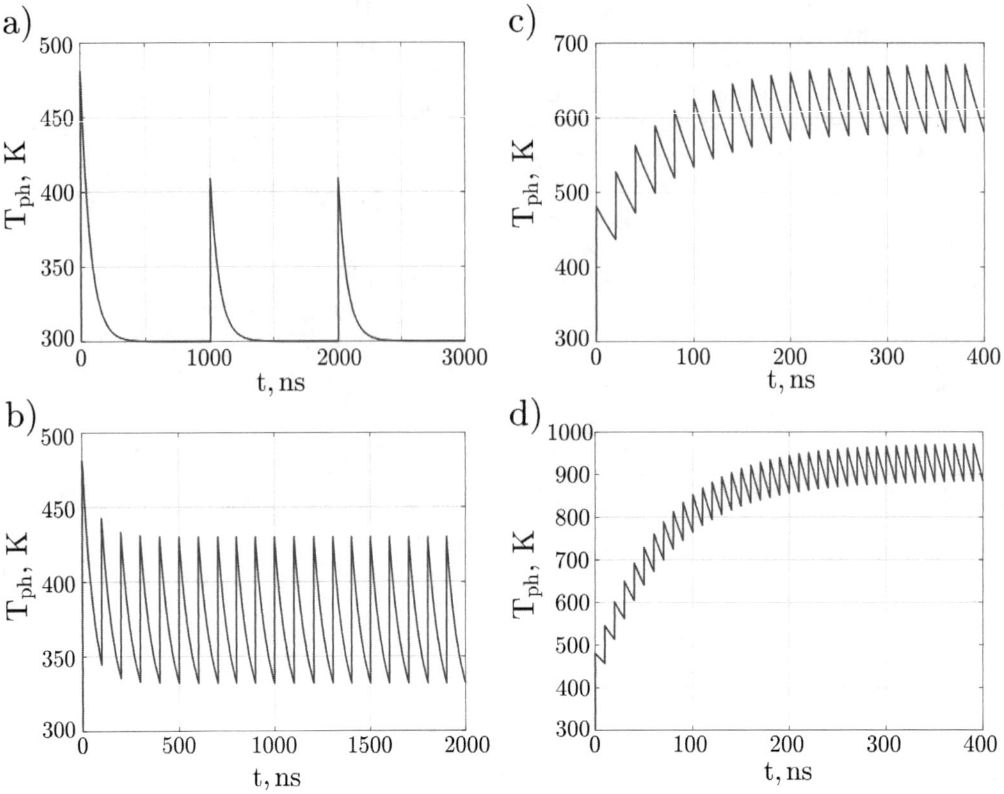

Figure 2. Dependence of the phonon temperature upon heating by successive pulses of femtosecond laser with intensity I = 100 GW/cm^2 for different pulse repetition rates. The maximum temperature is reached during the first pulse for cases a) and b) corresponding to the repetition rates $\nu = 1$ MHz and $\nu = 10$ MHz, respectively. For c) and d) with $\nu = 50$ MHz and $\nu = 100$ MHz, the temperature is established over a long period of time and the maximum is reached only after a series of pulses.

The modelling results lead to the existence of two different cases for temperature evolution. In the first case, at relatively low frequencies (Fig. 2(a, b)), the phonon temperature reaches its maximum during the first pulse. At the scale of a single cycle in case of a low repetition rate, the temperature drops to initial values due to the drain of heat into the substrate. An interesting fact is that the concentration of free carriers decreases much slower than the temperature of the cylinder. Since that the heat capacitance of free carriers stays relatively high $C_{e-h} = 3nk_B$ and significantly exceeds the initial value before the first pulse came. Thus, the first and subsequent pulses are significantly different in the temperature increase as it is seen in Fig.2(a). In the case of higher repetition rate (see Fig. 2(c)), the next pulse comes before the temperature drops to initial and the averaged temperature increases with subsequent pulses coming. Therefore, with each pulse, the temperature rises until it stabilizes (Fig. 2(c, d)). Under stable temperature we understand the situation in which the maximum temperature of the structure stays the same for all further pulses. It occurs when the initial conditions for all the parameters of our problem coincide for the subsequent pulses.

So, we estimate the maximum temperature during the multi-pulse heating process as a function of the pulse repetition rate for different intensities (Fig. 1(b)). By comparing the obtained results with the melting point of silicon, we can predict the maximum allowable pulse repetition rate before the structure is damaged. It is remarkable that for typical power of 150 GW/cm^2 the thermal damage is achieved at 80 MHz repetition rate which corresponds to real experimental conditions.

4. Conclusion

In conclusion, we studied a self-consistent two-temperature model of semiconductor nanocylinders heating via pulsed laser excitation. We took into account both the heat diffusion into the substrate and the change in the concentration of free carriers due to the optical excitation. On the basis of this model, the temperature evolution of nanodisks over a series of pulses has been observed. We estimated critical repetition rate of the laser pulses for different incident intensities before the structure under consideration is damaged. We assume further development of this model in the case of other nanoparticles shape or some resonant features for application to specific problems, and we are convinced that described calculations will be useful for analytical evaluation of heating when observing various pulse-excited effects in nanophotonics.

5. Acknowledgments

The work was supported by Presidential Grant MD-5954.2021.1.3

References

[1] Won R 2019 Nature Photonics **13** 585–587 ISSN 17494893

[2] Koshelev K, Kruk S, Melik-Gaykazyan E, Choi J H, Bogdanov A, Park H G and Kivshar Y 2020 Science **367** 288–292 ISSN 10959203

[3] Anthur A P, Zhang H, Paniagua-Dominguez R, Kalashnikov D A, Ha S T, Maß T W, Kuznetsov A I and Krivitsky L 2020 Nano Letters ISSN 15306992

[4] Melik-gaykazyan E V, Shcherbakov M R, Shorokhov A S, Staude I, Brener I, Neshev D N, Kivshar Y S and Fedyanin A A 2017 Philosophical Transactions of the Royal Society A: Mathematical, Physical and Engineering Sciences **375** 20160281

[5] Karl N, Vabishchevich P P, Shcherbakov M R, Liu S, Sinclair M B, Shvets G and Brener I 2020 Nano Letters **20** 7052–7058 ISSN 15306992

[6] Shcherbakov M R, Lemasters R, Fan Z, Song J, Lian T, Harutyunyan H and Shvets G 2019 Optica **6** 1441–1442 ISSN 23318422 (*Preprint* 1909.13637)

[7] Zograf G P, Petrov M I, Makarov S V and Kivshar Y S 2021 1–92 (*Preprint* 2104.01964) URL http://arxiv.org/abs/2104.01964

[8] Zograf G P, Petrov M I, Zuev D A, Dmitriev P A, Milichko V A, Makarov S V and Belov P A 2017 Nano Letters **17** 2945–2952 ISSN 1530-6984 URL http://pubs.acs.org/doi/abs/10.1021/acs.nanolett.7b00183

[9] Duh Y S, Nagasaki Y, Tang Y L, Wu P H, Cheng H Y, Yen T H, Ding H X, Nishida K, Hotta I, Yang J H, Lo Y P, Chen K P, Fujita K, Chang C W, Lin K H, Takahara J and Chu S W 2020 Nature Communications **11** 1–9 ISSN 20411723 URL http://dx.doi.org/10.1038/s41467-020-17846-6

[10] Zhang T, Che Y, Chen K, Xu J, Xu Y, Wen T, Lu G, Liu X, Wang B, Xu X, Duh Y S, Tang Y L, Han J, Cao Y, Guan B O, Chu S W and Li X 2020 Nature Communications **11** 1–9 ISSN 20411723

[11] Tsoulos T V and Tagliabue G 2020 Nanophotonics **9** 3849–3861 ISSN 21928614

[12] Baffou G and Rigneault H 2011 Physical Review B **84** 035415

[13] Downer M and Shank C 1986 Physical review letters **56** 761

[14] Chen J K, Tzou D Y and Beraun J E 2005 Int. J. Heat Mass Transfer **48** 501–509 ISSN 0017-9310

[15] Cunha J, Guo T L, Della Valle G, Koya A N, Proietti Zaccaria R and Alabastri A 2020 Advanced Optical Materials **8** 2001225 (*Preprint* https://onlinelibrary.wiley.com/doi/pdf/10.1002/adom.202001225) URL https://onlinelibrary.wiley.com/doi/abs/10.1002/adom.202001225

[16] Sokolowski-Tinten K and von der Linde D 2000 Phys. Rev. B **61** 2643–2650 ISSN 2469-9969

[17] Zograf G P, Ryabov D, Rutckaia V, Voroshilov P, Tonkaev P, Permyakov D V, Kivshar Y and Makarov S V 2020 Nano Letters **20** 5786–5791 pMID: 32579376 (*Preprint* https://doi.org/10.1021/acs.nanolett.0c01646) URL https://doi.org/10.1021/acs.nanolett.0c01646

DFT analysis of crystal polarity on graphene surface

Alexander Pavlov[1,2], Alexey Mozharov[1], Yury Berdnikov[3], Camille Barbier[4], Jean-Christophe Harmand[4], Maria Tchernycheva[4], Roman Polozkov[5,1], Ivan Mukhin[1,5]

[1] Alferov University (Former St Petersburg Academic University), St. Petersburg, 194021, Russian Federation
[2]Peter the Great St.Petersburg Polytechnic University, Polytechnicheskaya 29, 195251 St.Petersburg Russia
[3] St. Petersburg State University, Saint Petersburg, 199034, Russian Federation
[4] Centre de Nanosciences et de Nanotechnologies (C2T), Univ. Paris-Saclay, UMR 9001 CNRS, 10 Boulevard Thomas, Gobert, 91120 Palaiseau, France
[5] ITMO University, Saint Petersburg, 197101, Russian Federation

E-mail: a.pavlov@physics.spbstu.ru

Abstract. We report an *ab-initio* study of the preferred polarity for wurtzite GaN nanostructures on virtual graphene substrates. By means of the density functional theory analysis we show that N-polar nanostructures on graphene are energetically favorable in comparison to Ga-polar. These finding are in agreement with experimentally observed N-polarity of wurtzite GaN nanowires grown on graphene substrate. We believe that the revealed polarity preference is of importance for piezoelectric and optoelectronic device design.

1. Introduction

Gallium nitride (GaN) nanostructures can be employed as active components for a wide range of device applications such as light-emitting diodes [1, 2], solar cells [3–7], photodetectors [8] and piezogenerators [9–11].

GaN nanostructures usually form wurtzite crystal structure which lacks the inversion symmetry of [0001] and [000$\bar{1}$] directions [12] resulting in a spontaneous polarization along the c-axis (in the [0001] direction). Thus the Ga-polar structures growing in [0001] direction can be distinguished from the N-polar ones growing in [000$\bar{1}$] direction.

The polarity of the growth front is widely known to influence the GaN optical [13] and electronic [14] properties and therefore impacts their use in piezoelectric [10] and light-emitting [15] devices.

While the nanosized GaN crystals can be formed epitaxially on a number of different substrates, the growth on virtual graphene substrate is promising in terms of the so-called van-der-Waals epitaxy. In this case the nanostructure can be lifted off the growth substrate and transferred to any other arbitrary substrate. However, unlike the growth on conventional substrates such as silicon or sapphire, the polarity of GaN grown on graphene is not well-explored. In this work we employ the density functional theory (DFT) analysis to study the polar properties of GaN clusters forming on graphene.

2. Results and discussion

We perform the DFT calculations to determine the most energetically preferable geometry of the GaN nanocluster on graphene. Such a clusters can represent the nucleus of GaN nanoisland or nanowire. To reproduce the hexagonal symmetry of the wurtzite structure at least ten gallium and ten nitrogen atoms should be considered. In addition sixty more carbon atoms form the monolayer graphene lattice.

This model system system allows eight possible geometric configurations of GaN nanocluster: four Ga-polar and four N-polar. However, only two out of eight configurations are found to be stable during the energy optimization [16] within DFT calculations in the plane-wave basis set for projector augmented-wave method implemented in GPAW package [17]. The calculations was carried out using the Perdew–Burke–Ernzerhof generalized gradient approximation for exchange-correlation functional, with the plane wave energy cutoff set to 450 eV and k-space sampling in $4 \times 4 \times 4$ Monkhrost-Pack grid. To reduce computation time, the graphene atoms were fixed the during optimisation procedure.

The evolution of the two stable configurations, initially one Ga-polar and one N-polar, is illustrated in Figure 1 (a). We observe the initially Ga-polar cluster to change the Ga-N dipole orientation during the optimization procedure. Meanwhile, the initially N-polar nanocluster adjusts shape while keeping the initial arrangement of the atomic planes and the Ga-N dipole orientation. Thus, both stable configurations evolve to N-polar geometry corresponding to minimal cluster energy.

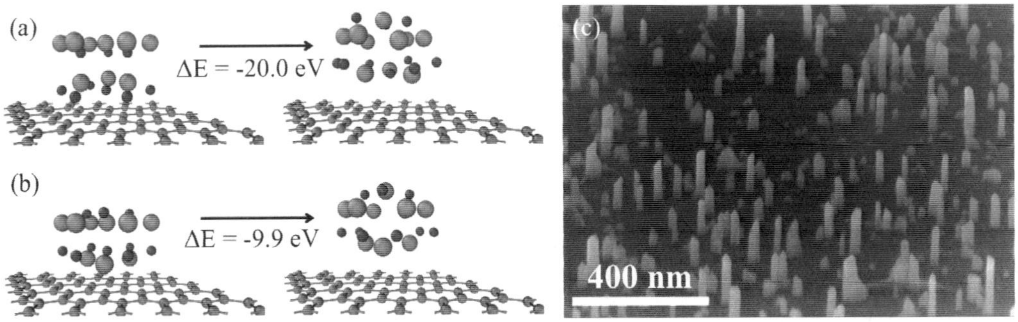

Figure 1. Snapshots of GaN NCs geometries before and after optimisation (a) for Ga-polar NC, (b) for N-polar NC. The ΔE show the energy difference between initial and final states. The energy difference between two systems in final states is less 0.5 eV. (c) 25° tilted SEM image of the system after the KOH etching.

Our model results are consistent with the experimentally observed preference of N-polarity in the growth of GaN nanowires on grahene/SiO_x/Si substrates. The details of GaN nanowire epitaxial growth can be found in our previous works [18, 19].

Figure 1 (c) shows a scanning electron microscopy (SEM) image of the grown GaN NWs on graphene and treated by KOH attack. Treatment in KOH solution leads to sharpening of NW tips due to crystal plane selective etching, which is a characteristic property of the N-polar structures [20]. The subsequent SEM imaging reveals that all GaN NWs grown on graphene possess N-polarity.

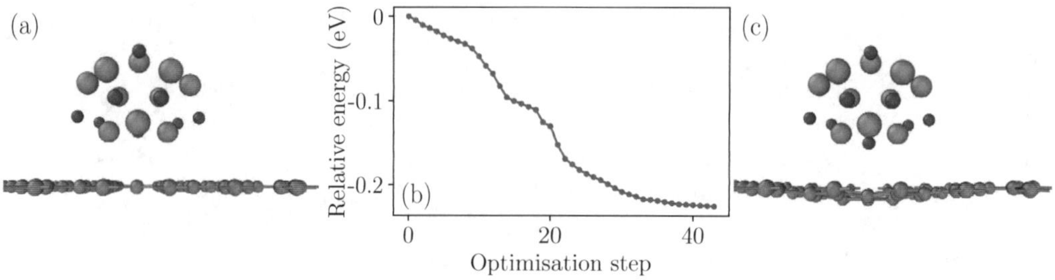

Figure 2. Snapshots of the N-polar GaN nanocluster (a) before optimisation procedure, (c) after optimisation procedure, (b) dependence of system potential energy on optimisation step.

Figure 2 shows the influence of all atom, including the graphene ones, optimisation on the final geometry of the system. For that we took the optimised geometry of N-polar GaN NC and include the graphene atoms into optimisation procedure. It is seen that the total energy of the system drops by $\approx$ 200 meV and the graphene layer sags aunder the NC. But, more importantly, there was no significant changes in the NC geometry.

3. Conclusions

The presented DFT study have shown that N-polar GaN structures on graphene substrates are energetically preferable to Ga-polar ones. Among different possible configurations of Ga and N atoms in the cluster only the N-polar one is stable while the initially Ga-polar structure demonstrates the change of polarity during the *ab-initio* optimization. The DFT modeling results are consistent with the experimental observations of exclusive N-polarity of GaN NWs grown on graphene layer on top of silicon substrate.

Acknowledgements

The work was supported by the Ministry of Science and Higher Education of the Russian Federation (FSRM-2020-0005). I.M. acknowledges the Russian Federation President Council for grants MK-2428.2020.2 and SP-2169.2021.1). Y.B. acknowledges Saint-Petersburg State University for a research grant 75746688. C.B. and M.T. acknowledge the financial support from the EU ERC project NanoHarvest (grant no. 639052). We acknowledge the Supercomputing Center of Peter the Great Saint-Petersburg Polytechnic University (SPbPU) for providing the opportunities to carry out large-scale simulations. The molecular structures in this article were rendered using Ovito software package [21].

References

[1] Guo W, Zhang M, Banerjee A and Bhattacharya P 2010 *Nano Lett.* **10** 3355–3359
[2] Hersee S, Fairchild M, Rishinaramangalam A, Ferdous M, Zhang L, Varangis P, Swartzentruber B and Talin A 2009 *Electron. Lett.* **45** 75–76
[3] Tian B, Kempa T J and Lieber C M 2009 *Chem. Soc. Rev.* **38** 16–24
[4] Tang Y, Chen Z, Song H, Lee C, Cong H, Cheng H, Zhang W, Bello I and Lee S 2008 *Nano Lett.* **8** 4191–4195
[5] Mozharov A, Bolshakov A, Cirlin G and Mukhin I 2015 *Phys. Status Solidi RRL* **9** 507–510
[6] Neplokh V, Ali A, Julien F H, Foldyna M, Mukhin I, Cirlin G, Harmand J C, Gogneau N and Tchernycheva M 2016 *Mater. Sci. Semicond. Process* **55** 72–78
[7] Shugurov K Y, Reznik R, Mozharov A, Kotlyar K, Koval O Y, Osipov A, Fedorov V, Shtrom I, Bolshakov A, Kukushkin S *et al.* 2019 *Mater. Sci. Semicond. Process* **90** 20–25
[8] Rigutti L, Tchernycheva M, De Luna Bugallo A, Jacopin G, Julien F, Zagonel L F, March K, Stephan O, Kociak M and Songmuang R 2010 *Nano Lett.* **10** 2939–2943
[9] Jamond N, Chrétien P, Houzé F, Lu L, Largeau L, Maugain O, Travers L, Harmand J C, Glas F, Lefeuvre E *et al.* 2016 *Nanotechnology* **27** 325403
[10] Gogneau N, Jamond N, Chrétien P, Houzé F, Lefeuvre E and Tchernycheva M 2016 *Semicond. Sci. Technol.* **31** 103002
[11] Lu L, Charron E, Glushkov E, Glushkova N, Bonello B, Julien F, Gogneau N, Tchernycheva M and Boyko O 2018 *Appl. Phys. Lett.* **113** 161903
[12] Morkoç H 2009 *Handbook of Nitride Semiconductors and Devices, Electronic and Optical Processes in Nitrides* vol 2 (John Wiley & Sons)
[13] Hite J, Twigg M, Mastro M, Freitas J, Meyer J, Vurgaftman I, O'Connor S, Condon N, Kub F, Bowman S *et al.* 2012 *Opt. Mater. Express* **2** 1203–1208
[14] Stutzmann M, Ambacher O, Eickhoff M, Karrer U, Lima Pimenta A, Neuberger R, Schalwig J, Dimitrov R, Schuck P and Grober R 2001 *Phys. Status Solidi B* **228** 505–512
[15] Carnevale S D, Kent T F, Phillips P J, Sarwar A, Selcu C, Klie R F and Myers R C 2013 *Nano Lett.* **13** 3029–3035
[16] Nocedal J and Wright S 2006 *Numerical optimization* (Springer Science & Business Media)
[17] Enkovaara J, Rostgaard C, Mortensen J J, Chen J, Dułak M, Ferrighi L, Gavnholt J, Glinsvad C, Haikola V, Hansen H *et al.* 2010 *J. Phys. Condens. Matter* **22** 253202
[18] Kumaresan V, Largeau L, Madouri A, Glas F, Zhang H, Oehler F, Cavanna A, Babichev A, Travers L, Gogneau N *et al.* 2016 *Nano Lett.* **16** 4895–4902
[19] Barbier C, Zhou T, Renaud G, Geaymond O, Le Fevre P, Glas F, Madouri A, Cavanna A, Travers L, Morassi M *et al.* 2020 *Cryst. Growth Des.*
[20] Largeau L, Galopin E, Gogneau N, Travers L, Glas F and Harmand J C 2012 *Cryst. Growth Des.* **12** 2724–2729
[21] Stukowski A 2009 *Model. Simul. Mater. Sci. Eng.* **18** 015012

Ablation-Free Laser Printing of Structural Colors in Reflection at 25,000 DPI

D Pavlov[1,2], V Lapidas[1], A Zhizhchenko[1,2], D Storozhenko[2] and A Kuchmizhak[1,2]

[1]Far Eastern Federal University, Vladivostok 690061, Russia
[2]Institute of Automation and Control Processes, Far Eastern Branch of the Russian Academy of Sciences, Vladivostok 690041, Russia

Abstract. Using direct femtosecond laser patterning of metal-insulator-metal (MIM) sandwich designed to support Fabry-Perot mode in the visible spectral range we demonstrate new practically relevant strategy for high-resolution color printing. Irradiation of the MIM sandwich by tightly focused laser pulses allows to produce unique 3D surface nanostructures – hollow nanobumps and nanojets - locally modulating surface reflectivity. Laser processing parameters control the 3D shape of such nanostructures allowing to gradually tune the reflected color from reddish brown to pure green. Up-scalable ablation-free laser fabrication method paves the way towards various applications ranging from large-scale structural color printing to optical sensors and security labeling at a lateral resolution of 25,000 dots per inch.

Abstract

1. Introduction

Unlike organic and inorganic pigments and dyes, structural colors originating from the interaction of broadband visible radiation with optically resonant nanostructures are non-fading and stable against UV radiation and thermal treatment [1,2]. Structural color technology is rather cheap in recycling, sustainable and durable, allowing to obtain vibrant (as compared to organic dyes) color tones with an extended gamut. Furthermore, utilization of optically resonant nanostructures paves the way toward high-quality imaging with the lateral resolution up to 100,000 dots per inch (DPI) that corresponds to the Abbe diffraction limit and makes the boundaries between neighboring pixels indistinguishable even with best dry microscope objectives. Lateral resolution of structural colors technology is two orders of magnitude higher comparing to those for best commercial printers. In the near future, structural colors are expected to revolutionize several technologically relevant areas as optical filters, next-generation displays, color marking and anti-counterfeiting of goods, etc. Moreover, a wide range of available materials for realization of optically resonant nanostructures makes the structural color technology potentially CMOS-compatible.

Despite the well-known physics of structural colors derived from standard bulk optics as prisms, thin films, photonic crystals or diffraction gratings, the interest in structural coloring of surfaces has recently refreshed. In part, this can be attributed to recent advances in such areas as nanophotonics, plasmonics and meta-optics, as well as the spreading of high-resolution planar nanofabrication technologies as electron- and ion- beam lithography that allowed to shrink the size of the structural colored pixel well below optical diffraction limit. Using semiconductor materials with high refractive index and low dissipative losses as well as plasmon-active metals [2,3], it is possible to fabricate sub-wavelength nanostructures with resonant optical response that can be tailored by varying characteristic geometry of nanostructures and their arrangement. Along with spectrally adjustable optical response, such nanoresonators can be also fabricated to have polarization anisotropy and angular dependence allowing to realize various schemes of optical information encryption [5,6].

In spite of a new look borrowed from meta-optics, majority of works used rather classical schemes based on thin-film Fabry-Perot filters [5], percolated films [7] and diffraction gratings [8] being combined with standard planar nano-resonators like nanodiscs, and nanoholes, whose planar geometry and arrangement are determined by planar fabrication technologies. Additional functionality and controllability over the single-pixel color and brightness appears to be achieved using 3D optical nanostructures. However, the ability to vary the height of the optically resonant nanostructures in each individual pixel of a color-coded picture cannot be implemented using the time- and money-consuming planar lithographic fabrication technologies.

Pulsed laser radiation is known to drive ultrafast solid-liquid phase transition on the surface of the exposed material that allows its sculpturing and subsequent resolidification in the form of various surface nano-morphologies like nanoparticles, nanoripples and nanobumps [9,10]. The latter type of structures is of particular interest as the 3D geometry of the nanobumps can be easily controlled by varying only the laser fluence and focusing conditions. Such structures can be imprinted on the surface of all plasmon-active metal films without ablation (ejection of the nanoparticles) thus ensuring ultra-clean and reproducible fabrication. Previously, by taking advantage of geometry-dependent resonant scattering of isolated nanobumps we demonstrated the direct laser writing of plasmonic colors that can be observed in the dark-field back-scattering regime [9]. Here, we show that ablation-free laser patterning of the metal-insulator-metal (MIM) sandwich with nanobumps allows to realize high-resolution (up to 25,000 DPI) structural coloring in more practically relevant reflection regime at high lateral resolution.

2. Material and Methods

For preliminary estimation, structural colors in reflection were generated by patterning a specially designed with FDTD numerical simulations MIM film containing 50-nm thick Au top layer (transmission ~ 6.5%), 100-nm thick Al_2O_3 middle layer and 500-nm thick Ag bottom layer thermally evaporated above silica glass slide (inset in the Figure 1a). The as-fabricated MIM film has reddish (as compared to the ordinary 50-nm thick Au on glass) color and represents a Fabry-Perot cavity designed to have a sharp resonant feature at 540 nm in reflection spectrum (Figure 1a).

The top layer of the MIM cavity was directly patterned using second-harmonic (515 nm) 200-fs laser pulses generated by regeneratively amplified Yb:KGW laser system (Pharos, Light Conversion). Laser pulses were focused onto the surface of MIM top layer using dry microscope objective having the numerical aperture (NA) of 0.42. The laser system was synchronized with a PC-driven attenuator (Standa) and 3D nanopositioning stages (Aerotech Gmbh.) allowing to process the surface according to programmable trajectory and vary the incident pulse energy. Morphology of the laser-printed nanostructures was correlated with laser-processing parameters using scanning electron microscopy (SEM, Ultra 55+, Carl Zeiss).

Reflectivity of the laser-patterned areas was studied using home-built microspectroscopy setup containing a bright-field optical microscope confocally aligned with a grating type spectrometer (Shamrock 303i, Andor) equipped with a thermo-electrically cooled CCD camera (Newton, Andor). Adjustable pinhole in the image plane was used to control the signal acquisition area. The reflection spectra were collected with an optical objective with NA=0.42 and normalized onto the signal from the bulk silver mirror. Reflection spectra were converted into the chromaticity coordinates of the hue-saturation-value (HCV) color space using the software package developed in [9].

3. Results and Discussions

Single-pulse ablation of "thermally" thin glass-supported Au films was extensively studied showing the formation of parabola- or cone-shaped hollow protrusions (nanobumps) at pulse energy E below the film ablation threshold [11]. Such behavior can be explained by a sequence of laser-induced phenomena as fast melting, acoustic relaxation at the film-substrate interface and resolidification [12]. Completely similar behavior was found for top Au layer of the MIM sandwich irradiated by tightly focused femtosecond laser pulses. In particular, similar nanobumps appeared upon laser irradiation with their lateral size and the height increases with a pulse energy E, as it is illustrated by a series of side-view SEM images (Figure 1b).

Remarkably, observation of the laser-patterned areas covered with nanobumps with an ordinary bright-field optical microscope revealed their structural colors (Figure 1c). More importantly, the color was found to be tuned from reddish brown to pure green upon increase of the nanobumps size as illustrated by a series of bright-field optical images collected with microscope objective at NA=0.42. The nanobumps in all the arrays were printed at a periodicity of 1 μm that ensured the homogeneously colored surface where the isolated nanostructures can not be resolved. Reflection spectra of the laser-patterned areas allowed to correlate the change of the structural color with corresponding variation of the nanobump morphology (see Figure 1d). As can be see, the increase of the nanobumps height causes the gradual decrease of the reflectivity in the yellow-red part of the optical spectrum. In contrast, blue part of the spectrum shows slower decrease of the reflectivity versus the nanobump size that explains the resulting green color of the textured surfaces printed at pulse energy of 1.8 nJ.

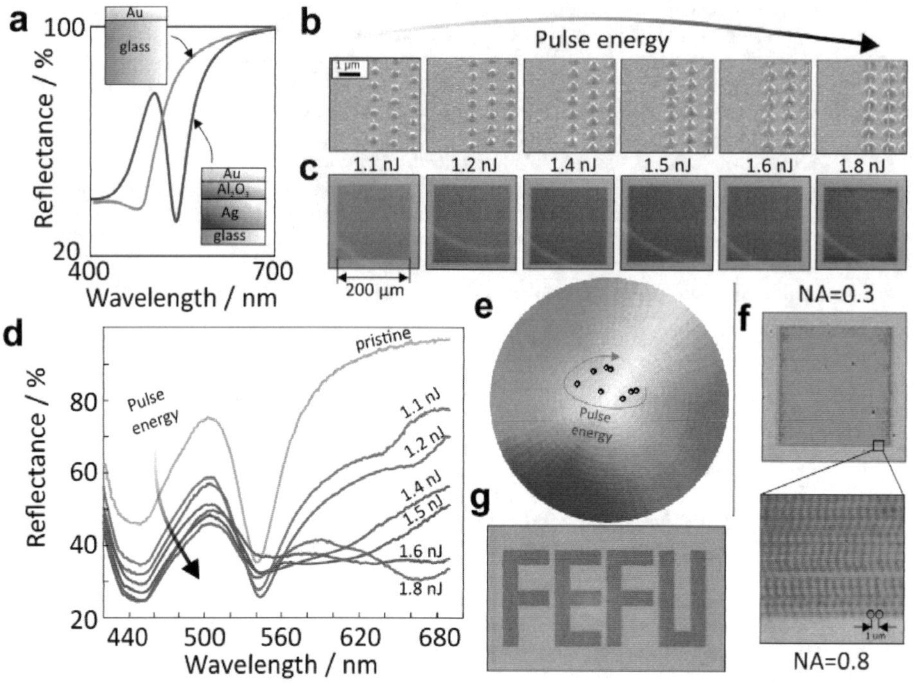

Figure 1. (a) Diffuse reflectance spectra of MIM sandwich (red curve) and bare 50-nm thick Au film on a glass substrate (yellow curve). (b) Series of side-view SEM images of the MIM surface textured at elevated pulse energy (from 1.1 to 1.8 nJ). (c) Corresponding bright-field optical image of the square-shape laser-patterned areas (200 x 200 μm^2) processed at elevated pulse energies E as well as reflectance spectra (d) of these areas. (e) HCV color space with markers indicating the colors of the pristine MIM sandwich and the patterned areas from (c). (f) Bright-field optical image of the laser-patterned area (E=1.8nJ) visualized with microscope objective at NA=0.3 and 0.8. (g) Bright-field optical image of the patterned area (400 x 80 μm^2) arranged to form "FEFU" letters. The nanobumps were imprinted at pulse energy E=1.8 nJ and periodicity of 1 μm.

Reflection spectra were further analyzed by converting them into the coordinates on the HCV color space to reveal the tuning range of the colors than can be reproduced using the developed laser-printing technique strategy. This data is summarized in Figure 1e illustrating that various rather pure colors can be reproduced via direct laser printing by changing the only experimental parameter – applied pulse energy. Noteworthy, imprinted structural colors of the laser-textured areas appeared similarly upon visualization with microscope objectives having variable NA that defines the collection angle of the reflected signal. Figure 1f compares the same laser-textured area produced at E=1.8 nJ upon its observation with low- and high-NA optics. As can be seen, visualization at NA=0.8 allowed to identify separate nanobumps that appeared as the green spots similarly to the whole laser-textured areas visualized at NA=0.3. In particular, this clearly illustrates possibility of the developed approach to imprint optical information and colored images at a lateral resolution up to 25,000 DPI. Finally, as an illustration of such possibility we imprinted several micro-scale surface areas arranged to form "FEFU" letters visualized in the reflection mode of the optical microscope at NA=0.42 (Figure 1g).

4. Conclusion

In conclusion, we demonstrated a novel approach to structural coloring of surface using the ablation-free direct laser patterning of metal-insulator-metal films. Using the proposed method, we prepared a test sample and performed a preliminary evaluation of the spectral properties of the fabricated structures. The structured surfaces demonstrate colors from reddish brown to pure green in reflection spectra depending on the shape of the single nanobamps.

Overall, the tailoring of the laser pulse parameters, as well as the thicknesses and materials of metal and dielectric films, provides a potential opportunity to shift the Fabry-Perot mode supported by the MIM sandwich in the full visible spectral range, and, consequently, to obtain colors from violet to red. The relative simplicity of the proposed approach, as well as the high lateral resolution of 25,000 dots per inch, pave the way for applications in optical sensors, security labeling, etc.

Acknowledgments
The work was supported by Russian Science Foundation 21-72-20122.

References

[1] Kristensen, A.; Yang, J.K.W.; Bozhevolnyi, S.; Link, S.; Nordlander, P.; Halas, N.J.; Mortensen, N. A. Plasmonic colour generation. Nat. Rev. Mater. 2016, 7, 16088

[2] Gu, Y.; Zhang, L.; Yang, J.; Yeoa, S.; Qiu, C.-W. Color generation via subwavelength plasmonic nanostructures. Nanoscale 2015, 7, 6409-6419.

[3] Olson, J.; Manjavacas, A.; Liu, L.; Chang, W.-S.; Foerster, B.; King, N. S.; Knight, M. W.; Nordlander, P.; Halas, N. J.; Link, S. Vivid, full-color aluminum plasmonic pixels. Proc. Natl. Acad. Sci. 2014, 111, 14348-14353.

[4] Zhu, X.; Yan, W.; Levy, U.; Mortensen, N.; Kristensen, A. Resonant laser printing of structural colors on high-index dielectric metasurfaces. Sci. Adv. 2017, 3, e1602487.

[5] Hail, C.U.; Schnoering, G.; Damak, M.; Poulikakos, D.; Eghlidi, H. A plasmonic painter's method of color mixing for a continuous RGB palette. ACS nano, 2020, 14(2), 1783-1791.

[6] Yokogawa, S.; Burgos, S.; Atwater, H. Plasmonic color filters for CMOS image sensor applications. Nano Lett. 2012, 12, 4349-4354

[7] Yang, C.; Shen, W.; Zhou, J.; Fang, X.; Zhao, D.; Zhang, X.; Ji, C.; Fang, B.; Zhang, Y.; Liu, X., Angle robust reflection/transmission plasmonic filters using ultrathin metal patch array. Adv. Opt. Mater 2016, 4, 1981-1986

[8] Kumar, K.; Duan, H.; Hegde, R.; Koh, S.; Wei, J.; Yang, J. Printing colour at the optical diffraction limit. Nat. Nanotechnol. 2012, 7, 557-561

[9] Wang, X.; Kuchmizhak, A.; Storozhenko, D.; Makarov, S.; Juodkazis, S. Single-step laser plasmonic coloration of metal films. ACS applied materials & interfaces. 2018, 10(1), 1422-1427.

[10] Sergeev, A. A.; Pavlov, D. V.; Kuchmizhak, A. A.; Lapine, M. V.; Yiu, W. K., Dong; Y., Rogach, A. L. Tailoring spontaneous infrared emission of HgTe quantum dots with laser-printed plasmonic arrays. Light: Science & Applications, 2020, 9(1), 1-10.

[11] Korte, F., Koch, J., Chichkov, B. N. Formation of microbumps and nanojets on gold targets by femtosecond laser pulses. Applied Physics A, 2004, 79(4), 879-881.

[12] Wang, X. W.; Kuchmizhak, A. A.; Li, X.,; Juodkazis, S.; Vitrik, O. B.; Kulchin, Y. N.; Inogamov, N. A. Laser-induced translative hydrodynamic mass snapshots: noninvasive characterization and predictive modeling via mapping at nanoscale. Physical Review Applied, 2017, 8(4), 044016.

Fluorescence lifetime-based intracellular thermometry for photothermal therapy

Oleksii O Peltek[1], Eduard Ageev[1], Pavel M Talianov[1], Mikhail V Zyuzin[1],

[1]Department of Physics and Engineering, ITMO University, Lomonosova 9, 191002, St. Petersburg, Russian Federation

*corresponding author: peltek.oleksii@gmail.com

Abstract. Photo-thermal therapy (PTT) is a rapidly developing approach for cancer treatment, that has greatly benefited from the tremendous advances in the synthesis of plasmonic nanoparticles (NPs), which can be used as light sensitive agents. This approach is based on the local induction of hyperthermia via light irradiation of plasmonic NPs in order to thermally kill cancerous cells. Such an approach ensures a high precision of treatment and the mechanism of action makes it as a valid alternative for treatment of malignant neoplasms with a multi-drug resistance. However, in order to minimize the adverse effects of PTT it is necessary to precisely measure and control the achieved temperatures at the tumor foci during light irradiation. Herein we report a novel technique to monitor intracellular temperatures during PTT based on temperature sensitive fluorescent dye, Rhodamine B. These findings may improve the quality of treatment and reduce unwanted adverse effects.

1. Introduction

Photo-thermal therapy (PTT) is a minimally invasive approach for cancer treatment that has gained a lot of attention in recent years as an additional therapy, which can be combined with traditional approaches of cancer treatment to improve therapy [1]. The main operating principle of this therapy is to convert the energy of electromagnetic radiation using photosensitizer to heat. The excessive heat leads to local hyperthermia, which causes cancer cell death. Recent works reported on the use of various photosensitizers, such as nanographene sheets [2], carbon nanodots [3], and a large number of organic molecules [4]. Plasmonic nanoparticles are the most extensively studied due to their superb optical properties, great tunability, and bio-targeting capabilities [5]. Owing to the effect of plasmon resonance, plasmonic NPs possess enhanced near-infrared (NIR) absorption, which is crucial for effective PTT. Depending on the temperatures reached during the PTT, cancerous cells can die via either apoptosis or necrosis pathway [6]. The difference is crucial for the effectiveness of cancer treatment since a large number of necrotic cells can cause a cytokine storm that may lead to various complications during cancer treatment. Therefore, despite all the advantages of PTT, it is required to precisely control the temperatures reached during the light irradiation.

Different studies have used various techniques to determine temperatures reached during PTT, which include infrared thermometers and thermocouples. However, these methods allow detecting of changes in temperature of not only cells but also the surrounding media, which significantly decreases the accuracy of the measurements. Thus, to accurately assess the temperatures reached in the cancerous cell during the irradiation another, more precise thermometry approach is required. For this purpose, certain fluorescent dyes can be utilized since their optical properties, such as fluorescence intensity and fluorescence lifetime depends on the temperature changes in the surrounding media [7]. Therefore, by detecting these changes it is possible to precisely measure temperatures reached in certain parts of cells, which were previously described in a number of works [8]. However, there are almost no studies that have utilized fluorescent probes for temperature monitoring during PTT.

In the present work we have developed an easy and precise method for intracellular thermometry based on fluorescence lifetime microscopy during PTT using Rhodamine B as a fluorescent temperature sensor.

2. Materials and methods

2.1. Materials

Gold (III) chloride trihydrate ($HAuCl4 \cdot 3H2O$, ≥99.9%, Sigma-Aldrich), ascorbic acid (AA, ≥99.0%, Sigma-Aldrich), sodium borohydride (NaBH4, 98%, Sigma-Aldrich), cetyltrimethylammonium bromide (CTAB, ≥99%, Sigma-Aldrich), sodium oleate (NaOL, > 97.0%, Sigma-Aldrich), Rhodamine B (RhB, ≥95%, Sigma-Aldrich).

2.2. Au NRs synthesis

Au NRs with absorbance peak in the NIR region were synthesized using a seeded growth approach and binary surfactant mixture [9]. The initial seed solution was prepared as follows: 5 mL of 0.2 M CTAB solution was added to 5mL of 0.5 mM HAuCl4. Afterward, freshly prepared 0.6 mL of 0.01 M NaBH4 was diluted to 1 ml and then added to Au(III)-surfactant solution, which changed color to brownish. Stirring was stopped after 2 minutes and the seed solution was left undisturbed for 30 minutes prior to further use. Meanwhile, the growth solution was prepared. For this 1.234 g of NaOL and 9 g of CTAB were added to 250 mL of water and then heated up to 50°C until they dissolved. Afterward, the solution was cooled down to 30°C and 36 mL of 4 mM AgNO3 were added. The solution was gently mixed and left undisturbed for 15 minutes. Then 250 mL of 1 mM HAuCl4 solution was added and was left under stirring (700 rpm) for 90 minutes. After this 3 mL of HCl was injected to adjust pH and stirred for another 15 minutes. Finally, 1.25 mL of 0.064 M AA was added to the solution, quickly stirred for 30 seconds, and injected with 0.2 mL of previously prepared seed solution. The obtained solution was left for 12 hours at 30°C for Au NRs to grow. On the following day, the solution was centrifuged at 7,000 rpm for 30 min to remove the supernatant and consequently stabilized using H2N-PEG-SH via ligand-exchange procedure [10].

2.3. Cells

Murine melanoma cell line (B16-F10 cells) was purchased from the American Type Culture Collection. Cells were cultivated in AlphaMEM with 10% of vol. fetal bovine serum and 2mM UltraGlutamine I at 37°C in a sterile humidified atmosphere with 5% CO_2.

2.4. Fluorescence lifetime temperature dependence

In order to obtain a calibration curve, describing the relation between fluorescence lifetime of RhB and various temperatures, 100 μM water solution of RhB was used. Values of fluorescence lifetime were obtained using Picoquant Picoharp 300 time-correlated single-photon counting system. The heating of the RhB solution was realized using a custom temperature control system, in which the temperature of the solution was monitored via a thermocouple.

2.5. Intracellular temperature measurements

For the temperature measurements, cells were seeded into confocal cell imaging dishes (d = 35 mm, Eppendorf) at the amount of 1.0 x 10^4 per dish. On the next day, Au NRs were added to the cells, and once again cells were left overnight. On the following day, cells were stained with 15 μL of 1 mM RhB solution. After 30 min, cells were washed twice with PBS and left in PBS supplemented with 4%

glucose. Fluorescence lifetime imaging microscopy (FLIM) was performed with a custom setup, which consisted of Picoquant PDL 800-D picosecond pulsed diode laser driver, Picoquant laser head LDH-FA 530XL, single-photon avalanche diode (MPD PDM PD-100-CTC-FC), and Picoquant Picoharp 300 time-correlated single-photon counting system.

3. Results

Synthesized Au NRs were characterized using Scanning Transmission Electron Microscopy (STEM) (Figure 1A). As determined from STEM pictures, the average width of Au NRs was around 20±1.1 nm and the length was approximately 124±5.4 nm. The obtained Au NRs had an absorbance maximum at 1173 nm (Figure 1B).

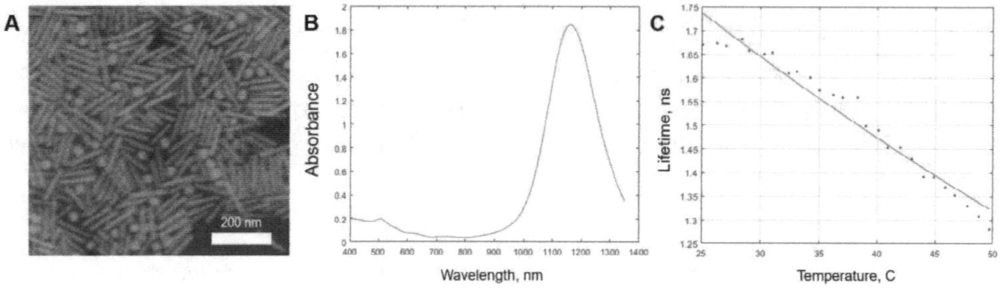

Figure 1. A. STEM images of Au NRs. B. Absorbance spectrum of Au NRs. C. Calibration curve of RhB fluorescence lifetimes depending on the solution temperature.

In order to estimate the relation between temperatures of the RhB solution and fluorescence lifetime, a calibration curve was obtained (Figure 1C). Afterward, FLIM of cells associated with Au NRs under NIR laser irradiation was performed. Depending on the laser intensity, the heating from Au NRs changed, which led to different intracellular temperatures and therefore to different fluorescence lifetimes detected via FLIM (Figure 2). Using the previously obtained calibration curve, we matched values of lifetime, which we detected with their corresponding values of intracellular temperature. Finally, based on the obtained microphotographs and calibration curve, the optimal parameters for achieving an intracellular temperature of 45.2±1.4 °C was power density equal to 0.32 mW/mm2 using a 1064 nm fiber laser with 100MHz repetition rate. According to the published works, this temperature may be considered optimal for the effective PTT, as it maximizes the apoptosis of cancer cells [6,11].

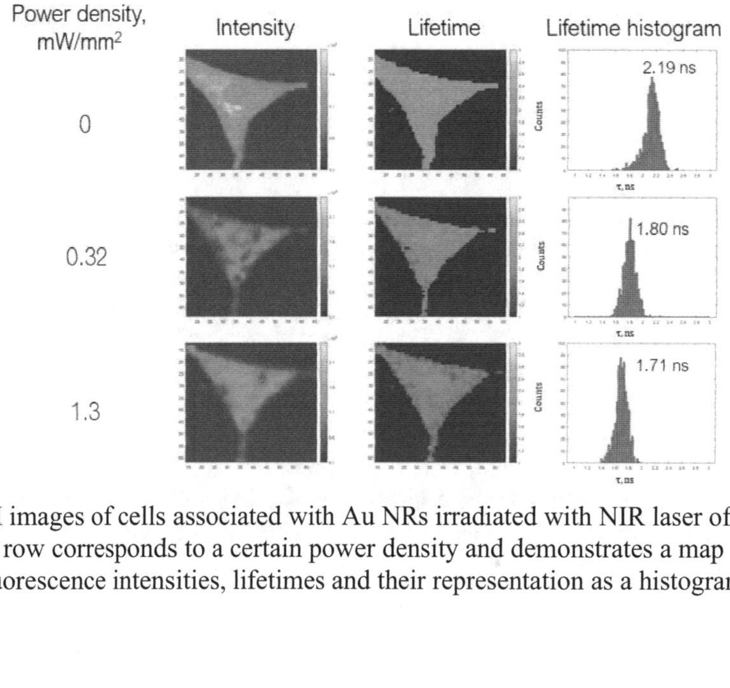

Figure 2. FLIM images of cells associated with Au NRs irradiated with NIR laser of various power densities. Each row corresponds to a certain power density and demonstrates a map of intracellular fluorescence intensities, lifetimes and their representation as a histogram.

4. Conclusion

In this work, we developed a novel technique that allows one to precisely estimate temperatures achieved during laser irradiation using fluorescence lifetime imaging microscopy and fluorescent dye – Rhodamine B. This approach can be used to select optimal parameters of laser irradiation to achieve the most efficient therapy. This method allows to control temperatures in an intracellular environment and, therefore, paves the way for the broader introduction of PTT in clinics.

5. Acknowledgments

This work was supported by the Russian Science Foundation (21-72-30018).

References

[1] Hu J-J, Cheng Y-J and Zhang X-Z 2018 *Nanoscale* **10** 22657–72.
[2] Yang K, Zhang S, Zhang G, Sun X, Lee S-T and Liu Z 2010 *Nano Lett.* **10**(9) 3318–23.
[3] Ge J, Jia Q, Liu W, Lan M, Zhou B, Guo L, et al. 2016 *Adv Healthcare Mater.* **5**(6) 665–75.
[4] Jung HS, Verwilst P, Sharma A, Shin J, Sessler JL and Kim JS. 2018 *Chem Soc Rev.* **47**(7) 2280–97.
[5] Ali MRK, Wu Y and El-Sayed MA. 2019 *J Phys Chem C.* **123**(25) 15375–93.
[6] Zhang Y, Zhan X, Xiong J, Peng S, Huang W, Joshi R, et al. 2018 *Sci Rep.* **8**(1) 8720.
[7] Chauhan VM, Hopper RH, Ali SZ, King EM, Udrea F, Oxley CH, et al. 2014 *Sensors and Actuators B: Chemical.* **192** 126–33.
[8] Moreau D, Lefort C, Burke R, Leveque P and O'Connor RP. 2015 *Biomed Opt Express.* **6**(10) 4105.
[9] Ye X, Zheng C, Chen J, Gao Y and Murray CB. 2013 *Nano Lett.* **13**(2) 765–71.
[10] Muslimov AR, Timin AS, Bichaykina VR, Peltek OO, Karpov TE, Dubavik A, et al. *Biomaterials Science.* **8**(4) 1137–47.
[11] Pattani VP, Shah J, Atalis A, Sharma A and Tunnell JW. 2015 *J Nanopart Res.* **17**(1) 20.

Angular Goos-Hanchen shift in subwavelength gratings enhanced by surface plasmon resonance

N I Petrov[1], V A Danilov[1] and B A Usievich[2]

[1] Scientific and Technological Centre of Unique Instrumentation of the Russian Academy of Sciences, 15 Butlerova str., Moscow 117342, Russia
[2] Prokhorov General Physics Institute of the Russian Academy of Sciences, Moscow, Russia

Corresponding author: petrovni@mail.ru

Abstract. The angular Goos-Hanchen shift dependence on the subwavelength grating parameters and the incident Gaussian beam width is investigated theoretically. High sensitivity of the angular Goos-Hanchen shift to the incident angle of a light beam near the surface plasmon resonance is demonstrated. Splitting of the reflected beam into two angularly separated beams is shown for strongly focused beam incident at the surface plasmon resonance angle.

1. Introduction

Plane waves of infinite width are widely used to describe the optical properties of diffraction gratings (see, for example, [1-3]). However, the fields associated with laser optical systems and other sources are spatially limited. Moreover, the fields of reflected and transmitted beams demonstrate diffraction phenomena that are not found in a plane wave. A variety of interesting phenomena emerges from the consideration of a bounded incident beam. If we consider the finite beam, it becomes obvious the excitation of surface plasmons, Goos-Hanchen (GH), focal and angular shifts [1, 4, 5]. In [6], the shift of the trajectory of sagittal rays propagating in a graded-index fiber was studied. In [7, 8] positive and negative GH shifts are demonstrated for the reflected beams from subwavelength gratings near the surface plasmon resonance. In [9] the theory of Goos-Hanchen shift and reshaping shift are developed and the GH shift for surface plasmons experimentally examined. In [10] the angular shift of a Gaussian beam reflected by a dielectric interface near the Brewster angle is analyzed. Angular shift and significant oscillations in the intensity distribution of the beam reflected by a dielectric interface was demonstrated for strongly focused incident beam in [11]. In [12] experimental demonstration of this angular deviation at reflection of an optical beam from an air-glass interface has been reported. A description of the spatial and angular Goos-Hänchen and Imbert-Fedorov shifts occurring at reflection of a light beam from plane interface is presented in [13]. GH shift has been observed experimentally in reflection from an air-gold plane interface [13]. In [14] an angular GH shift near the surface plasmon resonance (SPR) excitation is demonstrated.

In this paper, we investigate the angular GH shift when the surface plasmon resonance is excited in a subwavelength silver grating with a period $\Lambda = 400$ nm. The angular splitting of the reflected beam into two separated beams is shown for specific values of the width of the incident Gaussian beam and the depth of the grating.

2. Angular GH shift

A rigorous electromagnetic theory based on the C- method is used for calculations [2, 3, 7, 8, 15]. Note that an *s*- wave does not excite surface plasmon waves, so we consider the GH shift for the *p*-polarized incident wave only.

In Fig. 1 the intensity profiles as function of reflection angle are presented for different incident beam waists w_0. The wavelength of the incident beam is $\lambda = 632.8$ nm, the incident angle corresponding to SPR is $\theta_i = 33.25°$, the silver grating depth h = 30 nm and the sinusoidal grating period $\Lambda = 400$ nm.

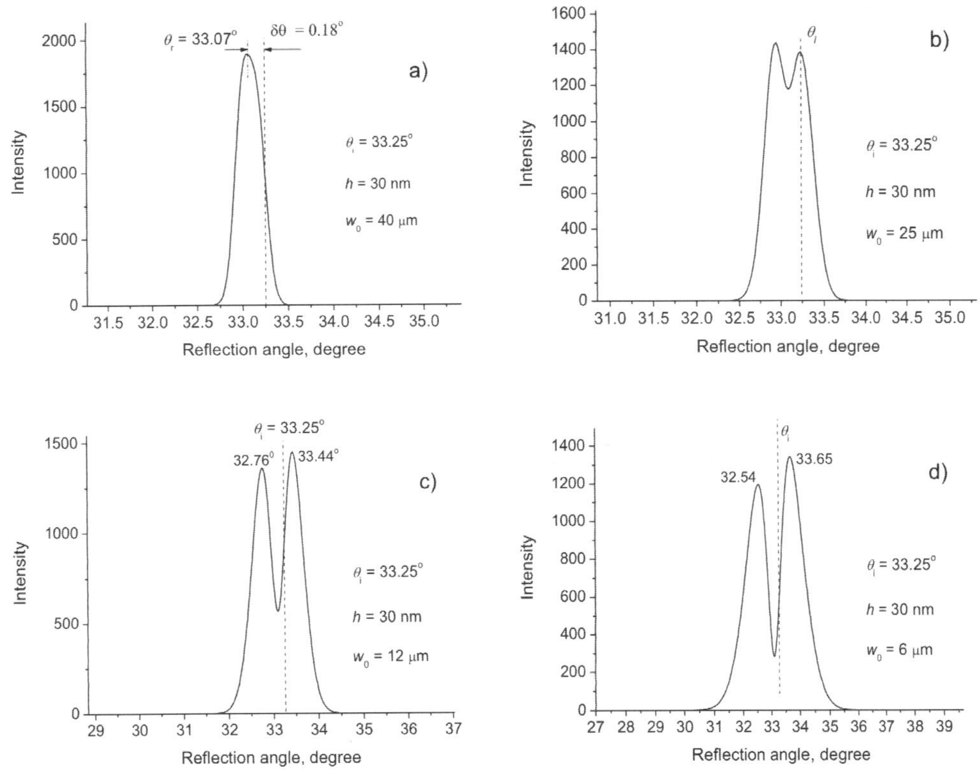

Figure 1. Intensity profiles as function of the reflection angle for different incident beam waists w_0: (a) $w_0 = 40$ μm; (b) $w_0 = 25$ μm; (c) $w_0 = 12$ μm; (d) $w_0 = 6$ μm.

It follows that negative angular GH shift of the order of the angular beam width occurs near the resonance angle of incidence. The reflected beam is divided into two beams when the incident beam is tightly focused. The spatial shift in this case becomes negligible and the angular shift becomes dominated.

In Fig. 2 the intensity distributions as function of the reflection angle are presented for the grating depth h = 20 nm. Note that this value of the depth corresponds to the minimum reflection of the plane wave at the resonance incidence angle [8]. Other parameters are the same as in Fig. 1.

We found that the angular shift of the reflected beam is sensitive to the angle of incidence and the depth of the grating. For tightly focused incident beams, there is a significant change in the shape of the reflected beam. In this case, two reflected beams appear, separated by an angular distance. At a certain value of the beam width, the amplitudes of these reflected beams become the same. This indicates that the subwavelength grating can work as a beam splitter.

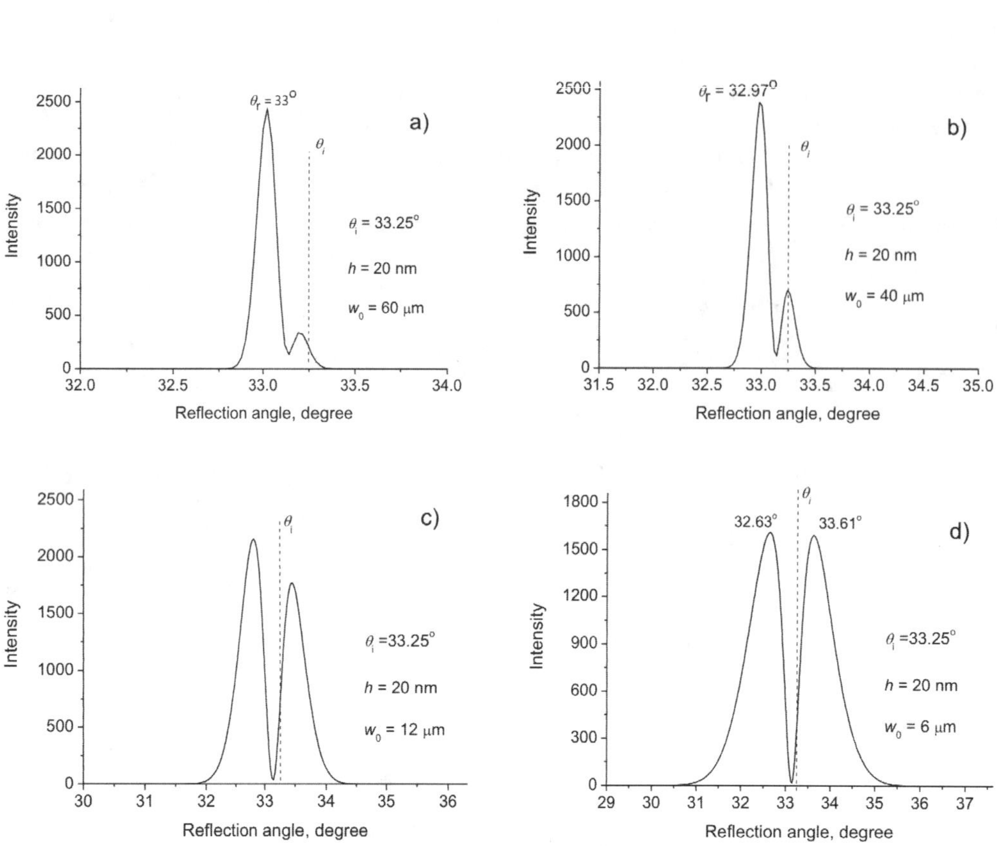

Figure 2. Intensity profiles as function of the reflection angle for different incident beam waists w_0: (a) $w_0 = 60$ μm; (b) $w_0 = 40$ μm; (c) $w_0 = 12$ μm; (d) $w_0 = 6$ μm.

It follows from the simulation that there is an effect of splitting the reflected beam into two parts with positive and negative angular shifts, respectively, when the width of the incident beam decreases. Such angular shifts are caused by energy flows propagating forward and backward along the grating-air interface at resonance. Indeed, in this case, a sharp change in the value of the phase and its sign of the reflected beam takes place. Noticeable distortions of the reflected beam appear near the critical grating depth $h_{cr} \approx 20$ nm even for incident beams with large waists (Fig. 2 (a, b)). However, at depths other than the critical depth of the grating, the distortions also disappear for highly focused beams. Note that positive and negative lateral beam shifts were observed on the flat surface silver–air interface near the surface plasmon resonance in [16]. The existence of forward and backward surface plasmon waves propagating along the metal–air interface at resonance is responsible for such positive and negative spatial shifts.

3. Conclusions

Thus, the angular GH shift near the SPR is investigated for different grating depths and the incident beam widths by using rigorous electromagnetic calculation. For the reflected beam, an angular shift is demonstrated, enhanced by a surface plasmon resonance of the order of the angular width of the beam. It is shown that the reflected beam splits into two angularly separated beams with large negative and positive angular shifts. The angular distance between the intensity peaks increases with the decrease of the incident beam waist.

The results can be useful in the development of different sensors, as well as in angular metrology and cantilever-based surface microscopies. They can be also used in the angular alignment of LIGO or LISA gravitational wave detectors [17].

Acknowledgements: This work was supported by the Russian Foundation for Basic Research (project No. 19-29-11026) and the Ministry of Science and Higher Education of the Russian Federation under the State contract No. 0069-2019-0006.

References

[1] Kou E F Y and Tamir T 1989 *Applied Optics* **28** 1169
[2] Petrov N I, Nikitin V G, Danilov V A, Popov V V and Usievich B A 2014 *Appl. Opt.* **53** 5740
[3] Petrov N I, Danilov V A, Popov V V and Usievich B A 2018 *Quantum Electronics* **48** 537
[4] Goos F and Hänchen H 1947 *Annalen der Physik* **436** 333
[5] Tamir T and Bertoni H L 1971 *J. Opt. Soc. Am.* **61** 1397
[6] Petrov N I 2013 *J. Opt.* **15** 014011
[7] Petrov N I, Danilov V A, Popov V V and Usievich B A 2020 *J. Phys.: Conf. Ser.* **1461** 012131
[8] Petrov N I, Danilov V A, Popov V V, Usievich B A 2020 *Optics Express* **28**(5) 7552
[9] Saito H and Tomita M 2021 *JOSA B* **38** 1048
[10] Chan C C, Tamir C 1985 *Opt. Lett.* **10** 378
[11] Petrov N I 2005 *J. Mod. Opt.* **52** 1545
[12] Merano M, Aiello A, van Exter M P and Woerdman J P 2009 *Nature Photonics* **3** 337
[13] Aiello A, Merano M and Woerdman J P 2009 *Phys. Rev. A* **80** 061801(R)
[14] Olaya C M, Hayazawa N, Hermosa N and Tanaka T 2021 *J. Phys. Chem. A* **125** 451
[15] http://www.mcgrating.com/
[16] Yin X, Hesselink L, Liu Z, Fang N and Zhang X 2004 *Appl. Phys. Lett.* **85** 372
[17] Centrella J M 2003 *Am. J. Phys.* **71** 520

Thin-Film Frustrated Total Internal Reflection Filter with Plasmonic Nanoparticle Inclusions in the Layers

N I Petrov

Scientific and Technological Centre of Unique Instrumentation of the Russian Academy of Sciences, 15 Butlerova str., Moscow 117342, Russia

Corresponding author: petrovni@mail.ru

Abstract. The influence of plasmonic nanoparticles embedded in the central and side layers of the frustrated total internal reflection filter on the resonant transmission of light is analyzed. It is shown that the frequency dispersion causes the splitting of the filter bandwidth and the angular splitting of the incident beam into several output beams.

1. Introduction

Resonant effects caused by wave interference occur when tunneling through a system of barriers. These effects are an optical analogy of quantum mechanical tunneling. They have found application in various areas of physics, including electron transfer in solid-state superlattices, integrated and fiber optics, and frustrated total internal reflection (FTIR) spectroscopy.

The FTIR filter was first proposed by Leurgens and Turner in 1947 [1]. It is known that FTIR filter uses resonant tunneling of light through a planar dielectric optical waveguide placed between two low refractive index thin films that act as potential barriers. Both integral and differential methods are used to solve Maxwell's equations to describe the FTIR effect in layered structures [2-5]. Currently, dielectric thin films with high thickness accuracy and low absorption and scattering losses are available. In [6] FTIR effect was considered in a circular polarization beam splitter [6]. High performance FTIR based thin-film linear polarizing beam splitter has been demonstrated in [7]. In [8], a FTIR based device was proposed for the spatial separation of an incident white light beam into three color beams. In [9] the effect of frequency dispersion in the resonator layer on the FTIR process has been investigated.

In this paper, the influence of the frequency dispersion caused by metal nanoparticles embedded in the layers of the FTIR filter on the resonant transmission of light is theoretically investigated. In comparison with the system considered in [9], here we analyze the effect of plasmonic nanoparticles embedded in the central and lateral layers. This allowed us to demonstrate the splitting of the incident beam into five spectrally and angularly separated output beams.

2. FTIR filter

The schematic view of the filter is shown in Fig. 1. The filter consists of a three-layer structure placed between two prisms with refractive indices n_p and n'_p. A high-index central layer with a high refractive index and thickness d_2 is placed between two films with low refractive indices n_1 and n'_1, and thicknesses d_1 and d_1', respectively. Angular and chromatic filtering of the incident light beam occurs due to the resonant diffraction effect when light propagates through a layered structure, i.e., the FTIR effect.

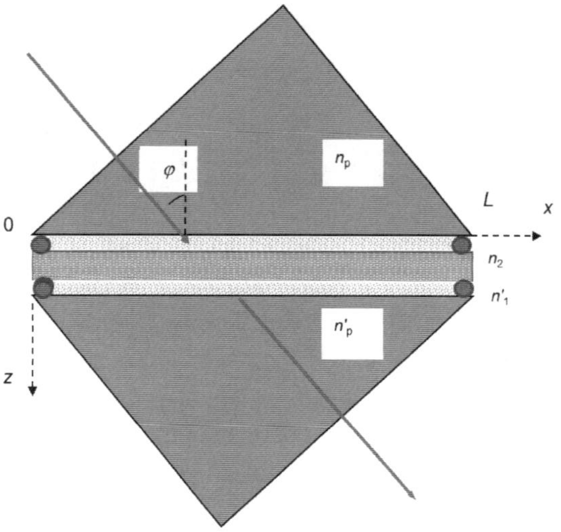

Figure 1. FTIR filter with nanoparticle inclusions in the layers.

The resonant condition for s-polarization is determined by the equation [8, 9]:

$$k_z' d_2 = -\arctan\left[\frac{k_z'(q_z + q_z')}{q_z q_z' - k_z'^2}\right] - Q_3, \quad (1)$$

where $Q_3 = 2q_z k_z'(q_z^2 - k_z^2)\exp(-2q_z d_1)/[(q_z^2 + k_z^2)(q_z^2 + k_z'^2)]$, $q_z = (\omega/c)\sqrt{n_p^2 \sin^2\varphi - n_1^2}$, $k_z = (\omega/c)\sqrt{n_p^2 - n_p^2 \sin^2\varphi}$, $q_z' = (\omega/c)\sqrt{n_p'^2 \sin^2\varphi - n_1'^2}$, $k_z' = (\omega/c)\sqrt{n_2^2 - n_p'^2 \sin^2\varphi}$ are the wavenumbers.

In the case of p- polarization, similar expressions for the resonance condition can be obtained.

For a fixed frequency of incident radiation, the dependence of the incidence angle on the dispersion is described by the expression:

$$\varphi(\omega) = \varphi_0(\omega) + \frac{d_2^*}{d_{2eff}} \frac{\omega/c}{n_p^2 \sin 2\varphi_0} \Delta\varepsilon_2' + \left(1 - \frac{d_2^*}{d_{2eff}}\right) \frac{2\Delta\varepsilon_1'}{n_p^2 \sin 2\varphi_0}, \quad (2)$$

where $\varphi_0(\omega)$ is the incident angle in the absence of dispersion, $\Delta\varepsilon_2'$ and $\Delta\varepsilon_1'$ are the changes of the real part of the permittivity in the central and side layers, respectively, $d_2^* = d_2 + 2q_z/\left(q_z^2 + k_z'^2\right)$.

Let's consider the layers with embedded metallic nanoparticles. Such a medium, in the framework of the Maxwell-Garnett model, is described by the effective dielectric permittivity. For spherical nanoparticles we have the relation [10, 11]:

$$\varepsilon_{eff} = \varepsilon_m + \frac{3\eta\left(\varepsilon_p - \varepsilon_m\right)\varepsilon_m}{3\varepsilon_m + \left(1-\eta\right)\left(\varepsilon_p - \varepsilon_m\right)}, \tag{3}$$

where η is the volume fraction of nanoparticles, ε_p is the dielectric permittivity of the nanoparticles. ε_m is the dielectric constant of the central layer.

In Fig. 2 the resonance incidence angle as function of the wavelength of *s*-polarized beam is shown for silver and gold nanoparticles. It follows that the resonant spectral lines shift when the angle of incidence of light changes. It is possible to get several resolved spectral lines at the output by changing the angle of incidence within certain limits. It can be seen that for a given angle of incidence, there are several resonant bands at once. This indicates that the resonant condition in the resonator is satisfied for several wavelengths at once. It follows from the simulation that the angular splitting of the beam occurs for a given wavelength, and for a certain wavelength of the incident beam, up to five output beams can be observed, separated by angles.

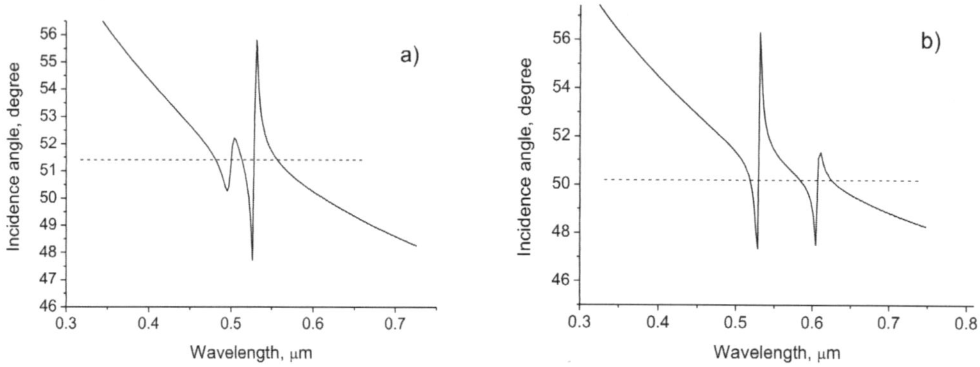

Figure 2. Resonance incident angle of *s*-polarized beam as function of wavelength. $n_p = 2.0$, $n_1 = 1.38$, $n_2 = 2.0$, $d_2 = 70$ nm; $\eta = 10^{-3}$. (a) $d_1 = 300$ nm; Ag nanoparticles in the central layer and Au nanoparticles in the side layers; (b) $d_1 = 500$ nm; Au nanoparticles in the central and side layers.

In Fig. 3 the spectral lines of transmitted light of FTIR filter with gold nanoparticle inclusions in the central and side layers are presented for the *s*- polarized incident beam. It follows that the spectral bandwidths decrease with the increase of the low-index layer thickness d_1 and the spectral line width $\Delta\lambda = 27$ nm was obtained at $\lambda = 518$ nm and $d_1 = 500$ nm. Five resonant bands exist at once for a given angle of incidence due to the fulfillment of the resonance condition for five wavelengths simultaneously.

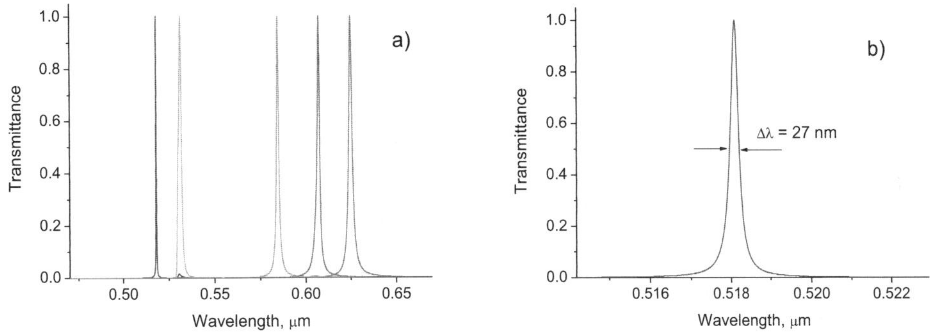

Figure 3. Spectral lines of transmitted light: $n_p = 2.0$, $n_1 = 1.38$, $n_2 = 2.0$, $d_1 = 500$ nm, $d_2 = 70$ nm, $\eta = 10^{-3}$. $\varphi = 50.18^{\circ}$.

It should be noted that resonant structures based on nanoparticle inclusions open up opportunities for creating infrared (IR) and terahertz technology devices that are inaccessible to conventional materials. In [11] acousto-optic filters with inclusions of metamaterials which provide ultra-narrow spectral lines due to resonant Bragg diffraction were considered. Large Goos-Hanchen (GH) shifts enhanced by the surface plasmon resonance in subwavelength gratings were demonstrated in [12]. In [13, 14] the resonance tunneling and GH and Imbert–Fedorov shifts of terahertz beams from graphene plasmonic metasurfaces were investigated.

3. Conclusions

Thus, the influence of photonic and plasmonic effects on the transmission spectrum of the FTIR filter with nanoparticle inclusions in the central and lateral layers is analyzed. The splitting of the filter bandwidth into several narrow-band spectral lines for a given incidence angle is shown. It is shown that an incident beam of a given wavelength splits into five angularly separated output beams. The considered thin-film beam splitter can be useful in many applications, such as color visualization systems, color display devices, sensors, spectroscopy, in spectral regions extending from the UV to the far IR range.

Acknowledgements: This work was supported by the Ministry of Science and Higher Education of the Russian Federation under the State contract No. 0069-2019-0006 and by the Russian Foundation for Basic Research, project No. 19-29-11026.

References

[1] Leurgens P and Turner A F 1947 *JOSA* **37** 983
[2] Iogansen L V 1961 *Sov. Phys. JETP* **13** 1291
[3] Bergstein L and Shulman C 1966 *Appl. Opt.* **5** 9
[4] Baumeister P W 1967 *Appl. Opt.* **6** 897
[5] Ulrich R 1970 *JOSA* **60** 1337
[6] Azzam R M A and De A 2003 *Opt. Lett.* **28** 355
[7] Li L and Dobrowolski J A 2010 *Opt. Express* **18** 3784
[8] Petrov N I 2007 *Opt. Lett.* **32** 2744
[9] Petrov N I 2020 *OSA Cont.* **3**(9) 2591
[10] Spanier J E and Herman I P 2000 *Phys. Rev. B* **61** 10437
[11] Petrov N I and Pustovoit V I 2019 *JETP Lett.* **109** 18
[12] Petrov N I, Danilov V A, Popov V V and Usievich B A 2020 *Opt. Exp.* **28**(5) 7552
[13] Chen J, Ban Y, Zhu Q B and Chen X 2016 *Opt. Lett.* **41** 4468
[14] Farmani A, Miri M and Sheikhi M H 2017 *J. Opt. Soc. Am. B* **34**(6) 1097

Direct laser writing method for micro-lens fabrication on the fiber facet

A.V. Pisarenko, D.A. Kolymagin, R.D. Zvagelsky, D.D. Merkushev, D. A. Shcherbakov, D.A. Chubich, A.G. Vitukhnovsky

[1]Moscow Institute of Physics and Technology (National Research University), Institutskii per. 9, Moscow Region, Dolgoprudny, 141700, Russian Federation

mipt.3D@gmail.com

Abstract. We present a way for micro-lens and micro-lens array fabrication on the fiber facet by direct laser writing (DLW) method. The proposed setup for DLW printing on the fiber facet can protect objective lens and makes it possible to safely using a wide range of resists. Micro-lens fabricated on the single-mode optical fiber facet is demonstrated. Based on lens-on-fiber system we proposed concept of the micro-lenses on baseplate array printed on the multi-core fiber, where each lens is rigidly aligned with the core. One of the possible micro-lenses on the baseplate array is presented and its optical properties such as focal spot size and resolution are investigated. The possible application of the proposed micro-lens array is complex optical elements, such as micro-objectives with optimized optical design. Moreover, suggested freeform lens array can find application in high-accuracy wavefront sensing.

1. Introduction

Nowadays, direct laser writing (DLW) method is commonly used for micro-optical elements fabrication. This state-of-the-art technology gives sub-micrometer accuracy and makes it possible to fabricate complex freeform 3D polymer structures. DLW opens large possibilities for fabrication high quality freeform micro-lenses and micro-objectives for wavefront transformation, such X-ray parabolic compound refractive lenses with curvature radius up to 5 μm at the paraboloid apex [1]. Also, the efficient single-photon source consisting of QD with a micro-lens, whose emission is collimated by the on-chip micro-objective has been demonstrated recently [2]. Moreover, DLW enables to combine polymer structures with optical fiber [3] or to create elements directly on the fiber facet [4], which is promising for sensing and imaging.

Here we demonstrate free-form micro-lens array on baseplate for MCF fiber and micro-lens on single-mode fiber facet fabricated by DLW and perform its optical characterization.

2. Experimental section

For the micro-lenses fabrication by direct laser writing method we used commercial 3D two-photon lithography system (Photonic Professional, Nanoscribe GmbH). The development for both micro-lens on the cover glass and micro-lens on the fiber facet were performed according to standard protocol (PGMEA 25 min, IPA 5 min).

2.1. Single micro-lens design and realization

For lens printing with DLW method we prepared the adaptive model with variable slicing and hatching distances in order to reducing RMS roughness. The different design (spherical and aspherical) with radius in the range from 10 to 150 μm were used for lenses fabrication. In order to reduce the focal spot size and improve the quality of light coupling we optimized the micro-lens shape to aspherical using optical design software Zemax. Also the geometry optimization tooks into account the dispersion of the refractive index of the polymer. It should be noted that combining micro-lenses which consists of a wide range of photosensitive compositions with different dispersion characteristics makes it possible to optimize such parameters as a chromatic aberration. As a transparent material for lenses fabrication, commercial IP-L resist (refractive index 1.52 @ 532 nm, [5]) and custom resists, such as acrylate with imidazole photoinitiator [6] were used. After fabrication we performed micro-lens characterization using setup that includes light source (white light or green laser), micropositioner and CCD camera. This setup makes it possible to examine the focal plane as well as to obtain an image of USAF target in order to estimate the resolution of micro-lens. The image of the USAF target (25 lp/mm) for optimized aspherical micro-lens is presented on Fig.1 (a). Also shape control of spherical and aspherical micro-lenses was conducted with atomic force microscopy (AFM) method. Besides lateral dimensions parameters measurements, AFM technique enables to control height, radius of curvature and surface roughness of the micro-lens.

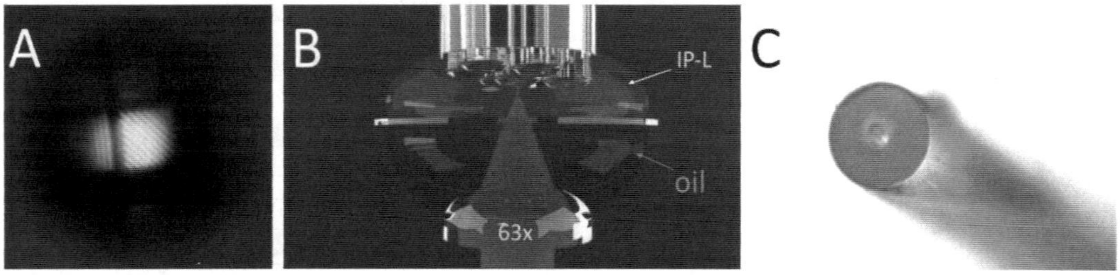

FIGURE 1. (a) The image of USAF target obtained with printed micro-lens. (b) Set-up for DLW printing on the facet of optical fiber (not to scale). (c) Micro-lens fabricated on the single-mode optical fiber rigidly alignment with the core (optical microscopy image)

2.2. Printing micro-lens on the fiber facet
For DLW printing on the optical fiber facet we used custom-made fiber holder combined with Nanoscribe (Fig.1 b). This set-up makes it possible to print elements on the optical fiber facet throw cover glass that protects the objective lens from mechanical damage, provides easy way to find the resin-glass interface and ability to use chemically aggressive resists. Before the printing process, the single mode telecom fiber were cleaved, cleaned and mounted into a holder. The image of the fabricated lens on the fiber facet is presented on fig.1 (c).

2.3. Lens on baseplate concept for multi-core fiber tools
In order to improve the quality of light coupling in optical waveguide we present an idea of baseplate with different controllable height (fig. 2 a). Holes in baseplate are made to allow the resist to flow out of the structure during development and reduce the fabrication time. Proposed approach of the micro-lens on baseplate is promising for fabrication a micro-lens array combined with multi-core fiber, where each lens is rigidly aligned with its core with sub-micron precision (Fig. 2 b). Microlens-array can significantly increases light coupling efficiency by focusing light directly into each single core of multicore fiber (as a commercially available MCF with 7 cores and cladding diameter of 125 μm). Moreover, the efficiency of coupling can be controlled by varying the shape of both sides of the lens and height of the baseplate (i.e. lens-facet distance). As a demonstration we fabricated the proposed multi-lens system and investigated its optical properties (Fig. 2 c).

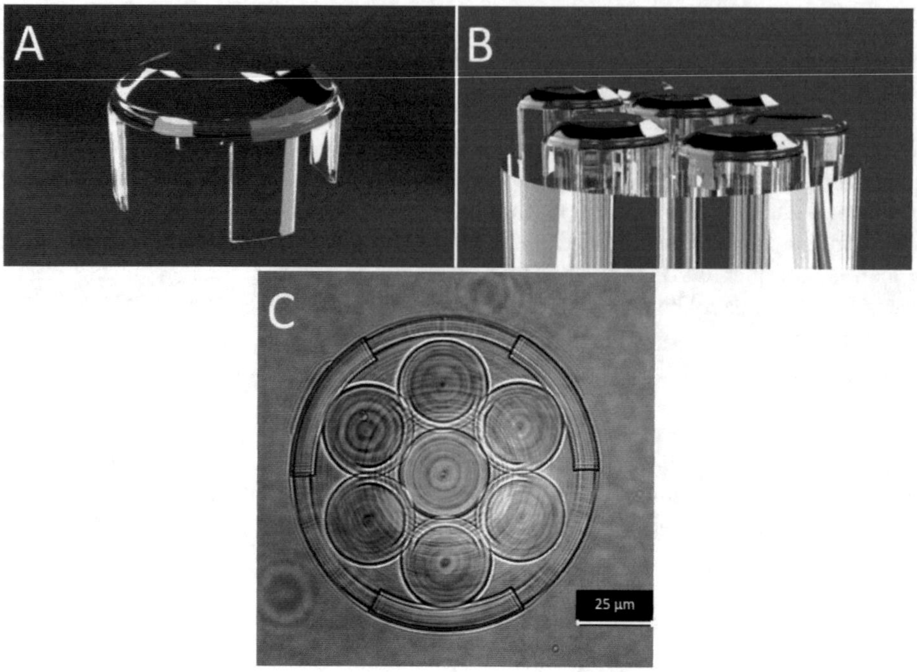

FIGURE 2. (a) Model of the single micro-lens on the baseplate. (b) Concept of micro-lenses on baseplates for light coupling in MCF. (c) Massive of micro-lenses on the baseplate fabricated with DLW method (optical microscope image).

3. Results and discussion

We designed and fabricated such optical components as single micro-lens, micro-lens array and micro-lens on the optical fiber facet. Also we presented the concept of the micro-lens on the baseplate array printed on the fiber facet, which makes it possible to control intensity and other characteristics of light coupling. Also this approach combined with methods of computer vison and correlation analysis can find application for improving the resolution of image systems (e.g. increasing depth of field). The system of such micro-lens array with 7 lens on baseplate is fabricated and its optical properties are measured. The obtained typical size of the focal spot is about 2 µm and the cross-section is Gaussian. It can be seen that the FWHM of the focal spot of single micro-lens is less than the typical core size of single-mode fiber (5-10 µm), which indicates the possibility of using proposed micro-lenses system for multi-core fiber assisting. DLW fabrication method on baseplate makes it possible to fabricate both plano-convex and convex-concave micro-lenses in order to improve quality of the wavefront conversion. In this way, one of the possible application of the proposed micro-lens array is complex optical elements, such as micro-objectives with optimized optical design.

4. Conclusion

The presented concept of micro-lens and micro-lens array fabrication on the fiber facet by direct laser writing (DLW) method is of great interest. The concept could be used for devices similar to fibre bundle for quantum correlation enhanced super-resolution microscopy. Suggested freeform lens array can find application in high-accuracy wavefront sensing, endoscopic systems and spatially-resolved spectroscopy.

5. Acknowledgments

This research was funded by Russian Foundation for Basic Research (project No.18-29-20129, micro-lens fabrication) and partially Russian Science Foundation (project No. 19-73-10173, new photocomposition synthesis). Also we are grateful to Alice Merkusheva for helping with design and illustrations.

References

[1] Sharipova M.I., Baluyan T.G., Abrashitova K.A., Kulagin G.E., Petrov A.K., Chizhov A.S., Shatalova T.B., Chubich D., Kolymagin D.A., Vitukhnovsky A.G., Bessonov V.O., Fedyanin A.A., "Effect of pyrolysis on microstructures made of various photoresists by two-photon polymerization: comparative study", Optical Materials Express (2021), V.11 (2), 371-384, DOI: 10.1364/ome.416457

[2] L. Bremer, K. Weber, S. Fischbach, S. Thiele, M. Schmidt, A. Kaganskiy, S. Rodt, A. Herkommer, M. Sartison, S. L. Portalupi, P. Michler, H. Giessen, and S. Reitzenstein, "Quantum dot single-photon emission coupled into single-mode fibers with 3D printed micro-objectives," APL Photonics 5, 106101 (2020)

[3] Pisarenko, A.V., Zvagelsky, R.D., Kolymagin, D.A., Katanchiev, B. V., Vitukhnovsky, A. G., Chubich, D. A. "DLW-printed optical fiber micro-connector kit for effective light coupling in optical prototyping". OPTIK (2020), V. 201, 163350, DOI: 10.1016/j.ijleo.2019.163350

[4] Malinauskas, M. et al. Femtosecond laser polymerization of hybrid/integrated micro-optical elements and their characterization. J. Opt. 12, 124010 (2010).

[5] M. Schmid, D. Ludescher, and H. Giessen, "Optical properties of photoresists for femtosecond 3D printing: refractive index, extinction, luminescence-dose dependence, aging, heat treatment and comparison between 1-photon and 2-photon exposure", Opt. Mater. Express 9(12), 4564 (2019).

[6] Yulia E. Begantsova, Roman Zvagelsky, Evgeny V. Baranov, Dmytro A. Chubich, Yuri V. Chechet, Danila A. Kolymagin, Anastasia V. Pisarenko, Alexey G. Vitukhnovsky, Sergey A. Chesnokov, "Imidazole-containing photoinitiators for fabrication of sub-micron structures by 3D two-photon polymerization", European Polymer Journal, V. 145, 110209 (2021).

Nontrivial optical response of silicon triangular prisms

M Poleva[1], K V Baryshnikova[1], K Frizyuk[1], and A B Evlyukhin[2,1]
[1]The Faculty of Physics, ITMO University, St. Petersburg 197101, Russia
[2]Leibniz Universitat Hannover, Hannover 30167, Germany

Abstract. The electromagnetic response of silicon triangle nanoprisms in the near-infrared region is investigated. It is revealed that the bianisotropic dipole approximation is insufficient for this geometry since the direct application of the Onsager-Casimir symmetry rule to the dipole response leads to a contradictory conclusion. We show that to resolve this contradiction, it is necessary to take into account the nonlocal contributions of higher orders to the excited electric and magnetic dipole moments of the prisms. However, the inclusion in the consideration of nonlocal corrections to the dipole moments leads to the need to take into account the excitation of multipoles of a higher order than dipoles.

1. Introduction

These days the key target of researchers in the field of modern technology is to increase the speed of transmitting and processing the information and the computational capacity. Since most of the devices work via electrical impulses, they possess severe limitations due to the finite mass and non-zero charge of the electron. A possible solution to overcome these difficulties is the usage of photons instead of electrons as the main data carriers. The science that investigates the interaction of light with nanostructures, which such devices consist of, is called nanophotonics [1, 2, 3].

Usually, nanostructures are based on the numbers of so-called meta-atoms – specific elements, from which the whole structure is constructed of. The properties of nanostructure depend both on the features of each meta-atom and their relative positions. If cumulative effects are not excited in the structure, the properties of each meta-atom (the size, the shape and the material) play a major role. That is why the investigation of individual nanoparticles and their clusters is of so high interest to the nanophotonics community [4]. To explain, manage and predict optical properties of nanoparticles several methods were proposed, including CDA (coupled dipole approximation) method [5]. It shows their high applicability to a lot of different nanophotonics tasks. However, the last researches show that some scattering effects could not be explained with this approximation, and need to take into account the high order multipoles [6] or non-local response [7]. It was shown that in the general case, the electric response of the nanoparticle can be associated with both electrical and magnetic component of the incident field and their derivatives, and vice versa, the magnetic response is not only with magnetic, but also electric field and their derivatives [8]. In this work, we consider a very specific case of light interaction with a single prismatic shaped dielectric building block with the equilateral triangle at the base and show that it requires explanation with both non-local and high-order multipoles consideration.

2. Results and Discussion

To investigate the electromagnetic response of the prismatic shaped dielectric particle with the equilateral triangle at the base we considered a dipole bianisotropic approximation, i.e. the electric and magnetic dipole moments are determined only by the incident fields in a certain point (ordinary this point corresponds to the particle centre of mass). The location of the particle in respect to the coordinate axes is shown in Fig.1(a). In order to realize the dipole resonances in the near-infrared range we chose the following parameters for modelling in COMSOL Multiphysics™ simulation software: radius of the circumscribed circle of the base triangle is $R = 300$ nm, height along the z-axis is $h = 188$ nm, the scatter material is silicon, for the surrounding medium $\varepsilon_d = 1$ and $\mu_d = 1$, the operating wavelength range is 900-1200 nm. In this range, for crystalline silicon, we can assume the dielectric constant to be $\varepsilon_p = 12.67$. To make sure that the usage of the dipole approximation is accurate we evaluated scattering cross-section and multipole decomposition for both lateral and normal incidence using exact formulas from [9, 10]. In Fig.1(b,c), we see that the quadrupole contributions to the total cross-section are significantly smaller than the dipole ones and the quadrupole response can be neglected. Hence, we assume that only bianisotropic dipole approximation is sufficient for our study. Later, we will show, why this assumption may be failed.

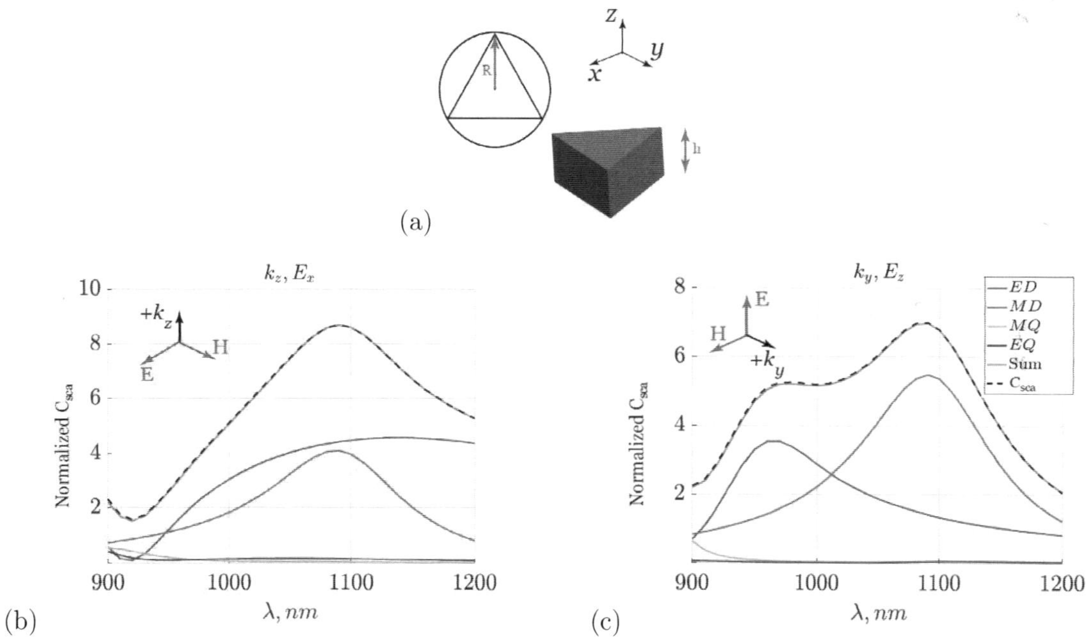

Figure 1. (a) Prismatic shaped nanoparticle with equilateral triangle at the base, R=300 nm, h=188 nm. Contributions to the scattering cross-sections of electric and magnetic dipole, electric and magnetic quadrupole, their sum, and scattering cross-section, calculated as the integral of the energy flux. Wave vector is directed along the (b) z-axis, electric field is x-polarized, (c) y-axis, electric field is z-polarized.

In bianisotropic dipole approximation the local electromagnetic response can be described with the following relations [11]:

$$\begin{aligned} \mathbf{p} &= \boldsymbol{\alpha}^{EE}\mathbf{D}_{in} + c_d^{-1}\boldsymbol{\alpha}^{EH}\mathbf{H}_{in} \\ \mathbf{m} &= \boldsymbol{\alpha}^{HH}\mathbf{H}_{in} + c_d\boldsymbol{\alpha}^{HE}\mathbf{D}_{in}, \end{aligned} \quad (1)$$

where $\mathbf{p}$ and $\mathbf{m}$ are the electric and magnetic dipole moments, respectively, excited by the electric displacement $\mathbf{D}_{in}$ and magnetic $\mathbf{H}_{in}$ incident fields, c_d is the speed of light in the ambient space, $\boldsymbol{\alpha}^{EE}, \boldsymbol{\alpha}^{EH}, \boldsymbol{\alpha}^{HE}, \boldsymbol{\alpha}^{HH}$ are the 3×3 blocks of polarizability tensor $\boldsymbol{\alpha}$, and each one of them represents a part of the dipole response to the incident fields. Since we consider the system without external time-odd bias field or external time modulation, i.e. reciprocal system, these blocks obey the Onsager-Casimir symmetry relations, which follow from the time-reversal symmetry of Maxwell's equations [11, 8] :

$$\boldsymbol{\alpha}^{EE} = (\boldsymbol{\alpha}^{EE})^T, \quad \boldsymbol{\alpha}^{HH} = (\boldsymbol{\alpha}^{HH})^T, \quad \boldsymbol{\alpha}^{EH} = -(\boldsymbol{\alpha}^{HE})^T \tag{2}$$

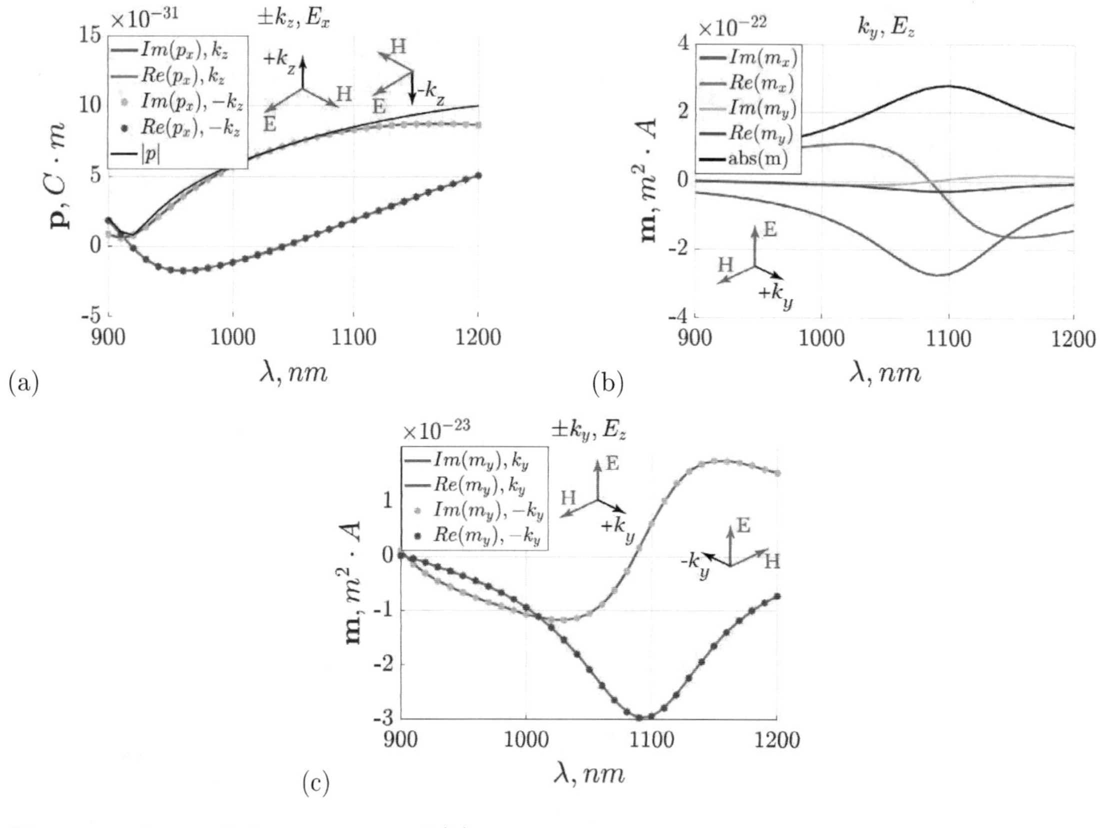

Figure 2. Spectral characteristics of (b) nonzero components of magnetic dipole for x-polarized wave along z and -z direction, (c) nonzero components of magnetic dipole for z-polarized wave from y direction, (d) bianisotropic electric dipole along y for z-polarized wave from y and -y directions calculated for nanoparticle of prismatic shape with equilateral triangle at the base.

Assuming that the dipole approximation (1) is valid for the prism with the equilateral triangle at the base, we check that symmetry rules (2) are satisfied using the approach presented in [12]. First, we calculate the electric dipole moments excited by external x-polarized plane wave propagating firstly from z and then $-z$ directions. As shown in Fig. 2(a), the electric dipole excites only in the direction of the incident electric field, i.e. oscillates along x-axis. For both incident directions with saving the polarization, the electric dipoles have the same absolute values and signs, which means that they are excited only by the incident electric field (since wave vector, electric and magnetic fields of the incident wave form the right-handed system, if we keep the direction of the electric field amplitude but rotate the wave vector by 180°

degrees, the direction of the magnetic field amplitude will be also rotated by 180° degrees, and the sign of magnetic field changes). In this case from Eqs. (1), we conclude that $\alpha_{zy}^{EH} = 0$. Hence, from (2), it is followed that $\alpha_{yz}^{HE} = 0$ as well. However, in Fig. 2(b), it is revealed that z-polarized wave from y-direction excites the magnetic dipole not only in the incident magnetic field x-direction but also in the incidence y-direction. Moreover, the magnetic dipoles oscillating along the incidence direction for both k_y and $-k_y$ have the same values and signs, which is shown in Fig. 2(c). This means that they are excited by the incident electric field, and $\alpha_{yz}^{HE} \neq 0$ for these irradiation conditions. Consequently, in the framework of the bianisotropic dipole approximation, we received a contradiction, which is as follows: on the one hand, from numerical calculations we obtained that $\alpha_{zy}^{EH} = 0$ and $\alpha_{yz}^{HE} \neq 0$, on the other hand, applying the Onsager-Casimir relations (2), we have $\alpha_{zy}^{EH} = -\alpha_{yz}^{HE}$. The last relation follows from the reciprocity property of the system with the triangular prism. Therefore, to resolve this contradiction we have to conclude that the bianisotropic dipole model (1) is not sufficient. To explain the phenomenon of dipoles exciting in the incidence direction we need to consider higher-order nonlocal contributions to the dipole moments stemmed from space derivatives of the incident fields. However, the inclusion in the consideration of nonlocal corrections to the dipole moments inevitably leads to the need to take into account the excitation of multipoles of a higher order than dipoles [8, 7].

3. Conclusion

It is been revealed that the bianisotropic dipole approximation is not sufficient to analyse the electromagnetic response of the considered dielectric prisms with the equilateral triangle at the base. Description of their dipole response must include nonlocal corrections higher than the bianisotropic terms. For an explanation of this feature, one can apply the symmetry theory and modal analysis. Our further investigations will be based on these approaches.

4. Acknowledgments

The work was supported by the Russian Science Foundation (project No. 21-72-20184).

References

[1] Krasnok A, Makarov S, Petrov M, Savelev R, Belov P and Kivshar Y 2015 *Metamaterials X* vol 9502 (International Society for Optics and Photonics) p 950203
[2] Evlyukhin A B, Tuz V R, Volkov V S and Chichkov B N 2020 *Phys. Rev. B* **101** 205415
[3] Zheludev N I and Kivshar Y S 2012 *Nat. Mater.* **11** 917–924 ISSN 1476-4660
[4] Terekhov P D, Baryshnikova K V, Artemyev Y A, Karabchevsky A, Shalin A S and Evlyukhin A B 2017 *Physical Review B* **96** 035443
[5] Evlyukhin A B, Reinhardt C, Seidel A, Luk'yanchuk B S and Chichkov B N 2010 *Physical Review B* **82** 045404
[6] Shamkhi H K, Baryshnikova K V, Sayanskiy A, Kapitanova P, Terekhov P D, Belov P, Karabchevsky A, Evlyukhin A B, Kivshar Y and Shalin A S 2019 *Physical review letters* **122** 193905
[7] Bobylev D A, Smirnova D A and Gorlach M A 2020 *Physical Review B* **102** 115110
[8] Achouri K and Martin O J F 2021 *arXiv:2102.08197* (*Preprint* 2102.08197) URL https://arxiv.org/abs/2102.08197v1
[9] Alaee R, Rockstuhl C and Fernandez-Corbaton I 2018 *Opt. Commun.* **407** 17–21 ISSN 0030-4018
[10] Evlyukhin A B and Chichkov B N 2019 *Phys. Rev. B* **100** 125415 ISSN 2469-9969
[11] Asadchy V S, Díaz-Rubio A and Tretyakov S A 2018 *Nanophotonics* **7** 1069–1094 ISSN 2192-8614
[12] Asadchy V S, Faniayeu I A, Ra'di Y and Tretyakov S A 2014 *Photonics Nanostruct. Fundam. Appl.* **12** 298–304 ISSN 1569-4410

Semitransparent visualizers of infrared lasers based on perovskite quantum dots

A S Polushkin[1], L E Zelenkov[1], D Khmelevskaia[1], D I Markina[1], A L Rogach[2] and S V Makarov[1]

[1]Department of Physics and Engineering, ITMO University, St. Petersburg 197101, Russia
[2]Department of Materials Science and Engineering, and Centre for Functional Photonics (CFP), City University of Hong Kong, Kowloon, Hong Kong SAR, P. R. China

E-mail: artem.polushkin@metalab.ifmo.ru
s.makarov@metalab.ifmo.ru

Abstract. These days halide perovskite is a very popular material to be applied both in photovoltaics and photonics due to its unique properties and low-cost methods of fabrication. High photoluminescence quantum yield and good nonlinear coefficients of perovskite material allow achieving multiphoton absorption in perovskite. Encapsulation of perovskite quantum dots in polymer matrix enables to prevent interaction with the environment and increase the material lifetime. In this paper, we present a semitransparent visualizer of infrared lasers working on two-photon absorption in halide perovskite nanocrystals encapsulated in the polymer matrix.

1. Introduction

Today high-power infrared lasers are used in many parts of human life, e. g., in medicine, in science, and for cutting and engraving metals. Infrared light is invisible for human eyes; as a result, different laser visualizers are used for visualizing the laser beam path. However, currently available visualizers possess some problems. Several of them work only in a narrow spectral range and are not applicable for any lasers. Other visualizers are "necessary to charge the active region with visible light" which means that they lose brightness of luminescence under the high power laser which makes them not very convenient to be used. Moreover, powerful laser emission can burn and destroy a visualizer. We have decided to develop an infrared laser visualizer that works in a wide spectral range, needs no "charging" and semitransparent which allows it to be used without laser beam path interruption.

After halide perovskite potential for solar cells was demonstrated, they attracted a lot of attention [1]. In addition, perovskite also shows some opportunities for the advanced nanophotonic device design due to its unique optical properties [2, 3]. Halide perovskites possess exciton at room temperature, a high refractive index, great optical gain, and large nonlinear coefficients. Perovskite quantum dots (QDs) synthesized by simple chemical methods can have a quantum yield up to 90% [4]. Furthermore, most laser visualizers are opaque, but active material can be used in a transparent polymer membrane for light up-conversion [5]. However, perovskite material can be damaged by water and oxygen, yet an encapsulation of perovskite QDs in a polymer matrix can protect them from harmful environment [6].

In this work, we synthesize $CsPbBr_3$ nanocrystals (NCs) encapsulated in polydimethylsiloxane (PDMS) matrix on a glass substrate and use it for infrared light up-conversion.

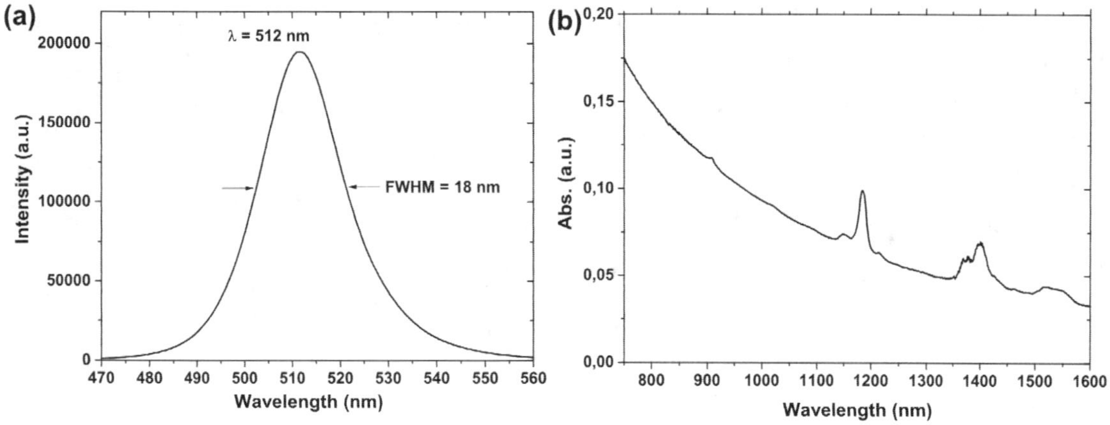

Figure 1. a) Photoluminescence and b) absorption spectrum of perovskite nanocrystals encapsulated in PDMS matrix

2. Results and discussion

$CsPbBr_3$ NCs were synthesized according to standard protocol [7, 8]. Perovskite NCs were separated by centrifugation and redispersed in toluene forming colloidally stable solutions. Dow Corning Sylgard 184 was used as an encapsulation matrix. A film curing of the resulting composite on a glass substrate was made by heating.

We have investigated obtained samples on two-photon absorption and photoluminescence (PL). Continuous PL from QDs in the polymer matrix was excited by a 365 nm UV mercury lamp and collected by Axio Imager A2m (Carl Zeiss) microscope with 50x objectives (Carl Zeiss EC Epiplan-NEOFLUAR). An optical fiber spectrometer (Ocean Optics QE Pro) connected with the microscope in a fluorescent regime was used for recording PL spectra. Fig. 1 represents obtained PL spectra and absorption spectra of perovskite NCs in the polymer matrix in the near-IR range. obtained by Shimadzu UV-3600 Plus UV-VIS-NIR Spectrophotometer. Peaks in fig. 1b correspond to the absorption of PDMS. A central wavelength of PL is 512 nm with full width at half maximum (FWHM) 18 nm. Photoluminescence quantum yield of perovskite QDs in PDMS was measured using Labsphere integrated sphere and appeared to be around 12%.

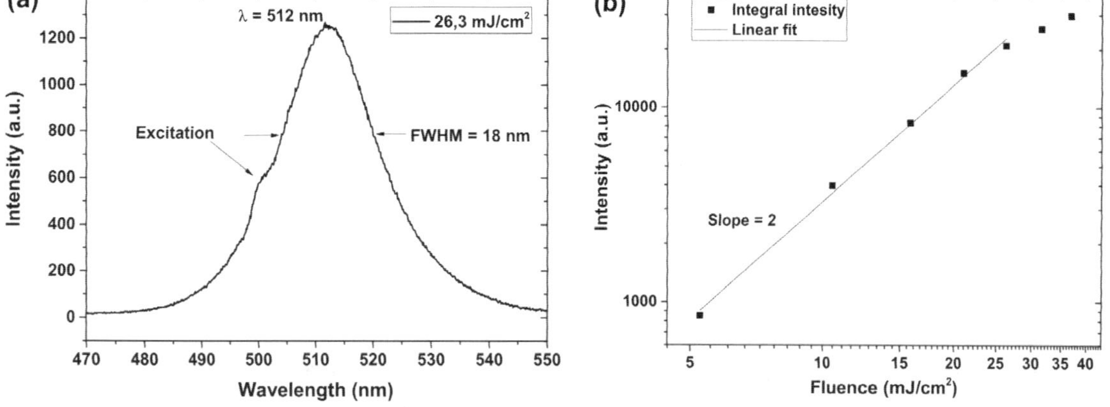

Figure 2. a) Up-conversion luminescence of perovskite quantum dots in PDMS b) dependence of luminescence integral intensity on pump fluence

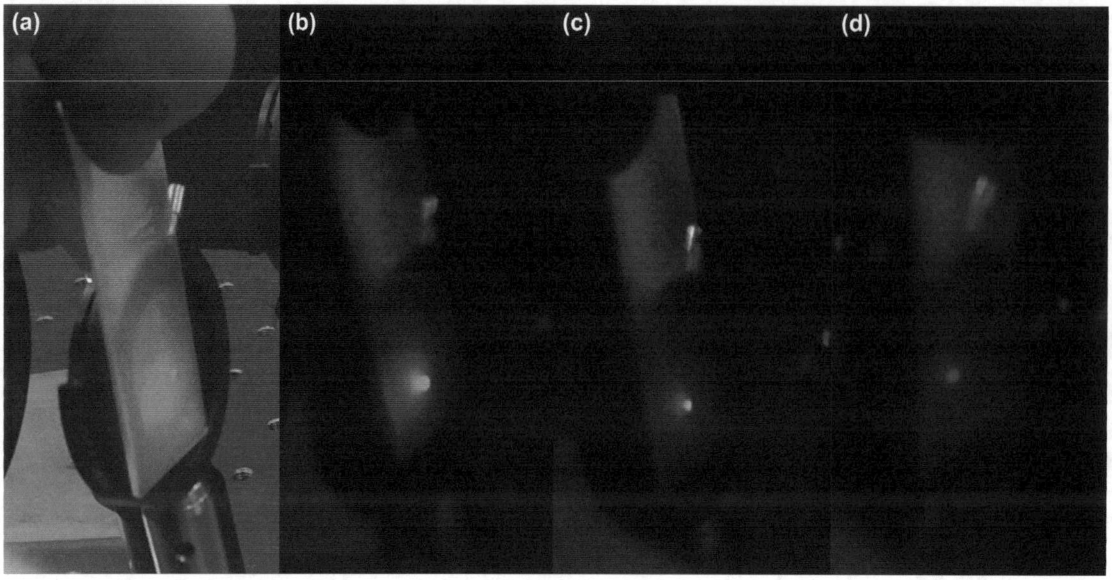

Figure 3. Visualization of IR laser beam a, b) 800 nm, c) 900 m, d) 1000 nm

Further, we investigated IR to visible light up-conversion. Pharos PH2-SP-20W-2mJ single-unit integrated femtosecond laser system and high power optical parametric amplifier Orpheus-FFor were used for excitation. A Laser beam was focused on the sample by 10x objective (Mitutoyo M Plan APO NIR, NA = 0.26) through the glass substrate and light was collected by 50x objective (Mitutoyo M Plan APO, NA = 0.55). Spectra were obtained by using the imaging spectrograph Andor Kymera 328i with CCD camera Andor iDus. Fig. 2a presents the luminescence spectrum of perovskite QDs in PDMS under the 1000 nm laser with pules duration 150 fs and repetition rate 100 kHz. The peak at 500 nm corresponds to excitation laser go to spectrometer due to no enough filtering. Fig. 2b shows the dependence of PL integral intensity on pump fluency in a log-log scale. A slope in fig. 2b equals two, which corresponds to the quadratic dependence, proving two-photon luminescence.

Moreover, we tested visualization of the infrared laser beam by our samples without any additional focusing. Fig. 3 demonstrates the luminescence under 800, 900, and 1000 nm laser with approximate beam diameter 2 mm. The visualizer is semitransparent and passes more than 50% of light. The sample also luminesces stably with a constant brightness under IR laser excitation and is not damaged by a 4 W laser beam.

In the future, we are planning to reveal the dependence of polymer on up-conversion by comparing samples with different polymers, and to study the dependence of polymer with QDs layer thickness on visualization quality. Creating a non-transparent sample can increase the efficiency of up-conversion by laser beam scattering but lose the semitransparency advantage. Another essential improvement of the visualizers can be the changing $CsPbBr_3$ NCs on $CsPbI_3$ NCs, which have a wider bandgap, that can increase spectral range and efficiency of up-conversion. However, $CsPbI_3$ NCs possess a lower quantum yield and less stability.

3. Conclusion

It has been demonstrated that $CsPbBr_3$ nanocrystals encapsulated in PDMS matrix have photoluminescence on 512 nm with FWHM 18 nm, and support IR to visible light up-conversion. We have also revealed that the dependence of integral intensity on excitation fluence is quadratic. Moreover, perovskite QDs in the polymer matrix can be used as semitransparent infrared laser beam visualizers.

4. Acknowledgments
This work is supported by the Russian Science Foundation (project #17-73-20336).

References
[1] Green M A, Ho-Baillie A and Snaith H J 2014 *Nature photonics* **8** 506–514
[2] Makarov S, Furasova A, Tiguntseva E, Hemmetter A, Berestennikov A, Pushkarev A, Zakhidov A and Kivshar Y 2019 *Advanced optical materials* **7** 1800784
[3] Berestennikov A S, Voroshilov P M, Makarov S V and Kivshar Y S 2019 *Applied Physics Reviews* **6** 031307
[4] Huang H, Bodnarchuk M I, Kershaw S V, Kovalenko M V and Rogach A L 2017 *ACS energy letters* **2** 2071–2083
[5] Fedorov V V, Bolshakov A, Sergaeva O, Neplokh V, Markina D, Bruyere S, Saerens G, Petrov M I, Grange R, Timofeeva M *et al.* 2020 *ACS nano* **14** 10624–10632
[6] Raja S N, Bekenstein Y, Koc M A, Fischer S, Zhang D, Lin L, Ritchie R O, Yang P and Alivisatos A P 2016 *ACS applied materials & interfaces* **8** 35523–35533
[7] Protesescu L, Yakunin S, Bodnarchuk M I, Krieg F, Caputo R, Hendon C H, Yang R X, Walsh A and Kovalenko M V 2015 *Nano letters* **15** 3692–3696
[8] Huang H, Chen B, Wang Z, Hung T F, Susha A S, Zhong H and Rogach A L 2016 *Chemical science* **7** 5699–5703

Second-harmonic generation from green printed selenium structures

E. Yu. Ponkratova[1], Y. Sun[1], Z. Zhang[2], M. Su[2], Y. V. Mikhailova[1], D. A. Zuev[1], Y. Song[2], S. Kumar[3], A. K. Ganguli[3] and P. A. Belov[1]

[1] Department of Physics, ITMO University, St. Petersburg 191002, Russia
[2] Institute of Chemistry, Chinese Academy of Sciences, Beijing 100190, China
[3] Department of Chemistry, Indian Institute of Technology, Delhi, New Delhi 110016, India

E-mail: ekaterina.grachkova@metalab.ifmo.ru

Abstract.
Plenary large-scale efficient second-harmonic generation sources remains a challenge in micro-/nanoscale research. In this work, we demonstrate the strong second-harmonic generation in arrays of selenium particles fabricated by green printing method at the pumping wavelength of 1047 nm. Second-harmonic generation mapping shows the strong signal over the whole area of selenium particle. We propose, that such structures can be a perspective platform for efficient frequency conversion systems and nonlinear sensing applications.

1. Introduction
Creation of frequency conversion systems at micro- and nano- scales plays an important role in modern photonics and optics. Particular interest lies in the fabrication of second-harmonic generation (SHG) sources which can be utilized for laser spectroscopy application [1], data recording, sensing [2] etc. For producing such structures, a variety of different methods are used: chemical [3], lithography [4], thermal [5] and many others. Green printing is a new and promising method for this goal due to its low cost, environmental friendliness and simplicity [6]. For instance, micro-/nanoscale pattern can be formed by micropillar array templates induced self-assembly of particles or deposited films. This method requires materials with deposition control and relatively low melting point [7]. One of suitable materials for this goal is selenium. Selenium is a semiconductor with a high refractive index and extremely low losses in the visible range that makes it an ideal system for providing interesting optical effects on printed micro- and nanoparticles [8].

In this work we produce arrays of selenium particles on silicon substrate by green printing method. We observe, that such particles exhibit a strong second-harmonic generation signal at the visible range. We propose that produced structures have a big potential for nanophotonic systems and devices.

2. Materials and methods
2.1. Se array fabrication
At the first step, polydimethylsiloxane (PDMS) template is replicated from the prepared Si groove shape stamp with width of grooves $w = 10\ \mu$m, height $h = 20\ \mu$m and period $p = 20\ \mu$m (see Fig. 1a). Firstly, 2 g of selenium power is loaded in the evaporation boat. Selenium film is evaporated on the PDMS template at the rate of 0.1 nm/s for a thickness of 2 μm for fluid-guided printing (see Fig. 1b). Then

the silicon substrate is covered by the PDMS template with selenium, and the fluid-guided printing is conducted at 100 °C. During this process, the selenium particles are printed on the substrate due to the Rayleigh instability (see Fig. 1c). Finally, the template is uncovered after cooling down to the room temperature and arrays of Se particles on silicon substrate are printed (see Fig. 1d).

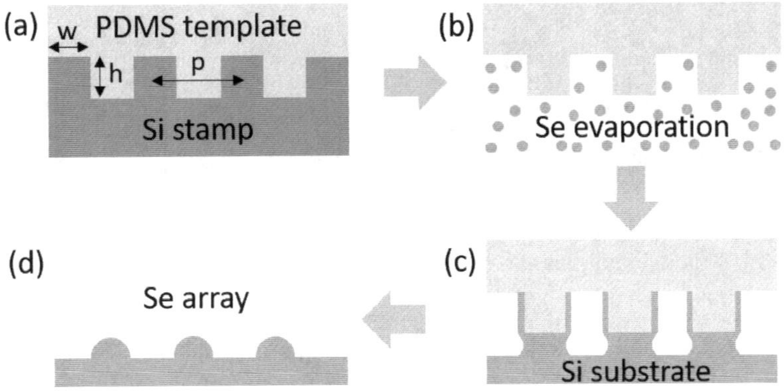

Figure 1. Green printing process. a) Creation of PDMS template by stamping from the groove-shaped silicon substrate; b) Se evaporation on the template; c) Narrowing and shaping of Se with the guidance of template; d) Printed Se particle arrays

2.2. Geometry characterization

External geometry characterization of structures with high accuracy was provided by Scanning Electron Microscopy (SEM). SEM images of the studied structures were obtained by Inspect (Field Electron and Ion Company, FEI) system in vacuum chamber under pressure 10^3-10^4 Pa and accelerating voltage 20 kV in mode of secondary electrons detection.

2.3. SHG signal measurements

SHG signal of the selenium structures were excited under the pumping of a femtosecond (fs) oscillator TEMA-150 (Avesta Project) generating at a wavelength of 1047$\pm$5 nm (pulse duration = 150 fs, repetition rate = 80 MHz). Laser beam was focused on the structures from the top by Mitutoyo Plan Apo NIR HR objective (100x, NA = 0.7). Excited SHG signal was collected by the same objective and directed to spectrometer Horiba LabRam HR800. The sample was placed on a piezo stage (AIST NT) with positioning accuracy of 10 nm during the two-dimensional mapping of SHG signal.

3. Results and discussion

Arrays of selenium particles on silicon substrate are fabricated by the green printing method which was proposed in [6]. SEM imaging is performed to investigate the configuration of the printed structures. SEM images of the whole structure and zoomed top and side images of a selenium semi-sphere are illustrated in Fig. 2. As it can be seen from Fig. 2a, the distance between each row is set to be 20 μm which is in a good agreement with period between grooves on Si stamp. The typical width and length of structures are 1-2 μm and several μm, respectively. It should be noted, that the width of the selenium array is determined by the evaporation procedure in Fig. 1b. For investigation of nonlinear properties, we choose the Se particle with a diameter around 1.2 μm which is indicated by a white dashed box in Fig. 2a and its magnified top view and side view SEM images are shown in Fig. 2b and Fig. 2c, respectively .

Measured SHG spectrum from the chosen particle is depicted in Fig. 3a. From the figure it can be observed, that fabricated Se structure generates quite strong SHG signal at the double frequency of the

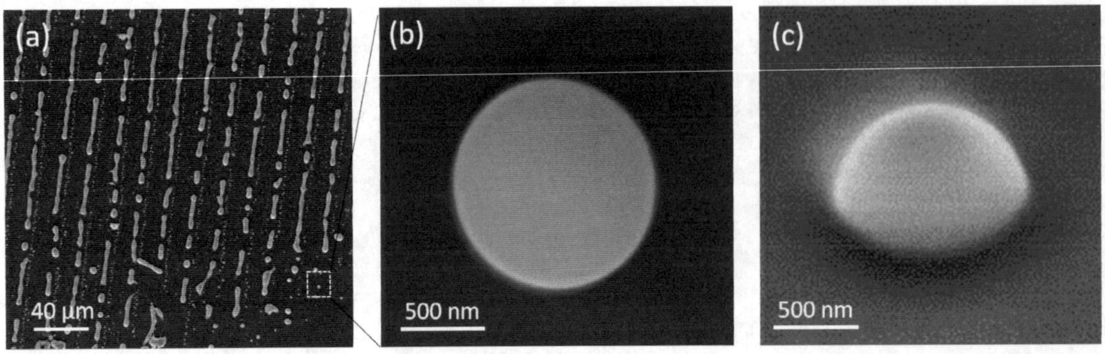

Figure 2. Geometry characterization of Se structures. a) SEM image of array; b) Top and c) side view of chosen area defined by white dashed box in a)

fundamental laser excitation. To investigate SHG intensity distribution over the Se particle area, a SHG map is recorded with a step size of 80 nm utilizing piezo stage (Fig. 3b). As it can be seen from the figure, the SHG signal is strong over the whole area of the particle. Thus, produced arrays of Se structures can be used as efficient SHG sources.

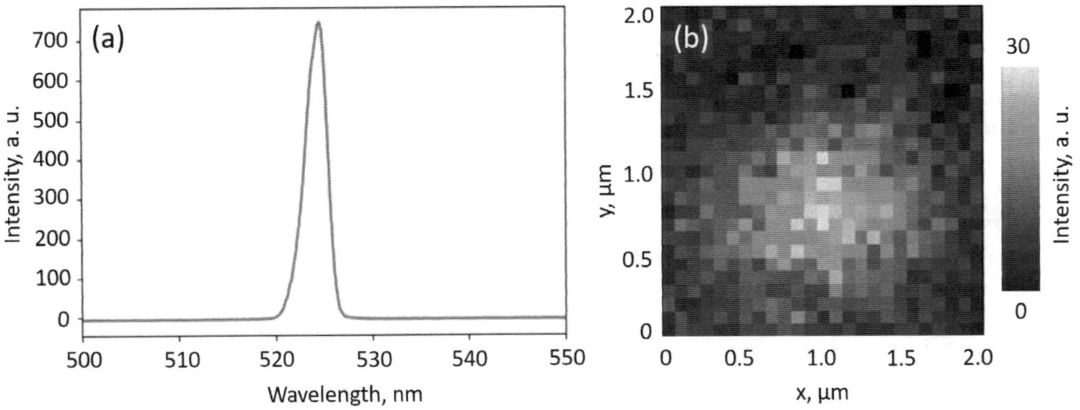

Figure 3. Measured nonlinear signal. a) SHG signal collected from one particle; b) SHG map of Se particle at the double frequency of the laser (524 nm)

4. Conclusion

In summary, we have fabricated Se arrays composed of microparticles on silicon substrate by the green printing method. These structures demonstrate strong SHG signal during irradiation by the beam of a fs-laser at the wavelength of 1047 nm. The SEM image presents homogeneous configuration of the Se particle. The SHG map provides strong signal over the Se particle area which demonstrates high potential in promising large-scale planar SHG source. Proposed green printing method can be applied for fabrication of both micro- and nanoarrays made of selenium which makes it a promising platform for creation of efficient nonlinear sensors and frequency conversion systems.

Acknowledgments

This work is supported by Russian Foundation for Basic Research (19-52-80036 (BRICS Project)), National Nature Science Foundation of China (Grant No. 51961145102 (BRICS Project)) and DST, Govt. of India (DST/ICD/BRICS/Pilot call3/GPTNOM/2019).

References

[1] Mund J, Fröhlich D, Yakovlev D R and Bayer M 2018 *Physical Review B* **98** 085203
[2] Ghirardini L, Baudrion A L, Monticelli M, Petti D, Biagioni P, Duò L, Pellegrini G, Adam P M, Finazzi M and Celebrano M 2018 *The Journal of Physical Chemistry C* **122** 11475–11481
[3] Xu Y, Zhao Y, Chen L, Wang X, Sun J, Wu H, Bao F, Fan J and Zhang Q 2015 *Nanoscale* **7** 6790–6797
[4] Black L J, Wiecha P R, Wang Y, De Groot C, Paillard V, Girard C, Muskens O L and Arbouet A 2015 *ACS photonics* **2** 1592–1601
[5] Makarov S V, Milichko V A, Mukhin I S, Shishkin I I, Zuev D A, Mozharov A M, Krasnok A E and Belov P A 2016 *Laser & Photonics Reviews* **10** 91–99
[6] Su M, Sun Y, Chen B, Zhang Z, Yang X, Chen S, Pan Q, Zuev D, Belov P and Song Y 2021 *Science Bulletin* **66** 250–256
[7] Gupta T D, Martin-Monier L, Yan W, Le Bris A, Nguyen-Dang T, Page A G, Ho K T, Yesilköy F, Altug H, Qu Y *et al.* 2019 *Nature nanotechnology* **14** 320–327
[8] Gampel L and Johnson F 1969 *JOSA* **59** 72–73

Journal of Physics: Conference Series **2015** (2021) 012114 doi:10.1088/1742-6596/2015/1/012114

Enhancement of the photocatalytic activity of nanoparticles due to the localization of electromagnetic fields

S R Ponomareva*, E A Gurvitz, K S Ladutenko

ITMO University, 49 Kronverkskiy pr., Saint Petersburg, 191002, Russia

E-mail: sofiya.ponomareva@metalab.ifmo.ru

Abstract. In this work, we consider the designs of nanoparticles based on titanium dioxide. We focus on localizing energy in the near-surface layer of particles in order to increase the generation of electron-hole pairs in this region. This should result in an enhancement in their photocatalytic activity.

1. Introduction

Photocatalysis is the process of acceleration in the rate of a chemical reaction that occurs due to the generation of electron-hole pairs under electromagnetic irradiation. It can be used in solving topical issues, for example, among applications of photocatalysis there are water and air purification in industry [1], sterilization in medicine [2], hydrogen production in energetics [3].

The efficiency of photocatalysis depends on many parameters. Some of them are determined by the material that is used as a photocatalyst. For example, recombination rate or diffusion length of electrons and holes. Other parameters depend on the geometry of the photocatalyst. The most important parameter among others is the contact area between the photocatalyst and the environment and the percentage of the light absorbed in the active region. One of the most popular materials to create a photocatalysist is titanium dioxide because it has a number of advantages: low cost and high accessibility, non-toxicity, and photoreactivity. At the same time, the bandgap width of TiO2 corresponds to the UV region of the spectrum, which limits its applicability under solar irradiation. Numerous approaches were proposed to increase its sensitivity to the visible range and to improve other parameters of the material such as crystal structure control, photosensibilization with dyes [4], metal and non-metal doping [5], creation of heterostructures [6], surface decoration [7] and many others. The materials science side of the problem has already been explored quite well. But there is another important and less studied drawback of photocatalysts, which is that the light is absorbed outside of the active region. If electrons are generated at the center of a particle, then there is a high probability that they will recombine and will not contribute to photocatalysis. For example, for TiO2 the diffusion length of minor carriers - holes is about 10 nm. Then it is logical to consider the generation of carriers in the area that is located at this distance from the surface.

There were published a few papers, which declare localization of electromagnetic fields connected with multipoles excitation for magnetic field separation [8], scattering maximization and amplification [9], localization of field in biological particles [10], light driven nanomixing [11].

Active all-dielectric nanophotonics researchers were also applied to photocatalysis. In the 2020 study [12] anapole excitation in TiO2 cylinders was suggested to increase the rate of Ag reduction. There were also studies related to the plasmonic enhancement of the fields [13].

However, localization of the energy inside of the particle doesn't lead to the photocatalytic activity enhancement if it occurs not in the active regions that we pointed out above. And that's why in this research we concentrated on the ability of particles to localize energy inside of the pre-surface layer with 10 nm width.

2. Results

In this work, we carried out numerical simulations for nanoparticles with the addition of titanium dioxide of various geometries. We started from the bulk TiO2 cylindrical nanoparticles as discussed in [12]. For the objective function we used the value of the electromagnetic energy inside of the 10 nm layer, divided by the value of the energy in the volume of the water with the same size:

$$W_{norm} = \frac{Re(\varepsilon_{TiO2})}{4} \frac{\int_V |\mathbf{E}(r)|^2 d^3r}{V_{10[nm]} \frac{|\mathbf{E}_0|^2}{2} \varepsilon_{water}} \tag{1}$$

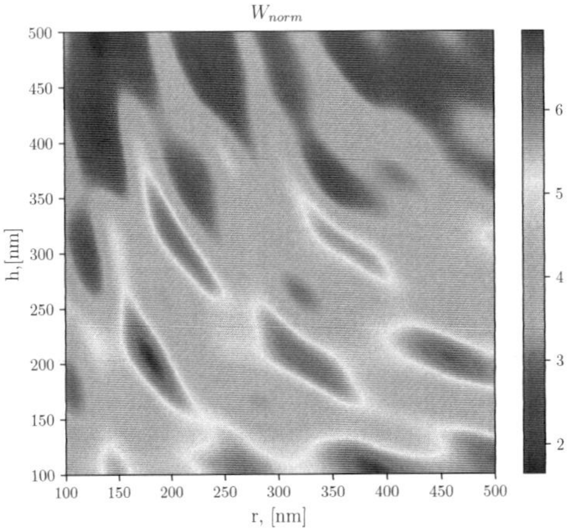

Figure 1. Destribution of normalized energy within 10nm pre-surface layer depending on the size parameters of TiO2 cylinder

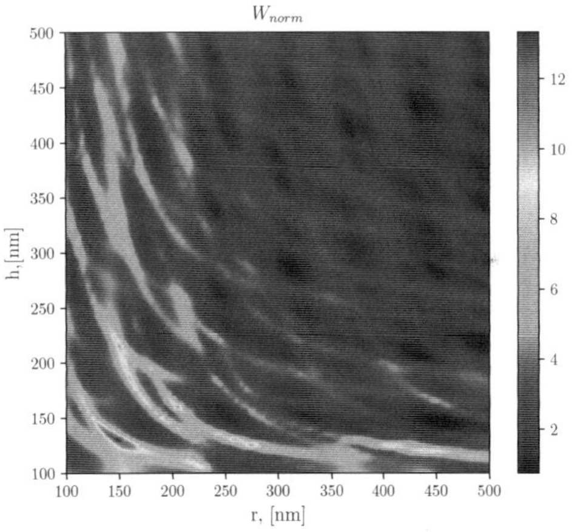

Figure 2. Destribution of normalized energy within 10nm pre-surface layer depending on the size parameters of Si@TiO2 cylinder

The obtained results are represented in figure 1. It shows multiple enhancement regions. For small height and large radius, electromagnetic filed configurations corresponding toroidal moments are localized in the vicinity of interface TiO2/solution. That rises photocatalysis enhancement and it was addressed in the paper [12] to the anapole states. It should be noted that the existence of an anapole state is not a requirement for the reaction acceleration. Photocatalitic reactions could be driven by basic multipoles (this definition was well described in the paper [14]. Significant field photocatalytic reaction acceleration is found for high cylinders with fairly low radius. Those regions are characterized by the combination of toroidal and basic moments in the scattering spectrum, which in sum shows better amplification rather than anapole-driven regions. In addition, layered strucrures with addition of Si were modeled (fig.2). It can be seen from obtained destribution that in the presence of a material with higher index the maximum value of the objective function has increased. A similar effect is observed for other considered geometries, for example, cones.

3. Conclusion

To conclude we examined the ability of nanoparticles to concentrate energy inside the subsurface layer to apply this effect in photocatalysis. It has been shown that it is possible to improve the existing results for anapole states, since the concentration should appear in certain regions. This work shows that photocatalysis is an attractive research area, where the use of already obtained knowledge about the optical responses of dielectric nanoparticles can lead to significant results.

References

[1] Ren H, Koshy P, Chen W F, Qi S and Sorrell C C 2017 *Journal of Hazardous Materials* **325** 340–366 ISSN 18733336 URL `http://dx.doi.org/10.1016/j.jhazmat.2016.08.072`

[2] Ohko Y, Utsumi Y, Niwa C, Tatsuma T, Kobayakawa K, Satoh Y, Kubota Y and Fujishima A 2001 *Journal of Biomedical Materials Research* **58** 97–101 ISSN 00219304

[3] Kumaravel V, Mathew S, Bartlett J and Pillai S C 2019 *Applied Catalysis B: Environmental* **244** 1021–1064 ISSN 09263373 URL `https://doi.org/10.1016/j.apcatb.2018.11.080`

[4] Jeon J P, Kweon D H, Jang B J, Ju M J and Baek J B 2020 *Advanced Sustainable Systems* **4** 1–19 ISSN 23667486

[5] Abbad S, Guergouri K, Gazaout S, Djebabra S, Zertal A, Barille R and Zaabat M 2020 *Journal of Environmental Chemical Engineering* **8** ISSN 22133437

[6] Gong Y, Wu Y, Xu Y, Li L, Li C, Liu X and Niu L 2018 *Chemical Engineering Journal* **350** 257–267 ISSN 13858947

[7] Yang J, Wang D, Han H and Li C 2013 *Acc. Chem. Res.* **46** 1900–1909 ISSN 0001-4842

[8] Baryshnikova K, Filonov D, Simovski C, Evlyukhin A, Kadochkin A, Nenasheva E, Ginzburg P and Shalin A S 2018 *Phys. Rev. B* **98** 165419 ISSN 2469-9969

[9] Terekhov P D, Shamkhi H K, Gurvitz E A, Baryshnikova K V, Evlyukhin A B, Shalin A S and Karabchevsky A 2019 *Opt. Express* **27** 10924–10935 ISSN 1094-4087

[10] Barhom H, Machnev A A, Noskov R E, Goncharenko A, Gurvitz E A, Timin A S, Shkoldin V A, Koniakhin S V, Koval O Yu, Zyuzin M V, Shalin A S, Shishkin I I and Ginzburg P 2019 *Nano Lett.* **19** 7062–7071 ISSN 1530-6984

[11] Valero A C, Kislov D, Gurvitz E A, Shamkhi H K, Pavlov A A, Redka D, Yankin S, Zemánek P and Shalin A S 2020 *Adv. Sci.* **7** 1903049 ISSN 2198-3844

[12] Hüttenhofer L, Eckmann F, Lauri A, Cambiasso J, Pensa E, Li Y, Cortés E, Sharp I D and Maier S A 2020 *ACS Nano* **14** 2456–2464 ISSN 1936086X

[13] Hou W and Cronin S B 2013 *Advanced Functional Materials* **23** 1612–1619 ISSN 1616301X

[14] Gurvitz E A, Ladutenko K S, Dergachev P A, Evlyukhin A B, Miroshnichenko A E and Shalin A S 2019 *Laser and Photonics Reviews* **13** 1–13 ISSN 18638899

Theoretical study of surface states excitation in one-dimensional photonic crystal by halide perovskite microstructures

A.A. Popkova[1], K.R. Safronov[1], D.I. Markina[2], A. Pushkarev[2], S.V. Makarov[2], V.O. Bessonov[1,3], and A.A. Fedyanin[1]

[1] Faculty of Physics, Lomonosov Moscow State University, Moscow 119991, Russia
[2] Department of Physics and Engineering, ITMO University, St. Petersburg 197101, Russia
[3] Frumkin Institute of Physical Chemistry and Electrochemistry, Russian Academy of Sciences, Moscow 119071, Russia

E-mail: popkova@nanolab.phys.msu.ru

Abstract. In this work we develop the concept of integrated light source based on halide perovskite microwires for Bloch surface waves (BSWs) in one-dimensional photonic crystals. We theoretically study modes supported by a $CsPbBr_3$ microwire placed on the surface of a photonic crystal. The modes with field distribution similar to BSW are observed. We study the coupling of microwire modes with BSWs depending on the geometric parameters of the wire. BSW excitation efficiency exceeding 40% is found for BSW-like modes of microwire. We also examine resonant modes of finite microwire and show that BSW-like modes play a main role in the excitation of BSWs with perovskite radiation. Our research paves the way for the implementation of complete integrated optical setups on the surface of photonic crystals.

1. Introduction

Bloch surface waves (BSW) are promising platform for integrated photonics. Due to a sensitive spectral-angular resonance, localization of the electromagnetic field near the surface and a long propagation length that can reach several millimeters [1]. BSWs are actively used in sensing [2], enhancement of luminescence emission [3] and nonlinear optics[4], micromanipulation [5] and others. The ability to control the excitation wavelength in a wide spectral range (from UV [6] to IR [7]) by changing the structure parameters allows one to combine the BSWs with various existing platforms.

At this moment, the main elements of integrated optics based on BSW, such as waveguides [8], lens [9], resonators [10], interferometers [11] have been demonstrated. However further development of BSW integrated photonics require on-chip BSW sources. Since the wave vector of the BSW is larger than the wave vector of light in vacuum, special schemes are required for BSW excitation [12]. Currently, the most efficient and simple scheme to implement on-chip BSW excitation is a miniaturized grating, which demonstrate the efficiency up to 18% [13]. Increased efficiency can be achieved by using lasing structures, such as perovskite microwires, on top of photonic crystals. Lead-halide perovskites have a high refractive index, a direct band gap and high quantum yield of luminescence at room temperature [14]. Due to relatively simple chemical fabrication methods [15], perovskites are actively employed for creating of solar cells [16], photodetectors [17], and light emitting devices [18]. The possibility of fabrication of microresonator structures with good quality on various substrates allows generating stimulated emission in the visible spectral range [19].

In this article, we propose to use of emitting perovskite microstructures to excite BSWs. The modes that can be excited in $CsPbBr_3$ microwires are studied numerically, and the efficiency of radiation coupling with BSWs is estimated. Our research demonstrates the possibility for the implementation of compact BSW sources on the surface of a photonic crystal, which is a new step in creating a complete integrated platform based on BSWs.

2. Results and discussion

Studied system is schematically shown in Fig.1a. The $CsPbBr_3$ microstructure is placed at the surface of one-dimensional photonic crystal (PC). The PC parameters are chosen to sustain TE-polarized BSWs in the spectral range of perovskite luminescense (510-550 nm). The parameters were optimized using the transfer matrix method as well as finite difference in the time domain (FDTD) method in the Lumerical Solutions software. The resulting structure is consisted of 6 pairs of alternating layers of SiO_2 ($n = 1.48$) and Ta_2O_5 ($n = 2.12$) with thicknesses of 143 nm and 101 nm, respectively, deposited on a glass substrate ($n = 1.52$). This number of layers provides high leakage radiation, making BSW to be easily visualized in experiments using leakage radiation microscopy [11].

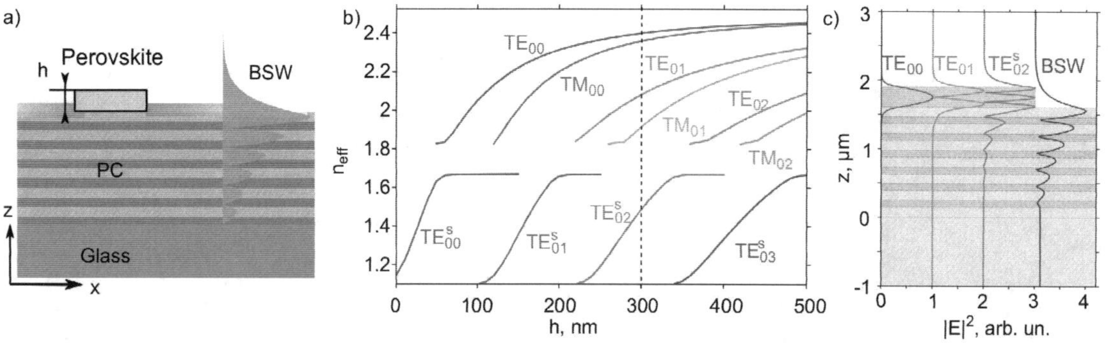

Figure 1. a) Schematic of the studied system. Perovskite microstructure is placed on the surface of a photonic crystal. Perovskite luminescence excites Bloch surface waves. b) Dependence of effective refractive index n_{eff} of modes inside perovskite film on the film thickness h at $\lambda = 535$ nm. c) Field distribution of TE_{00}, TE_{01}, TE^s_{02} modes inside the perovskite film with $h = 300$ nm and field distribution of Bloch surface wave on the bare photonic crystal at $\lambda = 535$ nm.

First, we numerically studied the modes supported by the perovskite film (with $n = 2.5$) with thickness h on the PC surface at $\lambda = 535$ nm. This wavelength corresponds to the center of the $CsPbBr_3$ gain spectral region [20], while n corresponds to refractive index of perovskite over lasing threshold. We used finite difference eigenmode solver realized in Lumerical FDTD. Our analysis (Fig. 1b) shows that, depending on the film thickness, the system may sustain from 1 (as h approaches zero) to 7 (at $h \approx 420$ nm) modes simultaneously. Modes can be divided into 2 classes according to the value of the effective refractive index n_{eff} equal to the ratio of the mode propagation constant and the vacuum wavenumber: high-index modes with $n_{eff} > 1.8$ and low-index modes with $n_{eff} < 1.7$. Range from $n_{eff} = 1.7$ to $n_{eff} = 1.8$ corresponds to transmission region of PC for $\lambda = 535$ nm. High-index modes can be TE- or TM-polarized, while all low-index modes are TE-polarized. As the bare PC supports the excitation of only TE-polarized BSWs, we consider only modes of the same polarization, since coupling between modes of different polarizations is inefficient. Fig.1c demonstrates the field distribution of TE-polarized modes supported by perovskite film with h = 300 nm. High-index modes (TE_{00} and TE_{01}) are conventional waveguide modes inside the film. The low-index mode (TE^s_{02}) is localized inside the film, like high-index modes, but attenuates inside PC with oscillations similar to those of BSW, so that we classified this mode as surface waveguide mode. Fig.1c also shows the field distribution of BSW on the bare PC surface. The field distribution of the TE^s_{02} surface waveguide mode is quite similar to BSW, while conventional waveguide modes are noticeably different. Thus, BSW should be excited more efficiently with surface waveguide modes.

To study the BSW excitation efficiency we launch a particular mode in a half-infinite wire with particular height h. The mode reaches the end of the wire, where it partially reflects, partially scatters, and partially couples to BSW on the bare surface. In order to differentiate scattering and BSW excitation, we calculated the field distribution across PC at a distance of 5 μm from the wire end and than calculated overlap integral of this field and the field distribution of BSW on the bare PC. So that we estimated BSW coupling efficiency η as the ratio of energy transferred to BSW and the energy of the launched mode. Surface waveguide modes demonstrate BSW coupling efficiency of the order of tens of percent in a narrow range of h (Fig.2a), while η is less than 0.3 for conventional TE_{00} mode and less than 0.1 for TE_{01} and TE_{02} modes. Thus, conventional waveguide modes excite BSW significantly weaker, since their field distributions differ from BSW field distribution. Fig.2b demonstrates the excitation of BSW with TE^s_{02} mode propagating inside the wire with h = 250 nm. We observe the field localization on the PC surface.

Then we consider finite microwire, which acts as Fabry–Perot cavity. In order to study resonant modes, we place a cloud of dipoles with random phases and positions inside microwire. Polarization of all the dipoles was the same (TE or TM). Our calculations confirm that depending on wavelength and polarization, one of the conventional or surface waveguide mode is excited inside the microwire. Quality factors of TE-polarized modes lies in the range from 200 to 500, while for TM-polarized modes, the value is around 200. We also studied BSW coupling efficiency for finite wire. Similar to the case of half-infinite wire, maximum η is observed when one of the surface waveguide modes is excited.

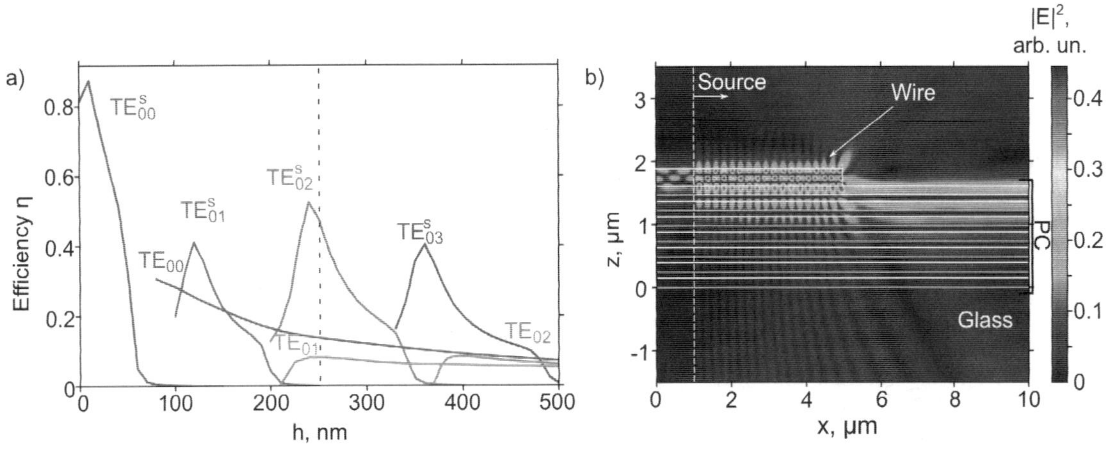

Figure 2. a) Dependence of BSW coupling efficiency for modes inside perovskite film on the film thickness h at $\lambda = 535$ nm. b) Excitation of BSW on the PC surface by TE^s_{02} mode inside the wire with $h = 250$ nm.

Conclusion

In this work, we propose integrated light source for BSW, based on halide perovskite microwires. We study the modes supported by microwires and efficiency of BSW excitation by them. The arrangement of the perovskite structure on the PC surface leads to the appearance of new modes that do not exist in the structure on the surface of the glass or ITO. These modes plays a major role in the excitation of BSW on the PC surface. Our research takes a significant step in the development of BSW-based integrated photonics.

Acknowledgements

The work was performed under financial support of the Russian Science Foundation (Grant No. 20-12-00371) and according to the Development program of the Interdisciplinary Scientific and Educational School of Lomonosov Moscow State University "Photonic and Quantum technologies. Digital medicine". Part of the research was supported by MSU Quantum Technology Centre. A.A.P. acknowledges support by Foundation for the Advancement of Theoretical Physics and Mathematics "BASIS" (Grant No. 19-2-6-28-1). S.V.M acknowledges the Presidential Grant (MD-5954.2021.1.3).

References

[1] Dubey R, Barakat E, Häyrinen M, Roussey M, Honkanen S K, Kuittinen M and Herzig H P 2017 *J. Eur. Opt. Soc., Rapid publ.* **13** 1–9
[2] Kuai Y, Chen J, Tang X, Xiang Y, Lu F, Kuang C, Xu L, Shen W, Cheng J, Gui H, Zou G, Wang P, Ming H, Liu J, Liu X, Lakowicz J R and Zhang D 2019 *Sci. Adv.* **5** eaav5335
[3] Stella U, Boarino L, De Leo N, Munzert P and Descrovi E 2019 *ACS Photonics* **6** 2073–2082
[4] Konopsky V N, Alieva E V, Alyatkin S Y, Melnikov A A, Chekalin S V and Agranovich V M 2016 *Light Sci. Appl.* **5** e16168
[5] Shilkin D A, Lyubin E V, Soboleva I V and Fedyanin A A 2015 *Opt. Lett.* **40** 4883–4886
[6] Badugu R, Mao J, Blair S, Zhang D, Descrovi E, Angelini A, Huo Y and Lakowicz J R 2016 *J. Phys. Chem. C* **120** 28727–28734
[7] Smolik G M, Descharmes N and Herzig H P 2018 *ACS Photonics* **5** 1164–1170
[8] Sfez T, Descrovi E, Yu L, Brunazzo D, Quaglio M, Dominici L, Nakagawa W, Michelotti F, Giorgis F, Martin O *et al.* 2010 *J. Opt. Soc. Am. B* **27** 1617–1625
[9] Yu L, Barakat E, Sfez T, Hvozdara L, Di Francesco J and Herzig H P 2014 *Light Sci. Appl.* **3** e124–e124
[10] Rodriguez G, Aurelio D, Liscidini M and Weiss S 2019 *Appl. Phys. Lett.* **115** 011101
[11] Safronov K R, Gulkin D N, Antropov I M, Abrashitova K A, Bessonov V O and Fedyanin A A 2020 *ACS Nano* **14** 10428–10437
[12] Kretschmann E 1972 *Opt. Commun.* **6** 185–187
[13] Kovalevich T, Boyer P, Suarez M, Salut R, Kim M, Herzig H P, Bernal M and Grosjean T 2017 *Opt. Express* **25** 5710–5715
[14] Makarov S, Furasova A, Tiguntseva E, Hemmetter A, Berestennikov A, Pushkarev A, Zakhidov A and Kivshar Y 2019 *Adv . Opt. Mater.* **7** 1800784
[15] Pushkarev A P, Korolev V I, Markina D I, Komissarenko F E, Naujokaitis A, Drabavicius A, Pakstas V, Franckevicius M, Khubezhov S A, Sannikov D A *et al.* 2018 *ACS Appl. Mater. Interfaces* **11** 1040–1048
[16] Berhe T A, Su W N, Chen C H, Pan C J, Cheng J H, Chen H M, Tsai M C, Chen L Y, Dubale A A and Hwang B J 2016 *Energy Environ. Sci.* **9** 323–356
[17] Ramasamy P, Lim D H, Kim B, Lee S H, Lee M S and Lee J S 2016 *Chem. Comm.* **52** 2067–2070
[18] Zhang X, Lin H, Huang H, Reckmeier C, Zhang Y, Choy W C and Rogach A L 2016 *Nano Lett.* **16** 1415–1420
[19] Sutherland B R and Sargent E H 2016 *Nat. Photonics* **10** 295
[20] Zhang Q, Su R, Du W, Liu X, Zhao L, Ha S T and Xiong Q 2017 *Small Methods* **1** 1700163

Comparison of different wireless coils for 1.5 T bilateral breast MRI

Viktor Puchnin[1,a], **Anton Nikulin**[1,b], **Anna Hurshkainen**[1,c], **Georgiy Solomakha**[1,d], **Anna Andreychenko**[2,e] **and Alena Shchelokova**[1,f]

[1]School of Physics and Engineering, ITMO University, Saint Petersburg, Russia
[2]Research and Practical Clinical Center of Diagnostics and Telemedicine Technologies, Department of Health Care of Moscow, Moscow, Russia

E-mail:
a) viktor.puchnin@metalab.ifmo.ru
b) a.v.nikulin@live.ru
c) a.hurshkainen@metalab.ifmo.ru
d) g.solomakha@metalab.ifmo.ru
e) anna.andreychenko@metalab.ifmo.ru
f) a.schelokova@metalab.ifmo.ru

Abstract.
This work compares three types of transceive wireless coils: Helmholtz-type coil, metamaterial-inspired coil, and quadrature coil based on their combination. Each transceive coil is electromagnetically coupled to a body "birdcage" type coil of a 1.5 T magnetic resonance scanner and improves bilateral breast imaging performance. While Helmholtz-type coil and metamaterial-inspired coil based on coupled split loop resonators are linearly polarized, their combination allows to couple with both linear components of the radiofrequency magnetic field, providing a more significant effect of a local boosting of the body coil's transmit efficiency and radiofrequency safety in comparison with birdcage coil only.

1. Introduction

Magnetic resonance imaging (MRI) is becoming an increasingly common and affordable method. The main advantages of this method are safety, noninvasiveness, and the diagnostic value of the examination. The most common systems in clinics with a field strength of 1.5 T generally use the transceive radiofrequency (RF) body coil built into the magnetic resonance (MR) scanner to excite electromagnetic signal and dedicated local coils to receive MR signal from the area of interest. However, this approach does not fully match the modern requirements for the quality of the obtained MR images for specific tasks, e.g., for early diagnosis of breast cancer. Therefore, specialized MRI of individual organs is now gaining popularity when the signal is transmitted and received only in a certain area where the human organ of interest is located. Nevertheless, mostly local transceive coils for head and extremities imaging are developed and used at clinical field strength (i.e., 1.5 and 3 T). These coils have proven themselves to investigate fine structures, i.e., knee cartilage; however, their design restricts their application for other parts of the body. Moreover, its construction includes RF cables positioned close to the patient and can potentially breach the RF safety of the MRI study during high-current transmission mode.

Recently, the targeted MRI concept was proposed [1]. This concept includes a passive focusing of the body coil's RF magnetic field to maximize its efficiency for small areas using the resonant structure located around the area of interest. In our previous work [2] it was demonstrated that the metamaterial-inspired resonant structure could be realized by using a system of coupled split-loop resonators, so-called metasolenoid coil (M-coil) [3], that could improve the transmit and receive efficiency of the 1.5 T body birdcage coil more than 4 times. However, this wireless coil supports linear polarization only, interacting only with the H_y-component of the transverse RF magnetic field produced by the body coil, which may limit its efficiency by a factor of $\sqrt{2}$ [4].

In this work, we propose a new quadrature bilateral breast coil design based on the combination of a Helmholtz pair and metasolenoid pair. Also, via electromagnetic simulation with a realistic female voxel model, we compare its performance in terms of transmit efficiency and RF safety with the linearly polarized Helmholtz pair and metasolenoid pair, respectively.

2. Methods

All electromagnetic simulations were performed in CST Studio Suite 2020. Fig. 1(a) demonstrates designs of the considered wireless coils: Helmholtz pair (HH-coil), metasolenoid pair (MM-coil), and quadrature coil based on their combination - HHMM-coil. Each Helmholtz coil consists of connected pair of three turns of copper wire forming three loops with the following dimensions: 124×105 mm^2, 114×95 mm^2, and 104×85 mm^2. A distance of 134 mm separates two pairs of turns. These dimensions lead to tuning an HH-coil to 63.68 MHz - a Larmor frequency of 1.5 T MR scanner. Each of the coils of the MM-system consisted of 10 split-loop resonators containing two parallel telescopic brass tubes (to perform resonant frequency fine-tuning to a specific load) connected by two PCB capacitors at their ends. The capacitors are implemented as the copper strips printed on the top and bottom layers of the dielectric substrate (Rogers RO4250B material) with $\varepsilon = 3.5$, $\tan\delta = 0.0004$ at 64 MHz with the following dimensions $164 \times 114 \times 1$ mm^3. The outer strips dimensions are 9×160 mm^2 and the inner ones 9×58 mm^2. These dimensions were also chosen to tune MM-coil's fundamental eigenmode with homogeneous field distribution to the Larmor frequency. A quadrature HHMM-coil was implemented as MM-coil and HH-coil combination, which do not electrically connect. Each of the four wireless coils was tuned to the same resonant frequency.

Due to the electromagnetic coupling between two resonant structures in each pair, these systems have several eigenmodes in the operating frequency range. The set of the modes for each design was evaluated using a non-resonant loop probe [schematically shown with blue lines in Fig. 1(b)] placed above one of the M-coils (in xz-plane) and between Helmholtz coils (in yz-plane). Each setup was loaded to the breast voxel model. We have used a realistic female body model based on Ella from the Virtual Family [5] and modified it by adding breast phantom no. 1 from the UWCEM Numerical Breast Phantom Repository [6].

To estimate the transmit efficiency and RF safety, we used a shielded high-pass body birdcage coil with 16 legs, 350 mm inner diameter, and 650 mm length. We compared four setups: (1) a voxel model placed inside the body coil without any wireless coil; the same system with HH-coil (2), MM-coil (3), and HHMM-coil (4) placed around the breast. The calculated B_1^+-field and specific absorption rate (SAR) averaged over 10g of tissue mass distributions were normalized to 1 W of total accepted power. Transmit efficiency, and RF safety was evaluated as the root mean squared value (RMS) of the transmit magnetic field – $|B_1^+|_{RMS}$ in the breast area per 1 W of accepted power, and $|B_1^+|_{RMS}$ in the area of interest per square root of maximum local SAR (maxSAR$_{av.10g}$), respectively. Transmit efficiency was calculated in the breast tissues limited by the volume of each wireless coil, respectively.

3. Results

Figure 1b shows the numerically obtained spectra of the magnetic field in the range of 40-80 MHz. The HH-coil has a single peak in this range, which corresponds to a co-directional magnetic field in both coils polarized in the x direction. This is a symmetric fundamental mode. It is associated with the greatest uniformity of the magnetic field within the coil volume. The anti-symmetric mode and the higher-order modes are outside the frequency range. As for MM-coil, two maxima correspond to the excitation of anti-symmetric and symmetric modes within the operating frequency range. They differ in

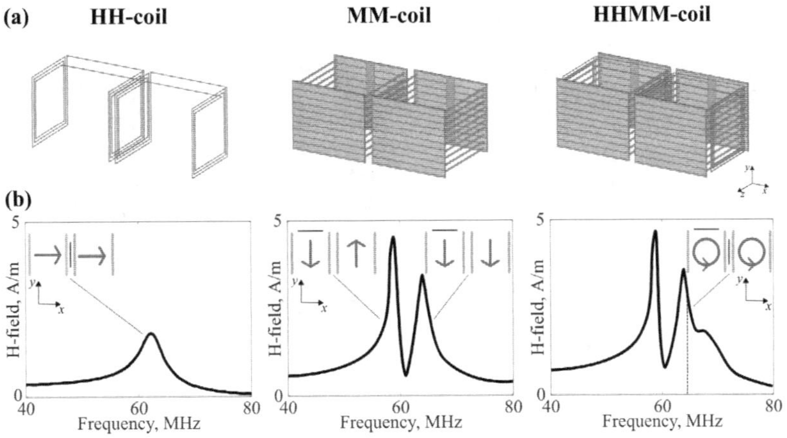

Figure 1. a) Schematic view of the Helmholtz pair (HH-coil), metasolenoid pair (MM-coil), and quadrature coil based on their combination - HHMM-coil. b) Numerically calculated spectra of the magnetic field in the middle of the left breast of the female voxel model on the frequency for each coil type correspondingly. The insets schematically demonstrate the magnetic field orientation within each coil: H_x-component for HH-coil; H_y-component for MM-coil, the opposite direction of red arrows indicates anti-symmetric and symmetric modes; circular arrow shows interaction with both H_x- and H_y-components for HHMM-coil. Blue lines depicted the position of the loop coils used for excitation.

the anti-directional and co-directional magnetic fields in separate M-coils polarized in the y direction. The symmetric mode is optimal because it is characterized by a more uniform transverse RF magnetic field distribution in the area of interest. The third spectrum, for the quadrature HHMM-coil, shows three modes. These are the symmetric and anti-symmetric MM-coil modes and the symmetric HH-coil mode. Due to the magnetic fields' orthogonality created by different coils, the electromagnetic coupling is minimal. Therefore, there is no additional mode splitting. When each type of coil's symmetric mode is adjusted to the operating frequency of 1.5 T MRI, a magnetic field with circular polarization is induced in the volume.

Since the HHMM-coil interacts with both components of the body coil's RF magnetic field, a greater transmit efficiency gain is observed than HH-coil and MM-coil. Fig. 2 demonstrates simulated $|B_1^+|_{RMS}$ and $SAR_{av.10g}$ maps inside the body coil without and with the proposed wireless coils designs placed around the breasts. Strong localization of the $|B_1^+|$-field in the breast area in the presence of the wireless coil leads to a 3.7, 4.3, and 9.7-fold enhancement in body coil's transmit efficiency for HH-, MM- and HHMM-coils, respectively. Although the magnetic field distribution in all cases is almost completely symmetric, $SAR_{av.10g}$ maps have an asymmetric distribution in the human body for the wireless coils that support linear polarization only. This could be explained by the complex interaction of the electric field produced by two feeding ports (with 90-degree phase shift) of the body with the HH- and MM-coils; it is not fully focused within the wireless coil's volume, and due to the arbitrary shape of the body model has asymmetric distribution. The maximum of $SAR_{av.10g}$ (depicted as a red circle in Fig. 2b) for body coil used alone located in the muscles of the back, while in the presence of wireless coils, it occurs in the breast or abdominal muscle areas. At the same time, the maximum $SAR_{av.10g}$ in the body model for the same 1 W total accepted power is increased by 3.8, 1.8, and 3 times for HH-, MM- and HHMM-coils, correspondingly, in comparison with a reference case (body coil used alone). Table 1 summarizes the results of transmit efficiency and RF safety for all the cases. The highest RF safety improvement (5.6-fold) compared to the reference case was achieved for the HHMM-coil.

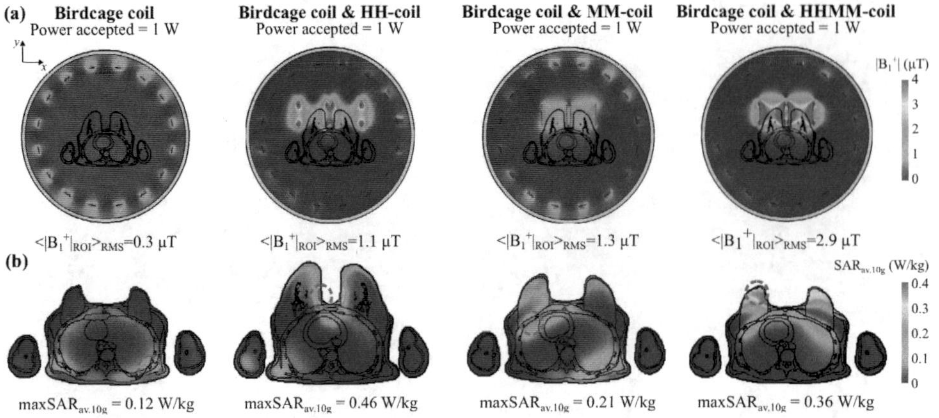

Figure 2. Numerically calculated $|B_1^+|_{RMS}$ maps (a) and $SAR_{av.10g}$ maps (b) for the voxel model placed inside the birdcage coil without and with the WLC. The $|B_1^+|_{RMS}$ value was calculated only for the volume of the resonator in the breast area (ROI). $SAR_{av.10g}$ distributions are build up through the local maximum plane. Local $SAR_{av.10g}$ maxima are indicated with red circles.

Table 1. Numerically calculated results of transmit and SAR efficiencies for different types of wireless coils.

	Body coil	HH-coil	MM-coil	HHMM-coil	Units
$\lvert B_1^+\rvert/\sqrt{P_{acc}}$	0.3	1.1	1.3	2.9	$[\mu T/\sqrt{W}]$
$maxSAR_{av.10g}$	0.12	0.46	0.21	0.36	[W/kg]
$\lvert B_1^+\rvert/\sqrt{maxSAR_{av.10g}}$	0.86	1.62	2.84	4.83	$[\mu T/\sqrt{W/kg}]$

4. Conclusion

In this work, we proposed and compared three designs of the wireless coil, which can be used for targeted bilateral breast imaging at 1.5 T. Two of them are linearly polarized coils based on pair of Helmholtz coils or volumetric metamaterial-inspired coils, another one is quadrature coil based on the combination of these linearly polarized coils. The combined coil interacts with the both (B_{1x} and B_{1y}) components of the transverse RF magnetic field generated by the body coil and focuses it in the region of interest within its volume with circular polarization. Due to this fact, electromagnetic studies with the female voxel model of the quadrature HHMM-coil have shown more than a 2.5-fold increase in transmit efficiency and a 1.7-fold increase in SAR efficiency compared to the linearly polarized MM-coil and HH-coil (Table 1). Future work includes the experimental realization of the HHMM-coil, *in vivo* imaging and comparison coil's receive performance with dedicated cable-connected coils.

Acknowledgments

This work was supported by the grant of the Russian Science Foundation (№ 18-75-10088).

References

[1] Shchelokova A *et al.* 2020 *Nat Commun.* **11** 3840
[2] Puchnin V *et al.* 2021 *J. Magn. Reson.* **322** 106877
[3] Maslovski S *et al.* 2005 *Prog. Electromagn. Res.* **54** 61–68
[4] Glover G *et al.* 1969 *J. Magn. Reson.* **64(2)** 255–270
[5] Christ A *et al.* 2010 *Phys. Med. Biol.* **55** 23–38
[6] Burfeindt M J *et al.* 2012 *IEEE Antennas Wirel. Propag. Lett.* **11** 1610–1613

Magneto-optical effects in second harmonic generation from W/Co/Pt nanofilms

V.V. Radovskaya[1], **E.A. Mamonov**[1], **I.Yu. Pashen'kin**[2], **N.S. Gusev**[2], **T.V. Murzina**[1]

[1] Department of Physics, Moscow State University, 119991, Moscow, Russia
[2] Institute for Physics of Microstructures RAS, 607680, Nizhny Novgorod, Russia

E-mail: radovskaia.vv16@physics.msu.ru

Abstract. Interfaces between ferromagnetic metals and nonmagnetic specimen attract much attention as they are very important for the formation of magnetic properties of nanostructures. Vice versa, specific magnetic ordering at such interfaces may provide new effects in their optical and nonlinear optical response. In this work we study the magnetization-induced effects in optical second harmonic generation (SHG) in W/Co/Pt-based thin films with the thicknesses of the Co layer of $2 - 10\ nm$. Besides common *odd* in magnetization effects in the SHG intensity, we observe additional one that is not expected for homogeneously magnetized ferromagnetic films, which consists in modulation of *p*-polarized SHG intensity under longitudinal magnetic field application. The phenomenological description of the observed effect is performed in terms of gradient and second-order in magnetization contributions to the SHG polarization, where gradient of magnetization along the normal to the structure plays the key role.

1. Introduction

Physics and technology of magnetic nanostructures is a hot research area known for a bunch of exciting effects originating from specific magnetic ordering at interfaces, which should appear as well in their optical and nonlinear optical response, first of all in surface-sensitive optical second harmonic generation (SHG) [1]. This is especially valid for the case of nanostructures composed of ferromagnetic and nonmagnetic materials, where such unique interface effects as perpendicular magnetic anisotropy, chiral magnetization states etc. have been demonstrated [2, 3, 4, 5]. Recently we studied magnetization-induced effects in SHG from bi- and trilayer films consisting of nanolayers of Co and nonmagnetic heavy metals (Pt, Ta). It was demonstrated that *odd* in magnetization intensity effect is observed for the p-polarized SHG in both reflection and transmission SHG schemes as the dc magnetic field was applied perpendicularly to the light, while it is symmetry-forbidden in dipole approximation for homogeneously magnetized structures. Moreover, it was found that the value of this "forbidden" magnetization-induced SHG effect depends on the symmetry of the nanostructures and is the largest in the case of asymmetric Pt/Co/W and Ta/Co/Pt films [6].

In order to investigate this effect in more detail, here we study the SHG in a nanostructure of the same nature, with the composition of W/Co/Pt, when applying *dc* magnetic field up to $1,5\ kOe$ in longitudinal geometry.

2. SECOND HARMONIC GENERATION

In the electric dipole approximation, the SHG polarization is expressed as: $\mathbf{P}_{2\omega} = \hat{\chi}^{(2)D} \vdots \mathbf{E}_\omega \mathbf{E}_\omega$, where $\hat{\chi}^{(2)D}$ is the electric dipole second-order susceptibility, $\mathbf{E}_\omega$ is the electric field of the fundamental light at the frequency ω. According to the Neumann's principle, $\hat{\chi}^{(2)D}$ is forbidden in the electric dipole approximation in the bulk of centrosymmetric materials, while it appears at the interfaces where the inversion symmetry is broken, thus giving rise to the interface SHG [7]. In the case of magnetic interfaces the magnetization-induced SHG (MSHG) in the most general form can be expressed as a series of the magnetization $\mathbf{M}$ [8]:

$$\chi_{ijk}(\mathbf{M}) = \chi^{(0)}_{ijk} + \chi^{(1)}_{ijkL} M_L + \chi^{(2)}_{ijkLM} M_L M_M + \chi^{(3)}_{ijklM} \nabla_l M_M + \chi^{(4)}_{ijkLmN} M_L \nabla_m M_N + \ldots, \quad (1)$$

where small letters in subindeces correspond to polar vectors ($\mathbf{E}_\omega$ and ∇), while the capital ones – to the axial vector $\mathbf{M}$. The first term $\hat{\chi}^{(1)}$ corresponds to nonmagnetic (crystallographic) dipole susceptibility, the second one is responsible for the common magnetooptical Kerr effect at the SHG wavelength in homogeneously magnetized media. The term $\hat{\chi}^{(2)}$ in Eq. (1) corresponds to the second-order (even) in $\mathbf{M}$ contribution to the nonlinear susceptibility, $\hat{\chi}^{(3)}$ and $\hat{\chi}^{(4)}$ describe the contributions induced by gradients of the magnetization; they are allowed in the bulk of centrosymmetric metals.

If the term $\hat{\chi}^{(1)}$ vanishes, which is realized for the case of p-polarized SHG and longitudinal or polar magnetic field application, *odd* in $\mathbf{M}$ MSHG effects are given by gradient terms described by $\hat{\chi}^{(3)}$ and $\hat{\chi}^{(4)}$; in what follows they are denoted as "forbidden" MSHG effect. It was explained in terms of "gradient" $\chi^{(4)}$ contribution, particularly given by the components like $\chi^{(4)}_{xxxXzZ}$, $\chi^{(4)}_{xzzXzZ}$ etc, all of them containing gradients of $\mathbf{M}$ along the normal to the surface (along the (OZ) direction); the coordinate frame is shown in Fig. 1. Besides, one can consider the second-order in $\mathbf{M}$ interface contribution described by $\chi^{(2)}$. Thus the SH intensity can be written as: $I_{2\omega} \approx |\mathbf{P}^{cr} + \mathbf{P}^{MM} + \mathbf{P}^{\nabla}|^2$. On the other hand, the SHG intensity can be expressed as a sum of *even* and *odd* in $\mathbf{M}$ contributions as

$$I_{2\omega} \approx |\mathbf{P}^{even} + \mathbf{P}^{odd}(\mathbf{M})|^{\mathbf{2}}; \quad (2)$$

these notation are convenient for the analysis of *odd* in magnetization effects in the SHG intensity and are used below when describing the experimental results.

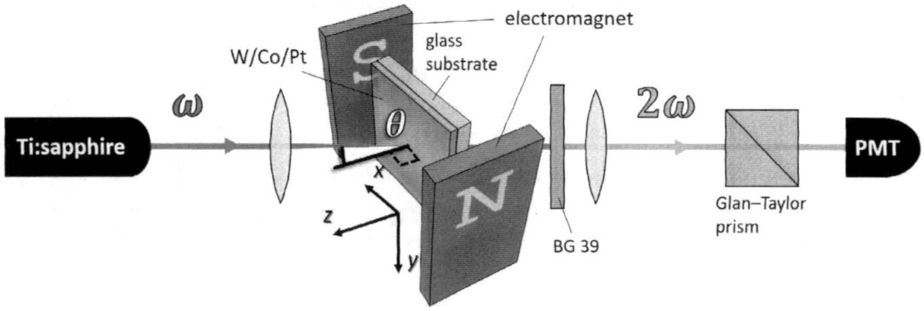

Figure 1. MSHG experimental geometry with the coordinate frame; θ is the angle of incidence.

3. RESULTS AND DISCUSSION

In this work we studied isotropic magnetic trilayer films of the composition W(3)/Co(x)/Pt(3) with various Co thicknesses ($x = 2, 3, 4, 10$ nm). The samples were grown by magnetron sputtering on glass substrates using a AJA 2200 multichamber system at the basic pressure of 10^{-5} Pa, the layers are denoted starting from the substrate.

The scheme of the MSHG experiment is presented in Fig. 1. As a pump we used the p-polarized radiation of the pulse Ti:sapphire laser with the following parameters: wavelength of 800 nm, pulse duration of approximately 20 fs, repetition frequency of 80 MHz. The pump was focused to the surface of the sample in a spot of 30 μm in diameter; the transmitted SHG light passed through the color BG39 filters, an analyser (Glan-Tailor prizm) and was detected by a PMT operating in the photon counting mode. The analyser was set at (i) 45° between p- and s-polarizations (the so called *mixed* polarization, "allowed" geometry), or (ii) p-polarization of the SHG radiation ("forbidden" geometry). The samples were placed in the magnetic fields formed by an electromagnet (H <1.5 kOe).

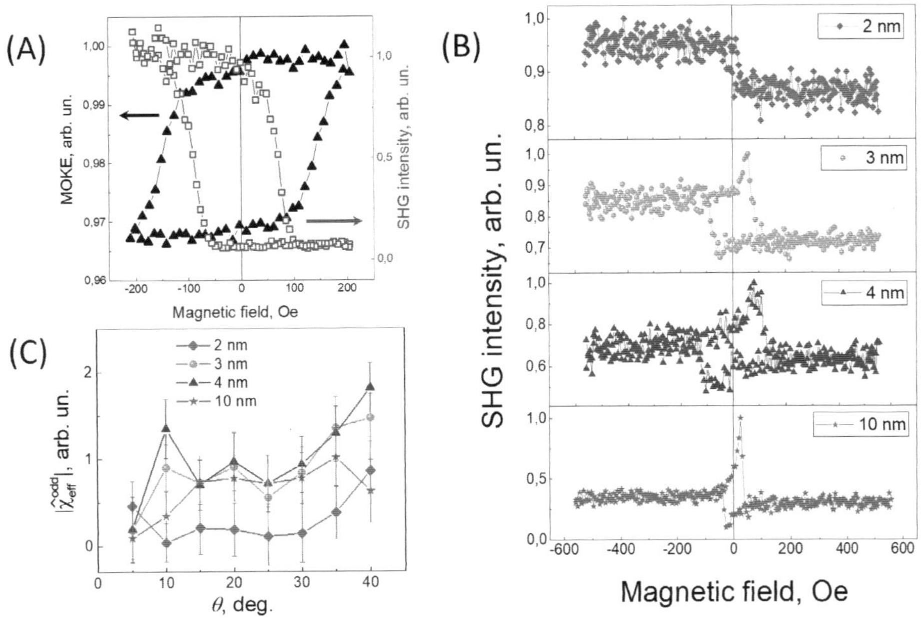

Figure 2. (A) Longitudinal MOKE hysteresis loop (triangular symbols) and analogous "allowed" SHG magnetic field dependence (open square symbols) measured for W(3)/Co(3)/Pt(3) film. (B) "Forbidden" SHG effect for structures W/Co/Pt with various Co thicknesses of $\theta = 30°$. (C) Estimated *odd* terms of the second-order susceptibility as a function of angle of incidence, $|\hat{\chi}^{odd}_{eff}|(\theta)$, for W/Co/Pt films with various Co thicknesses.

Figure 2(A) shows MOKE and MSHG magnetic hysteresis loops in the "allowed" geometry. Longitudinal magnetooptical Kerr effect (MOKE) was used for the characterization of the magnetic properties of the films, when using the probe radiation of He-Ne laser. As an example, MOKE hysteresis loop for W(3)/Co(3)/Pt(3) film is shown in Fig. 2(A) (black symbols). One can see that the width of the MOKE hysteresis is roughly twice as large as in the case of the MSHG. This may be related to different localization of the MOKE and MSHG sources, in the bulk of the cobalt film and at its interfaces, are their different magnetic properties. Using Eq. (1), one can estimate the ratio of $|\mathbf{P}^{odd}|$ and $|\mathbf{P}^{even}|$ of $\frac{|\mathbf{P}^{odd}|}{|\mathbf{P}^{even}|} \sim 0.56$.

Figure 2(B) shows magnetic hysteresis loops measured in the geometry of the "forbidden" MSHG effect for the angle of incidence $\theta = 30°$ for the samples with Co thicknesses $x = 2, 3, 4, 10$ nm; the magnetic field was applied in the longitudinal geometry, i.e. in plane of the samples and in the plane o incidence, (XOZ). Firstly, pronounced peaks in the SHG intensity for the magnetic fields lower than the saturation fields are seen, which can be explained as remagnetization of the structure by rotation of the average magnetization **M** in the plane of the sample through the transversal MOKE geometry. Secondly, one can observe the asymmetry of the SHG intensity in saturating magnetic fields (of up to 600 Oe) of the opposite directions. It can be characterized by the MSHG magnetic contrast $\rho_{2\omega} = \frac{I^+_{2\omega} - I^-_{2\omega}}{I^+_{2\omega} + I^-_{2\omega}}$, where $I^+_{2\omega}$ and $I^-_{2\omega}$ denote the SHG intensity for positive and negative magnetic fields, respectively. The existence of the MSHG contrast indicates the presence of *odd* with respect to **M** terms in the SHG polarization that appear in Eq. (2). As the common MOKE in SHG is forbidden under these conditions, we propose that this effect can be explained in terms of contribution of $\mathbf{P}^{MM}$ and $\mathbf{P}^{\nabla}$ from (1).

Measuring the "forbidden" effect for all W/Co/Pt samples for the incidence angles $\theta = 5°, 10°, \ldots, 45°$ revealed an increase of mean SHG intensity ($I^{mean} = 0.5(I^+_{2\omega} + I^-_{2\omega})$) with increasing θ. Similar phenomenon was observed for Pt/Co/W thin films (opposite order of layers) [6]. Such dependence represents the primary influence of z-component of nonlinear polarization on magneto-induced SHG. By calculating the mean SHG intensity and magnetic contrast, we estimated odd contribution as $|\mathbf{P}^{odd}| \sim |\hat{\chi}^{odd}_{eff}| \sim 0.5|\rho_{2\omega}|\sqrt{I^{mean}}$. Figure 2(C) demonstrates the dependence of this quantity on the angle of incidence $|\hat{\chi}^{odd}_{eff}|(\theta)$, which reveals quite similar values $|\hat{\chi}^{odd}_{eff}|$ for all x values at angles of incidence $\theta = 15° \ldots 30°$. A rise at $\theta \geqslant 30°$ can be explained as an increase of relative contribution of $\mathbf{P}^{MM}$ and $\mathbf{P}^{\nabla}$ with polar z indices.

4. Conclusion

Summing up, we studied experimentally magnetic field induced effects in optical second harmonic generation from W/Co/Pt nanolayered films. We demonstrate that when excluding the detection of common magnetooptical effects at the SHG wavelength in the longitudinal magnetic field, that is expected for homogeneous magnetic films, *odd* in magnetization changes of the intensity of the p-polarized SHG are observed. These studies confirm that gradient in magnetization terms of the second-order susceptibility should be necessarily taken into account when considering the MSHG effects in thin magnetic films.

Acknowledgments

This work was supported by Russian Science Foundation (grant No. 19-72-20103) and the Foundation for the Advancement of Theoretical Physics and Mathematics "BASIS" (grant No. 20-2-9-23-1).

References

[1] Boyd R W 2008 *Nonlinear optics, Third Edition* (USA: Elsevier)
[2] den Broeder F J A, Kuiper D, van de Mosselaer A P and Hoving W 1988 *Phys. Rev. Lett* **60** 2769
[3] Carcia P F, Meinhaldt A D and Suna A 1985 *Appl. Phys. Lett.* **47** 178
[4] Legrand W, Chauleau J-Y, Maccariello D, Reyren N, Collin S, Bouzehouane K, Jaouen N, Cros V and Fert A 2018 *Science Advances* **4** eaat0415
[5] Jiang W, Zhang S, Wang X, Phatak C, Wang Q, Zhang W, Jungfleisch M B, Pearson J E, Liu Y, Zang J, Cheng X, Petford-Long A, Hoffmann A and te Velthuis S G E 2019 *Phys. Rev.* B **99** 104402
[6] Murzina T V, Radovskaya V V, Pashen'kin I Yu, Gusev N S, Maydykovskiy A I and Mamonov E A 2021 *Optics Express* **29** 2106
[7] Shen Y R 1984 *The Principles of Nonlinear Optics* (Wiley, New York)
[8] Petukhov A V, Lyubchanskii I L and Rasing Th 1997 *Phys. Rev.* B **56**

Corrigendum: Non-collinear ground state and stable bimerons from four-spin chiral interactions in D_{3h} magnet (2021 *J. Phys.: Conf. Ser* **2015** *012118*)

G R Rakhmanova, A N Osipov, D I Ilin, I V Shushakova and I V Iorsh
School of Physics and Engineering, ITMO University, Saint Petersburg, 197101, Russia

gulnaz.rahmanova@metalab.ifmo.ru

Page 1:
In the Introduction section 1, the following text appears:
In this Letter, we discuss another possible origin of noncolliner magnetic textures in ferromagnets with the D_{3h} point group symmetry.
This should read:
The Letter is presented based on the results of the article [11] and will discuss possible origin of noncolliner magnetic textures in ferromagnets with the D_{3h} point group symmetry.

Page 4:
In the References section, the following text should appear:
"[11] Ado I A, Rakhmanova G R, Zezyulin D, Iorsh I V, Titov M 2021 arXiv:2105.14495 [Physical Review Letters]"

The authors acknowledge that it is worthwhile to mention the article "Alternative to Dzyaloshinskii Moriya interaction for monolayer Fe_3GeTe_2 and other two-dimensional ferromagnets with trigonal prismatic symmetry" written by Ado I A, Rakhmanova G R, Zezyulin D, Iorsh I V, Titov M (2021 arXiv:2105.14495 [Physical Review Letters]) as reference 11 due to huge contribution to the current work and through this corrigendum this is addressed.

Journal of Physics: Conference Series **2015** (2021) 012118 doi:10.1088/1742-6596/2015/1/012118

Non-collinear ground state and stable bimerons from four-spin chiral interactions in D_{3h} magnet

G R Rakhmanova, A N Osipov, D I Ilin, I V Shushakova and I V Iorsh

School of Physics and Engineering, ITMO University, Saint Petersburg 197101, Russia

gulnaz.rahmanova@metalab.ifmo.ru

Abstract. We demonstrate that four-spin interactions in crystals with D_{3h} point group of symmetry can cause a phase transition from a collinear state to a non-collinear magnetic ground state (such as magnetic vortices or magnetic skyrmions), while all anti-symmetric chiral terms are forbidden by symmetry in such crystals. Moreover, D_{3h} point group rather common among two dimensional magnets. Taking into account possible four-spin chiral exchange interactions is important for understanding noncollinear magnetic order in these systems. We also address a possible stabilization of bimerons by the same contribution.

1. Introduction

Electron-mediated indirect asymmetric exchange between a pair of localized spins in lattices with broken inversion symmetry or with external symmetry breaking is commonly referred to as the Dzyaloshinsky-Moriya interaction (DMI) [1]. The DMI leads to a variety of magnetic structures from the collinear state (ferromagnetic or anti-ferromagnetic) to complex, non-collinear spin textures [2]. Indeed, the presence of linear-in-gradient terms in micromagnetic energy, that DMI may induce, can make a collinear order unstable and create helix, cone, or skyrmion crystals. Those are important structures that are promising candidates for the role of information signal carrier in spintronics and also may be used to store and manipulate information. Such structures were observed in many conducting magnets or magnetic multilayers with broken inversion symmetry such as MnSi, FeGe, Ir/Co/Pt or Pt/CoFeB/MgO [3]. However, a little attention has been paid to the study of two-dimensional (2D) magnets. A breakthrough occurred in 2017, when long-range magnetic order was discovered in 2D van der Waals materials $Cr_2Ge_2Te_6$ [4] and CrI_3 [5]. Soon they were accompanied by a metallic ferromagnet Fe_3GeTe_2 [6, 7], where spin helical phase also has been detected [8]. These materials are characterized by D_{3h} point group of symmetry with honeycomb magnetic lattice, for which all Lifshitz invariant terms in micromagnetic functional are forbidden, while multi-spin chiral exchange interactions are allowed by symmetry [9]. That is why the origin of non collinear magnetic order in this material cannot be correctly explained by DMI [10]. It can affect magnetic order only at the sample boundaries. In this Letter, we discuss another possible origin of noncolliner magnetic textures in ferromagnets with the D_{3h} point group symmetry. As in D_{3h} all antisymmetric contributions vanish, it is thus worthwhile to consider terms of the fourth order with respect to the vector of local magnetization direction.

2. Helical phase from four-spin chiral interactions

Free energy functional of D_{3h} ferromagnet depends on a dimensionless magnetization vector $\boldsymbol{m}(r)$ of the unit length and reads:

$$F[m] = \int d^2r[A\sum_\alpha(\nabla m_\alpha)^2 + 8BI + Km_z^2 - \boldsymbol{Hm}], \quad (1)$$

where $\boldsymbol{H}$ stands for external magnetic field measured in energy units, $A > 0$, K, B are exchange, anisotropy and four-spin interaction constants, respectively. The group D_{3h} is characterized by two independent fourth order invariants with respect to the field $\boldsymbol{m}$(r). They can be written as

$$I_1 = m_x(m_x^2 - 3m_y^2)\nabla \mathbf{m}_\parallel, \quad I_2 = m_x(m_x^2 - 3m_y^2)\nabla \mathbf{m}_\perp. \quad (2)$$

Obviously, for a 2D system, the second one can be disregarded. Hence, we are left with a single fourth order term I_1.

A non-collinear ansatz for local magnetization field $\boldsymbol{m}(r)$, that has the biggest chance to become a ground state, can be written in a form:

$$\boldsymbol{m}(\boldsymbol{r}) = \boldsymbol{n}\cos\alpha + [\boldsymbol{m}_1\cos(\boldsymbol{kr}) + \boldsymbol{m}_2\sin(\boldsymbol{kr})]\sin\alpha, \quad (3)$$

where $\boldsymbol{n}$, $\boldsymbol{m_1}$ and $\boldsymbol{m_2}$ are mutually perpendicular unit vectors: $\boldsymbol{m}_1 = (\sin\phi, -\cos\phi, 0)^T$, $\boldsymbol{m}_2 = (\cos\phi\cos\theta, \sin\phi\cos\theta, -\sin\theta)^T$, $\boldsymbol{n} = \boldsymbol{m}_1 \times \boldsymbol{m}_2$. The angle $\alpha = 0$ corresponds to a collinear state, $\alpha = \pi/2$ – to a pure helix state. We assume that the ferromagnet is kept well below Curie temperature, hence $|\boldsymbol{m}| = 1$. The angle between the vectors $\boldsymbol{k}$ and $\boldsymbol{m}$ is an important characteristic of the state that can be measured in experiment.

Substituting (3) into (1) we minimize equation (1) with respect to k_x, k_y and get for I_1 the free energy as a function of three variables θ, ϕ, α:

$$\frac{F[m]A}{VB^2} = -\frac{9}{64}(\sin\alpha + 5\sin3\alpha)^2\sin^4\theta\cos^2\theta + \frac{KA}{B^2}\cos^2\theta\cos^2\alpha$$
$$+\frac{KA}{2B^2}\sin^2\theta\sin^2\alpha - \frac{A}{B^2}\sin\alpha\,\boldsymbol{Hm}, \quad (4)$$

where the four-spin interaction sets the energy scale A/B^2 that defines the non-collinear order.

It is interesting to note that states described by equation (4) are degenerate with respect to ϕ. It means the free energy does not depend on the azimuthal angle of the average magnetization vector $\boldsymbol{m}$. At the same time, for helical state with a finite wave vector, the angles between $\boldsymbol{k}$ and $\boldsymbol{m}$ are different for different values of ϕ:

$$\boldsymbol{k}\cdot\boldsymbol{m} \propto \sin3\phi. \quad (5)$$

It is seen in figure 1. Thus, by controlling the direction of the average magnetization it is also possible to control the propagation direction of the non-collinear state.

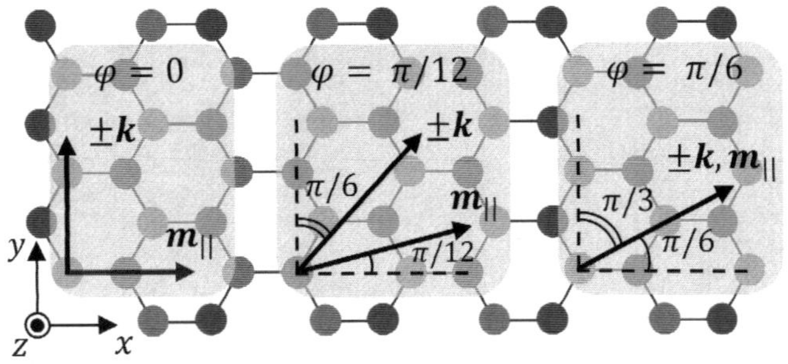

Figure 1. Geometry of the structure of the D_{3h} magnet. The direction of the helix propagation is defined by the wavevector $\boldsymbol{k}$. Panels show three directions of the wavevector and the in-plane component of $\boldsymbol{m}$.

Further we minimize equation (4) with respect to θ, ϕ, α numerically and the result of the energy minimization is illustrated in figure 2 by plotting the dependence α (an angle describing a transfer from collinear to non-collinear state) on both KA/B^2 and HA/B^2 at the absolute energy minimum for the cases of out-of-plane and in-plane magnetic field. The graph shows a phase transition from a collinear state ($\alpha = 0$) to non-collinear state ($\alpha \neq 0$) and back. In addition, the angle α smoothly decreases from

$\pi/6$ to almost 0 with H increase. It can be seen that the span of magnetic cone may, at best, only slightly exceed the value $\pi/6$, while the pure helix, $\pi/2$, is never reached. The corresponding jumps are also seen in the left panels.

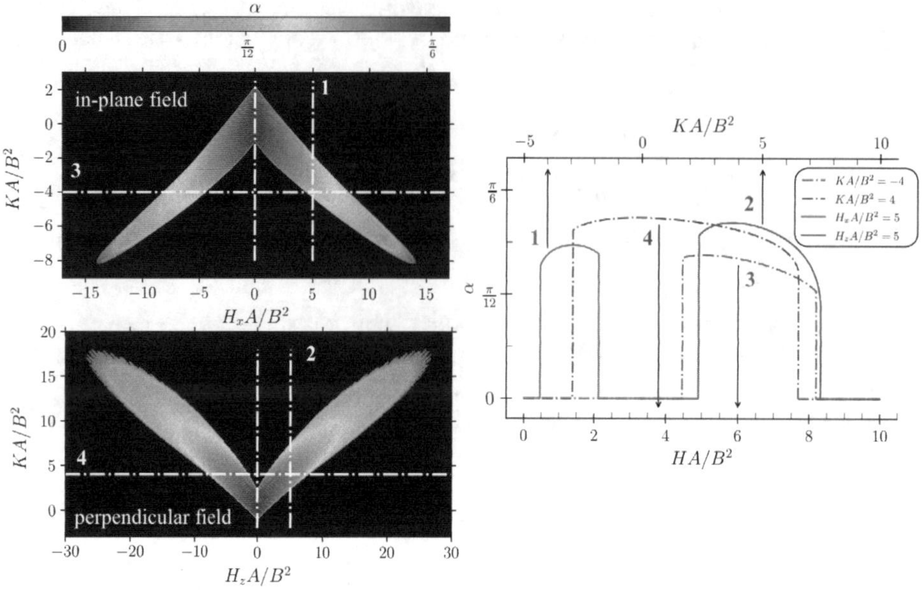

Figure 2. Phase diagram of the collinear and non-collinear phases as a function of anisotropy K and external magnetic field in lateral (upper panel) and perpendicular (lower panel) directions. The right panel shows the cross-sections of the phase diagrams along the lines marked with dashed lines.

3. Stabilization of bimerons

It is natural to explore the effect of four-spin interactions on skyrmions. It however appears that due to the antisymmetric nature of the term in equation 2 its contribution to the radially symmetric spin distributions such as skyrmion is zero. At the same time, topological excitations lacking the radial symmetry should be affected by the four-spin interactions. As an example, we consider a bimeron, an in-plane counter part of skyrmion with the magnetization profile given by $m = \hat{R}_z(\phi_0)m_0$, where $\hat{R}_z$ is the matrix of rotation with respect to z axis, and m_0 reads:

$$m_0 = [\cos\Theta(r), \sin\Theta(r)\cos\Phi(\phi), \sin\Theta(r)\sin\Phi(\phi)], \tag{6}$$

where $\Phi = Q\phi + \delta$ with integer Q being the topological charge, and an unknwon $\Theta(r)$ satisfies the boundary conditions $\Theta(0) = \pi$, $\Theta|_{r\to\infty} = 0$. In what follows we assume the positive exchange constant A>0. If we substitute the ansatz (6) for $Q = 1$ to the expression for the free energy we get the following Lagrange equation:

$$\Theta''(\rho) + \frac{\Theta'(\rho)}{\rho} - \frac{\sin 2\Theta(\rho)}{2\rho^2} - \frac{KA}{4B^2}\sin 2\Theta(\rho) + \frac{3B}{2|B|\rho}\sin(2\phi_0 + \delta)\left[5\cos^2\Theta(\rho) - 1\right]\sin^2\Theta(\rho) = 0. \tag{7}$$

The specific profile for Θ can be obtained by numerically solving the corresponding Euler-Lagrange equations. In figure 3(a) we show profiles of Θ for different values of the dimensionless parameter KA/B^2 and $2\phi_0 + \delta = \pi/2$.

The role of four-spin interaction can be revealed by a simple analysis. If we suppose that the specific radial bimeron profile $\Theta_{BM}(\rho) = \varphi_{BM}(\rho/\rho_0)$ is characterized by a single characteristic bimeron radius r_0 we can rewrite the free energy as

$$F = AF_{exch} - \tilde{B}\rho_0 F_{4spin} + \frac{KA}{2B^2}\rho_0^2 F_{anisotropy}, \tag{8}$$

where $\tilde{B} = (B/|B|)\sin(2\phi_0 + \delta)$ and $F_{exch}, F_{4spin}, F_{anisotropy}$ are scale-invariant contributions from exchange, four-spin interactions and anisotropy respectively. We can immediately note, that since

$F_{anisotropy} = \int \rho d\rho \sin^2 \varphi_{BM}(\rho) > 0$, the free energy has a minimum at finite ρ_0 only for $K>0$, i.e. for the case of easy-plane anisotropy. Further, let us employ a standard ansatz for the radial profile $\phi_{BM} = \pi - 2\arcsin(\tanh(\rho/\rho_0))$. Substituting this ansatz to the free energy functional and minimizing it with respect to ρ_0 we finally obtain the following equation for the $\tilde{\rho}_0$ corresponding to the minimal energy

$$\tilde{\rho}_0 \approx \frac{4.2B^2}{KA}\frac{B}{|B|}\sin(2\phi_0 + \delta). \qquad (9)$$

Thus, whenever $\tilde{B} = (B/|B|)\sin(2\phi_0 + \delta) > 0$ there exist a bimeron solution. The minimal energy corresponds to the condition $2\phi_0 + \delta = \pm\pi/2$ for positive (negative) B. In figure 3(a) we plot the ansatz solution with the proper radius ρ_0 with the dashed lines. It is seen that ansatz solution accurately describes the exact numerical solution at small radius but substantially discrepancies are seen at larger ρ. The scaling of the mean bimeron radius is also well described by the simple linear relation as can be seen in figure 3(b).

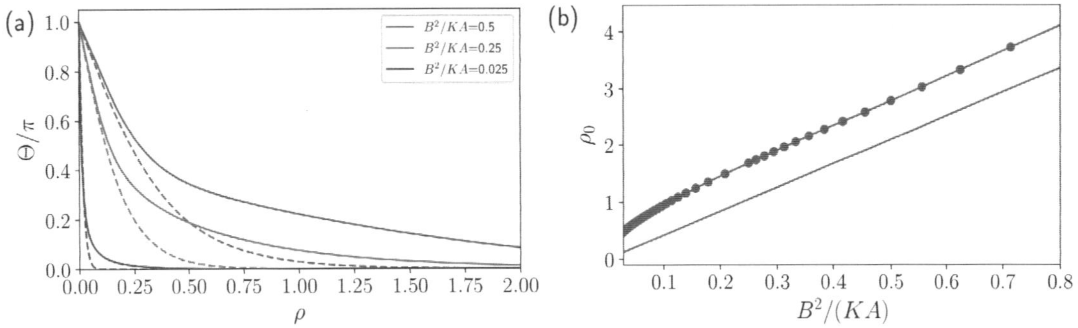

Figure 3. (a) Radial profile of the bimeron state for different values of KA/B^2 obtained via numerical solution of the Euler-Lagrange equation (solid lines) and with an ansatz. (b) Dependence of the mean bimeron radius on B^2/KA obtained numerically (blue line) and using the ansatz (green line).

4. Conclusion

In conclusion, we demonstrate that the DMI in a two-dimensional magnet with D_{3h} point group of symmetry does not contribute to the energy functional and cannot cause an instability of the collinear order, while the existence of the four-spin indirect magnetic interaction may be responsible for the appearance of a non-collinear magnetic order. We also address a possible stabilization of bimerons by the same contribution.

Acknowledgments

The work was supported by the Foundation for the Advancement of Theoretical Physics and Mathematics "BASIS" (№ 21-1-5-17-1).

References

[1] Dzyaloshinsky I 1958 *J. Phys. Chem. Solids* **4** 241
[2] Ishikawa Y, Tajima K, Bloch D and Roth M 1976 *Sol. Stat. Comm.* 19 525
[3] Muhlbauer S, Binz B, Jonietz F, Pfleiderer C, Rosch A, Neubauer A and Georgii R 2009 *Science* **323** 915
[4] Gong C, *et al* 2017 *Nature* **546** 265
[5] Huang B, *et al* 2017 *Nature* **546**, 270
[6] Deng Y, *et al* 2018 *Nature* **563** 94
[7] Fei Z, et al 2018 Nat. Mater. **17** 778
[8] Meijer M, Lucassen J, Duine R, Swagten H, Koopmans B, Lavrijsen R and Guimar M 2020 *Nano Lett.* **20** 8563
[9] Ado I, Qaiumzadeh A, Brataas A and Titov M 2020 *Phys.Rev.* B **101** 161403
[10] Laref S, Kim K-Wand Manchon A 2020 *Phys. Rev.* B **102** 060402(R)

Journal of Physics: Conference Series **2015** (2021) 012119 doi:10.1088/1742-6596/2015/1/012119

Temporal transition in parallel-plate waveguides: analysis of scattering and propagation at the temporal interface

L. Stefanini, D. Ramaccia, A. Toscano, and F. Bilotti

ROMA TRE University, Department of Engineering, via Vito Volterra 62, 00146, Rome, Italy

luc.stefanini@stud.uniroma3.it

Abstract. In this contribution, we present the analysis and numerical verification of the scattering phenomenon from a temporal interface in a parallel-plate waveguide realized by suddenly modifying the dimensions of the waveguide while the wave is propagating. As it is well known in guided wave theory, at the interface between two different waveguides there exists a change of the effective refractive index and wave impedance perceived by the propagating wave within the device, which inevitably scatters at the interface into a reflected and refracted wave. In analogous way, by suddenly changing the effective material properties within the whole waveguide, it is possible to realize the so-called temporal interface, as well. Here, we theoretically and numerically investigate on the scattering from a waveguide temporal interface induced by the abrupt change of the waveguide dimension, which in turn realize a change of the effective material properties perceived by the wave.

1. Introduction

In the last years, thanks to the recent achievements in electronics, in particular introducing the possibility to digitally control the material properties, there was a growing interest on space-time metamaterials [1,2]. Among them, temporal metamaterials are characterized by the possibility to modulate the refractive index in the time domain. An abrupt change of the refractive index in the whole space realizes a *temporal interface*, which leads to the generation of a reflected and a refracted wave. This phenomenon was originally investigated by Morgenthaler [3], who considered the scenario of a travelling wave in a medium whose characteristics are changed over time, he demonstrated that a change in the velocity, induced by a change in the material properties, will generate a forward and a backward waves in the new material. Moreover, respecting the duality with the spatial problem, the frequency of the new waves is changed rather than the wavelength. Combining more than one interface opens the door to realizing temporal devices, such as temporal Fabry-Perot cavities and matching slabs [4,5], antireflection coatings [6], broadband absorbers [7], just to name a few. Further advancing the degree of freedoms, introducing anisotropy, leads to singular behaviors such as an inverse prism [8] or temporal aiming [9]. However, the practical implementation of such a temporal metamaterial is not a straightforward task.

In this contribution, we consider a conventional waveguide whose geometry of a waveguide is altered in a certain instant in time, changing instantaneously the effective refractive index and wave impedance perceived by the propagating wave. In guided wave theory, the physical connection between two different waveguides realizes an abrupt change of these quantities in the spatial domain and, to satisfy the boundary conditions at the interface, the incident wave is split into a reflected and refracted wave that starts to propagate in opposite direction from the interface. A similar behavior is realized by introducing, in a certain instant in time, an abrupt change of the geometrical properties of the waveguide. The transition returns a response like the one just presented by the spatial interface. Here, we consider a conventional parallel-plate waveguide (PPWG), whose metallic plates are at a distance d_1 before the instant of time t_0, and d_2 after it, as illustrated in Fig. 1. We describe in detail the effect induced by the boundary modification in terms of induced frequency shift and coupling with the supported modes of the new waveguide. We report the analytically evaluated transmission and reflection coefficients and compare them with the numerical results.

2. Scattering at the temporal interface in a parallel-plate waveguide

Let us consider a PPWG infinitely extended in the yz-direction and with height d1 along the x-direction (Fig. 1), operating in a monomodal regime and supporting the propagation of a TM_1 mode. The electromagnetic wave is propagating in the z-direction with wavevector k_{z1} at the operative frequency f_0. In Fig. 2a, we report a graphical representation of the dispersion diagram at the frequency f_0 (orange solid circle) before the temporal boundary discontinuity. For a given height of the waveguide the x-component of k_0, *i.e.*, $\pm k_{x0}$, is set, defining, in turn, the value of k_{z1} via the intersection with the circumference. After the temporal discontinuity, the component k_x suddenly changes to k_{x1}, whereas the propagating wavenumber, k_{z1} ,dictated by wavelength is maintained. This forces the system to propagate with a different k_0, *i.e.*, k_{01} (orange solid circle), modifying instantaneously the temporal frequency of the propagating wave. Stressing the duality with a spatial interface, the incidence of a wave onto a space discontinuity will preserve the frequency which will in turn change the propagation constant k_z by the intersection with new k_{x1}.

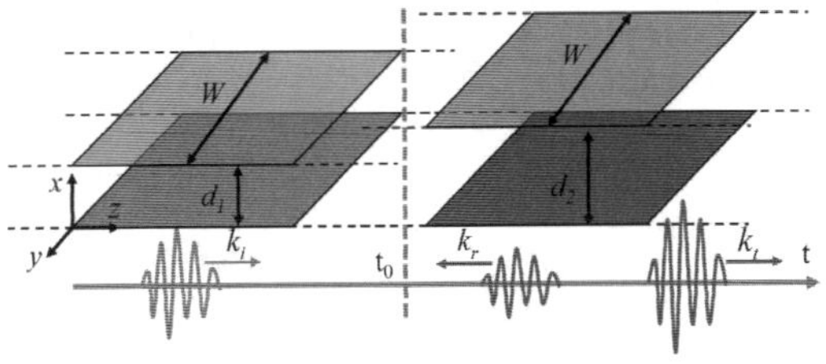

Figure 1. Boundary-induced temporal interface in a parallel plate waveguide. The height of the waveguide is suddenly increased from d1 to d2 at the time instant t0. The incident wave, orange, will be reflected and transmitted, blue, preserving its wavelength and changing its frequency.

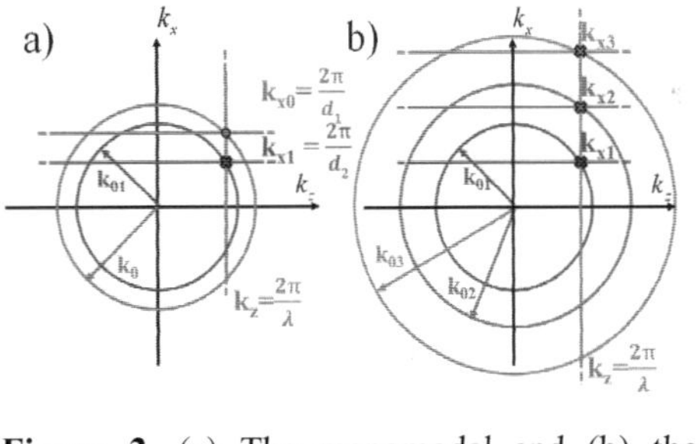

Figure 2. (a) The monomodal and (b) the multimodal case discontinuity induced by the temporal boundary discontinuity at t0. The time interface, represented by the line, intersects all the modes in the second structure, exciting them with their frequencies.

However, the waveguide is inherently a multi-modal structure, following the condition of $k_z = k_{z1}$, it is possible to represent the time interface as a vertical line which intersects all the modes of the second structure, as shown in Fig. 2b. There is not a mono-modal scenario, every mode can support a particular wavelength and every mode is intersected by the vertical line indeed. Moreover, at the transition, the initial spatial field distribution within the waveguide remains unaltered, but it is no longer an eigen solution for the new structure. Therefore, during the propagation the original mode starts decreasing exponentially and its energy is transferred to the new eigen-modes of the waveguide with the same k_z through a mode coupling mechanism, regardless being forward or backward propagating. Considering the analogy with a spatial interface between two different waveguides, we can say that the second device is always excited above its mono-modal bandwidth. In Fig. 2b the generic mth circumference in the dispersion diagram corresponds to the new frequency at which the mode is propagating, that is related to the original frequency ω_0 as follows:

$$\omega_m = \omega_0 \sqrt{1 - \frac{k_{x1}^2 - k_{xm}^2}{k_{02}^2}} \tag{1}$$

The exact amount of energy coupled to each mode from the original one is defined by the phase-matching condition and symmetry-matching between modes. By using coupled mode theory (CMT) together with the time discontinuity boundary conditions, we evaluated in closed form the transmission and reflection coefficients for each mode. The final expressions are reported below:

$$T_H = \frac{k_{xm}}{2k_{x1}} \kappa_{em} \left(1 + \frac{\omega_0}{\omega_m}\right)\left(\frac{\omega_0}{\omega_1}\right); \qquad R_H = \frac{k_{xm}}{2k_{x1}} \kappa_{em} \left(1 - \frac{\omega_0}{\omega_m}\right)\left(\frac{\omega_0}{\omega_1}\right) \tag{2}$$

where ω_1 is the frequency of the first mode in the new structure and κ_{em} is the coupling coefficient given by the overlap integral across the PPWG cross section. The factor ω_0 / ω_1 represent the time dilation coefficient which carries information on the energy's change during the transition; this in turn confirms the activeness of this phenomenon. Stressing duality with the waveguide-waveguide spatial interface is pointless, given that the space discontinuity is a passive phenomenon.

3. Numerical verification

In this section, we verify the temporal interface induced by the boundary change over time, evaluating numerically the frequency shift and the corresponding reflection and transmission coefficients for the induced modes after the temporal discontinuity. We developed a finite difference time domain (FDTD) numerical code [10]-[11] to simulate over time the propagation and transformation of the original mode into the final set of forward and backward propagating modes in the PPWG.
In the simulation, we consider a PPWG height of 2 cm, supporting the propagation of the lowest order TM mode at 10 GHz. Five different simulations have been performed, where the distance between the metallic plates was incrementally increased by a 20%, from 2 cm to 4.0 cm.

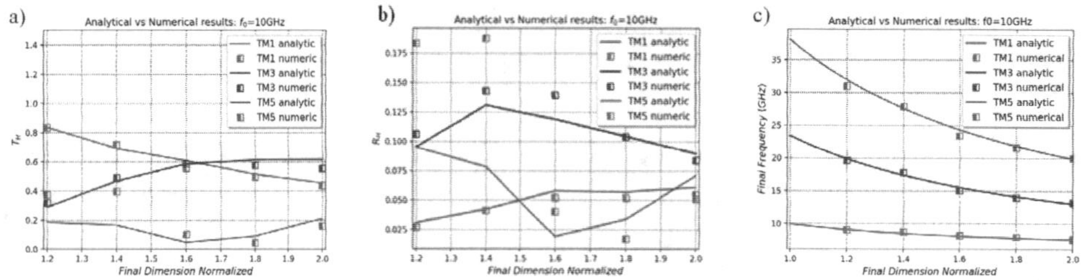

Figure 3. The source central frequency is set at 10 GHz and the frequency content spans 2 GHz in both directions; the initial height was 2 cm and the final height after the abrupt temporal change is sampled every 0.4 cm. On the horizontal axis we report the height normalized to the initial one. From top to bottom we plot (a) the transmission coefficients; (b) the reflection coefficients, and (c) the expected frequency shifts. Solid lines refer to the analytical results, while the squares to the numerical ones.

In Fig 3a, we report the comparison between the analytical (solid line) and numerical (squares) results in terms of the frequency shifts for the first three odd modes, *i.e.,* TM_1, TM_3, TM_5, showing that each mode has its frequency shifted due to the time interface.
In Figs. 3b-c, we report the amplitude of the reflection and transmission coefficients at the boundary-induced temporal interface for the modes after the waveguide discontinuity. The results show good agreement except for the TM_5 mode when small variations in the waveguide height are considered. This discrepancy is due to the finite resolution of the simulation which is not sufficient to resolve a mode having such a high frequency. However, for wider variations, which corresponds to lower frequencies, the results confirm the analytical model as all the other modes do.

Conclusions

To conclude, we have presented and analyzed a waveguide-to-waveguide transition in time domain. The theoretical analysis demonstrates that the abrupt transition of the waveguide dimensions induces a scattering like the one observed in time-modulated metamaterials, where the frequency changes to accommodate the wavelength. Closed form formulas of the frequency shifts and the energy distribution for each forward and backward mode are derived. Those preliminary results can clear the path for the implementation of a device based on a time interface, relaxing the need of directly modulating the material properties.

Acknowledgements

The work has been developed in the framework of the activities of the research contract MANTLES, funded by the Italian Ministry of Education, University and Research as a PRIN 2017 project (protocol number 2017BHFZKH).

References

[1] Caloz C and Deck-Léger Z-L 2020 Spacetime Metamaterials — Part I : General Concepts *IEEE Trans. Antennas Propag.* **68** 1569–82
[2] Caloz C and Deck-Léger Z-L 2020 Spacetime Metamaterials—Part II: Theory and Applications *IEEE Trans. Antennas Propag.* **68** 379–81
[3] Morgenthaler F R 1958 Velocity Modulation of Electromagnetic Waves *IEEE Trans. Microw. Theory Tech.* **6** 167–72
[4] Ramaccia D, Toscano A and Bilotti F 2020 Light propagation through metamaterial temporal slabs: Reflection, refraction, and special cases *Opt. Lett.* **45**
[5] Ramaccia D, Alù A, Toscano A and Bilotti F 2021 Temporal multilayer structures for designing higher-order transfer functions using time-varying metamaterials *Appl. Phys. Lett.* **118** 101901
[6] Pacheco-Peña V and Engheta N 2020 Anti-reflection temporal coatings *Optica* **7** 323–31
[7] Li H and Alù A 2021 Temporal switching to extend the bandwidth of thin absorbers *Optica* **8** 24
[8] Caloz C and Deck-Léger Z L 2019 Spacetime metamaterials *arXiv*
[9] Pacheco-Peña V and Engheta N 2020 Temporal aiming *Light Sci. Appl.* **9**
[10] Langtangen H P 2013 Finite difference methods for wave motion 1–108
[11] Strikwerda J C 2004 *Finite Difference Schemes and Partial Differential Equations, Second Edition* (Society for Industrial and Applied Mathematics)

Propagation and scattering effects in temporal metastructures

D. Ramaccia[1], A. Alù[2,3], A. Toscano[1], and F. Bilotti[1]

[1]ROMA TRE University, Department of Engineering, via Vito Volterra 62, 00146, Rome, Italy
[2] Photonics Initiative, Advanced Science Research Center, New York, NY 10031, USA
[3] Department of Electrical Engineering, City College, City University of New York, New York, NY 10031, USA

davide.ramaccia@uniroma3.it

Abstract. Electromagnetic scattering typically occurs when a change in the material properties is perceived by the propagating wave, that inevitably splits into a reflected and refracted wave to maintain the continuity of the field components at the interface between the two media. However, such a scattering phenomenon occurs also when the entire media suddenly switches its properties to other values at a certain instant of time, realizing the so-called temporal interface. After a temporal interface, a couple of waves, one reflected and one transmitted, starts to propagate in the new media with the same wavelength but at a different frequency. Exploiting the analogies and differences between spatial and temporal interfaces, in this contribution we present the temporal counterparts of conventional electromagnetic devices based on dielectric slabs and a cascade of them, *i.e.,* the multilayered structures. We discuss about the analysis and design strategies for synthetizing the desired scattering response in both transmission and reflection and present the possible families of devices based on multi-switched temporal metamaterials that can be conceived.

1. Introduction

Metamaterials and metasurfaces are a recognized platform for achieving unprecedented light-matter interactions. It has been demonstrated that spatial modulation of the constitutive parameters of a metamaterial, or the surface properties of metasurfaces, is able to tailor the phase-fronts of the interacting electromagnetic waves, i.e., the spatial frequencies of the wave, realizing some revolutionizing devices, such as invisibility cloaking [1,2], filtering or beam-steering [3–6] and focusing structures [3,7,8]. Introducing a modulation over time increasing the degrees of freedom, allowing to extend the control also to the temporal frequencies. It has been demonstrated that time-varying profiles of material properties can induce artificial non-reciprocity[9–14], generate/control harmonics [15], and control Doppler shift [16,17], thanks to the synthetic motion of a medium in one specific direction [11,12,18–20]. Time-varying metasurfaces also demonstrated to be able to induce frequency translation [19,21–23] metasurfaces.
Recently, metamaterials whose electromagnetic properties suddenly changes over time have been also investigated for inducing instantaneous frequency conversion of the propagating waves. The sudden change of the medium properties realizes the temporal counterpart of the well-known spatial interface

between two different media, generating two scattered waves, one backward- and one forward-propagating [24–26].

Here we present our recent investigation on propagation and scattering from a multi-switched temporal metamaterial, whose properties change several times during the wave propagation. According to the number of switching, the temporal duration and material properties, we can realize several different temporal devices, such as temporal impedance matching slabs, Fabry-Perot cavities, transparent multilayered structures, and amplifying devices, just to name a few.

2. Temporal metamaterial slab devices

A temporal metamaterial slab can be implemented by switching two times the properties of a uniform homogeneous medium in the whole space, as shown in Figure 1.

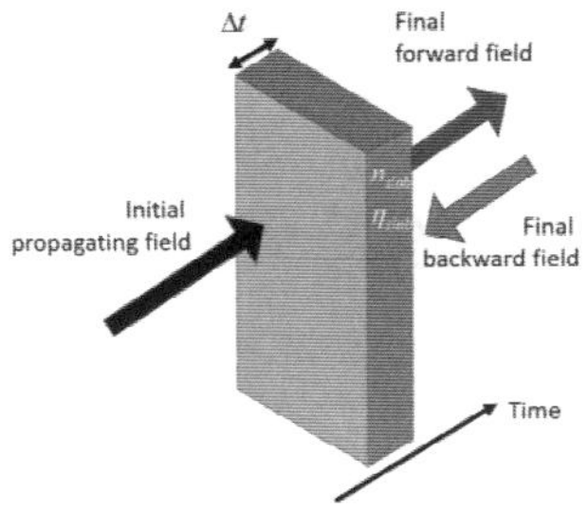

Figure 1. A temporal metastructure composed by a single temporal metamaterial slab applied for a period Δt.

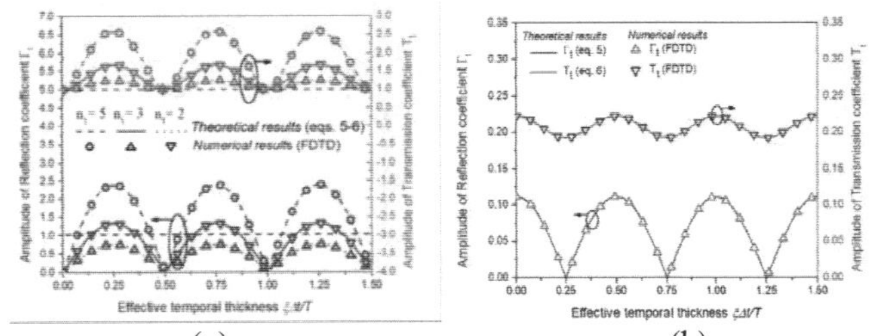

(a) (b)

Figure 2. Transmission and reflection coefficient as a function of the temporal thickness of slab when it is designed to act as (a) an impedance matching slab and (b) a Fabry-Perot resonator.

It realizes the temporal version of a dielectric slab between two semi-infinite media, *i.e.*, a temporal slab [27], which allows designing both a perfect matching temporal slab, as proposed in [28], and a temporal Fabry-Perot cavity [27]. In Fig. 1 we report a representation of a single temporal metamaterial slab of refractive index n_{slab} that is present in the whole space where the wave is propagating for the period Δt. Before and after such a temporal slab, the whole space is filled by a medium whose refractive indices are n_a and n_b, respectively. The two consecutive temporal interfaces, n_a-n_{slab} and n_{slab}-n_b, scatter four wave contributions whose amplitudes are defined by the Morgenthaler coefficients [16], [17]. They are combined into the overall reflection/transmission coefficients considering the phase delay introduced by the slab [18]:

$$\Gamma_t = e^{i\xi\omega_0\Delta t}\left(\tau_t\rho_t' + \tau_t'\rho_t e^{-i2\xi\omega_0\Delta t}\right)$$
$$T_t = e^{i\xi\omega_0\Delta t}\left(\tau_t\tau_t' + \rho_t'\rho_t e^{-i2\xi\omega_0\Delta t}\right) \quad (1)$$

where τ, τ', ρ, ρ' are the Morgenthaler coefficients at the two interfaces, ω_0 is the illumination frequency, and ξ that is nothing but the time-dilation factor related to the change in the speed of the wave after the first temporal boundary [17]. Equation (1) give us the possibility to design temporal Fabry-Perot cavities and impedance-matching coating for half-period and quarter-period temporal slabs, respectively, as shown in Fig. 2 where the scattering coefficients are reported.

3. Temporal multi-layered devices

Temporal multi-layer structures consist of a cascade of different media over time that, in analogy with what it can be achieved in conventional spatial multi-layer structures to realize several families of different devices.

In Fig. 3, we report some of our designs of multilayered temporal metamaterials that consists of m temporal slabs applied in cascade over time.

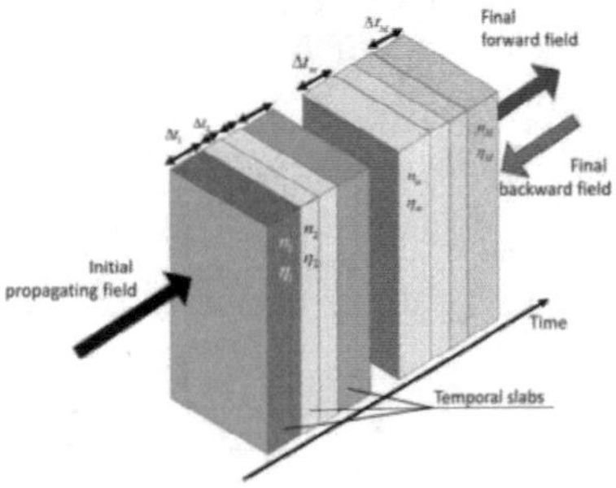

Figure 3. A temporal metastructure composed by a multilayered structure consisting of a cascade of M different temporal metamaterial slabs, each applied for a specific time.

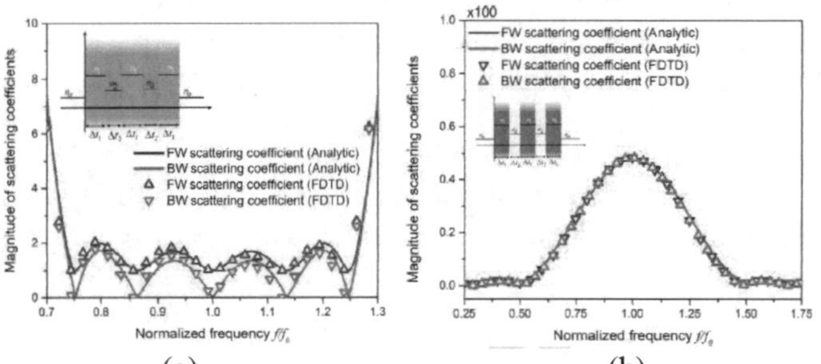

(a) (b)

Figure 4. Transmission and reflection coefficient as a function of the normalized frequency of (a) a transparent and (b) amplifying multilayered devices.

The higher number of dielectric slabs, each with its refractive index and application time, enables the design of further electromagnetic devices based on temporal metamaterials by engineering the scattering response from the structure. As shown in [19], the overall scattering coefficients can be derived in terms of transfer matrix TM that considers the scattering contribution of all temporal slabs:

$$\begin{bmatrix} E_b^+ \\ E_b^- \end{bmatrix} = \left(MM_1 \prod_{m=1}^{M} DM_m MM_{m+1} \right) \begin{bmatrix} E_a^+ \\ 0 \end{bmatrix} = TM_M \begin{bmatrix} E_a^+ \\ 0 \end{bmatrix} \tag{2}$$

where MM and DM identify the matching matrices and delay matrices, respectively. Again, by engineering the phase delay between two consecutive interfaces, we can design different classes of devices [19]. In Fig. 4, we report the scattering coefficient of two multilayered device: the first is a transparent device which exhibits zero reflection and unitary transmission at the design frequency (Fig. 4a); the second is an amplifying device for both the reflected and transmitted waves at the design frequency (Fig. 4b). More details on the design and numerical verification of the performances will be presented at the conference.

Conclusions

In this contribution, we have presented the possibilities offered by temporal metamaterial in realizing several types of devices. Exploiting the analogies and differences between spatial and temporal interfaces, in this contribution we have presented the temporal counterparts of conventional electromagnetic devices based on dielectric slabs and a cascade of them, i.e., the multilayered structures. We have discussed about the analysis and design strategies for synthetizing the desired scattering response in both transmission and reflection and present the possible families of devices based on multi-switched temporal metamaterials that can be conceived.

Acknowledgements
The work has been developed in the framework of the activities of the research contract MANTLES, funded by the Italian Ministry of Education, University and Research as a PRIN 2017 project (protocol number 2017BHFZKH).

References

[1] Pendry J B, Schurig D and Smith D R 2006 Controlling electromagnetic fields. *Science* **312** 1780–2
[2] Alù A 2009 Mantle cloak: Invisibility induced by a surface *Phys. Rev. B* **80** 245115
[3] Glybovski S B, Tretyakov S A, Belov P A, Kivshar Y S and Simovski C R 2016 Metasurfaces: From microwaves to visible *Phys. Rep.* **634** 1–72
[4] Li A, Singh S and Sievenpiper D 2018 Metasurfaces and their applications *Nanophotonics* **7** 989–1011
[5] Ramaccia D, Arcieri S, Toscano A and Bilotti F 2017 Core-Shell Super-Spherical Nanoparticles for LSPR-Based Sensing Platforms *IEEE J. Sel. Top. Quantum Electron.* **23** 380–7
[6] Ramaccia D, Toscano A and Bilotti F 2017 Scattering and absorption from super-spherical nanoparticles: analysis and design for transparent displays [Invited] *J. Opt. Soc. Am. B* **34** D62
[7] Ramaccia D, Bilotti F, Toscano A and Hrabar S 2013 Restoring the radiating performances of shortened horn antennas over a broad frequency range *2013 IEEE Antennas and Propagation Society International Symposium (APSURSI)* (IEEE) pp 964–5
[8] Ramaccia D, Scattone F, Bilotti F and Toscano A 2013 Broadband Compact Horn Antennas by Using EPS-ENZ Metamaterial Lens *IEEE Trans. Antennas Propag.* **61** 2929–37
[9] Ramaccia D, Bilotti F and Toscano A 2014 Angular Momentum-biased metamaterials for filtering waveguide components and antennas with non-reciprocal behavior *2014 8th International Congress on Advanced Electromagnetic Materials in Microwaves and Optics* (IEEE) pp 250–2
[10] Taravati S 2018 Giant Linear Nonreciprocity, Zero Reflection, and Zero Band Gap in Equilibrated Space-Time-Varying Media *Phys. Rev. Appl.* **9** 064012
[11] Yu Z and Fan S 2009 Complete optical isolation created by indirect interband photonic transitions *Nat. Photonics* **3** 91–4
[12] Lira H, Yu Z, Fan S and Lipson M 2012 Electrically Driven Nonreciprocity Induced by Interband Photonic Transition on a Silicon Chip *Phys. Rev. Lett.* **109** 033901
[13] Sounas D L and Alù A 2017 Non-reciprocal photonics based on time modulation *Nat. Photonics* **11** 774–83
[14] Galiffi E, Huidobro P A and Pendry J B 2019 Broadband Nonreciprocal Amplification in Luminal Metamaterials *Phys. Rev. Lett.* **123** 206101
[15] Taravati S and Kishk A A 2020 Space-Time Modulation: Principles and Applications *IEEE Microw. magaz* **21** 30–56
[16] Ramaccia D, Sounas D L, Alù A, Toscano A and Bilotti F 2017 Doppler cloak restores invisibility to objects in relativistic motion *Phys. Rev. B* **95** 075113
[17] Liu B, Giddens H, Li Y, He Y, Wong S-W and Hao Y 2020 Design and experimental demonstration of Doppler cloak from spatiotemporally modulated metamaterials based on rotational Doppler effect *Opt. Express* **28**
[18] Chamanara N, Taravati S, Deck-Léger Z-L and Caloz C 2017 Optical isolation based on space-time engineered asymmetric photonic band gaps *Phys. Rev. B* **96** 155409
[19] Ramaccia D, Sounas D L, Marini A, Toscano A and Bilotti F 2020 Electromagnetic Isolation Induced by Time-Varying Metasurfaces: Non-Reciprocal Bragg Grating *IEEE Antennas Wirel. Propag. Lett.* 1–1
[20] Ramaccia D, Sounas D L, Alù A, Bilotti F and Toscano A 2015 Nonreciprocal Horn Antennas Using Angular Momentum-Biased Metamaterial Inclusions *IEEE Trans. Antennas Propag.* **63**
[21] Zhang L, Chen X Q, Liu S, Zhang Q, Zhao J, Dai J Y, Bai G D, Wan X, Cheng Q, Castaldi G, Galdi V and Cui T J 2018 Space-time-coding digital metasurfaces *Nat. Commun.* **9** 4334
[22] Ramaccia D, Sounas D L, Alu A, Toscano A and Bilotti F 2020 Phase-Induced Frequency Conversion and Doppler Effect with Time-Modulated Metasurfaces *IEEE Trans. Antennas Propag.* **68** 1607–17

[23] Wu Z and Grbic A 2020 Serrodyne Frequency Translation Using Time-Modulated Metasurfaces *IEEE Trans. Antennas Propag.* **68** 1599–606

[24] Caloz C and Deck-Léger Z-L 2020 Spacetime Metamaterials—Part II: Theory and Applications *IEEE Trans. Antennas Propag.* **68** 379–81

[25] Morgenthaler F R 1958 Velocity Modulation of Electromagnetic Waves *IEEE Trans. Microw. Theory Tech.* **6** 167–72

[26] Xiao Y, Maywar D N and Agrawal G P 2014 Reflection and transmission of electromagnetic waves at a temporal boundary *Opt. Lett.* **39** 574–7

[27] Ramaccia D, Toscano A and Bilotti F 2020 Light propagation through metamaterial temporal slabs: Reflection, refraction, and special cases *Opt. Lett.* **45**

[28] Pacheco-Peña V and Engheta N 2020 Anti-reflection temporal coatings *Optica* **7** 323–31

Time-varying metamaterials and metasurfaces for antennas and propagation applications

D. Ramaccia, A. Toscano, and F. Bilotti

ROMA TRE University, Department of Engineering, via Vito Volterra 62, 00146, Rome, Italy

davide.ramaccia@uniroma3.it

Abstract. In this contribution we present the most recent results from our group about the opportunities offered by time-varying metamaterials and metasurfaces for conceiving antenna systems and devices exhibiting artificial non-reciprocity, frequency conversion, energy accumulation and temporal electromagnetic scattering. Such artificial metastructures are characterized by constitutive parameters (permittivity, permeability and/or surface impedance) that are modulated in time through an external control or requires modulated excitation signal for enabling anomalous scattering behaviour. Here, we briefly describe the physical insights of the unusual interaction arising between the electromagnetic field and such metamaterials and metasurfaces, and then we present some antennas and propagation applications, showing the performances of non-reciprocal antenna systems, magnet-less isolators, Doppler cloaks, temporal devices and metasurface-based virtual absorbers.

1. Introduction

Thanks to the advent of metamaterials and metasurfaces, the RF/microwave antennas and components have been redesigned to fully exploit the new opportunities offered by them in controlling almost at will the scattering [1–3], wavefronts [4], and frequency response [5–7] from such engineerable artificial structures. However, in the last few years a new degree of freedom has been exploited for increasing the possibilities offered by metamaterials and metasurfaces in the design of novel microwave/RF antenna systems and devices: the temporal modulation of the material and surface properties [8].

In this contribution, we review the most recent antenna systems and components proposed by our group based on time-varying metastructures. Here, we focus our attention on four classes of systems and devices: *magnet-less non-reciprocal antenna systems* [9–11], the *Doppler cloaks* [12–14], the *temporal devices* [9,19], and, finally, the metasurface-based *virtual electromagnetic absorbers* [20–22].

2. Time-varying metamaterials and metasurfaces applications

2.1. Non-reciprocal antennas and components

Electromagnetic reciprocity is one of the fundamental properties of any passive and linear electromagnetic system that forces the response in transmission to be always identical to the one in reception. However, it represents a strong limit in case an isolation between transmitter and receiver is needed. In Fig. 1a and Fig. 1e, we report two different strategies for achieving magnet-less non-reciprocal antenna systems based on space-time modulated metamaterials.

Journal of Physics: Conference Series **2015** (2021) 012121 doi:10.1088/1742-6596/2015/1/012121

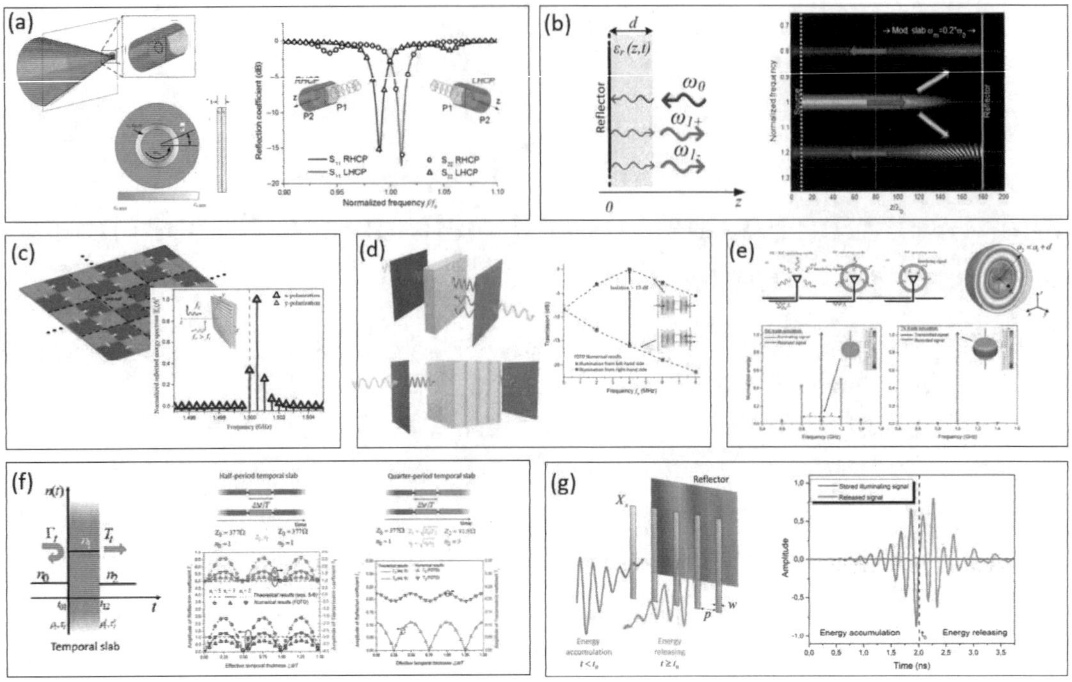

Figure 1. Devices based on space-time and time-varying metamaterials and metasurfaces: (a) non-reciprocal horn antenna with angular-momentum biased filtering inclusion [15,16]; (b) Doppler cloak realized though space-time modulated metamaterial [12]; (c) Time-varying reflective metasurface for frequency conversion and doppler cloaking [13,17]; (d) metasurface-based electromagnetic isolator [18]; (e) space-time modulated covers for antenna non-reciprocity [11]; (f) time-domain components achieved with temporal slabs and multilayered structures [9,19]; (g) metasurface-based virtual absorbers excited by complex frequencies [20–22].

The first one (Fig. 1a) is based on a filtering particle to be placed in the throat of a conventional horn antenna, whose filling material is modulated in both space and time for inducing an artificial Zeeman effect. The opposite propagating circular-polarized fields with the same handedness experience different transmission levels due to the particle modulation. The second one (Fig. 1e) is based on space-time modulated covers to be placed around the antenna. According to the modulation scheme of the material properties of cover, it can induce a frequency conversion for the received (transmitted) signals, while the transmitted (receiving) signals are let passing unaltered, realizing a non-reciprocal antenna system. Finally, in Fig. 1d we report an electromagnetic isolator based on time-varying metasurfaces with opposite frequency modulation capabilities. The two metasurfaces are located at the opposite side of a conventional and passive reciprocal filtering structure, realizing a local Doppler shifting in the volume where the filtering structure is located. This simply realizes an asymmetric transmission when illuminated from opposite sides.

2.2. Frequency conversion and Doppler cloaking

A space-time modulated metamaterial is an artificial material whose permittivity and/or permeability are modulated over space and time, realizing a propagating medium wave within the metamaterial. As shown in Fig. 1b, during the propagation of the electromagnetic wave within such a metamaterial, a frequency and wavevector shift is always induced and, according to the propagation direction of both electromagnetic and medium waves, the excitation of new modes can take place or not. This allowed to engineer the material properties to induces a desired frequency and wavevector shift such that the final modes emerging from the material can compensate the actual Doppler shifting induced by the motion. This allowed to design the first Doppler cloaking device based on space-time modulated metamaterial. More recently, a more compact version based on time-modulated metasurfaces has been designed (Fig. 1c). It exploits the temporal modulation of the reflection coefficient by exploiting the properties of a high-impedance surface modulated by a set of varactors. They shift periodically the resonant frequency of the high-impedance surfaces before and after its operative frequency, inducing an apparent Doppler shift for the reflected wave.

2.3. Temporal devices

Matching networks, Fabry-Perot resonators and multilayered electromagnetic devices exploits the scattering from their interfaces and the propagation delay within the materials for tailoring the overall scattering response in transmission and reflection. Temporal devices have been recently conceived by using the temporal dimension instead of the spatial one. The temporal device consists of a cascade of abrupt switching of the material properties where the wave is propagating, as shown in Fig. 1f. The material properties and timing of the switching must be properly selected to achieve the desired overall scattering response from the device, in full analogy with the spatial counterparts. In Fig. 1f, the transmission and reflection coefficient as a function of the temporal thickness of a temporal slab are reported showing that is possible to design temporal Fabry-Perot cavities and matching layers. Multilayered structures benefit of an higher number of degree of freedoms for the synthesis of more complex responses and devices.

2.4. Metasurface-based virtual absorbers

Finally, we present the design of metasurface-based virtual absorbers, which consist of a partially open cavity bounded by a metasurface and a metallic reflector. Such kind of structure always exhibits a reflection when illuminated by a monochromatic uniform plane wave, that can be interpreted through the strong mismatched between the equivalent reactive input impedance exhibited at the metasurface location and the real wave impedance of the free space where the wave is propagating. However, the system may exhibit zero-reflection when it is illuminated by a time-modulated plane wave with a specific complex frequency. As long as the excitation is applied, the system does not transmit or reflect, but store the electromagnetic energy within its cavity volume, realizing a virtual absorption effect, as shown in Fig. 1g.

Such a concept has been applied also to microwave circuits, showing not only that it is possible to achieve perfect virtual matching for reactive loads fed by lossless transmission lines, but we have derived the operative bounds for such a system for a given metasurface reactance and electrical dimension of the cavity.

Conclusions

To conclude, this contribution focused its attention on the most recent results from our group about the opportunities offered by time-varying metamaterials and metasurfaces for conceiving antenna systems and devices. In particular, we have presented some applications of time-varying metastructures, showing how it is possible to design non-reciprocal antenna systems, magnet-less isolators, Doppler cloaks, temporal devices and virtual absorbers.

Acknowledgements

The work has been developed in the framework of the activities of the research contract MANTLES, funded by the Italian Ministry of Education, University and Research as a PRIN 2017 project (protocol number 2017BHFZKH).

References

[1] Alù A 2009 Mantle cloak: Invisibility induced by a surface *Phys. Rev. B* **80** 245115
[2] Ramaccia D, Arcieri S, Toscano A and Bilotti F 2017 Core-Shell Super-Spherical Nanoparticles for LSPR-Based Sensing Platforms *IEEE J. Sel. Top. Quantum Electron.* **23** 380–

7
[3] Ramaccia D, Toscano A and Bilotti F 2017 Scattering and absorption from super-spherical nanoparticles: analysis and design for transparent displays [Invited] *J. Opt. Soc. Am. B* **34** D62
[4] Epstein A and Eleftheriades G V. 2016 Huygens' metasurfaces via the equivalence principle: design and applications *J. Opt. Soc. Am. B* **33** A31
[5] Ramaccia D, Di Palma L, Ates D, Ozbay E, Toscano A and Bilotti F 2014 Analytical Model of Connected Bi-Omega: Robust Particle for the Selective Power Transmission Through Sub-Wavelength Apertures *IEEE Trans. Antennas Propag.* **62** 2093–101
[6] Ramaccia D, Bilotti F, Toscano A and Vegni L 2013 Dielectric-free multi-band frequency selective surface for antenna applications *COMPEL - Int. J. Comput. Math. Electr. Electron. Eng.* **32** 1868–75
[7] Ramaccia D, Toscano A, Colasante A, Bellaveglia G and lo Forti R 2011 Inductive tri-band double element FSS for space applications *Prog. Electromagn. Res. C* **18** 87–101
[8] Engheta N, Kiasat Y, Mencagli M J, Nahvi E and Edwards B 2019 4D Metastructures : Merging Spatial and Temporal Metamaterials *2019 13th International Congress on Artificial Materials for Novel Wave Phenomena, METAMATERIALS 2019* pp 76–7
[9] Ramaccia D, Toscano A and Bilotti F 2020 Light propagation through metamaterial temporal slabs: Reflection, refraction, and special cases *Opt. Lett.* **45**
[10] Ramaccia D, Sounas D L, Alù A, Bilotti F and Toscano A 2015 Nonreciprocal Horn Antennas Using Angular Momentum-Biased Metamaterial Inclusions *IEEE Trans. Antennas Propag.* **63** 5593–600
[11] Ramaccia D, Sounas D L D L, Alu A, Bilotti F and Toscano A 2018 Nonreciprocity in antenna radiation induced by space-time varying metamaterial cloaks *IEEE Antennas Wirel. Propag. Lett.* **17** 1968–72
[12] Ramaccia D, Sounas D L, Alù A, Toscano A and Bilotti F 2017 Doppler cloak restores invisibility to objects in relativistic motion *Phys. Rev. B* **95** 075113
[13] Ramaccia D, Sounas D L, Alu A, Toscano A and Bilotti F 2020 Phase-Induced Frequency Conversion and Doppler Effect with Time-Modulated Metasurfaces *IEEE Trans. Antennas Propag.* **68** 1607–17
[14] Ramaccia D, Sounas D ., Alu A, Toscano A and Bilotti F 2017 Spatio-temporal modulated Doppler cloak for antenna matching at relativistic velocity *2017 11th International Congress on Engineered Materials Platforms for Novel Wave Phenomena (Metamaterials)* (IEEE) pp 277–9
[15] Ramaccia D, Sounas D L, Alù A, Bilotti F and Toscano A 2015 Nonreciprocal Horn Antennas Using Angular Momentum-Biased Metamaterial Inclusions *IEEE Trans. Antennas Propag.* **63**
[16] Ramaccia D, Bilotti F and Toscano A 2014 Angular Momentum-biased metamaterials for filtering waveguide components and antennas with non-reciprocal behavior *2014 8th International Congress on Advanced Electromagnetic Materials in Microwaves and Optics* (IEEE) pp 250–2
[17] Ramaccia D, Sounas D L, Alu A, Toscano A and Bilotti F 2018 Metasurface-based Doppler cloaks: Time-varying metasurface profile to achieve perfect frequency mixing *2018 12th International Congress on Artificial Materials for Novel Wave Phenomena (Metamaterials)* (IEEE) pp 331–3
[18] Ramaccia D, Sounas D L, Marini A, Toscano A and Bilotti F 2020 Electromagnetic Isolation Induced by Time-Varying Metasurfaces: Non-Reciprocal Bragg Grating *IEEE Antennas Wirel. Propag. Lett.* 1–1
[19] Ramaccia D, Alù A, Toscano A and Bilotti F 2021 Temporal multilayer structures for designing higher-order transfer functions using time-varying metamaterials *Appl. Phys. Lett.* **118** 101901
[20] Marini A V, Ramaccia D, Toscano A and Bilotti F 2020 Metasurface-bounded open cavities supporting virtual absorption: free-space energy accumulation in lossless systems *Opt. Lett.* **45** 3147–50

[21] Marini A V, Ramaccia D, Toscano A and Bilotti F 2021 Metasurface virtual absorbers: Unveiling operative conditions through equivalent lumped circuit model *EPJ Appl. Metamaterials* **8**
[22] Marini A, Ramaccia D, Toscano A and Bilotti F 2020 Perfect matching of reactive-loaded transmission lines through complex excitation *14th European Conference on Antennas and Propagation, EuCAP 2020* (Institute of Electrical and Electronics Engineers Inc.)

Highly Chiral Exceptional Point in Perturbed Coupled Resonators

S. Ramezanpour

Department of Physics and Engineering, ITMO University, Birzhevaya line, 199034, St. Petersburg, Russia

E-mail: ramezanpourshahab@gmail.com

Y. Ra'di

Advanced Science Research Center, City University of New York, Nicholas Terrace St., 10031, NY, USA

A. Alù

Advanced Science Research Center, City University of New York, Nicholas Terrace St., 10031, NY, USA

A. Bogdanov

Department of Physics and Engineering, ITMO University, Birzhevaya line, 199034, St. Petersburg, Russia

Abstract. Exceptional point (EP) is a non-Hermitian spectral degeneracy that has application in ultrasensitive sensors and laser mode selectivity. By employing strong chirality in an optical system, the direction of light propagation can be controlled and subwavelength particles can be detected. Here, we show that EP with high chirality can appear in the coupled resonators perturbed by a scatterer, in which both the distance and position of the scatterer can be tuned. We achieve strong chiral EP in two different distances between the resonators, with chirality around 0.99 in both cases.

1. Introduction

Exceptional point (EP) is a degeneracy in the space parameter at which two or more eigenvalues and their corresponding eigenfunctions coalesce, simultaneously. It has attracted many interests due to its exotic functionalities in such as unidirectional lasing and supersensitive sensors [1, 2, 3]. Highly directional emission can be achieved from a chaotic deformed microresonator [4], and from an elliptical resonator with a wavelength-size notch at the boundary [5]. Besides, chiral EP in a couple of asymmetry disk-like cavities have application in nanoparticle detection [6].

In this paper, we show that incorporating both gain/loss and scatterers in a coupled pair ring resonators EP with high chirality can be achieved. In this structure, with tuning the distances between the resonators, and the position of the scatterer, a strong chiral EP appears in two different distances between the resonators. It can be considered as a tunable device in which with tuning both distance between the resonators, and position of the scatterer the possible imperfections during the fabrication process can be compensated. Using tunable components such as scatterers and/or Kerr nonlinearity [7] can open a new way to achieve a regime extremely close to EP and chiral EP, enabling super-sensitivity/directionality.

Journal of Physics: Conference Series **2015** (2021) 012122 doi:10.1088/1742-6596/2015/1/012122

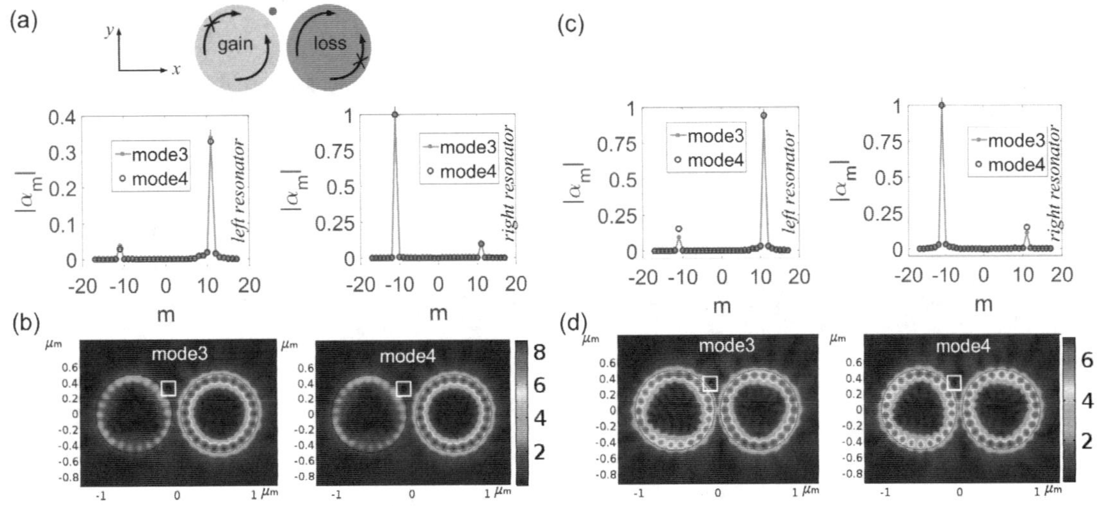

Figure 1. (a) Schematic of the WGM structure in perturbed coupled resonator at chiral EP, α_m components of left and right resonators, and (b) field distribution at chiral EP for the gap between them $g = 0.115\,\mu$m. (c) α_m components of left and right resonators, and (d) field distribution at chiral EP for the gap $g = 0.075\,\mu$m.

2. Coupled Ring Resonators

In a ring resonator, the WGMs reveal two degenerate states, even and odd modes. Incorporating two scatterers along the surface of a microcavity, and tuning their position, we can achieve unidirectional propagation, so-called chiral EP, related to the either clockwise (CW) or counter clockwise (CCW) direction, due to the destructive optical interference for CCW or CW propagation, respectively.

In a coupled ring resonators, the degeneracy of WGMs would be lifted due to the coupling between the rings. The Hamiltonian of this system is described by:

$$\hat{H} = \begin{pmatrix} \Omega_0 & \kappa \\ \kappa & \Omega_0 \end{pmatrix} \tag{1}$$

where κ is the coupling rate between two rings and Ω_0 is the resonant frequency of rings. Therefore, the eigenvalues of the coupled system are $\Omega_0 \pm \kappa$. In an ideal case, the system would have four eigenvalues, two of them are degenerated by the value $\Omega_0 - \kappa$, and the other two eigenvalues are degenerated by $\Omega_0 + \kappa$. However, the numerical result of (unperturbed) coupled resonator shows four different eigenvalues, with close values two by two. This slight difference between the analytical and numerical values can be attributed to the difference between the coupling rate of even and odd modes in the rings. These four eigenvalues can be analogous to a pair of two-level system (TLS).

Analogous to the chiral EP in a single resonator, in which two scatterers can bring the system to the chiral EP, in the coupled ring resonators, two mechanisms: 1) the scatterer which unevenly affects the resonators, and 2) coupling between the resonators, can create the conditions of chiral EP in the resonators. Our result shows that this kind of destructive interference can occur close to the eigenvalues related to the (inter-)EP in gain/loss coupled resonators. To this end, first, we bring the unperturbed coupled ring resonator to the EP (or close to it), by tuning the distance (coupling) and dissipation between them, and then we incorporate a scatterer, and tune its position to create the chiral EP. The schematic of such a mechanism is shown in Fig. 1(a) (top).

To calculate the field distribution, we solve the wave equation $-\nabla^2 \psi = n^2(x,y)\frac{\omega^2}{c^2}\psi$, numerically in Comsol, where ψ is associated to E_z for TM polarizaion, and $n(x,y)$, ω and c are the refractive index distribution, eigenfrequency and speed of light in vacuum, respectively. To calculate CW and

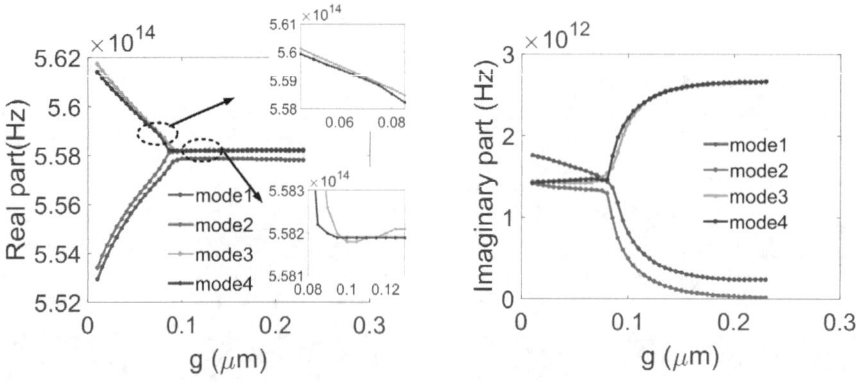

Figure 2. Real and imaginary parts of four eigenvalues versus the gap between the resonators

CCW components, we expand the wavefunction inside the cavity in cylinder harmonics as $\psi(\rho,\phi) = \sum_{m=-\infty}^{\infty} \alpha_m J_m(n\frac{\omega}{c}\rho)\exp(im\phi)$, where J_m is the mth order of first kind bessel function, and the positive (negative) angular momentum, m is corresponding to CCW (CW) component. The chirality can then be defined as

$$\alpha = 1 - \frac{\min\{\sum_{m=-\infty}^{-1}|\alpha_m|^2, \sum_{m=1}^{\infty}|\alpha_m|^2\}}{\max\{\sum_{m=-\infty}^{-1}|\alpha_m|^2, \sum_{m=1}^{\infty}|\alpha_m|^2\}} \quad (2)$$

[8]. We analyse a coupled resonators with the radii of rings $r = 0.5\,\mu m$, and the permittivity $\varepsilon = 6$. The dissipation is considered for the right resonator by adding an imaginary part to its refractive index equal to 0.012. The radius and gap of scatterer are tuned to $0.03\,\mu m$, and its angle is 35° with respect to the left resonator (here, the gap means center to center distance minus the radii of particles). The maximum chirality is obtained for two different gaps between the resonators $g = 0.115\,\mu m$ (Fig. 1(a)-1(b)), and $g = 0.075\,\mu m$ (Fig. 1(c)-1(d)).

Figure 1(a) depicts $|\alpha_m|$ components of both left and right resonators for the gap between them $g = 0.115\,\mu m$. From the $|\alpha_m|$ components, it can be inferred that WGM in the left (right) resonator is mainly in CCW (CW) direction, and the field is mostly distributed in the right resonator. Its field distribution is shown in Fig. 1(b). The vanished zero nodes of the WGM structure is the sign of unidirectional traveling wave characteristic. The chirality of the left (right) resonator related to the modes 3 and 4, $\alpha_{l,3}$, $\alpha_{l,4}$ ($\alpha_{r,3}$, $\alpha_{r,4}$) are estimated $\alpha_{l,3} = 0.9863$, $\alpha_{l,4} = 0.9916$, $\alpha_{r,3} = 0.9903$ and $\alpha_{r,4} = 0.9910$.

Figure 1(c) and 1(d) depicts $|\alpha_m|$ components of resonators and their field distribution for the gap $g = 0.075\,\mu m$. It shows that field distribution is distributed equally between the resonator, and WGM structure has mainly CCW (CW) direction in the left (right) resonator. The chirality is calculated $\alpha_{l,3} = 0.9890$, $\alpha_{l,4} = 0.9733$, $\alpha_{r,3} = 0.9879$ and $\alpha_{r,4} = 0.9781$.

In order to observe the interference of the energy levels at chiral EP, Fig. 2 shows real and imaginary parts of four eigenvalues related to the perturbed coupled resonators versus the gap between them. The zoomed regions of the dotted circle shows that the real parts of third and fourth eigenvalues depart and interfere locally, while their corresponding imaginary parts are pretty close to each other. It is similar to the chiral EP in a single resonator with two scatterers, where one of the scatterer lifts the degeneracy, and the another one creates the degenerated states.

3. conclusion

We study chiral EP in a coupled ring resonators containing scatterer(s) as a perturbation. The chirality around 0.99 could be achieved for two different distances between the resonators. Due to the tunability of both the distance between the resonators and the position of the scatterer, the proposed structure enables achieving high chirality, and can compensate the possible imperfections during the fabrication process.

ACKNOWLEDGMENTS
S. Ramezanpour and A. Bogdanov acknowledge support from the Russian Science Foundation (Project No. 21-12-00383).

References
[1] M.-A. Miri and A. Alu, "Exceptional points in optics and photonics", Science 363 (2019).
[2] H. Hodaei and et al., "Enhanced sensitivity at higher-order exceptional points", Nature 548 (2017).
[3] Z. Xiao and et al., "Enhanced sensing and nondegraded thermal noise performance based on pt -symmetric electronic circuits with a sixth-order", PRL (2019)
[4] C. Gmachl and et. al, "High-power directional emission from microlasers with chaotic resonators", Science 280 (1998).
[5] Q. J. Wang and et. al, "Whispering-gallery mode resonators for highly unidirectional laser action", PNAS 107 (2010).
[6] N. Zhang and et. al, "Far-field single nanoparticle detection and sizing," Optica 4 (2017).
[7] S. Ramezanpour and A. Bogdanov, "Tuning 2nd and 3rd order exceptional points with kerr nonlinearity", AIP Conference proceedings 2300 (2020).
[8] J. Wiersig, "Structure of whispering-gallery modes in optical microdisks perturbed by nanoparticles", Phys. Rev. A 84 (2011).

Light propagation and interaction modelling in ring fibre nonlinear microcavities

V A Razukov and L A Melnikov

Yuri Gagarin State Technical University of Saratov, 77 Politechnicheskaya street, Saratov, Russia, 410054

E-mail: razukov.vad@gmail.com, lam-pels@yandex.ru

Abstract. Usage of the "Cabaret" explicit-implicit numerical scheme is proposed to model field dynamics both in microcavities and fibre cavities. Two opposite running waves are investigated with a number of the nonlinear effects included in the model, with potential to include more if necessary.

1. Introduction

Modern communications network is a complex system that requires a lot of problems being solved to reach the required levels of the information being transferred per second, with the demand of it and thus the load on the network growing every year. The situation can basically be summarized by the requirement of the data transmission to be within the Tb/sec range, while the devices employed for that has to be as small as possible, preferably microscopic or even nanoscale level. And the biggest potential in the terabit communications seems to belong to the optical frequency combs [1]. Furthermore, generating Kerr combs in the nonlinear cavities [2, 3] could create the combs with tens of GHz in spacing, but that can also create a 'vicious circle' of sorts, because these combs are prone to strong phase noise [4, 5, 6], resulting in a new set of the data transmission problems. Experiments show [7] that, despite the mentioned difficulties, it is the Kerr combs that are the best suited to answer the modern communication demands.

So, optical connections capable of transmitting such data amounts are the most apparent way to get past the bottlenecks in the world communication networks and computational centres. One could use wavelength-division multiplexing (WDM) with tens or hundreds of channels together with the advanced data modulation formats to achieve the required terabit speeds, but with the energy consumption remaining quite optimal.

Distributed feedback lasers are a common method to create optical WDM line. However, it is quite problematic to make them fully silicon based, because they would then require multiple interfaces between various parts, leading to the total size growth. Moreover, spectral efficiency starts to suffer as well. One way to solve such problems would be to use the OFCs as a WDM source, because each of the spectral lines in a comb could be modulated on its own [1, 8, 9].

Hence it is obvious that being able to predict the field propagation inside of the cavities has an immense practical value. The task is made more difficult due to the high nonlinearity of the microcavities, resulting in the numerical only investigations being possible, and in the process it is common to look for the models that both are adequate in the process simulation, but which do not consume a lot of time during the calculations. The previous trend [10, 11, 12, 13, 14, 15, 16] was to use the modal approach, but the result is a system of up to hundreds of nonlinear bound equations, and solving those numerically is a daunting task. Modal method is one of the spectral ones, but there is an alternative way, which is to use a transport equations difference scheme, what have been already used to simulate Raman and SBS lasers [17, 18, 19]. This article further develops our model [17], and we discuss the results obtained with the approach.

2. Model introduction

The ultra-short pulse theory uses mostly two ways to solve problems of the kind: slowly varying amplitude and slow envelopes. As we have already state in our previous articles [17, 18], the left side of the transport equations is as follows:

$$\frac{\partial E}{\partial t} + v_g \frac{\partial E}{\partial z} - i\frac{\omega''}{2}\frac{\partial^2 E}{\partial z^2} + \ldots = 0, \tag{1}$$

Here v_g - group velocity, $\omega'' = -\beta'' v_g^3$, $\beta'' = -\frac{\lambda^2}{2\pi c} D_2$, where D_2 - is group velocity dispersion parameter [20]. Thus, the pulse equations within a microcavity are given as follows:

$$2i\left(\frac{\partial F}{\partial t} + v\frac{\partial F}{\partial z}\right) + D\frac{\partial^2 F}{\partial z^2} + 2\chi\left(|F|^2 + 2|B|^2\right)F = 0,$$

$$2i\left(\frac{\partial B}{\partial t} - v\frac{\partial B}{\partial z}\right) + D\frac{\partial^2 B}{\partial z^2} + 2\chi\left(2|F|^2 + |B|^2\right)B = 0. \tag{2}$$

The boundary conditions for the equations are:

$$F(0) = \sqrt{1-R}\sqrt{1-r}F(L) + \sqrt{R}\sqrt{A}\sqrt{1-r} + \sqrt{r}B(0),$$

$$B(L) = \sqrt{1-R}\sqrt{1-r}B(0) - \sqrt{r}(1-r)F(L) + \sqrt{Rr}\sqrt{1-R}\sqrt{A} \tag{3}$$

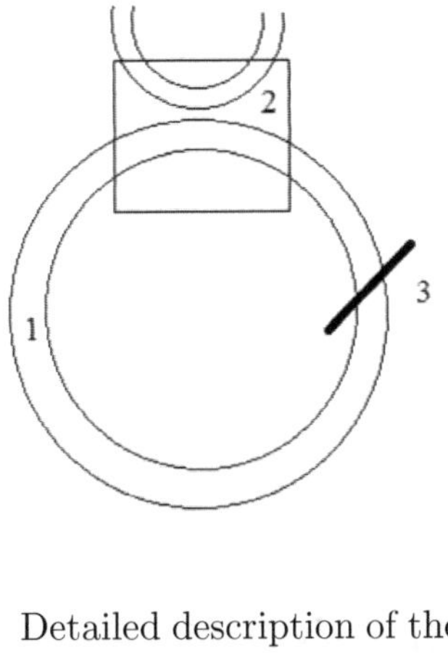

Figure 1. Model outline.

Detailed description of the model was provided by us before [21]. In the current investigation of two opposite running waves, we include in our model the ring cavity itself (1), nonlinear

Journal of Physics: Conference Series **2015** (2021) 012123 doi:10.1088/1742-6596/2015/1/012123

and dispersive; we also include the coupler (2), and a regular intracavity mirror (3) to simulate the light scattering. Modulation instability can also be present in the system [17, 20, 22], due to the negative GVD in the medium, resulting in the solution instability, while intensity stays time-constant. Their combined effect leads to the complicated dynamics under the investigation. The effective second order difference scheme "Cabaret" [23] is used to solve the problem, that we have already proven possible [21]. Computational algorithm stability was proven by having less than 1% of numerical losses after more than a 1000 cavity roundtrips.

3. Simulation results

Figures 2 and 3 show the unidirectional simulation results from the experiment start up to the appearing of the opposite wave born of the nonlinearity. It can be noted that a characteristic solitonic-like comb has been formed, with its spikes spaced roughly at the roundtrip time, resulting in a OFC appearing in the signal.

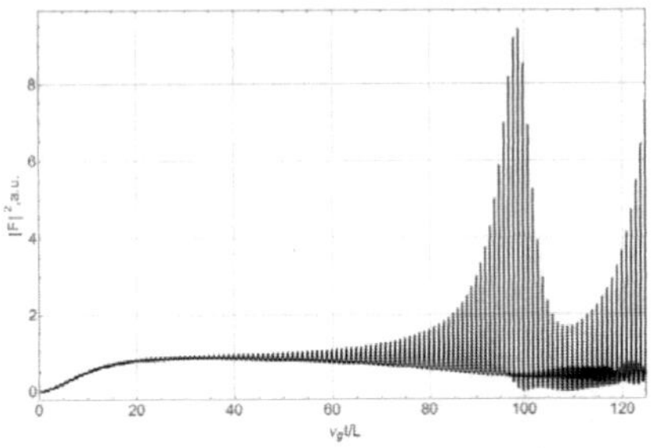

Figure 2. Pulse propagation in the single direction regime.

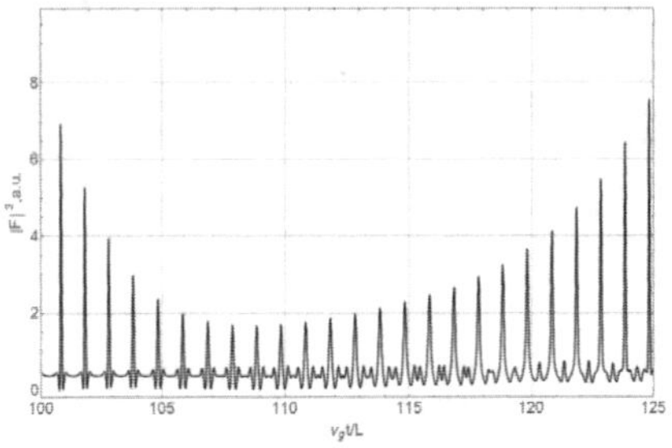

Figure 3. OFC appearence due to the nonlinear interaction. Phase modulation coefficient is extremely low ($\chi = 0,001$).

Next we introduce counterpropagating wave and simulate cavity rotation, which would lead to the increase in the nonreciprocal phase shift. Wave interface coefficient χ in (2) is increased from 0.001 to 0.25. That results in the OFC deformity and transition to the chaotic regime. Modeling results are presented in Figures 4 and 5. In both cases phase coefficient $\chi = 0,25$. Wave interface coefficient $r = 0.001$.

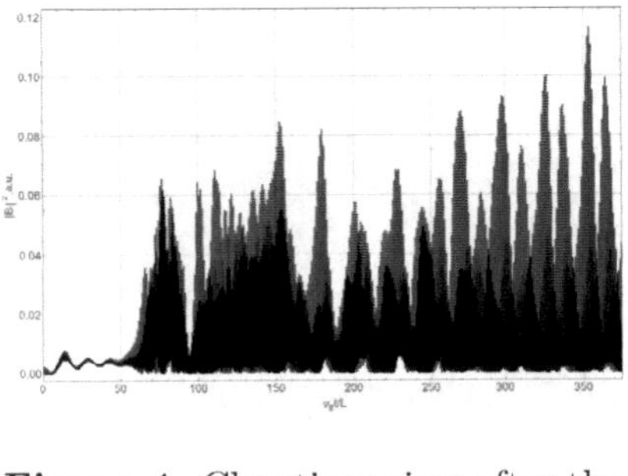

Figure 4. Chaotic regime after the second wave was introduced.

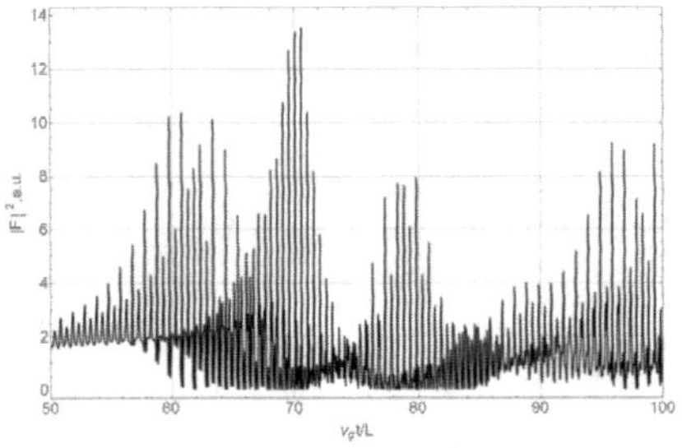

Figure 5. Detailed field profile of the clockwise wave interacting with the counterclockwise one.

Final step was to add Rayleigh scattering to the system. It is simulated by inserted mirrors in every Δz interval with the phase being given a new random value after the scattering occurs.

To deduce how it influences the fields, we increase the wave interface coefficient r from 0.001 to 0.05. The results are demonstrated in Figures 6 and 7. $\chi = 0,25$. Coupling coefficient $r = 0.001$. Rayleigh loss coefficient is always 0.0001.

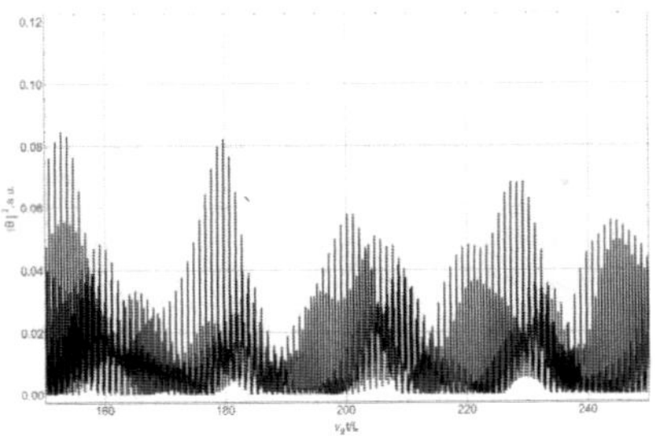

Figure 6. Several overlapping combs interfacing with each other and contributing phase noises into the signal.

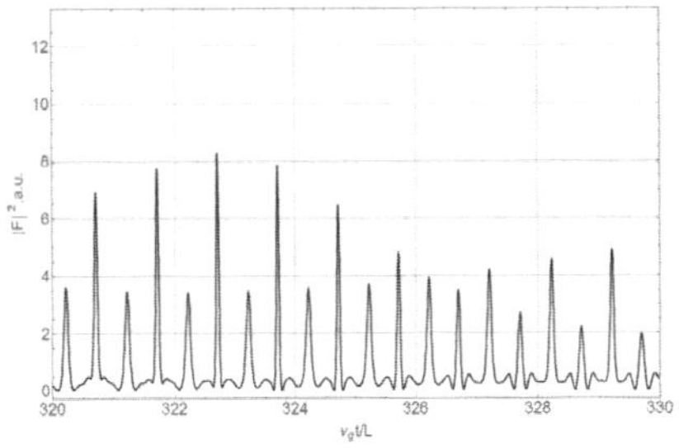

Figure 7. Detailed view of the field profile.

4. Conclusion

In conclusion we would like to state that our numerical model not just possesses increased stability, but it is very easy to add nearly any kind of a nonlinear effect in the investigated system without a noticeable growth in the computation times. Our further goals are to first make sure that our model fits the existing experimental data and refine it if necessary, and then attempt to find a promising regime that could have a significant practical value.

References

[1] Hillerkuss D, Schmogrow R, Schellinger T, Jordan M, Winter M, Huber G, Vallaitis T, Bonk R, Kleinow P, Frey F, Roeger M, Koenig S, Ludwig A, Marculescu A, Li J, Hoh M, Dreschmann M, Meyer J, Ben Ezra S, Narkiss N, Nebendahl B, Parmigiani F, Petropoulos P, Resan B, Oehler A, Weingarten K, Ellermeyer T, Lutz J, Moeller M, Huebner M, Becker J, Koos C, Freude W and Leuthold J 2011 *Nature Photonics* **5**(6) 364–371

[2] Del'Haye P, Schliesser A, Arcizet O, Wilken T, Holzwarth R and Kippenberg T J 2007 *Nature* **450**(7173) 1214–1217

[3] Levy J S, Saha K, Okawachi Y, Foster M A, Gaeta A L and Lipson M 2012 *IEEE Photonics Technology Letters* **24** 1375–1377

[4] Herr T, Hartinger K, Riemensberger J, Wang C Y, Gavartin E, Holzwarth R, Gorodetsky M L and Kippenberg T J 2012 *Nature Photonics* **6**(7) 480–487

[5] Pfeifle J, Weimann C, Bach F, Riemensberger J, Hartinger K, Hillerkuss D, Jordan M, Holtzwarth R, Kippenberg T, Leuthold J, Freude W and Koos C 2012 *Optical Fiber Communication Conference* (Optical Society of America) p OW1C.4 URL `http://www.osapublishing.org/abstract.cfm?URI=OFC-2012-OW1C.4`

[6] Wang P H, Ferdous F, Miao H, Wang J, Leaird D, Srinivasan K, Chen L, Aksyuk V and Weiner A 2012 *Optics express* **20** 29284–95

[7] Pfeifle J, Brasch V, Lauermann M, Yu Y, Wegner D, Herr T, Hartinger K, Schindler P, Li J, Hillerkuss D, Schmogrow R, Weimann C, Holzwarth R, Freude W, Leuthold J, Kippenberg T J and Koos C 2014 *Nature Photonics* **8**(5) 375–380

[8] Qian D, Huang M F, Ip E, Huang Y K, Shao Y, Hu J and Wang T 2011 *Optical Fiber Communication Conference/National Fiber Optic Engineers Conference 2011* (Optical Society of America) p PDPB5 URL `http://www.osapublishing.org/abstract.cfm?URI=OFC-2011-PDPB5`

[9] Witzens J, Baehr-Jones T and Hochberg M 2010 *Nature Photonics* **4**(1) 10–12

[10] Chembo Y K and Yu N 2010 *Phys. Rev. A* **82**(3) 033801 URL `https://link.aps.org/doi/10.1103/PhysRevA.82.033801`

[11] Maleki L, Ilchenko V S, Savchenkov A A, Liang W, Seidel D and Matsko A B 2010 *2010 IEEE International Frequency Control Symposium* pp 558–563

[12] Matsko A B, Savchenkov A A, Liang W, Ilchenko V S, Seidel D and Maleki L 2011 *Opt. Lett.* **36** 2845–2847 URL `http://ol.osa.org/abstract.cfm?URI=ol-36-15-2845`

[13] Herr T, Brasch V, Jost J D, Wang C Y, Kondratiev N M, Gorodetsky M L and Kippenberg T J 2014 *Nature Photonics* **8**(2) 145–152

[14] Herr T, Brasch V, Jost J D, Wang C Y, Kondratiev N M, Gorodetsky M L and Kippenberg T J 2014 *Nature Photonics* **8**(2) 145–152

[15] Chembo Y K and Menyuk C R 2013 *Phys. Rev. A* **87**(5) 053852 URL `https://link.aps.org/doi/10.1103/PhysRevA.87.053852`

[16] Gorodetsky M L 2011 *Opticheskiye Microrezonatory s gigantskoy dobrotnostyu. [Optical Microresonators with Huge Quality] [in Russian]* (Physical and Mathematical Literature)

[17] Razukov V A and Melnikov L A 2016 *Saratov Fall Meeting 2015: Third International Symposium on Optics and Biophotonics and Seventh Finnish-Russian Photonics and Laser Symposium (PALS)* vol 9917 ed Genina E A, Tuchin V V, Derbov V L, Postnov D E, Meglinski I V, Larin K V and Pravdin A B International Society for Optics and Photonics (SPIE) pp 788 – 792 URL `https://doi.org/10.1117/12.2229745`

[18] VA R, LA M, YuA M and SV S 2016 *Journal of Applied Spectroscopy* **83**(6–16) 344–345

[19] Melnikov L A and Mazhirina Y A 2017 *Quantum Electronics* **47** 1083–1090 URL `https://doi.org/10.1070/qel16560`

[20] Agraval G P 2007 *Nonlinear Fiber Optics* (Academic Press)

[21] Razukov V and Melnikov L 2020 *Izvestiya of Saratov University. New series. Series: Physics* **20** 64–71

[22] A M Y, LA M, SK T, DV C and NS T 2014 *Izvestyia Vuzov, Applied Nonlinear Dynamics* **22**(5) 73–82

[23] VM G and AA S 1998 *Matem. Mod.* **10**(1) 86–100

III-V nanowires with quantum dots: MBE growth and properties

R R Reznik[1-4], K P Kotlyar[1,2], V O Gridchin[2], I V Ilkiv[1,2], A I Khrebtov[1], Yu B Samsonenko[1,4], I P Soshnikov[1,4,5], N V Kryzhanovskaya[1], L Leandro[6], N Akopian[6] and G E Cirlin[1-4]

[1]Alferov University, Khlopina 8/3, 194021, St-Petersburg, Russia
[2]St. Petersburg State University, 199034 St. Petersburg, Russia
[3]ITMO University, Kronverkskiy pr. 49, 197101, St. Petersburg, Russia.
[4]Institute for Analytical Instrumentation RAS, Rizhsky 26, 190103, St-Petersburg, Russia
[5]Ioffe Institute, 194021 St. Petersburg, Russia
[6]DTU Fotonik, 2800 Kongens Lyngby, Denmark

E-mail: moment92@mail.ru

Abstract. We demonstrate growth of AlGaAs NWs with GaAs QDs and InP NWs with InAsP QDs on silicon substrates. Results of GaAs QDs optical properties study have shown that these objects are sources of single photons. In case of InP NWs with InAsP QDs, the results showed that ~ 100% of homogeneously oriented NWs were formed with good optical quality of this system on a Si(111). PL spectrum peak near 1.3 μm indicates that such system is promising for optoelectronic devices.

1. Introducrion

Semiconductor nanowires (NWs) and quantum dots (QDs) attract more and more interest of scientist because of their unique properties [1]. The most famous mechanism of QDs growth is Stranski - Krastanov method [2]. However, in case of QDs growth in body of NWs it is easier to control height, diameter and density of QDs. Diameter of such QDs is determinates by diameter of NWs, height - by growth time and surface density – by NWs density. Moreover, silicon is a key material for electronics now. Unique morphological properties of NWs allow to form these hybrid nanostructures on silicon platform

In this work we report about molecular-beam epitaxy (MBE) growth and physical properties of AlGaAs NWs with GaAs QDs and InP NWs with InAsP QDs on silicon substrates

2. Experiments

In all cases NWs with QDs were formed on Si(111) substrates by MBE. The growth techniques for GaAs/AlGaAs and InAsP/InP systems were similar, but the features of behavior during the growth of atoms of different materials were taken into account. First of all, Si substrates were cleaned using HF (10:1) and thermally at 850°C. On the next step thin Au film were deposed on silicon surface at 550°C For Au nanoparticles formation. On the next stage, we grew the AlGaAs and InP base of the NWs temperatures of substrates 510°C and 270°C, respectively. Then, the GaAs or InAsP nanoinsertions (or QDs) with lower bandgap were formed, which were covered again by the material of the NWs. The GaAs QDs formation time was varied in the range of 5-15 seconds. In case of InP NWs, the growth rate of InAs studied by a similar method was 0.4 ML/s. Two samples were grown with the As/P fluxes ratios 1/1.8 and 1/1.5.

The surface morphology is studied by scanning electron microscopy (SEM), structural properties of the NWs were studied by transmission electron microscopy (TEM). Optical properties of grown samples were studied by photoluminescence (PL) and micro-photoluminescence (μPL) techniques.

3. Results and discussion

Figure 1, (a) shows SEM picture of AlGaAs NWs (growth time – 20 minutes) with GaAs QDs on silicon. It seen that NWs were formed in 111 direction which indicates about epitaxiallity of the growth. The average height of NWs this QDs is ~ 2 μm which is enough for most applications. It is important to note the high surface density, which is determined by the density of the catalyst droplets. Controlling surface density of samples can have a significant impact in creating devices.

Figure 1, (b) shows typical SEM picture of the InP NWs (20 minutes - growth time) with InAsP QDs. As can be seen from the figure, the length of the NWs was approximately 4 micrometers. In turn, the dimater of the grown nanostructures turned out to be inhomogeneous in length and amounted to 100 nm at the base and 30 nm at the top of the NW. As in the case of AlGas NWs, InP NWs have a high surface density. It is important to note that the NWs were formed strictly in the direction of 111. In both cases, this happened due to the use of a unique technique for cleaning the substrates, sputtering the catalyst, and transferring the substrate to the growth chamber without losing ultrahigh vacuum conditions. Typically, the presence of an incompletely removed silicon oxide layer results in nonepitaxial growth in all directions. To achieve epitaxial growth, expensive methods of pregrowth processing of substrates, including lithography, are used [3].

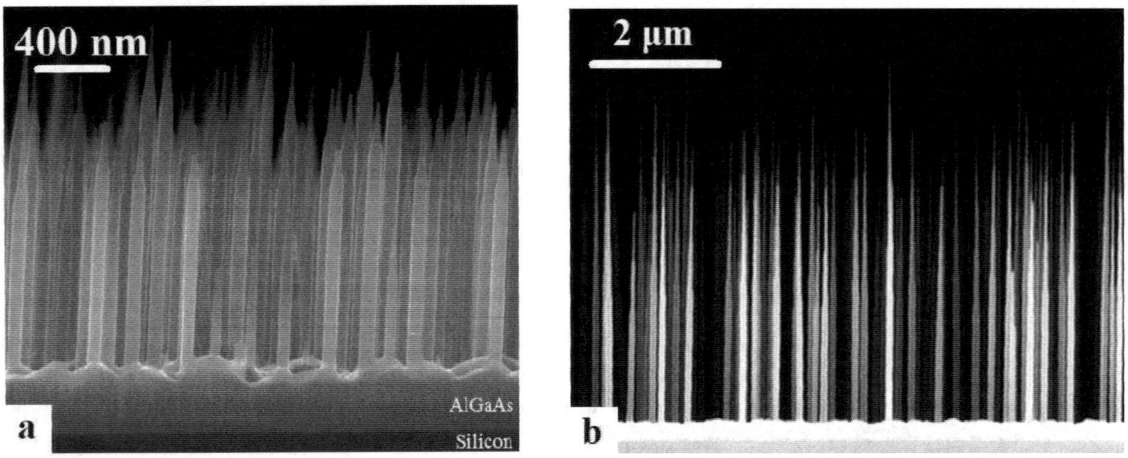

Figure 1. Typical SEM picture of NWs with QDs on silicon: (a) GaAs/AlGaAs; (b) InAsP/InP.

The optical properties of single AlGaAs NW with GaAs QD were studied using the low-temperature μPL technique. Figure 2, (a) shows typical μPL spectra at 4.2K from three samples containing QDs formed within various duration: 5 seconds, 7 seconds and 15 seconds. As expected, increased growth time increases the size of the QD, which decreases the emission energy (increases the radiation wavelength) of QD. As shown in insertion of Figure, the μPL spectra of quantum dots in the NWs have an ultra-small line width (Full Width on Half Maximum, FWHM) limited by the spectrometer resolution (30 μeV). The measurement results also confirm that we can control the changes in the emission wavelength of QD in NWs by choosing the QD growth time from 5 seconds (emission at 750 nm) to 15 seconds (emission at 780 nm).

Figure 2, (b) shows PL spectra for P-based samples at room temperature. Both samples demonstrate PL spectra at room temperature, which indicates the high quality of the grown structures. In addition to the peak corresponding to radiation from InP, both spectra contain two peaks corresponding to radiation from InAsP of different compositions. During the formation of an axial InAsP QD, a radial InAsP quantum wells (QWs) can also form as a result of lateral growth [4-6]. We correlate the higher-energy peak with emission from the QWs, and the lower-energy peak with emission from the QDs.

Figure 3 shows a typical TEM image of an InP NW with InAsP QD dispersed onto a carbon grid and the arsenic distribution (x) map for various parts of the NW. The results of TEM measurements confirmed the assumption about the formation of InAsP QDs and radial InAsP QWs in the body of InP NWs. It should be noted that the InP NWs under study were formed in a pure wurtzite crystallographic phase.

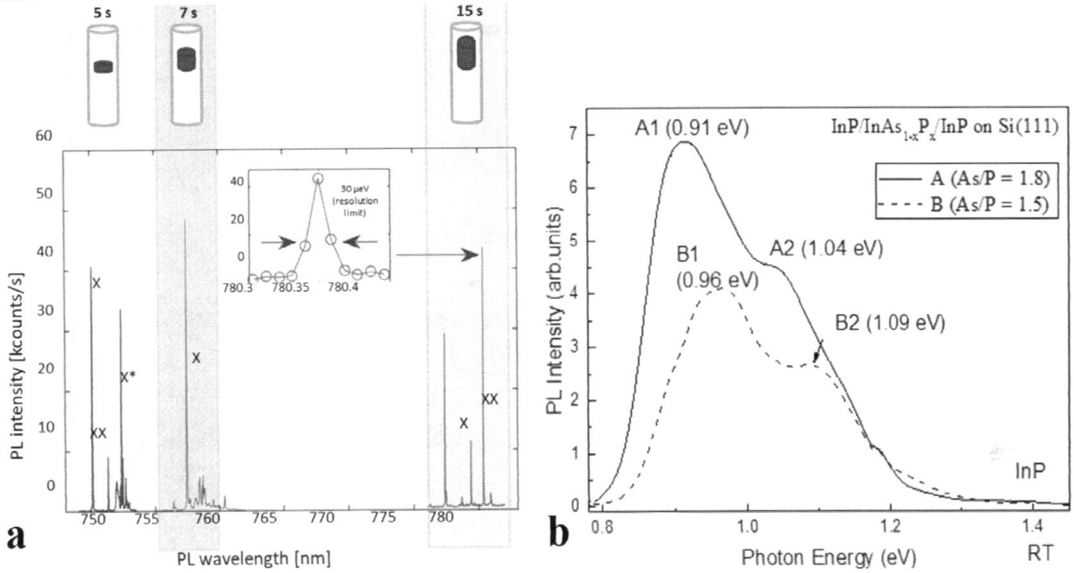

Figure 2. (a) PL spectra at T = 10 K for AlGaAs NWs with GaAs QDs with different Al contents and different growth times of GaAs QDs: x = 0.4, 12 s; x = 0.5, 15 s; (b) PL spectra at room-temperature for InP NWs with InAsP QDs grown with different ratios between the P and As fluxes.

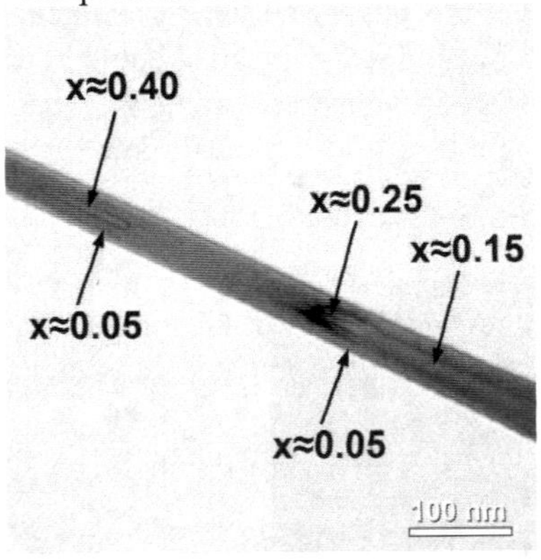

Figure 3. Typical TEM image of an InP NW with InAsP QD.

For registration of GaAs QDs autocorrelation function an optical setup was assembled, consisting of: a laser for exciting QD in single NW, a semitransparent mirror and two single photons detectors. After excitation by the laser, the radiation from the QD passed through the semitransparent mirror with a probability of 50% and registered by first detector, and with a probability of 50% it was reflected and registered by second detector. In fact, the measurement result was the multiplication of signals from both detectors. Figure 4 shows the autocorrelation measurements of the QD emission showing a characteristic minimum on the autocorrelation function at time zero $g^{(2)}$ ($\boldsymbol{\tau}$ = 0) = 0.19. The presence of the minimum on the autocorrelation function means that only one detector registered a photon at the initial moment of time. This fact characterizes the grown nanostructures as sources of single photons.

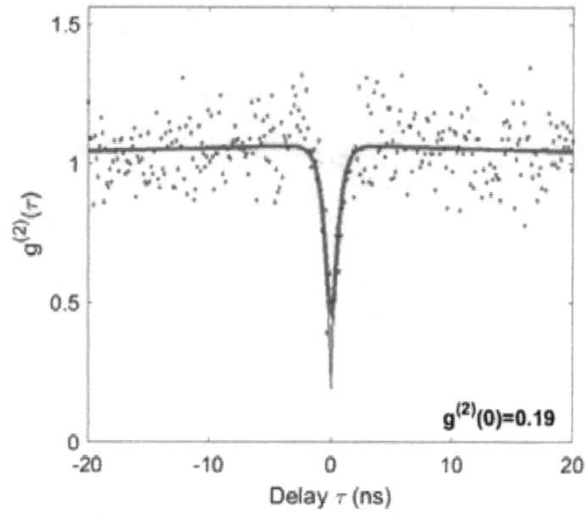

Figure 4. Autocorrelation measurements of an exciton (750 nm) in GaAs QD (5 seconds).

In addition, we would like to note that nanostructures based on AlGaAs NWs with InGaAs QDs were grown on silicon surface. Preliminary results of optical properties studies of the grown structures showed that PL spectra from InGaAs QDs are observed up to room temperature in a wide range of wavelengths 950-1200 nm depending on the composition of the QDs.

4. Conclusions

In conclusion, AlGaAs NWs with GaAs QDs and InP NWs with InAsP QDs have been grown by MBE on silicon substrates. Results of GaAs QDs optical properties studies have shown that these objects are sources of single photons. In case of InP NWs with InAsP QDs, The results of studies of the optical properties of the grown structures showed that nanostructures demonstrate PL spectra at room temperature. It is important to note that in all cases the NWs were formed strictly in the direction of 111 due to the use of a unique technique for cleaning the substrates, sputtering the catalyst, and transferring the substrate to the growth chamber without losing ultrahigh vacuum conditions. All of the above indicates that the grown nanostructures are promising for optoelectronic applications.

Acknowledgements

The samples were grown under support of Ministry of Science and Higher Education of the Russian Federation (state task № 0791-2020-0003). Optical study was conducted under RSF grant № 18-72-10047.

References

[1] Dubrovskii V G, Cirlin G E and Ustinov V M 2009 *Semiconductors* **12** 1539
[2] Dubrovskii V G, Cirlin G E and Ustinov V M 2003 *Phys. Rev. B* **68** 075409
[3] Munshi A M, Dheeraj D L, Fauske V T, Kim D C, Huh J, Reinertsen J F, Ahtapodov L, Lee K

D, Heidari B, van Helvoort A T J, Fimland B O and Weman H 2014 *Nano Lett. 14* 960
[4] Tchernycheva M, Patriarche G, Travers L, Zwiller V, Perinetti U and Harmand J-C 2007 *Nano Lett.* **7** 1500
[5] Harmand J-C, Jabeen F, Liu L, Patriarche G, Gauthron K, Senellart P, Elvira D and Beveratos A 2013 *J. Cryst. Growth* **378** 519
[6] Cirlin G E, Tchernycheva M, Patriarche G and Harmand J-C 2012 *Semiconductors* **46** 175

Retrieving the refractive index of a sphere from the phase spectrum of its light-scattering profile

A V Romanov[1, 2, 3] and M A Yurkin[1, 2]

[1] Voevodsky Institute of Chemical Kinetics and Combustion SB RAS, Institutskaya Str. 3, 630090, Novosibirsk, Russia
[2] Novosibirsk State University, Pirogova Str. 2, 630090, Novosibirsk, Russia
[3] Corresponding author: *avm.romanov@gmail.com*

Abstract. We studied the Fourier spectrum of the light-scattering profiles of single particles in the Rayleigh-Gans-Debye (RGD) and Wentzel–Kramers–Brillouin (WKB) approximations. In the case of a homogeneous sphere, we found the relationship between the key parameters of the spectrum (including its phase) and the sphere characteristics – both analytically and numerically in the framework of the approximations and the rigorous Lorentz–Mie theory, respectively. Based on these results, we have improved the existing spectral characterization method for spheres extending the applicability range to particles with a higher refractive index.

1. Introduction

Light scattering is one of the most common approaches to the non-invasive characterization of microparticles. Among multitude of existing methods, the most promising ones are those working with single particles, since they are more reliable than ensemble ones (the corresponding inverse problem is typically well posed). There are many approaches to solving such inverse light-scattering problems (Ref. [1] and references therein), one of which is the compression of information in the measured light-scattering profiles or patterns (LSP) into several parameters that determine the characteristics of the particle under study. An important common advantage of such methods is the resistance to various distortions introduced both by experiment and by the imperfection of the particle model. Several recent examples include spectral methods for determining the size and refractive index of spheres [2,3] and for non-sphericity estimation [4,5].

In particular, an accurate and robust characterization of spheres was demonstrated in a limited range of size and refractive index [2]. This method extracted two parameters from the Fourier spectrum of the one-dimensional LSP: the main peak position and the zero-frequency amplitude, which strongly correlated with the size and refractive index, respectively. However, the ambiguity of the latter parameter for high refractive indices limited its use for widespread polystyrene beads. In this work we show the reasons for this ambiguity and propose a way to improve the capabilities of the spectral method based on the usage of the phase spectrum.

2. Main analytical results

We focus on the versatility of our approach for variety of applications and mostly consider the scattered intensity of unpolarized light. However, one of the prominent examples, the scanning flow cytometer, measures the following combination of Mueller scattering matrices:

$$I(\theta) = \frac{1}{2\pi} \int_0^{2\pi} [S_{11}(\theta, \varphi) + S_{14}(\theta, \varphi)] \mathrm{d}\varphi, \tag{1}$$

where θ is polar angle from $\theta_1 = 10°$ to $\theta_2 = 65°$. Note that the integral of S_{14} is exactly zero for any axisymmetric particle [6]. We apply the FFT-based spectral transformation, described in [4], to the discrete measured data, in order to have a discrete representation of the following continuous transform:

$$F(v) = \frac{1}{2\pi} \int_{-\infty}^{\infty} w(\theta) I(\theta) \exp(-\mathrm{i} v \theta)\, \mathrm{d}\theta \tag{2}$$

where w(θ) is the Hann window function for the range [10°,65°].

Analyzing the light-scattering problem in the framework of the RGD approximation, we have managed to explain the origin of the main peak in the Fourier spectrum and its relation to the particle size. Furthermore, analyzing the WKB approximation, we found that the given Fourier transform of the scattering intensity has a phase factor linearly proportional to both the refractive index and the frequency. Such a transformation result can be expected already proceeding from the given formula for the intensity

$$|I(\theta)|^2 = \frac{1}{4\pi} \int_{\infty} \exp\left(i2 \sin\frac{\theta}{2} kp + i(m-1)kp\right) F(p, \theta)\, \mathrm{d}p \tag{3}$$

where $(m-1)kp$ is a phase factor, and F is some function defined by particle geometry, the spectrum of which we obtain as a result of the Fourier transformation. This will be discussed in more detail at the conference.

A factor of this type in the Fourier image indicates a shift of the original function (LSP) along the coordinate axis (θ) by a value proportional to the refractive index, which was observed and used earlier for spheres [7,8]

3. Numerical calculations and discussions

The above theoretical analysis suggests that the phase of the spectrum at peak frequency (further denoted as peak phase) strongly depends on the refractive index, so it is natural to combine it with the peak location, which strongly depends on size and only weakly on refractive index. In contrast to the zero-frequency amplitude, used in [4], the peak phase is equally applicable for any refractive index at the cost of being cyclic, i.e. inherently non-invertible in any wide range. Thus, its natural application niche is characterization of particles in a relatively narrow range of size and refractive indices. In particular, we further apply it to characterization of 4-μm polystyrene beads. Note, however, that similar calculations can be performed for other ranges of characteristics.

We used the Lorenz–Mie theory to calculate the LSPs in the ranges [3.7 μm, 4.3 μm] and [1.56, 1.62] for size and refractive index, respectively, and Fourier-transformed them according to Eq. (2). We assumed the wavelength 0.66 μm and host-medium refractive index equal to 1.333. In the resulting spectra, we determined the position of the main peak and the phase value at this point. The dependence of these parameters on the sphere characteristics are shown in Fig. 1. As can be seen from the figure, the phase has an almost linear dependence on the refractive index; however, there is some ripple associated with effects not accounted for in the WKB approximation.

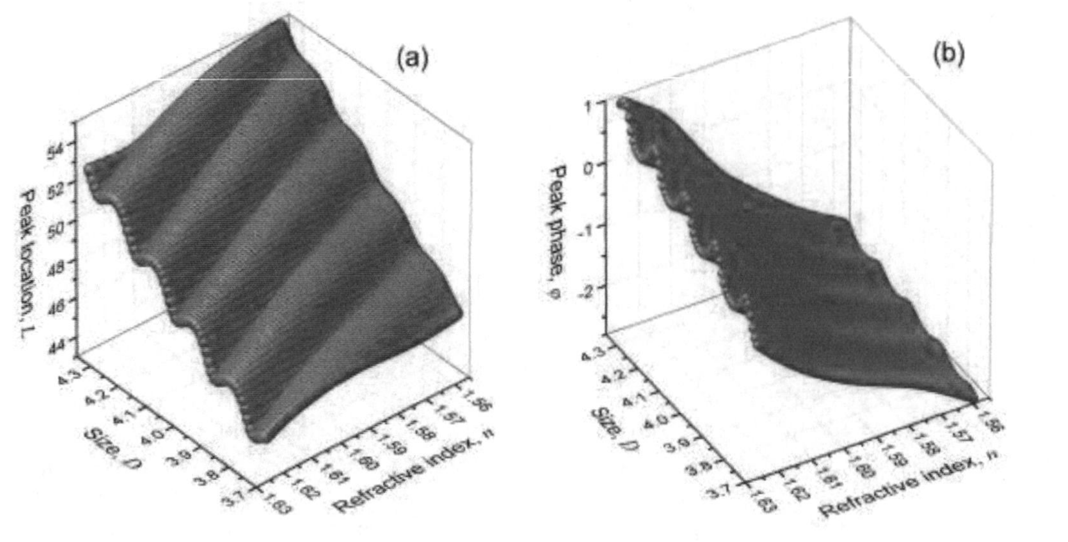

Figure 1. The dependence of peak location (a) and its phase value (b) on size and refractive index calculated by using the Lorentz–Mie theory.

Inverting this dependence by plotting the same data points in other coordinates (characteristics of the sphere as functions of the parameters of the spectrum), we solve the inverse problem using interpolation. Despite the underlying linear structure, ripples present significant problems by compromising uniqueness and affecting interpolation accuracy. Note that these ripples present problems for any other characterization methods as well, even for the least-square fit. For example, the LSPs of two spheres with size different by a ripple period (approximately 0.15 μm in this case) maybe more similar to each other than to LSP of spheres with sizes in between. To assess the influence of these ripples on our method, we tested it on synthetic LSPs of spheres, computed in the same range of characteristics but different from the interpolation grid nodes. While the maximum errors are 10 nm and $10^{-3}$for size and refractive index, the mean errors do not exceed 1 nm and 10^{-4}, respectively. The results of processing experimental LSPs of polystyrene beads will be presented at the conference.

4. Conclusion

We have analyzed the application of the phase spectrum of 1D LSP to the characterization of spheres, both analytically using the WKB approximation and numerically using the Mie theory. Moreover, we constructed an interpolant to characterize homogeneous spheres using the position and phase of the main peak in the Fourier spectrum. This interpolant works in the limited range of size and refractive indices (due to the cyclicality of the phase parameter), but proved to be fast, robust, and accurate for characterization of latex beads.

Acknowledgments

The research has been supported by the Russian Foundation for Basic Research (grant No. 19-32-90073).

References

[1] Romanov A V and Yurkin M A 2021 Single-particle characterization by elastic light scattering *Laser & Photon. Rev.* **15** 2000368

[2] Romanov A V, Konokhova A I, Yastrebova E S, Gilev K V, Strokotov D I, Chernyshev A V, Maltsev V P and Yurkin M A 2017 Spectral solution of the inverse Mie problem *J. Quant. Spectrosc. Radiat. Transf.* **200** 280–94

[3] Konokhova A I, Yastrebova E S, Strokotov D I, Chernyshev A V, Karpenko A A and Maltsev V P 2019 Ultimate peculiarity in angular spectrum enhances the parametric solution of the inverse Mie problem *J. Quant. Spectrosc. Radiat. Transfer* **235** 204–8

[4] Romanov A V, Konokhova A I, Yastrebova E S, Gilev K V, Strokotov D I, Maltsev V P and Yurkin M A 2019 Sensitive detection and estimation of particle non-sphericity from the complex Fourier spectrum of its light-scattering profile *J. Quant. Spectrosc. Radiat. Transf.* **235** 317–31

[5] Yastrebova E S, Dolgikh I, Gilev K V, Vakhrusheva I V, Liz E, Litvinenko A L, Nekrasov V M, Strokotov D I, Karpenko A A and Maltsev V P 2021 Spectral approach to recognize spherical particles among non-spherical ones by angle-resolved light scattering *Opt. Laser Technol.* **135** 106700

[6] Yurkin M A, Semyanov K A, Tarasov P A, Chernyshev A V, Hoekstra A G and Maltsev V P 2005 Experimental and theoretical study of light scattering by individual mature red blood cells with scanning flow cytometry and discrete dipole approximation *Appl. Opt.* **44** 5249–56

[7] Maltsev V P and Lopatin V N 1997 Parametric solution of the inverse light-scattering problem for individual spherical particles *Appl. Opt.* **36** 6102–8

[8] Shepelevich N V, Lopatin V V, Maltsev V P and Lopatin V N 1999 Extrema in the light-scattering indicatrix of a homogeneous sphere *J. Opt. A* **1** 448–53

Nonlinear TMOKE enhancement in 1D Au/Py magnetoplasmonic crystals

A.M. Romashkina, V.B. Novikov and T.V. Murzina

Department of Physics, Lomonosov Moscow State University, Moscow, 119991 Russia

E-mail: ram@shg.ru

Abstract. Resonant optical properties of the magnetoplasmonic crystals, which support propagation of surface plasmon polaritons (SPPs) accompanied by magnetooptical effects, have found success in magnetic field driven control of optical radiation. In this work we investigate the resonant magnetooptical effects in the second harmonic generation in the magnetoplasmonic crystal formed by gold/pemalloy bifilm covering dielectric grating. Strong transverse magnetooptical Kerr with the contrast up to 30% is revealed in the spectral vicinity of the SPP excitation.

1. Introduction

Magnetoplasmonic crystals (MPCs) are periodic metal-dielectric nanostructures, which bind together the excitation of surface plasmon polaritons (SPPs) and magnetooptical effects. Resonant optical properties of the SPP localized in close proximity to the interface of metal and dielectric make the optical response of the plasmonic crystals very sensitive to properties of media adjacent to the interface that was harnessed in sensing of molecules and gases [1]. Particularly, magnetization of media in MPCs can shift resonant frequencies of SPPs. For example, the interface of metal and magnetoactive dielectric semispaces supports propagation of SPPs, the wave vector magnitude of which is determined by both dielectric permittivities of neighboring media and the magnetization:

$$k_{spp} = k_0 \sqrt{\frac{\varepsilon_m \varepsilon_d}{\varepsilon_m + \varepsilon_d}}(1 + \alpha M), \tag{1}$$

where ε_m, ε_d are permittivities of metal and dielectric, respectively, M - magnetization perpendicular to the SPP propagation direction, $\alpha = a(-\varepsilon_m \varepsilon_d)^{-1/2}(1 - \varepsilon_m^2/\varepsilon_d^2)^{-1}$, $k_0 = \omega/c$ is the vacuum wave number, a is the gyration constant [2]. This sensitivity of SPP in MPCs to the applied magnetic field gives rise to the resonant manifold increase of intrinsically weak magneto-optical effects, e.g., transverse magneto-optical Kerr effect (TMOKE) consisting of the variation of the reflection coefficient of light under the magnetization of medium perpendicularly to the plane of light incidence. It has paramount importance in control of light routing and remote optical sensing of magnetic field strength [3].

It should be noted that the excitation of SPPs requires special efforts caused by the wavevector mismatch between k_{spp} and the vacuum wave number. In plasmonic crystals it is cured by the light diffraction involving the reciprocal lattice vector $\mathbf{G}$ of the periodic structure that leads to the quasi-phase matching condition

$$\mathbf{k}_{spp} = \mathbf{k}_{||} + m\mathbf{G}, \tag{2}$$

where $\mathbf{k}_{spp}$ is the wavevector of SPP, $\mathbf{k}_{||}$ is the tangential component of the wave vector of incident plane wave and m is the diffraction order.

In the field of investigation of magneto-optical effects in MPCs special attention is paid to the structures composed of noble and ferromagnetic metals [4, 5]. In this case both high-quality plasmon resonances supported by noble metals, like gold or silver, exhibiting lowest optical losses in visible spectral range, and high magneto-optical activity are combined in one structure. It results in a pronounced enhancement of magneto-optical effects, for example, TMOKE [6].

Being odd to the magnetization, the magnitude of TMOKE is characterized by the magnetic contrast given by

$$\rho = \frac{I(+M) - I(-M)}{I(+M) + I(-M)} \tag{3}$$

that is a relative change of reflected light intensity for opposite directions of the medium magnetization M. The magnetic contrast ρ has a typical value of about 10^{-3} that requires the intense efforts to detect it. In turn the nonlinear optical phenomena, particularly the second harmonic generation (SHG), can be harnessed to significantly intensify the effects of magnetic field in the optical response of MPCs [7]. Having nonlinear local field factors dependence, the resonant enhancement of the SHG in the spectral vicinity of the excitation of SPP at the fundamental frequency has been demostrated [8, 9, 10]. As well being sensitive to the optical properties of the media interface, pronounced nonlinear TMOKE was achieved via SPP excitation in periodically perforated Au/Co films [11].

This work is aimed at the investigation of the SHG in the MPCs based on one-dimensional dielectric lattice covered by thin gold and ferromagnetic permalloy films. Special attention is paid to the nonlinear TMOKE under the excitation of SPPs. Several tens of percent of the magnetic contrast in the SHG in the spectral vicinity of SPP excitation is revealed, that makes such structures promising for the magnetic field sensing.

2. Samples, experimental results and discussion

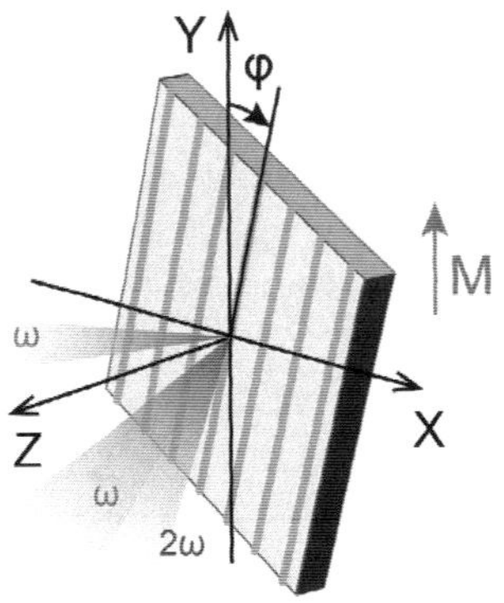

Figure 1. Experimental geometry. Light incidence plane XY, the sample is rotated around Z-axis, magnetic field is applied along Y-axis.

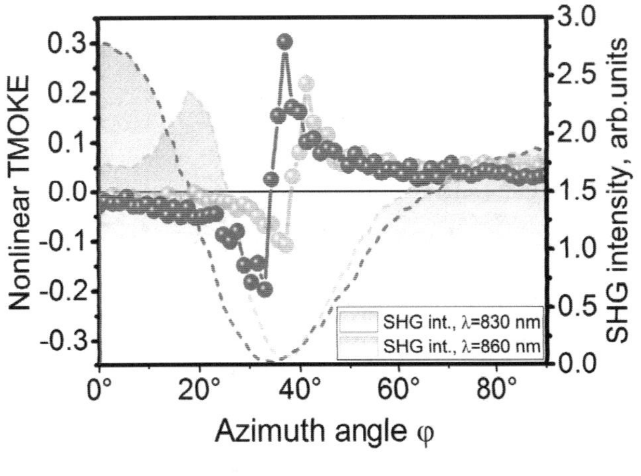

Figure 2. SHG (dotted lines) and nonlinear TMOKE (bubbles) azimuth dependence for pump wavelengths of 830 nm (yellow) and 860 nm (red).

The experimental studies were carried out for the MPC structure consisting of gold film with the thickness of 80 nm and ferromagnetic permalloy (Py) $Ni_{80}Fe_{20}$ film with the thickness of 10 nm, consistently sputtered on a one-dimensional polymer diffraction grating with a period of 600 nm located on the glass substrate.

The studies of the SHG and nonlinear TMOKE in MPCs were peformed using femtosecond titanium: sapphire laser tunable in the spectral range of 720 nm – 900 nm with the repetition rate of about 80 MHz, power of about 15 mW and the pulse duration of about 60 fs. The p-polarized light beam was incident at the angle of $\theta = 22.5^\circ$ and focused by lens with $f = 5$ cm on the MPC; the reflected light was collected by another lens and second harmonic (SH) signal selected by appropriate color filters was detected by a photomultiplier tube. The experimental studies were carried out under the azimuth rotation of the structure around its normal, while keeping the angle of incidence θ constant [Fig. 1]. The structure was located between two permanent neodymium magnets with the field strength of about 2 kOe, leading to the saturation magnetization of the permalloy layer in the MPC. The magnetic field was oriented along the y-axis that is perpendicular to the plane of the light incidence (xz), as shown in Fig. 1.

We revealed that considered structure supports the propagation of the SPPs, which appear as dips in the linear reflection spectrum of light that in our case takes place for both s- and p-polarizations of the incident light. We are focused on the SPP resonances located in the spectral vicinity of 860 nm, which correspond to the surface optical waves localized at the interface of Au/Py/Air and for the diffraction order $m = -1$ [see Eq. 2]. We demonstrated that the magnetic contrast of linear TMOKE is the most pronounced for these resonances compared to the SPPs localized at the interface SiO_2/Au. We revealed that the excitation of SPPs at the fundamental frequency in the considered structure is accompanied by the resonant Fano-type dependence of the SH intensity in the reflected light as the function the azimuth angle φ. It is shown by dashed curves in Fig. 2 for fundamental wavelengths of $\lambda = 830$ nm and $\lambda = 860$ nm. In this figure the SPP at the fundamental frequency corresponds to diffraction order $m = -1$ [see Eq. 2] and it is excited at the azimuth angle of about $\varphi = 30^\circ$ for $\lambda = 830$ nm and $\varphi = 20^\circ$ when $\lambda = 860$ nm. We suppose that specific form of the azimuth dependence of the SH intensity stems from the interplay of the local field enhancement of the fundamental light and the interference of SH light waves emitted by various parts of the MPC, where phase of nonlinear light sources changes at the variation of φ through the resonant value.

We revealed that the TMOKE in the SH intensity is resonantly enhanced in the close vicinity of the minimum of the SHG, that is close to the azimuth angle of the SPP excitation. This is illustrated by dots in Fig. 2 for both wavelengths. As seen at the variation of the azimuth angle, ρ abruptly changes with sign reversal. We achieved 20% TMOKE in SHG for fundamental wavelength of $\lambda = 830$ nm and 30% for $\lambda = 860$ nm, while nonresonant nonlinear TMOKE is about $2 - 5\%$. Obtained resonant values of TMOKE are significant for the structures where magneto-optical properties are introduced by the magnetoactive material film with thickness of only 10 nm. We emphasize that observed nonlinear magnetic contrast ρ is by 4 orders of the magnitude larger than the linear magnetic contrast that is about $3 \cdot 10^{-3}$ for considered structure. Obtained large values of magnetic contrast confirm superior possibilies of such type of MPC for the magnetic field sensing.

3. Conclusion

In this work we studied the SHG and nonlinear TMOKE in the MPC formed by a one-dimensional dielectric diffraction grating covered by continuous gold and ferromagnetic permalloy films that binds together diffraction-assisted excitation of the SPPs with high figure of merit and the magnetooptics. We revealed resonant effects of the excitation of the SPPs at fundamental frequency in the SHG, which appear as Fano-like shape of the angular dependence of the SH power. We observed that it is accompanied by manifold increase of the TMOKE in the SHG, which reaches the contrast of 30%, giving the opportunity of the application of considered structures for the magnetic field sensing.

Acknowledgments

This work was supported by the Russian Foundation for Basic Research, project no. 19-02-00826. Samples of plasmonic crystals were produced using the material and technical resources of the Common Use Center of the Research and Education Center "Functional Micro/Nanosystems" of the Bauman Moscow State Technical University (ID 74300). Authors are thankful to D. A. Ezenkova, I. A. Rodionov,

K. N. Afanas'ev and A. V. Baryshev for design and fabrication of MPCs. A.M.R. acknowledges the financial support of the Foundation for the advancement of theoretical physics and mathematics BASIS.

References

[1] Shaimanov A, Orlikovsky N, Khabushev E, Zverev A, Pishimova A, Sharonov G, Yankovskii G, Rodionov I and Baryshev A 2018 *Photonics and Nanostructures - Fundamentals and Applications* **32** 1 – 5 ISSN 1569-4410
[2] Maier S A 2007 *Plasmonics: Fundamentals and Applications* (Springer, New York, NY)
[3] Belotelov V, Akimov I, Pohl M, Kotov V, Kasture S, Vengurlekar A, Gopal A V, Yakovlev D, Zvezdin A and Bayer M 2011 *Nature Nanotechnology* **6** 370
[4] Borovkova O V, Hashim H, Kozhaev M A, Dagesyan S A, Chakravarty A, Levy M and Belotelov V I 2018 *Applied Physics Letters* **112** 063101
[5] Caballero B, García-Martín A and Cuevas J C 2016 *ACS Photonics* **3** 203–208
[6] Novikov V, Romashkina A, Ezenkova D, Rodionov I, Afanas'ev K, Baryshev A and Murzina T 2020 *Optics and Spectroscopy* **128** 1481–1486
[7] Chekhov A L, Razdolski I, Kirilyuk A, Rasing T, Stognij A I and Murzina T V 2016 *Phys. Rev. B* **93**(16) 161405
[8] Agarwal G S and Jha S S 1982 *Phys. Rev. B* **26**(2) 482–496
[9] Kauranen M and Zayats A V 2012 *Nature Photonics* **6** 737–748 ISSN 1749-4893
[10] Chen C K, de Castro A R B and Shen Y R 1979 *Opt. Lett.* **4** 393–394
[11] Razdolski I, Gheorghe D G, Melander E, Hjörvarsson B, Patoka P, Kimel A V, Kirilyuk A, Papaioannou E T and Rasing T 2013 *Phys. Rev. B* **88**(7) 075436

Topological edge states of anyon pairs emulated in electrical circuits

A D Rozenblit[1*], N A Olekhno[1], A A Dmitriev[1], P S Seregin[1], and M A Gorlach[1]

[1]ITMO University, 197101 Saint Petersburg, Russia

E-mail: *alina.rozenblit@metalab.ifmo.ru

Abstract. Recent advances in two-particle topological quantum states demonstrate resilience to geometrical imperfections and hold perspectives for robust quantum computations. In this context, particles with fractional quantum statistics, the so-called anyons, attract especial attention. In particular, topological edge states of anyon pairs in one-dimensional chains of coupled cavities were recently predicted to demonstrate localization at one or another edge of the array depending on details of the quantum statistics. In this paper, we propose an equivalent electrical circuit serving as a classical emulator of such topological states. Detailed numerical studies of resonances in the circuit fully support theoretical predictions, pointing towards future experimental realizations of anyonic states analogs in electrical circuits.

1. Introduction

Anyons are quasi-particles with fractional quantum statistics in-between bosons and fermions that exist in one- and two-dimensional systems and were demonstrated experimentally with a discovery of fractional quantum Hall effect [1]. However, the direct observation of this effect requires the use of high-quality semiconductor nanostructures, extremely low temperatures, and high magnetic fields [1, 2], thus being very challenging. On the other hand, anyons offer great potential for fault-tolerant quantum computations [1]. This led to active development of various strategies for their emulation, including photonic systems [3] and flux-tube models [4]. Despite considerable progress in realizing topological states of photon pairs [5] and their emulation [6], topological states of two anyons remain uncharted. Recently, it has been theoretically predicted that one-dimensional (1D) arrays of coupled cavities supporting anyonic excitations with on-site interaction possess topological edge states with localization depending on the quantum statistics [7].

In the present paper, we develop an equivalent electrical circuit allowing for a classical emulation of such two-particle quantum systems, and perform its numerical simulations with the help of Keysight Advanced Design System software.

2. Theoretical model

We consider a 1D array of cavities with nearest-neighbor tunneling couplings J, on-site interaction between two anyons with energy U, and additional two-particle tunneling links P corresponding to the simultaneous tunneling of two anyons between the neighboring cavities, as shown in Fig. 1(a). The system is described by the following Hamiltonian (considering energy units $\hbar = 1$):

$$\hat{H} = \omega_0 \sum_{m=1}^{N} \hat{n}_m + \sum_{m=1}^{N} U_m \hat{n}_m(\hat{n}_m - 1) - J \sum_{m=1}^{N-1} (\hat{a}_m^\dagger \hat{a}_{m+1} + \hat{a}_{m+1}^\dagger \hat{a}_m) + \\ + \frac{P}{2} \sum_{m=1}^{(N-1)/2} (\hat{a}_{2m-1}^\dagger \hat{a}_{2m-1}^\dagger \hat{a}_{2m} \hat{a}_{2m} + \hat{a}_{2m}^\dagger \hat{a}_{2m}^\dagger \hat{a}_{2m-1} \hat{a}_{2m-1}), \quad (1)$$

where N is the number of cavities in the array, ω_0 is the energy of one anyon in a cavity, $\hat{a}_m^\dagger$ and $\hat{a}_m$ are the creation and annihilation operators for the anyon at the cavity m, and $\hat{n}_m = \hat{a}_m^\dagger \hat{a}_m$ is a particle number operator. The first three terms correspond to the Hubbard model with tunneling links J and the on-site interaction energy U_m, while the latter term describing the simultaneous tunneling of a pair of anyons with the amplitude P is similar to the model considered in Ref. [6]. For the bulk sites ($m \neq 1$ and $m \neq N$) and edges ($m = 1$ or $m = N$), the interaction energy U_m equals U and $U + J^2/(2U)$, respectively. As anyons obey fractional statistics, their commutation relations involve the statistical exchange angle $\theta \in [0, \pi]$: $\hat{a}_l \hat{a}_k = \exp(i\theta \operatorname{sgn}(l-k)) \hat{a}_k \hat{a}_l$, $\hat{a}_l \hat{a}_k^\dagger = \delta_{lk} + \exp(-i\theta \operatorname{sgn}(l-k)) \hat{a}_k \hat{a}_l$. The limits of the statistical exchange angle $\theta = 0$ and $\theta = \pi$ correspond to the bosons and pseudo-fermions, respectively. The latter are characterized by the fermionic commutation relations in the case of different sites of the array ($l \neq k$), while at the same site ($l = k$) they feature bosonic commutation relations.

To find the system eigenmodes, we substitute the two-photon wave function in the form

$$|\psi\rangle = \frac{1}{\sqrt{2}} \sum_{m,n=1}^{N} \beta_{mn} \hat{a}_m^\dagger \hat{a}_n^\dagger |0\rangle, \quad (2)$$

to the eigenvalue problem $\hat{H}|\psi\rangle = E|\psi\rangle$, thus obtaining an equivalent two-dimensional (2D) tight-binding (TB) model with nearest-neighbor links J and diagonal links P, which is shown in Fig. 1(b). However, the vertical links next to the diagonal are replaced with the statistical angle-dependent couplings $Je^{i\theta}$, which describe the fractional quantum statistics. The TB eigenenergies ε are defined as $\varepsilon = E - 2\omega_0$. Then, we study such a single-particle problem of higher dimensionality numerically and obtain 2D maps of coefficients $|\beta_{mn}|^2$ characterizing the probability of detecting one anyon at the m-th cavity and the other one at the n-th cavity.

As seen from Fig. 1(c), for $\theta = 0$ there is a bandgap near energies $\varepsilon/J \approx 4$. The two bands above and below the bandgap (Fig. 1(d,f)) correspond to delocalized bound photon pairs, doublons, while the single in-gap mode represents a topological edge state localized at the first cavity, see Fig. 1(e). In the opposite case of $\theta = \pi$ shown in Fig. 1(g), the bands above and below the bandgap still represent doublons (despite the lower branch merges with the continuum of scattering states of anyons), see Fig. 1(h,j). However, the edge state localization pattern (Fig. 1(i)) changes, facilitating the presence of both anyons at the opposite edge of the cavity array.

3. Electrical circuit emulation

To support our theoretical results, we map the TB equations for the Hamiltonian Eq. (1) to a set of Kirchhoff's rules describing voltages and currents in the electrical circuit, following the lines of paper [6]. Then, on-site voltages in the circuit φ_{mn} represent wavefunction amplitudes β_{mn}, and frequencies of resonances f_j correspond to eigenenergies ε_j. The resulting electric circuits shown in Fig. 2(a,c) resemble the TB model (Fig. 1(b)) up to the replacement of tunneling links J and P with the capacitors C_J and C_P, respectively, and the statistical angle-dependent links

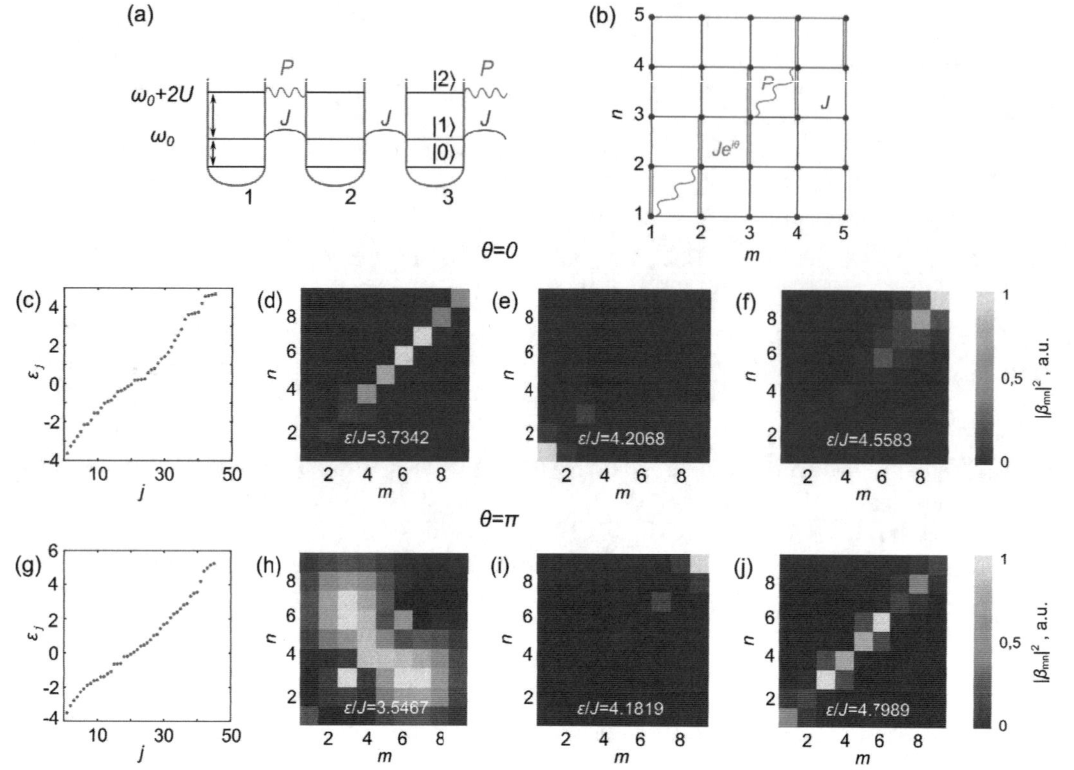

Figure 1. (a) Design of the considered system representing a 1D array of cavities coupled via tunneling links J and two-particle links P. (b) Equivalent 2D tight-binding model with the statistical exchange angle θ. The array of five cavities is considered as an example. Coordinates of the site (m, n) correspond to the location of the first and second anyons in the cavities m and n, respectively. (c) Spectrum of eigenvalues ε_j as a function of the eigenvalue number j for the model with the statistical exchange angle $\theta = 0$ and the size of 9×9 sites. The parameters of the model are $J = 1$, $U = 1.5$, and $P = -0.75$. (d-f) Maps of two-particle wave function amplitudes in the case of the bosonic statistics ($\theta = 0$). (g-j) The same as (c-f), but for the model supporting topological edge states of two pseudo-fermions ($\theta = \pi$).

$Je^{i\theta}$ by the complex impedances $C_\text{J}e^{i\theta}$. All nodes of the circuit are grounded with the inductors L, except for diagonal nodes ($m = n$) featuring the additional capacitors $C_\text{U} + 2C_\text{J}(1 - \cos(\theta))$ corresponding to the on-site interaction. Nodes neighboring to the diagonal ones are additionally grounded by the capacitors $C_\text{g} = 2C_\text{J}$. Edge nodes are also grounded with the capacitors C_J and C_P to compensate for the absent neighbors. The parameters of the circuit and the tight-binding model are related as follows:

$$J = 1, \qquad U = \frac{C_\text{P} + C_\text{U}}{2C_\text{J}}, \qquad P = -\frac{C_\text{P}}{C_\text{J}}, \qquad \varepsilon = \frac{f_0^2}{f^2} - 4, \qquad f_0 = \frac{1}{2\pi\sqrt{LC_\text{J}}}. \tag{3}$$

The impedance spectroscopy for $\theta = 0$ shown in Fig. 2(b) reveals two groups of resonant peaks corresponding to doublons, and a single peak in the bandgap corresponding to the edge state, in accordance with theoretical predictions. In the case of $\theta = \pi$, however, the tunneling links $Je^{i\theta}$ are represented by the negative capacitors $-C_\text{J}$. The latter can be approximated by the inductors L_J having the same absolute value of impedance as the capacitors C_J at the frequency of topological edge state, Fig. 2(c). Simulated spectra for this case, shown in Fig. 2(d),

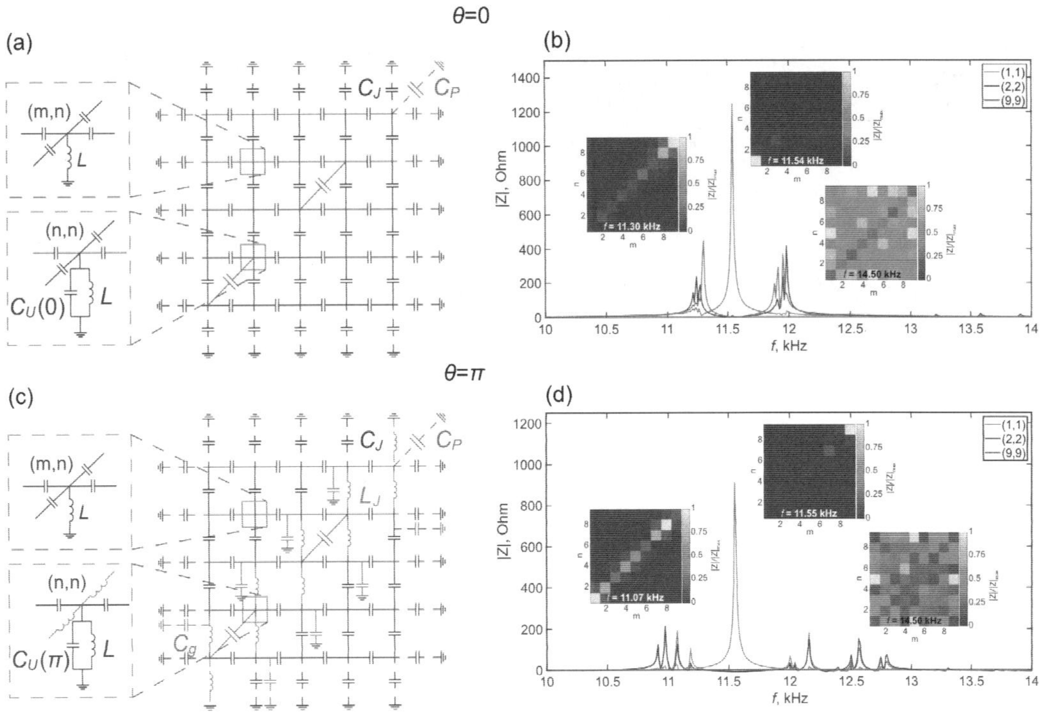

Figure 2. (a) Equivalent electrical circuit for the bosonic statistics $\theta = 0$. Capacitors C_{J} and C_{P} corresponding to the tunneling couplings J and P are shown in blue and red, respectively. (b) Numerically simulated spectra of the absolute value of impedance for the sites with coordinates $(1, 1)$, $(2, 2)$ and $(9, 9)$ in the circuit of Panel (a) with 9×9 nodes. Insets show normalized impedance maps at the indicated frequencies. (c) The same as (a), but for the pseudo-fermionic statistics $\theta = \pi$. Inductors L_{J}, replacing the appropriate capacitors C_{J}, are shown in green. (d) The same as (b), but for the circuit with $\theta = \pi$. Grounding capacitors C_{g} for the nodes neighboring to the diagonal ones are shown in pink.

again demonstrate the presence of two doublon bands and the in-gap edge state. Strikingly, this state localizes at the opposite corner of the circuit, corresponding to the localization of anyon pair at the opposite edge of the cavity array and facilitating the statistics-induced topological transition.

4. Conclusion

We have developed a rigorous electric circuit emulator of statistics-induced topological edge states of interacting anyons predicted previously [7]. Numerical simulations of electrical circuits fully support our theoretical findings, demonstrating the emergence of topological edge states.

Acknowledgments

This work was supported by the Russian Science Foundation (grant No. 20-72-10065). N.O. acknowledges partial support by the Foundation for the Advancement of Theoretical Physics and Mathematics "Basis".

References

[1] Nayak C, Simon S H, Stern Ady, Freedman M, and Das Sarma S 2008 *Rev. Mod. Phys.* **80**, 1083–1159

[2] Nakamura J, Liang S, Gardner G, and Manfra M 2020 *Nature Phys.* **16**, 931–936
[3] Longhi S, and Della Valle G 2012 *Opt. Lett.* **37**, 2160–2162
[4] Todorić M, Jukić D, Radić D, Soljačić M, and Buljan H 2018 *Phys. Rev. Lett.* **120**, 267201
[5] Blanco-Redondo A, Bell B, Oren D, Eggleton B J and Segev M 2018 *Science* **362** 568–571
[6] Olekhno N A, *et al.* 2020 *Nat. Commun.* **11**, 1436
[7] Olekhno N, Rozenblit A, Seregin P, Gorlach M 2020 *AIP Conference Proceedings* **2300**, 020093

A Facile Route of Manufacturing of Silicon-Based Nanostructures with Tuned Plasmonic Properties

Miroslava Flimelova[1, a)] **and Yury V Ryabchikov**[1,2, b)]

[1] HiLASE Centre, Scientific Laser Application Department,
Institute of Physics of the Czech Academy of Sciences, Dolni Brezany Czech Republic
[2] Department of Solid State Physics,
P.N. Lebedev Physical Institute of the Russian Academy of Sciences, Moscow, Russia

[a)] flimelova@fzu.cz, [b)] yury.ryabchikov@hilase.cz

Abstract. An environment-friendly method of pulsed laser ablation in liquids is successfully employed for structural modification of silicon nanoparticles leading to a considerable narrowing of their size distribution accompanied with a reduction of the mean size. Contamination-free conditions of synthesis ensure the chemical purity of formed nanostructures that may reduce toxicity issues. Such a laser-induced modification leads to the appearance of plasmonic properties in semiconductor-based nanomaterials. Their spectral position can easily be varied in the whole visible range. Combined in one nanoparticle properties of semiconductors and noble metals can strongly promote applications of composite laser-synthesized nanoparticles for biosensing (using their plasmonic-based surface-enhanced ability) and bioimaging (using their both optical and magnetic abilities) purposes.

1. Introduction

Semiconductor nanomaterials possess unique both linear and nonlinear optical properties that can result to various applications [1-7]. However, it is known that semiconductor nanostructures are characterized by the absence of any plasmonic features that are attributed to noble metal nanomaterials [8]. Such an issue can be overcome by formation of composite nanostructures containing both semiconductor and metal elements [9]. Among different methods of nanoparticle manufacturing, the most promising and facile one is a technique of environmentally-friendly pulsed laser ablation in liquids that provides contamination-free conditions of nanomaterial synthesis [10]. As it has been found lately, this approach is also an effective route of production of ligand-free composite semiconductor-based nanostructures with embedded metal impurities [11-13]. Nevertheless, this type of research is still very challenging being on its early stage requiring many additional studies.

In this work, different types of composite silicon-based nanoparticles with plasmonic properties were formed using femtosecond (fs) laser ablation. For this purpose, silicon wafers of different levels of purity were used for formation of silicon nanoparticle colloidal solutions. Prepared colloids were employed for further laser-induced structural modification by the ablation of various noble metal targets. Afterwards, structural, optical and electronic properties of prepared plasmonic nanocomposites were studied and their possible bioapplications are assessed.

2. Materials and methods

The preparation method is based on a fs laser source emitting light in the infrared spectral range. Its irradiation was focused by a galvanoscanner (162 mm focal distance) into a spot of 50 μm size (experimentally estimated). It also moved the laser beam by a surface of semiconductor and metallic wafers embedded in the deionized water (18 MΩ·cm, 8 mm water level) with 2 m/s velocity. In all experiments, the targets were exposed by the laser irradiation (10 kHz repetition rate and 150 μJ/pulse laser fluence) for 15 minutes. Commercially available wafers of "pure" and "metallurgical-grade" [14] silicon as well as of highly pure noble metals were used. Size distributions and chemical contents of composite nanoparticles were investigated using transmission/scanning electron microscope (TEM/SEM) equipped with energy dispersive X-ray (EDX) spectrometer. Structural properties were examined by means of X-ray methods such as X-ray diffraction (XRD) and X-ray photoemission spectroscopy (XPS). Optical properties were studied using a UV-VIS, Raman and FTIR spectroscopies. An electron paramagnetic resonance (EPR) method was employed to investigate their paramagnetic defect labels with unpaired electrons.

3. Results and discussion

In order to extend functionalities of semiconductor nanomaterials towards plasmonic-enhanced-based applications, a method of pulsed laser ablation was employed for their structural modifications. In this case, it was found that a laser impact on metal targets immersed in a colloidal solution of semiconductor nanoparticles provokes considerable changes of their properties. A typical TEM image of composite silicon-based nanoparticles is shown in Figure 1a. By turn, this treatment also leads to a significant variation of their chemical composition influencing structural and optoelectronic features of nanostructures [11]. In particular, considerable reduction of size distribution of composite semiconductor-metal nanoparticles as compared to initial ones is occurred [11]. One can see that a mean size of silicon-gold nanoparticles is approximately 6 nm with a narrow size dispersion (Figure 1a, inset). The presence of various content in nanoparticles is confirmed by EDX studies (Figure 1b) revealing several maxima corresponding to different elements. Their relative intensities determine ratios between these elements. In this case, signals from the copper can be neglected arising from a used TEM grid. It worth noting that in order to study Si-Cu NPs, nickel or gold grids were used. The responses attributed to gold and silicon justify the presence of both species in a nanoparticle reflecting a mass contribution of the corresponding elements. The aspect ratio between them strongly depends on used experimental conditions [11].

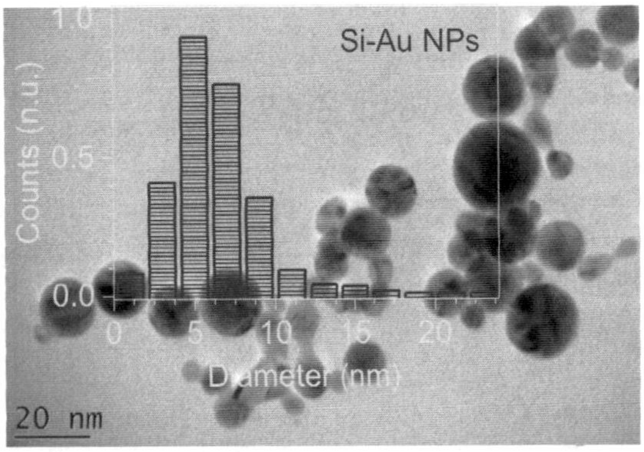

(a)

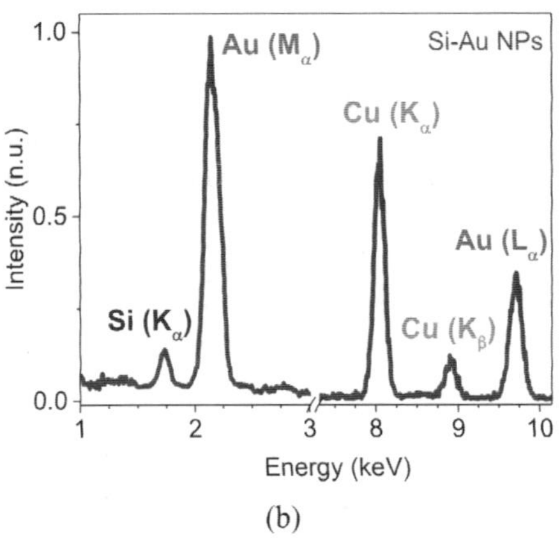

(b)

Figure 1. a) A transmission electron microscope image of composite nanoparticles formed by laser ablation of a gold target immersed in a colloidal solution of bare silicon nanoparticles. A corresponding size distribution is shown in inset. b) An energy-dispersive X-ray spectrum of composite silicon-gold nanoparticles.

The formation of composite nanoparticles can be caused due to an interaction between metal nanoclusters emitted from a target and silicon ones appeared due to laser-induced decomposition of silicon nanoparticles [15]. Possible mechanisms of the nanocomposite synthesis are described in more details in the work [12]. The presence of different types of nanoclusters in the laser plume leads not only to a significant reduction of their mean size but also to a change of the chemical composition of newly formed nanoparticles (Figure 1b).

Such a modification of the structure of nanoparticles provokes remarkable changes of their optoelectronic properties (Figure 2a and 2b) [10]. First of all, one can find significant alteration of optical transmission spectra caused by laser-induced actions on various metal targets in the presence of silicon nanoparticles (Figure 2a). In these cases, strong minima are appeared in the spectra of composite nanoparticles significantly depending on a used metal (Figure 2a). Indeed, each element has its own plasmonic frequency that allows us to tune spectral position in overall visible range (400 nm for Si-Ag NPs; 522 nm for Si-Au NPs; 613 nm for Si-Cu NPs). Recent studies show that laser-synthesized silicon-gold nanocomposites are promising for bacteria detection [16] using surface-enhanced Raman scattering (SERS) as well as for the identification of spoilage in the food industry [17] due to their significant plasmonic properties.

At the same time, composite silicon-gold nanoparticles reveal remarkable paramagnetic properties due to a great number of paramagnetic defects with unpaired electrons that can be detected using EPR spectroscopy. Figure 2b represents a typical experimentally observed EPR spectrum (derivative) of these kinds of nanoparticles. A performed calculation gives a following value of the defect states (in comparison with an etalon sample [12]) – $0.5 \cdot 10^{19}$ spin/g that is less than that of the initial bare silicon nanoparticles. Proposed mechanisms of these changes can be associated with the interaction between silicon and gold nanoclusters in the laser-induced plasma plume leading to charge redistribution. Additionally, possible strong localization of unpaired electrons may provoke fast dissipation of energy of the excited defects. A calculated value of g-factor (2.0055) corresponds to dangling bonds in amorphous silicon [12].

Thus, laser-synthesized composite nanoparticles combine in one nanoparticle modalities of semiconductor (silicon) and metallic (gold, silver, copper) nanostructures ensuring possibility to solve simultaneously several tasks in various applications. One of the most important one is related to nanomedicine where such nanoparticles can be served for bioimaging, biosensing, hyperthermia etc. at the same time using specific experimental conditions.

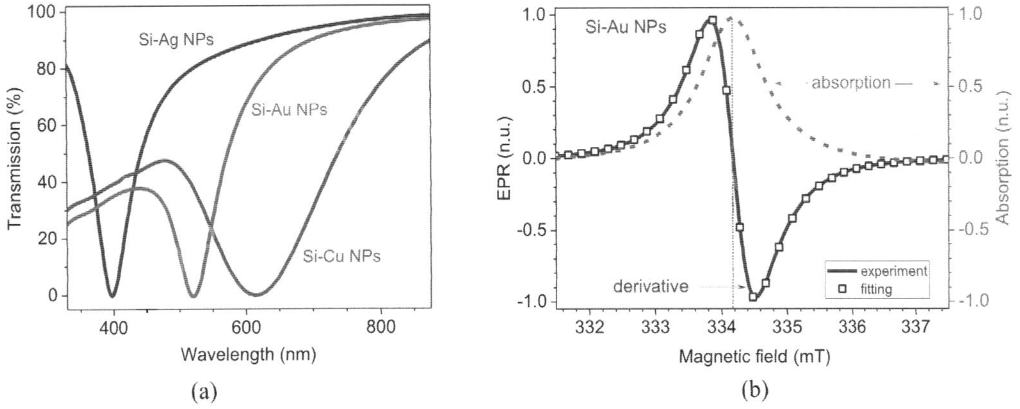

Figure 2. a) Optical transmission spectra of composite nanoparticles formed by laser ablation of the metallic targets (silver, gold, copper) immersed in a colloidal solution of bare silicon nanoparticles. b) Electron paramagnetic resonance response of composite nanoparticles formed by laser ablation of a gold target immersed in a colloidal solution of bare silicon nanoparticles: an experimental derivative curve (open points), its computer simulation (lines) and an absorption curve (closed points).

4. Conclusions

Pulsed laser ablation in liquids technique is shown to be a very effective and promising versatile route to synthesize composite nanostructures with various chemical content and different plasmonic and magnetic resonance properties. Laser-induced structural modification of silicon nanoparticles leads to a strong change of their size distribution and chemical content that can be controlled by experimental conditions. A combination of semiconductor and metallic materials in form of one nanoparticle provokes the appearance of considerable plasmonic features in silicon nanostructures with paramagnetic labels that can potentially be employed for biosensing and bioimaging (including magnetic resonance imaging) applications at the same time. The obtained results show high further perspectives of laser-synthesized composite nanostructures for solving multipurpose tasks in biomedical applications.

5. Acknowledgements

This project has received funding from the European Union's Horizon 2020 research and innovation programme under the Marie Skłodowska-Curie grant agreement No 897231 (LADENTHER). Authors also thank the European Regional Development Fund and the state budget of the Czech Republic (Project BIATRI: CZ.02.1.01/0.0/0.0/15_003/0000445), the Ministry of Education, Youth and Sports (Programs NPU I-Project no. LO1602). Yu.R. also acknowledges Freie Universität Berlin for support within the Excellence Initiative of the German Research Foundation (0503121810) as well as a COST project (ECOST-STSM-BM1205-120416-072252) and personally Prof. J. Behrends.

References

[1] Konstantinova E A, Demin V A, Vorontzov A S, Ryabchikov Yu V, Belogorokhov I A, Osminkina L A, Forsh P A, Kashkarov P K, Timoshenko V Yu 2006 J. Non–Cryst. Solids 352 1156–9
[2] Ryabchikov Yu V, Lysenko V, Nychyporuk T 2014 *J. Phys. Chem. C* **118** 12515–9
[3] Al–Kattan A, Nirwan V P, Popov A, Ryabchikov Yu V, Tselikov G, Sentis M, Fahmi A, Kabashin A V 2018 *Int. J. Mol. Sci.* **19(6)** 1563
[4] Belogorokhov I A, Ryabchikov Yu V, Tikhonov E V, Pushkarev V E, Breusova M O, Tomilova L G, and Khokhlov D R 2008 *Semicond.* **42(3)** 321–4
[5] Ryabchikov Yu V, Belogorokhov I A, Vorontsov A S, Osminkina L A, Timoshenko V Yu, Kashkarov P K 2007 *Phys. Status Solidi A* **204(5)** 1271–5
[6] Ryabchikov Yu V, Belogorokhov I A, Gongalskiy M B, Osminkina L A, Timoshenko V Yu 2011 *Semiconductors* **45(8)** 1059–63
[7] Ming T, Turishchev S, Schleusener A, Parinova E, Koyuda D, Chuvenkova O, Schulz M, Dietzek B, Sivakov V 2021 *Small* **17(8)** 2007650
[8] Lu X, Rycenga M, Skrabalak S E, Wiley B and Xia Y 2009 *Annu. Rev. Phys. Chem.* **60** 167–192
[9] Cortie M B, McDonagh A M 2011 *Chem. Rev.* **111** 3713–35
[10] Zhang B, Gökce S, Barcikowski S 2017 *Chem. Rev.* **117** 3990–410
[11] Ryabchikov Yu V 2019 *J. Nanopart. Res.* **21(4)** 85
[12] Ryabchikov Yu V, Behrends J 2020 *Opt. Quant. Electron.* **52** 117
[13] Kutrovskaya S, Gerke M, Osipov A, Shagurina A, Bukharov D, Kucherik A 2019 *J. Phys. Conf. Ser.* **1164(1)** 012012
[14] Pastushenko A, Lysenko V 2016 *RSC Adv.* **6(10)** 8093–4
[15] Ryabchikov Yu V 2019 *Phys. Status Solidi* A **216(2)** A1800685
[16] Kögler M, Ryabchikov Yu V, Uusitalo S, Popov A, Popov A, Tselikov G, Välimaa A L, Al–Kattan A, Hiltunen J, Laitinen R, Neubauer P, Meglinski I, Kabashin A B 2018 *J. Biophotonics* **11(7)** e201700225
[17] Uusitalo S, Popov A, Ryabchikov Yu V, Bibikova O, Alakomi H L, Juvonen R, Kontturi V, Siitonen S, Kabashin A, Meglinski I, Hiltunen J, Laitila A 2017 *J. Food Eng.* **212**, 47–54

Journal of Physics: Conference Series **2015** (2021) 012129 doi:10.1088/1742-6596/2015/1/012129

Optical heating of doped semiconductor nanocylinders supporting quasi-BIC modes

Daniil Ryabov, Olesia Pashina, George Zograf, Sergey Makarov and Mihail Petrov

The Department of Physics and Engineering ITMO University, Saint-Petersburg, Russia.

E-mail: daniil.ryabov@metalab.ifmo.ru

Abstract. In the recent years, semiconductor and dielectric nanophotonic structures attracted a lot of attention for their resonant optical properties finding applications in thermal tuning and optical heating. Exciting high quality optical modes of both electric and magnetic nature in nanoresonators of high-index materials, one can effectively enhance optical absorption in such structures. Another big advantage of semiconductor materials is the ability to finely control the level of optical losses in visible and near infrared (near-IR) range through varying the doping level. In this work, we show theoretically that by moderate carrier doping of silicon via donors from group V materials one can achieve effective heating of nanoresonators. We show that by tuning the doping level of crystalline silicon supporting high quality non-radiative modes based on quasi bound states in the continuum one can achieve strong heating in near-IR under continuous wave regime illumination. We believe that our finding will pave the way for an efficient semiconductor near-IR all-optical sensors and nanoheaters.

1. Introduction

The implementation of the concept of bound-states in the continuum (BIC) to photonics [1] attracted a lot of attention to a number of effects including harmonic generation [2, 3], lasing action [4] and many others. Although BIC requires infinite periodic structure for field localization, it was shown that single nanoresonators can support quasi-BIC mode through destructive interference of radiation in the far-field, manifested by the unprecedented increase of the quality factor of the optical mode [5]. Such prominent quality factor and field localization demonstrated record performance for SHG [6, 7], smallest optically-pumped laser [8, 9], third-harmonic generation [10], the ability for fine-tuning of the scattering response [11] and many others. Such tight field localization inside a single nanoobject is very profitable for efficient light absorption, however many semiconductor materials posses ultralow losses in visible and near-infrared range which prevents them from being utilized for optical heating applications[12]. This comes into sharp contrast with thermoplasmonics where the ohmic losses are inevitable. In this work we show theoretically that moderate doping of crystalline silicon with donor concentration of $n_d = 7.6 \cdot 10^{18}\,\mathrm{cm}^{-3}$ leads to significant increase of heating efficiency in nanocylinders supporting quasi-BIC resonance. We provide engineering of the optimal geometry and material parameters for achieving maximal optical heating in the near-IR optical range which is very attractive for bio-related photothermal applications and thermosensing.

2. Results

We start our study from consideration of the eigenmodes of a single cylindrical nanoresonator made of crystalline silicon. At the first stage we neglect the non-radiative losses which can be extremely low in undoped silicon in near-IR. The radiative losses which are inevitably present in any open resonant system can be significantly reduced by engineering the resonant configurations of the nanoparticles which results in high radiative Q-factors of the investigated geometries. The quasi-BIC modes are formed of strongly coupled modes of a cylinder providing the destructive interference of light in the far-field zone and, hence, better localization of light inside of the particle [13, 6]. Such a regime allows for significant increase of the modes' Q-factor, and opens a way for efficient heating of the particles with relatively low intensities of the incident field.

The high-Q quasi-BIC modes appeared in the nanodisk geometries posses cylindrical symmetry with zero azimuthal number $m = 0$ with respect to cylinder axis [6, 14, 15]. Figure 1 (a) shows the spectral positins and Q-factors of modes of a cylinder on top of glass substrate with $m = 0$. Red circle region highlights the anticrossing are where two modes of the same symmetry come into strong coupling regime (these modes branches are marked with thin black lines for the convenience).

High-Q mode branch in this region reveals the quasi-BIC configurations which are marked with red color dots. For the wavelength of 1400 nm declared quasi-BIC mode corresponds to the cylinder with radius $\rho = 619$ nm and height $h = 885$ nm. As shown in Fig.1(b) these mode can be accessed through a azimuthally polarized vector beam matching the symmetry of the mode.

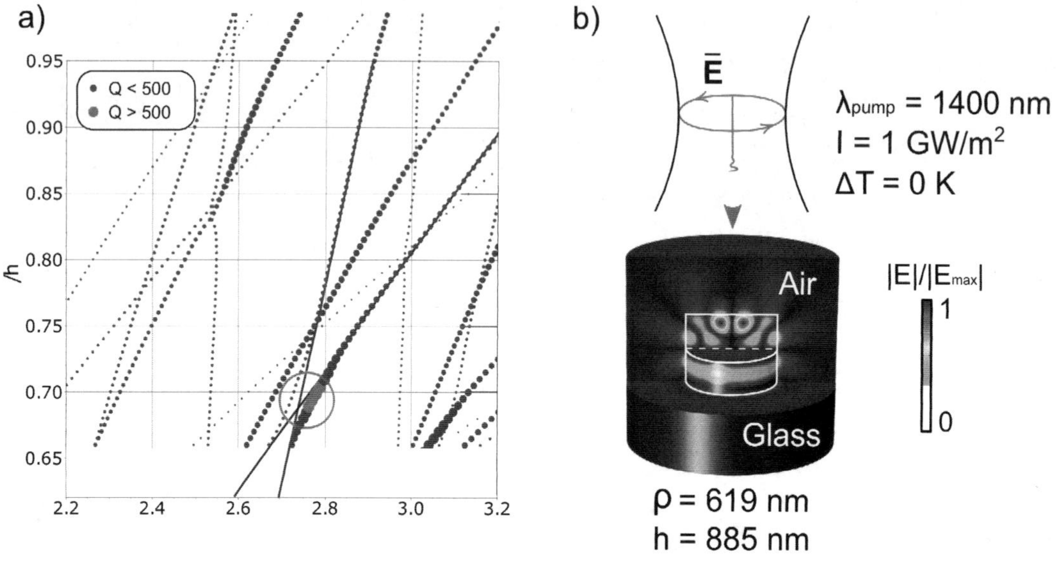

Figure 1. a) The Q-factor map of the cylinder eigenmodes depending on the dimensionless wavevector ($\kappa\rho = 2\pi\rho/\lambda$) and the ratio of the radius to the particle height. The size of the dots is proportional to the value of Q-factor. Points with a Q- factor greater than 500 are marked with red dots. The crossing thin black lines are shown for the convenience to highlight the anticrossing region. b) Quasi-BIC mode electric field distribution for the declared configuration of the nanocylinder (the field is shown in the slice of the resonator). Polarization of the excitation field (cylindrical vector beam with azimuthal polarization $m = 0$) with the intensity I = 1 GW/m^2 and incident wavelength of $\lambda_{\text{pump}} = 1400$ nm is shown schematically above the structure. The optical losses in silicon in near-IR range are close to zero, hence the heating of the particle is negligible (ΔT = 0K).

At the next step, we simulate the heating of the above-mentioned nanostructures supporting high-Q quasi-BIC mode with the commercial software COMSOL Multiphysics by direct solving of the heat conduction equation for the stationary case:

$$\nabla \cdot (\kappa(\mathbf{r})\nabla T(\mathbf{r})) + Q(\mathbf{r}) = 0. \tag{1}$$

Here, κ is the material coefficient of thermal conductivity; $\mathbf{r}$ is the spatial coordinate and T is the local temperature of the cylinder. In our calculations, the source of heating $Q(\mathbf{r})$ corresponds to electromagnetic power absorbed by the resonant structure from azimuthally polarized beam. Optical losses in silicon in the infrared range are close to zero, which results in almost negligible heating of the considered structure at near-IR. At the same time, heating efficiency can be increased by doping silicon with the donors from the group V materials (P, As, Sb), which results in a corresponding change in the carriers concentration in the conduction band. Since the typical ionization energy of donors for silicon is much less than the thermal energy at room temperature $E_d \ll kT$, we assume that all donors are ionized at 300 K. Thus, the concentration of charge carriers is completely determined by the concentration of the dopant n_d (see the inset of Fig. 2(a)). The temperature increase during heating is balanced by Ohmic and radiative losses [16]:

$$P_{\text{abs}} \sim \frac{\gamma_{ohm}}{(\gamma_{ohm} + \gamma_{rad})^2}. \tag{2}$$

In this formula γ_{ohm} and γ_{rad} are the material (Ohmic) and radiative losses respectively. In the case of fixed radiation losses, the maximum of this function is reached when the radiative and non-radiative losses are equal. Since we have already optimized the radiative losses in quasi-BIC regime, the maximum heating could be achieved with relatively low values of optical losses. We explicitly estimate the optimal value by simulating the temperature increase inside the nanostructure and obtain a maximal heating corresponding to the imaginary part of the complex refractive index k = 0.0031 (Fig. 2(a)). For further analysis of the quasi-BIC heating properties, we present the temperature increase dependence on the radius of the excited particles with the fixed height of 885 nm and optimal value of material losses (Fig. 2(b)). This graph confirms that resonant excitation of a high-Q mode leads to a drastic enhancement of the heating temperature reaching the value of 800 K for the intensity of 1 GW/m^2.

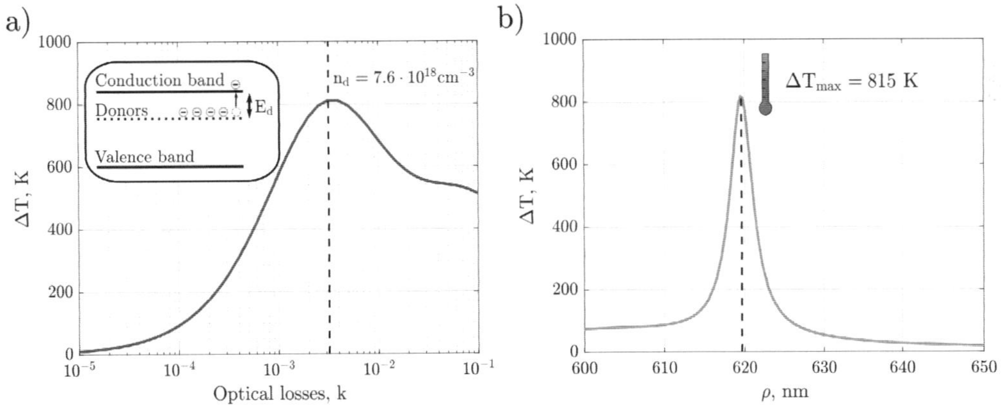

Figure 2. a) Optimization of quasi-BIC nanocylinder heating by adding optical losses k to the silicon. The inset shows the diagram of energy levels in a doped semiconductor and thermal transfer of the electrons from the donor level to the conduction band ($E_d \ll kT$). b) Temperature increase dependence of a doped nanocylinder on its radius at pump intensity of 1 GW/m^2. The optimal value shows by dotted line and corresponds to the previously calculated quasi-BIC mode.

Finally, we estimate the donors concentration necessary to achieve the optimal value of optical losses in silicon. In accordance with Drude model the imaginary part of dielectric permittivity change could be expressed as:

$$\Delta \text{Im}\varepsilon(n_d) = \frac{{\omega_p}^2(n_d)\tau_e}{\omega(1+\omega^2\tau_e^2)} = 2\tilde{n}k, \tag{3}$$

where ω is the incident light frequency; $\tau_e = 1$ fs - electron relaxation time associated with the electron-electron interaction; $\omega_p(n_d) = (n_d e^2/\varepsilon_0 m_{\text{eff}})^{1/2}$; $m_{\text{eff}} = 0.26 m_e$ is the effective mass of electons in the conduction band of c-Si; ε_0 is the permittivity of vacuum and e is the elemantary charge. Supposing that the real part of the refractive index $\tilde{n}$ does not change significantly with the free carrier concentration increase, we immediately obtain donors concentration $n_d = 7.6 \cdot 10^{18}$ cm^{-3} corresponding to the optimal value of $k = 0.0031$.

3. Conclusion

We theoretically investigated the strong optical heating of silicon cylindrical nanoparticles. It resulted from the efficient coupling of the incident azimuthally polarized laser beam with the quasi-BIC mode and optimization of material optical losses via free carriers doping. The described approach provides both high quality factor due to the tight localization of light in structures supporting quasi-BIC modes and non-zero losses necessary for effective thermooptic interaction between the laser radiation and the responding particles. We believe that our proposal will contribute to modern research on all-dielectric thermonanophotonics.

4. Acknowledgments
The work was supported by the Russian Science Foundation (project No. 18-72-10140).

References
[1] Hsu C W, Zhen B, Stone A D, Joannopoulos J D and Soljačić M 2016 Nature Reviews Materials **1** 1–13

[2] Hu H G, Yang M, Yue P, Bai Y T, Wang W J and Liu S D 2020 J. Phys. D: Appl. Phys. **53** 215101 ISSN 0022-3727

[3] Sautter J, Xu L, Miroshnichenko A E, Lysevych M, Volkovskaya I, Smirnova D A, Camacho-Morales R, Zangeneh Kamali K, Karouta F, Vora K, Tan H H, Kauranen M, Staude I, Jagadish C, Neshev D N and Rahmani M 2019 Nano Lett. **19** 3905–3911 ISSN 1530-6984

[4] Kodigala A, Lepetit T, Gu Q, Bahari B, Fainman Y and Kanté B 2017 Nature **541** 196–199

[5] Rybin M V, Koshelev K L, Sadrieva Z F, Samusev K B, Bogdanov A A, Limonov M F and Kivshar Y S 2017 Phys. Rev. Lett. **119**(24) 243901 URL `https://link.aps.org/doi/10.1103/PhysRevLett.119.243901`

[6] Koshelev K, Kruk S, Melik-Gaykazyan E, Choi J H, Bogdanov A, Park H G and Kivshar Y 2020 Science **367** 288–292 ISSN 0036-8075

[7] Carletti L, Koshelev K, De Angelis C and Kivshar Y 2018 Phys. Rev. Lett. **121**(3) 033903 URL `https://link.aps.org/doi/10.1103/PhysRevLett.121.033903`

[8] Tiguntseva E, Koshelev K, Furasova A, Tonkaev P, Mikhailovskii V, Ushakova E V, Baranov D G, Shegai T, Zakhidov A A, Kivshar Y et al. 2020 ACS nano **14** 8149–8156

[9] Mylnikov V, Ha S T, Pan Z, Valuckas V, Paniagua-Domínguez R, Demir H V and Kuznetsov A I 2020 ACS nano **14** 7338–7346

[10] Koshelev K, Tang Y, Li K, Choi D Y, Li G and Kivshar Y 2019 ACS Photonics **6** 1639–1644

[11] Melik-Gaykazyan E, Koshelev K, Choi J H, Kruk S S, Bogdanov A, Park H G and Kivshar Y 2021 Nano Letters **21** 1765–1771

[12] Zograf G P, Petrov M I, Makarov S V and Kivshar Y S 2021 1–92 (*Preprint* `2104.01964`) URL `http://arxiv.org/abs/2104.01964`

[13] Bogdanov A A, Koshelev K L, Kapitanova P V, Rybin M V, Gladyshev S A, Sadrieva Z F, Samusev K B, Kivshar Y S and Limonov M F 2019 Adv. Photonics **1** 016001 ISSN 2577-5421

[14] Sadrieva Z, Frizyuk K, Petrov M, Kivshar Y and Bogdanov A 2019 Physical Review B **100** 1–12 ISSN 24699969 (*Preprint* `1903.00309`)

[15] Gladyshev S, Frizyuk K and Bogdanov A 2020 Physical Review B **102** 75103 ISSN 24699969 URL `https://doi.org/10.1103/PhysRevB.102.075103`

[16] Zograf G P, Petrov M I, Zuev D A, Dmitriev P A, Milichko V A, Makarov S V and Belov P A 2017 Nano Lett. **17** 2945–2952 ISSN 1530-6984

Tetra-axial metamaterial

D. Sakhno[1], **E. Koreshin**[1], **P. Belov**[1]
[1]ITMO University, Kronverksky pr. 49, 197101, St. Petersburg, Russia
E-mail: `denis.sakhno@metalab.ifmo.ru`

Abstract. In this work we theoretically and numerically demonstrate that triple non-connected wire medium has four optical axes. The axes coinside with diagonals of the unit cell.

1. Introduction

In the general case, any homogeneous local dielectric medium can be described by symmetric effective permittivity tensor which can be diagonalized in some coordinate system. The principal elements ε_{ii} $(i = x, y, z)$ of the diagonal tensor are called principal permittivities [2] and the relations between them determine the shape of dispersion surfaces of the medium. There are three types of dielectric media ($\varepsilon_{ii} > 0$): 1) isotropic media ($\varepsilon_{xx} = \varepsilon_{yy} = \varepsilon_{zz}$) with isofrequency surfaces in the form of a sphere; 2) uniaxial media ($\varepsilon_{ii} \neq \varepsilon_{jj} = \varepsilon_{ll}$, for some $i \neq j \neq l$) with spherical isofrequency surface for ordinary waves and ellipsoid of revolution isofrequency surface for extraordinary waves which touch each other in direction of optical axis directed along i-axis; 3) biaxial media ($\varepsilon_{xx} \neq \varepsilon_{yy} \neq \varepsilon_{zz}$) with complex-shaped isofrequency surfaces with two axes in the plane il such that $\varepsilon_{ii} > \varepsilon_{jj} > \varepsilon_{ll}$ [3]. The presence of optical axes in biaxial media results in the effect of conical refraction which finds many applications in optics [4]. If not all principal permittivities of an uniaxial medium are positive then the extraordinary waves in the medium have hyperbolic isofrequency contours and such medium is called hyperbolic metamaterial [5]. The biaxial hyperbolic media with $\varepsilon_{ii} < 0, 0 < \varepsilon_{jj} \neq \varepsilon_{ll}$, for some $i \neq j \neq l$ are studied in [6]. Note, that since permittivity tensor elements depend on frequency, for some types of biaxial crystals the directions of main axes depend on the frequency. This effect is called the dispersion of the axes [2].

In the classification presented above the media are assumed to be local and no effects of spatial dispersion are taken into account. In the case of spatially dispersive materials, the nonlocal effects in the media may strongly influence on its dispersion properties. For example, in the crystals with cubic symmetry the spatial dispersion is destroying isotropy and leads to the effect called spatial-dispersion-induced birefringence [7, 8]. The shape of dispersion surfaces in nonlocal media is governed by the dependence of permittivity tensor $\varepsilon(\omega, \mathbf{k})$ on wave vector.

In this paper we study triple non-connected wire medium [1] and demonstrate that it has four optical axes.

2. Main part

In this work, we will consider a triple non-connected wire medium [1] with a unit cell shown in Fig. 1(a). The metamaterial consists of three two-dimensional arrays of parallel infinite straight

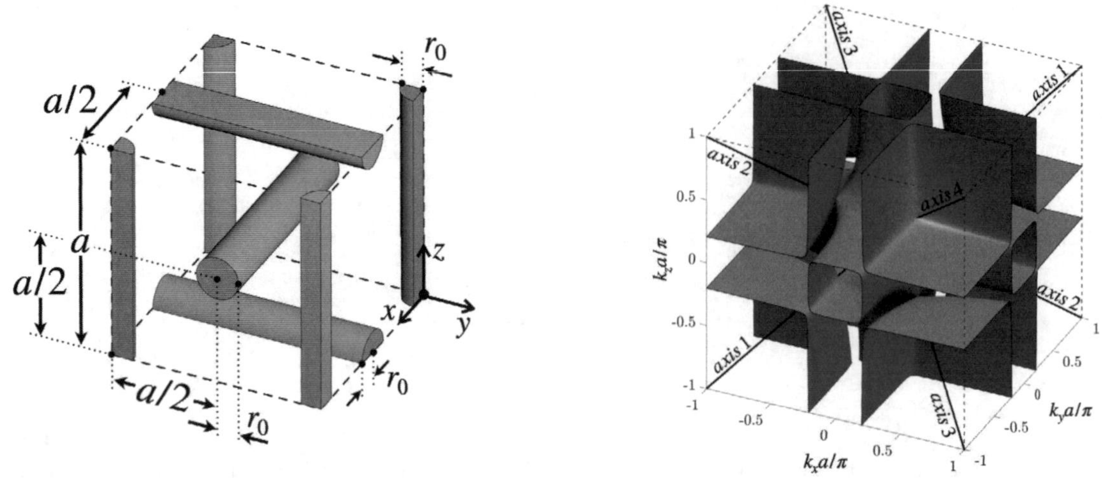

(a) (b)

Figure 1. (a) A single cubic unit cell of the structure with designation of all sizes: a – unit cell size, r_0 – wires radii (b) Isofrequency surface in k-space, according to the theory [1].

metal wires (along x, along y, or along z). In each of the arrays, the axes of the wires form a square lattice in a plane perpendicular to their direction. We will analyze a configuration of the medium in which the wires have the same radii r, the periodicity in all directions is the same and equal to a, and the distances between axes of nearest perpendicular wires is equal to half of the period $a/2$ (see Fig. 1(a)). Mathematically, the geometry can be described by defining coordinates of the wire axes by the following way [1]:

(i) the x-directed wires: $y = an + a/2$ and $z = al + a/2$,
(ii) the y-directed wires: $x = am + a/2$ and $z = al$,
(iii) the z-directed wires: $x = am$ and $y = an$,

The triple wire medium is a medium with strong spatial dispersion at low frequencies. Earlier it was shown that the effects of spatial dispersion (in quasi-static case, $ka << \pi$ and $k_0 a << \pi$) in this material are described by the permittivity tensor [1]:

$$\varepsilon_{ii} = 1 - \frac{k_p^2}{k_0^2 - k_i^2}, \quad i, \in \{x, y, z\}; \quad k_p^2 = \frac{2\pi/a^2}{\ln(a/2\pi r_0) + \pi/6} \tag{1}$$

where $\vec{k} = (k_x, k_y, k_z)^T$ is the wave vector in the medium of wires, k_0 is the wave vector in the host medium, k_p is the wave vector corresponding to the effective plasma frequency of the wire medium.

For an arbitrary anisotropic medium, the dispersion equation can be written in the following form [1, 2]:

$$\left(k_y^2 + k_z^2 - k_0^2\varepsilon_{xx}\right)\left(k_x^2 + k_z^2 - k_0^2\varepsilon_{yy}\right)\left(k_x^2 + k_y^2 - k_0^2\varepsilon_{zz}\right) - \\ - \left(k_y^2 + k_z^2 - k_0^2\varepsilon_{xx}\right)k_y^2 k_z^2 - \left(k_x^2 + k_z^2 - k_0^2\varepsilon_{yy}\right)k_x^2 k_z^2 - \left(k_x^2 + k_y^2 - k_0^2\varepsilon_{zz}\right)k_x^2 k_y^2 - 2k_x^2 k_y^2 k_z^2 = 0. \tag{2}$$

Now we can substitute the components of the tensor $\overline{\overline{\varepsilon}}$ from Eq. 1 into the dispersion equation (Eq. 2) and calculate the isofrequency surfaces in the space of wave vectors. In the previous work [1], the hyperbolic isofrequency contours in the $k_z = 0$ plane were mainly considered and the complete picture of the dispersion of the metamaterial under consideration was not discussed.

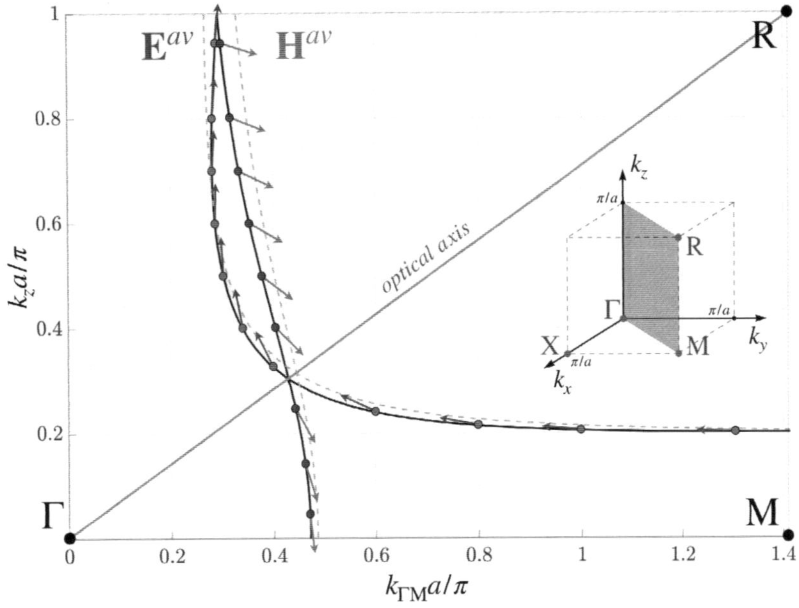

Figure 2. Isofrequency contour for $\omega a/2\pi c = 0.1$ in ΓMR section of the Brillouine zone (shown on the right side of the plot). Black lines are numerically confirmed isofrequency contours, pink dashed curves are determined by the Eq. 2, blue and red arrows show the direction of $\mathbf{E}^{av}$ and $\mathbf{H}^{av}$ averaged over the unit cell, diagonal ΓR gray line is an optical axis.

The figure 1(b) shows an isofrequency surface for the triple wire medium at $\omega a/2\pi c = 0.1$. It can be noticed that this surface has a rather unusual: namely, it has eight points of self-intersection (one in each of the eighth parts of the zone). These points are located at the on the main diagonals of the cubic Brillouin zone. This observation allows us to manifest the presence of four optical axes for the the wire medium.

We have performed a series of numerical simulations using commercial software package CST MWS to compute isofrequency surfaces for a triple wire medium (applying the periodic boundary conditions to the unit cell and using the eigenmode solver) in the ΓMR plane (plane $x = y$ in Fig. 1(b); according to the theory one of four optical axes lies in this plane) and to determine the polarization of the corresponding modes.

The simulation results are shown in Fig. 2 (the fourth part of the diagonal plane of the ΓMR zone section): black lines are numerically confirmed isofrequency contours, pink dashed curves – theoretical isofrequency contours according the dispersion equation in a quasistatic case (Eq. 2) where $ka << \pi$ and $k_0 a << \pi$, blue and red arrows show the direction of electric and magnetic fields averaged over the unit cell, respectively, at certain points of the contour [9]. Mode crossing at the boundary of the Brillouin zone is described by explicit dispersion equation [1].

This is a numerical confirmation of the presence of a self-intersection point of the isofrequency surface on the main diagonal of the cubic Brillouin zone ($k_{sip} = (k_d, k_d, k_d)^{\mathrm{T}}$), which coordinate in k-space depends on the frequency and changes according to the law (which can be deducted by applying to Eq. 2 a condition $k_x = k_y = k_z = k_d$):

$$k_d^2 = \frac{k_0}{3}\left(2k_0 + \sqrt{k_0^2 + 3k_{pl}^2}\right), \quad i \in \{x, y, z\} \tag{3}$$

Figure 3 shows this dependence, and also the points obtained from the simulation, which are in good agreement with Eq. 3 at low frequencies.

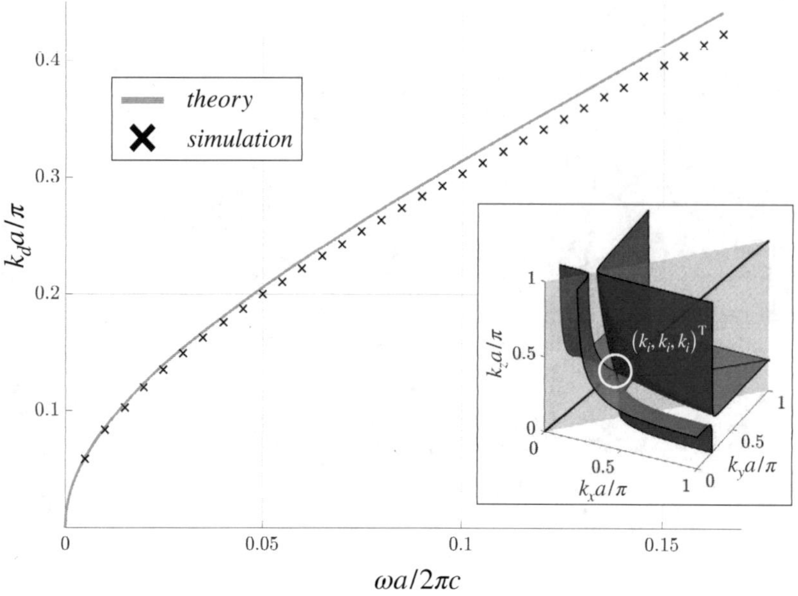

Figure 3. The diagonal self-intersection point (circled on the one eighth of the isofrequency surface (Fig. 1(b)) on the right side of the plot) coordinate k_d versus frequency ω plot. Pink line is given by the Eq. 3, black x-markers are obtained from numerical simulation.

3. Conclusion

The picture of mode polarization near the point of intersection of isofrequency surfaces convinces us of the presence of optical axes predicted by the theory for a given medium and, as a consequence, of the possibility of observing the effect of conical refraction for a given medium [4]. The latter effect has a very wide range of applications, from the creation of optical tweezers to trapping of Bose-Einstein condensates [4]. Thus, the triple wire medium is an easy-to-manufacture alternative to biaxial crystals with adjustable geometry parameters [1] and stable optical axes in a wide frequency range.

References

[1] C.R. Simovski, P.A. Belov, "Low-frequency spatial dispersion in wire media," *Physical Review E*, vol. 70(4), 046616, 2004.

[2] M. Born, E. Wolf, "Principles of optics: electromagnetic theory of propagation, interference and diffraction of light," *Elsevier*, 2013.

[3] G.H.C. New, "Biaxial media revisited", *European Journal of Physics*, vol. 34, pp. 1263-1276, 2013.

[4] A. Turpin, Y.V. Loiko, T.K. Kalkandjiev, J. Mompart, "Conical refraction: fundamentals and applications," *Laser & Photonics Reviews*, vol. 10(5), pp. 750-771, 2016.

[5] A. Poddubny, I. Iorsh, P. Belov, and Yu. Kivshar,"Hyperbolic metamaterial", *Nature Photonics*, Vol. 7, pp. 958-967, 2013

[6] K.E. Ballantine, J.F. Donegan, P.R. Eastham, "Conical diffraction and the dispersion surface of hyperbolic metamaterials," *Physical Review A*, vol. 90(1), 013803, 2014.

[7] A. V. Chebykin, M. A. Gorlach, P. A. Belov, "Spatial-dispersion-induced birefringence in metamaterials with cubic symmetry," *Physical Review B*, vol. 92(4), 045127, 2015.

[8] M. A. Gorlach, S. B. Glybovski, A. A. Hurshkainen, P. A. Belov, "Giant spatial-dispersion-induced birefringence in metamaterials," *Physical Review B*, vol. 93(20), 201115, 2016.

[9] M.G. Silveirinha, "Metamaterial homogenization approach with application to the characterization of microstructured composites with negative parameters," *Physical Review B*, vol. 75(11), 115104, 2007.

Journal of Physics: Conference Series **2015** (2021) 012131 doi:10.1088/1742-6596/2015/1/012131

Anomalous anti-Stokes Scattering in Amorphous Carbon Thin Films

S V Saparina[1], A R Gazizov[1,2], S S Kharintsev[1,2]

[1] Kazan Federal University, 18 Kremlevskaya str., 420008 Kazan, Russian Federation
[2] Tatarstan Academy of Science, 20 Bauman str., 420111 Kazan, Russian Federation

E-mail: sveta.saparina@yandex.ru

Abstract. This work is devoted to a study of temperature-dependent Raman scattering of amorphous carbon (a-C) thin films. An anomalous rise of the anti-Stokes intensity in respect to the Stokes intensity was observed. This result comes from the resonant enhancement of anti-Stokes scattering of defects of graphite-like crystals. The observed discrepancy is quantified through a coefficient, *a*, defined as a ratio of the anti-Stokes and Stokes scattering cross sections. Understanding the mechanism for anomalous enhancement of anti-Stokes scattering provides a way for correct probing the temperature of the a-C thin films exposed to cw laser illumination.

1. Introduction

Due to the flexible rearrangement of physical and chemical properties depending on the ratio of diamond and graphite-like phases, amorphous carbon thin films demonstrate the versatility in large area of applications: from protective coatings in highly sensitive fiber-optic sensors to functional elements in modern nanoelectronic devices [1]. However, an important problem limiting their widespread use is the chemical heterogeneity and structural instability [2] of the material under high temperatures and pressures. In particular, under such conditions the hermetic properties of carbon films degraded that leads to the inefficient operation of an electronic device. Therefore, it becomes an important task to control the evolution of the structural and morphological properties of a carbon film with temperature.

Raman thermometry demonstrates very unique advantages in thermal measurement than traditional methods, especially for measurements at the nanoscale. However, Raman-based temperature probing is material-specific. In this work, we consider the critical factors and define the coefficients for determination the temperature of an amorphous carbon film of via Raman spectroscopy measurement.

2. Experimental part

As the studied samples, we used curved glass substrates – in the form of cylinders with a diameter of 125 μm, covered with a carbon layer of various thicknesses – from 3 to 107 nm. The carbon coating was applied by chemical vapor deposition method with the same precursor-gas – propylene. The Raman spectrum was recorded using linearly polarized light with a wavelength of 632.8 nm. To prevent photo-heating of the samples of interest, the incident laser radiation power of 5.6 mW was set. A Raman spectrum was registered in the spectral range of 150–2000 cm^{-1}. The exposure time was 20 seconds. In this work was utilized an experimental setup that was realizing a reflection-based feedback mechanism. This mechanism allows one to reliably exclude side-effects due to thermal expansion and the inertia of surrounding constituents while doing the Raman measurement. The focus was set onto a specimen under study in an optimal way and measures the maximal response from a photomultiplier. Then, based on the feedback mechanism, a z-correction was made with a xyz piezoscanner in such a way that the focus size would not exceed that defined by the setpoint. To reduce its offset range of piezoscanner due to heating of the surroundings, the hot plate was covered with a thermal shield.

3. Results and discussion

Figure 1 shows a Raman spectrum of the 107 nm a-C coating at room temperature and decomposition into seven elementary peaks assigned to different species of graphite-like and disordered phases. Stokes and anti-Stokes component of Raman spectra were normalized to the intensity of the Stokes line. A blue solid curve is a sum of all peaks. It fits the experimental data of anti-Stokes and Stokes Raman spectra with a correlation coefficient of 0.996 and 0.998, respectively.

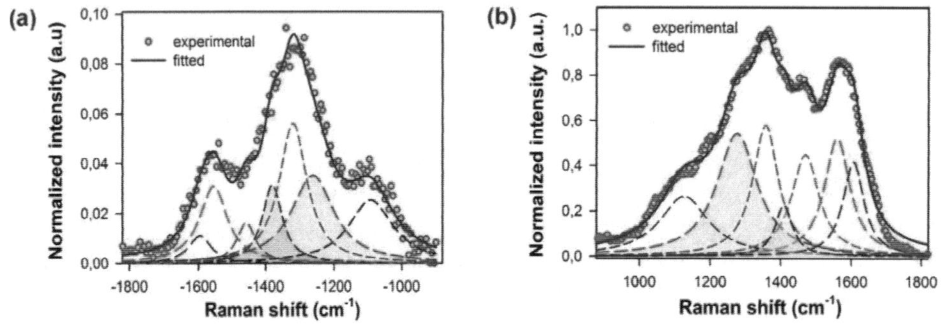

Figure 1. a) Anti-Stokes and b) Stokes Raman spectra of the a-C coating collected at room temperature and decomposed into elementary peaks.

It is known that the temperature of a carbon sample can be spectroscopic determined from the shift of G-mode, the linewidth broadening, the changes of the intensities of the anti-Stokes and Stokes components of the Raman spectrum. Due to the complex structure of Raman spectrum of amorphous carbon the intensity of lines is the most sensitive temperature indicator. Using the Boltzmann distribution for the mean number of phonons with energy $\hbar\Omega$ he temperature of sample can be determined using the following relationship [3]:

$$\boldsymbol{T} = \frac{\hbar c \Omega}{k_b \ln\left[a \frac{I_S\,(\omega_0+\Omega)^4}{I_{AS}(\omega_0-\Omega)^4}\right]}, \quad (1)$$

where I_S and I_{AS} are Stokes and anti-Stokes intensities; ω_0 is the excitation frequency; Ω is a phonon frequency; is the Planck constant; c is the speed of light; k_B is the Boltzmann constant; *a* is coefficient that is specific to material. In general case, the coefficient *a* in equation (1) contains the anti-Stokes to Stokes scattering cross sections ratio. In spontaneous Raman scattering of light these cross sections are equal to each other. However, under resonance conditions, the system can have different scattering cross sections for the ground and the excited vibration state that leads to an anomalous relationship between the Stokes and the anti-Stokes intensities.

Figure 2 (a) shows both the anti-Stokes and the Stokes Raman spectra of 31 nm thick film recorded at different temperatures. It can be seen that the Stokes part changes uniformly with temperature, while the anti-Stokes part grows anomalous in comparison with the predictions of Boltzmann's law. Two-dimensional (2D) temperature map calculated using the equation (1) shows the anti-Stokes to Stokes intensities ratio versus phonon mode frequency of carbon materials excited by the laser with a wavelength of 632.8 nm (figure 2 (b)). It can be seen that the maximum of the anti-Stokes to Stokes intensities ratio is achieved at low values of the vibration frequency and at elevated temperature. At room temperature the anti-Stokes to Stokes intensity ratio at characteristic vibrations frequencies of carbon materials should be only 1/600.

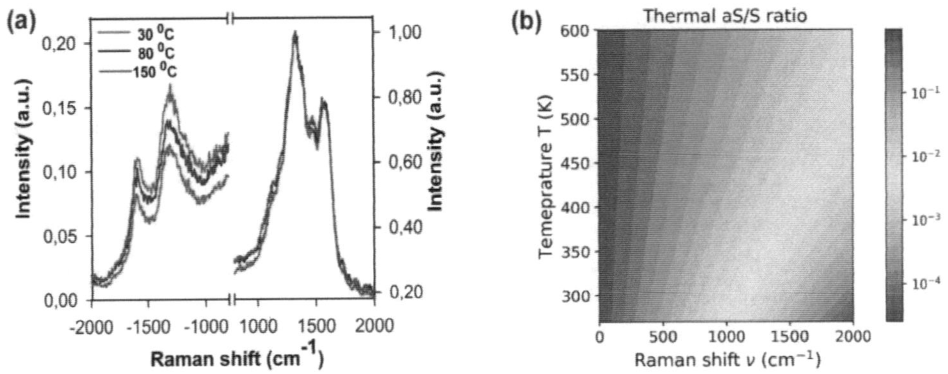

Figure 2. a) Temperature-dependent anti-Stokes and Stokes Raman spectra captured for a carbon film; b) 2-D temperature map of carbon materials.

In a number of articles, such anomalous behaviour is discussed within the framework of the vibration pumping [4], the correlated Stokes-anti-Stokes photon pair generation [5], and the charge transfer [6]. Usually, the contributions of the vibration pumping are reflected in the linear dependence of the anti-Stokes to Stokes intensities ratio on laser power. However, based on results of our spectroscopic investigations for the samples, the dependence was clearly approximated by an exponential function; therefore, the contributions of the vibration pumping are small. For the realization of the effect of correlated Stokes-anti-Stokes pairs the highly-oriented structures are necessary, such as highly crystalline graphite or nanodiamonds. However, the investigated carbon surface comprises crystalline and amorphous domains in concordance with results of atomic-force microscopy. Therefore, a phonon that generates photon pair quickly decay inhibiting the Stokes-anti-Stokes correlation. Also to estimate the contribution of heating under laser illumination of the carbon film to total temperature of the sample we investigated the dependence of *ln(Ias/Is)* on the incident laser power by using laser with wavelength of 632.8 nm. It was found that a significant photoinduced heating occurs in the carbon film at an illumination power of more than 6.0 mW. Samples of interest were exposed to a laser light with the wavelength of 632.8 nm and the power of 5.6 mW. So, contribution of the photoheating is negligible small. Therefore, we believe that anomalous behavior of anti-Stokes to Stokes intensity ratio caused by resonance anti-Stokes Raman scattering in disordered carbon film. Moreover, the resonance is the sensitive to the phonon mode (figure 3(a))

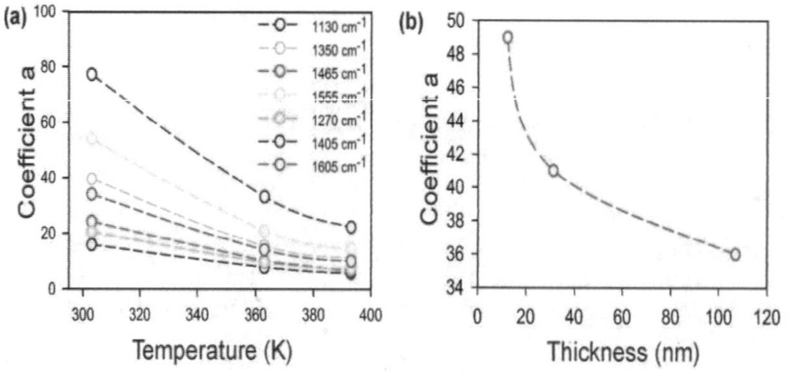

Figure 3. a) Dependence of the coefficient a on temperature for a 12 nm thick carbon film; b) Dependence of the coefficient a on thickness of carbon film.

The strength of this effect depends on the carbon film thickness. Figure 3 (b) shows a mean value of a coefficient a in formula (1) calculated for a graphite vibration mode. The accurate estimation of crystallinity degree in the highly disordered carbonaceous material containing a huge amount of structural defects and dopants consisted of the different phases by means of Raman spectroscopy is challenging [7]. Since it was shown in our previous study the smaller thickness results in larger concentration of the defects that related to the chemical inclusions, such epoxide, carbonyl, carboxyl functional groups [8]. That leads to the modification of potential surface of vibrations [9]. So, the possibility of the involvement of an electronic transition resonance in the anti-Stokes spectral domain increases. Tending toward thicker films (on the order of microns), this effect is waning and the temperature of the film will exactly correspond to the Boltzmann distribution with the coefficient a which has a value of unit.

Thus, in this work, we estimated the critical factors that contribute to the total temperature of the sample probed by means of Raman spectroscopy.

4. Conclusion

In this work, the temperature of the amorphous carbon films was probed by Raman spectroscopy. We found that the anomalous enhancement of the anti-Stokes Raman scattering of the a-C thin film. This effect is related to the resonant enhancement of the anti-Stokes scattering. This result opens a unique possibility for probing the temperature of a-C thin films exposed to cw laser illumination.

Acknowledgments

This work was supported by grant No. 19-12-00066 of the Russian Science Foundation. S.V.S. acknowledges the help from the Russian Foundation for Basic Research (grant No. 20-32-90086) for decomposing highly overlapped Raman bands into elementary components.

References

[1] Khan A and Jawaid M 2019 *Graphene Functionalization Strategies: From Synthesis to Applications* (Singapore: Springer)
[2] Chang H and Hongkai W 2013 *Energy Environ. Sci.* **12** 3483–3507
[3] Baibarac M, Baltog I, Matea A, Mihut L, Lefrant S 2015 *J. Raman Spectrosc.* **46** 32–38
[4] Ru E C Le and Etchegoin P G 2006 *Faraday Discuss.* **132** 63–75
[5] Jorio A, Kasperczyk M, Clark N, Novotny L 2015 *Phys. Status Solidi B.* **252** 2380–84
[6] Tao L et al *2020 Adv. Energy Mater.* **10** 1–19
[7] Kharintsev S, Saparina S, Fishman A, Stolov A, Li J 2020 *J. Phys.Chem. C.* **124** 15886–94
[8] Kharintsev S, Saparina S, Fishman A, Stolov A, Li J 2020 *J. Phys.Chem. C.* **124** 14752–8
[9] Larson D, Downing, K, Glaeser 2011 *J. Struct. Biol.* **174**. 420–3

Synthesis of prolate gold nanoparticles for use in plasmon-enhanced overtone near-infrared spectroscopy

A A Sapunova, D R Dadadzhanov*, A Y Dubovik, M A Baranov, N A Toropov and T A Vartanyan

Faculty of Photonics, Center of Information Optical Technologies, ITMO University, 49 Kronverksky pr., St. Petersburg 197101, Russia

Corresponding author: daler.dadadzhanov@gmail.com

Abstract. Gold nanoparticles were obtained by the method of three-stage growth from a seed solution. Scanning electron microscopy images as well as comparison of extinction spectra to the results of numerical simulations prove the formation of prolate nanoparticles. Such particles with localized plasmon resonance in the NIR region are much-needed for the plasmon-enhanced overtone spectroscopy.

1. Introduction

Near-infrared (NIR) spectroscopy is a type of vibrational spectroscopy that studies the interaction of light with matter in the wavelength range from 750 nm to 2500 nm. As a practical application of NIR spectroscopy, it is possible to conduct quantitative and qualitative analyses of organic molecules in a non-invasive and non-destructive way. This method can successfully detect substances containing functional groups such as CH, NH, SH or OH. Moreover, the spectra in the NIR region can be used to distinguish compounds from each other, since the overtones of the valence vibrations of the hydrogen atom that is part of these groups depend on which atom the hydrogen atom is bound to. NIR spectroscopy has received a new impetus due to the emergence of intensive sources and reliable radiation receivers, which have been intensively developed in recent years for applications in telecommunications based on optical fiber.

A significant limitation in the use of NIR spectroscopy is the relatively low intensity of the absorption bands of dipole vibrational transitions in this spectral region compared to those in the mid-infrared region. One of the promising ways to increase the probabilities of vibrational transitions is associated with their enhancement in the near field of metal nanoparticles with plasmon resonances. To do this, it is necessary to fulfill the resonant condition of the spectral overlap between the absorption band of the plasmon resonance and the absorption band of the overtone vibrational transitions [1,2]. Since the plasmon resonance in gold nanospheres is located in the visible range of the spectrum, to create chemical sensors that work on the basis of registering of optical transitions characteristic of organic substances in the near infrared region gold nanoparticles of a different shape, such as gold nanorods, are needed.

Currently, insufficient attention is paid to the synthesis of nanoparticles with plasmon resonances in the near-infrared region of the spectrum. Most of the papers is devoted to the synthesis and study of nanoparticles with plasmon resonances at wavelengths smaller than 1000 nm. Therefore, the creation of

Journal of Physics: Conference Series **2015** (2021) 012132 doi:10.1088/1742-6596/2015/1/012132

such nanoparticles, whose plasmon resonance is located at wavelengths of about 1500 nm, which are characteristic of the first overtone of the valence vibrations of the hydrogen atom, is an urgent task.
The position of the plasmon resonance of a nanorod depends on the ratio of its length to its diameter, which is called the aspect ratio. Depending on the magnitude of the aspect ratio, the position of the plasmon resonance in the gold nanorods can vary from the visible wavelength range to the infrared. Therefore, the aim of this work is to select a method for the synthesis of gold nanorods, in which the peak of their absorption will fall in the near infrared range.

2. Experimental section

The method we used for the synthesis of gold nanorods was that small spherical nanoparticles were first grown, which served as a seed in the subsequent stages of synthesis, in which the length of the nanorods increased without changing their diameter [3,4]. Growth control was achieved by using $HAuCl_4•3H_2O$ (hydrogen tetrachloroaurate (III)) and CTAB (cetyltrimethylammonium bromide).

To synthesize the gold nanorods a gold seed solution from hydrogen tetrachloroaurate, trisodium citrate, and sodium borohydride were prepared. To achieve the nanoparticles with an aspect ratio greater than 10, a three-stage method of nanoparticle growth was employed. In this case, solutions of $HAuCl_4$, CTAB, and ascorbic acid were added to three test-tubes. Further, these test-tubes were designated as "A", "B" and "C", respectively. Next, a seed solution was added to each test-tube. From the resulting solutions, 1 ml of solution A was added to solution B, and then 1 ml of solution B was added to solution C.

The purification of nanoparticles from the synthesis by-products was carried out in several stages. To begin with, 1 ml of the solution was taken from the test-tube C and centrifuged to deposit the nanoparticles on the bottom of the test-tube. The supernatant was selected and added to the sediment 500 ml of polyethylene glycol (PEG). The procedure was repeated 2-3 times.

3. Results

Figure 1 shows scanning electron microscopy (SEM) images of nanoparticles with the desired aspect ratio of 13, produced by the three-stage seeding method. Figure 1(a) clearly shows that nanoparticles of different geometries were prepared from simple spheres to nanowires of several hundred nm in length. The nanorods of interest are also present.

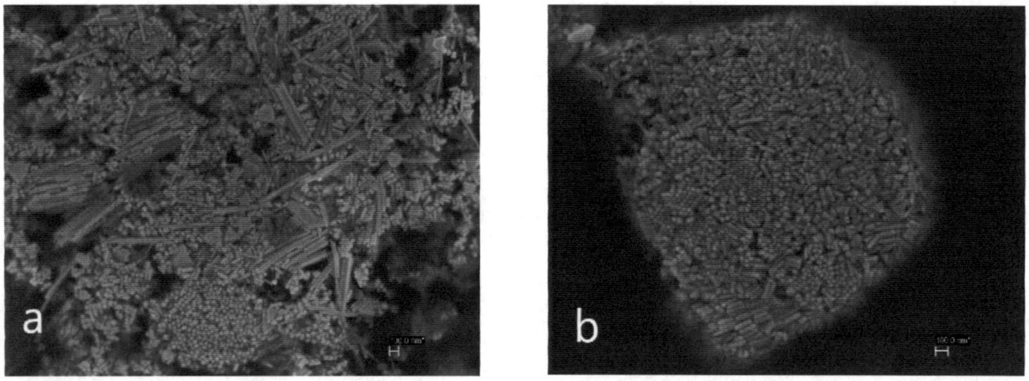

Figure 1. SEM-images of purified gold nanoparticles taken from solution C after the synthesis and deposited on the silicon substrate, where (a) and (b) are the SEM images from different spots. The scale bar is 100 nm.

The size and shape distribution of the prepared gold nanoparticles were determined from SEM-images using ImageJ software. Particularly, 100 nanowires, 100 nanorods, and 100 spheres were chosen to plot the histogram. The analysis of the histograms shows that the average length of the nanowires is 589 nm, the have the average length of the nanorods is 128 nm, and the average diameter of the nanospheres is 38 nm. Using these geometrical parameters, the absorption (ACS), scattering (SCS), and extinction (ECS) cross-sections of nanowires, nanorods, and nanospheres in water were numerically calculated in the COMSOL Multiphysics software. The resulting spectra are shown in Figures 2, 3 and 4.

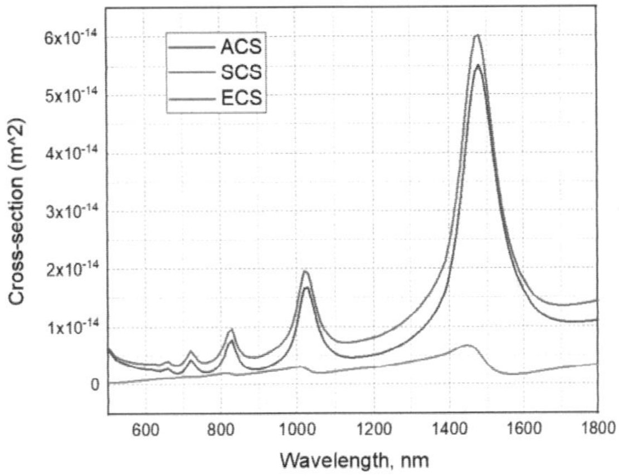

Figure 2. Absorption, scattering, and extinction cross-sections of the gold nanowire (590 × 18 nm) in the homogenous medium ($n_{water} = 1{,}33$).

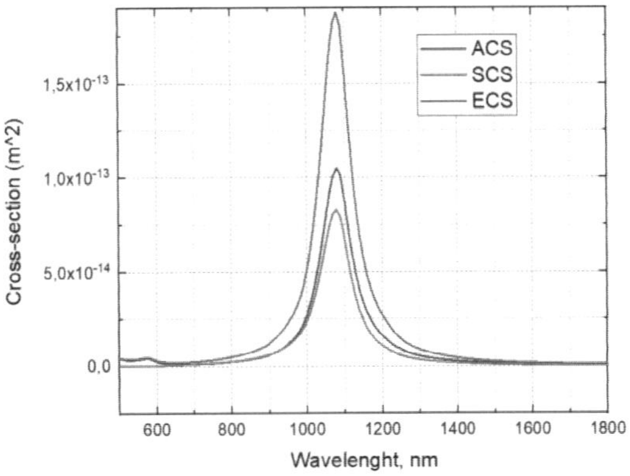

Figure 3. Absorption, scattering, and extinction cross-sections of a gold nanorod (128 × 28 nm) in the homogenous medium ($n_{water} = 1{,}33$).

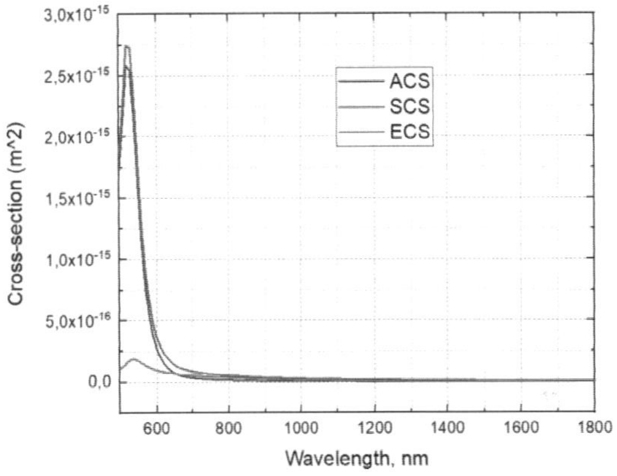

Figure 4. Absorption, scattering, and extinction cross-sections of the gold nanosphere (diameter of a 38 nm) in the homogenous medium ($n_{water} = 1{,}33$).

After the first synthesis, the nanoparticles were selected and purified from solutions B and C, and then the absorption spectra were measured using a Shimadzu UV Probe-3600 spectrophotometer. The measuring of the absorption spectra of the gold nanoparticles were done in deuterium water (D_2O) (Figure 5).

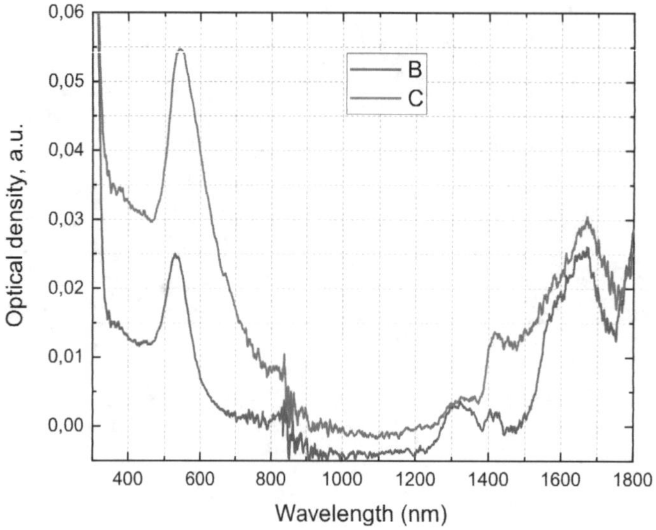

Figure 5. Absorption spectra of gold nanoparticles from solutions B and C.

The spectra have the pronounced plasmon resonance absorption band around 530 nm, which indicates the presence of spherical nanoparticles in solutions. It is worthy to notice, the appeared absorption band about 1700 nm, indicating the presence of the prolonged nanoparticles in the shape of nanorods, are located within the vibrational frequencies of CH and NH functional groups of many organic molecules.

4. Conclusion

The synthesized gold nanoparticles were examined using scanning electron microscopy. It turned out that the morphology of the nanoparticles is different, as there are both the desired nanorods and nanowires, as well as nanoparticles of other shapes, in particular, triangles and spheres. The average size of the obtained nanoparticles was determined. The obtained absorption spectra indicate that prolate gold nanoparticles with the plasmon absorption band in the near-infrared region have been synthesised. Thus, the results obtained make it possible to tune the plasmon resonance band, by tailoring the size and shape of the nanoparticles, for the development of label-free sensor based on surface-enhanced near-infrared absorption effect.

5. Acknowledgements

This work has been funded by Scholarship of the President of the Russian Federation SP-1654.2021.1

6. References

[1] Dadadzhanov, D. R., Vartanyan, T. A., & Karabchevsky, A. (2019). Differential extinction of vibrational molecular overtone transitions with gold nanorods and its role in surface enhanced near-IR absorption (SENIRA). Optics express, 27(21), 29471-29478.

[2] Dadadzhanov, D. R., Vartanyan, T. A., & Karabchevsky, A. (2020). Lattice Rayleigh Anomaly Associated Enhancement of NH and CH Stretching Modes on Gold Metasurfaces for Overtone Detection. Nanomaterials, 10(7), 1265.

[3] Khanal, B. and Zubarev, E. 2019 *ACS nano* **13(2)** 2370-2378.

[4] Nikoobakht, B. and El-Sayed, M. 2003 *Chemistry of Materials* **15(10)** 1957-1962.

Application of the SENSE Algorithm to Multimodal Switchable Metasurface Imaging

P Seregin, E Kretov, K Smolka, M Zubkov

Department of Physics and Engineering, ITMO University, Saint-Petersburg, Russia

E-mail: pavel.seregin@metalab.ifmo.ru

Abstract. This work aims to provide a way of performing parallel imaging with a single-channel variable-frequency resonant device. Different spatial sensitivity profiles required for SENSE reconstruction are achieved by switching between the device eigenmodes. A device capable of such switching is manufactured and several k-spaces are acquired using the device different eigenmodes. The k-spaces are then subject to downsampling to obtain the sensitivity profiles of each eigenmode and then - to perform SENSE-based reconstruction of the unaliased images with different acceleration factors.

1. Introduction

Advances in material science have found practical realization in the development of a variety of new material classes. The radiofrequency (RF) engineering part of the magnetic resonance imaging (MRI) community has paid particular interest to metamaterials and high dielectric permittivity ceramics. These have found use in the scanner hardware development via a series of resonant devices used as RF coils for high and ultra-high field MRI [1, 2, 3, 4].

A common metasurface- or dielectric-based RF coil is a fixed-geometry resonant device supporting a number of eigenmodes with different frequencies and spatial electromagnetic field distributions [1, 2, 5]. In order for such a resonant device to be used in MRI its geometry is optimized in such way that a suitable eigenmode frequency is tuned to the resonant frequency of a target nucleus at the selected magnetic field strength. This commonly results in a highly-efficient fixed-frequency device [1, 2, 3], metamaterial a number of variable-frequency devices have also been suggested [4, 6, 7].

Excitation of the metasurface or dielectric resonator eigenmode tuned to the scanner operating frequency is often performed in MRI via coupling the designed device to the scanner body coil and using the latter both in transmission and reception modes [1, 4, 8, 3]. One drawback of such approach is in the lack of multi-channel capabilities. This, in turn, leads to the inability to perform parallel imaging with the designed resonant devices. This drawback is common among the fixed-frequency resonant devices due to their implicit tuning to a lowest order mode, providing highest spatial field homogeneity in the whole region of interest, whereas parallel imaging techniques require the same region of interest to be covered by a set of different spatially inhomogeneous receive field profiles.

This work aims to overcome this drawback by providing a technique for parallel imaging with a single variable-frequency resonant device, allowing receiving the NMR signal with different spatial sensitivity profiles. This is achieved by switching between the device eigenmodes and acquiring several reduced k-spaces using each of the eigenmodes sensitivity profiles to perform a common SENSE-based reconstruction [9].

2. Materials and Methods

In order to perform the desired eigenmode switching, a tunable metasurface was assembled to be used as an RF coil. The metasurface was fabricated as a printed circuit board (PCB) on an 210×300 mm^2 FR-4 substrate (Relative permittivity=4.6; Loss Tangent=0.02 @ 40 MHz). The substrate hosts 10 straight 10 mm wide copper traces with 12 mm gaps between them (Fig. 1). Two breaks were introduced symmetrically into the mid-points of the metasurface edge traces: one used for the connection of the coaxial cable and the other for a matching 50 Ohm resistance. This solution allows avoiding additional matching circuits and reduces the change of the input impedance during eigenmode switching. The short coaxial cable at one of the mid-points is then used to connect the metasurface to the MR-scanner low noise amplifier via an RF-trap.

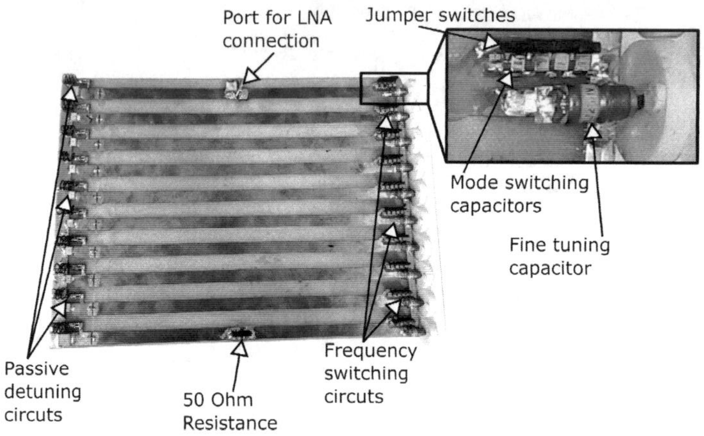

Figure 1. Photo of the switchable metasurface with the connection ports, detuning circuits and switching mechanisms labelled.

Each trace is also equipped with a passive detuning circuit located at the edge of the PCB. The opposite edge of each trace hosts a group of capacitors and jumper-type switches used for changing the metasurface eigenmode frequencies. Frequency fine-tuning is implemented via additional variable high-Q capacitors located at the same edge of the PCB.

Imaging was performed on a 1 T Siemens Magnetom Harmony clinical whole-body scanner operating at the 40.5 MHz Larmor frequency. Five images of a homogeneous MRI phantom were sequentially acquired using the gradient echo FLASH pulse sequence with a 300×300 mm^2 field of view, a 256×256 points matrix and TR/TE = 100/10 ms. After each of the acquisitions the metasurface resonant frequency was adjusted using the jumper switches so that each image was acquired using a sensitivity profile resulting from the magnetic field distribution of a different eigenmode. The sensitivity profiles provided by each eigenmode can be seen in Fig. 2, G and H.

The image processing pipeline comprised spatial sensitivity estimation, k-space downsampling and SENSE-based reconstruction. Each mode sensitivity profiles were estimated from a low-resolution phantom image, acquired via cropping the 256×256 full k-space to the central 256×$N_{profile}$ lines, where $N_{profile}$ was varied from 16 to 64 central lines. Next, the full 256×256 k-space was downsampled to 256×(256/R), where R is the acceleration factor. This was performed by retaining only every Rth k-space line. Finally, the reduced k-spaces and the low-resolution sensitivity maps were reconstructed using the SENSE algorithm, as implemented in the PULSAR toolbox [10] running under MATLAB 2020b. The reconstruction quality was assessed by computing the mean g-factor under various reconstruction conditions, particularly, under varying acceleration factors and number of eigenmodes (N) used for reconstruction.

3. Results

The SENSE method application to the aliased images resulting from the k-space downsampling provided a successful reconstruction of the full unaliased image with at all the tested acceleration factors and mode selection conditions (Fig. 2). A qualitative image improvement as well as the

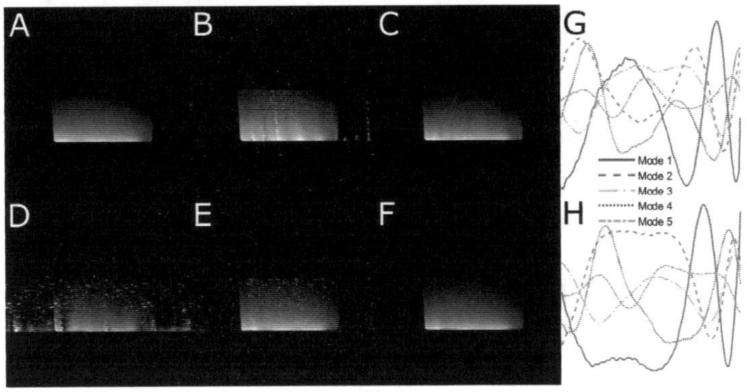

Figure 2. Sum of squares and SENSE reconstruction examples. A - sum of squares image reconstructed from full k-spaces of the 5 modes combined in image space. B - SENSE reconstruction using R = 2, N = 2 and $N_{profile}$ = 16. C - SENSE with R = 2, N = 5, $N_{profile}$ = 16. D - SENSE with R = 4, N = 4, $N_{profile}$ = 16. E - SENSE with R = 4, N = 4, $N_{profile}$ = 32. F - SENSE with R = 4, N = 5, $N_{profile}$ = 32. G - real part of the sensitivity profiles at the bottom of the phantom, provided by the modes in use. H - imaginary part of the same sensitivity profiles.

mean g-factor reduction was observed with the increasing number of modes used for SENSE reconstruction (Fig. 3). Similarly, an increase in image quality (manifesting also in the decrease

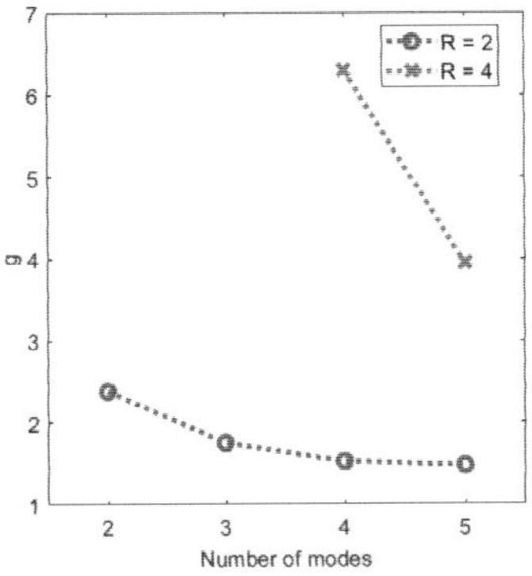

Figure 3. Mean g-factor dependence on the number of modes used for SENSE reconstruction. $N_{profile}$ = 16 was used for all reconstructions.

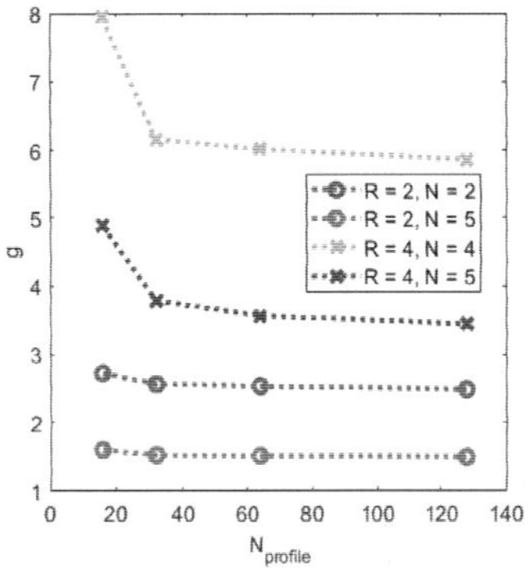

Figure 4. Mean reconstruction g-factor dependence on the number of phase encoding lines retained for sensitivity profile estimation.

in the mean g-factor) was observed with increasing sensitivity profile resolution (Fig. 4).

4. Conclusion

Current work shows a new principal way a single-channel limitation of the metamaterial or dielectric-based RF coils in MRI can be overcome, and parallel imaging can be performed. It was demonstrated, that acquiring the signal via the metasurface, serving as an RF coil, while tuning its different eigenmodes to the Larmor resonance frequency provides enough data for SENSE-based image reconstruction. While proof of principle has been demonstrated, true parallel imaging is yet to be performed, as, in the current work, images were acquired sequentially. The true parallel acquisition can nevertheless be attained due to a rather slow change in the acquired NMR signal relative to the modern electronic component switching times. Thus, in a single 1/(acquisition bandwidth) interval, the metasurface can be switched between different eigenmodes, and several k-space points might be recorded, so that different eigenmode k-spaces could be filled in an interleaved manner. This would allow the desired parallel imaging techniques to be applied to the data acquired in such a quasi-parallel interleaved technique.

Acknowledgments

This work was supported by the Russian Science Foundation (Project No. 21-79-30038).

References

[1] Mikhailovskaya A A, Shchelokova A V, Dobrykh D A, Sushkov I V, Slobozhanyuk A P and Webb A 2018 *Journal of Magnetic Resonance* **291** 47–52 ISSN 10907807
[2] Moussu M A C, Ciobanu L, Kurdjumov S, Nenasheva E, Djemai B, Dubois M, Webb A G, Enoch S, Belov P, Abdeddaim R and Glybovski S 2019 *Advanced Materials* **31** 1900912 ISSN 1521-4095
[3] Shchelokova A, Ivanov V, Mikhailovskaya A, Kretov E, Sushkov I, Serebryakova S, Nenasheva E, Melchakova I, Belov P, Slobozhanyuk A and et al 2020 *Nature Communications* **11** 3840 ISSN 2041-1723
[4] Brui E A, Shchelokova A V, Zubkov M, Melchakova I V, Glybovski S B and Slobozhanyuk A P 2018 *physica status solidi (a)* **215** 1700788 ISSN 18626300
[5] Kosulnikov S, Zalipaev V, Shchelokova A, Melchakova I, Glybovski S, Slobozhanyuk A and Belov P 2018 *Physical Review B* **98** ISSN 2469-9950, 2469-9969
[6] Hurshkainen A, Nikulin A, Georget E, Larrat B, Berrahou D, Neves A L, Sabouroux P, Enoch S, Melchakova I, Belov P, Glybovski S and Abdeddaim R 2018 *Scientific Reports* **8** 9190 ISSN 2045-2322
[7] Ivanov V A, Hurshkainen A A, Solomakha G A and Zubkov M A 2020 *Photonics and Nanostructures - Fundamentals and Applications* **38** 100747 ISSN 1569-4410
[8] Shchelokova A V, van den Berg C A, Dobrykh D A, Glybovski S B, Zubkov M A, Brui E A, Dmitriev D S, Kozachenko A V, Efimtcev A Y, Sokolov A V, Fokin V A, Melchakova I V and Belov P A 2018 *Magnetic Resonance in Medicine* **80** 1726–1737 ISSN 07403194
[9] Pruessmann K P, Weiger M, Scheidegger M B and Boesiger P 1999 *Magnetic Resonance in Medicine* **42** 952–962 ISSN 0740-3194, 1522-2594
[10] Ji J X, Son J B and Rane S D 2007 *Concepts in Magnetic Resonance Part B: Magnetic Resonance Engineering* **31B** 24–36 ISSN 1552-504X

Circularly polarized RF coil for energy harvesting in clinical MRI

P. S. Seregin[1], **O. I. Burmistrov**[1], **G. Solomakha**[1], **E.I. Kretov**[1], **N.A. Olekhno**[1], **A. Slobozhanyuk**[1]

[1]Department of Physics and Engineering, ITMO University, 197101 Saint Petersburg, Russia

E-mail: pavel.seregin@metalab.ifmo.ru

Abstract. Radiofrequency (RF) harvesting is a promising technology for the wireless power supply of various in-bore devices used in magnetic resonance imaging. However, current technical solutions in this area are based on the conversion of linearly polarized RF fields, and thus their efficiency is limited, as they interact only with a fraction of circularly polarized RF fields. In the present work, we introduce and experimentally realize a novel harvesting setup allowing for converting circularly polarized RF fields to direct current.

1. Introduction

Magnetic resonance imaging (MRI) offers great perspectives in medical diagnostics. Nevertheless, some specific MRI tasks, such as abdominal imaging, require gating and, as a result, patient parameter tracking. For this, various in-bore devices are often used, including respiration sensors and heart rate monitors, which in most cases need an external power source [1, 2, 3]. However, powering these devices with cables can create imaging artifacts and decrease the comfort of patients [4]. As a result, various strategies of wireless power supply are actively developed. RF harvesting is one of the wireless supply technology. It relies on the conversion of RF field B_1 that is readily present in the MRI scanner, thus not requiring additional transmitting coils [1].

The use of RF harvesting in MRI poses several difficulties. For example, currents flowing in the harvesting circuit potentially can distort the magnetic fields B_1 and B_0 [6], which can result in the image quality decrease. Moreover, the majority of harvesting setups for MRI exploit only one linear polarization component of the transmit field [5]. However, the B_1 field is circularly polarized, and thus the efficiency of such devices is limited. In the present paper, we summarize these main challenges and aim to overcome some of them by proposing a harvesting setup capable of utilizing a circularly polarized field. Here, we realize and experimentally test the harvesting coil for clinical MRI and study the effects of its location within the scanner bore.

2. Drawbacks in linearly polarized harvesting for MRI applications

The main disadvantage of MRI RF harvesting is its low efficiency. Despite the significant peak power of the radio transmitter (10-25 kW), the average power that RF Harvesting could collect is about 100 mW. This is because a significant part of power should be dissipated in the human body to provide excitation of nuclei. On the one hand, it is possible to increase the power of the received energy by increasing the area of the harvesting coil. However, an increase in the

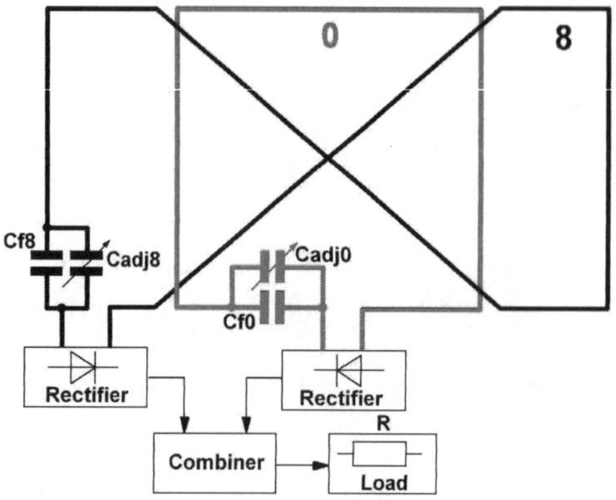

Figure 1. Design of the proposed circular field harvesting setup consisting of two coils, one in the shape of 0 (shown in red) and another one in the shape of 8 (shown in black). The setup includes two tuning capacitors and two rectifiers for each of the coils, along with a combiner. The grounding resistor emulates the load powered with the harvesting setup).

coil area can lead to distortion of the B_1 field and a rise in the specific absorption rate (SAR). Besides, DC after the rectifier may affect the B_0 field. Since it is a source of secondary constant magnetic fields.

Another factor that influences the power level of MRI RF harvesting is the pulse sequence used during the MRI examination. A powerful RF excitation pulse is present for a short time, usually no more than 10 ms, while the pause between the excitation pulses can be 100 ms.

3. Improving MRI harvesting efficiency with circular polarization conversion

To evaluate the impact of the pulse sequence, a series of experiments were performed with Siemens Espree 1.5 T MRI scanner. It was observed that the harvesting voltage for gradient-echo sequences (GRE) were small because of the low power examination (low flip angle (FA)) and only one pulse used for excitation. However, for some pulse sequences, such as the T2 turbo spin-echo (TSE) sequence, harvesting worked better due to the high-power examination (high flip angle), many pulses in the excitation phase, and low TR.

To get circularly polarized and, as a result, improve efficiency, we suggested a design with two coils with orthogonal B_1 fields. Design of coils are illustrated in Figure 1. Each coil receives the corresponding linear component of the circularly polarized B_1 field. Construction of each coil includes tuning capacitors to adjust the resonant frequency. The currents induced in the coils are rectified with the diodes and then combined together in the bridge-based combiner and go to the load. Its resistance was chosen to be 200 Ohms, which is equivalent to a DC load of a low noise amplifier (LNA) and a small RF transmitter to create a wireless communication channel. In experimental realization, we use the following coil parameters: length $L = 110$ mm, width $W = 67$ mm, and height $H = 32$ mm. Photo of the experimental realization is shown in Fig. 2(a).

It is essential to find an optimal position of the designed coil within the bore of the scanner to provide maximum energy harvesting and simultaneously eliminate the effect of the MRI RF harvesting setup on the B_0 and B_1 fields. To find the optimal location, the voltage dependence was evaluated, obtained with a harvesting setup as a function of the displacement of setup from the scanner isocenter. Since the region of the uniform RF field of excitation depends on the size of the body coil, a comparison was made between the Siemens Espree (red curve) and the Siemens Avanto (blue curve), which are equipped with different types of body coils. The output

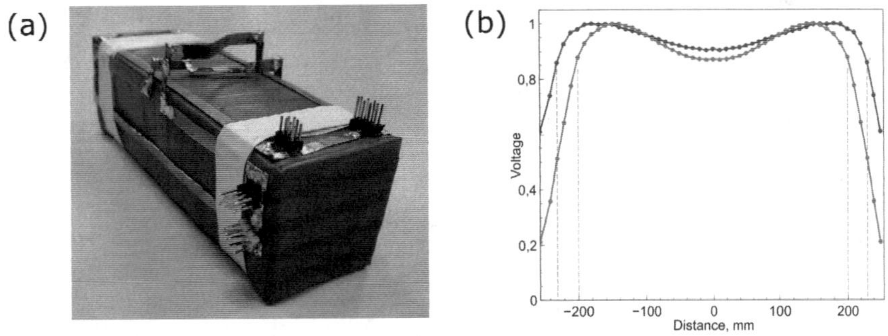

Figure 2. (a) Photo of the experimentally realized circular polarized harvesting setup. (b) The dependence of voltage obtained with a harvesting setup as a function of the displacement of setup from the scanner isocenter. Blue and red curves correspond to Siemens Avanto and Siemens Espree MRI scanners, respectively.

voltage was measured using Owon XDS3202A 14-bit oscilloscope attached by a coaxial cable with multiple RF cable traps to prevent the cable effect. We measured the output voltage with a single excitation pulse, as RF harvesting should provide voltage to the local coil circuitry during each phase encoding step. To this end, we applied a manual frequency adjustment procedure employing three separate pulses. However, only the first one was used to measure the harvesting output. To estimate the region in which the harvesting setup maintains stable functionality, we select a 15% drop in the output voltage as criteria. Therefore, the realized harvesting coil can be freely moved within ±20 cm range for Espree setup and within ±22.5 cm range for Avanto setup without considering a change in the output voltage.

To predict the effect of the harvesting setup to B_1 distribution, an experimental study was performed. We investigated different locations of the harvesting coil to ensure that the harvesting setup does not impact the MRI functionality. We use the method estimation B_1 map with two GRE sequences [7]. The experiment was performed using Siemens large spherical phantom. It has a diameter of 17 cm and part number 04761065. The phantom was placed at the isocenter $Z = 0$ for Espree MRI scanner. Three B_1 map were measured for two cases: without harvesting setup(Fig. 3(a)), with a harvesting coil at the distance of 1 cm (Fig. 3(b)) and 3 cm (Fig. 3(c)) from the edge of the phantom. It was found that for the case when harvesting is located at a

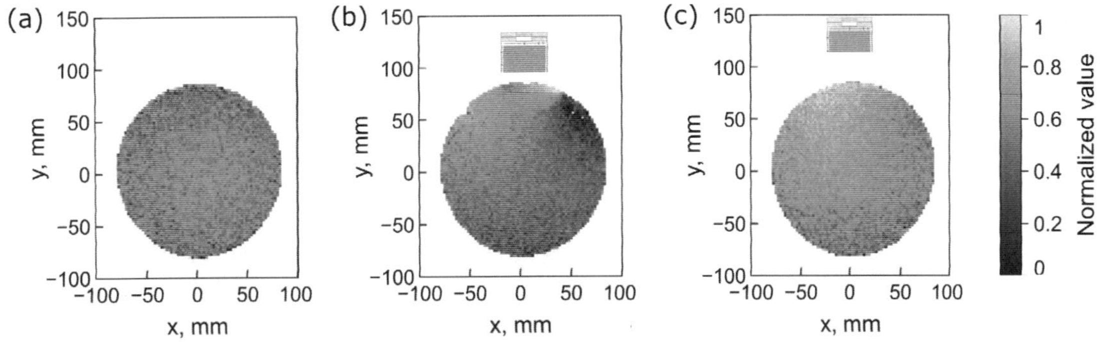

Figure 3. (a) Measured RF field B_1 inside the body phantom in Siemens Espree MRI scanner without a harvesting setup. (b) B_1 field with the harvesting setup placed at the distance 1 cm above the phantom. (c) B_1 field with the harvesting setup placed at 3 cm above the phantom.

distance of 1 cm, currents induced in the harvesting coil create local B_1 field inhomogeneity. However, if the distance between the harvesting coil and the phantom equals 3 cm, then the field again appears unperturbed. It is important to note that the given distance from the phantom depends on the size of the harvesting coil and the amplitude of the induced current.

4. Conclusion

This paper proposes a novel design of a RF harvesting setup based on two coils with an orthogonal magnetic field. Such a design is capable of converting a circularly polarized RF magnetic field to DC. Such design is promising for MRI applications due to the circular polarization of B_1 RF field in MRI scanners. As we have demonstrated with our experiments, the developed harvesting setup does not distort the B_1 field in the region of scan and can provide DC in a specific range of positions along MR-scanner isocenter.

The measurements demonstrated that it is possible to double voltage almost two times at the harvesting output with circular polarized RF coil. The reason for lower DC output could be in the losses in the voltage. The peak voltage during one excitation RF pulse reaches 100 V on the load 200 Ohm. Therefore the peak power is approximately 50 W. However, the duration of the exciting pulse is no longer than 5 mS, and the time repetition (TR) pulse time usually is 40-400 mS. The estimated average power at each step of the phase encoding cycle will be about 100-500 mW, and it depends on the selected pulse sequence. The proposed RF harvesting setup can be used to provide DC supply for cable-free receive arrays or wireless sensors.

Acknowledgments

This work was supported by the Russian Science Foundation (Project No. 21-79-30038).

References

[1] Nohava L, Ginefri JC, Willoquet G, Laistler E, Frass-Kriegl R. "Perspectives in Wireless Radio Frequency Coil Development for Magnetic Resonance Imaging" 2020 *Frontiers in Physics* **8**, 11

[2] Mandal R, Babaria N, Cao J, Lu KH, Liu Z. "MRI Powered and Triggered Current Stimulator for Concurrent Stimulation and MRI" 2019 *bioRxiv*, 715805

[3] Ganti A, Lin J, Wynn T, Ortiz T. "Achieving electromagnetic compatibility of wireless power transfer antennas inside MRI system" 2019 *Wireless Power Transfer* **6**, 138–153

[4] Shchelokova A, *et al.* "Ceramic resonators for targeted clinical magnetic resonance imaging of the breast" 2020 *Nature communications* **11**, 1–7

[5] Venkateswaran M, Kurpad K, Brown J E, Fain S, van der Weide D. "Wireless Power Harvesting During MRI" 2020 *2020 42nd Annual International Conference of the IEEE Engineering in Medicine & Biology Society (EMBC)*

[6] Twieg M, de Rooij MA, Griswold MA. "Active detuning of MRI receive coils with GaN FETs" 2015 *IEEE Transactions on Microwave Theory and Techniques* **63**, 4169–4177

[7] Insko E, Bolinger L. "Mapping of the radiofrequency field" 1993 *Journal of Magnetic Resonance, Series A* **103**, 82–85

A perturbation theory approach to the ground state exciton energy in the limit of a weak magnetic field in anomalous exciton Hall effect.

V A Shabashov[1,2], **V K Kozin**[1,3], **A V Kavokin**[4,5,6] **and I A Shelykh**[1,3]

[1] ITMO University, Kronverkskiy prospekt 49, Saint Petersburg 197101, Russia
[2] St. Petersburg Academic University of the Russian Academy of Sciences, 194021 St. Petersburg, Russia
[3] Science Institute, University of Iceland, Dunhagi-3, IS-107 Reykjavik, Iceland
[4] Westlake University, 18 Shilongshan Road, Hangzhou 310024, Zhejiang Province, China
[5] Institute of Natural Sciences, Westlake Institute for Advanced Study, 18 Shilongshan Road, Hangzhou 310024, Zhejiang Province, China
[6] School of Physics and Astronomy, University of Southampton, Highfield, Southampton SO17 1BJ, United Kingdom

Abstract.
The anomalous exciton Hall effect is a phenomenon that occurs in a quantum well in the presence of an external magnetic field applied perpendicular to the surface due to the interaction of the exciton dipole moment with an electric field, formed by the charged impurities. The effect was fully described in [1] for different magnetic field regimes. In this paper, we focus on the way the perturbation method was used for finding the ground state energy of an exciton in the limit of a weak magnetic field.

1. Introduction

Excitons, propagating in the presence of an external magnetic field orthogonal to their velocity, acquire an electric dipole moment perpendicular to both the magnetic field and their propagation direction [2]. The existence of a dipole moment causes the excitons to have a nonzero asymmetric scattering rate while interacting with charged impurities. The T-matrix formalism was exploited to describe the symmetric and asymmetric contributions to the scattering rates. The problem then boiled down to the calculation of T-matrix elements, which satisfy the Lippmann–Schwinger equation:

$$T_{\boldsymbol{k},\boldsymbol{k}'} = \hat{V}_{\boldsymbol{k},\boldsymbol{k}'} + \int \frac{d^2\boldsymbol{g}}{(2\pi)^2} \frac{\hat{V}_{\boldsymbol{k},\boldsymbol{g}} T_{\boldsymbol{g},\boldsymbol{k}'}}{E - \epsilon(\boldsymbol{g}) + i0}, \tag{1}$$

By virtue of the smallness of the magnetic field, one may desire to solve the Lippmann–Schwinger equation perturbatively:

$$T_{\boldsymbol{k},\boldsymbol{k}'} \approx \hat{V}_{\boldsymbol{k},\boldsymbol{k}'} + \delta T_{\boldsymbol{k},\boldsymbol{k}'}, \text{ where } \delta T_{\boldsymbol{k},\boldsymbol{k}'} = \mathrm{P}\int \frac{\mathrm{d}^2\boldsymbol{g}}{(2\pi)^2} \frac{\hat{V}_{\boldsymbol{k},\boldsymbol{g}} \hat{V}_{\boldsymbol{g},\boldsymbol{k}'}}{E - \epsilon(\boldsymbol{g})} - \frac{i\nu}{2} \int_0^{2\pi} \hat{V}_{\boldsymbol{k},\boldsymbol{g}} \hat{V}_{\boldsymbol{g},\boldsymbol{k}'} \, \mathrm{d}\phi_{\boldsymbol{g}}. \tag{2}$$

Here $\epsilon(\boldsymbol{g})$ is the ground state energy of an exciton with a wave vector $\boldsymbol{g}$ in the presence of a magnetic field; $\nu = |\partial\epsilon(\boldsymbol{k})/\partial k|^{-1}k/(2\pi)$; $\hat{V}$ is the exciton-impurity scattering potential.

However, the perturbation theory approach turned out to be invalid in this case, since the calculated first-order correction was of the same order as the main contribution. Nevertheless, being inapplicable for solving the Lippmann–Schwinger equation, the method can be used for the calculation of the ground state energy $\epsilon(\boldsymbol{g})$ and the matrix elements of the scattering potential.

In the following sections, the steps are presented for finding an analytical expression of the ground state energy of an exciton in the limit of a weak magnetic field using the perturbation method. During the calculations, several infinite sums appear, which are computed using an elegant method from [3].

2. Formulation of the problem

(i) We start by writing the unperturbed Hamiltonian of the relative electron–hole motion:

$$\hat{H}^{(0)} = -\frac{\hbar^2}{2\mu}\nabla_r^2 - \frac{e^2}{4\pi\varepsilon_0\varepsilon r} + \frac{\hbar^2 k^2}{2M} \tag{3}$$

Its wave functions $\Phi_{m_1,m_2}(\boldsymbol{r})$ $\left(= \Phi_{\boldsymbol{m}}(\boldsymbol{r}) = |\boldsymbol{m}\rangle\right)$ consist of a coordinate $(R_{m_1}(r))$ and an angular $(\Theta_{m_2}(\theta))$ parts and can be found in [4]. The ground state wave function and energy read ($a_B = 4\pi\varepsilon_0\varepsilon\hbar^2/(\mu e^2)$ is the Bohr radius):

$$\Phi_{\boldsymbol{m}_0}(\boldsymbol{r}) = \left(\frac{8}{\pi}\right)^{1/2}\frac{1}{a_B}\exp\left(-\frac{2r}{a_B}\right); \quad \epsilon_{\boldsymbol{m}_0}^{(0)}(\boldsymbol{k}) = -\frac{\mu e^4}{\hbar^2}\frac{1}{8\pi^2\varepsilon^2\varepsilon_0^2} + \frac{\hbar^2k^2}{2M} = \frac{\hbar^2k^2}{2M} - \epsilon_0 \tag{4}$$

(ii) The magnetic field affects the system through the perturbation potential $\hat{V}$, which assumed to be small:

$$\hat{V} = -\frac{i\hbar e B}{2}\left(\frac{1}{m_e} - \frac{1}{m_h}\right)[\boldsymbol{r}\times\nabla_r]_z + \frac{e^2B^2}{8\mu}r^2 + \frac{e\hbar}{M}\boldsymbol{B}\cdot[\boldsymbol{r}\times\boldsymbol{k}] \tag{5}$$

(iii) The goal of this paper is to find the ground state energy $\epsilon(\boldsymbol{k})$ using a perturbation theory approach.

3. Calculation of the ground state energy

(i) Let us expand the ground energy up to the second order in the magnetic field strength B:

$$\epsilon(\boldsymbol{k}) = \epsilon_{\boldsymbol{m}_0}^{(0)}(\boldsymbol{k}) + \langle\boldsymbol{m}_0|\hat{V}|\boldsymbol{m}_0\rangle + \sum_{\boldsymbol{m}\neq\boldsymbol{m}_0}\frac{|\langle\boldsymbol{m}|\hat{V}|\boldsymbol{m}_0\rangle|^2}{\epsilon_{\boldsymbol{m}_0}^{(0)}(\boldsymbol{k}) - \epsilon_{\boldsymbol{m}}^{(0)}(\boldsymbol{k})} \tag{6}$$

(ii) One may notice, that: $\langle\boldsymbol{m}_0|\hat{x}k_y - \hat{y}k_x|\boldsymbol{m}_0\rangle = 0$ and $\langle\boldsymbol{m}|\hat{L}_z|\boldsymbol{m}_0\rangle = 0$. Thus (we omit dependence on $\boldsymbol{k}$ in the energy for brevity; $\boldsymbol{B} \parallel Oz$):

$$(1) \quad \langle\boldsymbol{m}_0|\hat{V}|\boldsymbol{m}_0\rangle = \frac{e^2B^2}{8\mu}\langle\boldsymbol{m}_0|\hat{r}^2|\boldsymbol{m}_0\rangle = \frac{e^2B^2}{8\mu}\cdot\frac{3a_B^2}{8} = \frac{3e^2a_B^2B^2}{64\mu} \tag{7}$$

$$(2) \quad \sum_{\boldsymbol{m}\neq\boldsymbol{m}_0}\frac{|\langle\boldsymbol{m}|\hat{V}|\boldsymbol{m}_0\rangle|^2}{\epsilon_{\boldsymbol{m}_0}^{(0)} - \epsilon_{\boldsymbol{m}}^{(0)}} = \sum_{\boldsymbol{m}\neq\boldsymbol{m}_0}\frac{\left|\langle\boldsymbol{m}|\frac{e^2B^2}{8\mu}\hat{r}^2 + \frac{e\hbar}{M}\boldsymbol{B}\cdot[\hat{\boldsymbol{r}}\times\boldsymbol{k}]|\boldsymbol{m}_0\rangle\right|^2}{\epsilon_{\boldsymbol{m}_0}^{(0)} - \epsilon_{\boldsymbol{m}}^{(0)}}$$

$$= \left(\frac{e^2B^2}{8\mu}\right)^2\sum_{\boldsymbol{m}\neq\boldsymbol{m}_0}\frac{\left|\langle\boldsymbol{m}|\hat{r}^2|\boldsymbol{m}_0\rangle\right|^2}{\epsilon_{\boldsymbol{m}_0}^{(0)} - \epsilon_{\boldsymbol{m}}^{(0)}} + \left(\frac{e\hbar B}{M}\right)^2\sum_{\boldsymbol{m}\neq\boldsymbol{m}_0}\frac{\left|\langle\boldsymbol{m}|\hat{x}k_y - \hat{y}k_x|\boldsymbol{m}_0\rangle\right|^2}{\epsilon_{\boldsymbol{m}_0}^{(0)} - \epsilon_{\boldsymbol{m}}^{(0)}} \tag{8}$$

One can see that the analytical solution requires a calculation of several infinite sums. The smart way to perform that is presented in next subsections.

3.1. First auxiliary task

First, we want to concentrate on the following problem: how a hermitian operator $\hat{b}_1$, satisfying the relation $\mathrm{i}\mu[\hat{H}_0, \hat{b}_1] = \hat{x}$, acts on $\Phi_{\boldsymbol{m}_0}(\boldsymbol{r})$.

(i) Let us introduce $b_1(\boldsymbol{r})$ as $\hat{b}_1\Phi_{\boldsymbol{m}_0}(\boldsymbol{r}) = b_1(\boldsymbol{r})\Phi_{\boldsymbol{m}_0}(\boldsymbol{r})$. Then:

$$\begin{aligned} &\mathrm{i}\mu[\hat{H}_0, \hat{b}_1]\Phi_{\boldsymbol{m}_0}(\boldsymbol{r}) = \hat{x}\Phi_{\boldsymbol{m}_0}(\boldsymbol{r}) \\ &\Rightarrow -\frac{\mathrm{i}\hbar^2}{2}\left\{\left[\nabla_r^2 b_1(\boldsymbol{r})\right]\Phi_{\boldsymbol{m}_0}(\boldsymbol{r}) + 2\left[\nabla_r b_1(\boldsymbol{r})\right]\cdot\left[\nabla_r\Phi_{\boldsymbol{m}_0}(\boldsymbol{r})\right]\right\} = x\Phi_{\boldsymbol{m}_0}(\boldsymbol{r}) \end{aligned}$$

(ii) The solution for the equation above is:

$$b_1(\boldsymbol{r}) = -\mathrm{i}\frac{a_B^3}{\hbar^2}\left[\left(\frac{r^2}{4a_B^2} + \frac{3r}{16a_B}\right)\cos\phi + C\right]$$

with C being an arbitrary constant.

3.2. Second auxiliary task

Now let us focus on the similar problem: how a hermitian operator $\hat{b}_2$, satisfying the relation $\mathrm{i}\mu[\hat{H}_0, \hat{b}_2] = \hat{r}^2$, acts on $\Phi_{\boldsymbol{m}_0}(\boldsymbol{r})$.

(i) We introduce $b_2(\boldsymbol{r})$ as $[\hat{b}_2\Phi_{\boldsymbol{m}_0}(\boldsymbol{r}) = b_2(\boldsymbol{r})\Phi_{\boldsymbol{m}_0}(\boldsymbol{r})$. Then, following exactly the same steps, we obtain the equation on $b_2(\boldsymbol{r})$:

$$-\frac{\mathrm{i}\hbar^2}{2}\left\{\left[\nabla_r^2 b_2(\boldsymbol{r})\right]\Phi_{\boldsymbol{m}_0}(\boldsymbol{r}) + 2\left[\nabla_r b_2(\boldsymbol{r})\right]\cdot\left[\nabla_r\Phi_{\boldsymbol{m}_0}(\boldsymbol{r})\right]\right\} = \hat{r}^2\Phi_{\boldsymbol{m}_0}(\boldsymbol{r}) \tag{9}$$

(ii) The solution reads:

$$b_2(\boldsymbol{r}) = -\frac{\mathrm{i}a_B^4}{2\hbar^2}\left[\frac{1}{3}\left(\frac{r}{a_B}\right)^3 + \frac{3}{8}\left(\frac{r}{a_B}\right)^2 + \frac{3}{8}\frac{r}{a_B} + \frac{3}{32}\ln\left(\frac{r}{a_B}\right) + C\right] \tag{10}$$

with C being an arbitrary constant.

3.3. Calculation of the infinite sums

Using the results of the previous section, we can find analytical expressions for several sums, which appear during energy calculation.

(i) From the relations operators $\hat{b}_1$ and $\hat{b}_2$ satisfy, one obtains:

$$\mathrm{i}\mu\left(\epsilon_{\boldsymbol{m}_0}^{(0)} - \epsilon_{\boldsymbol{m}}^{(0)}\right)\langle\boldsymbol{m}_0|\hat{b}_1|\boldsymbol{m}\rangle = \langle\boldsymbol{m}_0|\hat{x}|\boldsymbol{m}\rangle; \quad \mathrm{i}\mu\left(\epsilon_{\boldsymbol{m}_0}^{(0)} - \epsilon_{\boldsymbol{m}}^{(0)}\right)\langle\boldsymbol{m}_0|\hat{b}_2|\boldsymbol{m}\rangle = \langle\boldsymbol{m}_0|\hat{r}^2|\boldsymbol{m}\rangle \tag{11}$$

(ii)

$$\begin{aligned} I_1 &= \sum_{\boldsymbol{m}\neq\boldsymbol{m}_0}\frac{|\langle\boldsymbol{m}_0|\hat{x}|\boldsymbol{m}\rangle|^2}{\left(\epsilon_{\boldsymbol{m}_0}^{(0)} - \epsilon_{\boldsymbol{m}}^{(0)}\right)^2} = -\sum_{\boldsymbol{m}}\frac{\langle\boldsymbol{m}_0|\hat{b}_1|\boldsymbol{m}\rangle\langle\boldsymbol{m}|\hat{b}_1|\boldsymbol{m}_0\rangle}{\left(\epsilon_{\boldsymbol{m}_0}^{(0)} - \epsilon_{\boldsymbol{m}}^{(0)}\right)^2}\cdot(\mathrm{i}\mu)^2\left(\epsilon_{\boldsymbol{m}_0}^{(0)} - \epsilon_{\boldsymbol{m}}^{(0)}\right)^2 \\ &= \mu^2\langle\boldsymbol{m}_0|\hat{b}_1^2|\boldsymbol{m}_0\rangle = \frac{159a_B^6\mu^2}{4096\hbar^4} \end{aligned}$$

$$I_2 = \sum_{\boldsymbol{m}\neq \boldsymbol{m}_0} \frac{|\langle \boldsymbol{m}_0|\hat{x}|\boldsymbol{m}\rangle|^2}{\epsilon^{(0)}_{\boldsymbol{m}_0} - \epsilon^{(0)}_{\boldsymbol{m}}} = \sum_{\boldsymbol{m}} \frac{\langle \boldsymbol{m}_0|\hat{x}|\boldsymbol{m}\rangle\langle \boldsymbol{m}|\hat{b}_1|\boldsymbol{m}_0\rangle}{\epsilon^{(0)}_{\boldsymbol{m}_0} - \epsilon^{(0)}_{\boldsymbol{m}}} \cdot (-\mathrm{i}\mu)\left(\epsilon^{(0)}_{\boldsymbol{m}_0} - \epsilon^{(0)}_{\boldsymbol{m}}\right)$$

$$= -\mathrm{i}\mu\langle \boldsymbol{m}_0|\hat{x}\hat{b}_1|\boldsymbol{m}_0\rangle = -\frac{21a_B^4\mu}{256\hbar^2}$$

$$I_3 = \sum_{\boldsymbol{m}\neq \boldsymbol{m}_0} \frac{|\langle \boldsymbol{m}_0|\hat{r}^2|\boldsymbol{m}\rangle|^2}{\epsilon^{(0)}_{\boldsymbol{m}_0} - \epsilon^{(0)}_{\boldsymbol{m}}} = \sum_{\boldsymbol{m}\neq \boldsymbol{m}_0} \frac{\langle \boldsymbol{m}_0|\hat{r}^2|\boldsymbol{m}\rangle\langle \boldsymbol{m}|\hat{b}_2|\boldsymbol{m}_0\rangle}{\epsilon^{(0)}_{\boldsymbol{m}_0} - \epsilon^{(0)}_{\boldsymbol{m}}} \cdot (-\mathrm{i}\mu)\left(\epsilon^{(0)}_{\boldsymbol{m}_0} - \epsilon^{(0)}_{\boldsymbol{m}}\right)$$

$$= -\mathrm{i}\mu\left(\langle \boldsymbol{m}_0|\hat{r}^2\hat{b}_2|\boldsymbol{m}_0\rangle - \langle \boldsymbol{m}_0|\hat{r}^2|\boldsymbol{m}_0\rangle\langle \boldsymbol{m}_0|\hat{b}_2|\boldsymbol{m}_0\rangle\right) = -\frac{105a_B^6\mu}{512\hbar^2}$$

3.4. Completion of the calculations

(i) Using the calculation of the sums from the previous section, 8 can be simplified:

(a) $$\left(\frac{e^2B^2}{8\mu}\right)^2 \sum_{\boldsymbol{m}\neq \boldsymbol{m}_0} \frac{\left|\langle \boldsymbol{m}|\hat{r}^2|\boldsymbol{m}_0\rangle\right|^2}{\epsilon^{(0)}_{\boldsymbol{m}_0} - \epsilon^{(0)}_{\boldsymbol{m}}} = -\left(\frac{e^2B^2}{8\mu}\right)^2 \cdot \frac{105a_B^6\mu}{512\hbar^2} \sim B^4 \Rightarrow \text{we neglect this term}$$

(b) $$\left(\frac{e\hbar B}{M}\right)^2 \sum_{\boldsymbol{m}\neq \boldsymbol{m}_0} \frac{\left|\langle \boldsymbol{m}|xk_y - yk_x|\boldsymbol{m}_0\rangle\right|^2}{\epsilon^{(0)}_{\boldsymbol{m}_0} - \epsilon^{(0)}_{\boldsymbol{m}}} = \left(\frac{e\hbar B}{M}\right)^2 k^2 \sum_{\boldsymbol{m}\neq \boldsymbol{m}_0} \frac{\left|\langle \boldsymbol{m}|x|\boldsymbol{m}_0\rangle\right|^2}{\epsilon^{(0)}_{\boldsymbol{m}_0} - \epsilon^{(0)}_{\boldsymbol{m}}}$$

$$= \left(\frac{e\hbar Bk}{M}\right)^2 \cdot \left(-\frac{21a_B^4\mu}{256\hbar^2}\right)$$

(ii) Having all the necessary calculations done, the analytical expression for the ground state energy in a weak magnetic field limit can be found, using the perturbation approach:

$$\epsilon(\boldsymbol{k}) = \epsilon^{(0)}_{\boldsymbol{m}_0}(\boldsymbol{k}) + \frac{3e^2a_B^2B^2}{64\mu} - \left(\frac{e\hbar Bk}{M}\right)^2 \frac{21a_B^4\mu}{256\hbar^2} \tag{12}$$

Introducing the magnetic length $l_B = \sqrt{\hbar/(eB)}$ and $l_2 = (3/128)^{1/4}a_B$, the energy can be represented in the form:

$$\epsilon(\boldsymbol{k}) = \frac{\hbar^2\mathbf{k}^2}{2M}\left[1 - 2|\kappa|\left(\frac{a_\mathrm{B}}{l_B}\right)^4\right] - \epsilon_0\left[1 - \left(\frac{l_2}{l_B}\right)^4\right] \tag{13}$$

4. Conclusion

A perturbation theory approach is not applicable for solving the Lippmann–Schwinger equation, but it allows us to find the ground state energy and the scattering matrix elements analytically. They in turn are substituted into the Lippmann–Schwinger equation, which was solved numerically in the main work.

It is interesting to note that the obtained expression for $\epsilon(\boldsymbol{k})$, having the same form as in [3], actually differs from the latter in the constant l_2.

References

[1] Kozin V K, Shabashov V A, Kavokin A V and Shelykh I A 2021 *Phys. Rev. Lett.* **126** 036801
[2] Thomas D G and Hopfield J J 1961 *Phys. Rev.* **124** 657
[3] Arseev P I, Dzyubenko A B 1998 *J. Exp. Theor. Phys.* **87** 200–9
[4] Yang X L, Guo S H, Chan F T, Wong K W and Ching W Y 1991 *Phys. Rev. A* **43** 1186

Generalized Kerker effect in dielectric antennas for enhanced backscattering modulation

Diana Shakirova[1], Dmitry Dobrykh[1,2], Sergey Krasikov[1], Anna Mikhailovskaya[1,2], Ildar Yusupov[1], Alexey Slobozhanyuk[1], Konstantin Ladutenko[1], Dmitry Filonov[3], Andrey Bogdanov[1], and Pavel Ginzburg[2]

[1] Department of Physics and Engineering, ITMO University, Saint Petersburg 197101, Russia
[2] School of Electrical Engineering, Tel Aviv University, Tel Aviv 69978, Israel
[3] Center for Photonics and 2D Materials, Moscow Institute of Physics and Technology, Dolgoprudny 141700, Russia

E-mail: `shakirovajjm@mail.ru`

Abstract. Enhancement of electromagnetic signal modulation is one of the key problems for modern contactless communication systems. Using resonance effects allows to achieve significant interaction between an electromagnetic wave and matter of an antenna, providing opportunity to control scattering. This work demonstrates efficiency of multipole engineering based on Mie theory for dielectric core-shell antennas, particularly we show that generalized Kerker effect is a useful tool for backscattering modulation magnification. Our approach allows to manipulate scattering properties of devices without increasing their size by using all-dieletric concept.

1. Introduction

The modulation of an electromagnetic signal determines the quality of wireless communication systems, which have a broad range of applications: submarine communications [1-2], optical communications [3], frontier space missions [4], etc. To narrow our problem, we focus on radio-frequency identification (RFID) applications, widely used in automated tracking, on-site inspection support systems, location of buried assets [5].

Resonance phenomena can be employed to enhance interaction between an electromagnetic wave and antenna. Single-resonant antennas have a significant drawback described by Chu-Harrington limit [6]: size reduction leads to the drop in operational bandwidth. A possible solution is to work with multi-resonant systems. To save the subwavelength scale of a structure, high permittivity dielectric materials should be used for device production. This method is linked to the concept of dielectric resonant antennas, which allows to decrease the size of a device without a significant bandwidth degradation [7-8].

2. Backscattering modulation enhancement in RFID applications

The quality of communication channel is determined by the power modulation of the signal scattered by an antenna in the direction of a reader [9]:

$$\Delta P^{RCS} = \frac{P_r \lambda^2 G_r^2 \Delta\sigma_{back}}{(4\pi)^3 r^4} \quad (1)$$

where P_r is power of the signal transmitted by a reader, λ is wavelength, G_r is gain of a reader, r is distance between antenna and reader, and $\Delta\sigma_{back}$ is backscattering modulation. All parameters are fixed for the certain application except $\Delta\sigma_{back}$, which becomes a key parameter for optimization. Backscattering modulation is given by the difference between backscattering cross-sections of an antenna in two states maintained by a system:

$$\Delta\sigma_{back} = \sigma_{back}^{max} - \sigma_{back}^{min} \quad (2)$$

To increase this difference we implement a multipole engineering approach by controlling interference between resonances supported by an antenna. Fig.1 presents the conceptual model of the antenna, which consists of the dielectric core with a wrapped split-ring resonator (SRR) and the shell supporting magnetic dipole (MD) and magnetic quadrupole (MQ) resonances, respectively. Core-shell geometry is chosen to apply Mie theory that provides a rigorous analytical solution to scattering problem on the sphere.

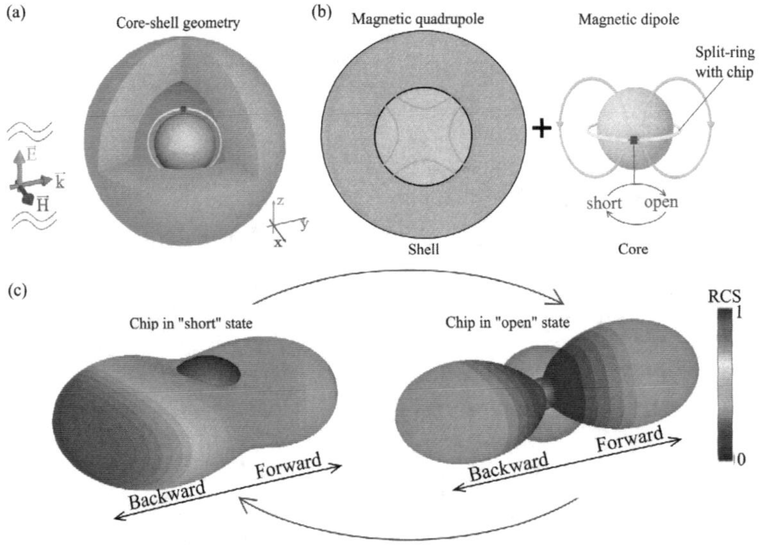

Figure 1. (a) A core-shell structure. Radius of core is 5.5 mm and inner and outer radiuses of the shell are 13.5 mm and 39.5 mm. (b) Principle of operation - magnetic quadrupole resonance in a shell, magnetic dipole in a core. The dipolar resonance is controlled by an impedance of a split ring resonator, wrapped around the sphere. (c) Modulation of radiation pattern for two states.

3. Multipole engineering with Mie theory

At the first step we consider a dielectric spherical particle consisting of the core with radius of 5.5 mm and shell with radius of 39.5 mm separated by an air layer with thickness of 8 mm.

According to Mie theory scattering and backscattering efficiencies are given as the following [10]:

$$Q_{sca} = \frac{2}{\chi_l^2}\sum_{n=1}^{\infty}(2n+1)(|a_n|^2 + |b_n|^2) \quad (3)$$

$$Q_{bk} = \frac{1}{\chi_l^2}\left|\sum_{n=1}^{\infty}(2n+1)(-1)^n(a_n - b_n)\right|^2 \quad (4)$$

where Q_{sca} is scattering efficiency, Q_{bk} - backscattering efficiency, $\chi_l = \frac{\pi d}{\lambda}$ is size parameter (d - diameter of the particle), n - multipole order, and a_n and b_n are scattering Mie coefficients. Implementation of these calculations is performed in PYTHON and results are shown in fig.2. There is a set of color maps, demonstrating the scattering efficiencies as functions of the system parameters. Permittivity of the core and shell are the subject for optimization. The operational frequency is chosen to be 900 MHz, consistently with RFID standards.

Here we consider both lossless and lossy systems. The ranges of permittivity values are 33-49 and 900-930 for the shell and core, respectively. The loss tangent of the materials is taken to be $tan\delta = 5 * 10^{-4}$, consistently with the experimental data reported [11]. The total and backscattering efficiencies are plotted on the maps, and the evolution of resonances is clearly seen. At the crossing of MD and MQ highlighted by a dashed rectangle in fig.2(c) one can observe the highest jump in the backscattering efficiency with a small change of the core permittivity. It corresponds to the different regimes of interference between MD and MQ in frame of generalized Kerker effect. This effect is valid for the lossy system in fig.2(d), which allows to take the MD-MQ intersection as a work point.

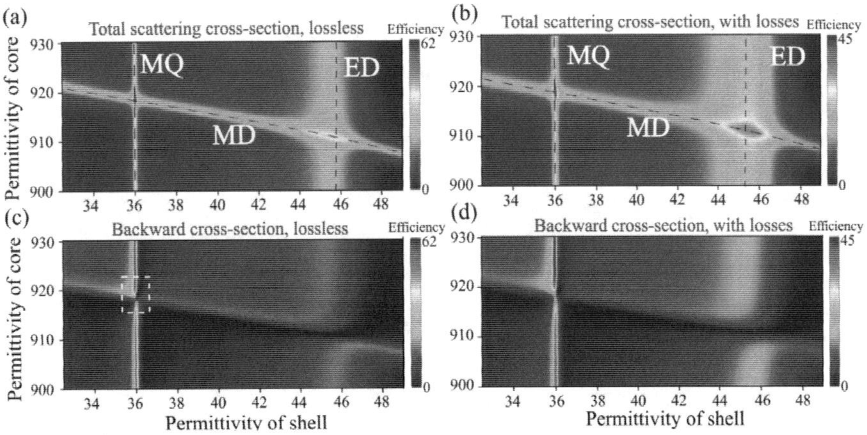

Figure 2. Scattering cross-section colormaps. Total scattering cross-section efficiencies for (a) lossless case, (b) lossy case. Backward scattering cross-section efficiencies for (c) lossless case, (d) lossy case.

4. Resonant antenna with an integrated circuit

The approach discussed is based on permittivity changing of the core, which is inapplicable to current devices. To make our model more practical, we introduce a metallic ring wrapped around the core as an integrated circuit (IC). This ring has a gap to maintain two states: open (state 1) and closed (short, state 2). Switching between these two states is equivalent to changing of the permittivity in the theoretical model.

Numerical calculations of backward cross-section efficiency are performed in CST Microwave Studio and presented in fig.3. The contribution of MD, MQ and ED is observed, and from the plotted difference between open and short states (yellow line) it is clearly seen, that the combination of MD and MQ is the most sensitive to switching between the states.

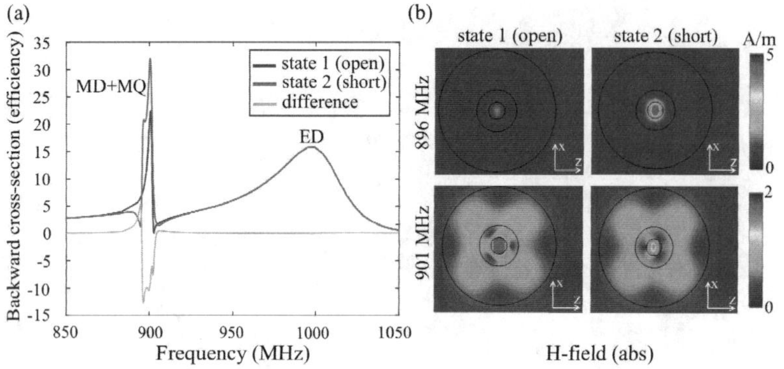

Figure 3. (a) Backscattering spectrum for open and short states and difference between states. (b) Magnetic near field maps - absolute values

Magnetic field amplitudes are calculated and shown in fig.3(b) to demonstrate the dependence of far-field changes on the inductive near-field coupling at the inner volume of the antenna. Two characteristic frequencies equal to 896 MHz and 901 MHz are chosen. At the first one the backscattering modulation achieves its maximal value. Now, it is obvious - the higher the field localization in the core is, the less energy is scattered in the far-field, and this localization is controlled by IC responsible for the switching mechanism.
Remarkable that the model with IC demonstrates less backscattering efficiency than the theoretical model. This effect is linked to the presence of SRR, which decreases the quality factor of MD resonance of the core.

5. Conclusion

We have applied the multipole engineering approach based on Mie theory to the enhancement backscattering modulation problem. Generalized Kerker effect is demonstrated on the dielectric core-shell antenna. We have shown that interference between higher-order resonances provides an increase of backscattering modulation efficiency, and this enhancement takes place for the lossy antenna model with SRR. The presented method allows to develop wireless communication system models, which demonstrate high efficiency without increasing the device size.

Acknowledgments

The research was supported by the Russian Science Foundation (Project No. 19-79-10232), ERC StG "In Motion" (802279), PAZY Foundation, and Israeli Ministry of Science and Technology (Project "Integrated 2D 3D Functional Printing of Batteries with Metamaterials and Antennas.")
D.D. and D.S. contributed equally to this work.

References

[1] Zheng Y, Xiao C, Liu X, Yang T, and Bin Yang W 2008 *Oceans'08 MTS/IEEE Kobe-Techno-Ocean'08—Voyage Toward The Future OTO'08* (Piscataway, NJ: IEEE), pp. 1–6
[2] Filonov D, Steinberg B Z, and Ginzburg P 2017 *Phys. Rev. B* **95** 235139

[3] Seeds A, Shams H, Fice M, and Renaud C 2015 *J. Light. Technol.* B **33** 579
[4] Li H, Tang X, Hang S, Liu Y, and Chen D 2017 *J. Appl. Phys.* **121** 123101
[5] Domdouzis K, Kumar B, Anumba C 2007 *Adv. Eng. Informatics* **21** pp. 350-355
[6] Harrington R 2001 *Time-Harmonic Electromagnetic Fields* 2nd ed. (New York: Wiley-IEEE Press)
[7] Petosa A 2007 *Dielectric Resonator Antenna Handbook* (Norwood, MA: ArtechHouse)
[8] Hopkins B, Filonov D, Miroshnichenko A, Monticone F, Alù A, and Kivshar Y 2015 *ACS Photonics* **2** 724
[9] Bolomey J, Capdevila S, Jofre L, and Romeu J 2010 *Proc. IEEE* **98** 1555
[10] Bohren C and Huffman D 1983 *Absorption and Scattering of Light by Small Particles* (Wiley, New York)
[11] Moussu M, Ciobanu L, Kurdjumov S, Nenasheva E, Djemai B, Dubois M, Webb A G, Enoch S, Belov P, Abdeddaim R, and Glybovski S 2019 *Adv. Matter* **31** 1900912

Supercattering Channels of Nonspherical structurers

H. K. Shamkhi[1], and A. Canós Valero[1]
[1]ITMO University, St. Petersburg, Kronverksky 49, 197101, Russia
E-mail: h.shamkhi@optomech.ifmo.ru

Abstract. A superscattering structure is an efficient energy-mapping device that of particular importance for various electromagnetic experiment methods, with potential sensing and energy harvesting applications. We study in this work the scattering cross-section of outgoing channels in the irreducible and singular basis for an arbitrary shape scatterer. The superscattering status is shown to occur within a single outgoing channel of an optimized cluster of cylinders, a forbidden mechanism in spherically symmetric Mie resonators.

1. Introduction

The expanded and outgoing investigation of the optical responses of high-index low-losses materials has resulted in elaborated design methods to enhance scattering significantly [1]–[8]. This is particularly important to increase efficiency of optical devices and technologies for various applications[5], [9]–[13]. For a given scattering problem, the far-field scattering is fully described by a set of incoming and outgoing spherical waves mapped through the S-matrix. It has been shown that the contribution of a single scattering channel is limited to $2(2n+1)\pi/k^2$ for scatterers with spherical symmetry, where n is the multipole order; and every Mie mode contributes to one multipolar channel [6], [14]. It is therefore established that if scattering exceeds (by far) the dipolar channel limitation ($6\pi/k^2$) it is regarded as superscattering[6]. However, any degradation of the special spherical symmetry allows for, along with the Mie modes, the excitation of the Fabry-Perot modes that oscillate transversely to the scatterer plane and scatter at various channels imposed by the new reduced symmetry. In this scenario, the limitation of the maximum power that can be carried out by a scattering channel (ex. spherical basis) can be determined by a collective of degenerate modes. We attempt in this work to obtain general symmetry-related limitation on these channels. Nevertheless, a structure with limited rotational or reflection symmetries allows for different approaches to enhance the overall scattering [15]. Considering the realization simplicity, we optimize a cylinder as a resonating building block in a symmetric cluster with a discrete point group symmetry (D_{4h}), achieving unrepresented values of superscattering.

2. Theory

To define a new basis for the scattering channels that is a property of the scatterer symmetry, we incorporate two methods acting on the S-matrix as defined in the spherical basis [16]: the singular value decomposition (SVD) and the irreducible representation. Starting from the SVD of the S-matrix, energy conservation requires a limitation on its singular values ($\lambda_w(\mathbf{S}) \leq 1$) [17]. Using the triangle inequality and submultiplicative properties of the SVD method, and the

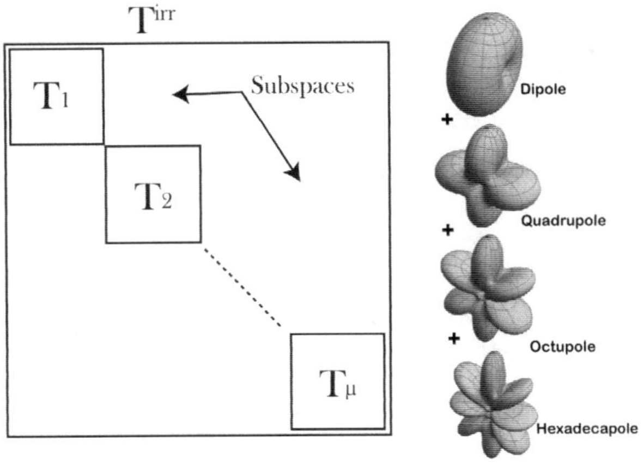

Figure 1. T-matrix block-diagonal form after being projected on the irreducible basis of D_{4h} symmetry with each (T_μ) represents a subspace. On the right side, the schematic illustrates that the differential scattering of each scattering channel in the irreducible basis is a combination of a set of spherical scattering channels.

relation of the S-matrix $\mathbf{S} = 2\mathbf{T} + \mathbf{I}$, we obtain the following limitation on the T-matrix singular values

$$\lambda_w(\mathbf{T}) = \frac{1}{2}\lambda_w(\mathbf{S} - \mathbf{I}) \leq \frac{1}{2}(1 + \lambda_w(\mathbf{S})), \tag{1}$$

which implies also that the T-matrix singular values are equal or less than unity. We then take a step further to exploit the symmetry properties of the scatterer with the projection of the T-matrix on the irreducible basis; which describes the transformation properties of a set of eigenfunctions and corresponds to a distinct energy eigenvalue. Transforming the reducible representation into the irreducible one breaks up the energy levels into fewer degenerate levels. For particles with discrete geometrical symmetries, the T-matrix can be made block diagonal $\mathbf{T}^{\text{irr}} = \mathbf{P}^{-1}\mathbf{T}\mathbf{P} = \text{diag}\left(\mathbf{T}^1, \ldots, \mathbf{T}^C\right)$ using group theory, where $\mathbf{P}$ is an orthogonal projector matrix [16], C is the number of the irreducible representations and it is identical to the number of the classes of the scatterer point group symmetry (see fig.1). Every irreducible representation is a regular matrix with $(w_\mu \times w_\mu)$, where $\mu = 1, \ldots, C$. Now we reapply the SVD for the T matrix in the irreducible basis $\mathbf{W}^{\text{irr}(\dagger)}\mathbf{T}^{\text{irr}}\mathbf{N}^{\text{irr}} = \Lambda^{\text{irr}}$, where $\Lambda^{\text{irr}} = \text{diag}\left(\lambda_{11}, \ldots, \lambda_{\mu w_\mu}\right) \in \mathbf{R}^{w \times w}$ since the projector matrix performs a similarity transformation, the singular values $\lambda_{\mu w_\mu}$ are identical to the singular values of the T-matrix in the reducible basis and thus subject to the aforementioned inequality. The left and right singular matrices $\mathbf{W}^{\text{irr}} \in \mathbf{C}^{w \times w}$ and $\mathbf{N}^{\text{irr}} \in \mathbf{C}^{w \times w}$ are unitary, orthonormal, and take the same block diagonal form of the T-matrix in the irreducible representation. Therefore, every submatrix of the degenerate subspace is, in fact, being decomposed separately from the other subspaces. We exploit this feature to derive the scattering cross section in the following form

$$\sigma_{sca} = \frac{2\pi}{k^2}||\mathbf{Ta}||^2 = \frac{2\pi}{k^2}||\mathbf{P}^{-1}\mathbf{W}^{\text{irr}}\Lambda^{\text{irr}}\mathbf{N}^{\text{irr}(\dagger)}\mathbf{Pa}||^2 = \frac{2\pi}{k^2}\sum_{\mu}\sum_{w_\mu}\lambda^2_{\mu w_\mu}\left|\mathbf{n}^{\mu(\dagger)}_{w_\mu} \cdot \hat{\mathbf{a}}\right|^2, \tag{2}$$

where the vector $\mathbf{a}$ represents the spherical expansion coefficients of the incident wave, the sub-vector $\hat{\mathbf{a}} = \mathbf{Pa} \in \mathbf{C}^{w_\mu \times w_\mu}$ refers to the projected input vector that is transferred by the

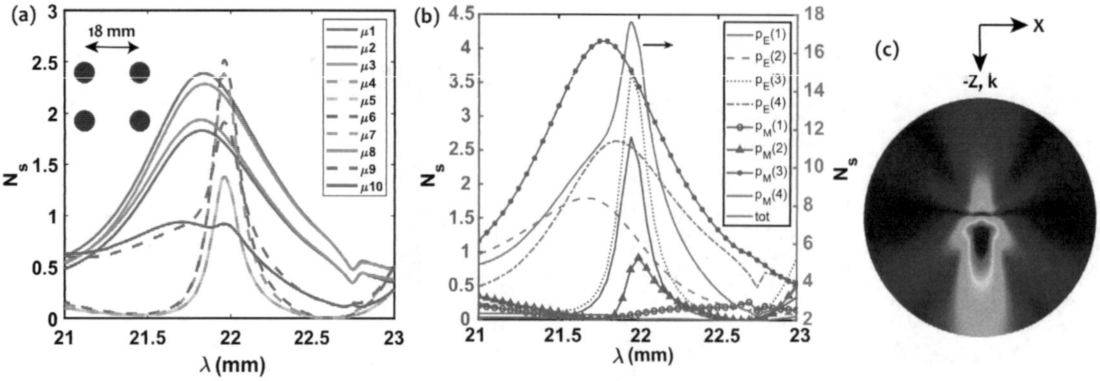

Figure 2. (a) Scattering cross section decomposed according to the power mapped by subspaces (μ) in the irreducible representation of the T-matrix [16]; and normalized to the dipolar-resonance maximum scattering in a spherical object $N_s = \sigma_{sca}/(6\pi/k^2)$. The optimized structure as shown in the inset is a D4h cluster of identical cylinders positioned in the xy-plane around the coordinate origin. The cylinders made from dielectric ceramic materials with height 9.82(mm), radius 3.778(mm), and permitivity 24.5. (b) Spherical multipole decomposition of scattering cross section. For clarity purposes the decomposition here is presented for only the first 4 mulltipoles orders, however to fully recover scattering 9 orders were considered [18]. The legend designate multipole order with either electric (E) or Magnetic polarity (M), and the red line is the total normalized scattering cross section. (c) Norm $|E_x|$ of the total electric field at wavelength maxima $21.95(mm)$. Notice the incident plane wave is x-polarized and propagating in the z-direction.

corresponding submatrix of the degenerate class μ, and $\mathbf{n}_{w_\mu}^{\mu(\dagger)}$ are vectors in the right singular matrix subspaces. Note that the matrices $\mathbf{P}$ and $\mathbf{W}^{\text{irr}}$ are assumed to preserve vector norms and thus are omitted. The scattering cross section of the decomposed T-matrix in the irreducible basis (Eq.2) limits one channel contribution to the energy that is mapped by one degenerate level. Every channel then carries power limited by the corresponding singular value and the inclination between the subvector $(\mathbf{Pa})^\mu$ and the right singular matrix basis.

3. Results

In Fig. 2, we present an optimized result of scattering by a dielectric cluster (see caption for details), these result has nearly tripled what has been shown in literature [6], [14]. Every element in the cluster is a resonating Huygens source with a highly asymmetric scattering pattern, which has reduced the destructive interference in the whole structure. As can be seen from Fig. 2(a), most of the degenerate levels transfer energy that is higher than the spherical dipolar level. However, the total scattering power is more evenly distributed between them if compared to the conventional spherical decomposition (Fig. 2(b)). We also note that the magnetic octupole channel carries four times the power of the spherical dipolar limit. Note that the octupolar channel remains in superscattering status if a normalization standard is considered relative to its maxima in an uncoupled spherical system $(14\pi/k^2)$. The near-filed profile in Fig. 2(c) illustrates the strong and directive concentrations of the input wave in the forward direction.

4. Conclusion

In conclusion, the incident wave and the symmetry of a scatterer have been formed in a rigorous way to determine a limit on a single channel scattering power for arbitrarily-shape scatterer. We also carried out a numerical optimization of a cluster of cylinders and obtain large superscattering values amount up to 18 times that of the maximum dipolar channel in a spherical scatterer.

5. Acknowledgement
The authors gratefully acknowledge the support of the RFBR Grant 20-52-00031 for the multipolar mode decomposition.

References

[1] A. García-Etxarri, R. Gómez-Medina, L. S. Froufe-Pérez, C. López, L. Chantada, F. Scheffold, J. Aizpurua, M. Nieto-Vesperinas, and J. J. Sáenz, *Opt. Express*, vol. 19, (6), (4815–4826), Mar. 2011.

[2] S. Jahani and Z. Jacob, *Nature Nanotechnology*, vol. 11, (23), Jan. 2016.

[3] R. M. Bakker, Y. F. Yu, R. Paniagua-Domínguez, B. Luk'yanchuk, and A. I. Kuznetsov, *Nano Letters*, vol. 17, (6), (3458–3464), Jun. 2017.

[4] K. Baryshnikova, D. Filonov, C. Simovski, A. Evlyukhin, A. Kadochkin, E. Nenasheva, P. Ginzburg, and A. S. Shalin, *Phys. Rev. B*, vol. 98, (165419), 16 Oct. 2018.

[5] P. D. Terekhov, K. V. Baryshnikova, Y. Greenberg, Y. H. Fu, A. B. Evlyukhin, A. S. Shalin, and A. Karabchevsky, *Scientific Reports*, vol. 9, (1), (3438), 2019.

[6] L. Verslegers, Z. Yu, Z. Ruan, P. B. Catrysse, and S. Fan, *Phys. Rev. Lett.*, vol. 108, (083902), 8 Feb. 2012.

[7] A. Mirzaei, A. E. Miroshnichenko, I. V. Shadrivov, and Y. S. Kivshar, *Applied Physics Letters*, vol. 105, (1), (011109), 2014.

[8] A. Mirzaei, I. V. Shadrivov, A. E. Miroshnichenko, and Y. S. Kivshar, *Nanoscale*, vol. 7, (42), (17658–17663), 2015.

[9] K. Koshelev and Y. Kivshar, *ACS Photonics*, vol. 8, (1), (102–112), Jan. 2021.

[10] V. Kozlov, D. Filonov, A. S. Shalin, B. Z. Steinberg, and P. Ginzburg, *Applied Physics Letters*, vol. 109, (20), (203503), 2016.

[11] N. Kostina, M. Petrov, A. Ivinskaya, S. Sukhov, A. Bogdanov, I. Toftul, M. Nieto-Vesperinas, P. Ginzburg, and A. Shalin, *Phys. Rev. B*, vol. 99, (125416), 12 Mar. 2019.

[12] C. R. Simovski, A. S. Shalin, P. M. Voroshilov, and P. A. Belov, *Journal of Applied Physics*, vol. 114, (10), (103104), 2013.

[13] P. D. Terekhov, H. K. Shamkhi, E. A. Gurvitz, K. V. Baryshnikova, A. B. Evlyukhin, A. S. Shalin, and A. Karabchevsky, *Optics Express*, vol. 27, (8), (10924), Apr. 2019.

[14] V. I. Shcherbinin, V. I. Fesenko, T. I. Tkachova, and V. R. Tuz, *Phys. Rev. Applied*, vol. 13, (024081), 2 Feb. 2020.

[15] A. C. Valero, H. K. Shamkhi, A. S. Kupriianov, V. R. Tuz, A. A. Pavlov, D. Redka, V. Bobrovs, Y. S. Kivshar, and A. S. Shalin, *arXiv*, vol. 2105.13119, 2021.

[16] M. Kahnert, *Journal of the Optical Society of America A*, vol. 22, (6), (1187), 2005.

[17] S. Molesky, P. Chao, W. Jin, and A. W. Rodriguez, *Physical Review Research*, vol. 2, (3), (033172), Jul. 2020.

[18] J. D. Jackson, *Classical Electrodynamics*, Third. New York: Wiley, Aug. 1998.

Polarized Raman spectroscopy of GaP nanowires under 5% elastic strain

Vladislav Sharov,[1, 2, a)] Vladimir Fedorov,[1] Prokhor Alekseev,[2] and Ivan Mukhin[1, 3]

[1)]*Alferov University, Saint-Petersburg 194021, Russia*

[2)]*Ioffe institute, Saint-Petersburg 194021, Russia*

[3)]*ITMO University, Saint-Petersburg 197101, Russia*

Abstract. Optical porperties of highly-strained gallium phosphide nanowires were investigated via polarized Raman spectroscopy. 5% elastic strain was created in individual nanowire lying on nickel substrate by the means of atomic force microscopy. Micro-Raman mapping along the nanowire cross section in parallel and perpendicular polarization was carried out. Strain-induced effects on transverse optical mode position and shape were analyzed. The pronounced splitting of the mode due to high level of strain was observed. It was found that in parallel polarization the mode shape is sensitive to the position of the pumping spot which can be attributed to enhanced light-nanowire coupling effects.

INTRODUCTION

Semiconductor nanowires (NWs) are prospective building blocks for next generation nanoscale electronic devices [1]. Due to high surface-to-volume ratio and surface strain relaxation, NWs can withstand large elastic strains (up to 15%) [2]. Such strains can significantly modify semiconductor properties due to the changes in symmetry and energy structure. Therefore, strain engineering, i.e. utilizing mechanical stress for NW-based device performance improvement, is of interest. Successful tuning or enhancement of NW optical and transport properties was recently shown in numerous papers. Thus, red shift of photoluminescence of ZnO [3] and GaAs [4] NWs under applied tension was reported. Also, the enhancement of carrier mobility in Si NWs up to 5 times was achieved under 2% strain [5]. Recently we utilized elastic strain for the enhancement of InGaAs NW conductivity [6] and GaAs NW solar cell efficiency [7]. Here we investigate stress-induced changes in Raman spectra of the bent GaP NWs. GaP NWs are prospective for nanophotonics as they demonstrate effective wave-guiding properties, exhibit one of the broadest transparency range, high values of the nonlinear refractive index and good thermal conductivity. Raman spectroscopy is conventionally used for strain analysis in semiconductors. Recently, some strain-induced effects in the Raman spectra of GaP NWs with 3.5% strain were revealed and explained [8, 9]. Higher strain level allows better understanding of strain-induced phenomena. In a logical continuation of the preceding works, here we provide data on the Raman spectra of the NWs with 5% strain.

RESULTS AND DISCUSSION

GaP NWs were grown in self-catalytic vapour-liquid-solid regime on (111) Si using solid-source molecular beam epitaxy, the main details about the growth process may be found elsewhere [10]. NW average length exceeded 25 μm (see SEM image on figure 1a). For subsequent study NWs were dispersed on nickel substrate by mechanical rubbing (see figure 1b). Figure 1c shows polarized Raman stupy of a straight NW. Optical measurements were carried out using Horiba LabRam HR 800 confocal Raman microscope equipped with 532 nm laser and aberration-corrected

[a)]Corresponding author: vl_sharov@mail.ru

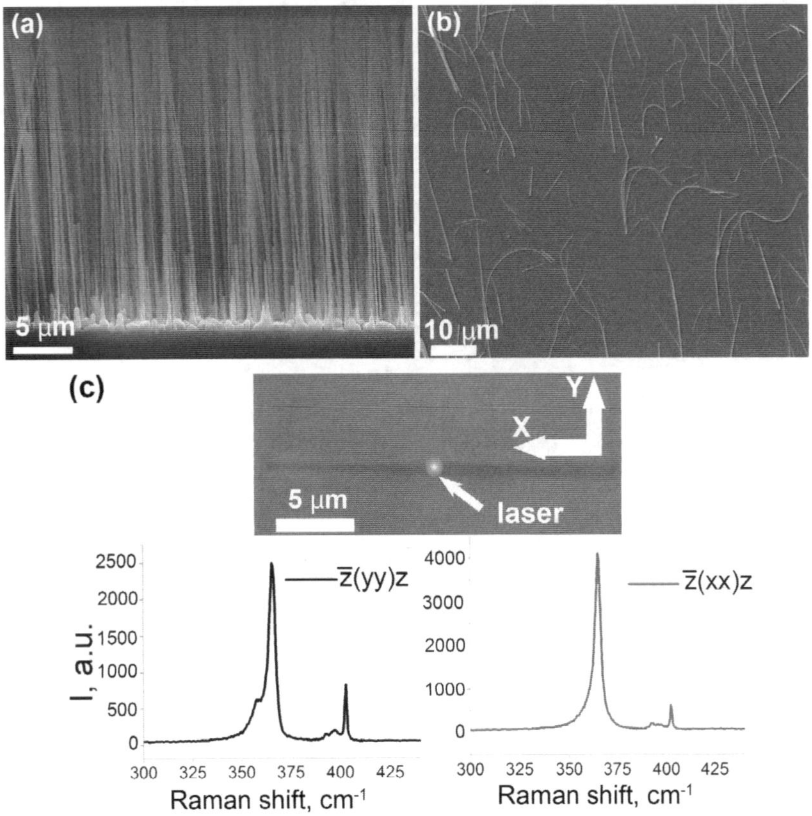

FIGURE 1. SEM image of studied GaP nanowire array (a) on the growth substrate, (b) on nickel substrate.

$\times$100 objective with 0.9 N.A. and optical resolution 280 nm. First, we conducted Raman study of a straight NW in two polarization geometries: ($\bar{z}(xx)z$ (parallel) and $\bar{z}(yy)z$ (perpendicular) accroding to Porto's notation [11]). Pronounced transverse optical (TO), surface optical (SO) and longitudinal optical (LO) modes appear at 366, 395 and 402 cm^{-1} respectively. Raman selection rules are not fully masked by antenna effect as NW diameter (around 120 nm) is large enough. Full width on half maximum (FWHM) of TO mode doesn't exceed 4 cm^{-1}.

Then, individual NWs were bent using atomic force microscope probe as a nanomanipulator. Due to the great NW length, the elastic energy was balanced with van der Waals forces between NW and the substrate. AFM topography of the bent NW is shown in figure 2a. The strain level can be estimated from the topography using the following expression: $\varepsilon = r/R$, where ε is elastic strain, r is the NW radius and R is the NW curvature radius. In our case, r=60 nm and R=1200 nm giving ε=5%. According to the latest study, Young modulus of zinc blende GaP NW is approximately 150 GPa [12] so the obtained strain level corresponds to 7.5 GPa stress which is much higher than the fracture toughness of bulk GaP single crystals (0.6 - 0.8 GPa) [13]. One-dimensional micro-Raman profiling along the cross section of the bent NW with the step size 90 nm in both polarizations was carried out (see figure 2b). The results for parallel and perpendicular polarizations are shown in figure 2c and d respectively. Y coordinate of the maps corresponds to the pumping spot coordinate. It can be seen, that in case of parallel polarization, the spectrum shape changes with this coordinate. On the contrary, in case of perpendicular polarization, the shape is constant. This is illustrated with figure 2e and f which show certain spectra from the maps. Spectra 1 correspond to the pumping spot coordinate near the tensed NW side. In case of spectra 2, the pumping spot is located at the center of NW cross section, and the impact from tensed and compressed regions is equal. Spectra 3 correspond to the case when the pumping spot is located near the compressed side.

The results are different as compared to the similar study that was recently carried out for bent GaP NW on glass

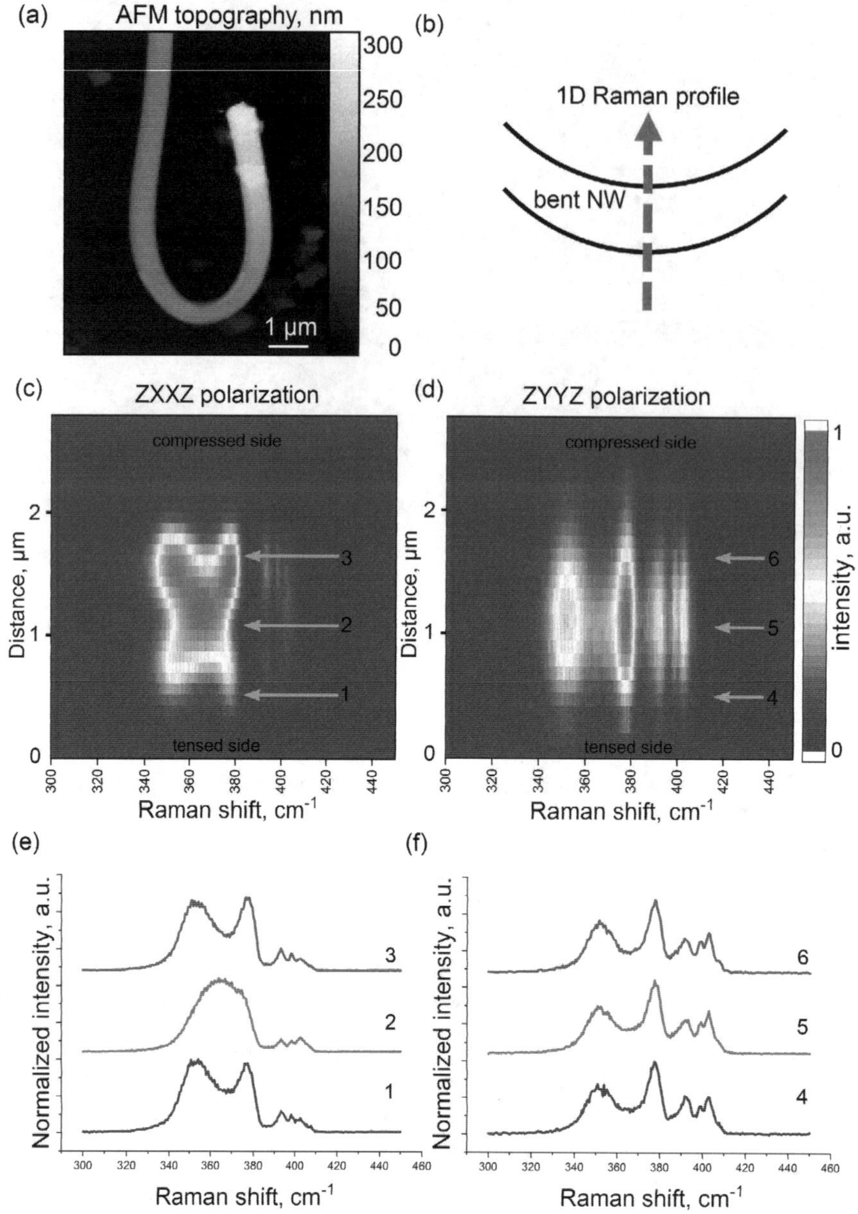

FIGURE 2. (a) AFM topography of the bent nanowire, (b) Scheme of Raman profiling, (c) Raman profile of the bent nanowire in parallel polarization, (d) Raman profile of the bent nanowire in perpendicular polarization, (e) The Raman spectra from figure 2c, (f) The Raman spectra from figure 2d.

[9] which showed only nonlinear broadening and linear shift of TO mode. According to deformation potential theory, tension and compression induce red and blue shift of the TO mode respectively [14]. Thus, bending deformation should lead to the uniform broadening. However, when the substrate material is changed to nickel, the broadening is masked due to substrate-induced effects. The uniform broadening only appears when the pumping coordinate is at the center of the NW cross section in parallel polarization (spectrum 2 in figure 2e). In other cases, the mode splitting is observed: TO mode splits into two components with maximums at 350 and 378 cm^{-1}. The spectra obtained in

perpendicular polarization were found not to be sensitive to the coordinate of pumping spot though showing a mode splitting. LO mode also undergoes strain-induced broadening and splitting in both polarizations, but its accurate analysis is hindered by the proximity to the SO mode. The obtained effects highlight the enhanced Raman signal from NW sides which may be connected with light-nanowire coupling effects leading to a specific field distribution inside the NW.

CONCLUSION

To conclude, we provided polarization-dependent Raman study of individual GaP nanowire under 5% elastic strain corresponding to 7.5 GPa stress. The studied NWs were lying on a nickel substrate. Strain-induced non-uniform broadening and splitting of TO and LO modes were observed. In perpendicular polarization, the spectrum shape is independent from pumping coordinate, the TO mode is split into two components with the maximums at 350 and 378 cm^{-1}. In parallel polarization, the TO mode is either split or not depending on the pumping coordinate. The results are different from previous studies and cannot be fully explained with deformation potential theory. The impact from local highly-strained NW regions is found to be more pronounced, which we attribute to substrate-induced effects. The comprehensive theoretical study of this phenomena needs to be done.

ACKNOWLEDGMENTS

The work was supported by the ministry of science and higher education of the Russian Federation (0791-2020-0005). V.S. and V.F. thank the Russian Foundation for Basic Research (RFBR, project number 19-32-90230) for the support of the optical measurements.

REFERENCES

1. M. Meyyappan and M. K. Sunkara, *Inorganic nanowires: applications, properties, and characterization* (CRC Press, 2018).
2. M. Yashinski, H. Gutiérrez, and C. Muhlstein, "On the origins of anomalous elastic moduli and failure strains of gap nanowires," Nanotechnology **28**, 065703 (2017).
3. R. Chen, Q.-L. Ye, T. He, T. Wu, and H. Sun, "Uniaxial tensile strain and exciton–phonon coupling in bent zno nanowires," Applied Physics Letters **98**, 241916 (2011).
4. G. Signorello, S. Karg, M. T. Björk, B. Gotsmann, and H. Riel, "Tuning the light emission from gaas nanowires over 290 mev with uniaxial strain," Nano letters **13**, 917–924 (2013).
5. Y.-M. Niquet, C. Delerue, and C. Krzeminski, "Effects of strain on the carrier mobility in silicon nanowires," Nano letters **12**, 3545–3550 (2012).
6. P. A. Alekseev, V. A. Sharov, M. S. Dunaevskiy, D. A. Kirilenko, I. V. Ilkiv, R. R. Reznik, G. E. Cirlin, and V. L. Berkovits, "Control of conductivity of in x ga1−x as nanowires by applied tension and surface states," Nano letters **19**, 4463–4469 (2019).
7. P. A. Alekseev, V. A. Sharov, B. R. Borodin, M. S. Dunaevskiy, R. R. Reznik, and G. E. Cirlin, "Effect of the uniaxial compression on the gaas nanowire solar cell," Micromachines **11**, 581 (2020).
8. H. S. Im, K. Park, J. Kim, D. Kim, J. Lee, J. A. Lee, J. Park, and J.-P. Ahn, "Strain mapping and raman spectroscopy of bent gap and gaas nanowires," ACS omega **3**, 3129–3135 (2018).
9. V. A. Sharov, A. D. Bolshakov, V. V. Fedorov, S. Bruyere, G. E. Cirlin, P. A. Alekseev, and I. S. Mukhin, "Deep-subwavelength raman imaging of the strained gap nanowires," The Journal of Physical Chemistry C **124**, 14054–14060 (2020).
10. V. Fedorov, A. Bolshakov, L. Dvoretckaia, G. Sapunov, D. Kirilenko, A. Mozharov, K. Y. Shugurov, V. Shkoldin, G. Cirlin, and I. Mukhin, "Self-catalyzed mbe-grown gap nanowires on si (111): V/iii ratio effects on the morphology and crystal phase switching," Semiconductors **52**, 2092–2095 (2018).
11. D. L. Rousseau, R. P. Bauman, and S. Porto, "Normal mode determination in crystals," Journal of Raman Spectroscopy **10**, 253–290 (1981).
12. P. A. Alekseev, B. R. Borodin, P. Geydt, V. Khayrudinov, K. Bespalova, D. Kirilenko, R. Reznik, A. Nashchekin, T. Haggren, E. Lähderanta, *et al.*, "Effect of crystal structure on the young's modulus of gap nanowires," Nanotechnology (2021).
13. K. Hayashi, M. Ashizuka, R. Bradt, and H. Hirano, "Cleavage of gallium phosphide," Mater. Lett. **1**, 116–118 (1982).
14. F. Cerdeira, C. Buchenauer, F. H. Pollak, and M. Cardona, "Stress-induced shifts of first-order raman frequencies of diamond-and zinc-blende-type semiconductors," Phys. Rev. B **5**, 580 (1972).
15. P. Alekseev, V. Sharov, M. Dunaevskiy, G. Cirlin, R. Reznik, and V. Berkovits, "Electromechanical switch based on in x ga 1−x as nanowires," Semiconductors **52**, 1833–1835 (2018).

Spatial mapping of optical modes in plasmonic nanoantenna by scanning tunneling microscopy

V A Shkoldin[1,2], D V Lebedev[1,3,4], A M Mozharov[1], D V Permyakov[2], L N Dvoretckaia[1], A O Golubok[4], A A Bogdanov[2], A K Samusev[2], I S Mukhin[1,2]

1 Alferov University, 8/3 Khlopina str., St. Petersburg, 194021 Russia
2 ITMO University, 9 Kronverksky pr., St. Petersburg, 197101 Russia
3 Saint Petersburg State University, 7/9 Universitetskaya nab., St. Petersburg, 199034 Russia,
4 Institute for analytical instrumentation RAS, 26 Rizhskii pr., St. Petersburg, 190103 Russia

shkoldin@spbau.ru

Abstract. Using of inelastic electron tunnelling is very promising approach to study of subwavelength photons and plasmons sources. Such sources are very important for improving of on-chip data processing. One of the ways for development of efficient and compact optical electrically-driven sources is using of nanoantenna placed into the tunnel junction. In this work, singe optical nanoantenna was investigated under ultra-high vacuum and ambient conditions. Photon maps of nanoantenna excited under scanning tunnel microscope tip was observed and the obtained results was compared with the theoretical predictions of electromagnetic near-field distribution.

1. Introduction

In recent years, the development of electrically controlled nanoscale photon sources has become very important in the field of integrated optoelectronics. Despite numerous advantages, conventional approaches in this field, such as Fabry-Perot semiconductor lasers, solutions based on nanowires or microdisc quantum dot lasers, have resonators with dimensions significantly exceeding the operating wavelength, which limits the density of elements on a microchip. A promising approach for the implementation of nanoscale light sources is the use of a tunnel contact between two metal surfaces positioned at a sub-nanometer distance. Light emission from a metal-insulator-metal (MIM) tunnel junction was first experimentally demonstrated in the pioneering work of Lambe and McCarthy [1]. The phenomenon of photon generation was interpreted in terms of inelastic (without energy conservation) tunneling of electrons (IET), excitation of surface plasmons and their transformation into photons due to scattering by surface roughness. However, MIM type systems have one significant disadvantage, which is the extremely low efficiency of photon generation in the process of inelastic electron tunneling, namely one generated photon per 10^4-10^6 electrons. Thus, the task of increasing the quantum efficiency of these processes is actual.

To solve this problem plasmon nanoantenna can be used. The nanoantenna placed into tunnel junction can increase internal quantum efficient of IET up to 10% [2, 3]. The increased local density of optical states (LDOS > 10^5) of metal nanostructure is responsible for this enhancement.

In this work the effect of a single nanoantenna on IET was investigated with scanning tunneling microscopy (STM) under ultrahigh vacuum (UHV) and ambient conditions. The used home-build STMs were coupled with optical systems to observe light emission from the tunnel junction. In addition, current-voltage measurements of tunnel junction were carried out.

2. Sample

The investigated nanoantennas were made by e-beam lithography and vacuum thermal evaporation. The gold nanoantennas had the form of cylinders and were located on the single-crystal gold film covering the mica substrate. Their height and width were equal to 50 nm and 180 nm, respectively. The crystallization of gold was made by annealing in UHV [4-6].

3. Experimental setup

The experiments were carried out with the use of two different STM setups. The first STM was UHV Omicron VT650 (Germany) with a vacuum not worse than 8.1e^{-9} mBar. Before the measurements STM chamber with loaded samples and tips was baked out for 8 hours at 150 °C. High quality cut Pt/Ir wires (DPT10, Bruker, USA) were used as tips. The coupled to STM optical system consisted of single photon counter IDQuantique ID120 (Switzerland) and allowed to perform STM-induced light emission mapping. The second STM was based on AIST-NT SmartSPM (USA) operating at ambient conditions. This setup allowed to collect light emission in high aperture objective, which was connected to a spectrometer Princeton Instruments SpectraPro-2500i (USA) equipped with PyLoN:400BR_eXcelon detector with liquid nitrogen cooling. For measurements at ambient conditions, we used tips made from Pt wire (300 mcm in diameter).

For UHV experiments the studied samples were loaded into the vacuum chamber and thermally annealed to remove a water film from surface. Then, STM tip was positioned under the single gold nanoantenna and spatial maps of tunnel electron-induced emission were obtained. However, the intensity of the collected light was not enough to obtain the spectrum. Additionally, current-voltage IV-measurements were caried out at numerous of points along the line crossed the nanoantenna.

To obtain STM light emission spectrum the measurements were performed at ambient conditions, where thin film of water is always present on the surface. This fact requires to use relatively high STM bias voltages (exceeding 2.5 V) in order to acquire STM-induced light maps. The STM scanning was carried out at relatively slow speed of 500 nm/sec and optical spectra was obtained with relatively short times (100 ms).

4. Experiment

At the first stage of experiments, the measurements were made in UHV STM. STM topography and maps of the emission intensity from the tunnel contact with single nanoantennas were acquired simultaneously. The results of mapping of the nanoantenna showed a nonuniform distribution of radiation pattern of the studied nanostructure. The maximum radiation intensity was observed at the edge of the disk (Figure 1a,b).

The numerical simulations show that two optical modes can be excited in the nanoantenna. With the probe positioned in the center of the nanoantenna, only one mode at 668 nm can be excited. When the STM probe is positioned at the edge of the nanoantenna, two modes can be efficiently excited at 668 nm and 566 nm (Figure 1c). Both modes have a «hot spot» at the edge of the nanoantenna, which leads to increasing LDOS at the periphery of the structure. This explains the increasing of the emission process efficiency from the tunnel contact, which was observed in the experiment (see Figure 1 a, b). IV-curves analysis also shows a change of LDOS at the edge of the nanoantenna.

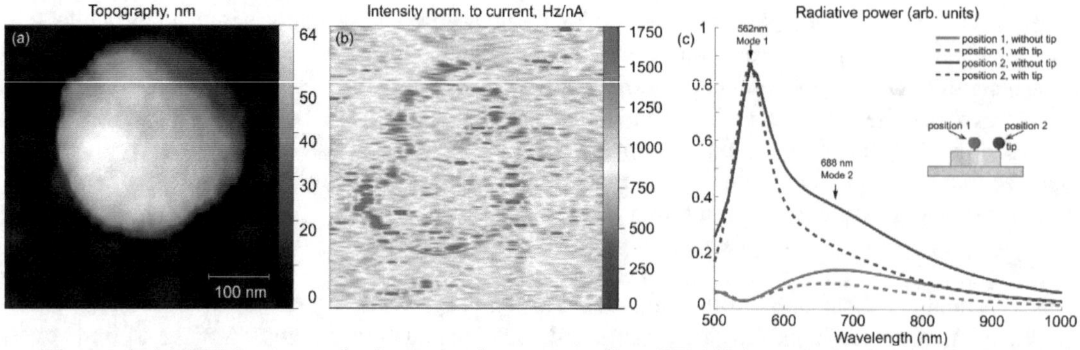

Figure 1. (a) STM topography image and corresponding STM light emission map of a single nanodisk on the surface of a gold film. (c) Simulated spectrum of radiative power of the system for different tip positions.

The optical spectra of the nanoantenna STM-induced emission contain two clear bands with maxima at 670 nm and 765 nm (Figure 2). The long-wavelength band dominates in the spectra obtained on the substrate (see green lines in Figure 2). And the short-wavelength band dominates in the spectra obtained on the nanoantenna (see blue lines in Figure 2). When the STM tip was located at the edge of nanodisk, optical spectra had the enhanced short-wavelength component. These experimental results qualitatively correspond to the results of numerical simulations, at the periphery of nanodisk the mode associated with the shorter wavelength component can be excited. We suggest that the difference of the peak positions predicted by modelling can be caused by the changing of metal work function at ambient conditions, as well as the imperfections of disk fabrication. The simulation was performed for an ideal cylinder in vacuum, while in air the sample surface is covered with a thin layer of water, which affects the parameters of the tunnel contact, including the energy position of Fermi level. The low selectivity of mode excitation is most likely related to the relatively large diameter of the STM tip, which is comparable with the size of the nanoantenna.

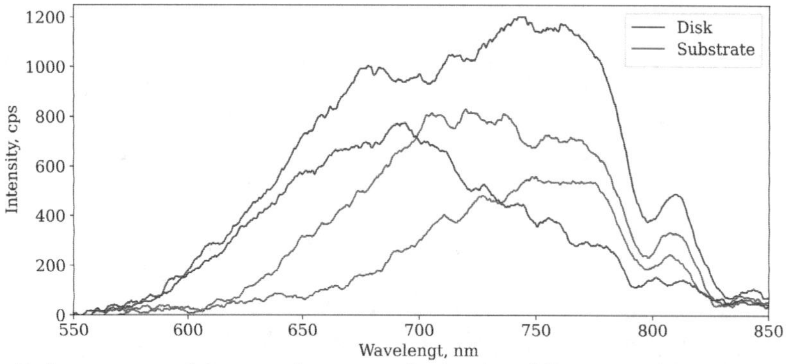

Figure 2. Radiation spectra of the tunnel contact obtained at different positions. The blue lines are spectra obtained from the tunnel contact with a nanoantenna. The green lines are typical spectra obtained from the gold substrate.

5. Conclusion

The single nanoantennas were investigated by STM at UHV and ambient conditions. The spatial distribution map of STM-induced light intensity was acquired. It was shown that radiation power depends on the tip position relative to the nanoantenna. The experimental results correspond well to the numerical simulations. Also, optical spectra of STM-induced light emission from gold nanoantenna were obtained, and they are governed by the optical resonances of the investigated nanostructures.

Acknowledgments
This work was done with the financial support of the Russian Federation President Council (MK-2428.2020.2) and the Russian Foundation for Basic Research (project 19-32-90028). D.V. Lebedev and A.O.Golubok thanks for financial support in the formation of gold samples the Russian Science Foundation (Project No. 21-79-10346)

References
[1] Lambe J and McCarthy S L 1976 *Phys. Rev. Lett.* **37** 923–5
[2] Doderer M and et al. 2019 Light Emission from a Waveguide Integrated MOS Tunnel Junction 2019 *Conf. Lasers Electro-Optics, CLEO 2019 - Proc.* 3–4
[3] Parzefall M and Novotny L 2019 Optical antennas driven by quantum tunneling: a key issues review *Reports Prog. Phys.* **82** 112401
[4] Lebedev D V, Mozharov A M, Bolshakov A D, Shkoldin V A, Permyakov D V, Golubok A O, Samusev A K and Mukhin I S 2020 Indirect Detection of the Light Emission in the Local Tunnel Junction *Phys. Status Solidi - Rapid Res. Lett.* **14**
[5] Lebedev D V, Shkoldin V A, Mozharov A M, Permyakov D V, Dvoretckaia L N, Bogdanov A A, Samusev A K, Golubok A O and Mukhin I S 2021 Scanning Tunneling Microscopy-Induced Light Emission and I(V) Study of Optical Near-Field Properties of Single Plasmonic Nanoantennas *J. Phys. Chem. Lett.* **12** 501–7
[6] Shkoldin V A and et al. S 2019 Crucial Role of Metal Surface Morphology in Photon Emission from a Tunnel Junction at Ambient Conditions *J. Phys. Chem. C* **123** 8813–7

A Mode Decoupling Method for Circular Patch Antennas

M. Siganov,[1] **S. Glybovski**[1], **D. Tatarnikov**[1,2,3]

[1)]Department of Physics and Engineering, ITMO University, St.Petersburg, Russia
[2)]Topcon Positioning Systems, Derbenevskaya nab. 7/22, Moscow, Russia
[3)]Moscow Aviation Institute, Moscow, Volokolamskoe shosse, 4

Abstract. In this contribution, we investigate the effect of mode decoupling for dual-feed circular patch antennas. We consider an antenna comprising a circular disc patch placed inside an annular-ring patch. Both patches are resonant radiators, which can be fed by two different vertical posts. Due to the field interaction of the patches, the feeds experience parasitic cross-talk limiting the receive diversity performance of the system. Here, we demonstrate the possibility to suppress the cross-talk at the operational frequency by including a capacitive decoupling element into the circular patch. The physical mechanism of the method consists in the excitation of two different eigenmodes (TM01 and TM11) in a certain combination canceling the mutual coupling. The presented decoupling method can be used for patches operating either at the same frequency or at two different frequencies. In the proposed dual-feed antenna it is possible to achieve an isolation level of better than -45 dB. In contrast to the method of decoupling and matching network used for the same type of antennas, one does not need couplers to feed the patches, which reduces losses.

INTRODUCTION

Patch antennas are widely used in satellite navigation and telecommunication systems. Dual-feed antennas are used in cases when the operation of several different services is required. The system may need to operate at two different frequencies with different radiation patterns (e.g.combining satellite navigation and mobile communication services). Alternatively, both antennas operate at the same frequency providing different patterns (the case of receive diversity improving the quality and reliability of a wireless link). Insufficient port isolation is one of the significant disadvantages of dual-feed antennas. Usually, it is hard to achieve a level of better than -15 dB, which is not enough for some applications. On the other hand, the need of miniaturization imposes limitations to the geometry and the size of the antenna making the isolation level even worse. Usually, it is possible to improve the isolation using decoupling and matching networks (DMNs), but DMNs require devices like -3 dB branchline-couples which introduce losses [1]. This approach allows to achieve a total efficiency level of 63.1%-85.1% [2]. In this work, we propose a way of isolating the ports of a dual-feed antenna composed of a circular disc patch surrounded by an annular-ring patch at the periphery both having the same ground plane as shown in Fig. 1(a). The design combining a ring and disc patches is known [3], but we introduce a new element, i.e. the decoupling capacitance connected between the disc patch and the ground plane at the antenna center.

RESULTS

The decoupling element can be designed as one or two circular metal plates planes connected to the centers of the disc patch and the ground plane by a metal post and introducing a structural capacitance between them as illustrated in Fig. 2(a). A layer of a dielectric is placed between the plates, and it is possible to adjust the capacitance of the element choosing the permittivity of the dielectric. The decoupling capacitance can be also realized as a surface-mount-device (SMD) as shown in Fig. 2(b) or have a tooth-wheel shape, shown in Fig. 2(c), which allows one to finely adjust the area of the capacitor plate after the antenna has been manufactured.

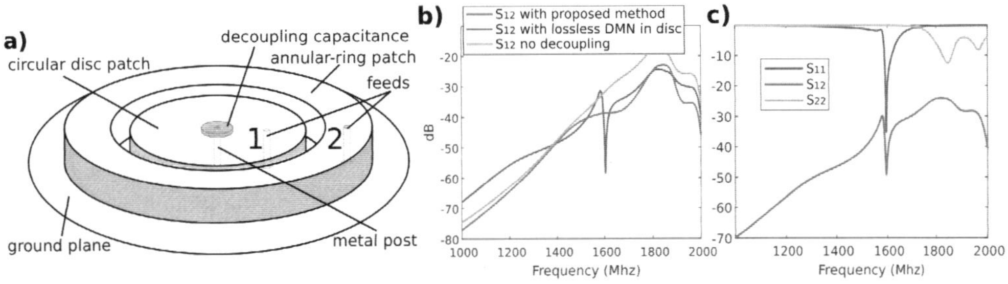

Fig 1: Design and characteristics of the proposed antenna: a) Geometry of the proposed antenna b) S_{12} without decoupling, with DMN and with the proposed decoupling capacitance; c) S-parameters of the proposed proposed antenna; all S-parameters for the antenna decoupled with our method.

Let us consider an example, in which the ring patch is tuned to the resonant frequency of TM01 mode to provide the monopole-like radiation pattern. However, the single feed excites also mode TM11 due to the asymmetry of its position relative to the center. Both modes are responsible for the coupling with the disc patch. However, TM01 mode produces the main contribution to the radiation pattern of the ring. For the disc patch, the situation is quite similar. If it is tuned to the resonance of TM11 mode, it is the dominant mode for the creation of far fields, while TM01 is a parasitic mode. Again, both modes are responsible for the coupling with the ring patch. The proposed decoupling capacitance affects the resonant frequency of TM01 mode of the disc and controls the strength of

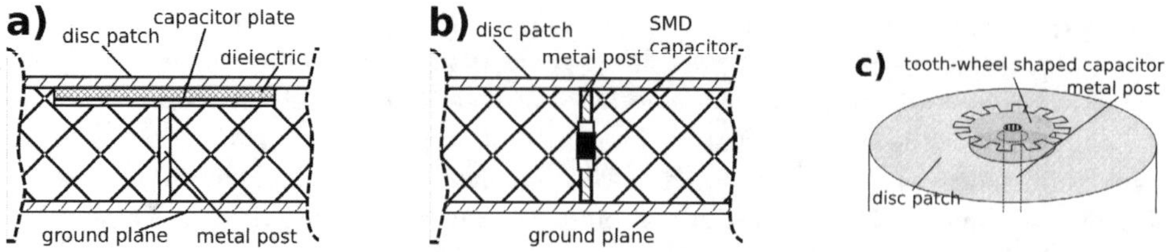

Fig 2: Possible implementations of decoupling capacitance

this mode in the excitation by the feed, without affecting mode TM11. By changing the capacitance, one can achieve mutual compensation of the two mechanisms of coupling between the disc and ring (through their TM01 modes and TM11 modes).

The selective tuning of TM01 mode of the disc is possible since that TM11 mode has zero electric field at the position of the central capacitance. As shown in the example below, when adjusting the capacitance, the decoupling frequency changes together with the resonance of TM01 mode of the disc. The method of selective tuning of mode TM01 using a central capacitance is known in the literature for combining two modes in the same patch resonator [4]. However, here we use it for decoupling between two different patches. In our method, when decoupled, both patches support a superposition of two exited modes (TM01 and TM11). Therefore, our method is an extension of the mode decoupling technique, previously applied to loop [5] and dipole [2] antennas, to patch antennas. The proposed method can be useful since the use of two separate patches gives more opportunities for choosing the operating frequencies of the antenna and the services with which the antenna works.

To demonstrate the achievable level of isolation, numerical simulations were made in CST Microwave Studio 2020. Fig. 1(b) shows the S_{12} parameter for a case without decoupling, with a DMN method (two feeds in the disc patch are driven with a phase shift of 180°) and with a decoupling capacitance. The properties were calculated for the antenna with the disc patch radius of 35 mm, the inner ring patch radius of 36.5 mm, and the outer ring patch radius of 45 mm. The values of the dielectric permittivity of the disc and ring radiators were 1.8 and 21, accordingly. The disc radiator was tuned and matched at the frequency of 1595 MHz (port 1), while the ring radiator was tuned and matched at the frequency of 1843 MHz (port 2) as it is seen from Fig. 1(c). It was numerically calculated that by using a decoupling element with the capacitance of 0.51 pF, it is possible to achieve a minimum of S_{12} at a 1595 MHz (resonant frequency of TM01 of the disc patch) with the improvement by more than 20 dB. For the comparison, the decoupling with a DMN improves the isolation over a wide frequency band, but only by 10 dB. Figures 3-4 show the radiation patterns for the case without decoupling, with a DMN, and with the proposed method. It can be seen that the proposed decoupling method leads to some asymmetry distortion in E-plane, which is, however, acceptable for receive diversity applications. This distortion is a price to pay for the coupling compensations and is due to the excitation of TM01 mode in the disk patch. The resulting radiation pattern is produced mostly

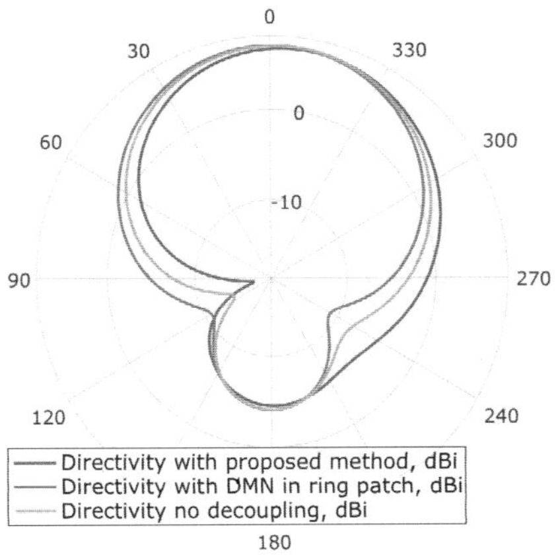

Fig 3: Radiation pattern in E-plane

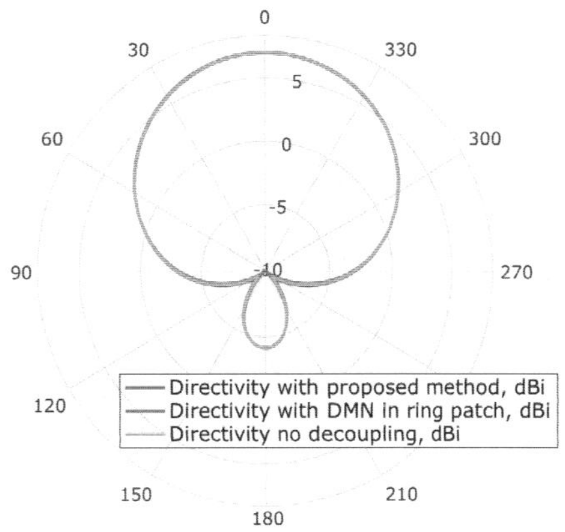

Fig 4: Radiation pattern in H-plane

by TM11, so the pattern shape compares to the case with no decoupling in the upper hemisphere, while having differences of 2-6 dB in the lower hemisphere (in directions of weak radiation). The TM11 mode of the disk has dominant contribution to the pattern, while the undesirable radiation of TM01 mode can be considered as minor.

CONCLUSION

In this work, the method of eigenmode decoupling was extended to circular patch antennas. This decoupling method improves the isolation between the annular-ring and disc patch antennas each one having only one feed. Unlike the existing DMN method, the proposed method does not require couplers, which implies lower losses. As shown via numerical simulations, the method achieves an isolation level of better than -45 dB but distorts the shapes of the antenna radiation patterns.

ACKNOWLEDGMENTS

The work was funded by Topcon Positioning Systems under contract № 42/768, PCT RU 2021/000088 March 02, 2021 (patent pending)

References

[1] Coetzee J C and Yu Y 2008 *IEEE Trans. Antennas Propag.* **56** 1587–1593

[2] Sun L, Li Y, Zhang Z and Wang H 2020 *IEEE Trans. Antennas Propag.* **68** 3423–3432

[3] Herscovici N, Christodoulou C, Rajo-Iglesias E, Quevedo-Teruel O and Sanchez-Fernandez M 2008 *IEEE Antennas and Propagation Magazine* **50** 197–205

[4] Wei K, Zhang Z, Chen W and Feng Z 2010 *IEEE Antennas Wirel. Propag. Lett.* **9** 562–565

[5] Yan X, Gore J C and Grissom W A 2018 *Nature Communications* **9** 3481

Photonic dualism of icosahedral quasicrystals

A D Sinelnik[1 a], I I Shishkin[1,2], X Yu[3], K B Samusev[1,4], P A Belov[1], M F Limonov[1,4], P B Ginzburg[2], M V Rybin[1,4]

[1]Department of Physics and Engineering, ITMO University, Saint Petersburg 197101, Russia
[2] Department of Electrical Engineering, Tel Aviv University, Ramat Aviv, Tel Aviv 69978, Israel
[3] Huazhong University of Science and Technology, Wuhan, 430074, China
[4] Ioffe Institute, St Petersburg 194021, Russia

[a)] E-mail artem.sinelnik@metalab.ifmo.ru

Abstract. Here we report on fabrication of 3D submicron-size dielectric icosahedral quasicrystals and show a photonic dualism of these structures. Relying on far-field measurements, we found pronounced patterns of unconventional Bragg diffraction, which indicates the existence of multiple photonic pseudogaps.

1. Introduction

An ordinary crystal possesses periodicity and translational symmetry – when the basis vectors are translated to any distance, the lattice transforms into itself. The translation of the unit cell allows you to completely fill the space without gaps and overlaps, thereby creating an endless crystal lattice. In addition to translational symmetry, crystals are characterized by rotation and / or reflection symmetry. Because of translational symmetry, the crystal can have symmetry axes of the 2nd, 3rd, 4th and 6th orders and cannot have the rotation symmetry of the fifth and any order higher than the sixth, since such rotations do not bring the crystal lattice into itself

Quasicrystals are a separate class of solids with quasuperiodicity, which does not imply filling the entire space during translation, as well as rotational symmetry axes of any order, in particular, those that are forbidden for ordinary periodic crystals. Quasicrystals, like ordinary crystals, can be one-dimensional, two-dimensional, and three-dimensional.

An example of two-dimensional quasicrystals are the Penrose mosaics, named after the mathematician Roger Penrose, who first considered such structures in his work "The Role of Aesthetics in Pure and Applied Mathematical Research."[1]. The Penrose mosaic family is built by sequentially adding building blocks consisting of a narrow and wide rhombus. The angle at the apex of a wide rhombus is $2\pi/5=72^\circ$, at the apex of a narrow rhombus $2\pi/10=36^\circ$.

An example of three-dimensional quasicrystals is the class of icosahedral structures. The family of icosahedrons is created by sequentially adding three-dimensional structural blocks according to a certain rule [2]. Icosahedrons have spatial axes of symmetry C_{2v}, C_{3v}, C_{5v}.

2. Discussion

For experimental studies of photonic dualism, icosahedral quasicrystals and a photonic crystal woodpile truncated along a sphere were created. A two-photon polymerization technique [3-4] was used to create these structures. Figure 1 shows SEM images of the generated structures.

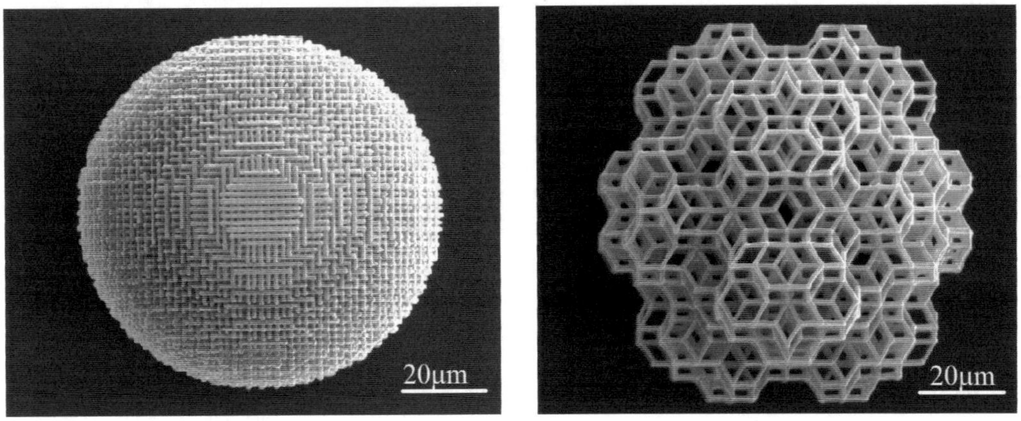

a b

Figure 1. SEM images of the fabricated photonic crystal icosahedral quasicrystal . (a) SEM image photonic crystal woodpile. (b) SEM image quasicrystal.

Next, optical diffraction by a photonic crystal and a quasicrystal was investigated. An Nd laser with a wavelength of 532 nm was used as a light source [3]. Figure 2 shows the diffraction patterns for a photonic crystal and a quasicrystal.

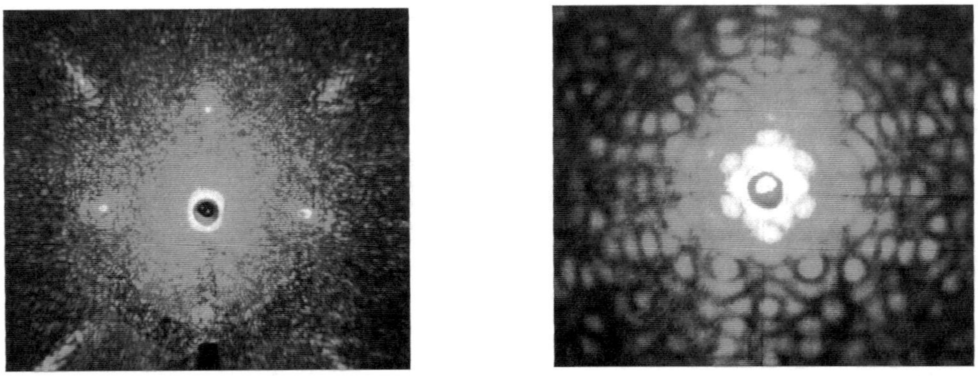

a b

Figure 2. Experimental diffraction patterns from photonic crystal (a) and photonic quasicrystal (b).

The diffraction pattern from a photonic crystal exhibits C4v symmetry and is a superposition of two components: Bragg diffraction in the form of bright ordered maxima associated with the crystal structure, and a background in the form of speckle-type diffraction due to scattering from individual rods [5]. The second scattering component was determined by the eigenmodes of the rods of finite length [6]. Due to the different lengths of the rods, the scattering is random, which leads to the formation of a speckle pattern. It is known that the presence of bright diffraction maxima on ordered periodic photonic structures is associated with the presence of stop bands. Although quasicrystals do not have translational symmetry, the presence of diffraction maxima is associated with the existence of a quasi-stop band in the band structure of a quasicrystal, since the diffraction pattern is closely related to reciprocal space [7]. The pronounced quasi-Bragg diffraction pattern in figure 2 indicate the existence of multiple photonic quasi-stop bands in quasicrystals.

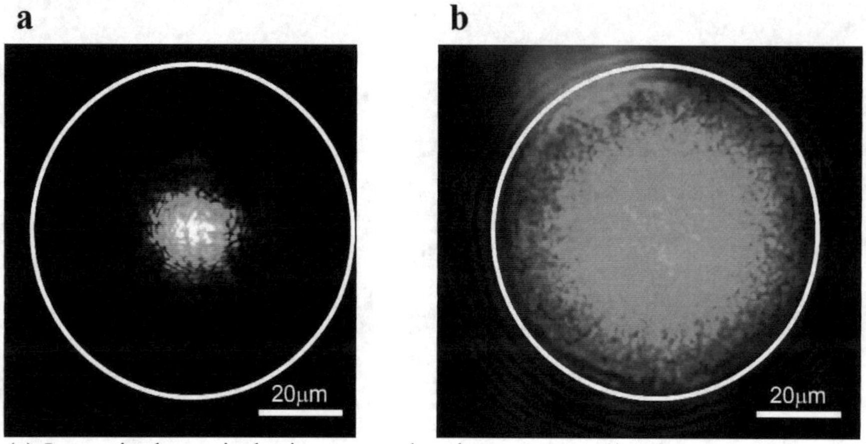

Figure 3. (a) Image in the optical microscope showing a propagating beam through woodpile. (b) Optical microscope image of icosahedral quasicrystal that is completely filled with scattered light due to intrinsic localization. The white circle shows the boundaries of the samples.

Figure 3 shows images of a photonic crystal and a quasicrystal that are irradiated with 530 nm laser radiation. Figure 3a demonstrates that when light passes through a photonic crystal, there is only a ballistic component and a weak wave structure. However, no light scattering is observed on the rest of the photonic crystal. For clarity, the sample boundaries are marked with a white line. In the case of icosahedral quasicrystals, the images are fundamentally different. From figures 3b it can be seen that the quasicrystal is "filled" with scattered light and it is impossible to separate the ballistic component. On the one hand, due to quasiperiodicity, quasicrystals have structural elements, which can generate quasi-Bragg scattering and, accordingly, energy quasi-stop bands, which is the main property of periodic photonic crystals. On the other hand, due to the absence of ideal periodicity, in quasicrystals there is an uncompensated component of multiple scattering into the total solid angle of 4π steradians, similar to scattering by disordered structures. This is precisely the picture that is observed when studying the scattering of light by icosahedrons. On the one hand, on the screen located behind the sample, a rather pronounced picture of numerous Bragg reflections is observed, figure 2. On the other hand, there is an unstructured "glow" of the entire icosahedral sample – figure 3 due to multiple rescattering of light in a quasicrystal, similar to the scattering of light in a turbid disordered medium.
Thus, it is because of this duality that photonic dualism is observed in quasicrystals. And such structures can behave both as ordered crystals and as disordered media, figure 4.

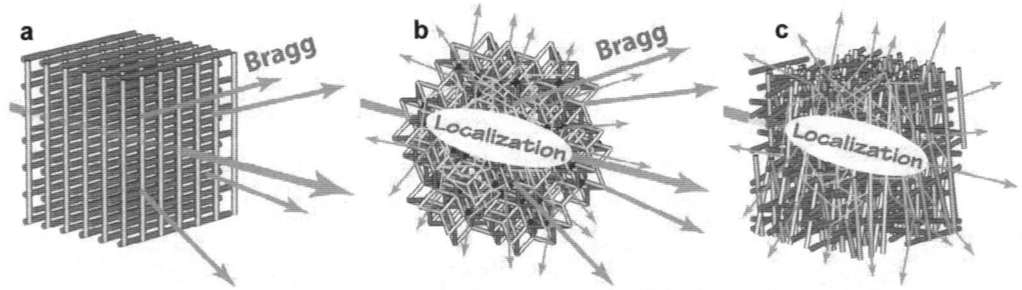

Figure 4. Schematic of light scattering in three states of solids. (a) Perfect crystals: Bragg scattering and a ballistic component of a transmitted laser beam. (b) Perfect 3D quasicrystals: Bragg scattering, multiple scattering of light within the structure, and a ballistic component. The light flux propagating in the direction of the incident laser beam can be separated into a ballistic component and a multiple scattering component, that is, light localized in a quasicrystal. (c) Disordered structures: multiple scattering of light localized within the structure, and the ballistic component, if observed.

3. Conclusions

In this work, we performed diffraction experiments characteristic of the study of crystalline samples. As a result, we demonstrated the photonic dualism of quasicrystals – both Bragg diffraction and multiple scattering owing to light localization in quasicrystals were observed. The photonic dualism of quasicrystals opens the way to variety applications in optics, from lasing and sensing to telecommunications.

Acknowledgements
This work was supported by the Russian Foundation for Basic Research (Grant No.20-02-00785a).

References

[1] Penrose R. *Bull. Inst. Math. Appl.* 10 8. 266–271 (1974).
[2] Jeon S. Y., Kwon H., Hur K. *Nat. Phys.* 13, 363-368 (2017)
[3] Sinelnik, A. D., Samusev, K. B., Rybin, M. V. & Limonov, M. F. *Phys. Rev. B*, 99, 174204 (2019).
[4] Farsari, M. & Chichkov, B. N. *Nat. Photon.* 3, 450 (2009).
[5] Oskooi A.F. et al. *Comput. Phys. Commun.* 181 3 687–702 (2010).
[6] Ioffe A.F., Regel A.R. *Prog. Semicond.* 4 237–291 (1960).
[7] Ledermann, A. et al. *Nat. mater.* 5, 942-945 (2006).

Strong light-matter coupling in topological metasurfaces integrated with transition metal dichalcogenides

Ivan Sinev[1], Mengyao Li[2,3,4], Fedor Benimetskiy[1], Tatiana Ivanova[1], Svetlana Kiriushechkina[2], Anton Vakulenko[2], Sriram Guddala[2,3], Dmitry Krizhanovskii[1,5] Anton Samusev[1], Alexander Khanikaev[2,3,4]

[1]Department of Physics and Engineering, ITMO University, St. Petersburg, Russia
[2]Department of Electrical Engineering, City College of New York, New York, NY, USA
[3]Physics Department, City College of New York, New York, NY, USA
[4]Physics Program, Graduate Center of the City University of New York, New York, NY, USA
[5]Department of Physics and Astronomy, University of Sheffield, Sheffield S3 7RH, UK
i.sinev@metalab.ifmo.ru

Abstract. Strong light-matter interactions enable unique nonlinear and quantum phenomena at moderate light intensities. Within the last years, polaritonic metasurfaces emerged as a viable candidate for realization of such regimes. In particular, planar photonic structures integrated with 2D excitonic materials, such as transition metal dichalcogenides (TMD), can support exciton polaritons – half-light half-matter quasiparticles. Here, we explore topological exciton polaritons which are formed in a suitably engineered all-dielectric topological photonic metasurface coupled to TMD monolayers. We experimentally demonstrate the transition of topological charge from photonic to polaritonic bands with the onset of strong coupling regime and confirm the presence of one-way spin-polarized edge topological polaritons. The proposed system constitutes a promising platform for photonic/solid-state interfaces for valleytronics and spintronics.

1. Introduction

Topological photonics has undergone a tremendous growth in the past years as various platforms from microwave to optical spectral ranges were used for implementation of topological phases of light[1,2]. One of the recently emerged research directions in this field is aimed at expanding the concepts of topological photonics towards systems with strong light-matter interaction, which opens the possibilities for creation and control of exotic quasiparticle systems. In particular, integration of topological photonic structures with active media has led to creation of topological lasers[3], topological polaritonic phases[4], and highly nonlinear[5] topological photonic devices.

Topological polaritons, not unlike their non-topological counterparts, represent coupled light-matter excitation that emerge due to strong coupling between electromagnetic and solid-state degrees of freedom. However, they also exhibit a plethora of useful phenomena due to their topological origin. In particular, one-way spin-polarized transport, topological protection against scattering characteristic for topological phases combined with high nonlinearity arising from light-matter hybridization may enable the observation of topological solitons and modulation instability and generation of squeezed topological light[6,7]. Furthermore, topological systems coupled to excitonic TMD materials represent a promising interface between spin- and valley-degrees of freedom due to unique properties of TMD.

So far, topolaritonic systems were mostly limited to 1D systems, or 2D systems characterized by 1D topological invariants. The first topolaritonic system characterized by a 2D topological invariant was demonstrated in the work by S. Klembt et al.[8] in GaAs quantum well lattices for the case of broken time-reversal symmetry induced by magnetic field. Here, we realize topolaritonic spin-Hall phase that does not require magnetic field in a topological metasurface based on planar Si photonic structure strongly coupled to excitons in TMD monolayers. We experimentally demonstrate the transfer of topological charge from photonic to polaritonic mode and one-way propagation of the edge topological polaritons.

2. Results and discussion

The topological structure that we use here is shown in Fig. 1a. It represents a Si photonic metasurface with a honeycomb shrink-expand lattice design that is based on the one proposed for a topological quantum optical interface[9]. We adjusted this design to support leaky topological edge modes near the exciton frequency in $MoSe_2$ (1.65 eV at 7K). These samples were fabricated by patterning Silicon on Insulator (SOI) substrates with the use of e-beam lithography followed by reactive ion etching.

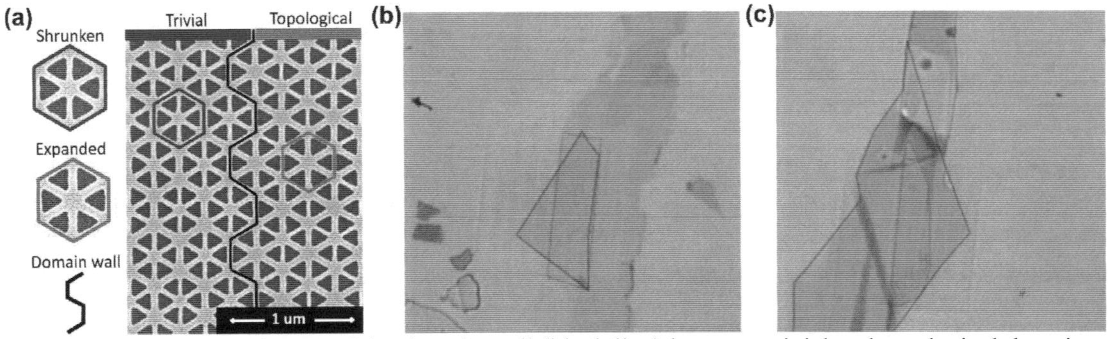

Figure 1. (a) SEM image of the domain wall (black line) between trivial and topological domains of an all-dielectric topological metasurface with hexagonal unit cells. (b,c) Optical images of topolaritonic metasurface with thin hBN layers (black) and $MoSe_2$ monolayer (red) transferred on top of the photonic topological metasurface with multiple domain walls.

To realize the transition to topological polariton regime, we transferred a MoSe2 monolayer on top of the metasurface together with a thin hBN layer. The latter enhanced the quality of the exciton and enabled the fine tuning of the spectral position of the photonic modes with respect to the exciton. Here, we study two samples with slightly different parameters of the metasurface unit cell (Fig. 1b,c) that provide different detuning of the lower topological band edge with respect to $MoSe_2$ exciton.

Leaky character of the metasurface modes allowed us to directly characterize the dispersion of photonic and polaritonic modes of the metasurfaces [10,11] using back focal (Fourier) plane imaging in our custom-built experimental setup. The setup was integrated with a closed-cycle helium cryostat and allows to capture the band diagrams in frequency-momentum space at temperatures down to 7K.

Fig. 2 shows the linearly polarized light reflectivity maps that illustrate the formation of topological polaritons with the onset of strong coupling regime with $MoSe_2$ exciton. We first studied the sample with a 10 nm hBN flake, but without the TMD monolayer. Fig. 2a,c reveal the topological band gap of the photonic lattice (~1.7-1.8 eV), with two gapless edge modes inside the bulk bandgap. These modes correspond to the spin-up and spin-down topological photonic edge states propagating along the domain wall in the opposite directions. The transfer of a $MoSe_2$ monolayer on top of hBN (Fig. 3b,d) leads to the appearance of the exciton in reflectivity maps crossing the lower photonic band of the topological metasurface approximately 0.07 eV below the band edge. These experimental spectra clearly reveal the anti-crossing behavior between the exciton and the photonic bands, manifesting the strong coupling regime and the formation of topological bulk states. Notably, Fig. 3b,d reveals that the edge modes asymptotically approach the newly formed polaritonic bulk band, which is indicative of the polaritonic nature of the edge states.

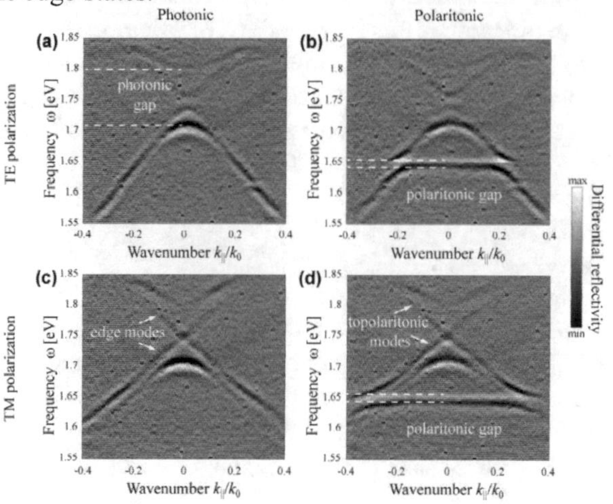

Figure 2. (a,c) Angle-resolved differential reflectivity measured from the vicinity of the domain wall of a metasurface with a 10nm hBN, but without MoSe2 for TE (a) and TM (c) polarized collection. (b,d) TE and TM polarized differential reflectivity maps of the domain wall with a MoSe2 monolayer on top hBN demonstrating strong coupling regime and the respective bulk and edge topolaritonic bands.

One of the most important properties of spin-Hall topological systems is the one-way spin-polarized character of their topological boundary states. While it was observed experimentally in photonic structures, a similar property should also emerge in spin-polarized topolaritonic boundary modes. Therefore, as a next step we measured the back focal plane reflectance spectra with left and right circularly polarized beams focused on the domain wall. The results shown in Figs. 3a,b that one of the edge modes is manifested in the reflectivity maps depending on the helicity of the incident light, which confirm the one-way nature of the edge topological polaritons.

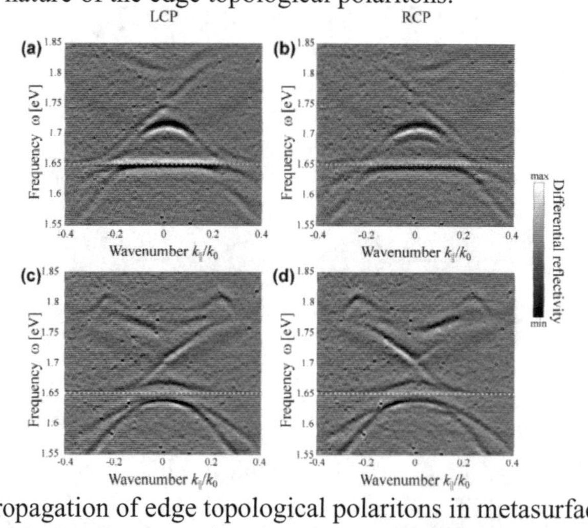

Figure 3. One-way propagation of edge topological polaritons in metasurfaces integrated with MoSe2: with exciton deep inside the photonic bulk mode (top row) and at its edge (bottom row). Differential reflectivity maps are shown for (a,c) left and (b,d) right circularly polarized light excitation.

We also performed the measurements on the second sample, with MoSe2 placed right on top of the metasurface and hBN transferred on top to provide larger overlap of photonic modes with TMD. The metasurface was scaled to ensure that the exciton resonance is aligned with the very edge of the lower energy photonic band at the Γ-point of the photonic Brillouin zone. The measured band diagrams for this sample, shown in Figs. 4c,d, once again reveal the formation of polaritonic bands; however, this sample demonstrates larger Rabi splitting between the polariton branches. Moreover, the topological edge modes in this case asymptotically approach the exciton frequency, which indicates their higher exciton fraction near the edge of the bulk band. As the data in Figs. 4c,d is measured for circularly polarized excitation, it also confirms the one-way propagation of the spin-polarized boundary states, with the forward and backward states that can be excited only with the circularly polarized light of proper helicity. We also note that even though MoSe2, as any other excitonic materials, is inherently absorbing, this absorption does not prevent the formation of the edge topological polaritonic modes and does not have a notable influence on the dispersion of the bulk photonic modes as can be seen in Figs. 2 and 3.

3. Conclusion

To conclude, we showed the formation of topological polaritonic phases with preserved time-reversal symmetry by engineering the strong coupling regime between silicon-based topological photonic metasurface and excitons in monolayers of transition metal dichalcogenides. We demonstrated experimentally the transfer of topological charge from photonic to polaritonic bulk modes and formation of one-way propagating spin-polarized edge topological polaritons. Through the combination of robustness of topological edge states and valley polarization in TMD, such platform holds great potential for future valleytronic and spintronic devices.

Acknowledgments
This work was supported by Ministry of Science and Higher Education of Russian Federation, goszadanie no. 2019-1246 and megagrant 14.Y26.31.0015. Optical measurements were funded by RFBR, project number 20-32-70185. IS acknowledges the support from The Russian Federation President Grant Council, project no. MK-4652.2021.1.2.

References

[1] L. Lu, J. D. Joannopoulos, and M. Soljačić, Nat. Photonics **8**, 821–829 (2014).
[2] A.B. Khanikaev and G. Shvets, Nat. Photonics **11**, 763–773 (2017).
[3] P. St-Jean, V. Goblot, E. Galopin, A. Lemaître, T. Ozawa, L. Le Gratiet, I. Sagnes, J. Bloch and A. Amo, Nat. Photonics **11**, 651–656 (2017).
[4] A.V. Nalitov, D.D. Solnyshkov, and G. Malpuech, Phys. Rev. Lett. **114**, 116401 (2015).
[5] D. Smirnova, D. Leykam, Y. Chong, and Y. Kivshar, Appl. Phys. Rev. **7**, 021306 (2020).
[6] D. Leykam and Y.D. Chong, Phys. Rev. Lett. **117**, 143901 (2016).
[7] J.-L. Tambasco, G. Corrielli, R.J. Chapman, A. Crespi, O. Zilberberg, R. Osellame and A. Peruzzo, Sci. Adv. **4**, eaat3187 (2018).
[8] S. Klembt, T. H. Harder, O. A. Egorov, K. Winkler, R. Ge, M. A. Bandres, M. Emmerling, L. Worschech, T. C. H. Liew, M. Segev, C. Schneider and S. Höfling, Nature **562**, 552–556 (2018).
[9] Y. Yang, Y. F. Xu, T. Xu, H.-X. Wang, J.-H. Jiang, X. Hu and Z. H. Hang, Phys. Rev. Lett. **120**, 217401 (2018).
[10] N. Parappurath, F. Alpeggiani, L. Kuipers and E. Verhagen, Sci. Adv. **6**, eaaw4137 (2020).
[11] M. A. Gorlach, X. Ni, D. A. Smirnova, D. Korobkin, D. Zhirihin, A. P. Slobozhanyuk, P. A. Belov, A. Alù and A. B. Khanikaev, Nat Commun **9**, 909 (2018).

Phase-Sensitive Optical Coherence Microscopy of Integrated Nanophotonics Devices

Maxim A. Sirotin,[1, a)] Evgeny V. Lyubin,[1] Kirill R. Safronov,[1] Daniil V. Akhremenkov,[1] Vladimir O. Bessonov,[1, 2] Irina V. Soboleva,[1, 2] and Andrey A. Fedyanin[1]

[1)] *Faculty of Physics, Lomonosov Moscow State University, Moscow 119991, Russia*
[2)] *Frumkin Institute of Physical Chemistry and Electrochemistry, Russian Academy of Sciences, Moscow 119071, Russia*

[a)] *Corresponding author: sirotin@nanolab.phys.msu.ru*

Abstract. We report on the development of a new approach for studying the internal structure of polymer integrated nanophotonics devices using phase-sensitive optical coherence microscopy. Visualization and flaw detection of devices and their internal structure was carried out using the example of coupling gratings and prisms for a miniature Otto configuration with a characteristic gap height of 50-300 nm.

INTRODUCTION

Optical coherence microscopy (OCM) is a method of optical nterferometry, which makes possible three-dimensional visualization of micro-objects. OCM is based on a combination of optical coherence tomography (OCT) and confocal microscopy [1]. OCM allows three-dimensional mapping of the refractive index of the sample and has a submicron resolution [2]. In addition, one of the methods, phase-sensitive OCM, has subnanometer axial sensitivity to changes in optical path-length [3]. The submicron resolution of the OCM makes it a powerful tool for studying semitransparent objects.

One of the areas of application of the OCM in recent years is flaw detection of microscopic devices and chips. OCM makes it possible to non-invasively study the internal structure of semitransparent objects, and therefore it can be used to search for defects in various polymer and crystalline structures without destroying them and allowing further use of the tested device. The method is already being used to study microscopic photonics and fiber optics devices [4].

The advantage of the method is its non-invasiveness and the ability to visualize the internal structure of semitransparent objects, which is important for flaw detection of integrated nanophotonics devices with a complex internal structure.

Our goal was to apply phase-sensitive OCM to study semitransparent polymer integrated nanophotonics devices fabricated by femtosecond laser polymerisation technique [5]: both for relatively simple devices and for devices with complex geometry and internal structure.

METHODS

Phase-sensitive OCM is based on the interference between two beams: reflected from the sample layer with intensity reflectance R_s located at a distance z_s from the beam splitter, and reflected from a reference mirror with reflectance R_r located at a distance z_r from the beam splitter. In our version of the method, the interference fringes appear on a wide spectrum ($\Delta\lambda$ = 300 nm) of the source depending on the wave vector k (Fig. 1).

Samples such as polymer microprisms and other integrated nanophotonics devices usually consist of several layers, differing in refractive index. Let's consider a device with two layers, the refractive indices of which are n_1 and n_2. At the boundary of these layers with coordinates z_{s1} and z_{s2}, reflection occurs (R_{s1} and R_{s2}).

Let us denote the thickness of the layer with the refractive index n_1 as d. As a result of the interference of rays reflected from the boundaries of the layer with the ray reflected from the reference mirror, characteristic interference fringes appear on a wide spectrum of the source. Then, from this interference pattern, using the Fourier transform, it is possible to extract the optical path that the ray travels in this layer, and from there — the layer thickness: $d * n_1 = \lambda_0 * \Delta\psi/4\pi$, where $\Delta\psi$ is the change in the Fourier phase when the ray passes through the layer with n_1, λ_0 is the center wavelength of the source. A polymer prism printed on a substrate (Fig. 2) or a gap between the prism and the substrate (Fig. 3) can act as such a layer.

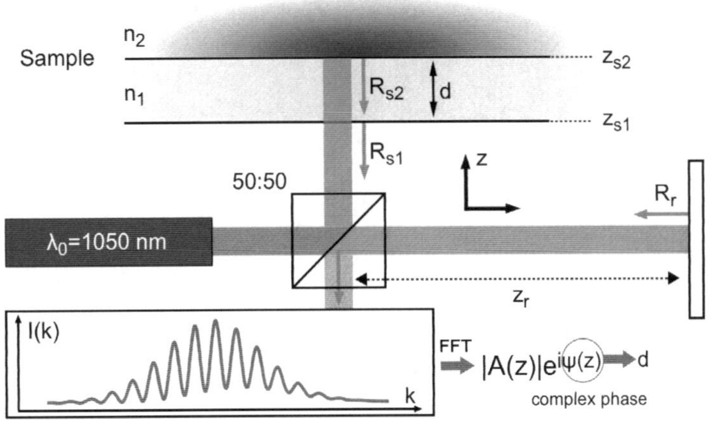

FIGURE 1. Schematic diagram of the phase-sensitive OCM method used for visualization of integrated nanophotonics devices.

A similar method is used to study the structure of biological cells [3]. However, the method was first used to study the internal structure of polymer integrated nanophotonics devices.

RESULTS AND DISCUSSION

At the beginning, we investigated polymer microprisms — prototypes of gratings for coupling surface electromagnetic waves on photonic chip with incident free-space radiation [5]. Figure 2 shows a comparison of images of a microprism obtained using an optical microscope (a), as well as using a phase-sensitive OCM — a scattering intensity map (b) and an optical path-length difference converted to heights (c).

Figure 2 clearly shows that the phase-sensitive OCM made it possible to determine the geometric dimensions of the prism, including its height (200 nm), which is difficult to do using conventional optical microscopy and other non-invasive methods, after which the object can be used.

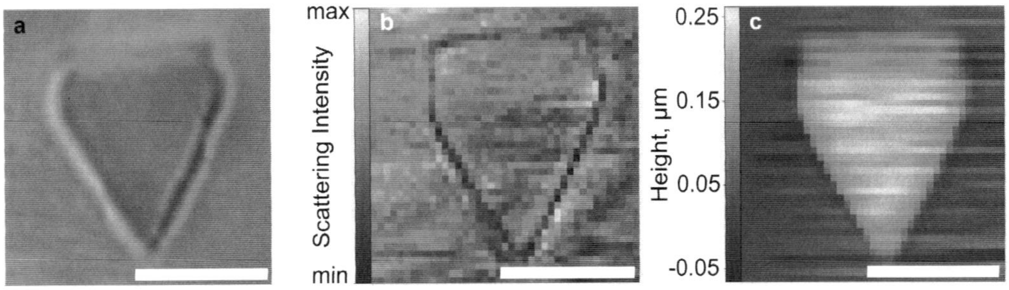

FIGURE 2. Image of a polymer microprism under an optical microscope (a) and its OCM scan: scattering intensity map (b) and an optical path-length difference converted to height (c). The bar represents 5 μm.

The next object of our study was miniature Otto configuration [6] devices, which have complex internal structure. These devices are more efficient and able to couple normally incident radiation. The prisms use the Otto scheme: excitation of a Bloch surface wave through a gap between the prism and the substrate of the order of 100 nm. The height and shape of the gap have a significant impact on the efficiency of coupling. The problems of studying the gaps are their small heights and their location inside the structure, which does not allow the use of contact visualization methods and methods for which the prism is opaque. This problem can be successfully solved by optical interferometry

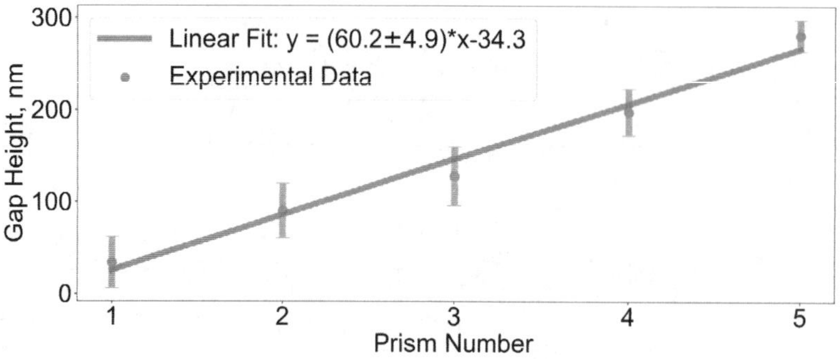

FIGURE 3. Dependence of the gap height of the miniature Otto configuration microprism on the sample number.

methods, such as phase-sensitive OCM.

Using the developed method for studying the gaps, we are able to determine the height of the gap in the prisms, depending on the manufacturing parameters (Fig. 3). The accuracy of determining the gap is up to 10 nm and varies depending on the print quality and the features of the prism shape. Figure 3 shows that the method makes it possible to reveal a linear dependence inherent in devices during manufacturing. In addition, the method also allows one to build a three-dimensional image of the gap — the dependence of the height on the transverse coordinates. These data can then be used in numerical simulations to explain the experimentally evaluated coupling efficiency and to determine the printing parameters that ensure the creation of the prism with the highest efficiency of the coupling.

CONCLUSION

Integrated nanophotonics devices are becoming more and more complex: devices with sophisticated internal geometry appear, for example, with gaps with a height of about 100 nm. Studying such devices, determining their characteristics, and detecting defects in them is important for predicting their optical properties and efficiency. However, there is a lack of non-invasive optical methods for studying such devices.

We proposed using the phase-sensitive OCM method for imaging and characterization of integrated nanophotonics devices, both relatively simple and devices with a complex internal structure. The method made it possible to construct images of objects and determine the geometric properties of the internal structure, which can be further used both for analyzing the efficiency of the device and for flaw detection.

ACKNOWLEDGMENTS

S.M.A. and S.I.V. thank MSU Quantum Technology Centre for the support. This work was also supported by the Russian Foundation for Basic Research (Grant No. 18-29-20097). This research was performed according to the Development program of the Interdisciplinary Scientific and Educational School of Lomonosov Moscow State University «Photonic and Quantum technologies. Digital medicine».

REFERENCES

1. M. Yamanaka, et al., Optical coherence microscopy in 1700 nm spectral band for high-resolution label-free deep-tissue imaging, Sci. Rep. **6**, 1-8 (2016).
2. A. Federici and A. Dubois, Full-field optical coherence microscopy with optimized ultrahigh spatial resolution, Opt. Lett. **40**, 5347-5350 (2015).
3. C. Joo, et al., Spectral-domain optical coherence phase microscopy for quantitative phase-contrast imaging, Opt. Lett. **30**, 2131 (2005).
4. D. R. Reyes, M. Halter, and J. Hwang, Dimensional metrology of lab-on-a-chip internal structures: a comparison of optical coherence tomography with confocal fluorescence microscopy, J. Microsc. **259**, 26-35 (2015).
5. K. A. Abrashitova, et al., Bloch surface wave photonic device fabricated by femtosecond laser polymerisation technique, Appl. Sci. **8**, 63 (2018).
6. K. R. Safronov, et al., Miniature Otto configuration implemented by two-photon laser lithography, J. Phys. Conf. Ser. **1461** (2020).

Scattering spectra of dielectric ring: microwave experiments

Nikolay Solodovchenko [a)] **and Mikhail Sidorenko**

Department of Physics and Engineering, ITMO University, Saint Petersburg 197101, Russia

[a)]E-mail: n.solodovechenko@metalab.ifmo.ru

Abstract. In recent years, dielectric ring resonators (RRs) have become an essential part of integrated optical circuits. This determines the growing interest in the study of the fundamental electromagnetic properties of these objects, which is far from complete. In particular, in the literature it is difficult to find information about light scattering spectrum, which should demonstrate a strong resonance character when the probe wavelength is comparable to the geometric dimensions of the RR, taking into account its dielectric constant. In this work, we present the results of an experimental study of the electromagnetic properties of a dielectric RR in the microwave range of the spectrum. The results of numerical calculations of the scattering spectrum are also presented, which demonstrate excellent agreement with the experimental data. In addition to the expected resonance character of the spectrum, we report the effects of strong light confinement, which are associated with Fano resonances between the eigenmodes and the component of the electromagnetic field scattered by the ring.

1. Introduction

Dielectric ring stands out among the most important structural elements of modern photonics, along with the sphere and the cylinder [1]. First of all, a RR is distinguished as the main element of an add-drop filter, on which modern integrated optical circuits are based [2, 3]. However, along with their use in optoelectronic devices, dielectric rings are a classical object of fundamental research. It should be noted that dielectric rings are used in a wide variety of areas of physics, among which the study of an interplay of two types of spontaneous symmetry breaking in optical RRs [4], strong nonlinear coupling in a Si3N4 RR [5], an observation of the multidimensional Purcell effect in an ytterbium-doped RR [6] and a generation of quantum-correlated photon pairs using electrically tunable ring resonators and passive Bragg reflectors [7] and many others.

Despite the great activity of researchers studying various effects associated with RR, as far as we know, neither the calculated nor experimentally measured spectra of light scattering by a dielectric ring in a wide spectral range were presented in the original articles or in reviews [3,8] or books [9]. Therefore, the goal of this work was to study experimentally the scattering spectra of electromagnetic waves from a dielectric ring in the far field, as well as to study the resonant eigenmodes of the ring using the near-field spectroscopy. The studies were carried out in the microwave spectral range, and taking into account the scaling of Maxwell's equations, the results obtained are valid for any spectral range.

2. Experimental setup

For our experiments, a RR was chosen with the dimensions of the inner diameter d = 93 mm and the outer diameter D = 115 mm and the height h = 7 mm and was made of microwave ceramics. The average dielectric constant $< \varepsilon >$ in the investigated frequency range (1-10 GHz) is 44. The radar cross-section (RCS) scattering measurements of the RR were performed in an anechoic chamber [10]. We use a rectangular horn antenna connected to a transmitting port of the vector network analyzer Rohde&Schwarz ZVB20 to approximate a plane-wave excitation. The transmitting antenna, fixed at the same level as the receiving antenna, illuminates the RR. The RR is located in the center of the measuring station on a stand made of polystyrene foam, interaction with which is neglected. It should be noted that the transmitting antenna is oriented so that the incident electric field is in the plane of the ring (TE polarization). The antennas are located at a distance of about 3 meters from the center.

In such an installation, both the magnitude and the phase are measured, which makes it possible to obtain a complex value of the electric field. To calculate the RCS of a ring, two measurements are needed: the electric field with the ring and empty space. Further, RCS is calculated by the optical theorem (Fig. 1).

The RR eigenmode sequence can be experimentally investigated in the near field. For operation in the near field, a radiating horn antenna is used, and a metal ring located close to the RR surface acts as a signal receiver, which fixes one component of the magnetic field at a given point in space. Thus, in three passes, you can get complete information (real and imaginary part) about the magnetic field near the RR surface (insert in Fig. 1).

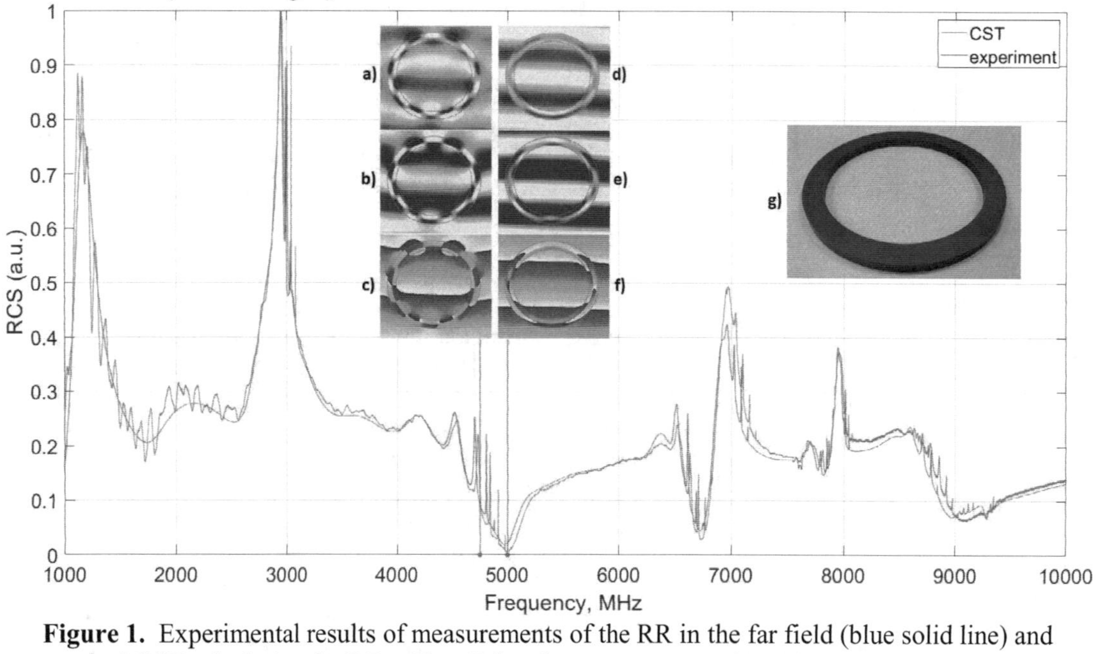

Figure 1. Experimental results of measurements of the RR in the far field (blue solid line) and numerical CST calculation for RR with a dielectric constant ε equal to 44 over the entire frequency range (orange solid line) of the ring resonator. The real part (a, d), imaginary part (b, e) and phase (c, f) of the z component of the magnetic field. Each insert (a-f) has its own intensity scale. The insert (g) is a photography of the sample

3. Experimental observation of light trapping in dielectric ring

Among the results obtained, we highlight the experimental observation of the effect of light trapping in several relatively narrow regions of the investigated spectrum. The scattering spectrum shown in Fig. 1 consists of background scattering, the approximate level of which is marked by a green dashed straight line, a number of resonance bands of different widths and intensities, as well as several regions in which the level of scattered light decreases sharply and, in particular, drops to almost zero in the 5 GHz frequency range. Since the incident electromagnetic wave is normalized in intensity over the entire spectral range, the absence of scattering means the capture of electromagnetic energy by the resonator at the corresponding frequencies. This effect can be associated with destructive interference in the far zone of two waves, one of which is associated with the intrinsic resonance of the ring, and the other with scattering at the ring as a whole. This effect is characteristic of the Fano resonance [10] observed in many of the simplest dielectric structures, in particular on a sphere [11] and a cylinder [12]. Another striking effect leading to the light confining by the resonant structure is bound states in the continuum (BIC) [13], which were also observed in the case of a separate dielectric cylinder [14]. However, the difference between the BIC regime and the effect observed in this work is that in the case of BIC, a sharp narrowing and decrease in the intensity of the resonance scattering line was observed, while in our case scattering does not occur at all.

The absence of scattered light when illuminating an object is equivalent to its invisibility. This phenomenon was studied theoretically earlier in the case of light scattering by a dielectric cylinder in the Fano resonance regime [15]. However, in this work, we succeeded to observe this phenomenon experimentally. Figure 1 shows the distribution of the electromagnetic field inside and around the ring obtained as a result of processing the experimental data: the spatial distribution of the real and imaginary magnetic components of the field and its phase for two frequencies - 4.75 and 5 GHz. It is clearly seen, especially by the example of the phase distribution, that the incident TE-polarized electromagnetic wave with a frequency of 5 GHz passes the ring practically without distortion, leaving the ring invisible from any angle of observation.

4. Conclusion

In this work, for the first time, as far as we know, an experimental study of light scattering by a dielectric ring was carried out. The far and near field scattering was investigated. In addition, the scattering spectra were calculated numerically, and excellent agreement between the calculated and experimental data was demonstrated.

The data obtained allow us to conclude that the eigenmodes of the dielectric ring are fundamentally different from the whispering gallery modes, since they do not form a quasi-equidistant sequence characteristic of the whispering gallery modes. In addition, we observed spectral regions in which there is no light scattering, that is, electromagnetic energy is confined by the resonator. This effect will be studied in detail in our next works.

Acknowledgments

The authors acknowledge fruitful discussions with Mikhail Limonov, Kirill Samusev, and Timur Seidov. This work was supported by the Russian Foundation for Basic Research (Grant No. 20-02-00785a).

References

[1] K. J. Vahala, *Nature* **424**, 839 (2003)
[2] A. Li and W. Bogaerts, *Laser Photonics Rev.* **13**, 1800244 (2019)
[3] P. Pintus, M. Hofbauer, C. L. Manganelli, M. Fournier, S. Gundavarapu, O. Lemonnier, F. Gambini, L. Adelmini, C. Meinhart, C. Kopp, F. Testa, H. Zimmermann, and C. J. Oton, *Laser Photonics Rev.* **13**, 1600219. (2019)
[4] F. Copie, M. T. M. Woodley, L. Del Bino, J. M. Silver, S. Zhang, and P. Del'Haye, *Phys. Rev. Lett.* **122**, 013905 (2019)
[5] S. Ramelow, A. Farsi, Z. Vernon, S. Clemmen, X. Ji, J. E. Sipe, M. Liscidini, M. Lipson, and A. L. Gaeta, *Phys. Rev. Lett.* **122**, 153906 (2019)
[6] D. Ding, L. M. C. Pereira, J. F. Bauters, M. J. R. Heck, G. Welker, A. Vantomme, J. E. Bowers, M. J. A. de Dood, and D. Bouwmeester, *Nat. Photonics* **10**, 385 (2016)

[7] N. C. Harris, D. Grassani, A. Simbula, M. Pant, M. Galli, T. Baehr-Jones, M. Hochberg, D. Englund, D. Bajoni, and C. Galland, *Phys. Rev. X* **4**, 041047 (2014)

[8] W. Bogaerts, P. De Heyn, T. Van Vaerenbergh, K. DeVos, S. K. Selvaraja, T. Claes, P. Dumon, P. Bienstman, D. Van Thourhout, and R. Baets, *Laser Photonics Rev.* **6**, 47 (2012)

[9] J. Heebner, R. Grover, T. Ibrahim, and T. A. Ibrahim 2008 *Optical microresonators: theory, fabrication, and applications* (New York: Springer Science & Business Media)

[10] C. Larsson, C. Sohl, M. Gustafsson, and G. Kristensson *2009 3rd European Conference on Antennas and Propagation* (Berlin: IEEE) pp. 3633-3636

[11] M. F. Limonov, M. V. Rybin, A. N. Poddubny, and Y. S. Kivshar, *Nat. Photonics* **11**, 543 (2017)

[12] M. I. Tribelsky and A. E. Miroshnichenko, *Phys. Rev. A* **93**, 053837 (2016)

[13] M. V. Rybin, K. B. Samusev, I. S. Sinev, G. Semouchkin, E. Semouchkina, Yu. S. Kivshar, and M.F. Limonov, *Opt. Express* **21**, 30107 (2013)

[14] C. W. Hsu, B. Zhen, A. D. Stone, J. D. Joannopoulos, and M. Soljačić, *Nat. Rev. Mater.* **1**, 16048 (2016)

[15] A. A. Bogdanov, K. L. Koshelv, P. V. Kapitanova, M. V. Rybin, S. A. Gladyshev, Z. F. Sadrieva, K. B. Samusev, Y. S. Kivshar, and M. F. Limonov, *Adv. Photonics* **1**, 016001 (2019)

[16] M. V. Rybin, D. S. Filonov, P. A. Belov, Yu. S. Kivshar, and M. F. Limonov, *Sci. Rep.* **5**, 8774 (2015)

Journal of Physics: Conference Series **2015** (2021) 012145 doi:10.1088/1742-6596/2015/1/012145

Resonant light scattering from dielectric ring of rectangular cross section

N S Solodovchenko[1 a)], **T Z Seidov**[1,2], **K B Samusev**[1,3] **and M F Limonov**[1,3]

[1]Department of Physics and Engineering, ITMO University, Saint Petersburg 197101, Russia
[2]St Petersburg University, 7-9 Universitetskaya Emb., St Petersburg 199034, Russia
[3]Ioffe Institute, St. Petersburg 194021, Russia

[a)]E-mail: `n.solodovechenko@metalab.ifmo.ru`

Abstract. In this paper we present the results of numerical calculations of electromagnetic properties for cylindrical ring resonators (RRs) with rectangular cross-section and a dielectric permittivity corresponding to silicon $\varepsilon = 12$. The calculation of the scattering spectrum (Radar Cross Section) and the field distribution of the modes were performed at in-plane polarized light excitation. The presence of four side walls in the RRs determines a richer spectrum of eigenmodes in comparison with cylindrical whispering gallery modes of disk resonators and creates more possibilities and diversity for their practical applications.

1. Introduction.

Well-known Whispering Gallery Modes (WGMs) can be considered as electromagnetic modes confined within the structure by total reflections from the curved side surface of the resonator with the proper phase condition after circling along the whole resonator. WGM resonators have been investigated in detail and are widely used in sensors and detectors because of their huge Q-factor and a narrow resonance line width [1–3]. In contrast, studies of RRs are far from complete, although they have significant advantages over WGM [4–9]. The presence of four sides in RRs with a square or rectangular cross section leads to strong light localization and the existence of volume resonance modes, which cannot be obtained in cylindrical resonator.

In the scattering spectrum of ring dielectric structure, we obtained a series of resonances, which we will hereinafter call the Ring Gallery Modes (RGMs) by analogy with the WGMs in disk dielectric microresonators. In this work, we found that, unlike WGMs, which form a continuous sequence of quasi-equidistant modes, RGMs are divided into several series, each consisting of one intense wide band with complex structure and several narrow high-Q lines, the intensity of which decreases rapidly according to the logarithmic law.

2. Scattering spectrum of a dielectric RR.

We consider here a dielectric RR with homogeneous dielectric permittivity $\varepsilon = 12$ (that corresponds to the dielectric constant of silicon) and zero attenuation. The ring is embedded in the air ($\varepsilon = 1$). The length of the center line $L = \pi(r_{out} + r_{in})$, the ratio of the inner radius r_{in} to the outer radius rout is 0.6, width $g = r_{out} - r_{in}$ and thickness h satisfy the relation $g/h = 5$. In our calculations, the electric field of the incident wave is linearly polarized in the plane of the ring (TE polarization). All the computations were performed based on finite-difference frequency-domain (FDFD) method using the commercial software COMSOL.

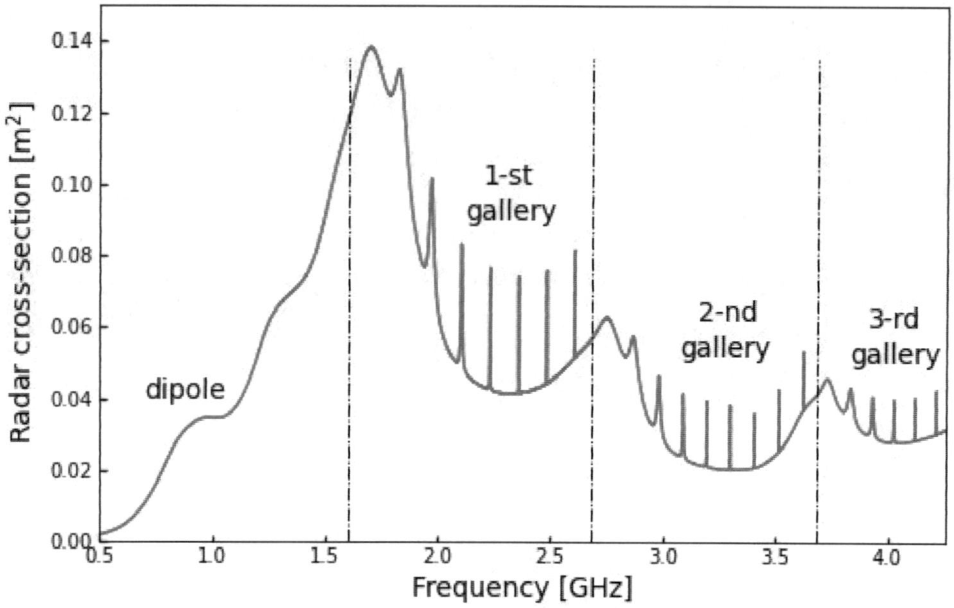

Figure 1. Total Radar cross-section spectrum of dielectric ring under normal incidence of an electromagnetic wave on the lateral face. The ring permittivity $\varepsilon = 12$. $L = 61.9\,cm$, $g/h = 5$. The ring is embedded in air, $\varepsilon_{air} = 1$.

A complete electromagnetic description of the ring resonance structure was obtained, containing information about the spectrum (eigenvalues) and the distribution of mode fields (eigenfunctions). This information made it possible to fully interpret the scattering spectrum using the standard notation of the eigenfunctions, which are characterized by the azimuthal (m), radial (r), and axial (z) mode indices, forming ordered triple (m, r, z).

Figure 1, which shows the calculated RCS spectrum for the dielectric constant $\varepsilon = 12$, allows us to analyze the main features of light scattering by a dielectric RR. The spectrum begins with a broad low-frequency band of low intensity, which corresponds to a dipole resonance. The dipole band is followed by a complex and, at first glance, chaotic set of wide and narrow resonance lines. However, a careful analysis of the spectrum clarifies the periodic sequence of line sets and their interpretation, which is confirmed by the shape of the calculated field patterns for each resonance, discussed below.

To interpret the results and discuss the spectra, we introduce the concept of RGM by analogy with the term WGM. Figure 1 shows the spectrum containing two full galleries and the beginning of the third gallery. The gallery begins with a wide band of complex shape and continues with a limited number of constantly narrowing symmetrical lines until it ends when the next wide band appears, belonging to the next gallery. Let us compare the mechanisms of the appearance of resonances in a ring resonator (RGM gallery) and a disk resonator (WGM gallery). WGM arise in a disk due to total reflections from the rim with the proper phase condition after circling along the entire resonator. In WGM - resonator the azimuthal index m has very large values;

Journal of Physics: Conference Series **2015** (2021) 012145 doi:10.1088/1742-6596/2015/1/012145

in real devices, m is usually in the range from about 102 to 105. In the case of RGM, which are considered in this work, the situation is completely different both with respect to the mechanism of the formation of resonances and in terms of the value of the azimuthal index. The boundary conditions are determined by the presence of all four walls and the electromagnetic field fills the entire volume of the ring resonator. The key to interpreting the spectrum in figure 1 is the presence of two side walls in the ring instead of one side wall in the disk. For simplicity, the calculations considered the case of a thin rectangular ring with the ratio of the width of the ring (the difference between the outer and inner radii) to its thickness equal to 5. Therefore, in the low-frequency region of the spectrum shown in figure 1, the axial mode index will not change (half the wavelength, $z = 0.5$), and the electromagnetic mode will be determined by a pair of azimuthal and radial indices (m, r).

However, our RR is thin, but not so narrow that with increasing frequency, i.e. with a decrease in the wavelength, the radial index r remained constant in the discussed spectral range. With decreasing wavelength, the boundary conditions in the radial direction, determined by the two side walls, change, leading to an increase in the radial index r. This is how the galleries of the RR are formed, each of which begins with a change in the radial index by 0.5, which corresponds to an increase in the transverse distribution of the electromagnetic field by half the wavelength. In this case, the boundary conditions on the side walls of the resonator remain unchanged and correspond to zero field intensity.

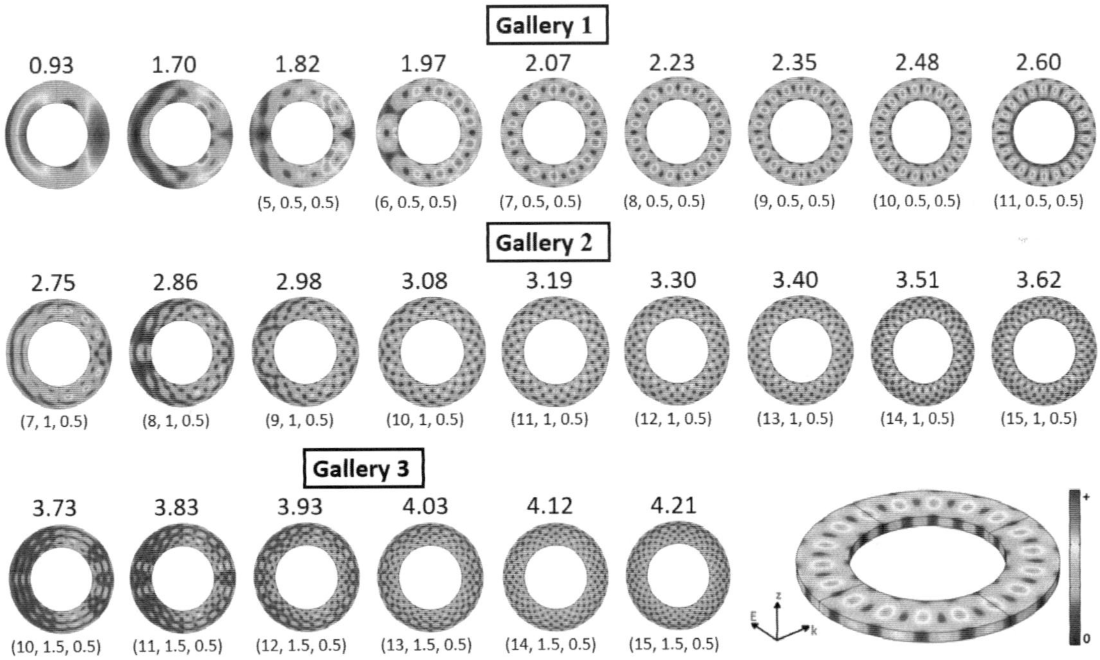

Figure 2. The calculated field patterns in the (x, y) cross sections (resonator top view) for ring permittivity $\varepsilon = 12$. The azimuthal, radial, and axial mode indices (m, r, z) are indicated for each resonance. The vertical scale corresponds to the field amplitude $|\mathbf{E}|$. The inset shows the scattering geometry: the incident wave is polarized in the plane of the ring.

Figure 2 confirms the above interpretation of the scattering spectrum in Figure 1. Detailed maps of the calculated resonant electromagnetic field distribution inside the ring depending on the wavelength are presented. The first gallery corresponds to the radial index $r = 0.5$. In the middle of the ring in the cross section, there is a single maximum $|\mathbf{E}|$ (yellow-red color) and at the side walls of the ring the field disappears (blue color) which corresponds to a half-wave. In the second gallery, in the cross section there are two field maxima $|\mathbf{E}|$ of and three minima (in the center and at the edges), which corresponds to one full wavelength, $r = 1$. And so on, with a successive increase in the radial index by $\triangle r = 0.5$.

3. Conclution

Devices based on silicon micro RRs are already indispensable for a wide range of applications and fundamental research. However, due to the limited design freedom of the RR (dielectric permittivity and roundtrip length), the functionality cannot always be fully explored and optimized. In our work, we have demonstrated that such resonators have a much more complex spectrum of eigenmodes in comparison with WGM. Structures with RGM contain both wide and narrow lines of various shapes, which will allow realizing new functionality in silicon ring resonators. Our results provide a new understanding of the phenomena of light localization and the relationship between the geometry and physical properties of microscopic photonic structures.

Acknowledgments

This work was supported by the Russian Foundation for Basic Research (Grant No. 20-02-00785a).

References

[1] Righini G, Dumeige Y, Feron P, Ferrari M, Nunzi Conti G, Ristić D and Soria S 2011 *La Rivista del Nuovo Cimento* **34** 435

[2] Dontsov A A, Monakhov A M and Averkiev N S 2013 *Quantum Electronics* **43** 414–417

[3] Baranov D, Samusev K, Shishkin I, Samusev A, Belov P and Bogdanov A 2016 *Optics letters* **41** **4** 749–52

[4] Kekatpure R 2011 *Journal of Lightwave Technology* **29** 253–259

[5] Bogaerts W, Heyn P D, Vaerenbergh T V, Vos K D, Selvaraja S, Claes T, Dumon P, Bienstman P, Thourhout D and Baets R 2012 *Laser Photonics Reviews* **6** 47–73

[6] Pintus P, Hofbauer M, Manganelli C, Fournier M, Gundavarapu S, Lemonnier O, Gambini F, Adelmini L, Meinhart C, Kopp C, Testa F, Zimmermann H and Oton C 2019 *Laser & Photonics Reviews* **13** 1800275

[7] Nojić J, Tabatabaei-Mashayekh A, Rahman T, Moscoso-Mártir A, Merget F, Sun X and Witzens J 2021 *IEEE Journal of Selected Topics in Quantum Electronics* **27** 1–12

[8] de Goede M, Dijkstra M, Chang L, Acharyya N, Kozyreff G, Obregón R, Martínez E and García-Blanco S M 2021 *Opt. Express* **29** 346–358

[9] Goswami K, Mondal H and Sen M 2021 *Optics Communications* **483** 126668

Wideband Coil Based on Microstrip Line for Multiheteronuclear Magnetic Resonance Spectroscopy

G. Solomakha[1], N. Egorova[1], R. Balafendiev[1], C. Simovski[1,2], S. Glybovski[1]

[1]Department of Physics and Engineering, ITMO University, 197101 Saint Petersburg, Russia
[2] Aalto University, School of Electrical Engineering, Department of Radio Science and Engineering, P.O. Box 13000, 00076 Aalto, Finland

E-mail: g.solomakha@metalab.ifmo.ru

Abstract. Magnetic Resonance Spectroscopy of X-nucleus is a method for mapping metabolite quantity *in-vivo* in the desired region of the human body. However, this method has a lot of difficulties. Since natural abundance of X-nucleus is much lower than for hydrogen, to improve SNR of the spectrum we need to work in higher static magnetic fields. But, even at such high fields, X-nuclei imaging still a very difficult process. One of the problems are RF-coils, which are required for transmission and reception of signals. Traditionally, for human body X-nuclei MRS multi-tuned loop coils are used. However, such coils suffer from additional losses in the circuits required for double-frequency tuning and high complexity. In this work, we propose an alternative approach, based on a recently introduced leaky-wave antenna for MRI, that allows creating wideband excitation which provides wideband transmit and receive. This wideband frequency range covering ^{13}C, ^{23}Na and ^{31}P Larmor resonant frequencies.

1. Introduction

The design of double and triple radio frequency (RF) coils for multi-nuclei spectroscopy is a challenging engineering problem. In the simplest case one loop are tuned to two resonant frequencies using pole-insertion method. This method is based on the addition of capacitance to the resonant LC-circuit. This method can be applied both to the surface and volume coils. However, such an approach leads to the significant decrease of coil's efficiency due to higher losses [1]. Another approach is based on using two separate loops, tuned to different frequencies. To decrease coupling usually high-impedance traps tuned to a particular frequency are used for decoupling. Another approach - using a coil with orthogonal magnetic field distributions. For example, combination of loop and butterfly coil [2] or combination of loop coil and stripline resonator [3].

In this work, we propose an alternative approach, based on the recently proposed leaky-wave antenna (LWA) for ultra-high field (UHF) magnetic resonance imaging (MRI) [5]. This antenna is based on a microstrip line with I-shaped slots, etched in the ground plane. The working principle of such an antenna is based on the fact that the wave propagating in the guiding structure radiates if the speed of propagation is higher than the speed of light [4]. Since the

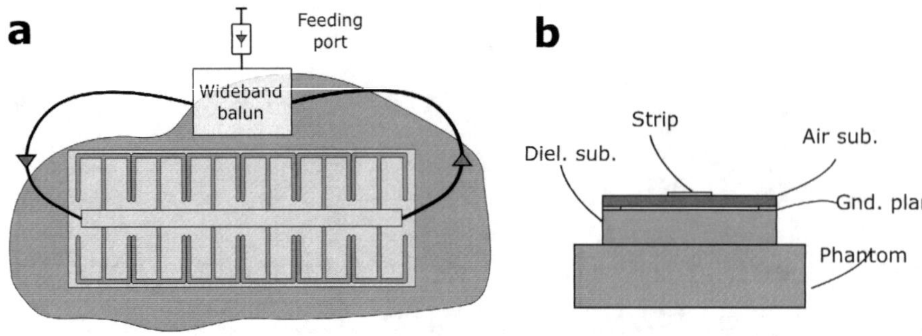

Figure 1. Schematic layout of the proposed leaky-wave antenna placed over a homogeneous phantom: (a) top view; (b) side view

human body has a high relative permittivity (around 60), the wave in the stripline is faster comparison to the wave in the media.

In this work, we showed by numerical simulations, that it's possible to achieve broadband field excitation, which covers the frequency range from 78 to 122 MHz. All simulations were performed in the case of LWA loading for the homogeneous phantom with properties ε=60, ς=0.6 S/m, which mimics human body averaged tissues over the 78 to 122 MHz frequency range.

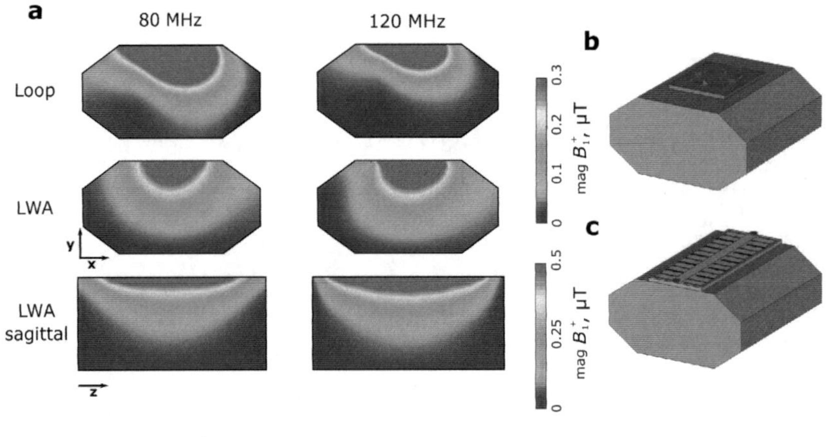

Figure 2. (a) Simulated B_1^+ magnitude distributions normalized by square root of accepted power. (b) CST model of investigated coils, loaded with homogeneous phantom.

2. Design and methods

In our work, we use the general approach based on unitcell optimization, described in the paper about 7T LWA [5], but with a different goal - to obtain a broadband excitation for heteronuclear spectroscopy. The proposed coil consisted of an array of six slots, placed at the top of the homogeneous phantom. The total length of the coil was 42 cm. Slots were etched in the common ground plane. The overall length of the slot was L=30.5 cm. I-shaped slots were used to increase the real part of the slot's input impedance and improve the loading of the antenna onto the phantom. Additional dielectric spacers were used to improve the loading of individual slots. Period of the slots was equal to 6,7 cm. This combination of slot length and period was chosen to provide good matching in the whole frequency range from 78 to 122 MHz. Slots had a period of 6 cm and were distributed equally along the z-axis. The width of antennas

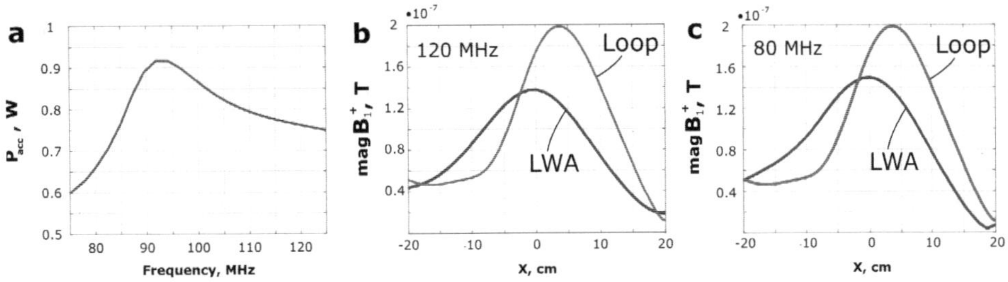

Figure 3. Results of numerical simulations: (a) Simulated power, dissipated in the phantom, over the wide frequency range for leaky-wave antenna. (b) Simulated B_1^+ magnitude at the 10 cm depth in the phantom for the LWA and loop coil at 120 MHz. (c) Simulated B_1^+ magnitude at the 10 cm depth in the phantom for the LWA and loop coil at 80 MHz.

was chosen to fit 8 channel configuration in a human body imaging application. The design of the proposed antenna is presented in Fig. 1.

The length and period of slots, as well as the permittivity of dielectric spacer were optimized to achieve the maximum efficiency of 10 cm while keeping a broadband excitation. Numerical simulation software CST Studio Suite 2020 (CST, Darmstadt, Germany) was used for the calculation of the RF magnetic field of the proposed antenna. Two loops, tuned to 80 MHz (^{23}Na resonant frequency) and 120 MHz (^{31}P resonant frequency) with four distributed capacitors were used as reference coils. Loop coils were simulated with the same homogeneous phantom loading. All results were obtained using Frequency Domain Solver with finite element meshing.

3. Results and discussion

The simulation has shown that the proposed coil design based on the leaky-wave approach is suitable for broadband RF-excitation of objects. Distributions of circular polarized magnetic RF-field B_1^+ for the LWA and loop coils are presented in Figure 2. Power accepted in the LWA in the broad frequency range is presented in Figure 3 (a). From this we can see that, in the whole frequency range, the power absorbed in the phantom is higher than 65 %. To compare leaky-wave antennas with loop coils more precisely B_1^+ fields measured in the depth of 10 cm at 80 at 120 MHz frequency along X-axis are presented at the Figure 3 (b) and (c). From these plots, we see that LWA creates a 30 % lower magnetic field at 120 MHz and 40 % lower at 80 MHz correspondingly. Such a decrease of field happens due to a substantially larger field of view of the leaky-wave antenna.

4. Conclusion

In this work, we propose a new RF-coil for heteronuclear 7 Tesla MRS based on the leaky-wave approach that could simultaneously receive and transmit signals of several frequencies. The results have shown that using this coil, one can create an RF-magnetic field comparable with conventional multi-tuned loop RF-coils.

5. Acknowledgment

This work was supported by the Russian Foundation for Basic Research (Grants No. 19-29-10038).

References

[1] M.D. Schnall, V. Harihara Subramanian, J. S. Leigh, B. Chance "A new double-tuned probed for concurrent ^{1}H and ^{31}P NMR," *Journal of Magnetic Resonance (1969)*, vol. 65, no. 1, pp. 122–129, 1985.

[2] G. Adriany and R. Gruetter, "A half-volume coil for efficient proton decoupling in humans at 4 Tesla," *Journal of Magnetic Resonance*, vol. 125, no. 1, pp. 178–184, 1997.

[3] O. Rutledge, T. Kwak, P. Cao, and X. Zhang, "Design and test of a double-nuclear RF coil for ^{1}H mri and (13)c mrsi at 7T," *Journal of magnetic resonance (San Diego, Calif. : 1997)*, vol. 267, p. 15"21, June 2016.

[4] D. R. Jackson, C. Caloz and T. !Itoh, "Leaky-Wave Antennas," in Proceedings of the IEEE, vol. 100, no. 7, pp. 2194-2206, July 2012.

[5] G. Solomakha, J. Svejda, C. van Leeuwen, A. Rennings, A. Raaijmakers, S. Glybovski, and D. Erni, "A self-matched leaky-wave antenna for ultrahigh-field magnetic resonance imaging with low specific absorption rate," *Nature Communications*, vol. 12, no. 1, pp. 1–11, 2021.

Active narrowband filter based on 2.5D metasurface from Ge2Sb2Te5

A I Solomonov[1,2 a)], **O M Kushchenko**[1], **D A Yavsin**[2], **M V Rybin**[1,2], **and A D Sinelnik**[1]

[1]Department of Physics, ITMO University, St. Petersburg 197101, Russia
[2]Ioffe Institute, St. Petersburg 194021, Russia

[a)]E-mail: aleksandr.solomonov@metalab.ifmo.ru

Abstract. We propose a new concept of an active narrowband filter based on a 2.5D metasurface from Ge2Sb2Te5 (GST). In this paper, we present a numerical calculation of the transmission spectrum from a structure of ellipsoids of revolution. For this 2.5D metasurface, modulation of narrow peaks in the IR range for s- and p-polarization is shown. A manufacturing technique using two-photon lithography and laser electrodispersion is proposed.

Introduction.

The phase transition between amorphous and crystalline state in the chalcogenide phase-change medium from germanium antimony telluride has long been used in rewritable optical storage discs [1]. During the last decade the possibility of using the chalcogenide materials for signal modulation by photonic metastructures in the optical range was demonstrated [2, 3, 4]. However, most available methods for fabrication such structures are not able to provide enough degrees of freedom to achieve a multifunctional response, which limit their efficiency. A possible solution is to combine a two-dimensional (2D) metasurface into a multilayer 2.5D structure [5, 6].

In this paper, we propose the concept of an active narrowband filter based on a 2.5D metasurface made of $Ge_2Sb_2Te_5$ (GST). For considered structures, numerical calculations were performed in the near-infrared spectral range, and the manufacturing procedure was presented.

1. Transmission spectra for 2.5D metasurfaces.

To obtain the reflection spectra, the commercial package CST Studio Suite 2020 was used, which is based on the finite-difference scheme in the frequency domain. The unit cell contains 2 identical ellipsoids of revolution from the GST with a polar radius of 200 nm and an equatorial radius of 250 nm. The period between the ellipsoids is 1.5 mkm. The height between the ellipsoids was a freely variable parameter. The complex dielectric constant of the GST was taken from [7]. The ellipsoids are embedded in the air. For our calculations, a linearly polarized plane wave was incident on a unit cell along the normal to the metasurface.

Figure 1 shows the reflection spectra of a 2.5D GST metasurface for two polarizations of the incident light. For s-polarization at a wavelength of 1645 nm, a narrow transmission dip is observed. For p-polarization, a doublet is observed with dips centered at 1614 and 1628 nm. The switching the GST phases between amorphous and crystalline the dips leads to disappearing of the dips. By varying the free parameter (i.e. the interlayer distance), the peak can be tailored to the operating range in the telecommunication wavelengths.

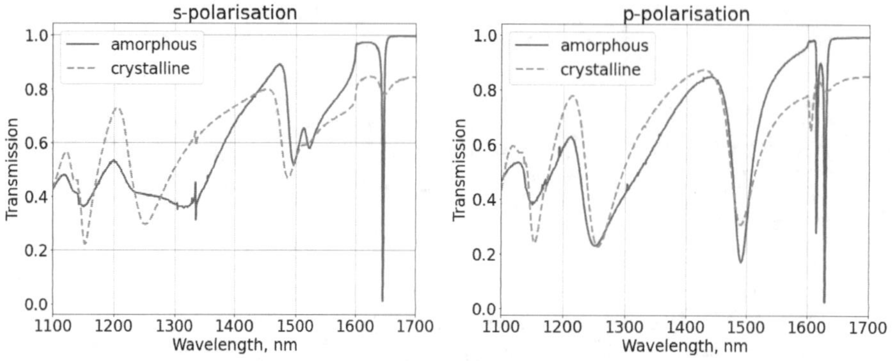

(a) (b)

Figure 1. The calculated transmission spectra for the 2.5D metasurface from GST for s-polarization (a) and p- polarization (b)

Figure 2. SEM images of 2.5D metasurface after GST film deposition.

2. Fabrication of 2.5D metasurface from GST.

We used following procedures for fabricating 2.5D metasurfaces corresponding to the simulated structures. At the first stage, polymer posts were made of zirconium propoxide with an Irgacure 369 photo-initiator by the method of two-photon lithography [8]. For this, the radiation of a 50-fc TiF-100F laser with a central wavelength of 780 nm was focused on the non-developed polymer film using a 40X microscope objective with numerical aperture NA = 0.75.

At the next stage, a GST layer is applied to the polymer posts by laser electrodispersion [9]. During deposition, the laser beam melted the GST target, pulling out submicron drops charged with plasma from the surface. When the force of the Coulomb repulsion exceeds the force of surface tension, it breaks up into many smaller droplets. The effect of the lack of repulsive forces can be seen in Fig. 2. White and shapeless spots are uncrushed GST droplets.

Conclusion

To summarize, the numerical simulations of the transmission spectra of 2.5D metasurfaces have been carried out in this work show the possibility of modulating the transmission in the IR range during the phase transition between the amorphous and crystalline states in GST. We have demonstrated that the proposed manufacturing technique using two-photon lithography and laser electrodispersion allows us to develop the design of a narrow-band filter and conduct an experimental comparison with the results of numerical calculations.

This work was performed with financial support from the Russian Science Foundation (grant 20-12-00272).

References

[1] Kolobov A V, Fons P, Frenkel A I, Ankudinov A L, Tominaga J and Uruga T 2004 *Nature materials* **3** 703–708
[2] Ahmadivand A, Gerislioglu B, Sinha R, Karabiyik M and Pala N 2017 *Scientific Reports* **7** 1–8 ISSN 20452322
[3] Sámson Z, MacDonald K, De Angelis F, Gholipour B, Knight K, Huang C, Di Fabrizio E, Hewak D and Zheludev N 2010 *Applied Physics Letters* **96** 143105
[4] Gholipour B, Karvounis A, Yin J, Soci C, MacDonald K F and Zheludev N I 2018 *NPG Asia Materials* **10** 533–539
[5] Fu X, Zhang Z, Jiang Y, Zhang W, Bi K, Sun K and Fan R 2020 *Advanced Engineering Materials* **22** 1901126
[6] Mansouree M, Kwon H, Arbabi E, McClung A, Faraon A and Arbabi A 2020 *Optica* **7** 77–84
[7] Dyakov S A, Gippius N A, Voronov M M, Yakovlev S A, Pevtsov A B, Akimov I A and Tikhodeev S G 2017 *Physical Review B* **96** ISSN 24699969
[8] Farsari M and Chichkov B N 2009 *Nature photonics* **3** 450–452
[9] Yavsin D, Kozhevin V, Gurevich S, Yakovlev S, Melekh B, Yagovkina M and Pevtsov A 2014 *Semiconductors* **48** 1567–1570

UV/Ozone Treatment and Open-Air Copper Plasmonics

Solovey V. R.[1, a)], Yakubovsky D. I.[1], Ermolayev G. A.[1], Lebedinskij Y. Y.[2], Markeev A. M.[2], Voronov A. A.[1], Zamboni F.[3], Popok V. N.[3], Arsenin A. V.[1], Volkov V. S.[1] and Novikov S. M.[1]

[1]*Center for Photonics and 2D Materials, Moscow Institute of Physics and Technology, 141700 Dolgoprudny, Russia*
[2]*Center of Shared Facilities in Nanotechnology, Moscow Institute of Physics and Technology, 141700 Dolgoprudny, Russia*
[3]*Departmnt of Materials and Production, Aalborg University Skjernvej 4A, 9220 Aalborg, Denmark*

[a)]Corresponding author: solovey.vr@phystech.edu

Abstract. Thin copper films with thickness ~28 nm deposited on SiO_2 substrate with the vacuum electron beam evaporation method and treated by UV-ozone are studied. It was found that a UV-ozone treatment of the copper film causes rapid formation of the thin ~3-4 nm oxide film. XPS analysis showed that CuO oxide predominates in this film. The formed oxide film effectively protects the copper against the following oxidation. The presented method of UV-ozone treatment is a simpler and cheaper approach compared to many other ways to form protective coatings of copper to preserve its functional properties. This method can be useful in nanoelectronic, nanooptical, and biosensors applications.

Introduction. Metals with plasmonic properties in the visible and near-infrared (NIR) ranges such as gold, silver, and copper can be used for optical, electronic, sensing and other applications, which are currently of high interest [1, 2]. The important issue is the stability of plasmonic properties, which often limits the use of some metals due to their chemical reactivity and the possibility of spurious effects. The most common material used for plasmonic is gold, which demonstrates excellent optical properties as well as resistance to oxidation. The limitations of gold in plasmonics include its high price and incompatibility with microelectronic technological processes. Silver shows superior performance due to low optical losses and also widely used [3-7] but generally considered to be less attractive due to lower chemical and, thus, plasmonic stability [8]. Copper is another metal, which has excellent optical properties. Compared with gold, it is inexpensive and has lower optical losses in the visible and NIR range. The advantages of copper in plasmonic applications were exploited for instance in the ultralow-loss copper plasmonic waveguides and biosensing applications [9-13]. Copper is prone to relatively fast surface oxidation upon exposure to an ambient atmosphere [14]. Under normal conditions, the dominant product is Cu_2O with a minor or no contribution of CuO. Thus, to use Cu films for plasmonic applications, one needs to protect the surface of structures against oxidation-induced degradation. It can be done by applying a protective shell/coating of SiO_2, Al_2O_3 or even graphene [10, 15]. In this work, we test a simple method of UV-ozone treatment, which leads to the rapid formation of a thin oxide layer on a copper film. This oxide layer effectively protects the copper against the following oxidation-related degradation of the plasmonic properties, which was recently proven for Cu nanoparticles [16]. We performed a complex analysis of the formed oxide layers. We anticipate that the presented results will be of interest as a simple and effective method for preserving the plasmonic properties of thin copper films to be used in non-linear optics or sensing applications.

Sample Fabrication. Copper films with thicknesses of 28 nm were deposited using electron beam evaporation in a NEE-4000 E-Beam Evaporation system. Clean silicon wafers capped with a top layer of 2 nm thick SiO_2 were placed in the vacuum chamber of the e-beam evaporator at a pressure of 3×10^7 Torr, at room temperature. As a material for deposition, granules of copper with a purity of 99.99% were used. The deposition rate was approximately 2 Å/s. 8 identical samples were fabricated simultaneously in one cycle. The quoted "thickness" of the cupper films is the average value measured by the quartz oscillator. The fabricated films were subsequently imaged by scanning electron microscope (SEM) JEOL JSM-7001F (Figure 1b).

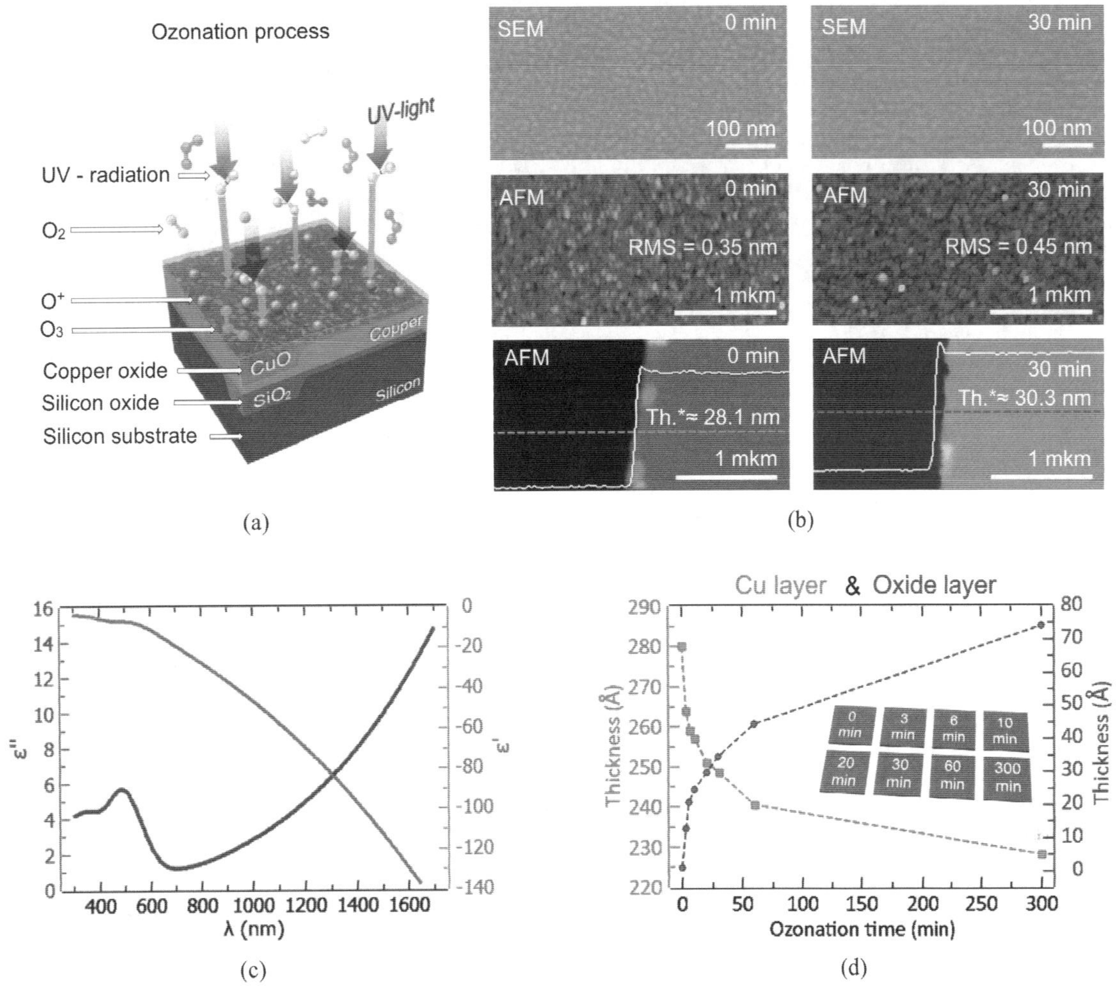

FIGURE 1. (a) The synthesis scheme of the CuO layer on the surface of a thin copper film using UV-ozonation, (b) SEM and AFM images of the surface of the deposited copper film (after 0 and 30 min ozonation); (c) Dependence of the real and imaginary parts of the dielectric function on wavelength for the pure copper film on a SiO2/Si substrate, (d) Dependence of the thickness of a copper film on the time of UV-ozonation and schematic representation of samples in the inset.

Right away after deposition, 7 out of 8 substrates were treated by the UV Ozone Cleaner (Ossila Limited). During UV-cleaning, the ultraviolet radiation breaks the oxygen-oxygen double bond resulting in the formation of two free radicals of oxygen O. These free radicals can react with molecular oxygen producing ozone molecules O_3. The mixture of O and O_3 can lead to a rapid oxide film formation on the copper surface (Figure 1a). In order to estimate the oxide film growth versus time, substrates were UV treated with different duration of time as 3, 6, 10, 20, 30, 60, and 300 min, see Figure 1d (each sample has only one own processing time). One Cu film was not treated by UV and was used as a reference sample for the measurement of Cu film optical properties.

Sample Characterization. The optical properties of copper films (reference and exposed to UV-cleaning) were studied with the V-VASE Ellipsometer (Woollam Co.). The spectral range was 300-1700 nm, and the angles of incidence of the light beam were 65°, 70°, and 75°. The thickness of the films was accurately measured by atomic force microscopy (AFM) at a step of the film on the substrate and it was found to be 28.0 ± 0.5 nm (Figure 1b). These measurements of the thickness are necessary to carry out a correct fitting of the ellipsometry data. The spectra of real

Journal of Physics: Conference Series **2015** (2021) 012148 doi:10.1088/1742-6596/2015/1/012148

and imaginary components of the dielectric function of the copper film (reference sample) were found from data measured by a variable-angle spectroscopic ellipsometer (Figure 1c). The multilayered model consisted of Si substrate, Cu film and CuO layer (data based on XPS analysis) was applied [17]. It was found that the copper thickness decreases with the increase of ozonation time, which can be an indication of oxide layer formation at the expense of metallic copper (see Figure 1d). It should be noted that the above-mentioned model of thickness estimation has drawbacks, because it accounts only for CuO but does not consider the possible formation of other copper oxides or hydroxides. Also, AFM data shows that the thickness of Cu film increases after UV-ozone treatment in comparison with the reference Cu film, non-treated by UV-ozone. (Fig 1b).

Additionally, we measured the reflectance spectra (Figure 2a) and performed XPS analysis (Figure 2b). Figure 2a shows the dependence of the reflectance spectra of all samples on the ozonation time. The chemical state and composition of the samples were analyzed using the Theta Probe tool (Thermo Scientific) under high-vacuum conditions with a monochromatic Al Kα X-ray source (1486.6 eV). The XPS analysis was carried 4 hours after the film fabrication and the samples were kept in an ambient atmosphere. The XPS spectra of the reference sample presented in Figure 2b show the characteristic bands at 932.6 and 952.4 eV. Unfortunately, Cu and Cu_2O have nearly the same binding energy, [18, 19], and the bands can be an indication of both Cu and Cu_2O. From the narrow bandwidth of $Cu2p_{3/2}$ one can suggest that the signal comes from the metal but we must allow the Cu_2O formation during the time the samples were kept in the air.

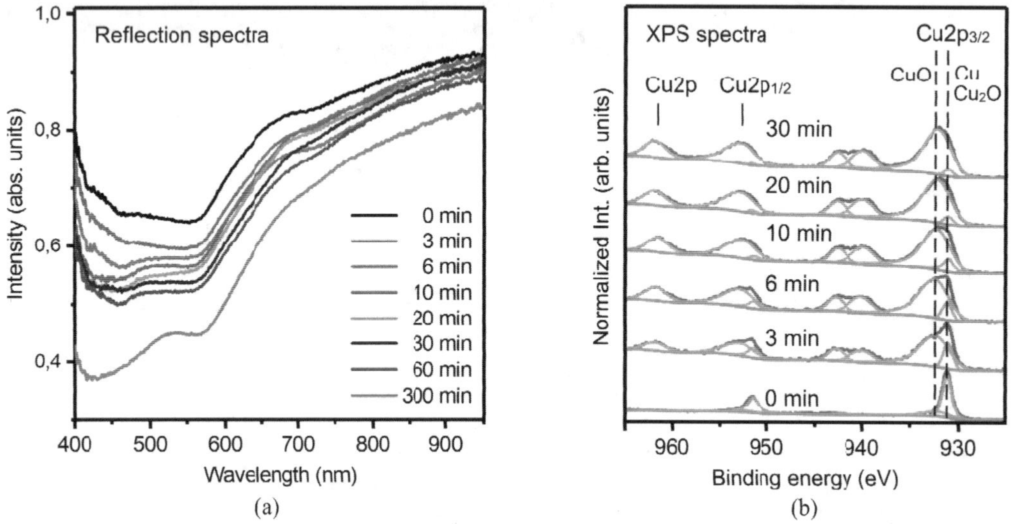

FIGURE 2. Reflection spectra (a) and XPS data analysis of the surface of a thin copper film in an air atmosphere and after UV-cleaning for 0, 3, 6, 10, 20 and 30 minutes (b).

The XPS spectra of the copper films ozonated for 0, 3, 6, 10, 20 and 30 min, are presented in Figure 2b. They show a significant increase of the $Cu2p_{3/2}$ and $Cu2p_{1/2}$ shoulders at 934.7 and 954.5 eV, respectively. The latter can be assigned to CuO, while the former one can correspond to copper dihydroxide, $Cu(OH)_2$ [19]. Two new strong satellites at 941 and 963 eV, which are typical for CuO spectra, are also observed [19]. The obtained results show a significant difference in the copper oxidation under the ambient conditions and UV-ozone treatment. While for the films oxidized in the ambient atmosphere the Cu_2O predominates with a minor or no contribution of CuO [20], for the UV-ozone treated copper film, the oxidation undergoes predominantly through the CuO formation. Based on the ellipsometry and XPS measurements (depth of XPS analysis is low, ≤ 3 nm) it is possible to estimate the thickness of formed oxide as 3-4 nm.

Summary. It was demonstrated that a UV-ozone treatment of the copper film causes rapid formation of the thin oxide layer. XPS analysis showed that CuO oxide predominates. This oxide layer can protect the metallic copper against the following oxidation. The presented method of UV-ozone treatment is a simpler and cheaper approach compared to many other ways utilizing protective coatings to preserve the metallic copper against chemical degradation. Thus, it can be recommended for the protection of copper surfaces to be used in plasmonic or sensing applications.

ACKNOWLEDGMENTS

We gratefully acknowledge financial support from the Ministry of Science and Higher Education of the Russian Federation (No. 0714-2020-0002) and the Russian Foundation for Basic Research (20-07-00475 and 20-07-00840). We thank the Shared Facilities Center of the Moscow Institute of Physics and Technology for the use of their equipment.

REFERENCES

1. S. Maier Plasmonics: Fundamentals and Applications. Springer (New York, 2007).
2. J. Langer, S.M. Novikov, L.M. Liz-Marzán, Nanotechnology **26**, 322001 (2015).
3. Y.V. Stebunov, O.A. Aftenieva, A.V. Arsenin, V.S. Volkov, ACS Applied Materials & Interfaces **7**, 21727-21734 (2015).
4. K.V. Voronin, Y.V. Stebunov, A.A. Voronov, A.V. Arsenin, V.S. Volkov, Sensors **20** (1), 203 (2020).
5. N. A. Brazhe, A. B. Evlyukhin, E. A. Goodilin, A. A. Semenova, S. M. Novikov, S. I. Bozhevolnyi, B. N. Chichkov, A. S. Sarycheva, A. A. Baizhumanov, E. I. Nikelshparg, et al. Sci. Rep. **5**, 13793 (2015).
6. K. M. McPeak, S. V Jayanti, S. J. P Kress, S. Meyer, S.Iotti, A.Rossinelli, D. J. Norris, ACS Photonics **2**, 326-333 (2015).
7. X. Zhang, E. M. Hicks, J. Zhao, G. C. Schatz, R. P. Van Duyne, Nano Lett. **5**, 1503 (2005).
8. S. M. Novikov, V. N. Popok, A. B. Evlyukhin, M. Hanif, P. Morgen, J. Fiutowski, J. Beermann, H.-G. Rubahn, S. I. Bozhevolnyi, Langmuir **33**, 6062 (2017).
9. V.S. Volkov, Yu.V. Stebunov, D.I. Yakubovsky, D.Yu. Fedyanin, A.V. Arsenin, AIP Conference Proceedings **1874**, 020008 (2017).
10. Y. V. Stebunov, D. I. Yakubovsky, D. Yu. Fedyanin, A. V. Arsenin, V. S. Volkov, Langmuir **34**, 4681-4687, (2018).
11. D. I. Yakubovsky, Y. V. Stebunov, R. V. Kirtaev, K. V. Voronin, A. A.Voronov, A. V. Arsenin, V. S.Volkov, Nanomaterials **8**, 1058, (2018).
12. Y.V. Stebunov, D.I. Yakubovsky, V.G. Leiman, A.V. Arsenin, V.S. Volkov, Journal of Physics: Conference Series **1092**, 012143 (2018).
13. V. Mkhitaryan, K. March, E. N. Tseng, X. Li, L. Scarabelli, L. M. Liz-Marzán, S. Y. Chen, L. H. G. Tizei, O. Stéphan, J. M. Song, M. Kociak, F. Javier García de Abajo, A. Gloter, Nano Lett. **21**(6), 2444–2452 (2021).
14. J. H. Par, P. Ambwani, M. Manno, N. C. Lindquist, P. Nagpal, S.-H. Oh, C. Leighton, D. J. Norris, Adv. Mater. **24**, 3988-3992 (2012).
15. V. G. Kravets, R. Jalil, Y.-J. Kim, D. Ansell, D. E.; Aznakayeva, B. Thackray, L.Britnell, B. D. Belle, F. Withers, I. P. Radko, Z. Han, S. I. Bozhevolnyi, K. S. Novoselov, A. K. Geim, A. N. Grigorenko Sci. Rep. **4**, 5517 (2014).
16. V.N. Popok, S.M. Novikov, Yu.Yu. Lebedinskij, A.M. Markeev, A.A. Andreev, I.N. Trunkin, A.V. Arsenin, V.S. Volkov, Plasmonics **16**, 333-340 (2021).
17. D. I. Yakubovsky, A. V. Arsenin, Y. V. Stebunov, D.Y. Fedyanin, V. S. Volkov, Optics Express **25**, 25574 –25587 (2017).
18. S. Poulston, P. M. Parlett, P. Stone, M. Bowker, Surf. Interface Anal. **24**, 811 (1996).
19. C. Biesinger, L.W.M. Lau, A. R. Gerson, R. St.C. Smart, Appl. Surf. Sci. **257**, 887 (2010).
20. J. H. Kim, S. H. Ehrman, T. A. Germer, Appl. Phys. Lett. **84**, 1278 (2004).

Fabrication and characterization of aluminum, gold-tin and titanium-tungsten alloys for mid-infrared plasmonic applications

J Spettel[1,2,4,*], G Stocker[1,4], T D Dao[2], R Jannesari[3], A Tortschanoff[2], P Saeidi[3], G Pühringer[3], F Dubois[2], C Fleury[2], C Consani[2], T Grille[1], E Aschauer[1], M Moridi[2] and B Jakoby[3]

[1] Infineon Technologies Austria AG, 9500 Villach, Austria
[2] Silicon Austria Labs GmbH, 9524 Villach, Austria
[3] Institute for Microelectronics and Microsensors, Johannes Kepler University Linz, 4040 Linz, Austria
[4] These authors contributed equally to this work.
*corresponding author, E-mail: Jasmin.Spettel@silicon-austria.com

Abstract. We numerically and experimentally investigate aluminum metal (Al), gold-tin (AuSn) and titanium-tungsten (TiW) metallic alloys as plasmonic materials in the mid-infrared (MIR) region using spectroscopic ellipsometry and reflection measurements of gratings. The angle dependence of the specular reflectance of shallow gratings is investigated using a free-beam measurement setup and compared to simulations. It is shown that the deep and narrow resonances observed for all three materials match the associated prediction from simulations.

1. Introduction

Due to the great potential applications of plasmonics and photonics in the mid-infrared (MIR) spectral range, the research interest within these fields have increased rapidly in recent years. Potential applications are for example waveguides [1], photonic integrated circuits [2], selective thermal emitters [3] or IR detectors [4]. Even though many groups already searched for alternative plasmonic materials, such as silicides, transparent conducting oxides, doped semiconductors, carbides and many more, most plasmonic applications are still based on gold (Au) and silver (Ag) [5,6]. Within this work, we present aluminum (Al), gold-tin (AuSn) and titanium-tungsten (TiW) as alternative plasmonic materials in the MIR regime. Al is very cheap and a thin native oxide layer serves as protection from chemical reactions with the environment [7]. AuSn is a very soft material with a low melting point [8] and thus can be used for molding nanoimprint lithography or 3D printing. TiW on the other hand is resistant to high temperatures [9] and hence is suitable for thermal devices, such as thermal emitters [10]. In addition, it is a common material for electrodes in electronics, leading to the advantage of prospective standard fabrication.

In this study, we first show spectroscopic ellipsometry measurements of the proposed materials. Then, the materials are tested regarding their usability for surface plasmon polariton (SPP) excitation using a grating configuration. The experimental data are also compared to simulations and show good agreement. All three materials exhibit strong and narrow resonances [11].

2. Experimental

First, a spectroscopic ellipsometry study of the proposed plasmonic materials was performed. Therefore, a sample with a 100 nm thick sputtered layer for each material was prepared. Details of the fabrication process are given in [11]. The ellipsometry measurement was performed at Semilab (SE-2000 combined with IRSE extension for full UV-MIR spectroscopic ellipsometry) and the retrieved data were modeled with a Drude-Lorentz model. In Figure 1a) and b) the real and imaginary part of the retrieved relative permittivity of Al, AuSn and TiW are compared to literature values of gold (Au) and silver (Ag) [12]. Within the investigated spectral range (wavelength of 0.25 μm – 20 μm) all three materials show a negative real part of permittivity, indicating them as promising candidates for plasmonics. The values of Al are even comparable with those of Au and Ag. The imaginary part of the relative permittivity is an indicator for the losses of the material. Similarly, Al is comparable with Au and Ag, whereas AuSn and TiW are expected to have lower losses.

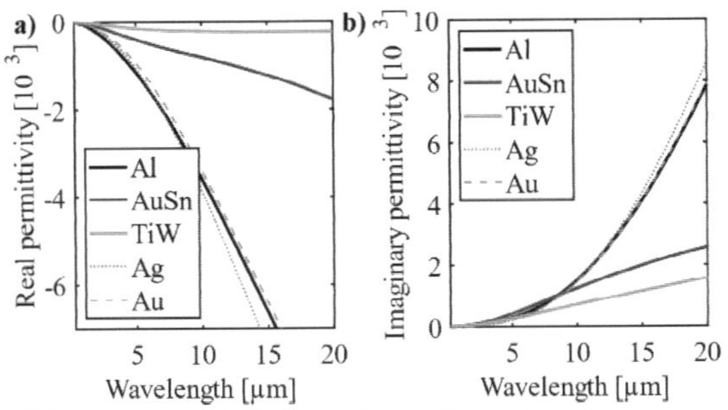

Figure 1: Comparison of a) real part and b) imaginary part of complex permittivity of the experimental data of Al, AuSn and TiW retrieved by spectroscopic ellipsometry with literature values of Au and Ag, taken from [12]. This figure is taken from [11].

The proposed plasmonic grating structures were fabricated on 8-inch silicon (Si) substrates in the cleanroom facilities of Infineon Technologies Austria AG. A schematic of the layer stack is shown in Figure 2a). Further details about the fabrication process are provided in our published paper [11]. In order to observe a resonance around an incidence angle of 30°, the grating period was chosen to be 2.8 μm (1.6 μm groove width) and the wavelength close to 4.2 μm (limited by the tunability of the laser). For characterization, a free-beam reflection measurement was carried out. Figure 2b) shows a schematic of the measurement setup, which is described in [11].

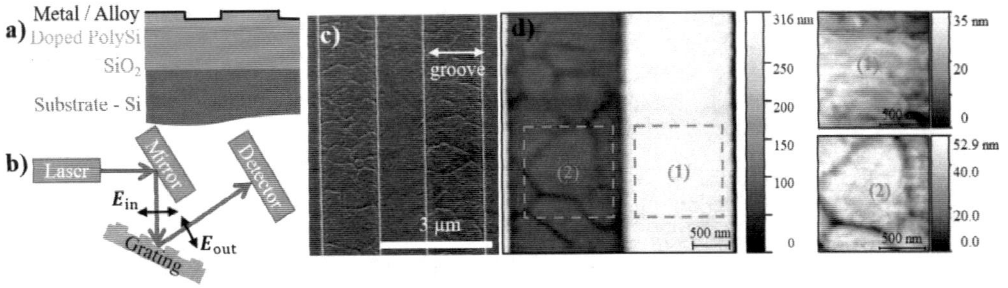

Figure 2: a) Layer stack of the gratings (adapted from [11]). b) Free beam measurement setup. The double-arrows indicate the p-polarization of the laser light. c) Top view SEM image of a polycrystalline Si grating with 100 nm Al coating. The grooves are indicated by the double-arrow. d) AFM measurement of one period of an Al coated grating with 225 nm depth.

Furthermore, the surface roughness of the gratings was investigated using a scanning electron microscope (SEM, Thermo Scientific Helios G4) and an atomic force microscope (AFM, Veeco Dimension 3100, Digital Instruments). Figure 2c) depicts an SEM image in top view showing the surface morphology of a polycrystalline Si grating coated with 100 nm Al. The large grains in the grooves (grooves are indicated by the double-arrow in the figure) have their origin in the structure of the polycrystalline Si layer after the etching process [11]. Figure 2e) depicts an AFM measurement of one period of an Al coated grating. Accordingly, the roughness is about 53 nm within the grooves (area 2) and 35 nm elsewhere (area 1).

In addition to the free-beam reflection measurement, the reflectance spectra of the gratings were simulated using a rigorous coupled-wave analysis software (RCWA) (DiffractMOD, Synopsys' RSoft package). Thereby, orders -3 through +3 were considered.

3. Results and discussion

Figure 3a), b) and c) depict the angle dependence of the specular reflectance at gratings with a depth of 150 nm and a coating of 100 nm sputtered Al, AuSn and TiW, respectively. Each left panel shows the measurement result, whereas the right panels depict the specular reflectance retrieved from simulations. In addition, the CO_2 absorption band is indicated by the shaded rectangle in the left panel in Figure 3a). As expected from the grating equation $k_{SPP} = k_0 \sin(\theta) \pm mG_x$ [13] the resonance dip moves to longer wavelengths for larger angles of incidence θ. Here k_{SPP} and k_0 are the wavenumber of the SPP and the incident wave, respectively, m is an integer and $G_x = \frac{2\pi}{P}$ is the wavenumber of the grating with period P. Furthermore, the resonance dip is asymmetric (Fano-like) due to the coupling between the surface wave, which is a bound state, and the diffraction from the periodic grating, which is a leaky wave [14]. Thus, the specular reflection towards shorter wavelengths regarding the resonance dip is attenuated in comparison to the specular reflectance towards longer wavelengths. In order to determine the width of the resonance (full width at half maximum – FWHM) the data at an incidence angle of 26° were fitted using a Fano-formula [15]:

$$R \propto \frac{(F\gamma + \lambda\lambda_0)^2}{(\lambda - \lambda_0)^2 + \gamma^2} + B\,. \tag{1}$$

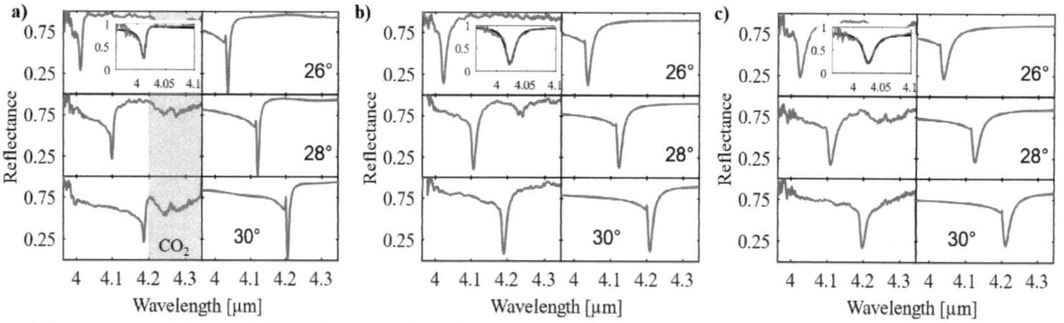

Figure 3 Comparison of the angle dependent specular reflectance retrieved from measurements (left) and simulation (right) of gratings with a depth of 150 nm coated with 100 nm of a) Al, b) AuSn and c) TiW at different angles of incidence (26°, 28°, 30°). The insets show the Fano-fit in addition to the measured data. The shaded rectangle in a) indicates the CO_2 absorption band. Adapted from [11].

Here R is the reflectance, F is the Fano-parameter, γ the resonance width, λ the wavelength, λ_0 the center wavelength of the resonance dip and B the offset. For Al a width of 4.7 nm, for AuSn 9.7 nm and for TiW 11.6 nm was determined (the simulations yielded widths of 5.5 nm, 11.5 nm and 16.9 nm, respectively). The fitted curve together with the measurement data is shown in the inlets of Figure 3. The broadening of the resonance from Al to AuSn to TiW is consistent with the retrieved relative permittivities, shown in Figure 1. Overall a good agreement between the measurements and the simulations was obtained. Small deviations are explained by the uncertainty of the zero-angle (±1°), leading to a slight shift in the position of the resonance, not perfectly p-polarized light during the measurements as well as by small variations in the fabrication process leading to slight deviations of the grating dimensions (period, groove width, side wall steepness). In addition, the surface roughness is not included in the simulations. See [11] for more details and the results of the gratings with 225 nm and 375 nm depth.

4. Conclusions

We have investigated Al, AuSn and TiW both numerically and experimentally as alternatives to Ag and Au for plasmonic applications in the MIR spectral range. A spectroscopic ellipsometry measurement was performed to retrieve the optical properties of the proposed materials. The materials show good plasmonic properties in the MIR region. As a demonstration, we fabricated a set of plasmonic gratings using Al, AuSn and TiW. Strong SPP resonances were observed from the fabricated gratings at about 4.2 μm resonance. FWHMs were obtained with 4.7 nm for Al, 9.7 nm for AuSn and 11.6 nm for TiW. A good agreement between experimental data and simulations was observed.

Acknowledgements

This work was performed within the PICASSO-project funded by the Bundesministerium für Klimaschutz, Umwelt, Energie, Mobilität, Innovation und Technologie in the framework of the program "Produktion der Zukunft" (Prj. Nr. 871417) and partly supported by the COMET K2-Center Symbiotic Mechatronics. Furthermore, the authors want to thank the involved staff working at the cleanroom of Infineon Technologies Austria AG, Villach, for their kind support with the fabrication of the structures, T. Ostermann for the mask layouts and L. Makai and A. Ertl at Semilab for their kind support in spectroscopic ellipsometry measurement.

References

[1] Tetienne J-P et al. I 2010 Injection of midinfrared surface plasmon polaritons with an integrated device *Appl. Phys. Lett.* **97** 211110

[2] Ono M, Taniyama H, Xu H, Tsunekawa M, Kuramochi E, Nozaki K and Notomi M 2016 Deep-subwavelength plasmonic mode converter with large size reduction for Si-wire waveguide *Optica* **3** 999

[3] Miyazaki H T, Ikeda K, Kasaya T, Yamamoto K, Inoue Y, Fujimura K, Kanakugi T, Okada M, Hatade K and Kitagawa S 2008 Thermal emission of two-color polarized infrared waves from integrated plasmon cavities *Appl. Phys. Lett.* **92** 141114

[4] Chang C-C, Sharma Y D, Kim Y-S, Bur J A, Shenoi R V, Krishna S, Huang D and Lin S-Y 2010 A Surface Plasmon Enhanced Infrared Photodetector Based on InAs Quantum Dots *Nano Lett.* **10** 1704–9

[5] West P R, Ishii S, Naik G V, Emani N K, Shalaev V M and Boltasseva A 2010 Searching for better plasmonic materials *Laser Photonics Rev.* **4** 795–808

[6] Zhong Y, Malagari S D, Hamilton T and Wasserman D 2015 Review of mid-infrared plasmonic materials *J. Nanophotonics* **9** 093791

[7] Gerard D and Gray S K 2014 Aluminium plasmonics *J. Phys. Appl. Phys.*

[8] Lee C H, Wong Y M, Doherty C, Tai K L, Lane E, Bacon D D, Baiocchi F and Katz A 1992 Study of Ni as a barrier metal in AuSn soldering application for laser chip/submount assembly *J. Appl. Phys.* **72** 3808–15

[9] Marx D R, Turn J C and Shi J 1999 *Tungsten titanium targets for VLSI device fabrication* (Materials Research Corporation)

[10] Laroche M, Arnold C, Marquier F, Carminati R, Greffet J-J, Collin S, Bardou N and Pelouard J-L 2005 Highly directional radiation generated by a tungsten thermal source *Opt. Lett.* **30** 2623

[11] Spettel J et al. 2021 Aluminium, gold-tin and titanium-tungsten alloys for mid-infrared plasmonic gratings *Opt. Mater. Express* **11** 1058

[12] Rakić A D, Djurišić A B, Elazar J M and Majewski M L 1998 Optical properties of metallic films for vertical-cavity optoelectronic devices *Appl. Opt.* **37**

[13] Sambles J R, Bradbery G W and Yang F 1991 Optical Excitation of Surface Plasmons: an introduction *Contemp. Phys.* **32** 173–83

[14] Miroshnichenko A E, Flach S and Kivshar Y S 2010 Fano resonances in nanoscale structures *Rev. Mod. Phys.* **82** 2257–98

[15] Luk'yanchuk B, Zheludev N I, Maier S A, Halas N J, Nordlander P, Giessen H and Chong C T 2010 The Fano resonance in plasmonic nanostructures and metamaterials *Nat. Mater.* **9** 707–15

Water-stable halide perovskite nanocrystals in biological environment

Pavel M. Talianov[1, a)], Oleksii O. Peltek[1], Mikhail A. Masharin[1], Soslan Khubezhov[2], Mikhail A. Baranov[3], Audrius Drabavičius[4], Alexander S. Timin[5,6], Lev E. Zelenkov[1], Anatoly P. Pushkarev[1], Sergey V. Makarov[1], Mikhail V. Zyuzin[1]

[1]Department of Physics and Engineering, ITMO University, Lomonosova 9, 191002 St. Petersburg, Russian Federation
[2]Department of Physics, North Ossetian State University, 362025 Vladikavkaz, Russian Federation
[3]Faculty of Photonics and Optical Information, Center of Information Optical Technologies ITMO University, 197101 St Petersburg, Russian Federation
[4]Center for Physical Sciences and Technology, LT-10257 Vilnius, Lithuania
[5]Research School of Chemical and Biomedical Engineering, National Research Tomsk Polytechnic University, 634050 Tomsk, Russian Federation
[6]RM Gorbacheva Research Institute for Pediatric Oncology, Hematology and Transplantation, Pavlov University, 197022 St. Petersburg, Russian Federation

[a)] Corresponding author: pavel.talianov@metalab.ifmo.ru

Abstract. Tetramethyl orthosilicate and triethoxyphenylsilane, which contains hydrophobic phenyl groups, were used as a silica (SiO_2) source in a modified ligand-assisted reprecipitation synthesis approach for the fabrication of water-stable perovskite nanocrystals. Hydrolysis-condensation reaction of tetramethyl orthosilicate and triethoxyphenylsilane results in a formation of 3D siloxane network. Employing triethoxyphenylsilane in the synthesis enhances the hydrophobic properties of the SiO_2 shell, which increases the stability of perovskites in aqueous medium. The stability of the $CsPbBr_3@SiO_2$ nanocrystals was estimated after 24 h of water exposure by the photoluminescence measurements at different time points. The synthesized $CsPbBr_3@SiO_2$ nanocrystals were visualized during *in vitro* experiments with murine melanoma B16-F10 cells. Hence, the potential of $CsPbBr@SiO_2$ nanocrystals for bioimaging purposes was observed.

1. Introduction

Recently, perovskite nanostructured materials have attracted significant interest due to their excellent photophysical properties, such as narrow full width at half maximum (FWHM), high photoluminescence (PL) quantum yield (PLQY), tunable emission spectra, etc. There are a number of reports dedicated to the fabrication of optoelectronic devices based on perovskite materials, e.g., light-emitting diodes [1], scintillators [2], lasers [3], solar cells. Moreover, not only the linear optical properties of perovskite nanocrystals (NCs) are worthy of attention, but also their nonlinear optical properties, such as near-infrared-triggered photon upconversion, which can be potentially applied for *in vivo* bioimaging due to an increased penetration depth of near-infrared light [4]. Nevertheless, the instability of perovskites in aqueous medium is still a challenge, which prevents from applying them in bioimaging. This study is aimed at obtaining lead halide perovskite ($CsPbBr_3$) nanocrystals coated with a silica shell exhibiting an improved stability in water.

2. Materials and methods

Synthesis of $CsPbBr_3$@SiO_2 NCs:
The perovskite nanocrystals covered with silica shells ($CsPbBr_3$@SiO_2 NCs) were synthesized using a modified ligand-assisted reprecipitation (LARP) method [5]. Briefly, lead halide perovskite NCs precursor was obtained by mixing PbBr, CsBr and stabilizers (oleylamine (OAm) and oleic acid (OA)) in dimethylformamide (DMF). Subsequently, the colloidal solution was stirred at 90 °C for 2 hours to obtain a clear, slightly yellow solution. Afterwards, a small amount of ammonia was added to the precursor solution, and then the resulting mixture was injected to the excess of toluene solution that contained a tetramethyl orthosilicate (TMOS) and triethoxyphenylsilane (phTEOS) (molar ratio= 50:50) for the hydrolysis-condensation reaction under vigorous stirring at room temperature. After 10 s, the stirring speed was decreased and kept constant overnight. The obtained $CsPbBr_3$ NCs were washed with toluene by centrifugation (8000 rpm for 5 min).
A control experiment with non-passivated NCs was performed by Qixuan Z et al [5]. The obtained perovskite sample demonstrated poor stability, decomposing in less than 16 minutes.

Characterization of $CsPbBr_3$@SiO_2 NCs:
To check the water resistance, a fresh sample of $CsPbBr_3$ NCs covered with SiO_2 shell ($CsPbBr_3$@SiO_2 NCs) was sustained in water during 24 h, and then the PL analysis of $CsPbBr_3$@SiO_2 NCs was performed using CARY 60 UV-Vis spectrophotometer (excitation $\lambda = 400$ nm). The high quality of the obtained $CsPbBr_3$@SiO_2 NCs was confirmed with scanning transmission electron microscopy (STEM) on Merlin 42-37 microscope, and with high-resolution transmission microscopy (HRTEM) on Tecnai F20 X-TWIN microscope. The elemental analysis was performed using energy-dispersive X-ray (EDX) measurements.

Cell experiments with $CsPbBr_3$@SiO_2 NCs:
B16-F10 cells were cultured in AlphaMEM medium (37°C, aseptic atmosphere with 95% air and 5% CO_2). $CsPbBr_3$@SiO_2 NCs incubated with B16-F10 cells were visualized using confocal laser scanning microscopy (CLSM). The cells were seeded into cell imaging dishes. Afterwards, the cells were stained with Rhodamine 800, and the cell nuclei with propidium iodide (PI). Then, $CsPbBr_3$@SiO_2 NCs were added to the cells and incubated for 24 hours. The stained cells were excited using a HeNe laser (λ = 633 nm, emission filter: LP λ = 650 nm). Perovskite NCs were excited using an argon laser (λ = 488 nm, emission filter: BP λ = 500-550 nm).

3. Results

$CsPbBr_3$ NCs covered with SiO_2 shells were visualized with STEM (Fig. 1) and HRTEM (Fig.2). According to the obtained data, the core size distribution of the individually coated $CsPBBr_3$@SiO_2 NCs is in the range of 20-30 nm, and their shell thicknesses are in the range of to 8-20 nm. HRTEM image analysis confirms the good quality and the high crystallinity of the synthesized $CsPbBr_3$ cores.

Journal of Physics: Conference Series **2015** (2021) 012150 doi:10.1088/1742-6596/2015/1/012150

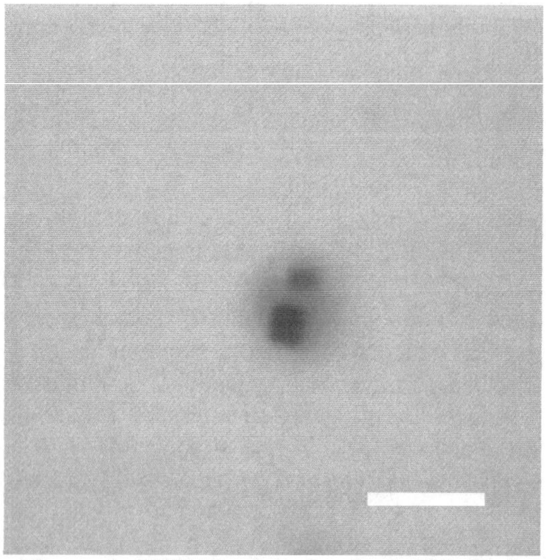

Figure 1. STEM image of a single NC. Scale bar corresponds to 100 nm

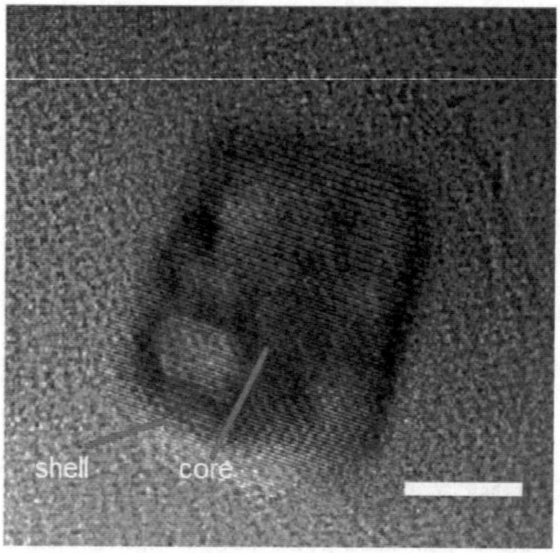

Figure 2. HRTEM image of single NC. Scale bar corresponds to 10 nm

EDX analysis was performed for a single $CsPbBr_3@SiO_2$ NC (Fig. 3). According to the EDX spectrum, it contains the peaks corresponding to Cs, Pb, Br (perovskite core) and Si (silica shell). To reveal stability of the obtained NCs in aqueous solutions, $CsPbBr_3@SiO_2$ NCs were sustained in water for 24 h, and the PL measurements were carried out at different time points (0, 1, 4, 24 h) (Fig. 4). According to the obtained PL data, the PL from the perovskite NCs decreased to 25% after 24 h of water exposure.
The water resistance of the silica shell formed around $CsPbBr_3$ NCs can be explained as follows. To obtain a silica shell, TMOS and phTEOS were added to the reaction at 50:50 molar ratio. It should be noted that phTEOS contains phenyl groups, which provide hydrophobic properties and, thus, the improved stability in water.

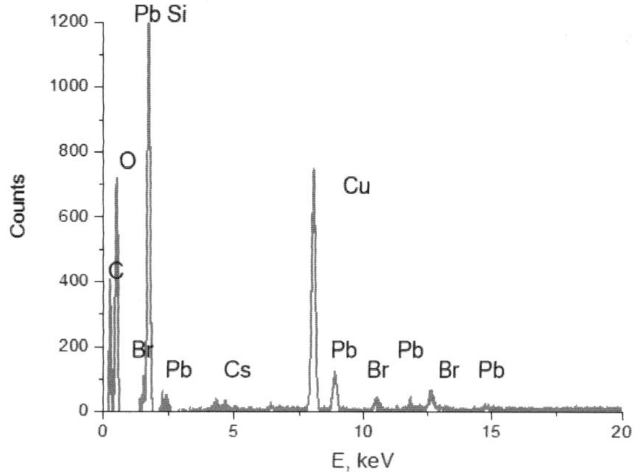

Figure 3. EDX analysis of a single NC

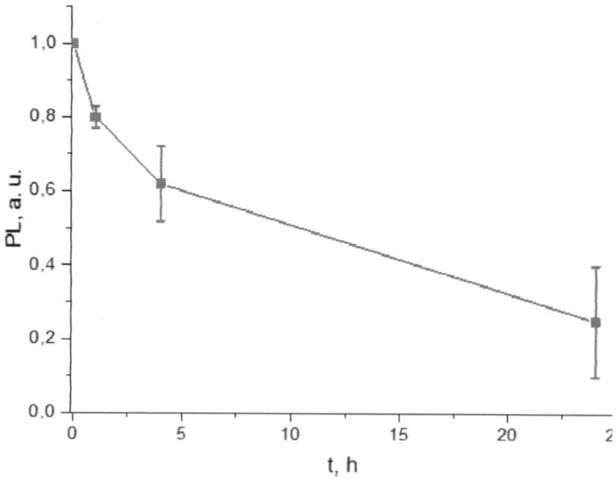

Figure 4. Relative PL intensity in water during 24 h

Further, the stability of $CsPbBr_3@SiO_2$ NCs was tested in a biological environment; apart from water, it contained organic compounds (ions, enzymes, and proteins), which can induce dissolution of nanomaterials. As a proof of concept, the synthesized $CsPbBr_3@SiO_2$ NCs were incubated with B16-

F10 cells and visualized with CLSM (Fig. 5). Bright green fluorescence from $CsPbBr_3@SiO_2$ associated with B16-F10 cells (stained with tetramethylrhodamine) (TRITC) confirmed the stability of perovskite nanocrystals in biological environment.

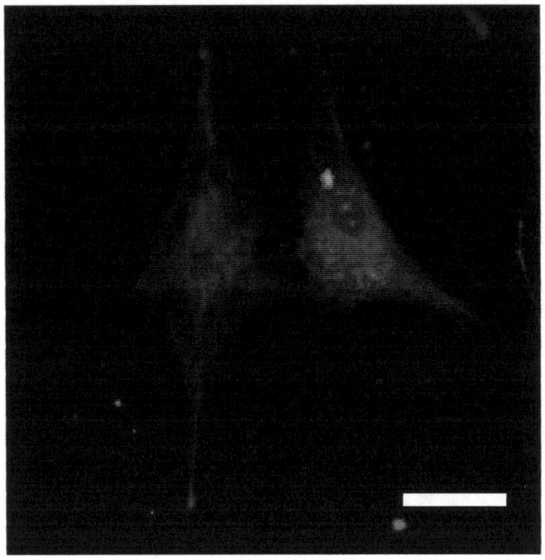

Figure 5. CLSM image of the association of $CsPbBr_3@SiO_2$ NC (green) with a stained B16 cell (red). Excitation $\lambda = 488$ nm. Scale bar corresponds to 20 μm.

4. Conclusions

To conclude, a synthesis approach to obtain water-stable $CsPbBr_3@SiO_2$ NCs was developed. For this, TMOS and phTEOS were added in molar ratio 50:50 to the reaction within LARP method. The used ratio of phTEOS and TMOS allows obtaining stable shells, which prevent water penetration to the perovskite core. The water stability of the sample was estimated with PL measurements. The crystal morphology, quality and elemental composition were verified by EDX, STEM, and HRTEM analyses. Moreover, lead halide perovskites were visualized in living cells using confocal laser scanning microscopy. The obtained results prove the good stability of $CsPbBr_3@SiO_2$ in water and biological environment.

5. Acknowledgments

This work was supported by Russian Science Foundation (Project 21-73-20189).

References

1. Veldhuis, S.A., Boix, P.P., Yantara, N., Li, M., Sum, T.C., Mathews, N. and Mhaisalkar, S.G., *Adv. Mater.*, 2016, 28: 6804-6834.
2. Chen, Q., Wu, J., Ou, X. et al., *Nature*, 2018, 561, 88–93.
3. Zhu, H., Fu, Y., Meng, F. et al. *Nature Mater,* 2015, 14, 636–642.
4. Liang, L., Chen, N., Jia, Y. et al. *Nano Res.*, 2019, 12, 1279–1292.
5. Qixuan, Z., Muhan C., Huicheng H.,et al., *ACS Nano,* 2018, 12, 8, 8579–8587.

Tuneable photoluminescence of TBAPY-based Metal-Organic Complex

Maria V. Timofeeva[1,a)], Yuliya A. Kenzhebaeva[1], Semyon V. Bachinin[1], Valentin A. Milichko[1,2]

[1]ITMO University, St. Petersburg 197101, Russia
[2] Institut Jean Lamour, Université de Lorraine, Nancy F-54011, France

[a)]Corresponding author: maria.timofeeva@metalab.ifmo.ru

Abstract. We report the synthesis of crystals of an organic metal complex based on a large redox-active ligand (tbapy) exhibiting luminescent properties. We have demonstrated the tuning of the luminescence of a Zn-based metal-organic complex, soaked in advance in dimethylformamide, with pumping light.

1. Introduction

Metal-Organic Frameworks (MOF) are porous crystalline materials in which metal ions are bonded to each other with organic ligands. MOFs can change their configuration reversibly under the influence of light, temperature, pressure, electric field, or chemical reaction. One of the explanations for the change in configuration is the removal of "guest" molecules from rather large nanosized pores of the MOF. This effect allows MOF to be used as a storage material.

2. Experimental section

Metal-Organic Complex(MOC) based on large ligands with π-π interactions have been attracting an increasing attention as potential favorites for luminescence studies. However, the synthesis of such crystals is usually a complex and multistep process. Here we show that TBAPY-based Metal-Organic Complex 'figure 1' crystals can be formed under the conditions of a solvothermal process[1],[2] and also demonstrate the change in the PL of the MOC by laser radiation.

2.1. Synthesis

Yellow single crystals of TBAPY- based Metal-Organic Complex were synthesized via solvothermal method: 0,018 g (0.06 mmol) of $[Zn(NO_3)_2*6(H_2O)]$ and 0,020 g (0.03 mmol) of H_4TBAPY were mixed together in 2:2:1 DMF-Dioxane-H_2O solvent mixture and 20μL of concentrated HCl. The mixture was kept for 3 hours at 120°C.The SEM of synthesized nanocrystals of MOC is shown on the 'figure 2'.

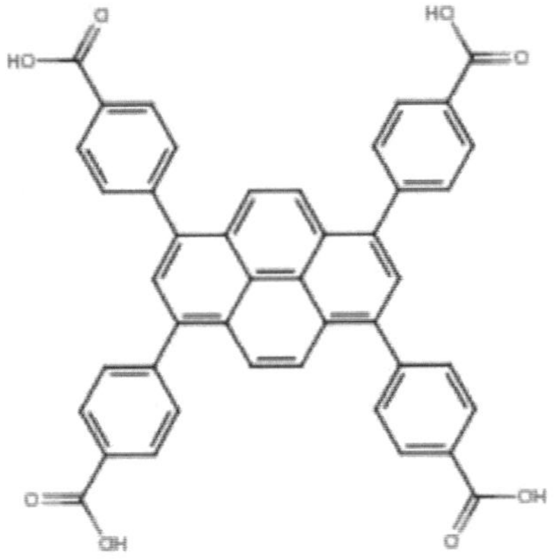

Figure 1. H_4TBAPY Ligand

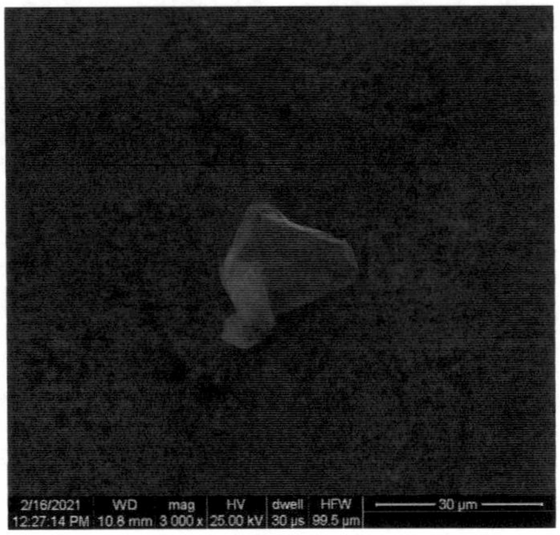

Figure 2. SEM images of TBAPY-based Zn-Complex.

2.2. Optical experiments

It is known that large ligands that are associated with zinc can exhibit changes in their luminescent properties[3]. When the coordination bond of the metal-ligand is broken, luminescence shifts can be observed. Here we represent an optical experiment on the change in the PL of a TBAPY-based MOC upon pulsed radiation of different photon energies It is shown on the 'figure 3,4'. We can observe a change in the shape of the PL spectrum after increasing the laser power to 60 µw. This is due to the rearrangement of the electronic configuration of the Metal-Organic Complex, since under prolonged irradiation or high powers of laser, the solvent molecules of the "guest" evaporate from the complex, which affects the change in the structure.

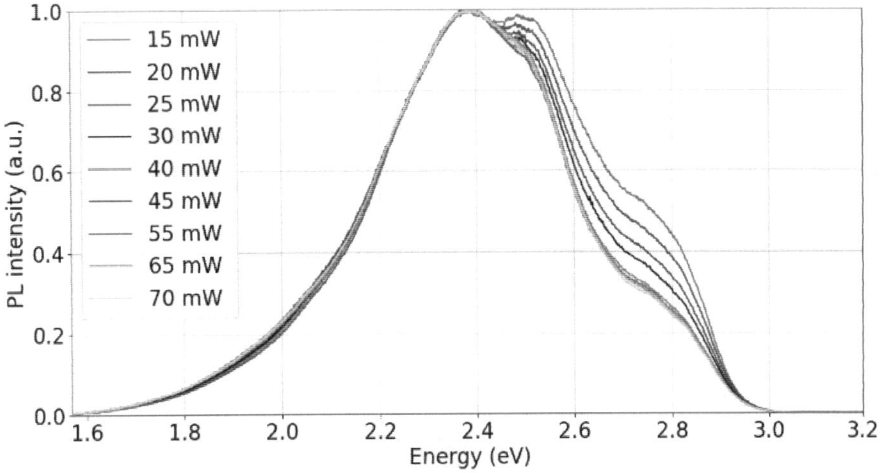

Figure 3. Image of PL spectrum of TBAPY-based Metal-Organic Complex excited by pulsed radiation with different photon energies.

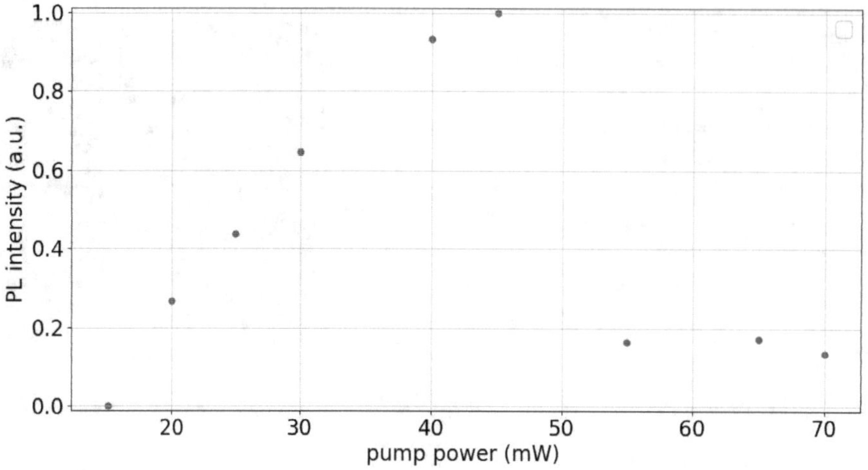

Figure 4. Normalized PL intensity of MOC as a function of pump power.

3. Conclusion

Thus, a synthesis of an organic metal complex exhibiting tunable luminescence properties is shown. The following materials were chosen for the synthesis: a d^{10} metal atom with a closed shell (Zn) and a redox-active ligand (tbapy), since they have high coordination plasticity and interesting photoluminescent properties. Also the intrinsic hybrid nature of these materials gives them the opportunity to experiment different charge transfer processes that enable various emission phenomena. In turn the porous nature of MOCs supports the uptake of guests that may also influence to the emission in several and useful ways, for example, tuning the emission colour and/or the selective detection of particular ions or molecules.

References

[1] K. C. Stylianou , J. Rabone , S. Y. Chong , R. Heck , J. Armstrong , P. V. Wiper , K. E. Jelfs , S. Zlatogorsky , J. Bacsa , A. G. McLennan , C. P. Ireland , Y. Z. Khimyak , K. M. Thomas , D. Bradshaw , M. J. Rosseinsky , J. Am. Chem. Soc. 2012 , 134 , 20466 –20478

[2] K. C. Stylianou, R. Heck, S. Y. Chong, J. Bacsa, J. T. A. Jones, Y. Z. Khimyak, D. Bradshaw, M. J. Rosseinsky, J. Am. Chem. Soc. 2010, 132, 4119

[3] V.A. Milichko, S.V. Makarov, A.V. Yulin, A.V. Vinogradov, A.A. Krasilin, E. Ushakova, V.P. Dzyuba, E. Hey-Hawkins, E.A. Pidko and P.A. Belov, Advanced Materials **29**, 1606034 (2017)

Total angular momenta quantization of dielectric sphere modes

Ivan Toftul
Department of Physics and Engineering, ITMO University, Saint-Petersburg, Russia
E-mail: `itoftul@itmo.ru`

Kristina Frizyuk
Department of Physics and Engineering, ITMO University, Saint-Petersburg, Russia

Mihail Petrov
Department of Physics and Engineering, ITMO University, Saint-Petersburg, Russia

Abstract. Spherical particles both dielectric and metallic are essential building blocks in nanophotonics. During the recent rapid development of *Mie-tronic* — nanophotonics devices heavily using various features of the Mie-resonances — the deep fundamental investigation of the eigenmodes of such particles by using the novel tools is still relevant and currently important. Moreover, eigenmodes of a sphere are closely related to the *Vector Spherical Harmonics* (VSH) which are widely used in the multipolar decomposition to analyze less symmetric structures. In this work, we study in detail the *canonical spin and angular momenta* (AM), helicity, and other properties of the eigenmodes of dielectric (nondispersive) and metallic (dispersive) spheres. We show that the canonical momentum *density* of the AM is quantized and has a close relation to the quantum picture of a single photon. Our work provides a solid platform for future studies and applications of the AM transfer from near fields of spherical particles to the matter in its vicinity.

1. Introduction

In recent years *Mie-tronics* started to gain popularity [1, 2, 3]. One of the key players in this branch of science are the Mie resonances. The first solution of the scattering of a plane by an arbitrary sphere was done by Gustav Mie in 1908 [13]. The scattered field was decomposed into the series of *vector spherical harmonics (VSH)*, and weights in the decomposition contained all the information about all possible resonances in the particle. These ideas are heavily used today in a wide number of topics in modern optics such as metasurfaces [1], metalenses [14], second harmonic generation [5], and more.

In parallel, the Abraham-Minkowski controversy gained its momenta, especially in the terms of dispersive medium [7, 8]. Such approach was successfully implemented for the dielectric and metallic cylindrical waveguides [4]. In this work, we find new opportunities of this theory in the scope of spherical particles. Multipolar decomposition into the series of VSH shows its great power when it comes to the symmetry analyses of the studied structures [5, 6].

Journal of Physics: Conference Series **2015** (2021) 012152 doi:10.1088/1742-6596/2015/1/012152

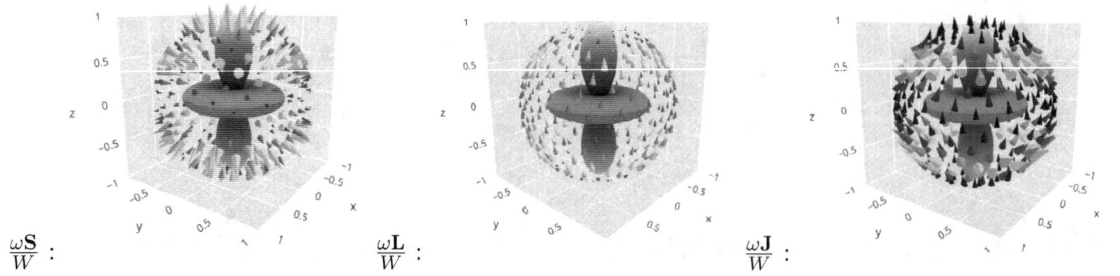

Figure 1. Distribution of the total angular momenta for the $(n = 2, m = 1)$ electric mode. The green surface shows the energy density W of the mode. 3D arrows shows vector fields of spin AM $\mathbf{S}$, orbital AM $\mathbf{L}$, and total AM $\mathbf{J}$. The color of arrows and its size represents the magnitude of the vectors: the closer to yellow the hiegher the magnitude.

Spherical particles both dielectric and metallic are essential building blocks in nanophotonics. During the recent rapid development of *mietronics* — nanophotonics devices heavily using various features of the Mie-resonances — the deep fundamental investigation of the eigenmodes of such particles by using the novel tools is still relevant and currently important. Moreover, eigenmodes of a sphere are closely related to the VSHs which are widely used in the multipolar decomposition to analyze less symmetric structures. In this work, we study in detail the *canonical spin and angular momenta* (AM) of the eigenmodes of dielectric (nondispersive) spheres. We show that canonical momentum *density* of the AM is quantized and has a close relation to the quantum picture of a single photon. Our work provides a solid platform for future studies and applications of the AM transfer from near fields of spherical particles to the matter in its vicinity.

2. Eigenmodes of a sphere and canonical momenta of electromagnetic field

We consider a dielectric particle of radius a placed at the origin which can be formally defined as

$$\varepsilon(\mathbf{r},\omega) = \begin{cases} \varepsilon_{\text{p}}(\omega), & \text{for } r < a \\ \varepsilon_{\text{host}}, & \text{for } r > a \end{cases} \quad \text{and} \quad \mu = 1. \tag{1}$$

The eigenmodes of a sphere is defined by a Helmholz equation

$$\nabla \times \nabla \times \mathbf{E} - k^2 \varepsilon(\mathbf{r},\omega)\mathbf{E} = 0. \tag{2}$$

The solution is going to give the set of eigenmodes with corresponding *complex* eigen frequencies $\omega = \omega_{\text{Re}} - i\gamma$, where γ is the dissipation rate.

We analyze the eigenmodes using canonical properties of the field introduced recently in [7] for the, generally, dispersive media, which is crucial for the case of metallic particles. The Brillouin energy W, linear momentum $\mathbf{P}$, spin $\mathbf{S}$, orbital $\mathbf{L}$, and total $\mathbf{J}$ angular momentum are given by [7, 8, 9]

$$W = \frac{1}{4}\left(\tilde{\varepsilon}\varepsilon_0|\mathbf{E}|^2 + \tilde{\mu}\mu_0|\mathbf{H}|^2\right), \tag{3}$$

$$\mathbf{P} = \frac{1}{4\omega}\operatorname{Im}\left(\tilde{\varepsilon}\varepsilon_0\mathbf{E}^* \cdot (\nabla)\mathbf{E} + \tilde{\mu}\mu_0\mathbf{H}^* \cdot (\nabla)\mathbf{H}\right), \tag{4}$$

$$\mathbf{S} = \frac{1}{4\omega}\operatorname{Im}\left(\tilde{\varepsilon}\varepsilon_0\mathbf{E}^* \times \mathbf{E} + \tilde{\mu}\mu_0\mathbf{H}^* \times \mathbf{H}\right), \tag{5}$$

$$\mathbf{L} = \mathbf{r} \times \mathbf{P}, \qquad \mathbf{J} = \mathbf{L} + \mathbf{S}. \tag{6}$$

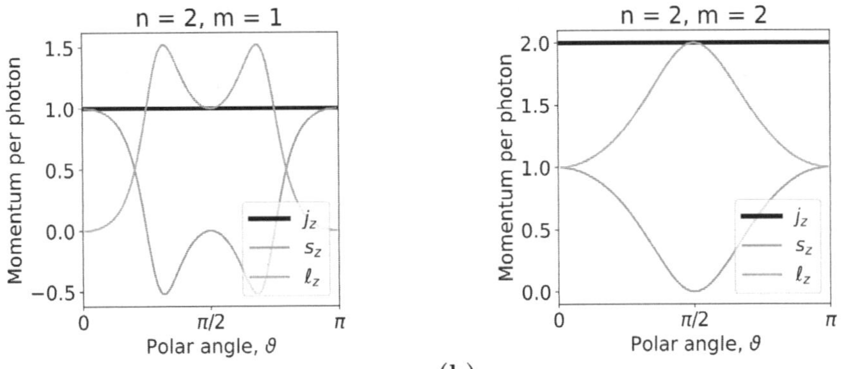

(a) (b)

Figure 2. Normalized angular momentum density as a function of polar angle ϑ for ($n = 2, m = 1$) mode **(a)** and for ($n = 2, m = 2$) mode **(b)**. Due to the azimuthal symmetry, there is no dependence on the polar angle φ. Here one can see that total angular momentum per one photon is quantized and equal m.

Here $(\tilde{\varepsilon}, \tilde{\mu}) = (\varepsilon, \mu) + \omega\partial_\omega(\varepsilon, \mu)$. Each of the components can be naturally decomposed into the electric and magnetic pars.

3. Angular momenta of eigenmodes of a dielectric sphere

Electromagnetic fields of the eigenmodes of the sphere are appeared to be connected with the electric field of a single photon. Akhiezer and Berestetsky showed in [10].

One of the key results of this work is that the total AM *density* normalized to the energy density per one photon is quantized separately for the electric and magnetic modes. We can define electric (TM) modes ($H_r = 0$) and magnetic (TE) modes ($E_r = 0$) through the complex VSH as

$$\mathbf{E} = \mathbf{N}, \qquad \mathbf{H} = -\frac{i}{Z_0}\mathbf{M} \qquad \text{TM modes} \tag{7}$$

$$\mathbf{E} = \mathbf{M}, \qquad \mathbf{H} = -\frac{i}{Z_0}\mathbf{N} \qquad \text{TE modes} \tag{8}$$

The vector structure of the spin, orbital, and total angular momenta is shown in Fig. 1. After some analytic calculations, we get

$$\boxed{j_z = \frac{\omega J_z}{W} = m.} \tag{9}$$

We illustrate this in Fig. 2. Otherwise it is non-integer. However, the absolute value does not follow the quantum limit

$$\mathbf{j}^2(\theta = 0) = \underbrace{m^2}_{\text{classical limit}} \neq \underbrace{n(n+1)}_{\text{quantum limit (1 photon)}}, \tag{10}$$

Which is explicitly shown in [10] and the exact resolution is given in [11], where the problem of angular momenta of N photons in free space was solved rigorously

$$\mathbf{j}^2 = \frac{N^2m^2 + Nn(n+1) - m^2}{N^2}. \tag{11}$$

4. Conclusion

To summarize, in this work we have applied the canonical momenta analyses to the eigen modes of a dielectric sphere. We showed analytically that the z-component of the total angular momenta of light per one photon is quantized which structures a bridge between AM in classics electrodynamics and in quantum theory of photons. We also demonstrated the vector field structure of spin $\mathbf{S}$, orbital $\mathbf{L}$, and total $\mathbf{J} = \mathbf{S} + \mathbf{L}$ angular momenta of the eigen modes. Our work provides a solid platform for future studies and applications of the AM transfer from near fields of spherical particles to the matter in its vicinity.

Acknowledgments

The reported study was funded by RFBR, project number 19-32-90237.

References

[1] Bonod, N. & Kivshar, Y. All-dielectric Mie-resonant metaphotonics. Comptes Rendus. Physique 21, 425–442 (2020).
[2] Koshelev, K. & Kivshar, Y. Dielectric Resonant Metaphotonics. ACS Photonics 8, 102–112 (2021).
[3] Tseng, Ming Lun, et al. "Dielectric Metasurfaces Enabling Advanced Optical Biosensors." ACS Photonics (2020).
[4] Picardi, M. F., Bliokh, K. Y., Rodríguez-Fortuño, F. J., Alpeggiani, F. & Nori, F. Angular momenta, helicity, and other properties of dielectric-fiber and metallic-wire modes. 5, (2018).
[5] Frizyuk, K., Volkovskaya, I., Smirnova, D., Poddubny, A. & Petrov, M. Second-harmonic generation in Mie-resonant dielectric nanoparticles made of noncentrosymmetric materials. Physical Review B 075425, 1–15 (2018).
[6] Gladyshev, S., Frizyuk, K. & Bogdanov, A. Symmetry analysis and multipole classification of eigenmodes in electromagnetic resonators for engineering their optical properties. Phys. Rev. B 102, 075103 (2020).
[7] Bliokh, K. Y., Bekshaev, A. Y. & Nori, F. Optical Momentum, Spin, and Angular Momentum in Dispersive Media. Physical Review Letters 119, 1–6 (2017).
[8] Bliokh, K. Y., Bekshaev, A. Y. & Nori, F. Optical momentum and angular momentum in complex media: From the Abraham-Minkowski debate to unusual properties of surface plasmon-polaritons. New Journal of Physics 19, (2017).
[9] Jackson, J. D. Classical Electrodynamics. vol. 1 (1998).
[10] A. I. Akhiezer and V. B. Berestetsky, Quantum electrodynamics, Vol. 2876 (US Atomic Energy Commission, Techical Information Service Extension, 1957).
[11] De Witt, C. M. & Jensen, J. H. D. Uber den Drehimpuls der Multipolstrahlung. Zeitschrift für Naturforschung A 8, (1953).
[12] Novotny, L. & Hetch, B. Principles of Nano-Optics. vol. 1 (2010).
[13] Bohren, C. F. & Huffman, D. R. Absorbtion and scattering of light by Small Particles. vol. 1 (1984).
[14] Chen, W. T. & Capasso, F. Will flat optics appear in everyday life anytime soon? Appl. Phys. Lett. 118, 100503 (2021).

Control of spontaneous emission rate in lead halide perovskite film on hyperbolic metamaterial

Pavel Tonkaev and Sergey Makarov

Saint-Petersburg National Research University of Information Technologies, Mechanics and Optics, Kronverksky av. 49 197101 St. Petersburg, Russia

E-mail: pavel.tonkaev@metalab.ifmo.ru

Abstract. Hyperbolic metamaterials represent a class of nanophotonic architectures with the possibility of controlling density of optical states. Due to this property, hyperbolic metamaterials can be employed as meta-electrodes in optoelectronic devices. On the other hand, lead halide perovskites have several promising properties for application in light-emitting devices. Moreover, a perovskite film is easily deposited on a hyperbolic metamaterial surface. Here, we theoretically show how to accelerate radiative recombination in a perovskite film with a hyperbolic metamaterial. This effect can be applied in light-emitting devices, where radiative recombination is extremely important.

1. Introduction

One of the structures, which supports high density of photonic states is a hyperbolic metamaterial (HMM)[1, 2]. Due to hyperbolic dispersion, these metamaterials maintain states with extremely high wavelength vectors[1] that allows to accelerate spontaneous emission[3]. This effect is useful for such nanophotonic applications as lasing [4, 5] and charge transfer[6]. On the other hand, acceleration of radiative recombination can be vastly useful in lead halide perovskite optoelectronic devices where high quantum yield is a crucial parameter[7]. Moreover, since perovskite emission wavelength is easily tuned[8], it can be matched to HMM resonance, as it was demonstrated for various nanophotonic designs [9, 10, 11].

In this work, we present a theoretical study of radiative properties of a dipole source with various orientations located at different position inside a perovskite film on the HMM. We revealed the optimal wavelength at which a significant enhancement of radiative recombination and the maximum available value of the Purcell factor are expected.

2. Results

By definition, HMMs are anisotropic media exhibiting hyperbolic isofrequency contours[1]. For periodic metal-dielectric structures with permittivities ϵ_m, ϵ_d, and thicknesses t_m, t_d, respectively, the valid effective medium approximation predicts a uniaxial permittivity tensor with the following components:

$$\varepsilon_{\|} = p\varepsilon_{\mathrm{m}} + (1-p)\varepsilon_{\mathrm{d}} \tag{1}$$

$$\varepsilon_{\perp} = \left(\frac{p}{\varepsilon_{\mathrm{m}}} + \frac{1-p}{\varepsilon_{\mathrm{d}}}\right)^{-1}, \tag{2}$$

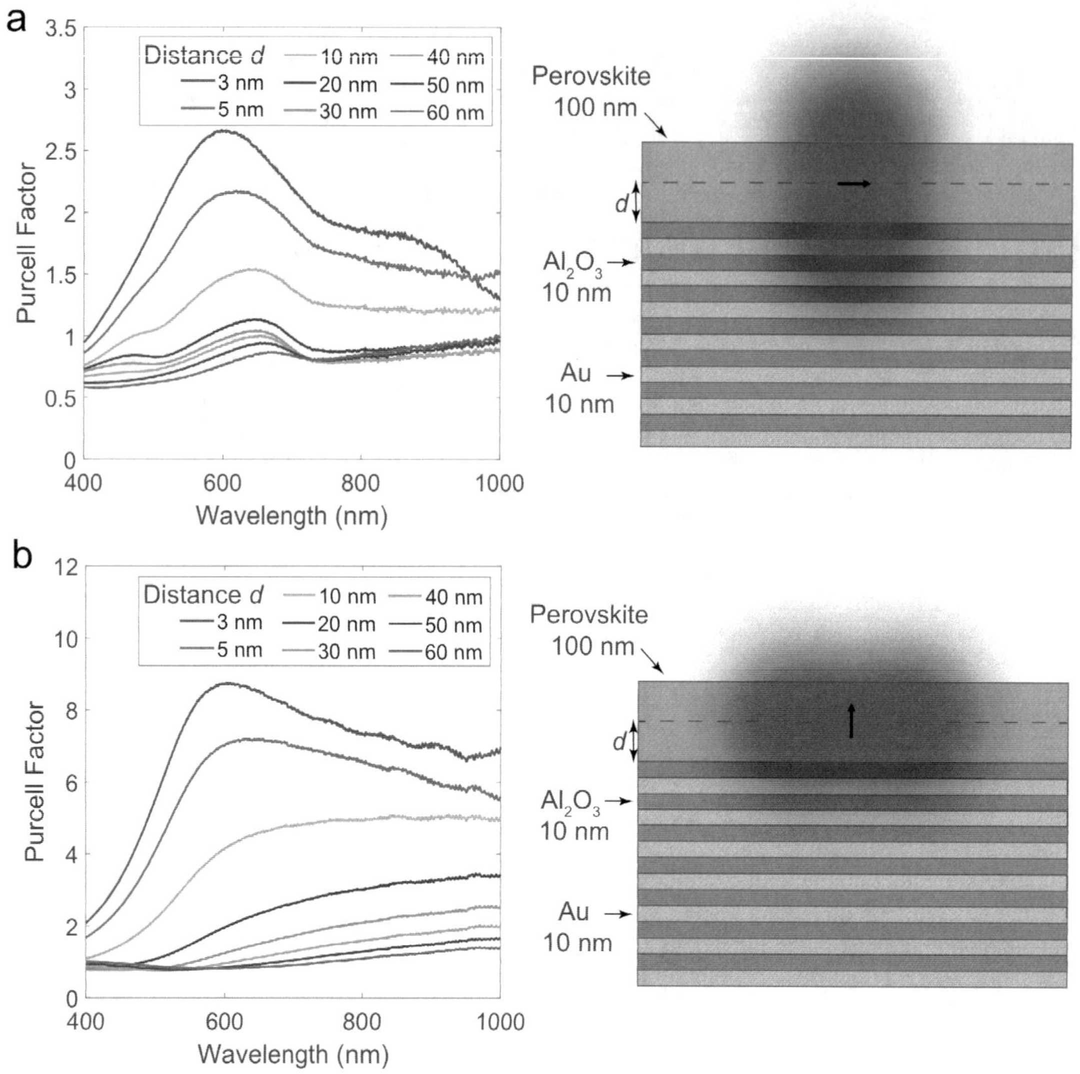

Figure 1. Schematic model design and theoretical dependence of the Purcell factor on wavelength for emitter located at different distance from HMM surface with parallel (a) and perpendicular (b) orientations.

where $p = t_m/(t_m + t_d)$ is the metal fill fraction in the unit cell. To achieve the hyperbolic regime, parallel $\varepsilon_{\parallel}$ and perpendicular $\varepsilon_{\perp}$ components of the tensor (relatively to the anisotropy axis) must satisfy the condition: $\varepsilon_{\parallel} \cdot \varepsilon_{\perp} < 0$.

In our study, we examine the HMM as a multilayered periodic structure with a period of 10 nm gold and 10 nm alumina layers. The fabrication method of this kind of structure is well developed[12] We assume that Al_2O_3 permittivity equals to 3.06 and Au has dispersion from[13]. Considering these parameters, our estimation based on Equations 1 and 2 reveals that for a wavelength above 500 nm HMM of type II ($\varepsilon_{\parallel} < 0$, $\varepsilon_{\perp} > 0$) is expected.

Numerical procedure to calculate the Purcell factor was conducted applying the impedance approach[14]. All our calculations were performed with the CST Microwave Studio. The size of a dipole emitter was considered with dimension of about $\sim \lambda/100$, that is equal to 3 nm. On the top of the hyperbolic metamaterial with 7 periods a 100 nm thick perovskite film with permittivity of 6.3 was placed. Open boundaries and two symmetry planes were applied to reduce the size of the computational domain. The calculations are made for two different orientations of the dipole: parallel (right Figure 1a) and perpendicular (right Figure 1b) to the surface. The dipole emitter was located in the perovskite film at distance d from 3 nm to 60 nm from the HMM.

According to our simulations, the spontaneous emission rate of the dipole emitter near the HMM is enhanced. The calculated Purcell factor for parallel and perpendicularly oriented dipoles located at various distances from the HMM are shown in Figures 1a (left) and 1b (left), respectively. For both orientations, the highest values of the Purcell factor is observed at the wavelength of 600 nm for the closest to the HMM emitter. The value of the Purcell factor for a perpendicular to the surface dipole is more than three times larger than for a parallel one. Based on electric field distribution, the coupling between high-k mode and dipole is stronger at this orientation.

With the shifting of the dipole emitter from the HMM surface, the value of the Purcell factor diminished drastically due to decrease of the coupling. At distance more than 20 nm, the value of spontaneous emission rate was practically equal to the value in a perovskite film with the HMM absence. The average value of the Purcell factor can be calculated as an arithmetical mean of the three different dipole orientations. Its highest value equals to 4.8 at 600 nm wavelength for the dipole near the HMM surface. However, due to strong distance dependence, to estimate the Purcell factor in a real perovskite film, averaging over the film thickness is required.

3. conclusion

We have studied Purcell factor in lead halide perovskite film on the hyperbolic metamaterial and revealed that for a perovskite with emission at wavelength 600 nm it is possible to reach acceleration of radiative recombination. Our numerical simulations are relevant for wide range of lead halide perovskite materials $APbX_3$ (where A is cation, and X is Cl^-, Br^-, I^- or their mixture), and their agree with recent experimental studies of perovskite emission interaction with HMM [15, 16, 17]. This approach can be applied for creation of efficient light emitting devices [18, 19] and solar cells [7, 20], where acceleration of radiative recombination plays one of the key roles [21].

Acknowledgments

This work was supported by the Russian Science Foundation (project 19-73-30023).

References

[1] Poddubny A, Iorsh I, Belov P and Kivshar Y 2013 *Nature Photonics* **7** 948–957
[2] Simovski C R, Belov P A, Atrashchenko A V and Kivshar Y S 2012 *Advanced Materials* **24** 4229–4248
[3] Noginov M, Li H, Barnakov Y A, Dryden D, Nataraj G, Zhu G, Bonner C, Mayy M, Jacob Z and Narimanov E 2010 *Optics Letters* **35** 1863–1865
[4] Chandrasekar R, Wang Z, Meng X, Azzam S I, Shalaginov M Y, Lagutchev A, Kim Y L, Wei A, Kildishev A V, Boltasseva A *et al.* 2017 *ACS Photonics* **4** 674–680
[5] Lin H I, Shen K C, Liao Y M, Li Y H, Perumal P, Haider G, Cheng B H, Liao W C, Lin S Y, Lin W J *et al.* 2017 *ACS Photonics* **5** 718–727
[6] Lee K J, Lee Y U, Fages F, Ribierre J C, Wu J W and D'aléo A 2018 *Nano Letters* **18** 1476–1482
[7] Richter J M, Abdi-Jalebi M, Sadhanala A, Tabachnyk M, Rivett J P, Pazos-Outón L M, Gödel K C, Price M, Deschler F and Friend R H 2016 *Nature Communications* **7** 1–8
[8] Sutherland B R and Sargent E H 2016 *Nature Photonics* **10** 295
[9] Tiguntseva E, Zograf G, Komissarenko F, Zuev D, Zakhidov A, Makarov S and Kivshar Y S 2018 *Nano letters* **18** 1185–1190

[10] Makarov S, Furasova A, Tiguntseva E, Hemmetter A, Berestennikov A, Pushkarev A, Zakhidov A and Kivshar Y 2019 *Advanced optical materials* **7** 1800784
[11] Berestennikov A S, Voroshilov P M, Makarov S V and Kivshar Y S 2019 *Applied Physics Reviews* **6** 031307
[12] Sukham J, Takayama O, Mahmoodi M, Sychev S, Bogdanov A, Tavassoli S H, Lavrinenko A V and Malureanu R 2019 *Nanoscale* **11** 12582–12588
[13] Edward D P and Palik I 1985 Handbook of optical constants of solids
[14] Greffet J J, Laroche M and Marquier F 2010 *Physical Review Letters* **105** 117701
[15] Adl H P, Gorji S, Habil M K, Suarez I, Chirvony V S, Gualdron-Reyes A F, Mora-Sero I, Valencia L M, de la Mata M, Hernandez-Saz J *et al.* 2020 *ACS photonics* **7** 3152–3160
[16] Li Z, Smalley J S, Haroldson R, Lin D, Hawkins R, Gharajeh A, Moon J, Hou J, Zhang C, Hu W *et al.* 2020 *ACS Photonics* **7** 1754–1761
[17] Tonkaev P, Anoshkin S, Pushkarev A, Malureanu R, Masharin M, Belov P, Lavrinenko A and Makarov S 2021 *Applied Physics Letters* **118** 091104
[18] Lin K, Xing J, Quan L N, de Arquer F P G, Gong X, Lu J, Xie L, Zhao W, Zhang D, Yan C *et al.* 2018 *Nature* **562** 245–248
[19] Gets D, Alahbakhshi M, Mishra A, Haroldson R, Papadimitratos A, Ishteev A, Saranin D, Anoshkin S, Pushkarev A, Danilovskiy E *et al.* 2021 *Advanced Optical Materials* **9** 2001715
[20] Furasova A, Calabró E, Lamanna E, Tiguntseva E, Ushakova E, Ubyivovk E, Mikhailovskii V, Zakhidov A, Makarov S and Di Carlo A 2018 *Advanced Optical Materials* **6** 1800576
[21] Masharin M A, Berestennikov A S, Barettin D, Voroshilov P M, Ladutenko K S, Di Carlo A and Makarov S V 2021 *Nanomaterials* **11** 45

Laser induced tunable Ge2Sb2Te5 phase-change gratings

Pavel I. Trofimov[1], **Irina G. Bessonova**[1], **Petr I. Lazarenko**[2], **Demid A. Kirilenko**[3], **Nikolay A. Bert**[3], **Sergey A. Kozyukhin**[4], **Ivan S. Sinev**[1]

[1]Department of Physics and Engineering, ITMO University, St. Petersburg, Russia
[2]National Research University of Electronic Technology, Zelenograd, Moscow, Russia,
[3]Ioffe Institute, St. Petersburg, Russia,
[4]Kurnakov Institute of General and Inorganic Chemistry of the Russian Academy of Sciences, Moscow, Russiap.trofimov@metalab.ifmo.ru

Abstract. Periodic photonic nano- and microstructures are routinely used for light manipulation at the nanoscale. However, their fabrication process is demanding in terms of time, cost and facilities. Here we demonstrate a rapid laser-assisted method for fabrication of gratings in $Ge_2Sb_2Te_5$ (GST) thin films, based on the formation of laser induced periodic surface structures (LIPSS). LIPSS formation mechanisms dependent on the wavelength of the operating laser, lead to high flexibility of the process, producing gratings with tunable period and orientation with respect to the initial laser polarization. The phase-change properties of GST, on the other hand, allows to fabricate phase gratings with strong modulation of refractive index, which are rewritable in nature.

1. Introduction

Formation of laser induced periodic surface structures (LIPSS) is a well-studied phenomena. The formation of such structures is triggered by the irradiation of the material with ultrashort laser pulses. The incident and scattered light forms the interference picture which is then imprinted in the sample material through the absorption which is usually followed by ablation [1]. However, as ablation implies removal of the material from the surface, such periodic structures are not rewritable in nature.

On the other hand, well known family of chalcogenide phase-change materials (PCM) allows for such flexibility [2]. These materials, for example $Ge_2Sb_2Te_5$ (GST), possess two different phase states – amorphous and crystalline – which are stable at normal condition and show significantly different physical properties, in particular, the refractive index. Moreover, GST can be reversibly switched between these states at the nanosecond time scale by applying electrical or optical pulses [3]. Therefore the concept of LIPSS formation in PCM is very promising for simple and rapid fabrication of tunable and rewritable periodic photonic structures [4].

2. Result & Discussion

In this work, we demonstrate the laser-assisted fabrication of high-quality phase-change gratings based on the formation of LIPSS in GST thin films. During the process we irradiate 50 and 100 nm GST film on a sapphire substrate with femtosecond laser pulses using a laser system with tunable wavelength (Light Conversion Pharos and Orpheus optical parametric amplifier, 1MHz repetition rate, 300 fs pulse duration) to induce a partial crystallization process. We scan the laser beam waist over the 50x50 μm^2 areas of the GST film with a piezoscanner to create the phase-change gratings. We then characterize the

Journal of Physics: Conference Series **2015** (2021) 012154 doi:10.1088/1742-6596/2015/1/012154

quality of the gratings and their periods using atomic force microscopy and study their internal structure with transmission electron microscopy (TEM).

The observed gratings are manifested periodic surface relief in the measured AFM maps, which is formed due to different densities of amorphous and crystalline states of GST. By changing the wavelength of the excitation laser within the range from 800 to 2000 nm, we observed the formation of gratings with periods varied from 110 to 700 nm. AFM image of the structure formed at 1400 nm is shown in Fig. 1a. This type of periodic structures with lines oriented parallel to the direction of laser polarization is usually attributed to the "dielectric" regime of LIPSS formation [5].

In turn, for shorter wavelengths (600–800 nm) we also observed the formation of the gratings perpendicular to the laser polarization which is usually characterized as "metallic" regime [6]. Within a narrow spectral region around 760 nm, we also observed the formation of structures with more complicated geometry that manifested bidirectional phase modulation (Fig. 1b).

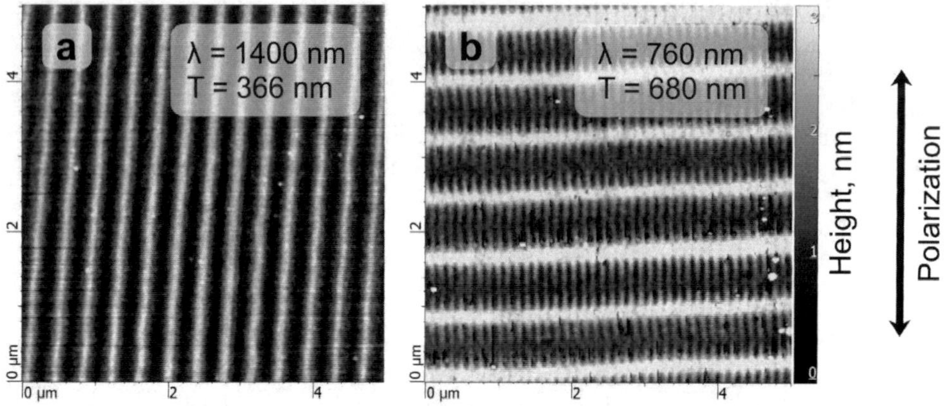

Figure 1. AFM images of the gratings fabricated in 50 nm GST film using different laser wavelengths, which are formed: (a) Parallel to the polarization of the incident light, (b) Perpendicular to the polarization of the incident light. The polarization of the incident light is indicated on the right with an arrow.

To confirm the phase–change nature of GST gratings, we fabricated thin lamellas from our samples and studied the distribution of amorphous and crystalline areas inside the GST films using TEM. TEM images for structures fabricated with femtosecond laser pulses of 1850 and 2000 nm wavelength in 50 and 100 nm thick GST films, respectively, are shown in Fig. 2. The images confirm the modulation of the phase state in the films. Importantly, while we observed nearly full depth crystallization of GST for 50 nm film (Fig. 2a), for the 100 nm film the crystallized areas did not reach the sapphire substrate. Therefore to fabricate homogeneous throughout the entire depth GST gratings at these laser wavelength it is better to use thin GST films with thickness less than 100 nm.

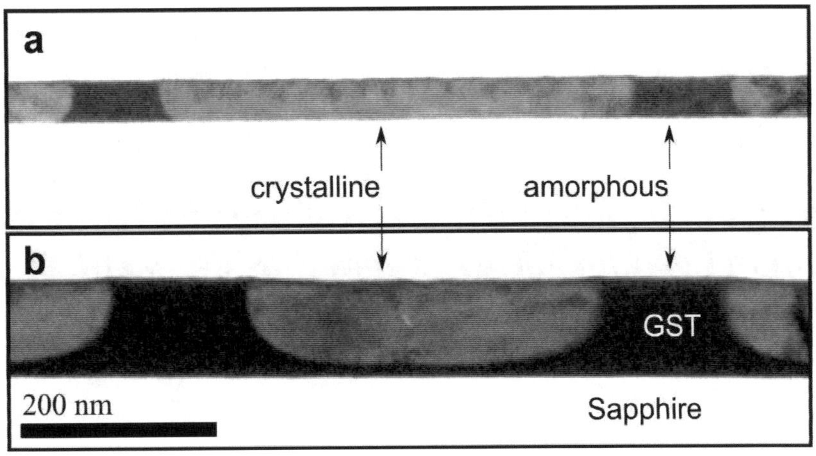

Figure 2. TEM images of the thin lamellas prepared from the phase change gratings. The cuts were made perpendicularly to the grating lines: (a) The cross section of grating created in 50 nm GST film on sapphire with a laser wavelength of 1850 nm; (b) The cross section of grating created in 100 nm GST film on sapphire with a laser wavelength of 2000 nm.

In conclusion, we demonstrated the formation of phase-change gratings in GST thin films. The laser fabricated gratings were formed with tunable period and orientation with respect to laser polarization. The formation of high quality gratings which are rewritable in nature opens up new possibilities for fast, and low-cost fabrication of tunable devices for light manipulation, such as beam steerers and grating couplers.

Acknowledgement
The authors acknowledge the support from Russian Science Foundation (project no. 19-72-10086)

References

[1] Bonse, J.; Graf, S. Maxwell Meets Marangoni — A Review of Theories on Laser-Induced Periodic Surface Structures. Laser & Photonics Reviews 2020, 2000215.

[2] Wuttig, M.; Taubner, T. Phase-change materials for non-volatile photonic applications. Nature Photonics 2017, 11, 465-476.

[3] Au Y.-Y., Bhaskaran H. & Wright C. D. Phase-change devices for simultaneous optical-electrical applications. Scientific Reports 2017, 7, 9688

[4] Kozyukhin, S. ; Smayev, M.; Sigaev, V.; Vorobyov, Y.; Zaytseva, Y.; Sherchenkov, A.; Lazarenko, P. Specific Features of Formation of Laser-Induced Periodic Surface Structures on Ge2Sb2Te5 Amorphous Thin Films under Illumination by Femtosecond Laser Pulses. Phys. Status Solidi B 2020, 257, 1900617.

[5] Bonse, J.; Hohm, S.; Kirner, S. V.; Rosenfeld, A.; Kruger, J. Laser-induced periodic surface structures — A scientific evergreen. IEEE Journal of selected topics in quantum electronics 2016, 23.

[6] Fuentes-Edfuf, Y.; S´anchez-Gil, J. A.; Florian, C.; Giannini, V.; Solis, J.; Siegel, J. Surface Plasmon Polaritons on Rough Metal Surfaces: Role in the Formation of Laser-Induced Periodic Surface Structures. ACS Omega 2019, 4, 6939–6946.

Tungsten disulfide nanoparticles produced by femtosecond laser ablation in water for nanophotonic applications

G I Tselikov[1], A A Popov[2], G A Ermolaev[1], A V Syuy[1], A V Kabashin[2], A V Arsenin[1] and V S Volkov[1]

[1]Center for Photonics and 2D Materials, Moscow Institute of Physics and Technology, Dolgoprudny 141700, Russia
[2]MEPhI, Institute of Engineering Physics for Biomedicine (PhysBio), Bio-nanophotonics Laboratory, 115409, Moscow, Russia.

celikov@physics.msu.ru

Abstract. We demonstrate nearly spherical nanoparticles of tungsten disulfide (WS_2) produced by femtosecond pulsed laser ablation of bulk target in deionized water. Structural and optical analysis reveals that produced nanospheres preserve the crystalline structure, high refractive index and support strong excitons and Mie resonances in the spectral range 400-700 nm, resulting in enhanced photothermal response probed by Raman spectroscopy.

1. Introduction

An emerging field of all-dielectric nanophotonics implies the utilization of dielectric nanoparticles with a high refractive index exhibiting a resonant optical response due to Mie resonances [1]. Currently, traditional semiconductor materials such as silicon [2], germanium [3] and others are mainly used for the implementation of optical devices based on nanostructures with Mie resonances. However, a fixed value of the dielectric constant of these materials imposes fundamental restrictions on the development of optical devices operating at a strict selectivity of resonant frequencies. One of the ways to improve the properties of modern photonic devices is to use materials with anisotropic dielectric properties. Recent results on the optical properties of nanoparticles synthesized from transition metal dichalcogenides (TMDC) [4] demonstrate their unique applied potential associated with their internal layered structure [5]. An important feature of these materials is their strong dielectric anisotropy and high refractive index [6]. Therefore, the high index TMDC materials are increasingly considered as a promising platform with rich excitonic physics and numerous nanophotonic applications. Thus, nanodisks made of TMDC were shown to provide significantly enhanced light-matter interaction due to excitation of so-called "anapole" states and its strong coupling to exciton resonances [4], resulting in a strong enhancement of second and third harmonic generation in such nanostructures [7]. However, despite the remarkable progress in nanopatterning of TMDC, the development of spherical nanoantennas from TMDC still poses a significant challenge.

In this work we demonstrate spherical nanoparticles of tungsten disulfide (WS_2) with variable size produced by femtosecond pulsed laser ablation of bulk target in deionized water followed by size separation step. Structural and optical investigations have shown that the produced nanoparticles are crystalline, have a narrow size distribution with an average diameter in the range from 5 to 150 nm, and support size-dependent Mie-resonant modes in extinction spectra along with exciton resonances in the spectral range 400-700 nm.

2. Results and discussion

First, for the production of WS_2 nanoparticles (NPs), we adapted methods of ultra-short pulsed laser ablation and fragmentation in liquid ambiance, which were earlier used for the preparation of bare Si and Au nanoparticles [8, 9]. We observed a red-brownish coloration of solution within a few minutes of the ablation process, indicating the formation of WS_2 NPs. The concentration of NPs increased linearly at the beginning of the process and then came to a saturation. In particular, for 100 μJ pulse energy and 10 kHz repetition rate, saturation time was 30 min, and the saturation concentration was about 100 μg mL^{-1}. Once the ablation was completed, the colloidal aqueous solutions of WS_2 were subjected to subsequent steps of centrifugation at increasing rotation speed 250-8000 min^{-1} (rpm) (45-2000g) to separate produced nanoparticles by size. Fig. 1a shows SEM image of the biggest nanoparticles separated at 250 rpm. It is seen that WS_2 NPs have nearly spherical shape and an average size of 100 nm. The TEM image of the smallest WS_2 NPs selected from the solution at 8000 rpm is shown at Fig. 1b. Insets show HR-TEM image of selected area and size distribution. It is seen that the smallest WS_2 with an average size of 5 nm are crystalline. The interlayer distance is about 0.67 nm, which corresponds to the d-spacing of WS_2 (002).

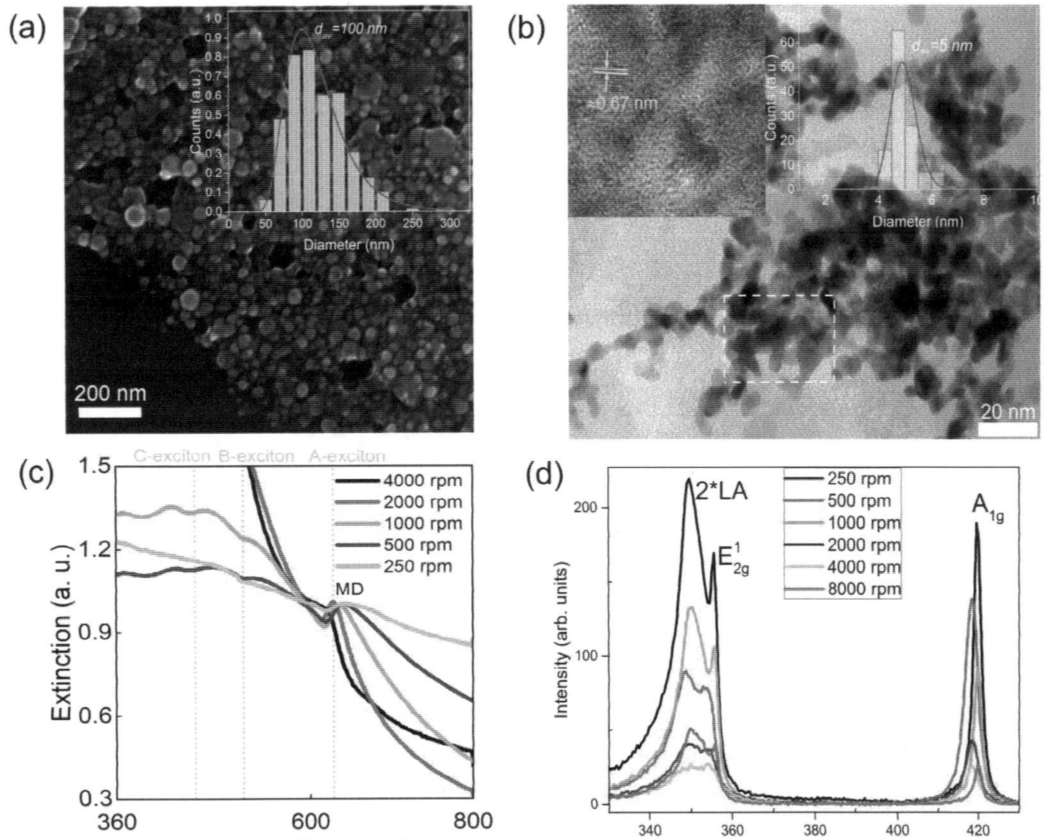

Figure 1. (a) SEM image of laser-ablated WS_2 NPs selected from a colloidal solution at 250 rpm. Inset shows the corresponding size distribution of WS_2 NPs. (b) TEM image of WS_2 NPs selected at 8000 rpm. Insets show HR-TEM image of selected area and size distribution. (c) Extinction spectra of WS_2 NPs selected at different rotation speed. (d) Raman scattering spectra of WS_2 NPs selected at different rotation speed.

Journal of Physics: Conference Series **2015** (2021) 012155 doi:10.1088/1742-6596/2015/1/012155

The extinction spectra of colloidal solutions of WS_2 NPs with different sizes selected at a different rotation speed of centrifuge were experimentally measured using standard UV-VIS technique calculated numerically using extended Mie theory. It is seen from Fig. 1c that the spectra are composed of several peaks in a broad spectral range from 400 to 700 nm (the excitonic transitions A, B and C are indicated on the graph). It was found that its spectral behavior can be well explained by taking into account the excitonic properties of WS_2 and Mie-scattering of light by high-refractive-index NPs with the size defined be SEM-measurements. The peak centered at 620-700 nm corresponds to magnetic-dipole (MD) resonances while the broad peak at 500 nm corresponds to electric-dipole resonance (ED). The position of these peaks shifts towards longer wavelengths with increasing the size of NPs in agreement with Mie-theory. Raman scattering spectra of WS_2 NPs with variable size (selected at different rotating speed) excited at 532 nm are presented in Fig. 1d. The spectra are in good agreement with previously reported data for bulk WS_2 and WS_2 flakes [10]. In particular, peaks at 420 and 356 cm^{-1} are attributed to $A_{1g}(\Gamma)$ and $E^1_{2g}(\Gamma)$ modes correspondingly. The most intense peak of Raman spectra centered at 561 cm^{-1} corresponds to scattering process involving two longitudinal acoustic phonons from the edge of the Brillouin zone 2LA(M).

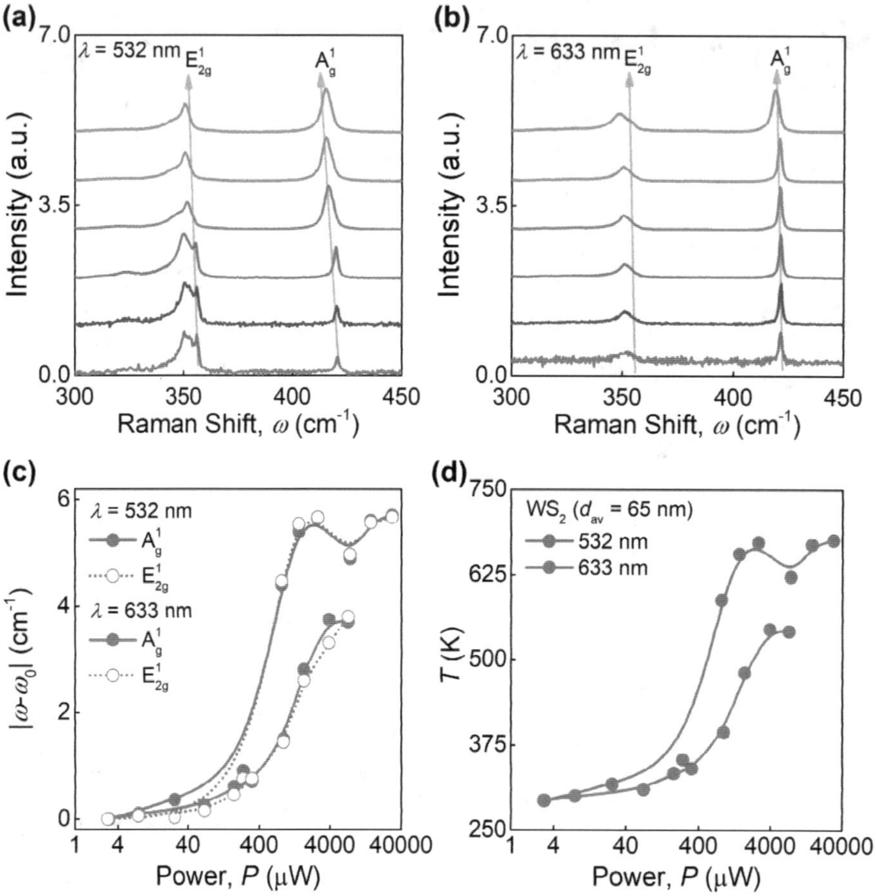

Fig. 2 Raman spectra of WS_2 NPs with an average size of 65nm on glass substrate recorded under different laser excitation power and at two laser wavelengths: (a) 532 nm and b 633 nm correspondingly. (b) Shift of the frequencies corresponding to A_{1g} and E^1_{2g} Raman scattering modes. (d) Dependence of WS_2 NPs local temperature on the power of exciting radiation for λexc = 532 nm and λexc = 633 nm.

One of the most promising therapeutic applications of biocompatible resonant dielectric nanoparticles is photothermal therapy (PTT) of cancer cells that implies heating induced destruction of cells under optical or ultrasonic excitation [11]. To test the photothermal response of studied resonant WS_2 NPs, we investigated the power-dependent Raman scattering spectra of laser-ablated WS_2 NPs selected by the centrifugation at 1000 rpm and having the average diameter of 65 nm assuming the linear dependence of the frequencies of both A_{1g} and E^1_{2g} modes from the temperature $\omega(T)=\omega_0+\chi_1\cdot\Delta T$, where ω_0 is the peak position of vibration A_{1g} or E^1_{2g} modes at room temperature and χ_1 is the first-order temperature coefficient of the A_{1g} or E^1_{2g} modes which was recently demonstrated for thin WS_2 on different substrates as well as for suspended layers of WS_2 [10]. Knowing the values of (A_{1g}) and (E^1_{2g}), we were able to recalculate the thermal shift of WS_2 Raman modes during the power-dependent Raman measurements into the local temperature of deposited on glass substrate WS_2 NPs. It should be noted that for the same values of the excitation power resonant WS_2 NPs demonstrate nearly 1.5 times higher local temperature rise when being excited at 532 nm then at 633 nm. It was explained by higher values of local electro-magnetic fields under the simultaneous excitation of ED and excitonic resonances under 532 nm excitation. This high photothermal response of WS_2 NPs opens the way to use them as new biocompatible agents for theranostics, providing high optical contrast and photothermal response.

3. Conclusions

We demonstrated spherical high-refractive nanoparticles made from bulk crystalline WS2 by versatile laser-assisted technique. Structural and optical analysis reveals that produced nanospheres preserve the crystalline structure, high refractive index and support strong excitons and Mie resonances in the spectral range 400-700 nm, resulting in enhanced photothermal response probed by Raman spectroscopy.

4. Acknowledgements

The work was performed under financial support of the Russian Science Foundation (grant 19-79-00350).

5. References

[1] A.B. Evlyukhin, S. M. Novikov, U. Zywietz, R. L. Eriksen, C. Reinhardt, S. I. Bozhevolnyi, and B. N. Chichkov, Nano Lett. 12, 3749–3755 (2012)
[2] I. Staude and J. Schilling, Nature Photonics, 1, 274 (2017).
[3] G. Grinblat, Y. Li, M.P. Nielsen, R.F. Oulton, S.A. Maier Nano Lett. 2016, 16, 4635−4640.
[4] R. Verre, D.G. Baranov, B. Munkhbat, J. Cuadra, M. Käll, and T. Shegai Nature Nanotechn. 14, 679-684 (2019).
[5] Manzeli, S., Ovchinnikov, D., Pasquier, D. et al. 2D transition metal dichalcogenides. Nat Rev Mater 2, 17033 (2017).
[6] Ermolaev, G. A., Grudinin, D. V., Stebunov, Y. V., Voronin, K. V., Kravets, V. G., Duan, J., Mazitov, A. B., Tselikov, G. I., Bylinkin, A., Yakubovsky, D. I., Novikov, S. M., Baranov, D. G., Nikitin, A. Y., Kruglov, I. A., Shegai, T., Alonso-González, P., Grigorenko, A. N., Arsenin, A. V., Novoselov, K. S., & Volkov, V. S. (2021). Nature Communications, 12(1), 854.
[7] Busschaert S, Reimann R, Cavigelli M, Khelifa R, Jain A and Novotny L 2020 ACS Photonics 7, 2482–2488.
[8] Blandin, P., Maximova, K. A., Gongalsky, M. B., Sanchez-Royo, J. F., Chirvony, V. S., Sentis, M., Timoshenko, V. Y., & Kabashin, A. V. (2013) Journal of Materials Chemistry. B, Materials for Biology and Medicine, 1(19), 2489–2495.
[9] Bailly, A.-L., Correard, F., Popov, A., Tselikov, G., Chaspoul, F., Appay, R., Al-Kattan, A., Kabashin, A. V., Braguer, D., & Esteve, M.-A. (2019) Scientific Reports, 9(1), 12890.
[10] Huang, X., Gao, Y., Yang, T., Ren, W., Cheng, H.-M., & Lai, T. (2016). Scientific Reports, 6, 32236.
[11] A.V. Kabashin and V. Yu. Timoshenko. Nanomedicine 11, 2247–2250 (2016).

Efficient Third Harmonic Generation with THz Graphene Metasurfaces

O Tsilipakos[1], A Theodosi[1,2], C M Soukoulis[1,3], E N Economou[1,4] and M Kafesaki[1,2]

[1]Institute of Electronic Structure and Laser (IESL), Foundation for Research and Technology - Hellas (FORTH), Heraklion, Crete GR-70013, Greece
[2]Department of Materials Science and Technology, University of Crete, Heraklion GR-70013, Greece
[3]Ames Laboratory—U.S. DOE and Department of Physics and Astronomy, Iowa State University Ames, Iowa 50011, USA
[4]Department of Physics, University of Crete, Heraklion GR-70013, Greece

Email: otsilipakos@iesl.forth.gr

Abstract. We theoretically study graphene-based metasurfaces for efficient third-harmonic generation in the THz regime. The graphene sheet is judiciously patterned into patch and cross geometries, in order to support sharp metasurface resonances at the fundamental and third harmonic frequencies. The reported conversion efficiencies reach -19dB (1.2%).

1. Introduction

Graphene is a two-dimensional (2D) material that possesses attractive properties for linear and nonlinear photonic applications [1], since it is characterized by a strong third-order nonlinearity and its linear and nonlinear conductivities can be tuned electrically. In addition, graphene can support tightly-confined surface plasmons in the technologically important THz regime. In particular, the short wavelength of propagating graphene plasmons enables the design of metasurfaces (MSs) supporting higher-order resonances, while the lattice constant remains subwavelength even for higher harmonic frequencies, avoiding diffraction effects (i.e., higher diffraction orders becoming propagating).

In this work, we design graphene-based MSs for efficient third harmonic generation (THG) in the THz regime. The design strategy is based on exploiting multiple resonances of the metasurface [2] and placing sharp resonances in the spectral vicinity of the fundamental (FF) and third harmonic (TH) frequencies, in order to harness the associated spatiotemporal energy confinement and boost the conversion efficiency (CE). This strategy has been termed double-resonant enhancement in Ref. [3]. Having as a reference the CE of -26dB (at input intensity of 0.1 MW/cm^2) reported for graphene ribbons on a metal-backed dielectric substrate in Ref. [4], we investigate and propose structures with 2D patterning, in order to further confine the plasmons and produce stronger local fields, as well as to have the possibility for polarization insensitive response in the x and y axes. By careful alignment of the fundamental and third-harmonic frequencies with MS resonances, we achieve efficiencies as high as -19dB (1.2%).

2. Graphene metasurfaces for third harmonic generation

We first study rectangular patches with size $w_x \times w_y$ in a square lattice [figure 1(a)]. The lattice constant is $a = 3.88$ μm. The patches reside on a dielectric substrate of thickness $h = 6.3$ μm, backed by a gold backreflector. A parametric study is conducted with respect to the length and width of the

Journal of Physics: Conference Series **2015** (2021) 012156 doi:10.1088/1742-6596/2015/1/012156

resonating patches, $w_x = r_x \times a$ and $w_y = r_y \times a$, with r_x and r_y varying from 0.1 up to 0.95. The conversion efficiency is calculated under CW conditions with the frequency-domain finite element method (FEM) by using two linear simulations at the FF and TH frequencies [4, 5] (see Methods). Control over the underlying resonant frequencies is exerted by varying w_y and consequently engineering the dispersion of the propagating plasmon; modifying w_x changes the length of the finite waveguide segment and specifies the frequencies satisfying the resonant condition. The optimum point in the w_x - w_y map results in a CE of -19dB (1.2%).

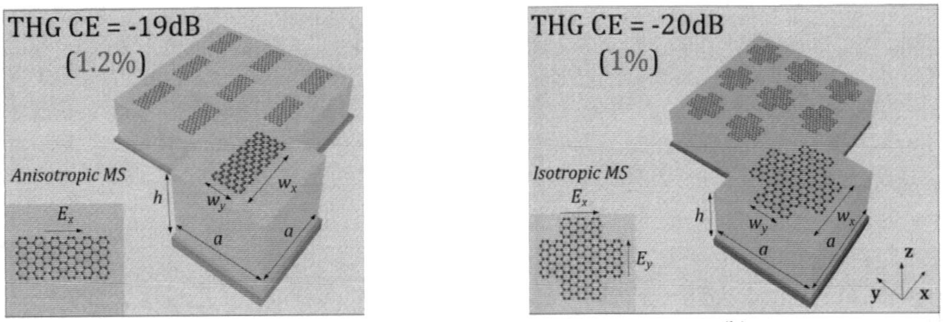

(a) (b)

Figure 1. Graphene-based metasurfaces for efficient third harmonic generation. (a) Rectangular graphene patches reside on a metal-backed dielectric substrate. A conversion efficiency of 1.2% is attained for E_x incident polarization. (b) Cross geometry of patches for isotropic response. The attained conversion efficiency is 1% for either E_x or E_y incident polarization.

However, this performance is obtained only for the E_x polarization, since after optimization we find $r_x \neq r_y$ meaning that the resulting patch is rectangular. To achieve isotropic response, we investigate different geometries such as the cross geometry in figure 1(b), which, following a similar parametric optimization procedure, leads to a THG conversion efficiency of -20dB (1%). To unveil the physical mechanism underlying the optimum point we plot the linear reflection and absorption spectrum (power coefficients for plane-wave incidence), as shown in figure 2. The operating frequency ω_{FF} is positioned at the first absorption peak (2.2 THz marked in figure 2 with a solid arrow), associated with the fundamental resonance of the structure. The TH frequency $\omega_{TH} = 3\times\omega_{FF}$ (6.6 THz marked in figure 2 with a solid arrow) is also well aligned with an absorption peak of the MS. These absorption peaks are associated with underlying resonances of the metasurface, as verified through eigenvalue simulations [6]; the respective eigenfrequencies are marked in figure 2 with dashed (red) arrows.

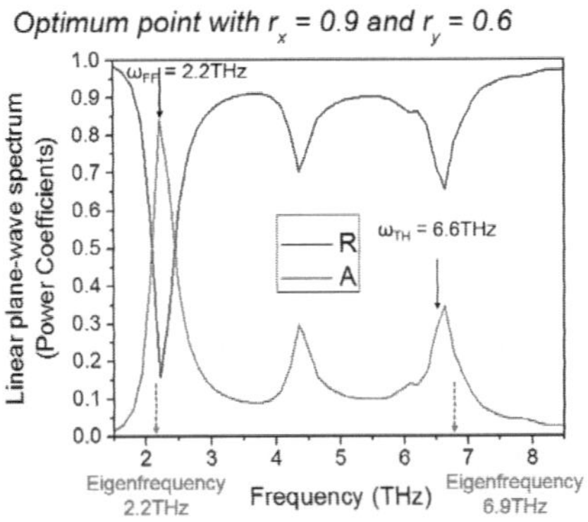

Figure 2. Linear plane-wave scattering spectrum (power coefficients) for cross geometry. Plotted data correspond to the optimum structure of the cross geometry with $r_x = 0.9$ and $r_y = 0.6$. Two resonances of the metasurface are in close proximity to the fundamental and third harmonic frequencies. The resonances are identified by the absorption peaks and verified through eigenvalue simulations (dashed arrows).

Journal of Physics: Conference Series **2015** (2021) 012156 doi:10.1088/1742-6596/2015/1/012156

3. Nonlinear modelling framework for continuous wave operation

3.1. Simulation details
For assessing the nonlinear response of the metasurfaces under study we perform full wave 3D electromagnetic simulations using COMSOL Multiphysics. A single unit cell is simulated with periodic boundary conditions at the x and y boundaries and input/output ports at the z boundaries. In order to simulate the THG nonlinear response under continuous wave (CW) illumination, we can decouple the nonlinear problem into two linear frequency domain simulations at the fundamental (FF) and third harmonic (TH) frequency [4,5]. In the first step, the metasurface is illuminated at the fundamental frequency, inducing currents on the graphene patches. Via frequency mixing mediated by the third order conductivity, a nonlinear current is produced oscillating at the third harmonic. This acts as a source for the second simulation radiating a plane wave at 3ω. The THG conversion efficiency (CE) is given by P_{TH}/P_{FF}, where P_{TH} is the power outflow at the third harmonic frequency and P_{FF} is the injected power at fundamental frequency.

3.2. Graphene properties
Graphene is naturally modelled as an infinitesimally thin material via a surface current boundary condition. It is a highly dispersive material in the THz regime; its linear conductivity can be described by a Drude dispersion model:

$$\sigma^{(1)} = -\frac{iD}{\left(\omega - \frac{i}{\tau}\right)}, \tag{1}$$

$$D = \frac{q^2\mu_c}{\hbar^2}, \tag{2}$$

where $\tau = 500$ fs is the electron relaxation time, q is the elementary charge, $\mu_c = 0.3$ eV is the chemical potential and $\hbar$ is the reduced Planck's constant.

The nonlinear current density that develops on graphene and oscillates at the third harmonic frequency is given by:

$$J_{3\omega} = \sigma^{(3)} \cdot \mathbf{E}_{||} \cdot \mathbf{E}_{||} \cdot \mathbf{E}_{||} e^{j3\omega t}. \tag{3}$$

Graphene's third order conductivity formula is taken from Cheng et al. [7]

$$\sigma^{(3)} = \frac{i\sigma_0(\hbar u_F q)^2}{48\pi(\hbar\omega_{FF})^4} T\left(\frac{\hbar\omega_{FF}}{2\mu_c}\right), \tag{4}$$

where

$$\sigma_0 = q^2(4\hbar)^{-1}, \tag{5}$$

$$T(x) = 17G(x) - 64G(x) + 45G(3x), \tag{6}$$

$$G(x) = ln\left|\frac{1+x}{1-x}\right| + (i\pi\theta|x| - 1), \tag{7}$$

in which u_F is the Fermi velocity and $\theta(x)$ the Heaviside step function.

4. Conclusion

We have proposed nonlinear THz metasurfaces based on 2D-patterned graphene patches atop a metal-backed dielectric substrate for efficient third harmonic generation. The two-dimensional patterning tightly confines the supported graphene plasmons and results in strong local fields, boosting the nonlinear conversion process. Both a rectangular and a cross graphene patch geometry have been studied, and are found to exhibit high conversion efficiencies of approximately -20dB (1%). The cross geometry leads, moreover, to polarization-independent (isotropic) response. Toward future work, additional metasurfaces combining graphene with metallic structures have been investigated and preliminary results indicate even higher CEs, reaching values as high as -9dB (12%).

References

[1] Hafez H, Kovalev S, Tielrooj K J, Bonn M, Gensch M, Turchinovich D 2020 *Adv. Opt. Mat.* **8** 1900771
[2] Tsilipakos O, Kafesaki M, Economou E N, Soukoulis C M, Koschny T 2020 *Adv. Opt. Mater.* **8** 2000942
[3] You J W, You J, Weismann M, Panoiu N C 2017 *Philos. Trans. R. Soc. A Math. Phys. Eng. Sci.* **375** 20160313
[4] Jin B, Guo T, Argyropoulos C 2017 *J. Opt. (United Kingdom)* **19** 094005
[5] Christopoulos T, Tsilipakos O, Sinatkas G, Kriezis E E 2018 *Phys. Rev. B* **98** 235421
[6] Christopoulos T, Tsilipakos O, Sinatkas G, Kriezis E E 2019 *Opt. Express* **27** 14505
[7] Cheng J L, Vermeulen N, Sipe J E 2014 *New J. Phys.* **16** 053014

Acknowledgments

This work was supported by the Hellenic Foundation for Research and Innovation (H.F.R.I.) under the "2nd Call for H.F.R.I. Research Projects to support Post-doctoral Researchers" (Project Number: 916, PHOTOSURF).

Microwave realization of multiresonant metasurfaces for achromatic pulse delay

Odysseas Tsilipakos[1], Lei Zhang[2], Maria Kafesaki[1,3], Costas M. Soukoulis[1,2], Thomas Koschny[2]

1. Institute of Electronic Structure and Laser (IESL), Foundation for Research and Technology – Hellas (FORTH), Heraklion, Crete GR-70013, Greece
2. Ames Laboratory—U.S. DOE and Department of Physics and Astronomy, Iowa State University Ames, Iowa 50011, USA
3. Department of Materials Science and Technology, University of Crete, Heraklion GR-70013, Greece

Email: otsilipakos@iesl.forth.gr

Abstract. We propose a microwave realization of a metasurface that can delay broadband pulses without distortion in reflection. In order to obtain large and broadband pulse delay, we harness the synergetic phase delay of five sharply-resonant meta-atoms. More specifically, three electric-LC and two split ring resonators, supporting electric and magnetic dipole resonances, respectively, are combined in a subwavelength unit cell. The resonances are spectrally interleaved and specifically designed to provide a spectrally-constant reflection amplitude and group delay according to the prescription in [ACS Photonics 5, 1101, 2018]. The designed metasurface is electrically ultrathin ($\lambda_0/19$), since it relies on resonant phase delay exclusively, instead of phase accumulation via propagation. We show delay of 700-MHz Gaussian pulses centred at 11 GHz by 1.9 ns, corresponding to approximately 21 carrier cycles. Our results highlight the practical potential of metasurfaces for broadband dispersion control applications.

1. Introduction

Metasurfaces, the two dimensional counterparts of metamaterials, promise to revolutionize electromagnetic wave control and replace conventional bulky components [1,2]. However, they have found limited use in broadband applications due to their resonant, narrowband nature. An approach that can increase the aggregate bandwidth is to combine multiple resonant meta-atoms in the unit cell. By properly designing the resonances, the respective phase delays stemming from each resonance (that are by themselves dispersive) can combine into an achromatic, spectrally-constant group delay that can be exploited for achromatic pulse delay [3] and wavefront manipulation [4] applications. The achieved group delay is enhanced with respect to that of the single resonance, due to the synergetic action of the resonance ensemble.

In this work, we demonstrate a microwave realization of a broadband time delay metasurface operating in reflection, by using five meta-atoms to implement five spectrally-interleaved sharp resonances (two magnetic and three electric), specifically designed according to the prescription in Ref. [3]. The designed metasurface is electrically ultrathin ($\lambda_0/19$), since it relies on resonant phase delay exclusively, instead of phase accumulation via propagation. We show delay of broadband, 700-MHz Gaussian pulses centred at 11 GHz by 1.9 ns (21 times the carrier cycle $T_0=1/f_0$).

2. Results

The underlying concept of this work is shown in Fig. 1. A singly-resonant metasurface can be described by a homogenized Lorentzian electric surface conductivity $\tilde{\sigma}_{se} = i\tilde{\kappa}_e\omega/(\omega^2 - \omega_0^2 + i\Gamma_e\omega)$, where $\tilde{\sigma}_{se} = \eta\sigma_{se}/2$ is the dimensionless conductivity, ω_0 the resonant frequency, $\tilde{\kappa}_e$ the coupling strength, and Γ_e the dissipation. A sharp resonance results in a narrow bandwidth of high reflection, as shown in Fig. 1(a); at the same time the supplied group delay rolls off rapidly (the peak is obtained on resonance and can be calculated analytically: $\tau_g(\omega_0) = 1/[\pi(\tilde{\kappa}_e + \Gamma_e)]$. A crude approach to extend the metasurface bandwidth would be to make the resonance wider [Fig. 1(b)]. However, this decreases the supplied group delay considerably. An alternative option is to utilize multiple sharp resonances; then, we can harness the strong phase delay of the constituent resonances and at the same time obtain a broad aggregate bandwidth, extending in principle arbitrarily the delay-bandwidth limit. By using the prescription in Ref. [3], the decaying tails of the respective group delay contributions combine in a flat-top region with increased group delay. An example with five resonances is shown in Fig. 1(c). Such a metasurface can delay broadband pulses in reflection without distortion, as schematically shown in Fig. 1(d). Note that we have allowed for both electric and magnetic polarizability, in order to obtain control over unidirectional scattering; operation in reflection requires interleaved electric and magnetic resonances, whereas operation in transmission requires overlapped resonances. Scaling of the metasurface bandwidth for a larger number of resonances, N, is depicted in Fig. 1(e), (f), referring to the half-power bandwidth (HPBW) of the reflection amplitude, BW_R^N, and the bandwidth of the group delay, BW_τ^N, measured at the 90% points, respectively. These quantities are normalized to the respective quantities for a single resonance, BW_R^1 and BW_τ^1. As can be seen in Fig. 1(f) for example, the constant-group-delay bandwidth follows a linear dependence $BW_\tau^N \sim 0.8(BW_\tau^1)N$. The slope of 0.8 indicates the contribution of each resonance to the aggregate bandwidth.

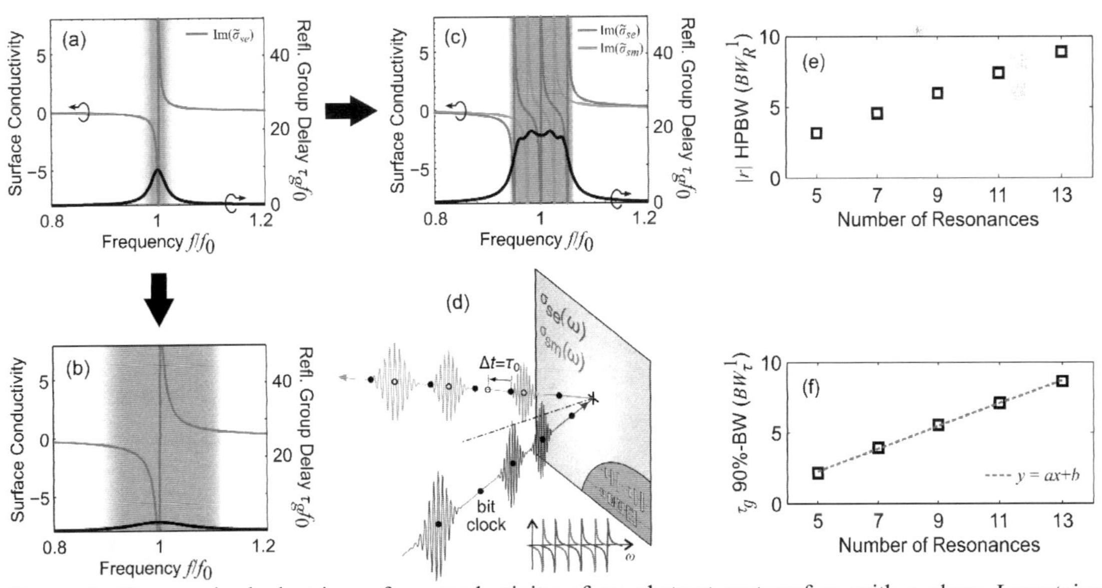

Figure 1. Homogenized electric surface conductivity of an abstract metasurface with a sharp Lorentzian resonance. The group delay reaches $9.5T_0$ but for a very narrow bandwidth that cannot accommodate real-world broadband signals. The gray shading denotes the bandwidth of high reflection amplitude. (b) The bandwidth can increase with a brighter (broader) resonance but at the expense of weak group delay. (c) Properly arranged multiple sharp resonances can provide a spectrally-constant reflection amplitude and group delay that can be used to delay broadband pulses without distortion. (e) Schematic illustration of pulse delay operation in

reflection. (e,f) Scaling of available bandwidth with number of resonances. (e) Half-power bandwidth (HPBW) of reflection amplitude, BW_R^N, normalized to $BW_R^1 = \tilde{\kappa}_e + \Gamma_e$. (f) Group delay bandwidth measured at the 90% points, BW_τ^N, normalized to $BW_\tau^1 = \tilde{\kappa}_e + \Gamma_e$. The slope of the linear fit is 0.8 indicating the contribution of each resonance to the achieved aggregate bandwidth.

We now focus on the physical implementation of the scenario in Fig. 1(c) for the 10-GHz regime. We use the most straightforward strategy of one meta-atom per resonance. Electric-LC (ELC) and split-ring resonator (SRR) meta-atoms implement the electric and magnetic dipole resonances, respectively, as seen in the generic schematic of Fig. 2(a). A three-metallization-layer symmetric printed circuit board (PCB) is used; the patches of the SRRs reside in the top and bottom layers, whereas the ELCs in the middle metallization layer. Resorting to the theoretical recipe in Ref. [3], we require electric and magnetic resonances with a frequency spacing of 500 MHz, a dissipation coefficient of ~35 MHz, and a coupling strength of 320 MHz. The resonances can be seen in the electric and magnetic surface conductivities depicted in Fig. 2(b). The conductivities have been retrieved by applying the expressions in Ref. [3] to plane-wave scattering coefficients obtained via full-wave simulations of a single unit cell in the periodic structure (periodic boundary conditions are applied at *x*- and *y*-boundaries). This specific combination of resonances, can theoretically provide a spectrally constant group delay of 2 ns over a 700-MHz bandwidth, while providing a high, flat reflection amplitude of $|r| = 0.8$. Indeed, the simulated response (full wave solution of Maxwell's equations) depicted in Fig. 1(c) indicates a very good approximation of the targeted response with only minor ripples of the reflection amplitude and group delay inside the designed reflection band. Further details on the physical implementation (meta-atom dimensions and exact placement in unit cell, etc.) and an experimental verification of the proposed structure can be found in Ref. [5].

The spectrum of a 700-MHz Gaussian pulse is also included in Fig. 2(c) with a thick black line; the entire pulse bandwidth nicely fits inside the reflection band. The pulse delaying properties of the designed metasurface are demonstrated in Fig. 2(d). The output pulse is delayed by 1.9 ns (~$21T_0$) without broadening nor distortion. The output peak intensity is ~0.65, since the amplitude reflection coefficient in Fig. 2(c) is ~0.8. For comparison, if a broad single resonance was utilized, as in Fig. 1(b), the pulse would be delayed by only 0.35 ns. Using five resonances we were able to gain approximately a five-fold improvement in the attained group delay. In addition, using a single resonance would lead to a pulse broadening of ~10%, due to the group delay dispersion inside the pulse bandwidth [cf. Fig. 1(b)]; in contrast, pulse broadening in Fig. 1(d) is below 1%.

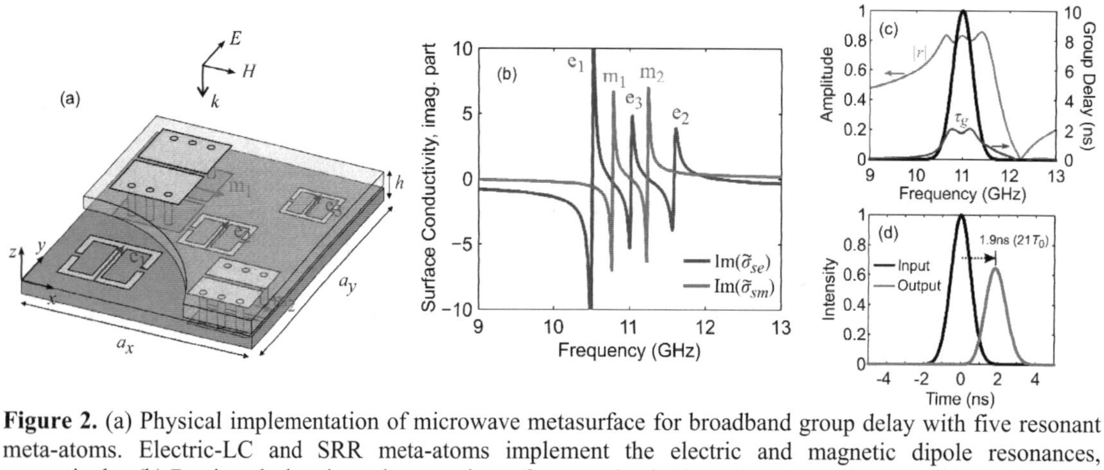

Figure 2. (a) Physical implementation of microwave metasurface for broadband group delay with five resonant meta-atoms. Electric-LC and SRR meta-atoms implement the electric and magnetic dipole resonances, respectively. (b) Retrieved electric and magnetic surface conductivities (dimensionless), as obtained from full wave simulations of plane-wave scattering using the expressions in Ref. [3]. (c) Metasurface response: reflection amplitude and group delay. The spectrum of a 700-MHz Gaussian pulse positioned at the centre of the reflection band is overlaid (thick black line). (d) Demonstration of broadband pulse delay operation for the 700-

MHz Gaussian pulse in the temporal domain. The pulse is delayed by 1.9 ns (~$21T_0$) without distortion.

3. Conclusion

We have designed a microwave physical implementation of an achromatic metasurface that can delay broadband pulses in reflection. Our approach relies on fitting five sharply-resonant meta-atoms inside a subwavelength unit cell. The metasurface is electrically ultrathin ($\lambda_0/19$), since it relies exclusively on resonant phase delay. The proposed metasurface has been fabricated in the form of two boards (0.787-mm thick) made of Rogers RT/duroid 5880 ($\varepsilon_r = 2.2$ and $tan\,\delta = 9 \times 10^{-4}$ at 10 GHz). One board is implementing the top SRR metallization layer and the ELC meta-atoms and the second is implementing the bottom SRR metallization layer. The boards have been assembled with chip resistors [instead of plated through vias schematically depicted in Fig. 2(a)] to complete the electrical continuity of the SRR meta-atoms and measurements with horn antennas and a vector network analyzer have been performed [5]. Our work highlights the practical potential of metasurfaces for dispersion control applications that rely on large and broadband phase delays.

4. References

[1] Glybovski, S B, Tretyakov, S A, Belov, P A, Kivshar, Y S and Simovski, C R 2016, *Phys. Rep.* **634** 1-72
[2] Sun, S; He, Q; Hao, J; Xiao, S; Zhou, L 2019, *Adv. Opt. Photonics* **11** 380-479
[3] Tsilipakos O, Koschny T, and Soukoulis C M 2018, *ACS Photonics* **5** 1101-1107
[4] Tsilipakos O, Kafesaki M, Economou E N, Soukoulis C M, Koschny T 2020, *Adv. Opt. Mater.* **8** 2000942
[5] Tsilipakos O, Zhang L, Kafesaki M, Soukoulis C M, Koschny T 2021, *ACS Photonics* **8**, 1649-1655

Acknowledgments

This work was supported by the Hellenic Foundation for Research and Innovation (H.F.R.I.) under the "2nd Call for H.F.R.I. Research Projects to support Post-doctoral Researchers" (Project Number: 916, PHOTOSURF). Work at Ames Laboratory was supported by the Department of Energy under Contract No. DE-AC02-07CH11358.

Comparison of angular-selective metasurfaces as tools for sub-THz single-frequency sensing

M A Tumashov[1], J P del Risco[2], S B Glybovski[1], A D Sayanskiy[1], S A Kuznetsov[3], J D Baena[4]

[1]Department of Physics and Engineering, ITMO University, St. Petersburg, Russia
[2]Department of Mathematics, Universidad Sergio Arboleda, Bogota, Colombia
[3]Physics Department, Novosibirsk State University, Novosibirsk, Russia
[4]Physics Department, Universidad Nacional de Colombia, Bogota, Colombia

E-mail: [1] s.glybovski@metalab.ifmo.ru

Abstract. We demonstrate and numerically characterize a set of single-layer metasurfaces based on capacitively-coupled bent metal strips deposited on a thin polypropylene film operating as band-stop filters at 139.1 GHz. Due to a high inter-digital coupling capacitance between neighboring inclusions formed by the strips, the resonance of the metasurface becomes highly sensitive to the incidence angle of a plane wave. At the same time, a narrow-band resonance is achieved by reducing the conductor length along the vertical polarization direction of the incident wave. We propose to use both properties for designing a sensor, which analyses the thickness of a thin slab of a known substance deposited over the structure by tracking the angle at which the minimum of transmission can be measured at a single frequency. We compare three versions of the proposed geometry by numerical calculation of the transmission coefficient depending on both the incidence angle and analyte slab's thickness. We select the most suitable structure for sensing.

1. Introduction

The sub-terahertz range starting from 100 GHz enables many important applications including astronomy, security, and spectroscopy [1]. Due to the possibility of relatively strong interaction between non-ionizing sub-THz waves and thin material slabs, spectroscopy is particularly useful for label-free sensing of various substances and biological samples [2]. However, the strong interaction of sub-THz waves with materials slabs is impossible to reach when the thickness is small compared to the wavelength, i.e. for micron-thick or even thinner slabs.

Recently, to improve sensing capabilities of thin films it was proposed to use metasurfaces, the two-dimensional and typically resonant periodic structures with subwavelength periodicity. Unit cells of such structures consist of resonant inclusions that under illumination with plane waves produce tiny electric-field hotspots with dimensions much smaller than the wavelength. Due to this resonant field confinement, metasurfaces provide a high sensitivity of films with deep-subwavelength thicknesses [3, 2]. The presence of an analyzed substance (analyte) slab deposited on a metasurface typically shifts the resonant frequency that can be detected using spectroscopy techniques. Based on this principle, many configurations of sub-THz and THz metasurfaces have been proposed for thin-film sensing (see e.g. in [4, 5, 6, 7]). A recent review of the topic is available in [8].

Conventional THz thin-film sensors based on resonant metasurfaces with high quality factor require spectroscopy quasi-optical setups being complex and expensive. In [9] a metasurface sensor with an alternative operational principle was proposed, in which an angular sensitivity of the metasurface properties is employed. The corresponding metasurface is designed so that the resonance frequency becomes dependent on the incidence angle. Also, it is sensitive to the presence of an analyte slab as for any metasurface sensor. Hence the thickness of the slab with a known permittivity can be measured by tracking the angle of incidence corresponding to the maximum/minimum of the transmission coefficient at a single operational frequency. The metasurface with a deposited analyte should be placed between a single-frequency source and detector and mechanically rotated. The main advantage of this technique concerning others is that it doesn't require expensive equipment such as a table-top spectrometer to measure the thickness of thin layers.

In this work, we propose and study a new set of metasurfaces for single-frequency thin-film sensing at 139.1 GHz having pronounced angular dependence of the stop-band resonance. This frequency corresponds to a commercial IMPATT-diode-based monochromatic radiation source from Terasense Group Inc., USA.

To increase angular sensitivity of the resonance we develop the idea of using chains of interconnected high-quality-factor resonators [10]. We propose and compare three single-layer metal patterns with a new shape of the subwavelength resonators implemented as bend capacitively-coupled metal strips and select one of them to be the most suitable for sensing films with thicknesses of less than 2 μm.

2. Metasurface design and characterization

Each of the three proposed unit cells consists of meandered strips etched in a single aluminum-based layer, which is placed on top of a polypropylene (PP) thin film with a thickness of 30 μm and a relative permittivity of 2.25. The aluminum layer has a conductivity of $1.5{\cdot}10^7$ S/m with a thickness of 0.4 μm.

The unit cells of three proposed structures are depicted in Fig. 1. Structure (a) is a starting one. It consists of two S-shaped bent metal strips. Each one has a vertical part in the middle, which is assumed to be parallel to the polarization direction of the incident wave. There are also relatively long horizontal parts responsible for making the unit cell resonant and for the strong capacitive coupling to similar shapes in neighboring unit cells. Differences in the metalization strips number of studied designs are justified by the fact that we must suggest the design which has the narrow-band resonance at the operation frequency and high aperture efficiency simultaneously to achieve the required performance. One of the adjusted parameters is the distance between the top and bottom trains. It was chosen for each design separately to provide a resonance behavior on an operating frequency of 139.5 GHz. The metasurfaces belong to the class of band-stop frequency-selective surfaces where high capacitive coupling between unit cells increases the sensitivity of the resonance frequency to the angle of incidence [10]. The width of all strips as well as the width of gaps between them was chosen to be 10 μm. the minimum possible value within the requirements of the presumed photolithographic manufacturing technology. With this limitation, the vertical strips were made as short as possible to maximize the quality factor of the resonance. Two mirror-reflected S-shaped strips were combined into one unit cell to cancel out the cross-polarization effect so that the metasurface interacts only with vertical polarization.

As shown by numerical simulations, the electric field is localized in the gaps between the capacitive horizontal strips. With the aim to enlarge the area covered by the electric-field hotspots two other structures were introduced. Structure (b) differs from (a) by having two parallel meandered strips in each S-shape. Structure (c) has ramified strips with further increased coupling capacitance. Note that to keep the resonant frequency at the normal incidence at the

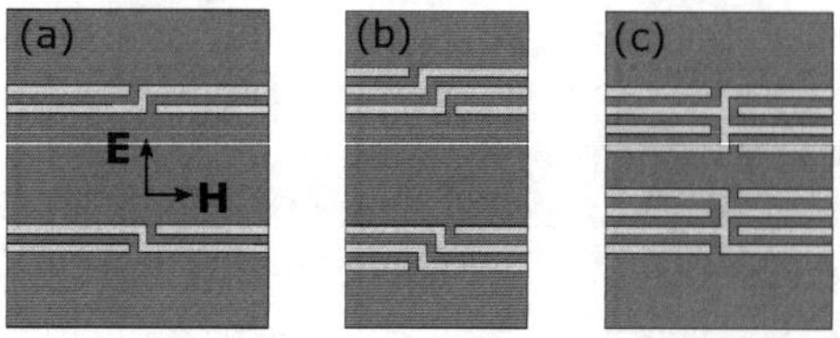

Figure 1. Three suggested unit-cell designs of a metasurface with angular sensitivity based on capacitively-coupled bent metal strips (yellow - aluminum, red - polypropylene).

same position of 139.1 GHz, all three structures have the same vertical periodicity of 340 μm, but different horizontal periodicity: 291, 200, and 250 μm for (a), (b) and (c), respectively.

All three structures were modeled and numerically analyzed with CST Microwave Studio 2020 (Frequency Domain Solver was used) by calculating transmission coefficients in the range of incidence angles. TE-polarization was kept with respect to the direction of vertical strips. The incidence angle is equivalent to the rotation angle of the metasurface in the proposed optical scheme of the sensor. The transmission properties were calculated for the analyte's relative permittivity of 2.65 and the slab thickness varied from 0.4 to 1.6 μm. It was also observed in the numerical study that the presence of an imaginary part in the permittivity has no much effect on the efficiency of the sensor. Therefore, it was decided to consider the analytes with the real part of permittivity only.

Initially, each metasurface was tuned to have a minimum of the transmission coefficient at 139.1 GHz in the absence of the analyte for the normal incidence (rotation angle of zero degrees). As the analyte of a given thickness of 1 μm appears, the minimum can be observed at another incidence angle defined as Θ. Therefore the structure should be rotated by the same angle to recover the minimum of transmission between the source and detector at 139.1 GHz.

For each design, we introduce a set of parameters to compare: the required rotation angle Θ (for the analyte slab's thickness of 1 μm), the width of the frequency stop band at the normal incidence in the presence of the analyte, and the angular accuracy (defined as the ratio of Θ to the half-width of the angular dependence of the transmission coefficient taken at the level of half maximum of spectrum characteristic). The calculated values of these parameters are summarized in Table 1. The angular accuracy is the most important parameter for our consideration. The higher the angular accuracy, the more precisely the properties of the analyte slab can be found by mechanical rotation because a small increment of thickness will lead to the abrupt increment of the required angle of rotation. The difference in angular accuracy for the proposed designs is illustrated in Fig. 2.

While the angle of rotation Θ is similar for all the structures and stays within the range of 24-30 deg, the angular width of the curves is very different. As can be seen, structure (c) is the most suitable for angular sensing since it provides the highest angular accuracy. Also, its value of Θ is slightly larger than for (a) and (b). In contrast, structure (b) has the minimum angular accuracy among the considered structures.

For two structures with low (b) and high (c) angular accuracy we investigated the dependence of the required rotation angle Θ on the deposited analyte slab's thickness. The results are presented in Fig. 3. As can be seen, the two curves are almost similar, so the range of mechanical rotation for the sensors based on metasurfaces (b) and (c) is the same.

Next, we estimated how technical fabrication features could affect the parameters of the selected designs. The minimum width of the metal strips is 10 μm, while the tolerance of the presumed technological procedure is about 1 μm. This was found to have almost no effect on the parameters. Due to the strong miniaturization of unit cells, dissipative losses take place around the resonance frequency (mostly conductive losses in the aluminum strips).

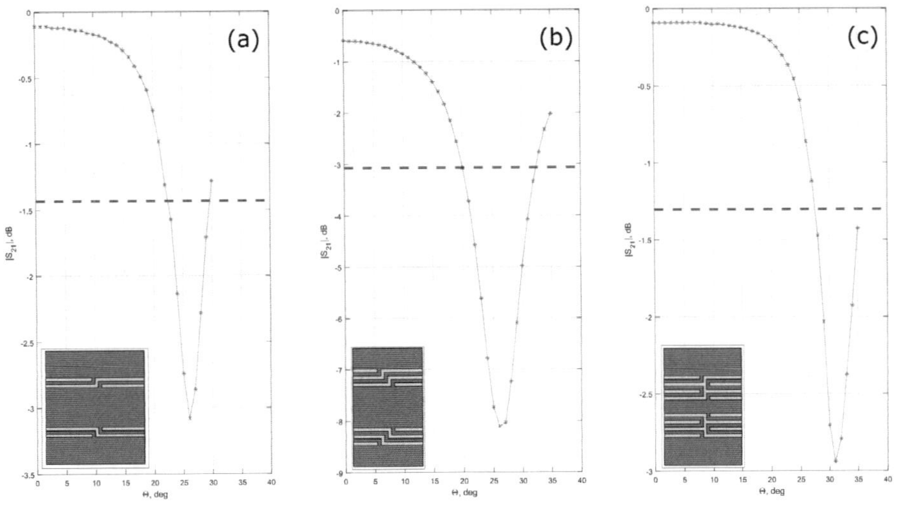

Figure 2. Angular dependence of transmission coefficient magnitude at 139.1 GHz for three proposed metasurfaces with a deposited 1-μm-thick analyte slab. Dashed lines indicate the levels at which angular accuracies calculated.

Table 1. Comparison of parameters of three suggested designs.

Parameter/Design	(a)	(b)	(c)
Stop-band width, GHz	2.49	3.96	1.79
Rotation angle Θ, deg, (for 1-μm analyte)	26	26	31
Angular accuracy	7.4	4.3	8.8

Therefore, for structures (b) and (c) we investigated the effect of the conductivity deviation from the nominal value of $1.5{\cdot}10^7$ S/m. Thus during the metal deposition and consequent photolithography procedure, the conductivity may drop to $7.5{\cdot}10^6$ S/m. Fig. 4 shows the spectra of the transmission coefficients with the nominal and reduced conductivity for structures (b) and (c). The results are shown for the normal incidence in the presence of a 1-μm-thick analyte slab. It can be checked that the conductance deviation also has a little effect on the metasurface properties. Indeed, for the transmission coefficient minimum, a frequency shift of fewer than 450 MHz for both structures is observed.

3. Conclusions

In this work, we proposed and numerically compared three versions of an angular-sensitive resonant metasurface operating in the transmit regime at 139.1 GHz. It was shown that the version with the shape having ramified strips with increased capacitive coupling with neighboring unit cells (structure (c)) had the most sensitive resonance to the angle of incidence. This property results in the highest accuracy of determination of the analyte slab's thickness of the sensor based on this metasurface. In particular, for the analyte's relative permittivity of 2.65, the sensor covers the range of thicknesses from 0.4 to 1.6 μm with the required mechanical rotation by the angle from 15 to 35 degrees. At the same time, the angular accuracy was found to be

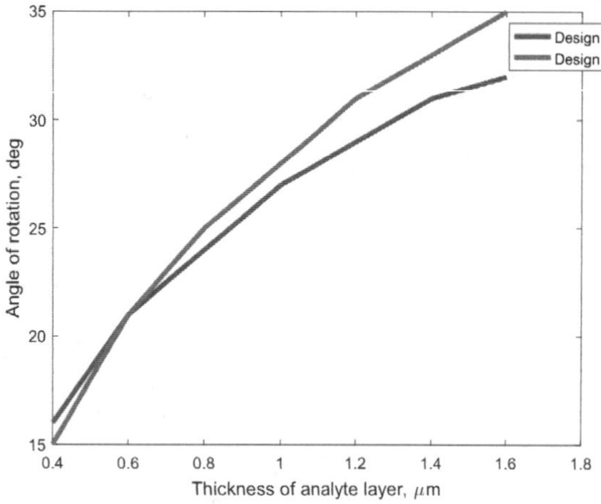

Figure 3. Dependence of the required rotation angle Θ on the analyte slab's thickness for metasurfaces (b) and (c).

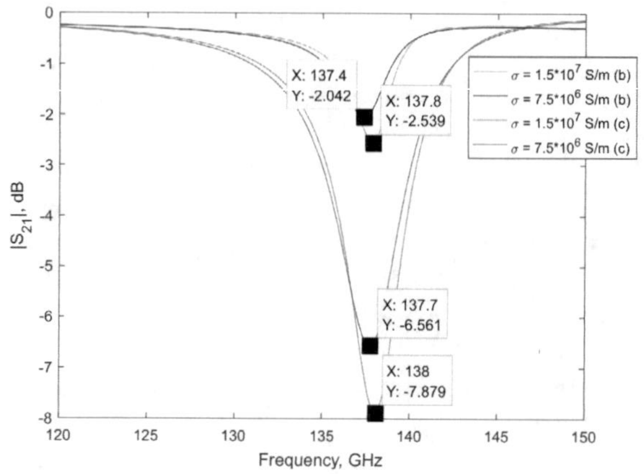

Figure 4. Frequency spectra of the transmission coefficient magnitude with the nominal and reduced conductivity of metal strips for structures (b) and (c).

around 2 degrees, i.e. better than 0.1 μm for the accuracy of thickness measurement.

Acknowledgments

This work was supported by the Russian Science Foundation (Project 19-79-10260).

References

[1] S. Dhillon, M. Vitiello, E. Linfield, A. Davies, M. C. Hoffmann, J. Booske, C. Paoloni, M. Gensch, P. Weightman, G. Williams *et al.*, "The 2017 terahertz science and technology roadmap," *Journal of Physics D: Applied Physics*, vol. 50, no. 4, p. 043001, 2017.

[2] J. F. O'Hara, W. Withayachumnankul, and I. Al-Naib, "A review on thin-film sensing with terahertz waves," *Journal of Infrared, Millimeter, and Terahertz Waves*, vol. 33, no. 3, pp. 245–291, 2012.

[3] T. Driscoll, G. O. Andreev, D. N. Basov, S. Palit, S. Y. Cho, N. M. Jokerst, and D. R. Smith, "Tuned permeability in terahertz split-ring resonators for devices and sensors," *Applied Physics Letters*, vol. 91, no. 6, p. 062511, 2007.

[4] M. Islam, S. J. M. Rao, G. Kumar, B. P. Pal, and D. R. Chowdhury, "Role of resonance modes on terahertz metamaterials based thin film sensors," *Scientific reports*, vol. 7, no. 1, pp. 1–8, 2017.

[5] I. Jáuregui-López, P. Rodriguez-Ulibarri, S. A. Kuznetsov, N. A. Nikolaev, and M. Beruete, "Thz sensing with anomalous extraordinary optical transmission hole arrays," *Sensors*, vol. 18, no. 11, p. 3848, 2018.

[6] I. Jáuregui-López, P. Rodríguez-Ulibarri, A. Urrutia, S. A. Kuznetsov, and M. Beruete, "Labyrinth metasurface absorber for ultra-high-sensitivity terahertz thin film sensing," *physica status solidi (RRL) – Rapid Research Letters*, vol. 12, no. 10, p. 1800375, 2018.

[7] "Ultra-thin subwavelength film sensing through the excitation of dark modes in thz metasurfaces," *Optics Communications*, vol. 453, p. 124366, 2019.

[8] M. Beruete and I. Jáuregui-López, "Terahertz sensing based on metasurfaces," *Advanced Optical Materials*, vol. 8, no. 3, p. 1900721, 2020.

[9] N. Nikolaev, S. Kuznetsov, and M. Beruete, "Terahertz thin-film sensing with angle-susceptable metasurface," in *2018 43rd International Conference on Infrared, Millimeter, and Terahertz Waves (IRMMW-THz)*, 2018, pp. 1–2.

[10] J. D. Ortiz, J. D. Baena, V. Losada, F. Medina, and J. L. Araque, "Spatial angular filtering by fsss made of chains of interconnected srrs and csrrs," *IEEE microwave and wireless components letters*, vol. 23, no. 9, pp. 477–479, 2013.

Dissipative Kerr solitons at the edge state of the Su-Schrieffer–Heeger model

A. K. Tusnin, A. M. Tikan, T. J. Kippenberg

Institute of Physics, Swiss Federal Institute of Technology Lausanne (EPFL), Lausanne, Switzerland

E-mail: tobias.kippenberg@epfl.ch

Abstract. We investigate analytically and numerically dynamics of dissipative Kerr solitons (DKS) at the edge state of the Su-Schrieffer–Heeger model. We demonstrate that four-wave mixing processes can lead to the formation of DKSs in the edge state of the resonator chain which subsequently initiates photon transfer to the bulk states. We discuss how the edge state soliton can be stabilized by limiting its width within the band gap. Our results contribute to advanced dispersion engineering via mode hybridization in chains of resonators — one of promising ways to achieve broadband frequency combs generation on chip.

1. Introduction

Photonics often serves as a primary platform for observation of effects found in various areas Physics ranging from materials in the solid-state to Bose–Einstein condensate, etc. Topological states, initially found in solids, have triggered in photonics a remarkable interest due to fundamental aspects as well as to their potential applications [1]. However, up to now the majority of studies have considered linear and single (or few) mode(-s) operation regime of photonic devices. In this work we theoretically analyze the nonlinear dynamics of the edge state in the Su-Schrieffer–Heeger model [2], constructed of dimerized optical high-Q microcavities with Kerr nonlinearity. On the single resonator level, the four-wave mixing processes lead to the generation of coherent and broadband frequency combs [3] which are related to the formation of dissipative Kerr solitons (DKS) [3].

2. Homogeneously coupled resonators

In the simplest (and for now trivial, i.e. not topological) case — a chain of linearly coupled resonators with periodical boundary conditions — we assume the field envelope $A_\ell(\phi)$ of an uncoupled ℓ-th resonator to be described by the Lugiato-Lefever equation (LLE) [3], and linear coupling to the neighbor resonators expressed by j. In the meanfield approximation, this system is described by the set of coupled LLEs [4] studied in details in Ref. [5]. Here we briefly recap the main features arising due to the nonlinear coupling between frequency modes of a single resonator ω_μ (μ is integer) and spatial supermodes k. Introducing the collective excitation amplitude $\psi_{\mu k}$ which is connected to the field envelope amplitudes $A_\ell(\phi)$ via the Fourier transform

$$\psi_{\mu k} = \frac{1}{2\pi\sqrt{N}} \int \sum_{\ell=1}^{N} A_\ell e^{2\pi i(\ell k/N + \mu\varphi)} d\varphi,$$

we obtain the governing equation for $\psi_{\mu k}$:

$$\frac{\partial \psi_{\mu k}}{\partial \tau} = -(1+i\zeta_0)\psi_{\mu k} - i\big[d_2\mu^2 - 2j\cos\frac{2\pi k}{N}\big]\psi_{\mu k} + \frac{i}{N}\sum_{\substack{k_1,k_2,k_3\\ \mu_1\mu_2\mu_3}} \psi_{\mu_1 k_1}\psi_{\mu_2 k_2}\psi^*_{\mu_3 k_3}\delta_{\mu_1+\mu_2-\mu_3-\mu}\delta_{k_1+k_2-k_3-k} + \delta_{k,k_0}\delta_{\mu,0}\tilde{f}_{k_0}, \quad (1)$$

where μ is the comb index, k is the spatial supermode index, ζ_0 stands for normalized laser detuning, d_2 is normalized group velocity dispersion ($d_2 > 0$ corresponds to anomalous dispersion case), j is normalized intraresonator coupling, $\tilde{f}_{k_0}$ is normalized pump to given supermode k_0. The term in square brackets depicts two-dimensional hybridized dispersion surface with different regions [e.g. anomalous-anomalous (anomalous-normal) in the vicinity $k = 0$ ($k_0 = N/2$)], the nonlinear term incorporates conservation laws in both frequency μ and spatial k dimensions. Therefore, the four-wave mixing processes occur in μ-k space, leading to the effective two-dimensional dynamics on the system.

3. Staggered coupling

Now, we can perform the first step towards the SSH model. We need to open a gap in the supermode dispersion which is realized by dimerazing the chain. By creating unit cells made of two resonators, we introduce intra-cell J_{intra} and inter-cell couplings J_{inter} as shown in Fig. 1(a) keeping the periodic boundary conditions. Now, we can introduce collective excitation amplitudes $\partial\psi_{\mu k}$ and $\phi_{\mu k}$ for both sublattices

$$\psi_{\mu k} = \frac{1}{2\pi\sqrt{N}}\int \sum_{\ell=1,3,5...} A_\ell e^{2\pi i(\ell k/N+\mu\varphi)}d\varphi,\ \phi_{\mu k} = \frac{1}{2\pi\sqrt{N}}\int \sum_{\ell=2,3,6...} A_\ell e^{2\pi i(\ell k/N+\mu\varphi)}d\varphi,$$

where N is the number of unit cells. In contrast to the previous case, now the dynamics is be described by a set of coupled equations on $\psi_{\mu k}$, $\phi_{\mu k}$

$$\frac{\partial \psi_{\mu k}}{\partial \tau} = -(1+i\zeta_0)\psi_{\mu k} - id_2\mu^2\psi_{\mu k} + i\big[J_{\text{intra}} + J_{\text{inter}}e^{-2\pi i k/N}\big]\phi_{\mu k} + \frac{i}{N}\sum_{\substack{k_1,k_2,k_3\\ \mu_1\mu_2\mu_3}} \psi_{\mu_1 k_1}\psi_{\mu_2 k_2}\psi^*_{\mu_3 k_3}\delta_{\mu_1+\mu_2-\mu_3-\mu}\delta_{k_1+k_2-k_3-k} + \delta_{k,k_0}\delta_{\mu,0}\tilde{f}^1_{k_0}, \quad (2)$$

$$\frac{\partial \phi_{\mu k}}{\partial \tau} = -(1+i\zeta_0)\phi_{\mu k} - id_2\mu^2\phi_{\mu k} + i\big[J_{\text{intra}} + J_{\text{inter}}e^{2\pi i k/N}\big]\psi_{\mu k} + \frac{i}{N}\sum_{\substack{k_1,k_2,k_3\\ \mu_1\mu_2\mu_3}} \phi_{\mu_1 k_1}\phi_{\mu_2 k_2}\phi^*_{\mu_3 k_3}\delta_{\mu_1+\mu_2-\mu_3-\mu}\delta_{k_1+k_2-k_3-k} + \delta_{k,k_0}\delta_{\mu,0}\tilde{f}^2_{k_0}. \quad (3)$$

According to Eq. (2), this system possesses two-dimensional dynamics as well, and the supermode dispersion can be obtained by diagonalizing the linear part of the system. The dynamics now incorporates interactions between two sublattices. Kerr nonlinearity acts within each of them, resulting into four-wave mixing processes on each of the dispersion surfaces. The coupling term can also lead to intraband mixing as shown in Ref. [4]. This property remains valid for open boundary conditions, where edge states can be found.

4. Nonlinearity at the edge state

We continue with qualitative analyses of a 20 resonator SSH chain. If $J_{\text{intra}} > J_{\text{inter}}$ edge states with eigenvalues in the center of the gap can appear. Such an example is presented in Fig. 1(a)

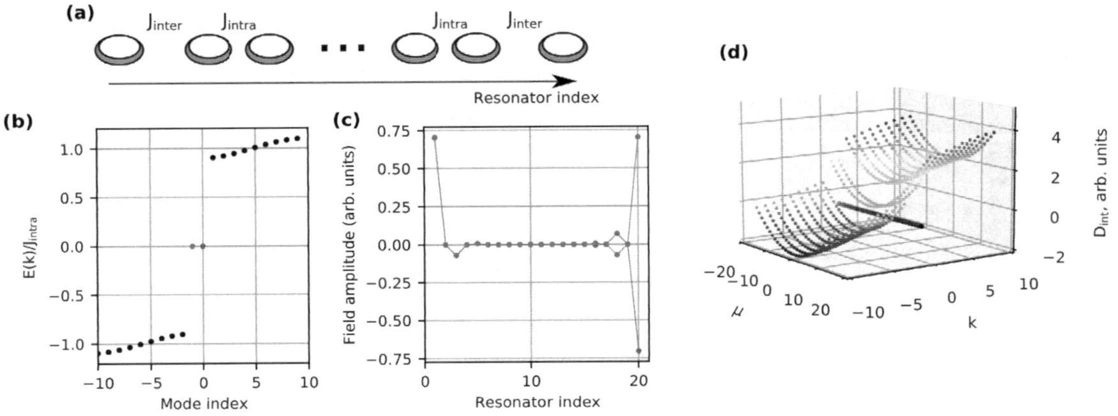

Figure 1. SSH model. (a) Sketch of a SSH chain consisting of optical microresonators. (b) Created band structure of spatial supermodes in case of 20 resonators. Red and green dots represent the edge states. (c) Spatial distribution of the edge modes. (d) Hybridized integrated dispersion of a 20 resonator chain with each resonator having anomalous dispersion. The parabolas in the gap correspond to the edge states. The black line schematically represents a soliton solution.

with corresponding single-mode band structure in Fig. 1(b) and spatial profile of the edge states in Fig. 1(c). Taking into account the frequency modes ω_μ and the group velocity dispersion, we obtain the hybridized dispersion to be of the form presented in Fig. 1(d). Since the edge parabolas are always at the center of the band gap, the four-wave mixing processes along k axes for fixed μ are suppressed and one can expect the nonlinear dynamics to be similar to single resonator case, leading to soliton solution to be just a line under the edge parabolas [6]. However, once a soliton state is achieved, the dispersionless line under the edge parabola [black line in Fig. 1(d)] crosses the lower band and the phase matching condition between the edge and bulk modes can be satisfied, leading to presence of 2D dynamics. We investigate this effect numerically.

We consider a chain of 10 identical resonators with physical parameters accessible in experiments: internal linewidth $\kappa_0/2\pi = 50$ MHz, group velocity dispersion $D_2/2\pi = 40$ MHz, coupling coefficients $J_{\text{intra}}/2\pi = 5$ GHz, $J_{\text{intra}}/2\pi = 1$ GHz. We numerically solve the set of linearly coupled LLEs similar to [4] by scanning the resonance from blue to red detuned sides. For the pump power in the bus waveguide 800 mW we observe rich nonlinear dynamics presented in Fig. 2. In the transmission trace (Fig. 2(a)) one can see dynamics qualitatively similar to single resonator dynamics. Indeed, most of the optical power is concentrated in the first resonator, and the transmission traces represent typical regions: Turing rolls, chaotic MI, soliton steps. However, the soliton step has irregular deviations of power due to generation of dispersive waves whose presence can be seen in other resonators. The corresponding spatio-temporal diagrams indeed show that the solitons in the first resonator experience a random walk [Fig. 2(b)], which is accompanied by the generation of light in other resonators [Fig. 2(c,d)].

In order to understand the four-wave mixing processes with regard to the linear band structure, it is useful to reconstruct the nonlinear dispersion relation (NDR) [6]. NDR for a given resonator reveals presence of collective linear and nonlinear structures [5]. We investigate closely the soliton regime corresponding to the step in Fig. 2(a) for 1st, 2nd and 10th resonators [Fig. 2(e-g)]. We observe ladder of tilted lines in the first resonator (Fig. 2(e)), which correspond to the breathing of the solitons and their random walk. In 2nd and 10th resonators we observe

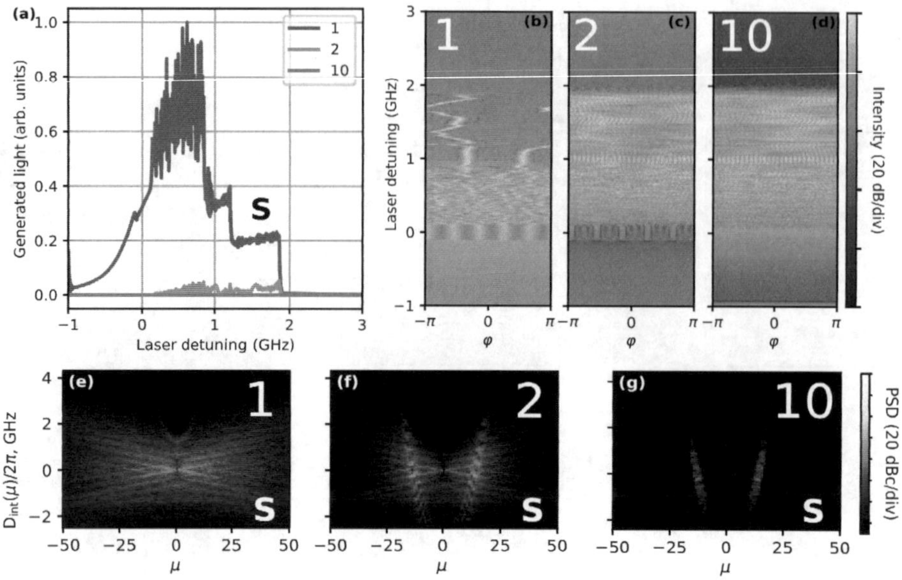

Figure 2. Scan of the edge state. (a) Transmission trace of the edge state. (b-d) Spatio-temporal dynamics with increased detuning in 1st, 2nd and 10th resonators. (e-g) Nonlinear dispersion relation reconstruction of a soliton state.

the predicted effect: the lines cross the lower bulk parabolas, achieving efficient phase matching, giving rise to photon transfer to the corresponding supermodes. Moreover, since the parabolas are symmetric, and there is a huge number of modes in the bulk which can be crossed, the soliton line can be tilted in both directions, leading to soliton random walk at the spatio-temporal diagram [Fig. 2(b)].

5. Conclusion

We have instigated nonlinear dynamics and dissipative Kerr soliton formation at the edge state of the Su-Schrieffer–Heeger model. The edge state dynamics in this system is similar to the single resonator case, but the presence of the bulk modes can perturb the solitonic state. If the gap width is such that the soliton spectrum does not overlap with the bulk modes, the soliton stability will not be perturbed. Alternatively, this can be achieved by choosing a resonator with a larger group velocity dispersion. If the spectral overlap is possible, the soliton provides a photon transfer to the bulk parabolas whose dynamics has two-dimensional nature. Our results reveal insights in nonlinear interactions in the simplest topological model in the nonlinear and multimode case, that can be further used and generalized for more complex models.

References

[1] Lu L, Joannopoulos J D and Soljačić M 2014 *Nature photonics* **8** 821–829
[2] Asbóth J K, Oroszlány L and Pályi A 2016 *The Su-Schrieffer-Heeger (SSH) Model* (Cham: Springer International Publishing) pp 1–22
[3] Kippenberg T J, Gaeta A L, Lipson M and Gorodetsky M L 2018 *Science* **361**
[4] Tikan A, Riemensberger J, Komagata K, Hönl S, Churaev M, Skehan C, Guo H, Wang R N, Liu J, Seidler P *et al.* 2021 *Nature Physics* 1–7
[5] Tusnin A K, Tikan A M, Komagata K and Kippenberg T J 2021 *arXiv preprint arXiv:2104.11731*
[6] Leisman K P, Zhou D, Banks J W, Kovačič G and Cai D 2019 *Phys. Rev. E* **100**(2) 022215

Light focusing by silicon nanosphere structures under conditions of magnetic dipole and quadrupole resonances

N Ustimenko[1], K V Baryshnikova[1], R Melnikov[1], D Kornovan[1], V Ulyantsev[1], and A B Evlyukhin[2,1]
[1]ITMO University, St. Petersburg 197101, Russia
[2]Leibniz Universitat Hannover, Hannover 30167, Germany

E-mail: nikita.ustimenko@metalab.ifmo.ru

Abstract. Metalens is a planar device for light focusing. In this work, we design and optimize c-Si nanosphere metalenses working on the magnetic dipole and quadrupole resonances of the c-Si nanoparticle. Resonant optical response of c-Si nanostructures is simulated by the multipole decomposition method along with the zero-order Born approximation. Limitations of this approach are investigated. The obtained results of optimization are verified by simulation via the T-matrix method.

1. Introduction

The structures of many subwavelength particles for light focusing, namely metalenses, are considered by researchers as the perspective elements for future optical devices [1, 2]. However, to design widely-functional metalenses with the desired characteristics, advanced optimization techniques are needed [2, 3]. In this case, speed and efficiency of optimization are determined by a physical model describing a collective response of metalens elements, typically arranged in rings. In this work, we consider metalenses composed of c-Si nanosphere rings. We suggest the model when the optical response of the nanoparticle is associated with several multipoles while the collective response of nanoparticles structure is described in the zero-order Born approximation [4]. We find the limitations of our approach and apply an evolutionary algorithm [5] for optimization of ultra-thin Si nanosphere metalenses working at the magnetic dipole and quadrupole resonances of the nanoparticle [6, 7]. The accuracy of the results obtained in the Born approximation is verified by comparison with the rigorous T-matrix method [8].

2. Results and Discussion

We consider the structures consisting of N c-Si nanospheres with a diameter of 200 nm. According to Fig. 1(a), the optical response of single particle can be associated with the dipoles and quadrupoles of electric and magnetic types. The interaction of multipoles corresponding to the different nanoparticles can be taken into account under the coupled multipole model (see Eq. (25) in [11]):

$$\mathbf{Y} = \mathbf{Y}_0 + \hat{\mathbf{V}}\mathbf{Y}, \tag{1}$$

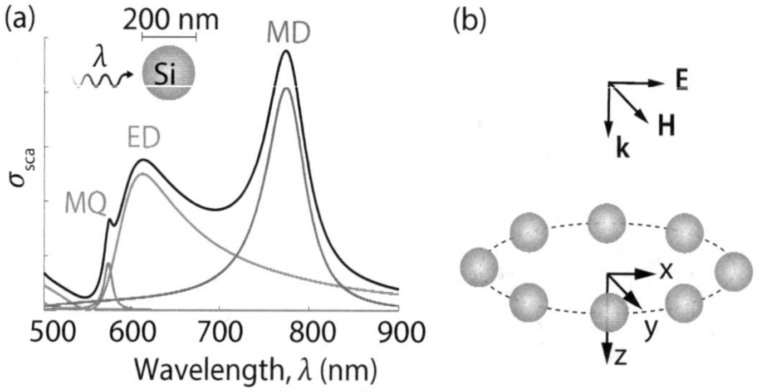

Figure 1. (a) Scattering spectrum of a silicon nanosphere having a diameter of 200 nm calculated from the Mie-theory [9]. Red, orange, blue, and green lines correspond to the MD, ED, MQ, and EQ contributions, respectively. Dispersion of c-Si is taken into account [10]. (b) Schematic view of a single nanosphere ring with the shown incident wave and Cartesian coordinates system.

where $\mathbf{Y}$ is the vector of self-consisted multipole moments:

$$\mathbf{Y} = \left[p_x^1, ..., p_z^N, m_x^1, ..., m_z^N, Q_{xx}^1, ..., Q_{zz}^N, M_{xx}^1, ..., M_{zz}^N\right]_{24N}^T, \tag{2}$$

$\mathbf{Y}_0$ is the vector of moments excited only by the incident wave:

$$\mathbf{Y}_0 = \left[p_{0,x}^1, ..., p_{0,z}^N, m_{0,x}^1, ..., m_{0,z}^N, Q_{0,xx}^1, ..., Q_{0,zz}^N, M_{0,xx}^1, ..., M_{0,zz}^N\right]_{24N}^T, \tag{3}$$

$\hat{\mathbf{V}}$ is the $24N \times 24N$ interaction matrix composed of Green's tensors blocks.

Solving Eq. (1), we can find the self-consisted multipole moments of nanoparticles:

$$\mathbf{Y} = (\hat{\mathbf{U}} - \hat{\mathbf{V}})^{-1}\mathbf{Y}_0, \tag{4}$$

if $\det(\hat{\mathbf{U}} - \hat{\mathbf{V}}) \neq 0$, where $\mathbf{U}$ is the $24N \times 24N$ unity matrix. If the electromagnetic coupling between particles is weak enough, it can be neglected, which corresponds to the zero-order Born approximation (ZBA). The ZBA multipole moments of nanoparticles

$$\mathbf{Y} = \mathbf{Y}_0 \tag{5}$$

connect with the incident fields $\mathbf{E}_0$ and $\mathbf{H}_0$ through the Mie-polarizabilities [11]:

$$\mathbf{p}_0^j = \alpha_p \mathbf{E}_0(\mathbf{r}_j), \tag{6}$$

$$\mathbf{m}_0^j = \alpha_m \mathbf{H}_0(\mathbf{r}_j), \tag{7}$$

$$\hat{Q}_0^j = (\alpha_Q/2)\left[\nabla \mathbf{E}_0(\mathbf{r}_j) + \mathbf{E}_0(\mathbf{r}_j)\nabla\right], \tag{8}$$

$$\hat{M}_0^j = (\alpha_M/2)\left[\nabla \mathbf{H}_0(\mathbf{r}_j) + \mathbf{H}_0(\mathbf{r}_j)\nabla\right], \tag{9}$$

Knowing the multipole moments, we can calculate the total electric $\mathbf{E}$ and magnetic $\mathbf{H}$ fields in the system (using Eqs. (13)-(16) from [11]), and the electromagnetic intensity $\mathrm{I}(\mathbf{r})/\mathrm{I}_0 = (|\mathbf{E}(\mathbf{r})|^2 + Z^2|\mathbf{H}(\mathbf{r})|^2)/2E_0^2$ where $\mathbf{r}$ is the observation point, Z is the free-space impedance, E_0 is the amplitude of electric field in the incident plane wave.

Using the ZBA, we study light focusing by a single ring of Si nanospheres shown in Fig. 1(b). Due to the axial symmetry, the ring has a focus on its axis (here z-axis). Hence, ring focal

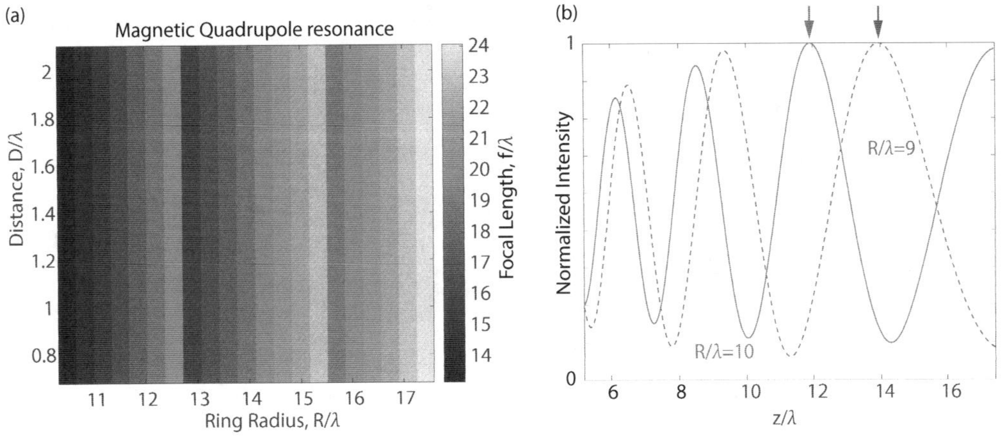

Figure 2. Features of light focusing at the MQ resonance $\lambda = 574$ nm. (a) Focal length of single nanosphere ring as a function of the ring radius and center-center distance between particles. (b) Normalized (on the maximum) electromagnetic intensity distribution along the z-axis for the ring radius $R/\lambda = 9$ (blue dashed line) and $R/\lambda = 10$ (red solid line). The number of particles in both rings $N = 32$. The arrows indicate the focal lengths of rings. The results are obtained in the ZBA.

length is defined as the distance between the ring plane and the position of the highest intensity peak on the ring axis. Dependence of the ring focal length on the ring parameters for the MQ resonance is shown in Fig. 2(a). The ring focal length is almost completely determined by its size. Moreover, this dependence is non-monotonic due to the interference origin of the intensity peaks. As a result of constructive interference between the incident wave and the scattered waves, the bigger ring may have a shorter focal length [see focal lengths in Fig. 2(b)]. Similar behavior of the focal length is observed at the MD resonance.

The focal length and focal intensity are the basic parameters of metalens. The error of calculating these quantities in the ZBA was studied for the single ring at the MD and MQ resonances. This error was defined as follows: $\Delta^{(V)} = 100\% \times |V^{\mathrm{CMM}} - V^{\mathrm{ZBA}}|/V^{\mathrm{CMM}}$, where V^{CMM} and V^{ZBA} are values of V calculated using moments (4) and (5), respectively. We calculate errors of the ring focal length $\Delta^{(f)}$ and intensity $\Delta^{(\mathrm{I})}$ as functions of ring radius 6 μm $\leq R \leq 10$ μm and distance between centers of particles 200 nm $\leq D \leq$ 1200 nm. We obtain that the focal length error $\Delta^{(f)} \leq 2.5\%$ at the both resonances for all ring parameters; the focal intensity error $\Delta^{(\mathrm{I})} \lesssim 10\%$ when the inter-particle distance $D \geq 0.66\lambda$ and $D \geq 0.92\lambda$ at the MQ and MD resonances, respectively. We will call these distance values limiting. Hence, to simulate Si nanosphere metalenses via the ZBA-based approach, the inter-particle distance should be higher than limiting.

Based on the ZBA calculation scheme is directly comparable with the optimization algorithm. To optimize metalenses, we choose an evolutionary SEMO algorithm [5]. We use this algorithm to optimize metalenses at the MQ and MD resonances. The objective functions of optimization are both desired focal length (here $f = 5$ μm) and high focal intensity. During the optimization, the algorithm varies the number of rings, their radii, and the number of particles in the ring. The fixed parameters: diameter of Si nanoparticle is 200 nm, metalens size to 20 μm, minimal inter-particle distance is equal to $1.11\lambda_{\mathrm{MQ}}$ and $1.07\lambda_{\mathrm{MD}}$. The optimized metalenses with $f = 5$ μm and their intensity profiles are shown in Fig. 3.

We compare intensity profiles, obtained in ZBA, with the ones calculated by the T-matrix method [8] and obtain a good agreement. Both methods determine the same focal length equalled to 5 μm and 4.9 μm for MQ and MD resonances, while the error of focal intensity: 5.7% and 3%, respectively.

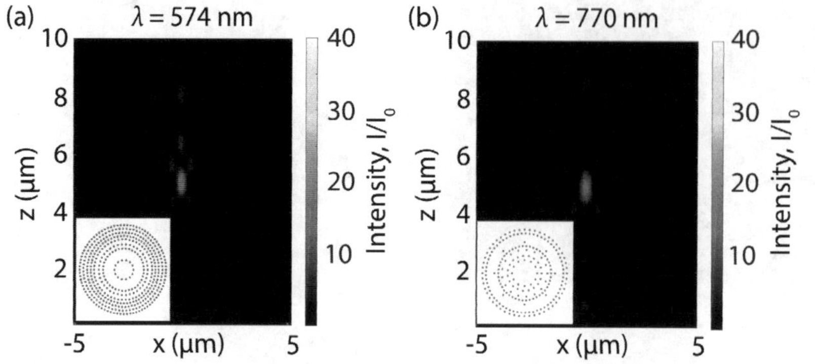

Figure 3. Normalized electromagnetic intensity distribution in the xz-plane for metalenses optimized at the MQ (a) and MD (b) resonances (I_0 is the intensity of incident wave). Both metalenses have the focal length $f = 5$ μm. In insets: distribution of nanoparticles in the xy-plane for optimized metalenses. The results are obtained in the ZBA.

3. Conclusion

We studied light focusing by c-Si nanosphere rings. The optical response of such structures was described using the dipole-quadrupole model along with the zero-order Born approximation. We showed that this approach provides a low-error simulation of the resonant optical response if the distance between particles $\gtrsim \lambda_0$. We optimized positions of Si nanospheres, using an evolutionary algorithm, and obtained effective metalenses with the desired focal lengths, working at the magnetic dipole and quadrupole resonances of the single nanosphere. The results of optimization in ZBA demonstrate a good agreement with the T-matrix method simulation.

4. Acknowledgments

K.V.B., and D.K. acknowledge support from the Foundation for the Advancement of Theoretical Physics and Mathematics "BASIS".

References

[1] Lalanne P and Chavel P 2017 *Laser Photonics Rev.* **11** 1600295
[2] Chen W T, Zhu A Y and Capasso F 2020 *Nat. Rev. Mater.* **5** 604–620
[3] Yao K, Unni R and Zheng Y 2019 *Nanophotonics* **8** 339–366
[4] Keller O, Xiao M and Bozhevolnyi S 1993 *Surf. Sci.* **280** 217–230
[5] Kalyanmoy D 2002 *Multi-Objective Optimization using Evolutionary Algorithms* (John Wiley & Sons, LTD, New York, USA)
[6] Kuznetsov A I, Miroshnichenko A E, Brongersma M L, Kivshar Y S and Luk'yanchuk B 2016 *Science* **354** aag2472
[7] Evlyukhin A B, Novikov S M, Zywietz U, Eriksen R L, Reinhardt C, Bozhevolnyi S I and Chichkov B N 2012 *Nano Lett.* **12** 3749–3755
[8] Egel A, Pattelli L, Mazzamuto G, Wiersma D S and Lemmer U 2017 *J. Quant. Spectrosc. Radiat. Transf.* **199** 103–110
[9] Bohren C F and Huffman D R 1983 *Absorption and Scattering of Light by Small Particles,* (Wiley, New York)
[10] Aspnes D E and Studna A A 1983 *Phys. Rev. B* **27** 985
[11] Babicheva V E and Evlyukhin A B 2019 *Phys. Rev. B* **99** 195444

Application of Born series for modeling of Mie-resonant nanostructures

N Ustimenko[1], D Kornovan[1], K V Baryshnikova[1], A B Evlyukhin[2,1], and M Petrov[1]

[1]School of Physics and Engineering, ITMO University, St. Petersburg 197101, Russia
[2]Leibniz Universitat Hannover, Hannover 30167, Germany

E-mail: nikita.ustimenko@metalab.ifmo.ru

Abstract. Born series formalism is a widely-used approach to solve a scattering problem in quantum mechanics and optics, including a problem of electromagnetic scattering on the ensembles of Mie-resonant nanoparticles. In the latter case, the Born series formalism can be used when the electromagnetic coupling between nanoparticles is weak. This can be violated near the multipole Mie-resonance of the nanoparticle. In this work, we analyze the applicability of the Born series approach for modeling the resonant optical response of Mie-nanoparticle ensembles and formulate quantitative criteria of Born series convergence and, subsequently, the applicability of this approach.

1. Introduction

The Born series formalism is a method to simulate scattering by a group of many nanoparticles in a perturbative manner [1, 2, 3, 4, 5, 6]. This method is based on constructing a convergent Born series and replacing it with a finite sum that successively approximate the interaction between particles where the accuracy depends on the number of terms included in the sum (i.e., on the Born approximation order). Born approximations of different orders have been used to simulate tip-substrate interaction [1, 2], calculate polarizability of a non-spherical particle [3, 4], and optimize metalens design [6]. All these problems require a simulation of the interaction in non-periodic structures of a large number of particles which is non-effective to perform, using full-wave simulation techniques; therefore, the Born series method, along with the coupled multipole model, [7, 8] was chosen. The applicability of the Born series method, as well as its convergence, are determined by the strength of electromagnetic coupling in the system. To the best of our knowledge, there is no exhaustive physical analysis of Born series applicability to simulate Mie-resonant nanostructures [9, 10] with quantitative criteria of the series convergence. In this work, we try to give such analysis and criteria and consider a nanosphere dimer having a simple analytical solution in the framework of the coupled dipole model.

2. Results and Discussion

We investigate the resonant scattering of a monochromatic electromagnetic plane wave $\mathbf{E}_0 e^{i\mathbf{kr}-i\omega t}$ on a dimer of identical nanospheres supporting dipole Mie-resonances at the optical range (as shown in Fig. 1(a)) [9, 10]. We use the electric dipole approximation of nanoparticle's optical response admissible at the electric dipole resonance of nanoparticle. The self-consisted dipole moments of nanoparticles can be calculated using the coupled dipole model (CDM) [7]:

$$\begin{aligned}\mathbf{p}^1 &= \alpha(\omega)\mathbf{E}_0 + \alpha(\omega)k^2\varepsilon_0^{-1}\hat{\mathbf{G}}^{12}\mathbf{p}^2,\\ \mathbf{p}^2 &= \alpha(\omega)\mathbf{E}_0 + \alpha(\omega)k^2\varepsilon_0^{-1}\hat{\mathbf{G}}^{21}\mathbf{p}^1.\end{aligned} \tag{1}$$

where $\alpha(\omega)$ is the dipole polarizability, ε_0 is the vacuum permittivity, $k = 2\pi/\lambda$ is the free-space wavenumber, $\hat{\mathbf{G}}^{ij} \equiv \hat{\mathbf{G}}(\mathbf{r}_i, \mathbf{r}_j)$ is the dyadic dipole Green's function [7]. Here and further, we omit the time-dependence of fields and dipole moments. In accordance to Fig. 1(a), the incident wave propagates along the z-axis with a polarization along the x-axis $\mathbf{E}_0 = [E_0, 0, 0]^T$ (T-polarization) or y-axis $\mathbf{E}_0 = [0, E_0, 0]^T$ (L-polarization). The nanoparticles are placed at the positions $\mathbf{r}_1 = [0, +D/2, 0]$ and $\mathbf{r}_2 = [0, -D/2, 0]$, therefore dot product $\mathbf{k}\cdot\mathbf{r}_i = 0$ $(i = 1, 2)$ in Eq. (1).

In Eq. (1), the dipole moment of the nanoparticle determines by the external field (the first term) and the dipole field generated by another particle (the second term). Due to the structure symmetry, both dipole moments have only one non-zero component along the incident field:

$$p_l^i = \frac{\alpha(\omega)E_0}{1 - \alpha(\omega)k^2\varepsilon_0^{-1}G_{ll}^{12}}, \quad i = 1, 2. \tag{2}$$

where $l = x$ for T-polarization, and $l = y$ for L-polarization. The corresponding elements of Green's function:

$$G_{xx}^{12} = \frac{e^{ikD}}{4\pi D}\left[1 + \frac{\mathrm{i}}{kD} - \frac{1}{k^2D^2}\right], \quad G_{yy}^{12} = \frac{e^{ikD}}{4\pi D}\left[-\frac{2\mathrm{i}}{kD} - \frac{4}{k^2D^2}\right]. \tag{3}$$

Dipole moments calculating by Eq. (2) rigorously takes into account the electromagnetic coupling between two dipoles. We can also approximate the coupling expanding moments (2) in the Born series. In this case, the Born series is an expansion of the denominator of (2) in geometric series:

$$p_l^i = \alpha(\omega)E_0\sum_{n=0}^{\infty}\left(\alpha(\omega)k^2\varepsilon_0^{-1}G_{ll}^{12}\right)^n, \quad i = 1, 2. \tag{4}$$

The solution of Eq. (1) in the Born approximation of m-th order is given by transition from an infinite series to a finite sum in Eq. (4), i.e., $\sum_{n=0}^{\infty} \to \sum_{n=0}^{m}$.

We can use the Born series approach if the Born series converges. The convergence criteria of geometric series such as Eq. (4) is well-known:

$$|\alpha(\omega)\cdot k^2\varepsilon_0^{-1}G_{ll}^{12}| < 1. \tag{5}$$

Electric-dipole polarizability of a homogeneous nanosphere can be expressed through the Mie-coefficient $a_1(nkR)$ [7, 11], depending on the particle refractive index n and radius R, and incident wavelength:

$$\alpha(\omega) = \mathrm{i}\frac{6\pi\varepsilon_0}{k^3}a_1, \tag{6}$$

then the convergence criteria (5) can be formulated as follows:

$$6\pi\cdot|a_1|\cdot|k^{-1}G_{ll}^{12}| < 1. \tag{7}$$

Non-absorptive particles ($\varepsilon'' = 0$). In this case, the resonance condition provides [12]:

$$a_1 = 1, \tag{8}$$

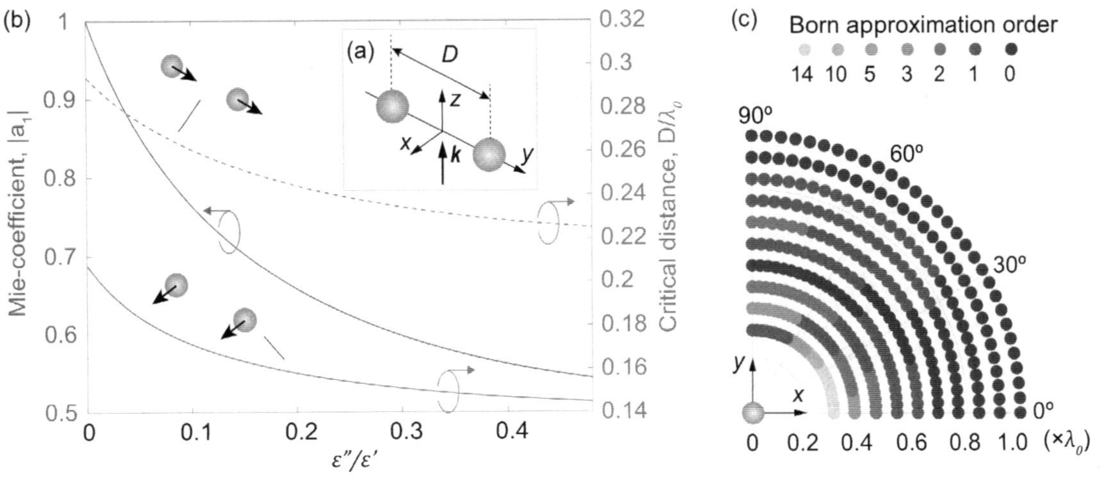

Figure 1. Analysis of the applicability of the Born series approach. (a) Schematic view of identical nanosphere dimer embedded in a vacuum with the marked wavevector of the external incident wave. Dipole moments with the orientation along the y-axis or x-axis are excited by y-polarized (L-polarized) or x-polarized (T-polarized) external wave, respectively. (b) The absolute value of Mie-coefficient (blue line) and critical center-center distance between particles (red lines) as functions of particle material ε'' (where $\varepsilon' = 12.5$). Red solid and red dashed lines correspond to T- and L-polarization of dipole moments, respectively. The Born series converges at the dipole Mie-resonance $\lambda = \lambda_0$ (where $\lambda_0 = 555$ nm) if the distance between particles D is higher than the critical one. (c) The Born approximation order as a function of the second particle position while the first particle is placed at the coordinates' origin. The position of the circle point indicates the second particle position, and its color corresponds to the order of Born approximation that allows calculating the extinction cross section in this approximation with an error (10) of less than 10%. The distance between particles is measured in resonant wavelength $\lambda_0 = 1140$ nm. The minimal inter-particle distance is $D = 0.3\lambda_0$, radius of particles $R = 100$ nm, permittivity of particle material $\varepsilon' = 64$, $\varepsilon'' = 0$. The normally incident plane wave is x-polarized.

corresponding to the resonance wavelength λ_0. By substitution of resonance condition (8) and expressions (3) in inequality (7), we obtain at-resonance convergence criteria of the Born series:

$$6\pi \cdot |k_0^{-1} G_{ll}^{12}| < 1, \tag{9}$$

explicitly independent on the particle refractive index and size (but the resonant wavelength depends). From (9), we find the *critical* distances for L- and T-polarization: $D^{\mathrm{L}} = 0.3\lambda_0$ and $D^{\mathrm{T}} = 0.21\lambda_0$, respectively. A word *critical* highlights that the Born series diverges when the center-center distance between particles is less than critical.

Absorptive particles ($\varepsilon'' \neq 0$). For the particles with absorption (Ohmic) losses, a condition (8) is not valid. The criteria of Born series convergence (7) depends on the absolute value of the Mie-coefficient $|a_1|$. As shown in Fig. 1(b), at-resonance value $|a_1|$ decreases as imaginary part of permittivity increases. Solving the inequality (7) for varying ε'', we find the critical distances and also show their in Fig. 1(b). It is seen that, for both T- and L-polarizations, the critical distances decrease as the absorption losses increase.

Knowing the critical distances between particles, we investigate the accuracy of calculating resonant optical response via the Born series approach. We put the first particle in the origin of the Cartesian coordinate system and vary the position of the second particle. The distance between two particles is higher than $D = 0.3\lambda_0$. The error of the m-order Born approximation is defined as the relative error of the extinction cross section (ECS):

$$\Delta = \frac{|\sigma^{(m)} - \sigma^{(\mathrm{CDM})}|}{\sigma^{(\mathrm{CDM})}} \times 100\%, \tag{10}$$

where $\sigma = k_0\varepsilon_0^{-1}|E_0|^{-2}\mathrm{Im}\left[\sum_{i=1}^{2} \mathbf{E}_0^*(\mathbf{r}_i) \cdot \mathbf{p}^i\right]$ is the ECS of resonant dimer in the dipole approximation [7], $\sigma^{(\mathrm{CDM})}$ and $\sigma^{(m)}$ is ECS calculated using the rigorous dipole moments (2) and dipole moments in the m-th Born approximation (4), respectively. We set a limit of 10% on the error value Δ and find the Born approximation order when a condition $\Delta \leq 10\%$ is satisfied. The results are presented in Fig. 1(c) for the dimer of non-absorptive nanoparticles. We see in Fig. 1(c) that particles separated higher than λ_0 can be considered as non-interacting because their collective response is accurately described in the zero-order Born approximation. This because a strong near-field coupling between particles becomes weak for inter-particle distances $D \gtrsim \lambda_0$.

3. Conclusion

We analyzed when and how the Born series approach can be used to accurately simulate a resonant optical response of nanoparticle structures. Studying a dimer of the nanospheres under the electric dipole approximation, we derive convergence criteria of the Born series, independent of the particle size and refractive index for non-absorptive particles.

4. Acknowledgments

The work was supported by the Russian Science Foundation (project No. 21-79-10190).

References

[1] Labani B, Courjon D, Labeke D V and Girard C 1990 *J. Opt. Soc. Am. B* **7** 936
[2] Keller O, Xiao M and Bozhevolnyi S 1993 *Surf. Sci.* **280** 217–230
[3] Fan Z H, Wang D X, Chen R S and Yung E K N 2006 *Microw. Opt. Techn. Lett.* **48** 1741–1746
[4] Singham S B and Bohren C F 1988 *J. Opt. Soc. Am. A* **5** 1867–1872
[5] Bereza A S, Nemykin A V, Perminov S V, Frumin L L and Shapiro D A 2017 *Phys. Rev. A* **95** 063839
[6] Ustimenko N A, Baryshnikova K V, Melnikov R V, Kornovan D F, Ulyantsev V I, Chichkov B N and Evlyukhin A B 2021 Multipole optimization of light focusing by silicon nanosphere structures (*Preprint* `arXiv:2103.01482`)
[7] Evlyukhin A B, Reinhardt C, Seidel A, Luk'yanchuk B S and Chichkov B N 2010 *Phys. Rev. B* **82** 045404
[8] Babicheva V E and Evlyukhin A B 2019 *Phys. Rev. B* **99** 195444
[9] Evlyukhin A B, Novikov S M, Zywietz U, Eriksen R L, Reinhardt C, Bozhevolnyi S I and Chichkov B N 2012 *Nano Lett.* **12** 3749–3755
[10] Kuznetsov A I, Miroshnichenko A E, Brongersma M L, Kivshar Y S and Luk'yanchuk B 2016 *Science* **354** aag2472
[11] Bohren C F and Huffman D R 1983 *Absorption and Scattering of Light by Small Particles,* (Wiley, New York)
[12] Tribelsky M I and Luk'yanchuk B S 2006 *Phys. Rev. Lett.* **97** 263902

Problems of sub-THz astronomy, radars and telecommunications: development of equipment and methods

Gregory Bubnov, Aleksandra Gunbina, Fedor Kovalev, Ilya Lesnov, Kirill Mineev, Sergey Vdovichev and Vyacheslav Vdovin

IAP RAS, 46, Ulyanov st., Nizhniy Novgorod, Russia, 603600

vdovin@ipfran.ru

Abstract. Some original results of development methods and equipment for subterahertz astronomy, radars and telecommunications are presented in this paper. There are problems of extremely low sensitivity of receivers, high power transmitters, atmosphere propagation and development of key solid state components for this band: detectors arrays, cooling systems, antennas and so on.

1. Introduction

In recent years, there has been a boom in the development of the terahertz range. To be more precise we are talking about its long-wave part from 0.1 to 1.0 THz called sub - THz waves. Sub - THz astronomy, radar and telecommunications have been widely developed last decade [1]. The talk presents the main approaches to solving the most critical problems of sub-THz waves: influence of the atmosphere and low signal / noise ratio in majority of sub-THz applications

2. Results

Among problems of sub-THz waveband development, a key place is occupied by the significant atmospheric absorption of these waves and the study of atmospheric absorption or microwave astroclimate [2-4], as well as the selection of adequate sites for the placement of radio telescopes and antennas for deep space communications, as well as radars effective for diagnosing of space debris using very high power gyrotrons and low noise receivers [5]. Some of the results of astroclimate research, as well as specially developed equipment and methods, are presented: solid state components for sub - THz band including superconducting detectors arrays, cryogenically cooling systems and interfaces, components of Sub THz antennas and so on.

The main task here was solution of critical problems listened above for relatively large antennas (13-70 m) assumed to be fabricated for Suffa observatory [2]. Suffa project now is multipurpose observatory with astronomy, telecon and radar applications. The development of extremely low noise receivers and high power transmitters is presented, cooled to deep cryogenic temperatures including subK levels, is presented in the talk [6]. Some technical approaches how to solve a problem of phase manipulation for sub-THz communications are also presented.

3. Conclusion

As a result of studies based on direct 2 and 3 mm measurements, as well as comparisons with the results of GPS and optical measurements, we've selected most promising sites for Suffa antennas and came to the conclusion that it is necessary to fabricate the receiver for direct measurements of the astroclimate at wavelengths of 1.3 and 0.8 mm and the total futility of subsequent transparency windows of 0.55 mm and further tasks of ground astronomy and radars. There is certain perspective for short-range communication (mobile 7G) with a distance between cell elements up to hundreds of meters up to 700 GHz waves. Methods for parrying a high level of absorption of sub-THz waves for communication, radar and astronomy problems through the use of powerful generators and superconducting receivers are proposed and investigated. A prototype of a 3 mm receiver based on commercial LNA from Farran was demonstrated, and a comparison was made with a two-frequency cooled receiver of 13 and 8 mm range. The development of microwave technologies has pushed the frontier of the promising application of cryogenic receivers to the sub-THz wave boundary. At lower frequencies, ground-based radar, communications and astronomy systems are dominated by the atmosphere and there is no point in deep cryogenic cooling. Despite the fact that the atmosphere at sub THz waves is less transparent than at long mm waves, cryogenic cooling of the equipment gives a more significant effect due to a priori higher noises of the less perfect uncooled receiving equipment of sub THz waves, which are still far from the quantum limits. The level of the quantum limit here is higher.

4. Acknowledgements

This work is supported by RSF grant #19-19-00499

References

[1] Vdovin V.F., Lesnov I.V. et al. Development and research of astronomical and telecommunication equipment of the Sub-terahertz frequency range. V International Conference on Metamaterials and Nanophotonics (METANANO2020) 14 - 18 September, 2020, AIP Conference Proceedings. Scopus, WoS. https://metanano.itmo.ru/sites/default/files/call-for-papers/METANANO2020_fullprogram.pdf

[2] Artemenko, Y.N., Balega, Y.Yu., Baryshev, A.M., Zinchenko, I.I., Vdovin, V.F. New stage of the Suffa submm observatory in Uzbekistan project ISSTT 2019 - 30th International Symposium on Space Terahertz Technology, Proceedings Book, 2019, c. 196-201, http://www.scopus.com/inward/record.url?eid=2-s2.0-85081100409&partnerID=MN8TOARS

[3] Grigoriy Bubnov, Vyacheslav Vdovin, Peter Zemlyanukha, and Igor Zinchenko Observations of astro-climate with the broad band radiometer using the atmospheric dip method Proc. of 5th International Conference on Metamaterials and Nanophotonics METANANO 2020, AIP Conference proc. 14 - 18 September, 2020, AIP Conference Proceedings. https://metanano.itmo.ru/sites/default/files/call-for-papers/METANANO2020_fullprogram.pdf

[4] Grigoriy M. Bubnov, Andrey S. Marukhno, Marat G. Mingaliev, Anastasia P. Markova, Nikolay I. Shatsky, Olga V. Voziakova, Vyacheslav F. Vdovin, Peter M. Zemlyanukha, Igor I. Zinchenko Millimeter-wave astroclimate studies in the North Caucasus: Expedition and first results Proc. of Terahertz and Microwave Radiation: Generation, Detection and Applications (TERA-2020), Tomsk Russia, 24-26 Aug 2020, p.54 The Proc. of SPIE Scopus, Ei Compendex, Inspec, International Aerospace Abstracts, Index to Scientific & Technical Proceedings, Physics Abstract: http://tera2020.tsu.ru/wp-content/uploads/2020/08/TERA_2020_abstracts.pdf

[5] Vyacheslav Vdovin; Ilya Lesnov; Grigoriy Bubnov Sub-terahertz Data Channels: Evaluation of Perspectives ICECCE 2020 doi: 10.1109/ICECCE49384.2020.9179204 Date of Conference: 12-13 June 2020

[6] Parshin V., Bubnov G, Gunbina A. et al. Terahertz Reflectivity of $YBa_2Cu_3O_{7-\delta}$ at Cryogenic Temperatures. IEEE Transactions on Applied Superconductivity, Vol. 30, No. 8, December 2020 DOI: 10.1109/TASC.2020.2994964

Laser-assisted formation of luminescent domains in metal- or semiconductor-doped silicate and phosphate glasses

A.S. Lipatiev, G.Yu. Shakhgildyan, M.P. Vetchinnikov, S.V. Lotarev, V.N. Sigaev

Mendeleev University of Chemical Technology, Miusskaya Sq. 9, Moscow, 125047, Russia

E-mail: vetchinnickov.maxim@yandex.ru

Abstract. In this study, silicate and phosphate glasses doped with Ag or CdS were exposed to femtosecond laser pulses and photoluminescence properties of the laser-written domains were investigated. Laser writing in phosphate glass doped with CdS was found to induce very weak photoluminescence, while laser-written domains in silicate glass had a comparatively high photoluminescence intensity, that was assigned to the formation of the sulphur vacancies in the CdS nanocrystals precipitated under the ultrafast laser pulses. Observed photoluminescence bands in Ag-containing glasses we assigned to the formation of different silver nanospecies which provide photoluminescence bands with the maxima at 685 and 600 nm in Ag-doped silicate and phosphate glasses, respectively.

1. Introduction

Glasses doped with noble metals as well as semiconductor nanocrystals have been known for a long time but still play a significant role in the development of materials with enhanced optical properties for light-emitting diodes, optical fiber amplifiers, modulators, sensors, etc. [1, 2]. Moreover, space-selective modification of optical properties in these glasses using ultrafast laser pulses is a very attractive technique paving the way for the development of optical memory based on photoluminescent emission from the laser-written domains. Domains induced by the stationary focused ultrafast laser beam are typically represented as pipe-shaped structures elongated along the laser writing direction formed during multi-photoionization and subsequent rearrangement of the glass structure [3, 4].

Many studies were performed in the field of direct laser writing in glasses by functional additives. For example, the possibility of laser-induced Ag cluster precipitation, as well as Ag nanoparticles formation in silver-doped phosphate glass, was proposed to use for the optical data recording [5] and single-mode waveguide fabrication [6, 7]. The space-selective precipitation of CdS or CdSe quantum dots in glasses under femtosecond laser irradiation was also studied [8, 9]. Nevertheless, the development of a reliable functionalized glass suitable for efficient direct laser writing is still remaining a challenge and a search for the best combination of a glassy matrix and a dopant is thus going on.

Here, we studied and compared the photoluminescence (PL) properties of laser-induced domains in two different glass matrices (silicate and phosphate) doped with Ag or CdS.

Journal of Physics: Conference Series **2015** (2021) 012163 doi:10.1088/1742-6596/2015/1/012163

2. Experimental

Two glass matrices were selected for doping including silicate glass in the K_2O-ZnO-B_2O_3-SiO_2 system (labelled as OS) and phosphate glass in the ZnO-P_2O_5 system (labelled as PZ). Both silicate and phosphate glasses were doped separately with Ag or CdS and hereafter are referred to, respectively, as OS-Ag, OS-CdS, PZ-Ag and PZ-CdS depending on glass matrix and dopant (Ag_2O – 0.03 mol.% for OS glass and 1 mol.% for PZ glass; CdS – 1 mol.% for OS and PZ glasses). High purity grade reagents (SiO_2, H_3BO_3, ZnO, K_2CO_3, P_2O_5, CdS, $AgNO_3$) were used for glass batch preparation. Glass samples were fabricated using the melt-quenching process according to the technique described elsewhere [10]. It is notable that we did not succeed in producing transparent OS glass with the content of silver over 0.03 mol.%.

Glass samples of 10 x 10 x 3 mm size were polished into plane-parallel plates for laser irradiation experiments. An experimental setup based on the Pharos SP femtosecond laser system with a central wavelength of 1030 ± 2 nm tuned to a pulse repetition rate of 100 kHz and a pulse duration of 180 fs was used for laser writing experiments. The laser beam was focused into the glass sample by an Olympus microscope objective (20X, N.A. = 0.45) at a depth of ~150 μm under the surface. The pulse energy was set to 200 nJ, which provided the most luminous domains in all investigated glasses. 10^6 laser pulses were deposited to each irradiated domain. The glass samples were positioned by means of a three-coordinate motorized air-bearing translation stage (Aerotech ABL1000) synchronized with the laser.

PL spectra and integral intensity distribution maps of the laser-written domains were recorded using an NTegra Spectra nanolaboratory (NT-MDT) with the 488 nm argon laser beam focused with Olympus microscope objective (40X, N.A. = 0.75) as excitation source and a scanning step of 0.8 μm.

3. Results and discussion

Fig. 1 presents typical PL integral intensity distribution maps of the domain laser-written in OS-CdS glass sample in the top view configuration. In general, observed PL appears due to the formation of Ag or CdS nanoclusters. A ring shape of the PL zone originates from the Gaussian distribution of the incident energy in the laser writing beam: the highest energy is in the middle, and it decreases to the edges. Therefore, one can see the highest PL intensity in the edges of the ring where the conditions of metal or semiconductors luminescent nanospecies are the most favorable [5, 8]. In order to compare PL spectra from the four domains in different glasses, we extracted PL spectra from the places with the highest PL intensity.

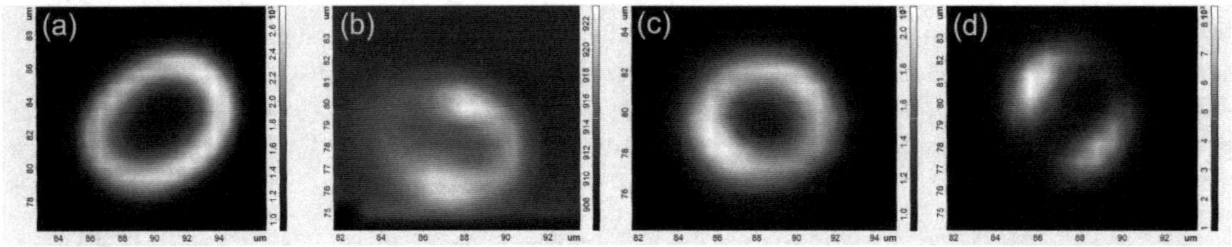

Figure 1 – Integral intensity map of PL in the range 520-800 nm registered for the top view of the domains laser-written in OS-CdS (a), PZ-CdS (b), OS-Ag (c) and PZ-Ag (d) glass samples

Fig. 2(a) presents PL spectra recorded from the irradiated domains in the PZ-CdS and OS-CdS glass samples. PL of the domain in CdS-doped PZ glass is almost imperceptible, which is expressed in high noise of its normalized PL spectrum and is evident from the corresponding PL intensity map (Fig. 1). However, a wide PL band peaking at ~600 nm can be extracted. This can be explained by very low concentration of the luminescent CdS nanocrystals in the domain laser-written in PZ glass. Domains obtained in OS-CdS glass are characterized by intense PL with a maximum at 675 nm. It is known that a PL band at 675 nm emerging under the 488 nm excitation is attributable to the recombination of an electron trapped in a sulfur vacancy with a hole in the valence band of small CdS nanocrystals (about 2-3 nm in size) [11, 12]. Comparing the results obtained for different glass matrices we can assume that the possible low concentration of CdS nanocrystals and therefore almost complete absence of the PL in domains laser-written in CdS-doped PZ glass may be associated with the structure of glass. For PZ glass where ZnO content higher than 50 mol.% the latter play an intermediate role of glass network former and modifier [13] because the behavior of Cd is similar to Zn. Thus, if cadmium and zinc are well embedded in structural units of the glass network then the equilibrium of laser modification process is shifted away from the formation of the sufficient quantity of CdS or ZnS.

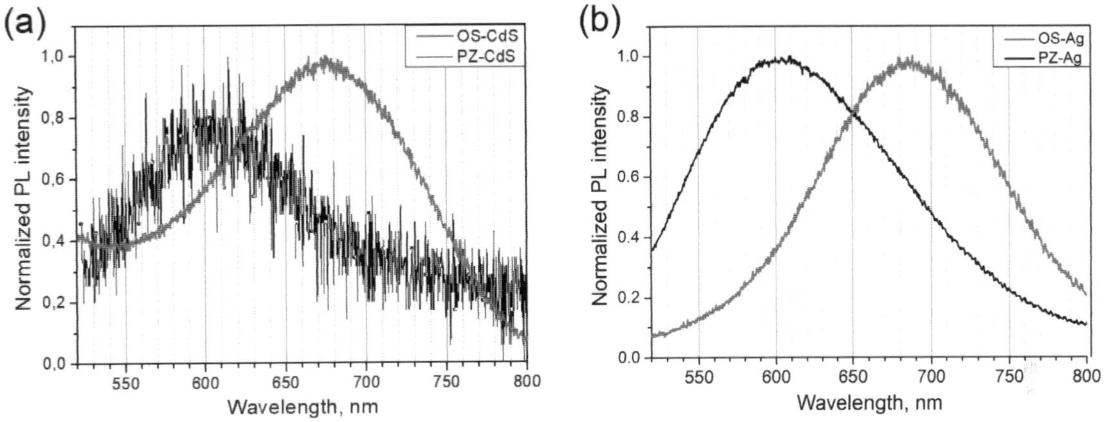

Figure 2 – PL emission spectra recorded from the laser-irradiated domains in PZ and OS glasses doped with CdS (a) or Ag (b)

In Ag-doped glasses, laser-written domains in both OS-Ag and PZ-Ag samples manifested relatively strong PL. The evident difference in the centers of the emission bands can be seen PL emission spectra for OS and PZ glass doped with Ag (Fig. 2(b)). A PL band peak of the domain in OS-Ag glass sample is located at 685 nm. In Ag-doped PZ glasses, the laser-written domains show a maximum of PL band at ~600 nm.

PL emission of silver clusters strongly depends on the excitation wavelength [14]. According to the available data [15], PL band peaking at 570 nm under the excitation of 480 nm in the oxyfluoride glass can be assigned to the electronic transitions in the Ag_m^{x+} clusters with m=4. In our case, excitation wavelength was 488 nm that is close to the mentioned work, thus we can assume that in PZ-Ag glass samples clusters of 4-5 atoms can be precipitated. The red-shift of the PL band registered in the domain in Ag-doped OS glass may indicate an increase of the m value, that is, bigger silver aggregates are formed [16]. A further study of PL properties using different excitation wavelengths as well as the optical absorption study is required and in progress in order to precisely determine all kinds of the silver nanospecies precipitated during the laser irradiation.

4. Conclusions

We studied PL properties of silicate and phosphate glasses separately doped with Ag or CdS induced by the femtosecond laser irradiation. In all cases, laser irradiation leads to the formation of domains exhibiting PL. In CdS-doped phosphate glass, PL was found to be almost imperceptible that may be due to the hampered formation of the sufficient quantity of CdS nanocrystals in Zn-rich glass. On the other hand, the laser-written domains in silicate glass with CdS are characterized with intense PL with a maximum at 675 nm presumably attributed to the formation of the sulfur vacancies in the CdS nanocrystals precipitated upon the laser heating. In Ag-containing PZ glass, we found PL assigned to the formation of Ag_m^{x+} clusters. Maxima of PL spectra differ in silicate and phosphate systems. In silicate glasses, PL bands have a peak at 685 nm, while in Ag-doped phosphate glasses, a maximum of PL bands is located at 600 nm. The observed red shift of PL in silicate glass as compared to phosphate glass may indicate a size increase of the precipitated silver nanospecies. The obtained results are essential for the future development of optical memory and the laser-assisted fabrication of functional waveguides in metal- or semiconductor-doped glasses.

Acknowledgments
This study was financially supported by the Council for Grants of the President of Russian Federation (grant MK-1497.2020.3) and the Russian Foundation for Basic Research (grant 21-53-12026).

References

[1] Wei Y, Ebendorff-Heidepriem H and Zhao J 2019 Recent advances in hybrid optical materials: integrating nanoparticles within a glass matrix *Adv. Opt. Mat.* **7** 1900702
[2] Xue J, Wang X, Jeong J H and Yan X 2020 Fabrication, photoluminescence and applications of quantum dots embedded glass ceramics *Chem. Eng. J.* **383** 123082
[3] Petit Y, Danto S, Guérineau T, Abou Khalil A, Le Camus A, Fargin E and Cardinal T 2018 On the femtosecond laser-induced photochemistry in silver-containing oxide glasses: mechanisms, related optical and physico-chemical properties, and technological applications *Adv. Opt. Technol.* **7** 291-309
[4] Mardilovich P, Fletcher L B, Troy N W, Yang L, Huang H, Risbud S H and Krol D M 2013 Ultrafast laser fabrication of hybrid micro-and nano-structures in semiconductor-doped borosilicate glasses *Int. J. Appl. Glass Sci.* **4** 87-99
[5] Shakhgildyan G Y, Lipatiev A S, Vetchinnikov M P, Popova V V, Lotarev S V, Golubev N V and Sigaev V N 2018 One-step micro-modification of optical properties in silver-doped zinc phosphate glasses by femtosecond direct laser writing *J. Non-Cryst. Solids* **481** 634-642
[6] Royon A, Bourhis K, Bellec M, Papon G, Bousquet B, Deshayes Y and Canioni L 2010 Silver clusters embedded in glass as a perennial high capacity optical recording medium *Adv. Mater.* **22** 5282-5286
[7] Abou Khalil A, Bérubé J P, Danto S, Cardinal T, Petit Y, Canioni L and Vallée R 2019 Comparative study between the standard type I and the type A femtosecond laser induced refractive index change in silver containing glasses *Opt. Mater. Express* **9** 2640-2651
[8] Vetchinnikov M P, Lipatiev A S, Shakhgildyan G Y, Golubev N V, Ignat'eva E S, Fedotov S S and Sigaev V N 2018 Direct femtosecond laser-induced formation of CdS quantum dots inside silicate glass *Opt. Lett.* **43** 2519-2522
[9] Bell G, Filin A I, Romanov D A and Levis R J 2016 Direct growth of CdSe semiconductor quantum dots in glass matrix by femtosecond laser beam *Appl. Phys. Lett.* **108** 063112
[10] Vetchinnikov M P, Shakhgildyan G Y, Lipatiev A S, Shakhgildyan A Y, Lotarev S V and Sigaev V N 2019 Femtosecond direct laser writing in silicate glasses doped with silver and cadmium sulfide *J. Phys. Conf. Ser.* **1199** 012028
[11] Xu G Q, Liu B, Xu S J, Chew C H, Chua S J and Gana L M 2000 Luminescence studies of CdS spherical particles via hydrothermal synthesis *J. Phys. Chem. Solids* **61** 829-836
[12] Kamat P V, Dimitrijevic N M and Fessenden R W 1987 Electron-transfer reactions of small cadmium sulfide colloids in acetonitrile *J. Phys. Chem.* **91** 396-401
[13] Suzuya K, Itoh K, Kajinami A and Loong C K 2004 The structure of binary zinc phosphate glasses *J. Non-Cryst. Solids* **345** 80-87
[14] Tikhomirov V K, Rodríguez V D, Kuznetsov A, Kirilenko D, Van Tendeloo G and Moshchalkov V V 2010 Preparation and luminescence of bulk oxyfluoride glasses doped with Ag nanoclusters *Opt. Express* **18** 22032-22040
[15] Kolobkova E V, Nikonorov N V, Sidorov A I and Shakhverdov T A 2013 Luminescence of molecular silver clusters in oxyfluoride glasses *Opt. Spectrosc.* **114** 236-239
[16] Klyukin D A, Dubrovin V D, Pshenova A S, Putilin S E, Shakhverdov T A, Tsypkin A N and Sidorov A I 2016 Formation of luminescent and nonluminescent silver nanoparticles in silicate glasses by near-infrared femtosecond laser pulses and subsequent thermal treatment: the role of halogenides *Opt. Eng.* **55** 067101

Journal of Physics: Conference Series **2015** (2021) 012164 doi:10.1088/1742-6596/2015/1/012164

Omnidirectional Photonic Bandgap in Two-dimensional Photonic Quasicrystal Made of Near-Transparent Dielectric Material

Ruslan Yafasov[1, a)] , Mikhail Sidorenko[1, b)] , Mikhail Rybin[1, c)], Alexander Petrov[1,2, d)] and Andrey Sayanskiy[1, e)]

[1] ITMO University, St. Petersburg 197101, Russia
[2] Institute of Optical and Electronic Materials, Hamburg University of Technology, Germany

[a)] Corresponding author: ruslan.yafasov@metalab.ifmo.ru
[b)] mikhail.sidorenko@metalab.ifmo.ru
[c)] m.rybin@mail.ioffe.ru
[d)] a.petrov@tuhh.de
[e)] a.sayanskiy@metalab.ifmo.ru

Abstract. Complete bandgap for all-dielectric photonic crystals in the microwave region can be obtained only by using high-contrast materials. This requires the usage of dielectric materials with high relative permittivity coefficient. In this paper, we study, both numerically and experimentally, a two-dimensional all-dielectric photonic quasicrystal made of polyurethane foam, which is considered in all microwave applications as a transparent material. The quasicrystal structure having an omnidirectional two-dimensional bandgap is mathematically generated by the direct inscription of Bragg's peaks of the structure in the reciprocal space. The sample of the quasicrystal was manufactured on CNC (computer numerical controlled) milling machine out of foam with very low dielectric permittivity of 1.254. The numerical simulations and the experimental study are in good agreement with the theoretical model.

INTRODUCTION

Over the past 30 years, photonic crystals have become an integral part of photonics. The main property, which makes them useful in many applications, is their ability to stop the propagation of electromagnetic waves in a certain wavelength range referred as to *photonic band gap* [1]. Although photonic band gaps exist for certain directions even at low refractive index contrast, the omnidirectional photonic band gap can be observed only for relatively high contrast [2-3].

One of the possible ways to overcome the symmetry limitations of photonic crystals is to consider photonic quasicrystals [4–5]. The quasicrystals possess the properties of both ordered structures and structures with disorder. In this paper, we consider the method first proposed in [6], which allows one to obtain 2D and 3D photonic quasicrystals with omnidirectional bandgaps. We consider here a two-dimensional quasicrystal, which depends on two coordinates only. For a two-dimensional structure, the omnidirectional bandgap is defined for all in-plane directions.

We present for the first time the results of numerical and experimental study of this model. A sample of the crystal is fabricated by means of computer numerical controlled (CNC) milling machine with polyurethane foam used as the material. The photonic density of states is measured experimentally in a parallel-plane waveguide for single TM polarization.

Journal of Physics: Conference Series **2015** (2021) 012164 doi:10.1088/1742-6596/2015/1/012164

THEORETICAL MODEL

In this paper a method of building a quasicrystal as a superposition of several 1D gratings with sinusoidal distribution of dielectric permittivity, which was first proposed in [6], is considered. The normals to these gratings are chosen to be uniformly distributed over a unit half-circle in the two-dimensional case. The formula for the refractive index variation for the superposition of N gratings reads:

$$\Delta n_g(\boldsymbol{r}) = \sum_{i=1}^{N} \Delta n \sin(\boldsymbol{g}_i \cdot \boldsymbol{r} + \varphi_i).$$

Here $\boldsymbol{g}_i$ is the direction of the normal to the grating with number i, Δn_i is the amplitude of the i-th grating refractive index variation, φ_i is the random phase term of each grating, introduced to obtain an isotropic structure. The Fig. 1, a) shows an example for the 22 Bragg peaks distribution in a reciprocal space, which corresponds to the sum of 11 gratings. On Fig. 1, b) an example of the sum of 11 gratings is plot as a function of coordinates.

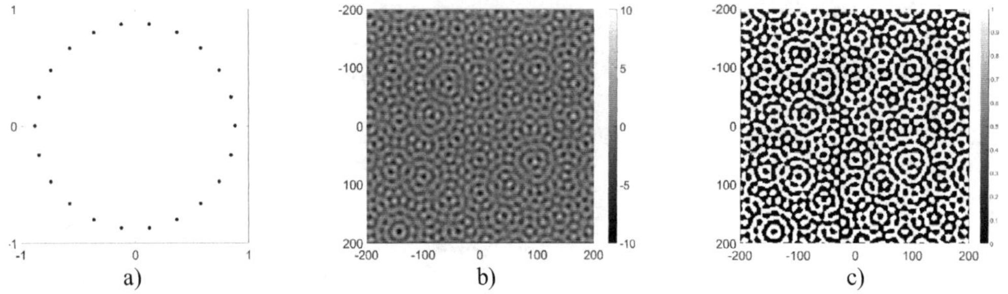

FIGURE 1. a) Vector directions for 11 gratings (both positive and negative directions are shown) b) An area of the 2D quasicrystal structure continuous refractive index variation for 11 gratings. The coordinate units is mm, the structure is designed to have a bandgap at the frequency of 10 GHz c) An area of the 2D quasicrystal structure after the binarization.

To fabricate the structure out of two materials, we apply the binarization process to the model, where one material takes the volume assigned to positive and the other to negative $\Delta n_g(\boldsymbol{r})$. A result of the binarization of the structure shown on Fig. 1, b) is shown on Fig. 1, c).

NUMERICAL SIMULATION

We consider a sample of the proposed quasicrystal in the numerical simulation. Two materials for the binarized structure have relative permittivity of 1 and 1.254. We set the number of gratings equal to 11, see [6] for the formula. We design the quasicrystal structure to have a complete photonic bandgap at the frequency of 10 GHz. The size of the structure is 1200x1200x5 mm^3. The goal is to simulate the local photonic density of states as the function of the frequency (see [7]). We put a small electric dipole inside the structure and simulate the ratio of power radiated by this dipole to the power radiated by the same dipole in the isotropic medium with dielectric permittivity equal to average value of 1.13. The simulations were performed in commercial software CST Microwave Studio. For TM polarization, the boundary conditions on Z axis are set to $E_\tau = 0$, and the dipole is aligned with Z axis. The simulated ratio P/P_0 is presented on Fig. 2, a), solid line. Strong suppression in the radiated power corresponds to the low density of states, which indicates the photonic bandgap.

EXPERIMENT

We fabricate an experimental sample with the parameters as above by using a CNC milling process. The sample of total 5 mm thickness is carved from the commercially available polyurethane foam panel. We scale down the model by the factor 1.75 to fit the sample to the CNC milling machine working area. The picture of the fabricated sample is presented on the Fig. 2, b). The bandgap frequency for the sample is then expected to be 17.5 GHz instead of 10 GHz. The material has the relative dielectric permittivity of 1.254 at 20 GHz. This value was obtained by using a special setup for dielectric properties measurement. The experimental sample was enclosed between two copper plates, which form the parallel plane waveguide, which allows us to consider TM polarization. The local photonic density of states was measured by putting a small electric dipole antenna inside the structure via the hole in one of the metallic planes. Then the S_{11} parameter was measured by means of a VNA and recalculated into the radiation impedance, which is presented on the Fig. 2, a), dashed line. The obtained results are in a good agreement with CST simulation.

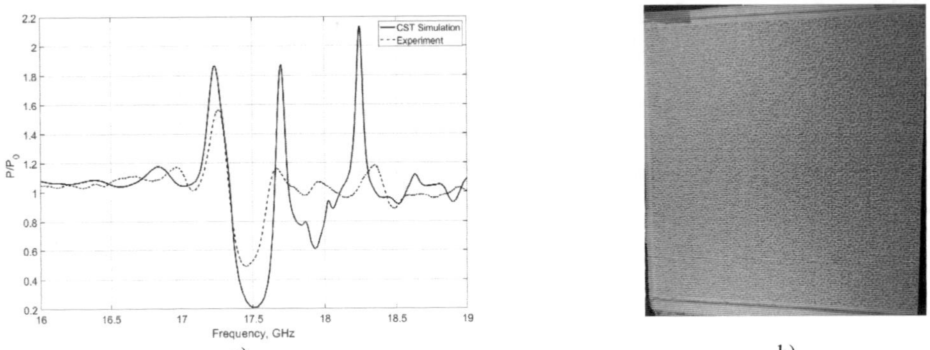

a) b)

FIGURE 2. a) The ratios P/P_0 for TM polarization for quasicrystal simulated in CST (solid line) and obtained from the experiment (dashed line). b) An experimental sample fabricated of polyurethane foam

CONCLUSION

In this paper we considered an all-dielectric photonic quasicrystal made of almost-transparent material. The quasicrystal structure was obtained by the method of summation of gratings. This allows obtaining the desired microwave properties of the final structure. The quasicrystal has the omnidirectional bandgap. To check the results, we measure the local photonic density of states. The results were proven both numerically and experimentally.

ACKNOWLEDGMENTS

This work was supported by the Russian Science Foundation (Project No. 20-79-10316)

REFERENCES

1 J. D. Joannopoulos, S. G. Johnson, J. N. Winn, R. D. Meade, Photonic crystals: Molding the flow of light (Princeton University Press, Princeton, ed. 2, 2008).

2 H. Men, K. Y. K. Lee, R. M. Freund, J. Peraire, S. G. Johnson, Robust topology optimization of three-dimensional photonic-crystal band-gap structures. Opt. Express 22, 22632–22648 (2014).

3 A. Cerjan, S. Fan, Complete photonic band gaps in supercell photonic crystals. *Phys. Rev. A* **96**, 051802(R) (2017).

4 W. Man, M. Megens, P. J. Steinhardt, P. M. Chaikin, Experimental measurement of the photonic properties of icosahedral quasicrystals. *Nature* **436**, 993–996 (2005).

5 P. N. Dyachenko, Y. V. Miklyaev, Band structure calculations of 2D photonic pseudoquasicrystals obtainable by holographic lithography. *Proc. SPIE* 6182, 618221 (2006).

6 L. Maiwald, T. Sommer, M. Schulz, M. Eich, A. Yu. Petrov, The limits for complete photonic bandgaps in low-contrast media, *Preprint: arXiv:1909.09521 [physics.optics]*.

7 Krasnok, A., Slobozhanyuk, A., Simovski, C. *et al.*, An antenna model for the Purcell effect. *Sci Rep* **5**, 12956 (2015)

Dielectric metasurface for emission control of magnetic dipole in the near-IR wavelength range

Vitaly Yaroshenko, Marina Obramenko, Artem Larin, Eduard Ageev, Dmitry Zuev

Department of Physics and Engineering, ITMO University, 197101 Saint-Petersburg, Russia

E-mail: v.yaroshenko@metalab.ifmo.ru

Abstract. Developing active dielectric systems is in high demand due to growing the internet of things. Such systems can control the optical properties of nanoemitters which leads to an increase in the performance of the telecommunication networks. Here we numerically investigate metasurface consists of all-dielectric erbium-doped silicon nanocylinders. We demonstrate that such a structure can effectively control and enhance 320-folds spontaneous emission in the near-IR wavelength range. The results of this paper can be used for creating new telecommunication systems.

1. Introduction

All-dielectric nanostructures can enhance and control the directivity of spontaneous emission[1, 2]. Recently, it was demonstrated that active dielectric nanoantennas[3, 4, 5] and metasurface[6] can work in the visible wavelength range. Hereby the development of nanostructures working in the optical telecommunication wavelength range is very promising and attractive.

Here we offer a conceptual design of metasurface consists of all-dielectric erbium-doped silicon nanocylinders. This material attracts attention due to several reasons. The first one is that silicon is lossless in the wavelength region higher than 1300 nm[7]. The second one is due to erbium has a transition at the wavelength of 1540 nm which corresponds to the standard telecommunication wavelength[8]. Mentioned above reasons make silicon doped by erbium a very prospective material in optical telecommunication wavelength.

In this work, we numerically study all-dielectric erbium-doped silicon metasurface for the near-IR wavelength range. We compute reflection and transmission spectra, estimate Purcell factor and directivity patterns of metasurfaces.

2. Results and discussion

In Figure 1, the designs of finite metasurfaces consisted of 5x5, 9x9, 15x15, and 31x31 unit cells are shown. The unite cell is represented by a silicon cylinder in a vacuum. We carry out numerical simulations to compute geometrical parameters of metasurface with magnetic dipole (MD) resonance at the wavelength of 1540 nm.

Firstly, we calculate the scattering of the single cylinder to tune the MD response at the wavelength of 1540 nm, using commercially available software CST Studio Suite and frequency-domain solver. In Figure 2 the scattering cross-section spectrum for the silicon nanocylinder with equal diameter and height of 380 nm is demonstrated. The electric and magnetic dipole

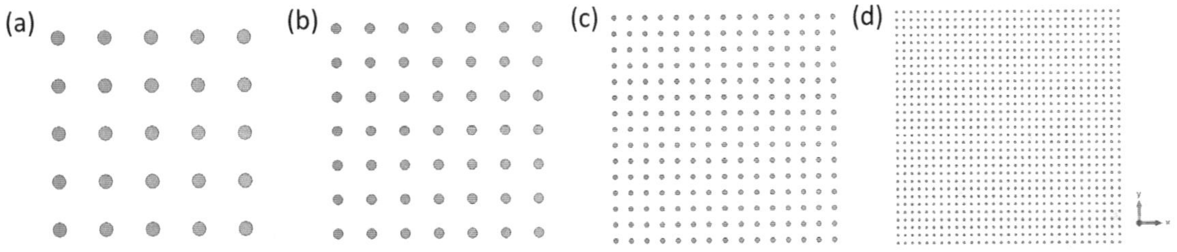

Figure 1. The designs of the finite metasurface consist of: (a) 5x5, (b) 7x7, (c) 15x15 and (d) 31x31 silicon nanocylinders

responses are observed at wavelengths of 1220 nm and 1540 nm, respectively. Highly important to notice that the concentration of erbium is negligibly low and has no influence on the optical properties of silicon.

Next, we perform the numerical simulation for the infinite metasurface using Floquet periodic ports and calculate the unit cell period. The transmission(T) and reflection(R) spectra are shown in Figure 2b. For the unit cell period of 1300 nm, the MD response is set up at the wavelength of 1540 nm, the ed response shifts to 1400 nm due to the near-field interaction.

Finally, we consider the finite metasurface, estimate the Purcell factor and compute directivity patterns. The electric dipole with a length of 10 nm is placed in the center of the central nanocylinder of the finite metasurface along y-axes (see Fig.1). Since the emission of erbium at the wavelength of 1540 nm described by magnetic dipole transition [9], we apply the duality of Maxwell's equations[10]. Using this technique, we replace the permeability by the permittivity and simulate the magnetic dipole by the electric dipole. Due to silicon is lossless in the wavelength range from 1400 to 2000 nm[7] we calculate only the radiative Purcell factor by the following equation[11]:

$$F_p = \frac{P}{P_0}, \tag{1}$$

where P is the power radiated by a dipole in the presence of the metasurface and P_0 - in a vacuum.

In Figure 2c calculated Purcell factor is shown. One can see that the Purcell factor achieves the value of about 340 and does not depend on the number of unit cells in the finite metasurface. However, the directivity patterns have some dependence on the number of unit cells. In Figure 3

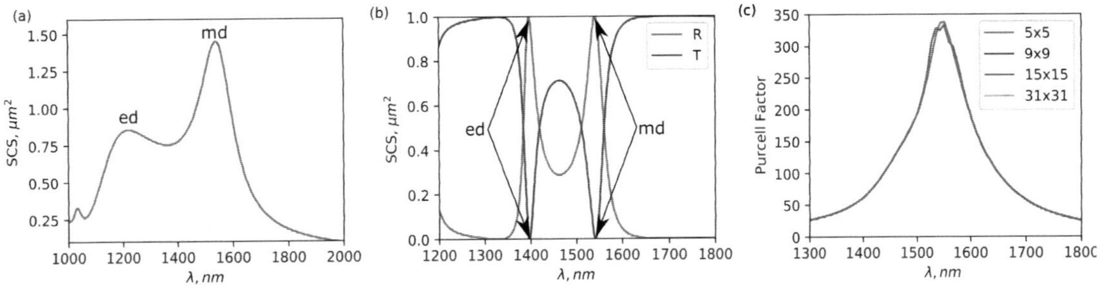

Figure 2. Numerically calculated: (a) scattering cross-section for the single silicon nanocylinder, (b) reflection(R) and transmission(T) coefficients for the infinite metasurface, and (c) the radiative Purcell factor for the finite metasurfaces. The legend shows the number of unit cells.

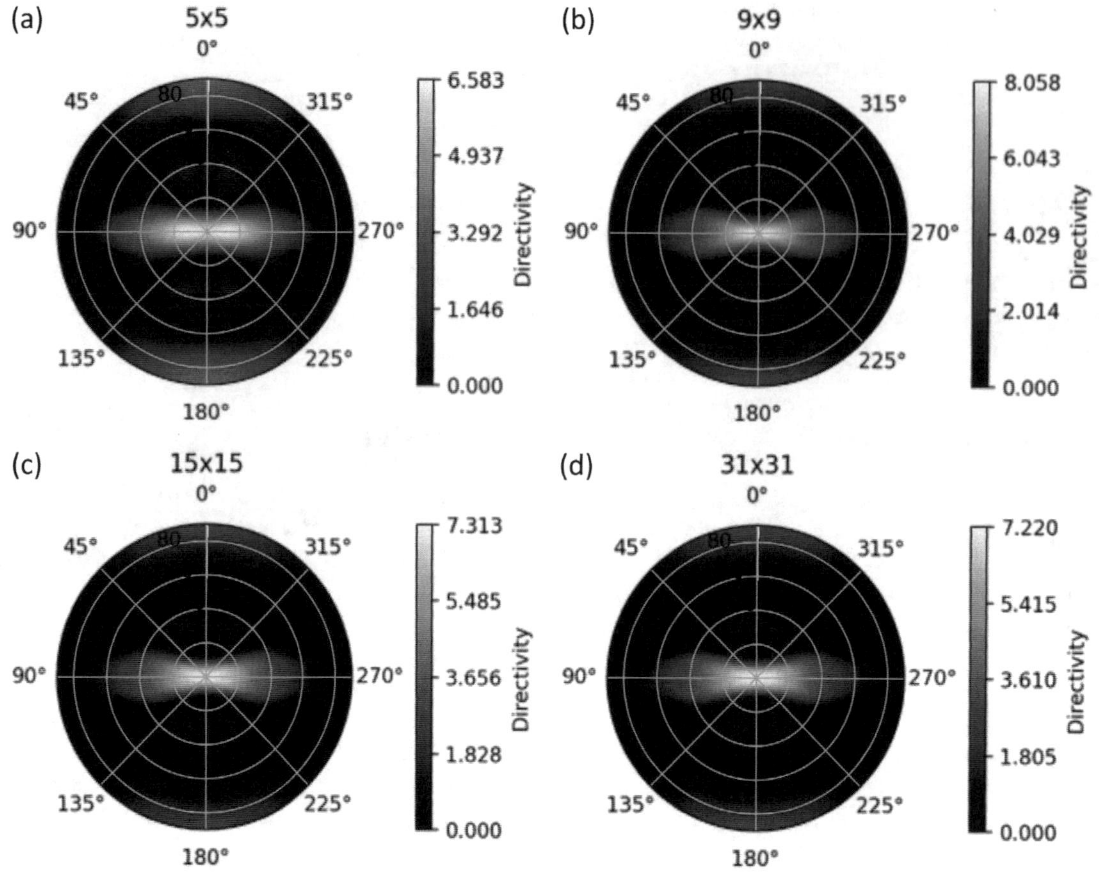

Figure 3. Directivity patterns for the finite metasurface consists of: (a) 5x5, (b) 7x7, (c) 15x15 and (d) 31x31 unit cells

the directivity pattern only in the upper-hemisphere is represented since the considered system is symmetrical relative to the z-axis (also to x- and y-axes). For the case of 5x5 unit cells(see Fig.3a) more power emits in x-direction compare to other cases(see Fig.3b-d). We observe an increase of the power radiated in z-direction for 9x9 unit cells(see Fig.3b) and patterns stay quite the same for 15x15 and 31x31 unit cell cases(see Fig.3c and d). Consequently, if dipoles are placed in each unit cells the directivity will have quite the same pattern and the emitters will radiate more power perpendicularly to the surface along the z-axis.

3. Conclusion

In this work, we demonstrate the results of a numerical study of all-dielectric metasurface consists of silicon nanoparticles doped by erbium. We compute the Purcell factor for finite metasurfaces and its value can be up to 320. We show that the emission of the dipole placed in the center of the central unite cell in the finite metasurface radiates mostly along the z-axis and an increasing number of the unit cell leads to decreasing of the emission in the plane of the metasurface.

The results presented in this paper can be useful in developing modern telecommunication systems and can improve the performance of existing ones.

Acknowledgments
This work was supported by Russian Foundation for Basic Research according to the research project No.18-29-20107

References

[1] Isabelle Staude, Thomas Pertsch, and Yuri S. Kivshar. All-dielectric resonant meta-optics lightens up. *ACS Photonics*, 2019. DOI: 10.1021/acsphotonics.8b01326.

[2] Aleksandr Vaskin, Radoslaw Kolkowski, A. Femius Koenderink, and Isabelle Staude. Light-emitting metasurfaces. *Nanophotonics*, 8(7):1151–1198, 2019.

[3] E. Y. Tiguntseva, G. P. Zograf, F. E. Komissarenko, D. A. Zuev, A. A. Zakhidov, S. V. Makarov, and Yuri. S. Kivshar. Light-emitting halide perovskite nanoantennas. *Nano Letters*, 18(2):1185–1190, 2018. PMID: 29365259.

[4] A. S. Zalogina, R. S. Savelev, E. V. Ushakova, G. P. Zograf, F. E. Komissarenko, V. A. Milichko, S. V. Makarov, D. A. Zuev, and I. V. Shadrivov. Purcell effect in active diamond nanoantennas. *Nanoscale*, 10:8721–8727, 2018.

[5] Viktoriia Rutckaia, Frank Heyroth, Alexey Novikov, Mikhail Shaleev, Mihail Petrov, and Joerg Schilling. Quantum dot emission driven by mie resonances in silicon nanostructures. *Nano Letters*, 17(11):6886–6892, 2017. PMID: 28968505.

[6] Aleksandr Vaskin, Soheila Mashhadi, Michael Steinert, Katie E. Chong, David Keene, Stefan Nanz, Aimi Abass, Evgenia Rusak, Duk-Yong Choi, Ivan Fernandez-Corbaton, Thomas Pertsch, Carsten Rockstuhl, Mikhail A. Noginov, Yuri S. Kivshar, Dragomir N. Neshev, Natalia Noginova, and Isabelle Staude. Manipulation of magnetic dipole emission from eu3+ with mie-resonant dielectric metasurfaces. *Nano Letters*, 19(2):1015–1022, 2019. PMID: 30605616.

[7] Carsten Schinke, P. Christian Peest, Jan Schmidt, Rolf Brendel, Karsten Bothe, Malte R. Vogt, Ingo Kröger, Stefan Winter, Alfred Schirmacher, Siew Lim, Hieu T. Nguyen, and Daniel MacDonald. Uncertainty analysis for the coefficient of band-to-band absorption of crystalline silicon. *AIP Advances*, 5(6):067168, 2015.

[8] Morito Matsuoka and Shun-ichi Tohno. 1.54 mkm wavelength emission of erbium-doped silicon films grown by ion beam epitaxy using sputtering-type metal ion source. *Applied Physics Letters*, 66(15):1862–1864, 1995.

[9] Denis G. Baranov, Roman S. Savelev, Sergey V. Li, Alexander E. Krasnok, and Andrea Alù. Modifying magnetic dipole spontaneous emission with nanophotonic structures. *Laser & Photonics Reviews*, 11(3):1600268, 2017.

[10] Ivan Fernandez-Corbaton and Gabriel Molina-Terriza. Role of duality symmetry in transformation optics. *Phys. Rev. B*, 88:085111, Aug 2013.

[11] Alexander E. Krasnok, Alexey P. Slobozhanyuk, Constantin R. Simovski, Sergei A. Tretyakov, Alexander N. Poddubny, Andrey E. Miroshnichenko, Yuri S. Kivshar, and Pavel A. Belov. An antenna model for the purcell effect. *Scientific Reports*, 5:12956 EP –, Aug 2015. Article.

Journal of Physics: Conference Series **2015** (2021) 012166 doi:10.1088/1742-6596/2015/1/012166

Dielectric nanoresonator for enhancement of 2D material photoluminescence

Vitaly Yaroshenko, Pavel Kustov, Dmitry Zuev

Department of Physics and Engineering, ITMO University, 197101 Saint-Petersburg, Russia

E-mail: v.yaroshenko@metalab.ifmo.ru

Abstract. Nowadays the volume of transmitted information exponentially grows and requires the development of new telecommunication systems. Dielectric nanoresonators can be considered as a basic part of such systems to control the emission of the nanoscale source. Here we numerically investigated resonant dielectric nanoresonators for emission enhancements of 2D nanomaterials. We show that the radiative Purcell factor can achieve the value of up to 21 and 12 for the magnetic quadrupole and dipole responses, respectively. Also, we compare the directivity patterns for magnetic dipole and quadrupole resonances. The results obtained in this work can be applied in the development of optical chips and interfaces.

1. Introduction

The development of dielectric resonators as a basic element of optical interfaces and chips to control the emission of nanoscale sources is on-high demand due to the increasing volume of transmitted information in the last decade[1]. Such interfaces can find their applications in optical information and telecommunication systems. One of the most important requirements for nanoscale sources is the ability to operate at room temperature[2]. Therefore two-dimensional material such as hexagonal boron nitride(hBN) is considered to satisfy this requirement

The emitters based on defects of hBN have such optical properties as an optical readout of the electron spin state[3] and single photon emission at room temperature[4]. It should be noted that the zero-phonon line (ZPL) of hBN is characterized by rather narrow emission lines at room temperature[5]. At the same time, ZPL wavelength various from deep ultraviolet to near-infrared due to the presence of various defects[6]. According to the above-mentioned reasons, hBN is a very promising candidate for the nanoscale emitter.

To control the emission of nanoscale emitters plasmonic and dielectric structures can be used. However, plasmonic systems have several problems such as complex directivity patterns and high dissipative losses in the visible range[7]. Otherwise, dielectric systems are devoid of these drawbacks and have both electrical and magnetic responses (Mie resonances) which lead to the rapid growth of emission control by all-dielectric nanophotonics[8].

In this paper, we offer and numerically investigate dielectric nanoresonator for enhancement of the spontaneous emission of the hBN flake. We study and compare the impact of the magnetic dipole and quadrupole resonances on the total and radiative Purcell factors.

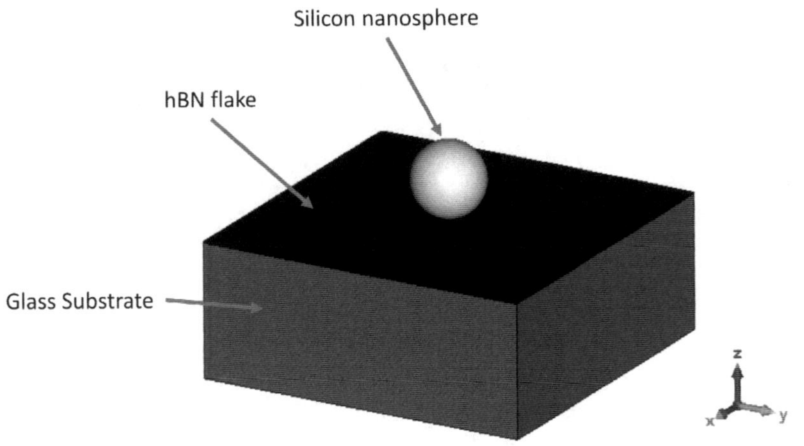

Figure 1. Design of the dielectric resonator on the hBN flake placed on the glass substrate.

2. Results and discussion

The dielectric nanoresonator consists of the silicon nanosphere placed on the glass substrate with the hBN flake. The diameter of the nanosphere depends on the excitation wavelength of defects and the resonance responses of the nanoparticle. It varies from 140 nm to 286 nm. The thickness of the hBN flake is 1 nm, which corresponds to several layers[4].

We start our consideration from the plane wave scattering on the silicon nanosphere placed on the infinite glass substrate. Using commercially available software CST Studio Suite and integral equation solver we calculate the diameters(see fig.2 legends) of the nanosphere which corresponds to the magnetic dipole(MD) and quadrupole(MQ) tuned to the three (min, mean and max) experimentally measured wavelengths[4]: 576 nm, 652 nm and 762 nm. The dispersions of silicon and hBN are taken from references [9] and [10], respectively. The refractive index of glass is constant and equals 1.45.

Next, we put 10 nm dipole in the center of the hBN flake along the x-axis precisely under

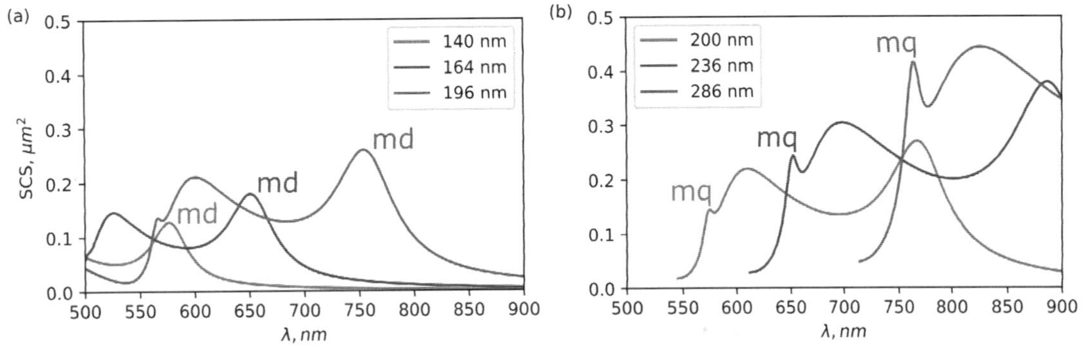

Figure 2. Numerically calculated scattering cross sections for the silicon nanosphere on the glass substrate for (a) magnetic dipole (MD) and (b) magnetic quadrupole (MQ) tuned on the hBN emission wavelengths. The legend shows the diameter of the silicon nanosphere.

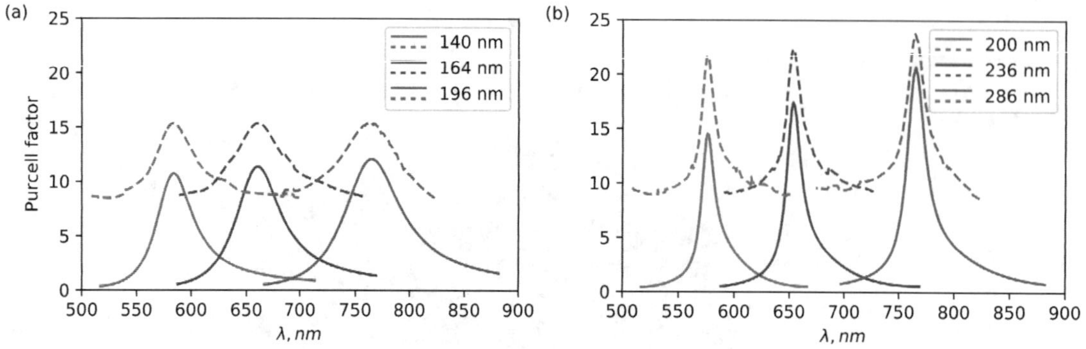

Figure 3. The total(dashed lines) and radiative(solid lines) Purcell factors for the silicon nanosphere on the glass substrate for (a) the magnetic dipole (MD) and (b) the quadrupole (MQ) tuned on the hBN emission wavelengths. The legend shows the diameter of the silicon nanosphere.

the silicon nanosphere to estimate the Purcell factor. Since silicon has losses in the considered wavelength range (500-900 nm) the total and radiative Purcell factors are calculated using the following equations[11]:

$$F_p^{Tot} = \frac{Re(Z)}{Re(Z_0)} = \frac{P_{rad} + P_{nonrad}}{P_0} \tag{1}$$

$$F_p^{rad} = \frac{P_{rad}}{P_0} \tag{2}$$

where Z is the impedance of the dipole in the presence of the system, Z_0 – the impedance of the dipole in a vacuum, P_{rad} – the power radiated by the dipole in the presence of the system, P_{nonrad} – the power dissipated in the system and P_0 – in a vacuum. We numerically computed radiated powers and impedances using a frequency-domain solver. On the figure3 calculated total and radiative Purcell factors for MD and MQ tuned at the chosen excitation wavelengths of hBN are represented. Considering the MD case the total Purcell factor achieves the values of about 15 for all considered cases (see fig.3a, dashed curves). For the radiative Purcell factor we observe a slow increase from 10 at the wavelength of 576 nm to 12 at 762 nm (see fig.3a, solid curves). At the same moment for the MQ case, the total and radiative Purcell factors increase from 22 to 24 and from 14 to 21, respectively (see fig.3b). One can see that the radiative Purcell

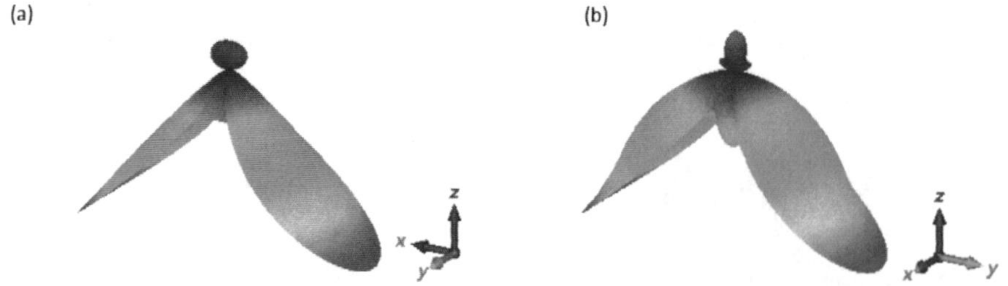

Figure 4. Directivity patterns for the excitation at the wavelengths tuned on (a) magnetic dipole (MD) and (b) magnetic quadrupole.

factor increases faster than the total one due to decreasing losses in silicon with increasing the wavelength.

The directivity patterns are calculated for the dipole emission in the considered system. It is important to stress out at the point that directivities differ for MD and MQ cases but do not depend on the tuned wavelength. Therefore in both cases, the emission mostly radiated in the substrate (see fig.4).

3. Conclusion

In this paper, we demonstrate the results of the numerical investigation of the silicon nanoparticle for enhancement of the hBN flake emission. We compare Purcell factors for MD and MQ resonance nanospheres. We show that for the MD case the radiative Purcell factors vary from 10 to 12, but the total one remains almost the same and equals 15 in the considered wavelength range. However, for the MQ case, we observe a fast increase of radiative Purcell factor from 14 to 21 and a slow increase of total Purcell factor from 22 to 24. For both cases, power radiates mostly in the substrate.

Investigated dielectric nanoresonator can find its application and telecommunication interfaces as a basic part of all-optical chips.

Acknowledgments

This work was supported by Russian Foundation for Basic Research according to the research project No.20-37-90147

References

[1] Isabelle Staude, Thomas Pertsch, and Yuri S. Kivshar. All-dielectric resonant meta-optics lightens up. *ACS Photonics*, 2019. DOI: 10.1021/acsphotonics.8b01326.

[2] Ali W. Elshaari, Wolfram Pernice, Kartik Srinivasan, Oliver Benson, and Val Zwiller. Hybrid integrated quantum photonic circuits. *Nature Photonics*, 14(5):285–298, May 2020.

[3] Annemarie L. Exarhos, David A. Hopper, Raj N. Patel, Marcus W. Doherty, and Lee C. Bassett. Magnetic-field-dependent quantum emission in hexagonal boron nitride at room temperature. *Nature Communications*, 10(1):222, Jan 2019.

[4] Toan Trong Tran, Christopher Elbadawi, Daniel Totonjian, Charlene J. Lobo, Gabriele Grosso, Hyowon Moon, Dirk R. Englund, Michael J. Ford, Igor Aharonovich, and Milos Toth. Robust multicolor single photon emission from point defects in hexagonal boron nitride. *ACS Nano*, 10(8):7331–7338, 2016. PMID: 27399936.

[5] A. Dietrich, M. Bürk, E. S. Steiger, L. Antoniuk, T. T. Tran, M. Nguyen, I. Aharonovich, F. Jelezko, and A. Kubanek. Observation of fourier transform limited lines in hexagonal boron nitride. *Phys. Rev. B*, 98:081414, Aug 2018.

[6] Sherif Abdulkader Tawfik, Sajid Ali, Marco Fronzi, Mehran Kianinia, Toan Trong Tran, Catherine Stampfl, Igor Aharonovich, Milos Toth, and Michael J. Ford. First-principles investigation of quantum emission from hbn defects. *Nanoscale*, 9:13575–13582, 2017.

[7] Nina Jiang, Xiaolu Zhuo, and Jianfang Wang. Active plasmonics: Principles, structures, and applications. *Chemical Reviews*, 118(6):3054–3099, 2018. PMID: 28960067.

[8] Arseniy I. Kuznetsov, Andrey E. Miroshnichenko, Mark L. Brongersma, Yuri S. Kivshar, and Boris Luk'yanchuk. Optically resonant dielectric nanostructures. *Science*, 354(6314), 2016.

[9] Carsten Schinke, P. Christian Peest, Jan Schmidt, Rolf Brendel, Karsten Bothe, Malte R. Vogt, Ingo Kröger, Stefan Winter, Alfred Schirmacher, Siew Lim, Hieu T. Nguyen, and Daniel MacDonald. Uncertainty analysis for the coefficient of band-to-band absorption of crystalline silicon. *AIP Advances*, 5(6):067168, 2015.

[10] Seong-Yeon Lee, Tae-Young Jeong, Suyong Jung, and Ki-Ju Yee. Refractive index dispersion of hexagonal boron nitride in the visible and near-infrared. *physica status solidi (b)*, 256(6):1800417, 2019.

[11] Alexander E. Krasnok, Alexey P. Slobozhanyuk, Constantin R. Simovski, Sergei A. Tretyakov, Alexander N. Poddubny, Andrey E. Miroshnichenko, Yuri S. Kivshar, and Pavel A. Belov. An antenna model for the purcell effect. *Scientific Reports*, 5:12956 EP –, Aug 2015. Article.

Open-source implementation of the discrete-dipole approximation for a scatterer in an absorbing host medium

M A Yurkin[1,2,4] and A E Moskalensky [1,2]

[1] Voevodsky Institute of Chemical Kinetics and Combustion SB RAS, Institutskaya Str. 3, 630090, Novosibirsk, Russia
[2] Novosibirsk State University, Pirogova Str. 2, 630090, Novosibirsk, Russia
[4] Corresponding author: yurkin@gmail.com

Abstract. Theoretical description of light scattering by single particles is a well-developed field, but most of it applies to particles located in vacuum or non-absorbing host medium. Although the case of absorbing host medium has also been discussed in literature, a complete description and unambiguous definition of scattering quantities are still lacking. Similar situation is for simulation methods – some computer codes exist, but their choice is very limited, compared to the case of vacuum. Here we describe the extension of the popular open-source code ADDA to support the absorbing host medium. It is based on the discrete dipole approximation and is, thus, applicable to particles with arbitrary shape and internal structure. We performed test simulations for spheres and compared them with that using the Lorenz-Mie theory. Moreover, we developed a unified description of the energy budget for scattering by a particle in a weakly absorbing host medium, relating all existing local (expressed as volume integrals over scatterer volume) and far-field scattering quantities.

1. Introduction

Light scattering by small particles is a ubiquitous phenomenon in nature and is widely used in science and technology. Theoretical description of this process is well developed (at least for single particles), allowing one to calculate any scattering quantities. For instance, the Lorenz-Mie theory, applicable to spheres, was developed more than 100 years ago [1]. Later, the T-matrix method [2] and the discrete-dipole approximation (DDA) [3,4], among other methods, enabled simulation of light scattering by arbitrarily shaped particles, even in inhomogeneous environment (for instance, placed onto a flat substrate).

Surprisingly, the case of a homogeneous but absorbing environment (host medium) is still often omitted both in theoretical descriptions and computational codes. Importantly, this case is relevant for many practical applications, e.g., for particles submerged in water and droplets in oil. For instance, the Lorenz-Mie theory was extended to the case of absorbing host medium only in 1996 [5], although some far-field results were obtained earlier [6]. The Maxwell's equations themselves are perfectly applicable for complex wavenumber k, and many light-scattering methods allow straightforward extension (by analytic continuation of all formulae). However, the extension of the existing codes has technical issues, e.g., with using special mathematical functions of complex arguments. Nevertheless, publicly available Lorenz-Mie codes are available for that case [7,8].

The situation is similar for the DDA. The underlying volume-integral equation (VIE) is valid for complex k [9] and the first DDA simulations for absorbing host medium has been recently reported [10]. However, the corresponding code is not publicly available. We have independently extended the popular open-source code ADDA [11] to the case of an absorbing host medium. In this paper we report its validation by comparison with the Lorenz-Mie theory.

A more fundamental issue, often overlooked, is the definition of the (measureable) scattering quantities. In the case of vacuum (or non-absorbing host medium), the analysis of the energy budget [12] leads to the classical definition of the intrinsic particle quantities: various cross sections, Mueller and amplitude scattering matrices [13]. They need to be calculated once for a given particle and real m_{h} (refractive index of the host medium), after which the measurement of any detector in the far-field zone can be easily simulated, both for a single particle and for a collection of many well-separated particles. Unfortunately, such theoretical description is lacking for the case of complex m_{h}, despite existing efforts in this direction [14–16]. The main problem is that the transformation of the integration surface for the flow of the Poynting vector from the near- to far-field region leads to the additional term associated with medium absorption. Thus, our second goal is to develop such description for relatively weak absorption, including the definition of the extinction cross section based on the energy-budget arguments.

2. Volume-integral equation

We consider a non-magnetic medium with the dielectric permittivity $\varepsilon_1 \stackrel{\text{def}}{=} \varepsilon_1' + \mathrm{i}\varepsilon_1''$, where $\varepsilon_1'' \geq 0$. The dielectric permittivity of the particle is $\varepsilon_2(\mathbf{r}) \stackrel{\text{def}}{=} \varepsilon_2'(\mathbf{r}) + \mathrm{i}\varepsilon_2''(\mathbf{r})$, $\varepsilon_2''(\mathbf{r}) \geq 0$. The VIE follows directly from the Maxwell's equation and can be written as [9,12]:

$$\frac{1}{3\varepsilon_1}\frac{\varepsilon_2(\mathbf{r}) + 2\varepsilon_1}{\varepsilon_2(\mathbf{r}) - \varepsilon_1}\mathbf{P}(\mathbf{r}) = \mathbf{E}_{\mathrm{inc}}(\mathbf{r}) + \omega^2\mu_0 \lim_{V_0\to 0}\int_{\mathbb{R}^3\backslash V_0} \mathrm{d}^3\mathbf{r}'\bar{\mathbf{G}}(\mathbf{r},\mathbf{r}')\cdot\mathbf{P}(\mathbf{r}'), \tag{1}$$

where $\mathbf{P}(\mathbf{r}) \stackrel{\text{def}}{=} [\varepsilon_2(\mathbf{r}) - \varepsilon_1]\mathbf{E}(\mathbf{r})$ is the polarization density, $\mathbf{E}(\mathbf{r})$ is the internal field, $\mathbf{E}_{\mathrm{inc}}(\mathbf{r})$ is the incident field, ω is the frequency, V_0 is the spherical exclusion volume centered at $\mathbf{r}$, and $\bar{\mathbf{G}}(\mathbf{r},\mathbf{r}')$ is the Green's dyadic:

$$\bar{\mathbf{G}}(\mathbf{r},\mathbf{r}') \stackrel{\text{def}}{=} \frac{\exp(\mathrm{i}kR)}{4\pi R}\left[\left(\bar{\mathbf{I}} - \frac{\mathbf{R}\otimes\mathbf{R}}{R^2}\right) + \frac{\mathrm{i}kR - 1}{k^2R^2}\left(\bar{\mathbf{I}} - 3\frac{\mathbf{R}\otimes\mathbf{R}}{R^2}\right)\right], \tag{2}$$

where $\mathbf{R} = \mathbf{r} - \mathbf{r}'$, $R = |\mathbf{R}|$, and $k \stackrel{\text{def}}{=} \omega^2\mu_0\varepsilon_1$ is the complex wavenumber:

$$k = k' + \mathrm{i}k'', \qquad k' = \omega^2\mu_0\varepsilon_1', \qquad k'' = \omega^2\mu_0\varepsilon_1'', \tag{3}$$

Eq. (1) can be discretized into a set of voxels (dipoles) which constitute a particle of interest, and their polarizations can be determined numerically. This is the basis of the DDA and its implementations, including the code ADDA. Importantly, it remains valid in the case of absorbing medium.

3. Energy budget and cross sections

The proper definition of power integrals and cross sections (absorption, scattering and extinction) in an absorbing medium is widely discussed in the literature [14–16]. In this case the flow of the incident Poynting vector does not vanish and the flow of any Poynting vector depends on the specific position of the integration surface. In particular, the integrals over the particle's surface differ from those taken in the far-field, the difference being the energy absorbed in the medium. Still, the power absorbed inside the particle has a clear physical meaning and can be calculated once $\mathbf{E}(\mathbf{r})$ or $\mathbf{P}(\mathbf{r})$ is known:

$$W_{\mathrm{abs}} \stackrel{\text{def}}{=} \frac{\omega}{2}\int_{V_{\mathrm{int}}} \mathrm{d}^3\mathbf{r}\,\varepsilon_2''(\mathbf{r})|\mathbf{E}(\mathbf{r})|^2, \tag{4}$$

where V_{int} is the particle volume.

Certain scattering and extinction powers can potentially be defined similarly (integrating corresponding components of the Poynting vector over the particle surface), however the physical meaning of such integrals is ambiguous. A more practical approach is to define these quantities through the far-field measurements [16–18]. For sufficiently large distances, the scattered wave is the outgoing spherical one:

$$\mathbf{E}_{\mathrm{sca}}(\mathbf{r}) \xrightarrow[r\to\infty]{} \exp(-k''r)\frac{\exp(\mathrm{i}k'r)}{r}\mathbf{F}\left(\frac{\mathbf{r}}{r}\right), \tag{5}$$

where $\mathbf{F}$ is the scattering amplitude. While Eq. (5) is valid for any ε_1'', it is practically relevant only for relatively small ε_1'' so that there exist r, such that it is large enough (for far-field limit) but the scattered field is still measureable. Obviously, the damping factor $\exp(-k''r)$ affects the scattered power passing through the far-field sphere with radius r, but it can be factored out to define the following normalized scattering power:

$$W_{\mathrm{sca}}^{\mathrm{n}} \overset{\mathrm{def}}{=} \frac{k'}{2\omega\mu_0}\oint \mathrm{d}^2\mathbf{n}\,|\mathbf{F}(\mathbf{n})|^2. \tag{6}$$

If we assume that the incident field is a plane wave with amplitude $\mathbf{E}_0$ (in the origin) propagating in the direction $\mathbf{n}_0$, one can similarly define the normalized extinction cross-section [16]:

$$W_{\mathrm{ext}}^{\mathrm{n}} \overset{\mathrm{def}}{=} \frac{2\pi}{\omega\mu_0}\mathrm{Im}[\mathbf{F}(\mathbf{n}_0)\cdot\mathbf{E}_0^*]. \tag{7}$$

However, this definition is meaningful (related to far-field power flow) only if $\exp(2k''r) \ll k'r$, i.e. the absorption is weak and r is not too large. Both Eqs. (6) and (7) are convenient because they are independent of r and can be further used to assess the energy budget for arbitrary far-field sphere. For a plane incident wave it is convenient to define cross sections dividing the powers by the irradiance of the incident wave in the origin $I_0 = k'|\mathbf{E}_0|^2/(2\omega\mu_0)$ [16,17]. The corresponding cross sections are computed by existing Lorenz-Mie codes, and we have implemented the same in the ADDA code.

The above theory is self-contained but has two important limitations. First, the far-field integration surface is limited to be a sphere (otherwise the normalized powers cannot be used), which is inconvenient for consideration of, e.g., extinction of a plane wave by a slab with many well-separated scatterers (the classical extinction problem for a non-absorbing host medium). Second, $W_{\mathrm{sca}}^{\mathrm{n}}$ requires integration of the scattered field, which is much slower in the framework of the VIE than a single volume integral to be computed for either $W_{\mathrm{ext}}^{\mathrm{n}}$ or W_{abs}.

We have managed to address both these issues. In short, using the generalized Jones lemma [16] we proved that the integral of the extinction (cross) part of the Poynting vector is the only part, which is relatively independent of the integration surface. Its value is determined by the interference of the incident and scattered fields in the forward direction, hence the surface can be freely deformed as long as it remains smooth and fully inside the intermediate region of r, described above, and the forward-direction part stays intact (at a fixed distance from the scatterer). Next, for weakly absorbing host medium we expressed $W_{\mathrm{sca}}^{\mathrm{n}}$ as an integral over V_{int} similarly to that in vacuum [12]. The details of these derivations will be presented at the conference.

4. Numerical results

We modified the ADDA code to support arbitrary m_{h}. The internal code modifications are mostly limited to changing a number of variables and functions from real to complex type. We also had to modify functions calculating the cross sections according to the developed theory. From user's point of view, the main change is the new command-line option `-mhost` setting m_{h} (any complex number). Previously the simulations with ADDA were possible only for real m_{h} through the manual scaling of wavelength and particle refractive index (the scattering problem depends only on relative refractive index and wavenumber, except for trivial scaling of some scattering quantities [19]). We verified that the enhanced code obeys this scaling rule (data not shown).

To validate ADDA for absorbing host medium (complex m_{h}), we compared the simulation results with the Lorenz-Mie theory using the publicly available Fortran code [7]. The results for a sphere with size parameter 10 and refractive index 1.5 + 0.1i are shown in Fig. 1. We tested three values of m_{h} (1 + 0i, 1 + 0.05i, and 1 + 0.1i) and two discretization levels (15 and 30 dipoles per the wavelength, corresponding to 24 and 48 dipoles per the particle diameter). One can see that ADDA results converge to the exact solution with increasing discretization, both for angle-resolved scattering intensity and cross sections. Moreover, the default discretization of 15 dipoles per the wavelength (for this particle refractive index) already gives reasonable results.

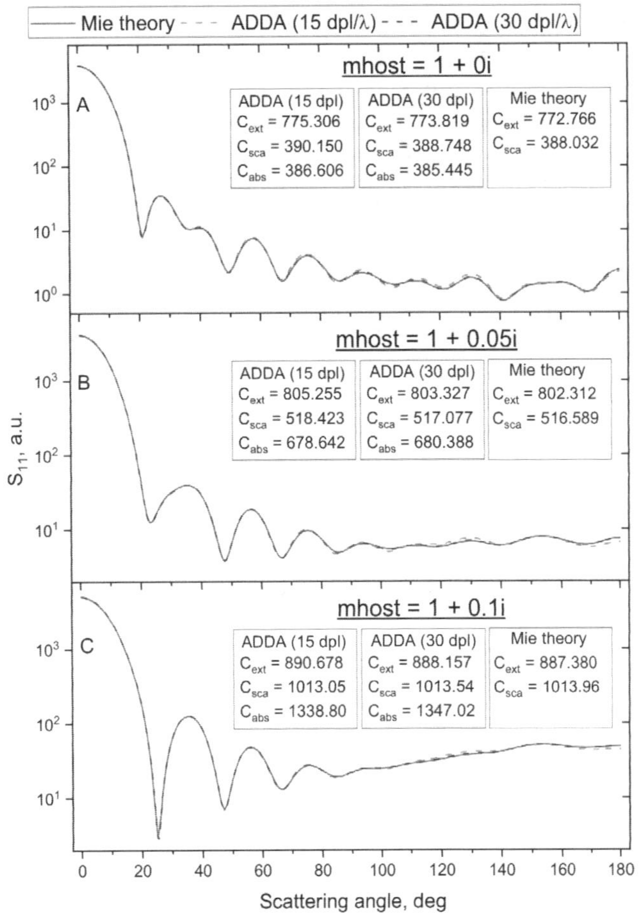

Figure 1 Comparison of ADDA simulations with the Lorenz-Mie theory for a sphere with size parameter 10 and refractive index 1.5 + 0.1i embedded in the medium with refractive index 1 + 0i (A), 1 + 0.05i (B), or 1 + 0.1i (C). Values of cross sections are shown in the insets.

5. Conclusion

We have implemented the support for the absorbing host medium in the popular open-source code ADDA and validated it by comparison with the exiting Lorenz-Mie code for spheres. The extended code is currently available at the separate branch: https://github.com/Sunmosk/adda-complex_freq, and will be incorporated into future releases of ADDA. Moreover, we extended the energy-budget analysis of the scattering problem to the case of weakly absorbing host medium, relating local (volume integrals over scatterer volume) and far-field scattering quantities.

Acknowledgments

The work was supported by the Russian Science Foundation (Grant No. 18-12-00052).

References

[1] Horvath H 2009 Light scattering: Mie and more – Commemorating 100 years of Mie's 1908 publication J. Quant. Spectrosc. Radiat. Transfer 110 783–6

[2] Mishchenko M I, Travis L D and Mackowski D W 1996 T-matrix computations of light scattering by nonspherical particles: A review J. Quant. Spectrosc. Radiat. Transfer 55 535–75

[3] Draine B T and Flatau P J 1994 Discrete-dipole approximation for scattering calculations J. Opt. Soc. Am. A 11 1491–9

[4] Yurkin M A and Hoekstra A G 2007 The discrete dipole approximation: an overview and recent developments J. Quant. Spectrosc. Radiat. Transfer 106 558–89
[5] Quinten M and Rostalski J 1996 Lorenz-Mie Theory for Spheres Immersed in an absorbing host medium Part Part Syst Charact 13 89–96
[6] Mundy W C, Roux J A and Smith A M 1974 Mie scattering by spheres in an absorbing medium* J. Opt. Soc. Am., JOSA 64 1593–7
[7] Mishchenko M I and Yang P 2018 Far-field Lorenz–Mie scattering in an absorbing host medium: Theoretical formalism and FORTRAN program J. Quant. Spectrosc. Radiat. Transfer 205 241–52
[8] Mishchenko M I, Dlugach J M, Lock J A and Yurkin M A 2018 Far-field Lorenz–Mie scattering in an absorbing host medium. II: Improved stability of the numerical algorithm J. Quant. Spectrosc. Radiat. Transfer 217 274–7
[9] Yurkin M A and Mishchenko M I 2018 Volume integral equation for electromagnetic scattering: Rigorous derivation and analysis for a set of multilayered particles with piecewise-smooth boundaries in a passive host medium Phys. Rev. A 97 043824
[10] Dong J, Zhang W and Liu L 2021 Discrete dipole approximation method for electromagnetic scattering by particles in an absorbing host medium Opt. Express 29 7690–705
[11] Yurkin M A and Hoekstra A G 2011 The discrete-dipole-approximation code ADDA: capabilities and known limitations J. Quant. Spectrosc. Radiat. Transfer 112 2234–47
[12] Moskalensky A E and Yurkin M A 2019 Energy budget and optical theorem for scattering of source-induced fields Phys. Rev. A 99 053824
[13] Bohren C F and Huffman D R 1983 Absorption and Scattering of Light by Small Particles (New York: Wiley)
[14] Chylek P 1977 Light scattering by small particles in an absorbing medium J. Opt. Soc. Am. 67 561–3
[15] Mishchenko M I 2007 Electromagnetic scattering by a fixed finite object embedded in an absorbing medium Opt. Express 15 13188–202
[16] Mishchenko M I and Yurkin M A 2020 Co- and counter-propagating wave effects in an absorbing medium J. Quant. Spectrosc. Radiat. Transfer 242 106688
[17] Yang P, Gao B-C, Wiscombe W J, Mishchenko M I, Platnick S E, Huang H-L, Baum B A, Hu Y X, Winker D M, Tsay S-C and Park S K 2002 Inherent and apparent scattering properties of coated or uncoated spheres embedded in an absorbing host medium Appl. Opt., AO 41 2740–59
[18] Mishchenko M I, Yurkin M A and Cairns B 2019 Scattering of a damped inhomogeneous plane wave by a particle in a weakly absorbing medium OSA Continuum 2 2362–8
[19] Mishchenko M I 2006 Scale invariance rule in electromagnetic scattering J. Quant. Spectrosc. Radiat. Transfer 101 411–5

Temperature sensor tag based on supercavity mode of dielectric resonator

Ildar Yusupov[1,5*], Dmitry Filonov[2,5], Pavel Ginzburg[3], Mikhail Rybin[1,4], Alexey Slobozhanyuk[1,5]

[1] Department of Physics and Engineering, ITMO University, Saint Petersburg 197101, Russia.
[2] Center for Photonics and 2D Materials, Moscow Institute of Physics and Technology, Dolgoprudny 141700, Russia
[3] School of Electrical Engineering, Tel Aviv University, Tel Aviv 69978, Israel
[4] Ioffe Institute, Saint Petersburg 194021, Russia.
[5] Sirius University of Science and Technology, Sochi 354340, Russia

E-mail: ildar.yusupov@metalab.ifmo.ru

Abstract. This paper presents a wireless temperature sensor design based on the excitation of a high-Q supercavity mode in a dielectric resonator. Narrow resonance bandwidth improves sensor performance enabling accurate temperature measurements. The sensor consists of a half split ceramic cylinder attached to a metal sheet. The resonator parameters which lead to the excitation of a supercavity mode were obtained numerically. When the ambient temperature increased continuously from 23 to 120°C the notable shift of the resonant frequency was experimentally demonstrated.

1. Introduction

Wireless sensing techniques are widely used in modern industrial and civil technology systems. Current devices make it possible to determine with high accuracy the parameters of the environment, or the parameters of specific components as part of a complex industrial system [1]. In harsh environmental conditions, the use of chipped technologies requires complicated solutions due to the presence of silicon components. Thus, chipless design of tags promises greater perspectives [2]. The absence of a chip allows chipless sensors to be integrated into systems where the use of silicone components is impossible. Chipless sensors are known to use to determine pressure [3], humidity [4], structural health monitoring systems [5], and many others. Such sensors represent a resonator or a set of resonators, which response allows assessing the change in the state of the object on which the chipless tag is placed. Chipless sensors are especially common in temperature monitoring systems. They can be based on dielectric substrates [6] or with the use of ceramic resonators. Ceramic materials were shown to be effective as RFID tags, which can be miniaturized and can have a relatively high reading range [7].

Nowadays there are several examples of exploiting ceramic resonators as a chipless tag for temperature sensing [8]. The operation principle of such sensors is based on the shift of resonant frequency of dielectric resonator fundamental mode due to heating. High-quality (Q) resonances can significantly increase the efficiency of the sensing tag. Narrow resonance bandwidth simplifies the reader operation and allows high accuracy measurements.

The proposed temperature sensor concept is based on a similar principle reported earlier in Refs. [9] and [10]. The main difference is the use of dielectric resonators supporting a supercavity mode. The observation of supercavity modes has been already shown for optics [11], and recently it was experimentally demonstrated for microwave range [12]. This property of a ceramic cylindrical resonator allows us to obtain responses with Q factor above 10^4. The structure of the proposed chipless sensor is shown in figure 1a. It consists of half a dielectric cylinder placed on a metal screen. The cylinder parameters are optimized to maintain the supercavity mode. The resonator is excited by a loop antenna connected to a vector network analyzer. The loop antenna is oriented along the x-axis. The obtained reflectance spectra are used to measure the ambient temperature.

2. Supercavity mode of dielectric resonator

Supercavity modes are realistic counterpart of an ideal mathematical concept of bound states in the continuum [13]. Supercavity modes can be sustained even by a single dielectric resonator due to the destructive interference of a pair of leaky cylinder modes with the same far-field profiles [11]. Adjusting for the destructive interference condition occurs by changing the aspect ratio of the cylinder resonator.

A numerical simulation using CST Microwave Studio was carried out to determine the parameters of the high-quality resonator. The requirement for the appearance of a supercavity mode is the ratio of the cylinder radius to its height as $r/h = 0.7$. As a result, the obtained parameters of the resonator: radius $r = 15$ mm, height $h = 21.42$ mm, permittivity $\varepsilon = 506$, loss tangent of ceramics $\tan\delta = 0.0004$. The optimized cylinder was cut in half and placed on the perfect electric conducting boundary. The result of numerical simulations is shown in figure 1b. The spectrum of the reflectance obtained by the loop antenna has a narrow resonance of -23 dB at 859.6 MHz. The inset shows the distribution of the absolute value of the magnetic field amplitude at the resonance. This distribution of the electromagnetic field is typical for the supercavity mode [12]. A change in the dielectric constant of the ceramic material leads to a shift in the resonance frequency of the mode under study. As the radius to height ratio remains constant, such a change of permittivity does not affect the supercavity mode parameters.

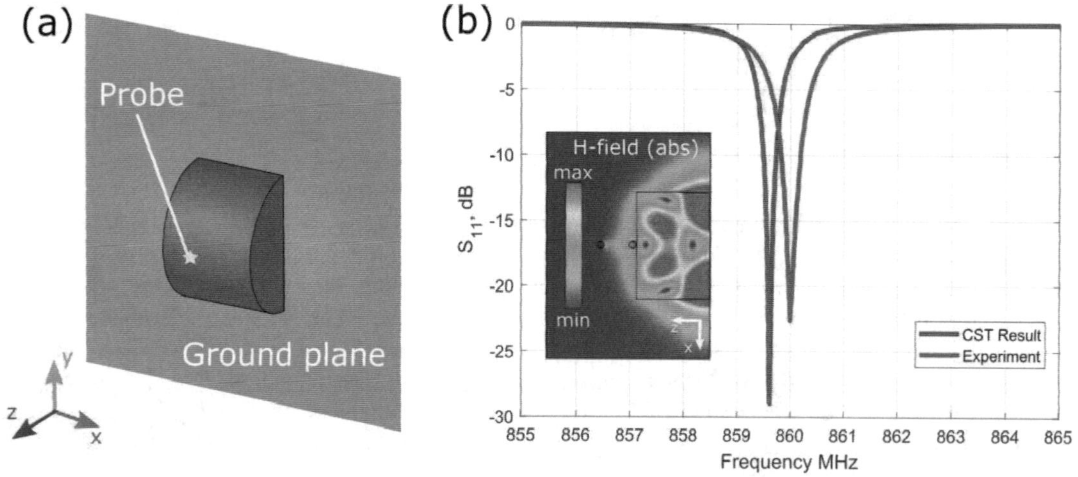

Figure 1. (a) Configuration of the proposed temperature sensor. Dielectric resonator with the shape of half-cylinder attached to a metallic ground plane. The resonator is excited by a loop probe connected to a vector network analyzer. The probe position is marked as a yellow star. The probe axis is oriented along the x-axis. (b) Numerically simulated and measured reflection coefficient of the resonator. Inset shows the magnetic field amplitude of supercavity mode.

3. Experimental demonstration

We performed experimental studies to demonstrate the temperature sensing effect. The experimental setup is shown in figure 2a. Half of the ceramic resonator with the parameters specified above was placed on a metal sheet with dimensions of 30x30 cm^2. The loop probe connected to the vector network analyzer was placed near the ceramic resonator, and the reflection coefficient was measured. Figure 1b shows a comparison of the numerically obtained and measured reflection coefficients. Both spectra are in an excellent agreement; the small deviation is explained by the inaccuracy of determining the exact parameters of the cylinder.

Let us describe the results on the demonstration of the temperature dependence of the resonant frequency. The ceramic resonator was heated with an air solder, and the temperature was controlled by a thermocouple sensor connected to the multimeter. Figure 2b depicts selected spectra corresponding to temperatures of 23, 40 and 80°C. The complete responsibility of the resonant frequency shift to ambient temperature from 23 to 120°C is shown in figure 2c. The resonant frequency gradually increases from 856 to 1115 MHz almost with linear temperature dependence.

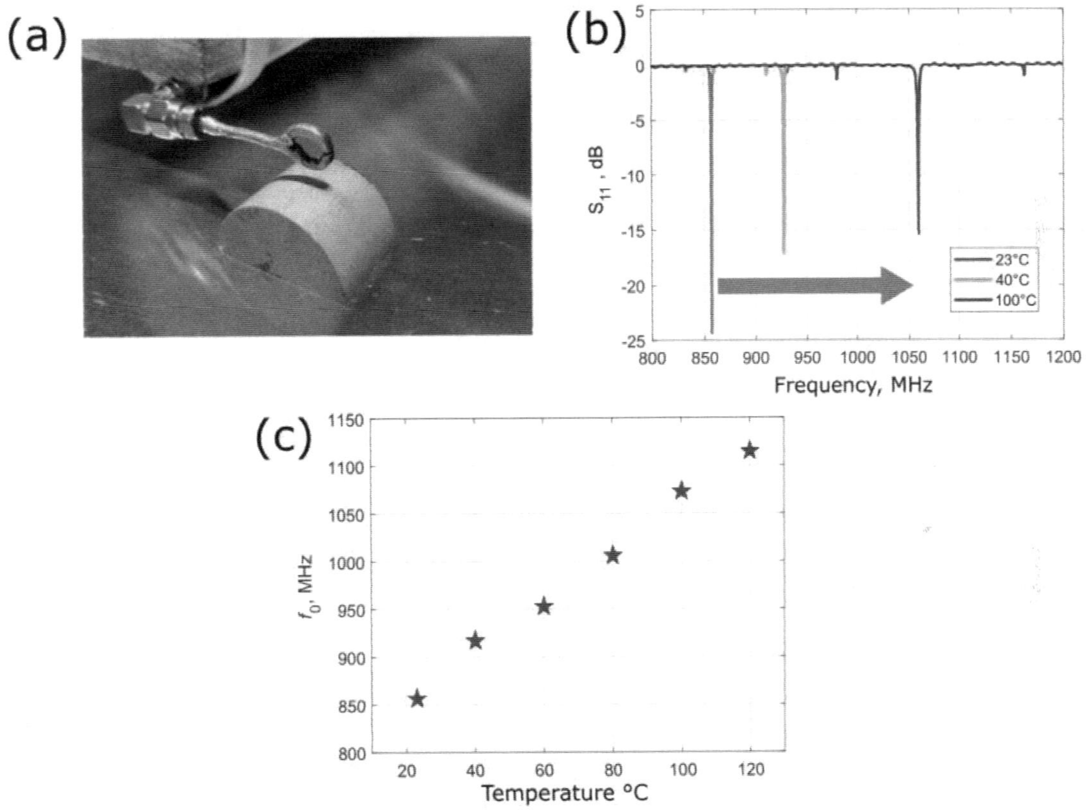

Figure 2. Experimental demonstration of resonance frequency shift due to increased ambient temperature. (a) Photo of the experimental setup. (b) Reflection spectra for temperatures of 23 (red curve), 40 (green) and 80°C (blue). The arrow shows direction of the resonance shift. (c) Measured resonance frequency versus temperature.

4. Conclusion

We have demonstrated the passive wireless temperature sensor's operation based on the dielectric resonator supercavity mode's resonant frequency shift. The proposed sensor has simple construction consisting of the half split ceramic cylinder attached to the metal sheet. The supercavity mode has been excited by the loop antenna and the spectra of the reflection coefficient have been measured. In the experiment we have confirmed the shift of the resonant frequency with the ambient temperature increasing continuously from 23 to 120°C. The future work will demonstrate the operation of the proposed sensor tag over broader temperature ranges. The far-field interrogation of the sensor using a directional antenna will also be investigated in the future.

Acknowledgments
The numerical and experimental results were supported by the Russian Foundation for Basic Research (Grant No. 20-37-51011). Materials for the experiment were supported by Russian Science Foundation (Grant No. 21-79-30038).

References
[1] L. Da Xu, W. He, and S. Li, "Internet of things in industries: A survey," *IEEE Trans. Ind. Informatics*, vol. 10, no. 4, pp. 2233–2243, 2014
[2] V. Mulloni and M. Donelli, "Chipless RFID sensors for the internet of things: Challenges and opportunities," *Sensors (Switzerland)*, vol. 20, no. 7, 2020
[3] C. Schuster, P. Schumacher, M. Schusler, A. Jiménez-Sáez, and R. Jakoby, "Passive chipless wireless pressure sensor based on dielectric resonators," *Proc. IEEE Sensors*, vol. 2017-Decem, pp. 1–3, 2017
[4] Y. Feng, L. Xie, Q. Chen, and L. R. Zheng, "Low-cost printed chipless RFID humidity sensor tag for intelligent packaging," *IEEE Sens. J.*, vol. 15, no. 6, pp. 3201–3208, 2015
[5] S. Dey, R. Bhattacharyya, S. E. Sarma, and N. C. Karmakar, "A Novel 'Smart Skin' Sensor for Chipless RFID Based Structural Health Monitoring Applications," *IEEE Internet Things J.*, vol. 8, no. 5, pp. 1–1, 2020
[6] X. Y. Guo, G. C. Wan, Y. M. Gao, and M. S. Tong, "A Novel Temperature Sensor Based on Chipless RFID Tags," no. 4, pp. 1–3, 2021
[7] D. Dobrykh *et al.*, "Long-range miniaturized ceramic RFID tags," *IEEE Trans. Antennas Propag.*, 2020
[8] M. Schüßler, C. Mandel, B. Kubina, and R. Jakoby, "Realization concepts for chipless wireless temperature sensing," *Sensoren und Messsyst. 2014 - 17. ITG/GMA-Fachtagung*, pp. 1–6, 2013.
[9] C. Mandel *et al.*, "Dielectric ring resonators as chipless temperature sensors for wireless machine tool monitoring," *2017 11th Eur. Conf. Antennas Propagation, EUCAP 2017*, pp. 3912–3916, 2017
[10] B. Kubina, M. Schusler, C. Mandel, A. Mehmood, and R. Jakoby, "Wireless high-temperature sensing with a chipless tag based on a dielectric resonator antenna," *Proc. IEEE Sensors*, pp. 4–7, 2013
[11] M. V. Rybin *et al.*, "High- Q Supercavity Modes in Subwavelength Dielectric Resonators," *Phys. Rev. Lett.*, vol. 119, no. 24, pp. 1–5, 2017
[12] M. Odit, K. Koshelev, S. Gladyshev, K. Ladutenko, Y. Kivshar, and A. Bogdanov, "Observation of Supercavity Modes in Subwavelength Dielectric Resonators," *Adv. Mater.*, vol. 33, no. 1, pp. 1–7, 2021
[13] M. Rybin and Y. Kivshar, "Supercavity lasing inflation identified," *Nature*, vol. 541, pp. 164–165, 2017.

Screening effect for excitonic spectrum of Coulomb coupling between two Dirac particles.

V. Zalipaev[2], **V. Kuidin**[1]

[1]Department of Physics and Engineering, ITMO University, 197101 Saint Petersburg, Russia
[2]Department of Photonics and Optoinformation, ITMO University, 197101 Saint Petersburg, Russia

Abstract. The properties of screening effect for energy spectrum of excitons in monolayer transition metal dichalcogenides are investigated using a multiband model. The excitonic Hamiltonian in the product base of the Dirac single-particle is used. The corresponding energy eigenvalue system of the first order ODE (radial equations) was solved using the finite difference method. This enables to determine the energy eigenvalues of the discrete excitonic spectrum and the wave functions. We compare the results for the energy spectrum and the corresponding eigen-functions forms for WS_2 and WSe_2 computed for two different potentials: pure Coulomb and screened Coulomb (Keldysh potential). It is demonstrated that excitonic energy levels for unscreened potential lie deeper, and the corresponding eigen-functions' forms differ from those obtained for screened one.

1. Introduction

In this paper, the properties of excitonic energy spectrum in transition metal dichalcogenide (TMD) monolayers are investigated using a multiband model. The starting point of our analysis is the excitonic matrix Hamiltonian in the product base of Dirac single-particle states constructed in [1]. Such Hamiltonian accounts for spin-orbit coupling. Following the separation of variables with respect to polar coordinates, we decouple the corresponding matrix eigenvalue problem rigorously and solve the resulting second order ODE by means of finite difference method (FDM). It is worth remarking that similar reduction for the matrix Hamiltonian energy eigenvalue problem was performed in papers [2], [3], [1], [4]. However, in these papers the reduction was approximate. In our approach the reduction to the second order resulting ODE for radial wave function is exact.

We determine excitonic energy levels and corresponding wave functions. Analysis the resulting radial ODE in both case when Coulomb or Keldysh potential is applied, we demonstrate that the discrete excitonic spectrum lies exactly below the value of the energy gap. In this paper we present a comparison analysis of the results for the energy spectrum and the corresponding eigen-functions for WS_2 and WSe_2. These were computed by means of FDM for case of two different potentials: pure Coulomb and screened Coulomb (Keldysh potential). We demonstrate that Coulomb potential levels of excitonic energy spectrum are located lower compare to eigen-levels for the Hamiltonian with Keldysh interaction potential.

According to [1], the exciton Hamiltonian is constructed in the basis $(|\phi_c^e\rangle \otimes |\phi_c^h\rangle, |\phi_c^e\rangle \otimes |\phi_v^h\rangle, |\phi_v^e\rangle \otimes |\phi_c^h\rangle, |\phi_v^e\rangle \otimes |\phi_v^h\rangle)^T$, i.e. the basis spanning the total Hilbert space that is given by the set of all possible tensor products combinations for the atomic orbital states of individual particles at the conduction and valence band edges. Using the orthonormality of the basis functions, the total exciton Hamiltonian (the symbol of the operator) could be written in the form introduced in [1] (see formulae (4) and (5)) with the Keldysh potential $-V(r)$ in polar coordinates that is given by

$$V(r) = V_0 \frac{\pi}{2}\left(H_0\left(\frac{r}{r_0}\right) - Y_0\left(\frac{r}{r_0}\right)\right), \qquad V_0 = \frac{e^2}{4\pi\varepsilon_0\chi r_0 t_h}, \quad \chi = \frac{\varepsilon_1+\varepsilon_2}{2}, \tag{1}$$

where $Y_0(r)$ and $H_0(r)$ are Newmann and Struve functions, respectively, r_0 is the screening length (the characteristic space scale of the problem), t_h is the hopping parameter, $\varepsilon_{1,2}$ is the dielectric constant of the environment above and below the TMD monolayer. It is worth noting that $at_h = v_F\hbar$, where v_F is the Fermi velocity and a is the lattice constant. For the case of Coulomb potential we have $V(r) = V_0(r/r_0)^{-1}$.

The eigenvalue problem for this Hamiltonian $H^{exc}_\alpha(\boldsymbol{k}^e, \boldsymbol{k}^h, r_{eh})|\Psi_{\alpha,n}\rangle = E^{exc}_{\alpha,n}(\boldsymbol{k}^e, \boldsymbol{k}^h)|\Psi_{\alpha,n}\rangle$ defines the exciton energy $E^{exc}_{\alpha,n}(\boldsymbol{k}^e, \boldsymbol{k}^h, r_{eh})$ and the exciton eigenstate $|\Psi_{\alpha,n}\rangle$, where α is a notation for the set of the spin and the valley indices s_e, τ_e, s_h, τ_h with the values ±1 for the electron-hole Dirac particles, $\boldsymbol{k}^e, \boldsymbol{k}^h$ are wave vectors, n means a set of quantum numbers of the excitonic states of discrete spectrum ([1]). The above eigenvalue problem is a matrix equation which, following a procedure analogous to [1] and [3], can be decoupled to a single ODE of the second order.

2. Excitonic Hamiltonian and separation of variables.

Let us consider the excitons with zero center-of-mass momentum $\boldsymbol{K} = \boldsymbol{k}_e + \boldsymbol{k}_h = 0$. Introducing $\boldsymbol{k} \equiv \boldsymbol{k}_e = -\boldsymbol{k}_h$ and using transformation to dimensionless polar coordinates $\partial_x = \cos\varphi\,\partial_r - \frac{\sin\varphi}{r}\,\partial_\varphi$, $\partial_y = \sin\varphi\,\partial_r + \frac{\cos\varphi}{r}\,\partial_\varphi$, we can write exciton Hamiltonian (see Ref. [1]) in the following way:

$$\begin{pmatrix} -V(r) & -he^{i\tau_h\varphi}D_h^- & -he^{-i\tau_e\varphi}D_e^+ & 0 \\ -he^{-i\tau_h\varphi}D_h^+ & \Delta_h - V(r) & 0 & -he^{-i\tau_e\varphi}D_e^+ \\ -he^{i\tau_e\varphi}D_e^- & 0 & -\Delta_e - V(r) & -he^{i\tau_h\varphi}D_h^- \\ 0 & -he^{i\tau_e\varphi}D_e^- & -he^{-i\tau_h\varphi}D_h^+ & \Lambda - V(r) \end{pmatrix}. \tag{2}$$

We define here differential operators $D^\pm_{e,h} \equiv i\tau_{e,h}\partial_r \pm \partial_\varphi/r$ and quantities $\Delta_{e,h} \equiv \Delta - \lambda s_{e,h}\tau_{e,h}$ (the effective band-gap, see [1]) with λ being the spin-orbit coupling strength, $\Lambda \equiv \lambda(s_e\tau_e - s_h\tau_h)$, and $h = a/r_0 = v_F\hbar/(r_0 t_h)$ is the dimensionless parameter. We have also expressed energy quantities (Δ, λ, $V(r)$ etc.) in the hopping parameter t units and the spatial variables in the screening length r_0 ones.

Due to spin splitting of the valence band, there are effectively two band gaps and, as a consequence, two different types of excitons (see [1]). These are commonly referred to as A and B excitons. The method presented in this paper can be applied with both values of $s_h\tau_h = \pm1$. When $s_h\tau_h = 1$ the Hamiltonian (2) describes the A exciton, while $s_h\tau_h = -1$ relates to the B exciton. The A excitons in the K and K' valley have $s_e = 1$, $\tau_e = 1$, $s_h = -1$, $\tau_h = -1$ and $s_e = -1$, $\tau_e = -1$, $s_h = 1$, $\tau_h = 1$, respectively. The B exciton has $s_e = -1$, $\tau_e = 1$, $s_h = 1$, $\tau_h = -1$. Hence, we obtain $\Lambda = 0$. For the A exciton in the K and K' valley (see [1]) we get $\Delta_h = \Delta_e = \Delta - \lambda = \Delta_{s,\tau}$. For the B exciton in the K valley (see [1]) we get $\Delta_h = \Delta_e = \Delta + \lambda$. Therefore, we can express Δ_h and Δ_e in terms of $\Delta_{s,\tau}$.

In order to separate variables for the A and B excitons in the K valley, we use the ansatz

$$\psi(r,\varphi) = e^{il\varphi}\left(\psi_l^{(1)}(r)e^{-i\varphi},\ \psi_l^{(2)}(r),\ \psi_l^{(3)}(r),\ \psi_l^{(4)}(r)e^{i\varphi}\right)^T,$$

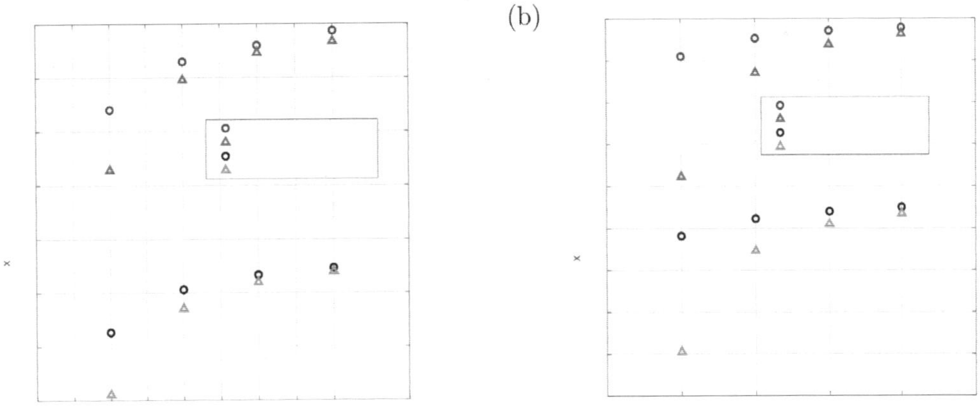

Figure 1. The excitonic spectrum E_{nl} for the s- and p-states ($l = 0, 1$) shown in (a) and (b), correspondingly, with $n = 0, 1, 2, 3$ for two examples of WS_2 and WSe_2, computed for Keldysh potential (circles) and for Coulomb potential (triangles).

where $l \in \mathbf{Z}$ is an orbital quantum number. Thus, the Hamiltonian (2) acting on $\psi_l(r) = \left(\psi_l^{(1)}(r), \psi_l^{(2)}(r), \psi_l^{(3)}(r), \psi_l^{(4)}(r)\right)^T$ takes the form $\hat{H}_l\psi_{nl}(r) = E_{nl}\psi_{nl}(r)$, $n = 0, 1, 2, ...$, with the radial quantum number n, and can be represented as the following system of four first-order ODE:

$$\begin{cases} V_E\psi_l^{(1)} + hi\left(\partial_r + \frac{l}{r}\right)\psi_l^{(2)} - hi\left(\partial_r + \frac{l}{r}\right)\psi_l^{(3)} = 0, \\ \left(\partial_r - \frac{l-1}{r}\right)\psi_l^{(1)} - ih^{-1}(\Delta_{s,\tau} + V_E)\psi_l^{(2)} - \left(\partial_r + \frac{l}{r}\right)\psi_l^{(4)} = 0, \\ \left(\partial_r - \frac{l-1}{r}\right)\psi_l^{(1)} + ih^{-1}(-\Delta_{s,\tau} + V_E)\psi_l^{(3)} - \left(\partial_r + \frac{l}{r}\right)\psi_l^{(4)} = 0, \\ -ih\left(\partial_r - \frac{l-1}{r}\right)\psi_l^{(2)} + ih\left(\partial_r - \frac{l-1}{r}\right)\psi_l^{(3)} + V_E\psi_l^{(4)} = 0, \end{cases} \tag{3}$$

where $V_E = -V(r) - E$. Transforming vector components $\psi_l(r)$ into $\varphi_l^{(1,4)} = \psi_l^{(1)} \pm \psi_l^{(4)}$, $\varphi_l^{(2,3)} = \psi_l^{(2)} \pm \psi_l^{(3)}$, we obtain a new system of ODE for the components of a new vector $\varphi_l(r) = (\varphi_l^{(1)}, \varphi_l^{(2)}, \varphi_l^{(3)}, \varphi_l^{(4)})^T$. This system is reduced into a scalar second-order ODE:

$$\left(\varphi_l^{(3)}\right)'' + \left(\frac{1}{r} - \frac{V'}{V_E}\right)\left(\varphi_l^{(3)}\right)' + \left(\frac{V_E^2 - \Delta_{s,\tau}^2}{4h^2} - \frac{l^2}{r^2}\right)\varphi_l^{(3)} = 0. \tag{4}$$

As soon as we solved spectral problem (4) for $0 < E < \Delta_{s,\tau}$ and found $\varphi_l^{(3)}(r)$, one could evaluate other components of $\varphi^l(r)$ and, finally, the components $\psi^l(r)$.

3. Screening effect for A exciton in K valley - numerical analysis

We consider the special case $s_e = \tau_e = 1$, $s_h = \tau_h = -1$ which corresponds to A exciton in K valley (see [1]). We solve our spectral problem for (4) using FDM. We take into account that the eigenfunction must decay exponentially. Fig. 1 shows comparison between the excitonic spectra E_{nl} for the s- and p-states ($l = 0, 1$) with $n = 0, 1, 2, 3$ obtained for Keldysh and Coulomb potentials in WS_2 with the parameters $\Delta_{s,\tau} = 2.4$ eV, $r_0 = 3.789$ nm, $a = 0.3197$ nm, $t_h = 1.25$ eV, and WSe_2 with $\Delta_{s,\tau} = 1.97$ eV, $r_0 = 4.511$ nm, $a = 0.3317$ nm , and $t_h = 1.13$ eV. Both examples involve the SiO_2 substrate with $\varepsilon = 3.9$. In Fig. 2, the probability distributions data for the case $n = 0$ and $l = 0$ are shown for Keldysh potential - (a) and for the Coulomb potential - (b), presenting $|\psi_{00}^{(1,2,3)}|^2$ dependence on r for WS_2.

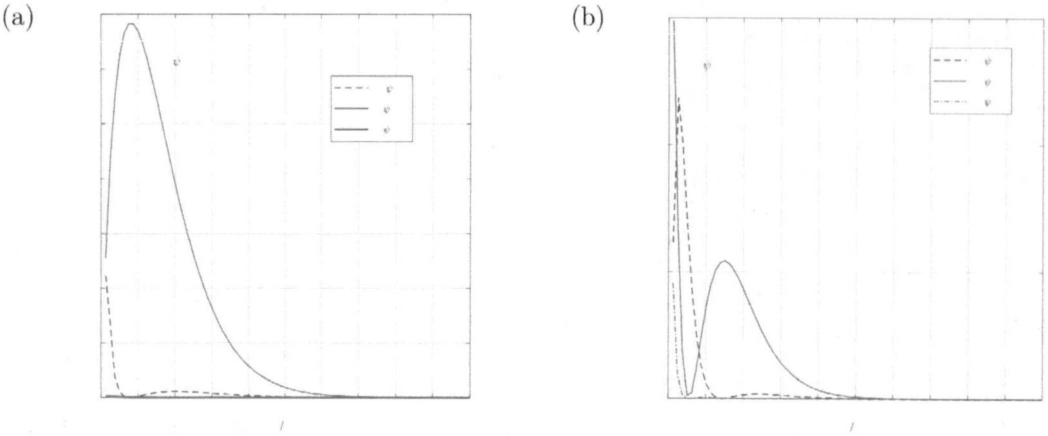

Figure 2. The probability distributions data for the case $n = 0$ and $l = 0$, shown in (a) for Keldysh and in (b) for Coulomb potentials, presenting $|\psi_{00}^{(1,2,3)}|^2$ dependence on r for WS_2.

We would like to remark that the binding energy $E_b = \Delta_{s,\tau} - E_{00}$, computed from experimental data for WS_2 and WSe_2 (see [5], [6]) is more efficiently described by Coulomb potential. Namely, we have $E_b = 0.29$ eV (see [5]), and the values for Coulomb and Keldysh potentials are $E_b = 0.272$ eV and $E_b = 0.18$ eV, respectively. Regarding the experimental data for WSe_2 we obtain $E_b = 0.31$ eV (see [6]), and the values for Coulomb and Keldysh potentials are $E_b = 0.259$ eV and $E_b = 0.16$ eV, respectively. Thus, one could see that the Coulomb potential model is more realistic for the discussed examples.

4. Conclusion

The energy spectrum of excitons in monolayer TMD has been calculated using a multiband model. Following the separation of variables, we decoupled the corresponding system of the first order ODE for the radial eigen-vector components rigorously and solved the resulting second order ODE by means of FDM. We compared the results for the energy spectrum and the corresponding eigen-functions for WS_2 and WSe_2 computed for two different potentials: Coulomb and Keldysh. It was shown that the Coulomb levels of excitonic energy spectrum are located lower, and the corresponding eigen-functions look differently. The energy levels perturbation appears stronger for the lowest states. The screening contributes most to the ground state shift.

5. Acknowledgments

The authors would like to thank Dr A. Vagov, Dr D. Gulevich for helpful discussions. This work is supported by the Russian Science Foundation under the grant 18-12-00429.

6. References

[1] Van der Donck M, Zarenia M and Peeters F 2017 *Physical Review B* **96** 035131
[2] Berman O L, Kezerashvili R Y and Ziegler K 2013 *Physical Review A* **87** 042513
[3] Berman O L and Kezerashvili R Y 2016 *Physical Review B* **93** 245410
[4] Trushin M, Goerbig M O and Belzig W 2016 *Physical Review B* **94** 041301
[5] Chernikov A, Berkelbach T C, Hill H M, Rigosi A, Li Y, Aslan O B, Reichman D R, Hybertsen M S and Heinz T F 2014 *Physical review letters* **113** 076802
[6] He K, Kumar N, Zhao L, Wang Z, Mak K F, Zhao H and Shan J 2014 *Physical review letters* **113** 026803

Numerical Study of Non-Radiating Near-Field Wireless Power Transfer System

E Zanganeh[1], M Song[2,1], M Korobkov[1], A Evlyukhin[3,1], A Miroshnichenko[4], P Kapitanova[1]

[1]Department of Physics and Engineering, ITMO University, Saint Petersburg, 199034, Russia

[2]College of Information and Communication Engineering, Harbin Engineering University, Harbin, 150001, China

[3]Institute of Quantum Optics, Leibniz Universität Hannover, Hannover, 30167, Germany

[4]School of Engineering and Information Technology, University of New South Wales Canberra, ACT, 2600, Australia

E-mail: esmaeel.zanganeh@metalab.ifmo.ru

Abstract. The main challenge in near-field wireless power transfer systems is the increase of power transfer efficiency. It can be achieved by reducing ohmic or radiation losses of the resonators included in the system. In this paper, we propose and investigate numerically a non-radiating near-field wireless power transfer system based on transmitter and receiver implemented as dielectric disk resonators. The transmitter and receiver geometrical parameters are numerically optimized to operate at the frequency of non-radiating state of high refractive index dielectric resonators instead of magnetic dipole mode. Under the non-radiating state, we determine the frequency with almost zero radiation to the far-field. We numerically study the wireless power transfer efficiency as a function of operation distance between the transmitter and receiver and demonstrate that the higher efficiency compared to magnetic dipole mode can be achieved at non-radiating state for a fixed distance due to suppression of the radiation loss.

1. Introduction

Improving power transfer efficiency and transmission range in wireless power transfer (WPT) systems by reducing the loss has attracted much attention during the last decade. Recently, WPT systems based on dielectric resonators have been proposed and studied [1-2]. The main reasons to use dielectric resonators instead of metal ones are (i) the minimization of ohmic loss, and (ii) minimization of radiation loss due to operation at higher-order modes. Thus, the power transfer efficiency of the WPT system operating at magnetic quadrupole mode can reach 83%, while the WPT efficiency at magnetic dipole (MD) mode operation is only 60 % [1]. Soon after that, a WPT system based on dielectric disk resonators with colossal permittivity operating at MD mode was demonstrated [2] and revealed 90%

efficiency at a distance of 4 cm between the transmitter and the receiver. However, there is still room to improve further the power transfer efficiency of WPT systems based on dielectric resonators.

Non-radiating (NR) sources have been recently introduced and studied [3-4]. In [3] the authors demonstrated an NR source based on four high permittivity dielectric cylinders excited by an electric dipole antenna. The NR source is formed from the radiation suppression through the destructive interference between electric toroidal dipole excited inside the cylinders and electric dipole radiation from the dipole antenna. In another work, Zurita-Sánchez [4] theoretically studied an NR source formed by electric dipole radiation cancellation between an electric point-like antenna placed near a subwavelength sphere and induced electric dipole radiation inside the sphere.

In this paper, we propose and study numerically non-radiating WPT system where the transmitter and receiver are implemented as dielectric disk resonators excited by a metal loop. The main novelty is the suppression of far-field radiation of the transmitter and receiver due to the operation at the frequency of NR state. The numerical optimization of the transmitter (receiver) parameters is done to find the frequency of the NR state characterized by the cancelation of far-field emission. The WPT efficiency as a function of the distance between the transmitter and the receiver is numerically evaluated and compared with the WPT system efficiency operating at MD mode.

2. Design of non-radiating transmitter and receiver

Let us first consider a single dielectric disk resonator made of ceramic with permittivity of ε=1000 and loss tangent tanδ= 2.5×10^{-4}. The thickness and radius of the disk are set as h=4.1 mm and R=42 mm, respectively. First, we studied the eigenmodes of the disk resonator in CST Microwave Studio and revealed two resonances at 278 MHz and 408.6 MHz. In order to excite these modes from near-field one can use a metal loop as shown in Figure 1 (a). This structure we now consider as a transmitter (receiver) of the WPT system.

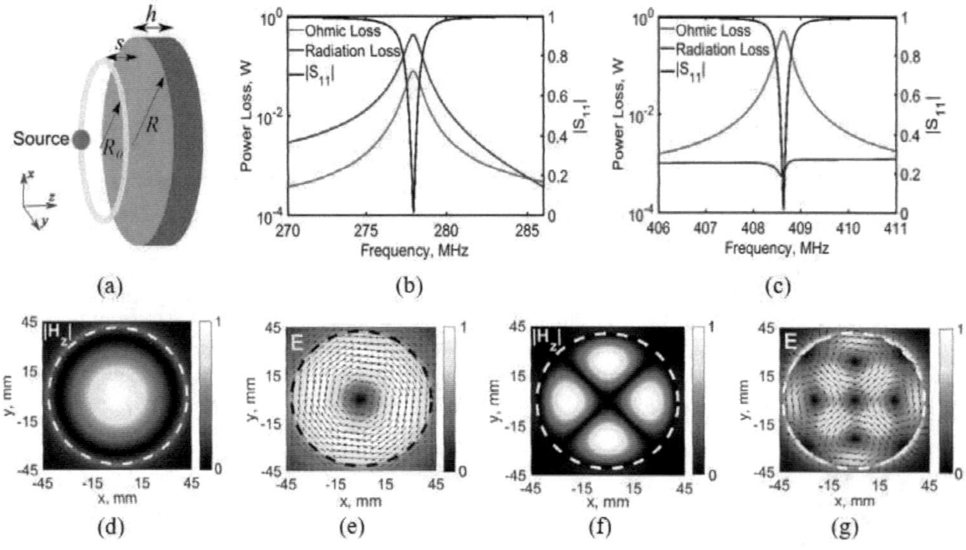

Figure 1. (a) Schematic view of the proposed transmitter (receiver). Simulated reflection coefficient, ohmic loss and radiation loss at MD mode resonance (b) and the frequency of NR state (c). Simulated normal component of the magnetic (d) and electric (e) fields at the frequency f= 278 MHz. Simulated normal component of the magnetic (f) and electric (e) fields at the frequency f=408.6 MHz. The white (black) dashed circle lines represent the border of the dielectric disk.

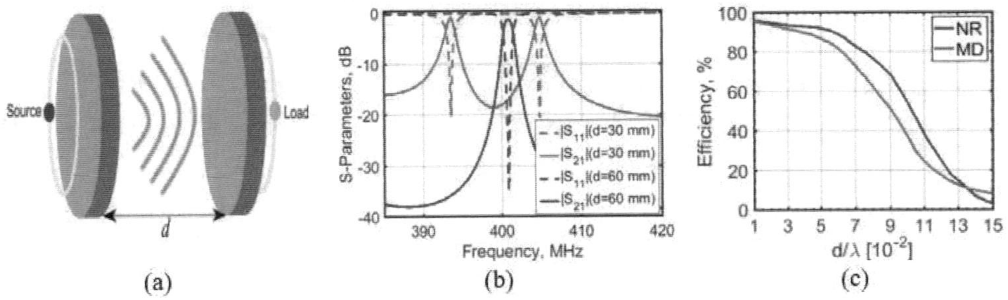

Figure 2. f (a) Schematic view of the WPT system. (b) S-parameters of the WPT system operating at the frequency of NR state for different separation distances d. (c) Results of numerical simulations of the WPT system efficiency operating at NR state and MD mode as functions of the separation distance. The normalization of distance to the corresponding wavelength to compare two WPT.

3. WPT system efficiency calculation

The revealed non-radiating property of the transmitter (receiver) at the frequency 408.6 MHz may help to improve the power transfer efficiency of the WPT system by getting rid of the radiation loss in the transmitter and receiver. Thus, we numerically studied the WPT efficiency of the system based on transmitter and receiver operating at MD mode and the frequency of NR state. The transmitting and the receiving parts are separated by a distance *d*, as shown in Figure 2(a). The power transfer efficiency can be written in terms of S-parameters as follows:

$$\eta = \frac{|S_{21}|^2}{1-|S_{11}|^2},$$

where the $|S_{21}|$ and $|S_{11}|$ are the transmission and reflection coefficients of the WPT system, which are obtained in CST Microwave Studio with help of frequency domain solver.

Figure 2 (b) shows simulated S-parameters of the WPT system operating at the frequency of NR state for two specific separation distances *d*. For *d*=60 mm, the optimal coupling conditions are satisfied and we observe single maximum of the transmission coefficient at 401 MHz. For *d*=30 mm, we observe the frequency splitting phenomena when the transmission coefficient has two maximums around 394 MHz and 404 MHz. It can be explained by the strong coupling between the transmitter and receiver. This phenomenon has also been observed earlier for WPT systems operating at MD and magnetic quadrupole modes [[2]], [[3]]. The WPT system efficiency is calculated at the frequency corresponding to the maximum of the transmission coefficient. For the strong coupling regime, we fix the frequency of the first maximum of the transmission coefficient.

The calculated efficiencies of the systems operating at MD mode and NR state are shown in Figure 2 (c) as a function of the separation distance *d*. We normalized the distance by the corresponding wavelength to compare the efficiencies of the WPT systems working at different frequencies. The system with the operational frequency of the MD mode, provides 90% efficiency within a short distance. This value drops with the increasing of separation distance. The WPT system operating at the NR state, reaches 96% efficiency. The efficiency decreases slower over the separation distance, which demonstrates the advantage of NR state towards the MD mode.

4. Conclusion

We have proposed and studied numerically the non-radiating WPT system. The transmitter and receiver of the system are based on dielectric disk resonator excited by metal loop which are optimized to resonate at NR state and provide almost zero radiation loss at the same time. We have demonstrated that, by getting rid of the transmitter and receiver radiation loss, the 96% of efficiency can be obtained which is higher compared to WPT system operating at MD mode. To better understand the physics of NR state described in this paper, multipole analysis is needed, which will be done later and published elsewhere. The experimental investigation of the WPT system characteristics is in the progress.

5. Acknowledgment

This work was supported by the Russian Science Foundation (Project No. 21-79-30038).

References

[1] A. Karalis, J. D. Joannopoulos, and M. Soljacic, Ann. Phys. 323(1), 34–48 (2008).
[2] M. Song, I. Iorsh, P. Kapitanova, E. Nenasheva, P. Belov, Appl. Phys. Lett. 108, 023902, (2016).
[3] M. Song, P. Kapitanova, P. Belov, Appl. Phys. Lett., 109, 223902, (2016).
[4] N.A. Nemkov, I.V. Stenishchev, and A.A. Basharin, Scientific Reports7, 1 (2017).
[5] J.R. Zurita-S´anchez, Physical Review Research1,033064 (2019).
[6] A.B. Evlyukhin, T. Fischer, C. Reinhardt, and B.N. Chichkov, Physical Review B94, 205434 (2016).
[7] E. A. Gurvitz, K. S. Ladutenko, P. A. Dergachev, A. B. Evlyukhin, A. E. Miroshnichenko, and A. S. Shalin, Laser & Photonics Reviews, 13,1800266 (2019).

Magnetic resonance imaging with a multi-tunable metamaterial-inspired radiofrequency coil

V Puchnin[1], V Ivanov[1], M Gulyaev[2] and M Zubkov[1]

[1] Department of Physics and Engineering, ITMO University, Saint Petersburg, Russia
[2] Faculty of Fundamental Medicine, Lomonosov Moscow State University, Moscow, Russia

E-mail: m.zubkov@metalab.ifmo.ru

Abstract. We present the initial experimental results obtained using a two-part receive/transmit (Rx/Tx) radiofrequency (RF) coil design for small animals magnetic resonance imaging at 7 T. The assembly uses a butterfly-type coil tuned to 300 MHz for scanning the ^{1}H nuclei and a non-resonant antenna with a metamaterial-inspired resonator tunable over wide frequency range for X-nuclei. ^{1}H, ^{31}P, ^{23}Na and ^{13}C are selected as test nuclei in this work. Coil simulations show the two parts of the RF-assembly to be efficiently operating at the required frequencies. Simulations and phantom imaging show sufficiently homogeneous transverse transmit RF fields and tuning capabilities for the pilot heteronuclear experiments.

1. Introduction

Small-animal MRI is a powerful noninvasive tool for preclinical research. It provides critical information on disorders or novel substances by studying in detail the physiological changes in various organs of the animal. Such research benefits from image quality improvement and scanning methods diversity expansion, although the two often lead to one another. This, in turn, results in the preclinical small-animal imaging being predominantly performed in high and ultra-high magnetic fields.

On the other hand, the common trend of increasing the static magnetic field [1] in MRI and the associated sensitivity growth has recently led to the possibility of routinely obtaining *in-vivo* non-hydrogen nuclei images without isotope enrichment. Nevertheless, imaging more than one non-hydrogen nucleus, while obtaining reference ^{1}H images requires a multi-tunable coil design [2, 3, 4]. A principle of such operation using a metamaterial-inspired design has recently been demonstrated [5] along with the comparison of metamaterial-based device performance to some classical radiofrequency (RF) coil solutions [6].

2. Methods

Here, a multi-tuned RF coil design previously suggested for 11.7 T imaging [5] was adapted for 7 T imaging and experimentally tested. The design comprised two separate Rx/Tx elements placed on the opposite sides of the object to be imaged. The two parts corresponded to the butterfly-type RF coil for ^{1}H imaging and a metamaterial-inspired multi-tunable coil for X-nucleus imaging (Fig. 1).

The ^{1}H 40 × 60 mm^2 copper butterfly coil was implemented on a PCB substrate with dimensions 65 × 41 × 1.5 mm^3 made from FR-4 material with $\varepsilon = 4.3$ and $\tan\delta = 0.033$ at

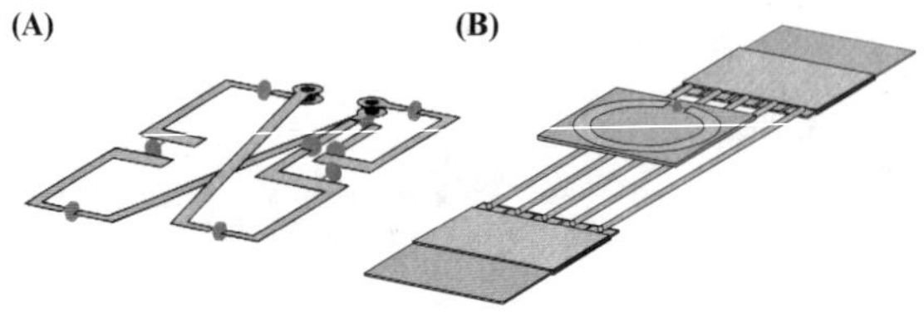

Figure 1. A – ^{1}H butterfly-type coil model with tuning/matching capacitors. B – X-nuclei tuneable coil model. Feeding of the X-nuclei coil is provided via inductive coupling to a small loop coil.

100 MHz (Fig. 1, A). Adaptation of the ^{1}H-imaging part from the original design was performed by selecting the necessary capacitors (six SMD capacitors 8.2 and 33 pF, and two variable capacitors SGNMA3T10005 1–10 pF).

The metamaterial-inspired X-nuclei coil was assembled using 5 telescopic brass wires connected at their ends to metallized PCBs substrates made from Arlon AD1000 with $\varepsilon = 10$ and $\tan\delta = 0.002$ in the operating frequency range (70-300 MHz) with dimensions $71 \times 49 \times 0.5$ mm^3. The metallized part consisted of 5 patches with dimensions of 30×10 mm^2. The capacitive load was provided by overlapping the patches with two metalized Arlon PCB substrates with dimensions $34.5 \times 49 \times 0.5$ mm^3 (Fig. 1, B). This structure supported a number of different eigenmodes [7], with the lowest frequency eigenmode having been selected for X-nucleus imaging. The mode tuning was implemented similarly to the original design, i.e., by changing the length of the metal wires and the overlap of patches with small PCB boards [5]. The mode excitation was provided via a small nonresonant loop antenna.

The proposed two-coil design was simulated with the load comprising a simple homogeneous rectangular phantom ($\varepsilon = 59$, $\sigma = 0.79$ S/m) with dimensions $154 \times 44 \times 37$ mm^3. All electromagnetic simulations were performed in CST Studio Suite 2020 in frequency domain solver.

Next, the assembled coil S-parameters were bench-measured using vector network analyzer. For the S_{11} parameter measurements the VNA was connected using singe port which powered either the feeding loop coil of the meatasurface-inspired part, or the butterfly coil.

In order to assess the RF-assembly imaging capabilities it was tested in a horizontal 300-mm-bore 7T scanner. The coil magnetic field distribution was measured and compared with the simulation results. Imaging was performed on a multi-nuclear phantom produced according to a 7 T muscle phantom recipe [8]. The experimental field maps were acquired after tuning the ^{1}H coil part to the 300.8 MHz and the X-coil part to the 79.6 MHz for sodium imaging.

Imaging capabilities were assessed on a 3-section 3D-printed phantom with different solutions in each section for ^{1}H and ^{23}Na imaging. The solutions in 3 sections were variations of the 7T muscle phantom recipe above with NaH_2PO_4 being substituted in the 2nd and 3rd sections by 2:1 mixture of $MnCl_2$ and NaCl, and pure $MnCl_2$ respectively. ^{1}H-imaging was done using a FLASH pulse sequence with TE = 3 ms, TR = 150 ms, 1 ms block excitation pulse, 100×100 mm^2 coronal plane FoV, a 100×100 matrix and a 2 mm-thick slice. ^{23}Na-imaging was done using a FLASH pulse sequence with TE = 3.28 ms, TR = 10.45 ms, 0.45 ms block excitation pulse, 100×100 mm^2 coronal plane FoV, a 128×128 matrix, 8192 averages and a 40 mm-thick slice.

3. Results

The S-parameter spectra were obtained for the operating ranges of X-coil and H-coil in the ranges of 60-140 MHz and 250-350 MHz, respectively. The results showed the possibility of

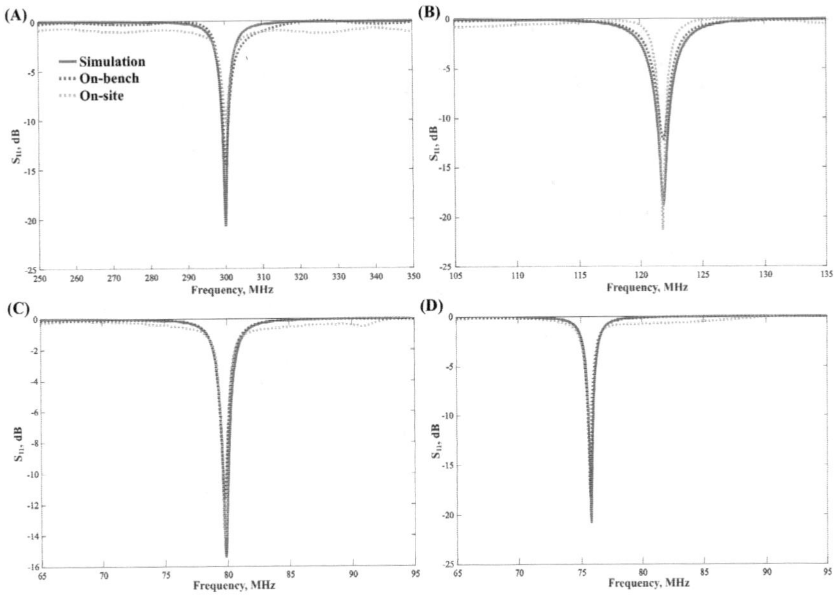

Figure 2. S_{11}-parameters of the coil assembly. A – ^{1}H coil tuning (at 300.3 MHz) calculated in simulation and measured on-bench and after the coil was installed in the scanner. B – X-coil tuning, when tuned to the ^{31}P resonant frequency (at 121.8 MHz). C - X-coil tuning, when tuned to the ^{23}Na resonant frequency (at 79.7 MHz). D - X-coil tuning, when tuned to the ^{13}C resonant frequency (at 75.7 MHz).

configuring a multiresonance coil for the reception and transmission on some of the the nuclei of interest frequencies (Fig. 2). For all spectra, the reflection coefficient at the work frequency was less than -10 dB. It confirms the possibility of tuning the metamaterial-inspired resonator to the Larmor frequencies of some nuclei of interest (^{31}P - 121.8 MHz (Fig. 2,B), ^{23}Na - 79.7 MHz (Fig. 2,C) and ^{13}C - 75.7 MHz (Fig. 2,D)).

Axial and coronal field profiles of the used coils (Fig. 3) demonstrated the simulated $|B_1^+|_{RMS}$ and the $|B_1^+|$ distribution maps measured in experiment to qualitatively match, confirming the correct practical design implementation.

The multi nuclear 3-section phantom images were successfully obtained by using ^{1}H-coil and X-coil, tuned to the Larmor frequency of ^{23}Na (Fig. 4).

4. Discussion

The RF-assembly assessment at 7 T has experimentally shown its possibility to be tuned to a number of frequencies corresponding to a set of biologically-relevant nuclei, including ^{1}H, ^{31}P, ^{23}Na and ^{13}C. Field mappings provide a confirmation of the ability of the coil to generate the desired imaging field of view in the coronal and axial planes.

Imaging data shows the possibility of heteronuclear imaging with the proposed coil design. Still, further protocol optimization is required for finding the optimal scanning conditions for the ^{23}Na imaging.

5. Conclusion

The presented data confirms the possibility of heteronuclear imaging with the suggested double-coil assembly, which opens up a perspective of expanding the experimental capabilities of a small-imaging MR-scanner.

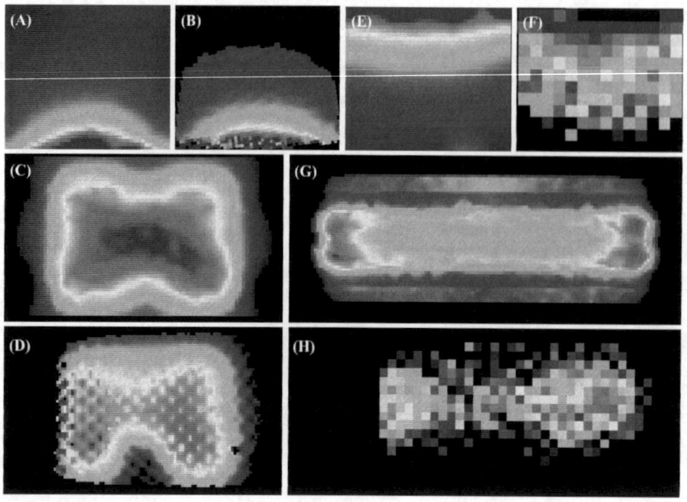

Figure 3. A-D – ^{1}H-coil field maps. A – simulation, axial plane; B – measured, axial plane; C – simulation, coronal plane; C – measured, coronal plane. E-H – X-coil field maps at the ^{23}Na resonant frequency. E – simulated, axial plane; F – measured, axial plane; G – simulated, coronal plane; H – measured, coronal plane. All fields are normalized to the same dynamic range.

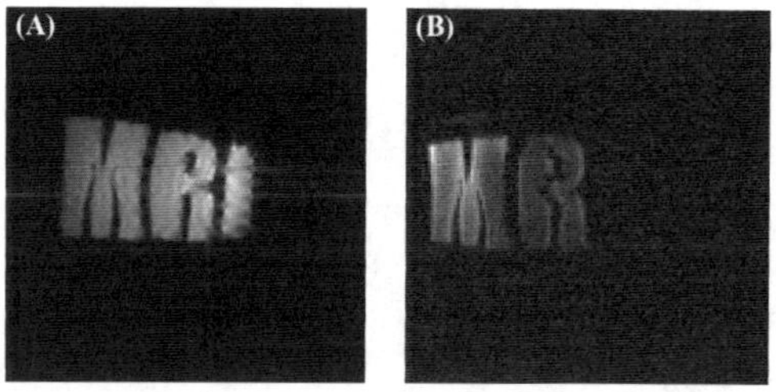

Figure 4. A - ^{1}H image of a 3-section phantom filled with uniformly with aqueous NaH_2PO_4 solution, B – ^{23}Na image of a 3-section phantom. Section 1, "M" contains aqueous NaH_2PO_4, section 2, "R", contains a solution of NaCl and $MnCl_2$. Section 3, "I", is filled with the $MnCl_2$ solution.

Acknowledgments

This work was funded by the Russian Science Foundation (Project 19-75-10104).

References

[1] Schick F 2005 *Eur. Radiol.* **15** 946
[2] Choi C H, Hong S M, Felder J and Shah N 2020 *Magn. Reson. Imaging* **72** 103
[3] Kirsch S, Augath M, Seiffge D, Schilling L and Schad L R 2010 *NMR Biomed.* **23** 592
[4] Kim K N, Seo J H, Han S D, Heo P, Im G H and Lee J H 2015 *Scanning* **37** 361
[5] Ivanov V, Hurshkainen A, Solomakha G and Zubkov M 2020 *Photonics Nanostruct.* **38** 100747
[6] Zubkov M, Hurshkainen A A, Brui E A, Glybovski S B, Gulyaev M V, Anisimov N V, Volkov D V, Pirogov Y A and Melchakova I V 2018 *NMR Biomed.* **31** e3952
[7] Hurshkainen A, Nikulin A, Georget E, Larrat B, Berrahou D, Neves A L, Sabouroux P, Enoch S, Melchakova I, Belov P, Glybovski S and Abdeddaim R 2018 *Sci. Rep-UK* **8** 9190
[8] Duan Q, Duyn J H, Gudino N, de Zwart J A, van Gelderen P, Sodickson D K and Brown R 2014 *Med. Phys.* **41** 102303

Journal of Physics: Conference Series **2015** (2021) 012172 doi:10.1088/1742-6596/2015/1/012172

Second harmonic generation in hybrid GaP/Au nanocylinders

Dmitry Pidgayko[1], Ilya Deriy[1], Vladimir Fedorov[2], Alexey Mozharov[2], Ivan Mukhin[2], Yuriy Zadiranov[3], Mihail Petrov[1], Anton Samusev[1], Andrey Bogdanov[1]

[1] Department of Physics and Engineering, ITMO University, 49, Kronverksky pr., 197101, Saint-Petersburg, Russia

[2] Alferov University (formerly St. Petersburg Academic University), 194021 St. Petersburg, Russia

[3] Ioffe Physical-Technical Institute of the Russian Academy of Sciences, Saint Petersburg 194021, Russia

E-mail: d.pidgayko@metalab.ifmo.ru

Abstract. To date, nanoscale dielectric and plasmonic systems with a nonlinear response are of great interest to researchers. This is due to a wide range of their potential applications in nonlinear optical converters and optical communication systems. The fundamental problem of nanoscale frequency converters is the low efficiency of nonlinear optical generation. The reason for this is that the main mechanism for increasing the efficiency of nonlinear signal generation via phase matching is not available at scales smaller than the wavelength. Here, we experimentally investigate the generation of the second optical harmonic in hybrid GaP/Au nanoparticles resonantly enhanced with plasmonic and Mie resonances. Using dark-field spectroscopy, it is shown that nanoantennas support a series of optical resonances in the visible range, the spectral position of which is in good agreement with the numerical simulation. We measured the second harmonic generation spectrum, with sharp resonances which is in accordance with linear scattering. Finally, the dependence of the second harmonic optical signal on the polarization is measured.

1. Introduction

To date, one of the main sources for of the optical harmonic generation at the nanoscale are plasmonic systems based on periodic arrays of particles [1] or single nanostructures [2], where plasmonic structures have resonance both at the fundamental and at the doubled frequency. Nevertheless, the generation efficiency of such systems is extremely low and does not exceed the value of 10^{-8} – 10^{-9}. Dielectric structures are also used for this purpose. For example, third harmonic generation (THG) is demonstrated in silicon–based nanostructures [3], [4]. The works where second harmonic generation (SHG) is amplified by the Mi resonances in resonant nanoparticles based on III-V semiconductors [5] and perovskites [6] are also of great interest. Finally, in [7], a record high value of the second harmonic generation efficiency (4.8×10^{-4}) in the near-IR spectral range was achieved due to its amplification by the so-called bound state in the continuum. An alternative approach to controlling the nonlinear optical properties of nanostructures is the use of hybrid systems consisting of a high-index resonant nanoparticle and a metal nanoparticle supporting localized plasmonic resonance. Such systems have recently shown very high efficiency THG based on coaxial silicon-gold systems [8]. But to date, investigation of hybrid nanoantennas for SHG generation have not been reported.

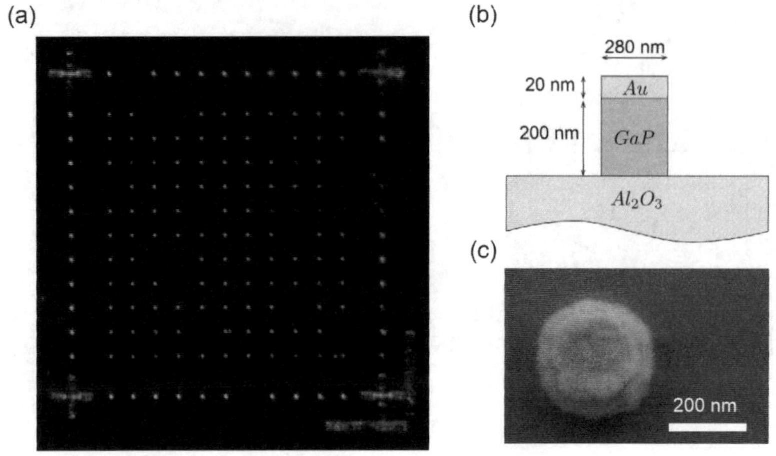

Figure 1. (a) Dark field image of GaP/Au nanoparticles array. (b) Schematic image of the hybrid nanoantenna. (c) SEM image of the nanoantenna.

Here we experimentally investigate hybrid GaP/Au nanoantennas (see Figure 1). The unit cell of gallium phosphide does not have an inversion center, so this material provides SHG in the bulk. Due to the high refractive index and zero losses, GaP nanocylinders maintain pronounced resonances in the visible range, which makes it possible to increase SHG when pumping a particle in the near-IR range. Finally, the metallic component, which gives a plasmon resonance in the near-IR region, also provides SHG generation amplification.

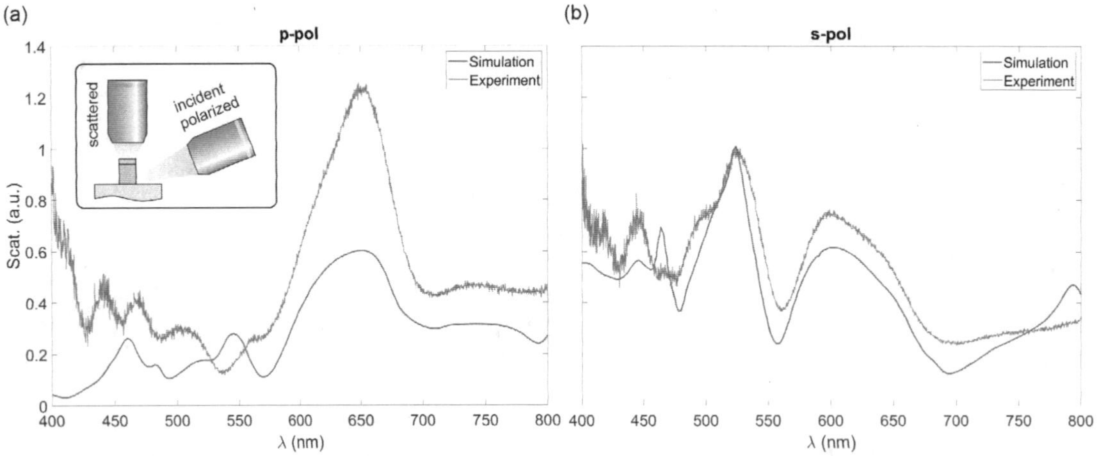

Figure 2. Experimental vs simulated scattering spectra for (a) p- and (b) s- polarised excitatino. Inset: scheme of dark-field experimental setup.

2. Results

2.1. Linear scattering

For nanocylinders fabrication the thin GaP layer (200 nm), grown by MBE on sapphire (0001) substrate was used. After coating with a thin layer of gold (20 nm), the nanocylinders were made by e-beam lithography. First, we studied optical properties of nanoparticles in linear regime – see Fig. 2. For numerical simulation we used The COMSOL Multiphysics package. The scattering spectra in were measured by dark-field spectroscopy, with an Ocean Optics HL-2000-LL light source, an 10x 0.26 N.A. Mitutoyo Plan Apo lens that illuminates particle arrays at an incident angle of 65 degrees, a 50x 0.42 N.A. Mitutoyo Plan Apo collection lens, and a LabRam HR 800 spectrometer. It is clearly seen that the particles support resonances in the visible part of the spectrum. Thus, SHG can be enhanced when pumping in the range of 1000-1600 nm due to the proper resonances of nanoparticles.

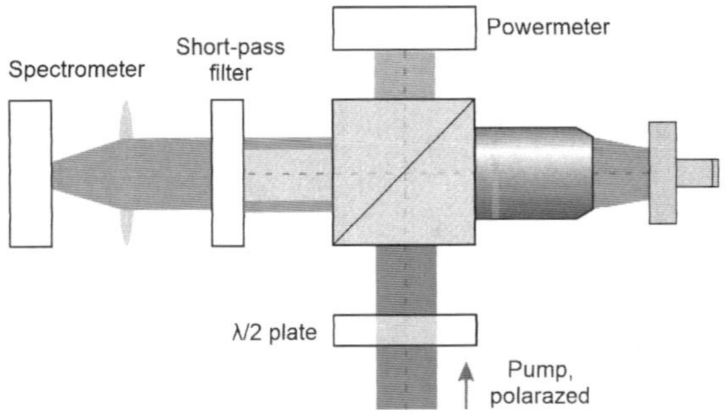

Figure 3. Experimental setup for SHG measurements.

2.2. SHG spectra and dependence on polarization direction.

The second harmonic generation was studied in the reflection geometry (see Fig. 3). As a radiation source, we used PHAROS Light Conversion (1053 nm, 1 MHz, 290 fs) with an optical parametric amplifier, which allowed us pump tunning in a wide spectral range of 1020-1500 nm.

For focusing and collecting, we used 50x 0.42 N.A. Mitutoyo Plan Apo NIR lens. We measured the pump power with use of high sensitivity thermal sensor OPHIR 3A. The second harmonic signal was analyzed using a Princeton SP 2558 spectrometer with a PyLoN 400BR eXcelon camera. The second harmonic spectra (see Fig. 4) exhibit pronounced resonant features that

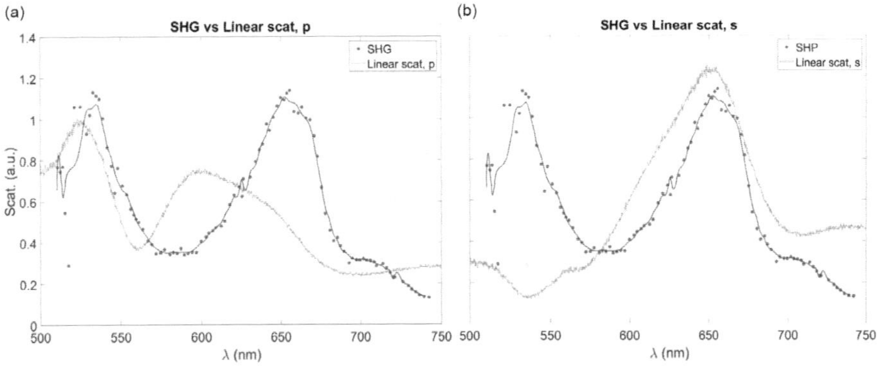

Figure 4. SHG vs linear scatering specta.

Journal of Physics: Conference Series **2015** (2021) 012172 doi:10.1088/1742-6596/2015/1/012172

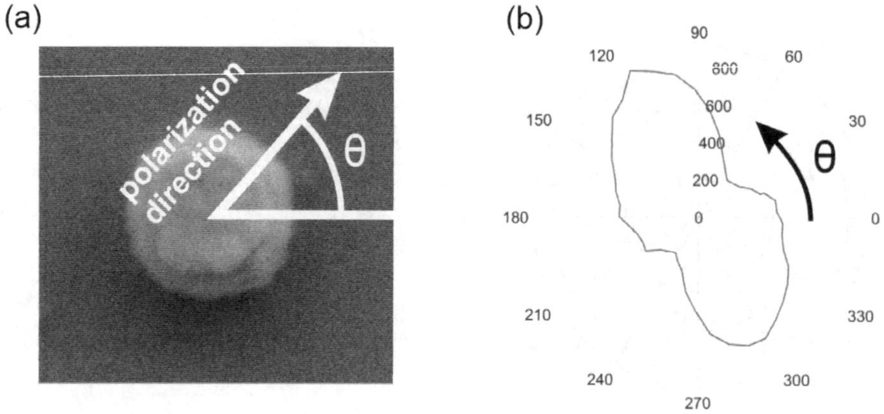

Figure 5. (a) SEM image of the nanoparticle and scheme of the pump polarization direction. (b) Dependence of SHG intensity on polarization direction in polar coordinaes.

are in good agreement with the linear scattering spectra. To study the dependence of SHG on the polarization, we fixed the wavelength at 1200 nm and placed a $\lambda/2$ plate in the pump channel, the direction of the fast axis of which was rotated in the experiment. As a result, we registered the dependence on the polarization, which is shown in the diagram (see Fig. 5). Thus, it can be concluded that the gallium phosphide from which the nanocylinders are made has a polycrystalline structure, despite the fact that the crystal lattices of Al_2O_3 and GaP are not consistent.

3. Conclusion

Here, we designed a hybrid plasmonic-semiconductor nanoantenna for the efficient second harmonic generation. Plasmonic part provide a resonance at the fundamental frequency while dielectric nanocylinder with intrinsic second-order susceptibility supports set of resonances at the doubled frequency. We measured second harmonic spectrum that demonstrates enhancement of nonlinear signal at wavelengths of nanocylinder eigenmodes. Finally, we experimentally showed the polarisation dependence of the second harmonic signal.

4. Acknowledgement

The authors Acknowledge Kristina Frizuk and Anna Nikolaeva for valuable discussions. Theoretical and technological part of the work is supported by Russian Foundation for Basic Research, project number 20-32-90243. Numerical and experimental part is supported by Russian Science Foundation project number 21-72-30018.

References

[1] Metzger B, Schumacher T, Hentschel M, Lippitz M and Giessen H 2014 *ACS Photonic* **6** 471
[2] Yang K Y ,Butet J, Yan C, Bernasconi G and Martin O J F 2017 *ACS Photonics* **6** 1522
[3] Shcherbakov M et al. 2014 *Nano Lett.* **14** 6488
[4] Smirnova D,Khanikaev A, Smirnov L and Kivshar Y 2016 *ACS Photonics* **3** 1
[5] Camacho-Morales R et al. 2016 *Nano Lett* **16** 7191
[6] Timpu F, Sergeyev A, Hendricks N, and Grange R 2017 *ACS Photonics* , **4**] 76
[7] Koshelev K, Kruk S, Melik-Gaykazyan E, Choi J H, Bogdanov A, Park H G and Kivshar Y 2020 *Science* **367(6475)** 288
[8] Shibanuma T, Grinblat G, Albella P, and Maier S 2017 *Nano Lett* **17** 2647

Corrigendum: Non-collinear ground state and stable bimerons from four-spin chiral interactions in D_{3h} magnet (2021 *J. Phys.: Conf. Ser* **2015** *012118*)

G R Rakhmanova, A N Osipov, D I Ilin, I V Shushakova and I V Iorsh
School of Physics and Engineering, ITMO University, Saint Petersburg, 197101, Russia

gulnaz.rahmanova@metalab.ifmo.ru

Page 1:
In the Introduction section 1, the following text appears:
In this Letter, we discuss another possible origin of noncolliner magnetic textures in ferromagnets with the D_{3h} point group symmetry.
This should read:
The Letter is presented based on the results of the article [11] and will discuss possible origin of noncolliner magnetic textures in ferromagnets with the D_{3h} point group symmetry.

Page 4:
In the References section, the following text should appear:
"[11] Ado I A, Rakhmanova G R, Zezyulin D, Iorsh I V, Titov M 2021 arXiv:2105.14495 [Physical Review Letters]"

The authors acknowledge that it is worthwhile to mention the article "Alternative to Dzyaloshinskii Moriya interaction for monolayer Fe_3GeTe_2 and other two-dimensional ferromagnets with trigonal prismatic symmetry" written by Ado I A, Rakhmanova G R, Zezyulin D, Iorsh I V, Titov M (2021 arXiv:2105.14495 [Physical Review Letters]) as reference 11 due to huge contribution to the current work and through this corrigendum this is addressed.

AUTHOR INDEX